2025 IEEE International Conference on Electron Devices and Solid-State Circuits (EDSS 2025)

Yinchuan, China
13-15 June 2025

IEEE Catalog Number: CFP25754-POD
ISBN: 979-8-3315-2209-4

**Copyright © 2025 by the Institute of Electrical and Electronics Engineers, Inc.
All Rights Reserved**

Copyright and Reprint Permissions: Abstracting is permitted with credit to the source. Libraries are permitted to photocopy beyond the limit of U.S. copyright law for private use of patrons those articles in this volume that carry a code at the bottom of the first page, provided the per-copy fee indicated in the code is paid through Copyright Clearance Center, 222 Rosewood Drive, Danvers, MA 01923.

For other copying, reprint or republication permission, write to IEEE Copyrights Manager, IEEE Service Center, 445 Hoes Lane, Piscataway, NJ 08854. All rights reserved.

*** This is a print representation of what appears in the IEEE Digital Library. Some format issues inherent in the e-media version may also appear in this print version.

IEEE Catalog Number:	CFP25754-POD
ISBN (Print-On-Demand):	979-8-3315-2209-4
ISBN (Online):	979-8-3315-2208-7

Additional Copies of This Publication Are Available From:

Curran Associates, Inc
57 Morehouse Lane
Red Hook, NY 12571 USA
Phone: (845) 758-0400
Fax: (845) 758-2633
E-mail: curran@proceedings.com
Web: www.proceedings.com

Table of Contents

Oral Session1 : Information functional materials and devices

001-S10174 **Defects engineering for stable halide perovskite solar cells** *1*
Prabakaran Selvaraj, Tapa Arnauld Robert, Haibing Xie

002-S10104 **Joint Process Optimization of IWZO Thin-Film Transistor for High-Resolution** *4*
Active-Matrix Displays
Bin Wang, Feilian Chen, Yan Yan, Meng Zhang

003-S10098 **Low-loss and Nonvolatile Multilevel Phase Shifter for Optical In-memory** *8*
Computing
Qiqi Wei, Jigeng Sun, Shaolin Zhou, Yi Zou

004-S10087 **Eliminating Hysteresis in Dual-Gate a-IZO TFTs via N_2O Plasma Passivation of** *11*
Back-Gate Insulator Interface Defects
Yuhan Zhang, Jinwen Liu, Huan Yang, Lei Lu, Shengdong Zhang

005-S10079 **Study of Flexible Tactile Sensors Based on Graphene/PDMS Composite Film** *15*
Yujun Ye, Weibin Lin, Shuangmei Xue, Qijun Sun, Yan Yan

006-S10023 **MoS_2-based RRAM with Asymmetric Contacts** *19*
Han Yan, Zichao Ma

007-S10041 **Two-dimensional $Mo_{0.5}W_{0.5}S_2$ based memristive artificial neuron** *22*
Hang Li, Wanhong Luan, Wei Zeng, Jiyu Zhao, Guanglong Ding, Su-Ting Han, Ye Zhou

008-S10040 **A near-infrared responsive P3HT synaptic transistor doped by upconversion** *25*
quantum dots
Wanhong Luan, Hang Li, Minglin Zheng, Zherui Zhao, Wei Zeng, Wenbin Zhang, Jiyu Zhao, Guanglong
Ding, Su-Ting Han, Ye Zhou

009-C0021 **Polarization Assisted n-AlGaN Electron Injection Layer in 229 nm far-UVC LEDs** *28*
for Hole Blocking and Current Density Enhancement
M. Nawaz Sharif, M. Ajmal Khan, Hiromitsu Sakai, Yuya Nagata, Kohei Fujimoto, Amina Yasin,
Mitsuhiro Muta, Hiroyuki Yaguchi, Yasushi Iwaisako, Yuhuai Liu, and Hideki Hirayama

Poster Session1: Information functional materials and devices

010-S10155   **Polarization-Modulated Elliptical Plasmonic Synapses for Optical Neural Network Applications**   *32*
Chengyan Zhong, Xiaobo She, Xiang Wang, Yu Liu

011-S10150   **Analysis of Photonic Crystal-Based Switching for Optical Neural Networks**   *35*
Yu Liu, Chengyan Zhong, Xiaobo She, Xiang Wang

012-C0020   **Near infrared spatial light modulator by metasurface-based pixel control**   *38*
Guozheng Lin, Zexuan Chen, Shanri Chen, Shaolin Zhou

013-S10099   **Alignment mark of meta-grating for sub-10 nanometer overlay inspection**   *42*
Zhaokai Qiu, Rui Liu, Shaolin Zhou, Xugang Ma

014-S10078   **Study of Negative Bias Illumination Stress Stability on IGZO/IZO Based Thin Film Transistors**   *46*
Le Bian, Jinxuan Wu, Yuxuan Shen, Siyuan He, Shuangmei Xue, Meng Zhang, Yan Yan

015-S10025   **A 2D Covalent Organic Framework (COF) Based Memristor**   *49*
Yaoli Guo, Yifan Zheng, Yan Yan, Jiaxing Yang, Yanlin Li, Zherui Zhao, Guanglong Ding, Shuangmei Xue

Oral Session2: Wide bandgap semiconductor materials and devices

016-S20147   **Magnesium intercalation in gallium nitride for enhanced p-type doping and device performance**   *52*
Jia Wang, Hiroshi Amano

017-S20058   **Control of 4H − silicon carbide (0001) oxidation rate via Argon ion-implantation**   *55*
Tao Zhu, YingFeng He, Rui Wang, Decai Liu, ZheYang Li, Rui Jin

Poster Session2: Wide bandgap semiconductor materials and devices

018-S20131   **Ga_2O_3 Photodetector for Single-Pixel Imaging Applications**   *58*
Yu Liu, Xiaobo She, Xiang Wang, Yufeng Guo

019-S20125   **Integrated 3×3 β-Ga_2O_3 Optoelectronic Synaptic Array for Efficient Edge Detection in Neuromorphic Computing**   *62*
Xiang Wang, Xiaobo She, Yingxu Wang, Yufeng Guo, Yu Liu

020-S20043 **The Regulating Effect of Electric Field on h-BN/ WTe$_2$ Heterostructure** 64

Jun Zhang,Hang Xu,Jiping Hu,Xiaotian Yang,Shipei Ji,Fang Wang,Juin J.Liou,YuhuaiLiu

021-S20038 **Optimization of AlGaN-based deep ultraviolet light emitting diodes with step-shaped hole blocking layer** 68

Haokai Jing

022-S20039 **Optimization of AlGaN-Based Deep-Ultraviolet Laser Diodes with Trapezoidal Quantum Barriers** 71

Wenlan Ma

023-S20036 **Enhanced optical confinement factor of DUV LD by optimizing the p-Cladding Layer** 75

Zhongqiu Xing,Yongjie Zhou, Fang Wang,Juin Liou,Yuhuai Liu

024-S20034 **Enhanced performance in AlGaN deep-ultraviolet light-emitting diodes without an electron blocking layer by using Graded Decreasing Quantum Barrier structure** 78

Xien Sang

025-S20029 **Optimizing Laser Diode Performance with Al-Composition Graded Last Quantum Barriers** 81

Xin Wang

026-S20019 **Electronic properties and optical modulation of heterostructures based on InP and h-BN** 85

Xiaotian Yang,Hang Xu,Jiping Hu,Jun Zhang,Shipei Ji,Yuhuai Liu

Oral Session3： Millimeter wave terahertz multibeam chip

027-S30169 **A W-Band 2.1 dB Insertion Loss Passive Attenuator with Magnetically Switchable Double-Layer Coupled-Lines in 65-nm Bulk CMOS** 89

Zhuang Miao,Guiyue Mao,Nengxu Zhu,Fanyi Meng,Kiat Seng Yeo

028-S30097 **A 100MHz Heterogeneous GaN/Si-CMOS 12 V-to-24 V Boost Converter with Soft-Switching Technique** 93

Qingsong Zhao,Zenglong Zhao,Fanyi Meng

029-S30016 **A 2-18GHz High Gain Power Amplifier Using Dual Current-Reuse Topology with ±0.3dB Gain Ripple** 97

Jianbing Liu,Fanyi Meng

030-S30054 **A 12-41 GHz Current-Reuse Composite-Cascode-Stack Power Amplifier in 65-nm** *100*
CMOS SOI
Guiyue Mao,Xuan Li,Yang Liu,Fanyi Meng

031-S30060 **A 1.36 ‒ 2.25 GHz Digitally Controlled Oscillator with Dynamic Element Matching** *104*
for NB-IoT Application
Binchen Wang,Mahalingam Nagarajan,Bharatha Thangarasu,Kaixue Ma,Fanyi Meng,Zhenghao Lu,Kiat Seng Yeo

032-S30134 **A Fully-Differential Bi-Directional 3.5-18 GHz 6-bit Active Phase Shifter in 0.18- μ** *108*
m SiGe BiCMOS
Lize Wang,Fanyi Meng

033-S30037 **A 5.1-33.5 GHz Variable Gain Low-Noise Amplifier With High Linearity in 0.13 μ** *111*
m BiCMOS
Shuai Li,Bharatha Thangarasu,Nagarajan Mahalingam,Kaixue Ma,Fanyi Meng,Zhenghao Lu,Anqing Chen,Hai Ye,Kiat Seng Yeo

034-S30033 **An 18-40 GHz 5-bit Passive Vector-Modulated Phase Shifter With an X-Type** *114*
Attenuator in 0.18 μ m SiGe BiCMOS
Yang Liu,Xuan Li,Guiyue Mao,Fanyi Meng,Kiat Seng Yeo

035-S30026 **A 10-40 GHz Phase Invariant Low Noise Variable Gain Amplifier with Pole** *117*
Staggering Bandwidth Extension Technology
Xuan Li,Fanyi Meng,Yang Liu,Guiyue Mao

036-S30045 **A 127-157 GHz Power Amplifier with 12dBm Saturated Output Power and 9.5%** *121*
Power-Added Efficiency in 65nm CMOS
Jiawei Chen,Yi Wu,Yuwen Long,Zongyang Zhang,Guangyin Feng

Poster Session3: Millimeter wave terahertz multibeam chip

037-S30168 **An Adaptive Dead-Time Controlled Hybrid Dual-Path Buck Converter With** *124*
Inductor Current Reduction Technique and High Power Density
Zishuo Li,Zhen Lin,Qingsong Zhao,Jian Wang,Fanyi Meng

038-S30162 **A Low-Loss Hybrid Power Combiner Designed for Ku-Band 32-Way Power** *128*
Combining Amplifiers
Yiting Zhang,Nengxu Zhu,Fanyi Meng

039-S30129 **A W-band 6-Bit Vector-Modulated Phase Shifter With Impedance-Invariant Technique** 131
Yue Zhang, Fanyi Meng

040-S30130 **A 13-17.4GHz Triple-Stacked Power Amplifier With 27 dBm P_{SAT} in 130-nm SOI Technology** 135
Shiyuan Fu, Fanyi Meng

041-S30161 **A W-Band Coupler-Based Differential Power Amplifier With 19.5 dB Power Gain and 19.2dBm PSAT in 130nm SiGe** 138
Jiaming Zhao, Pengfei Li, Fanyi meng

Oral Session4: Application of New Semiconductor Materials and Devices

042-S40171 **A Study on Thickness-Dependent Performances of High-k Gated Tellurium-Based Field-Effect Transistors** 142
Yang Hui Xia, Zi Chun Liu, Shu Ming Qi, Yu Hang Zheng, Yu Meng Wang, Xiao Long Xu, Hui Xia Yang, Yuan Xiao Ma, Yeliang Wang

043-S40158 **Comparative Analysis of Canny and Watershed Edge Detection for Line-Edge Roughness (LER) Quantification in SEM Images** 144
Shuyan He, Michail Michailow, Yufeng Jin

044-S40154 **Impact of Channel Geometry and Gate Dielectric Properties on GAAFET Device Performance** 148
Yue Zhang, Yongjie Zhao, Zhaonian Yang, Yufei Wang

045-S40111 **Field-effect transistor biosensor for ultra-high sensitivity detection of ssDNA** 151
Mengran Chen, Ya Li, Haoliang Li, Bing Chen, Wenchang Zhang, Xiaonan Yang

046-S40109 **A Study on Short-Circuit Parasitic Conduction Failure of 1200V SiC VDMOSFETs** 154
YaDong Zhou, HengYue Gong, YangHui Xia, YuMeng Wang, HuiXia Yang, HuiPing Zhu, YuanXiao Ma, Yeliang Wang

047-S40095 **Multilayer Nitride-based Memristors with AlScN Interlayer for Memory and Neuromorphic Computing Applications** 157
Yang Yang, Qilin Hua

048-S40070 **Performance Assessment of Stacked Monolayer WSe$_2$ GAA NSFETs for Sub-5 nm Nodes**   *159*

Ran Huo, Shijun Ou, Yihong Sun, Zichao Ma, Mansun Chan, Changjian Zhou

049-S40053 **A 6048-PPI High Performance OLEDoS Pixel Circuit with Self-Charging Capacitor Coupling Method**   *163*

Fan Guo, Yuanbo Sun, Congwei Liao, Lu Chang, Xin Zheng, Shengdong Zhang

050-S40032 **Using Nanophotonic Circuit to Calculate 2D Affine Transformation**   *166*

Jinzhi Mu, Jigeng Sun, Shaolin Zhou

Poster Session4: Application of New Semiconductor Materials and Devices

051-S40170 **Enhancement of Electrical Performance and Stability in p-Type SnO Thin-Film Transistors via Ultrathin Al$_2$O$_3$ Capping Layer**   *169*

Ruyu Liang, Lei Xu, Shi Zong

052-S40163 **Enhanced the reliability of InSnMgO thin-film transistors via in-situ sputtering of 2 nm MgO layer**   *172*

Shi Zong, Lei Xu

053-S40100 **Voltage-dividing Capacitor Sharing Enabled High PPI Pixel Circuit with Extended Data Voltage Range for OLED on Silicon Micro-display**   *174*

Yuanbo Sun, Lu Chang, Congwei Liao, Chen Chen, Xin Yuan, Zhiwei Ye, Xiufeng Zhou, Guangsheng Li, Shengdong Zhang

054-S40059 **Stress-Induced Band Structure Modulation in h-BN/HfSe$_2$ Heterostructures**   *178*

Yunhao Liu

055-S40047 **Band Structure Modulation of h-BN/HfSe$_2$ Heterostructures Under an External Electric Field**   *181*

Shipei Ji

056-S40042 **Electronic properties and energy band alignments of h-BN/TiS$_2$ heterostructure**   *185*

Cheng Li

Oral Session5: *Power semiconductor devices*

057-S50173 **1200V SOI N-channel LDMOS Adopting Partial Separation by Implantation of Oxygen** 188
Mingxin Sun, Chengwu Pan, Shipeng Chang, Siyang Liu, Weifeng Sun, Long Zhang

058-S50157 **Influence of the Charge-Imbalance Condition on the Short-Circuit Characteristics of the 4H-SiC Superjunction MOSFET** 191
Huan Ning, Da Wang, Zhi Lin

059-S50148 **Grounded Isolation Trenches in GaN-on-Si Power Integrated Circuits: An Electromagnetic Study for Trench Filling Considerations** 195
Rui Yao, Zijin Jiang, Miao Cui, Zhao Wang, Sang Lam, Stephen Taylor

060-S50143 **Theoretical and Numerical Investigation on the On-state Characteristics of a 1200V Multi-Gate 4H-SiC LDMOS** 199
Da Wang, Huan Ning, Zhi Lin

061-S50127 **A Novel Hybrid Boost Converter With Continuous Output Current and RHP Zero Elimination** 203
Zuyue Pang, Chenguang Lv, Ziyi Cui, Chi Zhang, Shuhai Chen, Shaowei Zhen

062-S50132 **A Novel Low Loss SOI-LIGBT with Carrier Stored Layer and P-drift** 207
Haoru Wang, Yifan Shu, Jun Zhang, Jialei Tan, Kaiwei Dai, Pei Guo, Jiayuan Wang, Jie Wei

063-S50074 **Study on Mechanisms of Heavy Ion Induced Leakage Current and Single Event Burnout of FRD** 211
Ailin Qiu, Huan Li, Xiaoping Dong, Mingmin Huang, Yao Ma, Qiang Yu

064-S50056 **Effect of Electron Irradiation and Post-Irradiation Annealing on Reverse-Conducting IGBT** 215
Pengwei Chen, Huan Li, Xiaoping Dong, Mingmin Huang, Yao Ma, Chang Chen, Qiang Yu

065-S50092 **1500V High-Voltage GaN HEMT Device with Multiple Field Plates for High Power Applications** 219
Moufu Kong, Yaowen Zhang, Yingzhi Luo, Kangxiang Zhao, Bingke Zhang, Bo Yi, Hongqiang Yang

066-S50083 **A Novel SiC Trench MOSFET With Integrated Junction Barrier Schottky for Improved Performances** 222
Bo Yi, JunFeng Duan, XinYi Wu, Tao Zhu, JunJi Cheng, HongQiang Yang

067-S50076   **A Novel SiC Trench MOSFET with Integrated N⁺- PolySi/SiC Heterojunction Diode**   **224**
Bo Yi,XinYi Wu,JunFeng Duan,Tao Zhu,JunJi Cheng,HongQiang Yang

068-S50057   **Self-Heating Switching for Recovering Threshold Voltage of SiC MOSFETs after Electron Irradiation**   **226**
Huanshi Guo,Huan Li,Mingmin Huang,Yao Ma,Nuoya Yang,Chang Chen,Yun Li,Qiang Yu

069-S50055   **Study on Single Event Effects in High-Voltage Silicon Power Semiconductor Devices**   **230**
Mingmin Huang,Huan Li,Ailin Qiu,Yichu Qin,Xiaoping Dong,Dongyang Wang,Yao Ma,Qiang Yu

070-S50048   **Electron Irradiation Effects on *I-V* Characteristics in p-GaN Gate HEMTs**   **234**
Dongyang Wang,Yifei Huang,Shuting Wang,Mingmin Huang,Qimeng Jiang,Yao Ma

Poster Session5 : Power semiconductor devices

071-S50123   **A Wide Input Range With A Small Area Bandgap Reference Voltage Source With Self-biased Cascode**   **238**
Sini Wu,Bharatha Kumar Thangarasu, Nagarajan Mahalingam, kaixue ma,fanyi meng,zhenghao lu,kiatseng yeo

072-S50091   **A Novel High-Voltage Lateral GaN Diode with Hybrid p-GaN Anode**   **242**
Kaijun Ding,Yimeng Tang,Yaowen Zhang,Yingzhi Luo,Bingke Zhang,Moufu Kong

073-S50022   **A Robust P-Type Bubble+Shield-Gate Terminal Ring for A 750V IGBT**   **245**
Kui Xiao,Shen Xu,Zheng Bian,Wei Yao

Oral Session6 : Digital and Memory Circuits

074-S60176   **Moore's Law at 60 - Still in Good Shape or Already Ailing?**   **249**
Frank Schwierz

075-S60017   **A Hybrid Clause Deletion Framework for Enhanced SMT Solving in Hardware Formal Verification**   **253**
Meihua Liu, WenDa Leng, Yufeng Jin

076-S60118   **A SOT-MRAM Based In-Memory Computing Macro with Enhanced Read-Current Stability and Multi-Bit MAC Operations**   **257**
Saiya Wang,Junzhan Liu,Guangyao Wang,Wang Kang

Poster Session6: Digital and Memory Circuits

077-S60172   **Design of Low-power Spike-Time Dependent Plasticity Synaptic and Neuron Circuits on a SOI process**   *261*
Zhuojun Chen,Qiaoyi Fu

078-S60112   **A Digital LDO Designed to Enhance the Side Channel Security with Random Noise Injection Technique**   *265*
Xiaodan Gu,Jiaji He,Yu Long,Mao Ye

Oral Session7: Artifical intelligence and its applications

079-S70081   **Sparse Diffusion Accelerator with Pattern Pruning,Dynamic Detection and Quantization**   *268*
Boran Cao,Sheng Zhang,Chen Tang,Xinyuan Lin,Leran Huang,Wentao Zhao,Yongpan Liu

080-S70046   **40 nm Core, MV, HV Devices Integration Database for Future AI Development**   *272*
Yi-Chuen Eng,Daiki Qin,Levi Chen,Wenhao Wu,Peter Liu,Cloud Wang,Chia-Yen Li,Mark Zheng,Nelson Yang,Junun Zhu,RR Zhu,XF Guan,Yuri Ma,Le Li,Hayato Wang,Leo Kuai,Carl Ma,Bo Yang,Jihong Yin,Kyle Xu,Rumeng Qiu

081-S70044   **Utilizing AI to Address the Poor Correlation of BEoL 1XDD VIA R_c and Its Inline Data at the 40 nm Node**   *276*
Peter Liu,Daiki Qin,Levi Chen,Yi-Chuen Eng,Wenhao Wu,Cloud Wang,Chia-Yen Li,Mark Zheng,Nelson Yang,Carl Ma,Bo Yang,Jihong Yin,Kyle Xu,Rumeng Qiu,Leo Kuai,Junun Zhu,RR Zhu,XF Guan,Yuri Ma,Le Li,Hayato Wang

Poster Session7: Artifical intelligence and its applications

082-S70156   **A sleep posture detection method for pregnant women based on AI algorithms and a flexible sleep monitoring belt with a MEMS IMU**   *280*
Chunhua He,Shangle Ye,Jian Zhan,Jing Lin,Heng Wu,Maojin Liang,Songqing Deng

083-S70152   **Underwater Acoustic Target Localization Method Based on Frequency-Domain Feature Enhancement**   *283*
Daoguang Zhang,Xianyou Zeng,Kai Yuan,Jing Lin,Heng Wu, Qinwen Huang, Chunhua He

084-S70105   **Moire-based Overlay Metrology Enhanced By Deep Learning**   *287*
Rui Liu,Zhaokai Qiu,Shaolin Zhou,Xugang Ma

085-S70035 **A Hybrid Algorithm for Automated Placement of CMOS Standard Cells** 291
Weiteng Hu, Junlang Yu, Bin Li, Zhaohui Wu

086-S70187 **Design of Ecological Environment Monitoring System Based on RT-Thread and Cloud Platform** 295
Jingyang Wen, Ling Li, Liangjing Bai, Xiyuan Wu, Wei Zhang, Hongxing Ma

Oral Session8: Communications

087-S80010 **Content Caching-oriented Popularity Forecast Algorithm Design** 297
Qi Chen, Wenze Gao, Takayuki Nakachi, Yitu Wang, Juinjei Liou

088-S80012 **Energy Efficient Scheduling with Forecast Information** 300
Xuying Zhou, Bijiao Yang, Takayuki Nakachi, Yitu Wang, Juinjei Liou

089-S80011 **Delay and Power Trade-off with Dynamic Lyapunov Function** 303
Qi Chen, Yang Yu, Takayuki Nakachi, Yitu Wang, Juinjei Liou

090-S80159 **Codebook based SCMA with Message Passing for Multi-User Detection** 306
Qixian Zheng, Hang Li, Lizhe Liu, Yashan Pang, Yangyang Guan, Zhiqun Cheng, Qinghua Guo

091-S80135 **A 12-bit 320MS/s Current-steering DAC With 70dBc SFDR in 22nm CMOS** 309
Tinghua Chen, Qiji Huang, Xinpeng Xing, Haigang Feng

092-S80030 **Robust Hybrid Holographic Beamforming Scheme Under the RHS Based UAV IoT System** 313
Mingcheng Shen, Keer Chen, Jichong Guo, Chunxia Su, Chang Ding

093-S80160 **LS and MMSE Channel Estimation in Dual Pulse Shaping Systems** 316
Moting Deng, Hang Li, Lizhe Liu, Yashan Pang, Yangyang Guan, Zhiqun Cheng, Qixian Zheng, Qinghua Guo

094-S80144 **A Background Calibration in Pipelined-SAR ADCs using Capacitor Flipping and Gain Verification** 319
Pengfei Ye, xinpeng xing, haigang feng

095-S80146 **An 18-bit 1-MS/s SAR ADC With a Continuous-Time SAR-Based Pre-Quantization** 323
Weifeng Qiao, Zengqing Liang, Haigang Feng, Xinpeng Xing

Poster Session8: Communications

096-S80186 **Study on the properties of composite dou-ble-layer films of anthocyanin agar and**   *327*
different concentrated TiO$_2$
Yuanzhan Sheng, Shiyu Zuo, Yan Liu, Pao-Hsun Huang

097-S80185 **A Negative Voltage Convertor Using Switch Capacitor Structure**   *330*
Shumin You, Zhipeng Liu, Ronglin Yang, Xixian Wang, Pao-Hsun Huang, Yan Liu

098-S80183 **A Method of Locusts Detection**   *334*
Hongxing Ma, Yingfei Wang, Xuan Liu, Jintai Chi, Fuyuan Wang, Junjie Liu

099-S80182 **Detection, Tracking, and Activity Monitoring of Small and Micro Insects Based on**   *338*
Deep Learning
Hongxing Ma, Haobo Jia, Jintai Chi, Yun Ma, Xuan Liu, Xiaobin Ren

100-S80179 **Locust Object Detection in Complex Backgrounds Using SPD-Conv and PPA**   *342*
Fusion in YOLOv11
Hongxing Ma, Xuan Liu, Kaiwen Chen, Lin Li, Haobo Jia, Wei Sun

101-S80180 **Research and Application of Mushroom Growth Status Monitoring Technology**   *346*
Hongxing Ma, Yun Ma, Fuyuan Wang, Haobo Jia, Kaiwen Chen, Xiaobin Ren

102-S80107 **RSMA Receiver Detection based on Neural Network Supervised Learning**   *350*
Dou Pei, Zhigang Zhou, Jian-gong Ni, Yejiang Lin, Gang Chen, Jian Zhou

103-S80080 **A Efficient Reed-Solomon Codes Recognizer Based on Galois Field Fourier**   *353*
Transform
Weiran Cao, Wei Zhang, Yihan Wang

104-S80188 **Effects of *3d* electron/hole doping on the static and high-frequency magnetic**   *356*
properties of *c*-oriented *hcp*-(CoIr) thin films with easy-plane magnetocrystalline
anisotropy
Tianyong Ma, Ao Han, Qi Liu, ZiSheng Li, Sha Zhang

105-S80189 **Investigation on the static and dynamic magnetic properties of *hcp*-(CoIr)$_{100-x}$M$_x$**   *361*
(M=B, SiO$_2$) thin films
Tianyong Ma, Qi Liu, Ao Han, ZiSheng Li, Sha Zhang

Oral Session9: MEMS/Sensors

106-S90149 **High-performance Piezoresistive Pressure Sensor Based on MEMS Technology** *366*
Zhikang Lan, Weiping Li, Tongqing Liu, Yizhou Ye, Xiaodong Huang

107-S90089 **Improved Plasmonic Scattering Imaging based on super-resolution algorithm for** *369*
image reconstruction
Zhaochen Huo, Bing Chen, Yu Li, Ya Li, Xiaonan Yang

108-S90096 **A Novel Active-Pixel Readout Circuit Based on Bias Current Cancellation for** *373*
Dynamic X-ray Imaging
Jiangbo Hu, Yuhan Zhang, Congwei Liao, Shengdong Zhang

Poster Session9: MEMS/Sensors

109-S90175 **Design of 10MHz Isolated Current Sense Amplifier Based on FDDA and Frequency** *377*
Modulation
Hongwei Shen, Xinghong Chen, Shaowei Zhen, Jingying Sun, Zupei Gu, Yongwang Ma, Yidong Yuan, Bo Zhang

110-S90138 **Magnetic Field Measurement Using Resonance Frequency-Optimized Fluxgate** *380*
Sensors
Wenbo Wang, Shanglin Yang

111-S90137 **A Narrow Band Surface Acoustic Wave Filter Based On 112° XY-LiTaO3** *384*
Yusuo Wang, Shanglin Yang, Zhijuan Zhao, Shunjing Lei, Hong Zhang, Juin Jei Liou

112-S90101 **A 1280×1024 DI ROIC with On-chip ADC and LVDS For Infrared Focal Plane** *388*
Arrays
Jia Shen, Yan Dong, Yaoxu An, Yao Li, Mao Ye

113-S90090 **POCT system based on lens-less imaging for complete blood counting** *392*
Yu Li, Zhaochen Huo, Jianing Li, Ya Li, Bing Chen, Xiaonan Yang

114-S90108 **A 320×256 15 μm-Pitch Readout Circuit with Column-Parallel 84dB-DR 167 KS/s** *396*
Incremental Sigma-Delta ADCs for IRFPA
Chenxu Zhao, Yao Li, Mao Ye, Qiuwei Wang, Jun Du, Yiqiang Zhao

115-S90102 **Two-Step Current Integration Enabled High Dynamic Range X-Ray Image Sensing** *400*
Yuezhong Duan, Tengyan Huang, Congwei Liao, Shengdong Zhang

116-S90064 **A Low-Noise Charge-Balanced Readout Circuit for MEMS Accelerometer** 404
Wenting Wang, Yao Li, Cheng Yuan, Mao Ye

117-S90050 **A 18.3-ENOB 160.2- μ W Fully Dynamic Discrete-Time Delta-Sigma ADC Using** 408
Floating Inverter Amplifiers
Shipeng Zhang, Jixiang Zhang, Yao Li, Mao Ye

118-S90049 **RCS-SAR: A 12-bit 20MS/s Secure ADC with Random Capacitor Switching Against** 412
Power Side-Channel Attacks
Jixiang Zhang, Shipeng Zhang, Jiaji He, Mao Ye

119-S90052 **A high integration and high stability design of analog driving ASIC for MEMS** 415
Gyroscopes
Hetian Sun, Mao Ye

120-S90031 **Design of a 12-bit 200MS/s asynchronous SAR ADC in 12nm FinFET technology** 419
Jingwen Li, Shaohui Pan, Yukun Fu, Huachao Xu, Bin Li

Oral Session10: RF & Microwave devices and circuits

121-S100139 **A 100.3dB-SNDR 10.8mW Pipeline-SAR ADC with second-order gain error shaping** 422
and one-time foreground calibration
Haowen Wu, Weihui Liu, Haigang Feng, Xinpeng Xing

122-S100153 **Electromagnetic Investigation of Substrate Coupling for Monolithic Microwave** 426
Integrated Circuits in GaN-on-Si Technology
Rui Yao, Miao Cui, Zhao Wang, Sang Lam, Stephen Taylor

123-S100136 **A multi-parameter tunable bandstop filter adopting parallel RF switch** 430
*Xiang LI, Xingkun You, Kuangyong Gao, Huafei Cheng, Huanghao Ying, Zhen Shi, Haizhen
Guo, Zhonghai Zhang*

124-S100117 **Design Method of Ultra-Wideband Power Amplifier Based on Optimization** 433
Approach
Xiang Chen, Guohua Liu, En Hong

125-S100061 **A Wide Band Reconfigurable PVT Robust high precision quarter phase detector for** 436
clock multiplier
*Anqing Chen, Nagarajan Mahalingam, Bharatha Kumar Thangarasu, Kaixue Ma, Fanyi Meng,
Zhenghao Lu, Kiat Seng Yeo*

126-S100190 **An OTA-C bandpass filter for infrared receiver chip** 439
Huilin Huang, Hongyi Liu, Shuxin Xu, Wanghui Zou

127-S100093 **Desig of Dual Band Synchronous Rectifier With a Novel Matching Network** 443
Guoqing Chen,Guohua Liu,Rui Zhang

128-S100113 **An Optimization Design for Planar Microwave Sensors Based on Slow-Wave** 445
Transmission Line
Huayi Wu,Guohua Liu

129-S100088 **A High-Gain and Low profile ESPAR Antenna Base on Alford Loop Antenna** 449
JiaYuan Fan,Chao Gu,Zhiwei Zhang,Dengfa Zhou

130-S100075 **A New Scalable De-embedding Method With a Distributed Pad Model and Lumped** 452
Compensation for Pad-Line Discontinuities in RF On-Wafer Characterization
Hongfei Su,Bharatha Kumar Thangarasu,Nagarajan Mahalingam,Kaixue Ma,Fanyi Meng,Zhenghao Lu,Kiat Seng Yeo

131-S100073 **A Thru-Line De-Embedding Method With Double-Type Pad Model for On-Wafer** 456
Device Characterization
Wu Yutong,Bharatha Kumar Thangarasu,Nagarajan Mahalingam,Kaixue Ma,Fanyi Meng,Zhenghao Lu,Kiat Seng Yeo

132-S100071 **An Adaptive Digital Loop Filter For Fast-Locking TDC-Based ADPLL** 460
Kaiqiang Qin, Mahalingam Nagarajan, Bharatha Thangarasu, Kaixue Ma, Fanyi Meng, Zhenghao Lu, Kiat Seng Yeo

133-S100068 **A 2.4 GHz Dual-Path Rectifier With Wide-Dynamic-Range for RF Energy** 463
Harvesting
Wansi Ge,Bharatha Kumar Thangarasu,Nagarajan Mahalingam,Kaixue Ma,Fanyi Meng,Zhenghao Lu,Kiat Seng Yeo

134-S100063 **A 2 - 9 GHz SiGe Broadband High-Gain Low Noise Amplifier Using Resistive** 467
Feedback Technology
Xiaozheng Guo, Bharatha Kumar Thangarasu, Nagarajan Mahalingam, Kaixue Ma, fanyi Meng, zhenghao Lu, Kaitseng Yeo

Poster Session10: RF & Microwave devices and circuits

135-S100126 **A superconducting tunable filter with fixed inputoutput position** 470
Mingchao Li, Kuangyong Gao, Xingkun You, Xiang Li, Huanghao Ying, Huafei Cheng, Haizhen Guo, Zhen Shi, Zhonghai Zhang

136-S100133 **A 5.5-6-GHz, 4.8mW, 5.7dBm IIP3 LNA in 22-nm CMOS Technology for Wi-Fi Application** 473
Yingqi Liu, Kaiyun Deng, Haoyu Dong, Bozhi Qiu, Haigang Feng

137-S100106 **High-Sensitivity Displacement Sensor Based on Frequency Selective Surface** 477
Bo Qi, Mi Lin, Guohua Liu

138-S100115 **Design of a wireless charging system for airborne sensor of a tunnel boring machine** 481
Zhenxuan Zhang, Xiaolong Wei, Zhen Huang, Yao Wang

139-S100084 **An Ultra-Wideband Power Amplifier based on 130nm SiGe Process** 485
Yu Hongshi, Bharatha Kumar Thangarasu, Nagarajan Mahalingam, Kaixue Ma, Fanyi Meng, Zhenghao Lu, KiatSeng Yeo

140-S100067 **A SAW Filter with 3-dB FBW of 15.5% on $LiNbO_3/SiO_2$/Quartz Substrate** 488
LiJun Feng, Changjian Zhou, Xiuyin Zhang

141-S100069 **A 130nm SiGe BiCMOS Ultra-Wideband High-Gain Power Amplifier With 86% Fractional Bandwidth and Digital Gain Control** 492
Yuqing Liu, Bharatha Kumar Thangarasu, Nagarajan Mahalingam, Kaixue Ma, Fanyi Meng, Zhenghao Lu, Kiat Seng Yeo

142-S100062 **Accurate Gummel Parameter Extraction for SiGe HBTs with Temperature-Dependent Saturation Current and Ideality Factor Models** 495
Xudong Cai, Guofang Yu, Jun Fu

143-N0014 **A Capacitorless Fully Synthesizable Digital-LDO Using scalable APR-friendly Power MOS Cell with 13.71A/mm2 current density** 500
Yun Xin Wang, Xu Liang Wang, Bo Ran Zhang, Ren Wei Chen, Xu Chen Men, Xiao Sen Liu

Oral Session 11: Electrical Static Discharge and Electrical Over Stress

144-S110151   **A Novel LVDDSCR Structure for Improved Latch-Up Immunity and Reduced Transient Overshoot Voltage**   *502*

Yilidana Mamuti, Hongyi Li, Zhiyuan He, Zihan Zheng, Jiyuan Huang, Yimu Yang, JiXiang Shang, JiZhi Liu, Ruibo Chen, Zhiwei Liu

145-S110120   **3D TCAD Optimization of Segmented SCRs for High-Voltage ESD Protection**   *505*

Tianyi Zhang, Qiang Cui, Boris Dobrichkov, Dimitar Nikolov, Vladimir Garistov

Poster Session 11: Electrical Static Discharge and Electrical Over Stress

146-S110164   **Degradation Behaviors of Bridged-Grain Polycrystalline Silicon Thin-Film Transistors under Dynamic Gate Voltage Stress with Fast Transition Time**   *509*

Ming Guo, Meng Zhang, Yunyang Wang, Lanrong Zou, Yan Yan, Lei Lu, Man Wong, Hoi-Sing Kwok

147-S110128   **A Novel Super Junction SCR with High Holding Voltage for On-chip ESD Protection**   *513*

Xinyu Zhu, Yipeng Chen, Shipeng Chen, Shurong Dong

148-S110145   **A novel fast-triggering DTSCR ESD protection device for low-voltage applications**   *516*

Gongtang Yu, Zhihua Zhu, Shanglin Yang, Juin Jei Liou

149-S110124   **Silver Film/CPSA ESD Failure Mechanism and Robustness**   *520*

Shipeng Chen, Zhencheng Xu, Yipeng Chen, Jiabei Pan, Ling Zhang, Shurong Dong

150-S110122   **Design of Low-Leakage ESD Power Clamp Circuit with Multi-level Control Network**   *523*

Gaoxiang Kai, Jiahao Xu, Liyao Wei, Hejiu Zhang, Yiqun Liu, Zhaonian Yang

151-S110121   **Design of a Compact Clamp Circuit Based on Current Mirror Structure**   *526*

Liyao Wei, Hejiu Zhang, Yiqun Liu, Zhaonian Yang

152-S110119   **Microcontroller Unit ESD-Induced Soft Failure Study**   *529*

Jiabei Pan, Yipeng Chen, Xinyu Zhu, Shipeng Chen, Ling Zhang, Shurong Dong

153-S110082   **Improved Snapback TVS Transient Behavior Model and Application to SEED**   *532*

Yipeng Chen, Xinyu Zhu, Jiabei Pan, Shipeng Chen, Ling Zhang, Shurong Dong

Oral Session12: Design, Simulation and Packaging of Chiplet IOs

154-S120140 **Electrical Performance of Chiplet Interconnect Channels under thermal conditions**   *536*
Tingyi Shi,Xiang Wang,Yuefeng He,Xiaofang Gao,Zhiqiang Zhu,Xujuan Wang,Xinnan Lin

155-S120103 **A Low-power 16 Gb/s Single-Ended Voltage-Mode Transmitter With Two-Tap FFE**   *540*
in 55-nm CMOS
Haitao You,Qiang Cui,Xiaofang Gao,Boris Dobrichkov,Dimitar Nikolov,Vladimir Garistov

Poster Session12: Design, Simulation and Packaging of Chiplet IOs

156-S120142 **AION-AMI：Chiplet System Simulation Platform**   *543*
Qing Qian,Xiang Wang,Yuefeng He,Xiaofang Gao,Haofeng Zhu,Xujuan Wang,Xinnan Lin

157-S120141 **Optimization Algorithm for High-Speed IO Systems**   *546*
Mengyuan Chu,Xiang Wang,Yuefeng He,Bo Zhang,Xinnan Lin

Proceedings of

2025 IEEE International Conference on Electron Devices and Solid-State Circuits (EDSS)

June 13-15, 2025

Yinchuan, Ningxia, China

Proceedings of

2025 IEEE International Conference on Electron Devices and Solid-State Circuits (EDSS)

Organizing Committee

Steering Committee

Charles Surya, Hong Kong Institute of Technology (HKIT)

Juin J. Liou, North Minzu University

Mansun Chan, The Hong Kong University of Science and Technology

Paul Yu, University of California San Diego

Honorary Chair

Junjie Li, North Minzu University

General Co-Chairs

Juin J. Liou, North Minzu University

Jiandong Mao, North Minzu University

Hu Zhao, North Minzu University

Feng Feng, Ningxia University

Conference Secretaries

Yuan Yang, North Minzu University

Yang Jiao, North Minzu University

Wenbo Wang, North Minzu University

Hesen Su, North Minzu University

Financial Chairs

Shanglin Yang, North Minzu University

Shoujun Tuo, North Minzu University

Technical Program Co-Chairs

Wanghui Zou, Changsha University of Science & Technology

Hao Liu, Ningxia University

Yitu Wang, North Minzu University

Bo Yi, University of Electronic Science and Technology of China

Long You, Huazhong University of Science and Technology

Long Zhang, Southeast University

Guifang Li, Northwestern Polytechnical University

Changjian Zhou, South China University of Technology

Local Arrangement Co-Chairs

Shanglin Yang, North Minzu University

Yang Jiao, North Minzu University

Yuan Yang, North Minzu University

Hesen Su, North Minzu University

Shoujun Tuo, North Minzu University

Wenbo Wang, North Minzu University

Publication Chair

Mengqi Zhou, Cambridge International Education

Publication Co-Chair

Jingyang Wen, North Minzu University

Registration Chairs

Zhaonian Yang, Xi'an UniversityofTechnology

Yansong Yang, The Hong Kong University of Science and Technology

International Advisory Committee Co-Chairs

Juin J. Liou, North Minzu University

Yang Chai, The Hong Kong Polytechnic University

Preface

On behalf of the Organizing Committee of the 16th IEEE International Conference on Electron Devices and Solid-State Circuits (EDSSC 2025), we are very glad to welcome all of you to Yinchuan, Ningxia, China for this exciting event together with many scientific researchers and professionals worldwide.

EDSSC 2025 is the 16th in a series of this successful conference jointly sponsored by the IEEE Hong Kong ED/SSC Joint Chapter, IEEE Beijing EDS Section, and North Minzu University. EDSSC 2025 continues the EDSSC tradition as a multidisciplinary forum for the exchange of ideas, research results, and industry experience in the broad areas of electron devices and solid-state circuits and systems. The technical program includes invited talks by famous scientists and contributed papers. Accepted papers will be submitted for inclusion into IEEE Xplore subject to meeting IEEE Xplore's scope and quality requirements.

EDSSC 2025 received more than 200 papers, which were reviewed by the technical committee and other experts. Such papers cover all major topics on theoretical research and empirical study, and include the study of information functional materials and devices, wide bandgap semiconductor materials and devices, millimeter wave terahertz multibeam chip, application of new semiconductor materials and devices, power semiconductor devices, digital and memory circuits, artificial intelligence and its applications, communications, MEMS/sensors, RF & microwave devices and circuits, electrical static discharge and electrical over stress, design, simulation and packaging of Chiplet IOs, and etc. As the sponsors, we would like to express our sincere thanks to IEEE Hong Kong ED/SSC Joint Chapter, IEEE Beijing EDS Section, and North Minzu University for their sponsorship and organization. Besides, we sincerely thank numerous volunteers for their dedicated service to the conference.

We would also like to express our deepest gratitude to the members of the Conference Committees and all reviewers for their professional reviews of the papers. Their expertise guaranteed the high quality of the technical program of the EDSSC 2025.

We look forward to seeing all of you in Yinchuan, Ningxia, China on June 13-15, 2025.

Organizing Committee of EDSSC 2025

Keynote Speaker

Jamal Deen

Professor , IEEE Fellow , Fellow of the Royal Society of Canada , Fellow of the Chinese Academy of Sciences

McMaster University, Canada

Title of the Keynote Report

Compact Modeling of Organic Thin Film Transistors For Flexible Electronics – Opportunities and Challenges

Dr. M. Jamal Deen is Distinguished University Professor and Director of the Micro-and Nano-Systems Laboratory at McMaster University. His current research interests are nanoelectronics, optoelectronics, nanotechnology, data analytics and their emerging applications to health and environmental sciences. As an educator, he won the Ham Education Medal from IEEE Canada, the McMaster University President's Award for Excellence in Graduate Supervision, and MSU Macademics' Lifetime Achievement Award for his exceptional dedication to teaching and significant contribution to student life, academia, and the community at large.

As an undergraduate student at the University of Guyana, Dr. Deen was the top ranked mathematics and physics student and the second ranked student at the university, winning the Chancellor's gold medal and the Irving Adler prize. As a graduate student, he was a Fulbright-Laspau Scholar and an American Vacuum Society Scholar. His awards and honors include the Callinan Award as well as the Electronics and Photonics Award from the Electrochemical Society; a Humboldt Research Award from the Alexander von Humboldt Foundation; the Eadie Medal from the Royal Society of Canada; and the McNaughton Gold Medal IEEE Canada. In addition, he was awarded the five honorary doctorate degrees in recognition of his exceptional research, scholarly and education accomplishments, exemplary professionalism and valued services.

Dr. Deen has been elected by his peers as Fellow/Academician of thirteen national academies and professional societies including The Royal Society of Canada - The Academies of Arts, Humanities and Sciences (the highest honor for academics, scholars and artists in Canada), the Chinese Academy of Sciences (China's highest national honor in the area of science and technology and highest academic title), the 4National Academy of Sciences India, the Canadian Academy of Engineering, IEEE, APS (American Physical Society) and ECS (Electrochemical Society). He served as the elected President of the Academy of Science, The Royal Society of Canada in 2015-2017. Most recently (Nov 2022), he was elected the inaugural Vice President (North) of The World Academy of Sciences, representing the developed countries. He was also elected to the Order of Canada, the highest civilian honor awarded by the Government of Canada.

Abstract

In the past few decades, the field of flexible organic/polymeric electronics has advanced significantly. This has been primarily because of the developments of new materials, improvements in the quality organic/polymeric materials, as well as the processing techniques, technologies and device designs. For example, roll-to-roll, sheet-to-sheet or printing technologies are being proposed as suitable manufacturing candidates because they can be carried out at room temperature, do not require the kind of clean room environment needed for traditional semiconductor manufacturing, and are very suitable for very low-cost, high volume production. Further, these advances are mostly stimulated by the promise of lighter and more robust devices and systems for applications that include large-area electronics, active matrix large-area displays, large-area solar cells, interactive displays, and conformable sensors and actuators. However, despite these advances, there remain challenges in the large-scale transfer of research prototypes into manufactured products. Furthermore, a major the limitation is the lack of accurate and computationally efficient compact models for organic/polymeric thin film transistors with associated parameter extraction techniques. In this presentation, we will discuss recent compact models and illustrate the merits and limitations of several of them as part of the electronic design automation platform. In this presentation, we will discuss our progress in developing industry-viable static and dynamic compact models for flexible transistors with predictable performance and the associated parameter extraction schemes including evolutionary computation for parameter extraction. Finally, we will present several on-going modeling challenges including illumination, hysteresis and contacts effects, as well as models that can predict stability, reliability, and lifetime.

Yi Luo

Professor, Chinese Academy of Engineering

Tsinghua University, China

Title of the Keynote Report

Optoelectronic information technology: review & prospect

Dr. Luo Yi, born in February 1960, is a professor in the Department of Electronic Engineering at Tsinghua University. He obtained a bachelor's degree from Tsinghua University in 1983, and master's and doctoral degrees from the University of Tokyo, Japan, in 1987 and 1990, respectively. Since 1992, he has been employed as a professor in the Department of Electronic Engineering at Tsinghua University. In 1995, he was funded by the National Science Fund for Distinguished Young Scholars, and in 1999, he was appointed as a Distinguished Professor of the "Changjiang Scholars Award Program" of the Ministry of Education. From 1997 to 2012, he served as the Director of the State Key Joint Laboratory of Integrated Optoelectronics for three consecutive terms. Currently, he is the Deputy Director of the Beijing National Research Center for Information Science and Technology, and the Convener of the Discipline Appraisal Group for Electronic Science and Technology of the Academic Degrees Committee of the State Council. In 2021, he was elected as an academician of the Chinese Academy of Engineering.

His main research focuses on compound semiconductor optoelectronic devices and their integrated application technologies, including lasers, LEDs, optical modulators, photodetectors, and their applications in fields such as optical fiber communication, broadband information perception, and semiconductor lighting. He has published 367 academic papers and authorized 34 invention patents. He has won 3 Second Prizes of the National Technical Invention Award and 1 Second Prize of the National Science and Technology Progress Award.

Abstract

Optoelectronic technology has developed rapidly since the 1980s. Achievements such as blue LEDs and semiconductor heterojunction lasers have won the Nobel Prize in Physics, and high-speed optical fiber networks have become the infrastructure of the information society. At present, with the growth of data, the gap between chip processing capability and the required computing power continues to widen, and high-speed optoelectronic chips are key elements for computing power construction. The team from Tsinghua University has made fruitful achievements in the field of optoelectronics. It took the lead in the research and development of gain-coupled distributed feedback (DFB) lasers in the world, achieving excellent spectral purity and high single-mode yield. The team has a clear goal: by 2025, it will break through the key technologies of high-speed optoelectronic chips, realize domestic substitution and high-end breakthroughs, and build an industrial ecosystem to cope with external restrictions. In addition, China's semiconductor lighting industry has developed rapidly, with its output value increasing by 152 times from 2003 to 2018. The global market for photodetection-related industries is expected to grow significantly from 2015 to 2025, and optoelectronic technology has a broad prospect.

Kiat Seng Yeo

Professor, IEEE Fellow, Academician of the Singapore Academy of Engineering

Tianjin University, China

Title of the Keynote Report

The Semiconductor Evolution: From 4 C's to 4 I's

Professor Kiat Seng YEO (M'00–SM'09–F'16) received his B.Eng. (EE) in 1993, and Ph.D. (EE) in 1996 both from Nanyang Technological University (NTU), Singapore. Currently, he is full professor, Advisor (Global Partnerships) and Director for Innovation and Enterprise (China) at Singapore University of Technology and Design (SUTD) and a distinguished professor at Tianjin University. He was Chairman of the University Research Board, Member of the Academic Council, Associate Provost for Research, Founding Associate Provost for Graduate Studies and Founding Associate Provost for International Relations at SUTD. He has over 30 years of experience in industry, academia, and consultancy. Before joining SUTD, he was full professor at NTU; and spent 13 years in management positions as Associate Chair (Research), Head of Circuits and Systems and Sub-Dean (Students Affairs). Professor Yeo was also a Fellow of the Renaissance Engineering Programme (REP) and served as Senator and Advisory Board Member at NTU. He has made many outstanding contributions to advance Singapore's education and research ambitions over the course of his career. As the Founding Director of VIRTUS, a S$52 million IC Design Centre of Excellence jointly set up by NTU and Singapore Economic Development Board (EDB), he contributed extensively to the economic development of integrated circuit design in Singapore by leading multidisciplinary research, with a focus on industry collaboration. In 2016, he initiated the FIRST (Fostering Industrial Research Success Together) Industry Workshop at SUTD. Today, it is a flagship event with an attendance of over 1,000 professionals from the industry.

Since 1996, Professor Yeo has been providing consultancy services to statutory boards, local SMEs, and multinational corporations in the areas of electronics and IC design. As Principal Investigator, he secured over SGD70 million in research funding from various funding agencies and industry. He is the author of 14 books, 7 book chapters and has published over 650 top-tier refereed journal and conference papers in his area of research and holds 55 patents, including 2 patents for the world's smallest integrated transformer, a patent for the world's smallest integrated filter for 60GHz standard, the inventor of several high Q-factor RF spiral inductors and co-inventor of quite a few novel circuit techniques for 5G/6G wireless communication and 7RF/mm-wave IC applications.

Professor Yeo is a world-renowned expert in low-power RF/mm-wave integrated circuit design and a recognized expert in CMOS technology. He holds/held positions such as advisor, chair, co-chair, and technical chair at many international conferences. He was awarded the Public Administration Medal (Bronze Award) by the President of the Republic of Singapore on National Day 2009 and was awarded the Nanyang Alumni Achievement Award in 2009 in recognition of his outstanding contributions to the University and society. Professor Yeo is an Academician of Singapore Academy of Engineering, an Academician of Singapore National Academy of Science, a Fellow of ASEAN Academy of Engineering and Technology, a Fellow of IEEE, a Fellow of the International Artificial Intelligence Industry Alliance (AIIA), and a Fellow of the Asia-Pacific Association for Artificial Intelligence (AAIA). He is the principal author of Integrated Circuit Design Research Ranking for Worldwide Universities 2008 and World University Research Rankings (WURR) 2020. Professor Yeo was recognized among the World's top 2% Scientists by Stanford University from 2020 to 2024, World's AI Top Scientist by AIIA in 2023, and Top Scholar by ScholarGPS in 2024.

Abstract

The semiconductor industry market will grow from US$600 billion today to over US$1 trillion by 2030. Traditionally, the growth of semiconductor industry is dominated by the 4 C's: Computers, Communications, Consumers, and Cars. But its growth will now come from the 4 I's: Intelligence, Integration, Innovation, and Interdisciplinary. The combination of these I's and C's will result in more ICs. Therefore, in addition to new energy, drones and smart cars, ICs will also be widely used in the human body, robots, future communications, biomedical, aerospace and many other fields. The invention of the transistor in 1948 at Bell Laboratories was a turning point in the history of electronics and IT. The transistor promises to revolutionize electronics and IT; indeed, it has become an integral and essential part of our life. This keynote begins with an overview of the semiconductor industry market, an introduction to early computers, the invention of integrated circuits and how it changes the electronics and IT industry. As integrated circuits continue to evolve, it is important to know the forces that have driven it along its historical trajectory, and to discover how much further it could go. Can integrated circuits keep pace with Moore's Law? Is it still evolving or has it come to an end? What is the next big thing? How is it going to affect us? This keynote will attempt to answer these questions. As a conclusion, the other trends to watch for the next revolution will be presented.

Defects engineering for efficient halide perovskite solar cells

Prabakaran Selvaraj, Tapa Arnauld Robert, Haibing Xie*
Institute for Advanced Study
Shenzhen University
Shenzhen 518060, P. R. China
xhbxal2021@szu.edu.cn

Abstract—The power conversion efficiency of single-junction perovskite solar cells has passed 26%, approaching their Shockley-Queisser theoretic limit. It is necessary to develop tandem solar cells for higher efficiency. Wide bandgap perovskite solar cells (1.65-1.7 eV) are ideal top cells for perovskite/Si and perovskite/Cu(In,Ga)Se$_2$ (CIGS) tandem solar cells. Therefore, in this work, we developed efficient wide bandgap perovskite solar cells through defects engineering via interface and bulk optimization. This work will pave a way for the development of highly efficient and stable perovskite based tandem solar cells with a low cost.

Keywords—Perovskite solar cells, wide bandgap, defects engineering

I. INTRODUCTION

Photovoltaic technology has significantly advanced in the laboratory and field in the last two decades. According to National Renewable Energy Laboratory (NREL), the power conversion efficiency (PCE) of the world-record perovskite solar cells of 26.95% is approaching to the Shockley-Queisser (S-Q) limit (~33%)[1,2]. New technology routes are urgently needed to overcome the S-Q limit of single-junction cells. Constructing a tandem solar cell with minimal thermalization losses may offer a theoretical PCE over 40%[3]. For instance, tandems with one wide bandgap ideal top sub-cell with a bandgap of 1.65-1.7 eV (perovskite) and another bottom ideal sub-cell with a narrow bandgap of 1.1 eV (Si or CIGS) can harvest a wide range of solar spectrum. The stacking of two different sub-cells can realize the sum of the individual open-circuit voltage (V_{OC}) and preserve high short-circuit current (J_{SC}) of photoactive materials. Here, we focus on the wide bandgap perovskite (WBG) solar cells and try to achieve maximum efficiency by defect engineering via modifying the passivation layer (PL) on perovskite surface, the additives inside perovskite, and the hole transporting layer self-assembled monolayers (SAMs). The appropriate functional groups of the passivation layer can target specific defect types of perovskite surface, leading to enhanced V_{OC} [4, 5]. The additives can reduce the ion migration, mitigate defects, and improve perovskite film quality via coordination bonding and Lewis acid-base interaction[6]. The few layers of organic molecule SAMs potentially anchor to the Indium-doped tin oxide (ITO) substrate's surface, following a favorable thermodynamic self-assembly process for defects passivation and hole extraction[7].

II. EXPERIMENTAL DETAILS

A. Materials

Cesium iodide (CsI), Formamidine iodide (FAI), Lead iodide (PbI$_2$), Lead bromide (PbBr$_2$), Phenyl-C61-butyric acid methyl ester (PCBM), isopropanol (IPA), chlorobenzene (CB), Dimethyl sulfoxide (DMSO), N, N-dimethylformamide (DMF) and nickel oxide (NiO$_x$) were purchased from the Advanced Election Technology. Ethane-1,2-diammonium iodide (EDAI$_2$), [4-(3,6-Dimethyl- 9H-carbazol-9-yl)butyl] (Me-4pacz), (4-(7H-dibenzo[c,g]carbazol-7-yl)butyl) phosphonic acid (4PADCB) and (aminomethyl)phosphonic acid (AMP) were purchased from Xi'an polymers. All chemicals were used as received without further purification.

B. Wide-bandgap perovskite precursor preparation

1M of FA$_{0.8}$Cs$_{0.2}$Pb(I$_{0.8}$Br$_{0.2}$)$_3$ perovskite precursor solution was prepared by dissolving 165 mg of FAI, 62.3 mg of CsI, 387.2 mg of PbI$_2$ and 132.1 mg of PbBr$_2$ in 1 mL mixed solvent of anhydrous N, N-dimethylmethanamide (DMF) and anhydrous dimethyl sulfoxide (DMSO) with a volume ratio of 3:1. Then 7.76 mg Pb(SCN)$_2$ and 1 mg of (aminomethyl)phosphonic acid (AMP) was added into the solution. The perovskite precursor solution was thoroughly stirred for 12 hours before use.

C. Device fabrication

The ITO glass substrates were cleaned with soap followed by sonicated in DI water and ethanol for 15 min in sequence. Then, we treated the dried substrates with UV-Ozone for 15 min before use. Then, the NiO$_x$ ink (10 mg/mL in DI water) was spin-coated at 2000 rpm for 60 s and annealed at 150 °C for 20 min in air. Then, we transferred the NiO$_x$ coated substrates to the glove box after the substrates cooled to room temperature. 1 mg/mL Me-4pacz, or 1 mg/ml 4PADCB, or mixed SAMs (dissolved in IPA) were spin-coated at 4000 rpm for 40 s and annealed at 100 °C for 10 min. The WBG perovskite solution was spin-coated at 500 rpm for 2 s and 4000 rpm for 60 s with 200 µL chlorobenzene (CB) dripped at 35 s in the second spinning step. We annealed the as-deposited films at 65 °C for 5 min and 100 °C for 15 min. The 1 mg/mL EDAI$_2$ solution (dissolved in IPA and stirred at 80 °C for 4 hours before use) was spin-coated on the perovskite top surface at 4000 rpm for 60 s and then annealed at 65 °C for 5 min. 20 mg/mL PCBM (dissolved in CB) was spin-coated on the perovskite top surface

979-8-3315-2209-4/25 $31.00 © 2025 IEEE

at 1500 rpm for 35 s and then annealed at 100 °C for 10 min. 0.5 mg/mL BCP solution (dissolved in IPA) was spin-coated on the PCBM layer at 4000 rpm for 30 s. Finally, 100 nm Ag was deposited using thermal evaporation.

D. Characterizations

I-V curves were tested at room temperature using a Keithley2400 source meter under AM 1.5 G at 100 mW cm^{-2} by a solar simulator (SS-300A-D, Zolix). The active area of the tested solar cells is 0.04 cm^2, and applied bias voltage ranges from 0 to 1.3 V for WBG perovskite solar cells. IPCE spectra were measured using a commercialized QE system (Zolix SCS600 series), the monochromatic light intensity of which was calibrated with a standard silicon diode.

III. RESULTS AND DISCUSSION

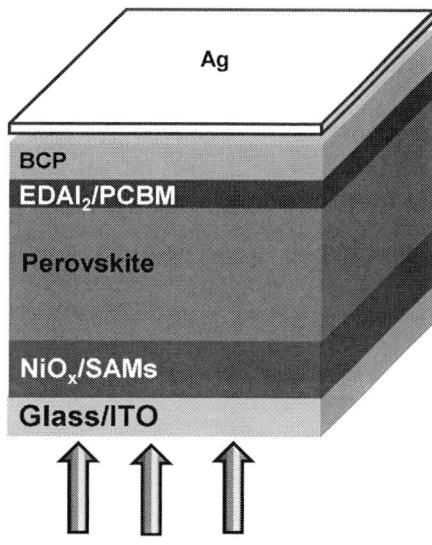

Fig. 1. Schematic diagram of wide-bandgap perovskite solar cells

Here, we selected the FA$_{0.8}$Cs$_{0.2}$Pb(I$_{0.8}$Br$_{0.2}$) compound as the wide bandgap perovskite (1.65 eV), and the device structure is Glass/ITO/NiO$_x$/SAMs/Perovskite/PL/PCBM/BCP/Ag (Fig. 1). The first step is fabricating the WBG perovskite solar cells and checking the efficiency using the above device structure. We achieved the maximum efficiency of 15.8% in the control device (without passivation layer), and 19.3% efficiency in the ethane-1,2-diammonium iodide (EDAI$_2$) surface passivated device (Fig. 2).

The EDAI$_2$ passivated device performed better than the control device could be due to the passivation of uncoordinated Pb^{2+} and field passivation of the perovskite surface [8]. The second optimization stage involves additives to improve crystallization and reduce the perovskite's defect formation. Here, we selected the (aminomethyl)phosphonic acid (AMP) as an additive in the perovskite [9].

Fig. 3 shows the efficiency of control devices and devices with different concentrations of AMP. The AMP concentrations

of 0.5 and 1.0 mg increased the efficiency from 18.0% to 19.8 % and 20.2% by improving the V$_{oc}$ and FF. Further increasing the

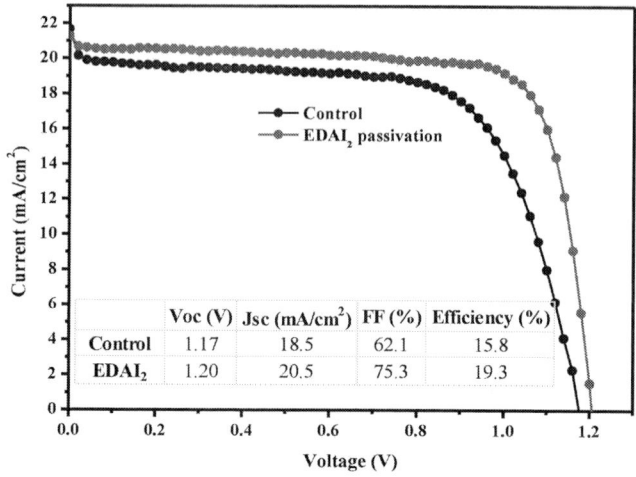

	Voc (V)	Jsc (mA/cm^2)	FF (%)	Efficiency (%)
Control	1.17	18.5	62.1	15.8
EDAI$_2$	1.20	20.5	75.3	19.3

Fig. 2. I-V curves of control and EDAI$_2$ passivated perovskite solar cells

concentration of AMP to 3.5 mg and 5.0 mg decreased the efficiency to 18.5% and 16.9%, respectively.

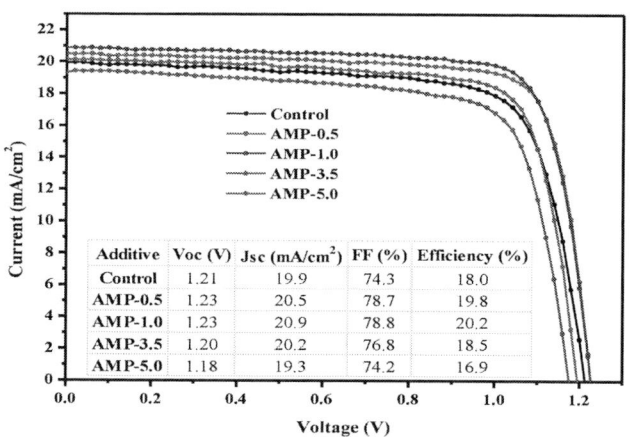

Additive	Voc (V)	Jsc (mA/cm^2)	FF (%)	Efficiency (%)
Control	1.21	19.9	74.3	18.0
AMP-0.5	1.23	20.5	78.7	19.8
AMP-1.0	1.23	20.9	78.8	20.2
AMP-3.5	1.20	20.2	76.8	18.5
AMP-5.0	1.18	19.3	74.2	16.9

Fig. 3. I-V curves of control and AMP additive modified perovskite solar cells

The third optimization stage is the SAMs layer, which improves the hole transport and reduces the interface recombination between the NiO$_x$ and perovskite layers. We selected two different SAMs, [4-(3,6- Dimethyl-9H-carbazol-9-yl) butyl] (Me-4pacz), (4-(7H-dibenzo[c,g]carbazol-7-yl)butyl) phosphonic acid (4PADCB). Me-4Pacz leads to higher V$_{OC}$ while 4PADCB provides better FF. Once we mixed Me-4pacz and 4PADCB with different ratios such as 1:1, 2:1, and 5:1, the device's overall efficiency improved due to enhanced V$_{OC}$ and FF. Fig. 4 shows the IV curves of Me-4pacz, 4PADCB, and different ratios of mixed SAMs. Among all, the 5:1 ratio of mixed SAMs showed the best efficiency of 21.6%. Fig. 5 shows the Incident Photon-to-Current Efficiency (IPCE) curves of Me-4pacz, 4PADCB, and mixed SAMs; among them, the ratio of

5:1 shows better IPCE and the integrated current matches well with the J_{SC}. The calculated bandgap of the perovskite solar cells is 1.65 eV using the absorption edge of the IPCE curves.

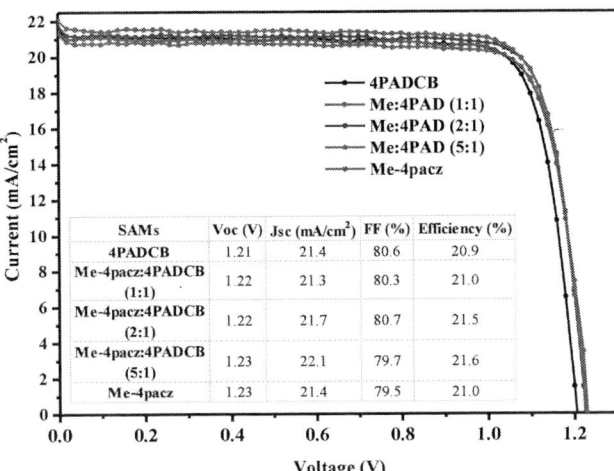

SAMs	Voc (V)	Jsc (mA/cm^2)	FF (%)	Efficiency (%)
4PADCB	1.21	21.4	80.6	20.9
Me-4pacz:4PADCB (1:1)	1.22	21.3	80.3	21.0
Me-4pacz:4PADCB (2:1)	1.22	21.7	80.7	21.5
Me-4pacz:4PADCB (5:1)	1.23	22.1	79.7	21.6
Me-4pacz	1.23	21.4	79.5	21.0

Fig. 4. I-V curves of SAMs modified perovskite solar cells

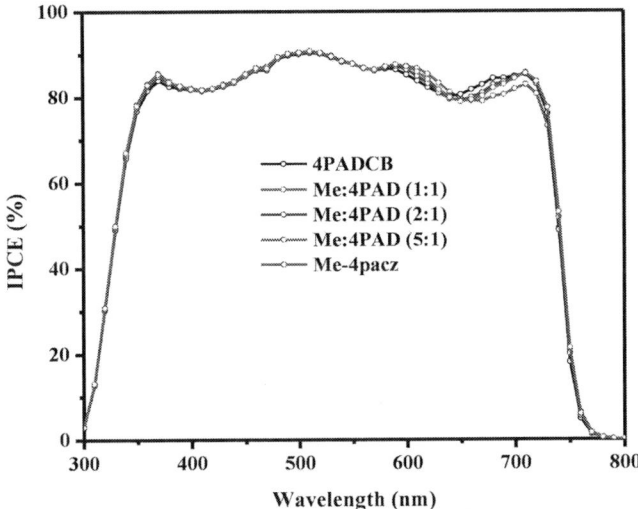

Fig. 5. IPCE of SAMs modified perovskite solar cells

IV. CONCLUSION

In conclusion, we have developed highly efficient wide bandgap perovskite solar cells with efficiency more than 21% by defects engineering via optimization of passivation layer, additives and SAMs. Further work is needed to improve the V_{OC} of WBG perovskite solar cells by minimizing interfacial and bulk defects and reducing non-radiative recombination. Deployment of WBG perovskite solar cells in perovskite/Si and perovskite/CIGS tandems is necessary to surpass the S-Q limit of single-junction perovskite solar cell.

ACKNOWLEDGMENT

H.X. thanks the start-up fund of Pengcheng Peacock Project and the start-up fund of Shenzhen University (no. 000002112129).

REFERENCES

[1] NREL, Best Research-Cell Efficiency Chart, https://www.nrel.gov/pv/cellefficiency.html.

[2] W. Shockley and H.J. Queisser, "Detailed balance limit of efficiency of p-n junction solar cells," J. Appl. Phys., vol. 32, pp. 510-519, 1961.

[3] T. Leijtens, K. A. Bush, R. Prasanna and M. D. McGehee, "Opportunities and challenges for tandem solar cells using metal halide perovskite semiconductors," Nat. Energy, vol. 3, pp. 828-838, 2018.

[4] J. Xu, H. Chen, L. Grater, C. Liu, Y. Yang, S. Teale, et al., "Anion optimization for bifunctional surface passivation in perovskite solar cells," Nat. Mater., vol. 22, pp. 1507-1514, 2023.

[5] H. Zhang, L. Pfeifer, S. M. Zakeeruddin, J. Chu and M. Grätzel, "Tailoring passivators for highly efficient and stable perovskite solar cells," Nat. Rev. Chem., vol. 7, pp. 632-652, 2023.

[6] F. Zhang and K. Zhu, "Additive engineering for efficient and stable perovskite solar cells," Adv. Energy Mater., vol. 10, p. 1902579, 2020.

[7] Z. Yi, X. Li, Y. Xiong, G. Shen, W. Zhang, Y. Huang, et al., "Self-assembled monolayers (SAMs) in inverted perovskite solar cells and their tandem photovoltaics application," Interdiscip. Mater., vol. 3, pp. 203-244, 2024.

[8] S. Li, Z. Zheng, J. Ju, S. Cheng, F. Chen, Z. Xue, et al., "A generic strategy to stabilize wide bandgap perovskites for efficient tandem solar cells," Adv. Mater., vol. 36, p. 2307701, 2024.

[9] Y. Zheng, C. Tian, X. Wu, A. Sun, R. Zhuang, C. Tang, et al., Dual-interface modification for inverted methylammonium-free perovskite solar cells of 25.35% efficiency with balanced crystallization, Adv. Energy Mater., vol. 14, p. 2304486, 2024.

Joint Process Optimization of IWZO Thin-Film Transistor for High-Resolution Active-Matrix Displays

Bin Wang, Feilian Chen, Yan Yan, *Meng Zhang
State Key Laboratory of Radio Frequency Heterogeneous Integration and College of Electronics and
Information Engineering, Shenzhen University, Shenzhen, China
Phone: 0086 755 26925737, *Email: zhangmeng@connect.ust.hk

Abstract—**This study performed a combined optimization of the processing conditions for indium tungsten zinc oxide (IWZO) thin film transistors (TFTs), optimizing multiple parameters, including the argon-oxygen ratio, active layer thickness, annealing temperature, and annealing time. As a result, high-performance IWZO TFTs were successfully fabricated. The device has a mobility of 85.01 cm^{2}V^{-1}s^{-1}, a threshold voltage of –2.8 V, and a subthreshold swing of 0.129 V/dec. After negative bias stress and positive bias stress testing, the threshold voltage shifts were –3.4 V and +0.5 V, respectively. The results demonstrate that IWZO TFTs are a viable thin-film transistor material, offering a promising direction for the fabrication of high-resolution active-matrix display devices.**

Keywords—**IWZO, joint process optimization, thin-film transistors**

I. INTRODUCTION

Metal oxide (MO) thin-film transistors (TFTs) are widely used in active matrix (AM) displays due to their high mobility and excellent stability [1], [2], [3], [4], [5], [6], [7], [8], [9], [10], [11], [12], [13]. However, the increasing demand for high-quality display panels has placed higher requirements on the mobility and reliability of TFTs [9], [14], [15], [16], [17], [18], [19]. High-resolution displays require TFTs with high mobility to ensure fast response times and high refresh rates, while also demanding excellent stability to ensure long-term, reliable operation [18], [20], [21], [22], [23]. Thus, it becomes critical to balance high mobility with long-term stability in device performance.

Traditional IGZO TFTs have been extensively used due to their superior stability. However, as the demand for high-resolution displays grows, the relatively low electron mobility of IGZO has become increasingly apparent, coupled with its higher production costs compared to alternative materials [8], [17], [23], [24], [25], [26]. IZO TFTs, on the other hand, offer higher mobility, good transparency, and photonic response, but their stability is relatively poor and they are susceptible to environmental light, making them less suitable for high-resolution applications [3], [27], [28], [29], [30], [31]. ITZO TFTs, which incorporate tin, show some improvement in stability, but studies indicate that ITZO is sensitive to water vapor and other environmental factors [32], [33], [34], [35], [36], [37]. As a result, finding new active layer materials has become an important trend in the field. IWZO TFTs, by introducing tungsten, significantly improve the device stability while maintaining high electron mobility, and the fabrication cost of IWZO is relatively low [38], [39], [40].

This study focuses on the process optimization of IWZO TFTs. By optimizing various parameters such as the argon-oxygen ratio, active layer thickness, annealing temperature, and annealing time, high-performance IWZO TFTs were successfully fabricated. The devices exhibited a mobility (μ_{FE}) of 85.01 cm^{2}V^{-1}s^{-1}, a threshold voltage (V_{th}) of –2.8 V, and a

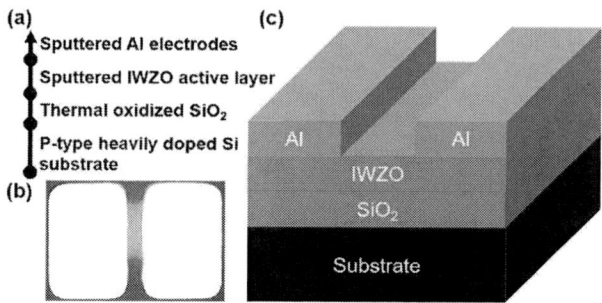

Fig. 1 (a) Process flow, (b) top view, and (c) schematic diagram of IWZO TFTs.

subthreshold swing (SS) of 0.129 V/dec. Under Negative Bias Stress (NBS) and Positive Bias Stress (PBS) testing, the threshold voltage shifts (ΔV_{th}) were –3.4 V and +0.5 V, respectively. This study offers a hopeful path for the production of high-resolution AM display devices.

II. EXPERIMENTAL

Fig. 1 illustrates the process flow, device top view, and device structure used in this study. RF magnetron sputtering technology was employed to deposit a 40 nm thick IWZO film on a heavily doped p-type silicon substrate coated with a 100 nm SiO$_2$ layer, using co-sputtering targets of IZO (In$_2$O$_3$:ZnO = 90:10 wt%) and WO$_3$ (99.99% purity). The film thickness was achieved by precise control of the sputtering time and characterized by a step meter. During the sputtering process, the powers of IZO and WO$_3$ were 60 W and 5 W, respectively, with an argon/oxygen (Ar/O$_2$) flow ratio of 20/15 sccm. Subsequently, a 100 nm thick aluminum source/drain electrode was deposited using DC magnetron sputtering in a pure Ar gas environment. Finally, the devices were annealed for 10 minutes in ambient air at 200 °C.

The width/length (W/L) for the TFT channel is 500 μm/200 μm. The μ_{FE} is derived from the following equation [41]:

$$\mu_{FE} = (W/L)(g_m/(V_{ds}C_{OX})) \qquad (1)$$

where g_m and C_{ox} represent the maximum device transconductance at V_{DS} = 0.1 V and gate insulator capacitance per unit area, respectively. The V_{th} is defined as the V_{gs} corresponding to I_{ds} of 10 nA when V_{ds} is equal to 0.1 V. The reliability of the device is measured by comparing the extent of V_{th} shift under NBS and PBS conditions, denoted as ΔV_{th}. ΔV_{th} represents the difference between V_{th} values on the transfer characteristic curve before and after stress. For NBS and PBS, the stress V_g is fixed at –20 V and +20 V, respectively, and the source and drain are grounded. Measurements and stress tests of the devices were conducted using a Keysight B1500A semiconductor parameter analyzer connected to a dark probe station. To ensure the reliability of the data, we conducted five replicate tests under each experimental condition.

979-8-3315-2209-4/25 $31.00 © 2025 IEEE

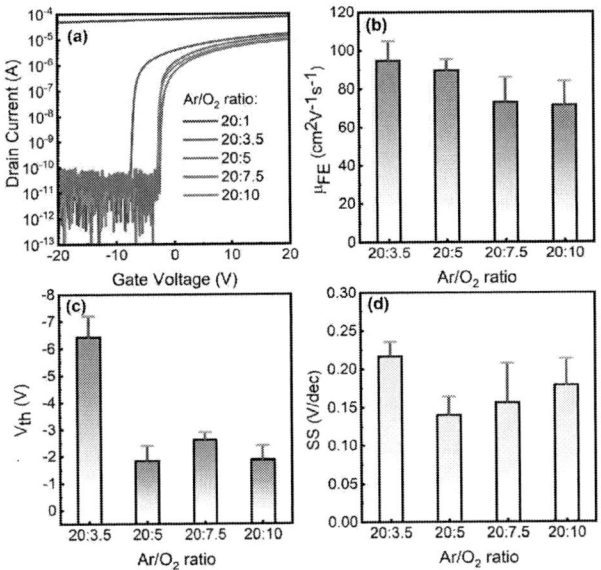

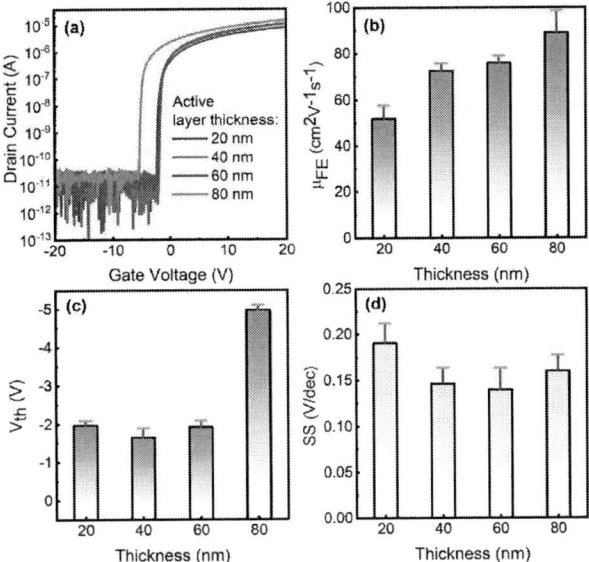

Fig. 2 (a) Transfer curves, (b) μ, (c) V_{th}, (d) SS of IWZO TFTs with different argon to oxygen ratios.

Fig. 3 (a) Transfer curves, (b) μ, (c) V_{th}, (d) SS of IWZO TFTs with active layer thicknesses.

III. RESULTS AND DISCUSSION

Fig. 2a shows the transfer characteristics of IWZO TFTs fabricated under different argon-oxygen ratios, with the ratio ranging from 20:1 to 20:10. The results indicate that the variation in the argon-oxygen ratio significantly affects the device's μ_{FE}, V_{th}, and SS. Fig. 2b illustrates the μ_{FE} of IWZO TFTs at different argon-oxygen ratios. As the oxygen content increases, the μ_{FE} gradually decreases. This trend can be attributed to the increased oxygen content, which fills more oxygen vacancies and consequently reduce the number of carriers. Fig. 2c and d show the changes in V_{th} and SS for IWZO TFTs under different argon-oxygen ratios. It can be observed that when the argon-oxygen ratio is 20:5, both V_{th} and SS reach their minimum value, while the μ_{FE} remains relatively high.

Based on an argon-oxygen ratio of 20:5, IWZO TFTs with different active layer thicknesses were fabricated, as shown in Fig. 3a. The active layer thickness varied from 20 nm to 80 nm. Fig. 3b presents the μ_{FE} of IWZO TFTs with different active layer thicknesses. As the active layer thickness increases, the μ_{FE} gradually improves. Within a certain range, a thicker active layer provides more carriers, thereby enhancing μ_{FE}. Figs. 3c and d show the V_{th} and SS of IWZO TFTs with varying active layer thicknesses. It is observed that both V_{th} and SS initially decrease and then increase with the increase in active layer thickness, reaching optimal values at 40 nm and 60 nm, respectively. Considering the combined effects on μ_{FE}, V_{th}, and SS, it can be concluded that the device performance is optimal with an active layer thickness of 40 nm.

Based on an argon-oxygen ratio of 20:5 and an active layer thickness of 40 nm, the influence of annealing temperature on IWZO TFTs was investigated. The annealing process was conducted in air, with temperatures ranging from 100 °C to 400 °C, and a fixed annealing time of 10 minutes, as shown in Fig. 4a. At an annealing temperature of 400 °C, the device loses its switching characteristics. As the annealing temperature increases, the μ_{FE} gradually improves, but the V_{th} exhibits a negative shift, as shown in Fig. 4b. Figs. 4c and d

show the V_{th} and SS of IWZO TFTs at different annealing temperatures. With increasing annealing temperature, V_{th} shifts negatively, while SS initially decreases and then increases. The optimal values for μ_{FE}, V_{th}, and SS are achieved at an annealing temperature of 200 °C.

Based on an argon-oxygen ratio of 20:5, an active layer thickness of 40 nm, and an annealing temperature of 200 °C, the influence of annealing time on the performance of IWZO TFTs was investigated. Fig. 5a shows the transfer characteristics for annealing times of 10, 20, 30, and 60 minutes. After considering the data presented in Figs. 5a, b, and c, it is concluded that the optimal performance, in terms of μ_{FE}, V_{th}, and SS, is achieved with an annealing time of 10 minutes.

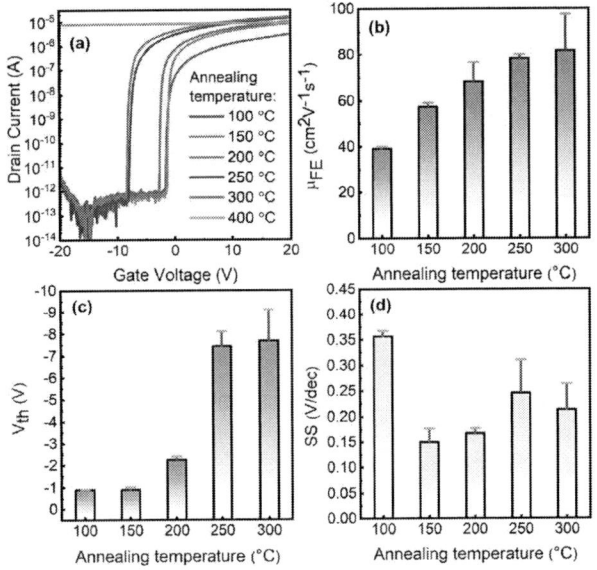

Fig. 4 (a) Transfer curves, (b) μ, (c) V_{th}, (d) SS of IWZO TFTs with different annealing temperatures.

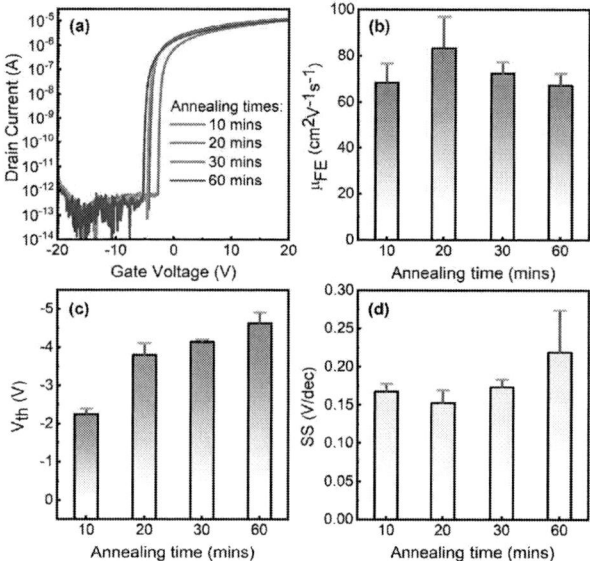

Fig. 5 (a) Transfer curves, (b) μ, (c) V_{th}, (d) SS of IWZO TFTs with different annealing times.

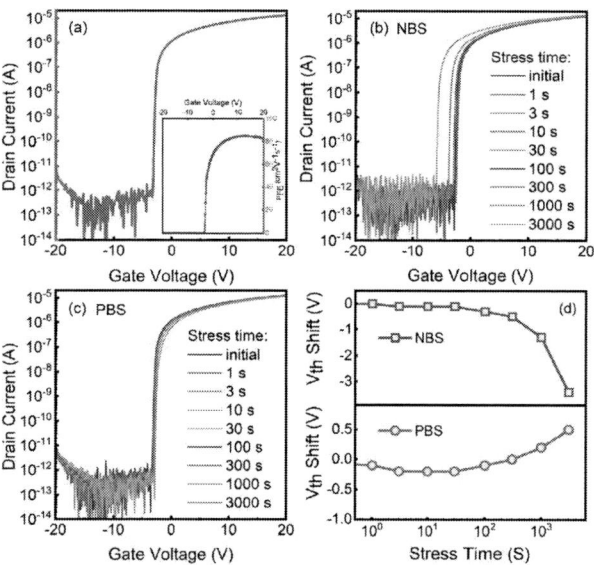

Fig. 6 (a) Optimized transfer curves for IWZO TFTs. Degradation of transfer curves of IWZO TFTs under (b) NBS and (c) PBS. (d) V_{th} shift dependent on stress time in IWZO TFTs under NBS and PBS.

The device with optimal performance was obtained through the aforementioned combined optimization, and its transfer characteristics are shown in Fig. 6a, with the inset displaying the calculated field-effect μ_{FE} at different gate voltages. The device exhibited a μ_{FE} of 85.01 cm²V⁻¹s⁻¹, a V_{th} of −2.8 V, and a SS of 0.129 V/dec. Figs. 6a and b show the test results of IWZO TFTs under NBS and PBS. Fig. 6d illustrates the variation of V_{th} with stress time. Under NBS and PBS testing, the ΔV_{th} was −3.4 V and +0.5 V, respectively.

IV. CONCLUSION

In summary, this study successfully fabricated high-performance IWZO TFTs through combined optimization. The reliability of the IWZO TFTs was evaluated through NBS and PBS tests. The research results show that IWZO TFTs are a new type of thin film transistors with good performance and high application potential. This research provides a promising direction for the fabrication of high-resolution AM displays devices.

V. ACKNOWLEDGMENTS

This work was financially supported by Shenzhen Science and Technology Program (JCYJ20240813141301003, JCYJ20230808105806014), Natural Science Foundation of China (62274111, 62374110), and Independent Scientific Research Program from State Key Laboratory of Radio Frequency Heterogeneous Integration (2024015).

REFERENCES

[1] K. Nomura, H. Ohta, A. Takagi, T. Kamiya, M. Hirano, and H. Hosono, "Room-temperature fabrication of transparent flexible thin-film transistors using amorphous oxide semiconductors," *Nature*, vol. 432, no. 7016, pp. 488–492, Nov. 2004, doi: 10.1038/nature03090.

[2] H. Hosono, "How we made the IGZO transistor," *Nat Electron*, vol. 1, no. 7, pp. 428–428, Jul. 2018, doi: 10.1038/s41928-018-0106-0.

[3] D. C. Paine, B. Yaglioglu, Z. Beiley, and S. Lee, "Amorphous IZO-based transparent thin film transistors," *Thin Solid Films*, vol. 516, no. 17, pp. 5894–5898, Jul. 2008, doi: 10.1016/j.tsf.2007.10.081.

[4] M. R. Moon *et al.*, "Microstructure and electrical properties of XInZnO (X=Ti, Zr, Hf) films and device performance of their thin film transistors—The effects of employing Group IV-B elements in place of Ga," *Journal of Alloys and Compounds*, vol. 563, pp. 124–129, Jun. 2013, doi: 10.1016/j.jallcom.2012.12.105.

[5] F. Chen, M. Zhang, Y. Wan, X. Xu, M. Wong, and H.-S. Kwok, "Advances in mobility enhancement of ITZO thin-film transistors: a review," *J. Semicond.*, vol. 44, no. 9, p. 091602, Sep. 2023, doi: 10.1088/1674-4926/44/9/091602.

[6] M. Zhang *et al.*, "Hot Carrier Degradation Accompanied by Recovery in InSnZnO Thin-Film Transistors," *IEEE Electron Device Lett.*, vol. 44, no. 7, pp. 1124–1127, Jul. 2023, doi: 10.1109/LED.2023.3277823.

[7] B. Wang, F. Chen, G. Zhu, L.-C. Chen, and M. Zhang, "Influence of N₂ Gas Flow on the Reliability of Indium Zinc Oxide Thin-Film Transistors," in *2024 IEEE International Symposium on the Physical and Failure Analysis of Integrated Circuits (IPFA)*, Singapore, Singapore: IEEE, Jul. 2024, pp. 1–5. doi: 10.1109/IPFA61654.2024.10690973.

[8] X. Xu *et al.*, "Realization of Visible-Light Detection in InGaZnO Thin-Film Transistor via Oxygen Vacancy Modulation through N2 Treatments," *Advanced Electronic Materials*, vol. 9, no. 10, p. 2300351, 2023, doi: 10.1002/aelm.202300351.

[9] G. Zhu *et al.*, "Significant Degradation Reduction in Metal Oxide Thin-Film Transistors via the Interaction of Ionized Oxygen Vacancy Redistribution, Self-Heating Effect, and Hot Carrier Effect," *IEEE Trans. Electron Devices*, vol. 70, no. 8, pp. 4198–4205, Aug. 2023, doi: 10.1109/TED.2023.3283940.

[10] L. Lu, J. Li, Z. Feng, H. S. Kwok, and M. Wong, "Elevated-Metal–Metal-Oxide Thin-Film Transistor: Technology and Characteristics," *IEEE Electron Device Letters*, vol. 37, no. 6, pp. 728–730, Jun. 2016, doi: 10.1109/LED.2016.2552638.

[11] F. Chen, M. Zhang, Y. Wan, X. Xu, M. Wong, and H.-S. Kwok, "Advances in mobility enhancement of ITZO thin-film transistors: a review," *J. Semicond.*, vol. 44, no. 9, p. 091602, Sep. 2023, doi: 10.1088/1674-4926/44/9/091602.

[12] Z. Jiang. M. Zhang, S. Deng, Y. Yang, M. Wong, and H.-S. Kwok, "Evaluation of Positive-Bias-Stress-Induced Degradation in InSnZnO Thin-Film Transistors by Low Frequency Noise Measurement," *IEEE Electron Device Lett.*, vol. 43, no. 6, pp. 886–889, Jun. 2022, doi: 10.1109/LED.2022.3165558.

[13] Z. Wang *et al.*, "Enhancement of Device Uniformity in IWO TFTs via RF Magnetron Co-Sputtering of In2O3 and WO3 Targets," *IEEE ELECTRON DEVICE LETTERS*, vol. 45, no. 12, 2024.

[14] Z. Chen *et al.*, "Effect of Moisture Exchange Caused by Low-Temperature Annealing on Device Characteristics and Instability in InSnZnO Thin-Film Transistors," *Advanced Materials Interfaces*, vol. 9, no. 14, p. 2102584, 2022, doi: 10.1002/admi.202102584.

[15] Z. Jiang, M. Zhang, S. Deng, M. Wong, and H.-S. Kwok, "Degradation of InSnZnO Thin-Film Transistors Under Negative Bias Stress," *IEEE Trans. Electron Devices*, vol. 70, no. 12, pp. 6381–6386, Dec. 2023, doi: 10.1109/TED.2023.3327975.

[16] X. Sun *et al.*, "High Performance Indium-Tin-Zinc-Oxide Thin-Film Transistor with Hexamethyldisilazane Passivation," *ACS Appl. Electron. Mater.*, p. acsaelm.4c00100, Apr. 2024, doi: 10.1021/acsaelm.4c00100.

[17] X. Xu *et al.*, "P-1.2: Realization of Visible Detection of InGaZnO Thin-Film Transistors by Coating (PEA)2PbI4 Light Absorbing Layer," *SID Symposium Digest of Technical Papers*, vol. 54, no. S1, pp. 441–444, 2023, doi: 10.1002/sdtp.16326.

[18] C. Peng *et al.*, "High performance amorphous IGO TFTs grown at low temperature," *physica status solidi (RRL) – Rapid Research Letters*, vol. n/a, no. n/a, doi: 10.1002/pssr.202300457.

[19] Y. Magari, T. Kataoka, W. Yeh, and M. Furuta, "High-mobility hydrogenated polycrystalline In2O3 (In2O3:H) thin-film transistors," *Nat Commun*, vol. 13, no. 1, p. 1078, Feb. 2022, doi: 10.1038/s41467-022-28480-9.

[20] K. Ide, K. Nomura, H. Hosono, and T. Kamiya, "Electronic Defects in Amorphous Oxide Semiconductors: A Review," *Physica Status Solidi (a)*, vol. 216, no. 5, p. 1800372, Mar. 2019, doi: 10.1002/pssa.201800372.

[21] Jiang B. *et al.*, "Research progress on crystalline IGZO thin film transistor," *Chinese Journal of Liquid Crystals and Displays*, vol. 38, no. 8, pp. 1031–1046, 2023, doi: 10.37188/CJLCD.2023-0121.

[22] C.-Y. Kim, "Achieving nearly 70 cm2/V·s field-effect mobility in single amorphous Ga-In-ZnO thin-film transistors deposited by sputtering process via intermediate patterned metal layer insertion," *Applied Surface Science*, 2025.

[23] H. Lee, S. Lee, A. B. Siddik, J. Lee, and J. Jang, "Improvement of Stability and Performance of Amorphous Indium Gallium Zinc Oxide Thin Film Transistor by Zinc-Tin-Oxide Spray Coating".

[24] X. Ji, Y. Yuan, X. Yin, S. Yan, Q. Xin, and A. Song, "High-Performance Thin-Film Transistors With Sputtered IGZO/Ga2O3 Heterojunction," *IEEE TRANSACTIONS ON ELECTRON DEVICES*, vol. 69, no. 12, 2022.

[25] Y.-S. Kim, T. Hwang, H.-J. Oh, J. S. Park, and J.-S. Park, "Reliability Engineering of High-Mobility IGZO Transistors via Gate Insulator Heterostructures Grown by Atomic Layer Deposition," *Advanced Materials Interfaces*, vol. n/a, no. n/a, p. 2301097, doi: 10.1002/admi.202301097.

[26] W. Cai *et al.*, "Significant Performance Enhancement of Very Thin InGaZnO Thin-Film Transistors by a Self-Assembled Monolayer Treatment," 2020.

[27] N. Ito, Y. Sato, P. K. Song, A. Kaijio, K. Inoue, and Y. Shigesato, "Electrical and optical properties of amorphous indium zinc oxide films," *Thin Solid Films*, vol. 496, no. 1, pp. 99–103, Feb. 2006, doi: 10.1016/j.tsf.2005.08.257.

[28] P.-T. Liu, Y.-T. Chou, and L.-F. Teng, "Environment-dependent metastability of passivation-free indium zinc oxide thin film transistor after gate bias stress," *Applied Physics Letters*, vol. 95, no. 23, p. 233504, Dec. 2009, doi: 10.1063/1.3272016.

[29] S. Y. Park *et al.*, "Improvement in Photo-Bias Stability of High-Mobility Indium Zinc Oxide Thin-Film Transistors by Oxygen High-Pressure Annealing," *IEEE Electron Device Lett.*, vol. 34, no. 7, pp. 894–896, Jul. 2013, doi: 10.1109/LED.2013.2259574.

[30] J. Huang *et al.*, "P-7: Enhanced Visible Light Response of Amorphous InZnO Thin-Film Transistors by Hydrogen Doping via Al2O3/SiO2 Gate Dielectric," *SID Symposium Digest of Technical Papers*, vol. 54, no. 1, pp. 1802–1805, 2023, doi: 10.1002/sdtp.16955.

[31] M. Zhang *et al.*, "Performance Enhancement of Indium Zinc Oxide Thin-Film Transistors Through Process Optimizations," *IEEE J. Electron Devices Soc.*, vol. 12, pp. 868–874, 2024, doi: 10.1109/JEDS.2024.3466956.

[32] Z. Jiang, M. Zhang, S. Deng, M. Wong, and H.-S. Kwok, "Degradation of InSnZnO Thin-Film Transistors Under Negative Bias Stress," *IEEE Trans. Electron Devices*, vol. 70, no. 12, pp. 6381–6386, Dec. 2023, doi: 10.1109/TED.2023.3327975.

[33] W. Zhong *et al.*, "Effect of Sc2O3 Passivation Layer on the Electrical Characteristics and Stability of InSnZnO Thin-Film Transistors," *IEEE TRANSACTIONS ON ELECTRON DEVICES*, vol. 68, no. 10, 2021.

[34] S. Lee *et al.*, "Thermal Dehydrogenation Impact on Positive Bias Stability of Amorphous InSnZnO Thin-Film Transistors," *ACS Appl. Mater. Interfaces*, p. acsami.4c03689, Jul. 2024, doi: 10.1021/acsami.4c03689.

[35] Z. Jiang, M. Zhang, S. Deng, M. Wong, and H.-S. Kwok, "Long-Term Recovery Behavior in InSnZnO Thin-Film Transistors after Negative Bias Stress," in *2023 IEEE International Symposium on the Physical and Failure Analysis of Integrated Circuits (IPFA)*, Pulau Pinang, Malaysia: IEEE, Jul. 2023, pp. 1–5. doi: 10.1109/IPFA58228.2023.10249075.

[36] Y. Wan *et al.*, "P-1.9: Optimization of InSnZnO Thin-Film Transistors Based on Ultra-Low-Temperature Annealing Process," *SID Symposium Digest of Technical Papers*, vol. 54, no. S1, pp. 464–467, 2023, doi: 10.1002/sdtp.16333.

[37] F. Chen, Y. Wan, P. Balasubramanian, and M. Zhang, "Influence of radio-frequency magnetron sputtering power on electrical characteristics and positive bias stress stability of indium tin zinc oxide thin-film transistors," *Semicond. Sci. Technol.*, vol. 40, no. 1, p. 015009, Dec. 2024, doi: 10.1088/1361-6641/ad9947.

[38] H. Tsuji *et al.*, "Development of Back-Channel Etched In-W-Zn-O Thin-Film Transistors," *J. Display Technol.*, vol. 12, no. 3, pp. 228–231, Mar. 2016, doi: 10.1109/JDT.2015.2445321.

[39] H. Li, M. Qu, and Q. Zhang, "Influence of Tungsten Doping on the Performance of Indium–Zinc–Oxide Thin-Film Transistors," *IEEE Electron Device Lett.*, vol. 34, no. 10, pp. 1268–1270, Oct. 2013, doi: 10.1109/LED.2013.2278846.

[40] Y.-J. Lee *et al.*, "Thin Transparent W-Doped Indium-Zinc Oxide (WIZO) Layer on Glass," *j nanosci nanotechnol*, vol. 12, no. 7, pp. 5604–5608, Jul. 2012, doi: 10.1166/jnn.2012.6296.

[41] F. Chen, Y. Wan, P. Balasubramanian, and M. Zhang, "Influence of radio-frequency magnetron sputtering power on electrical characteristics and positive bias stress stability of indium tin zinc oxide thin-film transistors," *Semicond. Sci. Technol.*, vol. 40, no. 1, p. 015009, Jan. 2025, doi: 10.1088/1361-6641/ad9947.

Low-loss and Nonvolatile Multilevel Phase Shifter for Optical In-memory Computing

Qiqi Wei*, Jigeng Sun*, Shaolin Zhou*, and Yi Zou*†
*School of Microelectronics, South China University of Technology, Guangzhou, China
zouyi@scut.edu.cn

Abstract—**We present a low-loss and nonvolatile multilevel phase shifter, designed to achieve precise and energy-efficient phase modulation. By incorporating both coarse and fine tuning blocks, our phase shifter offers 121 distinct phase-shifting states, with an exceptionally low insertion loss of 0.039 dB/π as demonstrated in simulations. To validate the performance of our proposed phase shifter, we utilize it in Mach-Zehnder Interferometer networks and base on this to construct optical convolution kernels for a Convolutional Neural Network,which is tested using the MNIST handwriting digit database and achieves an outstanding accuracy rate of over 97% in handwriting digit recognition tasks.**

Index Terms—**Silicon Photonics, Phase Shifter, Optical In-memory Computing, Phase Change Material**

I. INTRODUCTION

The rapid development of deep learning and artificial intelligence is continuously driving technological innovation. However, according to data from the International Energy Agency (IEA) [1], the total electricity consumption of data centres and artificial intelligence was approximately 460 *terawatt hours* (TWh) globally in 2022, with projections indicating that it may reach 1000 TWh by 2026. This highlights that the current computing architecture may lead to significant energy concerns. Optical in-memory computing integrates data storage and processing functions within the photonic domain, utilizing the inherent parallelism and high-speed characteristics of light to overcome the energy consumption and latency bottlenecks caused by data movement between processors and memory, thereby achieving high energy efficiency and ultrafast computation.

Phase shifters are essential components in optical in-memory computing. However, traditional phase shifters, which rely on *micro-electro-mechanical systems* (MEMS) [2], thermo-optical effect [3], electro-optical effect [4], and free-carrier-dispersion effect [5], suffer from large device footprints and high static power consumption. Moreover, their volatility limits their application in optical in-memory computing. Fortunately, due to the high refractive index contrast between the crystalline and amorphous states and the long-term nonvolatility of phase-change materials (PCMs), phase shifters utilizing nonvolatile phase change materials offer a promising solution to overcome these limitations.

---

† Corresponding author. This research is supported partly by the Guangzhou GJYC Fund No. 2024D01J0010,as well as partly by the SCUT Research Fund No. K3200890.

Representative optical phase-change materials include $Ge_2Sb_2Te_5$(GST) [6], $Ge_2Sb_2Se_4Te_1$(GSST) [7], Sb_2S_3 [8], and Sb_2Se_3 [9]. Among them, GST exhibits a large refractive index contrast during the phase transition, but it has a high extinction coefficient in the near-infrared range, leading to significant losses. GSST, which replaces Te with Se based on GST, shows a slightly reduced refractive index contrast but reduces the extinction coefficients in both the amorphous and crystalline states. Sb_2S_3 and Sb_2Se_3 both have lower absorption coefficients and feature multiple intermediate states. Furthermore, Sb_2S_3 has at least two crystalline phases with different dielectric properties and optical contrasts, and the possibility to switch between these different crystalline phases gives extra freedom for multilevel operation [8]. The ultralow-loss Sb_2S_3 stands out among these materials as an ideal PCM for non-volatile phase shifters [10]. Although multilevel phase shifters based on Sb_2S_3 have been demonstrated [11], [12], further exploration of more modulation levels for nonvolatile phase shifters can further enhance optical in-memory computing.

In this paper, we propose a multilevel, low-loss phase shifter based on the phase change material Sb_2S_3. By utilizing the intermediate states of Sb_2S_3 loaded onto a silicon waveguide, the phase shifter presents multilevel phase states. Inspired by the focus knobs in modern optical microscopes, we divide the phase modulation into coarse and fine tuning blocks. We demonstrate 121 distinct phase states by combining the intermediate states of two blocks. Additionally, our phase shifter features a low insertion loss of 0.039 dB/π.

Finally, we use our phase shifters to construct 4 convolution kernels of size 2×2 based on the Clements [13] and Reck [14] mesh for *handwriting digit recognition*. The multilevel phase states and low-loss characteristics of our phase shifter are critical to achieving accuracy rates of 97.03% and 97.32% for the Reck and Clements mesh, respectively.

II. PHASE SHIFTER DESIGN AND OPTIMIZATION

Fig. 1 depicts the structure of the proposed multilevel and low-loss phase shifter. It comprises three modules: a single mode Si waveguide, the coarse tuning block, and the fine tuning block.

To achieve the whole π phase change, we set the length of the tuning blocks according to (1),

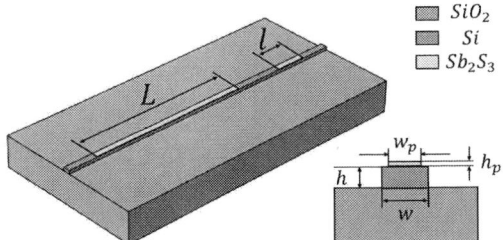

Fig. 1. The schematic illustration of the multilevel and low-loss phase shifter. The size of Si waveguide is given by the height $h = 220nm$ and the width $w = 450nm$.

$$L + l = \frac{\lambda\pi}{2(n_c - n_a)} \quad (1)$$

The variables L and l represent the length of the coarse and fine tuning block, respectively. n corresponds to the effective index of the phase shifter at the wavelength λ. The subscripts a and c denote the amorphous and crystalline Sb$_2$S$_3$, respectively. Moreover, referring to microscope operation, the coarse tuning block sets a rough phase-shifting state initially. The fine tuning block then refines the state more precisely. Therefore, the relationship between L and l can be expressed as follows,

$$L\Delta n_{L,\min} = l(n_c - n_a). \quad (2)$$

When implementing the intermediate states of Sb$_2$S$_3$, $\Delta n_{L,\min}$ represents the minimum effective refractive index difference of the waveguide between the two adjacent intermediates in the coarse tuning block. Equation (2) indicates that the minimum refractive index modulation step achievable by the coarse tuning block is equal to the maximum refractive index modulation achievable by the fine tuning block, ensuring the collaborative phase control between the coarse and fine tuning blocks.

Furthermore, assuming that the Sb$_2$S$_3$ in the coarse and fine tuning blocks possess i and j different phase states respectively, we describe the phase-shifting in our device as follows:

$$\theta = \frac{2\pi}{\lambda}\left[L(n_i - n_a) + l(n_j - n_a)\right]. \quad (3)$$

which enables a total of $i \cdot j$ different phase-shifting states.

To achieve multilevel phase-shifting and low transmission loss, we fix the thickness of Sb$_2$S$_3$ at $h_p = 20$ nm and set the width to $w_p = 350$ nm, 400 nm, and 450 nm to investigate the impact on loss. Due to the excellent low-loss property of Sb$_2$S$_3$, the PCM length has a negligible influence on the insertion loss of the device. A longer Sb$_2$S$_3$ facilitates to achieve more intermediate states in modulation [2], [3]. Therefore, we choose the 350 nm × 20 nm Sb$_2$S$_3$ for our phase shifter, and the insertion loss of our phase shifter is 0.039 dB/π. We assume that both coarse and fine blocks possess 9 intermediates and obtain the lengths of the two blocks as $L = 33.6$ μm and $l = 2.4$ μm.

III. SIMULATION

A. Multilevel Phase Modulation in MZI

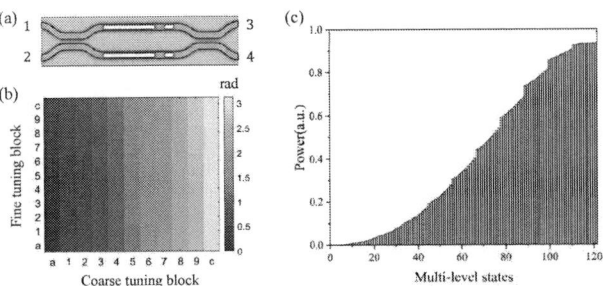

Fig. 2. (a) A balanced Mach-Zehnder Interferometer (MZI) with the phase shifters. (b) The achievable phase-shifting states under all combinations between the two phase modulation blocks. (c) The multi-level outputs of MZI at $1550nm$.

Fig. 2(a) shows a balanced Mach-Zehnder Interferometer (MZI) with both arms equipped with our phase shifter. We assume that both the coarse and fine blocks possess 9 intermediates states, and Fig. 2(b) illustrates all phase-shifting states generated by the cooperation of the two phase modulation blocks. We apply all phase-shifting states in one phase shifter and numerically calculate the MZI optical behaviors using the *finite-difference time-domain* (FDTD) method. When we modulate the top (bottom) phase shifter, the bottom (top) phase shifter always remains in its original state. TE polarized light with a wavelength of $1550nm$ is input into the device from port 1, and we detect the output power from port 3.

Fig. 2(c) illustrates the output transmission power at different phase-shifting states. As the phase difference between the two arms varies from 0 to π, the output power of port 3 increases step by step, presenting 121 different multilevel states. Since we define the functions of the coarse tuning and fine tuning blocks, we observe gradations in the multilevel output. There are 11 primary levels, each containing 11 smaller sub-levels, which depend on the available intermediate states of the PCM. Furthermore, the differences between primary levels become more pronounced as the output level increases.

B. Handwriting Digit Recognition Simulation

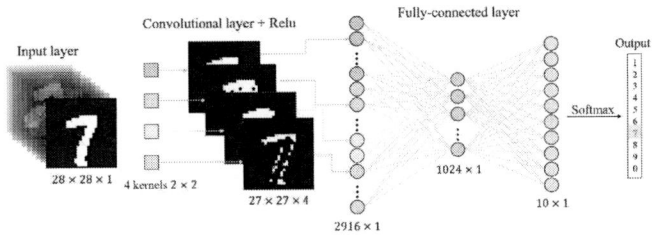

Fig. 3. A CNN for handwriting digit recognition.

To demonstrate the capability of the phase shifter in optical in-memory computing, we perform a simulation of *handwriting digit recognition*. A convolutional neural network (CNN) is constructed with one convolutional layer and two fully

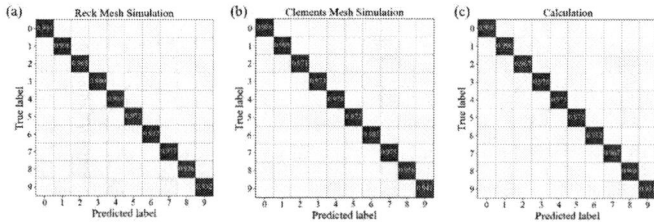

Fig. 4. (a-c) the confusion matrix of prediction results. (a) and (b) are the results for Reck and Clements mesh, respectively, and (c) is the calculated result.

connected layers, as shown in Fig. 3. After training the CNN, we utilize our phase shifter to reconstruct the 4 convolution kernels of size 2×2 based on the commonly used Clements and Reck mesh, respectively. Reck mesh is triangular, while Clements mesh is rectangular with a shorter optical depth, lower insertion loss, and higher phase error tolerance [13] .The reconstructed kernels are then employed to extract features from the input images.

Considering the phase error and loss, the elements of the reconstructed convolution kernel are complex numbers rather than real numbers. In a practical experiment, the pixel value is encoded in the light amplitude. To ensure consistency with this configuration, after performing the convolution separately with the real and imaginary parts of the reconstructed kernels, we sum the squares of the two results and then compute the square root. Since the imaginary part has a very small value, the sign of the result after taking the square root remains consistent with the convolution result from the real-part kernel. The filtered results are then activated by the *ReLU* function for subsequent recognition.

It should be noted that the phase-shifting values used in this study are extracted from Fig. 2(c). Although these values are discrete, the collaboration between the two phase shifters in the MZI still provides suitable phase-shifting functionality. Consequently,the confusion matrices of the prediction results are shown in Fig. 4(a-c). The simulated prediction accuracy achieves 97.03% for the Reck mesh and 97.32% for the Clements mesh respectively, which agree with the calculated accuracy of 97.38%. This demonstrates the potential of the proposed non-volatile multilevel phase shifter. Compared to the Reck mesh, the Clements mesh demonstrates higher prediction accuracy, primarily due to its better error tolerance and shorter optical depth.

IV. CONCLUSION

In this study, we present a multilevel, low-loss phase shifter based on the phase-change material Sb_2S_3. The device utilizes the intermediate states of Sb_2S_3 loaded on a silicon waveguide to achieve multilevel phase modulation, incorporating both coarse and fine tuning blocks to generate 121 distinct phase states. With a low insertion loss of 0.039 dB/π, the phase shifter demonstrates its potential in optical applications. We further demonstrate the effectiveness of the phase shifter by using it to construct convolution kernels for *handwriting digit*

recognition, achieving accuracy rates of 97.03% for the Reck mesh and 97.32% for the Clements mesh. This work highlights the capabilities of Sb_2S_3-based multilevel phase shifters for high-precision, low-loss optical computing applications.

REFERENCES

[1] IEA, "Electricity 2024," Paris, 2024, licence: CC BY 4.0. [Online]. Available: https://www.iea.org/reports/electricity-2024

[2] L. Midolo, A. Schliesser, and A. Fiore, "Nano-opto-electro-mechanical systems," *Nature Nanotechnology*, vol. 13, no. 1, pp. 11–18, 2018.

[3] M. Jacques, A. Samani, E. El-Fiky, D. Patel, Z. Xing, and D. V. Plant, "Optimization of thermo-optic phase-shifter design and mitigation of thermal crosstalk on the SOI platform," *Optics Express*, vol. 27, no. 8, pp. 10456–10471, 2019.

[4] V. Yoshioka, J. Jin, H. Zhou, Z. Tang, R. H. Olsson III, and B. Zhen, "CMOS-compatible, AlScN-based integrated electro-optic phase shifter," *Nanophotonics*, vol. 13, no. 18, pp. 3327–3335, 2024.

[5] A. Rahim, A. Hermans, B. Wohlfeil, D. Petousi, B. Kuyken, D. Van Thourhout, and R. Baets, "Taking silicon photonics modulators to a higher performance level: state-of-the-art and a review of new technologies," *Advanced Photonics*, vol. 3, no. 2, pp. 024003–024003, 2021.

[6] S. Cueff, A. Taute, A. Bourgade, J. Lumeau, S. Monfray, Q. Song, P. Genevet, B. Devif, X. Letartre, and L. Berguiga, "Reconfigurable flat optics with programmable reflection amplitude using lithography-free phase-change material ultra-thin films," *Advanced Optical Materials*, vol. 9, no. 2, p. 2001291, 2021.

[7] D. Sahoo and R. Naik, "GSST phase change materials and its utilization in optoelectronic devices: A review," *Materials Research Bulletin*, vol. 148, p. 111679, 2022.

[8] Y. Gutiérrez, A. P. Ovvyan, G. Santos, D. Juan, S. A. Rosales, J. Junquera, P. García-Fernández, S. Dicorato, M. M. Giangregorio, E. Dilonardo et al., "Interlaboratory study on Sb_2S_3 interplay between structure, dielectric function, and amorphous-to-crystalline phase change for photonics," *iScience*, vol. 25, no. 6, 2022.

[9] K. Murali, L. Thekkekara, M. A. Rahman, S. Sen, V. Shvedov, Y. Izdebskaya, C. Zou, S. A. Tawfik, I. Shadrivov, S. Sriram et al., "Tunability of Sb_2Se_3 phase change material for multi-domain optoelectronics," *Applied Materials Today*, vol. 40, p. 102338, 2024.

[10] M. Delaney, I. Zeimpekis, D. Lawson, D. W. Hewak, and O. L. Muskens, "A new family of ultralow loss reversible phase-change materials for photonic integrated circuits: Sb_2S_3 and Sb_2Se_3," *Advanced functional materials*, vol. 30, no. 36, p. 2002447, 2020.

[11] C. Ríos, Q. Du, Y. Zhang, C.-C. Popescu, M. Y. Shalaginov, P. Miller, C. Roberts, M. Kang, K. A. Richardson, T. Gu et al., "Ultra-compact nonvolatile phase shifter based on electrically reprogrammable transparent phase change materials," *PhotoniX*, vol. 3, no. 1, p. 26, 2022.

[12] R. Chen, Z. Fang, C. Perez, F. Miller, K. Kumari, A. Saxena, J. Zheng, S. J. Geiger, K. E. Goodson, and A. Majumdar, "Non-volatile electrically programmable integrated photonics with a 5-bit operation," *Nature Communications*, vol. 14, no. 1, p. 3465, 2023.

[13] W. R. Clements, P. C. Humphreys, B. J. Metcalf, W. S. Kolthammer, and I. A. Walmsley, "Optimal design for universal multiport interferometers," *Optica*, vol. 3, no. 12, pp. 1460–1465, 2016.

[14] M. Reck, A. Zeilinger, H. J. Bernstein, and P. Bertani, "Experimental realization of any discrete unitary operator," *Physical review letters*, vol. 73, no. 1, p. 58, 1994.

Eliminating Hysteresis in Dual-Gate a-IZO TFTs via N₂O Plasma Passivation of Back-Gate Insulator Interface Defects

Yuhan Zhang, Jinwen Liu, Huan Yang, Lei Lu, *Member, IEEE,* and Shengdong Zhang, *Senior Member, IEEE*

***Abstract*—Dual-gate (DG) oxide thin-film transistors (TFTs) have demonstrated superior driving capability and stability compared to conventional self-aligned top-gate (SATG) TFTs. In this work, the effects of nitrous oxide (N₂O) plasma treatment on bottom-gate insulator (BGI) of DG a-IZO TFTs were investigated. The results indicate that N₂O plasma treatment can effectively eliminate I-V hysteresis, shift the threshold voltage positively, improve the subthreshold swing and enhance the channel-length scalability. These improvements are attributed to the defect passivation at the BGI surface induced by N₂O plasma.**

***Keywords*—dual-gate, oxide semiconductor, thin-film transistor, N₂O plasma treatment, bottom-gate insulator.**

I. INTRODUCTION

In recent years, amorphous oxide semiconductor (AOS) thin-film transistors (TFTs), particularly exemplified by a-InGaZnO (a-IGZO) TFT, have garnered considerable attention, due to their high carrier mobility, low fabrication temperature, good large-area uniformity, ultra-low off current and steep subthreshold swing (*SS*) [1]-[3]. Compared to a-IGZO TFTs, a-InZnO (a-IZO) TFTs with a higher Indium (In) content exhibit higher mobility, along with tunable bandgap and high photosensitivity, making them more promising for some emerging applications, such as micro-LED displays [4], photodetectors [5] and flexible electronics [6].

However, single-gate (SG) AOS TFTs still face several challenges, including insufficient mobility and unsatisfactory stability [7], [8]. To solve these issues, dual-gate (DG) AOS TFT structure with two independent gate electrodes located on both sides of the active layer has been proposed [9]. It has been reported that DG a-IGZO TFTs with a sufficiently thin active layer can accumulate carriers within the bulk, leading to a higher mobility and a near-zero turn-on voltage compared to SG TFTs [8]. Additionally, DG TFTs show superior stability against bias voltage and environment factors, thanks to better channel

All authors are with the School of Electronic and Computer Engineering, Peking University, Shenzhen 518055, China. Shengdong Zhang is also with the School of Integrated Circuits, Peking University, Beijing 100871, China. (Corresponding author: Shengdong Zhang, email: zhangsd@pku.edu.cn)

This work was carried out at Guangdong Provincial Center for Oxide Semiconductor Devices and ICs, and supported financially by Ministry of Science and Technology Key Research and Development Program Grant 2022YFB3607200, and Shenzhen Municipal Scientific Program Grant JCYJ20220818100808019.

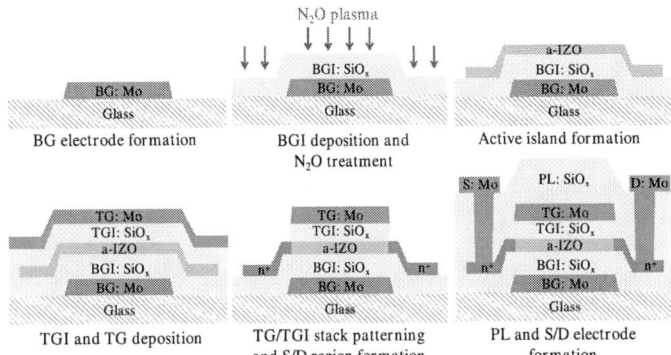

Fig. 1. The schematic process flow of the fabricated DG a-IZO TFT.

protection and reduced vertical electrical fields [10]. Moreover, the threshold voltage (V_{TH}) of DG TFTs during main gate scanning can be modulated by applying a constant bias to the control gate, enabling more flexible circuit designs [11]-[13].

To fabricate high-performance DG AOS TFTs and fully harness the advantages of the DG structure, it is essential to optimize the properties of the gate insulators (GI) and their interfaces with the active layer [14]. A study has shown that modulating the SiH_4/N_2O flow rate and deposition power can improve the quality of the SiO_x insulator and suppress hydrogen (H) diffusion into the active layer [15]. Moreover, N₂O plasma treatment has been reported to be an effective method for improving interface quality between the GI and the active layer, which has been widely used in the fabrication of self-aligned top-gate (SATG) TFTs [16], [17]. Compared to SATG TFTs, the bottom-gate insulator (BGI) bulk and its interface with the active layer also require elaborate optimization for DG TFTs.

In this work, the effects of N₂O plasma treatment on the electrical performance of DG a-IZO TFTs were investigated. The N₂O plasma was applied to the BGI surface. It is shown that N₂O plasma treatment can eliminate I-V hysteresis, positively shift the V_{TH}, and improve the channel-length scalability. The plasma treatment duration is further optimized to fabricate high-performance DG a-IZO TFTs.

II. EXPERIMENTS

The process flow of the fabricated DG a-IZO TFT is illustrated in Fig. 1. Firstly, a 100-nm-thick molybdenum (Mo) film was deposited and patterned to form the bottom-gate (BG)

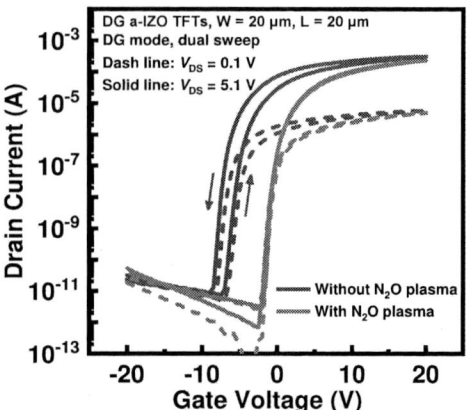

Fig. 2. The dual-sweeping transfer characteristics of the fabricated DG a-IZO TFTs with and without N₂O plasma treatment at BGI surface.

electrodes. Subsequently, a 100-nm-thick SiO$_x$ layer was deposited by PECVD at 300 °C to serve as the BGI. N₂O plasma treatments at 300 °C and 50 W with varying durations were then applied to the BGI surface to enhance transistor performance. Following that, a 20-nm-thick a-IZO active layer was deposited by direct-current (DC) sputtering at room temperature. Once the active islands were defined, another 100-nm-thick SiO$_x$ layer was deposited as the top-gate insulator (TGI), followed by thermal annealing in O₂ at 300 °C for 90 min. The top-gate (TG) Mo electrodes were then formed, and the TG/TGI stack was subsequently patterned. Using the Mo/SiO$_x$ gate stack as an in-situ mask, the self-aligned n⁺-IZO source/drain (S/D) regions were formed via an argon (Ar) plasma treatment [18]. After the deposition of the passivation layer (PL), the contact holes were opened, and S/D Mo electrodes were formed using the lift-off process.

The electrical properties of the DG a-IZO TFTs were measured using an Agilent B1500 semiconductor parameter analyzer at room temperature in a dark environment.

III. RESULTS AND DISCUSSIONS

Fig. 2 presents the transfer characteristics of the fabricated DG a-IZO TFTs with and without N₂O plasma pretreatment at BGI surface. The channel length (L) and channel width (W) are both 20 μm. In DG mode, the BG and TG electrodes are tied together, receiving the same sweeping voltage. In TG or BG mode, the gate sweeping voltage is applied only to the respective TG or BG electrode, while the other gate electrode remains grounded. The transfer curves in Fig. 2 are measured in DG mode. The field-effect mobility (μ_{FE}) is calculated from the transconductance (g_m), and the V_{TH} is determined at $I_{DS} \times L/W = 10^{-9}$ A. The SS is extracted as the minimum value of $dV_{GS}/d\log(I_{DS})$ with I_{DS} ranging from 10^{-11} A to 10^{-9} A.

As shown in Fig. 2, the DG TFT without N₂O plasma pretreatment exhibits a high μ_{FE} of 33.9 cm²/Vs, but a relatively negative V_{TH} of -5.7 V and a poor SS of 0.44 V/dec. These issues can be attributed to the diffusion of residual hydrogen (H) from BGI to a-IZO, generating a large number of defects at both the interface and within a-IZO bulk [15], [19]. In addition, it

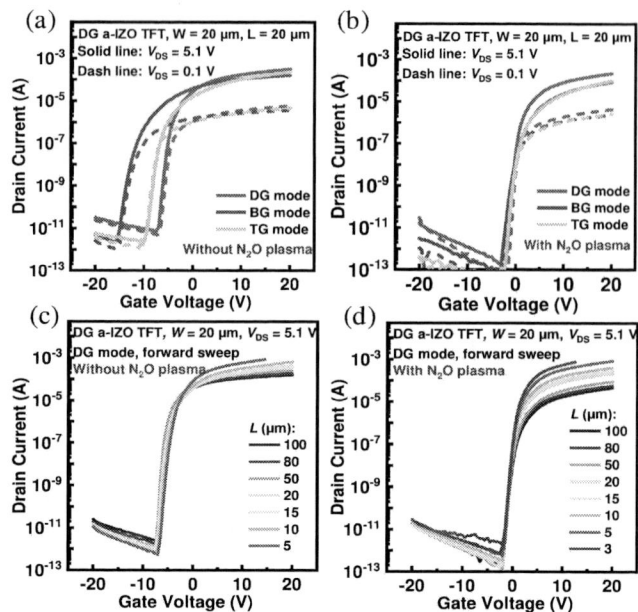

Fig. 3. The transfer characteristics and the L scalabilities of the fabricated DG a-IZO TFTs (a) (c) without and (b) (d) with 100-sec N₂O treatment at BGI surface.

exhibits a significant counterclockwise hysteresis of 1.69 V, indicating the presence of mobile charges and electrical dipoles within the BGI bulk and at the interface [20]. In contrast, the DG TFT with N₂O plasma pretreatment shows a significant positive shift in V_{TH} from -5.7 V to -1.0 V and an improvement in SS from 0.44 V/dec to 0.28 V/dec without any degradation in μ_{FE}. Most importantly, the counterclockwise hysteresis is almost completely eliminated. Generally, the interface trap state density (N_{it}) at the GI/channel interface can be estimated by the SS value using the following relationship:

$$ N_{it} = \left(\frac{SS}{\ln 10} \frac{q}{k_B T} - 1 \right) \frac{C_{ox}}{q} \tag{1} $$

where k_B is the Boltzmann constant, T is the absolute temperature, q is the elementary electron charge, and C_{ox} is the capacitance per unit area of the GI [21]. For DG TFTs in DG mode, the C_{ox} is the sum of the TGI capacitance and BGI capacitance. The calculated N_{it} for the DG a-IZO TFTs with and without N₂O plasma pretreatment is 1.88×10^{12} eV⁻¹cm⁻² and 3.25×10^{12} eV⁻¹cm⁻², respectively. The much smaller N_{it} for the N₂O-treated devices indicates that N₂O plasma can passivate the defects at the BGI interface [16], [21].

Fig 3 (a) and (b) show the transfer curves of the fabricated DG TFTs under three operation modes. For the devices without N₂O pretreatment, the V_{TH} in BG mode is more negative than in TG and DG modes, and the SS in BG mode is more gradual. This also indicates that the BGI interface quality is worse than that of the TGI interface. With N₂O plasma pretreatment on the BGI surface, the transfer curve in BG mode shows a significant positive shift and an improvement of SS. Moreover, both the SS and V_{TH} in BG mode become similar with those in TG mode. These results further suggest that N₂O plasma treatment can suppress H diffusion from BGI and passivate defects at the BGI

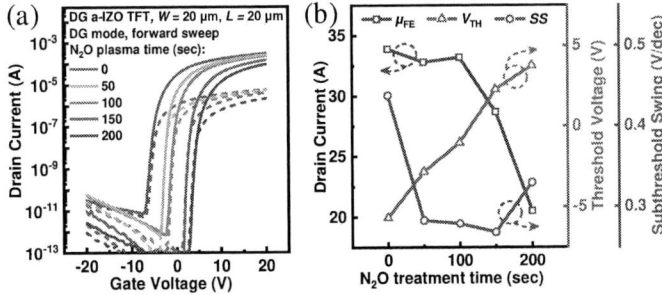

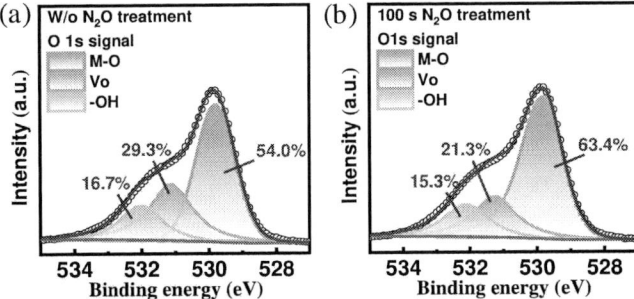

Fig. 4. (a) The transfer curves of the DG a-IZO TFTs with different BGI N_2O treatment times. (b) The dependence of the μ_{FE}, V_{TH} and SS on N_2O treatment time.

Fig. 5. XPS spectra of O 1s signal for a-IZO (a) without and (b) with 100-sec N_2O treatment.

surface. In addition, the L scalability of the DG TFTs with and without N_2O plasma treatment is shown in Fig. 3 (c) and (d). It is observed that the I_{DS} of the devices with N_2O plasma treatment increases in the expected order of magnitude as L decreases from 100 μm to 3 μm. In contrast, the I_{DS} of the devices without N_2O plasma treatment exhibits an abnormal dependence on L, which only increases slightly and is crowded with the variation of L. Thus, the N_2O plasma treatment at the BGI interface can also improve the L scalability of the DG a-IZO TFTs.

Fig. 4 illustrates the time dependence of the transistor performance with respect to the N_2O plasma treatment. As shown in Fig. 4(a), the transfer curves shift positively, and the extracted V_{TH} increases from -5.7 V to 3.6 V as the N_2O treatment duration increases from 0 sec to 200 sec. In addition, the μ_{FE} is well maintained at around 33 cm^2/Vs for N_2O treatment times ranging from 0 sec to 100 sec, but it begins to decrease as the treatment time is prolonged. The SS significantly improves after a 50-sec N_2O treatment and remains nearly constant from 50-sec to 150-sec. However, the SS worsens when the treatment time is extended to 200 sec. It is well known that the mobility is highly influenced by interface scattering, while SS is related to the interface trap density. Therefore, it can be inferred that N_2O pretreatment to the BGI can form a high-quality interface and reduce the interface trap density. The degradation of μ_{FE} and SS with a 200-sec treatment duration can be attributed to interface damage caused by excessive plasma bombardment [16].

To further investigate the influence of N_2O treatment on the SiO_x/a-IZO interface, the x-ray photoelectron spectroscopy (XPS) analysis and atomic force microscope (AFM) measurement were performed. Fig. 5(a) and (b) show the XPS spectra of O 1s peaks for the DG TFTs without and with N_2O treatment at the BGI interface, respectively. The broad peak of O 1s can be deconvoluted into three subpeaks centered at 529.5 eV, 531.0 eV and 532.0 eV, which are respectively associated with metal-oxygen (M-O) bonds, oxygen vacancies (V_o), and hydroxyl bonds (-OH). Compared to the untreated sample, the sample with N_2O treatment at the SiO_x interface exhibits a noticeable decrease in the relative area of V_o and –OH components, from 29.3% to 21.3% and from 16.7% to 15.3%, respectively, while a significant increase in M–O bonds from 54.0% to 63.4%. These results confirm that N_2O pretreatment can suppress V_o defect formation and H diffusion from BGI to

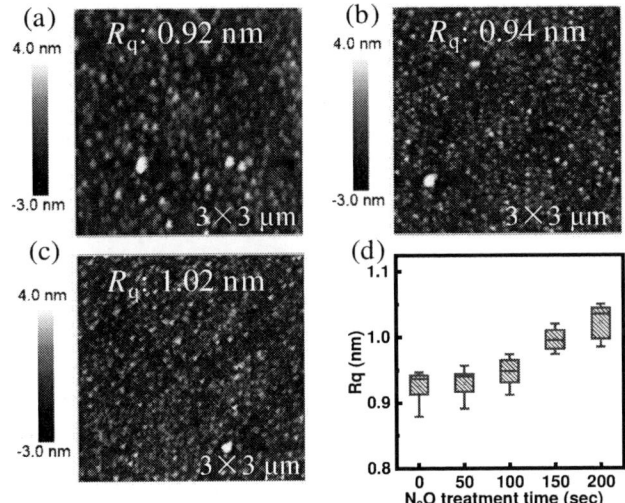

Fig. 6. AFM images of BGI SiO_x with N_2O treatment time of (a) 0 s, (b) 100 s, (c) 200 s, (d) Box plots of root-mean-square (RMS) roughness (R_q) as a function of N_2O treatment time.

a-IZO [15].

Fig. 6 shows the AFM images and the box plots of the BGI SiO_x with different N_2O treatment duration. After 100 sec of N_2O treatment, the root-mean-square (RMS) roughness (R_q) slightly increases from 0.92 nm to 0.94 nm. However, with a prolonged treatment time of 200 sec, the R_q significantly increases to 1.02 nm, which indicates the treatment duration should be carefully optimized to avoid damaging the roughness of the SiO_x/a-IZO interface. These results further validate that N_2O treatment not only chemically regulates the V_o concentration of the a-IZO active layer, but also physically modifies the interfacial roughness through bombardment effects.

IV. CONCLUSION

This work demonstrates that N_2O plasma treatment at the BGI surface can comprehensively enhance DG TFT performance by eliminating hysteresis, shifting V_{TH} positively, and improving L scalability. According to the XPS analysis, these improvements are primarily attributed to the oxidizing effect of N_2O plasma, which can passivate the interface defects and suppress H diffusion from BGI to a-IZO. Additionally, the results reveal that the effect of N_2O plasma is time-dependent.

979-8-3315-2209-4/25 $31.00 © 2025 IEEE

Excessive treatment time leads to degradation of μ_{FE} and SS, due to the interface damage caused by plasma bombardment, as confirmed by the AFM surface roughness measurements. The optimized treatment duration is determined to be 100 sec for DG a-IZO TFTs, under which high-performance metrics are achieved, including a high μ_{FE} of 33.2 cm^2/Vs, a steep SS of 0.28 V/dec, negligible hysteresis, and good L scalability down to 3 μm.

References

[1] S. Y. Lee, "Comprehensive review on amorphous oxide semiconductor thin film transistor," *Trans. Electr. Electron. Mater.*, vol. 21, no. 3, pp. 235-248, 2020, doi: 10.1007/s42341-020-00197-w.

[2] Y. Zhu, Y. He, S. Jiang, L. Zhu, C. Chen, and Q. Wan, "Indium–gallium–zinc–oxide thin-film transistors: materials, devices, and applications," *J. Semicond.*, vol. 42, no. 3, 2021, doi: 10.1088/1674-4926/42/3/031101.

[3] J. Troughton and D. Atkinson, "Amorphous InGaZnO and metal oxide semiconductor devices: an overview and current status," *J. Mater. Chem. C*, vol. 7, no. 40, pp. 12388-12414, Oct. 2019, doi: 10.1039/c9tc03933c.

[4] C. Liao, Y. Liu, and S. Zhang, "High performance A-PWM μLED pixel circuit design using double gate oxide TFTs," *Displays*, vol. 86, 2025, doi: 10.1016/j.displa.2024.102894.

[5] S. Jeon, I. Song, S. Lee, B. Ryu, S. E. Ahn, E. Lee, Y. Kim, A. Nathan, J. Robertson, and U. I. Chung, "Origin of high photoconductive gain in fully transparent heterojunction nanocrystalline oxide image sensors and interconnects," *Adv Mater*, vol. 26, no. 41, pp. 7102-9, Nov 5. 2014, doi: 10.1002/adma.201401955.

[6] J. Song, X. Huang, C. Han, Y. Yu, Y. Su, and P. Lai, "Recent developments of flexible InGaZnO thin-film transistors," *physica status solidi (a)*, vol. 218, no. 7, 2021, doi: 10.1002/pssa.202000527.

[7] T. Kamiya and H. Hosono, "Material characteristics and applications of transparent amorphous oxide semiconductors," *NPG Asia Mater.*, vol. 2, no. 1, pp. 15-22, 2010, doi: 10.1038/asiamat.2010.5.

[8] M. Mativenga, S. An, and J. Jang, "Bulk accumulation a-IGZO TFT for high current and turn-on voltage uniformity," *IEEE Electron Device Lett.*, vol. 34, no. 12, pp. 1533-1535, 2013, doi: 10.1109/led.2013.2284599.

[9] S. Kyoung-Seok, J. Ji-Sim, L. Kwang-Hee, K. Tae-Sang, P. Joon-Seok, C. Yun-Hyuk, P. KeeChan, K. Jang-Yeon, K. Bonwon, and L. Sang-Yoon, "Characteristics of double-Gate Ga–In–Zn–O thin-film transistor," *IEEE Electron Device Lett.*, vol. 31, no. 3, pp. 219-221, 2010, doi: 10.1109/led.2009.2038805.

[10] X. He, L. Wang, W. Deng, X. Xiao, L. Zhang, C. Leng, M. Chan, and S. Zhang, "Improved electrical stability of double-gate a-IGZO TFTs," *SID Symp. Dig. Tech. Pap.*, vol. 46, no. 1, pp. 1151-1154, 2015, doi: 10.1002/sdtp.10035.

[11] Y. Zhang, L. Chang, L. Lu, C. Liao, and S. Zhang, "Asymmetric double-gate (ADG) oxide thin-film transistor technology for medium- and small-sized AMOLED displays," *IEEE Electron Device Lett.*, pp. 1-1, 2025, doi: 10.1109/led.2025.3553527.

[12] Y. Chen, D. Geng, M. Mativenga, H. Nam, and J. Jang, "High-speed pseudo-CMOS circuits using bulk accumulation a-IGZO TFTs," *IEEE Electron Device Lett.*, vol. 36, no. 2, pp. 153-155, 2015, doi: 10.1109/led.2014.2379700.

[13] J. Hu, Y. Zhang, C. Liao, and S. Zhang, "A 1.33 μV/e$^-$ voltage mode active pixel sensor with threshold voltage compensation for dynamic X-ray imaging," *J. Soc. Inf. Disp.*, 2025, doi: 10.1002/jsid.2053.

[14] G. Baek, K. Abe, A. Kuo, H. Kumomi, and J. Kanicki, "Electrical properties and stability of dual-gate coplanar homojunction DC sputtered amorphous Indium–Gallium–Zinc–Oxide thin-film transistors and its application to AM-OLEDs," *IEEE Trans. Electron Devices*, vol. 58, no. 12, pp. 4344-4353, 2011, doi: 10.1109/ted.2011.2168528.

[15] K.-J. Zhou, T.-C. Chang, P.-Y. Yen, Y.-A. Chen, Y.-T. Chien, B.-S. Huang, P.-Y. Lee, T.-H. Juan, S. M. Sze, Y.-S. Fan, C.-S. Huang, and C.-H. Tsai, "Enhancing reliability of short-channel dual gate InGaZnO thin film transistors by bottom-gate oxide engineering," *IEEE Electron Device Lett.*, vol. 45, no. 4, pp. 593-596, 2024, doi: 10.1109/led.2024.3363131.

[16] T. Liang, X. Zhang, X. Zhou, L. Zhang, H. Lu, H. Lu, and S. Zhang, "Effects of N$_2$O plasma treatment time on the performance of self-aligned top-gate amorphous oxide thin film transistors," *SID Symp. Dig. Tech. Pap.*, vol. 48, no. 1, pp. 1299-1302, 2017, doi: 10.1002/sdtp.11880.

[17] Y. Zhang, J. Li, J. Li, T. Huang, Y. Guan, Y. Zhang, H. Yang, M. Chan, X. Wang, L. Lu, and S. Zhang, " 3-masks-processed sub-100 nm amorphous InGaZnO thin-film transistors for monolithic 3D capacitor-less dynamic random access memories," *Adv. Electron. Mater.*, vol. 9, no. 8, 2023, doi: 10.1002/aelm.202300150.

[18] Y. Zhang, J. Li, Y. Zhang, H. Yang, Y. Guan, M. Chan, L. Lu, and S. Zhang, "Deep sub-micron self-aligned bottom-gate amorphous InGaZnO thin-film transistors with low-resistance source/drain," *IEEE Electron Device Lett.*, vol. 44, no. 8, pp. 1300-1303, 2023, doi: 10.1109/led.2023.3287865.

[19] W. Pan, Y. Wang, Y. Wang, Z. Xia, F. S. Y. Yeung, M. Wong, H. S. Kwok, X. Wang, S. Zhang, and L. Lu, "Multiple effects of hydrogen on InGaZnO thin-film transistor and the hydrogenation-resistibility enhancement," *J. Alloys Compd.*, vol. 947, 2023, doi: 10.1016/j.jallcom.2023.169509.

[20] Z. Ye, Y. Yuan, H. Xu, Y. Liu, J. Luo, and M. Wong, "Mechanism and origin of hysteresis in oxide thin-film transistor and its application on 3-D nonvolatile memory," *IEEE Trans. Electron Devices*, vol. 64, no. 2, pp. 438-446, 2017, doi: 10.1109/ted.2016.2641476.

[21] J. Li, Y. Guan, J. Li, Y. Zhang, Y. Zhang, M. Chan, X. Wang, L. Lu, and S. Zhang, "Ultra-thin gate insulator of atomic-layer-deposited AlOx and HfOx for amorphous InGaZnO thin-film transistors," *Nanotechnology*, vol. 34, no. 26, Apr 12. 2023, doi: 10.1088/1361-6528/acc742.

Study of Flexible Tactile Sensors Based on Graphene/PDMS Composite Film

Yujun Ye   Weibin Lin   Shuangmei Xue   Qi-jun Sun   Yan Yan*

Abstract—**Pressure sensors are one of the core components of external sensing technologies. In applications such as robotic tactile sensing, with the continuous expansion of application scenarios, there are increasingly higher demands for flexible tactile sensors in perceiving complex 3D external information. This paper proposes a piezoresistive tactile sensor with a semi-cylindrical protrusion structure, which achieves a sensitivity of 4514.46 kPa^{-1} within the range of 0 to 2.4 kPa and exhibits strong stability. The sensor maintains highly stable output current waveforms under various pressure and frequency conditions over extended reliability tests. Furthermore, the sensor demonstrates the ability to dynamically recognize different texture types and can effectively measure vertical height information below 100 μm. This research provides new insights and references for the design of 3D high-precision and highly reliable flexible tactile sensors.**

Index Terms—**Pressure sensor, high sensitivity, graphene/PDMS composite, 3D perceiving**

I. INTRODUCTION

Human skin serves not only as an effective barrier protecting our bodies from external environmental damage but also as a crucial medium for perceiving external information such as texture, hardness, temperature, and humidity. Electronic skin (e-skin), which mimics the functions of human skin, holds significant application potential in fields such as robotics, healthcare, and wearable devices. In recent years, research on e-skin has focused on both the development of material properties, such as stretchability [1] and bending stability [2] and the expansion of functional characteristics, such as tactile sensing [3] and temperature/humidity detection [4]. Flexible tactile sensors are one of the core components enabling texture recognition in e-skin. Based on their sensing mechanisms, flexible tactile sensors can be categorized into capacitive [5], piezoresistive [6], piezoelectric [7], and triboelectric types [8]. Among these, piezoresistive sensors have gained popularity due to their advantages such as ease of signal acquisition, simple manufacturing processes, and strong anti-interference capabilities.

Previous research on piezoresistive sensors has primarily focused on improving sensing performance, such as sensitivity, detection range, and durability [9]. However, with the continuous optimization of sensor performance, the perception and recognition of externally measured objects have gradually garnered attention. Flexible tactile sensors typically measure current response signals while sliding over object surfaces and use the fast fourier transform (FFT) to obtain characteristic frequencies, enabling the differentiation of various texture morphologies [10]. Additionally, introducing microstructures on the sensor surface can further amplify the friction between the device surface and external textures, enhancing recognition accuracy [11]. Notably, previous studies on flexible tactile sensors have primarily focused on distinguishing different texture structures. In practical applications, however, there is a need for the quantitative recognition of more complex 3D information.

In this paper, we present a piezoresistive flexible tactile sensor based on a semi-cylindrical protrusion structure. The functional layer of the sensor, a pressure-sensitive film, is composed of a mixture of polydimethylsilylsiloxane (PDMS) prepolymer and graphene, forming a graphene-polydimethylsiloxane composite (GPC). This sensor not only exhibits exceptional pressure-sensing performance, with a sensitivity of 4514.46 kPa^{-1}, a response time of 73.53 ms, and a recovery time of 90.65 ms, but also demonstrates outstanding stability and pressure resistance. It maintains excellent performance under various measurement conditions, including different pressures, frequencies, and durations. Furthermore, the sensor is capable of dynamically recognizing 3D texture information. It can not only distinguish between different types of periodic texture widths but also measure and identify texture height information below 100 μm.

II. EXPERIMENTAL SECTION

The PDMS prepolymer solution was fabricated by mixing PDMS with curing agent (mass ratio of 5:1) and stirring well for 5 minutes. Then, graphite powder was added to the PDMS prepolymer and then mechanically stirred for 30 minutes to obtain GPC ink (graphite 30 wt%). Two PET films were pasted on the edges of the ITO-PET film to pattern the GPC film. After that, as shown in Figure 1a, the GPC ink was cast onto the ITO-PET substrate, and the sandpaper (180#) was pressed on the GPC ink as shown in Figure 1b. A glass rod was used to press the sandpaper surface to form a uniform GPC ink layer between the sandpaper and ITO-PET, following with a

---

Y. Yan, Y. J. Ye, W. B Lin, S. M. Xue are with College of Electronics and Information Engineering & the State Key Laboratory of Radio Frequency Heterogeneous Integration, Shenzhen University, Shenzhen, China. Q. J. Sun is with School of Physics and Optoelectronic Engineering & Guangdong Provincial Key Laboratory of Sensing Physics and System Integration Applications, Guangdong University of Technology, Guangzhou, P. R. China. This work is supported by the National Natural Science Foundation of China (Grant No. 62374110), and the Shenzhen Science and Technology Program (Grant No. JCYJ20230808105806014), and the State Key Laboratory of Radio Frequency Heterogeneous Integration (Independent Scientific Research Program No. 2024014). Corresponding author: (Yan Yan, email: yanyan@szu.edu.cn)

annealed treatment (90°C, 3h). Finally, as shown in Figure 1c, the sandpaper was peeled off the GPC film to obtain a GPC film with a microstructure on the surface of an ITO-PET substrate.

In this paper, two types of flexible tactile sensors were prepared. As shown in Figure 1d, a planar flexible pressure sensor (Figure 1d) consists of two layers of GPC pressure-sensitive films containing ITO-PET substrate electrodes symmetrically stacked, encapsulated with PI tape, and led out with copper tape [12]. Different from planar structures, in the case of the flexible pressure sensor based on a semi-cylindrical structure, a PDMS flexible substrate needs to be fabricated. The PDMS prepolymer is dripped into a mold with semi-cylindrical pits and spin-coated to form a film. Subsequently, a PET film is pressed onto the surface of the PDMS film and heated for curing. Finally, as shown in Figure 1e, the heated sample is peeled off from the mold to obtain a PDMS flexible substrate with semi-cylindrical convex structures. As illustrated in Fig. 1f, the substrate with semi-cylindrical convex structures is then integrated into the planar structure pressure sensor.

The surface topography and cross-section of the device structure of the GPC film were measured by scanning electron microscopy (SEM, JEOL, SU3500). The electrical performance of the tactile sensor is tested by a pressure sensing performance test system consisting of a computer, a source measure unit (Keithley 2614B), a force gauge (Mark-10), a motorized force tester (Mark-10), and a precision motion stage (JPLY-X140).

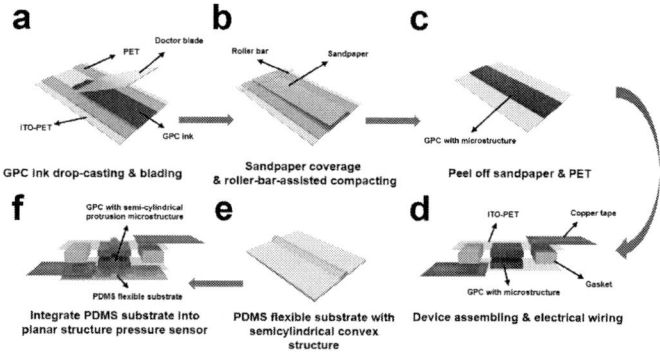

Fig. 1. Schematic diagram of the manufacturing process of the planar and semi-cylindrical flexible pressure sensor.

III. RESULTS AND DISCUSSION

Figure 2 presents the surface morphology and cross-sectional view of the GPC film microstructure under SEM. It can be observed that the surface of the GPC film replicates the morphology of the sandpaper template (Figure 2a), forming microstructures with irregular pore shapes. Additionally, the cross-sectional view of the GPC film (Figure 2b) reveals that these microstructures are predominantly concentrated on the surface of the GPC film, effectively reducing its elastic modulus. When subjected to external pressure, the GPC film with these microstructures is more prone to deformation, which

significantly enhances the sensitivity and detection range of the pressure sensor.

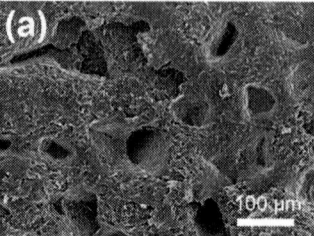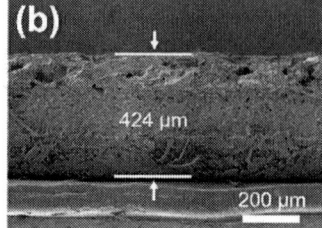

Fig. 2. SEM images (a) Surface topography of GPC thin film with microstructures; (b) Cross-section of a GPC film with microstructures.

Sensor sensitivity S is defined as:

$$S = d(\Delta I / I_{off}) / d(\Delta P) \tag{1}$$

where $\Delta I = (I_P - I_{off})$ is the value of the change in current, I_P is the current of the pressure sensor under external pressure, I_{off} is the initial current without external pressure, and ΔP represents the amount of change in external pressure. Figure 3a shows the sensitivity test curve of a flexible pressure sensor, and it is evident that both sensors exhibit high sensitivity. Specifically, in the low-voltage region (0 - 2.4 kPa), the sensitivity of planar structure devices is as high as 7726.67 kPa⁻¹, and the sensitivity of semi-cylindrical structure devices is also 4514.46 kPa⁻¹. Due to the introduction of a semi-cylindrical convex structure, the resistance of the pressure-sensitive film to deformation in the vertical direction is enhanced, resulting in a decrease in ΔI at the same pressure, which makes the sensitivity of the tactile sensor lower than that of the planar structure, but even so, its performance is still good. Figures 3b and 3c show that the response time is 69.81 ms and 85.81 ms for the planar structure, while the response time is 73.53 ms and 90.65 ms for the semi-cylindrical structure. Both show fast response and recovery speeds.

In addition to possessing exceptional pressure-sensing performance, the stability of the pressure sensor under various working conditions is equally crucial. Figures 3d to 3f present the reliability test results for both the planar structure and the semi-cylindrical protrusion structure pressure sensors. In the experiments, we applied different pressures at the same frequency, varied the frequency under the same pressure, and conducted continuous loading/unloading tests for 2000 seconds under specific pressure and frequency conditions. The results demonstrate that the output current waveforms of both sensors remain highly stable, with minimal peak variations, thereby fully validating their excellent stability and outstanding pressure resistance.

Through static testing, both the planar and semi-cylindrical flexible pressure sensors have demonstrated exceptional sensitivity, a wide detection range, and excellent durability. To achieve comprehensive perception of external 3D complex information, flexible tactile sensors need to possess dynamic recognition capabilities. Figures 4a and 4b display the current

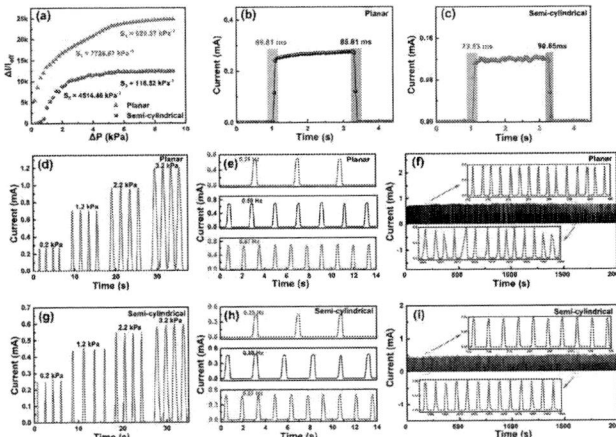

Fig. 3. (a) Relative current variation of planar and semi-cylindrical pressure sensors at different pressures; (b) Current response of a planar pressure sensor to a 2g weight. (c) The current response of a semi-cylindrical pressure sensor to a 2g weight; Reliability test of planar structure pressure sensor (d) cyclic test under different pressures (f=0.67Hz), (e) cyclic test at different frequencies (p=1.0 kPa) and (f) continuous test for 2000 seconds (p=1.0 kPa, f=0.67 Hz); Reliability test of semi-cylindrical pressure transducer (g) cyclic test at different pressures (f=0.67 Hz), (h) cyclic test at different frequencies (p=2.2 kPa) and (i) 2000 second cycle test (p=2.2 kPa, 0.67 Hz).

response curves of the two sensors as they slide over a specific stripe (with a width of 250 μm and a height of 80 μm).The observations reveal that when the planar structure sensor slides over the stripe, there is no obvious current response. In contrast, the semi-cylindrical structure sensor exhibits a significant change in current as it slides over the stripe, reaching a peak of 0.675 mA at the highest point of the stripe and quickly returning to its initial state after passing the stripe. This performance fully demonstrates its excellent potential for dynamically sensing texture information.

As shown in Figures 4c to 4f, we further utilized the semi-cylindrical tactile sensor to identify more complex periodic texture information through dynamic testing. By applying the FFT to the current signals, we obtained frequency spectra containing characteristic frequencies. According to the formula,

$$f = v / \lambda \qquad (2)$$

where f is the characteristic frequency, v is the velocity, and λ is the periodic size of the texture being measured. As shown in Figure 4d, the f of the current signal gliding through three periodic textures (Figure 4c, λ =700 μm, 800 μm, 900 μm) through the FFT of the semi-cylindrical pressure sensor are 1.42 Hz, 1.22 Hz, and 1.06 Hz (the calibrated displacement velocity in the measurement is 944.78 μm/s). Compared with the theoretical values of 1.35 Hz, 1.18 Hz, and 1.05 Hz, the error rates are 4.9%, 3.3%, and 0.9%, respectively, and it can be found that the sensor has excellent recognition ability of

periodic texture size.

$$R^2 = 1 - \frac{SSR}{SST} = 1 - \frac{\sum_{i=1}^{n}(y_i - \widehat{y})}{\sum_{i=1}^{n}(y_i - \overline{y})} \qquad (3)$$

In statistics, the coefficient of determination R^2 of the model is generally used to measure the predictive ability of the model, where SSR is the sum of squares of the residuals, y_i is the actual value, $\widehat{y}$ is the predicted value, and SST is the sum of the total squares, $\overline{y}$ is the average of the actual values. The maximum value of R^2 is 1, and the closer R^2 is to 1, the better the fit curve fits the observations.

The measured values and fitting curves of f under different λ (700-1400 μm) are shown in Figure 4e, and the R^2 of the fitting curve is 0.99, indicating that the fitting curve has a very high degree of fit, and the f of periodic textures of different sizes can be predicted by the fitting curve. The fitting formula is that the relative velocity between the sensor and the texture can be deduced from equation (1) to be 958.66 μm/s, and the error rate is 1.47% compared with the set displacement velocity of 944.78 μm/s.

To further study the impact of sliding speed on recognition accuracy, we passed the tactile sensor over the texture at different speeds and conducted 10 repetitions of feature frequency measurements at each speed. The summarized data is shown in Figure 4f. We found that within the range of displacement speed not exceeding 12 mm/s, the fluctuations of the feature frequency measured by the tactile sensor were very small, demonstrating excellent measurement stability. Notably, the device manifests a critical stability threshold at 18 mm/s sliding velocity, beyond which the GPC film exhibits marked deterioration. This may be the reason for the weakening of the microstructure interlocking between the two layers of GPC films, and the gradual amplification of the transverse shift, which leads to the microstructure separation.

The ability of tactile sensors to perceive longitudinal height information is also crucial. Figure 5a shows the fitting curve of the semi-cylindrical pressure sensor to identify the fringe summary data at different heights (10 measurements per sample). It is clear from the graph that the pressure sensor is able to distinguish the four different heights with very little fluctuation in the measurement data, indicating good stability and repeatability. To verify the performance of the tactile sensor in recognizing texture height, the tactile sensor was slid through a measured platinum wire, and the semi-cylindrical tactile sensor made 10 measurements and identified a fringe height of 64.28 ± 1.29 μm according to the fitted straight line analysis in Figure 5a (Figure 5b). Figure 5c shows the topography of the platinum wire under SEM, measured at a height of 60 μm. We can conclude that the measurement error rate of the tactile sensor with a semi-cylindrical structure is 7.1%, which shows excellent height information recognition ability.

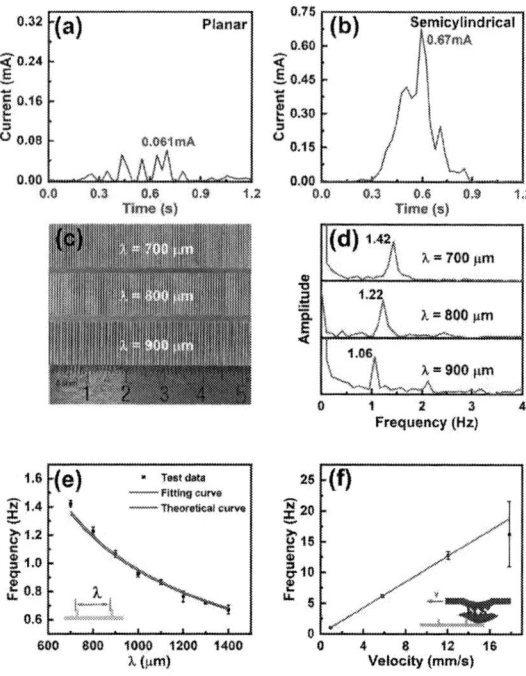

Fig. 4. (a) the current response curves of a planar tactile sensor and (b) a semi-cylindrical tactile sensor sliding through a single fringe process; (c) Photographs of stencils with periodic textured structures; The semi-cylindrical tactile sensor (d) glides through the template spectrogram and f, (e) glides through the f summary and fitting curve of ten fringe intervals, and (f) slides through the template 10 times at different speeds to summarize and fit the straight line.

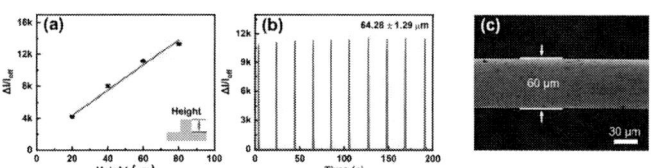

Fig. 5. (a) Relative current changes of the semi-cylindrical structure pressure sensor at different stripe heights; (b) Relative current changes with time when the semi-cylindrical structure pressure sensor slides over the platinum wire; (c) SEM photo of the platinum wire.

IV. CONCLUSION

We propose a novel flexible tactile sensor structure, which consists of two layers of GPC conductive films and a PDMS substrate with a semi-cylindrical protrusion structure. Research shows that this device not only exhibits excellent static pressure performance, with a sensitivity of 4514.46 kPa^{-1}, a response time of 73.53 ms, and a recovery time of 90.65 ms, but also demonstrates remarkable stability in pressure resistance tests. Furthermore, in sliding measurement and recognition, this flexible sensor showcases outstanding dynamic recognition capabilities, accurately distinguishing periodic textures and height information of different shapes. We believe that this work can provide valuable insights for the fabrication of flexible tactile sensors with high sensitivity and stability.

REFERENCES

[1] S. Hamaguchi, T. Kawasetsu, T. Horii, H. Ishihara, R. Niiyama, K. Hosoda, and M. Asada, "Soft Inductive Tactile Sensor Using Flow-Channel Enclosing Liquid Metal," Ieee Robotics and Automation Letters, vol. 5, no. 3, pp. 4028-4034, Jul, 2020.

[2] Y. D. Gu, J. Li, C. Y. Zheng, J. Huang, Z. D. Chen, T. Zhang, and S. B. Li, "A Flexible Pressure/Bending Bimodal Sensor for Hand Movements Monitoring," Ieee Sensors Journal, vol. 23, no. 24, pp. 30241-30248, Dec, 2023.

[3] T. C. Ma, W. J. Zhang, H. X. Zhao, and M. Zhang, "A Flexible Multifunctional Electronic Skin for Intelligent Tactile Perception," Ieee Sensors Journal, vol. 23, no. 15, pp. 17407-17414, Aug, 2023.

[4] Y. Kumaresan, O. Ozioko, and R. Dahiya, "Multifunctional Electronic Skin With a Stack of Temperature and Pressure Sensor Arrays," Ieee Sensors Journal, vol. 21, no. 23, pp. 26243-26251, Dec, 2021.

[5] S. Akbarzadeh, H. Ghayvat, C. Chen, X. Zhao, S. Hosier, W. Yuan, S. H. Pun, and W. Chen, "A Simple Fabrication, Low Noise, Capacitive Tactile Sensor for Use in Inexpensive and Smart Healthcare Systems," Ieee Sensors Journal, vol. 22, no. 9, pp. 9069-9077, May, 2022.

[6] Y. Zhu, Y. Liu, Y. N. Sun, Y. X. Zhang, and G. F. Ding, "Recent Advances in Resistive Sensor Technology for Tactile Perception: A Review," Ieee Sensors Journal, vol. 22, no. 16, pp. 15635-15649, Aug, 2022.

[7] S. Bonam, K. A. Bhagavathi, J. Joseph, S. G. Singh, and S. R. K. Vanjari, "An Ultra-Flexible Tactile Sensor Using Silk Piezoelectric Thin Film," Ieee Sensors Journal, vol. 23, no. 16, pp. 18656-18663, Aug, 2023.

[8] Y. Zhong, J. Q. Wang, L. G. Wu, K. S. Liu, S. P. Dai, J. Hua, G. G. Cheng, and J. N. Ding, "Dome-Conformal Electrode Strategy for Enhancing the Sensitivity of BaTiO$_3$-Doped Flexible Self-powered Triboelectric Pressure Sensor," Acs Applied Materials & Interfaces, vol. 16, no. 1, pp. 1727-1736, Dec, 2023.

[9] Z. M. Li, D. Feng, B. Li, W. B. Zhao, D. L. Xie, Y. Mei, and P. J. Liu, "Ultra-Wide Range, High Sensitivity Piezoresistive Sensor Based on Triple Periodic Minimum Surface Construction," Small, pp. 14, 2023 May, 2023.

[10] Y. Abbass, C. Gianoglio, H. A. Ali, M. Saleh, and M. Valle, "Texture Perception Using Tactile Sensing Glove Based on PVDF Sensors and Machine Learning," Ieee Sensors Letters, vol. 8, no. 7, pp. 4, Jul, 2024.

[11] X. G. Sun, T. Z. Liu, J. Zhou, L. Yao, S. L. Liang, M. Zhao, C. X. Liu, and N. Xue, "Recent Applications of Different Microstructure Designs in High Performance Tactile Sensors: A Review," Ieee Sensors Journal, vol. 21, no. 9, pp. 10291-10303, May, 2021.

[12] Y. Yan, R. X. Qiu, K. Y. Hu, H. H. Sun, S. M. Xue, M. Zhang, J. L. Xu, Y. Zhou, and Q. J. Sun, "Performance Optimization and Bending Stability Study of Flexible Pressure Sensors Based on Graphene/PDMS Composite Film," IEEE Transactions on Electron Devices, vol. 71, no. 12, pp. 7948-7954, Dec, 2024.

MoS$_2$-based RRAM with Asymmetric Contacts

Han Yan   Zichao Ma[*]

Abstract—**This work investigates the influence of contact asymmetry on the resistance-switching performance of MoS$_2$-based RRAM devices. The devices fabricated using combinations of Ag and Au electrodes with either van der Waals (vdW) contact or evaporated contacts are investigated. The device featuring vdW Ag contact and evaporated Au electrodes demonstrates notably reduced set/reset voltages, improved stability, and a 100 times higher on/off ratio. This is attributed to the vdW Schottky contact, which facilitates efficient control of conductive filament formation.**

Keywords—**RRAM, MoS$_2$, asymmetric contact**

I. INTRODUCTION

Resistive random-access memory (RRAM) based on two-dimensional (2D) materials has attracted significant attention for its potential in compute-in-memory technologies [1]. By leveraging the atomically thin, monocrystalline structure of 2D layered materials, these RRAM devices can be designed to exhibit a wide switching window, low operating voltages, and fast switching speed, effectively addressing the limitations of conventional memory devices [2]–[5]. Previous studies have primarily focused on device area scaling, electrode material selection, and enhancing material quality to reduce the operation power and improve device reliability [6]–[8]. The contact electrode plays a crucial role in determining the carrier injection, ion migration, and filament formation, all of which are essential to RRAM functionality [9]. However, despite being one critical factor that significantly influences many electronic devices, limited attention has been given to the effect of Schottky contact within the RRAM devices and their impact on device performance and reliability. This paper investigates the impact of work function and contact interface on RRAM performance, with a particular focus on the role of van der Waals (vdW) contacts in improving the resistive switching characteristics of MoS$_2$-based RRAM devices.

II. DEVICE FABRICATION AND MEASUREMENT METHODS

Polycrystalline MoS$_2$ film was synthesized via thermal decomposition using ammonium tetrathiomolybdate (ATTM, 99.95%, trace metal basis) as a precursor, following a previously reported method. The vertical grain boundaries of the MoS$_2$ provide channels for metal migration, which is important for the stable performance of RRAMs [10]. An optical image of the MoS$_2$ film prepared on a sapphire substrate at 700°C is shown in Fig. 1a. The film shows a thickness of 3.977 nm (6 layers) and a surface roughness of 3.93 Å, measured by an atomic force microscope. The Raman spectrum, displayed in the upper corner of Fig. 1a, exhibits characteristic E$_{2g}$ and A$_{1g}$ peaks of MoS$_2$, with full-width at half maximum (FWHM) of 7.56 cm^{-1} and 7.18 cm^{-1}, respectively, confirming the high quality of the film. The MoS$_2$ film was transferred as the resistive switching layer using the PMMA-assisted wet-transferred method onto an array of electrodes prepared on SiO$_2$/Si substrate, forming van der Waals (vdW) contacts. The electrodes were fabricated via photolithography, metal evaporation, and lift-off processes. The top electrodes were subsequently fabricated using the same method, forming the MoS$_2$ RRAM device array shown in Fig. 1b.

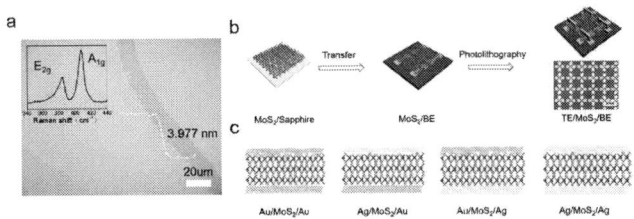

Fig. 1. (a) Optical image of MoS$_2$/sapphire, inset shows the Raman spectrum and the thickness profile measured by AFM; (b) Fabrication process and optical image of RRAM array; (c) Device structure of the MoS$_2$-based RRAM with asymmetric contact.

For most RRAM devices utilizing MoS$_2$, the commonly used electrode is currently Au and Ag due to their excellent conductivity, chemical stability, and compatibility with various fabrication processes. Fig. 1c shows four different types of MoS$_2$-based RRAM devices fabricated with combinations of evaporated top contacts and bottom vdW contacts using active Ag and inert Au metals. The fabricated devices exhibit asymmetric contacts in both the contact interface and the work function. The MoS$_2$ film transferred onto the bottom electrode forms a vdW gap that ensures a sharp interface and minimal defects, which is critical for achieving near-ideal Schottky contact. In contrast, the top electrode deposited via thermal evaporation of metal can result in a rough interface with the MoS$_2$ film. The Au/Au devices (Au as both the top and bottom electrodes) and Ag/Ag devices were tested by applying voltage biases to the top electrode while grounding the bottom electrode. For the Au/Ag devices (Au as the top electrode and Ag as the bottom electrode) and Ag/Au devices (Ag as the top electrode and Au as the bottom electrode), voltage biases were applied to the Ag electrode, with the Au electrode grounded.

Han Yan and Zichao Ma are with *School of Microelectronics, (South China University of Technology.)*, Guangzhou, P.R. China. This research was funded by the Guangdong Basic and Applied Basic Research Foundation 2025A1515010391, and the Guangzhou Science and Technology Program SL2024A04J3755. Corresponding author: (Zichao Ma, email: chauson@scut.edu.cn)

979-8-3315-2209-4/25 $31.00 © 2025 IEEE

III. RESULTS AND DISCUSSIONS

The resistive switching characteristics of the MoS_2-based RRAM device are shown in Fig. 2a to 2d, each accompanied by an optical image and a schematic illustrating device operation. All devices exhibit behavior consistent with a reverse-biased Schottky diode before the full formation of a conduction filament. This preliminary diode-like response reflects the high-resistance state and suggests that the switching mechanism involves Schottky barrier modulation, which transitions to filament formation under higher bias conditions. The Au/Au device, shown in Fig. 2a, exhibits relatively high set voltages exceeding 1V, and a large off-state leakage, resulting in a small on/off current ratio of approximately 10^2. The corresponding schematic and energy band diagrams show the formation of conducting filaments. In the pristine state, the top electrode and the bottom electrode are in Schottky contact with the film. When a bias voltage is applied to the top electrode, the metal atoms(Au or Ag) migrate to form a conductive filament connecting the top and bottom electrodes, and the conductive filament breaks after the reverse bias voltage is applied again. Due to the migration of metal atoms, the top electrode forms a metallic ohmic contact with the film [11].

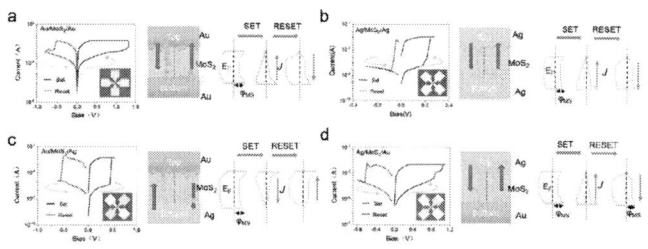

Fig. 2. Cyclic set-reset I-V characteristics for the RRAM devices with asymmetric contact using (a) Au/MoS$_2$/Au, (b) Ag/MoS$_2$/Ag, (c) Au/MoS$_2$/Ag, and (d) Ag/MoS$_2$/Au structure. Inset is an optical image of the fabricated device and illustrations of device operation mechanisms and band diagrams.

The high set/reset voltages are typical for RRAM devices using inert metal electrodes due to the higher activation energy required to form conductive filaments. The low on/off ratio indicates limited control over the conductivity between the high-resistance state and low-resistance state, which undermines the multi-bit capability. In comparison, the Ag/Ag devices, shown in Fig. 2b, display much lower set and reset voltages at 0.25V and -0.08V, respectively, and a much higher on/off ratio of more than 10^4. These low switching voltages are beneficial for low-power applications, but the symmetrical use of active Ag electrodes poses challenges in the formation and rupture of conductive filaments, as both electrodes contribute mobile ions, often resulting in uncontrolled and unstable filament formation.

Using one active and one inert electrode is a straightforward strategy to enhance device performance. Applying voltages to the active Ag electrode while grounding the inert Au electrode can stabilize the formation of conductive filaments, improving the resistive switching characteristics. Fig. 2c and 2d show the switching curves for the Ag/Au and Au/Ag devices and the corresponding energy band diagrams, where the Ag electrode serves either the top evaporated contact or the bottom vdW contact. Notably, the device with Ag as the bottom vdW contact exhibited set and reset voltages of 0.48 V and -0.4 V, respectively, compared to the device with evaporated Ag as the top electrode, which showed higher and asymmetric set and reset voltages at 1.05 V and -0.59 V. This discrepancy in performance is attributed to the Ag conductive filament in the Ag/Au structure, which induces defect state changes during each cycle, modulating the Schottky barrier at the Au/MoS$_2$ interface.

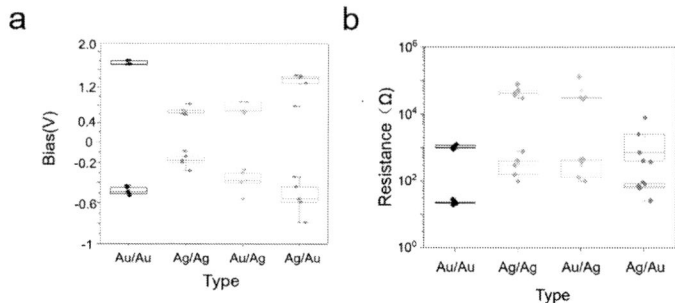

Fig. 3. Statistical analysis of (a) set and reset voltages, (b) high- and low-resistance states of the four types of RRAM devices.

As a result, the Ag/Au device exhibits desirable symmetric set and reset voltages compared to the Au/Ag device, along with an on/off ratio approximately 10 times larger, similar to that of the Ag/Ag devices. The consistency of the results was confirmed through statistical analysis of five devices for each configuration, as shown in Fig. 3a and 3b. This variation in switching characteristics can stem from differences in the interface properties between MoS$_2$ and the metal contacts of the RRAM devices. The I-V curves of the MoS$_2$-based RRAM devices in a high resistance state are summarized in Fig. 4. Under the same compliance current, devices with vdW Ag contacts exhibit a lower off current and reduced set voltage. This performance improvement can be attributed to the Schottky barrier at the MoS$_2$-Ag vdW interface. The barrier height was extracted by the Fowler-Nordheim tunneling (1),

$$I = \frac{q^3 E^2 m}{8\pi h \phi_{MS} m^*} \exp\left(\frac{8\pi\sqrt{2m^*}\phi_{MS}^{3/2}}{3heE}\right) \quad (1)$$

where E is the electric field across the MoS$_2$ layer, and ϕ_{MS} is the Schottky barrier height. The results show that the Au/Ag structure has the lowest metal-semiconductor barrier height (0.21 eV), which facilitates efficient charge injection and lowers the set voltage, as confirmed by the band diagram showing the synergistic effect of the low barrier at the Ag injection side and high stability at the Au grounding side. Combined with the lower activation energy required for Ag$^+$ ion migration, which demands lower voltage and current drive, the vdW Ag contact demonstrates the potential to significantly

979-8-3315-2209-4/25 $31.00 © 2025 IEEE

enhance the on/off-state resistance ratio and reduce operation power.

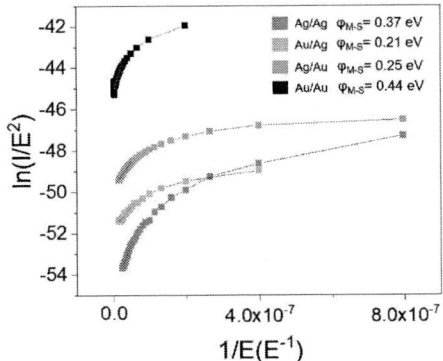

Fig. 4. The set I-V curves plotted in the 1/E versus $\ln(I/E^2)$ axis for Au/MoS$_2$/Au (black), Ag/MoS$_2$/Au (blue), Au/MoS$_2$/Ag (green), and Ag/MoS$_2$/Ag (red) structures.

IV. Conclusion

This study demonstrates the effectiveness of contact asymmetry in improving the resistive switching behavior of MoS$_2$-based RRAM devices. By examining both evaporated and vdW contact electrode structures using Au and Ag metals, it was found that devices with vdW Ag contacts and evaporated Au electrodes exhibited superior stability and lower switching voltages. This improvement is attributed to the vdW contact that efficiently regulates filament formation. In contrast, reversing the contact configuration led to increased off-state leakage, large set/reset voltages, and reduced stability, underscoring the importance of optimizing contact structures to enhance RRAM device performance.

References

[1] S. Yu, W. Shim, X. Peng and Y. Luo, "RRAM for compute-in-memory: From inference to training," *IEEE Transactions on Circuits and Systems I: Regular Papers*, vol. 68, no. 7, 2021, pp. 2753-2765.

[2] Q. A. Vu, et al., "A high-on/off-ratio floating-gate memristor array on a flexible substrate via CVD-grown large-area 2D layer stacking," *Advanced Materials*, vol. 29, no. 44, 2017, p. 1703363.

[3] L. Sun, et al., "Ultralow switching voltage slope based on two-dimensional materials for integrated memory and neuromorphic applications," *Nano Energy*, vol. 69, 2020, p. 104472.

[4] B. Tang, et al., "Wafer-scale solution-processed 2D material analog resistive memory array for memory-based computing," *Nature Communications*, vol. 13, no. 1, 2022, p. 3037.

[5] W. Banerjee, "Challenges and applications of emerging nonvolatile memory devices," *Electronics*, vol. 9, no. 6, 2020, p. 1029.

[6] Z. Li, P. Y. Chen, H. Xu and S. YU, "Design of ternary neural network with 3-D vertical RRAM array," *IEEE Transactions on Electron Devices*, vol. 64, no. 6, 2017, pp. 2721-2727.

[7] K. Liao, et al., "Memristor based on inorganic and organic two-dimensional materials: mechanisms, performance, and synaptic applications," *ACS Applied Materials & Interfaces*, vol. 13, no. 28, 2021, pp. 32606-32623.

[8] Q. Zhao, et al., "Current status and prospects of memristors based on novel 2D materials," *Materials Horizons*, vol. 7, no. 6, 2020, pp. 1495-1518.

[9] X. F. Wang, et al., "Interface engineering with MoS$_2$–Pd nanoparticles hybrid structure for a low voltage resistive switching memory," *Small*, vol. 14, no. 2, 2018, p. 1702525.

[10] P. Zhuang, et al., "Large-area multilayer molybdenum disulfide for 2D memristors," *Materials Today Nano*, vol. 23, 2023, p. 100353.

[11] Yan. H, et al., " Metal Penetration and Grain Boundary in MoS$_2$ Memristors.," *Advanced Electronic Materials* , vol. 11, 2025, p. 2400264.

Two-dimensional $Mo_{0.5}W_{0.5}S_2$ based memristive artificial neuron

Hang Li   Wanhong Luan   Wei Zeng   JiYu Zhao   Guanglong Ding   Ye Zhou   Su-Ting Han

Abstract—Currently, the "power wall" issue of traditional artificial neural networks (ANN) is becoming increasingly prominent. Designing spiking neural network (SNN) based on low-power characteristics of biological neural networks represents a highly effective solution to the problem of high-power consumption. The realization of SNN hinges on high-performance artificial neurons, and the development of such neurons is crucial for constructing low-power neural networks. In this study, we developed a threshold switching memristor (TSM) based on two-dimensional alloyed $Mo_{0.5}W_{0.5}S_2$ NS and constructed an artificial neuron according to the leaky integrate-and-fire (LIF) model, leveraging its threshold switching characteristics to successfully simulate the integration and firing characteristics of biological neurons. Our proposed two-dimensional alloyed $Mo_{0.5}W_{0.5}S_2$ NS based TSM offers a viable approach for the implementation of artificial neurons.

Index Terms—Threshold switching memristor $Mo_{0.5}W_{0.5}S_2$ NS

I. INTRODUCTION

As the scale of traditional ANN increases, energy consumption of ANN for training and inference rises sharply. This makes it difficult for traditional hardware circuits to further enhance computing performance without a significant increase in power consumption, leading to the so-called "power wall" problem. Compared with ANN, spiking neural network (SNN) more closely approximates the model of biological neural computing and features lower power consumption, which can effectively alleviate the "power wall" problem faced by ANN. With the aid of shunt capacitors, TSM can emulate the integration and firing characteristics of LIF neurons and serve as natural input stage for SNN. Therefore, the development of high-performance TSM is key to constructing artificial neurons and bionic low-power SNN.

As one of the two-dimensional materials, transition metal dichalcogenides (TMDs) have garnered extensive attention across various fields due to their unique optoelectronic properties. Memristors based on TMDs exhibit excellent performance. Cao et al. proposed a non-volatile memristor based on WS_2-PVP composite material and studied its switching mechanism.[1]. Zhang et al. proposed a flexible non-volatile memory with rewritable characteristics based on MoS_2-PVP composites, which exhibits write-once-read-multiple-times (WORM) feature.[2].

Because the electronic structure of alloyed TMDs can be precisely tuned by adjusting the composition ratio, Alloyed TMDs have been extensively studied in the fields of transistors and sensors in recent years. [3-5]. Preparation methods of 2D alloyed TMD $Mo_{1-x}W_xS_2$ have been well established, and they have been subsequently used in field effect transistors. [11]. This indicates that alloyed TMDs (such as $Mo_{0.5}W_{0.5}S_2$) have strong scientific value and application potential in the field of microelectronics. However, compared with traditional TMDs - based memristors, research on alloyed TMDs-based memristors is less extensive.

In this study, we prepared a TSM based on two-dimensional alloyed $Mo_{0.5}W_{0.5}S_2$ NS. The TSM exhibits a low threshold voltage (0.2V) and a high switching ratio (10^3). And based on TSM, we successfully constructed a neuron circuit with LIF model characteristics, thereby realizing the function of artificial neuron spike coding. This work provides a viable reference for the development of alloyed TMDs-based TSM.

II. EXPERIMENTAL DETAILS

Device fabrication: A dispersion of alloyed $Mo_{0.5}W_{0.5}S_2$ nanosheets at a concentration of 1 mg/ml (0.1 ml) was mixed with a PVP ethanol solution at a concentration of 20 mg/ml (0.9 ml). The mixture was then subjected to ultrasonication for 2 hours to obtain a well-dispersed PVP-nanosheet solution. The mixture is then spun onto the glass substrate with patterned ITO. After annealing at 120 °C for 3 hours, the metal top electrode was deposited using metal thermal evaporation (thickness Ag: 50 nm).

Electrical measurements: The electrical characteristics of the prepared devices were measured using the Keysight B2902A semiconductor analyzer, which also provided the input rectangular pulse (1 kHz, 8 V) during the testing of artificial neurons. The constructed LIF neuron circuit used an MR6000 memory recorder to measure the output voltage characteristics.

III. RESULTS AND DISCUSSION

Fig.1(a) shows the structure of TSM. The TSM has a vertical structure with ITO electrode at the bottom and Ag electrode at the top. The dielectric layer consists of Polyvinyl Pyrrolidone (PVP) doped with two-dimensional alloyed $Mo_{0.5}W_{0.5}S_2$

Hang Li, Wanhong Luan, Wei Zeng, JiYu Zhao and Ye Zhou are with Institute for Advanced Study, Shenzhen University, Shenzhen 518060, P. R. China. Guanglong Ding is with State Key Laboratory of Radio Frequency Heterogeneous Integration, Shenzhen University, Shenzhen 518060, PR China. Su-Ting Han is with Department of Applied Biology and Chemical Technology, The Hong Kong Polytechnic University, Hung Hom, Hong Kong SAR, P. R. China.

979-8-3315-2209-4/25 $31.00 © 2025 IEEE

NS. We chose PVP due to its excellent water solubility, mechanical properties, thermal stability, low dielectric constant, and chemical resistance. Fig.1(b) presents an atomic force microscope (AFM) image of the $Mo_{0.5}W_{0.5}S_2$ nanosheets. The thickness of the nanosheet is about 1.8 nm, which proves its two-dimensional structure. Fig.2 (c) shows 50 cycle IV characteristic curves of the same alloyed $Mo_{0.5}W_{0.5}S_2$ NS based memristor, indicating that the two-dimensional $Mo_{0.5}W_{0.5}S_2$ memristor exhibits a low threshold voltage (0.2V) and a high switching ratio (10^3). Fig.1(d) presents a statistical plot of the threshold voltage of the memristors, demonstrating that the memristor has good uniformity.

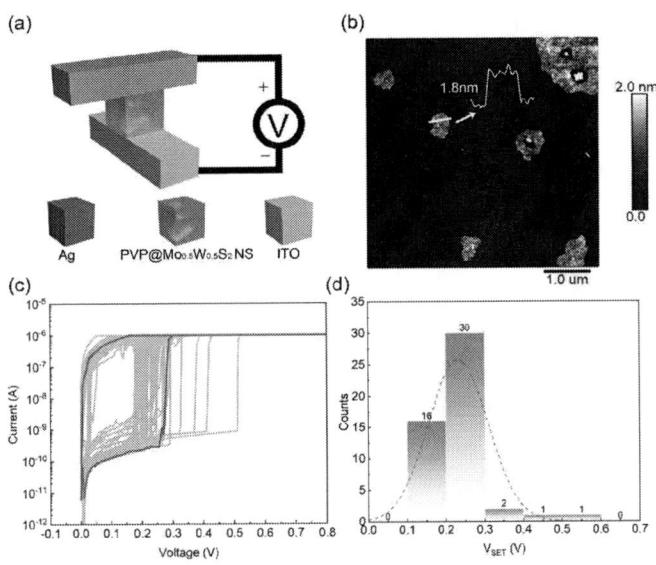

Fig. 1. (a) Schematic diagram of device structure. (b) The AFM image of $Mo_{0.5}W_{0.5}S_2$ NS. (c) The I-V curves (50 cycles) $Mo_{0.5}W_{0.5}S_2$ NS based memristor. (d) The threshold voltage (SET voltage) statistical analysis of $Mo_{0.5}W_{0.5}S_2$ NS based memristor.

As shown in Fig.2 (a), in biological systems, biological neurons receive input spikes from other neurons via synapses and transmit action potentials to downstream neurons when the membrane potential reaches a threshold. Fig.2(b) shows the memristive LIF artificial neuron circuit. [9-10]. In LIF artificial neuron circuit, when the accumulated capacitor potential reaches a threshold voltage higher than that of the TSM, the neuron fires and outputs a spike pulse. Its operation mechanism is divided into an integral process and a fire process. [6-8]. During the integration process, the input rectangular pulse waveform is applied to the capacitor, charging it until the TSM voltage reaches the threshold voltage, at which point the TSM switches from the high-resistance state (HRS) to the low-resistance state (LRS) and forms Ag filaments. During the fire process, the accumulated charge on the capacitor is released through the TSM in the low-resistance state, and a spike voltage across R2 is detected. When the TSM voltage drops below a certain level, the Ag filaments break, and the device reverts to the HRS. LIF artificial neuron circuit repeatedly charges and discharge capacitors, forming and breaking Ag

filaments in the active layer, thereby completing the neuronal spike coding function. Fig.2(c) shows the integration process on the capacitor. Fig.2(d) shows the fire process of LIF artificial neuron circuits.

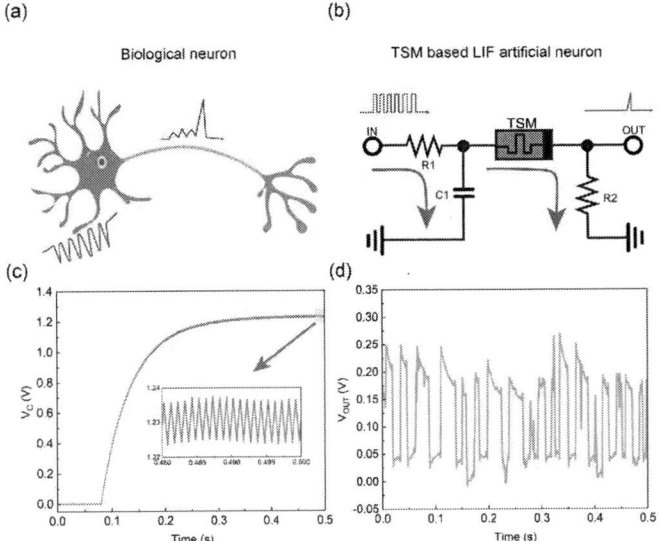

Fig. 2. (a) The schematic diagrams of biological neuron and (b) memristive artificial neuron circuit ($R_1 = 1M$, $C_1=100nF$, $R_2=100K$) . The integrating (c) and firing (d) behaviors of memristive LIF artificial neuron.

IV. CONCLUSION

In summary, we developed a TSM based on two-dimensional alloyed $Mo_{0.5}W_{0.5}S_2$ NS. And an LIF artificial neuron based on the TSM has been successfully constructed, which can realize integration and firing characteristics of biological neurons. And it should be noted that we proposed TSM based on two-dimensional alloyed $Mo_{0.5}W_{0.5}S_2$ NS features a low threshold voltage (0.2V), a high switching ratio (10^3), good repeatability, and stability, it provides a feasible scheme for the realization of artificial neurons.

V. ACKNOWLEDGMENTS

The Guangdong Basic and Applied Basic Research Foundation (Grant No. 2023A1515012479, 2024B1515040002), the Science and Technology Innovation Commission (Grant No. JCYJ20220818100206013 and JCYJ20230808105806014), the Frontier and Interdisciplinary Research Assessment Project, Academic Divisions of the Chinese Academy of Sciences (Grant No. XK2023XXA002) and Joint Research Program NTUT-SZU.

REFERENCES

[1] Qing Cao, Limiao Xiong, Xudong Yuan, Pengcheng Li, Jun Wu, Hailin Bi, Jun Zhang, "Resistive switching behavior of the memristor based on WS2 nanosheets and polyvinylpyrrolidone nanocomposite", Appl. Phys. Lett., 120,232105(2022).

[2] Zhang P, Gao C, Xu B, Qi L, Jiang C, Gao M, and Xue D, "Structural Phase Transition Effect on Resistive Switching Behavior of MoS2-Polyvinylpyrrolidone Nanocomposites Films for Flexible Memory Devices", Small.,15, pp.,2077-2084(2016).

[3] Haoxin Huang, Jiajia Zha, Shisheng Li, Chaoliang Tan, "Two-dimensional alloyed transition metal dichalcogenide nanosheets: Synthesis and applications", Chinese Chemical Letters.,33, pp.,163–176(2022).

[4] Li S, Hong J, Gao B, Lin Y.C, Lim H.E, Lu X, Wu J, Liu S, Tateyama Y, Sakuma Y, "Tunable Doping of Rhenium and Vanadium into Transition Metal Dichalcogenides for Two-Dimensional Electronics", Adv. Sci. 8, 2004438(2021).

[5] Jalouli A, Kilinc M, Marga A, Bian M, Thomay T, Petrou A, Zeng H, "Transition metal dichalcogenide graded alloy monolayers by chemical vapor deposition and comparison to 2D Ising model", J.Chem.Phys.,156,134704(2022).

[6] C. Chen, Y. He, H. Mao, L. Zhu, X. Wang, Y. Zhu, Y. Zhu, Y. Shi, C. Wan, Q. Wan, "A Photoelectric Spiking Neuron for Visual Depth Perception", Adv. Mater.,34,2201895(2022).

[7] Yan Wang, Yue Gong, Shenming Huang, Xuechao Xing, Ziyu Lv, Junjie Wang, Jia-Qin Yang, Guohua Zhang, Ye Zhou, Su-Ting Han, "Memristor-based biomimetic compound eye for real-time collision detection", Nature. Communications.,12,5979 (2021).

[8] G. Liu et al., "Experimental Demonstration of Coplanar NbOx Mott Memristors for Spiking Neurons," IEEE Electron Device Letters, 45, pp., 708-711(2024).

[9] H. Wang, Y. Xu, R. Yang and X. Miao, "A LIF Neuron with Adaptive Firing Frequency Based on the GaSe Memristor," IEEE Transactions on Electron Devices, 70, pp. 4484-4487(2023).

[10] D. Dev et al., "2D MoS2-Based Threshold Switching Memristor for Artificial Neuron," IEEE Electron Device Letters,41, pp. 936-939(2020).

[11] Jiandong Yao, Zhaoqiang Zheng, Guowei Yang, " Promoting the Performance of Layered-Material Photodetectors by Alloy Engineering," ACS Appl. Mater. Interfaces,8, pp. 12915-12924(2016).

A near-infrared responsive P3HT synaptic transistor doped by upconversion quantum dots

Wanhong Luan   Hang Li   Minglin Zheng   Zherui Zhao   Wei Zeng   Wenbin Zhang   Jiyu Zhao

Guanglong Ding   Su-Ting Han   Ye Zhou

Abstract—This study introduces a novel organic artificial synaptic device combining lanthanide upconversion quantum dots (UCQDs) and poly3-hexylthiophene (P3HT) for intelligent near-infrared (NIR) image recognition in lowlight conditions. The device leverages the unique properties of UCQDs to convert NIR light into visible light, enhancing the optical response of P3HT-based synaptic transistors. The synaptic transistor exhibits key features of biological synaptic behavior, such as excitatory postsynaptic current (EPSC) and pin-pulse facilitation (PPF), and offers excellent NIR detection sensitivity and stability. This research paves the way for the development of advanced intelligent NIR image recognition systems and provides valuable insights for researchers in related fields.

Index Terms—Near-infrared responsive, P3HT synaptic transistor, Upconversion quantum dots

I. Introduction

In autonomous driving technology, intelligent image processing is used to analyze road conditions and identify traffic signs and obstacles in real time. Through high-precision image processing, the automatic driving system can respond quickly to ensure driving safety. The exponential growth of machine vision information poses a major challenge to computing systems based on traditional Von Neumann architectures [1]–[3]. Von Neumann architecture is a classical computer system design scheme, its core lies in program storage and sequential execution, that is, computer programs and data are stored in the same memory, and through the central processing unit (CPU) sequential execution of the memory instructions [4]. However, with the rapid development of machine vision technology, this architecture has gradually exposed its limitations in terms of computational efficiency and power consumption. Data needs to be read and written frequently between the CPU and memory, creating a performance bottleneck [5]–[7]. However, the biological visual nervous system shows high accuracy and efficiency in perceiving and processing visual information. First, the efficient computing strategy of biological retina provides inspiration for highperformance intelligent image

recognition, and promotes the development of computing strategies within sensors. This strategy reduces data transfer requirements and improves system efficiency by integrating sensing, computing, and storage functions [8]. Near-infrared (NIR) light is an attractive alternative in low-light conditions where visible light recognition systems are inadequate. NIR can not only perform visual recognition tasks in low-light environments, but also penetrate obstacles such as smoke and dust to accurately transmit image information. However, the photon energy of NIR light is limited, which poses a challenge for excitation of traditional semiconductor materials. Rare earth upconversion quantum dots (UCQDs), with their excellent fluorescence properties, including long lifetime, narrow line width and tunable emission wavelength, have opened up more paths for the advancement of optoelectronic devices [9]. Among the many types of UCQDs, the lanthanide UCQDs is particularly prominent, which has significant advantages such as large Stokes shift, sharp luminescence, long life, high light stability and low environmental toxicity. Therefore, lanthanide quantum dots are regarded as promising candidate materials to expand the application field [10]. On the other hand, poly3-hexylthiophene (P3HT), as a typical P-type organic photoelectric semiconductor material, is favored for its easy solution spin coating and high carrier mobility, and is often used as the core channel material for high-performance neuromorphic devices/systems that enable insensor computing and visual recognition functions [11]. Notably, the main absorption spectral range of P3HT closely matches the main emission peak of LaF_3: Yb/Ho UCQDs at 980 nm excitation (about 540 nm). This spectral matching makes LaF_3: Yb/Ho UCQDs an ideal choice to broaden the optical response range of P3HT-based synaptic transistors [12]. In this study, the LaF_3: Yb/Ho UCQDs and P3HT are mixed initially, and the dual functions of charge capture and light sensing are realized through the synaptic transistor, thus creating a novel organic artificial synaptic device. The device has NIR light response and insensor computing ability. The synaptic transistor successfully reproduced key features such as excitatory postsynaptic current (EPSC) under near-infrared light excitation and pinpulse facilitation (PPF) phenomenon in the simulation of biological synaptic behavior. LaF_3: Yb/Ho UCQDS-doped P3HT synaptic transistors exhibit several advantages: excellent NIR detection sensitivity, efficient data processing, and excellent stability and sensitivity. These remarkable features not only

---

Wanhong Luan, Hang Li, Zherui Zhao, Wei Zeng, Wenbin Zhang, Jiyu Zhao and Ye Zhou are with *Institute for Advanced Study, Shenzhen University*, Shenzhen, P. R. China. Minglin Zheng and Guanglong Ding are with *College of Electronics and Information Engineering , Shenzhen University*, Shenzhen, P. R. China. Guanglong Ding and Ye Zhou are with *State Key Laboratory of Radio Frequency Heterogeneous Integration, Shenzhen University*, P. R. China. Su-Ting Han is with *Department of Applied Biology and Chemical Technology, The Hong Kong Polytechnic University*, Hung Hom, Hong Kong SAR, P. R. China. Corresponding author: (Guanglong Ding, email: dinggl@szu.edu.cn; Ye Zhou, email: yezhou@szu.edu.cn)

979-8-3315-2209-4/25 $31.00 © 2025 IEEE

provide a solid foundation for the development of the next generation intelligent NIR image recognition system in dark situation, but also bring valuable experimental verification and design inspiration for researchers in related fields.

II. EXPERIMENTAL DETAILS

Initially, a silicon wafer undergoes cleaning in isopropyl alcohol, followed by two subsequent cleanings in ethanol. The LaF_3: Yb/Ho UCQDs are dissolved in toluene and thoroughly mixed with a chlorobenzene solution of P3HT using ultrasonic waves in a container until fully dissolved. The resultant mixture is then spin-coated onto the silicon wafer, with the initial spin speed set at 500 r per minute for 5 s, followed by an increased speed of 2000 r per minute for an additional 25 s. Once the spin-coating process is complete, the coated wafer is annealed for 30 min. Subsequently, a mask is placed over the semiconductor film, and 50 nm of gold are deposited as electrodes through thermal evaporation.

III. RESULTS AND DISCUSSION

By sensing infrared light in a dark environment, the synaptic transistor can form an image recognition of the target, similar to the cheek sockets of infrared sensitive creatures such as snakes. Under NIR light, the infrared photosensitive performance of the synaptic transistor is improved by adding UCQDs (LaF_3: Yb/Ho) to the P3HT channel, which acts as the NIR sensing and charge capture site (Figure 1a). In order to verify the uniformity of the organic film, Figure 1b shows the scanning electron microscope (SEM) cross-section image of organic thin film doped with UCQDs on silicon wafer with SiO_2 as dielectric layer. The thickness of the film is uniform, and the fitness of the film with the silicon wafer is pretty good. To explore how doping concentrations affect device performance, the results are shown in Figure 1c. When the doping concentration is 0 mg/ml, the current ON/OFF ratio (I_{ON}/I_{OFF}) is 4×10^5. When the doping concentration increased to 0.3 mg/ml, the ratio decreased to 1.4×10^4, which had little effect on the performance of the device. When investigating the effect of the doping concentration on the light response, we investigated the absorption of light near 540 nm as the evaluation standard. When the doping concentration of UCQDs was 0.3 mg/ml, the absorption intensity was greater than that of the undoped one, which proved that the doping of UCQDs played a positive role in the NIR light absorption of organic thin films.

The photoresponse of the device to a single pulse under NIR light confirms that it has good NIR light response performance, which reflects its potential as an artificial photosynapse. PPF behavior that is double pulse facilitation, is usually used in synaptic neuroscience research to describe the phenomenon that when two stimuli are given in a short period of time, the second stimulus triggers a stronger response than the first stimulus, which proves the effectiveness of near infrared spectroscopy as a means of stimulation. In the application situation of NIR light as a form of stimulus, the successful simulation of biologically specific PPF behavior (Figure 2b).

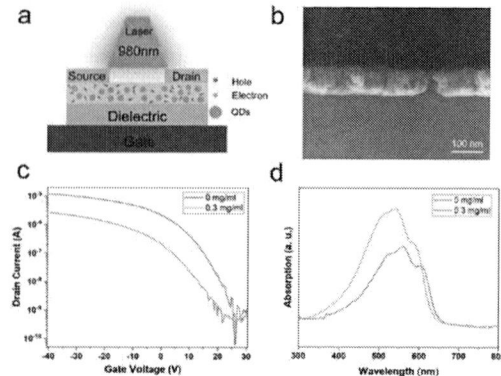

Fig. 1. (a) The device diagram of P3HT - based organic synaptic transistor. (b) The SEM cross-section image of organic thin film. (c) The transfer image of synaptic transistor with different UCQDs doping concentrations. (0 mg/ml; 0.3 mg/ml). (d) The absorption spectra of organic synaptic transistors with different UCQDs doping concentrations. (0 mg/ml; 0.3 mg/ml).

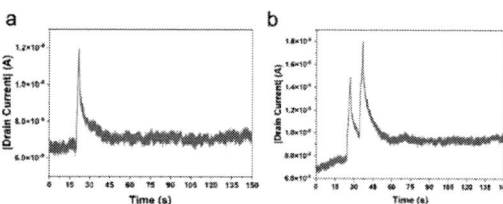

Fig. 2. (a) The current diagram of organic synaptic transistor triggered by a single light pulse (651 mW/cm^2). (b) The current diagram of organic synaptic transistor triggered by double light pulses (651 mW/cm^2).

REFERENCES

[1] Xu, K.; Peng, B. C.; Mao, H. W.; Wang, Z. P.; Gong, H. H.; Fu, C. Y.; Ren, F. F.; Yang, Y.; Wan, C. J.; Wan, Q.; Ye, J. D. Ga2O3 Bipolar Heterojunction-Based Optoelectronic Synapse Array with Visual Attention. J. Phys. Chem. Lett. 2024, 15, 556-564.

[2] Bhatnagar, P.; Patel, M.; Nguyen, T. T.; Kim, S.; Kim, J. Transparent Photovoltaics for SelfPowered Bioelectronics and Neuromorphic Applications. J. Phys. Chem. Lett. 2021, 12, 12426-12436.

[3] Dong, X. F.; Li, S. Y.; Sun, H.; Jian, L. J.; Wei, W. B.; Chen, J. B.; Zhao, Y.; Chen, J. T.; Zhang, X. Q.; Li, Y. Optoelectronic Memristive Synapse Behavior for the Architecture of Cu2ZnSnS4@BiOBr Embedded in Poly(methyl methacrylate). J. Phys. Chem. Lett. 2023, 14, 1512-1520.

[4] Wang, Y. F.; Sun, Q. J.; Yu, J.R.; Xu, N.; Wei, Y.C.; Cho, J.H.; Wang, Z.L. Boolean Logic Computing Based on Neuromorphic Transistor. Adv. Funct. Mater. 2023, 33, 2305791.

[5] Dong, Z. L.; Hua, Q. L.; Xi, J. G.; Shi, Y. H.; Huang, T. C.; Dai, X. H.; Niu, J. A.; 18Wang, B. J.; Wang, Z. L.; Hu, W. G. Ultrafast and Low-Power 2D Bi2O2Se Memristors for Neuromorphic Computing Applications. Nano Lett. 2023, 23, 3842-3850.

[6] Chen, C. S.; Zhou, Y. Q.; Tong, L.; Pang, Y.; Xu, J. B. Emerging 2D Ferroelectric Devices for In-Sensor and In-Memory Computing. Adv. Mater. 2024, 36, 2400332.

[7] Liao, C. R.; Liu, D.; Liu, Z.; Wang, J. C. Y.; Xie, X. S.; Li, J.; Zhou, G. D. Coexistence of the Negative Photoconductance Effect and Analogue Switching Memory in the CuPc Organic Memristor for Neuromorphic Vision Computing. 2024, 15, 6230-6236.

[8] Huang, P. Y.; Jiang, B. Y.; Chen, H. J.; Xu, J. Y.; Wang, K.; Zhu, C. Y.; Hu, X. Y.; Li, D.; Zhen, L.; Zhou, F. C.; Qin, J. K.; Xu, C. Y. Neuro-inspired optical sensor array for highaccuracy static image recognition and dynamic trace extraction. Nat. Commun. 2023, 14, 6736.

[9] Zhu, X. H.; Zhang, J.; Liu, J. L.; Zhang, Y. Recent Progress of Rare-Earth Doped Upconversion Nanoparticles: Synthesis, Optimization, and Applications. Adv. Sci. 2019,6, 1901358.

[10] Wu, S. W.; Han, G.; Milliron, D. J.; Aloni, S.; Altoe, V.; Talapin, D. V.; Cohen, B. E.; Schuck, P. J. Non-blinking and photostable upconverted luminescence from single lanthanide-doped nanocrystals. Proc. Natl. Acad. Sci. 2009, 106, 10917-10921.

[11] Xu, Z. H.; Chen, G. B.; Chen, S. B.; Xu, H. H. Mimicking Pain Conditioning Using an Electrolyte-Gated Organic Synaptic Transistor. Adv. Mater. Technol. 2024, 9, 2302047.

[12] Liu, C. J.; Deng, J. Y.; Gao, L.; Cheng, J. L.; Peng, Y. J.; Zeng, H. J.; Huang, W.; Feng, L. W.; Yu, J. S. Multilayer Porous Polymer Films for High-Performance Stretchable Organic Electrochemical Transistors. Adv. Electron. Mater. 2023, 9, 2300119.

Polarization Assisted n-AlGaN Electron Injection Layer in 229 nm far-UVC LEDs for Hole Blocking and Current Density Enhancement

1st M.Nawaz Sharif   2nd M.Ajmal Khan   3rd Hiromitsu Sakai   4th Kohei Fujimoto   5th Amina Yasin

6th Mitsuhiro Muta   7th Hiroyuki Yaguchi   8th Yasushi Iwaisako   9th Yuhuai Liu   10th Hideki Hirayama

Abstract—An Aluminum gallium nitride (AlGaN)-based (228-230 nm)-band far-ultraviolet-C (far-UVC) light-emitting diode (LED) is safe to be used in manned environments against viruses, bacteria, and fungi. However, the external quantum efficiency (EQE) of far-UVC LEDs is quite low due to hole leakage, electron leakage, and low carrier injection efficiency. Such a hole or electron leak can promote nonradiative recombination channels and pose major challenges in Al-rich far-UVC LEDs. In this work, a new polarization-assisted n-AlGaN electron injection layer (EIL) on the n-side (n-PA) of far-UVC LED was investigated both theoretically and experimentally to generate 3D electron density. The n-PA-based LED under study was grown by MOCVD on a 4 um-thick AlN template on c-plane sapphire. Subsequently, a 350 nm-thick Si-doped $n-Al_{0.87}Ga_{0.13}N$ buffer layer and a 30 nm-thin n-side polarization-assisted (n-PA) n $Al_{0.86 \to 0.97}Ga_{0.14 \to 0.03}N$ EIL were grown. The quantum well consists of a 4-fold QW (1 nm $Al_{0.94}Ga_{0.06}N$ barrier with 3.5 nm $Al_{0.86}Ga_{0.14}N$ well), a 6 nm $ud-Al_{0.94 \to 0.86}Ga_{0.06 \to 0.14}N$ final barrier, and a 9 nm thick p-AlN electron blocking layer. The LED is completed by a 75 nm $p-Al_{0.9}Ga_{0.1}N$ layer and finally a p-GaN layer (75 nm thick) on top, both layers heavily Mg-doped (10^{20} cm^{-3}). When Al-deposition is started from Ga-face crystals and graded from low Al composition (0.86) to high Al composition (0.97) in n-AlGaN EIL, the polarization-bound charge is positive, and it induces the generation of mobile 3D electrons. As a result, the EQE and light output power of the polarized far-UVC LED on-wafer under continuous-wave operation at RT were enhanced to 0.22% and 1.2 mW. More interestingly, the emission wavelength was pulled toward a shorter wavelength of 229 nm, and also the operating voltages were reduced in the n-PA LED. The new n-PA-based LED has been investigated theoretically using the SiLENSe simulation model. It was observed that the hole current density leakage toward the n-side was remarkably suppressed when compared to the conventional flat n-AlGaN EIL-based LED. Finally, a quite improved carrier injection efficiency (CIE) of 58% (n-PA LED) was observed when compared to the referenced structure, having a CIE of 29% (Flat n-AlGaN EIL LED).

Index Terms—AlGaN far-UVC LED, Polarization Assisted n-AlGaN Electron Injection Layer, Mobile 3D electrons

Author 1 Author 2 Author 5 and Author 10 are with *Cluster for Pioneering Research, RIKEN*, Saitama, Japan. Author 3 is with *Shin-Etsu Chemical Co., Ltd*, Tokyo, Japan. Author 4 Author 7 and Auther 8 are with *Graduate School of Science and Engineering, Saitama University*, Saitama, Japan. Author 6 is with *Nippon Tungsten Co., Ltd.*, Fukuoka, Japan. Author 9 is with *School of Electrical and Information Engineering, Zhengzhou University* Zhengzhou, China. Corresponding author: ((M.Nawaz Sharif, email: muhammadnawaz.sharif@riken.jp)

I. INTRODUCTION

We need a clean and green UV solution to protect human lives and nature from the adverse effects of viruses, bacteria, fungi, and other hazardous environmental pollutants, including Hg and CO2. According to a United Nations Organization (UNO) report, around 700,000 patients worldwide die every year from an infection with multidrug-resistant organisms (MROs) [1] [2]. Fungi such as Candida Auris have become more widespread throughout the USA, and deaths related to severe fungal infections are rising in the USA [3] [4]. Germicidal and virucidal inactivation of Severe Acute Respiratory Syndrome Coronavirus 2 (SARS-CoV-2), including MROs, i.e., Candida Auris and Methicillin-resistant Staphylococcus Aureus (MRSA), are possible in the manned environment using far-UVC light. Far-UVC light can be used for food, water, and air sterilization as well as for all kinds of surface disinfections in public spaces, including hospitals, city halls, airports, Schools, etc. Far-UVC light has the potential for clean refrigeration, air conditioning, safe food processing, and transportation [5] [6]. The expected risk of a new type of bacteria, pandemic, and viruses from outer space by astronauts also exists, and we can safely inactivate such new kinds of germs in the space station using light weight far-UVC LED module [7].

In recent years, the wavelength band below 230nm has been referred to as 'far-UVC', and the UVC region has gotten much attention due to its safe applications in the manned environment. The irradiated 230nm far-UVC light does not reach inside the cell and can only destroy the viruses attached to the surface without destroying the living human or animal cell. However, in the region above 230nm, the penetration length is sufficiently long, so the light penetrates the cornea of the eyes and stratum corneum on the cell surface and penetrates to the inside of the cell, destroying DNA and carrying the risk of developing cancer and other skin diseases [8]. The 222nm light source uses excimer lamps from USHIO. Inc requires a large excimer lamp body and a high-voltage power supply. A toxic halogen gas is used as a material source in the lamp, and the lamp itself also emits UV rays at wavelengths other than 222nm. A cut-off filter is used to obtain a specific wavelength, so there is a possibility of exposure to UV rays due to filter damage [9]. In contrast, group III-nitride semiconductors (AlGaN-based semiconductors) are used as materials for LEDs in the UV region because the band gap of AlGaN-based semiconductors can be changed in a wide range from 3.4 to 6.2 eV by changing their composition ratio [10]. Till now, the external quantum efficiency (EQE) of short-wavelength LEDs is much lower than that of long-wavelength LEDs. One reason for this is the decrease in internal quantum efficiency (IQE) in AlGaN-LEDs due to recombination without emission at non-emissive recombination centers results from the dislocation density and lattice mismatch between AlGaN and AlN [11]. One important factor in the decline in EQE is that the injection efficiency of carriers is significantly reduced with shorter wavelengths. Also, the activation energy of Si in the n-AlGaN EIL and Mg in the p-AlGaN hole injection layer (HIL) are quite high for high Al-contents, which also reduces the carrier injection efficiency (CIE) [12]. The light extraction efficiency (LEE) can be degraded due to low transmission if the Al content is not optimized in the n-AlGaN buffer layer (BL) and n-AlGaN EIL for the target emission wavelength [13]. Therefore, in this work, a new polarization-assisted n-AlGaN EIL on the n-side (n-PA) of far-UVC LED was investigated both theoretically and experimentally to generate 3D electron density and improve the CIE, which leads to higher EQE.

II. GROWTH METHOD

A 4 μm AlN film is deposited over a pseudo-substrate using an SR4000 MOCVD system manufactured by Taiyo Nippon Sanso [14]. The AlN template is inserted as a buffer layer. A high-temperature environment was maintained to promote surface diffusion of Al atoms and Al free radicals to grow a better AlN crystal. The horizontal type SR4000 can be operated at relatively high temperatures, and it is suitable for high-quality epitaxial growth of the AlN template on c-Sapphire [15]. The AlN template had a good crystallinity of about 150 arcsecs and 350 arcsecs of the full width at half maximum (FWHM) of the (0002) and (10-12) planes by ω-scanning of X-ray diffraction, which is a precious index of crystallinity. For sources of materials trimethylgallium (TMGa), trimethylaluminum (TMAl), ammonia (NH3), tetraethylsilane (TESi), and bis-cyclopentadienyl magnesium (Cp2Mg) were introduced for gallium (Ga), aluminum (Al), nitrogen (N), silicon (Si), and magnesium (Mg), respectively. As carrier gases, hydrogen gas (H2) and nitrogen gas (N2) were used. The growth condition of sample-Ref has been discussed elsewhere [16]. First, a flat 350nm n-$Al_{0.87}Ga_{0.13}N$ buffer layer was used in the reference LED as given in Fig 1. A 30nm new polarization-assisted n-AlGaN EIL on the n-side (n-PA) of far-UVC LED was introduced to exploit the 3D electron generation (see (1), as shown in Fig 1 2 3 4. The 3D electron generation in the n-AlGaN EIL through the polarization effect shall be achieved by growing compositionally graded Al in Si-doped n-AlGaN crystal instead of a sharp heterojunction.

The bound polarization-induced charge spreads to the bound 3D form by the well-known divergence law as:

$$\rho_\pi(z) = -\Delta.P(z) \qquad (1)$$

In Eq. (1) π is the volume charge density in the polar (z) direction, and Δ is the well-known divergence operator. When deposition is started from Ga-face crystals and graded from low Al composition to high Al composition, AlGaN, the polarization-bound charge is positive and induces the generation of mobile 3D electrons. Finally, the results of polarization assisted n-AlGaN EIL in 229 nm far-UVC LEDs for hole blocking and current density enhancement were discussed in the next section.

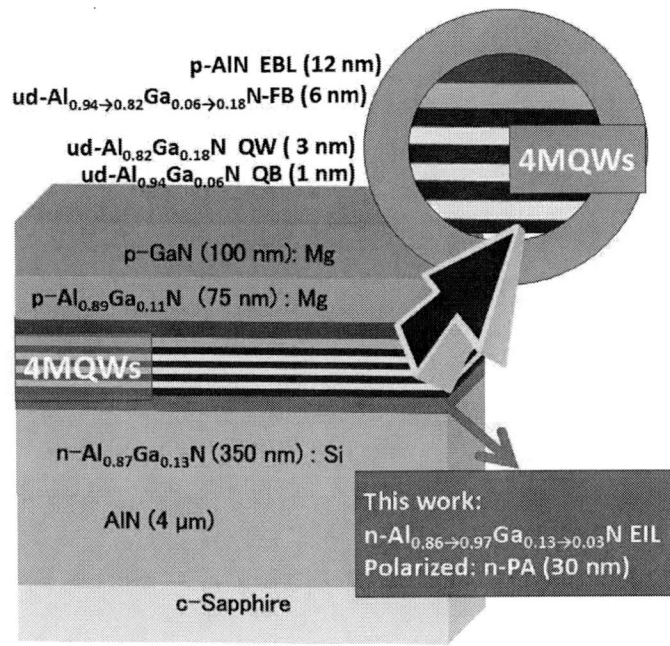

Fig. 1. Schematic of the AlGaN-based far-UVC LED structure (n-PA) grown on c-plane sapphire

III. RESULTS AND DISCUSSION

The performance evaluation (I-V-L) at room temperature (RT) under continuous-wave (CW) operation on wafer was attempted. The Ni/Au gold as a p-electrode on the top of p-GaN contact layer was evaporated and Indioum (In)-dot as n-electrode was pressed mechanically on the scribed part of n-AlGaN EIL. As a result, the EQE and light output power (LOP) of the polarized far-UVC LED (n-PA LED) on-wafer under CW operation at RT were enhanced to 0.22% as seen in Fig. 3 4. More interestingly, the emission peak wavelength was pulled toward a shorter side of 229 nm Fig. 4. The SiLENSe STR simulation model supported these results. In the device simulation, it was observed that the hole current density leakage toward the n-side was remarkably suppressed using a 30 nm-thin n-PA layer when compared to the reference flat n-AlGaN EIL-based LED as shown in Fig. 2. Similarly, the current density was greatly enhanced to 60 A/cm2 when compared to the flat n-AlGaN EIL-based LED (52 A/cm2). Finally, a quite improved CIE of 58% (for n-PA) was observed in the simulation compared to the referenced structure, having a CIE of 29%. The Al-rich based far-UVC LED has several crucial challenges of optical polarization, low carrier injection and donor–acceptor pair (DAP)-like emissions, etc [17] [18] [19] and we need to overcome such challenges in the future. These results open a new hope for promoting career injection efficiency and better hole blocking in the Al-rich far-UVC emitters, including LEDs and Laser diodes. It might be useful for peak emission wavelength control, too.

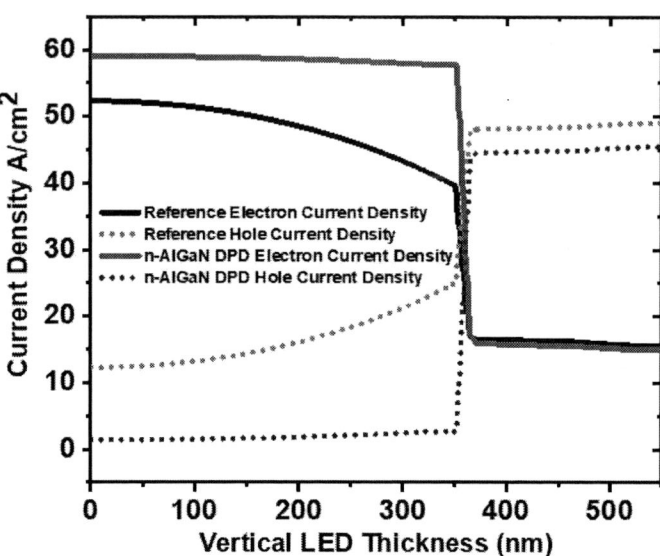

Fig. 2. Calculated current densities of the n-PA LED and Reference LED (flat type n-AlGaN EIL)

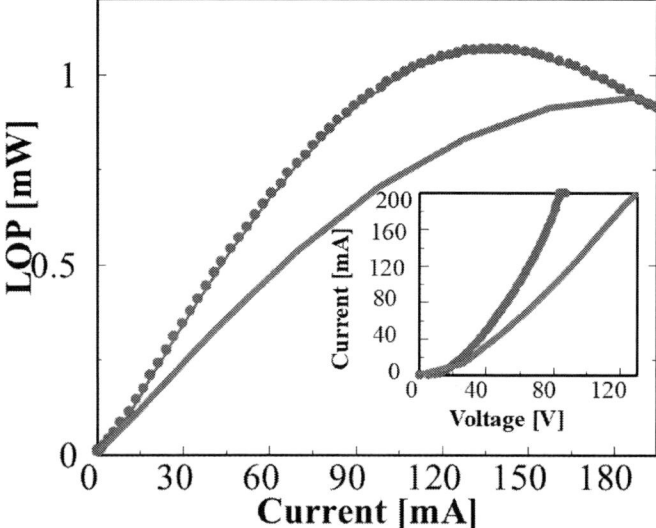

Fig. 3. Current vs output power (I–L) characteristics, just on bare-wafer level conditions under CW-operation at RT.

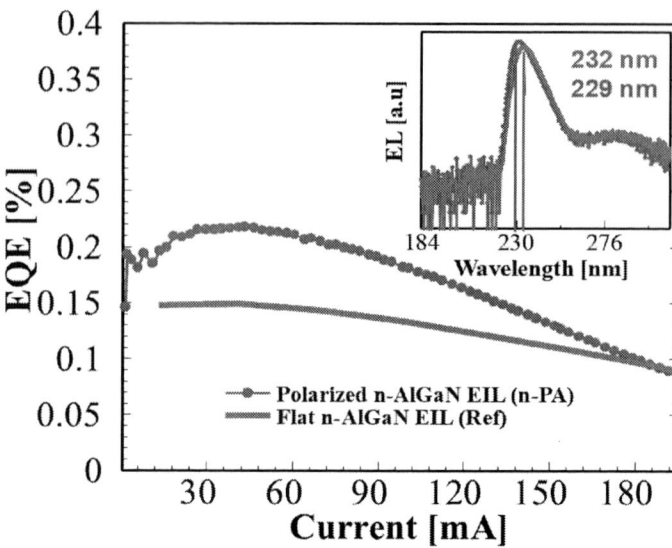

Fig. 4. Current vs EQEs (I-EQEs) characteristics, just on bare-wafer level conditions under CW-operation at RT. Electroluminescence (EL) spectral emission intensity under dc drive of 20mA at an emission wavelength of 229-232 nm is shown in the inset.

Summary

In summary, a new polarization-assisted n-AlGaN EIL on the n-side (n-PA) of far-UVC LED is used to exploit the 3D electron generation. As a result, the EQE and light output power of the polarized far-UVC LED on-wafer under continuous-wave operation at RT are enhanced to 0.22% and 1.2 mW. More interestingly, the emission wavelength is pulled toward a shorter wavelength of 229 nm, and the operating voltages were greatly reduced in the n-PA LED. The new n-PA-based LED has been investigated theoretically using the SiLENSe simulation model. It is observed that the hole current density leakage toward the n-side is remarkably suppressed when compared to the conventional flat n-AlGaN EIL-based LED. Finally, a quite improved CIE of 58% (n-PA LED) is observed when compared to the referenced structure, having a CIE of 29% (Flat n-AlGaN EIL LED).

References

[1] Glaab, J., et al., Skin tolerant inactivation of multiresistant pathogens using far-UVC LEDs. Scientific reports, 2021. 11(1): p. 14647.

[2] Assembly, U.G. and t. Committee, Financing of the United Nations Multidimensional Integrated Stabilization Mission in Mali: report of the 5th Committee: General Assembly, 70th session. 2016.

[3] O'neill, J., Antimicrobial resistance: tackling a crisis for the health and wealth of nations, in Rev. Antimicrob. Resist. 2014.

[4] Lyman, M., et al., Worsening spread of Candida auris in the United States, 2019 to 2021. Annals of Internal Medicine, 2023. 176(4): p. 489-495.

[5] Welch, D., et al., Far-UVC light: A new tool to control the spread of airborne-mediated microbial diseases. Scientific Reports, 2018. 8(1): p. 2752.

[6] Gomes, I., et al., Ultraviolet C irradiation: A promising approach for the disinfection of public spaces? 2023.

[7] Sharma, V.K. and H.V. Demir, Bright future of deep-ultraviolet photonics: Emerging UVC chip-scale light-source technology platforms, benchmarking, challenges, and outlook for UV disinfection. Acs Photonics, 2022. 9(5): p. 1513-1521.

[8] Yamano, N., et al., Long-term effects of 222-nm ultraviolet radiation C sterilizing lamps on mice susceptible to ultraviolet radiation. Photochemistry and photobiology, 2020. 96(4): p. 853-862.

[9] Kim, S.-S., et al., Application of the 222 nm krypton-chlorine excilamp and 280 nm UVC light-emitting diode for the inactivation of Listeria monocytogenes and Salmonella Typhimurium in water with various turbidities. Lwt, 2020. 117: p. 108458.

[10] Kaplar, R., et al., Ultra-wide-bandgap AlGaN power electronic devices. ECS Journal of Solid State Science and Technology, 2016. 6(2): p. Q3061.

[11] Ban, K., et al., Internal quantum efficiency of whole-composition-range AlGaN multiquantum wells. Applied physics express, 2011. 4(5): p. 052101.

[12] Khan, M.A., et al., Achieving 9.6% efficiency in 304 nm p-AlGaN UVB LED via increasing the holes injection and light reflectance. Scientific reports, 2022. 12(1): p. 2591.

[13] Maeda, N. and H. Hirayama, Realization of high-efficiency deep-UV LEDs using transparent p-AlGaN contact layer. physica status solidi (c), 2013. 10(11): p. 1521-1524.

[14] Hirayama, H., et al., 231–261nm AlGaN deep-ultraviolet light-emitting diodes fabricated on AlN multilayer buffers grown by ammonia pulse-flow method on sapphire. Applied Physics Letters, 2007. 91(7).

[15] Nagamatsu, K., et al., Reduction of parasitic reaction in high-temperature AlN growth by jet stream gas flow metal–organic vapor phase epitaxy. scientific reports, 2022. 12(1): p. 7662.

[16] Ajmal Khan, M., et al., Estimation of Junction Temperature in Single 228 nm-Band AlGaN Far-Ultraviolet-C Light-Emitting Diode on c-Sapphire Having 1.8 mW Power and 0.32% External Quantum Efficiency. physica status solidi (a), 2024: p. 2400064.

[17] Shahzeb Malik and M. Ajmal Khan et al., Polarization-dependent hole generation in 222 nm-band AlGaN-based Far-UVC LED: a way forward to the epi-growers of MBE and MOCVD. J. Mater. Chem. C, 2021,9, 16545-16557 (2021).

[18] Alexandra Ibanez et al. Excitons in (Al,Ga)N quantum dots and quantum wells grown on (0001)-oriented AlN templates: Emission diagrams and valence band mixings. J. Appl. Phys. 134, 193103 (2023).

[19] M. Ajmal Khan, et al. Milliwatt-Power AlGaN Deep-UV Light-Emitting Diodes at 254 nm Emission as a Clean Alternative to Mercury Deep-UV Lamps. Physica Status Solidi (a), 220 (1) (2023), p. 2200621. .

Polarization-Modulated Elliptical Plasmonic Synapses for Optical Neural Network Applications

Chengyan Zhong[1,2], Xiaobo She[1,2], Xiang Wang[1,2], Yu Liu[1,2*]

[1] College of Integrated Circuit Science and Engineering, Nanjing University of Posts and Telecommunications, Nanjing 210023, P. R. China
[2] National and Local Joint Engineering Laboratory of RF Integration and Micro-Assembly Technology, Nanjing 210023, P. R. China

* Corresponding author's email: liu_yu_24@163.com

Abstract—This work presents a plasmonic aluminum (Al) patch synapse with an elliptical geometry modulated by polarization angle for an optical neural network. Compared to conventional rectangular structures, elliptical patches offer superior fabrication advantages with reduced processing deviations and enhanced structural symmetry. The distribution and propagation characteristics of the electromagnetic field as they vary with polarization angle are investigated using finite-difference time-domain simulation. As the polarization angle increases from 0° to 90°, the synaptic weights are quantified within a range of 0.15 to 0.9. We capitalize on these polarization-dependent transmission characteristics to develop an optical computing platform tailored for Fashion-MNIST image classification tasks, achieving a remarkable classification accuracy of 86.8%. This approach paves the way for novel integration of plasmonic synaptic elements in optical neural network applications.

Keywords —Optical neural networks, nanophotonic, image classification, elliptical plasmonic synapses

I. INTRODUCTION

The fusion of optical computing and machine learning algorithms introduces an exciting new realm of possibilities. Conventional electronic neural networks face challenges with power consumption and processing speed, especially when handling large-scale datasets [1][2]. In contrast, optical neural networks show great potential due to their parallel processing capabilities and low energy consumption [3][4].

Recent advances have introduced various polarization-based approaches for optical neural networks. Previous work by Zhong et al. [5] demonstrated polarization-encoded neural networks using simplified grating patches with rectangular geometries, while Guo et al. [6] explored chromatic plasmonic polarizer-based synapses utilizing rectangular structures for all-optical convolutional neural networks. While these approaches established the viability of polarization-modulated weights, they relied on rectangular patches that present fabrication challenges including edge irregularities and structural asymmetry.

In this work, we explore a compact nanoscale optical neural network with an elliptical Al patch synapse. By manipulating the polarization angle to modulate the transmittance, we implement a novel weight modulation method for optical neural networks that overcomes the limitations of previous rectangular designs. The Fashion-MNIST classification challenge demonstrates the device's capability in machine learning applications, effectively connecting photonic engineering with computational intelligence.

II. RESULTS AND SIGNIFICANCE

The proposed plasmonic synaptic architecture is composed of an array of elliptical Al patches on a dielectric substrate. On the substrate, the Al patches are deposited at different angles as illustrated in Figure 1, simulating the process of retinal recognition of different polarization information. The polarization angle is the core parameter for regulating the spatial light field.

In the finite-difference time-domain (FDTD) simulation, we simulate the electromagnetic field distribution of the Al patch at different polarization angles. The simulation region is a 0.25 μm × 0.25 μm × 0.6 μm three-dimensional region. Periodic boundary conditions are set at 'x min', 'x max', 'y min' and 'y max', while perfectly matched layer (PML) boundary conditions are set at 'z min' and 'z max'. A plane wave source is used to excite the structure.

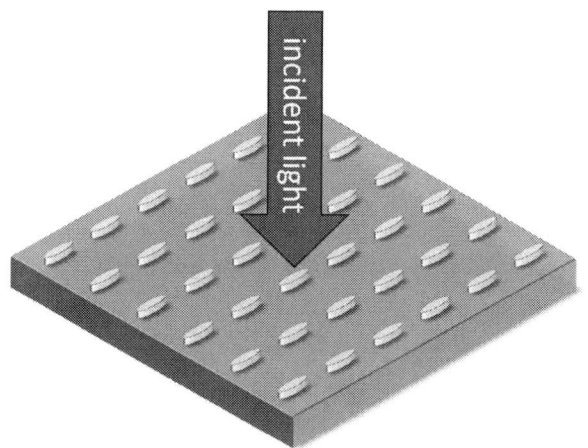

Fig.1 Schematic illustration of the proposed plasmonic synaptic architecture composed of an array of Al patches and a dielectric substrate.

Figure 2 shows the simulated electric field distribution (|E|) on the surface of the structure at three representative polarization angles (0°, 45°, and 90°). At a polarization angle of 0° (Figure 2(a)), the electric field distribution is primarily concentrated along the major axis of the ellipse, resulting in relatively low transmittance. As the polarization angle increases to 45° (Figure 2(b)), the electric field along the major axis gradually weakens while strengthening along the

979-8-3315-2209-4/25 $31.00 © 2025 IEEE

minor axis. At a polarization angle of 90° (Figure 2(c)), the electric field intensity along the minor axis reaches its maximum, resulting in maximum transmittance.

Figure 3 shows the spectra of transmittance, reflectance, and absorptivity at different polarization angles. Near wavelengths of 600 nm, the transmittance modulation range reaches its maximum, spanning from approximately 0.15 to 0.9. This wavelength band provides the optimal range for weight quantification, enabling maximum weight adjustment flexibility.

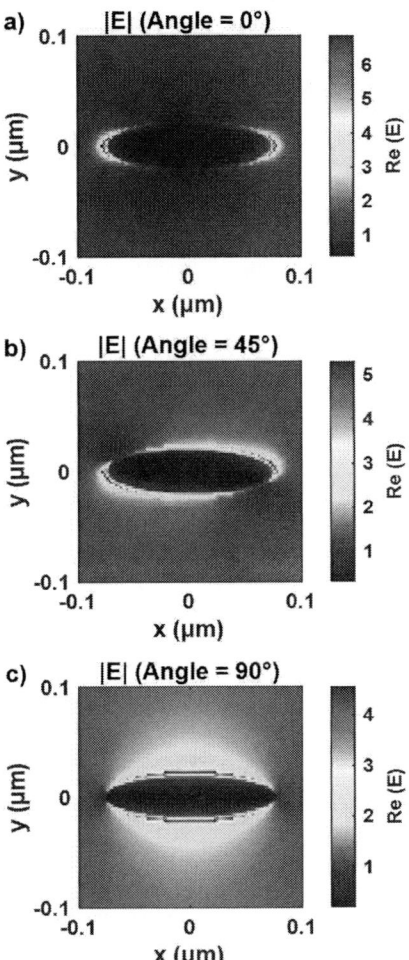

Fig.2 The electric field distributions (|E|) of the Al patch under different polarization angles: (a) Angle=0°, showing electric field concentration along the major axis of the ellipse; (b) Angle=45°, showing a gradual weakening of the electric field along the major axis and strengthening along the minor axis; (c) Angle=90°, showing electric field concentration along the minor axis of the ellipse.

Figure 4 presents the schematic overview of our optical neural network designed for Fashion-MNIST classification. It features a sample image as the network's input on the left. The architecture of the neural network, situated on the right, comprises fully connected layers. Within these layers, each connection represents a weight, which is realized by manipulating the transmission characteristics of the elliptical plasmonic elements. This design harnesses light signal

propagation to facilitate highly efficient parallel processing of complex image data.

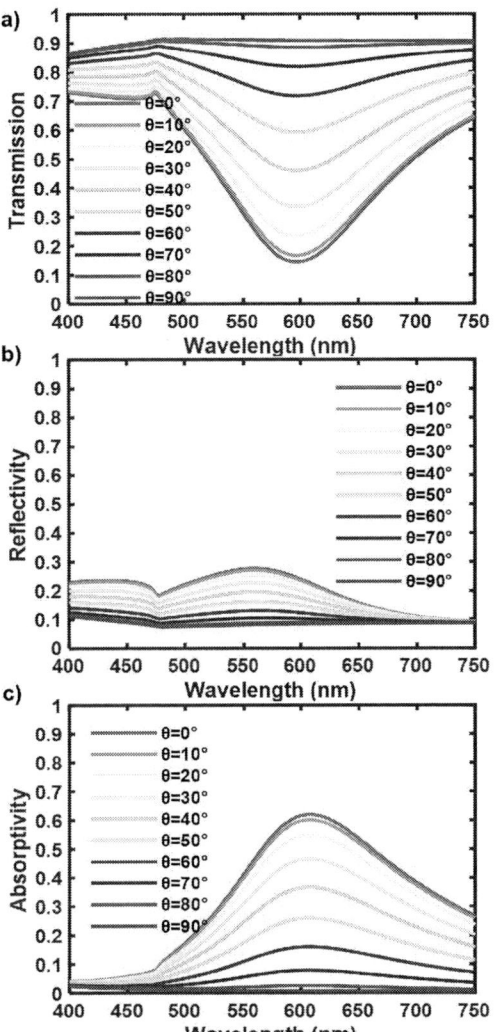

Fig.3 The (a) transmission, (b) reflectivity, and (c) absorptivity spectra for different polarization angles.

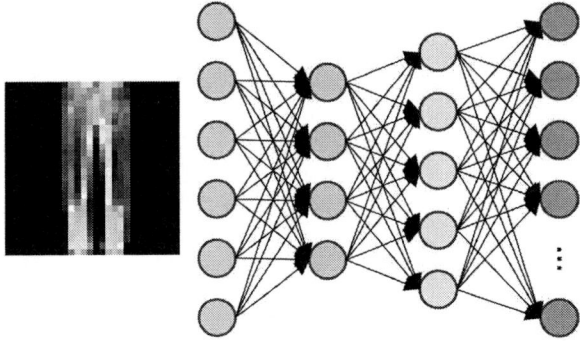

Fig.4 An illustration of the neural network architecture utilized for Fashion-MNIST classification, depicting the transformation of image inputs (shown on the left) into classification outputs via an optically implemented series of fully connected neural layers.

979-8-3315-2209-4/25 $31.00 © 2025 IEEE

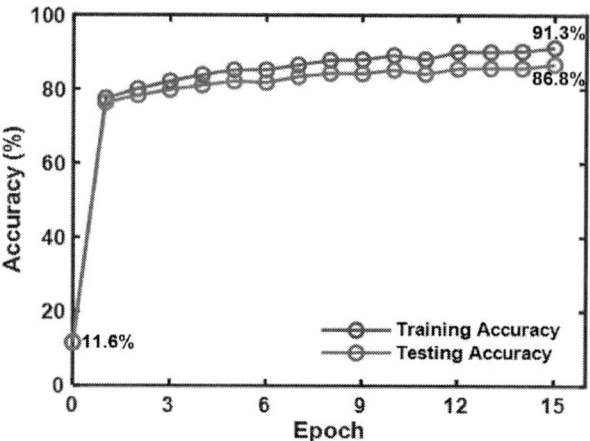

Fig.5 The classification accuracy of the Fashion-MNIST dataset increased rapidly from an initial 11.6% to over 75% after just one training epoch, ultimately achieving a final testing accuracy of 86.8%.

Figure 5 illustrates the classification performance of our model, showcasing the training and testing accuracy trends across 15 epochs. Starting with a baseline accuracy of approximately 11.6% (reflective of random guessing), the network exhibits rapid improvement, exceeding 75% accuracy after just one epoch. Throughout subsequent epochs, the training accuracy (depicted by the blue curve) methodically increases to a peak of 91.3% by the conclusion of the final epoch. Simultaneously, the testing accuracy (represented by the red curve) stabilizes at approximately 86.8%. This robust performance demonstrates the efficacy of our plasmonic synapse optical neural network in performing image classification tasks with high proficiency.

III. CONCLUSIONS

This work presents an elliptical Al patch plasmonic synapse for optical neural networks. Weight adjustment in image classification is achieved through polarization-dependent transmission characteristics. The simulation results show that weight quantization from 0.15 to 0.9 is realized as the polarization angle increases from 0° to 90°. Through 15 epochs of training, the classification accuracy of the Fashion-MNIST dataset reaches 86.8%. The elliptical geometry demonstrates significant advantages over previously explored rectangular structures, including easier fabrication, reduced structural deviations, and more uniform polarization response. Future work will focus on exploring more compact plasmonic synapse designs that provide higher integration density for fully integrated optical neural networks.

ACKNOWLEDGMENTS

This work was financially supported by the National Natural Science Foundation of China (grant nos. 62401276, U23B2042), Natural Science Research Start-up Foundation of Recruiting Talents of Nanjing University of Posts and Telecommunications (Grant No. NY223161) and in part by the Jiangsu Provincial Key Research and Development Program under Grant BE2022126.

REFERENCES

[1] Noda, S., Chutinan, A., & Imada, M. Trapping and emission of photons by a single defect in a photonic bandgap structure. Nature, 407(6804), 608-610 (2000).

[2] Soljačić, M., & Joannopoulos, J. D. Enhancement of nonlinear effects using photonic crystals. Nature Materials, 3(4), 211-219 (2004).

[3] Prather, D. W., Shi, S., Murakowski, J., Schneider, G. J., Sharkawy, A., Chen, C., & Metz, B. Photonic crystal structures and applications: Perspective, overview, and development. IEEE Journal of Selected Topics in Quantum Electronics, 12(6), 1416 - 1437 (2006).

[4] Lin, X., Rivenson, Y., Yardimci, N. T., Veli, M., Luo, Y., Jarrahi, M., & Ozcan, A. All-optical machine learning using diffractive deep neural networks. Science, 361(6406), 1004-1008 (2018).

[5] Zhong, C., Wang, X., Li, L., Cui, Y., Xiao, L., Song, D., Guo, J., Huang, W., Guo, Y. Polarization-encoded neural networks with simplified grating patch. Science China Technological Sciences 68(2), 1220901 (2022).

[6] Guo, J., Liu, Y., Lin, L., Li, S., Cai, J., Chen, J., Huang, W., Lin, Y., & Xu, J. Chromatic plasmonic polarizer-based synapse for all-optical convolutional neural network. Nano Letters 23(20), 9651-9656 (2023).

Analysis of Photonic Crystal-Based Switching for Optical Neural Networks

Yu Liu[1,2], Chengyan Zhong[1,2*], Xiaobo She[1,2], Xiang Wang[1,2*]

[1] College of Integrated Circuit Science and Engineering, Nanjing University of Posts and Telecommunications, Nanjing 210023, P. R. China

[2] National and Local Joint Engineering Laboratory of RF Integration and Micro-Assembly Technology, Nanjing 210023, P. R. China

* Corresponding author's email: b22030611@njupt.edu.cn, 1023223212@njupt.edu.cn

Abstract— **This work presents a photonic crystal (PhC) optical switching device that utilizes line defects for optical signal manipulation. Through finite-difference time-domain simulations, we characterized the electromagnetic field distributions and transmission properties as functions of rod radius. The device exhibits a transition from high-reflection to high-transmission states as rod radius increases. We harnessed these radius-dependent transmission characteristics to implement an optical computing system for Fashion-MNIST image classification. This approach establishes a foundation for employing photonic crystal structures in optical neural network applications.**

Keywords — *Optical neural networks, nanophotonic, image classification*

I. INTRODUCTION

The integration of optical computing with machine learning algorithms presents a particularly exciting frontier. Traditional electronic implementations of neural networks face challenges related to power consumption and computational speed, especially when processing large datasets [1][2]. Optical neural networks offer potential advantages through parallel processing capabilities and reduced energy requirements [3][4]. Recent developments have demonstrated promising approaches using various optical technologies, including surface plasmon polariton-based neural networks [5], polarization-encoded systems [6], and plasmonic synapses for convolutional neural networks [7]. However, implementing efficient weighting mechanisms required for neural network operations in optical systems remains challenging.

In this work, we explore a photonic crystal-based optical switching device with tunable transmission properties and demonstrate its application in image classification tasks. By leveraging the relationship between the rod radius in the PhC structure and the resulting transmission coefficients, we implement a weighting mechanism suitable for optical neural network operations. The Fashion-MNIST classification task serves as a practical demonstration of the device's potential in machine learning applications, bridging the gap between photonic engineering and computational intelligence.

II. RESULTS AND SIGNIFICANCE

Currently, we are working with theoretical simulations, and we plan to implement experimental fabrication in the future. The fabrication process for photonic crystals is shown in Fig. 1 First, we will use electron beam lithography to define the photonic crystal structure patterns on a silicon

substrate. Subsequently, reactive ion etching will be employed to transfer these patterns to the silicon layer, forming arrays of dielectric rods. Finally, the device fabrication will be completed through photoresist stripping, precision polishing and surface treatment techniques.

We utilized the finite-difference time-domain (FDTD) method to simulate the electromagnetic field distributions within the PhC structure for different rod radius. The simulation domain was set to 20 μm × 20 μm with perfectly matched layer (PML) boundary conditions to eliminate unwanted reflections. A continuous wave source was used to excite the structure.

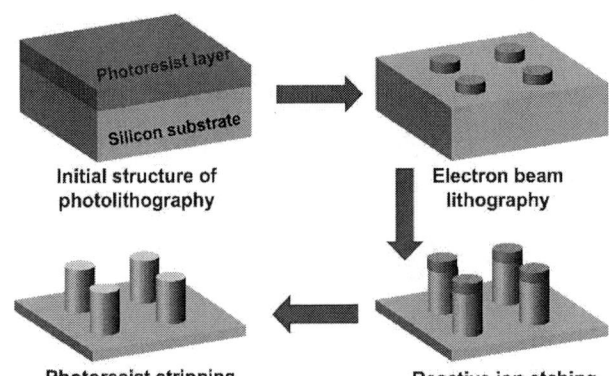

Fig.1 Process flow for photonic crystal fabrication.

Fig. 2 shows the electric field distributions (|E|) within the PhC structure for three representative rod radii. For small radius (r = 0.10 μm) in Fig. 2(a), the field predominantly shows light being reflected diagonally, with minimal transmission through the structure. As the radius increases to r = 0.30 μm in Fig. 2(b), a transition occurs with more light being channeled horizontally through the structure. At r = 0.35 μm in Fig. 2(c), a clear waveguiding effect emerges, with the majority of the incident light being transmitted along the horizontal waveguide path.

The optical computing implementation is illustrated in Fig. 3, which depicts an array of photonic crystal blocks. The red arrows indicate the input light signals entering from the left side of the structure, while the green arrows show the output signals emerging from the right side. Different colored blocks represent photonic crystals with varying rod radii, allowing for different transmission properties that function as programmable weights in the computational system.

Fig. 4 demonstrates the conceptual framework of our optical neural network for Fashion-MNIST classification. On the left, a sample image of a t-shirt is shown as input to the network. The neural network architecture on the right consists of fully connected layers where each connection represents a weight implemented through the photonic crystal's transmission properties. This optical implementation enables efficient parallel processing of image data through the propagation of light signals.

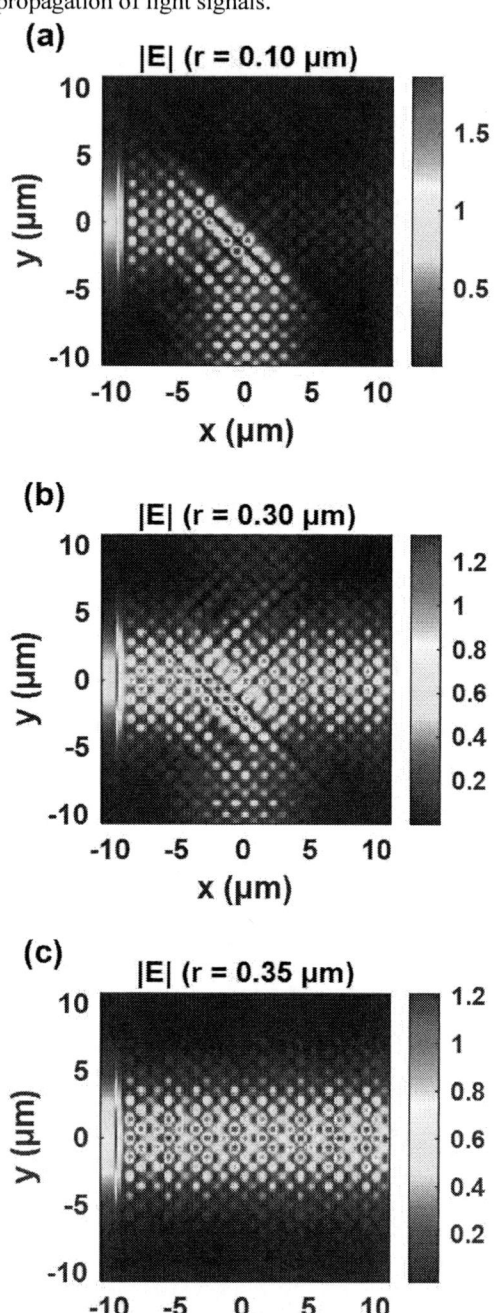

(a) |E| (r = 0.10 μm)

(b) |E| (r = 0.30 μm)

(c) |E| (r = 0.35 μm)

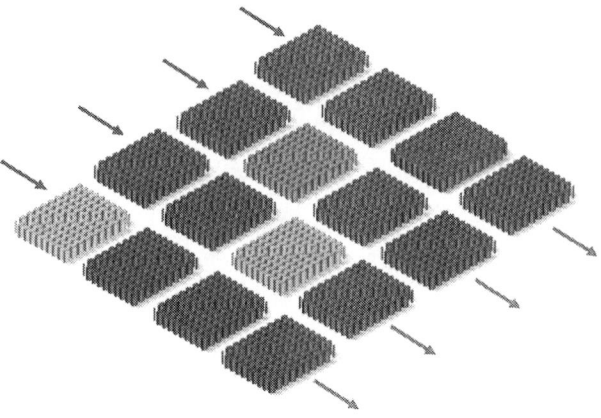

Fig.3 Schematic of the photonic crystal array configuration used for optical computing, with differently colored blocks representing varying rod radii that determine transmission properties for computational weighting.

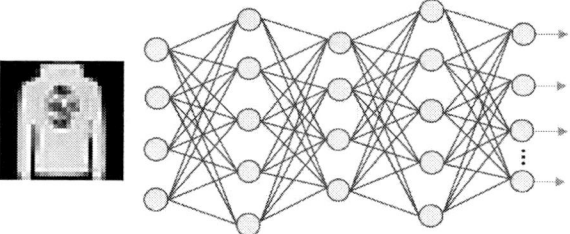

Fig.4 Neural network architecture for Fashion-MNIST classification, showing the conversion of image input (left) to classification output through an optical implementation of fully connected neural layers.

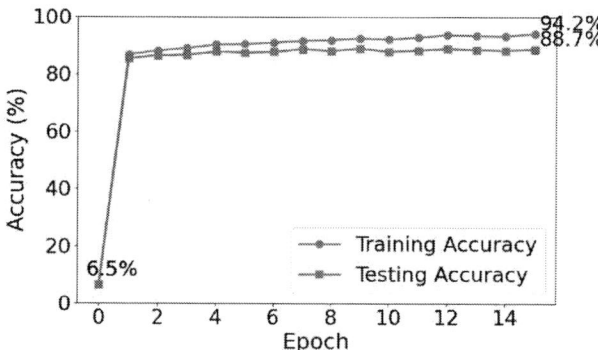

Fig.5 Fashion-MNIST classification accuracy versus training epochs, showing rapid improvement from initial 6.5% to over 80% after one epoch, with final testing accuracies of 88.7%.

Fig.2 Electric field distributions (|E|) in the photonic crystal structure with different rod radii: (a) r = 0.10 μm showing predominant reflection, (b) r = 0.30 μm showing the transition state, and (c) r = 0.35 μm showing enhanced horizontal transmission.

The classification performance is presented in Fig. 5, which displays the training and testing accuracy curves over 15 epochs. Starting from an initial accuracy of approximately 6.5% at epoch 0 (random chance), the network rapidly improves to over 80% accuracy after just one epoch. The training accuracy (blue curve) gradually increases to 94.2% by the final epoch, while the testing accuracy (red curve) stabilizes around 88.7%. This demonstrates the effectiveness of our photonic crystal-based approach for image classification tasks.

979-8-3315-2209-4/25 $31.00 © 2025 IEEE

III. CONCLUSIONS

This work presented a photonic crystal-based optical switching device utilizing line defects and demonstrated its application in image classification tasks. The simulation results revealed a strong dependence of the transmission properties on the rod radius, with a clear transition from a reflective to a transmissive state as the radius increased from 0.10 μm to 0.35 μm. The successful application for Fashion-MNIST image classification demonstrates the potential of photonic crystal structures in optical computing systems. Future work will focus on enhancing the device's functionality through active tuning mechanisms and exploring more complex PhC designs with multiple input and output channels for fully integrated optical neural network architectures.

ACKNOWLEDGMENTS

This work was financially supported by the National Natural Science Foundation of China (grant nos. 62401276, U23B2042), Natural Science Research Start-up Foundation of Recruiting Talents of Nanjing University of Posts and Telecommunications (Grant No. NY223161) and in part by the Jiangsu Provincial Key Research and Development Program under Grant BE2022126.

REFERENCES

[1] Noda, S., Chutinan, A., & Imada, M. Trapping and emission of photons by a single defect in a photonic bandgap structure. Nature, 407(6804), 608-610 (2000).

[2] Soljačić, M., & Joannopoulos, J. D. Enhancement of nonlinear effects using photonic crystals. Nature Materials, 3(4), 211-219 (2004).

[3] Prather, D. W., Shi, S., Murakowski, J., Schneider, G. J., Sharkawy, A., Chen, C., & Metz, B. Photonic crystal structures and applications: Perspective, overview, and development. IEEE Journal of Selected Topics in Quantum Electronics, 12(6), 1416 - 1437 (2006).

[4] Lin, X., Rivenson, Y., Yardimci, N. T., Veli, M., Luo, Y., Jarrahi, M., & Ozcan, A. All-optical machine learning using diffractive deep neural networks. Science, 361(6406), 1004-1008 (2018).

[5] Yang, C., Wu, Y., Zhong, C., Wang, X., Li, L., Guo, J., Huang, W., & Liu, Y. Compact and voltage-tunable surface plasmon polariton-based optical neural networks. Optics Letters 50(4), 1109-1112 (2023).

[6] Zhong, C., Wang, X., Li, L., Cui, Y., Xiao, L., Song, D., Guo, J., Huang, W., Guo, Y. Polarization-encoded neural networks with simplified grating patch. Science China Technological Sciences 68(2), 1220901 (2022).

[7] Guo, J., Liu, Y., Lin, L., Li, S., Cai, J., Chen, J., Huang, W., Lin, Y., & Xu, J. Chromatic plasmonic polarizer-based synapse for all-optical convolutional neural network. Nano Letters 23(20), 9651-9656 (2023).

Near infrared spatial light modulator by metasurface-based pixel control

Guozheng Lin*, Zexuan Chen*, Shanri Chen*, Shaolin Zhou*†

*South China University of Technology, Guangzhou 510640, China

†Email:eeslzhou@scut.edu.cn

Abstract—**The phase change materials (PCMs) of chalcogenide alloys exhibit a large permittivity contrast in different states and nonvolatile property, introducing one more degree of active control for the reconfigurable photonic device. In this paper, a metasurface-based near infrared spatial light modulator is proposed by sandwiching the film of Ge2Sb2Te5 alloy into a metal-insulator-metal (MIM) architecture for pixel-wise switchable control. For efficient electrical heating of GST film as well as the addressing of each metasurface pixel, the top and bottom electrode are customized for optimal triggering of GST phase transition. Both electrothermal and electromagnetic behaviors are verified by numerical simulations using finite element method.**

Keywords—track-following; metasurface; GST; electrothermal addressing

I. INTRODUCTION

With the exotic properties of electromagnetic metamaterials found since 2001[2], e.g., both negative permittivity and permeability[1], its complex 3D structure has been a difficulty for practical use and fabrication process. Therefore, a two-dimensional version of metamaterials (i.e. metasurfaces) emerges as feasible alternative for compact optoelectronic integrated devices[3]. Due to the merit of ultra-compact nature of metasurface (e.g. typically one-tenth the size of its operating wavelength), the manufacturing process is greatly eased. In such a manner, any desired response can be achieved by customizing the simply repeated meta-atoms.

Recently, one more degree of freedom is introduced for active control by integrating the chalcogenide phase change materials (PCMs) into metasurface-based devices. Upon the excitation of laser or electrical heating, PCMs undergo a phase transition and assume a large contrast of permittivity and optical constant. Therefore, the PCM-metasurface integrated devices enable viable reconfigurable devices by reversible phase change control of PCMs and the amplitude, phase, and polarization modulation of incident EM waves.

Herein, the commonly used vanadium dioxide (VO₂) suffers from the volatile nature[7], while the chalcogenide alloy such as GeSbTe exceeds in modulation speed, non-volatile and bistability[4][5][6]. To change the fully crystalline state of GST with regularly arranged lattice to amorphous state, it should be heated above the melting point at 900K (T_m)[4][9][12] with rapid cooling for quench (about 6~10°C/ns[9]), i.e. so called RESET. Conversely, the amorphous to crystalline phase transition called SET requires heating above the crystalline point at 450 K (T_c)[4][9] and holding for several microseconds[9]. As a result, we propose a type of near infrared spatial light modulator (SLM) by configuring GST film and metasurface into the MIM-based architecture[10][11] in this paper. The geometries of top metal and GST atoms are well-defined to customize the resonant wavelength at around 1064 *nm*, which is commonly used in laser area. Specifically, two separated bottom metal layers of MIM structure are used as electrodes of the entire SLM for GST heating with pixel-wise addressing along the column and row directions. In this manner, the pixel spacing and process complexities are minimized. In contrast, the metal wires for addressing are placed around the pixels in current mainstream schemes of routing[7], which inevitably lead to an expansion of pixel spacing that scales with the increasing number of pixels, ultimately resulting in pixel spacing exceed the physical dimensions of the pixels themselves. Further more, VO₂ is currently the mainstream material for addressable metasurfaces but its volatility leads a possibility of addressing failure during the addressing process. And GST based metasurfaces can avoid this problem.

II. ELECTROMAGNETIC RESPONSE AND ELECTROTHERMAL ADDRESSING

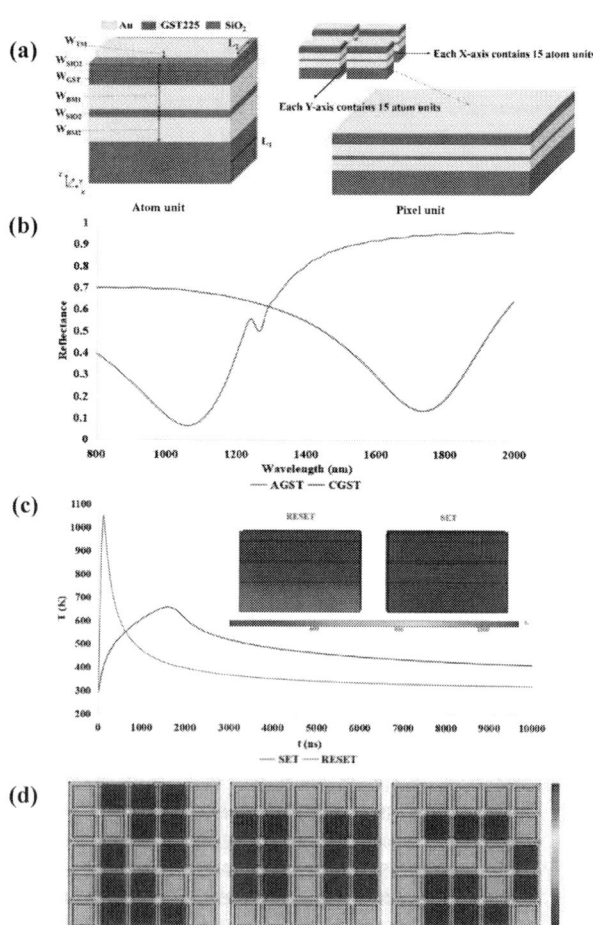

Fig. 1 . (a) The schematic of atom unit and pixel unit of proposed spatial light modulator. (b) A distinct reflection contrast at 1064 nm. (c) The temperature distribution of metasurface unit during SET and RESET process. (d) The contrast of reflected optical power by our metasurface SLM that is programmed to display the character of "N", "I" and "R". The incident light is linearly polarized along the top metal layer.

As illustrated in Fig. 1, each atom unit of the metasurface is configured into MIM architecture and each pixel of the SLM is composed of 15×15 atom units. First, two bottom layers (i.e. BM1 and BM2) of gold that are isolated by the SiO_2 layer sandwiched in-between, are used as the bottom reflector as well as the independent addressing lines for row and column operation. Then, GST film is deposited further between BM2 and the top SiO_2 passivation layer provides a Fabry-Pérot (F-P) cavity for the operating wavelength selection. For the amorphous state of GST, incident wave at the operation wavelength is fully trapped in the F-P cavity with large absorption and ideally zero reflection, i.e. the "off" or RESET state of our SLM. Upon the GST phase transition to crystalline state, the F-P cavity is deactivated and the pixel becomes maximally reflective, i.e. the "on" or SET stat of our SLM. Finally, the top metal (TM) thin layer of gold is used here for resonance mode adjustment. To confirm the validity of our SLM scheme, rigorous simulations of optical fields are performed using the finite difference time domain (FDTD) method. The boundary conditions along X-axis and Y-axis are periodic, while the Z-axis boundary condition is set as perfect matching.

After numerical analyses and optimization, as marked in the Fig. 1 (a), the parameters of our SLM pixel are customized as L_1=600 nm, L_2=200 nm, W_{SiO2}=10 nm, W_{GST}=40 nm, W_{BM1}=100 nm, W_{BM2}=100 nm and W_{TM}=10 nm. Therefore, the pixel size in the XOY plane is 9 μm×9 μm. As shown in Fig. 1(b), with the incidence of X-axis polarized light, a large modulation depth above 60% is obtained at the wavelength around 1064 nm, i.e. reflectance contrast of 0.67 versus 0.04. Herein, the electrothermal behaviors of GST phase transition are also verified by COMSOL to ensure the "RESET" and "SET" state switching, as shown in Fig. 1(c). Namely, two well-defined short voltage pulses of 100 ns at the row and column lines (V_BM1 and V_BM2) heat the GST film of chosen pixel unit above T_m for "RESET". While a voltage pulse of 1.5 μs only by the row line (V_BM2) heat the GST film of chosen unit above T_c for SET. As a result, a 5×5 array of SLM is constructed and simulated to show the programmable optical intensity profiles at SLM surface. As shown in Fig. 1(d), the letters of "N", "I" and "R" are distinctly programmed into our SLM.

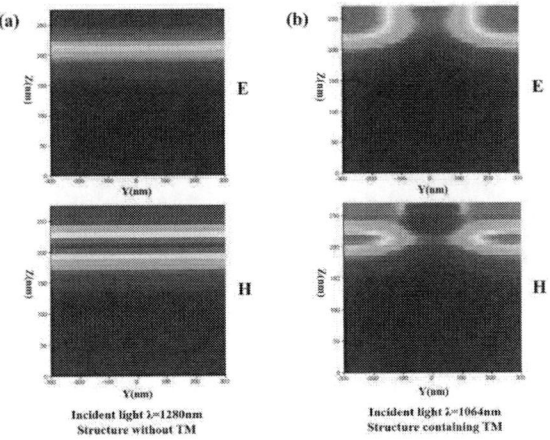

Fig. 2 . (a) YZ-plane electromagnetic field distributions at 1280nm of

structure without the TM layer. (b) YZ-plane electromagnetic field distributions at 1064nm of structure containing the TM layer.

In order to elucidate the mechanism of reflectance contrast, the electromagnetic field distributions at specific wavelengths of specific structures were proposed using FDTD method. As shown in Fig. 2(a), with an incidence of X-axis polarized light, an absorption peak at 1280 nm is achieved in a TM layer free structure of atom unit by employing a F-P cavity. The introduction of the TM layer induces magnetic resonance with the BM layer, as illustrated in Fig. 2(b). The magnetic resonance modifies the phase of incident light, resulting in a spectral shift of the absorption peak to 1064 nm.

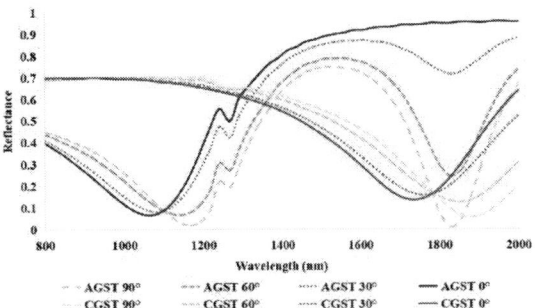

Fig. 3 The numerically calculated reflectance of our SLM at the incidence of varied directions of polarizations.

Further, to check the polarization dependency of our device, different reflectance are obtained by changing the polarization angle between the incident light and the X-axis, as shown in Fig. 3. Fortunately, polarization angle changes only cause minor resonance wavelength shift and a good reflectance contrast is still observable near 1064 nm, i.e. our SLM is almost polarization independent for a broadband device.

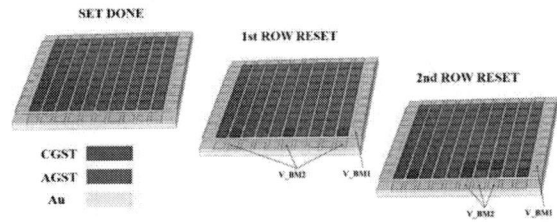

Fig. 4. A 10×10 pixel electrothermal addressing demo.

To elucidate the selective addressing and control of each pixel, a 10×10 array of SLM is illustrated in Fig. 4. Herein, the bottom metal layer of BM1 is connected with the row line along X-axis while the BM2 layer is connected with the column line along Y-axis. For independently selective control of each pixel, a addressing procedure is well defined below. Herein, two voltages, V_BM1 and V_BM2 can be applied to BM1 and BM2 individually. When V_BM1 is applied to the row line alone, pixels in the whole row can be instantly heated to high temperature within 100 ns, but it is still 100 K to 200 K away from 900 K (Tm)[4][9][12]). When V_BM2 is applied alone within 100 ns, all pixels in the selected column are only heated to a temperature below 450 K~500 K (Tc)[4][9], and no crystallization occurs unless the voltage is maintained to microseconds.

To start, a 1.5 μs voltage pulse of V_BM2 is applied to SET all columns of pixels, i.e. crystalline state of GST films. It takes ~20 μs for the device to cool down to room temperature when V_BM2 is removed. Then, to RESET as-selected pixels, a 100 *ns* voltage pulse of V_BM1 is applied to scan all rows sequentially followed by a 100 ns voltage pulse of V_BM2 applied to corresponding columns to access as-selected pixels, to which V_BM1 and V_BM2 are applied simultaneously. After RESET each column, it takes another 2 µs for the device to cool down to room temperature. Finally, the RESET process is repeated until all pixels are addressed. Herein, there might be several cases that need to be interpreted with detail. In the first case, for pixels expected to be RESET, V_BM1 and V_BM2 are applied simultaneously to address the row and column lines to heat the GST films above 900K (T_m). In this situation, for some pixels only heated by the column line (i.e. V_BM2), their temperatures might exceed T_c but only hold for tens of nanoseconds, which mean such pixels cannot be mistakenly SET and remain at their original state. In another case, for some pixels that are only heated by the row line voltage (i.e. V_BM1), and their temperatures are still more than 100 K away from T_m. They are not RESET and stay in crystalline state. Finally, for those pixels are not heated by the row or column lines, their temperatures are almost constant to maintain the original state.

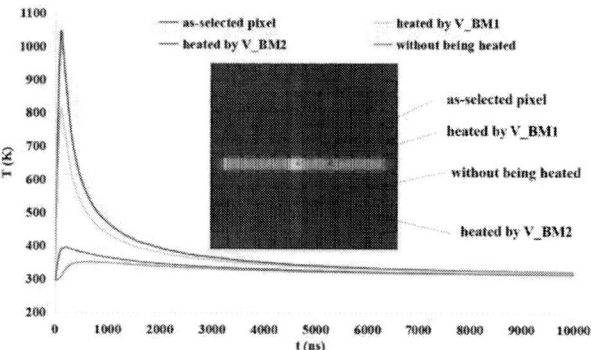

Fig. 5 The electrothermal behavior of our SLM during RESET process.

Herein, the voltage value of V_BM1 and V_BM2 can be reconfigured by changing the pixel spacing. For an example, the voltage of V_BM1=6.8V and V_BM2=1.6V are used for pixel spacing of 1 µm. For pixel spacing of 2 µm or 3 µm, the voltage of V_BM1=7.5V and V_BM2=2V or V_BM1=8.2V and V_BM2=2.2V are used instead.

III. Reliability of Electrothermal Addressing Scheme

To demonstrate thermal crosstalk behaviors between selected pixels and their neighboring pixels of our SLM, rigorous electrothermal simulations are also performed by COMSOL, as shown in Fig. 6.

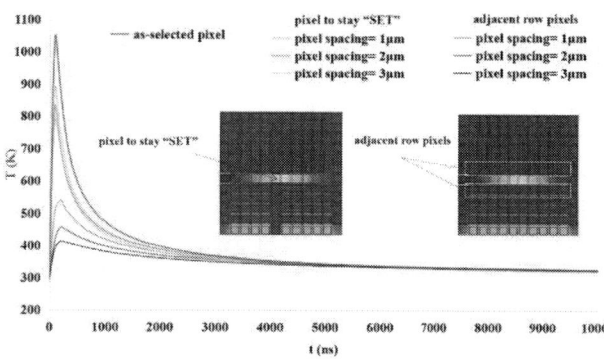

Fig. 6 The electrothermal behavior of our SLM where the undesired thermal crosstalk issues can be solved by slightly increasing the pixel spacings to change to addressing voltages of V_BM1 and V_BM2.

For severe thermal crosstalk for potential occurrence of addressing errors. The first case is that pixels that have just been RESET in the previous row might be SET unintentionally, i.e. the written data is erased by mistake. Similarly, for the second case where certain pixels in the selected row need to stay "SET", but thermal crosstalk from neighboring columns might RESET them undesirably. Thermal crosstalk arises from excessive heat transfer to neighboring pixels due to an insufficient pixel spacing. Herein, we propose a solution to mitigate thermal crosstalk. In our scheme, slight increase of pixel spacing can relieve the problem of thermal crosstalk. The thermal behaviors of our SLM with different pixel spacings of 1 µm, 2 µm and 3 µm are simulated and shown in Fig. 6. When the pixel spacing is 1 µm, the thermal crosstalk issue is quite serious and the above mentioned addressing errors might occur in certain situation. When the pixel spacing is increased to 2 µm, the thermal crosstalk issued is well resolved. When the pixel spacing is further increased to 3 µm, thermal crosstalk does not cause any problems even in the worst situation.

IV. Conclusion

In conclusion, we propose a metasurface-based SLM with a large reflectance contrast and large modulation depth in a broad wavelength range around 1064 nm. By integrating 15×15 units of atoms into a 9 µm×9 µm pixel, a metasurface SLM is configured. Electrothermal heating of each unit enables the switchable control of RESET and SET. Rigorous simulations confirm the validity of our SLM with for pixel-wise spatially programmable control. For selective control of each pixel as required by SLM programming, an addressing scheme is customized to optimize the electrothermal process of GST phase transition. Finally, by slightly tuning the pixel spacing, the potential thermal crosstalk issue can be well treated. Although the space utilization of each pixel is slightly sacrificed, it is insignificant compared with space cost by addressing lines in the conventional scheme of routing and wiring.

Acknowledgment

The authors gratefully acknowledge the assistance and guidance from Jigeng Sun, Rui Liu, Zhaokai Qiu and Jingxi Li for their suggestions and assistance to the project.

References

[1] Viktor G Veselago. REVIEWS OF TOPICAL PROBLEMS THE ELECTRODYNAMICS OF SUBSTANCES WITH SIMULTANEOUSLY NEGATIVE VALUES OF ε AND µ. American Institute of Physics. 10(4), (1968).

[2] D. R. Smith, Willie J. Padilla, Willie J. Padilla, et al. Composite Medium with Simultaneously Negative Permeability and Permittivity. PHYSICAL REVIEW LETTERS. 84(18), (2000).

[3] Ji C, Shanshan H, Shining Z, et al. Metamaterials: From fundamental physics to intelligent design. Interdisciplinary Materials. 2, 5-29, (2023).

[4] Zhihua F, Qinling D, Xiaoyu M and Shaolin Z. Phase Change Metasurfaces by Continuous or Quasi-Continuous Atoms for Active Optoelectronic Integration. Materials. 14(5), 1272, (2021).

[5] Shulin S, Qiong H, Jiaming H ,et al. Electromagnetic metasurfaces: physics and applications. Advances in Optics and Photonics. 11(2), 380-479, (2019).

[6] Niloufar R H,Junsuk R. Metasurfaces Based on Phase-Change Material as a Reconfigurable Platform for Multifunctional Devices. Materials. 10, 1046, (2017).

[7] Tingbiao G, Zhi Z, Zijian L, et al. Durable and programmable ultrafast nanophotonic matrix of spectral pixels. Nature Nanotechnology. (2024).

[8] Landy S, Sajuyigbe J, Mock D, et al. Perfect Metamaterial Absorber. PHYSICAL REVIEW LETTERS. 100, 20-23, (2008).

[9] Sajjad A, Omid H, Mohammad T, et al. Electrically driven reprogrammable phase-change metasurface reaching 80% efficiency. Nature Communications. 13, 1696, (2022).

[10] Santiago G C, Geoffrey R N, Hasan H, et al. The design of practicable phase-change metadevices for near-infrared absorber and modulator applications. Optics Express. 24(12), 13563-73, (2016).

[11] Tun C, Chenwei W, Lei Z, et al. Broadband Polarization-Independent Perfect Absorber Using a Phase-Change Metamaterial at Visible Frequencies. Scientific Reports. 4, 3955, (2014).

[12] Shaolin Z, Kezhou L, Yihan C, et al. Phase Change Memory Cell With Reconfigured Electrode for Lower RESET Voltage. 7, 1072-1079, (2019).

Alignment mark of meta-grating for sub-10 nanometer overlay inspection

Zhaokai Qiu*, Rui Liu*, Shaolin Zhou*‡, and Xugang Ma†
*School of Microelectronics, South China University of Technology, Guangzhou, China
†China Digital Industry Operation Group Co., Limited, Hong Kong, China
‡Email:eeslzhou@scut.edu.cn

Abstract—Overlay error of multi-layer lithographic process is a critical parameter to determine final pattern quality and critical dimension and yield of integrated circuits. Therefore, a fast and accurate overlay metrology technique is essential to support the process and yield control. This paper proposes a moiré-based overlay metrology scheme by using one type of composite meta-grating mark. By capturing the moiré fringe generated by composite meta-grating with phase analysis, overlay measurement at sub-10 nanometer level is achieved. Simulations and experiments demonstrate that our scheme can achieve accurate overlay without the need of movement for repeated focusing, offering valuable insights for further exploration of overlay metrology and the related high-accuracy measurement technologies.

Index Terms—Overlay, Overlay Metrology, Moiré Fringe, Meta-grating

I. Introduction

Photolithography process plays a vital role in the integrated circuit manufacturing industry. Overlay error is one of the most important performance metrics and the primary factor that determines the pattern quality, process defects or critical dimensions of multi-layer lithography. Along with the increasingly improved resolution of lithographic tool, new metrology scheme of overlay to intrinsically break the diffraction limit become highly demanding. Overlay error refers to the displacement between the pattern of the current layer and that of the reference layer. As required by the lithographic process, the maximum allowable overlay error is usually located at about 1/3-1/5 of the size of the smallest feature in the integrated circuit [1].

To ensure device performance, overlay metrology is required to verify the overlay accuracy between different steps or layers of pattern transfer. Traditional measurement techniques such as the scanning electron microscope (SEM), atomic force microscopy (AFM), and scanning tunneling microscope (STM) can achieve extremely high measurement accuracy. However, they often cause destructive damage to semiconductor devices [2], [3]. Compared to traditional measurement methods of overlay inspection, optical measurement techniques offer more advantages,such as high efficiency, noninvasive and non-destructive measurement. These advantages have led to their widespread application in control and yield management of the lithography-based integrated circuit manufacturing process.

Due to the structural complexity of semiconductor devices, it is difficult to directly measure the overlay error of multi-layer structures using optical methods. Therefore, optical measurement methods often resort to alignment marks of relatively simple geometrical patterns on different layers. These alignment marks for overlay inspection are fabricated simultaneously with the device features. In such a manner, the overlay error of internal structure of the semiconductor device is directly characterized by the misaligned displacement of overlay marks. Currently, the mainstream overlay error detection methods based on optical metrology can be roughly divided into two categories: Image-Based Overlay (IBO) [4], [5] and Diffraction-Based Overlay (DBO) [6], [7]. The IBO method is limited by the resolution of optical microscopes as the technology nodes continue to shrink. Additionally, continuous adjustments of focusing and wavelength are performed to improve the image contrast during detection [8]. Furthermore, the overlay measurement accuracy is sensitively susceptible to the patterning process of alignment marks used for IBO detection [9]. In advanced node processes, the DBO method has gained wider applications due to its lower sensitivity to process variations. However, as it has a shorter development history, there has been no traceable record measurement standards. As a result, the measurement results may have larger errors [10].

In this paper, we propose a moiré-based overlay technique by using the alignment marks of meta-gratings similar as the DBO method. By using a dual-layer meta-grating superposed configuration to diffract and modulate the incident light, moiré fringes are generated. The moiré fringe pattern is captured and the phase information of the fringes is extracted and analyzed to characterize the overlay offset. This allows the displacement offset between the upper and lower layers to be determined, thereby the overlay error can be obtained. This method eliminates the need of repeated focus adjustments during image acquisition, enabling fast and accurate overlay inspection. Experimental results effectively and feasibly demonstrate that our moiré-based method can achieve overlay inspection at sub-10 nanometer level.

II. Scheme and Principle

A. Moiré fringes by overlay inspection marks

When two gratings with different periods (P_1 and P_2, corresponding to frequencies f_1 and f_2) are superposed vertically,

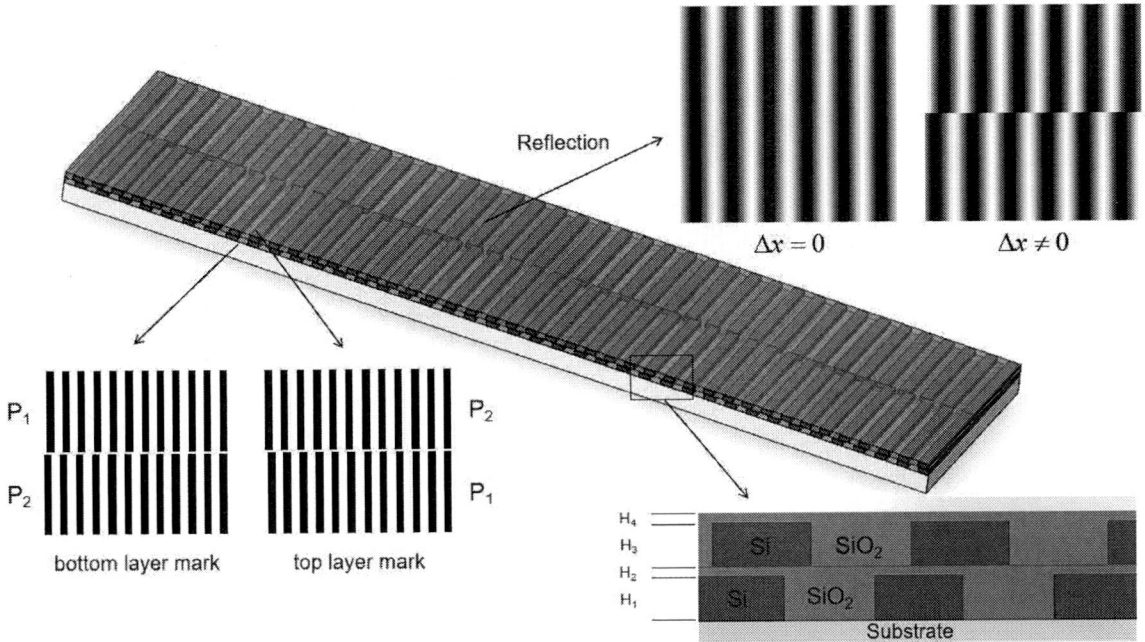

Fig. 1. The schematic of our scheme of overlay inspection by using the alignment mark of meta-grating

the incident light undergoes a double-layer diffraction. The diffraction field can be similarly regarded as linear recombination of all diffracted fields of each harmonic generated individually by the second grating, following the diffraction of incident light by the first grating. The complex amplitude distribution is given by (A_n and B_m are the diffraction coefficients):

$$E(x,y) = \sum_m B_m \exp(2\pi m f_2 x) \sum_n A_n \exp(j2\pi n f_1 x)$$
$$= \sum_{(n,m)} A_n B_m \exp\left[j2\pi(nf_1 + mf_2)x\right] \tag{1}$$

Clearly, the field distribution after the overlapping gratings contains abundant frequency information, most of which is indistinguishable. Only a few low-frequency components are easily distinguishable. When the periods of the two gratings are close, the lowest and most distinguishable fundamental frequency is $|f_1 - f_2|$. This component is referred to as the (1, -1) order moiré fringe, and its complex amplitude distribution is given by:

$$E_{(1,-1)}(x,y) = \sum_{n=-\infty}^{\infty} A_n B_n \exp\left[j2\pi n(f_1 - f_2)x\right] \tag{2}$$

B. Meta-grating mark for overlay inspection

The overlay error describes the positional deviation between adjacent circuit patterns in the lithography process. Therefore, both adjacent circuit patterns need overlay measurement target for detection to ensure that the overlay error meets the process requirements. The overlay measurement target used in this method are shown in Figure 1. Both the top layer and the bottom layer have two sets of gratings with periods of P_1 and P_2, respectively. The grating with period P_1 from the top layer is positioned above the grating with period P_2 from the bottom layer, and the grating with period P_2 from the top layer is positioned above the grating with period P_1 from the bottom layer.

The cross-sectional schematic of the meta-grating structure is shown in the figure. It consists of two layers of stacked silicon gratings, with the gaps between the gratings in each layer and the space above the gratings filled with silica as a dielectric impedance matching buffer layer. This design allows more light to be projected into the bottom structure and reflected back, thereby enhancing the contrast of the Moiré fringes. By choosing similar grating periods P_1 and P_2, the resulting Moiré fringe period becomes larger, which improves the measurement accuracy. The specific structural dimensions are as follows: the grating periods are $P_1 = 1\,\mu m$ and $P_2 = 1.1\,\mu m$, the grating duty cycle is 50%, the silicon grating heights are $H_1 = H_3 = 100\,nm$, and the silica filling layer heights are $H_2 = 20\,nm$ and $H_4 = 10\,nm$.

This overlay measurement target causes the light, after undergoing two diffractions, to generate two sets of Moiré fringes with identical periods. When there is a displacement Δx between the top layer and the bottom layer, the two sets of Moiré fringes will shift in opposite directions. The resulting relative phase shift $\Delta\varphi$ satisfies the following relationship:

$$\Delta x = \Delta\varphi / [2\pi(f_1 + f_2)] \tag{3}$$

979-8-3315-2209-4/25 $31.00 © 2025 IEEE

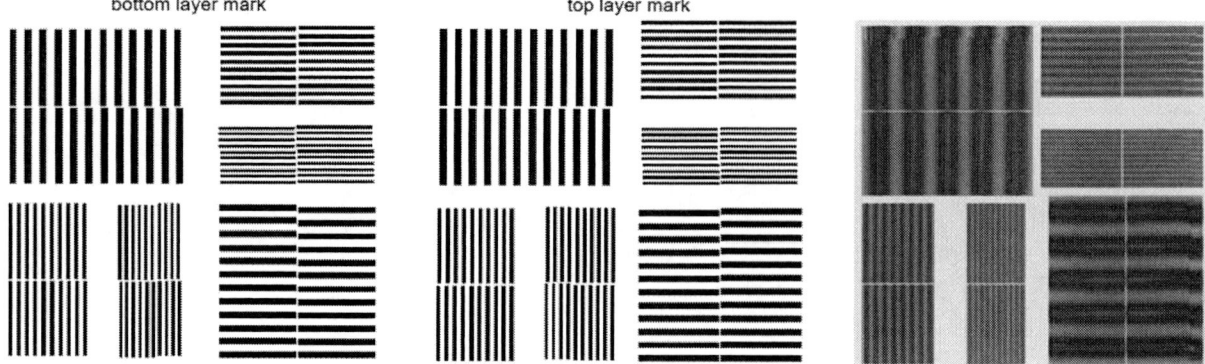

Fig. 2. The alignment marks composed by meta-gratings for overlay inspection in the X and Y directions

TABLE I
DIFFERENCE BETWEEN META-GRATING OVERLAY METROLOGY AND SEM MEASUREMENT

target number	target1	target2	target3	target4	target5	target6
Measurement Difference (nm)	5.4783	7.8486	7.6145	5.3506	1.9181	0.3875

By rotating the overlay measurement target pattern, overlay errors in different directions can be measured. In this paper, only the overlay error measurement result in the x-direction is used to validate its effectiveness.

III. SIMULATION AND EXPERIMENT

The meta-grating proposed in this paper is simulated using the FDTD (Finite-Difference Time-Domain) method. In the simulation, different offset values between the top and bottom layers are set as overlay errors. The near-field electric field intensity images of the reflection surface under different overlay errors are obtained through simulation. These images are then subjected to filtering and phase analysis to calculate the overlay measurement values. The overlay measurement values obtained through phase analysis have an average error of 0.254491 nm and a maximum error of 0.531 nm compared to the set overlay error values.

In the experiment, we fabricated several sets of meta-grating markers to validate the feasibility of the proposed method. Moiré fringe images generated by the markers were captured using an optical microscope, followed by phase analysis and offset prediction. Since overlay errors are inherently introduced during the fabrication process, the true value of the overlay error is difficult to determine. To verify the accuracy of the proposed overlay metrology method, one set of markers was measured using a high-resolution Scanning Electron Microscope (SEM) to determine the offset between the upper and lower layers. The average of these SEM measurements was used as the reference overlay error. The differences between the overlay measurements based on the meta-grating method and the SEM-measured overlay errors are shown in Table I. The average difference is 4.7663 nm, with a maximum difference of 7.8486 nm. The experimental results demonstrate that the meta-grating overlay metrology method can achieve sub-10 nanometer prediction accuracy.

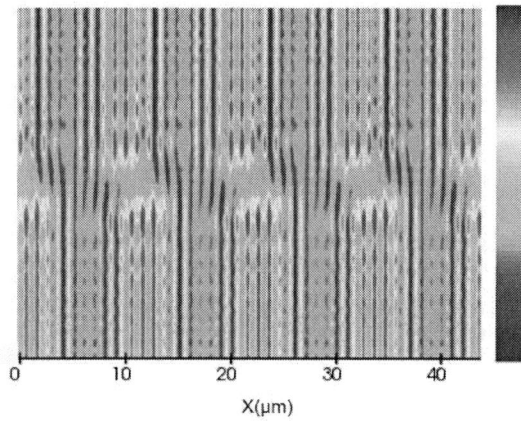

Fig. 3. The simulated near-field electric field intensity profile by superposition of meta-gratings

Fig. 4. The alignment marks of meta-gratings used for overlay inspection experiments

IV. CONCLUSION

In summary, we have designed a meta-grating structure and an overlay metrology method based on it. High-precision overlay measurement is achieved by analyzing the phase of moiré fringes generated by the dual-layer grating structure. By optimizing the specific structural dimensions, the measurement accuracy can be further improved. Simulation and experimental results demonstrate that this method can achieve sub-10 nanometer overlay measurement accuracy. This level of precision has great potential for application in advanced nanolithography processes such as DUV (Deep Ultraviolet), EUV (Extreme Ultraviolet) and nanoimprint. These findings highlight the accuracy and feasibility of this method for application in integrated circuit manufacturing.

REFERENCES

[1] H. Liu, J. Wang, J. Shi, G. Li, X. Chen, and W. Zhou, "Overlay error measurement method based on through-focus scanning optical microscopy," *Acta Optica Sinica*, vol. 44, no. 9, p. 0912004, 2024.

[2] C. G. Frase, E. Buhr, and K. Dirscherl, "Cd characterization of nanostructures in sem metrology," *Measurement Science and Technology*, vol. 18, no. 2, p. 510, jan 2007.

[3] T.-H. Hsu, H.-T. Lue, C.-C. Hsieh, E.-K. Lai, C.-P. Lu, S.-P. Hong, M.-T. Wu, F. H. Hsu, N. Z. Lien, J.-Y. Hsieh, L.-W. Yang, T. Yang, K.-C. Chen, K.-Y. Hsieh, R. Liu, and C.-Y. Lu, "Study of sub-30nm thin film transistor (tft) charge-trapping (ct) devices for 3d nand flash application," in *2009 IEEE International Electron Devices Meeting (IEDM)*, 2009, pp. 1–4.

[4] P. L. Rigolli, L. Rozzoni, C. Turco, U. Iessi, M. Polli, E. Kassel, P. Izikson, and Y. Avrahamov, "Aim technology for nonvolatile memories microelectronics devices," in *Metrology, Inspection, and Process Control for Microlithography XX*, vol. 6152. SPIE, 2006, pp. 1407–1419.

[5] B. Minghetti, T. Brunner, C. Robinson, C. Ausschnitt, D. Corliss, and N. Felix, "Overlay characterization and matching of immersion photoclusters," in *Optical Microlithography XXIII*, vol. 7640. SPIE, 2010, pp. 351–361.

[6] N. Wright, M. van der Schaar, A. d. Boef, P. Hinnen, M. Shahrjerdy, V. Wang, S. Lin, C. Wang, J. Huang, and W. Wang, "New approaches for scatterometry-based metrology for critical distance and overlay measurement and process control," *Journal of Micro/Nanolithography, MEMS and MOEMS*, vol. 10, no. 1, pp. 013013–013013, 2011.

[7] M. Ebert, P. Vanoppen, M. Jak, G. vd Zouw, H. Cramer, T. Nooitgedagt, and H. vd Laan, "New approaches in diffraction based optical metrology," in *Metrology, Inspection, and Process Control for Microlithography XXX*, vol. 9778. SPIE, 2016, pp. 834–839.

[8] M. S. Tamer, M. van der Lans, and H. Sadeghian, "Image-based overlay measurement using subsurface ultrasonic resonance force microscopy," in *Metrology, Inspection, and Process Control for Microlithography XXXII*, vol. 10585. SPIE, 2018, pp. 133–139.

[9] F. Dettoni, T. Shapoval, R. Bouyssou, T. Itzkovich, R. Haupt, and C. Dezauzier, "Image based overlay measurement improvements of 28nm fd-soi cmos front-end critical steps," in *Proc. SPIE 10145, Metrology, Inspection, and Process Control for Microlithography XXXI*, vol. 10145, 2017, p. 101450C, 28 March 2017.

[10] Y. Li, L. Yang, X. Wang, S. Shan, F. Deng, Z. He, Z. Liu, and X. Li, "Overlay metrology for lithography machine," *Laser & Optoelectronics Progress*, vol. 59, no. 9, p. 0922023, 2022.

Study of Negative Bias Illumination Stress Stability on IGZO/IZO Based Thin Film Transistors

Le Bian   Jinxuan Wu   Yuxuan Shen   Siyuan He   Shuangmei Xue   Meng Zhang   Yan Yan[*]

Abstract—The stability of thin film transistors (TFTs) under negative bias illumination stress (NBIS) is paramount in flat panel display applications. This work studies the NBIS stability of indium gallium zinc oxide (IGZO)/indium zinc oxide (IZO) bilayer-based TFTs annealed in various atmospheres. The findings reveal that TFTs annealed in an oxygen atmosphere exhibit optimal I-V characteristics and NBIS stability, with minimal threshold voltage shifts (ΔV_{TH}) of -0.61V, -1.75V, and -8.87V under red, green, and blue light, respectively. This improvement is attributed to the reduction of oxygen vacancies (V_O) and interface traps.

Index Terms—Thin Film Transistors, NBIS; Annealing, Metal Oxide

I. INTRODUCTION

Amorphous oxide semiconductor (AOS) materials are promising as channel materials in advanced display technologies due to their high carrier mobility, consistent large-scale production, and low cost. Currently, AOS pixel driver circuits have been successfully commercialized [1]–[3]. Thin film transistors (TFTs) play a core role as pixel switches and current driving devices in source matrix driving circuits. The stability of AOS TFTs directly impacts the opening of pixels and the display effect of the monitor. Among regular display designs, the active layers of TFTs are covered by blocking layers to prevent light infiltration, yet a portion of scattered light can incidentally penetrate the active layers [4]. To ensure that the pixels remain turned off when not displayed, a negative gate voltage must be applied to the device. Consequently, negative bias illumination stress (NBIS) testing is essential for predicting the long-term stability of devices by simulating stress conditions in actual usage environments.

In the research on optimizing TFTs performance, employing bilayer-based semiconductor thin film technology has been proven to be an effective strategy [5]–[7]. Additionally, annealing is a crucial approach to enhancing TFTs performance. During the preparation of TFTs, film stress may arise due to differences in the thermal expansion coefficients of the materials. The annealing process can effectively mitigate film stress and stabilize the thin film structure. In addition, the process promotes atomic recombination and eliminates weak

Le Bian, Jinxuan Wu, Siyuan He, Shuangmei Xue, Meng Zhang, and Yan Yan are with the College of Electronics and Information Engineering, Shenzhen University, Shenzhen, China. Yuxuan Shen is with College of Physics and Optoelectronic Engineering, Shenzhen University, Shenzhen, China. This work is supported by the National Natural Science Foundation of China (Grant No. 62374110), and the Shenzhen Science and Technology Program(Grant No. JCYJ20230808105806014), and the State Key Laboratory of Radio Frequency Heterogeneous Integration (Independent Scientific Research Program No. 2024014). Corresponding author: (Yan Yan, email: yanyan@szu.edu.cn)

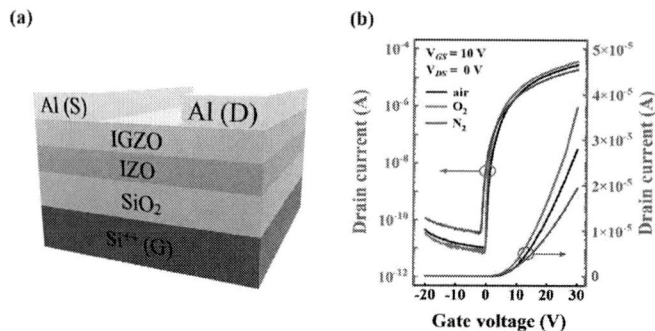

Fig. 1. (a) presents the structural diagram of IGZO/IZO TFTs, while (b) depicts the I-V curves of IGZO/IZO TFTs annealed in air, oxygen, and nitrogen atmosphere.

bonds and partial defects, sequentially improving the film's density and uniformity [3]. Therefore, studying the stability of bilayer-based metal oxide (MO) TFTs annealed under various environmental atmospheres in NBIS is highly significant.

In this work, IGZO/IZO bilayer-based thin films were produced as active layers for TFTs with various annealing atmospheres in air, oxygen, and nitrogen. The corresponding I-V characteristics were characterized. Subsequently, the NBIS stability of TFTs under red (650 nm), green (550 nm), and blue (450 nm) light was characterized. Threshold voltage shifts (ΔV_{TH}), as critical parameters, were extracted and analyzed. The results indicate that NBIS significantly reduces the electrical performance and stability of TFT, and the devices are significantly different under different annealing atmospheres. The TFT annealed in an oxygen atmosphere exhibits the best performance. This phenomenon can be attributed to the filling of oxygen vacancy (V_O) defects in the film by oxygen. The work can provide valuable insights for evaluating MO TFTs performance in practical display applications.

II. EXPERIMENTAL

Indium nitrate hydrate (99.999% trace metal basis, Sigma Aldrich), gallium nitrate hydrate (99.999% trace metal basis, Sigma Aldrich), and zinc nitrate hydrate (99.999% trace metal basis, Sigma Aldrich) were dissolved in 2-methoxy ethanol at ratios of In:Ga:Zn = 7:1:3 and In:Zn = 7:3, respectively. The solutions were sonicated for 30 minutes to ensure complete dissolution as precursor solutions. Bottom-gate, top-contact TFTs were fabricated using silicon as the substrate, with N-type highly doped silicon as the gate electrode and 100 nm SiO_2 film as the dielectric layer, as depicted in Fig. 1(a).

979-8-3315-2209-4/25 $31.00 © 2025 IEEE

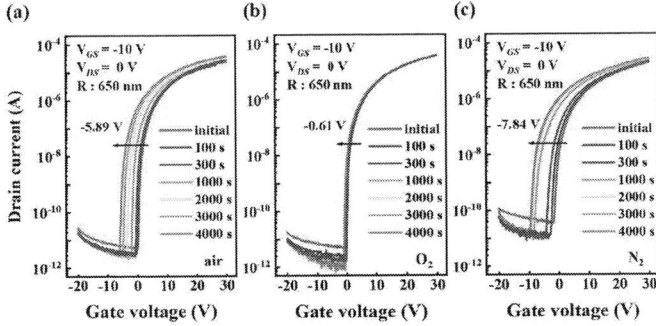

Fig. 2. The 650nm-NBIS of (a) IGZO/IZO-air TFTs, (b) IGZO/IZO-O$_2$ TFTs, (c) IGZO/IZO-N$_2$ TFTs device.

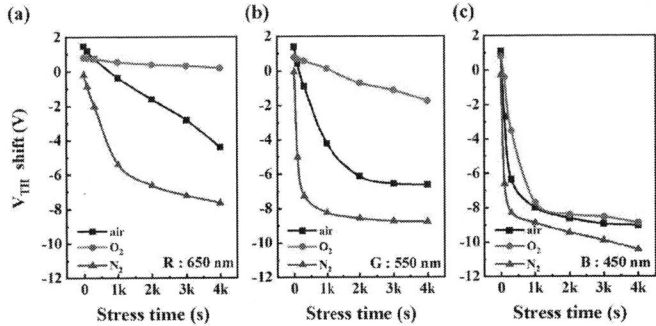

Fig. 3. The V$_{TH}$ shifts of IGZO/IZO TFTs under negative bias stress with different annealing atmospheres at wavelengths of (a) 650nm, (b) 550nm, and (c) 450nm.

IZO and IGZO precursor solutions were sequentially spin-coated onto the SiO$_2$ film at 4000 rpm for 40 s, dried at 120°C for 10 minutes, and then annealed at 300°C for 1 hour in air, O$_2$, or N$_2$. Then, 80 nm thick source (S) and drain (D) Al electrodes with (W/L =1000μm/100μm were deposited by vacuum thermal evaporation. Finally, the TFTs were annealed in air at 100°C to enhance crystal semi-contact. The I-V characteristics of the TFTs were measured using a probe station coupled with an Agilent B1500A semiconductor parameter analyzer.

III. RESULTS AND DISCUSSION

Fig. 1(b) displays the transfer curves of IGZO/IZO TFTs annealed in air, oxygen, and nitrogen, respectively designated as IGZO/IZO-air TFTs, IGZO/IZO-O$_2$ TFTs, and IGZO/IZO-N$_2$ TFTs in subsequent text. We can find that the mobilities of TFTs are 0.53 cm$^2 \cdot$V$^{-1} \cdot$s^{-1}, 0.79 cm$^2 \cdot$V$^{-1} \cdot$s^{-1}, and 0.37 cm$^2 \cdot$V$^{-1} \cdot$s^{-1} and the I$_{on}$/I$_{off}$ ratios of TFTs are 2.8 × 10^6, 5.8 × 10^6, and 6.2 × 10^6, respectively. Significantly, IGZO/IZO-O$_2$ TFTs exhibit the best electrical performance among them. This enhancement is due to the filling of unoccupied oxygen vacancies during annealing in an oxygen atmosphere, which reduces defects that act as electron traps and improves the transport capacity of the conductive channel [8].

To further evaluate the operational stability of MO TFTs, the NBIS stability of IGZO/IZO TFT annealed in three different atmospheres (air, oxygen, and nitrogen) is also studied. The test conditions were as follows: light wavelength = 650 nm, optical power density = 0.1 mW/cm², drain-source voltage (V$_{GS}$) = -10 V, and t = 4000 s.

As illustrated in Fig.2(a) to 2(c), IGZO/IZO-O$_2$ TFTs demonstrate lesser ΔV$_{TH}$, specifically -0.61V, whereas IGZO/IZO-air TFTs and IGZO/IZO-N$_2$ TFTs exhibit larger ΔV$_{TH}$, at -5.89V and -7.84V, respectively. Under NBIS, the negative shifts in the V$_{TH}$ are affected by several factors. Firstly, applying a negative gate voltage leads to the accumulation of the majority carrier in the N-type channel. This accumulation leads to an increase in electron concentration in the channel, making it impossible to restore balance in a very short period of time. Secondly, the negative gate voltage ionizes neutral V$_O$ into positively charged oxygen vacancy

ions, while generating new electrons [9]. In addition, under illumination, the active layer absorbs photons and generates electron hole pairs, causing a large number of electrons to transition to the conduction band and become free carriers [10]. The results indicate that annealing in an oxygen atmosphere can significantly improve the NBIS stability of the device, which reduces the concentration of V$_O$ at the interface between the active layer and the gate insulating layer, effectively reducing the number of ionized oxygen vacancies [11], [12]. The lower NBIS stability of IGZO/IZO-N$_2$ TFTs is due to the high concentration of oxygen vacancies in the channel layers, caused by the desorption of oxygen atoms during nitrogen annealing [13].

In order to investigate the effect of the annealing atmosphere on the stability of TFT under different types of illumination, we tested the NBIS reliability of IGZO/IZO TFT under three illumination wavelengths (650 nm, 550 nm, and 450 nm). As shown in Fig.3(a) to 3(c), the shorter the wavelength, the poorer the stability of NBIS of IGZO/IZO TFTs. To be more specific, IGZO/IZO TFTs annealed in N$_2$ exhibited the lowest stability, with ΔV$_{TH}$ reaching -7.64 V (red), -8.78 V (green), and -10.42 V (blue), respectively. IGZO/IZO TFTs annealed in air showed ΔV$_{TH}$ of -4.42 V (red), -6.63 V (green), and -9.02 V (blue). In contrast, IGZO/IZO TFTs annealed in O$_2$ demonstrated superior stability, with ΔV$_{TH}$ reaching -0.61 V (red), -1.75 V (green), and -8.87 V (blue), respectively.

Photon energy is inversely proportional to wavelength:

$$E = (h * c)/\lambda \tag{1}$$

Where E denotes photon energy, h represents Planck constant, c signifies the speed of light, and λ stands for the wavelength of light. As the wavelength decreases, the photon energy increases. Consequently, shorter wavelengths excite more photo-generated carriers, increasing the generation rate of electron-hole pairs in AOS materials. Additionally, under conditions of higher energy illumination, high-density neutral oxygen vacancies in deep energy levels above the valence band maximum (VBM) become more ionized into V$_o^+$ or V$_o^{2+}$, resulting in larger negative shifts in V$_{TH}$ [14].

IV. CONCLUSION

This study systematically investigated the effects of different annealing atmospheres on the I-V characteristics and NBIS stability of IGZO/IZO bilayer based TFTs. The experimental results show that compared with devices annealed in air or nitrogen atmosphere, TFTs annealed in an oxygen atmosphere not only exhibit the best electrical properties, but also exhibit smaller V_{TH} shifts under three different wavelengths of light compared to the devices annealed in air or nitrogen atmosphere. The effective reduction of V_O defects in the active layer under oxygen annealing conditions is considered as an internal mechanism for improving performance. This work shows reference significance for optimizing the electrical performance and the NBIS reliability of MO TFTs.

REFERENCES

[1] Y Shen, M Zhang, S He, L Bian, J Liu, Z Chen, S Xue, Y Zhou, and Y Yan, "Reliability Issues of Amorphous Oxide Semiconductor-Based Thin Film Transistors," J. Mater. Chem. C, vol. 12, no. 35, pp. 13707–13726, 2024.

[2] H Kim, S Maeng, S Lee, and J Kim, "Improved Performance and Operational Stability of Solution-Processed InGaSnO (IGTO) Thin Film Transistors by the Formation of Sn–O Complexes," ACS Appl. Electron. Mater., vol. 3, no. 3, pp. 1199–1210, Mar. 2021.

[3] Y Li, R Yao, J Zhong, Y Yang, Z Liang, Y Fu, X Zeng, G Su, H Ning, and J Peng, "The Hump Phenomenon and Instability of Oxide TFT Were Eliminated by Interfacial Passivation and UV + Thermal Annealing Treatment," ACS Appl. Electron. Mater., vol. 5, no. 9, pp. 4846–4862, Sep. 2023.

[4] GW Shim, W Hong, J Cha, JH Park, KJ Lee, and S Choi, "TFT Channel Materials for Display Applications: From Amorphous Silicon to Transition Metal Dichalcogenides," Advanced Materials, vol. 32, no. 35, p. 1907166, Sep. 2020.

[5] F He, Y Qin, L Wan, J Su, Z Lin, J Zhang, J Chang, J Wu, and Y Hao, "Metal Oxide Heterojunctions for High Performance Solution Grown Oxide Thin Film Transistors," Applied Surface Science, vol. 527, p. 146774, Oct. 2020.

[6] K Liang, Y Wang, S Shao, M Luo, V Pecunia, L Shao, J Zhao, Z Chen, L Mo, and Z Cui, "High-Performance Metal-Oxide Thin-Tilm Transistors Based on Inkjet-Printed Self-Confined Bilayer Heterojunction Channels," J. Mater. Chem. C, vol. 7, no. 20, pp. 6169–6177, 2019.

[7] J Lee and DS Chung, "Heterojunction Oxide Thin Film Transistors: A Review of Recent Advances," J. Mater. Chem. C, vol. 11, no. 16, pp. 5241–5256, 2023.

[8] SJ Park and TJ Ha, "Effects of Sn Doping on the Electrical Performance and Stability of Sub-V Operating Metal-Oxide Thin-Film Transistors Fabricated by Oxygen Annealing," IEEE Electron Device Lett., vol. 44, no. 4, pp. 642–645, Apr. 2023.

[9] P He, S Zuo, W Wang, R Hong, L Tang and X Zou, "Enhanced NBIS Stability of Oxide Thin-Film Transistors by Using Terbium-Incorporated Alumina," IEEE Electron Device Lett., vol. 45, no. 9, pp. 1594–1597, Sep. 2024.

[10] YF Tu, IN Lu, HC Chen, WC Su, YH Hung, KJ Zhou, YZ Zheng, LC Sun, YS Shih, JJ Chen, CY Lien, HC Huang, CH Lien, and TC Chang, "Improving a-InGaZnO TFTs Reliability by Optimizing Electrode Capping Structure Under Negative Bias Illumination Stress," IEEE Electron Device Lett., vol. 41, no. 8, pp. 1221–1224, Aug. 2020.

[11] J Zhang, G Lin, P Cui, M Jia, Z Li, and L Gundlach, "Enhancement-/Depletion-Mode TiO_2 Thin-Film Transistors via O_2 /N_2 Preannealing," IEEE Trans. Electron Devices, vol. 67, no. 6, pp. 2346–2351, Jun. 2020.

[12] N Choi, MJ Kim, H Hong, DY Shin, J Go and TG Weldemhret, "Quantitative Insight of Annealing Atmosphere-Induced Device Performance and Bias Stability in a Ga-Doped InZnSnO Thin-Film Transistors," IEEE Trans. Electron Devices, vol. 71, no. 9, pp. 5393–5400, Sep. 2024.

[13] HS Jeong, HS Cha, SH Hwang, and HI Kwon, "Effects of Annealing Atmosphere on Electrical Performance and Stability of High-Mobility Indium-Gallium-Tin Oxide Thin-Film Transistors," Electronics, vol. 9, no. 11, p. 1875, Nov. 2020.

[14] Z Yang, T Meng, Q Zhang, and HP D Shieh, "Stability of Amorphous Indium–Tungsten Oxide Thin-Film Transistors Under Various Wavelength Light Illumination," IEEE Electron Device Lett., vol. 37, no. 4, pp. 437–440, Apr. 2016.

A 2D Covalent Organic Framework (COF) Based Memristor

Yaoli Guo Yifan Zheng Yan Yan Jiaxing Yang Yanlin Li Zherui Zhao Guanglong Ding Shuangmei Xue

Abstract—In this paper, we successfully synthesized a high-quality, defect-free, free-standing two-dimensional (2D) covalent organic framework (COF) film for the development of high-performance memristors [Ag/COF/indium tin oxide (ITO)]. The resulting memristors exhibit low variability and superior reliability, with VSET values range from 0.87 to 0.95 V. The ON/OFF ratios of these devices range from 1.88×10^3 to 2.79×10^3.

Index Terms—covalent organic framework, memristor, low variability, high reliability

I. INTRODUCTION

Memristors, characterized by their compact size, simple structure, and low power consumption, are regarded as one of the most promising candidates for constructing brain-inspired computing systems, enabling the realization of high-energy-efficiency in-memory computing for addressing the von Neumann bottleneck [1]- [4]. In recent years, organic memristors have gained significant attention because of their simple fabrication process, large-scale production capabilities, and flexibility [5]- [8]. Two-dimensional covalent organic frameworks (2D COFs), with their highly ordered and, tunable chemical structures, tunable chemistry and high porosity, can significantly improve storage capacity and charge carrier separation efficiency of electronic devices [9]- [12]. Consequently, they can be applied as resistive switching (RS) layers for developing high-performance memristors. This study proposes a strategy to achieve high-quality COF-TPB/TPOC$_5$ films through interface-assisted method. Owing to the defect-free property, customizable and highly uniform porous structure and limited positional freedom, the obtained COF-TPB/TPOC$_5$ film endows the memristors with low variability and high reliability.

Yaoli Guo is with *State Key Laboratory of Radio Frequency Heterogeneous Integration (Shenzhen University), College of Physics and Optoelectronic Engineering (Shenzhen University)*, Shenzhen, China. Yifan Zheng and Zherui Zhao are with *Institute for Advanced Study (Shenzhen University)*, Shenzhen, China. Yan Yan, Jiaxing Yang, Yanlin Li, Guanglong Ding and Shuangmei Xue are with *State Key Laboratory of Radio Frequency Heterogeneous Integration (Shenzhen University)*, Shenzhen, China. This work is supported by the Guangdong Major Project of Basic and Applied Basic Research under Grant 2023B0303000008, the Shenzhen Science and Technology Program under Grants KQTD 20221101093555005, ZDSYS 20220527171402005 and 20231121155257002. Corresponding author: (Shuangmei Xue, email: sxue@szu.edu.cn)

II. EXPERIMENTAL SECTION

A. Preparation of the COF-based memristors

To prepare the COF-based memristors, the glass (2 cm × 2 cm) substrate coated with indium tin oxide (ITO, 185 nm) was cleaned by DI water. The COF-TPB/TPOC$_5$ films was prepared though interface-assisted method at the free oil-water interface. In brief, the aqueous solution containing 0.50 wt% 1,3,5-Tris(4-aminophenyl)benzene (TPB) and 0.15 wt% acetic acid was poured into a beaker. Then, the 2,5-bis(pentyloxy)terephthalaldehyde (TPOC$_5$) solution of 0.15 wt% in ethyl acetate was gently added into the above breaker. After the solution stabilized and stratified, the reaction was maintained at room temperature for 12 hours to facilitate film formation. Subsequently, the COF-TPB/TPOC$_5$ films were carefully transferred on the glass/ITO by stamping process. A 50 nm thick Ag electrode was deposited as the top electrode via thermal evaporation with the assistance of a shadow mask. The chemical structure, and device fabrication process are illustrated in Fig. 1 and 2. Keysight B2900A semiconductor parameter analyzers with a PRCBE probe station with a temperature controller was used to character the electrical properties of COF-based memristors. All the set-up was fixed on antivibration mounting.

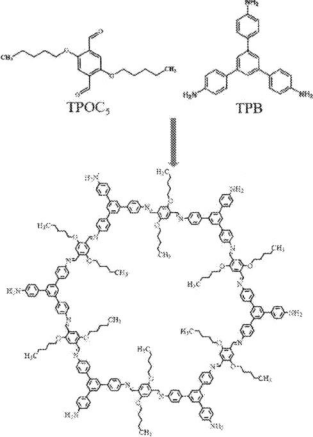

Fig. 1. Chemical Structures of TPB, TPOC$_5$ and COF-TPB/TPOC$_5$.

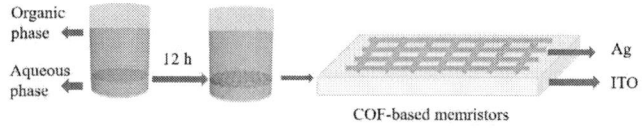

Fig. 2. The fabrication process of COF-based memristor.

III. RESULTS AND DISCUSSION

A. Structural caracterization of COF-TPB/TPOC$_5$ films

The chemical compositions of the COF-TPB/TPOC$_5$ films were characterized. Compared with the Fourier transform infrared spectra (FTIR) of the corresponding monomers, typical C=N imine stretch at 1592 cm^{-1} is emerged in that of the COF-TPB/TPOC$_5$ films, while the C=O and N-H stretches at 1681 and 3355 cm^{-1} are almost disappeared (Fig. 3), indicating the successful synthesis of COF-TPB/TPOC$_5$. The surface morphology of the COF-TPB/TPOC$_5$ film was characterized using atomic force microscopy (AFM), revealing a thickness of approximately 39 nm and a smooth, defect-free surface (Fig. 4).

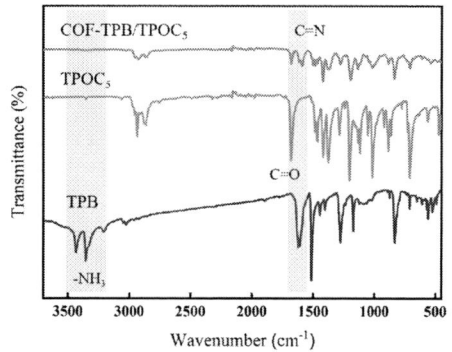

Fig. 3. FTIR spectra of TPB, TPOC$_5$ and COF-TPB/TPOC$_5$.

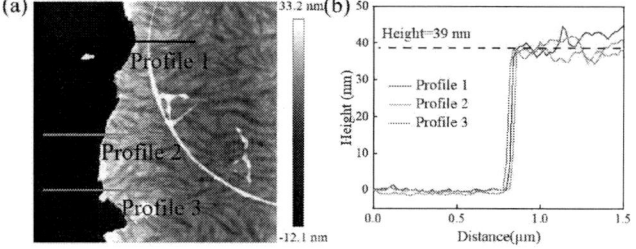

Fig. 4. AFM Height image (a) and thickness analysis (b) of COF-TPB/TPOC$_5$ film.

B. Electrical properties of the COF-based memristors

After the electroforming process at 1.50 V, the obtained COF-memristor exhibited stable volatile threshold (TS) switching under the voltage sweep of 0-2.00 V. According to the I-V curves in Fig. 5, the obtained device demonstrates relatively high temporal uniformity, with the average SET voltage (V$_{SET}$) of 0.88 V, ON/OFF ratio of 1.78×10^3, and relatively low coefficient of variation (CV, 0.159). To estimate the device spatial uniformity, the I-V curves of four devices were tested for 8 cycles, and the results about V$_{SET}$ and ON/OFF ratio of each device are shown in Table 1. The spatial CV of V$_{SET}$ and ON/OFF ratio are about 0.04 and 0.19.

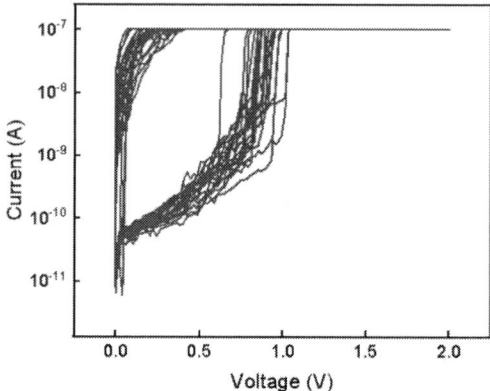

Fig. 5. The typical I-V curves (20 cycles) of COF-based memristors.

TABLE I
THE SUMMARIZE OF RS PROPERTIES OF COF-BASED MEMRISTORS

Device	SET voltage	ON/OFF ratio
1	0.87	1.99×10^3
2	0.89	1.88×10^3
3	0.88	2.03×10^3
4	0.95	2.79×10^3

IV. CONCLUSION

In this paper, we report the synthesis of highly dense, an defect-free 2D COF-TPB/TPOC$_5$ films using an interface-assisted method and their successful application in memristor. The VSET and ON/OFF ratio obtained Ag/COF/ITO devices are about 0.88 V and 1.78×10^3, respectively. At the same time, this memristor exhibit low variability and high reliability, with low temporal and spatial CV values.

REFERENCES

[1] M. R. Sarkar and C. Y. Yi, "An In-Memory Computing Architecture Utilizing Energy-Efficient VGSOT MRAM Device," IEEE Transactions on Circuits and Systems II: Express Briefs, vol. 71, no. 7, pp. 3258-3262, July 2024.

[2] S. -H. Park, J. Kim, J. Ko, J.Im, Y. Yang and J. -J. Kim, "Optimization of Programming Pulse Shape for Vertical NAND Flash Memory Using Neural Networks," IEEE Electron Device Letters, vol. 45, no. 11, pp. 2102-2105, Nov. 2024.

[3] Y. -J. Cho, Y. H. Kown, N. J. Seong, K. J. Choi, H. -O. Kim and J. -H. Yang, "Device Feasibility of 60-nm-Scaled Vertical-Channel Memory Transistors Using InGaZnO Channel and ZnO Charge-Trap Layers," IEEE Transactions on Electron Devices, vol. 71, no. 3, pp. 1839-1844, March 2024.

[4] P. Meihar, R. Srinu, S. Lashkare, A. K. Singh, H. Mulaosmanovic and V. Deshpande, "Ferroelectric MirrorBit-Integrated Field-Programmable Memory Array for the TCAM, Storage, and In-Memory Computing Applications," IEEE Transactions on Electron Devices, vol. 71, no. 5, pp. 2957-2962, May 2024.

979-8-3315-2209-4/25 $31.00 © 2025 IEEE

[5] Q. Ding, W. Jiao, H. Wang, X. Zhang and S. Gao, "Study of a Stretchable Polymer for Adjustable Flexible Organic Memristor," IEEE Transactions on Electron Devices, vol. 70, no. 7, pp. 3921-3927, July 2023.

[6] H. -T. Huang, J. Luo, J. -L. Wu, X. -E. Han, Z.-D. Zhang, J.-W. Cai, X. Gao, J.-L. Xu, Y. -N. Zhong, B. Dong, S. M. Morozova and S. -D. Wang, "Solution-Processed Organic Memristor Matrix With Behavior of Clustered Synaptic Plasticity," IEEE Electron Device Letters, vol. 44, no. 10, pp. 1724-1727, Oct. 2023.

[7] B. Shkodra, M. Petrelli, A. Douaki, M. Ahmad, A. Altana, L. Petti, S. Carrara and P. Lugli, "Spray-Coated Thin-Film Organic Memristor for Neuromorphic Applications," 2023 IEEE International Conference on Flexible and Printable Sensors and Systems, pp. 1-4, 2023.

[8] J. M. Fernandes, N. Xavier, S. Roy and S. Dutta, "Volatile Switching in Flexible Hybrid Halide Perovskite Memristors," 2024 IEEE International Flexible Electronics Technology Conference, pp. 1-4, 2024.

[9] C. Li, D. Li, W. Zhang, H. Li, G. Yu, "Towards High-Performance Resistive Switching Behavior through Embedding a D-A System into 2D Imine-Linked Covalent Organic Frameworks," Angewandte Chemie International Edition, vol. 60, pp. 27135, 2021.

[10] L. Liu, B. Geng, W. Ji, L. Wu, S. Lei, W. Hu, "A Highly Crystalline Single Layer 2D Polymer for Low Variability and Excellent Scalability Molecular Memristors," Advanced. Materials, vol. 35, pp. 2208377, 2023.

[11] Y. Yue, H. Li, H. Chen and N. Huang, "Piperazine-Linked Covalent Organic Frameworks with High Electrical Conductivity," Journal of the American Chemical Society, vol. 144, no. 7, 2022.

[12] A. Giri, G. Shreeraj, T. K. Dutta, A. Patra, "Transformation of an Imine Cage to a Covalent Organic Framework Film at the Liquid-Liquid Interface," Angewandte Chemie International Edition, vol. 62, Dec. 2023.

Magnesium intercalation in gallium nitride for enhanced p-type doping and device performance

1st Jia Wang

2nd Hiroshi Amano

Abstract—**P-type doping in gallium nitride (GaN) has traditionally relied solely on substitutional magnesium (Mg) incorporation. However, persistent limitations such as low hole mobility, limited hole concentration, and poor ohmic contact remain unresolved. Interstitial magnesium, long regarded as a deep-level compensating impurity detrimental to p-type doping, exhibits unexpected behavior at exceptionally high concentrations. This study demonstrates that, under such conditions, interstitial Mg atoms self-organize into aligned single-atomic layers, a phenomenon known as intercalation. Each Mg layer induces periodic polarity transitions in GaN, which may account for the experimentally observed increase in ionized acceptors and hole concentration post-intercalation. Furthermore, the uniaxial strain generated by these interstitial Mg layers modifies the valence band structure and enhances hole mobility. These synergistic advantages are attributed to "2D-doping" effects which offer fresh insights into semiconductor doping mechanisms. Crucially, achieving this 2D-doping regime requires only thermal annealing of a metallic Mg thin film deposited on GaN, offering a scalable and industrially viable pathway to overcome the constraints of conventional substitutional Mg doping. This discovery holds broad implications for optoelectronic and power device applications.**

Index Terms—**Gallium nitride, Magnesium, Intercalation, P-type doping, Ohmic contact**

I. INTRODUCTION

Gallium nitride (GaN) is a representative wide-bandgap semiconductor. The initial breakthrough in the realization of p-type conduction in gallium nitride (GaN) by substituting gallium (Ga) atoms with magnesium (Mg) and activating the Mg dopants marked the advent of white-light LEDs and catalyzed extensive research into GaN-based technologies [1] . Despite this progress, the fundamental understanding between GaN and Mg remains a fascinating puzzle. While Mg substituting Ga sites (Mg_{Ga}) is well-documented, interstitial magnesium (Mg_i) has proven far more challenging to detect and study [2]. Furthermore, p-type GaN still faces core challenges such as low hole mobility, insufficient hole concentrations, and bottleneck issues in achieving reliable ohmic contacts, all of which hinder its practical performance.

Jia Wang and Hiroshi Amano are with *Institute for Advanced Research, and Institute of Materials and Systems for Sustainability, Nagoya University,* Furo-cho, Chikusa-ku, Nagoya, 464-8601, Japan. This work is supported by Japan Science and Technology Agency (JST) as part of Adopting Sustainable Partnerships for Innovative Research Ecosystem (ASPIRE) program, Grant Number JPMJAP2311, and by the Japan Society for the Promotion of Science (JSPS) KAKENHI Grant Number JP24K17305. Corresponding author: (Jia Wang, email: wang@nagoya-u.jp)

II. SPONTANEOUS INTERCALATION OF MG INTO GAN

A. Intercalation in Semiconductor Materials

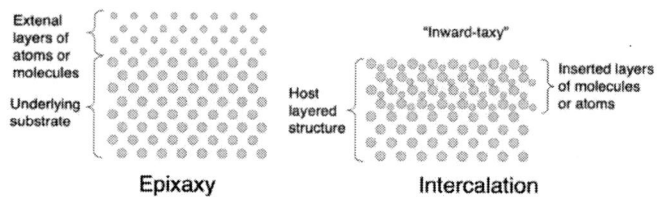

Fig. 1. Schematic illustration comparing the differences between epitaxy and intercalation.

Intercalation has been established as a versatile tool for developing a novel class of materials. Van der Waals materials are commonly selected for intercalation due to the weak attraction between their planar layers, which creates preferential sites for inserting intercalant atoms or molecules. In contrast, intercalation in non-van der Waals solids primarily involves MAX phases—a family of materials essential for MXene fabrication [3]. Recently, intercalation has also been reported in semiconductors [4]. From this perspective, intercalation can be understood as "inward epitaxy," where atomic layers are inserted into underlying crystals while maintaining lattice registry, similar to how external layers align in conventional epitaxy.

B. Mg-intercalated GaN superlattices (MiGs)

Recently, we identified a rare and unique instance of intercalation, in which monoatomic Mg_i layers are inserted into (0001) GaN, forming a 2D Mg-intercalated GaN superlattice (MiGs) structure [5] .

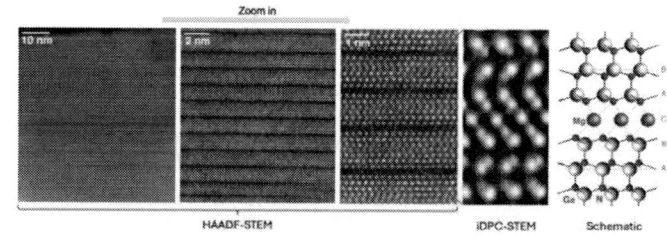

Fig. 2. Cross-sectional scanning transmission electron microscopy (STEM) images showing the typical structure of Mg-intercalated GaN superlattices [5].

As shown in Fig. 2, Mg_i layers are intercalated every few GaN layers, typically ranging from 5 to 8, though instances extending to several tens have also been observed. These Mg_i layers occupy the C site without disrupting the original ABAB hexagonal lattice symmetry. Moreover, Ga atoms are repelled from Mg_i atoms, while the positions of N atoms remain unchanged, resulting in contrasting polarity above and below each 2D Mg_i layer.

The fundamental requirement for two materials to form a superlattice is minimal interfacial energy and lattice mismatch [6]. Interestingly, GaN and Mg not only share the same hcp lattice structure, but their lattice constants—both the basal-plane a and out-of-plane c—are nearly identical.

As shown in TABLE I, the basal-plane lattice mismatch between GaN and Mg is significantly smaller than that of other materials commonly used for GaN heteroepitaxy, such as sapphire, Si, and SiC. This may also explain the tendency of Mg to segregate on (0001) GaN.

TABLE I
COMPARISON OF BASAL-PLANE LATTICE MISMATCH OF GAN WITH OTHER COMMON MATERIALS

Basal-plane lattice mismatch	$(a_f\text{-}a_s)/a_s$
GaN on Mg (0001)	-0.3%
GaN on Si (111)	+17%
GaN on sapphire (30° rot.)	-16%
GaN on SiC	+3.5%

C. Annealing of Metallic Mg Film on GaN

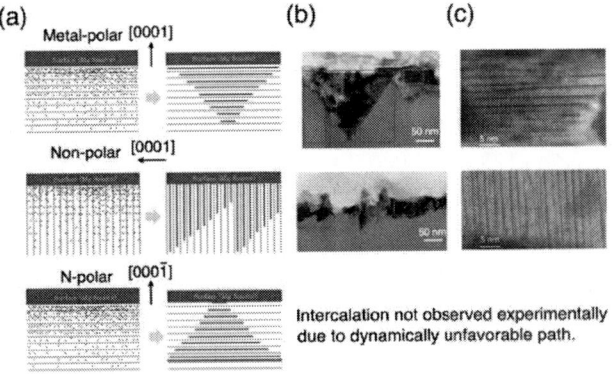

Fig. 3. (a) Diffusion (left) and intercalation (right) of Mg into GaN with different polarities. (b) High-angle annular dark-field scanning transmission electron microscopy (HAADF-STEM) images of the surface region, and (c) a detailed view of the region outlined in green in (b).

Interestingly, we initially observed these unique structures by simply depositing bare metallic Mg film onto GaN template or substrate via electron-beam evaporation or sputtering, followed by annealing at 550–900 °C in nitrogen under atmospheric pressure for a few minutes.

As shown in Fig. 3, when Mg is annealed on Ga-polar or non-polar GaN substrates (Fig. 3a), the 2D Mg_i layers intercalate laterally or vertically (Fig. 3c), respectively. This phenomenon fundamentally arises from the tendency of Mg to segregate on the basal plane of GaN. Furthermore, due to the polarity inversion effect of 2D Mg_i, the Mg layers arrange themselves into large pyramidal shapes to minimize total energy (red box in Fig. 3b).

Notably, we also repeated this process on N-polar GaN. However, no intercalation structure was observed in our STEM analysis. This result aligns with theoretical expectations, as a large pyramid in N-polar GaN would form with its tip facing upward, making top-down diffusion of Mg dynamically unfavorable.

III. 2D-MG DOPING EFFECTS

Since the monoatomic Mg_i layers are orderly arranged in a confined volume, they exert uniaxial compression on the GaN layers, creating an interstitial effect that subjects GaN to uniaxial strain as high as -10 %. This strain significantly alters the valence band structure of GaN, notably reversing the sign of crystal field splitting energy, which lifts the crystal field split-off hole band to become the predominant valence band [7]. Furthermore, it has been demonstrated that the MiGs structure greatly reduces specific contact resistivity (ρ_c) to p-GaN owing to the increased tunnelling probability. As described by *Eqn.* (1) [8]:

$$\rho_c = \exp\left(\frac{2\phi_B}{q\hbar}\sqrt{\frac{\epsilon_s m^*}{N_a}}\right) \quad (1)$$

where ϕ_B is the surface barrier height, q is the elementary charge, $\hbar$ is the reduced Planck constant, ϵ_s is the permittivity, m^* is the effective mass and N_a is the acceptor concentration. This reduction in results from a combination of effects, including an increase in N_a, a decrease in m^*, and a reduction in ϕ_B.

These unique characteristics enabled by intercalating monoatomic Mg_i layers can be regarded as novel 2D-Mg doping effects, distinct from the p-type doping achieved by conventional substitutional Mg on Ga sites, categorized as 0D-Mg doping. The intriguing features of the 2D-Mg_i doping process may provide fresh insights into the doping mechanism of nitride semiconductors and open avenues for the development of novel electronic devices.

IV. EXAMPLES OF 2D-DOPING OF MG IN GAN-BASED DEVICES

Figure 4 compares the *I-V* characteristics of GaN-based *p-n* junction diodes (PNDs) fabricated without and with the incorporation of a MiGs structure (i.e., 2D Mg doping). In Figure 4(b), the forward and reverse bias currents are plotted on the same semi-logarithmic scale. The diode containing the MiGs structure shows a forward-bias current enhancement by several orders of magnitude, while its reverse leakage current remains below the detection limit. Additionally, the current density and specific on-resistance are calculated and plotted on a semi-logarithmic scale, whereas the ideality factor (IF) and forward *I-V* characteristics for the diode with MiGs are presented on a linear scale. As shown in Figure 4(b), after 2D Mg doping, the GaN-based PND achieves a low turn-on voltage of 3 V, an ultra-low IF of 1.3 for GaN at 2.9 V, and a high current density (J) of 1 kA/cm² at 3.5 V (assuming a device diameter of 80 μm). These results were among the best reported for GaN-based PNDs [9].

These results confirm that 2D-Mg doping effectively restores an ideal Ohmic contact to *p*-type GaN that was either insufficiently doped during the epitaxial growth stage or subsequently compromised by plasma damage during device processing. The superior forward-bias *I-V* characteristics of the *p-n* junction diode further highlight the benefits of excellent Ohmic contact to *p*-GaN. The feasibility and effectiveness of this approach demonstrate its strong potential to overcome the longstanding challenge of forming low-resistance Ohmic contacts to *p*-type GaN.

[6] K.-N. Tu, "Surface and interfacial energies of CoSi$_2$ and Si films: Implications regarding formation of three-dimensional silicon-silicide structures," *IBM J. Res. Dev.*, vol. 34, no. 6, pp. 868–874, 1990.

[7] S. Poncé, D. Jena, and F. Giustino, "Route to high hole mobility in GaN via reversal of crystal-field splitting," *Phys. Rev. Lett.*, vol. 123, no. 9, p. 096602, 2019.

[8] A. Y. C. Yu, "Electron tunneling and contact resistance of metal-silicon contact barriers," *Solid-State Electron.*, vol. 13, no. 2, pp. 239–247, 1970.

[9] K. Nomoto, B. Song, Z. Hu, M. Zhu, M. Qi, N. Kaneda, T. Mishima, T. Nakamura, D. Jena, and H. G. Xing, "1.7-kV and 0.55-m$\Omega \cdot$ cm^2 GaN PN diodes on bulk GaN substrates with avalanche capability," *IEEE Electron Device Lett.*, vol. 37, no. 2, pp. 161–164, 2015.

[10] J. Wang *et al.*, "Ohmic contact to p-type GaN enabled by post-growth diffusion of magnesium," *IEEE Electron Device Lett.*, vol. 43, no. 1, pp. 150–153, 2021.

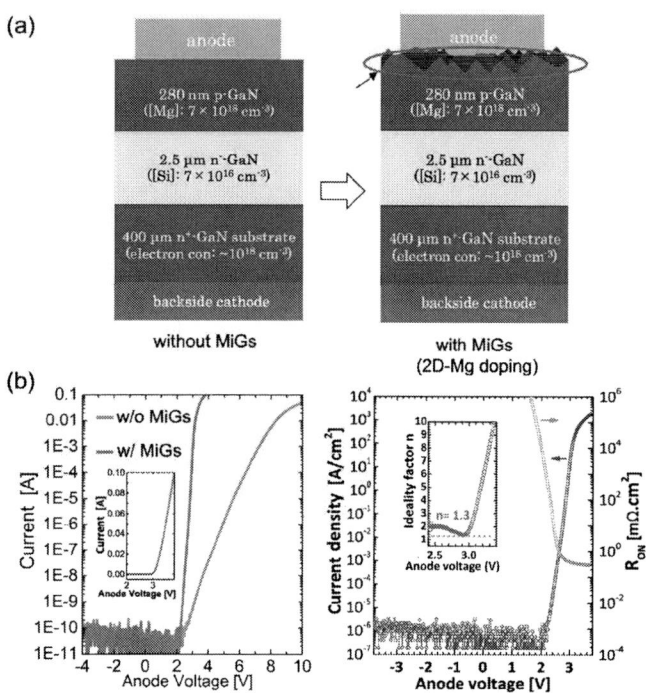

Fig. 4. (a) Cross-sectional schematic illustrations of GaN-based *p-n* junction diodes (PNDs) without (blank) and with the incorporated MiGs structure (indicated by the red circle). (b) Left: Forward and reverse *I-V* characteristics of both diodes plotted on a semi-logarithmic scale. The insets present the *I-V* characteristics of the diode with MiGs near the turn-on region on a linear scale. Right: Current density and specific on-resistance, both plotted on a semi-logarithmic scale as functions of the anode voltage for the diode with MiGs. The insets show the ideality factor as a function of anode voltage. [10]

REFERENCES

[1] H. Amano, "Nobel Lecture: Growth of GaN on sapphire via low-temperature deposited buffer layer and realization of p-type GaN by Mg doping followed by low-energy electron beam irradiation," *Rev. Mod. Phys.*, vol. 87, no. 4, p. 1133, 2015.

[2] U. Wahl *et al.*, "Lattice location of Mg in GaN: A fresh look at doping limitations," *Phys. Rev. Lett.*, vol. 118, no. 9, p. 095501, 2017.

[3] M. Naguib, V. N. Mochalin, M. W. Barsoum, and Y. Gogotsi, "25th anniversary article: MXenes: A new family of two-dimensional materials," *Adv. Mater.*, vol. 26, no. 7, pp. 992–1005, 2014.

[4] D. Zeng *et al.*, "Single-crystalline metal-oxide dielectrics for top-gate 2D transistors," *Nature*, vol. 632, no. 8026, pp. 788–794, 2024.

[5] J. Wang *et al.*, "Observation of 2D-magnesium-intercalated gallium nitride superlattices," *Nature*, vol. 631, no. 8019, pp. 67–72, 2024.

Control of 4H–silicon carbide (0001) oxidation rate via Argon ion-implantation

1st Tao Zhu   2nd YingFeng He   3rd Rui Wang   4th Decai Liu   5th ZheYang Li   6th Rui Jin

Abstract—The thermal oxidation of 4H–silicon carbide (SiC) pretreated by Ar^+ implantation has been studied. The thermal SiO_2 growth rates were determined through atomic force microscopy and Spectroscopic ellipsometer. The results prove that the oxidation of 4H-SiC (0001) was enhanced through Ar^+ implantation. The effect of the implantation dose on the growth rate and surface/interface morphology of SiO_2 was studied. The optical property of the samples was investigated via Spectroscopic ellipsometer. This work offers a new perspective on growth of thick SiO_2 on 4H-SiC by thermal oxidation.

Index Terms—SiC, oxidation, ion implantation, surface, damage degree

I. INTRODUCTION

Silicon carbide (SiC) is a promising semiconductor material for power devices applications due to its excellent characteristics such as wide-bandgap, high field breakdown, high saturation velocity for electrons and so on [1-3]. Among the various polytypes of SiC, 4H–SiC is the most suitable polytype for power device applications. The 600-1700 V 4H-SiC metal-oxide-semiconductor field effect transistors (MOSFETs) has been employed in various applications. For these 4H-SiC MOSFETs, common thicknesses of gate oxide range from 40 nm to 50 nm. However, for higher voltage 4H-SiC MOSFETs or trench 4H-SiC MOSFETs, a high-quality thick SiO_2 gate is needed. The growth rate of thermal oxide on SiC is low at typical oxidation temperature. Therefore, it is necessary to develop a method for enhancing thermal oxidation of SiC.

In this study, Ar^+ implantation was used to improve the 4H-SiC (0001) oxidation rate. It is found that high dose implantation leads to a large oxidation rate. However, high dose implantation can result in rough surface and interface of the as-grown SiO_2. Besides, the implantation dose will influence the refractive index (n) of the SiO_2.

TABLE I
THICKNESS OF SiO_2

Sample	SiO_2 thickness (AFM)	SiO_2 thickness (SE)
A	61.180 nm	60.152 nm
B	89.449 nm	89.500 nm
C	119.744 nm	114.800 nm

All authors are with *Beijing Institute of Smart Energy, Beijing Huairou Laboratory*, Beijing, China. This work is supported by project of Huairou Laboratory "High Voltage Silicon Carbide Electronic Devices". Corresponding author: (Author Rui Jin, email: jinrui@bise.hrl.ac.cn)

II. EXPERIMENT DETAILS

The N-type Si-face 4H-SiC substrates used for the study were supplied by CREE. The substrates were first cleaned using a standard RCA-clean. Then the substrates were divided into three groups: sample A was oxidized directly, while sample B and C were implanted with Ar^+ ions before thermal oxidation. Sample B and C were subjected to 30 keV Ar^+ irradiation with the fluence of 5×10^{14} cm^{-2} and 5×10^{15} cm^{-2}. All samples were oxidized under the same condition. (dry oxygen ambient, 1250°C).

Atomic force microscopy (Dimension ICON, Bruker, USA) was employed to investigate the surface and interface morphology of SiO_2. Refractive index of the SiO_2 was measured using Spectroscopic ellipsometer (SE). Both AFM and SE were used to determine the thickness of SiO_2.

The SRIM 2013 was used to simulate the nuclear energy loss of implanted Ar^+ to determine the damage degree of implanted SiC.

III. RESULTS AND DISCUSSION

Table 1 shows the SiO_2 thicknesses of the three samples. The results indicate that Ar^+ implantation pretreatment can enhance the oxidation of 4H-SiC (0001). The implantation dose has a significant effect on the oxidation rate. The oxidation rate was doubled under the condition of 5×10^{15} cm^{-2}. According to previous work, there existed a critical energy density which is necessary for creating a continuous amorphous layer. This value is between 2×10^{21} keV/cm^3 and 5×10^{21} keV/cm^3 [4].

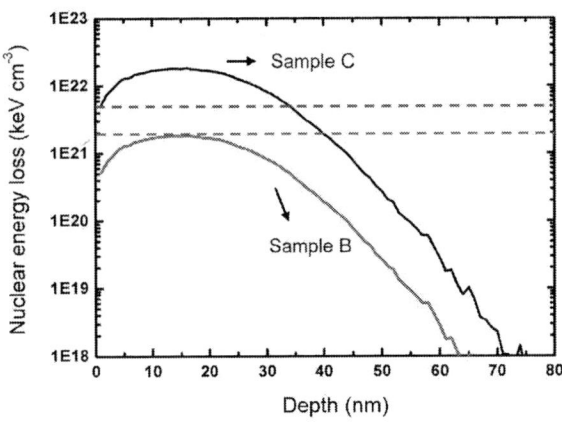

Fig. 1. SRIM simulation of nuclear energy loss of implanted Ar^+.

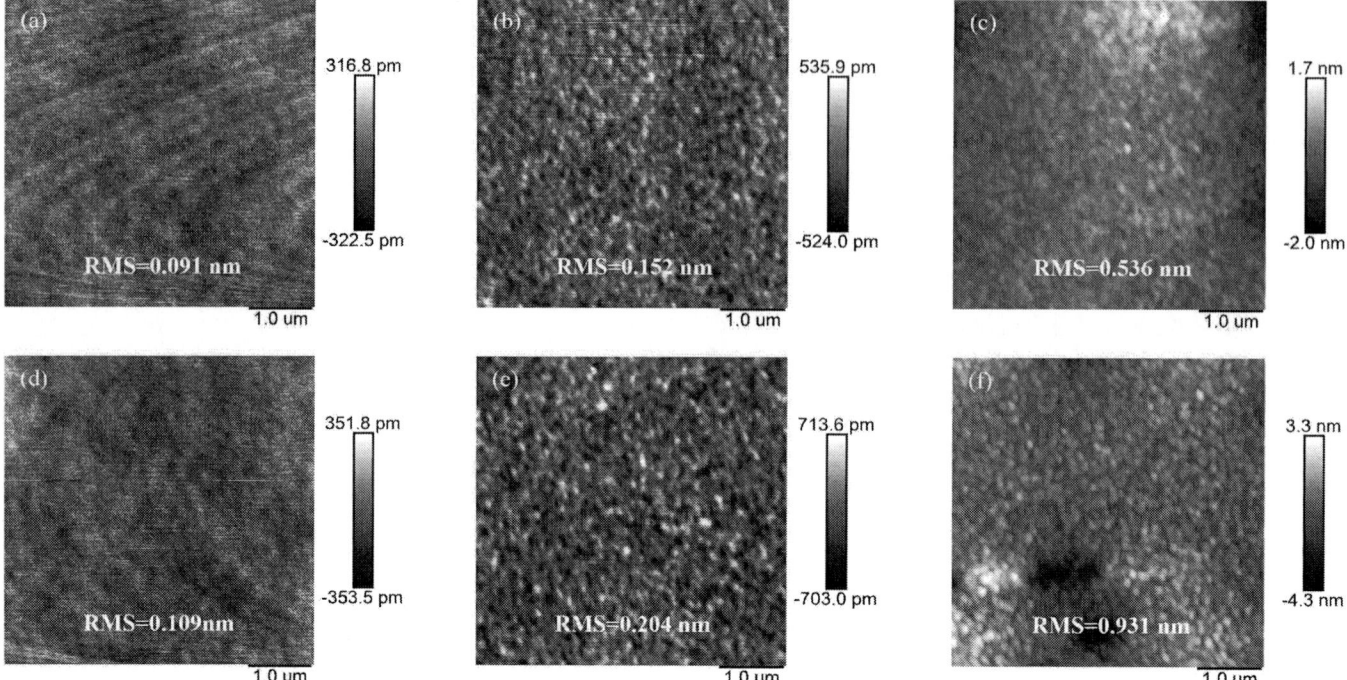

Fig. 2. The surface morphology of (a) sample A, (b) sample B, (c) sample C and the interface morphology of (d) sample A, (e) sample B, (f) sample C(Scan area is $5 \times 5 \ \mu m^2$ for all the images).

Our simulation (Fig. 1) shows that the condition of 5×10^{15} cm^{-2} is enough to offer the threshold energy density. Thus, sample C has a continuous amorphous layer. The previous work has shown that amorphous SiC layers have the highest oxidation rate [5]. Therefore, the oxidation of sample C is significantly enhanced. For sample B, although the energy density is lower than the critical energy density, the higher oxidation rate indicates that parts of the SiC substrate has been amorphized. Further study is needed to determine the effect of this incomplete amorphization.

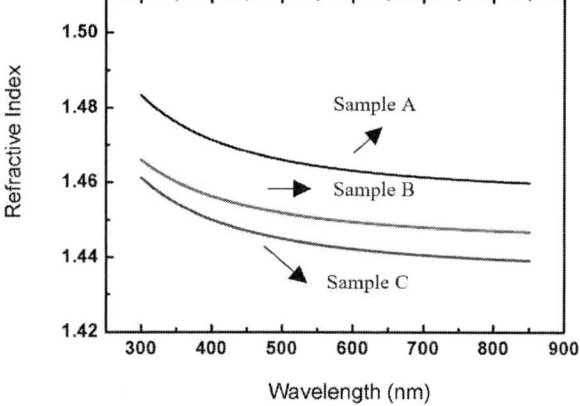

Fig. 3. Extracted refractive index of the SiO^+ for sample A, B and C as a function of wavelength.

The roughness of both SiO_2 and SiC substrate (after chemical removal of the SiO_2) surfaces was shown in Fig. 2. The root-mean-square (rms) roughness values of SiO_2 surface for sample A, B and C are 0.091 nm, 0.152 nm and 0.536 nm, respectively. The results reveal that the higher implantation dose will lead to a rougher surface. Incomplete amorphization (sample B) did not lead to an nonuniform oxidation. The rms roughness values of SiO_2 interface for sample A, B and C are 0.109 nm, 0.204 nm and 0.931 nm, respectively. The origin of the observed morphology changes is attributed to the crystallization of the amorphous layer underneath the oxide.

In Fig. 3, refractive index values (the wavelength ranges between 300 and 850 nm) of the samples were extracted from the SE data. The n values for three samples are from 1.44 to 1.49, which are within the range of values given by reference [6]. The spectral variation of the refractive index of SiO_2 in sample B and C follows that of sample A. However, the sample C n values are the lowest compared with sample A and B, indicating that SiO_2 in sample C has the least dense structure.

IV. CONCLUSION

We have demonstrated the enhancement of 4H-SiC (0001) thermal oxidation via Ar^+ implantation pretreatment. The AFM and SE results show that the oxidation rate was doubled when the implantation dose is 5×10^{15} cm^{-2}. Under the condition, the RMS surface and interface roughness of the sample is 0.536 nm and 0.931 nm, respectively. Spectroscopic ellipsometry data show that the samples' refractive index value at 633 nm was close to the standard value. Note, higher dose implantation makes the as-grown SiO_2 films have a less dense structure, rougher surface and interface. Therefore, the ion implantation condition used to enhance the oxidation should be carefully considered. This study provides an applicable approach to increase the 4H-SiC oxidation rate which is promising for applications in trench and high voltage SiC Power MOSFETs.

REFERENCES

[1] B. J.Baliga, SiliconCarbide PowerDevices. Singapore:World Scientific, 2005.

[2] L. Zhang, X. Yuan, X. Wu, C. Shi, J. Zhang, and Y. Zhang, "Performance evaluation of high-power SiC MOSFET modules in comparison to Si IGBT modules," IEEE Trans. Power Electron., 2019, 34(2), 1181–1196.

[3] X. She, A. Q. Huang, Ó. Lucía, B. Ozpineci, "Review of silicon carbide power devices and their applications," IEEE Trans. Ind. Electron., 2017, 64(10), 8193–8205.

[4] S. Leclerc, A. Declémy, M. F. Beaufort, C. Tromas, and J. F. Barbota, "Swelling of SiC under helium implantation," Appl. Phys. Lett., 2005, 98, 113506.

[5] N. Nipoti, A. Parisini, and A. Poggi, "Annealing kinetics of implantation-induced amorphous layer in 6H-SiC (0001)," Mater. Sci. Forum, 2002, 389, 1109.

[6] O. J. Guy, T. E. Jenkins, M. Lodzinski, A. Castaing, S. P. Wilks, P. Bailey and T. C. Q. Noakes, "Ellipsometric and MEIS studies of 4H-SiC/Si/SiO$_2$ and 4H-SiC/SiO$_2$ interfaces for MOS devices," Mat. Sci. Forum, 2007, 556-557, 509.

Ga₂O₃ Photodetector for Single-Pixel Imaging Applications

Yu Liu[1,2], Xiaobo She[1,2*], Xiang Wang[1,2], Yufeng Guo[1,2*]

[1] College of Integrated Circuit Science and Engineering, Nanjing University of Posts and Telecommunications, Nanjing 210023, P. R. China

[2] National and Local Joint Engineering Laboratory of RF Integration and Micro-Assembly Technology, Nanjing 210023, P. R. China

* Corresponding author's email: 1224228524@njupt.edu.cn, yfguo@njupt.edu.cn

Abstract— Gallium oxide (Ga₂O₃), with its wide bandgap of 4.9 eV, demonstrates exceptional responsivity to ultraviolet (UV) radiation while maintaining solar blindness, making it an ideal candidate for selective UV imaging. We developed a single-pixel imaging system utilizing a Ga₂O₃ photodetector coupled with compressive sensing algorithms to efficiently capture UV images with minimal measurements. Our experimental results demonstrate that reconstruction in the discrete cosine transform (DCT) domain significantly outperforms spatial domain reconstruction across various noise levels and compression ratios. At a noise level of 0.05, DCT domain reconstruction achieved a Peak Signal-to-Noise Ratio (PSNR) between 14.38-16.63 dB compared to 6.92-8.84 dB for spatial domain reconstruction. This work highlights the potential of Ga₂O₃ photodetectors in specialized UV imaging applications such as biological sample analysis, material inspection, and security systems.

Keywords —*Single-Pixel Imaging, Ga₂O₃, Discrete cosine transform*

I. INTRODUCTION

UV imaging has gained significant importance in various fields including biomedical analysis, material science, and security applications due to its ability to reveal features invisible under normal illumination.[1][2][3] Conventional UV imaging typically requires expensive array detectors that are sensitive in the UV range, presenting challenges for widespread adoption. Single-pixel imaging has emerged as a cost-effective alternative that enables imaging with just one detector by making sequential measurements with different spatial encoding patterns.[4][5][6]

Ga₂O₃ has recently attracted considerable attention as a promising material for UV photodetection due to its ultra-wide bandgap, excellent thermal stability, and intrinsic solar blindness.[7][8] These properties make Ga₂O₃ photodetectors particularly suitable for UV detection without interference from visible light. When integrated into single-pixel imaging systems, Ga₂O₃ photodetectors offer the potential to create specialized UV imaging systems at a fraction of the cost of traditional UV camera arrays. In this work, we investigate the application of Ga₂O₃ photodetectors in a compressive single-pixel UV imaging system and evaluate different reconstruction algorithms to optimize imaging performance.

II. RESULTS AND SIGNIFICANCE

Fig. 1(a) illustrates the device structure of our fabricated Ga₂O₃ photodetector, which features a metal-semiconductor-metal (MSM) architecture with Ti/Au electrodes deposited on a Ga₂O₃ thin film. Fig. 1(b) presents the photoresponse characteristics of the Ga₂O₃ detector, showing a peak

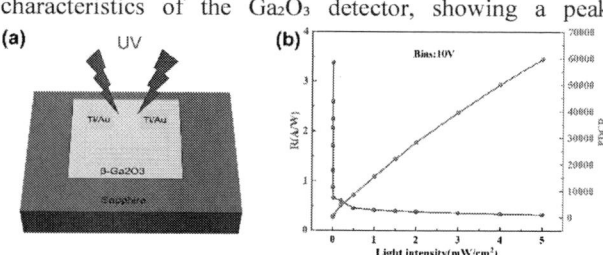

Fig.1 (a) Schematic diagram of gallium oxide structure. (b) R and PDCR under 10 V bias voltage.

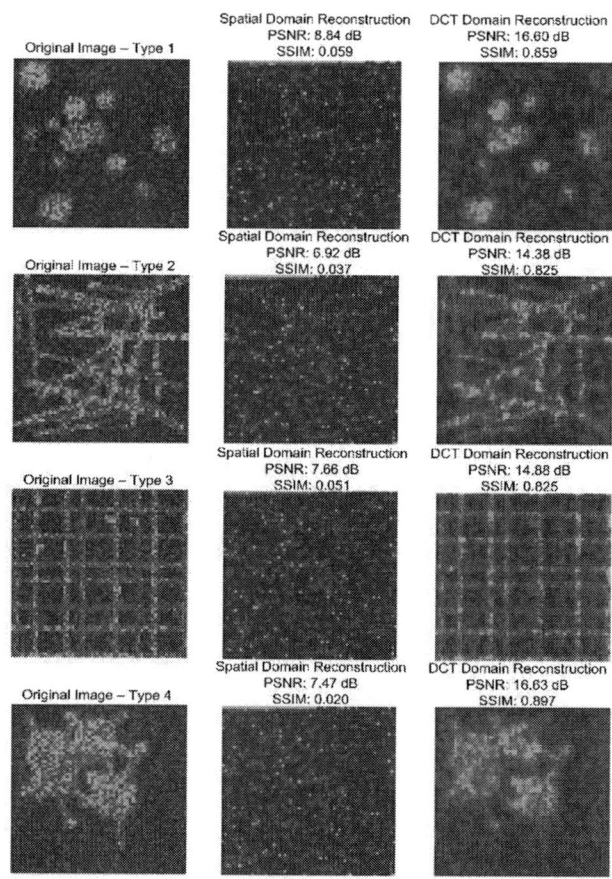

Fig.2 Reconstruction Comparison at Noise Level 0.01.

responsivity of 3.38 A/W at 254 nm, with a maximum photocurrent to dark current ratio of 59813.

For the imaging process, random binary patterns were generated and displayed sequentially on the digital micromirror device (DMD). For each pattern, the Ga₂O₃

979-8-3315-2209-4/25 $31.00 © 2025 IEEE

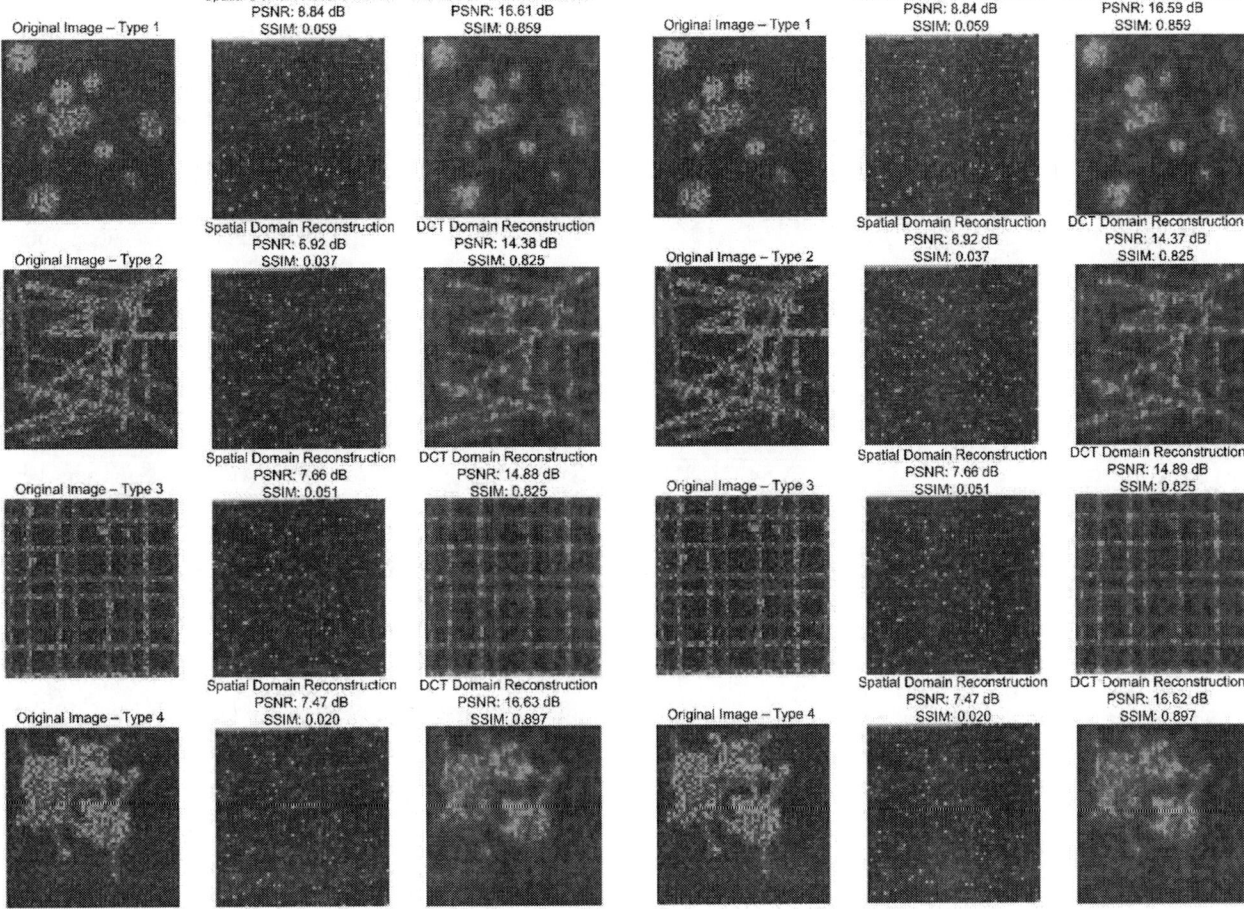

Fig.3 Reconstruction Comparison at Noise Level 0.05.

Fig.4 Reconstruction Comparison at Noise Level 0.1.

photodetector recorded a single intensity measurement representing the correlation between the pattern
and the UV image. These measurements, along with knowledge of the projection patterns, were then used to computationally reconstruct the original UV image.

We implemented and compared two reconstruction approaches: spatial domain reconstruction using direct L1-regularized least squares optimization (Lasso), and reconstruction in the discrete cosine transform (DCT) domain. Both methods solve the fundamental compressive sensing problem but differ in how they exploit image sparsity. We evaluated reconstruction performance across different measurement noise levels (0.01, 0.05, and 0.10) and compression ratios (0.1 to 0.8, representing the ratio of measurements to total pixel count).

Fig. 2, 3, and 4 comprehensively illustrate the reconstruction performance across varying noise levels (0.01, 0.05, and 0.1) for four distinct types of UV images. Fig. 2 demonstrates reconstruction results at the lowest noise level (0.01), where DCT domain reconstruction consistently outperforms spatial domain reconstruction across all image types. For Type 1 circular feature images, DCT domain reconstruction achieves a Peak Signal-to-Noise Ratio (PSNR) of 16.60 dB and Structural Similarity Index Measure (SSIM) of 0.859, compared to spatial domain's significantly inferior 8.84 dB and 0.059. This approximately 88% improvement in

PSNR and 14-fold improvement in SSIM underscores the DCT method's superior cap ability in preserving critical structural information.

Fig. 3 presents reconstruction results at a moderate noise level (0.05), where the performance difference remains pronounced. For Type 2 fibrous structures, DCT domain reconstruction maintains excellent quality with PSNR of 14.38 dB and SSIM of 0.825, while spatial domain reconstruction fails to preserve the critical linear features with PSNR of merely 6.92 dB and SSIM of 0.037. Similarly, for Type 3 grid patterns, DCT domain reconstruction achieves PSNR of 14.88 dB and SSIM of 0.825, compared to spatial domain's poor performance (PSNR: 7.66 dB, SSIM: 0.051). Type 4 images showing complex cluster patterns demonstrate the most dramatic performance gap, with DCT domain reconstruction achieving PSNR of 16.63 dB and SSIM of 0.897 versus spatial domain's 7.47 dB and 0.020.

Fig. 4 exhibits reconstruction performance at the highest noise level (0.1), revealing the DCT domain method's remarkable noise resilience. Even under these challenging conditions, DCT domain reconstruction maintains consistent performance across all image types, with PSNR values between 14.37-16.62 dB and SSIM values between 0.825-0.897. In stark contrast, spatial domain reconstruction continues to produce inadequate results with PSNR values between 6.92-8.84 dB and SSIM values between 0.020-0.059.

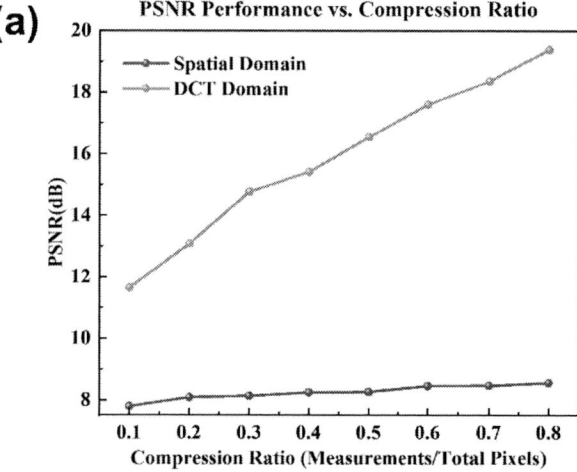

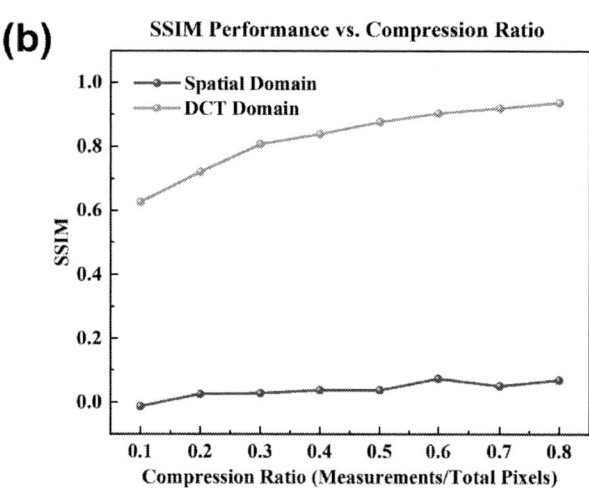

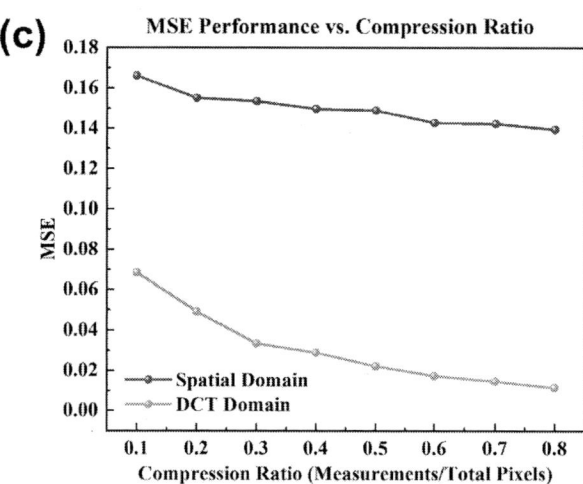

Fig.5 The quantitative performance across varying compression ratios: (a)PSNR. (b)SSIM. (c)MSE.

Most notably, the DCT domain reconstruction's performance remains virtually unchanged across all three noise levels, with less than 0.1% variation in quality metrics between noise levels 0.01 and 0.1, demonstrating exceptional robustness to measurement noise—a critical advantage for practical UV imaging applications in challenging environments.

The quantitative performance across varying compression ratios is presented in Fig. 5. The top panel illustrates PSNR performance, where DCT domain reconstruction exhibits a strong positive correlation with compression ratio, improving from 11.6 dB at 0.1 compression ratio to 19.4 dB at 0.8 compression ratio. This represents a 67.2% improvement in PSNR with increased sampling. In contrast, spatial domain reconstruction maintains a nearly flat response around 8 dB regardless of additional measurements, indicating only marginal improvement of approximately 8% across the entire range of compression ratios. The middle panel of Fig. 5 presents SSIM metrics, where DCT domain reconstruction demonstrates substantial quality enhancement as compression ratio increases, with SSIM values improving from approximately 0.6 at 0.1 compression ratio to over 0.9 at 0.8 compression ratio, representing a 50% improvement in structural fidelity. Spatial domain reconstruction, however, yields consistently poor SSIM values near zero across all compression ratios, confirming its fundamental inability to preserve structural features in UV images. The bottom panel shows Mean Squared Error (MSE) performance, where DCT domain reconstruction exhibits an inverse relationship with compression ratio, decreasing from approximately 0.07 at minimal sampling to around 0.01 at maximal sampling—a sevenfold reduction in reconstruction error. Spatial domain reconstruction maintains persistently high MSE values between 0.14-0.17 with minimal improvement across the compression ratio range.

III. CONCLUSIONS

This study demonstrates the effectiveness of Ga_2O_3 photodetectors in single-pixel UV imaging applications using compressive sensing techniques. Our findings conclusively show that reconstruction in the DCT domain substantially outperforms spatial domain reconstruction, achieving higher PSNR and SSIM values across all tested conditions. The DCT domain approach exhibits remarkable resilience to measurement noise and significant performance improvements with increased sampling rates. The combination of Ga_2O_3 photodetectors with optimized DCT domain reconstruction algorithms presents a promising approach for cost-effective, high-performance UV imaging systems. This technology has potential applications in diverse fields requiring specialized UV imaging, including biological fluorescence imaging, material inspection, forensic analysis, and security systems. Future work will focus on optimizing the Ga_2O_3 detector design for enhanced sensitivity and developing adaptive sampling strategies to further reduce acquisition time while maintaining image quality.

ACKNOWLEDGMENTS

This work was financially supported by the National Natural Science Foundation of China (grant nos. 62401276, U23B2042), Natural Science Research Start-up Foundation of Recruiting Talents of Nanjing University of Posts and Telecommunications (Grant No. NY223161) and in part by the Jiangsu Provincial Key Research and Development Program under Grant BE2022126.

REFERENCES

[1] S. Goossens, G. Navickaite, C. Monasterio et al., "Broadband image sensor array based on graphene-CMOS integration," Nature Photon, vol. 11, pp. 366-371, 2017.

[2] Z. Li, T. Yan, and X. Fang, "Low-dimensional wide-bandgap semiconductors for UV photodetectors," Nat Rev Mater, vol. 8, pp. 587–603, 2023.

[3] G. Li, D. Xie, Q. Zhang et al., "Interface-engineered non-volatile visible-blind photodetector for in-sensor computing," Nat Commun, vol. 16, no. 57, 2025.

[4] M. P. Edgar, G. M. Gibson, and M. J. Padgett, "Principles and prospects for single-pixel imaging," Nature Photon, vol. 13, pp. 13–20, 2019.

[5] N. Horiuchi, "Colour imaging with single-pixel detectors," Nature Photon, vol. 7, pp. 943, 2013.

[6] O. Graydon, "Retina-like single-pixel camera," Nature Photon, vol. 11, pp. 335, 2017.

[7] X. Wang et al., "Ultra-low Power In-sensor Computing β-Ga$_2$O$_3$ Ultraviolet Optoelectronic Synaptic Devices," IEEE Photonics Technology Letters, 2024.

[8] X. Wang et al., "Two-in-one functionality in a 28× 28 β-Ga$_2$O$_3$ array: bias-voltage switching between photodetection and neuromorphic vision," Optics Express, vol. 32, no. 22, pp. 39515-39524, Oct. 2024.

Integrated 3×3 β-Ga$_2$O$_3$ Optoelectronic Synaptic Array for Efficient Edge Detection in Neuromorphic Computing

Xiang Wang[1,2], Xiaobo She[1,2], Yingxu Wang[1,2], Yufeng Guo[1,2*], Yu Liu[1,2*]

[1] College of Integrated Circuit Science and Engineering, Nanjing University of Posts and Telecommunications, Nanjing 210023, P. R. China

[2] National and Local Joint Engineering Laboratory of RF Integration and Micro-Assembly Technology, Nanjing 210023, P. R. China

* Corresponding author's email: liu_yu_24@126.com, yfguo@njupt.edu.cn

Abstract— **Neuromorphic visual sensors, inspired by retinal neuron synapses, have garnered significant attention in the field of novel optoelectronic imaging and sensing. In this work, we fabricated a novel 3×3 β-Ga$_2$O$_3$ optoelectronic synaptic array by introducing additional oxygen vacancies to achieve persistent photoconductivity. This enables the device to emulate critical biological synaptic characteristics, such as short-term plasticity and learning-forgetting-relearning cycles, which are essential for dynamic data processing. We demonstrated the potential of a β-Ga$_2$O$_3$ optoelectronic synaptic array in convolutional image processing, where it successfully extracted various features using multiple convolutional kernels. Our work provides a promising hardware platform for efficient and scalable β-Ga$_2$O$_3$-based neuromorphic computing.**

Keywords — *Optoelectronic synaptic array, β-Ga$_2$O$_3$, Convolutional image processing*

I. INTRODUCTION

Inspired by the dynamic synaptic behavior of retinal neurons, neuromorphic visual sensors are driving significant innovation in optoelectronic imaging and sensing.[1],[2] By replicating the adaptive nature of biological vision systems, these sensors offer promising potential to redefine how visual information is acquired and processed in real-time applications.

β-Ga$_2$O$_3$, with its wide bandgap and distinctive optoelectronic properties, has been identified as a potential candidate for deep ultraviolet (DUV) photodetection applications.[3],[4] Emerging research has investigated the use of Ga2O3-derived neuromorphic devices for optimized image processing tasks, capitalizing on the material's inherent synaptic-like properties. These synaptic responses are linked to deep-level defects, particularly oxygen vacancies (Vo), which function as recombination centers and trigger persistent photoconductivity (PPC) effects.[5]

However, most current studies focus solely on fabricating devices to investigate synaptic plasticity performance, with little attention given to exploring their practical integration into visual sensing and image processing applications.[6],[7] To address these limitations, we developed a novel 3×3 β-Ga$_2$O$_3$ optoelectronic synaptic array to examine its utility in convolutional image processing. The demonstrated robust synaptic plasticity, along with its ability to dynamically adjust synaptic weights in response to varying input stimuli for feature extraction, suggests that the fabricated array device is highly suitable for neuromorphic hardware computing.

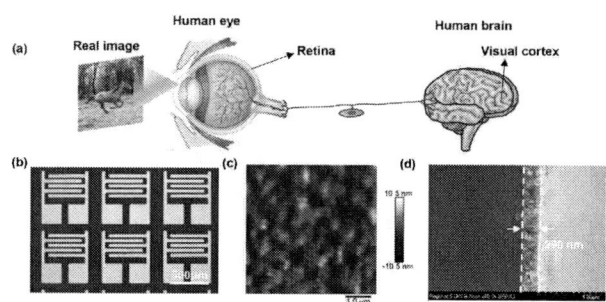

Fig.1 (a) Schematics of the human visual system. (b) Optical microscope image showing a magnified view of a portion of the array. (c) AFM image of the β-Ga$_2$O$_3$ thin film. (d) the SEM cross-section of the β-Ga$_2$O$_3$ device.

II. RESULTS AND SIGNIFICANCE

The human visual system, as shown in Figure 1(a), processes information in two stages: first, the retina captures and processes visual stimuli, and then additional information is extracted, followed by the transmission of this information via the optic nerve to the brain's visual cortex for processing. To mimic the retina's filtering function, we developed a gallium oxide optoelectronic synaptic array (shown in Figure 1(b)), which utilizes its capacity to dynamically adjust synaptic weights in response to varying input stimuli, enabling the detection of the original target image while simultaneously extracting its features.

β-Ga$_2$O$_3$ thin films were synthesized on a sapphire substrate using the metal-organic chemical vapor deposition (MOCVD) technique, with the reactor temperature set at 760°C. During the deposition process, the Vo content in the film was increased by raising the TEGa flow rate, while keeping the oxygen flow rate and other process conditions stable. After the deposition, the crystal quality of the film was evaluated using scanning electron microscopy (SEM) and atomic force microscopy (AFM). The array device was constructed through a series of fabrication techniques, including UV lithography, lift-off, electron beam evaporation, and annealing. The design incorporated interdigitated electrodes, consisting of a Ti/Au bilayer, which were applied directly onto the Ga2O3 thin film. The AFM, as depicted in Figure 1(c), yields a root-mean-square (RMS) roughness of approximately 2.16 nm, suggesting a relatively flat surface

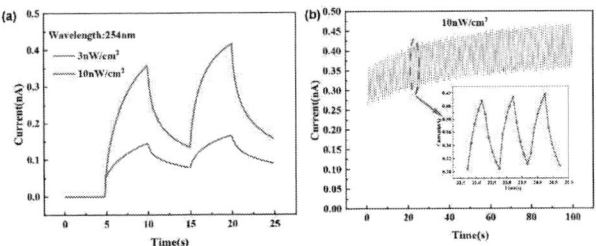

Fig.2 (a) Current responses to a pair of light pulses with different intensities. (b) Current response under consecutive 100 light pulses with an intensity of 10 nW/cm^2.

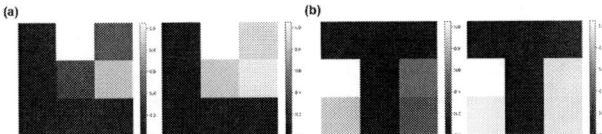

Fig.3 Demonstration of image memory and noise processing capability for the letters (a) L and (b) T, respectively.

Fig.4 (a)Original image. Convolutional results of (b) Sobel-x, (c) Sobel-y, (d) Prewitt-x, (e) Prewitt-y and (f) Embossing.

topography. Additionally, the SEM image in Figure 1(d) confirms that the β-Ga$_2$O$_3$ thin film possesses a smooth and uniform surface morphology.

As depicted in Figure. 2(a), two consecutive light pulses varying in intensity were administered. The second pulse triggered a larger current response compared to the first, mimicking the learning-forgetting-relearning cycle typical of biological entities, corresponding to the PPC phenomenon related to Vo defects present in the oxides. The stability and consistency of the light-triggered responses in our neuromorphic devices were confirmed through the application of 100 on/off cycles of light pulses, each with an intensity of 10 nW/cm^2, a duration of 0.5 s, and an interval of 0.5 s. A slight enhancement in the response was noted with each pulse, as depicted in the inset of Figure. 2(b), eventually plateauing after 100 cycles, as shown in the main part of Figure. 2(b).

Demonstrated in Figures 3(a) and 3(b), these devices successfully memorized the letters 'L' and 'T', respectively. Each memory cell represents an image pixel. Optical masks shaped as 'L' and 'T' were used in experiments involving 100 optical pulses at specified intensities for each letter. Post-exposure, the recorded images showed increased conductance. These findings demonstrate the devices' potential for robust image processing applications.

To demonstrate the practical utility of the β-Ga$_2$O$_3$ optoelectronic synaptic array, Figure 4 presents the results of performing various convolutional image processing tasks using this array. The original image, shown in Figure 4(a), serves as the input, while Figures 4(b)-4(f) display the outputs after processing with different convolutional filters for feature extraction: Sobel-x, Sobel-y, Prewitt-x, Prewitt-y, and Embossing, respectively. The data shows that these filters accurately capture specific features from the target image, effectively capturing both horizontal and vertical features. These findings underscore the capability of the β-Ga$_2$O$_3$ optoelectronic synaptic array in feature delineation, a critical aspect of image processing applications. By successfully extracting relevant features from the input image, the array demonstrates its potential for tasks such as edge detection, object recognition, and image segmentation.

III. CONCLUSIONS

In this study, we developed a 3×3 β-Ga$_2$O$_3$ optoelectronic synaptic array by introducing oxygen vacancies to achieve persistent photoconductivity. The device could successfully emulate biological synaptic characteristics and dynamically adjust synaptic weights in response to varying inputs. The array demonstrates excellent performance in convolutional image processing, effectively functioning for feature extraction and edge detection, thus providing an efficient hardware platform for β-Ga$_2$O$_3$-based neuromorphic computing.

ACKNOWLEDGMENTS

This work was financially supported by the National Natural Science Foundation of China (grant nos. 62401276, U23B2042), Natural Science Research Start-up Foundation of Recruiting Talents of Nanjing University of Posts and Telecommunications (Grant No. NY223161) and in part by the Jiangsu Provincial Key Research and Development Program under Grant BE2022126.

REFERENCES

[1] J. Li et al., "An Optoelectronic Synaptic Transistor with Autotuning Active Photoadaptation for Artificial Visual Perception," IEEE Electron Device Letters, vol. 44, no. 9, pp. 1516-1519, Sep. 2023.

[2] J. Guo, Y. Liu, L. Lin, S. Li, J. Cai, J. Chen, W. Huang, Y. Lin, and J. Xu, "Chromatic Plasmonic Polarizer-Based Synapse for All-Optical Convolutional Neural Network," Nano Lett, vol. 23, no. 20, pp. 9651-9656, Aug. 2023.

[3] X. Wang et al., "Ultra-Low Power In-Sensor Computing β-Ga2O3 Ultraviolet Optoelectronic Synaptic Devices," in IEEE Photonics Technology Letters, vol. 36, no. 23, pp. 1393-1396, 1 Dec. 2024.

[4] M. Orita, H. Ohta, M. Hirano, and H. Hosono, "Deep-ultraviolet transparent conductive β-Ga$_2$O$_3$ thin films, " Appl. Phys. Lett., vol. 77, no. 25, pp. 4166-4168, Dec. 2000.

[5] X. Wang, L. Li, H. Peng, Y. Wang, L. Zhang, Y. Gu, X. She, M. Zhang, Y. Guo, and Y. Liu, "Two-in-one functionality in a 28 × 28 β-Ga$_2$O$_3$ array: bias-voltage switching between photodetection and neuromorphic vision," Opt. Express, vol. 32, no. 22, pp. 39515-39524, Oct. 2024.

[6] M. L. Schneider et al., "Ultralow power artificial synapses using nanotextured magnetic Josephson junctions," Sci. Adv., vol. 4, no. 1, Jan. 2018, Art. no. e1701329.

[7] X. R. Han, Y. H. Mo, Y. B. Wang, X. Q. Yang, J. Y. Wang, B. C. Luo, "Amorphous Ga$_2$O$_3$/GaN heterostructure for ultralow-energy-consumption optically stimulated synaptic devices," Appl. Phys. Lett., vol. 124, no. 1, Jan. 2024.

The Regulating Effect of Electric Field on h-BN/WTe$_2$ Heterostructure

Jun Zhang   Hang Xu   Jiping Hu   Xiaotian Yang   Shipei Ji   Fang Wang   Juin J. Liou   Yuhuai Liu

Abstract—The regulating effect of the electric field on two - dimensional (2D) vertical van der Waals heterostructures (vdW) is crucial. So, in this paper, a 2D vertical h-BN/WTe$_2$ heterostructure was constructed. Using the first-principles method based on density functional theory, we studied how the electric field regulates the h-BN/WTe$_2$ heterostructure. The results show that the h-BN/WTe$_2$ heterostructure is a direct band gap semiconductor with a band gap of 0.831 eV. The electric field can affect the band gap size of the h-BN/WTe$_2$ heterostructure, but it won't change the band gap type or the band-alignment type. Also, it's important to note that when an electric field of -2V/Å is applied, both the conduction band minimum (CBM) and the valence band maximum (VBM) of the h-BN/WTe$_2$ heterostructure are larger than those without the electric field. And the VBM of the heterostructure exceeds the Fermi level, which means an ohmic contact will form within the heterostructure.

Index Terms—h-BN, WTe$_2$,Van der Waals heterostructures, 2D materials

I. INTRODUCTION

During the rapid development of condensed matter physics and materials science, two-dimensional materials are attracting a lot of attention. As a transition metal chalcogenide, WTe$_2$ has extraordinary properties. Its electrical transport shows high anisotropy, and the novel superconducting state at low temperatures is an excellent basis for exploring quantum phenomena [1]. h-BN with excellent mechanical stability, high

---

Jun Zhang, Hang Xu, Jiping Hu, Xiaotian Yang and Shipei Ji are with *National Center for International Joint Research of Electronic Materials and Systems, International Joint-Laboratory of Electronic Materials and Systems of Henan Province, School of Electrical and Information Engineering, Zhengzhou University, Zhengzhou, Henan 450001, P. R. China,* Fang Wang and Yuhuai Liu are with *National Center for International Joint Research of Electronic Materials and Systems, International Joint-Laboratory of Electronic Materials and Systems of Henan Province, School of Electrical and Information Engineering, Zhengzhou University, Zhengzhou, Henan 450001, P. R. China, International Joint Laboratory for Integrated Circuits Design and Application, Ministry of Education, School of Physics, Zhengzhou University, Zhengzhou 450052, China, Institute of Intelligence Sensing, Research Institute of Industrial Technology Co. Ltd., Zhengzhou University, Zhengzhou, Henan 450001, P. R. China,Zhengzhou Way Do Electronics Co. Ltd., Zhengzhou, Henan 450001, P. R. China.* Juin J. Liou is with *School of Electrical and Information Engineering, North Minzu University, Yinchuan, Ningxia 750001, P. R. China.* This work is supported by the [National Natural Science Foundation of China] under Grant [No.62174148]; the [National Key Research and Development Program] under Grant [No.2022YFE0112000 and 2016YFE0118400]; the [Key Program for International Joint Research of Henan Province] under Grant [No.231111520300]; the [Ningbo Major Project of 'Science, Technology and Innovation 2025'] under Grant [No.2019B10129]; and the [Zhengzhou 1125 Innovation Project] under Grant [No.ZZ2018-45]. Corresponding author: (Fang Wang, email:iefwang@zzu.edu.cn;Yuhuai Liu, email:ieyhliu@zzu.edu.cn)

---

thermal conductivity, and chemical inertness, often serves as a growth substrate for two - dimensional materials. As a growth substrate for two-dimensional materials, h-BN can effectively reduce interface scattering and help improve the performance of materials. When the two form a heterostructure, their advantages are complementar [2]. h-BN creates stable and high-quality growth conditions for WTe$_2$, allowing its intrinsic properties to be fully displayed. The interaction at their interface gives rise to new physical phenomena such as charge transfer and band reconstruction, greatly expanding the electronic state structure. The electric field, as a key external regulating factor, is of great significance in this heterostructure. When an electric field is applied, it can precisely regulate carrier concentration, mobility, and even trigger phase transitions [3]. This is beneficial for delving deep into the physical essence of two-dimensional material heterostructures and also opens the door for cutting-edge applications such as high-performance electronic devices and quantum bits. It is thus evident that the research on this heterostructure and the modulation of the heterostructure by the electric field is of far-reaching significance and broad prospects.

II. CALCULATION METHODS

All first-principles calculations in this paper are performed in the Pwmat software. The SG15 pseudopotential is used for all atomic calculations. The Generalized Gradient Approximation (GGA) - Perdew-Burke-Ernzerhof (PBE) is used for the exchange-correlation functional [4]. Since the research object is a two-dimensional material, periodic boundary conditions are applied to the structure boundaries along the X-axis and Y-axis, and a vacuum layer with a thickness of 15 Å is established along the Z-axis to ensure the decoupling between periodically repeated systems. The vdW interaction between h-BN and WTe$_2$ is described by DFT-D3 [5]. The cutoff energy of the plane wave is 50 Ryd, and all structures are fully relaxed until the energy and residual force on each atom converge to 10 eV and 0.05 eV/Å respectively. In addition, the Monkhorst-Pack method is used to generate K-points with symmetry. Meanwhile, K-point grids of 5×5×1 and 9×9×1 are used respectively for structural optimization and self-consistent calculations [6]. The structural model of the h-BN/WTe$_2$ heterostructure and the schematic diagram of the differential charge density are displayed through the VESTA software [7].

979-8-3315-2209-4/25 $31.00 © 2025 IEEE

III. RESULTS AND DISCUSSIONS

A. Geometric structure

First, the electronic structures and properties of single-layer h-BN and WTe$_2$ materials were calculated. The lattice constants of monolayer h-BN and WTe$_2$ are respectively 2.49 Å[8] and 3.56 Å [9]. The band gaps and energy band diagrams of monolayer h-BN [11] and WTe$_2$ [10] are shown in Fig. 1. Both monolayer h-BN and WTe$_2$ are direct band gap semiconductors, and their band gaps are 4.66 eV and 1.07 eV respectively. All of these are consistent with the reports of other researchers. Then, the h-BN was expanded into a 3×3×1 supercell and vertically stacked on the 2×2×1 WTe$_2$ supercell to form the h-BN/WTe$_2$ heterostructure as shown in Fig.2.(a). An important physical quantity for measuring whether the structure of a heterostructure can exist stably is the binding energy E_b of the heterostructure. The calculation formula of E_b is as follows:

$$E_b = \left(E_{h-BN/WTe_2} - E_{h-BN} - E_{WTe_2} \right)/n \qquad (1)$$

E_{h-BN/WTe_2} represents the total energy of the entire heterostructure, E_{h-BN} and E_{WTe_2} represent the total energies of the h-BN and WTe2 monolayers respectively. After calculation, the E_b of this heterostructure is -23.94 meV, indicating that this heterostructure can exist stably. As shown in Fig.2.(b) and (c), the h-BN/WTe$_2$ heterostructure is a direct band gap semiconductor with a band gap of 0.831 eV. Its CBM and VBM are mainly contributed by the W-5D orbital.

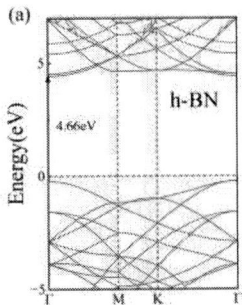 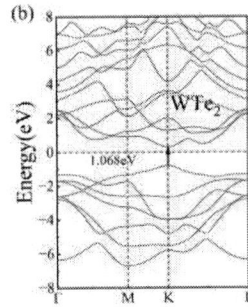

Fig. 1. (a) is the energy band structure of h-BN.(b) is the energy band structure of WTe$_2$.

B. External electric field effect

Finally, the regulatory effect of the electric field on the h-BN/WTe$_2$ heterostructure was explored. In the vertically stacked h-BN/WTe$_2$ heterostructure, we applied a positive electric field from the h-BN monolayer to the WTe$_2$ monolayer. As shown in Fig. 3, when a positive electric field is applied, the direction of this electric field is consistent with that of the built-in electric field, which will intensify the drift motion. The quasi-Fermi level of h-BN bends upward, resulting in an increase in CBO, a decrease in VBO, and a reduction in the forbidden band width, and the energy band

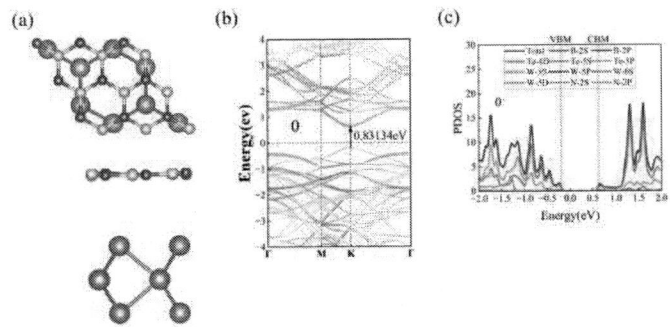

Fig. 2. (a) is the side view (lower layer) and top view (upper layer) of the h-BN/WTe$_2$ heterostructure. (b) is the energy band structure of the h-BN/WTe$_2$ heterostructure. (c) is the projected density of states of the h-BN/WTe$_2$ heterostructure.

alignment type is type I. Moreover, when the electric field increases, this change will be more obvious. When a negative electric field is applied, the direction of this electric field is opposite to that of the built-in electric field, which will intensify the diffusion motion. The quasi-Fermi level of h-BN bends downward, resulting in an increase in VBO, a decrease in CBO, and an increase in the forbidden band width, and the energy band alignment type is type I. Moreover, when the electric field increases, this change will be more obvious. In addition, when the electric field increases to 0.2 V/Å, the VBM of the heterostructure exceeds the Fermi level, and an ohmic contact will be formed within the heterostructure [12]. Through the above analysis, it can be obtained that when a positive electric field is applied to the h-BN/WTe$_2$ heterostructure, the band gap decreases, and when a negative electric field is applied to the h-BN/WTe$_2$ heterostructure, the band gap increases. As shown in Fig. 4, when the applied electric field changes from -3 V/Å to +3 V/Å, the band gap of the h-BN/WTe$_2$ heterostructure keeps decreasing, which is consistent with the above analysis, and the maximum band gap is 0.832 eV and the minimum band gap is 0.831 eV. In addition, the electric field does not change the direct-indirect band gap type of the h-BN/WTe$_2$ heterostructure. As shown in Fig. 4, when the applied electric field changes from -3 V/Å to +3 V/Å, the h-BN/WTe$_2$ heterostructure is always a direct band gap. In addition, as shown in Fig. 6, when the applied electric field is -2 V/Å, both the CBM and VBM of the h-BN/WTe$_2$ heterostructure are larger than those without an applied electric field, while the CBM and VBM of the h-BN/WTe$_2$ heterostructure under other electric fields are almost the same as those without an applied electric field. As shown in Figure 5, both the CBM and VBM of the h-BN/WTe$_2$ heterostructure without an applied electric field are mainly contributed by the W-5D state, and the application of an electric field does not change the contributors of the CBM and VBM of the h-BN/WTe$_2$ heterostructure. The above research provides guidance for semiconductor optoelectronic devices and ohmic contacts.

979-8-3315-2209-4/25 $31.00 © 2025 IEEE

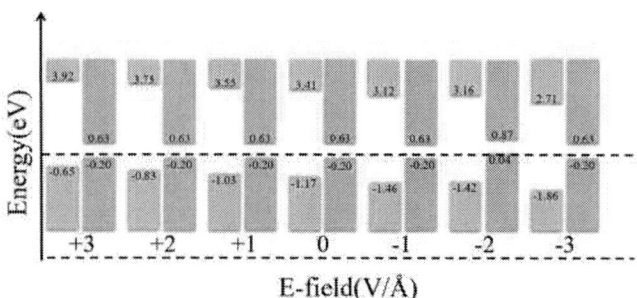

Fig. 3. The energy band alignment of the h-BN/WTe$_2$ heterostructure after applying an electric field ranging from -3 V/Å to +3 V/Å.

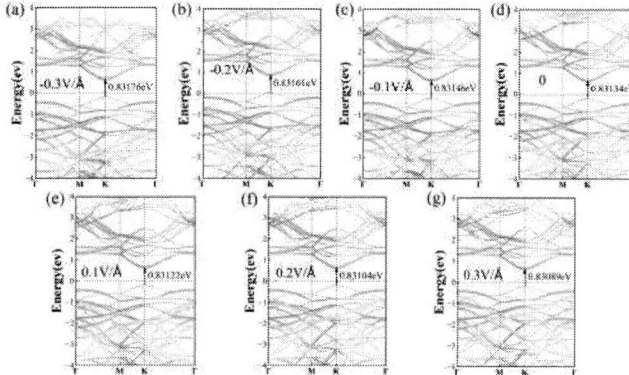

Fig. 4. (a)-(g) are the energy band structures of the h-BN/WTe$_2$ heterostructure after applying an electric field ranging from -3 V/Å to +3 V/Å respectively.

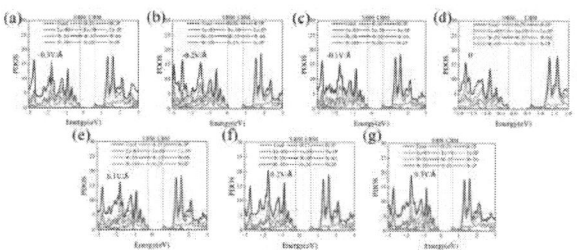

Fig. 5. (a)-(g) are the projected density of states of the h-BN/WTe$_2$ heterostructure after applying an electric field ranging from -3 V/Å to +3 V/Å respectively.

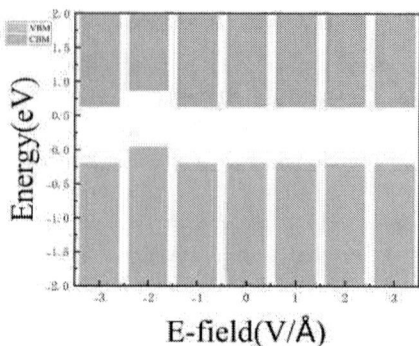

Fig. 6. The band edge positions of the h-BN/WTe$_2$ heterostructure after applying an electric field ranging from -3 V/Å to +3 V/Å.

IV. CONCLUSIONS

In conclusion, this study used the first-principles research method to investigate the electronic properties of monolayer h-BN, monolayer WTe$_2$ and the h-BN/WTe$_2$ heterostructure, as well as the regulatory effect of the electric field on the h-BN/WTe$_2$ heterostructure. The research results show that monolayer h-BN, monolayer WTe2 and the h-BN/WTe$_2$ heterostructure are all direct band gap semiconductors. The electric field will affect the band gap size of the h-BN/WTe$_2$ heterostructure but will not affect the band gap type and energy band alignment type of the h-BN/WTe$_2$ heterostructure. Additionally, it is worth noting that when the applied electric field is -2 V/Å, both the CBM and VBM of the h-BN/WTe$_2$ heterostructure are larger than those without an applied electric field, and the VBM of the heterostructure exceeds the Fermi level, and an ohmic contact will be formed within the heterostructure. The research results provide guidance for the exploration and application of the h-BN/WTe$_2$ heterostructure in electronic devices under different electric fields, and provide a simple theoretical basis for obtaining ohmic contacts of the h-BN/WTe$_2$ heterostructure.

V. ACKNOWLEDGMENTS

This work was supported by the [National Natural Science Foundation of China] under Grant [No.62174148]; the [National Key Research and Development Program] under Grant [No.2022YFE0112000 and 2016YFE0118400]; the [Key Program for International Joint Research of Henan Province] under Grant [No.231111520300]; the [Ningbo Major Project of 'Science, Technology and Innovation 2025'] under Grant [No.2019B10129]; and the [Zhengzhou 1125 Innovation Project] under Grant [No.ZZ2018-45].

REFERENCES

[1] J.B. Barot, S.K. Gupta and P.N. Gajjar. Optical properties of WTe2- a layered topological insulator: A DFT study[J]. Materials Today: Proceedings, 2023.

[2] J Wang, F Ma, W Liang, et al. Electrical properties and applications of graphene, hexagonal boron nitride (h-BN), and graphene/h-BN heterostructures[J]. Materials Today Physics, 2017, 2: 6-34.

[3] S. Brozzesi, C. Attaccalite, F. Buonocore, et al.Ab Initio Study of Graphene/hBN Van der Waals Heterostructures: Effect of Electric Field, Twist Angles and p-n Doping on the Electronic Properties, Nanomaterials 12 (2022) 2118.

[4] J.P. Perdew, K. Burke and M. Ernzerhof, Generalized gradient approximation made simple, Phys. Rev. Lett. 77 (1996) 3865–3868.

[5] S. Grimme, J. Antony, S. Ehrlich,et al. A consistent and cccurate ab initio parametrization of density functional dispersion correction (DFT-D) for the 94 Elements H-Pu, J. Chem. Phys. 132 (2010) 154104.

[6] H.J. Monkhorst, J.D. Pack. Special points for Brillouin-zone integrations, Phys. Rev. B 13 (1976) 5188.

[7] K. Momma, F. Izumi. VESTA 3 for three-dimensional visualization of crystal, volumetric and morphology data, J. Appl, Crystals 44 (2011) 1272–1276.

[8] G.S. Aga, G.S. Gurmesa, Q Zhang, et al. First-principles study of the electronic and optical properties of homo-doped 2D-hBN monolayer[J]. Computational Condensed Matter, 2022, 30: e00628.

[9] E Torun, H Sahin, S Cahangirov, et al. Anisotropic electronic, mechanical, and optical properties of monolayer WTe2[J]. Journal of Applied Physics, 2016, 119(7).

[10] M.R. Islam, M.R.H. Mojumder, B.K. Moghal, et al. Impact of strain on the electronic, phonon, and optical properties of monolayer transition metal dichalcogenides XTe2 (X= Mo and W)[J]. Physica Scripta, 2022, 97(4): 045806.

[11] T Liang, C Hu, M Lou, et al. High carrier mobility and controllable electronic property of the h-BN/SnSe2 heterostructure[J]. Langmuir, 2023, 39(31): 10769-10778.

[12] W Xiong, C Xia, X Zhao, et al. Effects of strain and electric field on electronic structures and Schottky barrier in graphene and SnS hybrid heterostructures[J]. Carbon, 2016, 109: 737-746.

Optimization of AlGaN-based deep ultraviolet light emitting diodes with step-shaped hole blocking layer

Haokai Jing   Xien Sang   Xin Wang   Fang Wang   Yuhuai Liu   Juin J. Liou

Abstract—To improve the luminous efficiency and optimize the performance of AlGaN-based deep ultraviolet light-emitting diodes (DUV-LEDs), a step-shaped hole blocking layer (HBL) was proposed. Using Crosslight software, the structure with the step-shaped HBL was analyzed in terms of hole concentration, radiative recombination efficiency, internal quantum efficiency (IQE), and P-I characteristics. Compared to the baseline structure, the radiative recombination rate increased by 93.7%, and the IQE reached 36.3%, representing a 65.6% improvement. The output power at a current of 100 mA reached 162.88 mW, which is a 66.6% enhancement over the baseline structure. The results indicate that the step-shaped HBL effectively suppresses the leakage of the hole from the active region, increases the hole concentration within the quantum wells, and improves the radiative recombination efficiency. Simultaneously, it enhances the IQE and output power of the DUV-LED. The step-shaped HBL provides valuable insights for addressing the severe hole leakage issue in DUV-LEDs.

Index Terms—AlGaN, deep ultraviolet light emitting diodes, internal quantum efficiency, hole blocking layer

I. INTRODUCTION

Devices based on AlGaN are widely used in electronic and optoelectronic devices that operate in high power and high frequency environments. AlGaN is an important wide-bandgap semiconductor material. It has excellent thermal

---

Haokai Jing and Xien Sang and Xin Wang are with *National Center for International Joint Research of Electronic Materials and Systems, International Joint-Laboratory of Electronic Materials and Systems of Henan Province, School of Electrical and Information Engineering, Zhengzhou University, Zhengzhou, Henan 450001, P. R. China*, Zhengzhou, Henan China. Fang Wang and Yuhuai Liu are with *National Center for International Joint Research of Electronic Materials and Systems, International Joint-Laboratory of Electronic Materials and Systems of Henan Province, School of Electrical and Information Engineering, Zhengzhou University, Zhengzhou, Henan 450001, P. R. China, International Joint Laboratory for Integrated Circuits Design and Application, Ministry of Education, School of Physics, Zhengzhou University, Zhengzhou 450052, China, Institute of Intelligence Sensing, Research Institute of Industrial Technology Co. Ltd., Zhengzhou University, Zhengzhou, Henan 450001, P. R. China, Zhengzhou Way Do Electronics Co. Ltd., Zhengzhou, Henan 450001, P. R. China*, Zhengzhou, Henan, China. Juin J. Liou is with *School of Electrical and Information Engineering, North Minzu University, Yinchuan, Ningxia 750001, P. R. China*, Yinchuan, Ningxia, China. This work is supported by the [National Natural Science Foundation of China] under Grant [No.62174148]; the [National Key Research and Development Program] under Grant [No.2022YFE0112000 and 2016YFE0118400]; the [Key Program for International Joint Research of Henan Province] under Grant [No.231111520300]; the [Ningbo Major Project of 'Science, Technology and Innovation 2025'] under Grant [No.2019B10129]; and the [Zhengzhou 1125 Innovation Project] under Grant [No.ZZ2018-45]. Corresponding author: (Fang Wang, email: iefwang@zzu.edu.cn; Yuhuai Liu, email: ieyhliu@zzu.edu.cn)

stability and electrical properties, making it especially suitable for extreme operating conditions. Due to its wide bandgap characteristics, AlGaN-based light-emitting devices can cover almost the entire ultraviolet spectrum (210-400 nm). This allows AlGaN-based DUV-LEDs to be applied in a variety of disinfection equipment, such as air purifiers, germicidal lamps, and medical instrument sterilizers [1]. Compared to traditional ultraviolet light sources, such as mercury lamps, DUV-LEDs do not contain harmful substances, have lower energy consumption, and longer lifespans, making them an environmentally friendly light source. In addition, DUV-LEDs are also used in the information storage field. DUV-LEDs can provide high-energy ultraviolet light for fast curing of inks and coatings, improving production efficiency, while also having lower energy consumption and environmental impact [2].

Although DUV-LEDs have great potential in the field of ultraviolet light sources, their development faces several significant challenges, including low hole concentration in the active region, low internal quantum efficiency, severe electron and hole leakage, and low radiative recombination rate. To address these issues and improve the performance of deep ultraviolet light-emitting diodes, numerous optimizations have been made to structures such as quantum wells, quantum barriers, and electron blocking layers. Qian et al. proposed a lattice-matched quaternary AlInGaN/AlGaN superlattice electron blocking layer [3]. Yang et al. integrated inverted V-shaped quantum barriers to mitigate the efficiency droop in deep ultraviolet light-emitting diodes [4]. However, research on the hole blocking layer in DUV-LEDs is still relatively limited. The design of multiple quantum barrier structures reduces the effective barrier height for holes in the valence band. After holes are injected into the active region, those that do not recombine with electrons in the quantum wells will leak into the n-type region, where they undergo non-radiative recombination with electrons. This reduces the electron concentration in the n-type region, which in turn lowers the radiative recombination rate within the quantum wells and decreases the device's efficiency.

The quantum barriers have limited confinement effects on holes. Introducing an AlGaN hole blocking layer with a higher Al composition can enhance the effective valence band barrier height for holes, effectively confining them within the quantum wells. This paper proposes a step-shaped hole blocking layer (HBL) structure. Simulations are conducted and compared for

DUV-LEDs without a hole blocking layer, DUV-LEDs with a rectangular hole blocking structure, and DUV-LEDs with a step-shaped hole blocking structure. The simulation results show that the step-shaped hole blocking layer structure can effectively reduce hole leakage into the n-type region, increase the hole concentration in the active region of the device, and improve the radiative recombination rate in the quantum wells. As a result, the internal quantum efficiency and output power of the device are enhanced, leading to a significant improvement in the device's performance.

II. SIMULATION STRUCTURE AND PARAMETERS

In this design, simulations of different DUV-LED structures are performed using Crosslight software. The design incorporates a 20 nm thick hole blocking layer. As shown in Figure 1(a), the basic structure of the LED starts with a 100 μm sapphire substrate, followed by a 3 μm thick n-$Al_{0.7}Ga_{0.3}N$ n-type region (dopant concentration of 5×10^{25} cm^{-3}). A 20 nm thick hole blocking layer (dopant concentration of 7×10^{25} cm^{-3}) is added between the n-type region and the active region. The active region consists of six 10 nm thick $Al_{0.7}Ga_{0.3}N$ quantum barriers and five 4 nm thick $Al_{0.6}Ga_{0.4}N$ quantum wells, alternating. The p-type region consists of a 20 nm thick $Al_{0.9}Ga_{0.1}N$ electron blocking layer (EBL, dopant concentration of 2×10^{24} cm^{-3}) and a 120 nm thick $Al_{0.7}Ga_{0.3}N$ contact layer (dopant concentration of 5×10^{24} cm^{-3}). In this structure, all n-type doping is done with Si, and all p-type doping is done with Mg. In this simulation study, the device simulations are carried out using the APSYS tool, which is an advanced semiconductor device physics modeling software. APSYS solves the Poisson equation and the current continuity equation for electrons and holes to calculate the electrical behavior of all the LEDs.

The reference structure is the one without a hole blocking layer (denoted as LED1), as shown in Figure 1(b). A hole blocking layer is added while ensuring that the total thickness and the average aluminum composition remain the same. Two additional structures are designed: a rectangular structure (denoted as LED2) and a step-shaped structure (denoted as LED3). Figure 1(c) shows schematic diagrams of the hole blocking layers for LED2 and LED3. The hole blocking layer of LED2 is made of 20 nm thick $Al_{0.90}Ga_{0.10}N$. The hole blocking layer of LED3 is composed of 7 nm of $Al_{0.94}Ga_{0.06}N$, 6 nm of $Al_{0.90}Ga_{0.10}N$, and 7 nm of $Al_{0.86}Ga_{0.14}N$, arranged sequentially. Both hole blocking layers are n-type doped.

To determine the carrier recombination and losses, important physical parameters such as the Auger recombination coefficient, radiative recombination coefficient, Shockley-Read-Hall (SRH) recombination lifetime, and the energy band offset ratio at the AlGaN/AlGaN interface were used. These parameters are set as follows: Auger recombination coefficient of 1×10^{-30}cm^{6}s^{-1}, radiative recombination coefficient of 0.5×10^{-16}m^{3}s^{-1}, SRH recombination lifetime of 10 ns, and an energy band offset ratio of 50:50 [5]. All calculations were performed at room temperature (T = 300 K).

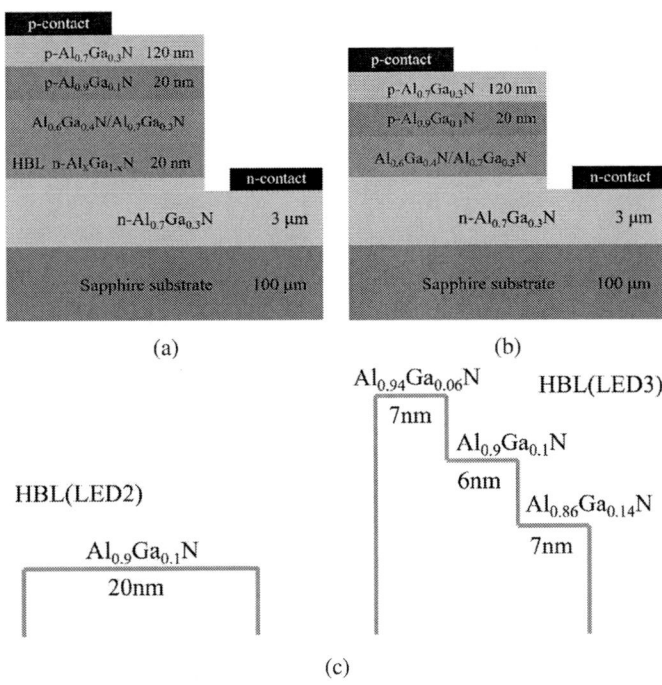

Fig. 1: (a) Schematic of the DUV-LED structure with HBL, (b) Schematic of the reference DUV-LED structure, (c) Schematics of HBL of LED2 and LED3.

III. RESULTS AND DISCUSSION

The reduction in hole leakage can be directly reflected in the increase of hole concentration in the MQW (multiple quantum wells). The hole concentration distribution in the MQW is shown in Figure 2(a). From the figure, it can be seen that both LED2 (with a rectangular hole blocking layer) and LED3 (with a step-shaped hole blocking layer) have higher hole concentrations in the MQW compared to LED1, which lacks a hole blocking layer. This indicates that the hole blocking layer effectively prevents hole leakage and increases the hole concentration in the active region. As a result of the increased carrier concentration in the active region, there is a greater chance for electrons and holes to recombine in the quantum wells. Figure 2(b) shows the radiative recombination rate distribution in the MQW. Compared to the conventional structure LED1, the radiative recombination rates for LED2 and LED3 have increased by 58.5% and 93.7%, respectively.

IQE (Internal Quantum Efficiency) is a crucial indicator of LED performance. A higher IQE means that more electrons and holes can recombine in the active region to generate photons, thereby improving the LED's light emission efficiency. In an LED, if electrons and holes are not effectively confined within the emission region and leak out, it significantly reduces the chances of recombination, thus lowering the IQE. Figure 3(a) shows the IQE of the three LEDs. It can be observed that after the addition of the hole blocking layer, hole leakage is reduced, resulting in an increase in IQE. The IQE of LED2, with a rectangular hole blocking layer, reaches 32.7% at a current density of 90 mA, representing a 49.3% improvement

979-8-3315-2209-4/25 $31.00 © 2025 IEEE

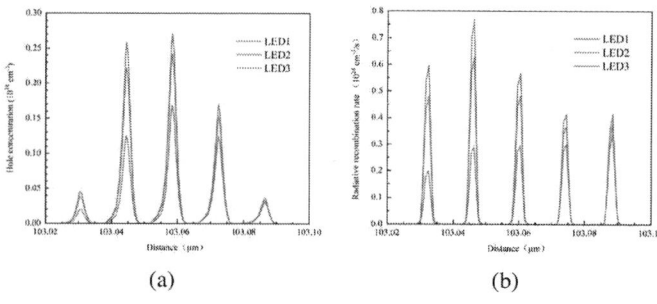

(a)　　　　　　　　(b)

Fig. 2: (a) The hole concentration of the three LEDs, (b) The radiative recombination rate of the three LEDs.

over LED1. The IQE of LED3, which has a step-shaped hole blocking layer, reaches 36.3% at a current density of 90 mA, a 65.6% improvement over LED1. As shown in Figure 3(b), with the increase in current, when the current reaches 100 mA, the power of LED2 reaches 146.02 mW, a 49.3% increase compared to the power of LED1 (97.78 mW). Meanwhile, the power of LED3 reaches 162.88 mW, which is a 66.6% increase compared to LED1. These results indicate that the step-shaped hole blocking layer can effectively optimize the optical and electrical characteristics of DUV-LEDs.

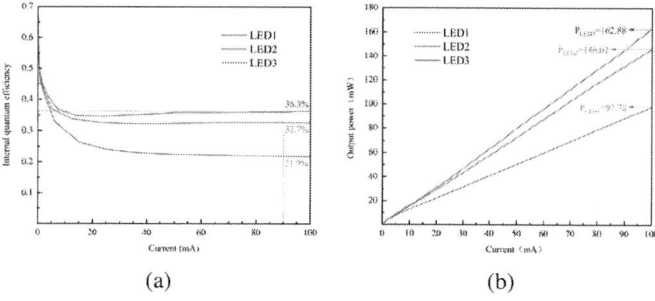

(a)　　　　　　　　(b)

Fig. 3: (a) The IQE of the three LEDs, (b) The Output power of the three LEDs.

IV. CONCLUSIONS

To reduce hole leakage in the n-type region of DUV-LEDs, rectangular and step-shaped hole blocking layers were added to the basic LED structure. Simulations were performed using Crosslight software to compare the optical and electrical performance of three structures: LED1 without a hole blocking layer, LED2 with a rectangular hole blocking layer, and LED3 with a step-shaped hole blocking layer, in deep ultraviolet light-emitting diodes. The comparison includes parameters such as the radiative recombination rate and hole concentration in the device's active region, internal quantum efficiency, and P-I characteristics. The results show that the step-shaped hole blocking layer effectively prevents hole leakage from the active region, optimizing the working performance of the DUV-LED.

V. FUNDING

This work was supported by the [National Natural Science Foundation of China] under Grant [No.62174148]; the

[National Key Research and Development Program] under Grant [Nos.2022YFE0112000 and 2016YFE0118400]; the [Key Program for International Joint Research of Henan Province] under Grant [No.231111520300]; the [Ningbo Major Project of 'Science, Technology and Innovation 2025'] under Grant [No.2019B10129]; and the [Zhengzhou 1125 Innovation Project] under Grant [No.ZZ2018-45].

REFERENCES

[1] Zefeng Lin, Lucheng Yu, Qicheng Zhou, Yehang Cai, Fawen Su, Shengrong Huang, Feiya Xu, Xiaohong Chen, Ling Li, Duanjun Cai. AlGaN-Based Deep-UV LED with Novel Transparent Electrodes and Integrated Array Device for Efficient Disinfection[J]. Laser & Optoelectronics Progress, 2024, 61(5): 0523002.

[2] Hofstetter, D.; Bour, D.P.; Beck, H. Proposal for Deep-UV Emission from a Near-Infrared AlN/GaN-Based Quantum Cascade Device Using Multiple Photon Up-Conversion. Crystals 2023, 13, 494.

[3] Dai, Q., Zhang, X., Wu, Z. et al. High Performance of a Non-Polar AlGaN-Based DUV-LED with a Quaternary Superlattice Electron Blocking Layer. J. Electron. Mater. 51, 5389–5394 (2022).

[4] Y. Kang et al., "Efficiency Droop Suppression and Light Output Power Enhancement of Deep Ultraviolet Light-Emitting Diode by Incorporating Inverted-V-Shaped Quantum Barriers," in IEEE Transactions on Electron Devices, vol. 67, no. 11, pp. 4958-4962, Nov. 2020, doi: 10.1109/TED.2020.3025523.

[5] H. Yu et al., Advantages of AlGaN-based deep-ultraviolet light-emitting diodes with an Al composition graded quantum barrier, Optics express, 27, 20, pp. A1544-A1553, 2019

Optimization of AlGaN-Based Deep-Ultraviolet Laser Diodes with Trapezoidal Quantum Barriers

Wenlan Ma   Xin Wang   Xien Sang   Fang Wang   Juin J. Liou   Yuhuai Liu

Abstract—To solve the issue of relatively low output power of AlGaN-based deep-ultraviolet laser diodes (DUV-LDs), this paper designs a multi-quantum-wells (MQWs) structure in which the Al composition of the last quantum barrier (QB) is trapezoidal. Simulation studies are carried out respectively on two laser structures, including the reference structure and the structure with a trapezoidal-Al-composition QB. Moreover, the characteristics of the energy band structure, P-I curve, electron and hole concentrations of two devices are compared and analyzed. The results show that changing the Al composition of the last QB to a trapezoidal structure reduces electron leakage and increases hole injection, effectively optimizing the output performance of DUV-LD. Compared with the traditional structure, the output power of the structure proposed has increased from 90.07 mW to 118.07 mW at a current of 80 mA.

Index Terms—AlGaN, deep ultraviolet laser diode, trapezoidal quantum barriers

I. INTRODUCTION

DUV-LDs have become a research hotspot in the optoelectronic field in recent years due to their advantages such as small size, light weight, simple structure, easy integration, and high reliability [1,2]. They have great application value in the fields of electronic manufacturing industry, optical data storage, biomedical research, water purification, and sterilization [3-5]. However, the current development of AlGaN-based DUV-LDs still faces many challenges. For example, the AlGaN material itself has a very strong polarization effect [6]. At the same time, due to lattice mismatch [7] and high dislocation density, problems such as low hole injection efficiency, electron leakage, and hole leakage occur [8], and these issues ultimately lead to low output power and low optical conversion efficiency of DUV-LDs. To address these problems and improve the performance of DUV-LDs, researchers have

attempted to conduct research from multiple aspects such as material growth and structure design [9,10]. For instance, the MQWs structure has been proposed to increase the radiative recombination rate [11]; high-Al-composition electron blocking layers (EBL) have been adopted to reduce electron leakage and optimize the structure of the EBL [12,13]; hole storage layers (HSL) have been added to improve the hole injection efficiency, and made improvements on the basic structure to achieve better results [14,15].

In this paper, for the basic AlGaN-based DUV-LD model, it is proposed to change the Al composition profile of the last QB in the MQWs of the LD to a trapezoidal structure, and a comparative simulation is carried out with the basic model. Simulation results indicate that the structure proposed in this paper increases the electron and hole concentrations in the active region. This is beneficial for enhancing the radiative recombination efficiency of carriers in the active region, thus improving the output power of the DUV-LD. The trapezoidal QB structure can effectively reduce the quantum-confined Stark effect. Moreover, it can mitigate the impact of the polarization effect at the hetero-interface between the last quantum barrier and the p-type region on the band bending.

II. SIMULATION MODEL AND PARAMETERS

Figure 1(a) shows the basic structure of a DUV-LD with a sapphire substrate. From bottom to top, it consists of a n-type region, an active region, and a p-type region. The n-type region is composed of a 3 nm n-type GaN contact layer, a 1 μm n-type $Al_{0.75}Ga_{0.25}N$ cladding layer (n-CL), and a 110 nm n-type $Al_{0.67}Ga_{0.33}N$ lower waveguide layer (n-LWG). The active region is alternately composed of three 8 nm $Al_{0.68}Ga_{0.32}N$ quantum barriers (QBs) and two 3 nm $Al_{0.62}Ga_{0.38}N$ quantum wells (QWs). The p-type region is composed of a 70 nm p-type $Al_{0.68}Ga_{0.32}N$ upper waveguide layer (p-UWG), a 10 nm $Al_{0.9}Ga_{0.1}N$ EBL, a 400 nm p-type $Al_{0.76}Ga_{0.24}N$ cladding layer (p-CL), and a 10 nm p-type GaN contact layer. Si and Mg are used as the doping materials for the n-type and p-type regions respectively.

The basic structure of the MQWs is shown in Figure 1(b). Taking the basic structure as the reference Structure A, this paper designs a special Structure B in which the Al composition of the last QB is changed to a trapezoidal structure that first increases and then decreases. As shown in Figure 1(c), the last 8 nm QB is divided into three parts. In the first part, the Al composition linearly increases from 0.68 to 0.74 with a thickness of 2 nm. In the second part,

---

Wenlan Ma and Xin Wang and Xien Sang are with *National Center for International Joint Research of Electronic Materials and Systems, International Joint-Laboratory of Electronic Materials and Systems of Henan Province, Zhengzhou, China.* Fang Wang and Yuhuai Liu is with *National Center for International Joint Research of Electronic Materials and Systems, International Joint-Laboratory of Electronic Materials and Systems of Henan Province, School of Electrical and Information Engineering, Zhengzhou University, International Joint Laboratory for Integrated Circuits Design and Application, Ministry of Education, School of Physics,Institute of Intelligence Sensing, Research Institute of Industrial Technology Co. Ltd., Zhengzhou University,Zhengzhou Way Do Electronics Co. Ltd.,* Zhengzhou, China. Juin J. Liou is with *School of Electrical and Information Engineering, North Minzu University,* Yinchuan, China. This work is supported by National Nature Science Foundation of China (Grant No. 62174148), National Key Research and Development Program (NKRDP Grant No. 2022YFE0112000), Key Program for International Joint Research of Henan Province (Grant No. 231111520300). Corresponding author: (Fang Wang, email:iefwang@zzu.edu.cn. Yuhuai Liu, email: ieyhliu@zzu.edu.cn)

979-8-3315-2209-4/25 $31.00 © 2025 IEEE

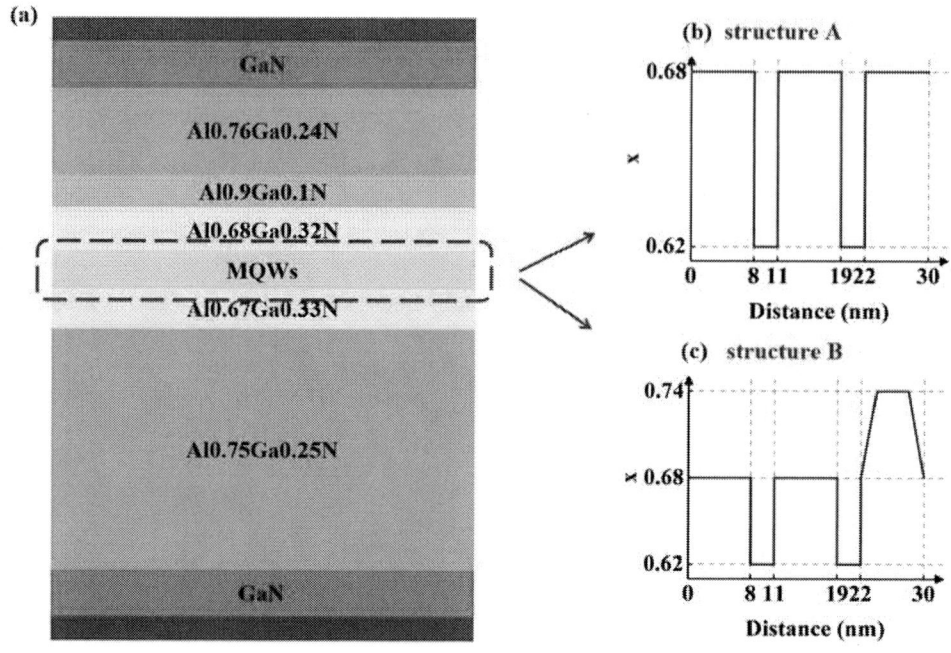

Fig. 1. (a) Schematic diagram of the DUVLD structure; (b) Basic MQWs structure A; (c) Structure B with a trapezoidal Al-composition QB.

the Al composition remains at 0.74 with a thickness of 4 nm. In the third part, the Al composition linearly decreases from 0.74 to 0.68 with a thickness of 2 nm. With Structure A serving as the reference structure, apart from the last QB, all other structures and parameters of Structure B are identical to those of Structure A. Laser performance was analyzed using Pics3D software. All simulations relevant to DUV-LDs were performed for room temperature.

III. RESULTS AND DISCUSSIONS

Figures 2(a) and 2(b) show the energy band diagrams and quasi-Fermi level diagrams of Structure A and Structure B. The effective barrier heights of electrons and holes are the energy differences between the conduction band maximum and the electron quasi-Fermi level, and between the valence band minimum and the hole quasi-Fermi level, respectively. The barrier height affects the carrier migration ability. The larger the barrier height, the stronger the ability of the active region to confine carriers. The effective electron barrier heights of Structure A and Structure B are 32.2 meV and 289.1 meV, respectively. The effective hole barrier heights of Structure A and Structure B are 386.4 meV and 296.8 meV, respectively. By comparing the effective barrier heights of the two structures, it can be found that the effective electron barrier height of Structure B is significantly increased compared with that of Structure A, indicating that Structure B has a stronger ability to block the electrons leak into the P-type region. At the same time, the effective hole barrier height of Structure B is lower than that of Structure A, indicating that Structure B also has an improved ability to inject holes into the active region. Figures

3 and 4 are the comparison diagrams of the hole concentration and electron concentration in the active regions of the two structures respectively. It can be seen from the figures 3 that the hole concentration in the first QW is increased by 2.8 %. The electron concentration of Structure B is also higher than that of Structure A. The figure 4 shows that electron concentration in the active region of Structure B is higher than that of Structure A, with increases of 3.8 % and 19.4 % in the two QWs respectively. Since Structure B has a higher effective electron barrier and a lower effective hole barrier, both the electron and hole concentrations in the active region of Structure B are higher, which is beneficial for improving the radiative recombination efficiency of carriers in the active region.

Figure 5 is the comparison diagram of the P-I curves of the two structures. It can be seen from the figure that the output powers of Structure A and Structure B at a current of 80 mA are 90.07 mW and 118.07 mW respectively. Compared with Structure A, the power of Structure B is increased by 31.1 % at a current of 80 mA. In the two structures, Structure B has a higher output power. The reason is that the change in the structure of the last QB hinders the electrons leak from the active region, increases the electron concentration in the MQWs, and promotes the recombination of electrons and holes.

IV. CONCLUSIONS

In this paper, a MQW structure is designed for AlGaN-based DUV-LD, in which the Al composition of the last QB is a trapezoidal structure. Simulation and modeling are carried out

979-8-3315-2209-4/25 $31.00 © 2025 IEEE

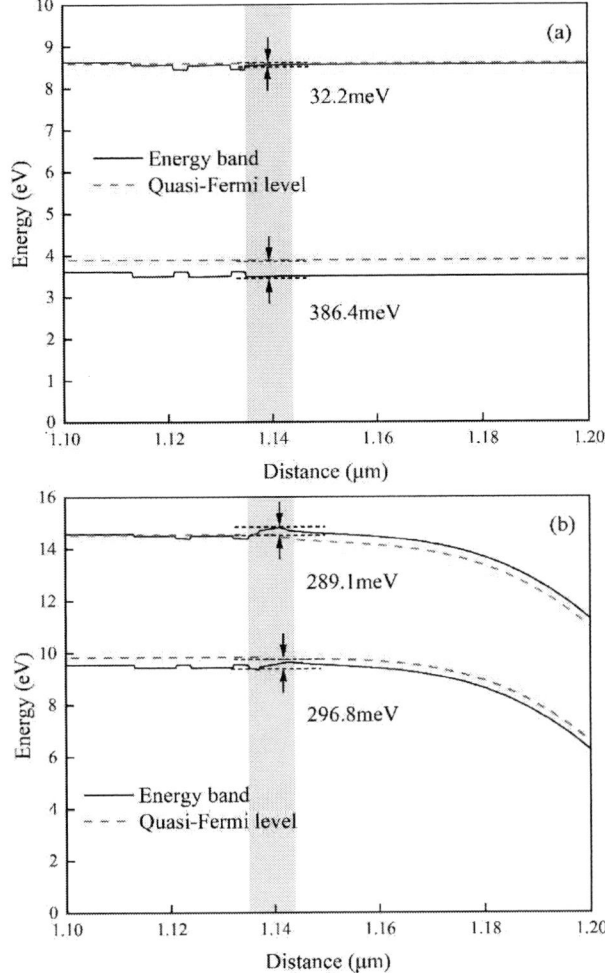

Fig. 2. Diagrams of Energy Bands and Quasi-Fermi Levels (a) Structure A (b) Structure B.

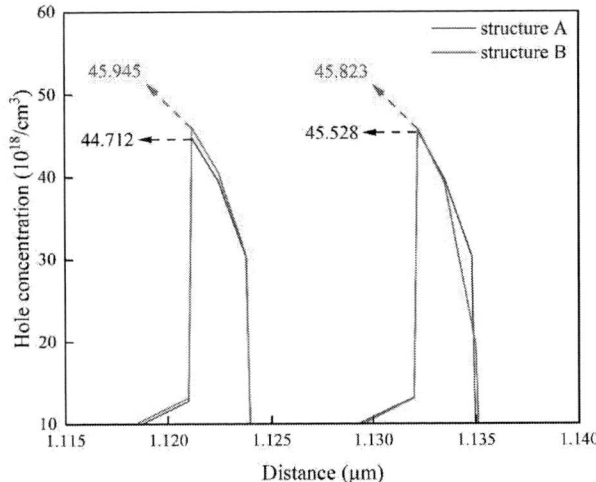

Fig. 3. Hole Concentration Diagrams in the Active Regions.

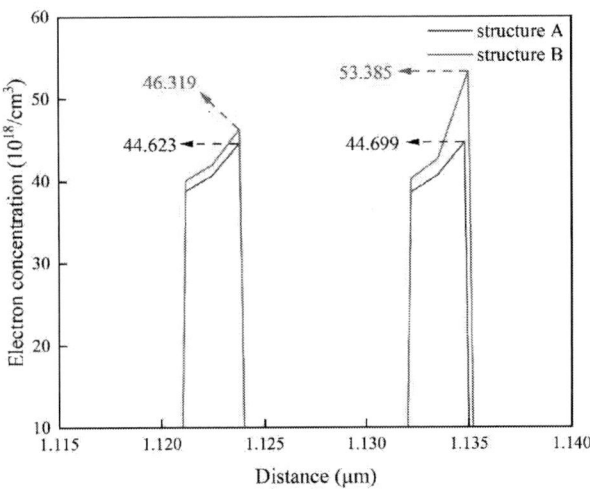

Fig. 4. Electron Concentration Diagrams in the Active Regions.

on two DUV-LDs with QBs of different structures. The results show that, compared with the basic structure A, in structure B where the Al composition of the last QB is trapezoidal, the electron leakage of the DUV-LD is significantly improved, and the output power is increased. The output power of structure B at a current of 80 mA is 119.09 mW, which represents a 31.1% increase compared to structure A. The excellent performance of structure B is attributed to the fact that the trapezoidal Al composition of the last QB increases the carrier concentration within the MQWs, improves the recombination of electrons and holes in the MQWs, and thus enhances the device performance.

V. ACKNOWLEDGMENTS

Supported by National Nature Science Foundation of China (Grant No. 62174148), National Key Research and Development Program (NKRDP Grant No. 2022YFE0112000), Key Program for International Joint Research of Henan Province (Grant No. 231111520300).

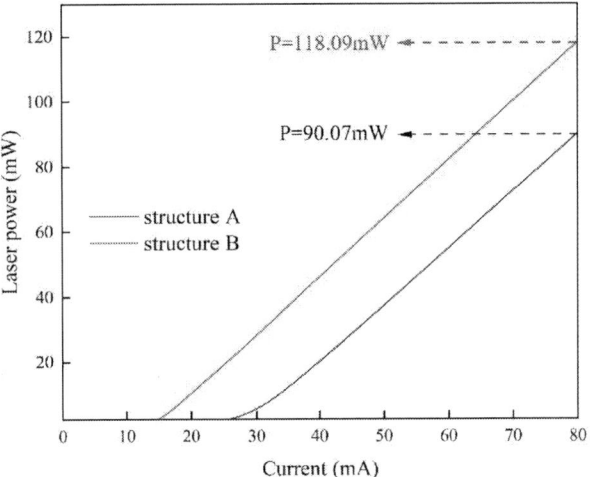

Fig. 5. P-I Curve Graphs.

979-8-3315-2209-4/25 $31.00 © 2025 IEEE

REFERENCES

[1] Nakamura, S. "Future technologies and applications of III-Nitride materials and devices". Engineering, 2015, 1, 161.

[2] Liu, X., Zhang, Y., Wang, Z., et al. "Recent progress in deep-ultraviolet light-emitting diodes and laser diodes". Chinese Physics B, 2020, 29(7), 074207.

[3] Zhang, M., Li, Y., Wang, Y., et al. "Deep-ultraviolet light-emitting diodes for biomedical applications". Optics Express, 2020, 28(15), 22121-22133.

[4] Chen, Y., Wang, Y., Li, Y., et al. "Deep-ultraviolet light-emitting diode-based gas sensors for environmental monitoring". Sensors and Actuators B: Chemical, 2021, 334, 129696.

[5] Zhao, Y., Zhang, Y., Wang, Z., et al. "Deep-ultraviolet laser-diode-based optical data storage". Optics and Lasers in Engineering, 2020, 129, 105973.

[6] Liu, S., Ye, C., Cai, X., et al. "Performance enhancement of AlGaN deep-ultraviolet light-emitting diodes with varied superlattice barrier electron blocking layer". Applied Physics A, 2016, 122, 1-6.

[7] Zhao, H., Liu, G., Zhang, J., et al. "Analysis of internal quantum efficiency and current injection efficiency in III-nitride light-emitting diodes". Journal of Display Technology, 2013, 9(4), 212-225.

[8] Huang, H. H., Chu, S. Y., Kao, P. C., et al. "Enhancement of hole-injection and power efficiency of organic light-emitting devices using an ultra-thin ZnO buffer layer". Journal of Alloys and Compounds, 2009, 479(1 - 2), 520-524.

[9] Liu, X., Wang, Z., Zhang, Y., et al. "Material growth and device fabrication of AlGaN-based deep-ultraviolet light-emitting diodes". Journal of Crystal Growth, 2020, 544, 125774.

[10] Wang, Y., Li, Y., Zhang, Y., et al. "Novel structure design for high-performance AlGaN-based deep-ultraviolet light-emitting diodes". Optics Express, 2020, 28(26), 38422-38433.

[11] Xing, Z., Zhou, Y., Chen, X., et al. "Increased radiative recombination of AlGaN-based deep-ultraviolet laser diodes with convex quantum wells". Optoelectronics Letters, 2020, 16(2), 87-91.

[12] Xing, Z. Q., Zhou, Y. J., Liu, Y. H., et al. "Reduction of Electron Leakage of AlGaN-Based Deep Ultraviolet Laser Diodes Using an Inverse-Trapezoidal Electron Blocking Layer". Chinese Physics Letters, 2020, 37(2), 67-71.

[13] Zhang, A. X., Ren, B. Y., Wang, F., et al. "Performance Optimization of AlGaN-Based Deep-Ultraviolet Laser Diodes with a Stepped Superlattice Electron-Blocking Layer and a Wedge-Shaped Hole-Blocking Layer Structure". Laser Optoelectronics Progress, 2023, 60(15), 339-346.

[14] Li, H., Kang, J., Li, P., et al. "Enhanced performance of GaN-based light-emitting diodes with a low-temperature p-GaN hole injection layer". Applied Physics Letters, 2013, 102(1).

[15] Sang, X. E., Xu, Y., Yin, M. S., et al. "Performance Optimization of Deep-Ultraviolet Laser Diodes Using a Mountain-Shaped Hole Storage Layer Structure". Journal of Atomic and Molecular Physics, 2023, 40(6), 123-128.

979-8-3315-2209-4/25 $31.00 © 2025 IEEE

Enhanced optical confinement factor of DUV LD by optimizing the p-Cladding Layer

Zhongqiu Xing   Yongjie Zhou   Fang Wang   Juin J. Liou   Yuhuai Liu

Abstract—Both low hole injection efficiency and high threshold current density reduce the lasing efficiency of the Deep ultraviolet (DUV) laser diode (LD) and affect the performance of the device. Therefore, the structural optimization of the p-type cladding layer is crucial for LDs. To solve this problem, deep-ultraviolet lasers with Al-composition gradient p-AlGaN cladding layers are proposed in this study to improve the optical confinement factor (OCF). The bulk-polarized charge of this cladding layer relieves the polarized electric field at the interface between the p-type waveguide layer and the p-type cladding layer, allowing more carriers to enter the quantum well, which leads to an increase in the carrier concentration and radiative complexity in the active region. Compared with LD with traditional CL layer, the optical limiting factor of this structure is increased by 8.9%.

Index Terms—Deep ultraviolet laser diode, optical confinement factor, p-type cladding layer

I. Introduction

Deep ultraviolet (DUV) laser diode (LD) has the advantages of short wavelength, high energy resolution and high photon flux [1], [2]. These properties make it of great value in various applications such as virus detection, UV curing, UV communication, materials science, chemical analysis, biomedical research, cutting of polymer materials, semiconductor detection and processing, and medical treatment [3]–[5]. The optical confinement factor (OCF) refers to the degree of relative coupling between the effective luminous region in the LD and the light in the entire device structure [6]. Higher OCF is enough to make the energy of light more concentrated in the active region, which means that for the carriers injected into the active region, more light will be released through the recombination process, which enhances the luminous efficiency of the laser [7]. The improvement of OCF depends on efficient carrier injection, because higher carrier injection efficiency allows more carriers to participate in the luminescence process

---

Zhongqiu Xing, Fang Wang, and Yuhuai Liu are with National Center for International Joint Research of Electronic Materials and Systems, International Joint-Laboratory of Electronic Materials and Systems of Henan Province, School of Electrical and Information Engineering, Zhengzhou University, Zhengzhou, China. Yongjie Zhou is with Basic Teaching Department, Zhengzhou University of Railway Engineering, Zhengzhou, China. Juin J. Liou is with School of Electrical and Information Engineering, North Minzu University, Ningxia, China. This work is supported by National Nature Science Foundation of China (Grant No. 62174148), National Key Research and Development Program (NKRDP Grant No. 2022YFE0112000), Key Program for International Joint Research of Henan Province (Grant No. 231111520300). Corresponding author: (Yuhuai Liu, email: ieyhliu@zzu.edu.cn)

[8]. Unlike LEDs, the design of the LD structure needs to ensure that the waveguide does not reduce the limit of light, while also effectively controlling the distribution of carriers.

II. Device architectures and physical parameters

Fig.1 shows a schematic of the structure of the proposed DUV LD with different EBLs of $Al_{0.9}Ga_{0.1}N$ and $B_{0.45}Ga_{0.55}N$. The structure is grown on a sapphire substrate, followed by a 1 µm thick $Al_{0.75}Ga_{0.25}N$ n-cladding layer (n-CL) and a 110-nm thick $Al_{0.68}Ga_{0.32}N$ n-waveguide layer (n-WG). The active region comprises of two $Al_{0.58}Ga_{0.42}N$ (6 nm each) quantum wells (QWs) and three $Al_{0.68}Ga_{0.32}N$ (10 nm each) QBs emitting at 267 nm. Above the last QB, there is a 10-nm thick $Al_{0.9}Ga_{0.1}N$ EBL, followed by a 110-nm thick $Al_{0.68}Ga_{0.32}N$ p-waveguide layer (p-WG). This is followed by a 400-nm thick $Al_{0.75}Ga_{0.25}N$ p-cladding layer (p-CL) and a 10-nm thick $Al_{0.68}Ga_{0.32}N$ p-buffer layer (p-BL). This structure is labeled LD1, and the p-CL of LD2 and LD3 are $Al_{0.7-0.8}GaN$ and $Al_{0.8-0.7}GaN$, respectively. The photoelectric properties of the two LDs were simulated using a photonic integrated circuit simulator (PIC3D). The electrical characteristics were computed by solving the Poisson equation, drift-diffusion model, current continuity equation, and polarization charge model, while the optical behavior was analyzed using the vector Helmholtz wave equation and Adachi refractive index model. Carrier transport and distribution were determined via the transmission matrix method and Schrödinger-Poisson self-consistent approach. Both samples shared identical growth conditions, with a width of 1.5 µm, cavity length of 534 µm, and a back loss of 2400 m^{-1}. The LDs emitted at approximately 267 nm, with all simulations conducted at room temperature ($T = 300$ K). The SP and PZ constants for BGaN and AlGaN were obtained from references [9], [10], with their accuracy verified in [11]. The doping configurations included n-type (Si, 1×10^{18} cm^{-3}) and p-type (Mg, 1×10^{19} cm^{-3}). The built-in interface charge was set to 40% of the theoretical value [12], [13], and the epitaxial layer exhibited a band offset ratio of $\Delta E_c / \Delta E_v = 0.7/0.3$. The front and rear cavity reflectivities were 85% and 75%, respectively.

979-8-3315-2209-4/25 $31.00 © 2025 IEEE

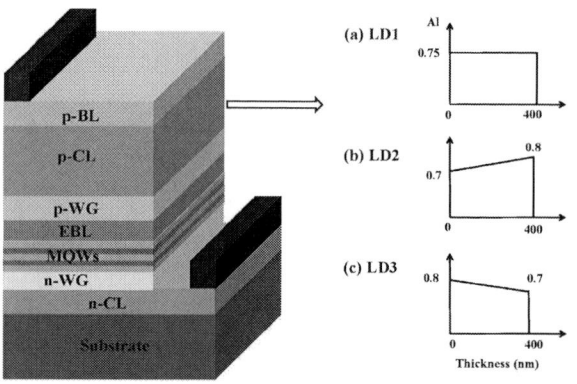

Fig. 1. Schematic cross-sectional structure of DUV LD p-CLs: $Al_{0.75}Ga_{0.25}N$ (LD1), $Al_{0.7-0.8}GaN$ (LD2) and $Al_{0.8-0.7}GaN$ (LD3).

III. Results and Discussions

As shown in Fig.2, in this paper, the true refractive index distributions and optical field distributions of the base structure LD1, the cladding layer LD2 with a gradual increase of Al composition, and the cladding layer LD3 with a gradual increase of Al composition are calculated. In the UVC LD, in order to confine the light field generated by carrier radiation compounding in the active region, the refractive indices of the materials on both sides of the active region of the multi-quantum well tend to decrease and the refractive index of the material in the active region is the highest, as shown in Fig. 1. This is to concentrate the distribution of light in the active region, thus improving the optical output performance of the laser. The OCF is related to the thickness of the active region, the refractive index of each layer, etc., whereas the reason for the increased optical confinement factor of LD2 is related to the refractive index of the structure's cladding layer using a gradient Al composition. The bulk-polarized charge of this cladding layer relieves the polarized electric field at the interface between the P-type waveguide layer and the P-type cladding layer, allowing more carriers to enter the quantum well, which leads to an increase in the carrier concentration and radiative complexity in the active region. Thus, the optical confinement factor of structure LD2 is increased as a result of the use of the gradient Al composition cladding layer. In addition, the optical confinement factor of structure LD3 is improved over both structure LD1 and structure LD2, and it is 8.9% higher than that of structure LD1, and most of the optical confinement factors of AlGaN deep-ultraviolet lasers are known to be reported to be less than 3% [14]. Therefore, such a large optical confinement value is more indicative of the fact that the continuous gradient P-type cladding layer provides better confinement of the optical field in the active region, which effectively suppresses the optical field leakage of the laser.

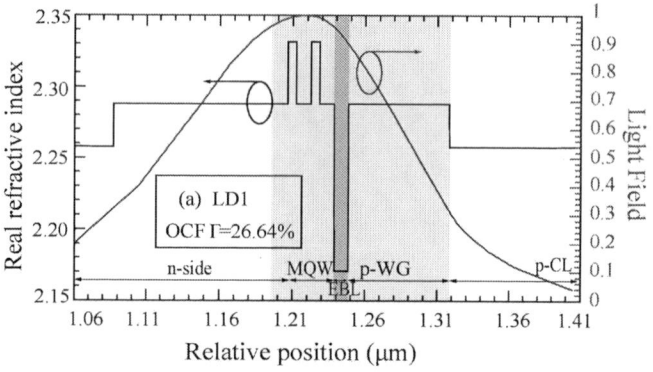

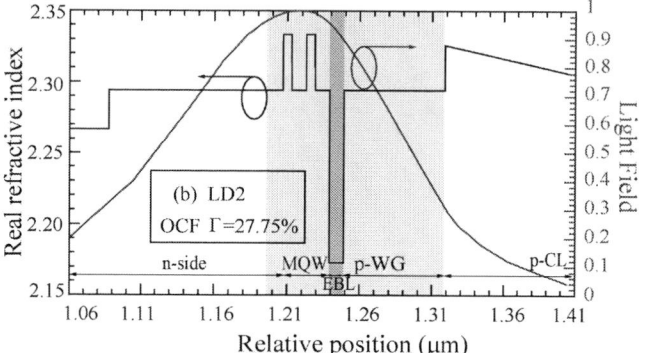

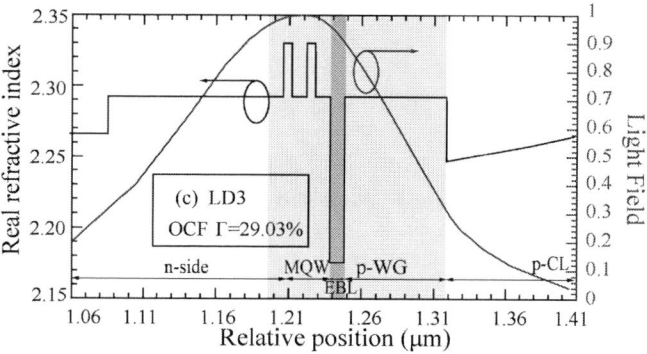

Fig. 2. Schematic cross-sectional structure of DUV LD EBLs: $Al_{0.9}Ga_{0.1}N$ (sample A) and $B_{0.45}Ga_{0.55}N$ (sample B).

Conclusion

In this study, we proposed and analyzed deep-ultraviolet laser diodes (DUV LDs) with Al-composition gradient p-AlGaN cladding layers to enhance the optical confinement factor (OCF). By optimizing the p-cladding layer design, the bulk-polarized charge effectively mitigated the polarization-induced electric field at the p-waveguide/cladding interface, thereby improving hole injection efficiency and carrier concentration in the active region. The gradient Al-composition structure (LD3) achieved an 8.9% increase in OCF (up to 29.03%) compared to conventional designs, significantly outperforming typical reported values (<3%) for AlGaN-based DUV

LDs. These findings highlight the critical role of band engineering and polarization management in DUV LD performance, offering a viable pathway for developing high-efficiency ultraviolet lasers for applications in spectroscopy, sensing, and biomedical technologies.

References

[1] L. Dabing, J. Ke J, and S. Xiaojuan S, "AlGaN photonics: recent advances in materials and ultraviolet devices," Advances in Optics & Photonics, 2018, 10(1), 43(2018).

[2] Xu and B. Sadler, "Ultraviolet Communications: Potential and State-Of-The-Ar," Communications Magazine IEEE 46(5):67-73(2008).

[3] Heilingloh, U. Aufderhorst , L. Schipper, U. Dittmer,O. Witzke, D. Yang, X. Zheng, K. Sutter, M. Trilling, M. Alt, E. Steinmann, and A. Krawczyk, "Susceptibility of SARS-CoV-2 to UV irradiation,". American Journal of Infection Control 48(10), 1273–1275(2020).

[4] Mehnke, M. Guttmann, J. Enslin, C. Kuhn, C. Reich, J. Jordan, S. Kapanke, A. Knauer, and M. Lapeyrade, "Gas Sensing of Nitrogen Oxide Utilizing Spectrally Pure Deep UV LEDs," IEEE Journal of Selected Topics in Quantum Electronics 23(2), 29-36(2017).

[5] G. Hechenblaikner, T. Ziegler, and I. Biswas, "Energy distribution and quantum yield for photoemission from air-contaminated gold surfaces under ultraviolet illumination close to the threshold," Journal of Applied Physics 111(12), 124914(2012).

[6] Xiang M., Zhang Y., Li G., et al, "Wide-waveguide high-power low-RIN single-mode distributed feedback laser diodes for optical communication," Optics Express, 30(17), 30187-30197(2022).

[7] Alahyarizadeh G., Amirhoseiny M., Hassan Z, "Effect of different EBL structures on deep violet InGaN laser diodes performance," Optics & Laser Technology, 76, 106-112(2016).

[8] Yang J., Zhao D., Zhu J., et al, "Effect of Mg doping concentration of electron blocking layer on the performance of GaN-based laser diodes," Applied Physics B, 125, 1-5(2019).

[9] Xing, Y. Wang, F. Wang, and Y. Liu, "Improvement of the optoelectronic characteristics in deep-ultraviolet laser diodes with tapered p-cladding layer and triangular electron blocking layer," Applied Physics B 128, 197(2022).

[10] L.Nam, M. Nakarmi, J. Li, J. Lin, and H. Jiang, "Mg acceptor level in AlN probed by deep ultraviolet photoluminescence," Applied physics letters 83(5), 878-880(2003).

[11] AA.Zhang, W. Liu, Z. Ju, S. Tan, Y. Ji, X. Zhang, L. Wang, Z. Kyaw, X. Sun, and H. Demir, "Polarization self-screening in [0001] oriented InGaN/GaN light-emitting diodes for improving the electron injection efficiency," Applied physics letters 104, 251108(2014).

[12] I.Piprek, "Efficiency droop in nitride-based light emitting diodes," Physica Status Solidi (A) Applications and Materials 207(10), 2217-2225(2010).

[13] Y.Li, Y. Huang, Y. Lai, "Investigation of Efficiency Droop Behaviors of InGaN/GaN Multiple-Quantum-Well μLEDs With Various Well Thicknesse," IEEE Journal of Selected Topics in Quantum Electronics 15(4), 1128-1131(2019).

[14] Kurnosov V., Kurnosov K, "Numerical simulation of the divergence and optical confinement factor of a semiconductor laser with an asymmetric periodic multilayer AlGaInAs/InP waveguide," Quantum Electronics, 50(9), 816-821(2020).

Enhanced performance in AlGaN deep-ultraviolet light-emitting diodes without an electron blocking layer by using Graded Decreasing Quantum Barrier structure

Xien Sang   Xin Wang   Wenlan Ma   Fang Wang   Juin J. Liou   Yuhuai Liu

Abstract—**With the continuous advancement of technology, the demand for deep ultraviolet (DUV) light sources has gradually increased across various fields. In order to enhance the light output power of AlGaN-based DUV light-emitting diodes (LEDs), this study proposes a quantum barrier (QB) structure with a graded Al composition decrease, based on the removal of the electron blocking layer. Through an analysis of the energy band diagram, it is demonstrated that this structure can increase the effective electron barrier height, effectively suppress electron leakage, and improve hole injection efficiency. This leads to an increase in carrier concentration in the active region, enhancing the recombination efficiency within the quantum wells. At a current of 100 mA, the LED's light output power reached 439 mW, with an internal quantum efficiency of 0.96. This approach provides a new perspective for improving the light-emitting efficiency and reliability of LEDs.**

Index Terms—**LED, AlGaN, Graded Decreasing Quantum Barrier,without an electron blocking layer**

I. INTRODUCTION

With the growing demand for DUV light sources, DUV-LEDs have demonstrated broad application potential in various fields such as water treatment, environmental monitoring, and medical disinfection due to their advantages of high efficiency, stability, and miniaturization [1]. However, despite the significant potential of DUV-LEDs in terms of applications, they still face numerous technical challenges. One of the key factors affecting the development of LEDs is the severe

---

Xien Sang and Xi Wang and Wenlan Ma are with *National Center for International Joint Research of Electronic Materials and Systems, International Joint-Laboratory of Electronic Materials and Systems of Henan Province, School of Electrical and Information Engineering, Zhengzhou University,*Zhengzhou, Henan, China. Fang Wang and Yuhuai Liu are with *National Center for International Joint Research of Electronic Materials and Systems, International Joint-Laboratory of Electronic Materials and Systems of Henan Province, School of Electrical and Information Engineering, Zhengzhou University, International Joint Laboratory for Integrated Circuits Design and Application, Ministry of Education, School of Physics,Institute of Intelligence Sensing, Research Institute of Industrial Technology Co. Ltd., Zhengzhou University,Zhengzhou Way Do Electronics Co. Ltd.*, Juin J. Liou is with *School of Electrical and Information Engineering, North Minzu University*, Yinchuan, Ningxia, China. This work is supported by National Nature Science Foundation of China (Grant No. 62174148), National Key Research and Development Program (NKRDP Grant No. 2022YFE0112000), Key Program for International Joint Research of Henan Province (Grant No. 231111520300). Corresponding author: (Fang Wang, email: iefwang@zzu.edu.cn;Yuhuai Liu, email: ieyhliu@zzu.edu.cn)

efficiency degradation with increasing current. To address this issue, researchers have proposed several approaches. For instance, Xing et al. suggested replacing the AlGaN electron blocking layer (EBL) with BGaN material to prevent electron leakage [2], while Wang et al. designed a Double-Tapered EBL [3]. Although the introduction of high-Al composition electron blocking layers can somewhat alleviate the electron leakage problem, increasing the Al content strengthens the ionization energy of the Mg-doped AlGaN layer, limiting the hole concentration in the p-region. Additionally, polarization charges can accumulate at the interface between the EBL and the QB, leading to the accumulation of holes at the last quantum well of the p-type layer and resulting in severe electron leakage. Studies have shown that removing the EBL structure can weaken the strong polarization effects caused by lattice mismatch between the QB and the EBL. In this paper, we design a QB structure with a graded decreasing Al composition. This new structure, by removing the traditional EBL, mitigates the band bending caused by strong polarization, reduces the quantum confinement Stark effect in the active region, increases the electron barrier height, preventing electrons from leaking through the active region to the p-type region, and lowers the hole barrier height, making it easier for holes to pass into the active region. Enhanced carrier injection into the active region results in effective recombination, thereby alleviating the efficiency degradation of the device.

II. DEVICE STRUCTURER

As shown in Figure. 1(a), the basic LED structure consists of a sapphire substrate, which is followed by a 200 nm Si-doped GaN buffer layer, a 3 μm thick n-GaN layer, MQWs composed of five 4 nm thick$Al_{0.6}Ga_{0.4}N$ quantum wells (QWs) and six 10-nm-thick $Al_{0.7}Ga_{0.3}N$ QB a 20 nm thick $Al_{0.9}Ga_{0.1}N$ electron barrier layer, a 12 nm Mg-doped (($n - doping = 5 * 10^{24} cm^{-3}$) GaN layer is used above the EBL, which is the underlying reference structure for LED1. The LED2 structure is based on LED1 with the electron blocking layer removed, while the LED3 structure features a graded-decreasing QB. The specific structure is shown in Figure 1(b). In this numerical study, device simulations were performed using the Advanced Physical Modeling of Semi-conductor Devices (APSYS) tool. APSYS software is used to calculate the electrical behaviour of all LEDs by solving the Poisson's equation and the current continuity equation for electrons and holes. In this study, the electrical and optical properties of the LED structure were analyzed in detail. Simulation parameters include energy band shift ratio, radiation recombination coefficient, Shockley-Read-hall (SRH) recombination lifetime, and Auger Recombination Coefficient is set to be 0.58, $0.5 * 10^{-16} m^3/s$, 100 ns, and $1 * 10^{-46} m^6/s$ respectively [4]. Other materials parameters of AlN and GaN such as lattice constant, deformation potential, elastic constant, etc. are listed elsewhere. The variation functions of electron and hole mobilities were calculated using the most commonly used Arora model. All simulations are performed by assuming that the LED devices operate at room temperature.

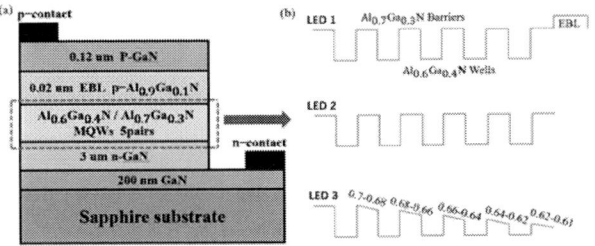

Fig. 1. Schematics of the DUV-LED structure, and LED1, LED2, and LED3 structure diagram.

III. RESULTS AND DISCUSSION

To study the performance of the proposed structure, the band diagrams of three different structures, LED1, LED2, and LED3, were obtained using the well-calibrated APSYS software. The principle of improving hole injection and electron leakage is to reduce the effective barrier height of the valence band and increase the effective barrier height of the conduction band. The effective barrier height is defined as the energy difference between an energy band and its corresponding quasi-Fermi level[5] and is a reliable parameter for evaluating the electron confinement ability and hole injection efficiency of a laser. As shown in Figure 2, compared to the reference structure LED1, both LED2 and LED3 exhibit increased electron effective barrier heights and decreased hole barrier heights, with the change in LED3 being the most significant. The electron barrier height in LED3 increases to 529 meV from 411 meV in LED1, effectively preventing electron leakage from the active region into the P region. Meanwhile, the hole barrier heights for LED1, LED2, and LED3 are 559 meV, 411 meV, and 264 meV, respectively. The hole barrier height in LED3 is lower compared to LED1, which enhances the efficiency of hole injection into the active region. This is mainly because for the graded decreasing AlGaN QB, the lattice mismatch between the QW and the barriers is reduced. Consequently, this design generates fewer piezoelectric fields, leading to lower band bending. Therefore,

the hole effective barrier height is reduced to improve hole injection efficiency, while the electron effective barrier height is increased to mitigate electron leakage.

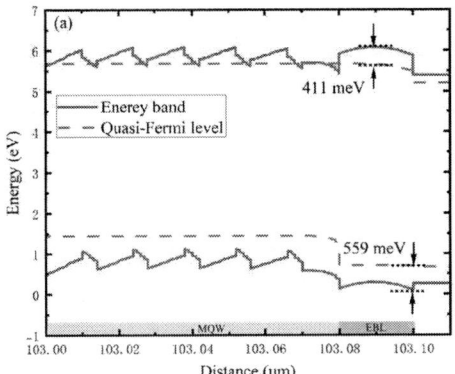

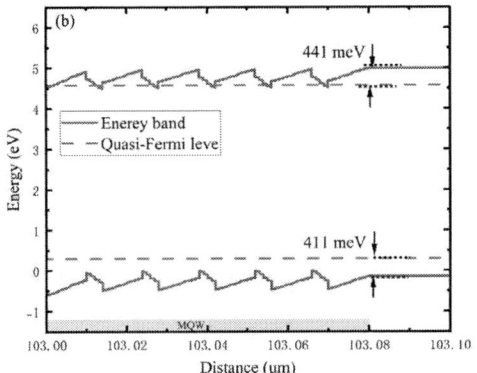

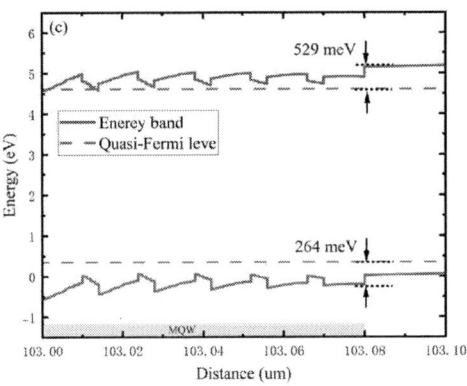

Fig. 2. Energy band diagram and quasi fermi level (a) LED1, (b) LED2, and (c) LED3.

To further validate the analysis of the LED energy band structures, the output optical power and internal quantum efficiency (IQE) of the three structures were investigated. As shown in Figure 3(a), our conclusion is confirmed: with increasing current, LED3 reaches a power of 439 mW at

979-8-3315-2209-4/25 $31.00 © 2025 IEEE

100 mA, which is a 352% increase compared to LED1 (97 mW). Due to the strong electron confinement within the MQW, electron leakage into the P-type region is suppressed, while holes can be efficiently injected into the active region. Consequently, the IQE of the designed LED3 is higher than that of traditional LEDs, reaching 95% at a current density of 80 mA, as shown in Figure 3(b). In comparison with traditional LEDs, the efficiency of LED3 gradually increases with increasing injection current, reaching a stable value at 95%, indicating that the designed LED exhibits higher radiative recombination within the MQW. Therefore, the designed LED3 can effectively address the issue of light efficiency degradation due to carrier overflow at high current injection, providing insights for further research on deep ultraviolet LEDs.

region, weakening the electron confinement ability, while simultaneously raising the hole barrier height, which severely impacts hole injection. Therefore, by removing the electron blocking layer and adopting a graded-decreasing QB structure, it was found that the electron effective barrier height increased by 28%, while the hole effective barrier height decreased by 52%. As a result, the light output power increased from 97 mW to 439 mW, the required threshold current decreased, electron leakage was effectively suppressed, and efficient hole injection was achieved, effectively addressing the issue of device efficiency degradation.

V. ACKNOWLEDGMENTS

Supported by National Nature Science Foundation of China (Grant No. 62174148), National Key Research and Development Program (NKRDP Grant No. 2022YFE0112000), Key Program for International Joint Research of Henan Province (Grant No. 231111520300).

REFERENCES

[1] S. Nakamura, Future technologies and applications of III-nitride materials and devices. Engineering. 1, 161 (2015).
[2] Z.Q. Xing, F. Wang, Y. Wang, J.J. Liou, Y.H. LIU, Enhanced performance in deep-ultraviolet laser diodes with an undoped BGaN electron blocking layer. Opt. Express. 30, 36446 (2022).
[3] Y.F. Wang, M.I. Niass, F. Wang, Y.H. Liu, Reduction of electron leakage in a deep ultraviolet nitride laser diode with a double-tapered electron blocking layer. Chin. Phys. Lett. 36, 057301 (2019).
[4] H.B. Yu, M.H. Memon, D.H. Wang, Z.J. Ren, H.C. Zhang, C. Huang, M. Tian, H.D. Sun, S.B. Long, AlGaN-based deep ultraviolet micro-LED emitting at 275 nm. Opt.Letters. 46 3271-3274 (2021).
[5] Z.Q. Xing, Y.J. Zhou, A.X. Zhang, Y.P. Qu, F. Wang, Y. Wang, J.J. Liou, Y.H. LIU, Non-heavy doped pnp-AlGaN tunnel junction for an efficient deep-ultraviolet light emitting diode with low conduction voltage. Opt. Express. 32. 10284-10294 (2024).

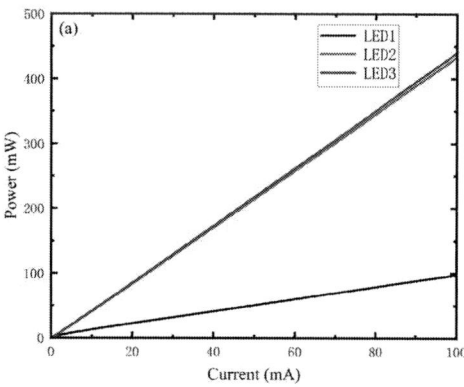

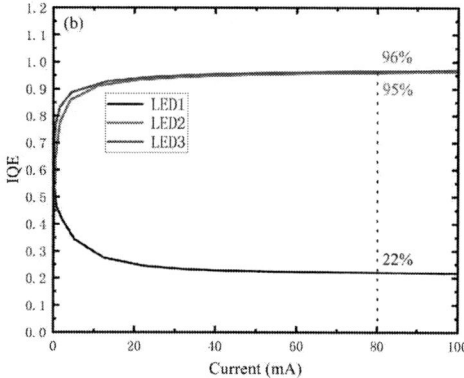

Fig. 3. (a) P-I curve of three structures, (b) Internal quantum efficiency of three structures.

IV. CONCLUSIONS

This paper proposes a graded-decreasing QB structure without an electron blocking layer (EBL) to improve the performance of DUV-LEDs. The cause of electron leakage is attributed to the strong polarization effect at the heterointerface between the EBL and LQB, which leads to energy band bending. This increases the electrostatic field in the active

Optimizing Laser Diode Performance with Al-Composition Graded Last Quantum Barriers

Xin Wang   Wenlan Ma   Xien Sang   Fang Wang   Juin J. Liou   Yuhuai Liu

Abstract—In this study, the effect of a linearly increasing Al composition in the last quantum barrier (LQB) on deep ultraviolet (DUV) laser diode (LD) was investigated. Simulation results show that, compared to traditional quantum barriers (QBs) with a constant Al composition, DUV LD with a linearly increasing Al composition LQB exhibit a reduced threshold current of 14.6 mA. The output power reaches 117.7 mW, and the slope efficiency increases from 1.64 W/A to 1.77 W/A. These results indicate that the use of a linearly increasing Al composition LQB structure effectively limits electron leakage, while also optimizing hole injection efficiency. As a result, the stimulated recombination rate in the active region is enhanced, leading to improved performance of the DUV LD.

Index Terms—Deep ultraviolet,laser diodes,AlGaN,carrier injection efficiency

I. INTRODUCTION

Deep ultraviolet (DUV) light is typically defined within the wavelength range of 200 to 300 nm. Compared to ultraviolet (UV) and visible light, DUV light has a shorter wavelength, higher energy, and unique properties when interacting with matter [1]. Deep ultraviolet (DUV) lasers represent a significant advancement in laser technology, with broad potential applications due to their unique characteristics. Among various DUV laser technologies, AlGaN-based DUV laser diodes (LDs) have attracted considerable attention due to their solid-state nature, ability to generate high optical power, and high energy conversion efficiency [2]. Traditional DUV light sources, such as gas discharge lamps and xenon fluoride lasers, suffer from issues such as large size, high operating costs, short lifespans, and low efficiency. In contrast, AlGaN-based DUV LDs address these issues by offering a smaller form factor, longer lifespan, and higher efficiency, making them

---

Xin Wang and Wenlan Ma and Xien Sang are with *National Center for International Joint Research of Electronic Materials and Systems, International Joint-Laboratory of Electronic Materials and Systems of Henan Province, School of Electrical and Information Engineering, Zhengzhou University,*Zhengzhou, Henan, China. Fang Wang and Yuhuai Liu are with *National Center for International Joint Research of Electronic Materials and Systems, International Joint-Laboratory of Electronic Materials and Systems of Henan Province, School of Electrical and Information Engineering, Zhengzhou University, International Joint Laboratory for Integrated Circuits Design and Application, Ministry of Education, School of Physics,Institute of Intelligence Sensing, Research Institute of Industrial Technology Co. Ltd., Zhengzhou University,Zhengzhou Way Do Electronics Co. Ltd.,* Juin J. Liou is with *School of Electrical and Information Engineering, North Minzu University,* Yinchuan, Ningxia, China. This work is supported by National Nature Science Foundation of China (Grant No. 62174148), National Key Research and Development Program (NKRDP Grant No. 2022YFE0112000), Key Program for International Joint Research of Henan Province (Grant No. 231111520300). Corresponding author: (Fang Wang, email: iefwang@zzu.edu.cn;Yuhuai Liu, email:ieyhliu@zzu.edu.cn)

---

an ideal alternative for industrial and scientific applications [3].However, the development of AlGaN-based DUV LDs still faces numerous challenges, including defects in the crystal quality of AlGaN material, lattice mismatch between AlGaN and substrates (such as sapphire or silicon carbide), and issues related to carrier injection efficiency [4]. Among these, carrier injection efficiency is one of the key factors affecting the performance of AlGaN-based DUV LDs. It is directly related to the LD's output power, threshold current, and overall stability [5]. Therefore, improving carrier injection efficiency has become a critical challenge. To address the issue of low carrier injection efficiency, the use of electron blocking layers (EBLs) and hole blocking layers (HBLs) can reduce carrier leakage, increase recombination rates, and optimize device performance. Xing et al. proposed an undoped BGaN EBL to reduce electron leakage in the p-type region [6]. Zhang et al. proposed two asymmetric compositional graded quantum well structures to improve the performance of DUV LDs [7]. Yin et al. investigated the structures of p-doped, n-doped, and n-p-doped quantum barriers (QBs) in the active region of a quaternary AlInGaN LQB LD without an EBL, which resulted in a reduced threshold current for the LD [8]. The quantum barrier layer, serving as the contact layer between the active region and the p-type region, primarily functions to limit electron overflow from the active region [9]. LQB provides bandgap continuity for the active region. Some studies have shown that the electron blocking capability of the LQB plays a crucial role in the radiative recombination rate. Issues such as electron leakage and low radiative recombination rates can be resolved by optimizing the LQB [10]. Therefore, we conducted simulations to investigate a DUV LD with a linearly increasing structure for the LQB. From the perspectives of threshold current, electron concentration, hole concentration, and radiative recombination efficiency, we verified that the linearly increasing structure of the LQB can improve the performance of the DUV LD.

II. DEVICE STRUCTURER

The DUV LD model studied in this paper is shown in Figure 1.The conventional model grown on a sapphire substrate is referred to as LD1. The n-side consists of an Si-doped GaN contact layer($n - doping = 1 * 10^{20} cm^{-3}$),a 1 μm thick Si-doped $Al_{0.75}Ga_{0.25}N$ ($n - doping = 1 * 10^{18} cm^{-3}$) cladding layer (n-CL),and a 0.11 μm thick Si-doped $Al_{0.67}Ga_{0.33}N$($n - doping = 1 * 10^{18}cm^{-3}$) lower waveguide layer (n-WG). The active region consists of two 3

979-8-3315-2209-4/25 $31.00 © 2025 IEEE   81

nm thick $Al_{0.62}Ga_{0.38}N$ quantum wells and three 8 nm thick $Al_{0.68}Ga_{0.32}N$ QBs. Above the active region, there is a 70 nm Mg-doped $Al_{0.68}Ga_{0.32}N(p-doping = 1*10^{19}cm^{-3})$ upper waveguide layer (p-WG). This is followed by a 10 nm Mg-doped $Al_{0.90}Ga_{0.10}N(p-doping = 1*10^{19}cm^{-3})$ electron blocking layer (EBL). The p-type region consists of a 0.4 µm thick Mg-doped $Al_{0.76}Ga_{0.24}N(p-doping = 1*10^{19}cm^{-3})$ cladding layer and a 10 nm thick Mg-doped $Al_{0.80}Ga_{0.20}N(p-doping = 1*10^{19}cm^{-3})$ contact layer. This device is inspired by the model proposed by Hameed et al [11]. In addition to the reference structure, the linearly increasing LQB structure studied in this paper is referred to as LD2. In LD2, the LQB is composed of $Al_{0.68-0.90}Ga_{0.32-0.10}N$. Except for the difference in the LQB structure, all other structural parameters are identical to those of LD1.

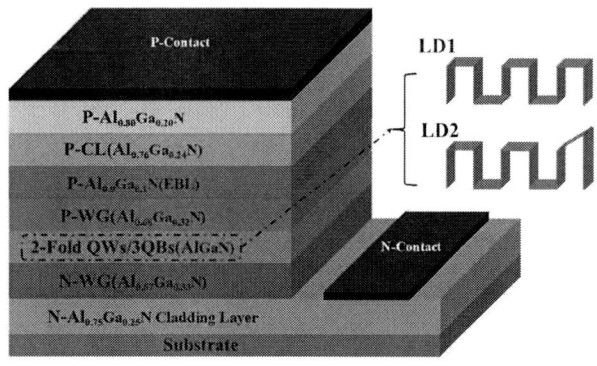

Fig. 1. Schematic structures for the studied LDs.

Simulations were performed using the Photonic Integrated Circuit simulator (PIC3D) to explore the optoelectronic characteristics of the two semiconductor lasers described above. PIC3D employs the Poisson equation, drift-diffusion model, current continuity equation, and polarization charge model to calculate the electrical stimulation of the laser diode (LD). Optical simulations of the LD are performed using the vector Helmholtz wave equation and the Adachi refractive index model. Carrier transport and distribution characteristics are calculated using the transfer matrix method and the Schrödinger-Poisson self-consistent approach. For both LD1 and LD2 samples, the same growth conditions are used: a width of 1.5 µm, cavity length of 534 µm, with the back loss assumed to be 2400 m^{-1}. The emission wavelength of the LD is approximately 267 nm, and all simulations are conducted at room temperature (T = 300 K).

III. RESULTS AND DISCUSSION

Figure 2 presents the *P-I* characteristic curves for the studied structures LD1 and LD2, showing the relationship between output power and injection current for both structures. The results indicate that, at the same injection current, the output power of LD2 is significantly higher than that of LD1. When the injection current is 80 mA, the output power of LD1 is 90.0 mW, while that of LD2 is 117.7 mW. Compared to LD1, the output power of LD2 increases by approximately 30.75%.The

threshold current for LD1 is 26.4 mA, while for LD2 it is 14.6 mA, resulting in a reduction of about 11.8 mA compared to LD1. The calculated slope efficiency increases from 1.64 W/A for LD1 to 1.77 W/A for LD2. This improvement is attributed to the linearly increasing LQB, which suppresses electron leakage from the active region to the p-type region, enhances hole injection efficiency in the active region, and increases the radiative recombination rate. The stimulated recombination

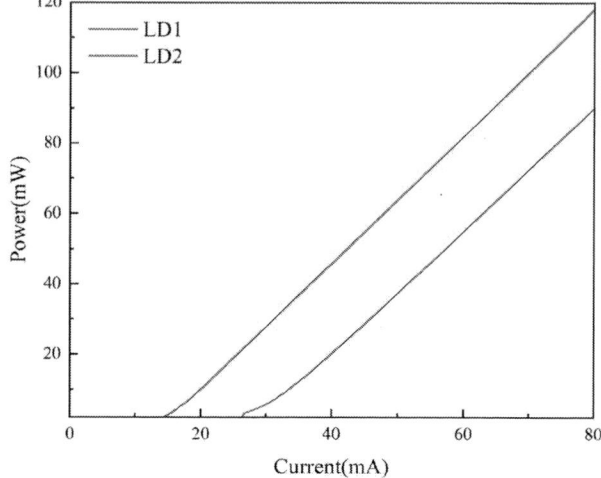

Fig. 2. The *P-I* curves of two LDs.

rate of the two DUV LDs is shown in Figure 3, where the stimulated recombination rate in the active region of LD2 is approximately 50.9% higher than that of LD1.This is because both the electron and hole concentrations in the active region have been increased, promoting the recombination probability of electrons and holes. Furthermore, the linearly increasing Al composition causes a gradual change in the band structure of the quantum well, improving the band alignment of electrons and holes. The wavefunctions of electrons and holes are more likely to overlap, effectively reducing non-radiative recombination. Figures 4 (a) and (b) show the electron concentration and hole concentration in the active region of LD1 and LD2, respectively. A comparison reveals that both the electron and hole concentrations in the quantum well of the active region of LD2 are higher than those in LD1. This indicates that the use of a linearly increasing LQB effectively limits electron leakage and enhances hole injection efficiency. This may be due to the gradual increase in barrier height as the Al content increases, which reduces the probability of electron escape from the quantum well. Additionally, the linearly increasing Al composition of the LQB can adjust the band structure near the quantum well, making hole injection more efficient and further improving hole injection efficiency. Figure 5 (a) shows the band structure near the active region of the conventional LD1 structure. From the figure, it can be seen that at the LQB, the effective electron barrier height of the LD1 structure is approximately 30 meV, while the effective hole barrier height is approximately 416 meV. Figure 5 (b) shows the band structure near the active region of LD2, with the

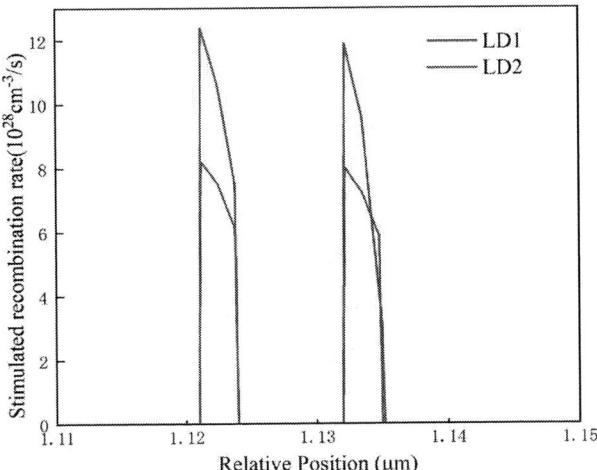

Fig. 3. Stimulated recombination rate.

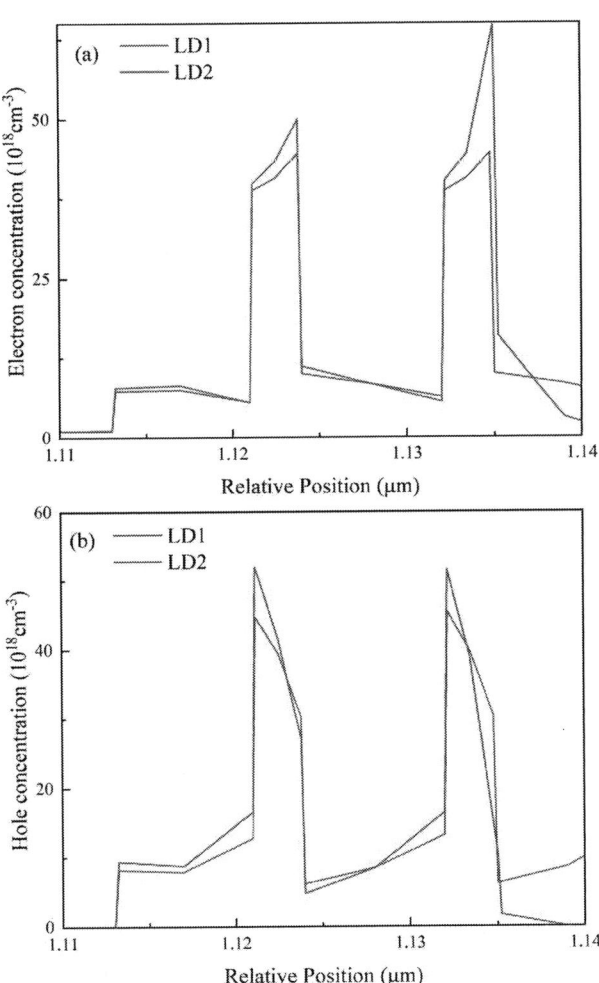

Fig. 4. (a)Electron concentration,(b)Hole concentration.

linearly increasing Al composition in the LQB. At the LQB, the effective electron barrier height of the LD2 structure is approximately 412 meV, while the effective hole barrier height is approximately 254 meV. A comparison reveals that the LD2 structure alters the band structure at the LQB,exhibiting a higher effective electron barrier height and a lower effective hole barrier height, which increases the carrier concentration in the active region. The higher carrier concentration enhances the recombination chance of electrons and holes, and the higher radiative recombination rate in the active region reduces the threshold current of the DUV LD, improving the output power, consistent with the P-I curve.Therefore, it indicates that LD2 has superior electron confinement and hole transport capabilities

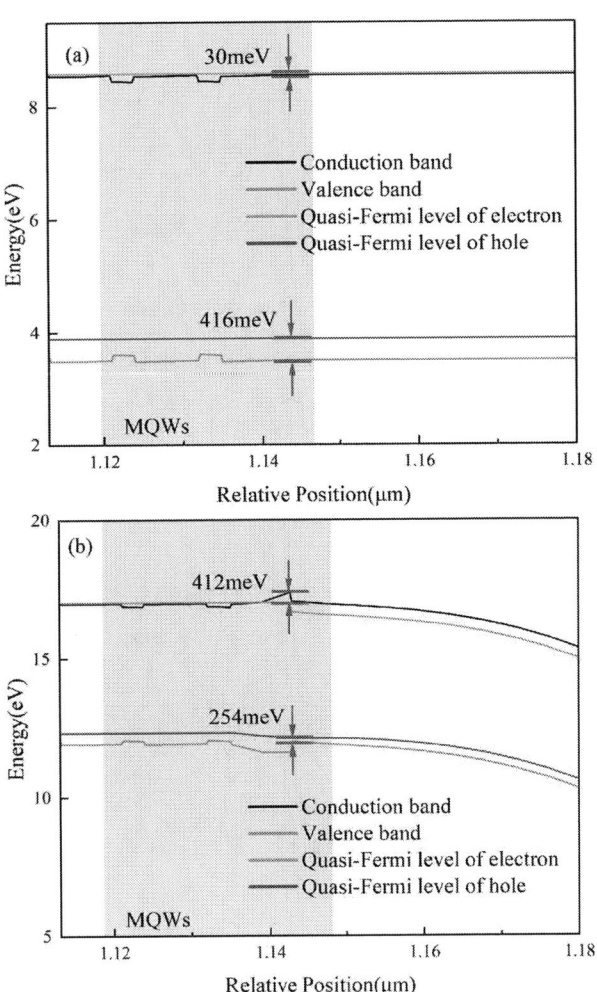

Fig. 5. Energy band diagram for (a) LD1, (b) LD 2

IV. CONCLUSIONS

This study conducts a simulation of LDs with a linearly increasing Al composition LQB structure, and compares them with traditional LD structures. The P-I curves, electron concentration, hole concentration, stimulated recombination rate, and band structure in the active region of the two structures are analyzed.The results show that for the LD2 with the linearly increasing Al composition LQB the threshold current decreases from 26.4 mA to 14.6 mA, the output power increases from 90.0 mW to 117.76 mW, and the radiative recombination rate is also improved. This improvement is mainly attributed to the linearly increasing Al composition LQB , which enhances the band structure, suppresses electron leakage, and improves hole injection efficiency. This provides valuable insights for the fabrication of high-efficiency DUV LDs.

V. ACKNOWLEDGMENTS

Supported by National Nature Science Foundation of China (Grant No. 62174148), National Key Research and Development Program (NKRDP Grant No. 2022YFE0112000), Key Program for International Joint Research of Henan Province (Grant No. 231111520300).

REFERENCES

[1] Kuo, Y-K, "Design and Optimization of Electron-Blocking Layer in Deep Ultraviolet Light-Emitting Diodes," IEEE Journal of Quantum Electronics, Vol. 56, No. 1 (2019), pp. 1-6.

[2] Guo, Y., Zhang, Y., Wang, J., et al., "Development of AlGaN-based deep ultraviolet light-emitting diodes and laser diodes," Proceedings of the 12th China International Forum on Solid State Lighting (SSLCHINA), IEEE, 2015, pp. 4-7.

[3] Xu, Y., Wei, S.-Q., Zhang, P.-F., et al., "Radiative recombination characteristics of deep ultraviolet laser diodes based on double-concave waveguide layers," Laser & Optoelectronics Progress, Vol. 60, No. 15 (2023), pp. 1514007.

[4] Liang, G., Ya-Nan, G., Jian-Kun, Y., et al., "Effect of barrier height on modulation characteristics of AlGaN-based deep ultraviolet light-emitting diodes," Chinese Journal of Luminescence, Vol. 43, No. 1 (2022), pp. 1-7.

[5] Guo, C., Sun, X., Guo, K., et al., "Recent progress of solar blind light emitting diodes for ultraviolet optical wireless communication use," Chinese Journal of Luminescence, Vol. 44, No. 10 (2023), pp. 1849-1861.

[6] Xing, Z., Wang, F., Wang, Y., et al., "Enhanced performance in deep-ultraviolet laser diodes with an undoped BGaN electron blocking layer," Optics Express, Vol. 30, No. 20 (2022), pp. 36446-36455.

[7] Zhang, P., Jia, L., Zhang, A., et al., "Composition-graded quantum barriers improve performance in AlGaN-based deep ultraviolet laser diodes," Optical Engineering, Vol. 61, No. 7 (2022), pp. 076113.

[8] Yin, M., Sang, X., Xu, Y., et al., "Doped effects of quaternary AlInGaN last quantum barrier for deep-ultraviolet laser diodes without electron blocking layer," Journal of Russian Laser Research, Vol. 44, No. 4 (2023), pp. 407-414.

[9] Sang, X., Xu, Y., Yin, M., et al., "InGaN multiple quantum well based light-emitting diodes with indium composition gradient InGaN quantum barriers," Optoelectronics Letters, Vol. 20, No. 2 (2024), pp. 89-93.

[10] Choubey, B., Ghosh, K., "Efficient and droop-free AlGaN-based UV-C LED using the inverted linearly graded active region and engineered hole source layer," Optik, Vol. 311 (2024), pp. 171941.

[11] Rehman, H. U., Bi, W., Rahman, N. U., et al., "A design and improvement in synergy between un-doped W-AlxGa1-xN/B-AlyGa1-yN quantum wells (QWs) and electron blocking layers (EBL) for DUV emission," Optics & Laser Technology, Vol. 181 (2025), pp. 112025.

979-8-3315-2209-4/25 $31.00 © 2025 IEEE

Electronic and optical properties modulation of heterostructures based on GeP and h-BN

1st Xiaotian Yang   2nd Jun Zhang   3rd Shipei Ji   3rd Yuhuai Liu*   4th Fang Wang

Abstract—**Indium phosphide (InP) has attracted attention due to its high carrier mobility and broad-spectrum absorption, making the maintenance of conductivity and flexible bandgap tuning research priorities. This paper employs density functional theory to study the stability, electronic, and optical properties of InP heterostructures incorporating an h-BN layer, as well as the effects of atomic vacancy defects. The results demonstrate that the InP/h-BN heterostructure adjusts the bandgap while preserving the material's inherent properties. Additionally, defects significantly alter charge transfer, thereby modulating thermoelectric performance. These findings provide guidance for researching InP/h-BN heterostructures in optoelectronic devices under various conditions.**

Index Terms—**electronic material, heterostructure, electronic structure, 2D materials, first-principles calculations**

I. INTRODUCTION

Two-dimensional materials play an important role in advanced fields because of their exceptional properties, and graphene, boron nitride, and indium phosphide are used in research and applications in this area [1]. However, monolayer indium phosphide is prone to degradation in humid or oxidative environments, necessitating surface passivation to enhance stability [2], as interface quality can easily affect device performance. Therefore, exploring various methods to modulate its bandgap and thermoelectric properties has become increasingly important [3]. Studies indicate that vertically stacked van der Waals (vdW) heterostructures can optimize performance and expand applications [4]. Hexagonal boron nitride (h-BN), with its lack of dangling bonds and ultra-wide bandgap, serves as an ideal material for adjusting the bandgap of InP [5]. Utilizing density functional theory, we selected monolayer and bilayer InP and optimized electronic properties through h-BN layers to investigate the electronic characteristics of InP-h-BN heterostructures [6]. The results demonstrate that different heterostructures can modulate the properties of InP, while thoroughly examining the impact of defect structures, thereby enhancing material performance and broadening potential applications [7].

II. CALCULATION METHODS

In this study, all first-principles calculations were conducted using pwmat, utilizing the Generalized Gradient Approximation (GGA) with the Perdew-Burke-Ernzerhof (PBE) exchange-correlation functional. Periodic boundary conditions were applied along the x and y directions, with an 18 Å vacuum layer in the z direction to prevent interactions between the top and bottom surfaces of the heterostructures. The plane-wave cutoff energy was set at 50 Ryd. The two-dimensional heterostructures underwent stepwise optimization, with full relaxation of geometric and atomic positions until the Hellmann-Feynman forces were reduced to below 0.02 eV/Å. The Monkhorst-Pack method was employed to generate K-points, ensuring symmetry, with a k-point grid of 551. Structural diagrams and charge density difference maps of the constructed heterostructures were obtained using VESTA software.

III. RESULT AND DISCUSSION

A. Geometric structure

Firstly, we calculated and verified the electronic structures and properties of monolayer and bilayer InP, as well as h-BN, to ensure the rigor of the study. The heterostructures were formed by relaxing a 3×3×1 InP rhombic supercell, vertically stacked on a 5×5×1 h-BN rhombic supercell, with lattice constants of 12.63 and 12.44, respectively. The lattice mismatch is minimal, approximately 1.52%. We used InP as the base structure for comparison with previous results and constructed multilayer InP by cleaving surfaces to explore the stacking properties of the materials. As shown in Figure 1(a) - (c), the monolayer properties of h-BN are stable and exhibit strong structural rigidity, making it a suitable novel insulating layer for constructing heterostructures to modulate the properties of intrinsic materials. Comparisons with previous studies show consistent results; hence, we included optimized lattice constants, bond lengths, bond angles, and binding energies in Table 1 for more detailed comparison. InP and h-BN both belong to the hexagonal crystal system.

Author 1, Author 2 and Author 3 are with *National Center for International Joint Research of Electronic Materials and Systems, International Joint-Laboratory of Electronic Materials and Systems of Henan Province. (School of Electrical and Information Engineering)*, ZhengZhou, China. Author 4 and Author 5 is with *National Center for International Joint Research of Electronic Materials and Systems, International Joint-Laboratory of Electronic Materials and Systems of Henan Province, School of Electrical and Information Engineering, Zhengzhou University, International Joint Laboratory for Integrated Circuits Design and Application, Ministry of Education, School of Physics*, ZhengZhou, China. *Institute of Intelligence Sensing, Research Institute of Industrial Technology Co. Ltd., Zhengzhou University*, ZhengZhou, China. *Zhengzhou Way Do Electronics Co. Ltd.*, Zhengzhou, ZhengZhou, China. This work is supported by [National Natural Science Foundation of China] under Grant [No.62174148]; the [National Key Research and Development Program] under Grant [Nos.2022YFE0112000 and 2016YFE0118400]; the [Key Program for International Joint Research of Henan Province] under Grant [No.231111520300]; the [Ningbo Major Project of 'Science, Technology and Innovation 2025'] under Grant [No.2019B10129]; and the [Zhengzhou 1125 Innovation Project] under Grant [No.ZZ2018-45]. Corresponding author: (Author 4, email: ieyhliu@zzu.edu.cn)

979-8-3315-2209-4/25 $31.00 © 2025 IEEE

Therefore, all subsequent structural optimizations and property calculations are based on the high-symmetry points of the hexagonal system, with their first Brillouin zone illustrated in Figure 2(d). For simplicity, we have abbreviated the structures as follows: monolayer InP is denoted as I, bilayer InP as I-Double, and the heterostructures as I/B, I-Double/B, I/B-VIn, and I/B-VP, respectively. As observed in Figures (a) - (c), (e), and (f), no visible deformation occurs during structural optimization without defect introduction. However, upon introducing defects, h-BN continues to exhibit good structural rigidity, whereas InP undergoes varying degrees of deformation. The deformation is significantly greater with In defects, closely related to charge transfer, which will be analyzed in detail in the differential charge density map in Figure 3.

To accurately assess the stability of each heterostructure, we introduce the binding energy Eb to evaluate its structural stability. The calculation formula is as follows:

$$E_b = (E_Hetero - E_InP - E_{(}h - BN))/n \qquad (1)$$

Where E Hetero represents the total energy of the heterostructure, while E InP and E h-BN denote the energies of the constituent components of the heterostructure. The Eb values for various heterostructures are presented in Table 1, ranging from -15.30 eV to -3.92 eV. These values indicate that the structures are readily formable, with a larger binding energy magnitude implying a stronger binding strength.

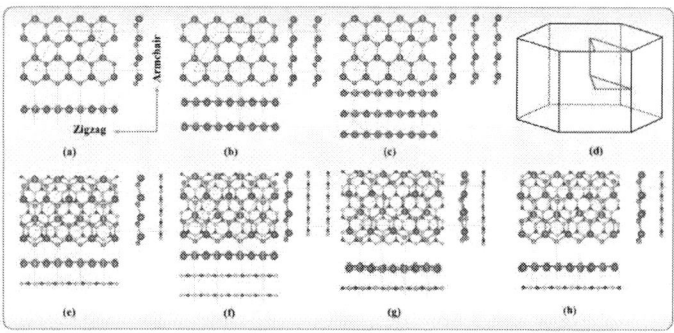

Fig. 1. (a) - (c) are the optimized crystal structures for monolayer, bilayer and trilayer InP, respectively. (d) is the first brillouin zone corresponding to the hexagonal system. (e) and (f) shows optimized crystal structures of InP combined with 1 to 2 layers of h-BN, respectively. (g) and (h) display the band structure and DOS of a monolayer InP and h-BN heterostructure with a single In / P defect.

B. Electric properties

The introduction of heterostructures significantly impacts the electronic properties of materials. Firstly, the bandgap of InP is greatly influenced by the number of stacking layers. Analyzing the band structure of a monolayer InP reveals that both the conduction band minimum (CBM) and the valence band maximum (VBM) are located at the point, exhibiting a typical direct bandgap. The bandgap value is sensitive to the number of stacking layers. As seen in Figures 3(a) - (c), the bandgap of InP is notably affected by the stacking layers. The

bandgap of monolayer InP is 1.134 eV, which sharply reduces to 0.584 eV in a bilayer stack and further decreases to 0.276 eV in a trilayer stack. Therefore, thicker InP exhibits higher conductivity, and the bandgap becomes zero when the number of stacking layers exceeds three. Secondly, the introduction of h-BN can maintain the intrinsic properties of the material while improving its environmental adaptability. As shown in the DOS diagrams in Figures 3(d) - (f), different h-BN stacking layers contribute only around +2 and -2 eV, with a slight impact on the bandgap of the InP monolayer. Thirdly, the adsorption between heterostructures is mainly driven by weak van der Waals forces, allowing the intrinsic characteristics of the constituent components to be preserved. The comparison of band structures shows that h-BN minimally affects the original material near the Fermi level and significantly influences the heterostructure only below -2 eV, preserving the intrinsic properties of the InP monolayer.

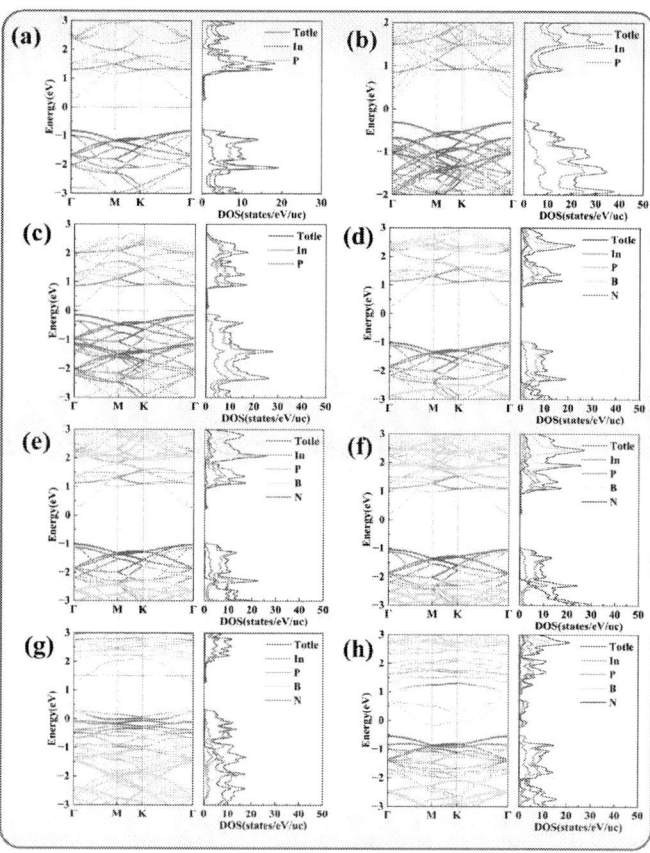

Fig. 2. (a), (b) and (c) are the calculated projected energy band structures and corresponding DOS for monolayer, bilayer and trilayer InP, respectively, and (d) - (f) are the same information for the heterostructures composed of single layer InP with 1-3 layers of h-BN. (g) and (h) display the band structure and DOS of a monolayer InP and h-BN heterostructure with a single In / P defect, respectively.

Currently, molecular dynamics is commonly used to observe the structural and thermal stability of two-dimensional materials. Therefore, we conducted 1000 fs AIMD simulations at 300 K on six different structures, incorporating dipole correction to

979-8-3315-2209-4/25 $31.00 © 2025 IEEE

study their thermal stability, as shown in Figure 4. The results indicate that all these structures exhibit good thermodynamic stability, with minimal changes in free energy, all less than 5 eV. Additionally, the introduction of In defects significantly reduces the sheet bandgap and causes an overall upward shift of the h-BN energy bands. The introduction of P defects results in the appearance of interface states and the creation of additional traversing bands. These introduced defects disrupt the periodicity of the lattice and introduce defect states in the band structure, which are located near the Fermi level, thus reducing the bandgap.

To comprehensively analyze the reasons for this effect, we further calculated the differential charge density of the materials. For a visual analysis of the charge transfer process, the charge density difference of the heterostructure along the z-direction was plotted using Vesta software. The formula for calculating the differential charge density is as follows:

The calculated differential charge density is depicted in Figure 3, where the yellow regions indicate electron accumulation, and the blue regions denote electron depletion. The illustration suggests that, within these heterostructures, electrons predominantly accumulate near the InP layer while depleting near the h-BN monolayer, indicating a transfer of electrons from the h-BN monolayer to the InP monolayer. Within the InP monolayer, electrons gather near the P atoms, whereas interlayer regions exhibit electron accumulation around In atoms. This observation corroborates the significant impact of defects on the properties of the material.

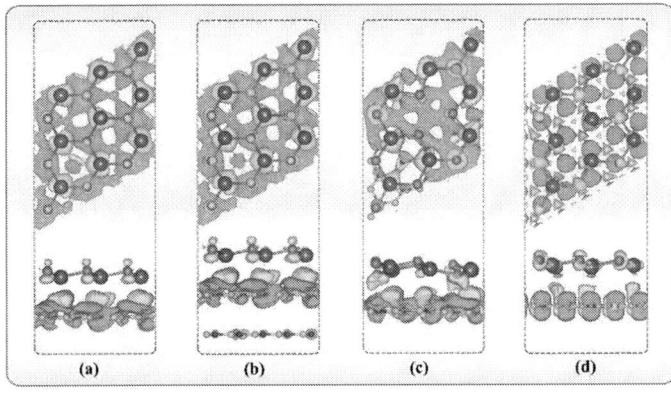

Fig. 3. Charge density difference with an isosurface of $0.9 \times 10 \exp(-3)$ e/Å of the heterostructures composed of monolayer and bilayer InP with h-BN (a, b), and a monolayer InP and h-BN heterostructure with a single In / P defect (c, d), respectively.

Consequently, defects in In atoms reduce the interlayer charge transfer capability, leading to a substantial upward shift in the h-BN energy bands, whereas P atom defects directly disrupt the original charge transfer scenario. As shown in Figure 3(d), electron accumulation almost envelops the P atoms, and the continuous charge depletion region in the lower layer is partitioned into multiple areas, markedly altering the distribution of electronic states near the Fermi level and thereby promoting the emergence of traversing bands in the heterostructure.

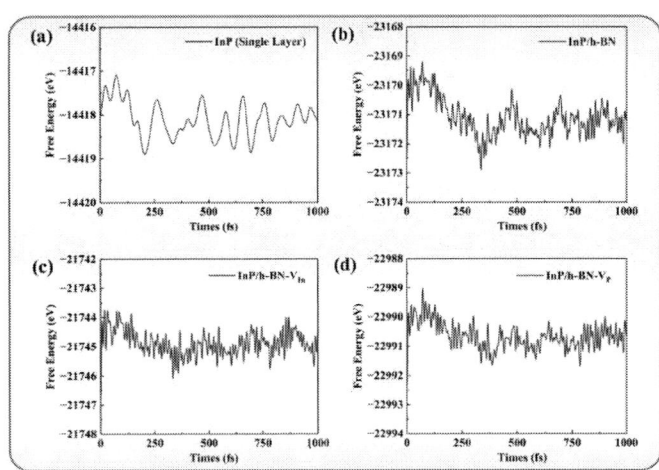

Fig. 4. (a) to (d) are the curves showing the change in free energy over time during AIMD simulations at 300 K for different structures.

TABLE I
OPTIMIZED STRUCTURAL PARAMETERS FOR H-BN, INP MONOLAYER, BILAYER, HETEROSTRUCTURES WITH H-BN: LATTICE MISMATCH (), BANDGAP, LATTICE PARAMETERS, BINDING ENERGY (EB)

Material	(%)	PBE gap (eV)	a / b (Å)	Eb (meV)
h-BN Monolayer(B)		4.569 Direct	12.43	
InP Monolayer (I)		1.134 Direct	12.627	
InP Bilayer (I2)		0.584 Direct		
Trilayer(I3)		0.276 Direct		
Hetero(I/B)	1.53%	1.298 Direct	12.532	-6.88
Hetero(I/B2)	1.53%	1.286 Direct	12.532	-11.13
Hetero(I/B3)	1.53%	1.279 Direct	12.532	-15.30
Hetero(I/B-VIn)	1.53%	0.974 Direct		-3.92
Hetero(I/B-VP)	1.53%		18.01	

[a]Sample of a Table footnote.

IV. CONCLUSIONS

In summary, this study investigates the structural and electronic-optical properties of heterostructures composed of monolayer InP, bilayer InP, and h-BN using first-principles calculations. Firstly, the heterostructure formed by monolayer InP and h-BN precisely tunes the material's bandgap while effectively preserving the intrinsic band structure of monolayer InP. Similarly, the InP heterostructure with a bilayer h-BN sandwich exhibits comparable properties. Secondly, the introduction of defects significantly alters the charge density distribution in the heterostructure, leading to substantial changes in the material's intrinsic electronic properties. P defects even result in the emergence of traversing bands and disrupt the continuous electron depletion trend in the h-BN layer. Thirdly, compared to monolayer InP without h-BN, the deformation due to defects in the I/B heterostructure is considerably smaller, indicating that h-BN significantly enhances the rigidity of the original material. Moreover, its chemical inertness greatly improves the environmental adaptability of the intrinsic material. The findings provide guidance for the exploration and application of various InP/h-BN heterostructures in electronic devices under different environmental conditions and offer a theoretical basis for their use as foundational materials in thermoelectric switches and memristors.

REFERENCES

[1] Lu, X., Li, L., Guo, X., Ren, J., Xue, H. and Tang, F., 2021. Effects of vertical strain and electric field on the electronic properties and interface contact of graphene/InP vdW heterostructure. Computational Materials Science, 198, p.110677.

[2] Chen, L., Zhou, X. and Yu, J., 2019. First-principles study on the electronic and optical properties of the ZnTe/InP heterojunction. Journal of Computational Electronics, 18(3), pp.749-757.

[3] Y. Xu, Z. Shi, X. Shi, K. Zhang, H. Zhang, Recent progress in black phosphorus and black-phosphorus-analogue materials: properties, synthesis and applications, Nanoscale 11 (2019) 14491-14527.

[4] T. Qi, Y. Gong, A. Li, X. Ma, P. Wang, R. Huang, C. Liu, R. Sakidja, J.Z. Wu, R. Chen, L. Zhang, Interlayer transition in a vdW heterostructure toward ultrahigh detectivity shortwave infrared photodetectors, Advanced Functional Materials 30 (2020) 1905687.

[5] H. Wang, S. Gao, F. Zhang, F. Meng, Z. Guo, R. Cao, Y. Zeng, J. Zhao, S. Chen, H. Hu, Y.J. Zeng, Repression of interlayer recombination by graphene generates a sensitive nanostructured 2D vdW heterostructure based photodetector, Advanced Science 8 (2021) 2100503.

[6] X. Dong, T. Chen, G. Liu, L. Xie, G. Zhou, M. Long, Multifunctional 2D g-C4N3/MoS2 vdW heterostructure-based nanodevices: spin filtering and gas sensing properties, ACS Sensors 7 (2022) 3450-3460.

[7] J. Zha, S. Shi, A. Chaturvedi, H. Huang, P. Yang, Y. Yao, S. Li, Y. Xia, Z. Zhang, W. Wang, H. Wang, Electronic/optoelectronic memory device enabled by tellurium-based 2D van der Waals heterostructure for in-sensor reservoir computing at the optical communication band, Advanced Materials 35 (2023) 2211598.

A W-Band 2.1 dB Insertion Loss Passive Attenuator with Magnetically Switchable Double-Layer Coupled-Lines in 65-nm Bulk CMOS

Zhuang Miao[^], Guiyue Mao[^], Nengxu Zhu, Fanyi Meng*, Kiat Seng Yeo
1.School of Microelectronics, Tianjin University, Tianjin, China
2.Tianjin Key Laboratory of Imaging and Sensing Microelectronic Technology, Tianjin, China
3.Key Laboratory of Organic Integrated Circuit, Ministry of Education, Tianjin University, Tianjin, China
*Email: mengfanyi@tiu.edu.cn

Abstract-This paper proposes a novel passive attenuator with magnetically switchable double-layer coupled-lines in 65-nm Bulk CMOS, which is suitable for millimeter-wave/terahertz applications. Compared to the traditional attenuator topology where the insertion loss (IL) of reference state is seriously affected by transistor on-resistance R_{on}, the proposed attenuator utilizes R_{on} as part of attenuator network to avoid its contribution on IL. Specially, thanks to the double-layer structure of the magnetically coupled-line, the designed attenuator can achieve higher attenuation on the same chip area. Finally, the measurement results shows that the proposed attenuator achieves a small insertion loss of 2.1-3.5 dB, a 0.14-0.5 dB RMS amplitude error and a return loss better than 10 dB from 85 to 105 GHz. The circuit only occupies a 0.066 mm² core chip area and consumes zero DC power.

Keywords-magnetically switchable double-layer coupled-lines, passive attenuator, W-band, millimeter-wave

I. INTRODUCTION

Passive attenuator is an important amplitude control module for millimeter-wave (mmW)/terahertz (THz) phased array transceivers, which can be applied to 5G/6G mmW base stations and terminals, satellite communication, phased array radar, and radio astronomy observations [1-2]. Compared to variable gain amplifier (VGA), it has the advantage of zero power consumption, wider operation bandwidth and higher power handing capability. In order to improve the link gain, sensitivity, and dynamic range of phased array transceiver systems, it is necessary to reduce the insertion loss of passive attenuators. However, due to the limitations of traditional attenuator topologies, the parasitic effects of transistors rapidly increase the insertion loss of attenuators, making it difficult for CMOS/SiGe attenuators to be applied in the mmW/THz systems.

Fig. 1 shows the topology of several traditional attenuators, such as π-type, T-type, bridge T-type and distributed attenuators [1-6]. For the first three types of attenuators, their common feature is the presence of switch transistors in the signal path, which makes the insertion loss will be affected by the switching characteristics of the transistors, i.e. $R_{ON} \times C_{OFF}$, especially in the mmW/THz frequency range. On the other hand, distributed attenuator is composed of transmission lines and switch transistors [3-6], which make it features a wider operating bandwidth, but also results in higher insertion loss as the frequency increases. Obviously, the above attenuator topology

[^]: equal contribution

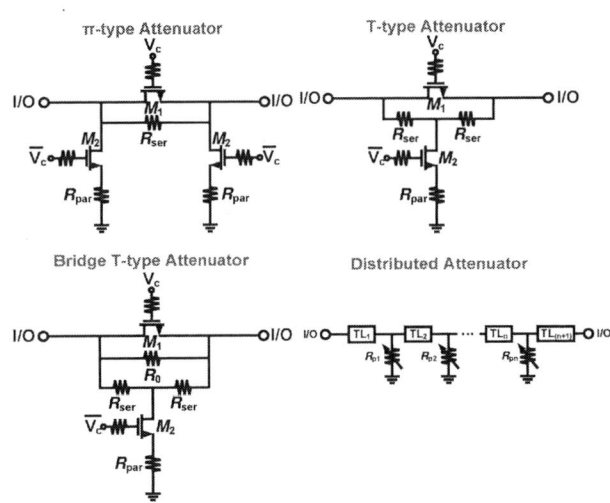

Fig. 1. The π-type, T-type, bridge T-type and distributed attenuators.

is difficult to meet the application requirements at mmW frequencies.

This paper proposes a novel passive attenuator with magnetically switchable double-layer coupled-lines (MSCL) in 65-nm Bulk CMOS, which cleverly places the switch transistors outside the signal path and makes R_{on} part of the attenuation network so as not to affect insertion loss. Meanwhile, the special double-layer structure of the coupled-line provides a physical basis for achieving higher attenuation compared to [9]. The measurement results show excellent performance from 85 to 105 GHz, with a remarkable insertion loss of 2.1-3.5 dB and 0.14-0.5 dB RMS amplitude error at operation frequencies.

II. PROPOSED ATTENUATOR DESIGN

Fig. 2 shows the schematic of attenuation cell implemented using the proposed attenuator topology. The double-layer coupled-line includes a main line and a double-layer auxiliary line, with coupling coefficient k. At reference state, the switch transistor M_1 is turned off, and its equivalent turn off-capacitance C_{off} prevents signal leakage and maintains a good signal path. In the attenuation state, the switch transistor M_1 is turned on, and its equivalent on-resistance R_{on} provides a low resistance signal to ground path, which can effectively attenuate

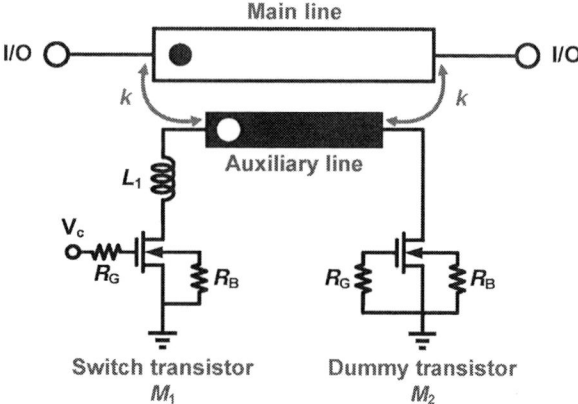

Fig. 2. The schematic of magnetically switchable double-layer coupled-line attenuation cell.

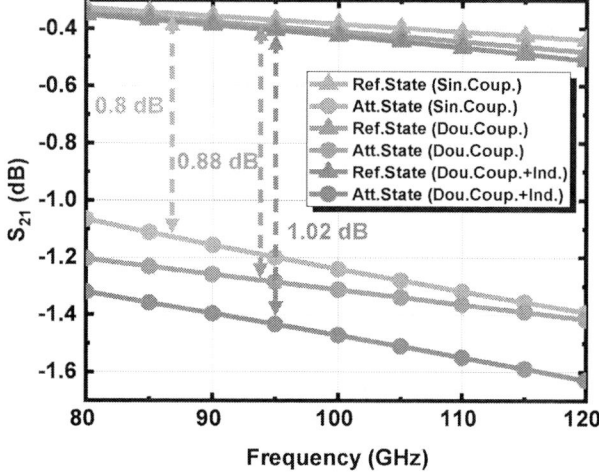

Fig. 3. The improvement of double-layer coupled-line and series inductance L_1 on attenuation (coupled-line length = 150 μm, width = 2 μm, switch transistor W/L = 12×0.6/0.13, dummy transistor W/L = 1×1/0.13, L_1 = 20 pH).

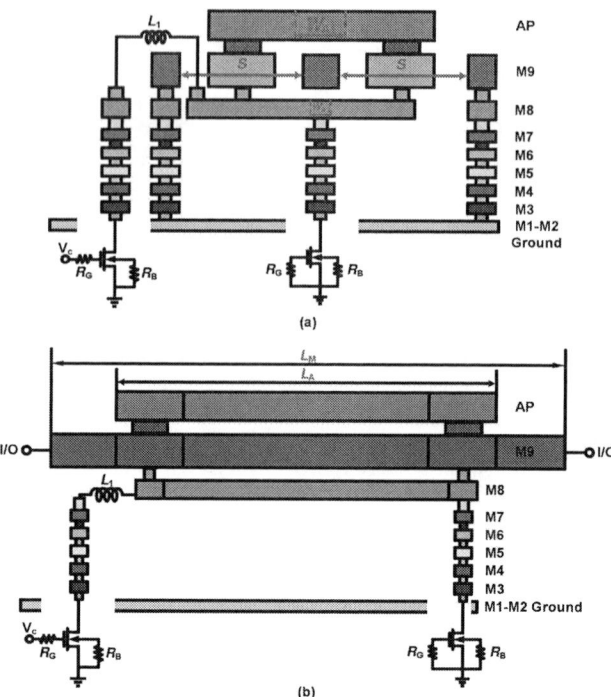

Fig. 4. The magnetically switchable double-layer coupled-line attenuation cell: (a) cross-sectional view (b) side view.

the signal amplitude. The dummy transistor M_2 is always in the off state, which can compensate for the phase error between states using its off capacitance. The series inductor L_1 is connected to the drain terminal of M_1, which can form an LC resonant network with the parasitic capacitance of M_1, thereby further improving the attenuation. The parasitic capacitance of M_1 and M_2 can provide good impedance matching for signal together with double-layer coupled-lines.

The improvement of double-layer coupled-line and the series inductance L_1 on attenuation as shown in Fig. 3. Although the two methods also have a certain impact on the insertion loss of the reference state, it is not obvious, and can significantly improve the attenuation. At 94 GHz, the double-layer structure of the coupled-line and the series inductance L_1 make the attenuation reach 1.02 dB, which is 0.14 dB higher than that without the series inductance L_1, and 0.22 dB higher than that of the single-layer coupled-line, with an increase of more than 25%.

Fig. 4 depicts the cross-sectional view and side view of magnetically switchable double-layer coupled-line attenuation cell. The main line using the top metal layer M9 is the signal path, and the auxiliary line composed of the sub top metal layer M8, aluminum (AP) and their connecting through holes are utilized to couple the radiation signal from the main line in the top and bottom directions, which can effectively enhance the coupling coefficient k of the coupled-line and achieve higher attenuation. One end of the auxiliary line on M8 is connected to the series inductor L_1 on M9, and then connected to the switch transistor through L_1. The other end is connected to the dummy transistor for phase compensation, with metal layers M1 and M2 as ground. The structures shown above all use full-wave electromagnetic (EM) simulation to extract parasitic parameters.

Fig. 5 shows the schematic and design parameters of W-band 5-bit magnetically switchable double-layer coupled-line attenuator. The attenuator consists of five cascaded attenuation cells with 0/0.5 dB, 0/1 dB, 0/2 dB, 0/4 dB, and 0/8 dB attenuations, with the attenuation step of 0.5 dB and attenuation range of 15.5 dB. Due to the small attenuation, the 0/0.5 dB attenuation unit doesn't use the series inductance L_1. The 0/8 dB attenuation cell is achieved by cascading two 0/4 dB attenuation cells. To obtain better port matching, attenuation cells with 0/0.5 dB and 0/1 dB are located at both ends of the entire attenuator.

III. RESULTS AND DISCUSSION

The proposed W-Band 5-bit passive attenuator is fabricated in 65-nm Bulk CMOS technology. The final chip occupies an area of 1×0.35 mm² (with pads), as shown in Fig. 6. The core chip area without pads is only 0.066 mm².

The S-parameters are measured using the R&S ZNA67 network analyzer and frequency extender modules from 75 to 110 GHz.

979-8-3315-2209-4/25 $31.00 © 2025 IEEE

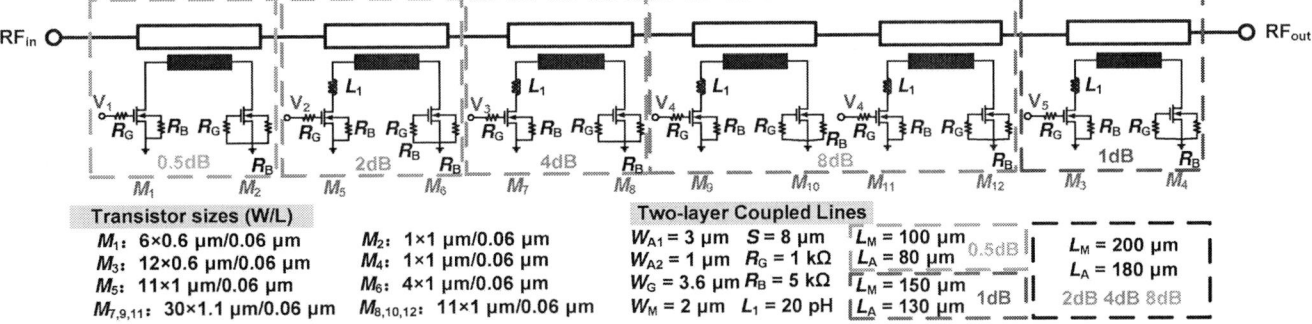

Fig. 5. The schematic and design parameters of W-band 5-bit magnetically switchable double-layer coupled-line attenuator.

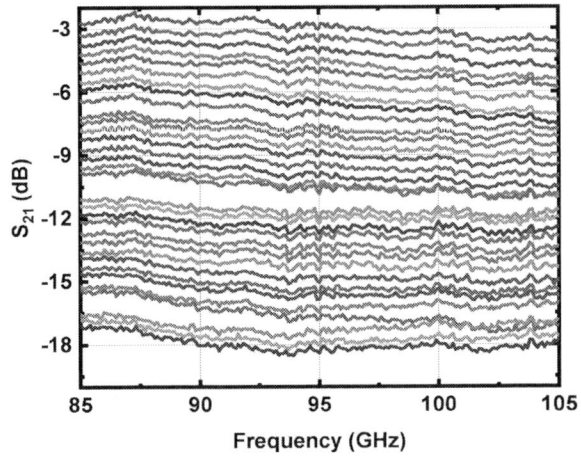

Fig. 6. The chip micrograph of W-band 5-bit passive attenuator.

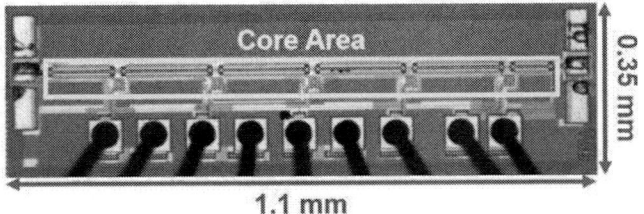

Fig. 7. The measured S_{21} curve in 32 operation states.

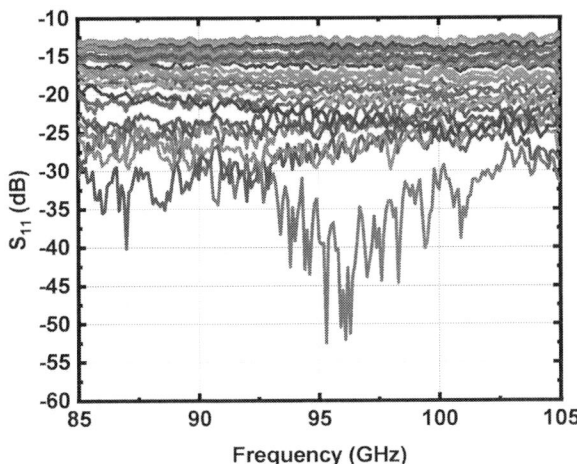

Fig. 8. The measured S_{11} curve in 32 operation states.

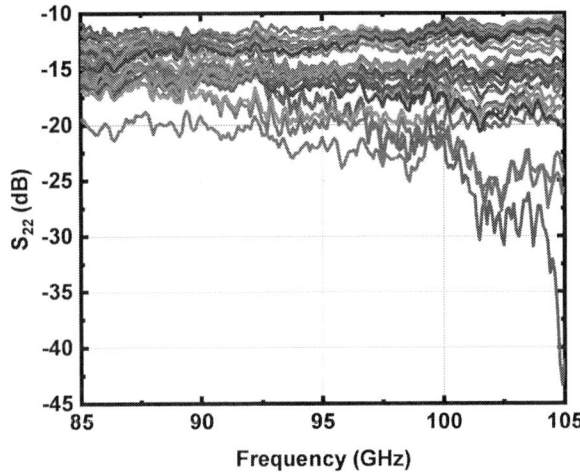

Fig. 9. The measured S_{22} curve in 32 operation states.

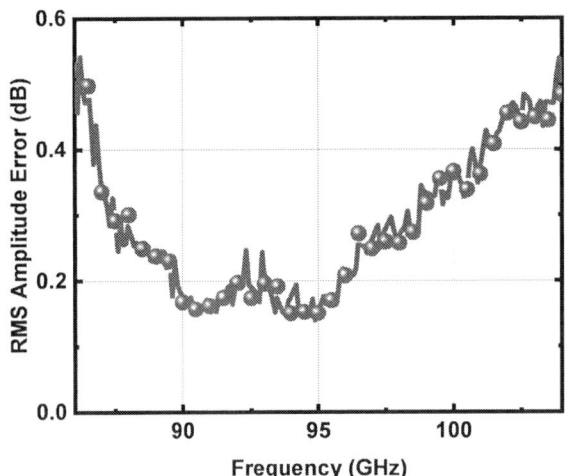

Fig. 10. The measured RMS amplitude error.

The final measurement results are shown in Fig. 7-Fig. 10. Fig.7 illustrates the S_{21} curve of proposed attenuator, which indicate it features 32 operation states with 5-bit digital control and 0.5 dB step-size. The IL of attenuator is only 2.1-3.5 dB from 85 to 105 GHz. The proposed attenuator achieves an input/output return loss better than 10 dB within the operation bandwidth, which is shown in Fig. 8 and Fig. 9. The calculated RMS amplitude error based on the results in Fig. 7 is shown in

979-8-3315-2209-4/25 $31.00 © 2025 IEEE

TABLE I PERFORMANCE SUMMARY AND COMPARISON WITH OTHER STATE-OF-THE-ART PASSIVE ATTENUATORS

Ref.	Tech.	Topology	Frequency (GHz)	Att. Range (dB)/ Step-size (dB)	Insertion Loss (dB)	RMS Amp. Error (dB)	Core Area (mm²)
2014 MWCL [3]	65-nm CMOS	Distributed	50-110	0.75/10	5.6-11.2	-	0.38
2020 TMTT [4]	65-nm CMOS	Distributed	10-50	1/14	2.6-6.2	0.3-2.2	0.19
2022 MWCL [6]	0.18-μm BiCMOS	Distributed	60-100	1/14.5	<3.5	0.7	0.201
2022 TCASII [7]	0.13-μm BiCMOS	π/T-type	DC-67	0.5/31.5	7.8	0.4	0.144
2023 MWTL [8]	65-nm CMOS	T-type	85-105	1/7	>10*	0.6	0.082
Thiis Work	**65-nm CMOS**	**MSCL**	**85-105**	**0.5/15.5**	**2.1-3.5**	**0.14-0.5**	**0.066**

*Estimated results from measurement figure

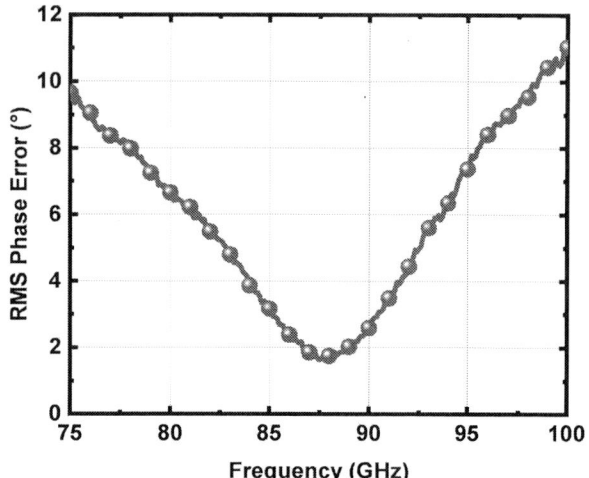

Fig. 11. The measured RMS phase error.

Fig. 10, where the minimum error reaches 0.14 dB and the error is less than 0.5 dB within the bandwidth. Finally, the measured RMS phase error is shown in Fig. 11, where the minimum error is less than 2° and the error is less than 11° from 75 to 100 GHz.

Table I benchmarks the performance of the proposed design and compares with other passive attenuators. Thanks to proposed topology, the proposed attenuator achieves the lowest IL and smaller core chip area compared to other W-band passive attenuators [3][6][8].

IV. CONCLUSION

This paper proposes a novel passive attenuator with magnetically switchable double-layer coupled-lines, which avoid the contribution of switch transistors to IL in conventional topologies, thereby greatly reducing IL. The proposed topology is analyzed and verified by EM simulation. According to the measurement results, it features a IL of 2.1-3.5 dB and RMS amplitude error of 0.14-0.5 dB from 85 to 105 GHz, with in-band return loss better than 10 dB. The proposed topology can provide a feasible solution for the amplitude control module in mmW/THz phased array transceiver systems.

ACKNOWLEDGMENT

This work was supported by the National Key Research and Development Program of China under Project 2023YFB4403200, the National Natural Science Foundation of China under Project 62271347 and U21A2045.

REFERENCES

[1] I. Song, M.-K. Cho, and J. D. Cressler, "Design and analysis of a low loss, wideband digital step attenuator with minimized amplitude and phase variations," *IEEE J. Solid-State Circuits*, vol. 53, pp. 2202–2213, Aug. 2018.

[2] T. N. Ross, K. T. Ansari, S. Tiller, and M. Repeta, "A 5-bit, 0.25 dB step variable attenuator at E-band," in *IEEE RFIC Symp.*, pp. 156–159, June 2018.

[3] K. Kim, H.-S. and B.-W. Min, "V-W band CMOS distributed step attenuator with low phase imbalance," *IEEE Microw. Wireless Compon. Lett.*, vol. 24, pp. 548–550, Aug. 2014.

[4] K. Park, S. Lee, and S. Jeon, "A new compact CMOS distributed digital attenuator," *IEEE Trans. Microw. Theory Techn.*, vol. 68, pp. 4631–4640, Nov. 2020.

[5] B. Suh, and B.-W. Min, "A 20-36-GHz voltage-controlled analog distributed attenuator with a wide attenuation range and low phase imbalance," *IEEE Trans. Microw. Theory Techn.*, vol. 69, pp. 2485 2493, May 2021.

[6] S. G. Rao, C. D. Cheon and J. D. Cressler, "A Millimeter-Wave, Transformer-Based, SiGe Distributed Attenuator," *IEEE Microw. Wireless Compon. Lett.*, vol. 32, no. 2, pp. 145-148, Feb. 2022.

[7] C. D. Cheon, S. G. Rao, W. Lim et al. "Design Methodology for a Wideband, Low Insertion Loss, Digital Step Attenuator in SiGe BiCMOS Technology," *IEEE Trans. Circuits Syst. II, Exp. Briefs*, vol. 69, no. 3, pp. 744-748, Mar. 2022.

[8] Q. Zhang et al., "Mechanism Analysis and Design of a Switched T-Type Attenuator With Capacitive Phase Compensation Technique," *IEEE Microw. Wireless Technol. Lett.*, vol. 33, no. 10, pp. 1438-1441, Oct. 2023.

[9] F. Meng and N. Zhu, "An MSCL-Based Attenuator With Ultralow Insertion Loss and Intrinsic ESD-Protection for Millimeter-Wave and Terahertz Applications," *IEEE Trans. Microw. Theory Techn.*, vol. 71, no. 1, pp. 240-249, Jan. 2023.

A 100MHz Heterogeneous GaN/Si-CMOS 12 V-to-24 V Boost Converter with Soft-Switching Technique

Qingsong Zhao[1,2,3], Zenglong Zhao[1,2,3] Fanyi Meng[1,2,3,*]

[1] School of Microelectronics, Tianjin University, Tianjin, China
[2] Tianjin Key Laboratory of Imaging and Sensing Microelectronic Technology, Tianjin, China
[3] Key Laboratory of Organic Integrated Circuit, Ministry of Education, Tianjin University, Tianjin, China
* Email: mengfanyi@tju.edu.cn

Abstract—This article presents a 100 MHz single-switch boost converter with heterogeneous integration of gallium nitride (GaN) and Si-CMOS technologies. GaN devices are used as power switches and freewheeling diodes, which are integrated with silicon-based Si-CMOS circuits. A soft-switching technique is proposed as an alternative to traditional resonant converters, leveraging the intrinsic parasitic capacitance of GaN devices to replace traditional off-chip capacitors for soft switching, thereby significantly improving efficiency and system integration. In addition, a closed-loop control circuit is designed in this work to ensure a stable output voltage under varying loads. Simulations show that the proposed boost converter achieves a peak efficiency of 89.7% at an output power of 12 W, operating at a 100 MHz frequency with a 12 V to 24 V voltage conversion ratio. The maximum output power can reach up to 15.94 W, while occupying an area of only 3.9×1.9 mm².

Index Terms—Boost converter, gallium nitride (GaN), heterogeneous integration (HI), soft-switching technique.

I. INTRODUCTION

In recent years, with the advancement of millimetre-wave (mmWave) and terahertz (THz) systems, the demand for higher switching frequencies and power densities in power converters for systems power supply has increased significantly. The size of a power converter is primarily determined by its passive components, which can be minimized by increasing the switching frequency, thereby enhancing power density. However, higher switching frequencies exacerbate parasitic effects, limiting the overall efficiency of the converter. Third-generation wide-bandgap semiconductors, such as gallium nitride (GaN), have been widely adopted in power converter designs due to their superior electrical characteristics. GaN high-electron-mobility transistor (HEMT) devices, with their lower on-state resistance and high electron mobility, have become a preferred alternative to silicon-based power switches, offering significant performance advantages [1], [2]. Nevertheless, the significant footprint of discrete GaN devices conflicts with the ongoing trend toward highly integrated power converters. As GaN devices play an increasingly critical role in power conversion, heterogeneous integration technology has emerged as a key solution for integrating GaN and silicon-based chips while minimizing chip area [3], [4], thereby enabling higher integration.

In addition to device selection, circuit topology plays a crucial role in determining the performance of a power converter. Traditional boost converters utilize power diodes as freewheeling components. However, hard-switching operation results in significant power losses, which escalate with increasing switching frequencies. Consequently, many studies [5] have explored the use of synchronous rectification with two power transistors to enhance efficiency. However, synchronous rectification requires additional bootstrap diodes and capacitors, increasing the overall circuit footprint. To address these challenges, some studies [6]–[8] have investigated the adoption of the E-class converter topology, which achieves zero-voltage switching (ZVS) prior to turn-on, thereby improving conversion efficiency. However, this approach necessitates numerous additional passive components, resulting in a substantial increase in circuit area.

To achieve high switching frequencies while reducing the passive component count and enhancing integration, this paper proposes a 100 MHz, 12–24 V boost converter designed for mmWave and THz systems power supply. The control and driver circuits are implemented using silicon-based CMOS technology, featuring a closed-loop feedback mechanism to regulate output voltage across varying load conditions, while GaN devices serve as power switches and freewheeling diodes. To enhance efficiency and integration, a soft-switching technique is introduced, eliminating the need for additional off-chip capacitors.

II. CIRCUIT DESIGN AND IMPLENTATION

A. Soft-Switching Technique

In traditional boost converters, especially under light load conditions, the switching losses caused by hard switching account for the majority of the total losses, and these losses increases as the switching frequency is raised. Consequently, many reported designs [6], [7] employ L-C resonant topologies to achieve soft switching. Fig. 1 shows the schematic of a Class-E DC-DC converter. With reasonably optimized design, the resonant frequency of the resonant network can be set slightly below the switching frequency, thereby achieving zero-voltage switching.

979-8-3315-2209-4/25 $31.00 © 2025 IEEE

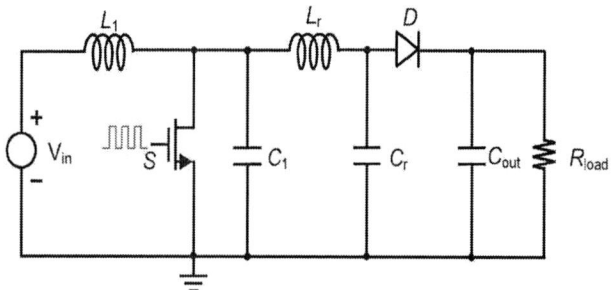

Fig. 1. Schematic of Class-E DC-DC boost converter.

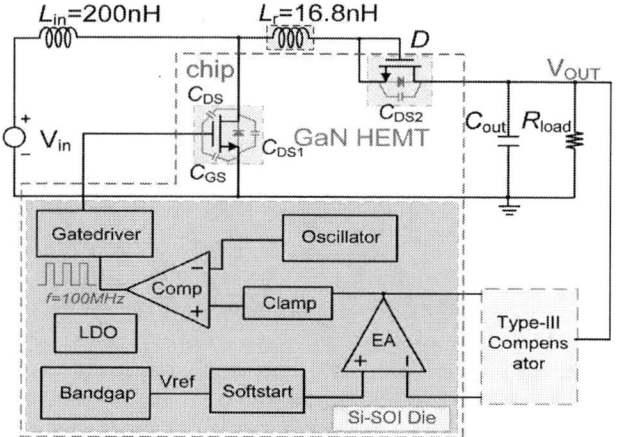

Fig. 2. Schematic diagram of the proposed boost converter.

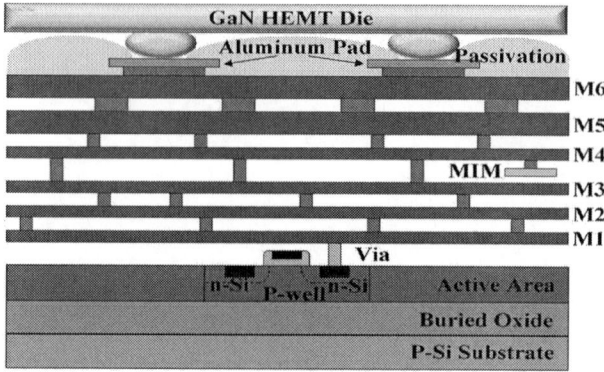

Fig. 3. Cross-sectional view of this work using SMIC 0.13um CMOS SOI process.

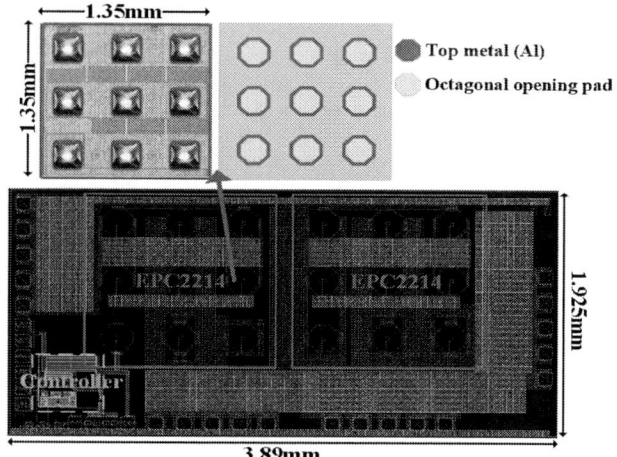

Fig. 4. Chip layout of the proposed boost converter.

To reduce the use of off-chip capacitors and improve integration, this paper proposes a soft-switching technique. Fig. 2 presents the overall schematic diagram of the proposed converter, where asynchronous rectification technology is employed, and the high-side GaN power switch acts as a diode by shorting the gate to the source, thereby reducing the additional area consumption associated with the bootstrap diode and bootstrap capacitor in synchronous control methods.

As shown in Fig. 2, the parasitic capacitance of the GaN HEMT is demonstrated, which is used to form a resonant network with the additional inductance L_r in this work. When the GaN switch is turned off, under an appropriately designed resonant frequency, this resonant network can extract charge from the parasitic capacitance, thereby enabling soft switching without the need for the additional capacitors required in traditional Class-E converters. The values of the inductor L_r and capacitor C_r are crucial in achieving soft switching, as they determine the voltage gain and can be chosen as follows:

$$\frac{V_{\text{OUT}}}{V_{\text{in}}} = \frac{8}{\pi^2} \cdot \frac{1}{\sqrt{\left(1 - \left(2\pi\sqrt{L_r C_{\text{DS2}}}f_s\right)^2\right)^2 + \left(2\pi C_{\text{DS2}}f_s R_{\text{D}}\right)^2}}$$

(1)

where f_s is the switching frequency, and R_D is the effective resistance as seen at the input of diode D. Due to the parasitic effects of the circuit components, unintended inductance and

capacitance may be introduced. Therefore, during the design process, the effects of these parasitics must be accounted for, and L_r and C_{DS2} should be carefully optimized to maximize efficiency.

B. Control Circuit Design and Implentation

To prevent the output voltage of the resonant converter from varying with load changes, a closed-loop control circuit is designed for this work to ensure a stable output voltage and a compensation network is introduced to stabilize the system. The chip generates a 100 MHz high-frequency signal internally to minimize the volume of external passive components. The control and driver modules are designed using silicon-based CMOS technology, and the GaN devices are integrated with control and drive circuits through heterogeneous integration technology, significantly reducing chip area.

This work is realized by heterogeneous integration technology through GLOBALFOUNDRIES GaN2BCD™ technology [3], which connects GaN devices to silicon-based chips vertically through small metal solder balls. Previous publications [4], [8] have demonstrated the feasibility of this technology. Fig. 3 shows a cross-sectional view of the SMIC

979-8-3315-2209-4/25 $31.00 © 2025 IEEE

TABLE I
PERFORMANCE COMPARISON OF RECENTLY REPORTED ON-CHIP DC–DC BOOST CONVERTERS

Design	2020 [10]	2019 [4]	2018 [7]	2023 [8]	This Work
Process	0.13 μm BCD	0.18 μm BCD + GaN	0.18 μm CMOS + GaN + IPD	0.13 μm SOI + GaN	0.13 μm SOI + GaN
Switching Frequency	1.5 MHz	50 kHz	300 MHz	500 MHz	100 MHz
$V_{\text{IN}}/V_{\text{OUT}}$	3.3 V / 12 V	3.3 V / 70 V	12 V / 18 V	6 V / 14 V-20 V	12 V / 24 V
Peak Efficiency	90.3 %	70.3 %	47.3 %	58 %	89.7 %
Max Out Power	5.25 W	1.68 W	4.16 W	4.19 W	15.94 W
Area	2.4 mm × 0.93 mm[*]	3.2 mm × 1.8 mm[*]	9.4 mm × 9.8 mm	3 mm × 3 mm	3.9 mm × 1.9 mm[*]

*The chip area does not include the inductors and filtering capacitors.

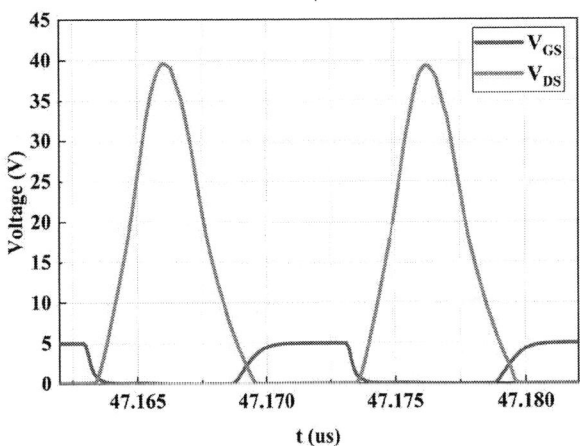

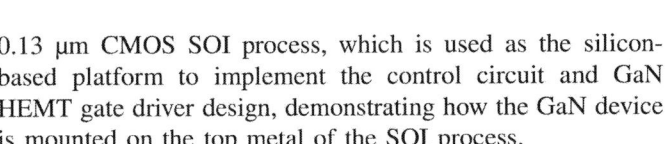

Fig. 5. The simulated V_{DS} and V_{GS} waveforms with soft-switching technique.

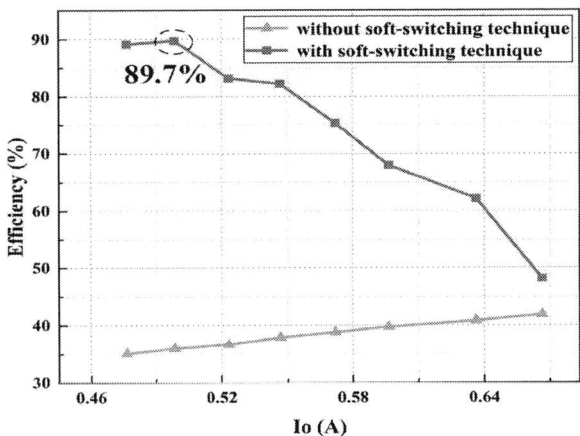

Fig. 6. The simulated efficiency versus output current.

0.13 μm CMOS SOI process, which is used as the silicon-based platform to implement the control circuit and GaN HEMT gate driver design, demonstrating how the GaN device is mounted on the top metal of the SOI process.

III. SIMULATION RESULT

Fig. 4 shows the overall chip layout of the proposed boost converter, with a total chip area of 3.9 × 1.9 mm². It also includes a micrograph of the 80-V commercial eGaN HEMT (EPC2214) [9] from Efficient Power Conversion (EPC) Corporation, along with the design of a regular octagonal pad in the SOI process for flip-chip mounting of the EPC2214. The chip's integration level is improved through heterogeneous integration technology. At a high switching frequency of 100 MHz, meanwhile, only two inductors—200 nH and 16.8 nH—are required as L_{in} and L_r, respectively, significantly reducing the size of the passive components.

Fig. 5 shows the gate drive and drain voltage waveforms of the GaN power switch. Thanks to the aid of soft-switching technology, the V_{DS} rapidly decreases before the switching transistor turns on, enabling the transistor to operate close to Zero-Voltage Switching (ZVS) and significantly reducing switching losses. Fig. 6 presents the conversion efficiency of the proposed power converter under different load conditions.

It is evident that under a 12 V voltage input, the chip can achieve a stable 24 V output within a certain load range. Even at a switching frequency as high as 100 MHz, the conversion efficiency can reach up to 89.7%. However, it is important to note that when the load deviates from the nominal load, the resonant frequency of the circuit may shift away from the switching frequency, resulting in the loss of ZVS operation. As a result, the conversion efficiency significantly decreases. Table I compares the proposed design with other boost converters. In comparison with previous designs, this design maintains high efficiency at high switching frequencies, delivering an output power of up to 15.94 W with an area of 3.9×1.9mm².

IV. CONCLUSION

In this work, a 100 MHz BOOST converter for converting 12 V to 24 V is designed and simulated. GaN HEMT devices are used as power switches and diodes, integrated with a gate driver and control circuitry, which are designed using a silicon-based SOI process and are integrated into a single chip using heterogeneous integration technology. Additionally, to reduce switching losses and improve integration, this work proposes a soft-switching technique, utilizing the parasitic capacitance of GaN devices to achieve soft switching without

979-8-3315-2209-4/25 $31.00 © 2025 IEEE

the need for additional off-chip capacitors. Even at a high switching frequency of 100 MHz, the converter achieves a peak efficiency of 89.7% with an output power of 12 W. The maximum output power reaches up to 15.94 W, and the chip area is only 3.9×1.9 mm^2.

ACKNOWLEDGMENT

This work was supported by the National Key Research and Development Program of China under Project 2023YFB4403200, the National Natural Science Foundation of China under Project 62271347 and U21A2045.

REFERENCES

[1] J. Das et al., "A 96% Efficient High-Frequency DC–DC Converter Using E-Mode GaN DHFETs on Si," *IEEE Electron Device Lett.*, vol. 32, no. 10, pp. 1370-1372, Oct. 2011.

[2] J. Choi, D. Tsukiyama, Y. Tsuruda and J. M. R. Davila, "High-Frequency, High-Power Resonant Inverter With eGaN FET for Wireless Power Transfer," in *IEEE Trans. Power Electron.*, vol. 33, no. 3, pp. 1890-1896, March 2018.

[3] D. Disney, F. Meng, X. Yi, and C. C. Boon, "Integrated DC–DC boost converter with gallium nitride power transistor," US Patent 15/648,105, Jul. 2017.

[4] F. Meng et al., "Heterogeneous Integration of GaN and BCD Technologies and Its Applications to High Conversion-Ratio DC–DC Boost Converter IC," *IEEE Trans. Power Electron.*, vol. 34, no. 3, pp. 1993-1996, March 2019.

[5] X. Ke, J. Sankman, Y. Chen, L. He and D. B. Ma, "A Tri-Slope Gate Driving GaN DC–DC Converter With Spurious Noise Compression and Ringing Suppression for Automotive Applications," *IEEE J. Solid-State Circuits.*, vol. 53, no. 1, pp. 247-260, Jan. 2018

[6] P. Choi, U. Radhakrishna, C. Boon, D. Antoniadis, and L. Peh, "A fully integrated inductor-based GaN boost converter with self-generated switching signal for vehicular applications," *IEEE Trans. Power Electron.*, vol. 31, no. 8, pp. 5365–5368, Aug. 2016.

[7] M. -J. Liu and S. S. H. Hsu, "A Miniature 300-MHz Resonant DC–DC Converter With GaN and CMOS Integrated in IPD Technology," *IEEE Trans. Power Electron.*, vol. 33, no. 11, pp. 9656-9668, Nov. 2018.

[8] Z. Liu, Z. Lin, J. Wang, K. Ma, D. Disney and F. Meng, "A Fully Integrated Heterogenous Si-CMOS/GaN 500 MHz 6 V-to-18 V Boost Converter Chip," *IEEE Trans. Power Electron.*, vol. 38, no. 5, pp. 5615-5618, May 2023.

[9] EPC, "EPC2214 datasheet," May. 2019. Accessed: Jan. 2025. [Online]. Available: https://epc-co.com/epc/Portals/0/epc/documents/datasheets/EPC2214_datasheet.pdf.

[10] D. Yun, H. Kim, D. Baek, S. Cho, J. Yoon and J. Lee, "A Fixed-Frequency Synchronous Boost Converter Based on Adaptive On-Time Control with a New Reverse Phase Ripple Injection Compensation," in *IEEE Energy Convers. Congr. Expo.*, Detroit, MI, USA, 2020, pp. 2244-2250.

A 2-18GHz High Gain Power Amplifier Using Dual Current-Reuse Topology with ±0.3dB Gain Ripple

Jianbing Liu, Fanyi Meng*

1.School of Microelectronics, Tianjin University, Tianjin, China
2.Tianjin Key Laboratory of Imaging and Sensing Microelectronic Technology, Tianjin, China
3.Key Laboratory of Organic Integrated Circuit, Ministry of Education, Tianjin University, Tianjin, China
*Email: mengfanyi@tju.edu.cn

Abstract—An ultra-wideband high flatness power amplifier (PA) for 2-18 GHz applications in 0.18-μm SiGe BiCMOS process is presented in this paper. An improved dual current-reuse topology has been proposed and applied to the PA driver stage. This design reduces power loss and provides two additional stages of gain, significantly improving gain flatness. De-Q (decrease the quality factor) resistors are used to trade off some power loss in exchange for maximizing bandwidth and flatness. The proposed PA achieves a gain of 40.5 dB with a gain ripple of less than ±0.3 dB across the entire 2-18 GHz band. The simulated output power at the 1-dB compression point (OP$_{1dB}$) reaches up to 12.5 dBm at 10 GHz. A saturated output power (P$_{sat}$) of 16.1 dBm is obtained with maximum power-added efficiency (PAE) of 20.5% at 8 GHz. The DC power consumption is 137.6 mW. The core area is 0.25 mm^2.

Index Terms—Power Amplifiers, Current Reuse, ultra-wideband, High flatness, SiGe BiCMOS.

I. INTRODUCTION

In recent decades, there has been a significant increase in data traffic for both mobile networks and data centers. PAs are critical components in radar systems and wideband wireless communication systems, significantly impacting the overall system performance, particularly in gain, output power, efficiency, and linearity. An ultra-wideband PA with minimal gain ripple is essential for ensuring signal accuracy and consistency in wireless, wireline, and radar applications [1]. Distributed amplifiers can achieve excellent bandwidth and flatness performance, but they have many disadvantages such as large area, low output efficiency, and complex design [2]. Previously published power amplifiers often employ cascode or multi-transistor stacked topologies in the output stage. These topologies offer high stable gain, excellent reverse isolation, and improved output power performance. However, in driver-stage circuits, the breakdown voltage constraint necessitates the use of a stacked structure, as the power supply is shared with the output stage. This design results in unnecessary current loss [3]. To mitigate this, a novel dual current-reuse topology is proposed. In this topology, the current from the third stage is split into two components: a large current flowing into the second stage and a smaller current flowing into the first stage. Additionally, due to resonance effects, the inductance value of an inductor varies significantly across the 2–18 GHz range.

To mitigate this, a De-Q resistor is employed to suppress the resonance of reactive components, thereby reducing

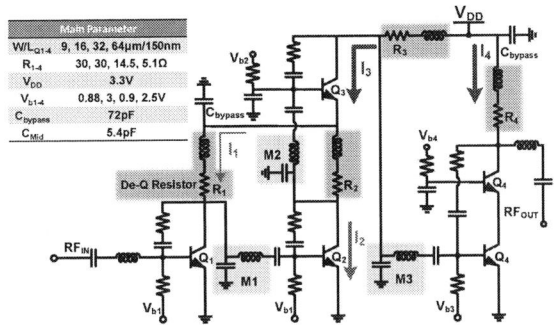

Fig. 1. Schematic of the proposed ultra-wideband high flatness power amplifier.

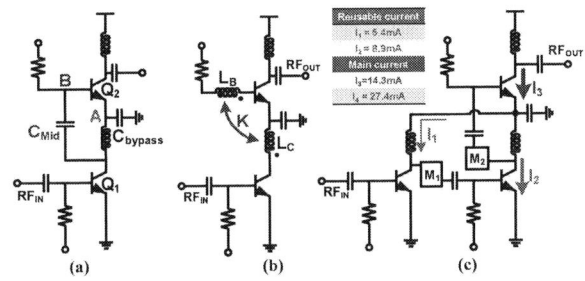

Fig. 2. (a) Classic capacitive-coupling. (b) Classic transformer-coupling. (c) Proposed dual current-reuse topology.

sharpness of impedance resonance and enhancing the overall performance of the circuit. The introduction of the De-Q resistor significantly improves bandwidth and flatness, with only a slight reduction in power efficiency.

II. CIRCUIT DESIGN

A. Schematic of PA

The proposed PA is constructed utilizing HHNEC SiGe 0.18-μm BiCMOS process. Fig. 1 illustrates the complete circuit diagram of the designed PA. A dual current-reuse topology is employed to split the single-stage cascode into a three-stage cascade. This structure provides two additional gain stages and poles, thereby improving efficiency and broadening the bandwidth while maintaining the drive stage current. The output stage utilizes a cascode structure to achieve a large, stable output impedance, resulting in high broadband gain and

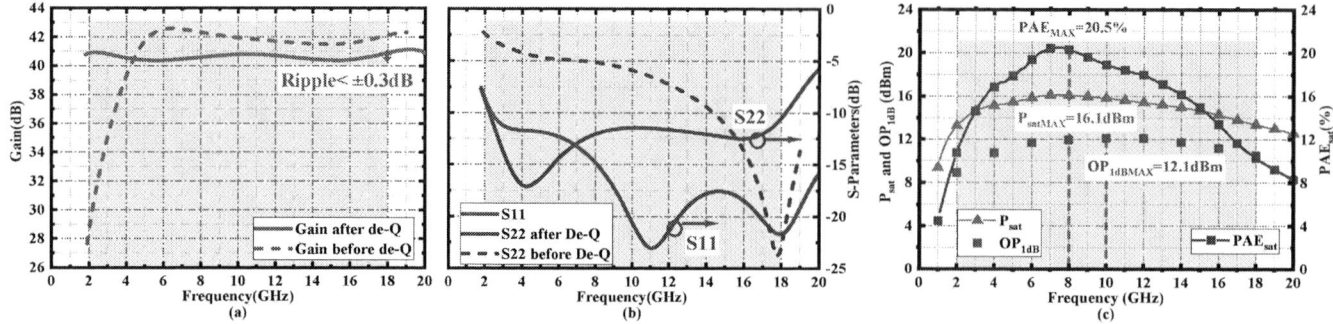

Fig. 3. (a) Simulated gain, (b) Small-signal S-parameters, (c) P_{sat}, OP_{1dB} and PAE versus frequency.

increased output power. De-Q resistors (R_1-R_4) are introduced to expand bandwidth and reduce gain ripple.

B. Dual Current Reuse Topology

Fig. 2 (a) (b) presents the topologies of classic current-reuse techniques [4]. In both structures, the inductor L_C is selected to be sufficiently large to block the RF signal within the desired frequency band. The bypass capacitor C_{bypass} is also chosen to be large enough to provide effective AC grounding at point A. The RF signal, after amplification by Q_1, is then directly coupled to point B either through C_{Mid} or by inductive coupling. The DC path behaves like a cascode amplifier, while the RF path forms a two-stage cascaded common-emitter (CE) amplifier, with both stages sharing the same supply current.

Compared to the traditional two-stage cascaded CE structure, the current-reuse topology provides better isolation and higher gain due to its increased output impedance. This structure has been widely adopted in low-noise amplifier (LNA) designs to achieve higher gain and a lower noise figure [3]. However, in PAs, the shared current introduces additional power loss in the driver stage, which also leads to degraded linearity. In this paper, an innovation to the aforementioned current-reuse topology is proposed, as shown in Fig. 2 (c).

For the DC path, the driving stage operates at a supply voltage of 3.3 V with a constant drive current I_3. For the RF path, a three-stage cascaded structure is implemented, where the current increases progressively at each transistor stage, with the condition $I_3 = I_1 + I_2$. This design minimizes power loss in the driver stage while providing additional gain. Furthermore, two inter-stage matching networks, M_1 and M_2, are introduced to adjust the high-frequency gain, thereby broadening the bandwidth and improving flatness.

C. Choke Inductance with De-Q resistor

The term "De-Q" refers to the process of decreasing the quality factor of a reactive component, such as an inductor or capacitor. This effectively broadens its bandwidth and reduces the sharpness of its resonant frequency. A De-Q resistor is typically placed in parallel with an inductor or integrated into a matching network. The resistor helps dampen the resonance of the reactive component, thereby reducing the sharpness of

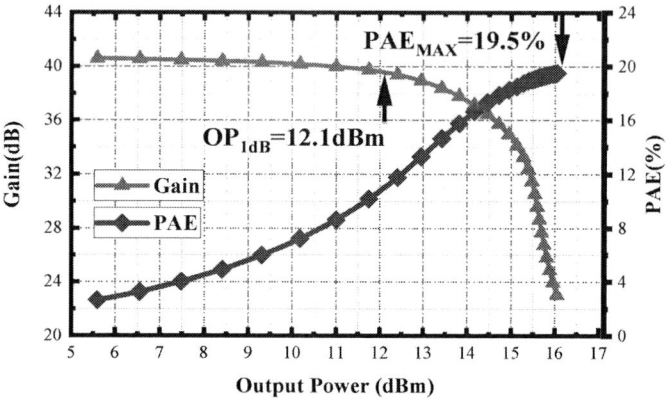

Fig. 4. Simulated large signal performance of the PA at 10 GHz.

impedance resonance and improving the overall performance of the circuit.

For high-frequency gain above approximately 8 GHz, the inter-stage matching networks M_1, M_2, and M_3 can be adjusted to flatten the gain. However, this adjustment has minimal effect in the low-frequency range especially at 2–4 GHz. To maintain the low-frequency gain, a choke inductor with De-Q resistors (R_1–R_4) is introduced. The DC power consumption associated with the de-Q resistors results in a reduction in efficiency [1]. Taking R_3 and R_4 as examples, they will contribute to a power loss of $P_{R3-4} = 6.74mW$, As a result, the output power decreases by approximately $P_{loss} = 1.3$ dBm and the efficiency drops by $PAE_{loss} = 0.98\%$ Although some efficiency and power are sacrificed, the De-Q resistor provides a more stable and flat gain profile. It also helps maintain better impedance matching by reducing sharp impedance mismatches and resonance peaks, thereby minimizing reflection losses and improving the output return loss (S22) across the operating bandwidth. As shown in Fig. 3 (a) and (b), the solid line represents the gain and S-parameters after decreasing the quality factor of the inductance, while the dashed line shows the gain and S22 without the addition of the De-Q resistor R_{3-4}. It is evident that the gain bandwidth is extended to 2 GHz, and the output return loss is significantly improved. This trade-off is fully acceptable.

TABLE I
PERFORMANCE COMPARISON WITH SIMILAR PAS

Ref.	ESSERC' 24 [1]	MWCl' 17 [2]	TMTT' 24 [3]	MWCL' 19 [5]	TMTT' 17 [6]	APMC' 23 [7]	This work
Process	28nm CMOS	0.18-μm CMOS	0.13-μm BiCMOS	65nm CMOS	0.13-μm CMOS	65nm CMOS	**0.18-μm BiCMOS**
Topology	De-Q.R$^{\$}$ / Inter-stage Inductance	4-stacked / DA$^{\#}$ / ATL$^{\&}$	Serial–parallel current-reuse	Cascode Transformer-coupled	4-stacked FET uniform DA$^{\#}$	3-stacked T-type network	**De-Q.R$^{\$}$ / Current-reuse**
Freq (GHz)	0.1-19.7	1-23.8	22.7-30.7	8-11.4	2-16	2-20	**2-18**
Gain (dB)	15.8-16.8	9.9-11.9*	28.9-31.9*	21.4-24.4*	9.8-10.4	15.4-17.2	**40.2-40.8**
Ripple (dB)	±0.5	±1*	±1.5*	±1.5*	±0.3	±0.9	**±0.3**
OP$_{1dB}$ (dBm)	4.5-5.6	8.9-14.5	<13.9	13-15.2*	13-15.5	10.5-12	**9-12.1**
P$_{sat}$ (dBm)	9.1-11.6	13-17*	15-16.2*	15-20.5*	15-18.5	13.9-16	**13-16.1**
PAE$_{max}$ (%)	7.7	10	20.5	24.5	17	20.9	**20.5**
Core Area (mm^2)	0.234	1.7	0.56	0.48	0.83	0.49	**0.25**

$^{\$}$De-Q resistor. $^{\#}$Distributed amplifier. $^{\&}$Artificial transmission line *Estimated from plots.

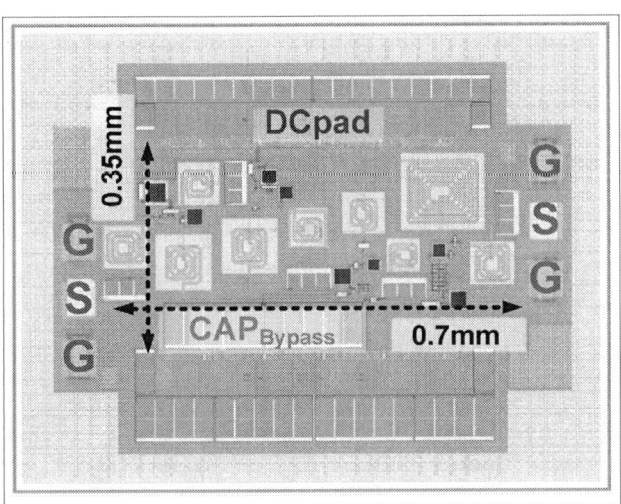

Fig. 5. Layout of proposed PA

III. SIMULATION RESULTS

Fig. 3 and Fig. 4 show the simulated small-signal S-parameters and large-signal performance of the PA across the 2–18 GHz frequency range. The proposed PA demonstrates a gain of 40.5 dB at 10 GHz, with a gain fluctuation of less than ±0.3 across the entire frequency band. The OP$_{1dB}$ reaches a maximum of 12.1 dBm at 10 GHz and remains above 9 dBm across the 2–18 GHz range. At 8 GHz, the simulation results show a maximum P_{sat} of 16.1 dBm with a peak PAE of 20.5%. The DC power consumption is approximately 137.6 mW with a 3.3 V supply. The core area is 0.25 mm^2 (0.7 mm× 0.35 mm) and the layout is shown in Fig. 5. A performance comparison is provided in Table 1. The proposed design simultaneously achieves high gain, a wide operating bandwidth, extremely low gain ripple, and a compact area.

IV. CONCLUSION

This paper presents a 2–18 GHz high-gain power amplifier with a ±0.3 dB gain ripple. A dual current-reuse topology is employed to minimize unnecessary current loss and provide two additional stages of gain and matching networks. De-Q resistors are used to reduce the sharpness of impedance resonance and enhance low-frequency gain. The simulation results demonstrate that these design approaches effectively broaden the bandwidth and improve flatness, while maintaining the desired output power.

REFERENCES

[1] An Sun, Haoqi Qin, Hao Xu, and Na Yan, "A Compact 0.1-19.7GHz Ultra-Wideband Power Amplifier with ±0.5dB Gain Ripple in 28 nm CMOS Process," *2024 IEEE European Solid-State Electronics Research Conference (ESSERC)*, pp. 388–391, 2024.

[2] Y. Zhang and K. Ma, "A 2–22 GHz CMOS distributed power amplifier with combined artificial transmission lines," *IEEE Microwave and Wireless Components Letters*, vol. 27, no. 12, pp. 1122–1124, Dec. 2017.

[3] Qingfeng Zhang, Chenxi Zhao, Wenhao Li, Yiming Yu, Yunqiu Wu, Huihua Liu, Wenquan Che, Quan Xue, and Kai Kang, "A Ka-Band SiGe High-Gain Power Amplifier With Stability–Efficiency–Reliability-Enhanced Serial–Parallel Current-Reuse Technique," *IEEE Transactions on Microwave Theory and Techniques*, vol. 72, no. 3, pp. 1657–1673, 2024.

[4] V. Giammello, E. Ragonese and G. Palmisano, "A Transformer-Coupling Current-Reuse SiGe HBT Power Amplifier for 77-GHz Automotive Radar," *IEEE Transactions on Microwave Theory and Techniques*, vol. 60, no. 6, pp. 1676–1683, June 2012.

[5] Van-Son Trinh, Hyohyun Nam, and Jung-Dong Park, "A 20.5-dBm X-Band Power Amplifier With a 1.2-V Supply in 65-nm CMOS Technology," *IEEE Microwave and Wireless Components Letters*, vol. 29, no. 3, pp. 234–236, 2019.

[6] Mohsin M. Tarar and Renato Negra, "Design and Implementation of Wideband Stacked Distributed Power Amplifier in 0.13-μm CMOS Using Uniform Distributed Topology," *IEEE Transactions on Microwave Theory and Techniques*, vol. 65, no. 12, pp. 5212–5222, 2017.

[7] K. Kumar, S. Kumar, and M. P. Gupta, "A L/S/C/X/Ku-Band Three-Stack, Two stages Fully Integrated CMOS Power Amplifier with 20.9% PAE Using T-Network," in *Proc. Asia-Pacific Microwave Conf. (APMC)*, Taipei, Taiwan, 2023, pp. 219–221.

979-8-3315-2209-4/25 $31.00 © 2025 IEEE

A 12-41 GHz Current-Reuse Composite-Cascode-Stack Power Amplifier in 65-nm CMOS SOI

Guiyue Mao, Xuan Li, Yang Liu, Fanyi Meng*

1 School of Microelectronics, Tianjin University, Tianjin, China
2 Tianjin Key Laboratory of Imaging and Sensing Microelectronic Technology, Tianjin, China
3 Key Laboratory of Organic Integrated Circuit, Ministry of Education, Tianjin University, Tianjin, China
*Email: mengfanyi@tju.edu.cn

Abstract—A two-stage wideband power amplifier in 65-nm CMOS SOI process is proposed. The power stage of the amplifier adopts a triple-stacked topology, and shunt inductors and feedback capacitors are used to improve the stability and power performance at high frequencies. In addition, the driving stage, which consists of a cascode structure with current reuse technique, is designed for high efficiency. To achieve a broadband operation, a *RLC*-structure based on the combination of resistive shunt feedback and parallel LC loads is introduced as an input matching network.The simulation results show that the 3-dB bandwidth of the amplifier covers 11.8-41.2 GHz. And the PA achieves 16.2-dBm saturated output power (*Psat*) , the peak power-added efficiency (PAE) is 18.2% with OP1dB of 13.2 dBm at 28 GHz. The 1-dB bandwidth of P*Psat* is from 18 to 37 GHz, and the relative bandwidth is 64%.

Index Terms—Broadband power amplifier (PA), millimeter wave, stacked, CMOS, current reuse technique

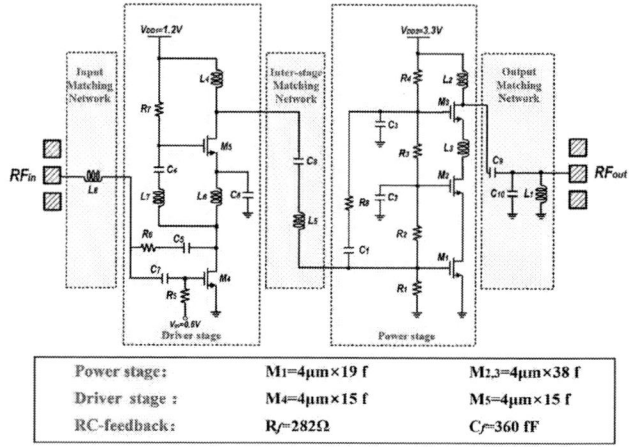

Fig. 1. Circuit schematic of the proposed PA.

I. INTRODUCTION

Ku/K/Ka bands are vital for radar detection, satellite communication, and 5G millimeter-wave systems due to their high resolution, low latency, and wide bandwidth. Phased-array antenna integrated circuits that cover multiple bands enhance transceiver integration and spectrum flexibility, offering cost advantages over multiple narrowband designs. The power amplifier (PA), as the final stage in the transmitter chain, dictates the linear frequency modulation bandwidth of the transmitted signal.

To achieve broadband power amplifiers, distributed structures with wide small-signal gain characteristics are widely used. However, since the bandwidth of the distributed structure is limited by transistor parasitics and inductive components, leading to frequency-dependent mismatches. The gain is also constrained by series configurations, introducing non-negligible signal attenuation. Balanced amplifiers offer high gain and flatness at the expense of single amplifier standing wave ratio, but their complex layout requires extensive orthogonal coupler area. Stacked power amplifiers increase the voltage swing that the transistor can carry under the silicon-based process and to increase the output power, increasing optimal impedance to facilitate broadband output matching.

In this brief , a two-stage power amplifier in 65-nm CMOS SOI is proposed, operating from 12 to 41 GHz, covering Ku/K/Ka bands. By carefully designing the LC matching network, the designed power amplifier achieves broadband characteristics with a small chip area.

II. CIRCUIT DESIGN

The schematic of proposed wideband power amplifier is shown in Fig.1. The power stage utilizes a triple-stacked topology to increase the output power of the PA, while RC negative feedback is employed to improve circuit stability. The high-frequency gain is further expanded by compensating for the shunt inductors. The drive stage employs a current reuse structure to reduce power consumption, thereby increasing PAE. Meanwhile, the input matching network expands the bandwidth by employing RLC technology.

A. Three-stack structure with internal node adjustment

Load-pull analysis is conducted on the three-stack circuit, considering output power and transistor voltage margin. The size of the first-layer transistor was finally determined to be $76\mu m$, while the second and third layers were each $152\mu m$.

To maximize output power, the self-bias network was carefully designed with voltage divider resistors to achieve equal voltage swing division for each stacked transistor. The gate of the stacked FET is connected to the grounded capacitors, which allows reliable transistor operation under large aggregate voltage swings. The tuned values of C2 and C3 were 455 fF and 261 fF, respectively. In the millimeter wave band, the internal parasitic capacitance in transistors can cause drain current oscillations. To mitigate this effect, a shunt inductor was introduced to align the voltage waveform phase by adjusting the imaginary part of the internal node impedance, thereby improving high-frequency gain and gain flatness.

B. Improved Output Matching With RC Feedback

The imaginary part of the optimal output impedance of the stacked PA increases with the number of stacked layers. As a result, it becomes difficult to match the output impedance with the optimal output impedance over a wide bandwidth, leading to significant return loss. In the proposed PA, the feedback loop composed of Cf and Rf is located between the drain of the third stacked transistor and the gate of the first transistor. The feedback capacitor Cf is connected from the PA output to the inputs with opposite phases, so the gate-drain parasitic capacitance Cds of the transistor can be offset by the Miller effect [1]. This corrects Z$_{out}$, which is affected by frequency changes, and reduces the output return loss S22. On the other hand, relatively complex circuit structure of the stacked structure introduces multiple undesired positive feedback loops, which deteriorate the stability of the circuit. The introduced RC feedback loop can effectively improve stability of the circuit.

C. Current Reuse Technique

The drive stage employs current reuse technique to reduce PA power consumption and enhance circuit efficiency while maintaining the same output power. The signal is amplified by the CS transistor and then coupled through a low-impedance path to the CG transistor for further amplification. Additionally, the two transistors have the same size to maintain consistent DC current. Compared to the traditional cascode structure, current reuse technology significantly reduces output power. However, in drive stage design, it is acceptable to sacrifice some output power for higher gain. The drive stage achieves a saturated output power of 12dBm, as shown by

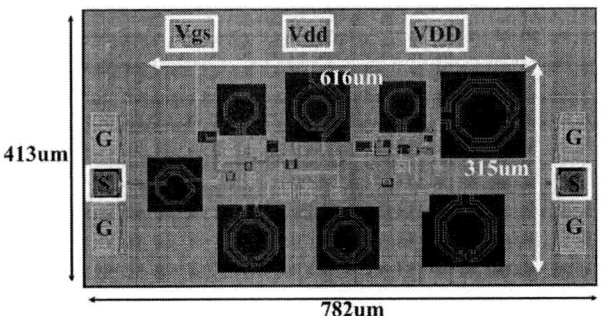

Fig. 3. The layout of the proposed PA.

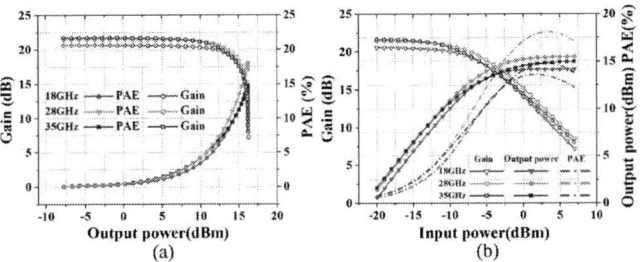

Fig. 4. Simulated and measured results of power gain and PAE vs (a) output power (b) input power at 18/28/35 GHz.

large-signal simulations, providing enough power to drive the power stage PA to saturation.

D. Wideband Input Matching Network with RLC-Branch

The input matching network of the proposed circuit is equivalent to two RLC branches, utilizing resistive shunt-shunt feedback with a parallel LC load. The schematic of the RLC feedback is shown in Fig.2(a) and the small-signal mode is shown in Fig.2(b). By adjusting the value of the passive device, the resonance points of the drive stage will be distributed at the low frequency and high frequency of the operating frequency, so the input return loss S11 will be significantly concave at the two resonance points, as shown in Fig.2(c), thereby broadening the matching bandwidth. In high-frequency matching, the quality factor of passive devices

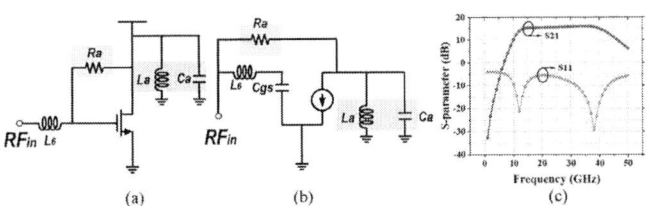

Fig. 2. RLC feedback branch: (a) schematic (b) small-signal model (c) S-parameter simulation results.

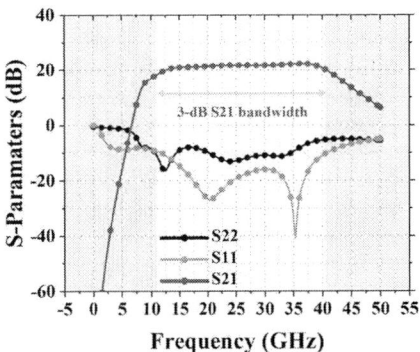

Fig. 5. Simulated S-parameters of the proposed PA.

979-8-3315-2209-4/25 $31.00 © 2025 IEEE   101

TABLE I
PERFORMANCE SUMMARY AND COMPARISON

Reference	MWCL'21 [2]	IMS'20 [3]	TMTT'20 [4]	MWCL'22 [5]	TCAS I'19 [6]	RFIT'22 [7]	**This work**
Technology	65nm CMOS	28nm Bulk	45nm CMOS	28nm CMOS	65nm CMOS	65nm CMOS	**65nm CMOS SOI**
Topology	1-stage,2-way	2-stage,CS	1-stage,satck	2-stage,CS	1-stage,CS	2-stage,CS	**2-stage,stack**
Gain (dB)	16.8	20.4	16.8	19.1	10	18.6	**21.4**
3-dB BW.(GHz)	26-41	20.8-41.6	26-36	19.7-38.9	25-35	23-39	**11.8-41.2**
3-dB BW.(%)	44	66	32	67	33	51	**111**
Psat(dBm)	25.5	15.1	18.2	16.9	14.75	14.5	**16.1**
OP$_{1dB}$(dBm)	21.5	12.4	15.6	12.8	13.2	13.5	**13.2**
1-dB Psat BW (%)	17	15.5	20	23.8	26	45.8	**64**
Peak PAE(%)	25.5	30.1	27	20.9	44.5	29.1	**18.2**
Area(mm^2)	0.28	0.1	0.37	0.1	0.12	0.17	**0.19**

greatly affects the actual matching effect. Increasing the R$_a$ value improves the quality factor and circuit stability of passive devices but also increases the input return loss. Therefore, it is crucial to balance matching characteristics, stability, and gain in the design process. After optimization, the resistance value was determined to be 180Ω.

III. POST-LAYOUT SIMULATION RESULTS

The chip fabricated with a standard 65-nm CMOS SOI technology occupies an area of $0.32mm^2$ including pads, and the core idea is $0.19mm^2$. The layout of the proposed PA is shown in Fig.3. As shown in Fig.5, the S-parameter simulation results indicate that the designed PA has a peak gain of 22 dB with a 3-dB BW from 11.8 to 41.2 GHz, which covers the full K/Ka/Ku-band with fractional BW of 111%. Besides, from 12 to 40 GHz, the input return loss is beyond 10 dB, illustrating broadband input matching.

Fig. 4 shows the large-signal performance of the proposed PA at 18/28/35 GHz . The proposed PA achieves 16.2-dBm saturated output power P$_{sat}$,the peak PAE is 18.2% with OP$_{1dB}$ of 13.2 dBm at 28 GHz. The PA achieves 15.5/14.9 dBm saturated output power and 13.6%/14.7% peak PAE at 18/35GHz.And its summary is shown in Fig.6.Also, the output power of the PA reports 64% P$_{sat}$ -1dB BW from 18 to 37 GHz.

The performance of this design and compares it with state-of-the-art mm-wave silicon-based PAs is summarized in Table 1. The table shows that the proposed PA has a significant advantage in gain, achieving the highest small-signal gain 3-dB bandwidth and output power 1-dB bandwidth. It offers a good balance between power gain, output power, PAE, and ultra-wideband.

IV. CONCLUSION

This paper presents a wideband power amplifier that covers the K, Ku, and Ka bands. The power stage of the amplifier utilizes parallel inductors and feedback capacitors to tune the intermediate node impedance of the stacked structure. It also employs an RC negative feedback loop to mitigate the mismatch caused by the positive feedback path due to parasitic capacitance during frequency changes, thereby optimizing the

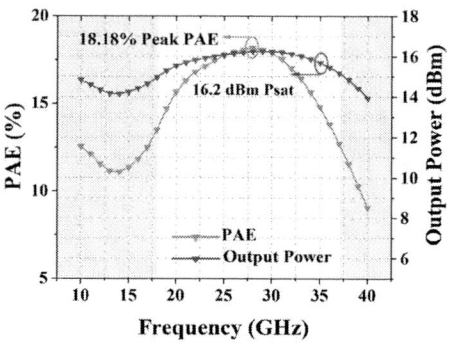

Fig. 6. Simulated large-signal performance versus frequency

output matching network. A current reuse structure is applied to the drive stage to achieve the desired output power while reducing power consumption. Additionally, the input matching network based on an RLC network enables the power amplifier to achieve extremely low return loss within the ultra-wideband. Simulation results show that the realized PA exhibits good broadband performance in terms of gain and output power.

ACKNOWLEDGMENT

This work was supported by the National Key Research and Development Program of China under Project 2023YFB4403200, the National Natural Science Foundation of China under Project 62271347 and U21A2045.

REFERENCES

[1] H. Jeong, H. D. Lee, B. Park, S. Jang, S. Kong and C. Park, "Three-Stacked CMOS Power Amplifier to Increase Output Power With Stability Enhancement for mm-Wave Beamforming Systems," *IEEE Trans. Microw. Theory Techn.*, vol. 71, no. 6, pp. 2450-2464, June.2023.

[2] T. -H. Fan, Y. Wang and H. Wang, "A Broadband Transformer-Based Power Amplifier Achieving 24.5-dBm Output Power Over 24–41 GHz in 65-nm CMOS Process," *IEEE Microw. and Wireless Compon. Lett.*, vol. 31, no. 3, pp. 308-311, Mar. 2021.

[3] C.-W. Wang, Y.-C. Chen, W.-J. Lin, J.-H. Tsai, and T.-W. Huang, "A 20.8–41.6-GHz transformer-based wideband power amplifier with 20.4-dB peak gain using 0.9-V 28-nm CMOS process," *IEEE MTT-S Int. Microw. Symp. Dig.*pp. 1323–1326, Aug. 2020.

[4] X. Fang, J. Xia, and S. Boumaiza, "A 28-GHz beamforming Doherty power amplifier with enhanced AM-PM characteristic," *IEEE Trans. Microw. Theory Techn.*, vol. 68, no. 7, pp. 3017–3027, Jul. 2020.

[5] T.-W. Huang, H.-C. Yen, J.-H. Tsai, W.-T. Bai, J.-C. Hung, and Y.-J. Liang, "A 19.7-38.9-GHz Ultrabroadband PA With Phase Linearization for 5G in 28-nm CMOS Process," *IEEE Microw. and Wireless Compon. Lett.* 2022.

[6] S. N. Ali, P. Agarwal, S. Gopal, S. Mirabbasi and D. Heo, "A 25–35 GHz neutralized continuous Class-F CMOS power amplifier for 5G mobile communications achieving 26% modulation PAE at 1.5 Gb/s and 46.4% peak PAE," *IEEE Trans. Circuits Syst. I*: Regular Papers, vol. 66, no. 2, pp. 834-847, Feb. 2019.

[7] C. Li, R. Wang, J. Zhang, W. Zhu and Y. Wang, "A Compact Broadband Power Amplifier Covering 23-39 GHz for 5G Mobile Communication," *Int. Symp. Radio-Freq. Integr. Technol.*, pp. 9-11, Republic of,2022.

A 1.36–2.25 GHz Digitally Controlled Oscillator with Dynamic Element Matching for NB-IoT Application

Binchen Wang*, Nagarajan Mahalingam*, Bharatha Kumar Thangarasu*, Kaixue Ma*
Fanyi Meng*, Zhenghao Lu†, Kiat Seng Yeo*‡

*School of Microelectronics, Tianjin University, Tianjin, China
†Soochow University, Suzhou, China
‡Singapore University of Technology and Design, Singapore

Abstract—This paper proposes a Class-B digitally controlled oscillator (DCO) with Dynamic Element Matching (DEM) module for NB-IoT Application. By using DEM module to control the finest capacitor array, the tuning linearity of DCO is significantly enhanced. This Class-B DCO have been implemented in 65 nm TSMC CMOS process. It has 21 bit frequency tuning elements for wide frequency tuning range from 1.36 to 2.25 GHz (49.3%). The achieved phase noise is -127.1 dBc/Hz@1MHz at 1.36 GHz and -126.4 dBc/Hz@1MHz at 2.25 GHz, while consuming 5.58-6.138 mW at 1.1V, resulting in a -184 dBc/Hz FoM and -198 dBc/Hz FoMT.

Index Terms—digitally controlled oscillator, linearity, tuning linearization, dynamic element matching, Internet of Things (IoT)

I. INTRODUCTION

Narrowband IoT (NB-IoT) technology is built upon cellular networks as its infrastructure and is compatible with the construction of new technological frameworks. This approach facilitates long-range communication, low data rates, low energy consumption, and a diverse range of services. Naturally, high-performance ADPLLs designed specifically for NB-IoT applications impose higher requirements on the Digitally Controlled Oscillator (DCO). As the most important component of the ADPLL, the DCO significantly influences the key performance such as frequency range, power consumption, phase noise, reliability, and area [1].

This paper implements a wide frequency tuning range (FTR) and good phase noise Class-B LC-DCO based on TSMC 65nm CMOS process. The FTR and frequency resolution of the presented DCO are improved and optimized by employing four capacitor arrays. Furthermore, a dynamic element matching (DEM) module has been configured for capacitor array that has the smallest frequency step size, effectively suppressing the mismatch of the frequency tuning units and significantly enhancing the tuning linearity [2]. The remainder of this paper

Binchen Wang, Nagarajan Mahalingam, Bharatha Kumar Thangarasu, Kaixue Ma, Fanyi Meng and Kiat Seng Yeo are with *the School of Microelectronics, Tianjin University*, Tianjin, China. Zhenghao Lu is with *the School of Electronic and Information Engineering, Soochow University*, Suzhou, China. Kiat Seng Yeo is also with *Engineering Product Development, Singapore University of Technology and Design*, Singapore. Corresponding author: (Binchen Wang, email: wbc@tju.edu.cn)

is organized in the following manner. Section II describes the presented DCO with DEM block. Simulation results are presented in Section III. Finally, the conclusion is drawn in Section IV.

II. CIRCUIT DESIGN

A. DCO Methodology

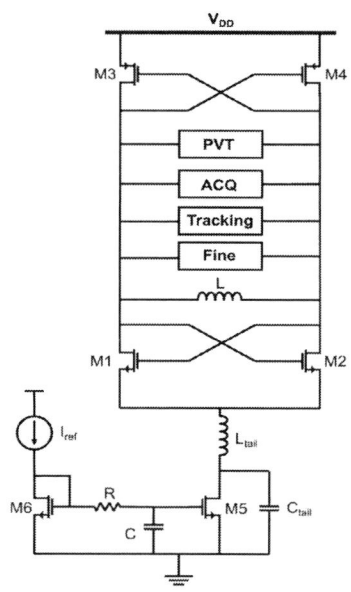

Fig. 1: DCO structure.

A CMOS cross-coupled structure is chosen for its better circuit symmetry, which help reduce flicker noise. Meanwhile, considering that the oscillator core needs to work in coordination with a large number of frequency tuning units, and to leave sufficient design margin for subsequent circuit modules, a Class-B DCO with higher oscillation amplitude and more stable start-up was selected here. Although this type of DCO may exhibit slightly inferior performance in terms of power consumption and phase noise compared to other oscillator types (such as Class-C and Class-F oscillators), its

979-8-3315-2209-4/25 $31.00 © 2025 IEEE

structure aligns more closely with the requirements of NB-IoT application [3].

The DCO designed in this paper is illustrated in Fig. 1. Cross-coupled pairs consist of M1-4, among them, the ratio of the $W/L_{(M3,M1)}$ and $W/L_{(M4,M2)}$ is set as:

$$\frac{(W/L)_{M3}}{(W/L)_{M1}} = \frac{(W/L)_{M4}}{(W/L)_{M2}} = 2.5 \qquad (1)$$

to ensure the better match between the upper and lower mos. The tail current source is composed of M5 and M6, the low-pass filter composed of R and C is used to filter the noise from the external current source. The filter network composed of L_{tail} and C_{tail} is used to block the propagation of the second harmonics in the resonator, so as to optimize the phase noise of DCO. The frequency tuning of DCO is mainly realized through four switched capacitor banks, namely PVT bank, ACQ bank, Tracking bank and Fine bank, which have different frequency step. Among them, the frequency step of Fine bank is the smallest, which determines the frequency resolution of DCO.

B. Switched Capacitor Array

The resonant frequency f_{DCO} of DCO can be tuned by the on-off of the switch in the switched capacitor array, and its expression is as follows:

$$f_{\text{DCO}} = \frac{1}{2\pi\sqrt{L\left\{\sum_{k=0}^{N-1}\left(C_{0,k} + d_k\Delta C_{k-1}\right) + C_p\right\}}} \qquad (2)$$

where $C_{0,k}$ represents the capacitance value when the switch is off, $d_k=0,1$ corresponds to the on and off state of switch, ΔC_k represents the difference of the capacitance value in different switch states, C_p represents the total parasitic capacitance [4].

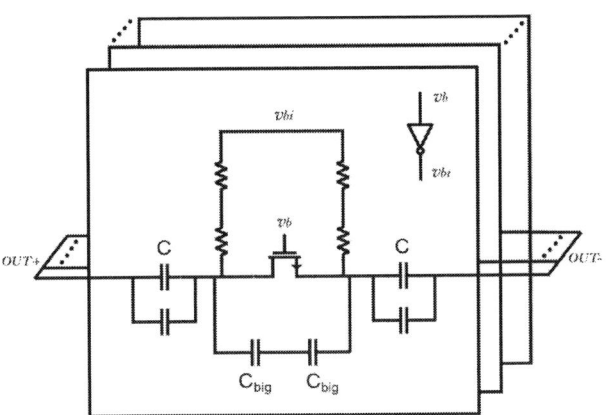

Fig. 2: DCO capacitor array.

The structure of the switched capacitor array is illustrated in Fig. 2. In order to cover the range of 1.4-2.1 GHz while reducing passive devices usage, all four banks are composed of binary weighted units. Among them, 7 bit Tracking bank and 5 bit Fine bank are shown in Fig. 2, while 4 bit PVT bank and 5 bit ACQ bank remain the same structure as the former except for their lacking of C_{big}. Considering that Tracking and Fine bank need finer frequency steps, a pair of large capacitors

are added in parallel on the basis of PVT bank, so that the capacitance difference between switch on and off is:

$$\Delta C_{\text{Tracking,Fine}} = \frac{C^2}{2\left(C + C_{\text{big}}\right)} \qquad (3)$$

For the fixed capacitor $C = 2.16fF$, if $\Delta C = 36aF$ is needed, then $C_{big} = 205fF$ can be calculated from (3). By simulating the curve of the equivalent capacitance versus frequency, the results shown in Fig. 4 can be obtained. It can be seen that when the frequency is 2GHz, the difference of switched capacitor ΔC is about 35.4aF, which basically corresponds with the calculated results [5].

C. Design of DEM Module

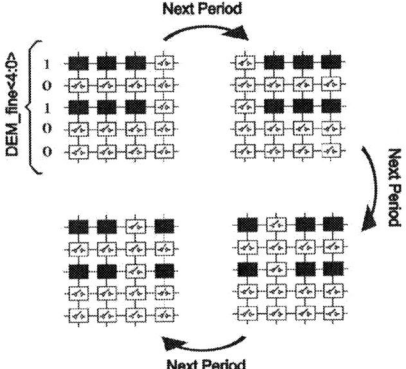

Fig. 3: DEM operation mode.

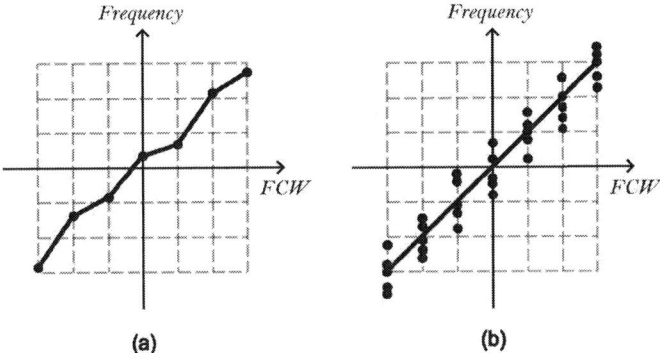

Fig. 4: Tuning curve of Fine bank (a) without DEM module and (b) with DEM module.

One of the notable limitations of a DCO compared to a voltage-controlled oscillator (VCO) is the presence of additional quantization noise, which arises from discontinuities in the DCO's frequency characteristic curve. This phenomenon makes the designer consider the tuning linearity especially, as poor linearity can degrade the overall performance of digital systems or even lead to functional errors. In particular, when applied in an all digital phase locked loop (ADPLL), inadequate linearity can significantly impair the loop's frequency locking capability.

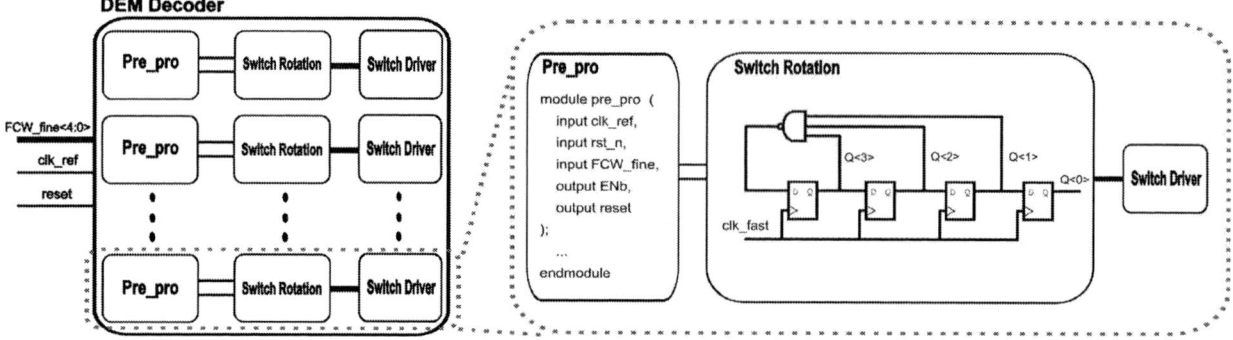

Fig. 5. Internal Structure of DEM.

Regarding the errors and mismatches in DCO capacitor array, the following factors should be taken into consideration:

- First, the equivalent capacitance of the unit varies with frequency, and this variation cannot be completely ignored for the capacitance difference of aF level.
- Second, parasitic effects within the circuit become increasingly non-negligible for such minute capacitance difference.
- Last, for an aF-level differences, mismatches of the same components due to process errors also need to be taken into account.

Consequently, the Fine bank exhibits inferior tuning linearity compared to the other three banks. To address this issue, a DEM module is employed to enhance the tuning linearity of the fine bank. As shown in Fig. 3, in order to match with the DEM module, Fine bank is expanded from the original 1×5 bit vector to a 5×4 bit matrix, where each matrix element is a switched capacitor. The input frequency control word (FCW) is processed by "DEM Decoder" block, and the output 20 bit "DEM_fine" is used to control the on-off of each unit in the new fine bank. When a certain bit of the FCW_fine is 1, the output of DEM Decoder will control the corresponding row of Fine bank, which will turn on three of the four switches in each cycle, and turn on the other three switches by shifting in the next cycle. Compared with the original Fine bank, each FCW will correspond to a group of frequency points instead of a single one. Even if there are various mismatch errors mentioned above, they will be converted into white noise which is unrelated to the input signal when calculating the average frequency, as shown in Fig. 4. Therefore, the linearity of the frequency tuning curve will be improved [6].

The internal structure of DEM module is shown in Fig. 5. It is composed of several identical signal processing modules (5 in this design). Fig. 5 also shows the composition of each signal processing module in detail. It is mainly composed of three sub-modules, which are the signal pre-processing module, switch rotation circuit and switch driver circuit. The signal preprocessing module works under the reference clock clk_ref. According to the input 1bit FCW, the module outputs the corresponding enable signals ENb and reset to control the subsequent switch rotation circuit. The switch rotation

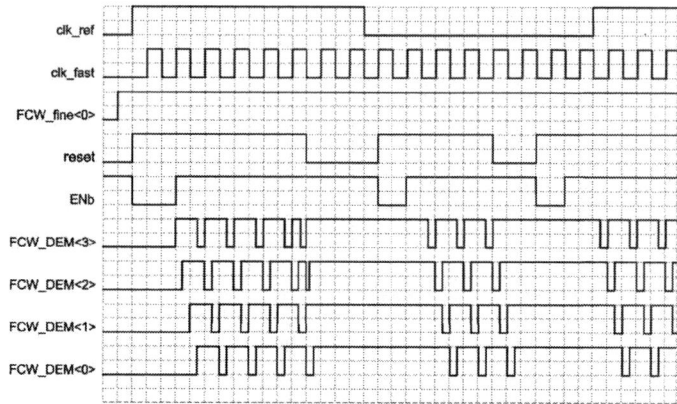

Fig. 6. Timing diagram of DEM.

circuit works under the higher frequency clock clk_fast, and is composed of a shift register and a NAND gate, which controls the periodic switching of the capacitor unit in the corresponding row according to the ENb and reset. The switch driver circuit is used to ensure the synchronization of each bit of control signal FCW_DEM output by DEM.

The timing diagram of DEM module is shown in Fig. 6. When FCW_fine is 1, signal "reset" is triggered to 1 at the rising edge of clk_ref, and signal "ENb" outputs a short period of 0. At this time, the 4 bit output signal "FCW_DEM" will periodically flip the level in a right-shift mode during reset=1. In one clk_fast period, only 1 of the 4 bit output signals remains low, and the rest are high. As can be seen from the timing diagram, the circuit realized its switch rotation function.

III. SIMULATION RESULTS

The proposed DCO and the switching rotation circuit in DEM are implemented on 65nm TSMC CMOS technology, while the other modules in DEM are mainly implemented in Verilog. The tuning range of DCO is 1.36-2.25 GHz, yielding a wide FTR of 49.3%. In order to ensure the full coverage of frequency band, the overlapping between different frequency bands are set aside at least 20% margin. Fig. 7 (a) (b) shows the frequency curve of the Fine bank without and with DEM. In Fig. 7 (b), each FCW corresponds to a group of at least

979-8-3315-2209-4/25 $31.00 © 2025 IEEE

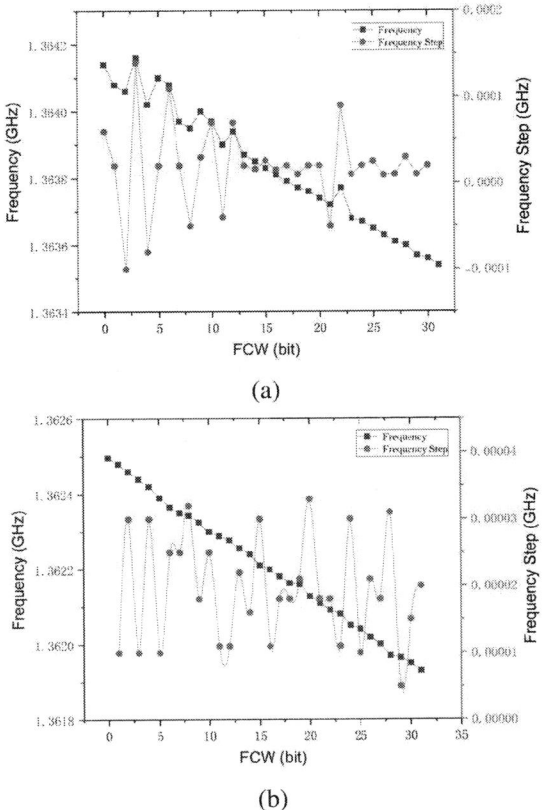

(a)

(b)

Fig. 7. Frequency, frequency step versus FCW curves of Fine bank (a) without DEM, (b) with DEM.

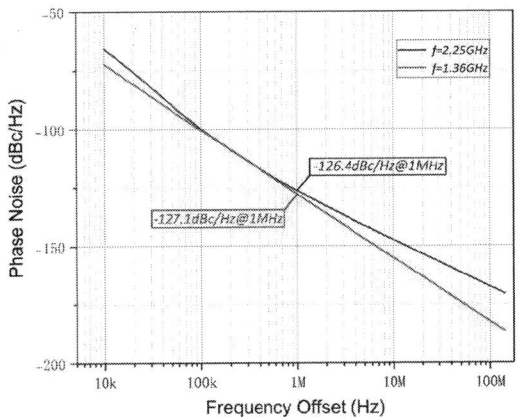

Fig. 8. DCO phase noise at minimum and maximum oscillation frequency.

TABLE I
COMPARISON WITH STATE-OF-THE-ART

	RFIC[7]	JSSC[8]	TMTT[9]	This work
Technology (nm)	28	65	65	**65**
TR (%)	24	9.1	40.25	**49.3**
V_{DD} (V)	1.2	1.2	1.2	**1.1**
Power (mW)	13	1.5	9.6	**5.58-6.138**
Resolution (Hz)	60k	10k	N/A	**40k**
PN(dBc/Hz)	-115.1	-117	-105.8~-108.7	**-127.1**
FoM(dBc/Hz)	-183	-173	-177.8~-184.2	**-184**
FoMT(dBc/Hz)	-190.6	-172	-196.3	**-198**

eight frequency points, and the frequency curve consists of the average frequency of each group. Comparing the frequency step of two fine banks in the same frequency range, it is evident that DEM module has a significant improvement on tuning linearity.

Fig. 8 shows the simulation results of phase noise of -127.1 dBc/Hz@1MHz at 1.36 GHz, and -126.4 dBc/Hz@1MHz at 2.25 GHz respectively. Table I gives a performance summary of the proposed DCO and compares it with state-of-the-art DCO.

IV. CONCLUSION

This paper presents a Class-B DCO with DEM module. With four optimized switched capacitor banks, the DCO operates from 1.36 to 2.25 GHz with 49.3% FTR, and has a frequency resolution of around 40kHz. Through the use of DEM module, a better tuning linearity is guaranteed. The achieved phase noise is -127.1 dBc/Hz@1MHz at 1.36 GHz and -126.4 dBc/Hz@1MHz at 2.25 GHz, while consuming only 5.58 mW at 1.1V voltage supply when all the switch are off and 6.138 mW when all the switch are on. The final FoM and FoMT are -184 dBc/Hz and -198 dBc/Hz, respectively.

REFERENCES

[1] Y. -P. E. Wang et al., "A Primer on 3GPP Narrowband Internet of Things," in IEEE Communications Magazine, vol. 55, no. 3, pp. 117-123, March 2017.

[2] I. Galton, "Why Dynamic-Element-Matching DACs Work," in IEEE Transactions on Circuits and Systems II: Express Briefs, vol. 57, no. 2, pp. 69-74, Feb. 2010.

[3] A. Hajimiri and T. H. Lee, "A general theory of phase noise in electrical oscillators," in IEEE Journal of Solid-State Circuits, vol. 33, no. 2, pp. 179-194, Feb. 1998.

[4] R. B. Staszewski, Chih-Ming Hung, D. Leipold and P. T. Balsara, "A first multigigahertz digitally controlled oscillator for wireless applications," in IEEE Transactions on Microwave Theory and Techniques, vol. 51, no. 11, pp. 2154-2164, Nov. 2003.

[5] V. K. Chillara et al., "9.8 An 860μW 2.1-to-2.7GHz all-digital PLL-based frequency modulator with a DTC-assisted snapshot TDC for WPAN (Bluetooth Smart and ZigBee) applications," 2014 IEEE International Solid-State Circuits Conference Digest of Technical Papers, San Francisco, CA, USA, pp. 172-173, 2014.

[6] K. L. Chan, J. Zhu and I. Galton, "A 150MS/s 14-bit Segmented DEM DAC with Greater than 83dB of SFDR Across the Nyquilst band," 2007 IEEE Symposium on VLSI Circuits, Kyoto, Japan, pp. 200-201, 2007.

[7] L. Wang et al., "An 8.2-10.2 GHz Digitally Controlled Oscillator in 28-nm CMOS Using Constantly-Conducting NMOS Biased Switchable Capacitor," 2022 IEEE Radio Frequency Integrated Circuits Symposium, Denver, CO, USA, pp. 207-210, 2022.

[8] T. Siriburanon et al., "A 2.2 GHz -242 dB-FOM 4.2 mW ADC-PLL Using Digital Sub-Sampling Architecture," in IEEE Journal of Solid-State Circuits, vol. 51, no. 6, pp. 1385-1397, June 2016.

[9] H. Liang, S. Liu, Y. Shen, J. Chang and Z. Zhu, "A 12.3–18.5-GHz Single-Core Oscillator Using a Dual-Mode Variable Inductor With a Tunable Self-Resonant Frequency Technique," in IEEE Transactions on Microwave Theory and Techniques, vol. 71, no. 3, pp. 1356-1365, March 2023.

979-8-3315-2209-4/25 $31.00 © 2025 IEEE

A Fully-Differential Bi-Directional 3.5-18 GHz 6-bit Active Phase Shifter in 0.18-μm SiGe BiCMOS

Lize Wang[1,2,3], Fanyi Meng[1,2,3,*]

1. School of Microelectronics, Tianjin University, Tianjin, China
2. Tianjin Key Laboratory of Imaging and Sensing Microelectronic Technology, Tianjin, China
3. Key Laboratory of Organic Integrated Circuit, Ministry of Education, Tianjin University, Tianjin, China
*Email: mengfanyi@tju.edu.cn

Abstract—This paper presents a fully-differential wideband 6-bit bi-directional active phase shifter in 0.18-μm SiGe BiCMOS technology for SHF-band phased arrays. To achieve precise quadrature signals from 3.5 to 18 GHz, the quadrature coupler adopts metal-via-metal structure to realize the enhanced-edge-coupling quadrature coupler. The quadrature signals are vector-modulated and amplified by bandwidth-extended bi-directional amplifiers. According to the post-layout simulation results, the RMS phase error is 2.25° to 3.5°, and the RMS gain error is 0.23 dB to 0.43 dB for the transmitter mode and the receiver mode within the frequency range from 3.5 to 18 GHz. The circuit consumes less than 145.26 mW from 1.8 V supply voltage and has an input P_{1dB} above -4.71 dBm in all states. The circuit core occupies 1.73 mm².

Index Terms—active phase shifter, bi-directional, edge coupling, fully-differential, SHF-band.

I. INTRODUCTION

Phased array system can realize the directional transmission of signals, which is essential for the development of active phased array antenna (APAA). Phase shifter is the core module of the SHF-band phased array system to realize beamforming, beam-scanning and other functions. The research on ultra-wideband phase shifters in the SHF-band is of great significance for the development of satellite communication, broadcasting, and cellular networks.

Recent research reports show different topological applications such as vector-modulated phase shifters [1], [2], [3], reflection-type phase shifters [4], and switch-based phase shifters [5]. Nevertheless, several drawbacks such as high insertion loss, narrow bandwidth, and linearity problems restrict their application in the SHF-band domain.

In order to address the aforementioned challenges, we propose a novel concept of a bi-directional active phase shifter based on bandwidth-extended bi-directional amplifier and enhanced-edge-coupling quadrature coupler (EECQ coupler). A prototype phase shifter was designed for the APAA and was manufactured by 0.18-μm SiGe BiCMOS technology, supporting a 6-bit, 360° bi-directional phase-shifting function over the frequency range of 3.5-18 GHz.

II. CIRCUIT PRINCIPLE AND DESIGN

The proposed 6-bit bi-directional active phase shifter consists of the bi-directional variable gain amplifier (Bi-VGA)

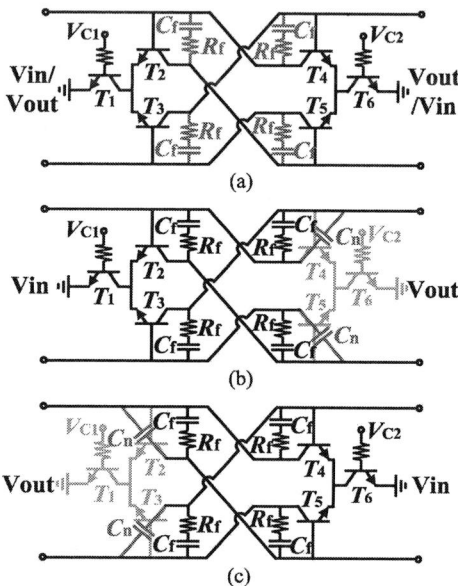

Fig. 1. (a)The core of the bandwidth-extended bi-directional amplifier in (b) transmitter mode and (c) receiver mode.

for amplitude adjustment, the EECQ coupler for quadrature signal conversion, and the bi-directional amplifier (Bi-AMP) for signal amplification, as shown in Fig.1 and Fig.2.

A. Bandwidth-extended Bi-directional Amplifier

As shown in Fig.1, the transmitter mode and the receiver mode of the amplifier can be achieved through the neutralized bi-directional technique [6]. In order to further facilitate its application in ultra-wideband systems, the bandwidth-extended bi-directional amplifier employs the proposed negative feedback structure consisting of the feedback resistor R_f and the feedback capacitor C_f.

By optimizing the value of the R_f, the bandwidth of the amplifier can be extended, as depicted in Fig. 3(a). And the function of the C_f (150 fF) is to cut off the direct current (DC) path between the collector and the base of the transistor, stabilizing the quiescent operating point of the amplifier. When V_{C1} is 1.3 V and V_{C2} is 0 V, the amplifier is in the transmitter mode, T_2 and T_3 are in the active region while T_4 and T_5 are in

979-8-3315-2209-4/25 $31.00 © 2025 IEEE

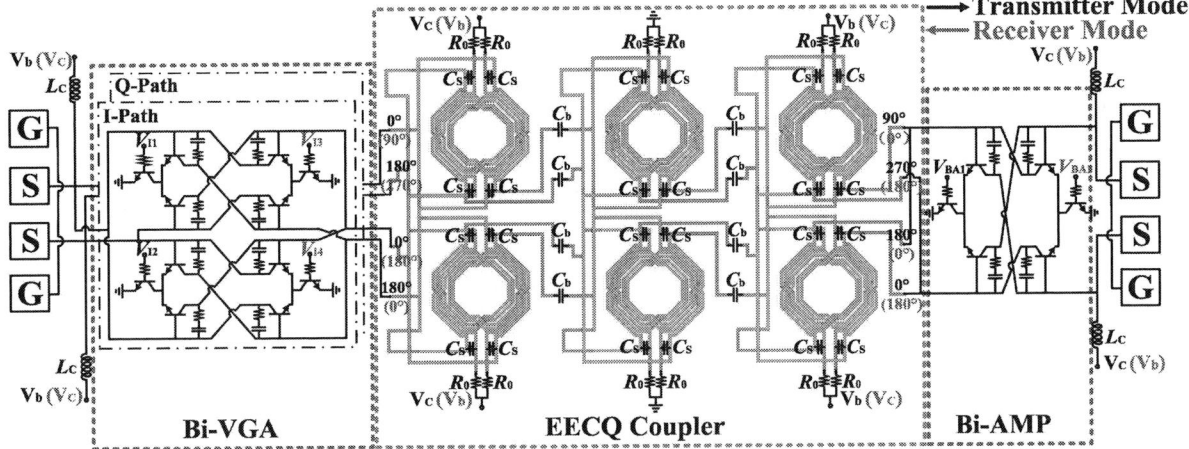

Fig. 2. Proposed phase shifter topology and bi-directional signal flow of the EECQ coupler.

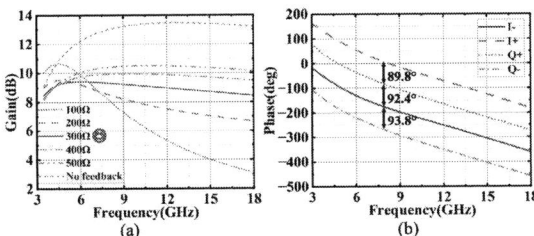

Fig. 3. (a) The relationship between the value of the R_f and the bandwidth and gain flatness of the amplifier; (b) The performance of the EECQ coupler.

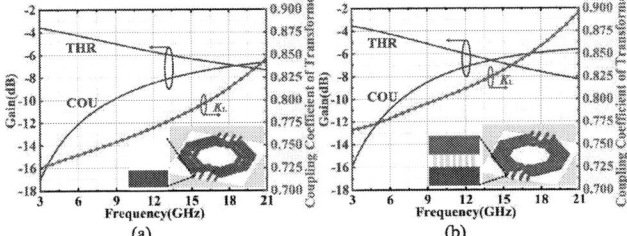

Fig. 4. Comparison of amplitude performance and coupling coefficient between (a) single mental layer structure and (b) proposed edge coupling enhancement structure.

the cutoff region. Most of the parasitic capacitance of T_2 and T_3 can be neutralized by the parasitic capacitance C_n of T_4 and T_5. When V_{C1} is 0 V and V_{C2} is 1.3 V, the amplifier is in the receiver mode and works in the same way as the transmitter mode. (Size width/length: T_1, T_6 are 1×150 nm/15 µm, T_2-T_5 are 1×150 nm/4.5 µm)

As illustrated in Fig. 2, the Bi-AMP consists of one bandwidth-extended bi-directional amplifier and the Bi-VGA consists of four cross-connected bandwidth-extended bi-directional amplifiers with control signals in the range of 0.6 to 1.2 V. The inductor required for collector power supply V_C (1.8 V) or base bias V_b (1.3 V) is provided by EECQ coupler and inductor L_C.

B. EECQ Coupler

The traditional quadrature signal generator [7] has the advantages of large bandwidth, low loss and high precision, but it needs to occupy a large area to achieve good orthogonal performance at low frequencies. The proposed EECQ coupler uses metal-via-metal structure to design a high-coupling coupler, which extends the bandwidth at low frequencies with the same area.

As shown in Fig.2, the proposed EECQ coupler is composed of 6 identical minimum units, which are equivalent to the lumped model of the quarter-wavelength transmission line. Through edge coupling enhancement technique, the coupling coefficient of the transformer is increased, and the performance of the minimum unit is improved at low frequencies, as shown

in Fig.4. In addition, the use of double layers of metal allows the interconnects between the minimum units to be divided into two layers, reducing the use of vias when the wires intersect, thereby reducing non-ideal effects.

In order to realize wideband quadrature signals, the minimum unit requires not only the inductive coupling, but also the capacitive coupling, so it is necessary to connect capacitors C_s (310 fF) in parallel between the ports to meet the capacitive coupling requirement. To reduce the circuit area and improve the utilization rate, the proposed EECQ coupler adopts a symmetrical structure, and the stages are connected by capacitor C_b (2 pF) to realize the function of supplying power to the amplifiers on both sides. As shown in Fig. 2, the isolation ports of the EECQ coupler are connected to the DC power supply or GND with R_0 (50 Ω) to achieve matching and isolation. Fig. 3(b) exhibits the performance of the EECQ coupler.

III. SIMULATION RESULTS

Fig. 5 depicts the layout of the proposed phase shifter with an area of 2068 µm × 836 µm. Fig. 6(a) and (b) reveal the simulated relative phase shifting and the gain in the receiver mode. Within the frequency range of 3.5-18 GHz, in the transmitter mode, the RMS phase error is 2.43-3.21° and the RMS gain error is 0.29-0.43 dB. In the receiver mode, the RMS phase error is 2.25-3.5° and the RMS gain error is 0.23-0.41 dB. The average gain is -2.94 dB at 6 GHz in the receiver mode. Fig. 6(c) and (d) show the simulated return loss and

979-8-3315-2209-4/25 $31.00 © 2025 IEEE

TABLE I
PERFORMANCE COMPARISON

Reference	Process	Frequency (GHz)	Resolution (bit)	Bi-directional	RMS phase Error (deg)	RMS gain Error (dB)	IP$_{1dB}$ (dBm)	Power (mW)	Area (mm²)
[3]	0.18μm BiCMOS	8-12	5	NO	<4.6	<0.6	N.A.	73.92	0.6
[5]	0.13μm SiGe HBT	8-12	6	YES	<2.2	<0.9	-15	195	3.9*
[8]	0.13μm BiCMOS	6-18	6	NO	<5.6	<1.05	5-8	93.06	2.44*
[9]	0.18μm BiCMOS	6.5-14.5	6	NO	<2.1	<0.5	11-26	133	0.64
This work	0.18μm BiCMOS	3.5-18	6	Yes	2.43-3.21** 2.25-3.5***	0.29-0.43** 0.23-0.41***	>-2.9** >-4.71***	<141.3** <145.26***	1.73

*: The overall chip size;   **: Transmitter Mode;   ***: Receiver Mode

Fig. 5. Layout of the proposed phase shifter.

the S_{21} at 10 GHz in the receiver mode, respectively. For the large-signal performance simulation, the input P$_{1dB}$ is higher than -2.9 dBm in the transmitter mode and -4.71 dBm in the receiver mode. A comparison with previous phase shifters is shown in Table I.

IV. CONCLUSION

This paper presents the design of a fully-differential 3.5-18GHz 6-bit bi-directional active phase shifter using edge coupling enhancement technique and bandwidth-extended bi-directional amplifier in 0.18-μm SiGe BiCMOS process. The metal-via-metal structure realizes a high-coupling quadrature coupler to provide wideband quadrature signals for the Bi-VGA, to achieve ultra-wideband bi-directional phase shift function.

ACKNOWLEDGMENT

This work was supported by the National Key Research and Development Program of China under Project 2023YFB4403200, the National Natural Science Foundation of China under Project 62271347 and U21A2045.

REFERENCES

[1] P. Gu, D. Zhao and X. You, "A wideband vector-modulated variable gain phase shifter for 5G NR FR2 in 40-nm CMOS," *IEEE Trans. Microw. Theory Techn.*, vol. 72, no. 9, pp. 5274-5284, Sept. 2024.

[2] G. Shi, Z. Li, Z. Dai, Z. Li, Z. Hou and L. Liu, "An X-band 6-bit active vector-modulated phase shifter with 0.29° rms phase error using linearization control technique," *IEEE Microw. Wireless Technol. Lett.*, vol. 33, no. 10, pp. 1482-1485, Oct. 2023.

[3] Z. Li, J. Qiao and Y. Zhuang, "An X-band 5-bit active phase shifter based on a novel vector-sum technique in 0.18μm SiGe BiCMOS," *IEEE Trans. Circuits Syst. II, Exp. Briefs*, vol. 68, no. 6, pp. 1763-1767, June 2021.

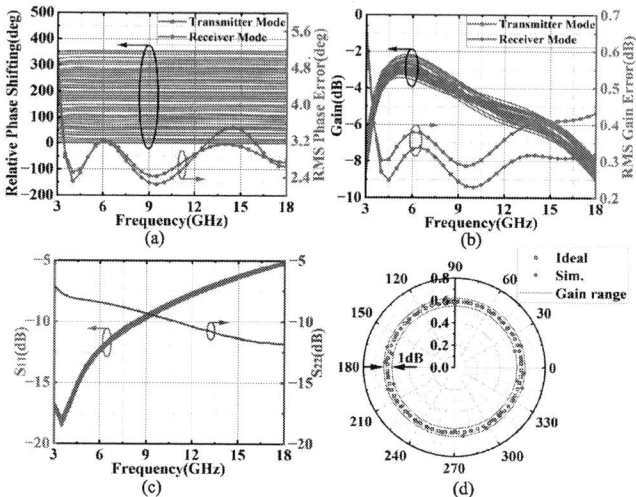

Fig. 6. (a) Simulated relative phase shifting in the receiver mode and the RMS phase errors; (b) Simulated gain in the receiver mode and the RMS gain errors; (c) Simulated return loss; (d) Simulated S_{21} at 10 GHz in the receiver mode.

[4] M. Hazer Sahlabadi, H. Yu, J. Xia and S. Boumaiza, "A digitally controlled bidirectional 24-32-GHz variable gain phase shifter in 45-nm SOI CMOS," *IEEE Trans. Circuits Syst. II, Exp. Briefs*, vol. 71, no. 8, pp. 3755-3759, Aug. 2024.

[5] Y. Gong, M. -K. Cho and J. D. Cressler, "A bi-directional, X-band 6-bit phase shifter for phased array antennas using an active DPDT switch," in *Proc. IEEE Radio Freq. Integr. Circuits Symp. (RFIC)*, June 2017, pp. 288-291.

[6] J. Pang et al., "21.1 A 28GHz CMOS phased-array beamformer utilizing neutralized bi-directional technique supporting dual-polarized MIMO for 5G NR," in *IEEE Int. Solid-State Circuits Conf. (ISSCC) Dig. Tech. Papers*, Feb. 2019, pp. 344-346.

[7] Y. Yu et al., "An 18~30 GHz vector-sum phase shifter with two-stage transformer-based hybrid in 130-nm SiGe BiCMOS," *IEEE Trans. Circuits Syst. I: Reg. Papers*, vol. 70, no. 12, pp. 5138-5151, Dec. 2023.

[8] Y. Yao, Z. Li, G. Cheng, L. Luo, W. He and Q. Li, "A 6-bit active phase shifter for X- and Ku-band phased arrays," in *Proc. IEEE Int. Conf. Integr. Circuits, Technol. Appl. (ICTA)*, Nov. 2018, pp. 124-125.

[9] T. Fujiwara and M. Shimozawa, "Broadband and highly accurate X-band vector-sum phase shifter using LC-type power splitter," in *Proc. 13th Eur. Microwave Integr. Circuits Conf. (EuMIC)*, Sept. 2018, pp. 122-125.

979-8-3315-2209-4/25 $31.00 © 2025 IEEE

A 5.1-33.5 GHz Variable Gain Low-Noise Amplifier With High Linearity in 0.13 μm BiCMOS

Shuai Li   Bharatha Kumar Thangarasu   Nagarajan Mahalingam   Kaixue Ma

Fanyi Meng   Zhenghao Lu   Anqing Chen   Hai Ye   Kiat Seng Yeo

Abstract—This paper presents a low-noise amplifier (LNA) designed using 0.13 μm BiCMOS technology. By incorporating additional poles through feedback loops, the LNA achieves a 3 dB bandwidth from 5.1 to 33.5 GHz. Key performance metrics include an input third-order intercept point (IIP3) of 1.7 dBm, an input 1 dB compression point (IP1dB) of -9.9 dBm, a peak gain of 14.3 dB, and a minimum noise figure (NF) of 3.3 dB. The LNA features an 8-step gain control via a 3-bit digital interface, along with an independent switch. The core area of the LNA is 0.075 mm², and its figure of merit (FoM) demonstrates a competitive advantage over existing designs.

Index Terms—BiCMOS, figure of merit (FoM), low noise amplifier (LNA), ultra-wideband (UWB), variable gain.

I. INTRODUCTION

As communication systems evolve, particularly with the advent of radar, satellite networks, and next-generation technologies, there is an increasing demand for components that can support ultra-high data rates. To meet these needs, systems must operate across wider frequency bands and higher bandwidths, which are essential for achieving faster data transmission and ensuring interoperability with future wireless standards [1]. Consequently, there is a growing need for ultra-wideband (UWB) low-noise amplifiers (LNAs) that can efficiently cover frequencies from the lower microwave ranges to the millimeter-wave spectrum [2].

Various techniques have been explored to extend the bandwidth (BW) of LNAs. Although distributed amplifiers can significantly increase bandwidth, they typically come at the expense of gain, noise performance, and power consumption [3]. Incorporating RC negative feedback and inductive peaking optimizes input matching for wideband performance but introduces additional noise [4]. In [5], the base-collector capacitance is included in the input matching network with degenerative inductance, avoiding extra passive components. This reduces device-induced noise but worsens gain flatness.

In this work, a 3-bit variable gain, switchable amplifier is presented, operating from 5.1 to 33.5 GHz, implemented using 0.13 μm BiCMOS technology. The LNA achieves UWB performance by introducing additional poles in the feedback loop, ensuring better gain flatness across the frequency range.

Shuai Li, Bharatha Kumar Thangarasu, Nagarajan Mahalingam, Kaixue Ma, Fanyi Meng, Anqing Chen, Hai Ye and Kiat Seng Yeo are with *the School of Microelectronics, Tianjin University*, Tianjin, China. Zhenghao Lu is with *the School of Electronic and Information Engineering, Soochow University*, Suzhou, China. Kiat Seng Yeo is also with *the Engineering Product Development, Singapore University of Technology and Design,*, Singapore. Corresponding author: (Shuai Li, email: ls3019232063@tju.edu.cn)

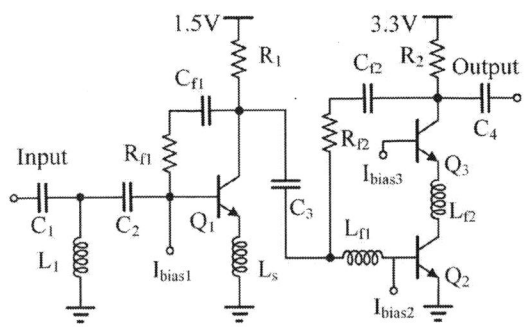

Fig. 1. Schematic of the core LNA circuit.

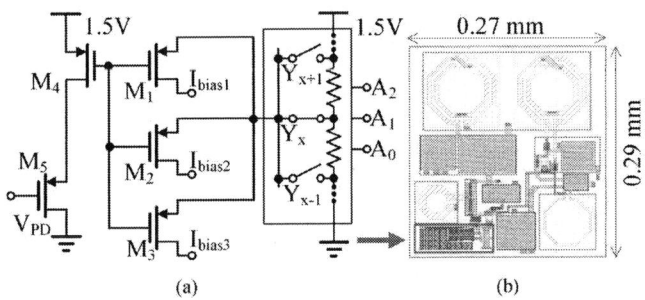

Fig. 2. (a) 3-bit bias circuit with PD mode, (b) layout of the proposed LNA.

II. CIRCUIT DESIGN OF LNA

A. Core LNA Design

Fig. 1 illustrates the schematic of the core LNA. The first stage achieves input matching through the combined use of an emitter degeneration inductor, RC feedback, and a T-type matching network. The emitter degeneration inductor effectively mitigates the high-frequency attenuation caused by the input capacitance of the transistor, thereby enhancing the amplifier's high-frequency response. Since the overall noise figure (NF) of the system is predominantly determined by the noise and gain of the first stage, the proposed feedback inductors were deliberately avoided in this stage to prevent the parasitic resistance of the inductors from degrading the NF. As will be demonstrated in subsequent analysis, the feedback inductance of the first stage can be equivalently modeled as a component of the load, specifically contributing to the input inductance of the second stage, thereby yielding a similar optimization effect.

979-8-3315-2209-4/25 $31.00 © 2025 IEEE   111

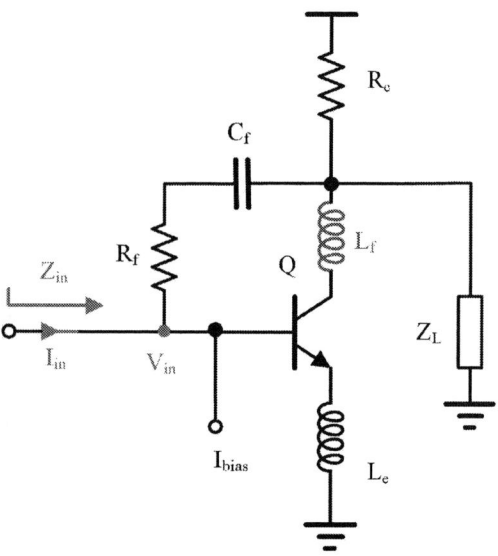

Fig. 3. Schematic of introducing inductance in negative feedback.

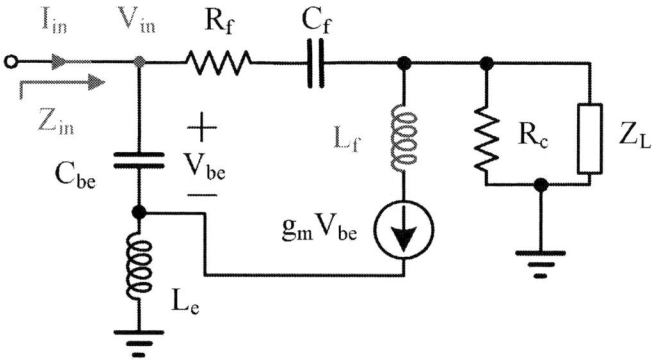

Fig. 4. Equivalent circuit of introducing inductance in negative feedback.

The second stage employs a cascode configuration, where additional feedback inductors (L_{f1}) and (L_{f2}) are incorporated into the feedback network. These inductors introduce additional poles, which effectively extend the bandwidth and ensure gain flatness across the operating frequency range. Since the system's linearity is primarily influenced by the linearity of the latter stages, the second stage is designed to operate under a higher supply voltage, enabling a larger output voltage swing. This approach achieves improved linearity at the cost of a modest increase in power consumption. Furthermore, the cascode topology, known for its high reverse isolation, compensates for the limited reverse isolation provided by the common-emitter (CE) configuration in the first stage. This enhancement not only improves overall stability but also mitigates the risks of potential feedback-induced oscillations, ensuring stable performance over a wide frequency range.

B. Bias Circuit Design

Fig. 2(a) shows the schematic of the 3-bit bias circuit with a power-down (PD) mode. This circuit integrates a 3 to 8 digital decoder with a resistive voltage divider, providing 8

discrete steps with a 1.5 dB increment. For instance, when the input values A_2, A_1, and A_0 are set to 101, the gate voltage of PMOS transistor Y_5 is set to 0, while the gate voltages of the remaining PMOS transistors are set to 1, where a '1' corresponds to a voltage level of 1.5 V and a '0' corresponds to a voltage level of 0 V. This configuration simultaneously controls the base currents of transistors Q_1, Q_2, and Q_3, achieving gain flatness across the wide frequency range for different control bit combinations.

In the operational state of the LNA, the gate voltage of transistor M_5, denoted as V_{PD}, is set to 0. In PD mode, V_{PD} is set to 1, which significantly reduces the current through the left section of the circuit to nearly zero. As a result, under the current mirror configuration, the gate voltages of M_1, M_2, and M_3 are shifted to the required levels, causing the base currents of Q_1, Q_2, and Q_3 to decrease as well. This reduction enables the PD mode operation [6].

C. Incorporating Inductance in Negative Feedback

In classical circuit structures, narrowband matching is typically achieved by using source or emitter degeneration inductance L_e to neutralize the gate-source capacitance or base-emitter capacitance C_{be}. Wideband matching, on the other hand, usually requires RC negative feedback, which controls the placement of the zero and pole in the feedback path, suppressing the gain rise at low frequencies while improving high-frequency response [4]. Consider the circuit shown in Fig. 3. By incorporating an inductor L_f in the feedback path formed by R_f and C_f, the frequency characteristics of the feedback path can be tuned, thereby optimizing the feedback performance.

Based on the small-signal equivalent circuit model in Fig. 4, where g_m is the transconductance of the transistor, R_c is the load resistance of the transistor, and Z_L is the input impedance of the subsequent stage, the input impedance can be derived as shown in equation (1). The expressions for Z_1 and Z_2 are given in equations (2) and (3), respectively. It can be observed that both L_e and R_f contribute to the real part of Z_{in}, while L_f and C_f exhibit frequency-selective characteristics. Within a certain broadband range, the imaginary part of Z_1 significantly decreases, leading to improved impedance matching.

It can be observed that the first-stage feedback inductance L_{f1}, when equivalently incorporated into the input impedance of the second stage, does not affect the pole and zero characteristics of Z_{in}. Meanwhile, the noise introduced by L_{f1} is filtered by the coupling capacitor C_3.

$$Z_{in} = \frac{Z_1 Z_2}{Z_1 + Z_2 + \frac{g_m R_c Z_L}{(R_c + Z_L) s C_{be}}}. \tag{1}$$

$$Z_1 = \frac{R_c Z_L}{R_c + Z_L} + R_f + s L_f + \frac{1}{s C_f}, \tag{2}$$

$$Z_2 = s L_e + \frac{g_m L_e}{C_{be}} + \frac{1}{s C_{be}}. \tag{3}$$

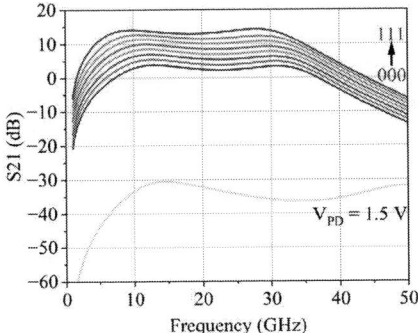

Fig. 5. S21 under 3-bit control and PD mode.

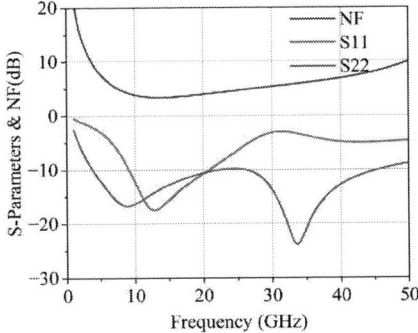

Fig. 6. NF, S11 and S22 at maximum gain.

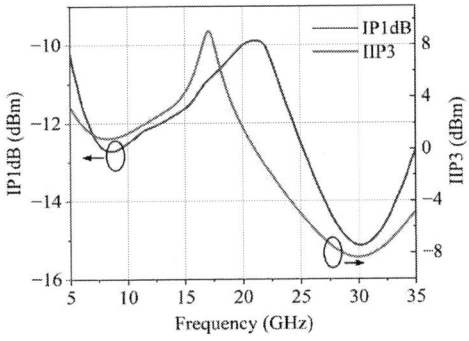

Fig. 7. IP1dB and IIP3 from 5 to 35 GHz.

TABLE I
PERFORMANCE SUMMARY AND COMPARISON

Reference	[4]	[5]	[7]	This work
Technology(SiGe)	130 nm	130 nm	130 nm	**130 nm**
BW (GHz)	1-27	8-12	0.3-15	**5.1-33.5**
FBW(%)	185.7	40	192.1	**147.2**
Gain (dB)	27.0	14.0	37.3	**14.3**
NF (dB)	1.4	1.3	1.8	**3.3**
IIP3 (dBm)	-8.0	0	-27.3	**1.7**
IP1dB (dBm)	-19.9	-7.5	-	**-9.9**
Pdc (mW)	85.0	48.5	52.0	**56.7**
FoM(GHz)[a]	1.2	1.2	0.065	**3.2**

[a] FoM[GHz] = Gain[abs.] * BW[GHz] * IIP3[mW] / ((F-1) * Pdc[mW]).

III. SIMULATED RESULTS AND DISCUSSION

Fig. 2 (b) shows the complete layout of the LNA, with the bias circuit located at the lower-left corner. The resistance introduced by the interconnects is included in the resistances of the voltage divider to minimize the mismatch in step increments while also minimizing the overall chip area., which is 0.27×0.29 mm². The achieved gain is shown in Fig. 5, where the peak gain and the valley gain within the bandwidth exhibit a flatness of $\pm$ 0.6 dB, with a maximum gain of 14.3 dB. The 3-bit digital control allows for 8 discrete steps with a 1.5 dB step size. The LNA operates in PD mode with a gain of -30 dB and a power consumption of 0.2 mW. Fig. 6 demonstrates that the LNA achieves a minimum NF of 3.3 dB, with S11 less than -8 dB from 8 to 25 GHz and S22 less than -8 dB from 3 to 50 GHz. Fig. 7 shows the peak values of input third-order intercept point (IIP3) at 9 dBm at 17 GHz and input 1dB compression point (IP1dB) at -9.9 dBm at 21 GHz. Table I shows that although the LNA has a comparable NF and gain compared to other designs, it offers a larger bandwidth and better figure of merit (FoM).

IV. CONCLUSION

In this work, a high-performance UWB LNA designed using 0.13 μm BiCMOS technology has been presented in this work. The design leverages feedback loops to introduce additional poles, achieving a 3 dB bandwidth ranging from 5.1 to 33.5 GHz. This LNA provides high gain and excellent gain flatness across the entire frequency range. Furthermore, it incorporates a 3-bit digital control for variable gain and independent switching for efficient power management, delivering a competitive FoM compared to other designs. This work aims to enhance wideband RF systems by offering an optimized LNA solution for future wireless communication and sensing applications.

REFERENCES

[1] M. El-Nozahi, E. Sánchez-Sinencio and K. Entesari, "A millimeter-wave (23–32 GHz) wideband BiCMOS low-noise amplifier", *IEEE J. Solid-State Circuits*, vol. 45, no. 2, pp. 289-299, 2010.

[2] J. K. Wang and H. Zhang, "A 22-to-47 GHz 2-stage LNA with 22.2 dB peak gain by using coupled L-type interstage matching inductors", *IEEE Trans. Circuits Syst. I, Reg. Papers*, vol. 67, no. 12, pp. 4607–4617, Dec. 2020.

[3] G. Nikandish and A. Medi, "Unilateralization of MMIC distributed amplifiers", *IEEE Trans. Microw. Theory Techn.*, vol. 62, no. 12, pp. 3041-3052, Dec. 2014.

[4] Z. Wang et al., "A 1–27 GHz SiGe Low Noise Amplifier With 27-dB Peak Gain and 2.85±1.45 dB NF," in *IEEE Trans. on Circuits Syst. II: Exp. Briefs*, vol. 71, no. 5, pp. 2629-2633, May 2024.

[5] C. Çalışkan, I. Kalyoncu, M. Yazici and Y. Gurbuz, "Sub-1-dB and Wideband SiGe BiCMOS Low-Noise Amplifiers for X -Band Applications," in *IEEE Trans. Circuits Syst. I, Reg. Papers*, vol. 66, no. 4, pp. 1419-1430, April 2019.

[6] B. K. Thangarasu, K. Ma and K. S. Yeo, "A Ku-Band Fully Differential Current-Reuse Stacked Low-Noise Amplifier in 0.18-μm SiGe BiCMOS Technology," in *IEEE Microw. and Wireless Techno. Lett.*, vol. 34, no. 4, pp. 407-410, April 2024

[7] S. Zeinolabedinzadeh, A. Ç. Ulusoy, M. A. Oakley, N. E. Lourenco and J. D. Cressler, "A 0.3–15 GHz SiGe LNA With >1 THz Gain-Bandwidth Product," in *IEEE Microw. and Wireless Compon. Lett.*, vol. 27, no. 4, pp. 380-382, April 2017.

An 18-40 GHz 5-bit Passive Vector-Modulated Phase Shifter With an X-Type Attenuator in 0.18 μm SiGe BiCMOS

Yang Liu, Xuan Li, Guiyue Mao, Fanyi Meng*, Kiat Seng Yeo
1.School of Microelectronics, Tianjin University, Tianjin, China
2.Tianjin Key Laboratory of Imaging and Sensing Microelectronic Technology
3.Key Laboratory of Organic Integrated Circuit, Ministry of Education, Tianjin University, Tianjin, China.
*Email: mengfanyi@tju.edu.cn

Abstract—This paper presents the design of an 18-40 GHz 5-bit passive vector-modulated phase shifter in 0.18 μm BiCMOS technology. The passive phase shifter comprises a quadrature all-pass filter (QAF) with an embedded LC compensation network, two 5-bit analog X-Type attenuators with complementary voltage control, and Marchand Baluns. The 360° full-span vector modulated phase shifter achieves a phase resolution of 11.25° while exhibiting an average insertion loss of 24 dB, with 0.5-3.8° RMS phase error and 0.2-2.2 dB RMS gain error. The core area of the circuit is 1.2mm × 0.4mm.

Index Terms—Vector-modulated phase shifter, X-type attenuator, ultra-wideband, quadrature all-pass filter

I. INTRODUCTION

The advancement of 5G millimeter wave communication, automotive radar, and satellite communication has created new bandwidth requirements for phased array systems. The phase shifter is essential in phased array systems, controlling the phase state to form beams. Its bandwidth and phase resolution directly affect the system's bandwidth and beam control resolution [1].

Phase shifters are classified into two types: active and passive. The passive phase shifter is superior to the active type in broadband capability, DC power consumption, and linearity, making it ideal for broadband phase shifter design. Switching and reflective phase shifters are the main types of passive phase shifters. However, Switching phase shifters require gain compensation, and reflective phase shifters have a limited phase shift range. In contrast, passive vector-modulated phase shifters are preferred in the millimeter wave frequency band for their wide bandwidth, bidirectional operation, and full quadrant coverage. As the gain control module, the X-Type attenuator is increasingly used in passive vector synthesis phase shifters for its excellent phase invariance and wide attenuation range.

This paper presents a 5-bit ultra-wideband vector synthetic passive phase shifter based on an X-Type attenuator operating at 18-40 GHz. The structure of this paper is organized as follows: Section II presents the design details of the VMPS. Section III discusses the simulated results and Section IV demonstrates the conclusion.

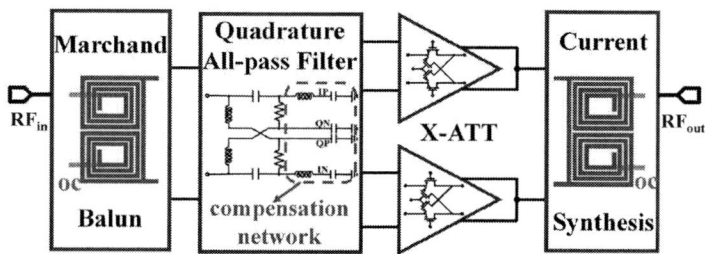

Fig. 1. Simplifed block diagram of the proposed passive vector synthesis phase shifter.

II. CIRCUIT DESCRIPTION AND IMPLEMENTATION

A. Architecture

Fig. 1 depicts the structure of a vector-modulated phase shifter, which typically includes a quadrature signal generator(quadrature all-pass filter), a gain control module (X-Type attenuator), and a power/current synthesis module. Firstly, the RF input signal is converted into a differential signal by a Marchand balun, which is then transformed into four IQ signals (IP IN QP QN) using a quadrature all-pass filter. The X-Type attenuator is employed to apply varying levels of attenuation to these signals. Finally, the attenuated signals are synthesized and output via a current-combining technique.

B. Broadband Quadrature All-Pass Filter

Generating an orthogonal signal at 18-40 GHz is challenging. Single-stage PPFs cannot produce the required IQ signal, and multistage cascade PPFs have unacceptable insertion loss. The quadrature all-pass filter is preferable for broadband signal generation due to its wide bandwidth and low insertion loss. Fig. 1 illustrates the schematic of the QAF, showcasing a traditional filter component on the left and an LC compensation network on the right. LC compensation network has two functions: it compensates IQ signals to reduce amplitude and phase mismatches and minimizes output impedance differences between the two IQ channels, allowing

979-8-3315-2209-4/25 $31.00 © 2025 IEEE

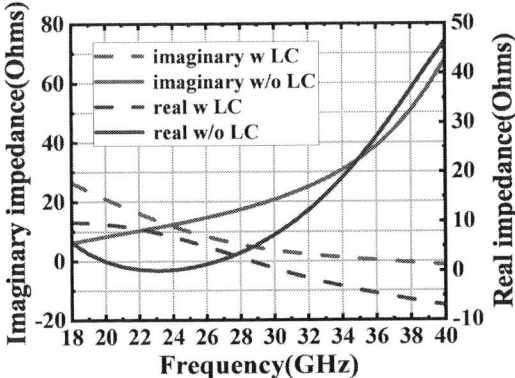

Fig. 2. Comparison of impedance discrepancies in IQ channels with and without LC compensation

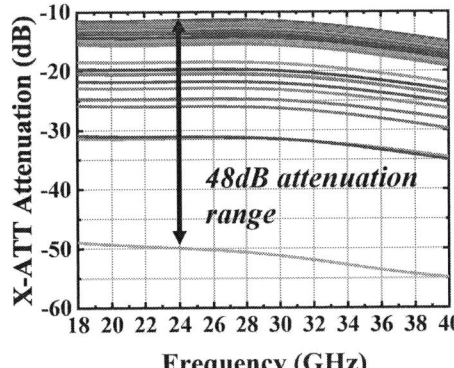

Fig. 3. Simulated attenuation of the analog-controlled X-ATT

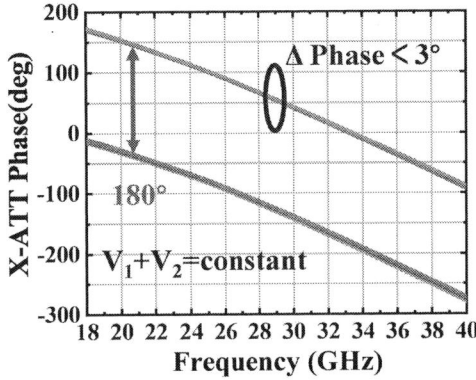

Fig. 4. Simulated phase of the analog-controlled X-ATT

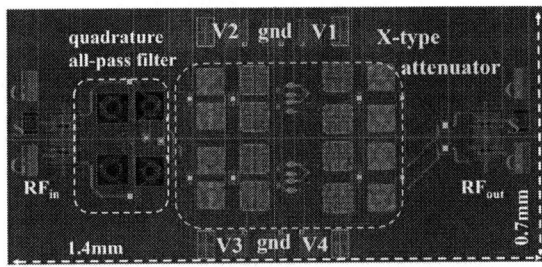

Fig. 7. Layout of the proposed wideband phase shifter

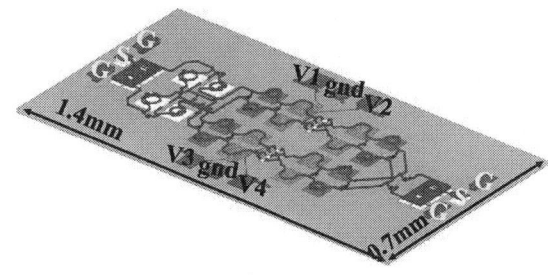

Fig. 8. EM model of the proposed wideband phase shifter

them to use the same matching network and further reduce discrepancies.

The simulatied results in Fig. 2 indicate a substantial decrease in impedance discrepancies between the real and imaginary components of the I and Q channels. Notably, the maximum discrepancy in the imaginary component decreased from 67.9 dB to 26.4 dB, whereas the maximum discrepancy in the real component dropped from 46.5 dB to 9.8 dB.

C. Complementary Voltage-Controlled X-Type Attenuator

X-Type attenuators are typically classified into two categories: analog-controlled and digital-controlled. The layout of digital-controlled X-Type attenuator necessitates a significant number of metal interconnects, which introduce parasitic inductances that adversely affect the amplitude and phase performance of the X-Type attenuator. Consequently, the X-Type attenuator controlled by complementary voltage functions as the gain control module, attenuating IQ signals to enable the synthesized output signal to cover the entire quadrant.

Fig. 1 depicts the schematic of the X-Type attenuator, consisting of four identically sized and cross-connected MOSFETs (M_1-M_4). Each MOSFET is modeled as a parallel resistance and capacitance [2]. These transistors act as voltage-controlled resistors, providing varying attenuation by canceling out out-of-phase signals. The grid control voltages for M_1/M_4 and M_2/M_3 are complementary pairs, with $V_1+V_2=$ 2V.

Fig. 3 and Fig. 4 showcases the simulated results of the X-Type attenuator, revealing an attenuation range of 48 dB and an additional phase shift of less than 3°.

III. SIMULATED RESULTS

Fig. 5 and Fig. 6 depicts the layout and EM model of the proposed 5-bit ultra-wideband phase shifter , with a core circuit area of 1.2 mm × 0.4 mm. Fig. 7 shows the simulated results of the phase shifter, which achieves 32 phase shift states, covering a 360° range with 11.25° steps and an RMS phase error of 1.1° at 29 GHz. Fig. 8 depicts an average insertion loss of -24 dB across the 32 phase-shift states, along with an RMS amplitude error of 0.2 dB at 29 GHz.

Table I compares the performance of the proposed vector synthesis phase shifter with other phase shifters, revealing the widest FBW, relatively low RMS phase error and zero power consumption.

TABLE I
PERFORMANCE COMPARSION OF PHASE SHIFTERS

Reference	A-SSCC [3]	IMS[4]	ESSCIRC[5]	MWCL[6]	TCAS-II[7]	TCAS-I[8]	**This Work**
technology	65-nm CMOS	40-nm CMOS	22-nm CMOS	0.18-μm CMOS	65-nm CMOS	130-nm SiGe	**0.18-μm SiGe**
Frequency(GHz)	27.5-30.5	26-32	24-29.5	27-33	30-32.5	18-30	**18-40**
FBW	10.4%	20.7%	20.56%	20%	8%	50%	**76%**
Topology	RTPS	P-VMPS	STPS	A-VMPS	VGPS	VMPS	**P-VMPS**
Resolution(°)	5.625	5.625	22.5	continuous	22.5	5.625	**11.25**
RMS phase error (°)	2.61	1.8-2.6	5	0.51-4	3.5	1.4-3.4	**0.78-3.5**
RMS gain error (dB)	1.26	1.2	N/A	0.86-1.3	0.4	1.2-1.4	**0.2-2.2**
Power(mW)	0	0	0	6.6	18	12.2	**0**

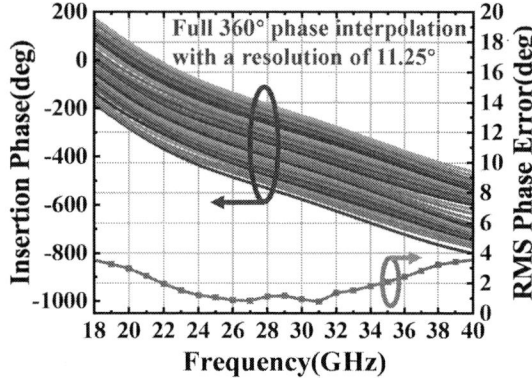

Fig. 5. Simulated results of the insertion phase and RMS phase error

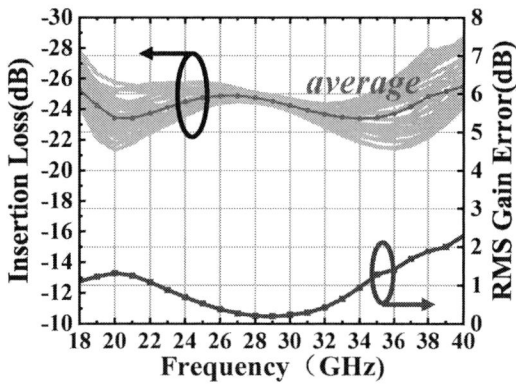

Fig. 6. Simulated results of the insertion loss and RMS gain error

IV. CONCLUSION

This paper presents an ultra-wideband 5-bit passive vector-modulated phase shifter in 0.18 μm SiGe technology. An QAF with an embedded LC compensation network for generating ultra-wideband IQ signals, a complementary voltage-controlled X-Type attenuator with a wide attenuation range and minimal phase shift serving as the gain control module and a Marchand balun utilized for the conversion between single-ended and differential signals. The simulated results demonstrate that the ultra-wideband phase shifter can achieve a phase shift range of 360° with a step size of 11.25° in 18-40 GHz. At 29 GHz, it exhibits a RMS phase error of 1.1° and a RMS amplitude error of 0.2 dB.

ACKNOWLEDGMENT

This work was supported by the National Key Research and Development Program of China under Project 2023YFB4403200, the National Natural Science Foundation of China under Project 62271347 and U21A2045.

REFERENCES

[1] G. -H. Park, C. W. Byeon and C. S. Park, "60 GHz 7-Bit Passive Vector-Sum Phase Shifter With an X-Type Attenuator," *IEEE Trans. Circuits Syst. II, Exp. Briefs*, vol. 70, no. 7, pp. 2355-2359, July 2023.

[2] P. Gu, D. Zhao and X. You, "Analysis and Design of a CMOS Bidirectional Passive Vector-Modulated Phase Shifter," *IEEE Trans. Circuits Syst. I, Reg Papers*, vol. 68, no. 4, pp. 1398-1408, April 2021.

[3] A. Li et al., "A Fully Integrated 27.5-30.5 GHz 8-Element Phased-Array Transmit Front-end Module in 65 nm CMOS," in *Proc. IEEE Asian Solid-State Circuits Conf. (A-SSCC)*, Macau, Macao, 2019.

[4] Y. Tian et al., "A 26-32GHz 6-bit Bidirectional Passive Phase Shifter with 14dBm IP1dB and 2.6° RMS Phase Error for Phased Array System in 40nm CMOS," in *IEEE MTT-S Int. Microw. Symp. Dig. (IMS)*, San Diego, CA, USA, 2023.

[5] E. Kobal, T. Siriburanon, R. B. Staszewski and A. Zhu, "A 28-GHz Switched-Filter Phase Shifter with Fine Phase-Tuning Capability Using Back-Gate Biasing in 22-nm FD-SOI CMOS," in *Proc. IEEE European Solid-State Circuit Conf. (ESSCIRC)*, Grenoble, France, 2021.

[6] Y. -T. Chang, Z. -W. Ou, H. Alsuraisry, A. Sayed and H. -C. Lu, "A 28-GHz Low-Power Vector-Sum Phase Shifter Using Biphase Modulator and Current Reused Technique," *IEEE Microw. Wireless Compon. Lett.*, vol. 28, no. 11, pp. 1014-1016, Nov. 2018.

[7] J. Park, G. Jeong and S. Hong, "A Ka-Band Variable-Gain Phase Shifter With Multiple Vector Generators," *IEEE Trans. Circuits Syst. II, Exp. Briefs*, vol. 68, no. 6, pp. 1798-1802, June 2021.

[8] Y. Yu et al., "An 18 30 GHz Vector-Sum Phase Shifter With Two-Stage Transformer-Based Hybrid in 130-nm SiGe BiCMOS," *IEEE Trans. Circuits Syst. I, Reg Papers*, vol. 70, no. 12, pp. 5138-5151, Dec. 2023.

A 10-40 GHz Phase Invariant Low Noise Variable Gain Amplifier with Pole Staggering Bandwidth Extension Technology

Xuan Li, Yang Liu, Guiyue Mao, Fanyi Meng*

1 School of Microelectronics, Tianjin University, Tianjin, China
2 Tianjin Key Laboratory of Imaging and Sensing Microelectronic Technology, Tianjin, China
3 Key Laboratory of Organic Integrated Circuit, Ministry of Education, Tianjin University, Tianjin ,China
*Email: mengfanyi@tju.edu.cn

Abstract—This paper presents an ultra-wideband (UWB) low noise variable gain amplifier (LNVGA) implemented in 0.18 μm SiGe BiCMOS process. Pole staggering bandwidth extension and peaking inductance techniques are used to achieve a flat broadband response. The phase invariance under different gain states is realized by the phase compensation network. The proposed LNVGA achieves a gain tunability range of 15 dB with a 0.5-dB fine-tuning step. The root mean square (RMS) phase error simulated in the 10-40 GHz range is less than 1.9°. The RMS gain error is less than 0.55 dB, with the minimum error being 0.13 dB at 27 GHz. At maximum gain, the minimum noise figure is 3.71 dB at 10 GHz. Additionally, the chip occupies a silicon area of 0.27 mm², including all pads, and consumes 53 mW.

Index Terms—LNVGA, SiGe BiCMOS, pole staggering, peaking inductance, phase compensation.

I. INTRODUCTION

Phased array technology is a crucial method for enhancing link robustness and effective isotropic radiated power (EIRP) in millimeter-wave wireless systems. In phased array systems, the signal amplitude is regulated by a variable gain amplifier (VGA) to minimize side lobes. Due to its capability to adapt to signals across various frequency bands, ultra-wideband (UWB) VGA has garnered significant attention in multi-standard and multi-frequency communication systems. To ensure system compatibility and flexibility, UWB phased array systems necessitate that the VGA maintains phase invariance while precisely controlling amplitude.

In many proposed VGA applications for millimeter-wave phased array systems, the current steering structure of the VGA can maintain input impedance while adjusting gain. However, the parasitic capacitance of the drain transistor introduces varying insertion phases during this tuning process. Although low phase error can be achieved through phase compensation in [1], this method typically has a narrow bandwidth. While a lower phase error was attained in the frequency range of 27-43 GHz in [2], the gain adjustment range remains quite limited. Additionally, due to the impact of noise on the high dynamic range of phased array systems, the VGA can be designed as a low-noise variable gain amplifier (LNVGA).

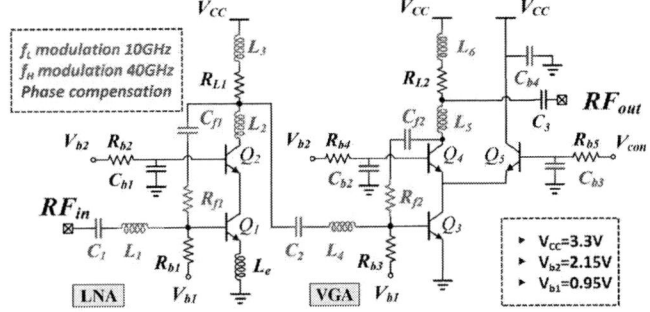

L_1	R_{f1}	C_{f1}	L_2	L_3	C_2
170pH	300Ω	90fF	300pH	180pH	200fF
R_{f2}	C_{f2}	L_4	L_5	C_{b1}	C_{b2}
2kΩ	90fF	120pH	90pH	200fF	2pF
C_{b3}	L_6	L_e	R_{L1}	R_{L2}	Q_{1-5}
150fF	150pH	30pH	70Ω	45Ω	4×2.5um

Fig. 1. Schematic of the proposed LNVGA.

In this work, a 10-40 GHz LNVGA is proposed. The structure of this paper is as follows. The circuit design of the proposed LNVGA is presented in Section II. Section III shows the simulated results. The conclusion is arranged in Section IV.

II. CIRCUIT DESIGN OF LNVGA

A. Structure of Proposed LNVGA

Fig. 1 shows the complete circuit schematic of the proposed LNVGA based on the 0.18 μm SiGe BiCMOS process. As the first stage, LNA uses cascode structure to enhance gain and isolation. In the second stage, the gain is regulated by adjusting the base bias of the transistor Q_5 in the current steering VGA. They all use resistance-capacitance (RC) feedback and peaking inductance to extend the bandwidth. Currently, the match between stages is facilitated by an LC network.

B. UWB Based on Pole Staggering Technology

Due to the frequency characteristics of the transistors and the influence of parasitic capacitance, the gain of amplifiers decreases as the frequency increases. The high-frequency gain

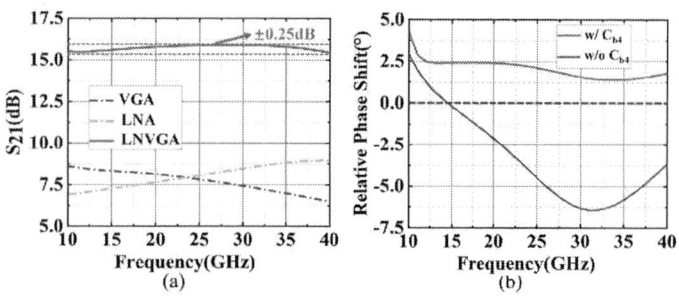

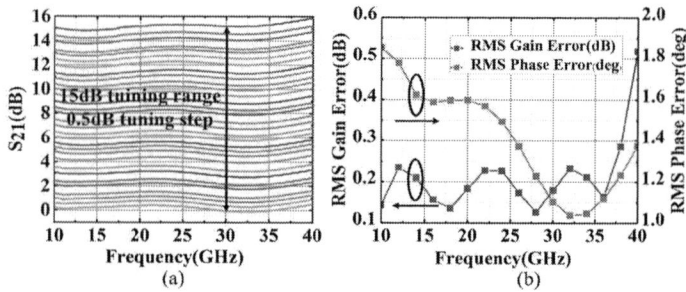

Fig. 2. (a) Simulated S_{21} for LNA, VGA, and LNVGA and (b) influence of C_{b4} on relative phase shift.

Fig. 4. (a) Simulated S_{21} and (b) RMS gain and phase error of the LNVGA.

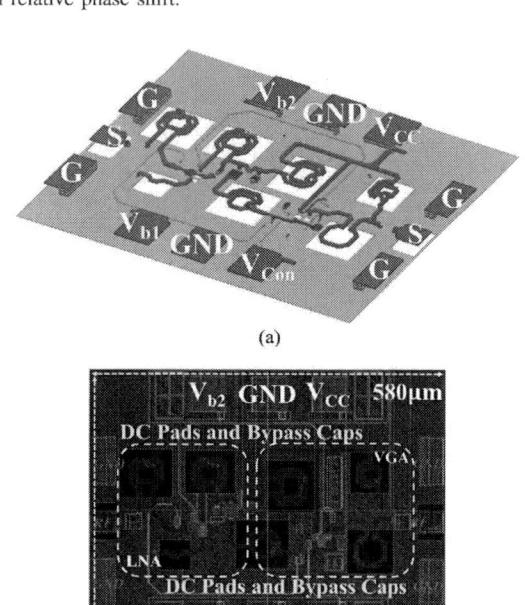

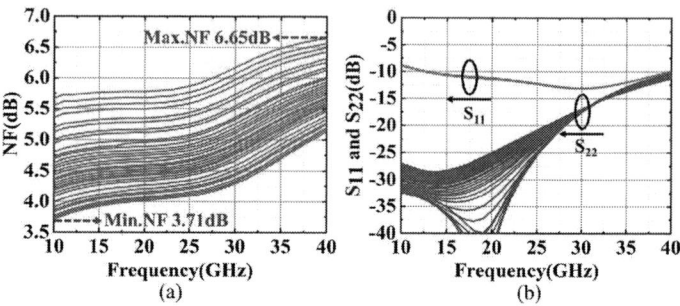

Fig. 5. (a) Simulated NF and (b) S_{11} and S_{22} of the LNVGA.

Fig. 3. (a) EM model and (b) Layout of the proposed LNVGA.

can be enhanced by employing peak inductance technology to improve the impedance of the high-frequency output, but this method alone cannot achieve a low gain error in UWB. As shown in Fig. 2(a), the UWB technology with pole staggering presented in this paper expands the bandwidth by introducing RC feedback to cancel parasitic capacitance and control cascaded LNA and VGA reach their maximum gain at 39 GHz and 10 GHz, respectively, thereby expanding the bandwidth.

In addition, utilizing input stage series inductance and emitter negative feedback inductance for input impedance matching in UWB applications helps to prevent noise performance degradation caused by resistance matching. Output impedance matching is accomplished by adjusting the peak inductance value.

C. Method of Phase Invariance

When tuning the gain by modifying the base bias of the discharge tube Q_5, the bias of Q_3 remains constant, thereby not affecting the phase. According to [2], variations in the parasitic capacitance of Q_4 and Q_5 induce a phase shift in

the current steering VGA during tuning. To achieve a low phase shift across a broad frequency range, smaller transistors are preferred to minimize parasitic capacitance. However, this approach has a limited impact on phase invariance.

To maintain a consistent phase across different gain states, the influence of the peaking inductance at the load of the current steering structure on the phase is analyzed in [1]. Additionally, a phase compensation using a cascode is proposed in [3]. However, these methods achieve low phase errors only within a limited bandwidth. This paper adopts a technique for neutralizing transistor parasitic capacitance by tuning C_{b2} and C_{b3}, and utilizes their opposite phase shifts to achieve low phase error. Furthermore, the error can be minimized by selecting appropriate values for L_5 and L_6 of the peaking inductor. In addition, incorporating C_{b4} effectively mitigates the impact of parasitic capacitance in the interconnect on the circuit phase, as illustrated in Fig. 2(b).

III. SIMULATED RESULT

In Fig. 3, the electromagnetic model (EM) model and layout of the proposed LNVGA is presented. The core area is 0.48 mm × 0.2 mm and the chip area size is 0.58 mm × 0.46 mm including pads. Fig. 4 shows that the proposed LNVGA achieves a gain range of 0 to 15 dB across a frequency spectrum of 10 to 40 GHz, with a root mean square (RMS) gain error of less than 0.55 dB and an RMS phase error of less than 1.9°. The minimum noise figure for this study is 3.71 dB, as shown in Fig. 5(a). The simulated S_{11} and S_{22} parameters, depicted in Fig. 5(b), and they are both below -8.7 dB and -10 dB, respectively, across the entire 10-40 GHz frequency range. Fig. 6 shows the fundamental and third harmonic power levels at 40 GHz. The input third-order intercept point (IIP3)

979-8-3315-2209-4/25 $31.00 © 2025 IEEE

TABLE I
PERFORMANCE SUMMARY AND COMPARISON

Reference	MWCL 2023 [1]	MWTL 2023 [2]	TMTT 2021 [4]	EuMIC 2022 [5]	TMTT 2020 [6]	IMS 2020 [7]	JSSC 2016 [8]	**This work**
Technology	65nm CMOS	65nm CMOS	55nm CMOS	130nm SiGe	45nm CMOS	120nm SiGe	SiGe HBT	**180nm SiGe**
Freq(GHz)	38-40	27-43	6.5-12	8-18	22-44	28-39	10-14.4	**10-40**
Relative BW(%)	5.1	45.7	59.4	76.9	66.7	32.8	36.1	**120**
Max. Gain (dB)	22	16.3	20.7	13.2	26.2	12.5	13	**15**
Gain range(dB)	16	8.4	18	30	16	18	22	**15**
Phase Error(°)	<2.67	<1.46	<4.5	<12.3	<6	<1	<2[a]	**<1.95**
Gain Error(dB)	N/A	<0.45	<0.6	N/A	<1.9	<1	N/A	**<0.55**
Min. NF(dB)	N/A	N/A	3.26	1.93	3	N/A	5.1	**3.71**
IIP3(dBm)	N/A	N/A	-6	1	-18	N/A	-3	**-5**
Pdc(mW)	38	27.6	75	24	112	45	83	**53**
Area(mm²)	0.37	0.25	0.98	0.58	1.86	0.08	0.7	**0.27**

[a]Relative phase shift

is -5 dBm. The simulated output power and power gain as a function of input power at 40 GHz are presented in Fig. 7.

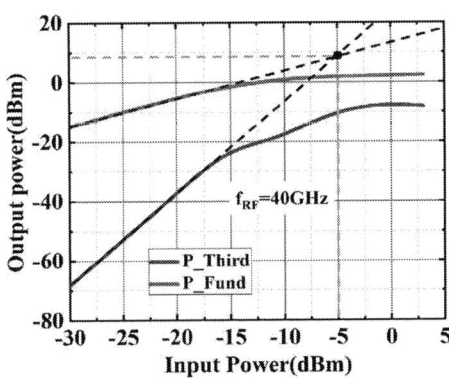

Fig. 6. Simulated fundamental and third order powers of the LNVGA.

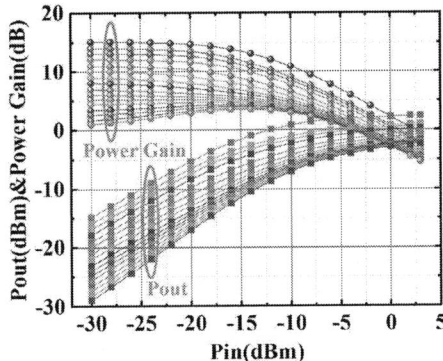

Fig. 7. Simulated power performances of the LNVGA.

The overall performance comparison with prior art is presented in Table I, where the proposed LNVGA exhibits the highest relative bandwidth and has extremely low RMS gain error and phase error in the frequency range of 10-40 GHz.

IV. CONCLUSION

This paper proposes an UWB LNVGA that operates in the 10-40 GHz frequency range and is based on a 0.18μm SiGe BiCMOS process. The LNVGA is composed of a cascaded cascode LNA and a current steering VGA. By employing pole staggering technology and adjusting the compensation capacitance to neutralize the parasitic capacitance, a flat UWB gain and extremely low phase error can be achieved. The peak inductor used can not only enhance high-frequency gain but also serve for phase compensation. Simulated results indicate the gain varies from 0 to 15 dB with a 0.5 dB gain tuning step. The RMS gain error is less than 0.55 dB, and the RMS phase error is less than 1.9°. S_{11} and S_{22} are both less than -8.7 dB across the entire 10-40 GHz frequency band, and less than -10 dB within the 12-40 GHz range. The proposed LNVGA with low phase and gain errors provides a practical solution for millimeter-wave UWB phased array systems.

ACKNOWLEDGMENT

This work was supported by the National Key Research and Development Program of China under Project 2023YFB4403200, the National Natural Science Foundation of China under Project 62271347 and U21A2045.

REFERENCES

[1] J. Tsai and C. Lin, "A 40-GHz 4-Bit Digitally Controlled VGA With Low Phase Variation Using 65-nm CMOS Process," *IEEE Microw. Wireless Compon. Lett.*, vol. 29, no. 11, pp. 729-732, Nov. 2019.

[2] J. -H. Tsai and Y. -T. Chen, "A 27–43 GHz CMOS Body-Biased Digital Current-Steering VGA With 4 Bit and Low Phase Shift," *IEEE Microw. Wireless Technol. Lett.*, vol. 33, no. 2, pp. 196-199, Feb. 2023.

[3] F. Ellinger, U. Jorges, U. Mayer and R. Eickhoff, "Analysis and Compensation of Phase Variations Versus Gain in Amplifiers Verified by SiGe HBT Cascode RFIC," *IEEE Trans. Microw. Theory Techn.*, vol. 57, no. 8, pp. 1885-1894, Aug. 2009.

[4] H. Gao et al., "A 6.5–12-GHz Balanced Variable-Gain Low-Noise Amplifier With Frequency-Selective Gain Equalization Technique," *IEEE Trans. Microw. Theory Techn.*, vol. 69, no. 1, pp. 732-744, Jan. 2021.

[5] K. Altintas, T. A. Ozkan, M. Yazici, M. Kaynak and Y. Gurbuz, "A 8–18 GHz Low Noise Variable Gain Amplifier with 30 dB Gain ControlRange," *in Proc. 17th Eur. Microw. Integr. Circuits Conf. (EuMIC)*, Milan, Italy, 2022, pp. 304-307.

[6] L. Gao and G. M. Rebeiz, "A 22–44-GHz Phased-Array Receive Beamformer in 45-nm CMOS SOI for 5G Applications With 3–3.6-dB NF," *IEEE Trans. Microw. Theory Techn.*, vol. 68, no. 11, pp. 4765-4774, Nov. 2020.

[7] R. B. Yishay and D. Elad, "A Compact Frequency-Tunable VGA for Multi-Standard 5G Transceivers," in *IEEE MTT-S Int. Microw. Symp. Dig. (IMS)*, Los Angeles, CA, USA, 2020, pp. 325-328.

[8] F. Padovan, M. Tiebout, A. Neviani and A. Bevilacqua, "A 12 GHz 22 dB-Gain-Control SiGe Bipolar VGA With 2° Phase-Shift Variation," *IEEE J. Solid-State Circuits*, vol. 51, no. 7, pp. 1525-1536, July 2016.

A 127-157 GHz Power Amplifier with 12dBm Saturated Output Power and 9.5% Power-Added Efficiency in 65nm CMOS

Jiawei Chen, Yi Wu, Yuwen Long, Zongyang Zhang, and Guangyin Feng[*]

School of Microelectronics, South China University of Technology, Guangzhou, China

[*]Email: gyfeng88@scut.edu.cn

Abstract—**This paper presents a broadband power amplifier (PA) operating in D-band, which consists of a three-stage driver amplifier with stagger tuning technique and one-stage PA with slot-balun-based output matching. The slot-balun-based output matching network presents a small insertion loss as well as a large bandwidth. Designed in a 65-nm CMOS process, the proposed broadband PA achieves a 3dB bandwidth (BW) from 127GHz to 157GHz, and S_{11} and S_{22} are less than -10dB within the 3dB bandwidth. In addition, ultimately achieves a peak gain of more than 24 dB, a saturated output power (P_{sat}) of 12dBm, and a maximum power-added efficiency (PAE_{max}) of 9.5%.**

Keywords—**Power Amplifier (PA), CMOS, D-band, Slot-Balun, Wideband Matching.**

I. Introduction

With the rapid development of communication technology, millimeter-wave frequency band has been paid attention to by academia and industry due to its rich spectrum resources and advantages of large broadband and high communication rate. Among them, the D band (110-170 GHz), which is known as the "atmospheric window", has a high research value due to its low group dispersion, low signal broadening, difficult signal distortion, and high transmission stability, which has a great potential for application in imaging, communication, and military industry [2]-[4].

CMOS process is widely used in millimeter-wave integrated circuit due to its advantages of easy integration and low cost. As the most energy-consuming module in the transmitting system, the power amplifier is crucial, and the performance of the PA directly affects the transmitted power as well as the energy consumption of the system. In order to realize the high rate of transceiver system communication, the amplifier is required to have a large operating bandwidth in the time domain [6]. Therefore, realizing PAs with large bandwidth and high output power is one of the most important challenges in the millimeter-wave band.

To achieve a larger bandwidth, there are several existing ways as shown below. One is the distributed power amplifier (DPA) structure, as shown in Fig.1(a). Conventional PA stages have reduced power gain and efficiency due to the large parasitic capacitance associated with their large transistor size, and the power gain is skewed towards low frequencies. In contrast, DPA shrink the transistor size and use a distributed matching network for each small sized CMOS transistor, which ultimately allows for ultra-wideband [1]. The disadvantage of this approach is that it requires a large area as well as small transistors resulting in limited output power. The second way is to use a higher order matching network, as shown in Fig.1(b), such as transformers. Due to the multi-pole nature of the higher order network, the poles can be designed to stay within the bandwidth, thus achieving a broadband effect [7]. The

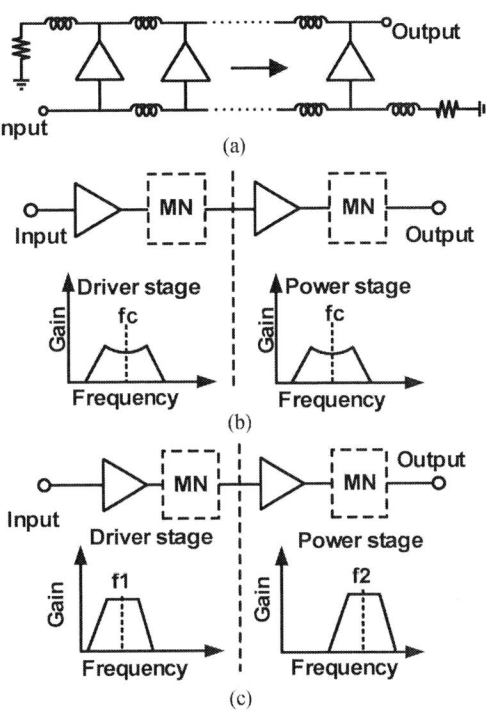

Fig. 1. Broadband amplifier types (a) Distributed power amplifier (b) High order matching network (c) Stagger tuning technique

disadvantage of this approach is that the gain ripple is large. The last one is the gain stagger tuning, as shown in Fig.1(c), where different stages are designed to have gains in different frequency bands, resulting in a large bandwidth [7]. However, it is more difficult to achieve a flat gain-frequency response.

In this paper, a broadband amplifier is designed by using stagger tuning technique and slot-balun-based output matching. It achieves a 3-dB BW of 30GHz with a peak gain of 24 dB and a P_{sat} of 12 dBm.

II. Circuit Design

The schematic of proposed PA is shown in Fig.2, which is a 4-stage amplifier structure with neutralization capacitor compensation technique for each stage to improve the power gain as well as reverse isolation. Since this design is used in a D-band transmitter, the input of PA is connected to a differential mixer with 100Ω output impedance, while the output is connect to a single-ended antenna with 50Ω input impedance. The first three DR stages of the circuit use low-k transformer matching networks and the PA stage uses a slot-balun for low-loss broadband matching.

979-8-3315-2209-4/25 $31.00 © 2025 IEEE

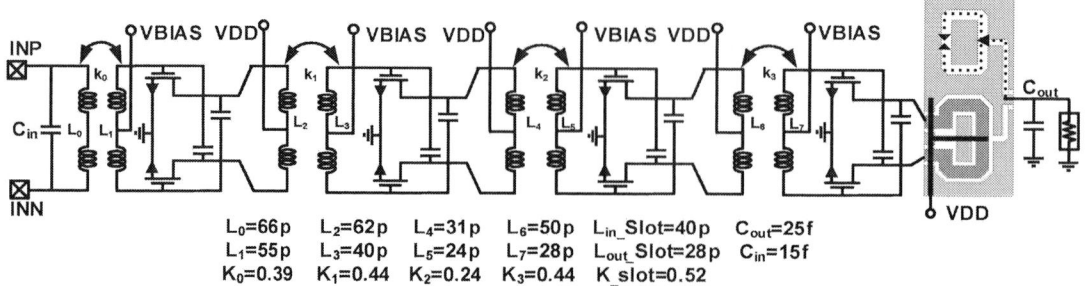

L₀=66p L₂=62p L₄=31p L₆=50p Lin_Slot=40p Cout=25f
L₁=55p L₃=40p L₅=24p L₇=28p Lout_Slot=28p Cin=15f
K₀=0.39 K₁=0.44 K₂=0.24 K₃=0.44 K_slot=0.52

Fig. 2. Schematic of the proposed PA.

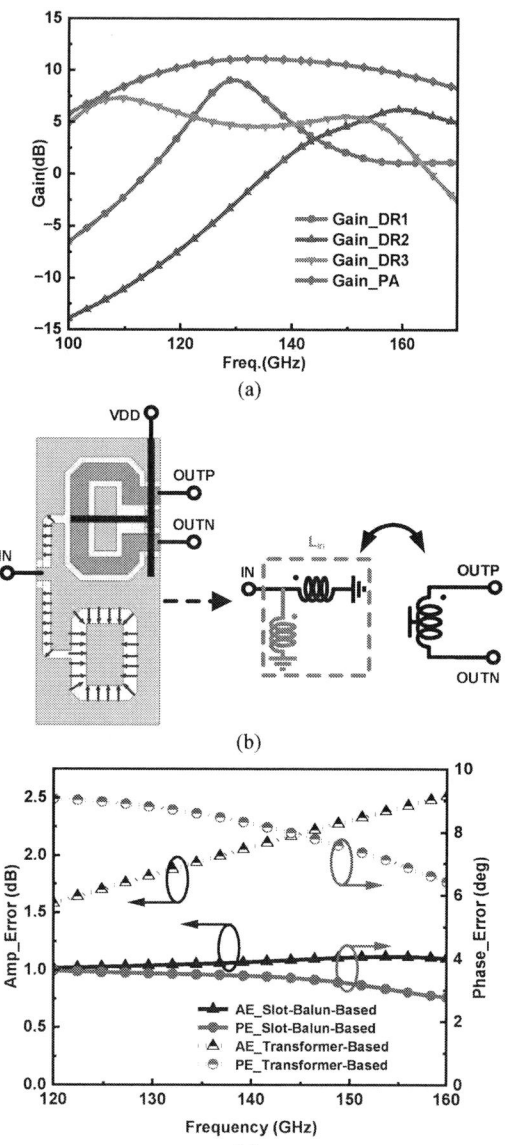

Fig. 3. (a) Gain distribution curve for each stage of the circuit, (b) Slot-balun and its equivalent lumped model, (c) Phase error and amplitude error comparison

A. DR Stage

The first three DR stage matching network of the designed circuit uses low-k transformer network, which makes the impedance curve of the transformer network more broadband and flat, and by designing the transformers for different DR stages, it makes the three DR stages have flat gain curves at different frequencies to achieve gain stagger tuning effect. As shown in Fig. 3(a), the peaks of DR stages gain curves are in different frequencies and the final output broadband effect. And the amplifier stages used in the circuit are differential pairs with neutralizing capacitors, which improves the power gain and reverse isolation of the amplifier stages.

B. PA Stage

In order to output larger power, the output stage transistors of conventional PAs are generally set larger, which brings a large parasitic capacitance. This leads to the fact that conventional transformer networks cannot achieve better power matching. The slot-balun network used in this paper has high-order matching characteristics and can better realize the power matching of the output stage under D-band.

Conventional PA designs use a transformer matching network as a balun for differential to single-ended conversion. In this case, the transformer single-ended side has a line to ground, resulting in less balanced phase and amplitude of the transformer network and a larger insertion loss. The slot-balun matching network used in the output stage is structured to achieve nearly balanced differential to single-ended conversion [8]. The structure and its equivalent lumped model are shown in Fig.3(b). Fig.3(c) shows the simulated phase error and amplitude error of transformers and slot-balun with the same inductance value and coupling factor. Compared to the conventional transformer balun, the slot-balun achieves a smaller insertion loss as well as a larger bandwidth.

III. SIMULATION RESULTS

Fig.4 shows the simulation results of S-parameters. The proposed PA achieves a simulated peak gain of 24.4dB, 3dB bandwidth of 30GHz. S_{11} and S_{22} are less than -10dB within the 3dB bandwidth. Large-signal simulation results are shown in Fig.5. The saturated output power is about 12dBm and the PAE_{max} is 9.5%. Fig.6 shows the saturated output power and maximum efficiency under different frequencies. It can be seen that proposed PA presents more than 11dBm output power with PAE greater than 8% within the 3dB bandwidth. Table I summarizes the performance of the proposed PA and compares with reported D-band broadband PAs in CMOS process. This

979-8-3315-2209-4/25 $31.00 © 2025 IEEE

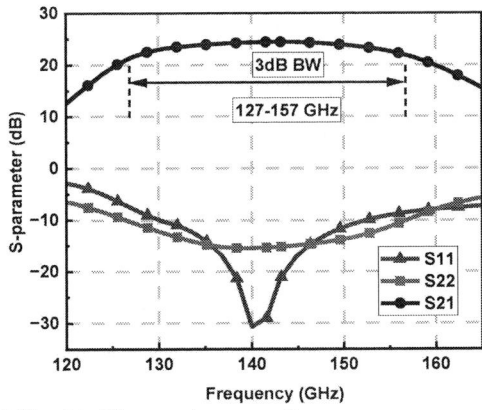

Fig. 4. Simulated S-parameters versus frequency

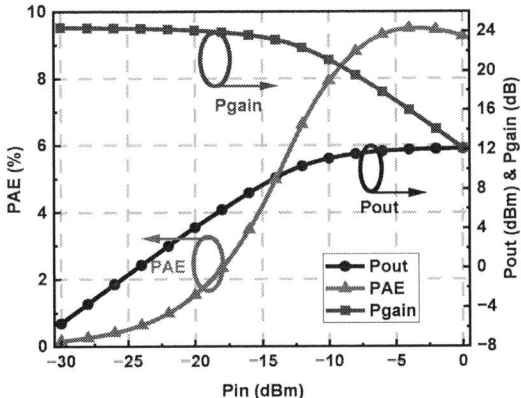

Fig. 5. Simulated P_{out}, gain, and PAE versus input power

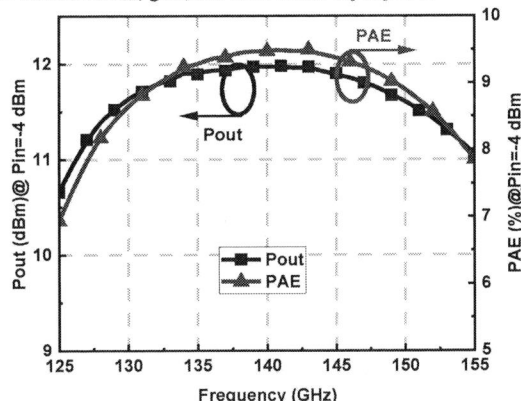

Fig. 6. Simulated P_{sat} and peak PAE versus frequency

TABLE I. PERFORMANCE SUMMARY AND COMPARISON OF D-BAND BROADBAND PAs IN CMOS PROCESS

Reference	This Work	RFIC'22 [3]	RFIC'18 [4]	TCAS-I'20[2]	RFIC'21 [6]
Process	65nm	28nm	40nm	28nm	45nm SOI
Topology	differential	4-way combing	4-way combing	differential	4-way combing
Frequency (GHz)	127-157	124-152	125-142	121-143	130-151
FBW(%)	21.1	20.3	12.8	16.7	15
P_{sat} (dBm)	12	16.2	14.8	8	16
PAE_{max} (%)	9.5	8.6	8.9	6.6	12.5
Gain(dB)	24.4	22.6	20.3	22.5	22.2

ACKNOWLEDGMENT

This work was supported in part by the National Key R&D Program of China under Grant 2023YFB4403200, in part by National Natural Science Foundation of China under Grant 62474069, and in part by Guangdong Basic and Applied Basic Research Foundation under Grant 2023A1515012311.

REFERENCES

[1] O. El-Aassar and G. M. Rebeiz, "A Compact DC-to-108GHz Stacked-SOI Distributed PA/Driver Using Multi-Drive Inter-Stack Coupling, Achieving 1.525THz GBW, 20.8dBm Peak P1dB, and Over 100Gb/s in 64-QAM and PAM-4 Modulation," in *IEEE International Solid-State Circuits Conference - (ISSCC)*, San Francisco, CA, USA, 2019, pp. 86-88.

[2] X. Tang et al., "Design of D-Band Transformer-Based Gain-Boosting Class-AB Power Amplifiers in Silicon Technologies," *IEEE Transactions on Circuits and Systems I: Regular Papers*, vol. 67, no. 5, pp. 1447-1458, May 2020

[3] J. Zhang, T. Wu, Y. Chen, J. Ren and S. Ma, "A 124–152 GHz > 15-dBm Psat 28-nm CMOS PA Using Chebyshev Artificial- Transmission-Line-Based Matching for Wideband Power Splitting and Combining," in *IEEE Radio Frequency Integrated Circuits Symposium (RFIC)*, Denver, CO, USA, 2022, pp. 187-190.

[4] D. Simic and P. Reynaert, "A 14.8 dBm 20.3 dB Power Amplifier for D-band Applications in 40 nm CMOS," in *IEEE Radio Frequency Integrated Circuits Symposium (RFIC)*, Philadelphia, PA, USA, 2018, pp. 232-235

[5] K. Katayama, M. Motoyoshi, K. Takano, L. C. Yang and M. Fujishima, "133GHz CMOS power amplifier with 16dB gain and +8dBm saturated output power for multi-gigabit communication," in *European Microwave Integrated Circuit Conference*, Nuremberg, Germany, 2013, pp. 69-72.

[6] S. Li and G. M. Rebeiz, "A 130-151 GHz 8-Way Power Amplifier with 16.8-17.5 dBm Psat and 11.7-13.4% PAE Using CMOS 45nm RFSOI," in *IEEE Radio Frequency Integrated Circuits Symposium (RFIC)*, Atlanta, GA, USA, 2021, pp. 115-118

[7] W. Wu, X. Bao, S. Chen, Y. Wang and L. Zhang, "A 67.8-to-108.2GHz Power Amplifier with a Three-Coupled-Line-Based Complementary-Gain-Boosting Technique Achieving 442GHz GBW and 23.1% peak PAE," in *IEEE International Solid-State Circuits Conference (ISSCC)*, San Francisco, CA, USA, 2024, pp. 526-528

[8] Y. Wu, J. Liu, J. Li, G. Feng, F. Meng and Y. Wang, "A 70.6–98.4GHz Power Amplifier With Equivalent Low-K Matching Network and Slot-Balun-Based Power Combiner in 40-nm CMOS," in *IEEE International Symposium on Radio-Frequency Integration Technology (RFIT)*, Chengdu, China, 2024, pp. 1-3

table shows that this PA achieves higher output power and PAE at a large bandwidth. In addition, this design achieves high saturated output power without multi-way power combining.

IV. CONCLUSION

In this paper, a broadband PA covering a bandwidth from 127GHz to 157GHz is designed in a 65nm CMOS process. A low-k transformer network is used for interstage matching to achieve the broadband effect through the complementary gain curves of each stage. The PA stage uses a slot-balun network to obtain low loss and phase imbalance. The final simulation results show that the design achieves a peak gain of 24dB, P_{sat} of 12dBm, and a PAE_{max} of 9.5%.

An Adaptive Dead-Time Controlled Hybrid Dual-Path Buck Converter With Inductor Current Reduction Technique and High Power Density

Zishuo Li[1,2,3], Zhen Lin[1,2,3], Qingsong Zhao[1,2,3], Jian Wang[1,2,3], Fanyi Meng[1,2,3,*]

1. School of Microelectronics, Tianjin University, Tianjin, China
2. Tianjin Key Laboratory of Imaging and Sensing Microelectronic Technology, Tianjin, China
3. Key Laboratory of Organic Integrated Circuit, Ministry of Education, Tianjin University, Tianjin, China
*Email: mengfanyi@tju.edu.cn

Abstract—In response to the increasing demand for compact, high-performance power management in mobile, wearable, and IoT devices, this paper introduces a dual-path buck converter (DPBC) with a high step-down conversion ratio. The proposed architecture leverages a flying capacitor to establish two parallel power paths, thereby effectively reducing the DC level of the inductor current and significantly lowering conduction losses. The reduction in inductor current loss directly contributes to enhanced power density, making the design particularly suited for highly integrated systems. In addition, an adaptive dead-time controller is implemented to achieve zero voltage switching (ZVS) over a wide load range, ensuring optimal performance under varying conditions. Fabricated in a 55 nm CMOS process and operating at 20 MHz, the proposed DPBC demonstrates a peak efficiency of 88.5% at V_{IN} = 3.3 V and I_{LOAD} = 300 mA. Overall, the design provides an attractive trade-off between efficiency, power density, and integration, marking it as a promising solution for next-generation power converters in compact electronic systems.

Index Terms—Dual-path buck converter, hybrid Converter, inductor current reduction, adaptive dead-time control, power density.

I. INTRODUCTION

Modern mobile, wearable, and IoT devices integrate sensing, computation, and communication functions into SoCs fabricated in nm-scale CMOS processes, operating at low voltages (0.6–1.2 V). Li-ion batteries, with their high energy density, are widely used, requiring high power density DC-DC converters to efficiently step down the 2.8–4.2 V supply while maintaining system reliability.

Conventional buck converters (CBCs) rely on a single current path through a bulky power inductor L, leading to excessive heat dissipation due to its high R_{DCR}. The demand for miniaturization further exacerbates this issue, as small inductors inherently exhibit high R_{DCR} in the hundreds of milliohms range.

The DPBC architecture leverages a flying capacitor C_{fly} to establish two parallel current paths for I_{LOAD}, reducing IL,dc and mitigating conduction losses, as shown in Fig.1. Acting as an SPDT switch, DPBC alternates between charging C_{fly} and discharging it in parallel with L, effectively lowering power loss. By redistributing current flow and alleviating heat

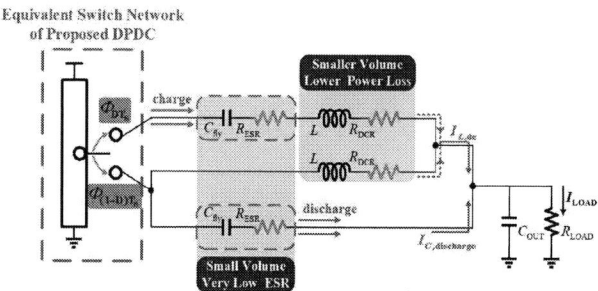

Fig. 1. Inductor current reduction strategy in DPBC.

dissipation from R_{DCR}, DPBC achieves higher efficiency than CBCs, making it a promising solution for compact, low- power applications.

Like most hybrid topologies, DPBC can be realized with a relatively smaller inductor and capacitor. Besides, the voltage swing at switching nodes is reduced in the DPBC structure as C_{fly} takes part of the voltage drop, contributing to lower voltage stress and switching loss of power switches, as well as smaller inductor ripple. Recently, various dual-path topologies have been explored [1]. A modified DPBC with an auxiliary switch [2] extends the output range but suffers from poor efficiency at high step-down ratios. While the topology in [3] meets this requirement, its complex driving scheme demands multiple off-chip components. Dual-path buck-boost converters in [4], [5] build on earlier DPBC designs but fail to achieve a high step-down ratio in buck mode.

This paper is organized as follows. Section II summarizes and compares several existing popular inductor power loss reduction strategies, then proposes a DPBC structure with high step-down ratio and analyzes its operation principles, including its inductor current reduction technique. The circuit implementation is given in Section III. Section IV presents simulation results and Section V concludes the paper.

979-8-3315-2209-4/25 $31.00 © 2025 IEEE

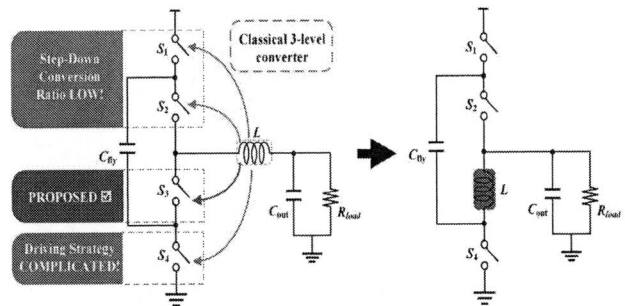

Fig. 2. Comparison among topology transformation strategies (left) and the proposed DPBC (right).

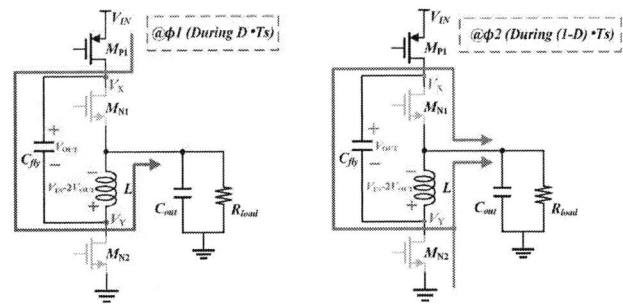

Fig. 3. Two-phase circuit operation of the proposed DPBC.

II. TOPOLOGY OF DPBC WITH HIGH STEP-DOWN CONVERSION RATIO

A. Comparison of inductor power loss reduction strategies

As I_{LOAD} increases, significant power loss occurs due to the high R_{DCR} of compact inductors. Although reducing R_{DCR} can decrease conduction loss, it results in larger inductor volume, which contradicts the trend toward compact portable devices. Instead, optimizing $I_{L,\text{DC}}$ and inductor current ripple ΔI_L is a more effective approach for minimizing power loss.

Multi-level converters utilizing flying capacitors effectively reduce ΔI_L by lowering inductor voltage stress, but their four-switch topology introduces additional switching and conduction losses [6]. Multi-phase interleaved converters mitigate $I_{L,\text{DC}}$ per inductor but increase system size and cost. Alternatively, increasing switching frequency enables on-chip integration, yet efficiency degrades due to the low Q factor of high-frequency inductors [7].

In fact, after appropriate structural transformation, the 3-level converter mentioned earlier will be very competitive. According to [3], there are four kinds of DPBC converter topologies generated through replacing one of the four switches of 3-level converter with an inductor respectively.

With S_1 or S_2 replaced by an inductor, it's easy to find that their VCRs are both higher than 1/2 based on inductor volt-second balance and capacitor charge balance, which has been reported in [1] and [2] respectively. [3] has suggested to take the place of S_4 with an inductor to obtain high step-down conversion ratio DPBC. While as depicted in Fig. 3, adopting an inductor to replace S_3 is more conductive to achieve high step-down conversion ratio under easier driving strategy.

B. Operation of the proposed DPBC

The proposed high step-down conversion ratio DPBC consists of three power switches (M_{P1}, M_{N1}, and M_{N2}), one flying capacitor (C_{fly}), one power inductor (L), and one output capacitor (C_{OUT}). As depicted in Fig.3, the proposed DPBC operates in two phases (Φ_1 and Φ_2).

During Φ_1, M_{P1} is turned on, while M_{N1} and M_{N2} are turned off. The input power supply V_{IN} is in series with C_{fly} and L, charging C_{fly} up to V_{OUT}. During this period, I_{OUT} equals I_L, and L is the sole provider of I_{LOAD}.

During Φ_2, M_{N1} and M_{N2} are turned on, while M_{P1} is turned off. L continues to supply I_{LOAD}, and the charge stored in C_{fly} during Φ_1 contributes to I_{LOAD}. In this phase, I_{OUT} is composed of I_L and I_C.

According to the inductor volt-second balance, the voltage conversion ratio (VCR) of the proposed DPBC is given by:

$$\frac{V_{OUT}}{V_{IN}} = \frac{D}{1 + D} \quad (2)$$

where V_{IN} and V_{OUT} are the input and output voltages, respectively, T_S is the switching period, and D is the duty cycle ($0 \le D \le 1$). Over the duty cycle range, the VCR remains below 1/2.

With two parallel power paths established by C_{fly} and L, the DC inductor current ($I_{L,dc}$) is reduced compared to a conventional buck converter (CBC) under the same load conditions, leading to lower inductor power loss. The derivation is as follows: Assuming that the direction of I_C flowing from V_X to V_Y is positive, during Φ_1, we have:

$$\int_0^{DT_S} I_{OUT}(t)dt = \int_0^{DT_S} I_C(t)dt = \int_0^{DT_S} I_L(t)dt = DT_S I_{L,dc} \quad (3)$$

Similarly, for Φ_2:

$$\int_{DT_S}^{T_S} I_{OUT}(t)dt = \int_{DT_S}^{T_S} I_L(t)dt + \left| \int_{DT_S}^{T_S} I_C(t)dt \right| \quad (4)$$

From capacitor charge balance, the charge stored in C_{fly} during Φ_1 is fully transferred to the output in Φ_2:

$$\int_0^{DT_S} I_C(t)dt = \left| \int_{DT_S}^{T_S} I_C(t)dt \right| \quad (5)$$

Combining (3) and (5), we derive (4) as:

$$\int_{DT_S}^{T_S} I_{OUT}(t)dt = (1 - D)T_S I_{L,dc} + DT_S I_{L,dc} = T_S I_{L,dc} \quad (6)$$

which simplifies to:

$$\int_0^{T_S} I_{OUT}(t)dt = (1 + D)T_S I_{L,dc} \quad (7)$$

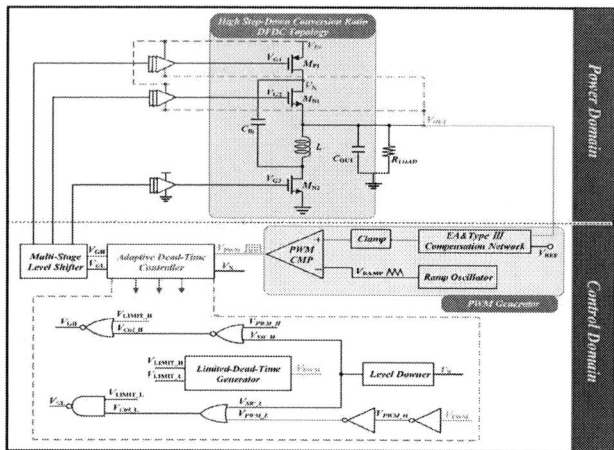

Fig. 4. Total block diagram of the proposed DPBC.

Thus, the load current is obtained as:

$$I_{LOAD} = \frac{\int_0^{T_S} I_{OUT}(t)dt}{T_S} = (1 + D)I_{L,dc} \qquad (8)$$

Finally, solving for the ratio of $I_{L,dc}$ to I_{LOAD}:

$$\frac{I_{L,dc}}{I_{LOAD}} = \frac{1}{1 + D} \qquad (9)$$

Since D ranges from 0 to 1, $I_{L,dc}$ in the proposed DPBC is always lower than I_{LOAD} across duty cycle variations. Additionally, with C_{fly} sharing part of the voltage drop, the voltage swing at switching nodes (V_X and V_Y) is reduced, leading to lower voltage stress and switching loss. Furthermore, based on $V_L = L\frac{dI_L(t)}{dt}$, the ripple current ΔI_L decreases with lower V_L, further mitigating inductor power loss.

III. CIRCUIT IMPLEMENTATION

The entire proposed DPBC system can be roughly divided into power domain and control domain as shown in Fig.4. The power domain includes gate drivers and power stage. The gate drivers are designed to provide enough driving capacities through buffers with relatively large size. And the power stage of the proposed DPBC takes a PMOS as the high side power switch. Therefore, based on flexible use of the input voltage and the output voltage depicted in Fig.4 (top), a fully integrated controller-driver system is completed through pure CMOS process, without any needs of bootstrap circuit or high voltage level shifter, which usually results in unnecessary voltage stress problem for process limitation.

For conduction power loss reduction, the non-overlap clock generator is usually designed to make sure that the two-phase power switches are not conductive at the same time. Different from this general strategy with a fixed dead-time introduced alone, an adaptive dead-time controller is suggested in this paper to realize zero voltage switching (ZVS) operation under almost the entire load range [8].

After the voltage level detection for the switching node V_X through the level downer, a sampling signal will be generated,

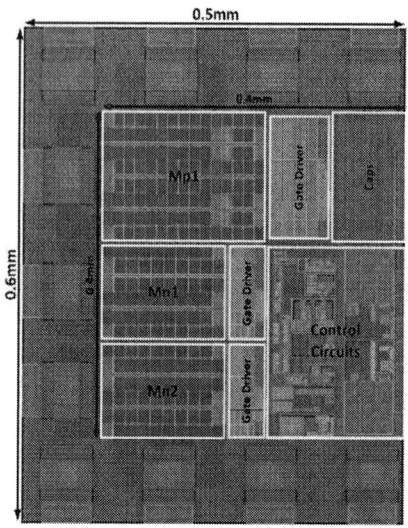

Fig. 5. The overall chip layout of the proposed DPBC.

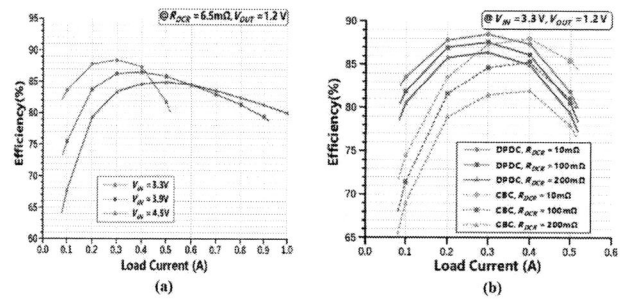

Fig. 6. Converter efficiency (a) efficiency variation with Load (b) comparison diagram between buck and the proposed converter.

containing the information about the rising and falling edge at V_X. Then, this sampling signal will be processed by different combinations of logic gates with the positive and reverse PWM signals, respectively. Later, two control signals, V_{Ctrl_H} and V_{Ctrl_L}, correspondingly aimed at high-side and low-side driving, will be generated.

It can be seen after a series of simulations that under heavy load conditions, the dead time generated only by V_{Ctrl_H} and V_{Ctrl_L} is longer than that really needed by the system. This situation deteriorates as I_{LOAD} increases continuously. To mitigate this issue, a conventional fixed dead-time generator is introduced to set the longest allowed dead time.

Through another logical operation for V_{Ctrl_H}, V_{Ctrl_L} and their own restrictive conditions V_{LIMIT_H} and V_{LIMIT_L}, the final dead-time control signals will be generated and used as the input signals of the multi-stage level shifter.

IV. SIMULATION RESULTS

The proposed DPBC is designed with a 1P8M 55nm CMOS process. Fig.5 shows the overall chip layout, occupying 0.5×0.6 mm^2. During the post-simulation process, it operates at a switching frequency of 20 MHz. For our simulation, a 47 nH external inductor with 6.5 mΩ R_{DCR}, along with two external

979-8-3315-2209-4/25 $31.00 © 2025 IEEE

TABLE I
PERFORMANCE SUMMARY AND COMPARISON

Design	This Work	2021 [1]	2022 [9]	2023 [10]
Process (nm)	55	65	65	65
Topology	DPBC	SIC	Tri-Path Buck	CPL-Buck
Max Load Current (A)	1	0.533	0.3	2
Peak Efficiency	88.5 %	78 %	84 %	92.9 %
Power Density (mW/mm^2)	381	150	21.6	300
Area (mm^2)	0.3	0.65	0.83	4

capacitors of 22 nF and 220 nF, are used for L, C_{fly}, and C_{OUT}, respectively.

Fig.6 (a) presents the power efficiency versus I_{LOAD} with R_{DCR} set to 6.5 mΩ. The simulation results for the proposed DPBC under $V_{\text{IN}} = 3.3$ V, 3.9 V, and 4.5 V are shown. Generally, it achieves high efficiency of over 80% across a relatively wide load range. At $V_{\text{IN}} = 3.3$ V and $I_{\text{LOAD}} = 300$ mA, the proposed DPBC achieves a peak efficiency of 88.5%. At $V_{\text{IN}} = 4.5$ V and $I_{\text{LOAD}} = 1$ A, the proposed DPBC reaches a full-load efficiency of 80%.

Fig.6 (b) illustrates the power efficiency comparison between the proposed DPBC and the CBC under several typical values of R_{DCR}. It can be observed that the proposed DPBC achieves better efficiency over a wide range of I_{LOAD} compared to the CBC. As R_{DCR} increases, the efficiency improvement due to the inductor current reduction technique becomes increasingly significant. For the typical case of $V_{\text{IN}} = 3.3$ V and $I_{\text{LOAD}} = 300$ mA, the efficiency of the proposed DPBC is improved by 1.1%, 2.9%, and 5% for R_{DCR} values of 10 mΩ, 100 mΩ, and 200 mΩ, respectively, compared to that of the CBC.

Finally, a performance comparison of the proposed DPBC with the state-of-the-art step-down converters is summarized in Table I. With ultra compact size consumed, the proposed DPBC in this paper achieves an overwhelming advantage in both power density and current density. Though it is not the best in terms of load range and efficiency among the mentioned designs, a maximum load current of 1 A and peak efficiency of 88.5% are both acceptable.

V. CONCLUSION

This paper presents a silicon-based hybrid dual-phase buck converter with an inductor loss reduction technique. By leveraging the DPBC topology, the converter effectively reduces $I_{\text{L,dc}}$ and ΔI_{L}, mitigating losses in compact inductors. Adaptive dead-time control ensures full ZVS operation across all load conditions. Operating under PWM voltage mode control, it supports an input voltage of 2.8-4.5 V, a 20 MHz switching frequency, and a 1.2 V output voltage. The chip occupies 0.3 mm², with a total system area of 2.52 mm², balancing efficiency and power density in a miniaturized design. Post-layout simulations confirm a peak efficiency of 88.5% in a

3.3 V-to-1.2 V application, aligning with recent high-quality designs. The converter achieves a maximum power density of 381 mW/mm², significantly outperforming similar designs.

REFERENCES

[1] N. Tang, W. Hong, B. Nguyen, Z. Zhou, J. -H. Kim and D. Heo, "Fully Integrated Switched-Inductor-Capacitor Voltage Regulator With 0.82-A/mm2 Peak Current Density and 78% Peak Power Efficiency," in *IEEE J. Solid-State Circuits.*, vol. 56, no. 6, pp. 1805-1815, June 2021.

[2] Yeunhee Huh, Sung-Wan Hong and Gyu-Hyeong Cho, "A Hybrid Structure Dual-Path Step-Down Converter With 96.2% Peak Efficiency Using 250-m Large-DCR Inductor", *IEEE J. Solid-State Circuits.*, vol. 54, no. 4, pp. 959-967, 2019.

[3] Shaowei Zhen, Rui Yang, Dongming Wu, Yufan Cheng and Ping Luo, "Design of Hybrid Dual-Path DC-DC Converter with Wide Input Voltage Efficiency Improvement", *2021 IEEE International Symposium on Circuits and Systems (ISCAS)*.

[4] A. Mishra, W. Zhu, B. Wicht and V. D. Smedt, "An All-1.8-V-Switch Hybrid Buck–Boost Converter for Li-Battery-Operated PMICs Achieving 95.63% Peak Efficiency Using a 288-m DCR Inductor," in *IEEE Trans. Power Electron.*, vol. 38, no. 3, pp. 3444-3454, March 2023.

[5] I. Park, J. Maeng, J. Jeon, H. Kim and C. Kim, "A Four-Phase Hybrid Step-Up/Down Converter With RMS Inductor Current Reduction and Delay-Based Zero-Current Detection," in *IEEE Trans. Power Electron.*, vol. 37, no. 4, pp. 3708-3712, April 2022.

[6] S. S. Amin and P. P. Mercier, "A Fully Integrated Li-Ion-Compatible Hybrid Four-Level DC–DC Converter in 28-nm FDSOI," in *IEEE J. Solid-State Circuits.*, vol. 54, no. 3, pp. 720-732, March 2019.

[7] Y. -W. Huang, T. -H. Kuo, S. -Y. Huang and K. -Y. Fang, "A Four-Phase Buck Converter With Capacitor-Current-Sensor Calibration for Load-Transient-Response Optimization That Reduces Undershoot/Overshoot and Shortens Settling Time to Near Their Theoretical Limits," in *IEEE J. Solid-State Circuits.*, vol. 53, no. 2, pp. 552-568, Feb. 2018.

[8] H. A. Zadeh, H. R. Kooshkaki, K. -Y. Lee and P. P. Mercier, "An Adaptive Constant-on-Time-Controlled Hybrid Multilevel DC–DC Converter Operating From Li-Ion Battery Voltages With Low Spurious Output," in *IEEE Trans. Power Electron.*, vol. 38, no. 5, pp. 5763-5776, May 2023.

[9] C. Wang, Y. Lu and R. P. Martins, "A Highly Integrated Tri-Path Hybrid Buck Converter With Reduced Inductor Current and Self-Balanced Flying Capacitor Voltage," in *IEEE Trans. Circuits Syst. I*: Regular Papers, vol. 69, no. 9, pp. 3841-3850, Sept. 2022.

[10] G. Cai, Y. Lu and R. Martins, "A Battery-Input Sub-1V Output 92.9% Peak Efficiency 0.3A/mm2 Current Density Hybrid SC-Parallel-Inductor Buck Converter with Reduced Inductor Current in 65nm CMOS," *2022 IEEE International Solid-State Circuits Conference (ISSCC)*, San Francisco, CA, USA, 2022, pp. 312-314.

A Low-Loss Hybrid Power Combiner Designed for Ku-Band 32-Way Power Combining Amplifiers

Yiting Zhang, Nengxu Zhu, Fanyi Meng
School of Microelectronics, Tianjin University, Tianjin, China
zhangyiting@tju.edu.cn

Abstract—This paper introduces a low-loss hybrid power combiner (HPC) designed for 32-way power combining amplifiers. The HPC integrates four-way transformers, strip-line transformers, and T-line networks, achieving reduced insertion loss within a compact footprint. Across the 9–16 GHz frequency range, the phase difference between S_{12} and S_{13} is tightly controlled, varying only between 182.7° and 183.2°, while the amplitude imbalance between S_{12} and S_{13} remains consistently low for four-way transformers. For the 32-way PA, the experimental results verify a 3-dB bandwidth of 10.5–14.5 GHz, 30.8 dBm saturated output power (P_{SAT}), and 19.9% peak power-added efficiency (PAE). This design offers a scalable solution for power-combining networks in 5G-Advanced and 6G FR3 power amplifiers.

Index Terms—5G-Advanced/6G, FR3 band, hybrid power combiner (HPC), PA.

I. INTRODUCTION

The shift toward 5G-Advanced and 6G wireless technologies demands power amplifiers (PAs) that can achieve three key objectives: high output power (exceeding 30 dBm), broad bandwidth (7–24 GHz for FR3), and seamless integration into compact phased-array transceivers [1]. Although III-V semiconductors are prevalent in high-power RF applications, CMOS SOI offers a more economical alternative, featuring improved breakdown voltage, reduced substrate coupling, and higher integration density. Recent advancements in CMOS PA design have primarily explored two approaches: multi-way power combining and stacked transistor configurations. While stacked-FET designs enhance voltage tolerance, they rely on large gate capacitors, which hinder scalability [2], [3], [4], [5]. These challenges underscore the importance of developing novel power combining techniques that maintain high performance while preserving compactness and efficiency.

This paper presents a novel low-loss hybrid power combiner (HPC) specifically engineered for 32-way power combining amplifiers, addressing the growing demand for efficient and compact power synthesis solutions in advanced communication systems. The proposed HPC architecture incorporates a sophisticated integration of four-way transformers, strip-line transformers, and T-line networks, which collectively contribute to a significant reduction in insertion loss while maintaining a highly compact form factor. This innovative design ensures optimal performance across a wide operational bandwidth, making it particularly suitable for high-frequency applications such as those required in 5G-Advanced and 6G FR3 systems. Within the frequency range of 9–16 GHz, the HPC demonstrates exceptional phase stability, with the phase difference between S_{12} and S_{13} tightly constrained to a narrow variation and the amplitude imbalance between S_{12} and S_{13} is consistently maintained at minimal levels, highlighting the design's robustness and precision. These performance metrics underscore the HPC's ability to deliver reliable and efficient power combining, even under demanding operational conditions. By addressing the limitations of traditional stacked-FET designs and offering a scalable, low-loss solution, this work bridges the gap between the stringent requirements of next-generation wireless standards and the capabilities of current PA technologies. The scalability and versatility of this HPC design make it an ideal candidate for power combining networks in next-generation communication technologies, particularly for 5G-Advanced and 6G FR3 power amplifiers. By offering a compact, low-loss, and high-performance solution, this work paves the way for enhanced efficiency and miniaturization in future wireless communication systems, addressing critical challenges in power amplification and signal integrity. The proposed HPC not only meets the demanding requirements of modern wireless standards but also provides a pathway for further advancements in RF power combining, enabling the development of more efficient and compact phased-array transceivers for emerging applications.

II. DESIGN OF A NOVEL LOW-LOSS HYBRID POWER COMBINING NETWORK

Fig. 1 illustrates the impedance transformation in a 32-way power-combining PA using the HPC network. The 1-to-16 path (PA pair 1-8) and 12-to-32 path (PA pair 9-16) are symmetrically aligned around reference plane A. Signals from eight differential unit PAs are first combined through a four-way transformer, then further merged via T-line power combining, and finally converted to a single-ended output using a strip-line transformer. This design utilizes a fully balanced series-parallel combining approach to maximize power and efficiency.

According to (1) [6], the transformer's impedance conversion ratio (r) and efficiency (η) are inversely related, meaning higher r values lead to lower η. To achieve an optimal balance between output power and efficiency, the strip-line transformers are designed with an r of 0.7 (35 Ω /50 Ω).

$$\eta = \frac{Q_{ind}^2 + 1}{Q_{ind}^2 + \dfrac{r + \sqrt{r^2 + 4Q_{ind}^2(r-1)}}{2}} \tag{1}$$

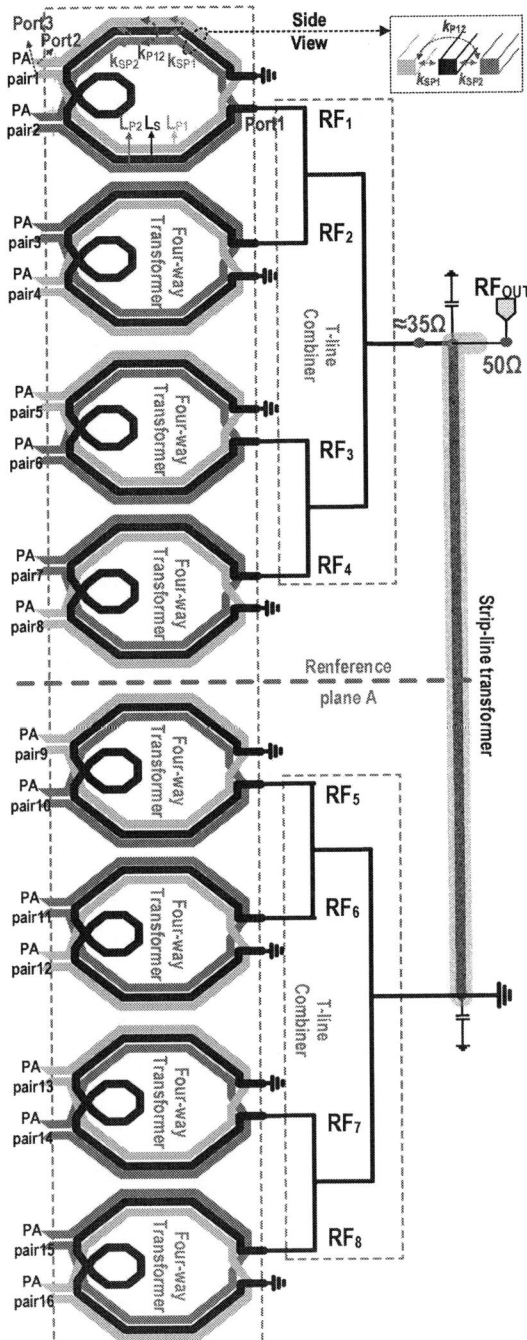

Fig. 1. Simplified schematic and Impedance transformation of the HPC power combiner in a 32-way PA.

The left part of Fig. 1 shows the layout of the four-way transformer, featuring two primary coils and a single secondary coil integrated within a unified inductor footprint. Fig. 2 shows the simulated passive efficiency of the four-way trans-

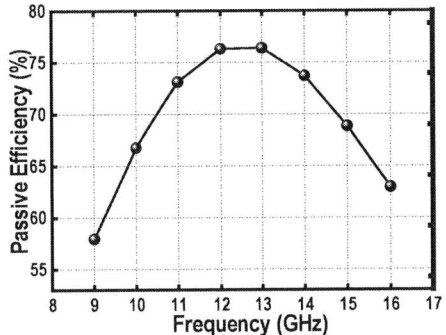

Fig. 2. Simulated passive efficiency of the four-way transformer.

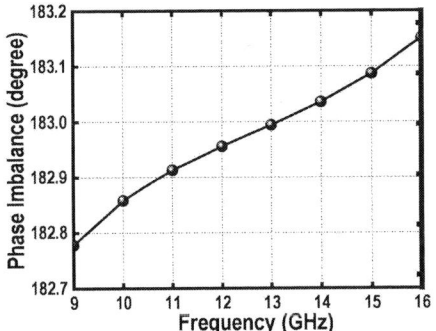

Fig. 3. Simulated phase imbalance of the four-way transformer.

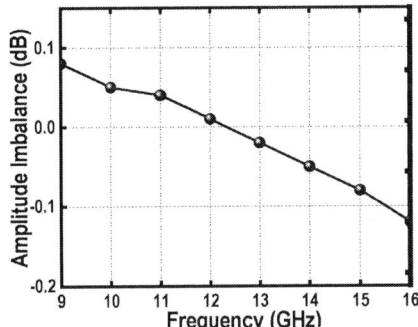

Fig. 4. Simulated amplitude imbalance of the four-way transformer.

former. Fig. 3 and Fig. 4 present the simulated phase and amplitude responses. Across the 9–16 GHz range, the phase difference between S_{12} and S_{13} remains tightly controlled, varying only between 182.7° and 183.2°, while the amplitude imbalance between S_{12} and S_{13} remains consistently low.

III. LAYOUT AND CONCLUSION

Fig. 5 shows the layout of the 32-way PA with the proposed HPC power combiner. Fig. 6 illustrates the simulated performance: a peak gain of 15.7 dB at 12 GHz, with a 3-dB bandwidth of 10.5–14.5 GHz. Fig. 7 shows the large-signal results at 12 GHz. The PA achieves a maximum P_{SAT} of 30.8 dBm

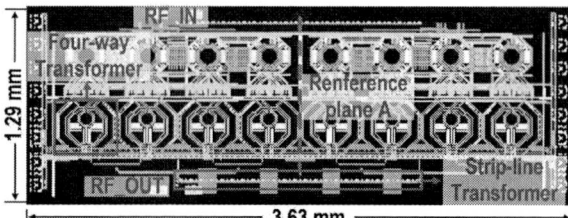

Fig. 5. Layout of the 32 way PA with the proposed HPC power combiner.

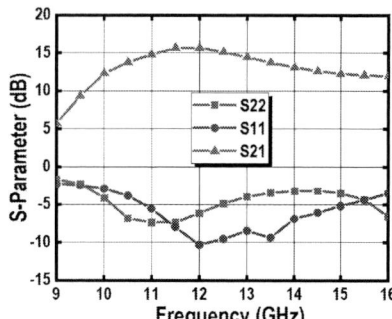

Fig. 6. Experimental results of *S*-parameters.

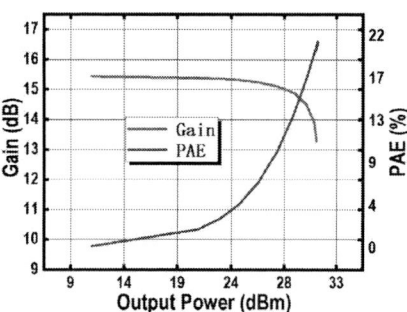

Fig. 7. Experimental results of large-signal performance.

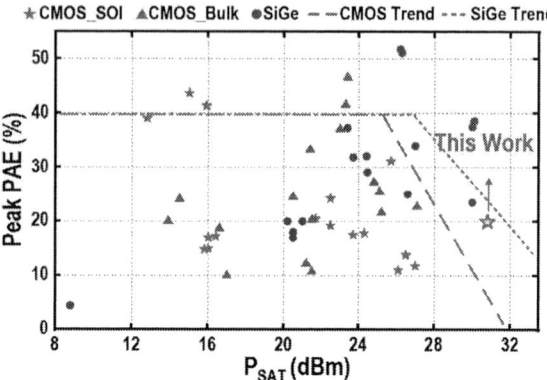

Fig. 8. Performance comparison (PAE and P_{SAT}) of the proposed PA with 8–18 GHz PAs published from December 2010 to June 2024 [7].

and a peak PAE of 19.9% at 12 GHz. As shown in Fig. 8, the PA achieves superior output power while maintaining competitive PAE among CMOS PAs. This paper presents a low-loss hybrid power combiner for 32-way power combining amplifiers, addressing the challenges of 5G-Advanced and 6G FR3 systems. The HPC integrates four-way transformers, strip-line transformers, and T-line networks, achieving minimal insertion loss, exceptional phase stability, and low amplitude imbalance within 9–16 GHz. Its compact, scalable design offers a high-performance solution for next-generation wireless communication, enabling efficient power amplification and miniaturization in advanced phased-array transceivers.

REFERENCES

[1] Prime Movers Lab. (2022). LEO SATCOM. [Online]. Available: https://www.primemoverslab.com/resources/ideas/leo-satcom.pdf

[2] C. Liu, Q. Li, Y. Li, X. Li, H. Liu and Y. -Z. Xiong, "An 890 mW stacked power amplifier using SiGe HBTs for X-band multifunctional chips," *2015 IEEE European Solid-State Circuits Conference (ESSCIRC)*, Sep. 2015, pp. 68–71.

[3] K. Kumar, S. Kumar and M. P. Gupta, "A L/S/C/X/Ku-Band Three-Stack, Two stages Fully Integrated CMOS Power Amplifier with 20.9 % PAE Using T-Network," *2023 Asia-Pacific Microwave Conference (APMC)*, Taipei, Taiwan, 2023, pp. 219-221.

[4] N. Rostomyan, J. A. Jayamon and P. M. Asbeck, "15 GHz Doherty Power Amplifier With RF Predistortion Linearizer in CMOS SOI," *IEEE Trans. Microw. Theory Techn.*, vol. 66, no. 3, pp. 1339-1348, 2018.

[5] B. Coquillas et al., "Ku Band SiGe Power Amplifier With High Output Power and SWR Robustness Up to 120 °C," *IEEE Trans. Circuits Syst. I, Reg. Papers*, vol. 70, no. 7, pp. 2744-2751, July 2023.

[6] Q. J. Gu, Z. Xu, and M.-C. F. Chang, "Two-way current-combining W-band power amplifier in 65-nm CMOS," *IEEE Trans. Microw. Theory Techn.*, vol. 60, no. 5, pp. 1365–1374, May 2012.

[7] Hua Wang, et al., "Power Amplifiers Performance Survey 2000- Present," [Online]. Available: https://ideas.ethz.ch/research/surveys/pa survey.htmls

979-8-3315-2209-4/25 $31.00 © 2025 IEEE

A W-band 6-Bit Vector-Modulated Phase Shifter With Impedance-Invariant Technique

Yue Zhang, Fanyi Meng*

1.School of Microelectronics, Tianjin University, Tianjin, China

2.Tianjin Key Laboratory of Imaging and Sensing Microelectronic Technology

3.Key Laboratory of Organic Integrated Circuit, Ministry of Education, Tianjin University, Tianjin, China.

*Email: mengfanyi@tju.edu.cn

Abstract—**This paper presents a 6-bit vector-modulated active phase shifter applying the impedance invariant technique for W-band phased-array systems. The circuit uses an impedance-invariant variable gain amplifier to reduce the phase and amplitude errors of the different states. The proposed phase shifter circuit implemented in 65-nm CMOS technology consumes 23 mW of dc power and its core occupies only 0.1 mm². Through simulation, the RMS phase error of the proposed phase shifter is 1.02° - 3.9°, while the RMS gain error is less than 0.51 dB over 89-102 GHz.**

Keywords—Phased Array, Active Phase Shifter, Vector Modulator, Mm-Wave, Impedance-Invariant

I. INTRODUCTION

The W-band demonstrates significant advantages in phased array systems due to its unique frequency characteristics. Firstly, the W-band offers extremely high bandwidth resources, enabling ultra-high-speed data transmission and meeting the urgent demands of modern communication systems for large data capacity and low latency. Secondly, the short wavelength of the W-band allows for a substantial reduction in the size of phased arrays. This not only facilitates the realization of highly integrated phased array systems but also enhances beam steering accuracy and flexibility[1].

The phase shifter is one of the core components of a phased array system, directly influencing the precision of beam control in the system. To achieve high-precision phase control, it is essential to generate accurate quadrature signals while effectively managing impedance matching with the VGA. Generally speaking, the input impedance of a VGA can vary significantly under different gain control states, leading to a degradation in impedance matching with the preceding quadrature signal generation stage. This is the primary culprit that hinders the improvement of phase shifter accuracy[2].

This work proposes a 6-bit 89-100 GHz vector-modulated active phase shifter using impedance-invariant technique. With the help of a specially designed VGA, the impedance variation between different gain states is significantly minimized, laying the foundation for the eventual improvement of the vector synthesis accuracy. The arrangement of this article is as follows. Section II states the circuit implementation. The

co-simulation results of the post-layout with EM are shown in Section III and the conclusions are drawn in Section IV.

II. CIRCUIT IMPLEMENTATION

Based on the vector-modulated architecture, the phase shifter mentioned in this paper consists of an IQ signal generator, two impedance-invariant variable gain amplifiers and a signal synthesis circuit. The detailed schematic of the proposed phase shifter is depicted in Fig. 1.

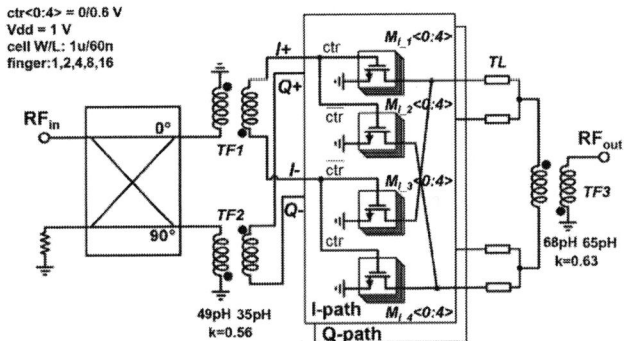

Fig. 1. Block diagram of the proposed phase shifter

A. IQ Signal Generator

The amplitude and phase of the synthesized signal are greatly limited by the quality of the IQ signal. Therefore, it is necessary to consider the phase and amplitude balance of the quadrature generator in a comprehensive manner, taking into account its insertion loss, bandwidth and chip area. In alternative implementations such as poly-phase filters, quadrature all-pass filters, etc., the 3-dB coupler exhibits performance metrics better suited for millimeter-wave band applications[3].

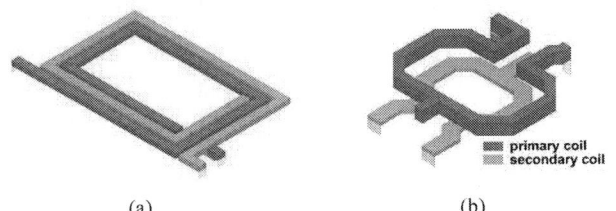

Fig. 2. EM models of the proposed quadrature coupler(a) and balun(b)

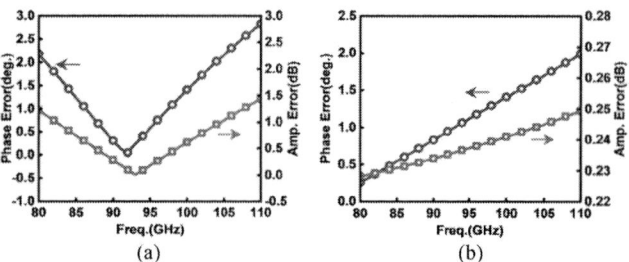

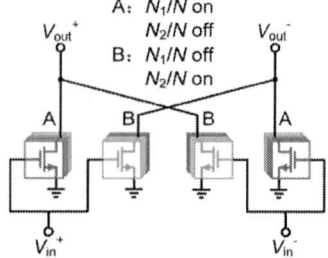

Fig. 3. Simulation phase and amplitude of the (a)IQ generator and (b)balun

Fig. 4. Schematic of the impedance-invariant VGA

In this paper, a 3-dB quadrature coupler is implemented using a rectangular spiral winding that occupies a small chip area. The EM simulation model of this coupler is shown in Fig. 2(a), and its overall size is 104*58 um^2. Simulation results are shown in Fig. 3(a). As evidenced by the performance of the coupler, it is effective in its intended function.

Subsequent to the generation of a quadrature signal, balun is necessary to facilitate the conversion of a single-ended signal to a differential signal. This paper utilizes a transformer structure to achieve this, and Fig. 2(b) shows the model. In order to optimize the phase balance of the balun, we have changed the length of the ground terminal of the primary coil. As you can see from the model, the primary coil is not symmetrical. At the expense of a tiny amount of amplitude balance, nearly perfect differential performance is achieved.

B. Impedance-Invariant Technique

A conventionally designed variable gain amplifier changes the gain of the circuit by adjusting the DC bias point of the transistor, but due to the parasitic effect of the transistor, this causes the amplifier circuit to generate an additional phase shift, which degrades the phase shift accuracy. To minimize the variation of the transistor parasitic parameters, the DC bias of the amplifier cannot be changed; instead, the gain adjustment should be achieved by controlling the size of the transistor. To this end, a transistor with a large width-to-length ratio can be subdivided into multiple cell transistors, thereby forming a numerically controlled transistor array. Consequently, the size of an activated transistor can be modified simply by determining the state of specific unit transistors. In this configuration, the operating point of the activated transistors remains constant during the gain control period.

Variable gain amplifiers also require a phase-shifted polarity selection function, which in conventional structures is usually achieved by complementary control of the gate voltage of the common source stage or common gate stage of the Gilbert cell. Similarly, in an amplifier in the form of a transistor array, the signal path is divided into two, and the unit transistors in the transistor array located in both paths are controlled by inverters and are turned on and off in a complementary manner. This configuration ensures that the effective width of the active or inactive state transistor located at the input node remains constant for all gain states.

In path A, there are N_1 transistors activated, while in path B, N_2 transistors are off. Since the transistors in the two paths are in complementary states, the sum of N_1 and N_2 remains constant. Similarly, by cross-connecting the output terminals of the positive signal path and the negative signal path, there are N active and inactive transistors at each output node. As a result, the output impedance does not vary with changes in gain. The gain of this circuit is determined only by the difference between the number of on-state transistors in the A and B paths. The magnitude of N_1 and N_2 determines the polarity of the phase.

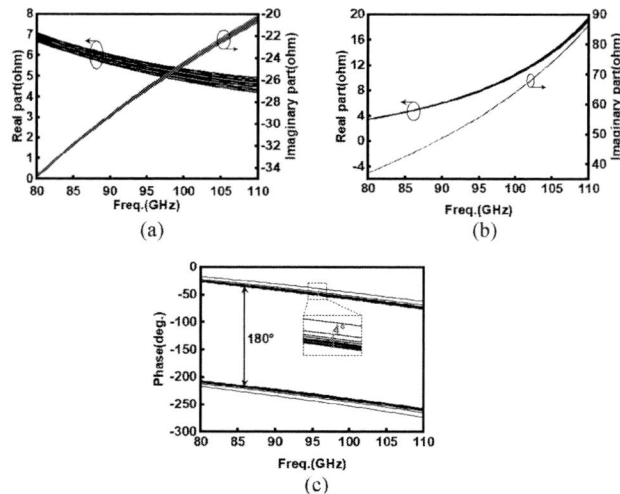

Fig. 5. Simulation of the impedance-invariant VGA (a)input impedance (b) output impedance (c)insertion phase of the VGA

The maximum change of the real part of the input impedance is only 0.7 ohm, and that of the imaginary part is 0.6 ohm. The real part of the output impedance changes by 0.7 ohm and the impedance part by 0.4 ohm. After abandoning some extreme states, the overall additional phase shift is less than 4° and the phase reversal effect caused by complementary control works well.

III. POST-LAYOUT SIMULATION RESULTS

The proposed 6-bit vector-modulated phase shifter is implemented in a 65nm CMOS technology. The whole design is modeled by HFSS and co-simulated with virtuoso. The layout of the phase shifter is shown in Fig. 6.

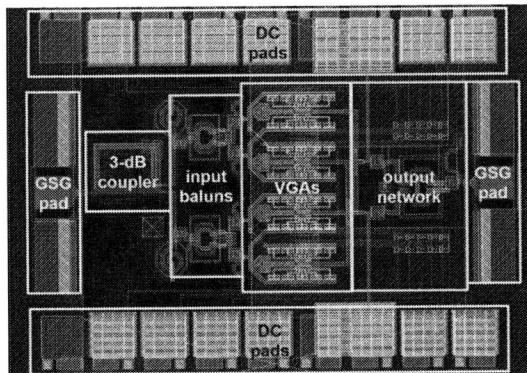

Fig. 6. Layout of the proposed phase shifter

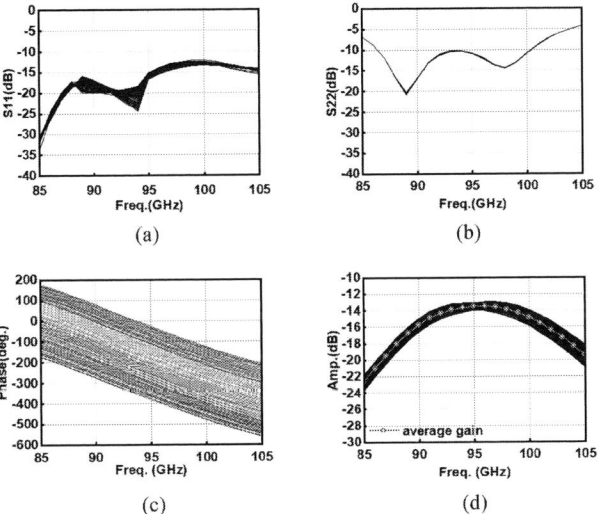

Fig. 7 Simulated (a) S11, (b) S22, (c) phase shift and (d) S21 for all phase states

Fig. 7(a) and 7(b) exhibit the S11 and S22 parameters in all states, respectively. The S11 parameters are superior to -12 dB for total 64 phase shifted states across the operating frequency band. The S22 parameters are below than -10 dB from 87 to 100 GHz. The relative phase shifting range is 360° with 6-bit accuracy as shown in Fig. 7(c). Fig. 7(d) depicts the gain curves. The phase shifter presented exhibits the 3-dB bandwidth from 89 to 102 GHz. The simulated RMS phase error is 1.02° at 95 GHz and is less than 3.9° in the 3-dB bandwidth. The RMS gain error has been calculated to be 0.22 dB at 95 GHz and is less than 0.51 dB from 89 to 102 GHz.

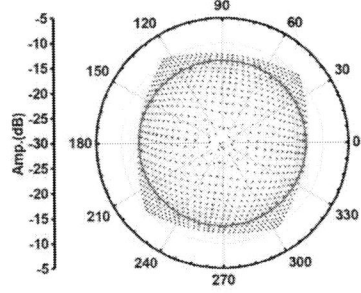

Fig. 8. All phase states on a polar plot at 94GHz

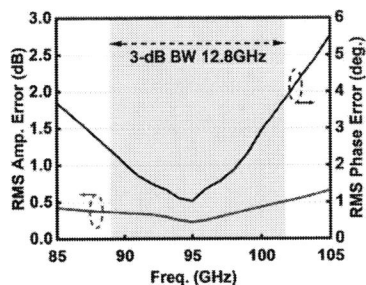

Fig. 9. RMS errors of the phase shifter

IV. CONCLUSION

This paper presents an active phase shifter, adopting impedance-invariant technique, with a 6-bit digital control developed in 65nm CMOS technology. The proposed phase shifter achieves a low RMS phase error level of 1.02°, and the RMS gain error is lower than 0.9 dB over the 3-dB bandwidth. The core of the phase shifter covers 0.44*0.23(0.1)mm² of the chip area. This design significantly reduces the amplitude error and fluctuation of the phase shifter while maintaining relatively small phase error.

TABLE I   Comparison With Other Phase Shifters

	[4]	[5]	[6]	**This Work**
Technology	40nm CMOS	65nm bulk CMOS	SiGe	**65nm CMOS**
Architecture	X-type attenuator	Vector Modulator	Vector Modulator	**Vector Modulator**
Resolution (bits)	6	6	6	**6**
Freq. (GHz)	70-90	91-100	86-106	**89-102**
Insertion Loss(dB)	-15.1	13.6*	2.3	**-13.5**
RMS Phase Error (deg.)	<2.4	<5.625	<6.6	**1.02-3.9**
RMS Gain Error (dB)	±0.6(abs)	<1	<1	**0.22-0.51**
DC Power (mW)	0	N/A	49	**23**
Core Area (mm²)	0.15	0.077	0.027	**0.1**

*with a PA

ACKNOWLEDGMENT

This work was supported by the National Key Research and Development Program of China under Project 2023YFB4403200, the National Natural Science Foundation of China under Project 62271347 and U21A20459.

REFERENCES

[1] J.-O. Plouchart, W. Lee, C. Ozdag, Y. Aydogan, M. Yeck, A. Cabuk, A. Kepkep, E. Apaydin, and A. Valdes-Garcia, "A fully-integrated 94-GHz 32-element phased-array receiver in SiGe BiCMOS, " in *Proc. IEEE Radio Freq. Integr. Circuits Symp. (RFIC)*, Honolulu, HI, USA, 2017, pp. 380-383.

[2] G. H. Park, C. W. Byeon, and C. S. Park, "A 60-GHz low-power active phase shifter with impedance-invariant vector modulation in 65-nm cmos, " *IEEE Trans. Microw. Theory Techn.*, vol. 68, no. 12, pp. 5395–5407, 2020.

[3] D. Pepe and D. Zito, "Two mm-wave vector modulator active phase shifters with novel IQ generator in 28 nm FDSOI CMOS, " *IEEE J. Solid-State Circuits*, vol. 52, no. 2, pp. 344–356, Feb. 2017.

[4] P. Gu, D. Zhao and X. You, "Analysis and Design of a CMOS Bidirectional Passive Vector-Modulated Phase Shifter, " *IEEE Trans. Circuits Syst. I Regul. Pap.*, vol. 68, no. 4, pp. 1398-1408, April 2021.

[5] T. Elazar and E. Socher, "A 90-100 GHz Vector Modulator 7-Bit Phase Shifter with Voltage Summation Topology, " in *IEEE MTT-S Int. Microw. Symp. Dig.*, Washington, DC, USA, 2024, pp. 251-253

[6] K. Smirnova, M. van der Heijden, X. Yang, K. Giannakidis, D. Leenaerts and A. Ç. Ulusoy, "W-Band 6-Bit Active Phase Shifter Using Differential Lange Coupler in SiGe BiCMOS, " *IEEE Microw. Wireless Technol. Lett.*, vol. 33, no. 7, pp. 1035-1038, July 2023.

A 13-17.4GHz Triple-Stacked Power Amplifier With 27 dBm P_{SAT} in 130-nm SOI Technology

Shiyuan Fu, Fanyi Meng*

School of Microelectronics, Tianjin University, Tianjin, China
Tianjin Key Laboratory of Imaging and Sensing Microelectronic Technology, Tianjin, China
Key Laboratory of Organic Integrated Circuit, Ministry of Education, Tianjin University, Tianjin, China
*Email: mengfanyi@tju.edu.cn

Abstract—A fully integrated broadband triple-stacked power amplifier (PA) in 130-nm SOI process is presented in this paper for wireless communication systems. A high-order broadband input matching network is employed to expand the bandwidth. Power combining techniques are utilized to enhance the overall output power, and compact three-winding coupled transformers are adopted for two-way power combining to reduce the chip area. Phase Compensation inductors are utilized to realize intermediate nodal impedance tuning of the stacked structure. The proposed PA shows simulated results of the 3-dB S_{21} bandwidth from 13 to 17.4 GHz. Simulation results show that the relative bandwidth is 29%, the maximum gain is 16.3 dB, the saturated output power reaches 27 dBm, and the maximum power-added efficiency at this point is 20.2%. The core area is only 0.78 mm².

Keywords—*power amplifier, radio frequency, triple-stacked, power combining, transformer, output power.*

I. INTRODUCTION

At present, the electronic information industry is thriving, and research on higher-frequency millimeter wave has increasingly become a focus of attention for various scientific research institutions and RF (Radio Frequency) development companies. Future technologies such as autonomous driving, the construction of smart cities, and derivative products of the sharing economy all rely on the development and advancement of millimeter wave communication technology [1]. Furthermore, the development of millimeter wave transceiver systems is a hot topic worldwide.

In the design of modern wireless communication systems, power amplifiers, as one of the key components, directly impact the transmission efficiency, coverage, and signal quality of the entire system [2]. With the continuous advancement of technology and the growing demand for ultra-wideband communications, the development of efficient and stable Ku-band ultra-wideband power amplifiers has become a hot research topic and challenge [3]. This paper designs a four-way combined differential triple-stacked power amplifier operating at 13-17.5 GHz based on the SMIC 130nm SOI process.

II. CIRCUIT DESIGN

This proposed PA comprises four differential power cells and each power cell is composed of one-stage three-stacked amplifier with neutralization capacitors as shown in Fig. 1. In this work, four-way current-combing transformer power com-

biner is used to increase the output current swing, and stacked FET technique is applied to increase the output voltage swing. Therefore, the output power can be optimized. Also, neutralization technique is used to improve overall stability.

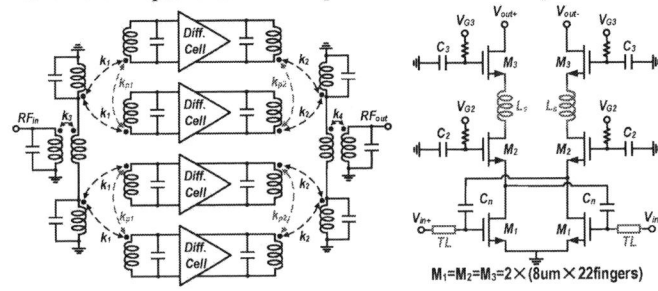

Fig. 1. Circuit schematic of the proposed PA.

A. Broadband Input Matching Network

In the radio frequency band, the input impedance of a common-source transistor is significantly influenced by parasitic capacitances, particularly the gate-source capacitance and the gate-drain capacitance. In differential circuits, the gate-drain capacitance, can be canceled out by using a neutralization capacitance [4]. Therefore, the gate-source capacitance plays a decisive role in the input impedance. Since the impedance of capacitors varies with frequency, it is not conducive to achieving broadband input matching.

As shown in Fig. 2, two series inductors L_S are added between the transformer and the gate. Since the two series inductors at the gate cancel out the input parasitic capacitance of the transistor, the input transformer only needs to perform the conversion of the real part of the impedance during impedance matching, thereby reducing the design complexity of the input transformer and facilitating broadband matching. In the actual layout design, to achieve a more compact layout, these two series inductors at the gate are replaced by two sections of transmission lines.

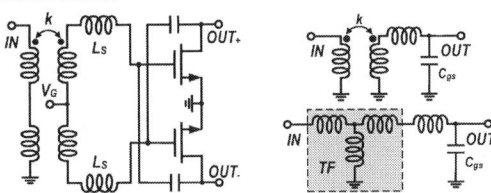

Fig. 2. Broadband input matching network and its equivalent circuit.

B. Phase Compensation Technique

To ensure the stable operation of stacked transistors, a grounding capacitor needs to be connected in parallel to the transistor gate as shown in Fig. 3 (a). Moreover, an appropriate gate capacitor can adjust the phase of the drain voltage of the stacked transistors. However, in the radio frequency band, the impact of parasitic capacitance at the intermediate nodes of stacked transistors becomes increasingly severe, leading to phase misalignment of the drain voltage. Due to the impedance mismatch, the signal transmission between stacked transistors cannot be guaranteed to be at maximum efficiency, thereby reducing the stacking efficiency of the transistors. Therefore, additional circuitry is required to counteract the effects of the parasitic capacitance at the intermediate nodes.

In this design, a series inductor L_S is adopted to make each stacked transistors reach an appropriate internal impedance, as shown in Fig. 3 (b). The series inductor can help to realize the phase alignment of the voltage waveform by adjusting the imaginary part of the impedance of the nodes $Z_{S,2}$ and $Z_{opt,1}$.

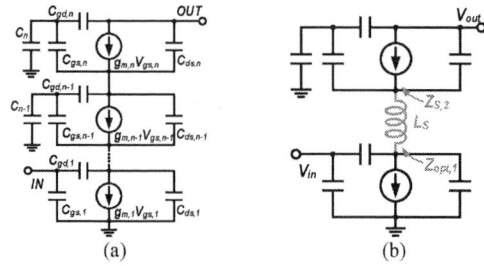

(a)　　　　　　(b)

Fig. 3.　The equivalent model of a triple-stacked transistor with parasitic capacitance, and phase compensation structure based on series inductance.

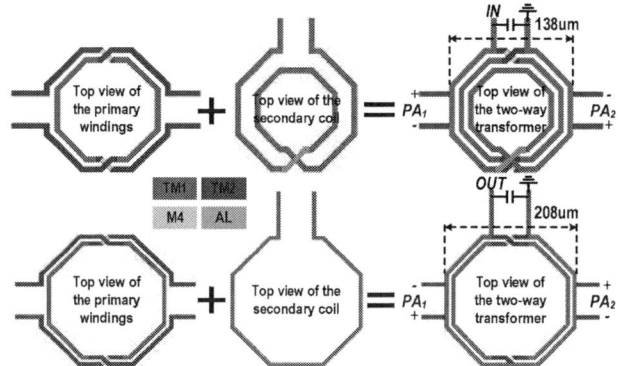

Fig. 4.　The three-winding coupled transformer for power division and combination of two-channel differential power amplifiers.

C. The Design of Power Division and Combination Networks

To minimize chip area, the three-winding coupled transformer is used to simultaneously perform power division and combination for two of the differential amplifiers. Finally, the overall power division and power combination are achieved through a conventional transformer. Furthermore, all transformers are involved in impedance matching at both the input and output terminals.

The overall structure of the three-winding coupled transformer consists of two primary windings and one secondary winding as shown in Fig. 4. In the non-overlapping sections of the three coils, top-layer double-thick metal traces are used for all wiring, which not only maintains the original coupling in the vertical direction but also enhances the coupling on the sides. This design allows for achieving a high coupling coefficient [8], with a coupling coefficient of 0.62 between the primary and secondary coils for both transformers. Fig. 5 presents the insertion loss of the entire output matching network, which is only 2.5dB near the center frequency.

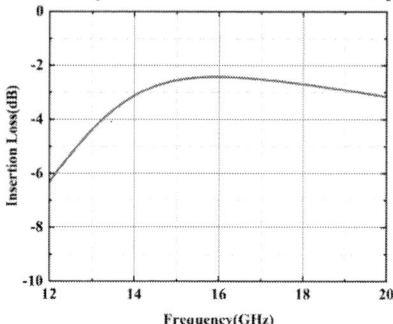

Fig. 5.　The insertion loss of the entire output matching network.

III. POST-LAYOUT SIMULATED RESULTS

Fig. 6 shows the overall layout of the PA, with a core area of 1.13mm×0.69mm. The power supply voltage of the PA is 3.6V. Fig. 7 shows the simulated S-parameters and stability factor of the PA, the peak gain is 16.3 dB at 15 GHz with a 3-dB bandwidth of 4.4 GHz from 13 GHz to 17.4 GHz. And this PA can operate stably across the entire frequency band.

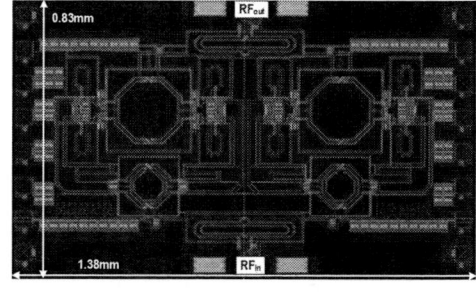

Fig. 6.　The layout of the proposed PA.

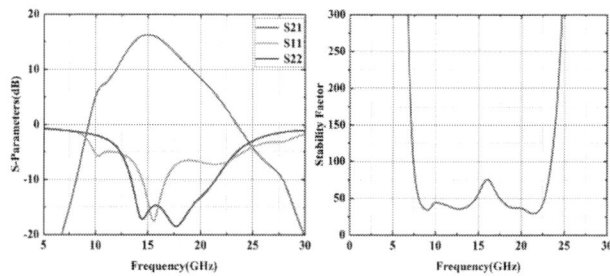

Fig. 7.　Simulated S-parameters of the PA.

979-8-3315-2209-4/25 $31.00 © 2025 IEEE

TABLE I
PERFORMANCE SUMMARY AND COMPARISON

Reference	[3]	[4]	[5]	[6]	[7]	[8]	[9]	**This Work**
Technology	130nm SiGe	65nm CMOS	180nm CMOS	65nm CMOS	130nm SiGe	130nm SiGe	90nm SiGe	**130nm SOI**
Freq. (GHz)	18	13.5-19	14-22	13.7-16.7	18	10.7-12.7	5.2-13	**13-17.4**
Gain (dB)	21.4	20.6	12.0	21.9	13.5	26.2	18.5	**16.3**
P_{sat} (dBm)	30.0	13.9	16.6	14.5	26.1	24.4	25.2	**27.0**
OP_{1dB} (dBm)	28.8	11.6	12.3	-	-	22.7	22.6	**24.1**
PAE_{max} (%)	23.5	20.0	18.7	24.1	27.2	32.1	21.6	**20.2**
Area (mm²)	1.13	0.62#	0.57#	0.30	1.20#	0.21	0.70	**0.78**

#include PADs.

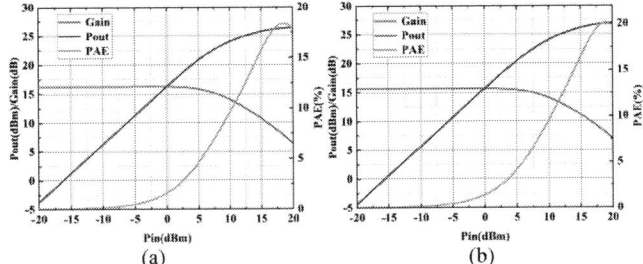

Fig. 8.　Simulated large-signal performance at 15 GHz and 16 GHz. (a) 15GHz. (b) 16GHz.

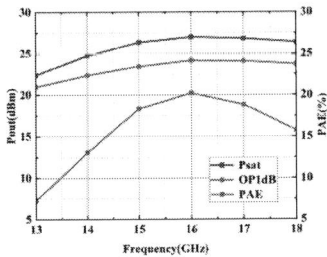

Fig. 9.　Simulated large-signal performance versus frequency of the proposed PA.

Fig. 8 shows the large-signal performance of the proposed PA at 15 GHz and 16 GHz. Fig. 9 illustrates the output power and the PAE of the PA in the frequency range of 13-18 GHz. The proposed PA achieves a saturated output power of 27 dBm and maximum PAE of 20.2% at 16 GHz. And in the frequency range of 15-18 GHz, the fluctuation in saturated output power is only 0.7 dBm.

Table I summarizes the performance of the proposed design and comparable works, demonstrating that this design achieves wider bandwidth and higher output power within a relatively compact area.

IV. CONCLUSION

A broadband triple-stacked PA is presented in this paper. A high-order broadband input matching network is employed to expand the bandwidth. Power combining techniques are utilized to enhance the overall output power, and compact three-winding coupled transformers are adopted for two-way power combining to reduce the chip area. Phase Compensation inductors are utilized to realize intermediate nodal impedance tuning of the stacked structure. And this PA is implemented in 130-nm SOI technology, with a core area of 0.78mm². The simulation results indicate that the proposed PA exhibits excellent performance and shows promising application prospects in the radio frequency band.

REFERENCES

[1]　W. Hong et al., "The Role of Millimeter-Wave Technologies in 5G/6G Wireless Communications," *IEEE J. Microw.*, vol. 1, no. 1, pp. 101-122, Jan. 2021.

[2]　H. Tao et al., "High-Power Ka/Ku Dual-Wideband GaN Power Amplifier With High Input Isolation and Transformer-Combined Load Design," *IEEE Microw. Wireless Compon. Lett.*, vol. 31, no. 1, pp. 49-51, Jan. 2021.

[3]　B. Coquillas et al, "Ku Band SiGe Power Amplifier With High Output Power and SWR Robustness Up to 120 °C," *IEEE Trans. Circuits Syst. I, Reg. Papers*, vol. 70, no. 7, pp. 2744-2751, July 2023.

[4]　B. Chen, L. Lou, K. Tang, Y. Wang, J. Gao and Y. Zheng, "A 13.5–19 GHz 20.6-dB Gain CMOS Power Amplifier for FMCW Radar Application," *IEEE Microw. Wireless Compon. Lett.*, vol. 27, no. 4, pp. 377-379, April 2017.

[5]　O. Z. Alngar, A. Barakat and R. K. Pokharel, "High PAE CMOS Power Amplifier With 44.4% FBW Using Superimposed Dual-Band Configuration and DGS Inductors," *IEEE Microw. Wireless Compon. Lett.*, vol. 32, no. 12, pp. 1423-1426, Dec. 2022.

[6]　J. Zhong, D. Zhao and X. You, "A Ku-Band CMOS Power Amplifier With Series-Shunt LC Notch Filter for Satellite Communications," *IEEE Trans. Circuits Syst. I, Reg. Papers*, vol. 68, no. 5, pp. 1869-1880, May 2021.

[7]　B. Coquillas et al., "A 27dBm Ku-Band SiGe Power Amplifier Working up to 90°C with High Robustness to the 2:1 SWR," in *Eur. Microw. Integr. Circuits (EuMIC) Conf.*, London, United Kingdom, 2022, pp. 152-155.

[8]　B. Yoon, I. S. Han, J. Kim and I. Ju, "A Compact, Highly Linear Ku-Band SiGe HBT Power Amplifier Using Shared Single Center-Tap Four-Way Output Transformer Balun for Emerging Low Earth Orbit SATCOM Phased-Array Transmitter," in *Proc. IEEE Radio Freq. Integr. Circuits (RFIC) Symp.*, Washington, DC, USA, 2024, pp. 295-298.

[9]　H. Wang, C. Sideris and A. Hajimiri, "A CMOS Broadband Power Amplifier With a Transformer-Based High-Order Output Matching Network," *IEEE J. Solid-State Circuits*, vol. 45, no. 12, pp. 2709-2722, Dec. 2010.

A W-Band Coupler-Based Differential Power Amplifier With 19.5-dB Power Gain and 19.2-dBm PSAT in 130-nm SiGe

Jiaming Zhao[*†‡], Pengfei Li[*†‡], Fanyi Meng[*†‡]

[*]School of Microelectronics, Tianjin University, Tianjin, China
[†]Tianjin Key Laboratory of Imaging and Sensing Microelectronic Technology, Tianjin, China
[‡]Key Laboratory of Organic Integrated Circuit, Ministry of Education, Tianjin University, Tianjin, China

Abstract—This paper introduces a W-band two-stage differential power amplifier (PA) in 130-nm silicon germanium (SiGe) technology. The proposed PA employs the coupler-based balun with low insertion loss as its input and output matching network, which also presents a compact layout and superior power-handling and efficiency. The PA achieves a peak gain of 19.5 dB with a 3-dB bandwidth (BW) from 81 GHz to 96 GHz. At 90 GHz, it delivers a saturated output power (Psat) of 19.2 dBm and demonstrates a maximum power-added efficiency (PAE) of 13.7%, in a compact core area of 0.12 mm².

Index Terms—W-band, coupler-based, balun, differential, power amplifier (PA), SiGe BiCMOS

I. INTRODUCTION

In recent years, W-band has been widely used in point-to-point multi-Gb/s communication, millimeter-wave (mm-wave) imaging, satellite communication, radar systems and other fields for its natural advantages of rich spectrum resources, high data transmission rate and low atmospheric attenuation. With the increase of f_T/f_{max} in silicon-based technology, the performance gap between silicon-based devices and III-V devices is also gradually narrowing. Low-cost and highly integrated silicon-based processes have become more attractive and are gradually being used in W-band circuit design.

From the development trend of mm-wave PA, the differential structure with transformers is more favored by designers [1], mainly because it can: (1) suppress the interference of common-mode signal; (2) provide a convenient DC bias path; (3) inherently form a 3-dB power combiner to improve the output power; (4) allow neutralization capacitance and other structures to improve the gain and stability of the PA. However, the folded layout of transformers causes current circular flow and creates negative coupling between opposite sides of the transformer. As the PA working frequency increases, the smaller transformer sizes will lead to stronger negative coupling and larger losses. In addition, the self-resonance

The authors are with the School of Microelectronics, Tianjin University, Tianjin 300072, China, and are also with the Tianjin Key Laboratory of Imaging and Sensing Microelectronic Technology, and the Key Laboratory of Organic Integrated Circuit, Ministry of Education, Tianjin University, Tianjin, China. Corresponding author: Fanyi Meng (email: mengfanyi@tju.edu.cn).

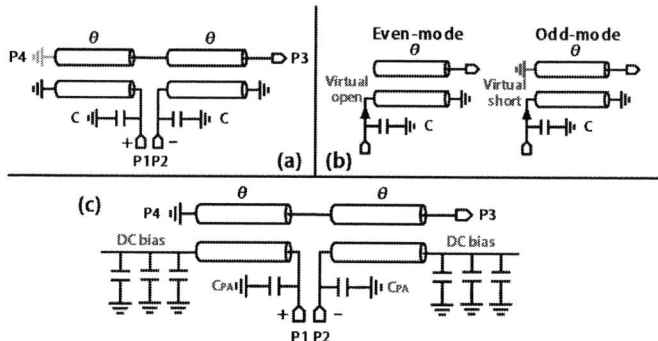

Fig. 1. (a) Coupler-based balun. (b) Even/odd circuit decomposition of coupler-based balun. (c) Practical coupler-based balun.

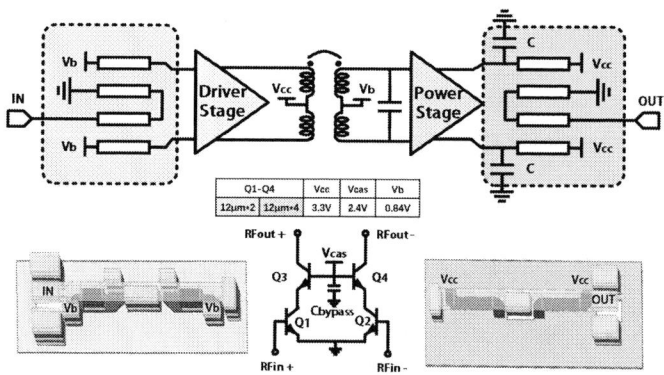

Fig. 2. The schematic of the proposed PA.

frequency also limits the application of transformers in high frequencies.

In this work, a W-band coupler-based differential PA is presented for radar systems. The proposed PA consists of two stages, each using a cascode structure. A coupler-based balun is used to realize the input and output matching network. Compared with the transformer structure, the insertion loss of the matching network is reduced, thus the gain, output power, and efficiency of the PA are improved.

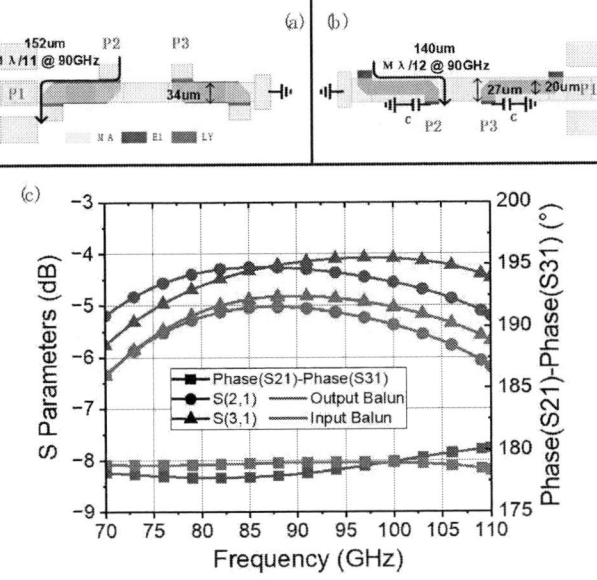

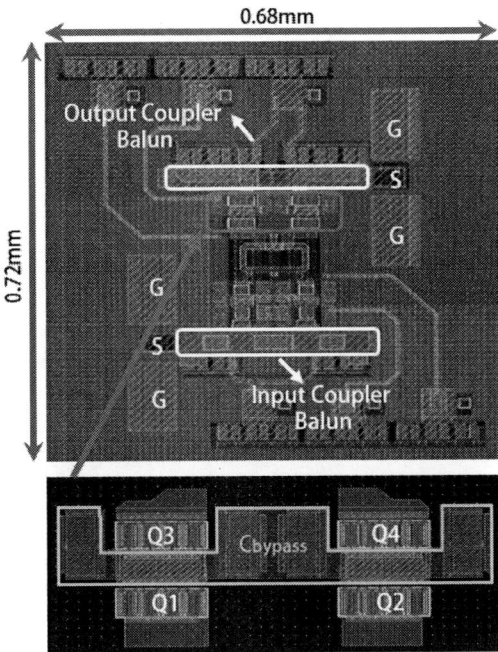

Fig. 3. (a) Input coupler-based balun structure. (b) Output coupler-based balun structure. (c) Simulation results.

Fig. 4. The layout of the proposed PA.

II. CIRCUIT DESIGN

A. Coupler-Based Balun

The conventional coupler-based balun consists of two distributed coupling lines and two lumped capacitors, as shown in Fig. 1(a). When an even-mode signal is input, it is equivalent to having an open terminal at the coupling port, as shown in Fig. 1(b). Even-mode transmission is always blocked, so it has theoretically perfect amplitude and phase balance [2], which allows its application in differential PA input and output matching networks.

According to [3], if the P4 is shorted to ground instead of open, the length of the transmission line can be reduced by 1/3, while the differential to single-ended conversion remains unchanged. Therefore, in this design, P4 is all shorted to the ground to reduce the area of the passive structure. As shown in Fig. 1(c), in the practical design, the two AC short-circuited terminals of the coupler-based balun provide a choke-free DC bias path, and the lumped capacitors C can absorb the parasitic capacitance C_{PA} of the PA core [1].

B. PA Design

Implemented in 8XP 130-nm SiGe technology, the proposed PA is shown in Fig. 2. The PA consists of two-stage differential cells, each of which adopts a cascode structure to ensure higher output power, higher gain, and better isolation. Due to the differential characteristics, the two bases of the common base (CB) in the differential cascode amplifier can be directly connected to form a perfect RF ground. However, in the mm-wave bands, long routing distance will introduce a parasitic inductance, so we connect a bypass capacitor here to provide better grounding characteristics.

In W-band PA design, the choice of transistor size is crucial. While large transistor sizes can increase output power, they can also increase the base resistance or parasitic capacitance, thus limiting the gain and high-frequency performance. Considering these trade-offs, at the output stage we chose transistors with 12 μm width and 4 multipliers. The driver stage is half the size of the output stage to both improve gain and provide about 3 dB margin to drive the output stage. The PA is biased in Class AB, with V_b set at 0.84 V, V_{cas} set at 2.4 V, and V_{cc} set at 3.3 V.

To increase the bandwidth of the PA, we utilize transformers for interstage matching and supply DC bias through the transformer's center tap. The input and output matching networks adopt the coupler-based balun, which greatly reduces the insertion loss of the matching networks and optimizes the key performance such as efficiency and output power of the PA, compared with transformers.

C. Input and Output Coupler-Based Balun

Fig. 3(a) and Fig. 3(b) show the electromagnetic (EM) model of the input and output balun of the proposed PA, which consists of the top three metal layers of MA, E1, and LY. To design a good matching network, we constantly change the width of the transmission line to adjust its characteristic impedance, thereby achieving corresponding impedance conversion. In addition, we also adjust the coupling coefficient of the coupling line to achieve good matching performance. At the differential port of the output balun, a capacitor of appropriate value needs to be added to achieve matching. It absorbs part of the capacitance of the output stage transistors

979-8-3315-2209-4/25 $31.00 © 2025 IEEE   139

TABLE I
PERFORMANCE SUMMARY AND COMPARISON OF W-BAND PAS

Reference	Technology	Topology	Freq. (GHz)	3-dB BW (GHz)	Gain (dB)	Psat (dBm)	PAE$_{MAX}$ (%)
[4] TCSI'18	130 nm SiGe	3-stage 2-way balanced	100	89-101	14.5	16.3	14.1
[5] TMTT'13	180 nm SiGe	2-stage 4-way combining	78	68-87	18.3	14	2
[6] TMTT'20	130 nm SiGe	3-stage 2-way combining	77	68.5-90	26.7	18.5	12.9
[7] RFIC'16	90 nm SiGe	5-stage class-E cascode	85	75-105	17	22	19.1
[8] TMTT'15	65 nm CMOS	3-stage CS 2-way combining	86	84-88.8	18.6	11.9	9
[9] MWCL'11	65 nm CMOS	3-stage diff. cascode	105	110-117	13.4	13.8	9.4
This work	130 nm SiGe	2-stage diff. cascode	90	81-96	19.5	19.2	13.7

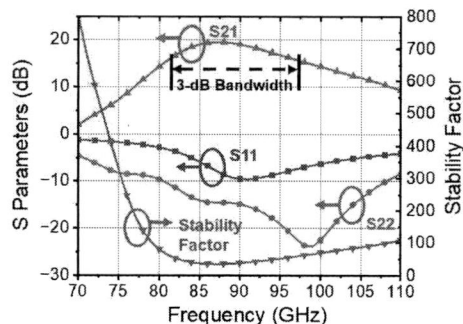

Fig. 5. Simulated small-signal S-parameters and stability factor of the PA.

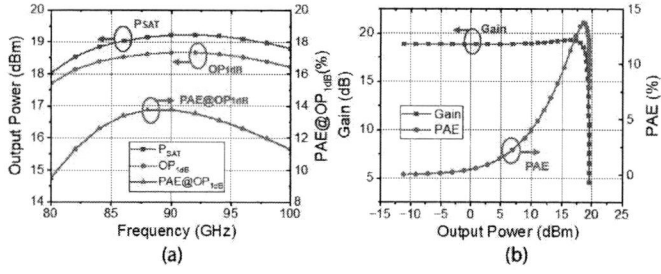

Fig. 6. (a) Simulated large-signal performance versus frequency of the proposed PA. (b) Simulated large-signal performance at 90GHz.

and finally, we choose a 46 fF metal-oxide-metal (MOM) capacitor.

The proposed balun provides two AC short-circuited terminals which can be used as DC bias paths. We directly connect bypass capacitors to these two ports to power the PA without consuming additional area.

The electrical lengths of the input and output baluns are $\lambda/11$ and $\lambda/12$ at 90 GHz. Fig. 3(c) demonstrates the phase and amplitude balance across 70–110 GHz. At 90 GHz, the input and output balun has about 1.9-dB and 1.2-dB insertion loss, respectively, and 0.2-dB and 0.1-dB amplitude imbalance. In the whole band, the phase difference between the two differential ports is less than 2 degrees.

III. SIMULATION RESULTS

The layout of the proposed PA is shown in Fig. 4. The total chip area, inclusive of pads, measures 0.49 mm^2, while the core area without pads is only 0.12 mm^2. Fig. 5 shows the simulated S-parameters and stability factor of the PA. The proposed PA shows a peak gain S_{21} of 19.5 dB at 88 GHz, and a 3-dB bandwidth (BW) of 15 GHz, covering from 81 to 96 GHz. During the 3-dB BW, the output return loss S_{22} is less than -10 dB. This PA is unconditionally stable as the stability factor > 1 is satisfied across 70–110 GHz.

Fig. 6(a) shows the simulated large-signal performance of the proposed PA between 80 GHz and 100 GHz. The curve is flat, and within the 3-dB BW, the changes in P_{sat} and output power of 1-dB compression point (OP1dB) are both less than 0.2 dBm. At 90 GHz, the proposed PA has the best large-signal performance, with 19.2 dBm P_{sat}, 18.7 dBm OP1dB, and 13.7% PAE, as shown in Fig. 6(b).

IV. CONCLUSION

A W-band two-stage cascode differential PA is presented in this letter. A coupler-based balun is used in the input and output matching network to reduce the loss of mm-wave passive devices and improve the performance of the PA. The proposed PA demonstrates notable characteristics, including a peak gain of 19.5 dB and a P_{sat} of 19.2 dBm at 90 GHz. Besides, within the 3-dB BW, the variations of the P_{sat} and OP1dB are less than 0.2 dBm. The simulation results indicate that the coupler-based balun has a good application prospect in mm-wave circuit design.

REFERENCES

[1] H. T. Nguyen and H. Wang, "A Coupler-Based Differential mm-Wave Doherty Power Amplifier With Impedance Inverting and Scaling Baluns," IEEE Journal of Solid-State Circuits, vol. 55, no. 5, pp. 1212-1223, May 2020.

[2] Kian Sen Ang, Yoke Choy Leong and Chee How Lee, "Analysis and design of miniaturized lumped-distributed impedance-transforming baluns," IEEE Transactions on Microwave Theory and Techniques, vol. 51, no. 3, pp. 1009-1017, March 2003.

[3] H. Jia, B. Chi, L. Kuang and Z. Wang, "A W-Band Power Amplifier Utilizing a Miniaturized Marchand Balun Combiner," IEEE Transactions on Microwave Theory and Techniques, vol. 63, no. 2, pp. 719-725, Feb. 2015.

979-8-3315-2209-4/25 $31.00 © 2025 IEEE

[4] Z. J. Hou et al., "A W-Band Balanced Power Amplifier Using Broadside Coupled Strip-Line Coupler in SiGe BiCMOS 0.13-um Technology," IEEE Transactions on Circuits and Systems I: Regular Papers, vol. 65, no. 7, pp. 2139-2150, July 2018.

[5] M. Thian, M. Tiebout, N. B. Buchanan, V. F. Fusco and F. Dielacher, "A 76–84 GHz SiGe Power Amplifier Array Employing Low-Loss Four-Way Differential Combining Transformer," IEEE Transactions on Microwave Theory and Techniques, vol. 61, no. 2, pp. 931-938, Feb. 2013.

[6] Y. Yu et al., "A 68.5 90 GHz High-Gain Power Amplifier With Capacitive Stability Enhancement Technique in 0.13 um SiGe BiCMOS," IEEE Transactions on Microwave Theory and Techniques, vol. 68, no. 12, pp. 5359-5370, Dec. 2020.

[7] K. Datta and H. Hashemi, "75–105 GHz switching power amplifiers using high-breakdown, high-fmax multi-port stacked transistor topologies," in Proc. IEEE Radio Freq. Integr. Circuits Symp. (RFIC), May 2016, pp. 306–309.

[8] H. Jia, B. Chi, L. Kuang and Z. Wang, "A W-Band Power Amplifier Utilizing a Miniaturized Marchand Balun Combiner," IEEE Transactions on Microwave Theory and Techniques, vol. 63, no. 2, pp. 719-725, Feb. 2015.

[9] Z. Xu, Q. J. Gu and M. -C. F. Chang, "A 100–117 GHz W-Band CMOS Power Amplifier With On-Chip Adaptive Biasing," IEEE Microwave and Wireless Components Letters, vol. 21, no. 10, pp. 547-549, Oct. 2011.

979-8-3315-2209-4/25 $31.00 © 2025 IEEE

A Study on Thickness-Dependent Performances of High-k Gated Tellurium-Based Field-Effect Transistors

Yang Hui Xia, Zi Chun Liu, Shu Ming Qi, Yu Hang Zheng, Yu Meng Wang, Xiao Long Xu,
Hui Xia Yang[*] Yuan Xiao Ma[*] and Yeliang Wang[*]
School of Integrated Circuits and Electronics, Beijing Institute of Technology, Beijing 100081, China
*Authors to whom the correspondence should be addressed: yanghuixia@bit.edu.cn, yxma@bit.edu.cn and yeliang.wang@bit.edu.cn

Abstract—In this work, p-type tellurium-based field-effect transistors (Te-FETs) have been successfully fabricated. A mixture high-k HfLaO is developed as gate dielectric in replacement of the traditional SiO2 to enhance the field effect, lowering the operation voltage from -50 V to -8 V for a high power efficiency. Moreover, the Te channel film is readily thinned by shortening the deposition duration from 90 seconds to 20 seconds. This can suppress off-state current due to the reduced amount of hole carriers in a thinner film. Furthermore, a low-temperature annealing ($\sim$ 150 °C) is adopted to additionally decline the off-state current because Te content can be decreased by oxygen compensation during the annealing in air, diminishing hole carriers. Interestingly, the on-state current is concurrently promoted, which should result from the atomic rearrangement by absorbing sufficient thermal energy during the annealing, improving the quality of Te film and Te/HfLaO interface. As a result, the on/off ratio of the annealed device has been successfully improved about 10 times as compared to the pristine device.

Index Terms—Te-FETs, high-k dielectric, annealing treatment

I. Introduction

Tellurium (Te), a promising p-type semiconductor, exhibits exceptional potentials due to its narrow bandgap and unique chain lattice structure, which enables high hole mobility [1]. Accordingly, Te-based field-effect transistors (Te-FETs) have been attracting significant attentions in optoelectronic devices [2] and logic devices [3]. The most of as-reported Te-FETs were fabricated on traditional SiO$_2$ gate dielectric [4], presenting a high operation voltage with undesirable power consumption due to its week field effect. One effective way to reduce the operation voltage is to adopt gate dielectric with high dielectric constant, namely high-k dielectric to enhance field effect. Although various high-k oxides, such as La$_2$O$_3$, HfO$_2$ and Ta$_2$O$_5$, have been extensively explored, these oxide dielectrics with single metal element are always associated with severe oxide defects of hygroscopicity and oxygen vacancies. Therefore, mixture high-k dielectrics have been developed to allow them to complement with each other, improving interfacial quality. In this work, Te FETs have been fabricated on traditional SiO$_2$ and mixture high-k HfLaO dielectrics. The operation voltage of the SiO$_2$-gated device can be dramatically reduced by adopting the mixture

HfLaO. Moreover, the electrical performances of the Te-FETs can be effectively improved by thinning the Te channel film and adopting a low-temperature post-deposition annealing treatment.

II. Experimental

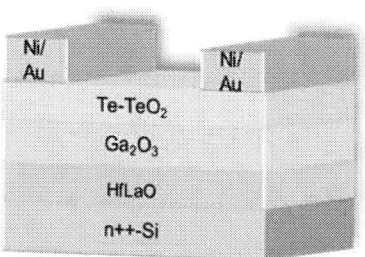

Fig. 1. Structure schematic of HfLaO-gated Te-FETs.

Fig. 1 illustrates the schematic structure of the Te-FETs gated by high-k HfLaO dielectric in this work. Initially, heavily-doped n-type silicon substrates (1$\sim$10 Ω·cm) were cleaned by successively dipping them in alcohol and deionized water under ultrasound each for 15 min. Then, HfLaO was deposited on the Si substrates by co-sputtering a La metal target at a DC 45 W and a Hf metal target at a DC 12 W, during which the ambience was Ar/O$_2$/N$_2$ = 24:6:1 sccm under a 1 Pa pressure. Next, a gallium oxide film was deposited as high-k buffer layer by sputtering a Ga$_2$O$_3$ ceramic target at a RF 100 W under an Ar/O$_2$ = 20:10 ambience at 0.2 Pa. Subsequently, Te channel film was deposited by co-sputtering a metallic Te target at DC 8 W and a ceramic TeO$_2$ target at a RF 25 W under an ambience of Ar = 30 sccm at 0.5 Pa. The thickness of the Te channel film was readily fabricated by modulating the deposition duration from 90 second to 20 second. Moreover, the counterpart SiO$_2$-gated samples were simultaneously fabricated on SiO$_2$/Si substrates at the same run of the deposition of 90s Te film. In the following, post-deposition annealing treatment was conducted on partial high-k gated samples in air at 150°C for 1 hour. Finally, the source/drain electrodes were formed by evaporating Ni/Au

979-8-3315-2209-4/25 $31.00 © 2025 IEEE

(10/50 nm) through a shadow mask onto the Te channel films by electron-beam method because Ni has a relatively high work function to reduce the contact resistance in p-type transistors. The patterned channel width (W) and length (L) were 300 and 40 μm, respectively.

III. RESULT AND DISCUSSION

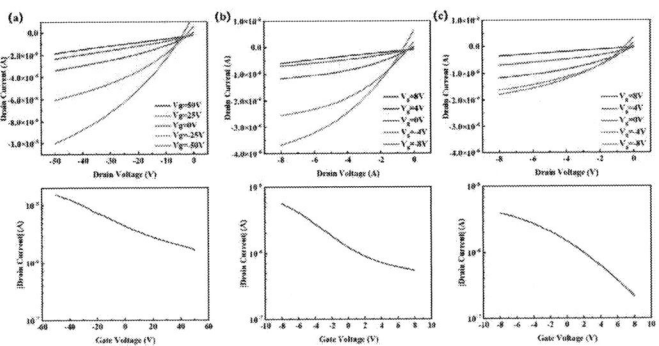

Fig. 2. Electrical characteristics of the Te-FETs: (a) 90s SiO$_2$-sample, (b) 90s HfLaO-sample and (c) 20s HfLaO-sample.

Fig. 2 shows the transfer and output characteristics of the 90s SiO$_2$-sample, 90s HfLaO-sample and 20s HfLaO-sample. As depicted in Fig. 2a, a normal p-type Te FET is successfully obtained with SiO$_2$ gate dielectric but its operation voltage is up to -50 V due to the weak field effect, which is not conducive to power consumption. The operation voltage can be dramatically reduced from -50 V to -8 V by replacing the SiO$_2$ with a mixture HfLaO high-k dielectric, as shown in Fig. 2b. However, the 90s HfLaO-sample presents a high off-state current because the bandgap of Te is inherently narrow ($\sim$ 0.3 eV) to produce excessive spontaneous hole carriers at room temperature, degrading the on/off current ratio of the Te FETs [5]. One way to reduce the off-state current of the Te FETs is to thin the Te channel film since the amount of hole carriers is expected to be declined in a thinner layer. As shown in Fig. 2c, although the off-state current can be slightly decreased in a thinner Te channel layer by shortening the deposition time from 90s to 20s, the on-state current is inevitably diminished due to the reduction of hole carriers.

Additionally, a post-deposition annealing treatment was conducted on the 20s HfLaO-sample. As displayed in Fig. 3, the off-state current is further reduced by the post-deposition annealing. This is attributed to the additional oxidization of the Te channel film during the annealing in air, declining the Te content and related hole carriers. Importantly, the on-state current is promoted by the annealing treatment, which should result from: (1) the lattice reconstruction of Te to annihilate grain boundaries and defects formed during the sputtering process; (2) improvement of Te/HfLaO interfacial quality due to atomic rearrangement during annealing. As a result, the on/off ratio of the annealed sample is about 10 times higher than that of the unannealed sample.

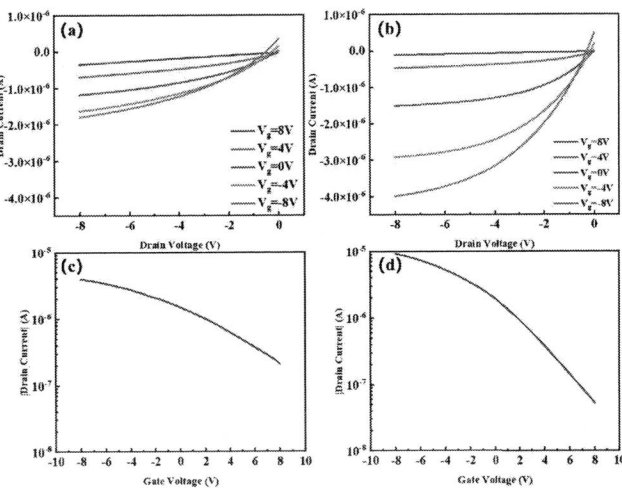

Fig. 3. Electrical characteristics of 20s HfLaO-sample: (a) and (c) non-annealed device. (b) and (d) annealed device.

IV. CONCLUSION

In this work, p-type Te-based FETs have been successfully fabricated at low temperature. Operation voltage of the Te-FETs is dramatically reduced by replacing SiO$_2$ with a mixture high-k HfLaO. The Te channel film is thinned by shortening the deposition duration from 90s to 20s, lowering the amount of hole carriers. Moreover, a 150 °C post-deposition annealing in air is conducted on the Te-FET to further reduce the off-state current due to the suppression of hole carriers by an extra oxidization. Meanwhile, the on-state current is raised thanks to the improvement of Te/HfLaO interfacial quality, promoting the on/off ratio nearly 10 times as compared to the pristine device.

REFERENCES

[1] Y. Wang et al., "Field-effect transistors made from solution-grown two-dimensional tellurene," Nat Electron, vol. 1, no. 4, pp. 228–236, Apr. 2018.

[2] U. Jeong et al., "Plasma-engineered high-performance tellurium field-effect phototransistors," Advanced Functional Materials, 2025, 2421140.

[3] Y. Ran et al., "Large-scale vertically interconnected complementary field-effect transistors based on thermal evaporation", Small 2024, 20, 2309953.

[4] T. Kim et al., "High-Performance Hexagonal Tellurium Thin-Film Transistor Using Tellurium Oxide as a Crystallization Retarder," IEEE Electron Device Lett., vol. 44, no. 2, pp. 269–272, Feb. 2023.

[5] G. Qiu, A. Charnas, C. Niu, Y. Wang, W. Wu, and P. D. Ye, "The resurrection of tellurium as an elemental two-dimensional semiconductor," npj 2D Mater Appl, vol. 6, no. 1, pp. 1–10, Mar. 2022.

979-8-3315-2209-4/25 $31.00 © 2025 IEEE

Comparative Analysis of Canny and Watershed Edge Detection for Line-Edge Roughness (LER) Quantification in SEM Images

Shuyan He   Michail Michailow   Yufeng Jin[*]

Abstract—**Line-edge roughness (LER) is a critical factor in nanoscale fabrication, affecting device performance and reliability. This study evaluates Canny and Watershed edge detection algorithms for LER quantification in SEM images by analyzing Ra, Rrms, Rad, Rrmsd, skewness, kurtosis, correlation length, and FFT-based spectral analysis. Results show that Watershed over-segments, inflating roughness values (Ra = 3.429 nm, Rrms = 3.979 nm) compared to better segmentation (Ra = 2.342 nm, Rrms = 2.836 nm), with increased correlation length (33.555 nm vs. 27.074 nm) and kurtosis (3.721 vs. 3.153). Canny provides precise edge localization but struggles with low-contrast transitions, causing roughness underestimation (Ra = 3.439 nm, Rrms = 3.952 nm) and higher kurtosis in poor segmentation (6.869 vs. 3.902). FFT and PSD analyses confirm that better segmentation exhibits dominant low-frequency peaks, while poor segmentation increases high-frequency noise. Canny is preferable for high-contrast SEM images, Watershed for complex, low-contrast structures, though both require parameter optimization. These findings provide quantitative insights for improving SEM-based LER measurements.**

Index Terms—**Watershed algorithm, Canny algorithm, Line-edge roughness (LER)**

I. INTRODUCTION

Line-edge roughness (LER) is a critical factor in nanoscale fabrication, particularly in CMOS manufacturing, where precise patterning is required to ensure device performance and reliability. Scanning Electron Microscopy (SEM) is widely used for LER characterization, but the extraction of accurate roughness parameters is dependent on effective edge detection methods. Traditional tools like Gwyddion are commonly applied for surface roughness analysis but are not optimized for line edge roughness measurement in SEM images [1].

This paper investigates two edge detection algorithms—Watershed and Canny, to assess their effectiveness in extracting LER from SEM images. We analyze roughness metrics including Ra, Rrms, Rad, Rrmsd, skewness, kurtosis, correlation length, and FFT, based spectral analysis to compare the impact of these methods. Our results demonstrate that the choice of edge detection significantly influences roughness quantification, providing insights into optimizing SEM-based LER measurements.

II. EDGE DETECTION ALGORITHMS FOR LER ANALYSIS

A. Importance of Edge Detection in SEM-Based LER Measurement

Edge detection plays a critical role in line edge roughness (LER) quantification, as it directly affects the accuracy of edge profiles extracted from scanning electron microscopy (SEM) images. The precision of this step influences the subsequent calculation of roughness parameters such as Ra (arithmetic mean roughness), Rrms (root mean square roughness), correlation length, and frequency domain characteristics.

Given the nanoscale resolution of the SEM images, effective edge detection must balance noise robustness, edge continuity, and computational efficiency. Different algorithms exhibit varying performance depending on factors such as image contrast, noise level, and edge complexity. Among the widely used methods, the Canny and Watershed algorithms are particularly relevant for SEM-based LER analysis because of their distinct advantages in edge extraction.

B. Canny and Watershed Edge Detection: Overview and Comparative Analysis

The Canny edge detector is a gradient-based algorithm optimized for precise edge localization and noise suppression. It follows a multi-stage process (Gaussian smoothing, gradient computation, non-maximum suppression, and hysteresis thresholding), making it effective for high-contrast edges in SEM images. However, its reliance on strong intensity transitions limits performance on low-contrast or gradual edges, requiring careful parameter tuning [4].

The Watershed algorithm is a morphological segmentation method that interprets an image as a topographical surface, where intensity gradients define "basins" and "ridges." It is well-suited for complex, low-contrast structures in SEM images but tends to over-segment, necessitating post-processing (e.g., marker-controlled Watershed). Unlike Canny, it considers regional intensity variations but has higher computational complexity [3].

The Laplacian operator, a second-order derivative-based method, enhances edge detection by capturing both fine-scale and large-scale intensity variations, making it effective for multicolored and moderately complex images [2]. However,

---

Shuyan He and Yufeng Jin are with School of Electronic and Computer Engineering (ECE), Peking University Shenzhen Graduate School, Shenzhen, China.Michailow, Michail is with Chair of Semiconductor Electronics and Institute of Semiconductor Electronics, RWTH Aachen University, Aachen, Germany. This work is supported by Shenzhen Science and Technology Program under Grant No. KJZD20230923115005009. It is also supported by the project under Grant No. KQTD20200820113105004. Corresponding author: Yufeng Jin (email: yfjin@pku.edu.cn)

979-8-3315-2209-4/25 $31.00 © 2025 IEEE

Criterion	Canny Edge Detector	Watershed Algorithm
Best suited for	High-contrast images with sharp edges	Low-contrast structures with complex textures
Edge continuity	Strong, but may miss weak edges	Can detect weak edges, but prone to over-segmentation
Noise robustness	Moderate (Gaussian filtering reduces noise)	High (considers regional intensity variations)
Computational cost	Lower (efficient for large images)	Higher (requires morphological preprocessing)
Parameter sensitivity	Requires careful tuning of thresholds	Less sensitive, but needs post-processing
Handling of overlapping structures	Limited	Strong (can separate closely spaced features)

Fig. 1: Comparison of different edge detection algorithms

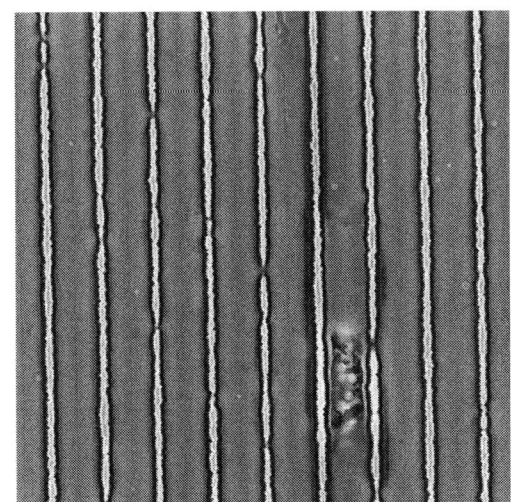

Fig. 2: Original SEM Image

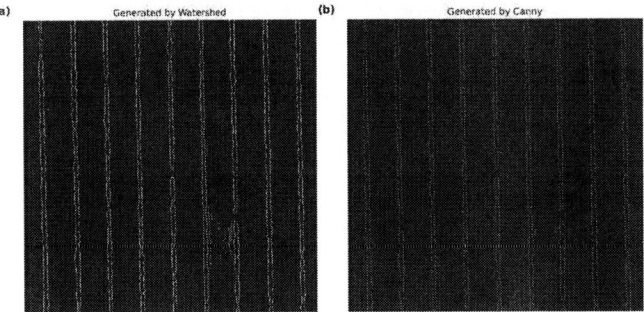

Fig. 3: (a)Generated by Watershed, (b)Generated by Canny

its sensitivity to noise often necessitates preprocessing techniques such as Gaussian smoothing or anisotropic diffusion. In contrast, the Sobel and Prewitt operators are first-order gradient-based methods that approximate intensity derivatives. The Sobel operator computes the absolute gradient magnitude, providing a simple yet effective edge detection approach, while the Prewitt operator incorporates directional weighting for improved sensitivity to specific edge orientations. Despite their computational efficiency, both methods exhibit lower precision and struggle with noise suppression compared to more advanced techniques like Canny or Watershed. Among the simplest edge detection techniques, the Roberts operator relies on diagonal gradient computation, offering high computational efficiency but limited noise resistance and lower edge localization accuracy. Consequently, while these methods provide efficient edge detection, their applicability in SEM analysis is constrained by their sensitivity to noise and precision limitations (figure 1).

Based on these characteristics, the Canny algorithm is preferable for SEM images with well-defined, high-contrast edges requiring precise boundary extraction, whereas the Watershed algorithm is more suitable for weak or complex edges where regional intensity variations are critical. A comparative study is essential to quantify their impact on LER measurements, particularly regarding roughness parameters such as Ra, Rrms, and correlation length.

Apply the Watershed and Canny edge detection algorithms to a post-etching SEM image by choosing two different segments which can be recognized the performance easily (figure 2) for boundary extraction. The comparison between Watershed-based (figure 3 (a)) and the Canny-based (figure 3 (b)) highlights the distinct characteristics of these two algorithms in detecting boundaries in SEM images. In Fig. 3(a), the Watershed algorithm successfully extracts closed contours, ensuring complete segmentation of structures. However, it exhibits a higher level of noise, particularly in regions with subtle intensity variations, leading to over-segmentation and the presence of extraneous edge artifacts. This oversegmentation effect is a known limitation of the Watershed method, as it

is highly sensitive to local intensity gradients and can introduce spurious edges in areas with noise or texture variations. In Fig. 3(b), the Canny edge detection algorithm produces cleaner and more precise edges, effectively suppressing noise and capturing major structural boundaries. However, a notable drawback is the presence of numerous discontinuities along the detected edges, indicating that Canny struggles with weak or low-contrast transitions. This results in fragmented edge representations, which could lead to an underestimation of roughness parameters in quantitative analysis.

The following section presents an experimental evaluation of both algorithms, assessing their performance in LER quantification.

III. ROUGHNESS ANALYSIS BASED ON CANNY AND WATERSHED EDGE EXTRACTION

Surface roughness analysis is essential in SEM image processing, as it provides critical insights into the morphological characteristics of material surfaces. Edge detection plays a fundamental role in quantifying roughness, as the contours extracted directly influence the accuracy of roughness parameters. To evaluate the impact of different edge detection algorithms on roughness characterization, we examine key roughness metrics derived from SEM images processed using

TABLE I: Comparison of different SEM images based on two algorithms

Roughness Methods Parameter	Different Algorithms in Comparing Edge Performance			
	Watershed (Better)	Watershed (Worse)	Canny (Better)	Canny (Worse)
Ra (nm)	2.342	3.429	2.338	3.439
Rrms (nm)	2.836	3.979	2.862	3.952
Rad (nm)	0.638	0.635	0.422	0.491
Rrmsd (nm)	0.835	0.859	0.715	0.806
Skewness	2.136	2.136	2.252	2.639
Kurtosis	3.153	3.721	3.902	6.869
Correlation Length (nm)	27.074	33.555	20.263	35.111

the Canny and Watershed algorithms. The selected roughness parameters include Ra (Arithmetic Mean Roughness), Rrms (Root Mean Square Roughness), Rad (Derivative of Arithmetic Mean Roughness), Rrmsd (Derivative of Root Mean Square Roughness), histogram skewness and kurtosis, correlation length, and FFT spectral analysis.

To evaluate the impact of different edge detection techniques on roughness quantification, we compare roughness parameters obtained from Canny and Watershed algorithms under varying segmentation conditions. The results, as summarized in Table 1, highlight differences in roughness characteristics between better and worse segmentation cases for both methods.

A. Arithmetic Mean Roughness (Ra) and Root Mean Square Roughness (Rrms)

The Ra and Rrms values indicate the overall roughness amplitude of the extracted edges. For the Watershed algorithm, the worse segmentation case results in a higher Ra (3.429 nm) and Rrms (3.979 nm) compared to the better segmentation case (Ra = 2.342 nm, Rrms = 2.836 nm). This suggests that over-segmentation in the worse case leads to an overestimation of roughness due to excessive contour fragmentation. Similarly, for the Canny algorithm, the worse segmentation case produces higher roughness values (Ra = 3.439 nm, Rrms = 3.952 nm) compared to the better case (Ra = 2.338 nm, Rrms = 2.862 nm). This trend is consistent across both methods, indicating that poorer segmentation quality inflates roughness estimates by introducing artificial edge variations.

B. Higher-Order Roughness Measures: Rad and Rrmsd

The derivative-based roughness parameters (Rad and Rrmsd) capture localized intensity variations, providing insight into the fine-scale texture of the extracted edges. For Watershed, the Rad values remain nearly identical in both cases (0.638 nm vs. 0.635 nm), whereas Rrmsd is slightly higher in the worse segmentation case (0.859 nm vs. 0.835 nm). This minor variation suggests that Watershed segmentation errors primarily affect large-scale roughness rather than localized fluctuations. In contrast, for Canny, both Rad and Rrmsd increase in the worse segmentation case, with Rad rising from 0.422 nm to 0.491 nm and Rrmsd increasing from 0.715 nm to 0.806 nm. This indicates that Canny's edge extraction is more sensitive to segmentation quality, with worse segmen-

tation leading to a greater amplification of local roughness fluctuations.

C. Analysis of FFT and PSD

The Fast Fourier Transform (FFT) is a mathematical tool that decomposes a signal into its constituent frequency components, allowing for the identification of dominant periodic features. The Power Spectral Density (PSD) represents the distribution of power across different frequencies, providing insights into structural roughness and periodicity. In edge detection applications, a high concentration of power at low frequencies indicates smoother and more continuous boundaries, whereas increased power at high frequencies suggests the presence of noise, fragmentation, or fine structural details.

In the case of Watershed-based segmentation, two SEM image edge profiles are analyzed: one yielding a better segmentation result and the other exhibiting poorer performance. In the better segmentation case (figure 4) compared with worse on (figure 5), the extracted edge profile appears more structured, with smoother transitions and fewer abrupt variations. The corresponding PSD shows a strong peak at low frequencies, indicating that the extracted edges are predominantly continuous with minimal fragmentation. The decay of the PSD at higher frequencies suggests limited high-frequency noise, confirming the effectiveness of the segmentation. The

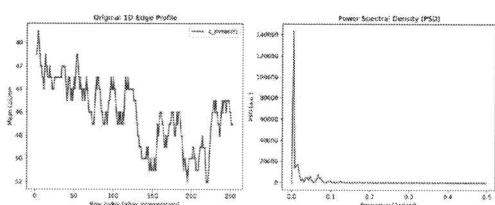

Fig. 4: Edge Profile and Power Spectral Density (PSD) Analysis of Well-Segmented Watershed Result

comparative analysis of Watershed and Canny-based edge detection in SEM images using FFT and PSD reveals distinct segmentation characteristics. The better Watershed segmentation exhibits a dominant low-frequency PSD peak, indicating smooth and continuous boundaries with minimal artifacts. Conversely, the poorer Watershed result shows increased high-frequency components, suggesting over-segmentation and artificial boundary artifacts. Compared to Watershed, Canny segmentation generally produces finer edges but is more sensitive

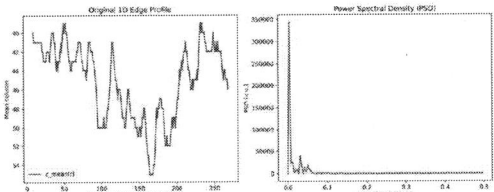

Fig. 5: Edge Profile and Power Spectral Density (PSD) Analysis of Over-Segmented Watershed Result

to noise, as reflected in the PSD distribution. While the better Canny result (figure 6) closely resembles the better Watershed segmentation in terms of spectral characteristics, the poorer Canny result (figure 7) introduces more mid-to-high frequency noise, reflecting edge fragmentation. Ultimately, FFT and PSD provide a quantitative measure of segmentation quality, where a strong low-frequency peak indicates optimal segmentation, while high-frequency dominance suggests excessive noise or over-segmentation artifacts.

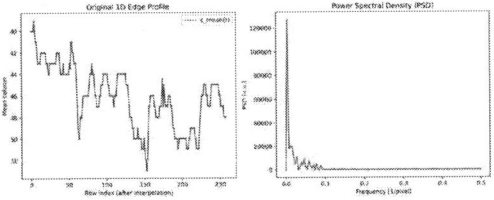

Fig. 6: Edge Profile and Power Spectral Density (PSD) Analysis of Well-Segmented Canny Result

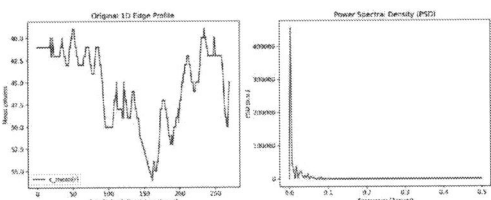

Fig. 7: Edge Profile and Power Spectral Density (PSD) Analysis of Noisy Canny Result

D. Histogram-Based Skewness and Kurtosis Analysis

The skewness and kurtosis metrics provide statistical insights into the distribution of roughness values. For Watershed, the skewness remains constant (2.136) in both cases, while the kurtosis increases from 3.153 to 3.721 in the worse segmentation case. This suggests that worse segmentation results in a sharper peak in the roughness distribution, likely due to excessive fine-scale segmentation artifacts. For Canny, the skewness increases from 2.252 in the better case to 2.639 in the worse case, indicating a shift in the roughness distribution toward more pronounced deviations. Furthermore, the kurtosis of Canny-extracted edges increases significantly from 3.902 to 6.869, suggesting that poor segmentation leads

to a highly peaked roughness distribution with more extreme variations. This highlights that Canny is more susceptible to segmentation-induced outliers compared to Watershed.

E. Correlation Length and Edge Continuity

Correlation length (Lc) measures the spatial coherence of extracted edges, with longer values indicating smoother and more continuous boundaries. For Watershed, the worse segmentation case results in a higher correlation length (33.555 nm) compared to the better case (27.074 nm), suggesting that over-segmentation introduces artificial continuity artifacts rather than true edge coherence. In contrast, for Canny, the correlation length increases significantly from 20.263 nm in the better case to 35.111 nm in the worse case, implying that poor segmentation distorts the natural edge structure, leading to artificially extended correlation lengths (table I). This further confirms that Canny is more affected by segmentation errors, as its gradient-based approach can misinterpret noise-induced variations as part of the true edge structure.

IV. Conclusions

This study examined the impact of Canny and Watershed edge detection on line-edge roughness (LER) quantification in SEM images. By analyzing roughness parameters such as Ra, Rrms, Rad, Rrmsd, skewness, kurtosis, correlation length, and FFT-based spectral analysis, we demonstrated that edge detection choice significantly affects roughness characterization.

The Watershed algorithm, while effective in extracting closed contours, tends to over-segment, introducing artifacts and overestimating roughness. In contrast, the Canny algorithm provides precise edge localization and better noise suppression but struggles with low-contrast transitions, leading to fragmented edges and roughness underestimation. FFT and PSD analyses confirmed that better segmentation results exhibit dominant low-frequency peaks, while poor segmentation increases high-frequency noise and artifacts.

Overall, Canny is preferable for high-contrast SEM images, while Watershed is more suitable for complex, low-contrast structures, though both require careful parameter optimization. Future work will explore hybrid approaches integrating morphological and gradient-based techniques to enhance segmentation accuracy and LER quantification.

References

[1] Wang J, Yao P, Liu W, et al. A hybrid method for the segmentation of a ferrograph image using marker-controlled watershed and grey clustering. Tribology Transactions, 2016, 59(3): 513-521.

[2] Paris S, Hasinoff S W, Kautz J. Local Laplacian filters: Edge-aware image processing with a Laplacian pyramid[J]. ACM Trans. Graph., 2011, 30(4): 68.

[3] Lu R, Shen Y, Wang Q, et al. Ultrasound Image Segmentation Based on the Early Vision Model and Watershed Transform[C]//2005 IEEE Instrumentationand Measurement Technology Conference Proceedings. IEEE, 2005, 3: 2016-2018.

[4] Ding L, Goshtasby A. On the Canny edge detector[J]. Pattern recognition, 2001, 34(3): 721-725.

Impact of Channel Geometry and Gate Dielectric Properties on GAAFET Device Performance

Yue Zhang[1*], Yongjie Zhao[1], Zhaonian Yang[1], Yufei Wang[2]

[1] School of Automation and Information Engineering, Xi'an University of Technology, Xi'an, China.
[2] Xi'an Microelectronics Technology Institute, Xi'an, China.
*email: zhangyue@xaut.edu.cn

Abstract—**Three-dimensional FETs are considered a significant innovation in advanced CMOS technologies. Notably, the introduction of the Gate-All-Around (GAA) structure obviously improved electrostatic performance and increased current drive. In this paper, the effects of device channel geometries and gate dielectric selections on the performance of GAAFET devices are investigated through simulation. The key performance metrics of FETs, including the on/off current ratio (I_{on}/I_{off}), subthreshold slope (SS), and drain-induced barrier lowering (DIBL), are analyzed in detail to provide a deeper understanding of the GAAFET structure and offer significant insights for device design optimization.**

Keywords—**CMOS, GAAFET, subthreshold slope, drain-induced barrier lowering**

I. INTRODUCTION

Over the past decades, the progress of CMOS integrated circuits (ICs) has been significantly driven by continuous innovations in transistor architectures. The emergence of GAAFETs offers superior electrostatic integrity and higher drive current compared to FinFETs, making them the most promising successor to the post-FinFET technology for advanced ICs [1-3].

With the gate electrodes fully surrounding the channel, GAAFETs exhibit significantly enhanced gate controllability due to the larger gate contact area and the capability for stacking, which enable more effective enhancement of device switching characteristics and short-channel effect suppression [4-5]. The typical GAAFET structures can be classified into three categories based on the cross-sectional shape of the channel: nanowire FET (NWFET), nanosheet FET (NSFET), and nanotube FET (NTFET) [6]. Advanced architectures in the nanometer regime pose unique design challenges that necessitate detailed evaluation of device performance to guarantee effective and reliable implementation for the next generation. The performance of GAAFETs is influenced by channel-related parameters, including channel shape and dimensions. Additionally, it is also associated with factors such as gate dielectric material, which require further research for the design of device structures.

In this paper, the impact factors for GAAFETs are investigated through simulation. Initially, simulations were conducted for three different types of GAA structures. Based on the results, the nanowire device was selected for further analysis. The impacts of parameters such as channel length, gate dielectric thickness, and dielectric constant on device

characteristics were investigated, including I_{on}/I_{off}, SS, and DIBL.

II. SIMULATION INFORMATION AND SETTINGS

The simulations of GAAFETs were performed using Silvaco TCAD tools. The device structures, featuring various GAA structures with different channel shapes, are illustrated in Fig. 1.

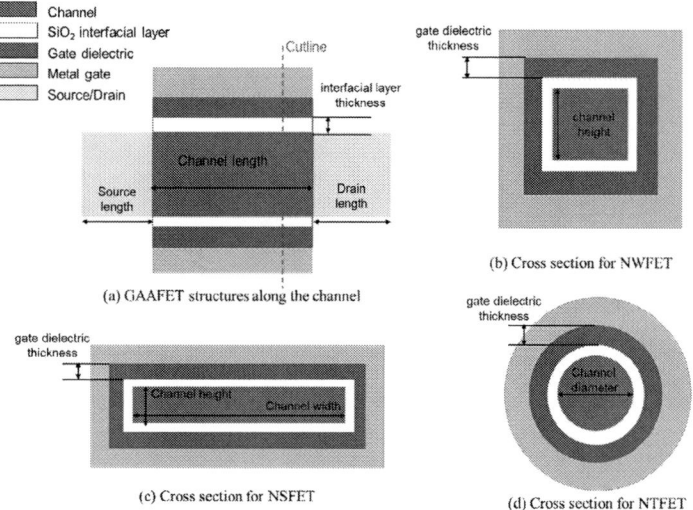

Fig. 1 Device structures for simulation

A comprehensive summary of the dimensional parameters is provided in Table I, which is based on the International Roadmap for Devices and Systems (IRDS) projected dimensions for the 2 nm technology node [7].

TABLE I.   DEVICE PARAMETERS SETTINGS FOR SIMULATION

Parameters	Values
Gate dielectric constant (SiO$_2$)	3.9
Gate dielectric thickness	1nm
Channel height	5nm
Channel length	10nm
Source/Drain length	10nm
Channel width	7nm
Channel Doping	1×10^{15}cm^{-3}
Source/Drain Doping	1×10^{20}cm^{-3}

III. RESULTS AND DISCUSSION

A. The effect of GAA structure on the characteristics of GAAFETs

The I-V transfer characteristics are simulated at $V_{ds} = 0.7V$ and $V_{ds} = 0.05V$ to encompass both the saturation and linear operating regimes. I_{on} and I_{off} are extracted from I-V curves when $V_{gs} = 0.7V$ and $V_{gs} = 0V$, respectively, with V_{ds} maintained at 0.7V.

The transfer and output characteristics of the three structures of GAAFETs are depicted in Fig. 2. Notably, NWFET exhibits a higher current driving capability.

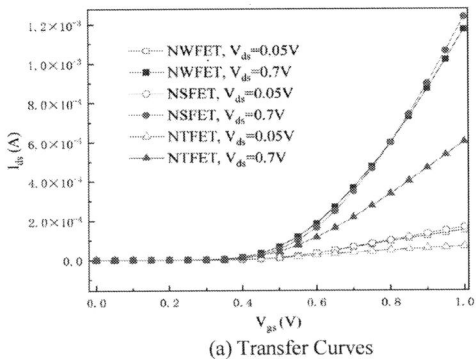

(a) Transfer Curves

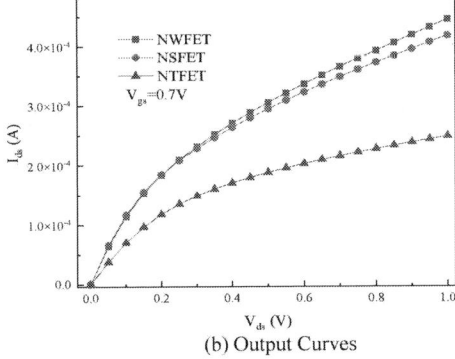

(b) Output Curves

Fig.2 I-V Characteristic Curves of Different GAA Structures

From the I-V characteristics, we can also extract the performance metrics, which are summarized in Table II. It reveals that the I_{on} and I_{off} of the three structures are comparable, lying within the same order of magnitude.

TABLE II.   PERFORMANCE PARAMETERS OF DIFFERENT GAA STRUCTURES

GAA structure	I_{on}(A)	I_{off}(A)	I_{on}/I_{off}	SS (mV/dec)	DIBL (mV/V)
NWFET	2.560×10^{-4}	2.709×10^{-11}	9.460×10^{6}	73.192	61.388
NSFET	3.486×10^{-4}	2.170×10^{-10}	1.606×10^{6}	79.272	66.356
NTFET	2.160×10^{-4}	6.000×10^{-11}	3.600×10^{6}	71.203	155.433

Among them, NTFET has the smallest SS but the largest DIBL value, while NWFET exhibits the smallest DIBL and a similarly low SS, only slightly higher than that of NTFET. Furthermore, NSFET has intermediate values for both SS and DIBL. Indeed, a smaller SS and DIBL values indicate superior gate controllability and device performance.

Moreover, in terms of I_{on}/I_{off}, NWFET demonstrates the highest value, which implies faster switching and lower power consumption. Consequently, considering the comparable I_{on}/I_{off}, SS and DIBL, NWFET is chosen for further investigation.

B. The effect of channel length on the characteristics of NWFETs

The simulations of NWFETs are performed with the channel lengths varied to 5 nm, 7 nm, 10 nm, 12 nm, and 15 nm, respectively, while other parameters are set according to the simulation settings detailed in Table I.

As indicated in Table III, based on the analysis of the I-V characteristics, the currents increase with decreasing channel length at the same V_{gs}. However, while I_{on} varies by only one order of magnitude across different conditions, I_{off} exhibits a variation of seven orders of magnitude. This significant disparity is attributed to the fact that I_{off} is highly sensitive to channel length. Smaller I_{off} are observed at larger channel lengths, thereby leading to an increase in I_{on}/I_{off}.

For SS and DIBL values, both increase with decreasing channel length, indicating weakened gate controllability and exacerbated short-channel effects.

Moreover, when the channel length exceeds 10 nm, the changes in device performance parameters become less pronounced when compared to those at shorter lengths of just a few nanometers. Thus, the channel length should be maintained around 10 nm to avoid excessive reduction.

TABLE III.   PERFORMANCE PARAMETERS AT DIFFERENT CHANNEL LENGTHS

Channel length (nm)	I_{on}(A)	I_{off}(A)	I_{on}/I_{off}	SS (mV/dec)	DIBL (mV/V)
5	1.251×10^{-3}	7.963×10^{-6}	1.571×10^{2}	169.421	219.465
7	5.466×10^{-4}	1.251×10^{-8}	4.369×10^{4}	95.312	125.432
10	2.560×10^{-4}	2.709×10^{-11}	9.460×10^{6}	73.192	61.388
12	1.818×10^{-4}	2.467×10^{-12}	7.369×10^{7}	67.771	49.161
15	1.227×10^{-4}	3.202×10^{-13}	3.832×10^{8}	63.961	37.246

C. The effect of gate dielectric on the characteristics of NWFETs

Given that the performance of GAAFETs is significantly influenced by the gate dielectric-related parameters, the simulations were conducted to investigate the impact of and dielectric constant and thickness using various dielectric materials. The gate dielectric materials selected include SiO_2, Si_3N_4, Al_2O_3, and HfO_2, which exhibit dielectric constants of 3.9, 7, 9.3, and 25, respectively. The corresponding oxide layer thicknesses are 1 nm, 2 nm, and 3 nm.

The performance metrics derived from the transfer characteristic curves are summarized in Tables IV. The data presented in all tables consistently indicate that an increase in the dielectric constant of the gate dielectric material correlates with enhanced device performance, characterized by higher I_{on}, lower I_{off}, improved I_{on}/I_{off}, and reduced SS and DIBL values.

As the thickness of the gate dielectric increases, both the SS and DIBL exhibit an upward trend. However, the magnitude of this deterioration varies among different materials. For instance, HfO_2 exhibits a relatively smaller increase in SS and DIBL values compared to SiO_2, which shows a more pronounced change. This difference can be attributed to the higher dielectric constant of HfO_2, which enhances gate controllability and mitigates short-channel effects, whereas SiO_2, with its lower dielectric constant, is more susceptible to these issues. The superior performance obtained with HfO_2 makes it an ideal choice for advanced semiconductor applications.

TABLE IV.   PERFORMANCE PARAMETERS AT DIFFERENT THICKNESSES WITH VARIED DIELECTRIC CONSTANT

Dielectric constant ε	Dielectric thickness (nm)	$I_{on}(A)$	$I_{off}(A)$	I_{on}/I_{off}	SS (mV/dec)	DIBL (mV/V)
3.9	1	2.560×10^{-4}	2.709×10^{-11}	9.460×10^6	73.192	61.388
	2	2.497×10^{-4}	2.354×10^{-9}	1.061×10^5	89.700	80.385
	3	2.565×10^{-4}	4.324×10^{-8}	5.932×10^3	107.33	99.694
7	1	2.935×10^{-4}	1.742×10^{-12}	1.685×10^8	66.384	58.711
	2	2.691×10^{-4}	4.082×10^{-11}	6.593×10^6	73.963	69.827
	3	2.700×10^{-4}	4.821×10^{-10}	5.600×10^5	81.913	76.599
9.3	1	3.106×10^{-4}	9.275×10^{-13}	3.348×10^8	65.130	56.431
	2	2.797×10^{-4}	1.421×10^{-11}	1.969×10^7	71.018	62.379
	3	2.778×10^{-4}	1.342×10^{-10}	2.070×10^6	77.22	75.706
27	1	4.024×10^{-4}	2.108×10^{-13}	1.909×10^9	62.480	46.789
	2	3.444×10^{-4}	7.716×10^{-13}	4.464×10^8	64.720	51.201
	3	3.328×10^{-4}	2.925×10^{-12}	1.124×10^8	67.175	51.867

IV. CONCLUSION

The effects of channel geometry and gate dielectric properties on the device performance of GAAFETs are investigated through simulation in this work. An initial comparative analysis among the three typical GAAFET structures—namely NWFET, NSFET, and NTFET—indicates that the NWFET device demonstrates relatively superior performance. Subsequently, for NWFETs, the effects of channel length, gate dielectric thickness, and dielectric constant on device performance metrics are simulated. The results reveal that shorter channel lengths weaken gate controllability and exacerbate short-channel effects. Thus, the channel length should be maintained around 10 nm to avoid excessive reduction. The results also demonstrate that higher dielectric constants and optimized thicknesses improve device performance by enhancing I_{on}/I_{off} while minimizing SS and DIBL values.

REFERENCES

[1] C. H. Kim, A. Castro-Carranza, M. Estrada, A. Cerdeira, Y. Bonnassieux, G. Horowitz, and B. Iniguez, "A compact model for organic field-effect transistors with improved output asymptotic behaviors," IEEE Transactions on Electron Devices, 2013, vol. 60, issue 3, pp. 1136-1141, 2018.

[2] Huang XJ, Lee WC, Kuo C, et al. Sub-50 nm P-channel Fin FET[J]. IEEE Transactions on Electron Devices, 2001, 48(5): 880-886.

[3] S. Sayeef, N. Kai, and D. Suman, "The era of hyper-scaling in electronics," Nature Electronics, vol. 1, pp. 442-450, 2018.

[4] A. Veloso, T. Huynh-Bao, P. Matagne, D. Jang, and D. Mocuta, "Nanowire & nanosheet FETs for ultra-scaled, high-density logic and memory applications," In 2019 Joint International EUROSOI Workshop and International Conference on Ultimate Integration on Silicon (EUROSOI-ULIS), Grenoble, France, April 1-3, 2020.

[5] M. Kobrinsky, J. D Silva, E. Mannebach, S. Mills, M. Abd El Qader, O. Adebayo, N. Arkali Radhakrishna, M. Beasley, J. Chawla, S. Chugh, A. Dasgupta, U. Desai, E. De Re, G. Dewey, T. Edwards, C. Engel, V. Gudmundsson, J. Hicks, B. Krist, R. Mehandru, I. Meric, P. Morrow, D. Nandi, P. Patel, R. Ramamurthy, D. Samanta, L. Shoer, A. St Amour, L. H. Tan, S. Yemenicioglu, X. Wang, and T. Ghani, "Novel cell architectures with back-side transistor contacts for scaling and performance," In 2023 IEEE Symposium on VLSI Technology and Circuits (VLSI Technology and Circuits), Kyoto, Japan, June 11-16, 2023.

[6] Qingzhu Zhang, Yongkui Zhang, Yanna Luo and Huaxiang Yin, "New structure transistors for advanced technology node CMOS ICs," National Science Review, vol. 11, nwae008, 2024.

[7] The International Roadmap for Devices and Systems, "2022 update: more Moore," IRDS 2022 white paper, IEEE, 2022.

Field-effect transistor biosensor for ultra-high sensitivity detection of ssDNA

Mengran Chen[1], Ya Li[2], Haoliang Li[3], Bing Chen[4], Wenchang Zhang[2], Xiaonan Yang[1]

1 School of Electrical and Information Engineering, Zhengzhou University, Zhengzhou 450001, China

2 Department of Gastroenterology, The First Affiliated Hospital of Zhengzhou University, Zhengzhou, 450052, China

3 State Key Laboratory of Fabrication Technologies for Integrated Circuits，Institute of Microelectronics, Chinese Academy of Sciences, Beijing, 100029, China

E-mail addresses: iexnyang@zzu.edu.cn

Abstract—A high-sensitivity and low-detection-limit field-effect transistor (FET) biosensor array, based on IGZO thin-film material, has been designed and fabricated. Through simulation, process optimization, and surface functionalization, the biosensor is capable of detecting single-stranded DNA (ssDNA) with high sensitivity. Firstly, the process parameters and fabrication process of the IGZO-FET are obtained through simulation design. The performance of the fabricated device was tested, revealing outstanding electrical properties; subsequently, the capture probe of ssDNA was firmly attached to the surface of the sensing area through surface functionalization, and the functionalization effect was characterized by both fluorescence and electrical signals. Finally, electrical signal detection of ssDNA solutions at different concentrations was performed to verify and analyze the detection limit and sensitivity of the field-effect transistor biosensor. The results demonstrate that the IGZO-FET can sensitively detect ssDNA over a concentration range of 10^{-6} to 10^{-20} mol/L.

Keywords--*IGZO-FET, ssDNA, biosensor, sensitivity*

I. INTRODUCTION

Recent advancements in genomics and molecular biology have led to an increasing demand for single-stranded DNA (ssDNA) detection. Detection and quantitative analysis of ssDNA are critical in biomedical applications such as disease diagnosis, genetic research, and environmental monitoring[1][2]. Existing DNA detection methods, such as polymerase chain reaction (PCR)[3] and fluorescence[4], are time-consuming, require complex equipment. In the field of genetic testing, FET sensors[5] offer advantages such as non-contact detection, excellent detection limits, and stability. IGZO, as a transparent oxide semiconductor, exhibits excellent conductivity, high electron mobility[6]. However, despite some research exploring the application of IGZO-FET in DNA detection[7], challenges remain in further improving DNA detection sensitivity and lowering detection limits detection. These challenges underscore the need for continuous advancements in sensor design, surface functionalization, and signal amplification techniques to enhance the performance of IGZO-FET-based sensors for ultra-sensitive ssDNA detection[8]. This study aims to develop a novel IGZO-FET-based sensor for the ultra-sensitive detection of ssDNA at low detection limits. Our goal is to significantly enhance the sensitivity and specificity of ssDNA detection by optimizing the structure of the IGZO-FET and functionalizing its surface, achieving high-precision detection at extremely low concentrations at extremely low concentrations.

II. DEVICE DESIGN AND METHODS

A. Design of IGZO-FET

The IGZO-FET was modeled and simulated using Silvaco TCAD and COMSOL, the specific design steps for the preparation process of the IGZO-FET sensor were established. The preparation process of the IGZO-FET sensor is shown in Fig. 1(a). Magnetron sputtering, electron beam evaporation and lithography are mainly used. To improve the device performance, we determined that the optimal thickness of IGZO is 7 nm, which enhances both the mobility and sensitivity of the device. The IGZO layer was annealed using a multi-step process, including nitrogen annealing, to increase the sub-threshold swing and enhance the sensing sensitivity of the device. The design of the IGZO-FET sensor unit is shown in Fig. 1(b). The DNA probe is modified onto the sensitive layer to enable the capture of target DNA.

B. Aptamer-functionalized biosensor

The IGZO-FET sensor was immersed in a 5% APTES solution and incubated at 180 rpm for 1 hour at 25°C.Then repeatedly rinsed with anhydrous ethanol and placed in an oven at 110°C for 30 minutes to undergo surface silanization.Sodium 4-(N-maleimide methyl)cyclohexane - 1 - carboxylate sulfosuccinimidate (sulfo-SMCC) was prepared as a 1 mg/mL solution in 10 mol/L PBS buffer (pH 7.4). The FET sensor was immersed in the prepared solution and incubated at 25°C for 2 hours. The amino groups in the sensing area of the device form stable amide bonds with sulfo-SMCC. The sensor was subsequently rinsed with PBS (pH 7.4) and dried quickly. The DNA probe was added dropwise to the sensing area of the device and incubated at 4°C for 12 hours. The thiol group at the 5' end of the DNA probe forms a stable thioether bond with sulfo-SMCC, thereby attaching to the sensing area.

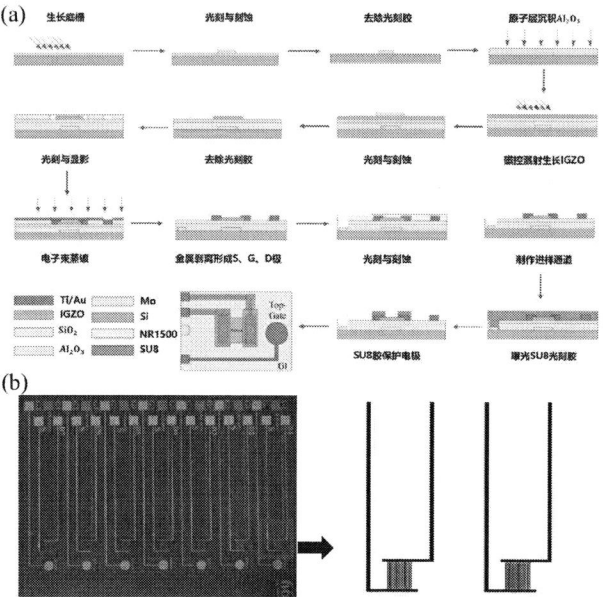

Fig. 1. (a)The preparation process of the IGZO-FET sensor were determined. (b) The design of the IGZO biosensor-FET.

C. Signal detection

The electrical characteristics of the IGZO-FET were measured using a semiconductor parameter analyzer and a probe station. The liquid gate method was employed for sample detection. The silver chloride paste was applied to the surface of the liquid gate. When measuring the output characteristic curve, the source is grounded, the drain voltage is scanned from 0 V to 1 V, with a step size of 500 mV.When measuring the transfer characteristic curve, the gate voltage is scanned from - 0.2 V to 1 V, with a step size of 500 mV. After surface functionalization, the detection sample was added to the sensing area. The source is grounded, and the drain voltage is set to 0.5 V. We set the Y-axis function to obtain the lg(Id)-Vg curve. The voltage was scanned from −0.5 V to 1.5 V, with a step size of 100 mV.

III. RESULTS AND DISCUSSION

A. Working Principile

After surface functionalization of the FET, the solution of the selected ssDNA sequence is dropped onto the sensing area and incubated for 5 minutes. The ssDNA of

samples with different concentrations can be detected by the change in drain current. Since the isoelectric point of DNA is between pH 4 and 5, and the pH of the DNA solution we used is 7.4, the ssDNA in the sample solution is negatively charged. The ssDNA will introduce additional electrons upon binding to the probe, thereby reducing the drain current，the increase in concentration results in a continuous decrease in the drain current . The study employs a liquid gate detection method to replace the top gate detection. The silver chloride slurry is applied to the surface of the liquid gate to reduce curve jitter and achieve a smoother curve in the linear region.

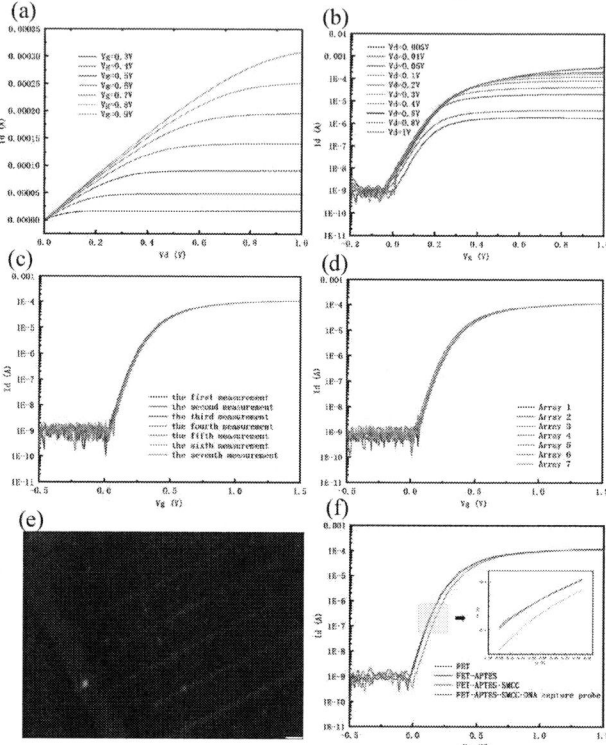

Fig. 2. (a) output characteristic curve of the IGZO-FET, (b) the characteristic curve of the IGZO-FET sensor, (c)the same sensor array repeatedly detects the Id-Vg curve for six times, (d) Id-Vg curves detected by different sensor arrays, (e)surface functionalized fluorescence characterization, (f)Surface functionalized electrical signal characterization.

B. Device characteristics

Fig. 2(a)(b) show the output characteristic curve and transform characteristic curve of the IGZO-FET sensor respectively.The device turns on near 0V and reaches saturation more quickly, demonstrating that the device exhibits a on/off ratio and good control over the conductive channel. Fig. 2(c)(d) show the repeatability and consistency test curves of the IGZO-FET sensor array, respectively. It can be observed that the transfer characteristic curve of the same sensing unit exhibits negligible offset after seven tests, and the transfer characteristic curves of seven different sensing units on the same sensor align well with each other, demonstrating that the device possesses excellent repeatability and consistency.

C. Surface functionalization

Fig. 2(e) shows the the fluorescence images of 6FAM-DNA capture probes incubated with APTES and sulfo-SMCC solution-treated FETs. After treatment with functional reagents, the DNA probe exhibited distinct fluorescence on the surface of the device in the sensing area following 12 hours of incubation. This confirms the successful surface functionalization of the device. Additionally, during the surface functionalization process, we also measured the transfer characteristic curve of the device. As shown in Fig. 2(f), the infiltration of solutions without DNA probe incubation did not result in a significant shift in the drain current. After DNA probe incubation, the current curve shifted to the right. This is because the DNA probe is negatively charged, introducing electrons into the sensing area, which also confirms successful surface functionalization.

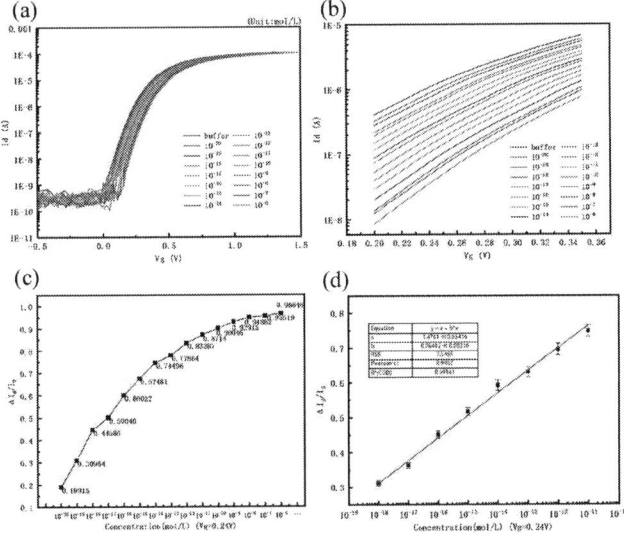

Fig. 3. (a)the Id-Vg concentration gradient curves of buffer and ssDNA solution diluted 10 times from 10^{-6} mol / L to 10^{-20} mol / L were plotted, (b)the linear region of the concentration gradient curve from Vg = 0.2 V to Vg = 0.35 V was characterized, (c)the current response of ssDNA solution at different concentrations was measured, (d)the current signals of ssDNA solutions from 10^{-18} to 10^{-11} mol / L were characterized by linear fitting.

D. Detection of ssDNA

Fig. 3 shows the Id-Vg curves measured for buffer and DNA solutions, with concentrations ranging from 10^{-6} to 10^{-20} mol/L, diluted tenfold for each step. As shown in Fig. 3(a)(b), as the sample concentration increases, the curve continuously shifts to the right, and the linear region of each curve remains completely separated without any overlap. The Id-Vg curve demonstrates that the ssDNA sequence can be successfully detected through binding to the capture probe. At the same time, since the sample is diluted by a factor of ten each time, and the current decreases as the concentration increases,

the linear region of the curve indicates that the current decreases linearly with concentration within a certain range. It can be seen from Fig. 3(c) that within the detection concentration range, the higher the concentration of the ssDNA , the greater the response of the drain current change. In this study, the current at Vg = 0.3 V and Vd = 0.5 V within a certain concentration range was selected for linear fitting. As shown in Fig. 3(d), when testing samples with concentrations from 10^{-11} to 10^{-18} mol/L, the current increased linearly with concentration, and $R^2 > 0.99$.

IV. CONCLUSIONS

We report an ultrasensitive biosensor array based on dual-gate IGZO-FETs for the detection of ssDNA. IGZO field-effect transistor arrays, with indium gallium zinc oxide (IGZO) thin films as the sensing material, were fabricated using magnetron sputtering, electron beam evaporation, atomic layer deposition, etching, and lithography. To enhance the sensitivity of the device, we controlled the thickness of the IGZO film to 7 nm. The IGZO-FET array we fabricated exhibits a large switching ratio and excellent electrical stability, enabling us to produce a consistent and reproducible high-performance biosensor. This also offers great potential for development in the detection of non-amplified molecules. By using stable chemical bonds to firmly attach DNA probes to the device's sensing area, we developed a high-performance IGZO-FET biosensor and evaluated its performance with DNA standards. The results demonstrate that the biosensor exhibits high sensitivity, with a detection limit of 0.01 aM. The standard concentration range can meet the low concentration sensitive detection of ssDNA.

REFERENCES

[1] Song, et al. "Limitations and opportunities of technologies for the analysis of cell-free DNA in cancer diagnostics." Nature biomedical engineering 6.3 (2022): 232-245.

[2] Dang, Donna K. "Circulating tumor DNA: current challenges for clinical utility." The Journal of clinical investigation 132.12 (2022).

[3] McDonald, Caitlin, and Adrian Linacre. "PCR in Forensic Science: A Critical Review." Genes 15.4 (2024): 438.

[4] Li, Zhenqing, et al. "High throughput DNA concentration determination system based on fluorescence technology." Sensors and Actuators B: Chemical 328 (2021): 128904.

[5] Wang, et al. "Aptamer-functionalized field-effect transistor biosensors for disease diagnosis and environmental monitoring." Exploration. Vol. 3. No. 3. 2023.

[6] Shin, Wonjun, et al. "Annealing Ambient and Film Thickness Dependent NO₂ Response and 1/f Noise Characteristics of IGZO Resistor-Type Gas Sensors." IEEE Transactions on Electron Devices (2023).

[7] Zhou, Yuying, et al. "Recent Advances in Enhancing the Sensitivity of Biosensors Based on Field Effect Transistors." Advanced Electronic Materials (2024): 2400712.

[8] Hwang, et al. "Wide-range and selective detection of SARS-CoV-2 DNA via surface modification of electrolyte-gated IGZO thin-film transistors." Iscience 27.3 (2024).

A Study on Short-Circuit Parasitic Conduction Failure of 1200V SiC VDMOSFETs

YaDong Zhou[1], HengYue Gong[1], YangHui Xia[1], YuMeng Wang[1], HuiXia Yang[1], HuiPing Zhu[2], YuanXiao Ma[1,*], and Yeliang Wang[1,*]

[1]Yangtze Delta Region Academy, and The School of Integrated Circuits and Electronics,
Beijing Institute of Technology, Beijing 100081, China

[2]Institute of Microelectronics, The Chinese Academy of Science, Beijing 100029, China

*Authors to whom the correspondence should be addressed: yxma@bit.edu.cn and yeliang.wang@bit.edu.cn

Abstract—In this study, the parasitic conduction failure of typical 1200V SiC VDMOSFETs under short-circuit conditions have been systematically investigated. Based on experimental and simulation results, the mechanism of parasitic conduction failure during short-circuit events is elucidated. The device operates in saturation mode when a short circuit occurs, during which current flows through the channel, JFET region, and N-drift region to cause a significant heating. As a result, the parasitic npn transistor is triggered as the temperature rises, leading to a second current rise and resultant thermal failure. Accordingly, a Multi-pillar structure featuring four pillars with various depths (0.5~0.2 μm) beneath the p-well region is proposed to mitigate the parasitic conduction failure. As verified by simulation, the peak short-circuit current is reduced by 3.16 %, and the short-circuit withstand time (SCWT) is increased by 18.18 %. This Multi-pillar structure partially blocks parasitic current path and inhibits the activation of the parasitic transistor.

Index Terms—SiC, VDMOSFET, Short-Circuit, Parasitic Conduction

I. INTRODUCTION

SiC MOSFETs have attracted many attentions due to their superior performances in high-voltage, high-frequency, and high-temperature applications, for which the reliability performance is a key characteristic. [1] As an important failure phenomenon, the short-circuit failure have been extensively investigated to improve the reliability of the SiC MOSFETs. However, the research on failure mechanisms has been primarily focused on gate oxide rupture [2] and metal electrode melting, [3] while the parasitic conduction failure under short-circuit conditions has been neglected to some extent. In this work, the thermal failure caused by the conduction of the parasitic npn transistor under short-circuit conditions has been studied for a typical SiC MOSFET (1200 V, 120 mΩ). The experimental and simulation analyses have been both conducted, based on which a Multi-pillar device structure has been proposed.

II. EXPERIMENTAL

A. Test Circuit

The test circuit diagram is shown in Figure 1, where the R_g, R_d, and R_s are the gate series resistance, drain series resistance, and source series resistance, respectively. Moreover, V_{bus} is the bus voltage, and L_d and L_s are the drain inductance and source inductance, respectively. The pulse source provides

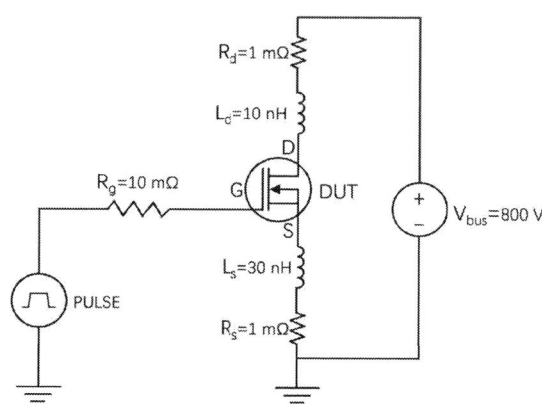

Fig. 1. Schematic diagram of short-circuit test.

gate pulse voltage signals, and the device under test (DUT) is a SiC VDMOSFET product (1200 V, 120 mΩ), namely H1M120F120.

B. Test Results

Figure 2 shows the measurement results of the SiC VD-MOSFET under short-circuit conditions with various gate pulse durations.

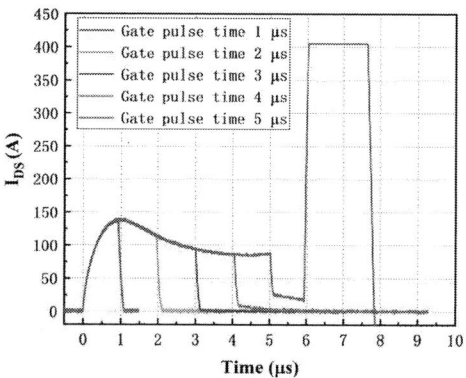

Fig. 2. Short-circuit current at a fixed V_{bus}=800 V and V_{GS}=18/–4 V with different gate pulse durations (1 μs, 2 μs, 3 μs, 4 μs and 5 μs).

979-8-3315-2209-4/25 $31.00 © 2025 IEEE

The short-circuit test results show that the peak current is 143 A. A significant turn-off tail current appears at 5 μs [4], which should result from the threshold voltage reduction due to heat accumulation [5] and conduction of the parasitic npn transistor formed by the N+ source, P-well, and N-drift regions [6]. After a 5 μs short-circuit, the device fails to completely turn off, and thermal failure occurs with a current surge at 6 μs.

III. SIMULATION

A. Modeling

As shown in Figure 3, the short-circuit failure behavior has been further studied based on experimental and TCAD simulation results.

Figure 3(a) shows the device structure of the SiC VDMOS-FET, and the corresponding simulation result for the 5 μs gate pulse duration has been shown in Figure 3(b), which matches the experimental result well to evidence the accuracy of the simulation model.

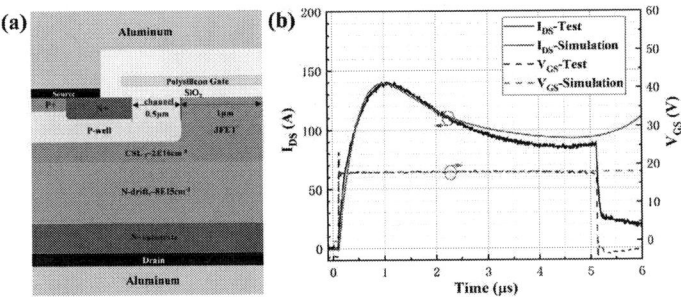

Fig. 3. (a) Schematic diagram of SiC VDMOSFET model. (b) Short-circuit current fitting (V_{DS} = 800 V; V_{GS} = 18 V; T_{CASE} = 25 °C).

B. Short-Circuit Failure Analysis

The simulation of the short-circuit failure process is shown in Figure 4(a). The short-circuit current goes through three stages: rising, falling, and rising again [7]. Points A, B, and C represent the short-circuit current peak (1 μs), the onset of parasitic conduction (4.6 μs), and the device failure point (5.5 μs).

Figure 4(b) shows the distribution of lattice temperature, hole current density, electron current density, and total current density at points A, B, and C. From the initial time to the point A, a large amount of current flows through the device and the short-circuit current reaches the first peak point, causing the device to heat up rapidly. Next, the short-circuit current exhibits a downward trend from point A to B to reach a minimum value, which is attributed to the elevated device temperature to strengthen lattice vibration and resultant phonon scattering. Oppositely, the short-circuit current starts to increase from point B to C due to the thinning depletion layer between the p-well and N-drift regions as shown in Figure 4(b). This thinning is attributed to the continuous heating accumulation, which increases the intrinsic carrier concentration and thereby weakens the built-in potential. Furthermore, the depletion

region is ruptured at point C to form a current transport path, which suggests the activation of the parasitic npn transistor. As a result, the current dramatically rises from the point C, which means the appearance of thermal failure.

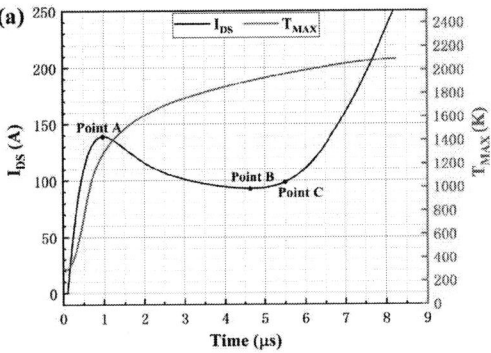

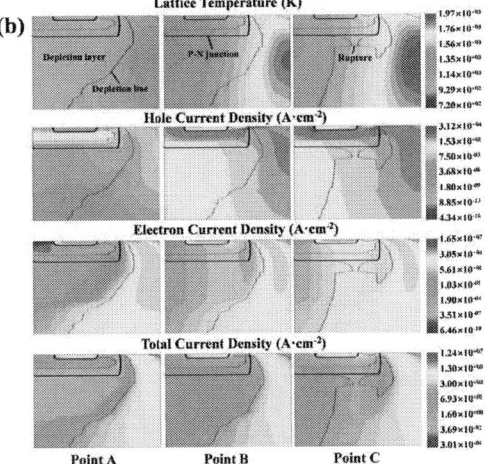

Fig. 4. (a) Continuous short-circuit simulation current and T_{MAX} (V_{DS} = 800 V; V_{GS} = 18 V; T_{CASE} = 25 °C);(b) Simulated distributions of lattice temperature, hole current density, electron current density, and total current density at point A point B, and point C.

The simulation results indicate that V_{bus} is directly applied to the device as short circuit appears, which quickly enforces the device to operate in the saturation region. A large amount of current flows through the channel, JFET region, and N-drift region to arrive at the drain terminal, which produces sufficient heat to raise the temperature for a electrothermal conversion in the JFET region. [8] As the temperature in the JFET region continuously rises, excessive electron and hole carriers are generated to forward bias the p-n junction between p-well and N+ regions, which can turn on the parasitic npn transistor [9] to eventually trigger its conduction. This causes the current to rise again, intensifying electrothermal conversion and resulting in thermal failure.

IV. DESIGNED STRUCTURE

One way to inhibit the short-circuit parasitic conduction failure is to improve the short-circuit withstand time (SCWT) of device, which can be achieved by suppressing the activation of the parasitic transistor and the short-circuit current. Figure

979-8-3315-2209-4/25 $31.00 © 2025 IEEE

5(a) shows the schematic of the as-proposed Multi-pillar structure, which features four pillars with various depths from 0.5 μm to 0.2 μm beneath the p-well region. As shown in Figure 5(b), the short-circuit current of the as-proposed Multi-pillar structure is compared to that of the original structure via simulation. As a result, the peak short-circuit current of the Multi-pillar device is 134.7 A, which is 3.16 % lower than the that of original structure (139.1 A). Moreover, the failure time as represented by the SCWT is improved to 1 μs, achieving a 18.18 % increase.

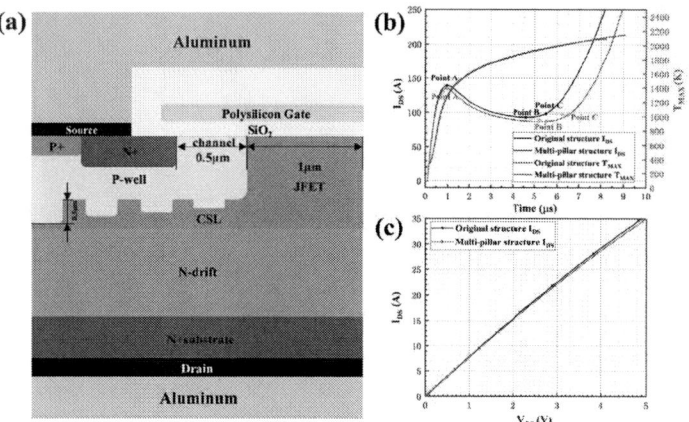

Fig. 5. (a) Schematic diagram of Multi-pillar structure;(b) Short circuit simulation current and T_{MAX} between Multi-pillar structure and original structure (V_{DS} = 800 V; V_{GS} = 18 V; T_{CASE} = 25 °C);(c) Transfer characteristic curve of original structure and Multi-pillar structure (V_{DS} = 0~5 V; V_{GS} = 20 V; T_{CASE} = 25 °C).

The simulated output characteristics of the original and Multi-pillar structures are shown in Figure 5(c). The current capability of the Multi-pillar structure is slightly degraded because the depletion region below the p-well region is expanded by the pillars, which can narrow the current path in the epitaxial region and thus minimally increases the on-state resistance.

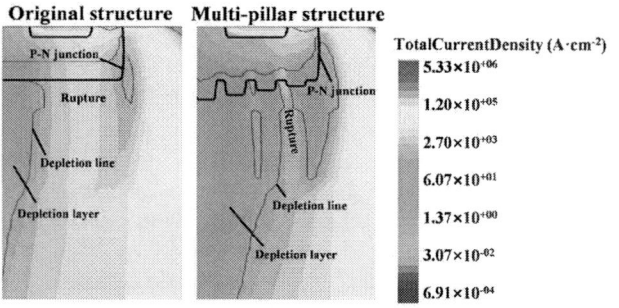

Fig. 6. Current density distribution at 6.5 μs under short-circuit between the original structure and the Multi-pillar structure (V_{DS} = 800 V; V_{GS} = 18 V; T_{CASE} = 25 °C).

Moreover, Figure 6 shows the short-circuit current density distribution of the original and Multi-pillar structures under the

same operating conditions at 6.5 μs. It can be observed that the Multi-pillar structure effectively blocks several conduction paths of the parasitic transistor.

V. CONCLUSION

The parasitic transistor conduction failure of 1200V SiC MOSFET under short-circuit conditions have been analyzed through experiments and simulations in this work. Accordingly, a novel Multi-pillar device structure is proposed, which improves the short-circuit withstand time by approximately 18.18 % with minimally affecting the conduction capability.

REFERENCES

[1] D. Xing et al., "1200-V SiC MOSFET Short-Circuit Ruggedness Evaluation and Methods to Improve Withstand Time," *IEEE J. Emerg. Sel. Topics Power Electron.*, vol. 10, no. 5, pp. 5059–5069, Oct. 2022.

[2] K. Yao, H. Yano, H. Tadano, and N. Iwamuro, "Investigations of SiC MOSFET Short-Circuit Failure Mechanisms Using Electrical, Thermal, and Mechanical Stress Analyses," *IEEE Trans. Electron Devices*, vol. 67, no. 10, pp. 4328–4334, Oct. 2020.

[3] T. Ziemann, A. Tsibizov, B. Kakarla, L. Bort, and U. Grossner, "Time-Resolved Short Circuit Failure Analysis of SiC MOSFETs," in *Proc. ISPSD*, 2019, pp. 219–222.

[4] G. Romano et al., "A Comprehensive Study of Short-Circuit Ruggedness of Silicon Carbide Power MOSFETs," *IEEE J. Emerg. Sel. Topics Power Electron.*, vol. 4, no. 3, pp. 978–987, Sept. 2016.

[5] K. Li, P. Sun, X. Ma, X. Huang, Q. Li, and L. Chen, "Impact of Short-Circuit Events on the Threshold Voltage Instability of SiC MOSFETs," in *Proc. PEAS*, 2023, pp. 1874–1878.

[6] B. Kakarla, A. Tsibizov, R. Stark, I. K. Badstübner, and U. Grossner, "Short Circuit Robustness and Carrier Lifetime in Silicon Carbide MOSFETs." in *Proc. ISPSD*, 2020, pp. 234–237.

[7] H. Chen et al., "Investigation on Short Circuit Test of 3300V SiC MOSFET," in *Proc. EDSSC*, 2019, pp. 1–3.

[8] C. Lin, N. Ren, H. Xu, and K. Sheng, "Performance and Short-Circuit Reliability of SiC MOSFETs With Enhanced JFET Doping Design," *IEEE Trans. Electron Devices*, vol. 70, no. 5, pp. 2395–2402, May 2023.

[9] X. Chen et al., "Investigation on Short-Circuit Characterization and Optimization of 3.3-kV SiC MOSFETs," *IEEE Trans. Electron Devices*, vol. 68, no. 1, pp. 184–191, Jan. 2021.

Multilayer Nitride-based Memristors with AlScN Interlayer for Memory and Neuromorphic Computing Applications

Yang Yang
Beijing Institute of Technology
Beijing, China
yangyang@bit.edu.cn

Qilin Hua
Beijing Institute of Technology
Beijing, China
huaqilin@bit.edu.cn

Abstract—**Nitride-based memristors are known for remarkable properties, including high on/off ratio, rapid switching speed, and seamless compatibility with CMOS processes. In this study, we introduce a multilayer nitride-based memristor with a stacked composition of AlN/AlScN/AlN, showcasing enhanced bipolar resistive switching characteristics. This novel design features AlScN as an interlayer within the AlN matrix, establishing a triple-stacked structure. The built-in electric field, generated by the ferroelectric polarization of the AlScN layer, effectively hinders the migration of electrons and ions within the AlN layer. Such memristors will exhibit promising potential applications in next-generation non-volatile memory and neuromorphic computing technologies.**

Keywords—Memristor, Nitride, AlScN, Multilayer

I. INTRODUCTION

Emerging memristors, particularly those utilizing Group III nitride materials, showcase exceptional scalability, speed, efficiency, and CMOS compatibility, positioning them as highly promising for next-generation non-volatile memory and neuromorphic computing applications [1-3]. AlN-based memristors are reported to demonstrate exceptional resistive switching characteristics including ultrafast switching (85 ps), low switching current, and low energy dissipation [4,5]. However, AlN-based memristors often show abrupt switching behavior, limiting their multilevel capabilities for memory and neuromorphic computing. Recently, a novel ferroelectric material AlScN is found to offer enhanced electromechanical coupling and superior resistive switching performance.

In this study, we introduce a multilayer nitride-based memristor with a stacked composition of AlN/AlScN/AlN. Additionally, we conducted a systematical investigation into the bipolar resistive switching characteristics of the memristor and analyzed the conduction mechanism utilizing AlScN as an interlayer for this novel multilayer nitride-based memristor.

II. DEVICE FABRICATION

A. Sputtered Ag/AlN/AlScN/AlN/Pt multilayer

The Ag/AlN/AlScN/AlN/Pt multilayer structure was fabricated through a series of precise deposition steps. First, a 20 nm titanium (Ti) adhesion layer and a 200 nm platinum (Pt) bottom electrode were deposited onto a SiO_2/p^+-Si substrate using DC magnetron sputtering. The Ti layer improved adhesion, preventing Pt detachment. This step was performed under an Ar flow rate of 30 sccm and a sputter power of 150 W for 15 minutes. Next, alternating layers of 20 nm AlN, 20 nm AlScN, and 20 nm AlN were deposited using RF magnetron sputtering with Al/AlSc metal targets in a N_2/Ar atmosphere (3:1 ratio). The deposition conditions included an Ar flow rate of 10 sccm, N_2 flow rate of 30 sccm, a sputter power of 250 W, and a duration of 3 minutes. Finally, a silver (Ag) top electrode was deposited via DC sputtering through a

150 μm diameter metal mask, with parameters set to 120 W power, 40 sccm Ar flow, and 15 minutes sputtering time. Fig. 1 provides a schematic of the multilayer nitride-based memristor.

B. Electrical Characterizations

The electrical characterization of multilayer nitride-based memristor was performed using a testing setup consisting of a probe station and a semiconductor device analyzer (Keysight B1500A), as illustrated in Fig. 1. For all current-voltage (I-V) measurements detailed in this study, the Ag layer served as the top electrode, while the Pt layer functioned as the bottom electrode and was grounded.

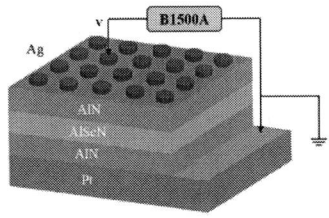

Fig. 1. Schematic structure of the multilayer nitride-based memristor.

III. RESULTS AND DISCUSSION

The fabricated memristor starts in a high-resistance state (HRS), requiring a higher voltage sweep to initiate resistive switching behavior. The device is tested at 1 mA current compliance, and exhibits a forming voltage of approximately 1.25 V. Fig. 2a illustrates the forming process of the multilayer nitride-based memristor.

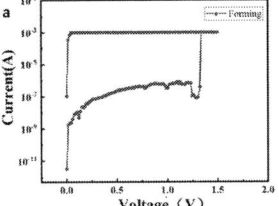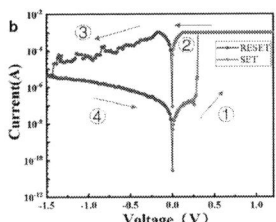

Fig. 2. (a) Forming process and (b) Typical I-V curve of the memristor.

In the I-V characterization, a current compliance of 1 mA was implemented during the SET operation to prevent the breakdown of the device, with the applied voltages sequentially set at 0 V → 1.5V → 0 V → -1 V → 0 V. Fig. 2b illustrates the electrical performance of the device, exhibiting a counterclockwise I-V hysteresis loop. Impressively, a gradual resistance switching behavior is observed during the RESET operation, distinguishing it from the abrupt transition typically seen in conventional AlN-based memristors [6].

979-8-3315-2209-4/25 $31.00 © 2025 IEEE

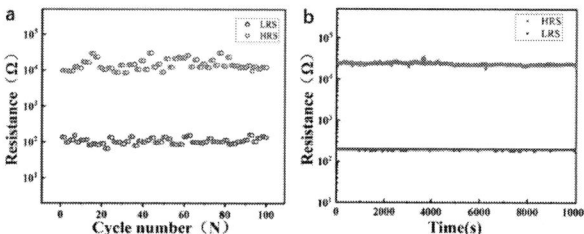

Fig. 3. (a) Endurance and (b) retention characteristics of the memristor.

To further assess the device performance, we measured the endurance and retention characteristics of the device, with the results illustrated in Fig. 3. The resistances of the device in its HRS and low resistance state (LRS) are depicted across 100 direct-current (DC) voltage sweeping cycles (see Fig. 3a), showcasing HRS values ranging from 10^4 to 10^5 Ω and LRS values between 10^2 and 10^3 Ω, with the latter exhibiting a more uniform distribution. Fig. 3b illustrates the retention characteristics, demonstrating that both HRS and LRS remain stable for over 10000 seconds at a read voltage of 0.1 V at room temperature, highlighting the device's excellent retention capability. Moreover, the on/off ratio consistently exceeds 100 with minimal variation, confirming the robust stability of the multilayer nitride-based memristor.

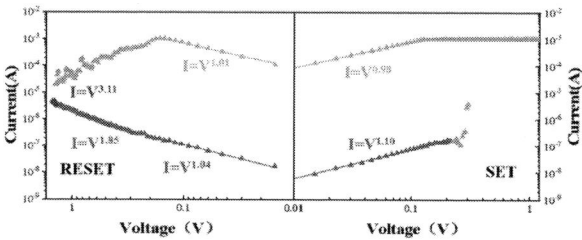

Fig. 4. Double-logarithmic I-V characteristics of the memristor at HRS and LRS with fitting results.

To investigate the resistive switching mechanism, we applied double logarithmic fitting on experimental I-V curves. Fig. 4 shows the linear fitting results at a RESET voltage of -1 V. In the LRS, slopes nearing unity in both voltage domains indicate the dominance of Ohm's law ($I \propto V$). In the HRS, the low-voltage region also follows Ohm's law, while higher voltages exhibit a quadratic relationship ($I \propto V^2$), consistent with Child's law. At even higher voltages, the current rises sharply until reaching the threshold, suggesting space-charge-limited conduction (SCLC) governs the HRS. Consequently, the LRS adheres to Ohm's law during SET and RESET operations, whereas the HRS aligns with the SCLC mechanism [7].

In Fig. 5, the device's energy band diagram unveils that the ferroelectric polarization of the AlScN interlayer suppresses electron/ion migration, creating a polarization-modulated energy barrier. During RESET, a positive voltage on the Ag top electrode drives Ag⁺ ions downward, forming conductive paths through atomic accumulation. However, the ferroelectric polarization of AlScN poses a substantial barrier [8], as its internal electric field (E_{int}) hinders Ag⁺ ion migration from the AlN layer to the Pt bottom electrode. The Ag⁺ ions generated in the AlScN layer during RESET further impedes electron/ion migration, enhancing the device's energy barrier. Additionally, during the RESET operation, the internal polarization field (E_{int}) of AlScN hampers Ag⁺ ion movement

from the AlN layer to the Ag top electrode, decelerating the RESET process and causing a gradual switching phenomenon.

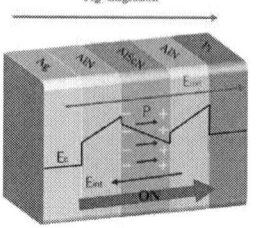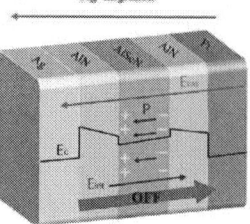

Fig. 5. Schematic illustration of working mechanism of the memristor.

IV. CONCLUSION

In summary, we developed a novel multilayer nitride-based memristor by incorporating an AlScN layer into AlN layers. Through DC I-V testing, we evaluated the device's resistive transitions unveiling exceptional resistive switching characteristics, including a low switching voltage, a high switching ratio, and prolonged retention time. Additionally, we delved into the physical mechanisms governing these resistive transitions. The LRS is primarily influenced by Ohm's law, while the HRS predominantly follows the space-charge-limited conduction (SCLC) mechanism. The implementation of a multilayer structural design presents considerable potential for the fabrication of nitride-based memristors exhibiting multilevel resistance states and linear and symmetric conductance modulation for non-volatile memory and neuromorphic computing applications.

ACKNOWLEDGMENT

This work was supported in part by the National Natural Science Foundation of China (62374018), Guangxi Key Laboratory of Brain-inspired Computing and Intelligent Chips (BCIC-23-K5), the Fundamental Research Funds for the Central Universities, Beijing Institute of Technology Research Fund Program for Young Scholars, and Xiaomi Young Scholars Program.

REFERENCES

[1] S. Wang, T. Yang, D. Zhang, Q. Hua, and Y. Zhao, "Unveiling Gating Behavior in Piezoionic Effect: toward Neuromimetic Tactile Sensing," *Adv Mater*, vol. 36, no. 36, p. 26, 2024.

[2] Y. Shi et al., "Neuro-inspired thermoresponsive nociceptor for intelligent sensory systems," *Nano Energy*, vol. 113, p. 108549, 2023/08/01/ 2023.

[3] Z. Dong et al., "Ultrafast and Low-Power 2D Bi2O2Se Memristors for Neuromorphic Computing Applications," *Nano Letters*, vol. 23, no. 9, pp. 3842-3850, 2023/05/10 2023.

[4] Z. Zhang et al., "All-Metal-Nitride RRAM Devices," *IEEE Electron Device Letters*, vol. 36, no. 1, pp. 29-31, 2015.

[5] B. J. Choi et al., "High‐Speed and Low‐Energy Nitride Memristors," *Advanced Functional Materials*, vol. 26, no. 29, pp. 5290-5296, 2016.

[6] H.-D. Kim, H.-M. An, E. Lee, and T. Kim, "Stable Bipolar Resistive Switching Characteristics and Resistive Switching Mechanisms Observed in Aluminum Nitride-based ReRAM Devices," *IEEE Transactions on Electron Devices - IEEE TRANS ELECTRON DEVICES*, vol. 58, pp. 3566-3573, 10/01 2011.

[7] E. W. Lim and R. Ismail, "Conduction Mechanism of Valence Change Resistive Switching Memory: A Survey," *Electronics*, vol. 4, no. 3, pp. 586-613. doi: 10.3390/electronics4030586.

[8] D. Wang et al., "Thickness scaling down to 5 nm of ferroelectric ScAlN on CMOS compatible molybdenum grown by molecular beam epitaxy," *Applied Physics Letters*, vol. 122, no. 5, p. 052101, 2023.

Performance Assessment of Stacked Monolayer WSe₂ GAA NSFETs for Sub-5 nm Nodes

Ran Huo   Shijun Ou   Yihong Sun   Zichao Ma   Mansun Chan   Changjian Zhou

Abstract—**2D materials have received great research interest as promising candidates to extend Moore's Law for sub-5 nm technology nodes. In particular, WSe₂ exhibits bipolar transport characteristics, making it highly suitable for constructing future Complementary FETs (CFETs) and stacked CMOS circuits. In this work, p-type stacked WSe₂ and Si GAA nanosheet (NS) FETs at N5 ~ N0.5 and ultra-scaled gate lengths are implemented and their performances are assessed by 3D TCAD simulations with careful experimental calibrations. WSe₂ NSFETs can outperform the Si counterparts in terms of exceptional ON-current (~ 893.2 μA/μm) with OFF-current tightly controlled at 0.1 ~ 1 fA/μm, near-ideal SS (~ 60.39 mV/dec), and negligible DIBL (~ 5.927 mV/V) for sub-5 nm technology nodes. These metrics reach or even exceed the high-density (HD) projections from IRDS 2023 and the performance remains consistent and remarkable under aggressive scaling of the gate length, which underscores the potential for future advanced CFETs based on the same material.**

Index Terms—**WSe₂, Sub-5 nm, GAA NSFET, TCAD**

I. INTRODUCTION

Scaling down field-effect transistors while boosting device performance has been the main driver of the IC industry over the last six decades. To extend Moore's Law to the sub-5 nm technology nodes, traditional Si MOSFETs experience issues including severe mobility degradation due to carrier scattering effect, and increased subthreshold swing (SS) because of the worse gate controllability, etc. Applying innovative structures and new materials can effectively improve the device's performance. We have witnessed the evolution of device structures from planar to FinFET and now gate-all-around (GAA) nanosheet FET (NSFET) to enhance gate controllability. Meanwhile, the advent of 2D materials (e.g. MoS₂ [1], [2] et al.) poses another choice to replace channel materials. 2D FETs and recently standard SRAM circuits are predicted to exhibit superior performance over Si counterparts. However, previous studies on FETs paid more attention to nFETs, while pFETs are overlooked. Unlike MoS₂ which is usually used for fabricating nFETs, the polarity of WSe₂ can be tuned by thickness modulation and contact engineering [3], which pave the way for pFETs and future Complementary FET (CFETs) implementation.

Ran Huo, Shijun Ou, Yihong Sun, Zichao Ma and Changjian Zhou are with *School of Microelectronics, South China University of Technology*, Guangzhou, China. Mansun Chan is with *Department of Electronic and Computer Engineering, The Hong Kong University of Science and Technology*, Hong Kong, China. This work is supported by Guangdong Provincial Key Field Research and Development Program (2022B0701180002) and GRF (16201223) from the Research Grant Council of Hong Kong. Corresponding author: zhoucj@scut.edu.cn (Changjian Zhou)

In this work, p-type monolayer (ML) WSe₂ GAA NSFETs were constructed and assessed using 3D TCAD simulations. In comparison, Si NSFETs with the same device structures are also simulated. For sub-5 nm technology nodes, different stacking channel layers are considered to reach the standards posed by the IRDS 2023 edition. The results suggest that the WSe₂ NSFETs exhibit superior performances over the Si NSFETs in terms of the main performance indicators (I_{ON} ~ 900 μA/μm, SS ~ 60 mV/dec, DIBL ~ 6 mV/V), and verified its potential for future transistors of sub-5 nm technology nodes by ultra-scaled gate lengths.

II. DEVICE STRUCTURE AND SIMULATION CONFIGURATION

Fig. 1 shows the typical device structure constructed using Synopsys Sentaurus TCAD. The geometry and structural sizes of simulation devices exactly follow the standard in IRDS 2023 [4]. The channel material is ML WSe₂, and various stacked nanosheets can be incorporated into the GAA NSFETs. The source (S) and drain (D) regions are heavily doped to reduce the contact resistance and also compensate for the lack of free carriers induced by bias voltage at the spacer regions. To enhance Schottky barrier tunneling, platinum is selected as S/D metal.

The simulation setup utilizes self-consistent solutions coupling equations of drift-diffusion transport, Poisson distribution, and carrier continuity. Moreover, carrier generation and recombination as well as the Schottky barrier tunneling model are involved. For ML WSe₂ stacks, ballistic mobility (μ_{bal}) is modified with channel length dependency. Using the Frensley rule, final mobility (μ) is obtained with drift-diffusion

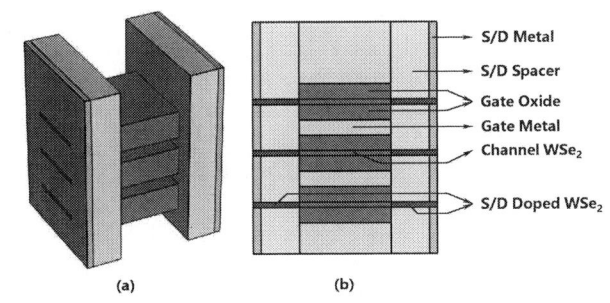

Fig. 1. Device geometry of ML WSe₂ GAA NSFETs (e.g. 3 channel stacks): (a) 3D schematic and (b) cross-sectional view.

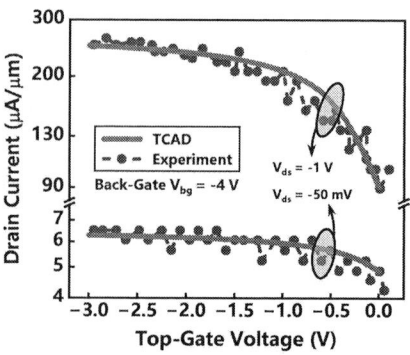

Fig. 2. Experimental calibration for simulation techniques.

TABLE I
PARAMETERS OF ML WSe$_2$ IN THIS SIMULATION

Parameter	Value
Atomic thickness (nm)	0.7
Bandgap (eV)	1.7
Relative dielectric constant	16
Electron affinity (eV)	3.85
Effective electron mass m$_e$ (m$_0$)	0.340
Effective hole mass m$_h$ (m$_0$)	0.454
Tunneling carrier mass m* (m$_e$)	1.05 [10]
Saturation velocity (cm/s)	3×10^6
Electron mobility (cm^{2}V^{-1}s^{-1})	30
Hole mobility (cm^{2}V^{-1}s^{-1})	100 [11]

TABLE II
STRUCTURAL GEOMETRIES OF NSFETs IN THIS SIMULATION

Parameter	Value (unit: nm)
Gate length L$_g$	18 (N5), 16 (N3), 14 (N2), 12 (N1), 12 (N0.5)
Nanosheet thickness	Si: 5, WSe$_2$: 0.7 (N5, N3, N2, N1, N0.5)
Nanosheet width	40 (N5), 35 (N3), 30 (N2), 15 (N1, N0.5)
Oxide thickness	1 (N5, N3, N2, N1, N0.5)
Spacer width	7 (N5), 6 (N3), 6 (N2), 5 (N1), 4 (N0.5)

TABLE III
CHARACTERIZATION CONDITIONS IN THIS SIMULATION

ON/OFF-Current (I$_{ON}$, I$_{OFF}$, unit: μA/μm)	
Parameter	Value
Drain voltage (V)	-0.75 (N5), -0.7 (N3), -0.65 (N2), -0.6 (N1, N0.5)
Gate voltage (V)	I$_{ON}$: -3, I$_{OFF}$: 0
Subthreshold Swing (SS, unit: mV/dec)	
Parameter	Value
Drain voltage (V)	-0.75 (N5), -0.7 (N3), -0.65 (N2), -0.6 (N1, N0.5)
Drain current (A/μm)	$10^{-7} \sim 10^{-8}$
Drain-Induced Barrier Lowering (DIBL, unit: mV/V)	
Parameter	Value
Drain voltage (V)	$-0.05 \sim -1.05$
Threshold voltage (V)	Gate voltage at I$_d$ = 10^{-6} A/μm

mobility (μ_{dd}, involving phonon scattering effect) under room temperature:

$$\frac{\mu}{\mu_{dd}} + \left(\frac{\mu}{\mu_{bal}}\right)^2 = 1 \qquad (1)$$

However, it has been reported that wide nanosheets of 2D materials can compensate performance degradation for channel length scaling, and even outperform against edge-related scattering due to the increasingly effective carrier density contributed by the increase of local electric field [5]. Thus, together with other promising reinforcement techniques (e.g. charge-transfer layer [6] et al.), mobility enhancement in the procedure of scaling should also be taken into consideration for projection. Considering the velocity saturation and strong interfacial scattering in the practice of ultra-scaled ML WSe$_2$ FETs, calibration using 100 cm^{2}V^{-1}s^{-1} [7] for phonon scattering limited mobility instead of bulk mobility (210 cm^{2}V^{-1}s^{-1}) should be more reasonable.

To ensure the reliability of our prediction, especially for the accuracy of material parameters and our model selection, Fig. 2 presents the calibration of the transfer curves of the WSe$_2$ FET with the previous experiments [8]. Here, the calibrated device parameters are identical to the experimentally reported devices and so does the configuration of bias voltage. Compared to the devices we construct, the calibrated device is undoped and the current completely depends on S/D tunneling current. Therefore, we need to consider contact resistance in calibration. However, to predict the best performance in our simulation, contact resistance should be neglected in our

NSFETs after heavily doping S/D. After cautious calibration, WSe$_2$ parameters are extracted and referenced in Table 1 [9].

With the accomplished 3D device structure, the main performance indicators of the GAA NSFETs are evaluated for 5 representative technology nodes, including N5, N3, N2, N1, and N0.5. The corresponding geometric parameters (e.g. gate length et al.) and characterization conditions for these technology nodes follow the suggestions from IRDS 2023 [4] and established rules, which are summarized in Table 2 and 3. For fairly comparative analysis, the geometric parameters of Si NSFETs are identical to ML WSe$_2$ (0.7 nm) devices, differing only in the implementation of a 5 nm silicon channel for each stack, following the fabricated device configuration. For each technology node, the number of stacked nanosheets varied from 1 to 5 to further evaluate its competence for high-performance applications. Besides, ultra-scaled WSe$_2$ devices based on N0.5 will be built to evaluate their scaling potential, which will be discussed later.

III. RESULTS AND DISCUSSIONS

A. NSFETs from N5 to N0.5

Fig. 3 shows the transfer characteristics of NSFETs with various technology nodes. The OFF-current for WSe$_2$ NSFETs ($\sim$ 0.1 fA/μm) is much lower than that of Si NSFETs ($\sim$ 10 nA/μm), which is far below the requirement for high-density (HD) devices from IRDS 2023 ($\sim$ 100 pA/μm). These superior characteristics can be attributed to the GAA structure. The circumferential gate configuration provides enhanced electrostatic control over the channel region. Additionally, the inherent atomically thin feature of WSe$_2$ facilitates further optimizing gate controllability.

979-8-3315-2209-4/25 $31.00 © 2025 IEEE

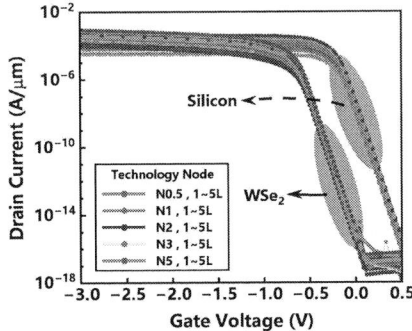

Fig. 3. Transfer characteristics of WSe$_2$ and Si NSFETs with various stacks at N5 $\sim$ N0.5.

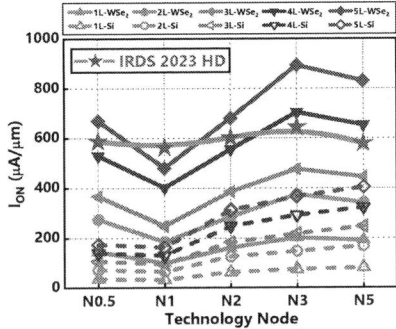

Fig. 4. ON-current of WSe$_2$ and Si NSFETs with various stacks at N5 $\sim$ N0.5 and projected values for HD devices from IRDS 2023.

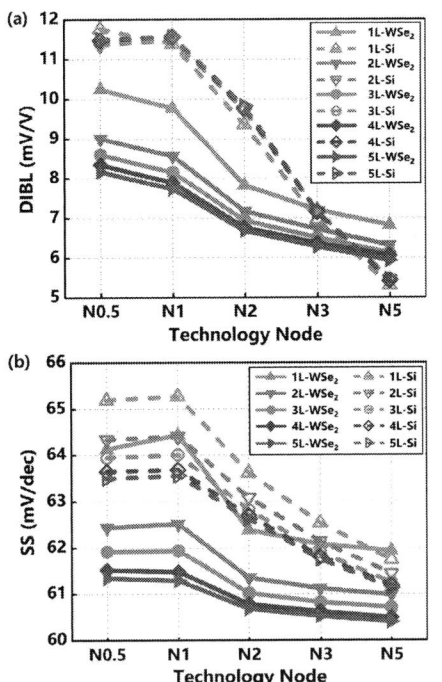

Fig. 5. Switching performance of WSe$_2$ and Si NSFETs with various stacks at N5 $\sim$ N0.5: (a) DIBL and (b) SS.

For all of the studied sub-5 nm technology nodes, the ON-currents of WSe$_2$ NSFETs are superior to Si NSFETs (Fig. 4). This advantage stems from the exceptional hole mobility exhibited by WSe$_2$, in contrast to silicon which typically experiences scattering-induced hole mobility limitations of around 50 cm^{2}V^{-1}s^{-1} at atomic-thickness levels [12]. WSe$_2$ lacks interfacial dangling bonds due to better van der Waals bonding, resulting in lower scattering probability and fewer interface trap states which contributes to a higher mobility of holes [3]. For each technology node, we have further investigated the effect of stacking more layers vertically for enhanced drive currents. As expected, the ON-current of the NSFETs exhibits a positive correlation with the number of stacked channels. When the stacking number reaches five, the WSe$_2$ NSFETs could be on par with or even outperform the performance projections for HD devices in IRDS 2023 at several sub-5 nm technology nodes except N1. At nodes N3 and N0.5, WSe$_2$ NSFETs achieve an extremely outstanding ON-current of approximately 893.2 µA/µm and 669.0 µA/µm while Si NS-FETs may need more complex processes to achieve the goal. Such outstanding ON-current should be attributed to the carrier compensation for Schottky barrier tunneling and modulation of S/D metal contact. Moreover, by modulating the trade-off of threshold voltage (V$_{TH}$), a higher ON-current adjustment for WSe$_2$ NSFETs can be achieved while maintaining the OFF-current within the maximum value specified by IRDS 2023. In contrast, Si NSFETs cannot benefit from it due to their

OFF-current being already near the specified limitations. These findings may provide valuable insights and potential pathways for future device scaling strategies in advanced semiconductor technologies.

The test conditions for threshold voltage, drain-induced barrier lowering (DIBL), and subthreshold swing (SS) are defined in Table 3, following the previously established criteria. Fig. 5 shows that the DIBL of the Si NSFETs ranges from 5.295 mV/V to 11.760 mV/V for different numbers of channel stacks. It is noteworthy that WSe$_2$ NSFETs with the same multiple stacked nanosheets exhibit a smaller DIBL (5.927 $\sim$ 10.240 mV/V) than silicon devices, which also reflects better suppression of threshold voltage roll-off especially in smaller devices. In terms of SS, stacked WSe$_2$ NSFETs exhibit more ideal characteristics approaching the limit of $\sim$ 60 mV/dec due to fewer interface states. Notably, as Si NSFETs scale down showing severe deteriorating performance for ON-current, WSe$_2$ devices are pleasingly getting back on track and exceeding IRDS 2023 projections of HD devices within height limits for stacking. These behaviors benefit tremendously from the excellent gate control capabilities of the GAA structure and the superiority of WSe$_2$ after carrier concentration optimization noted before.

B. NSFETs at Ultra-Scaled Gate Lengths

Owing to its superior ON-current characteristics and enhanced switching performance compared to silicon, WSe$_2$ emerges as a promising candidate for future scaling of pFETs. Therefore, we comprehensively investigated device performance at ultra-scaled technology nodes beyond those specified

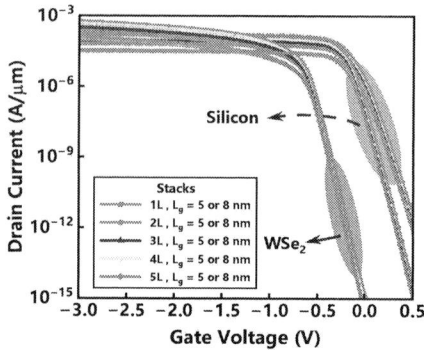

Fig. 6. Transfer characteristics of WSe$_2$ and Si NSFETs with various stacks at ultra-scaled gate lengths.

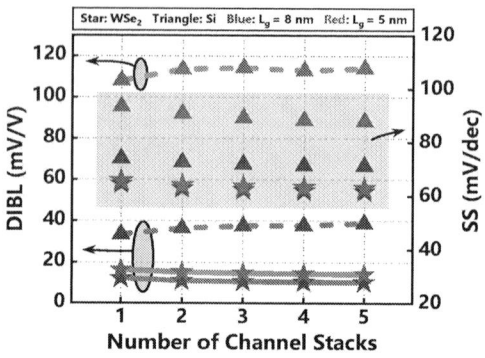

Fig. 7. Switching performance (DIBL and SS) of WSe$_2$ and Si NSFETs with various stacks at ultra-scaled gate lengths.

in the IRDS 2023. We implemented new NSFETs (ultra-scaled L_g = 8 or 5 nm, 1 ∼ 5 stacks) with gate lengths scaled to 8 nm and 5 nm respectively, and keep other structural design and test conditions consistent with N0.5.

As shown in Fig. 6, a comparative analysis of these ultra-scaled devices reveals that WSe$_2$ NSFETs with 5 channel stacks still demonstrate exceptional leakage current control (∼ 1 fA/μm) while achieving an ON-current of 638.2 μA/μm (L_g = 8 nm) and 489.2 μA/μm (L_g = 5 nm). These traits can be attributed to the weak Fermi pinning effect and low density of surface states in WSe$_2$ and the aforementioned compensation of wide nanosheets to a great extent.

Furthermore, Fig. 7 demonstrates that under identical conditions of 8 nm gate length, the SS and DIBL of WSe$_2$ NSFETs exhibit better characteristics than Si NSFETs, showing a negligible difference. However, upon scaling to a 5 nm gate length, Si NSFETs demonstrate a significant degradation in gate-to-channel control. Specifically, the average SS across various channel stacks with gate lengths of 8 nm and 5 nm respectively increases by 12.69% and 40.39% relative to N0.5. The DIBL characteristics show even more pronounced degradation, with multi-channel Si NSFETs experiencing an average increase of 220.28% (L_g = 8 nm) and 878.95% (L_g = 5 nm) compared to N0.5 counterparts. In contrast, WSe$_2$ NSFETs maintain near-ideal SS and almost negligible DIBL characteristics even at ultra-scale gate length. These characteristics position WSe$_2$ NSFETs as a promising solution for future applications that demand both high driving capability and superior switching performance while maintaining low standby power, which paves the way for advanced logic devices.

IV. CONCLUSION

We have compared the main performance indicators of Si and ML WSe$_2$ GAA NSFETs across various sub-5 nm technology nodes. WSe$_2$ exhibits remarkable advantages over silicon for NSFETs with 5 stacked nanosheets, achieving more near-ideal SS (< 61.4 mV/dec) and negligible DIBL (< 8.2 mV/V). Additionally, WSe$_2$ GAA NSFETs simultaneously deliver superior ON-current of up to 893.2 μA/μm, with ultra-low OFF-current of 1.47 fA/μm. Furthermore, we predict the performance of WSe$_2$ and Si NSFETs at ultra-scaled gate

lengths. Notably, these enhanced performance metrics remain robust even at aggressively scaled dimensions down to L_g = 8 or 5 nm, meeting or surpassing IRDS 2023 projections. Our simulations indicate the theoretical optimal performance of WSe$_2$ NSFETs and reveal the substantial promise of scalability for 2D materials for future scaled FETs in sub-5 nm regimes.

REFERENCES

[1] A. Rawat and B. Rawat, "Multichannel Two-Dimensional MoS$_2$ Nanosheet MOSFET for Future Technology Node," *IEEE Transactions on Electron Devices*, vol. 71, no. 6, pp. 3945-3951, June 2024.

[2] Y. C. Lu, J. K. Huang, K. Y. Chao, L. J. Li, and V. P. H. Hu, "Projected Performance of Si- and 2D-Material-Based SRAM Circuits Ranging from 16 nm to 1 nm Technology Nodes," *Nature Nanotechnology.*, vol. 19, pp. 1066–1072, June 2024.

[3] W. Liu et al., "Role of metal contacts in designing high-performance monolayer n-type WSe$_2$ field effect transistors," *Nano letters*, vol. 13, no. 5, pp. 1983-1990, March 2013.

[4] International Roadmap for Devices and Systems. (2023). MORE MOORE. [Online]. Available: https://irds.ieee.org/editions/2023/20-roadmap-2023-edition/130-irds%E2%84%A2-2023-more-moore

[5] F. Zhang, CH. Lee, J. A. Robinson, and J. Appenzeller, "Exploration of channel width scaling and edge states in transition metal dichalcogenides," *Nano Research*, vol. 11, pp. 1768-1774, March 2018.

[6] J. Park et al., "Charge-transfer contacts for the measurement of correlated states in high-mobility WSe$_2$," *Nature Nanotechnology*, vol. 19, pp. 948-954, July 2024.

[7] A. Pal, T. Chavan, J. Jabbour, W. Cao, and K. Banerjee, "Three-dimensional transistors with two-dimensional semiconductors for future CMOS scaling," *Nature Electronics*, vol. 7, pp. 1-11, December 2024.

[8] X. Xiong et al., "Top-Gate CVD WSe$_2$ pFETs with Record-High I_d∼594 μA/μm, G_m∼244 μS/μm and WSe$_2$/MoS$_2$ CFET based Half-adder Circuit Using Monolithic 3D Integration," *2022 IEEE International Electron Devices Meeting (IEDM)*, San Francisco, CA, USA, pp. 20.6.1-20.6.4, December 2022.

[9] K. H. Kim et al., "Tuning polarity in WSe$_2$/AlScN FeFETs via contact engineering," *ACS nano*, vol. 18, pp. 4180-4188, January 2024.

[10] D. H. Shin et al., "Microscopic Quantum Transport Processes of Out-of-Plane Charge Flow in 2D Semiconductors Analyzed by a Fowler-Nordheim Tunneling Probe," *Advanced Electronic Materials*, vol. 9, April 2023.

[11] K. P. O'Brien et al., "Advancing 2D Monolayer CMOS Through Contact, Channel and Interface Engineering," *2021 IEEE International Electron Devices Meeting (IEDM)*, San Francisco, CA, USA, pp. 7.1.1-7.1.4, December 2021.

[12] L. Donetti, F. Gamiz, N. Rodriguez, and A. Godoy, "Hole Mobility in Ultrathin Double-Gate SOI Devices: The Effect of Acoustic Phonon Confinement," *IEEE Electron Device Letters*, vol. 30, no. 12, pp. 1338-1340, October 2009.

979-8-3315-2209-4/25 $31.00 © 2025 IEEE

A 6048-PPI High Performance OLEDoS Pixel Circuit with Self-Charging Capacitor Coupling Method

Fan Guo, Yuanbo Sun, Congwei Liao, Lu Chang, Xin Zheng, and Shengdong Zhang*
School of Electronic and Computer Engineering
Peking University, Shenzhen, China
Email: zhangsd@pku.edu.cn

Abstract—**This paper presents a compact 5T-2C OLED microdisplay pixel circuit with PPI of 6048 and high luminance uniformity. A new compensation scheme is proposed to compensate the voltage threshold (V_{th}) deviation of the driving transistor, and this scheme suppresses emission current errors caused by body effect. Furthermore, two serial-connected capacitors are used to reduce the overdrive voltage of the driving transistor to extend the data voltage range. The performance of the proposed pixel circuit is verified using SMIC 40-nm CMOS process. The layout area of the proposed pixel circuit is only 4.2 μm× 1.4 μm, corresponding to 6048 PPI. The data voltage range is increased to 1.73V, which is 8.12 times larger than that of the conventional pixel circuit. The emission current error of the proposed pixel circuit varies from -0.97% to 0.24%, with the condition of $V_{th} \pm 16$ mV.**

Keywords—microdisplay, data range, Organic light-emitting diode on silicon (OLEDoS), pixel circuit, Vth compensation.

I. INTRODUCTION

Organic light-emitting diode on silicon (OLEDoS) displays have been widely used in virtual reality (VR) and augmented reality (AR) applications. Compared with LCDoS and μLEDoS microdisplay technologies, OLEDoS displays have the advantages of higher PPI, higher contrast, good brightness, and mature processing for mass production of near-eye display applications.

In OLEDoS displays, the voltage programing method is widely adopted in pixel circuit design due to its compact structure and low power consumption. However, achieving good display uniformity remains a significant challenge. The small size of the pixel circuits and the associated reduction in driving transistor dimensions contribute to increased deviations in threshold voltage (V_{th}), leading to variations in emission current and heightened luminance non-uniformities. Moreover, the driving transistor operates in the subthreshold region, where the current in OLEDoS pixel circuits typically ranges from picoamperes (pA) to nanoamperes (nA). Consequently, the corresponding data voltage range is limited to only a few hundred millivolts. While the most precise source drivers currently available can maintain an accuracy of approximately 5 mV [1], necessitating a data voltage range of at least 1.28 V to achieve an 8-bit color depth. Therefore, it is imperative to compensate for the V_{th} deviations of the driving transistors and to extend the data voltage range in order to enhance display uniformity.

Pixel circuits with compensation functions have been proposed to solve the afore-mentioned problems. Kimura et. al. [2] proposed a 4T-2C pixel circuit with self-discharging compensation method to achieve -2% to 2% emission current error with V_{th} variation of ± 50 mV, and the data voltage range is about 2-3 times larger than the conventional pixel circuit. But the data voltage extension ratio is restricted by the coefficient of the body effect. Kwak [3] reported a 4T-1C pixel circuit with parasitic capacitor voltage division scheme to extend the data voltage range and the emission current error ranges from -1.63% to $+1.15\%$, but it needs a large storage capacitor for V_{th} compensation. [4], [5] used source degeneration method to extend the data voltage range while the body effect and transistor size variations haven't be considered, and their emission current errors are larger than 4% . Cheng [6] used a 8T-2C dual data pixel circuit with V_{th} compensation to realize 10-bit gray levels, but this design endure large emission current error due to a part of data voltage is decided by the coupling capacitor, which endure deviations among channels.

To improve the display uniformity and achieve a higher resolution, this paper proposes a 5T2C pixel circuit with precise V_{th} compensation and ultra-high PPI for OLEDoS display.

II. PROPOSED PIXEL CIRCUIT

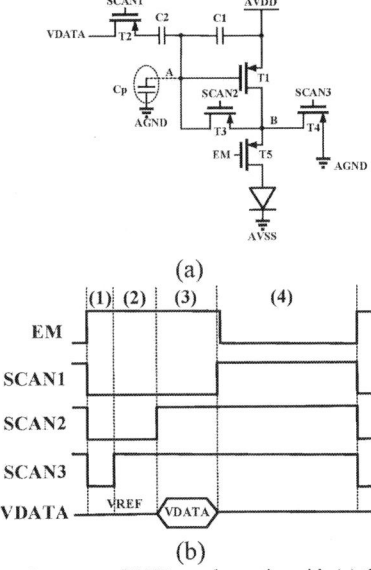

Fig. 1. The proposed compact OLEDos schematic, with (a) the 5T-2C pixel circuit, and (b) the timing diagrams of the circuit.

Fig. 1(a) and 1(b) respectively show the proposed pixel circuit and its timing diagram. The proposed pixel circuit consists of 5 p-channel MOSFETs and 2 storage capacitors (Cp is the parasitic capacitance between node A and the ground.), and operates in 4 phases.

(1) Initial Phase

The SCAN1, SCAN2, and SCAN3 signal lines are all maintained at low voltage to turn on T2, T3, and T4, respectively, and the gate and drain voltage of T1 is set to 0 V. The VDATA signal line provides the reference signal VREF, which is applied to the upper plate of the capacitor C2. The EM signal is high turn off T5 to prevent current from flowing through the OLED. The node voltage of A, $V_{A,1}$ is

$$V_{A,1} = V_{ref} \tag{1}$$

(2) V_{th} Extraction Phase

The SCAN3 signal becomes high to turn off T4. The SCAN1 and SCAN2 signals maintain low to turn on T2 and T3, respectively. T1 is in the diode connected mode, and the voltage stored in capacitor C1 and C2 equals VDD, which is larger than the threshold voltage of T1. In that case, T1 turns on and node A is charged by C1 and C2 through the path T1 and T3. The voltage of node A rises up until T1 is turned off. Therefore, the threshold voltage V_{th} is stored in C1 and C2. The node voltage of A, $V_{A,2}$ is

$$V_{A,2} = VDD - V_{th,T1} \tag{2}$$

Now the voltage difference of $V_{A,E}$ and V_{ss} is larger than the turn-on voltage of OLED, so signal EM maintains high to prevent current from flowing through OLED, thus enabling a high contrast ratio.

(3) Data Writing Phase

The SCAN2 signal becomes high to turn off T3, while SCAN1, SCAN3 and EM maintain the same as the V_{th} extraction phase. The data voltage V_{data} is applied to the upper plate of C1 through T2. V_{data} starts to charge node A through the coupling of C1 and C2. The voltage of the node A is

$$V_{A,3} = VDD - V_{th,T1} + \left(V_{data} - V_{ref}\right)\frac{C_2}{C_1+C_2+C_p} \tag{3}$$

The source to gate voltage of the driving transistor T1, which is stored in C1, is

$$V_{SG,T1,3} = V_{th,T1} - \left(V_{data} - V_{ref}\right)\frac{C_2}{C_1+C_2+C_p} \tag{4}$$

The V_{data} is divided by (C1+Cp) and C2, so the data voltage range is extended to $\frac{C_1+C_2+C_p}{C_2}$ times the original value.

(4) Emission Phase

The SCAN1 signal becomes high to turn off T2, the EM signal becomes low to turn on T5. SCAN2 and SCAN3 maintains the same as the data writing phase. The charge stored in C1 provides the overdrive voltage of T1. The voltage change on node B has little effect on the emission current. The emission current of T1 flowing through the OLED is

$$I_{em} = \frac{W}{L}I_0 e^{(V_{SG,T1,W} - V_{th})/(nV_T)}$$
$$= \frac{W}{L}I_0 e^{\frac{\left(V_{ref}-V_{data}\right)\frac{C_2}{C_1+C_2+C_p}}{nV_T}} \tag{5}$$

Where W/L, V_{th}, n, and V_T are the aspect ratio, threshold voltage, subthreshold slope factor, and thermal voltage, respectively. I_0 is the drain current at the threshold normalized for W/L. I_{em} is proportional to $(V_{ref} - V_{data})$ during the emission phase.

The proposed pixel circuit uses PMOS driving scheme. Compared with NMOS driving scheme, the source and body voltages of the driving transistor T1 are always consistent, so V_{th} extraction is more precise and the influence of body effect is avoided.

III. RESULTS AND DISCUSSIONS

Fig. 2 shows the layout of the proposed pixel circuit using SMIC 40-nm CMOS process with 2.5 V voltage devices. The proposed pixel circuit occupies a unit sub-pixel area of 4.2 μm× 1.4 μm, corresponding to a resolution of 6048 pixels per inch (PPI). The design parameters are summarized in Table I.

The maximum emission current is determined by using the white OLED efficiency, maximum luminance, and transmittance of the red, green and blue color filters, which is 4.52 nA in the experiment. Fig. 3. Shows the emission current (I_{em}) versus the gray levels.

TABLE I.   DESIGN PARAMETERS OF THE PROPOSED PIXEL CIRCUIT

Design parameter	Value
AVDD (V)	2.5
AVSS (V)	-2.5
W/L of T1	320 nm/2.7 μm
W/L of T2, T3, T4 and T5	320 nm/270 nm
C1/C2	5/1
Sub pixel area (μm²)	4.2 μm× 1.4 μm
Spatial resolution (PPI)	6048

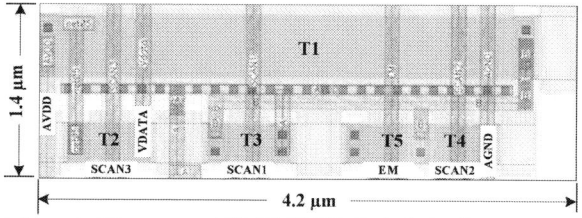

Fig. 2. Layout of the proposed 5T-2C OLEDoS pixel circuit

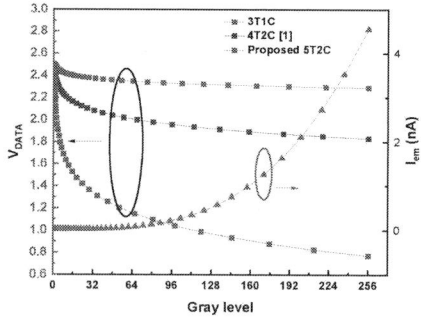

Fig. 3. Data voltage range of conventional 3T-1C, 4T-2C [1], the proposed 5T-2C pixel circuit and the emission current versus gray level.

979-8-3315-2209-4/25 $31.00 © 2025 IEEE

Fig. 3. shows data voltage range of different works, the data voltage range of conventional 3T-1C pixel circuit is only 213 mV, and the data voltage range of the 4T-2C pixel circuit [2] is 670 mV. They cannot satisfy the 8-bit color depth. The data voltage range of the proposed pixel circuit is 1.73 V, which is Consequently, the emission current can be more precisely controlled.

The pixel circuit of OLEDoS is designed to be small enough to ensure high PPI, so the size of the driving transistor T1 is also small, resulting in large V_{th} deviations of the T1. The V_{th} deviation, which is a function of the transistor size, has a Gaussian distribution, of which the one-sigma standard deviation is expressed as:

$$\delta \Delta V_{th,T1} = \frac{A_{Vth}}{\sqrt{W_{eff}*L_{eff}}} \qquad (6)$$

Where A_{Vth} is a constant of 4.96 mV·μm for 2.5 V PMOS in SMIC 40-nm process. W_{eff} and L_{eff} are the effective width and length of driving transistor T1, which are 0.32 μm and 2.7 μm in this design, respectively. The one-sigma standard deviation of $V_{th,T1}$ can be calculated to be 5.33 mV using (6). Thus, the deviation range of $V_{th,T1}$ is from -16 mV to 16 mV in the three-sigma standard deviation range.

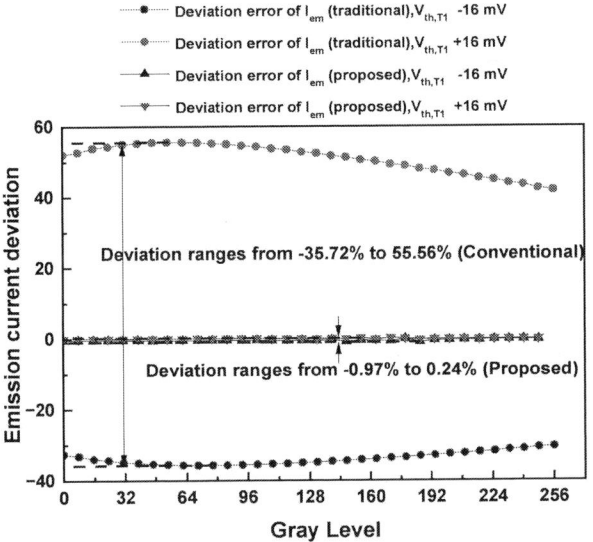

Fig. 4 Emission current deviation comparison between traditional 3T1C and proposed 5T2C pixel circuit.

To verify the performance of the proposed pixel circuit. Fig. 4 shows the emission current errors versus display gray levels. The deviation of traditional 3T1C circuit, which does not compensate for the V_{th} deviation of the driving transistor, ranges from -35.72% to 55.56% while the emission current deviation of proposed 5T2C pixel circuit ranges from -0.97% to 0.24% after V_{th} compensation, demonstrating that the proposed pixel circuit can improve the luminance uniformity efficiently. Table II summarizes the performance comparison of the proposed pixel circuit with previous works. Compare to these previous works, the proposed pixel circuit has the lowest emission current deviation error, thus achieving the highest display uniformity.

TABLE II. COMPARISON OF THE PROPOSED WORK WITH PRIOR WORKS FOR HIGH-PPI DISPLAYS

Index	[6]	[5]	[7]	[2]	This work
Process	55-nm	130-nm	180-nm	180-nm	40-nm
V_{th} **deviation**	*NAN*	*NAN*	±5 mV	±50 mV	±16 mV
PPI	6318	4670	3136	3256	6048
I_{em} **Deviation Error (%)**	-9.69 to 11.3	-19 to 19	-2.1to 2.08	-2 to 2	-0.96 to 0.24
Data Voltage Extension Ratio	*NAN*	3.44x	6.2x	2x	8.12x

IV. CONCLUSIONS

This paper proposes a high PPI OLEDoS pixel circuit with good uniformity and precise data voltage control. Compared with previous works, the proposed pixel circuit occupies a small unit sub-pixel area of 4.2 μm× 1.4 μm, corresponding to 6048 PPI. Moreover, the proposed pixel circuit has the least emission current deviation error ranges from -0.97% to 0.24% and thus achieves highest display uniformity. At the same time, the data voltage range is extended to 1.73 V to support 8-bit color depth. Therefore, the proposed OLEDoS pixel design is well suitable for high resolution and high display uniformity OLEDoS displays.

REFERENCES

[1] G. W. Lim, G.-G. Kang, H. Ma, M.-I. Jeong, and H. Kim, "A 10b Source-Driver IC with LSB-Stacked LV-to-HV-Amplify DAC Achieving 2688μm2/channel and 4.8mV DVO for Mobile OLED Displays," 2022 IEEE International Solid- State Circuits Conference (ISSCC), vol. 65, pp. 110-112, 2022.

[2] K. Kimura, Y. Onoyama, T. Tanaka, N. Toyomura, and H. Kitagawa, "New pixel driving circuit using self‐discharging compensation method for high‐resolution OLED micro displays on a silicon backplane," Journal of the Society for Information Display, vol. 25, no. 3, pp. 167-176, 2017.

[3] B.-C. Kwak and O.-K. Kwon, "A 2822-ppi resolution pixel circuit with high luminance uniformity for OLED microdisplays," Journal of Display Technology, vol. 12, no. 10, pp. 1083-1088, 2016.

[4] B. Liu, D. Ding, T. Zhou, and M. Zhang, "P‐53: A Novel Pixel Circuit Providing Expanded Input Voltage Range for OLEDoS Microdisplays," in SID Symposium Digest of Technical Papers, 2017, vol. 48, no. 1: Wiley Online Library, pp. 1438-1441.

[5] H.-J. Shin, Y.-D. Kim, S. Kim, and B.-D. Choi, "71-2: Distinguished Paper: 4,670-PPI OLEDoS Pixel Circuit Design for Wide Data Voltage Range in a 5V 0.13μm CMOS Process," SID Symposium Digest of Technical Papers, vol. 55, no. 1, pp. 979-982, 2024, doi: https://doi.org/10.1002/sdtp.17700.

[6] S.-S. Cheng and P. C.-P. Chao, "An Ultra-High 6318-PPI Pixel Circuit for Micro-OLED Displays with Vth Compensated up to 10-bit Gray Levels," IEEE Journal of Solid-State Circuits, 2024.

[7] X. Huo, C. Liao, M. Zhang, H. Jiao, and S. Zhang, "A Pixel Circuit With Wide Data Voltage Range for OLEDoS Microdisplays With High Uniformity," IEEE Transactions on Electron Devices, vol. 66, pp. 4798-4804, 2019.

Using Nanophotonic Circuit to Calculate 2D Affine Transformations

Jinzhi Mu, Jigeng Sun, Shaolin Zhou*

School of Microelectronics, South China University of Technology, Guangzhou 511442, China

202264711054@mail.scut.edu.cn; 202310192279@mail.scut.edu.cn; eeslzhou@scut.edu.cn

Abstract—As AAA games become increasingly popular, the demand for computational power of graphic cards keeps increasing. However, traditional electronic circuit is approaching the limitation of Moore's Law and this calls for a new solution. Optical matrix computation promises high parallel processing capability, high scalability and low power consumption. This work proposes a nanophotonic circuit to calculate 2D affine transformation, which is a basic image transformation method in graphics pipeline. Theory analysis shows that this nanophotonic circuit can finish a 2D affine transformation in only 10 to 20ns, or 1×10^8 to 2×10^8 2D affine transformations per second. It needs no more than 3.64nJ for a single 2D affine transformation and only has 8% deviation from ideal coordinate after each 2D affine transformation.

Keywords—*nanophotonic circuit, calculate, 2D affine transformation,*

I. INTRODUCTION

Affine transformation is a crucial yet time-consuming step in real-time rendering, widely applied in various image transformations and image detection tasks[1]. A 2D affine transformation is essentially a 3×3 matrix operation[2]. Optical matrix computation offers higher computational efficiency and lower power consumption compared to traditional electronic digital matrix computation. In this paper, we propose a 3×3 nanophotonic circuit based on a Mach-Zehnder interferometer topology-cascaded architecture to compute 2D affine transformations. We also numerically calculate its theoretical operational speed, power consumption and accuracy.

II. 2D AFFINE TRANSFORMATION

A. Basic 2D affine transformation

The basic 2D affine transformation is a 3×3 matrix, where the 2×2 portion in the top-left corner represents the fundamental 2D linear transformation, including Scaling, Shearing, Rotation, and Reflection. The rightmost column represents the 2D Translation. The general form of an 2D affine transformation is expressed as:

$$\begin{bmatrix} m_{11} & m_{12} & x_t \\ m_{21} & m_{22} & y_t \\ 0 & 0 & 1 \end{bmatrix} \qquad (1)$$

TABLE I.   BASIC 2D AFFINE TRANSFORMATION AND THEIR FORMS

Transformation Type	Form
Scaling: Change length and direction of images but not the coordinate.	$\begin{bmatrix} s_x & 0 & 0 \\ 0 & s_y & 0 \\ 0 & 0 & 1 \end{bmatrix}$
Shearing: Push images sideways.	Shear-x(s): $\begin{bmatrix} 1 & s & 0 \\ 0 & 1 & 0 \\ 0 & 0 & 1 \end{bmatrix}$
	Shear-y(s): $\begin{bmatrix} 1 & 0 & 0 \\ s & 1 & 0 \\ 0 & 0 & 1 \end{bmatrix}$
Rotation: Rotate images by an angle ϕ.	$\begin{bmatrix} \cos\phi & -\sin\phi & 0 \\ \sin\phi & \cos\phi & 0 \\ 0 & 0 & 1 \end{bmatrix}$
Reflection: Reflect images across either of the coordinate axes.	Reflect-x: $\begin{bmatrix} 1 & 0 & 0 \\ 0 & -1 & 0 \\ 0 & 0 & 1 \end{bmatrix}$
	Reflect-y: $\begin{bmatrix} -1 & 0 & 0 \\ 0 & 1 & 0 \\ 0 & 0 & 1 \end{bmatrix}$
Translation: Move images but not change its shape.	$\begin{bmatrix} 1 & 0 & x_t \\ 0 & 1 & y_t \\ 0 & 0 & 1 \end{bmatrix}$

B. Complex 2D affine transformation

At the same time, complex 2D affine transformations can be achieved by stacking multiple basic 2D affine transformations. If we suppose A_1, A_2, …, A_n are the basic 2D affine transformations to be applied and A_f is the combined matrix which represents the complex 2D affine transformation, we can obtain A_f by:

$$A_f = A_n \cdot … \cdot A_2 \cdot A_1 \qquad (2)$$

III. PHOTONIC DESIGN

As we know, any arbitrary m×n matrix M can be represented as $M=U\Sigma V^\dagger$ by singular value decomposition, where U is a m×m unitary matrix, Σ is a m×n diagonal matrix and $V^\dagger$ is a n×n unitary matrix[3]. Furthermore, the unitary matrix U and the unitary matrix $V^\dagger$ can be further decomposed into m(m-1)/2 2×2 unitary matrix blocks and n(n-1)/2 2×2 unitary matrix blocks respectively through unitary matrix decomposition. Thus the 3×3 2D affine transformation matrix talked above can be decomposed into a 3×3 unitary matrix U, a 3×3 diagonal matrix Σ and a 3×3 unitary matrix $V^\dagger$.

979-8-3315-2209-4/25 $31.00 © 2025 IEEE

A. Unitary Martix Blocks and 3×3 Unitary Matrix

The 3×3 unitary matrix can be further decomposed into three 2×2 unitary matrix blocks, which can be realized by two 3dB Mach–Zehnder interferometers and two nano-opto-electro-mechanical (NOEM) phase shifters[4] on silicon photonics integration platform with the transmission matrix:

$$\begin{bmatrix} E_{2a} \\ E_{2b} \end{bmatrix} = \frac{1}{2} \begin{bmatrix} e^{j\varphi}(e^{j\phi}-1) & je^{j\varphi}(e^{j\phi}+1) \\ j(e^{j\phi}+1) & -e^{j\phi}+1 \end{bmatrix} \begin{bmatrix} E_{1a} \\ E_{1b} \end{bmatrix} \quad (3)$$

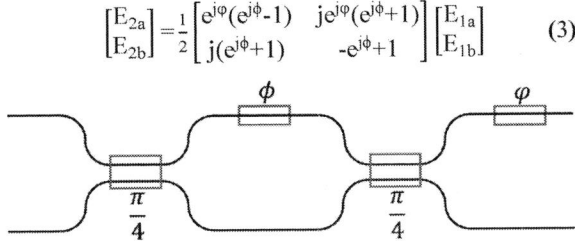

Fig. 1. 2×2 unitary matrix block. The blue rectangles represent the 3dB Mach–Zehnder interferometers and the red rectangles represent the NOEM phase shifters.

Then we can topologically cascade three 2×2 unitary matrixes to construct the 3×3 unitary matrix by either triangular structure or rectangular structure[5].

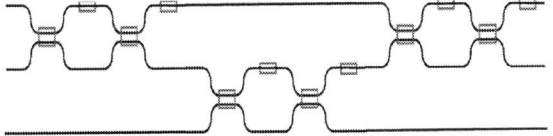

Fig. 2. Topologically cascade three 2×2 unitary matrixes to construct a 3×3 unitary matrix.

B. Diagonal Matrix Units

At the same time, the 3×3 diagonal matrix can be further decomposed into three diagonal matrix units which can be realized by two 3dB Mach–Zehnder interferometers and one nano-opto-electro-mechanical (NOEM) phase shifter[4] on silicon photonics integration platform with the transmission matrix:

$$\begin{bmatrix} E_{2a} \\ 0 \end{bmatrix} = \frac{1}{2} \begin{bmatrix} e^{j\phi}-1 & j(e^{j\phi}+1) \\ j(e^{j\phi}+1) & -e^{j\phi}+1 \end{bmatrix} \begin{bmatrix} E_{1a} \\ 0 \end{bmatrix} \quad (4)$$

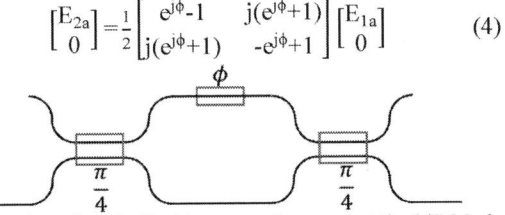

Fig. 3. Diagonal matrix unit. The blue rectangles represent the 3dB Mach–Zehnder interferometers and the red rectangle represent the NOEM phase shifter.

C. Nanophotonic Circuit Architecture and Working Priciple

Finally, we can get the nanophotonic circuit architecture: The three 2×2 unitary matrix blocks on the left are topologically cascaded to form the first 3×3 unitary matrix after the singular value decomposition of the 3×3 2D affine transformation. The three diagonal matrix units in the middle form the corresponding 3×3 diagonal matrix after the singular value decomposition. The three 2×2 unitary matrix blocks on the right are topologically cascaded to form the second 3×3 unitary matrix after the singular value decomposition.

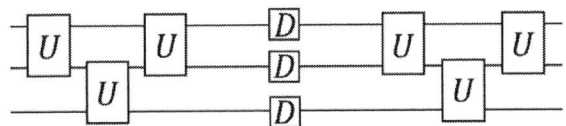

Fig. 4. The nanophotonic circuit architecture. The letter 'U' represents the 2×2 unitary matrix and the letter 'D' represents the diagonal matrix unit.

And the working principle can be represented by the Fig. 5: At first, the image is extracted as a series of coordinates represented in electronic vector form, which are then converted into light amplitude via DAC and electro-optic converters. Simultaneously, the required 2D affine transformation matrix is decomposed by singular value decomposition and unitary matrix decomposition, yielding the values of each 2×2 unitary matrix block and diagonal matrix unit. The corresponding complex phase is then loaded onto each NOEM phase shifter. Finally, the light carrying the coordinate information passes through the designed nanophotonic circuit, during which its amplitude changes accordingly. This results are in the coordinates in the optical domain after the 2D affine transformation, which are then converted into electronic coordinates via the electro-optic converter and DAC. These coordinates represent the final image after the 2D affine transformation.

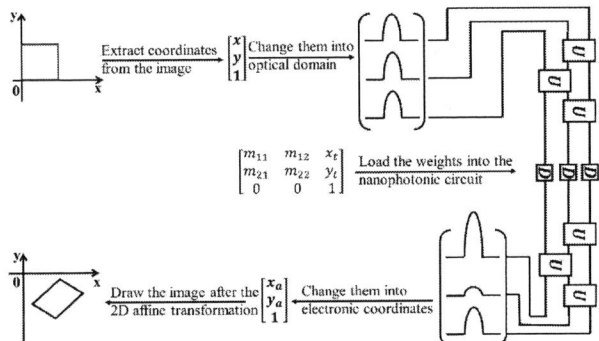

Fig. 5. Working principle of this nanophotonic circuit used for 2D affine transformation

IV. PERFORMANCE ANALYSIS

A. Processing Speed

Processing speed is determined by the processing time of the nanophotonic circuits, which consists of two parts: matrix weights loading time and net computation time. The matrix weights loading time basically depends on the speed of NOEM phase shifters, which have a modulation frequency of 100~200MHz[6]. The net computation time is the speed of light. It is obvious that the total processing speed is determined by the modulation frequency. So, this nanophotonic circuit can finish a 2D affine transformation within 10~20ns, or implement 1×10^8~2×10^8 2D affine transformations per second.

B. Power Consumption

The power consumption of this nanophotonic circuit is mainly determined by the fifteen NOEM phase shifters[4]. The power can be calculated by W=Pt=IVt, where 't' is the reciprocal of modulation frequency and 'I' can be obtained by the PN junction model:

$$I = I_0 \left[\exp\left(\frac{qV}{kT}\right) - 1 \right], \quad I_0 = Aqn_i^2 \left(\frac{D_p}{L_p N_d} + \frac{D_n}{L_n N_a} \right) \quad (5)$$

Since the voltage applied to each NOEM phase shifter is increasing with the phase shift we want to achieve and 850 mV will be enough to bring a π radian of phase shift. Then we can calculate that the total power consumption will be no more than 3.64nJ for every single 2D affine transformation.

C. Accuracy

2D affine transformation accuracy measures how accurate the result of the transformation is, which is basically determined by the insertion loss of the NOEM phase shifters of this nanophotonic circuit, which will introduce insertion loss of 0.04dB for every single one[4]. Thus, every coordinate calculated will suffer from total insertion loss of 0.36dB, which means that the coordinate obtained by using this nanophotonic circuit will have an 8% error compared to the ideal coordinate after 2D affine transformation.

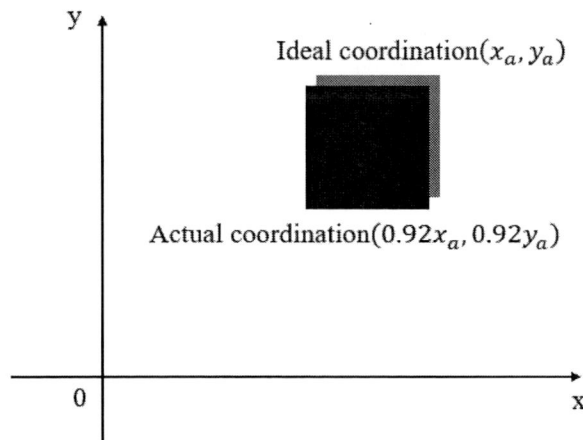

Fig. 6. Ideal coordinate and actual coordinate.

V. CONCLUSION

Using nanophotonic circuit to 2D affine transformation is proposed first time in this paper for accelerating real-time rending. A nanophotonic circuit structure based on Mach–Zehnder interferometer on silicon photonics integration platform is chosen to realize the 3×3 matrix multiplication. The performance analyses shows that this nanophotonic circuit can have a calculation speed of 10~20ns per 2D affine transformation or $1 \times 10^8 \sim 2 \times 10^8$ 2D affine transformations per second, with no more than 3.64nJ power consumption for a single 2D affine transformation and coordinate accuracy of 92%. Using nanophotonic circuit to calculate 2D affine transformation serves as a promising future real-time rendering that required heavy matrix transformation.

ACKNOWLEDGMENT

This work was supported by the 2025 Centistep Ladder Climbing Program of South China University of Technology (NO. j2tw202402005).

REFERENCES

[1] Lengyel, Jed, and John Snyder. "Rendering with coherent layers." Proceedings of the 24th annual conference on Computer graphics and interactive techniques. 1997.

[2] Shirley, Peter, Michael Ashikhmin, and Steve Marschner. "Transformation matrices." Fundamentals of computer graphics. AK Peters/CRC Press, 2009.

[3] Shen, Yichen, et al. "Deep learning with coherent nanophotonic circuits." Nature photonics 11.7 (2017): 441-446.

[4] Baghdadi, Reza, et al. "Dual slot-mode NOEM phase shifter." Optics Express 29.12 (2021): 19113-19119.

[5] Clements, W. R., Humphreys, P. C., Metcalf, B. J., Kolthammer, W. S., & Walmsley, I. A. (2016). Optimal design for universal multiport interferometers. Optica, 3(12), 1460-1465.

[6] Demirkiran, Cansu, et al. "An electro-photonic system for accelerating deep neural networks." ACM Journal on Emerging Technologies in Computing Systems 19.4 (2023): 1-31.Transl. J. Magn.

Enhancement of Electrical Performance and Stability in p-Type SnO Thin-Film Transistors via Ultrathin Al₂O₃ Capping Layer

1st Ruyu Liang   2nd Lei Xu   3rd Shi Zong

Abstract—With the rise of the electronic era, p-type oxide thin-film transistors (TFTs) have become essential components, driving advancements in integrated circuits for logic applications. Enhancing their performance and stability is therefore critical. This study focuses on p-type SnO TFTs, where a 4 nm ultrathin Al_2O_3 capping layer was deposited on the back-channel to suppress Sn valence state oxidation and reduce defect density near the channel. By adjusting the oxygen partial pressure during Al_2O_3 deposition, the regulatory effect of layer on oxygen vacancies (V_O) in the back-channel was explored. Optimal oxygen partial pressure enabled mild doping, filling oxygen sites and reducing V_O, thereby lowering overall defect density. At 20% O_2, the device achieved a field-effect mobility of 4.21 $cm^2V^{-1}s^{-1}$, alongside improvements in subthreshold swing and interface trap density. The on/off current ratio rose from 8.92×10^2 to 2.79×10^3, and the threshold voltage shift (ΔV_{th}) under positive bias stress from 6.55 V to 0.63 V. These results highlight dual role of the ultrathin Al_2O_3 layer in enhancing electrical performance and stability, paving the way for broader practical applications.

Index Terms—thin-film transistors, p-type SnO, ultrathin capping layer, stability

I. INTRODUCTION

Oxide semiconductors (OS) have great advantages in transparency, flexibility, and large-area fabrication, making them an important research topic, particularly in the field of thin-film transistors (TFT) used for various electronic applications, such as displays, sensors, flexible electronics, and emerging electronic devices. Among them, p-type oxide semiconductors have enormous application potential in complementary metal-oxide-semiconductor (CMOS) and various logic circuits, which will pave the way for the next generation of transparent electronic circuits[1]. Therefore, researching p-type oxide TFTs with good performance and stability is of significant importance. SnO TFTs have better development potential due to their high hole concentration and low-temperature fabrication.

The valence state content of Sn in p-type SnO films is crucial for device performance. For p-type SnO, the dominant divalent Sn is metastable and more likely to form SnO_2 in the back-channel[2], which is detrimental to the overall stability of the channel valence state and affects device performance. Therefore, capping the back-channel surface with a dense insulating layer is an effective strategy to further stabilize the content of Sn^{2+} based on the preparation of SnO TFTs. A large number of studies have applied the Al_2O_3 layer to SnO TFT[3-5]. However, these works all used passivation layers thicker

Ruyu Liang, Lei Xu and Shi Zong are with *the School of Electronic Engineering, North China University of Water Resources and Electric Power*, Zhengzhou, China. Corresponding author:(Lei Xu, e-mail xulei@ncwu.edu.cn).

than 10 nm for treatment. This may introduce additional fabrication processes, complicating the preparation costs.

In this work, a simple all-magnetic sputtering process was used to deposit an ultra-thin Al_2O_3 back-channel capping layer to improve the performance of SnO TFTs. After the addition of a 4 nm Al_2O_3 capping layer, the device's mobility and on/off current ratio(I_{on}/I_{off}) were improved. The oxygen partial pressure during the deposition of Al_2O_3 was compared to assess the situation of oxygen vacancy (V_O) defects near the SnO back-channel after capping, and XPS depth profiling confirmed slight doping of Al. Stability test results indicated that the stability of the Al_2O_3 capping layer prepared under 20% O_2 was significantly improved compared to the original uncapped SnO TFT. The enhancement in electrical performance and stability indicates the feasibility of the ultra-thin Al_2O_3 capping layer.

II. METHODS

As shown in Fig. 1(a), a 20 nm SnO film was deposited on an n+-Si substrate with 100 nm SiO_2 via DC sputtering using a Sn target in 2.1% O_2/Ar. A 4 nm Al_2O_3 layer was then RF-sputtered at 100 W under 0%-30% O_2, followed by air annealing at 225°C. Mo source/drain electrodes (100 nm) were DC-sputtered, with channel/electrode patterns defined by shadow masks (channel: 150 μm × 120 μm). The chemical composition and bonding states at the Al2O3/SnO interface were characterized by X-ray photoelectron spectroscopy (XPS, Thermo Scientific). The electrical properties and positive bias stress tests of all devices were performed at room temperature in the dark using a semiconductor parameter analyzer (Keithley 4200A).

III. RESULTS AND DISCUSSION

The back-channel surface is crucial for TFT performance. Depositing Al_2O_3 on SnO passivates surface states, while co-annealing introduces new bonding states influenced by Al_2O_3 deposition oxygen partial pressure (0%-30%). Fig. 1(b) and table 1 show transfer curves and electrical properties. At 0% O_2, the device exhibited ambipolar behavior due to defect reduction, enabling Fermi level shifts[6]. As oxygen pressure increased, ambipolar behavior vanished, and the 20% O_2 device showed improved mobility (2.85 to 4.21 $cm^2V^{-1}s^{-1}$) and on/off current ratio (8.92×10^2 to 2.79×10^3). The Al_2O_3 layer effectively passivates the defect state on the channel surface, and the subband gap state is suppressed. Due to the cascade effect, the interface trap density is decreased[7].

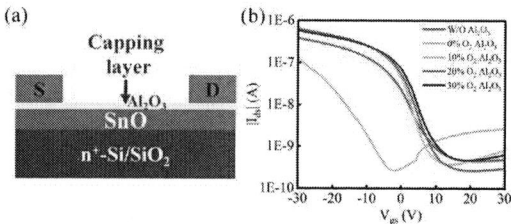

Fig. 1. (a) Device structure. (b) Transfer characteristic curves of SnO TFT with different oxygen partial pressure of Al₂O₃ layer.

TABLE I

ELECTRICAL PERFORMANCE PARAMETERS OF SnO TFT WITH DIFFERENT AT DIFFERENT OXYGEN PARTIAL PRESSURE OF Al₂O₃

Capping layer OPP	μ_{FE} (cm²V⁻¹s⁻¹)	I_{on}/I_{off}	V_{th} (V)	SS (V/decade)	N_{it} (cm⁻²eV⁻¹)
W/O Al₂O₃	2.85±0.28	8.92×10²	-1.69±0.26	6.85±0.27	2.39×10¹³
0% O₂	0.72±0.15	0.47×10²	-17.28±0.45	7.08±0.38	2.53×10¹³
10% O₂	4.73±0.22	1.78×10³	1.96±0.24	3.85±0.35	1.37×10¹³
20% O₂	4.21±0.26	2.79×10³	3.38±0.31	3.71±0.39	1.32×10¹³
0% O₂	4.45±0.33	1.60×10³	3.96±0.33	4.06±0.35	1.44×10¹³

To investigate the bonding states at the back-channel interface with the Al₂O₃ capping layer, XPS was performed on SnO films with Al₂O₃ deposited under varying oxygen partial pressures. Figure 2(a) shows the XPS elemental profile of a SnO film capped with 20% O₂ Al₂O₃. Based on Al₂O₃ and SnO thicknesses and etching rates, Al₂O₃ partially diffused into the SnO film, inducing doping effects. This interaction likely modifies the chemical bonding states near the back-channel interface.

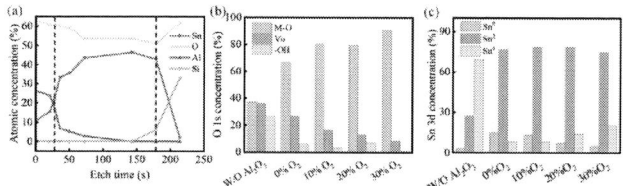

Fig. 2. (a) Variation of elemental content with 20%O₂ Al₂O₃ capped SnO film etching time. (b) relative contents of M-O, V_O and -OH peaks. (c) relative contents of Sn⁰、Sn²⁺ and Sn⁴⁺ peaks.

Fig. 2(b) displays deconvoluted O 1s spectra of the back-channel surface under varying conditions. The pristine device showed high V_O (36.12%) and -OH (26.70%) due to air exposure. Al₂O₃ capping reduced -OH, lowering band-tail states and confirming defect reduction[8]. V_O content decreased with higher O₂ partial pressure, as Al₂O₃ doping suppressed V_O formation. At 0% O₂, reduced subgap states and optimal V_O enabled ambipolar behavior. At 20% O₂, further V_O reduction enhanced mobility and on/off current ratio. At 30% O₂, off-state current stabilized, likely due to Sn valence state changes. Fig. 2(c) presents Sn 3d XPS spectra. The uncapped SnO film surface was heavily oxidized, with Sn⁴⁺ content reaching 69.64%. Al₂O₃ capping stabilized Sn²⁺, reduced Sn4+, and slightly increased Sn⁰. At 30% O₂, interfacial oxidation raised Sn⁴⁺ to 20.24%,

potentially introducing defect states, degrading subthreshold swing, and weakening p-type performance. Thus, the Al₂O₃ layer reduces back-channel defect states, significantly improving device performance.

To evaluate the effect of the Al₂O₃ capping layer on device stability, positive bias stress (PBS) tests were performed on a device with a 20% O₂ Al₂O₃ layer under V_{gs} = +30 V for 3600 s. As shown in Figure 3(a), the pristine device exhibited a threshold voltage shift (ΔV_{th}) of 6.55 V, while the Al₂O₃-capped device achieved a tenfold improvement, with ΔV_{th} reduced to 0.63 V. This improvement stems from the Al₂O₃ layer's ability to reduce defects, effectively suppressing carrier ionization and trapping. These findings highlight the crucial role of the Al₂O₃ layer in enhancing device reliability and offer key insights for optimizing device design and stability.

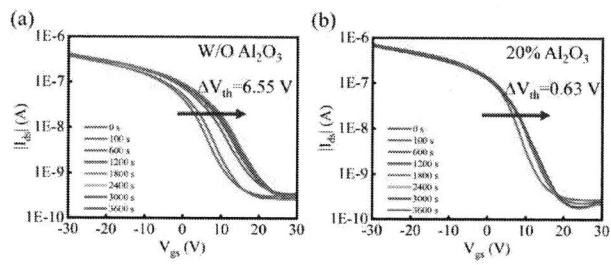

Fig. 3. Stability test curve of (a) original SnO TFT and (b) SnO TFT of 20% O₂ Al₂O₃ capping at positive bias stress of 30 V.

IV CONCLUSIONS

This work addresses the metastability of Sn²⁺ and detrimental defects in SnO by depositing a dense Al₂O₃ layer on the back-channel to regulate valence states and defect properties. The ultrathin Al₂O₃ capping layer isolates the channel from environmental oxygen and moisture, preventing further oxidation and adsorption, while reducing the formation of -OH and Sn⁴⁺, thereby lowering defect states. By varying the oxygen partial pressure during Al₂O₃ deposition, the oxygen content at the Al₂O₃/SnO interface is modulated, influencing V_O and Sn valence states. The 20% O₂ Al₂O₃ capping layer optimizes V_O and Sn valence states, enhancing device stability and performance. The field-effect mobility improves to 4.21 cm²V⁻¹s⁻¹, while the subthreshold swing and interface trap density are also enhanced. The on/off current ratio increases to 2.79×10³, and the positive bias stability improves tenfold, with the threshold voltage shift (ΔV_{th}) reduced from 6.55 V to 0.63 V. These results demonstrate the potential of the ultrathin Al₂O₃ layer to significantly enhance both performance and stability of the device.

REFERENCES

[1] Yan, A.; Wang, C.; Yan, J.; Wang, Z.; Zhang, E.; Dong, Y.; Yan, Z. Y.; Lu, T.; Cui, T.; Li, D.; Shen, P.; Jin, Y.; Liu, H.; Yang, Y.; Ren, T. L., Thin - Film Transistors for Integrated Circuits: Fundamentals and Recent Progress, Adv. Funct. Mater, vol. 34, Jun. 2023.

[2] Luo, H.; Liang, L. Y.; Liu, Q.; Cao, H. T., Magnetron-Sputtered SnO Thin Films for p-Type and Ambipolar TFT Applications, ECS Journal of Solid State Science and Technology, vol. 3, pp. Q3091, Sep. 2014.

[3] Bae, K.-H.; Shin, M. G.; Hwang, S.-H.; Jeong, H.-S.; Kim, D.-H.; Kwon, H.-I., Electrical Performance and Stability Improvement of p-Channel SnO Thin-Film Transistors Using Atomic-Layer-Deposited Al_2O_3 Capping Layer, IEEE Access, vol. 8, pp. 222410-222416, Dec. 2020.

[4] Kim, H. M.; Choi, S. H.; Lee, H. U.; Cho, S. B.; Park, J. S., The Significance of an In Situ ALD Al2O3 Stacked Structure for p‐Type SnO TFT Performance and Monolithic All‐ALD‐Channel CMOS Inverter Applications, Advanced Electronic Materials , vol. 9, Mar.2023.

[5] Mashooq, K.; Jo, J.; Peterson, R. L., Effect of Metal Capping Layer in Achieving Record High p-Type SnO Thin Film Transistor Mobility, IEEE Trans. Electron Devices vol. 71, pp. 574-580, Nov. 2024.

[6] Zhou, Y.; Song, Y.; Hong, R.; Liu, X.; Zou, X.; Iníguez, B.; Flandre, D.; Li, G.; Liao, L., Electrical Evolution of p-Type SnOx Film and Transistor Deposited by RF Magnetron Sputtering, IEEE Trans. Electron Devices, vol. 70, pp. 3100-3105, Apr. 2023.

[7] Luo, H.; Liang, L.; Cao, H.; Dai, M.; Lu, Y.; Wang, M., Control of Ambipolar Transport in SnO Thin-Film Transistors by Back-Channel Surface Passivation for High Performance Complementary-like Inverters, ACS Appl. Mater. Interfaces, vol. 7, pp. 17023-17031,Jul. 2015.

[8] Tu, Y. F.; Chiang, C. L.; Chang, T. C.; Hung, Y. H.; Sun, L. C.; Kuo, C. W.; Tu, H. Y.; Huang, H. C.; Lien, C. H., Improving Reliability of a-InGaZnO TFTs With Optimal Location of Al_2O_3 Passivation in Moist Environment, IEEE Trans. Electron Devices vol. 69, pp. 3181-3185, Jun. 2022.

Enhanced the reliability of InSnMgO thin-film transistors via in-situ sputtering of 2 nm MgO layer

Shi Zong, Lei Xu

Abstract—This study significantly enhances the electrical performance and environmental stability of InSnMgO thin-film transistors by introducing an in-situ deposited 2nm ultrathin MgO layer. The optimized device achieves a field-effect mobility of 52.79 cm²/Vs, an on/off ratio >10⁸, a subthreshold swing of 0.238 V/dec, and a threshold voltage of 0.11 V. The MgO layer exhibits dual functions of back channel passivation and oxygen vacancy suppression, reducing threshold voltage drift from -2.65 V to -0.53 V during positive bias stress testing under 85% humidity while maintaining stable performance after 45 days of ambient storage. This process provides an effective solution for industrial-grade high-reliability oxide thin-film transistors.

Index Terms—thin-film transistors, InSnMgO, Stability

I. INTRODUCTION

In recent years, amorphous oxide semiconductors (AOS) have been widely studied due to their high mobility, transparency, large-area uniformity, and promising applications in flexible electronic devices, gas sensors, new display technologies, and large-area integrated circuits.[1-3] Among them, amorphous indium gallium zinc oxide thin film transistors have successfully replaced silicon-based TFTs in the driving circuits of active matrix liquid crystal displays and active matrix organic light-emitting diode displays.[4-6] With the continuous deepening of research on the applications of TFTs in high resolution low power displays, flexible/curved displays, back-end-of-line compatible devices, logic and memory devices, the performance of a-IGZO TFT is increasingly challenged in these application domains. This study proposes the use of in-situ magnetron sputtering to deposit an MgO layer on top of the InSnMgO channel By controlling the sputtering time, a high quality MgO ultrathin layer is deposited to optimize the electrical performance and enhance the environmental stability of the device. The MgO ultrathin layer effectively isolates the device from moisture and oxygen in the air. During the PBS test under 85% ambient humidity, the threshold voltage shift is only -0.53 V, and after storage for 48 hours under 85% ambient humidity or 45 days in air, the electrical performance of the device remains almost unchanged. This work demonstrates the significant potential of in-situ deposited MgO ultrathin layers in optimizing threshold voltage and enhancing device environmental stability.

Shi Zong and Lei Xu are with the School of Electronic Engineering, North China University of Water Resources and Electric Power, Zhengzhou, China Corresponding author(Lei Xu e-mail xulei@ncwu.edu.cn)

II. Results and Disscussion

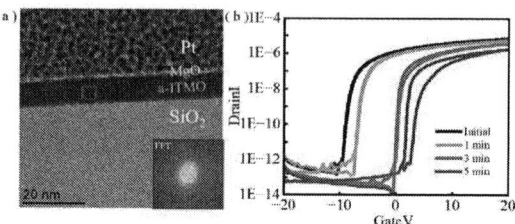

Fig. 1 (a) TEM cross sectional image of the layered thin film, inset: FFT diffraction pattern of the ITMO channel. (b) Transfer characteristics of devices with different in-situ sputtering times of magnesium oxide

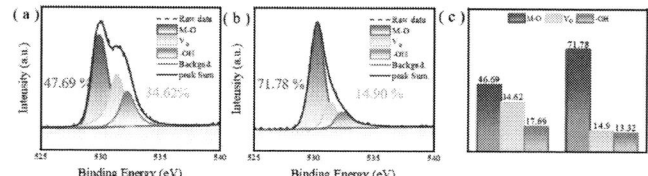

Fig. 2 (a), (b) O1s XPS spectra and curve fitting results of the Initial device and the In-situ MgO device. (c) The statistical diagram of M-O, Vo, and -OH in the channel.

Fig.1 (a) shows the cross ectional structure of the thin film characterized by transmission electron microscopy (TEM, FEI Tecnai G2 F20), with the inset presenting the fast Fourier transform (FFT) diffraction pattern of the channel layer. TEM analysis shows that the thicknesses of the MgO ultra-thin layer and ITMO channel layer are ~2 nm and ~8 nm, respectively. Fig. 1 (b) systematically studied the influence of in-situ sputtering time of MgO on the transfer characteristics of the device. When the sputtering time is 1 minute, the impact on the initial electrical performance of the device is relatively limited due to the inability to effectively form a high quality MgO ultra-thin layer. It is worth noting that when the sputtering time is extended to 5 minutes, the excessively thick MgO layer significantly increases the contact resistance between the electrode and the active layer, resulting in a significant decrease in drain current (I_{ds}) and exacerbating hysteresis.

Fig. 2 (a) shows the O 1s fine spectrum of a single layer ITMO film, and three characteristic peaks can be resolved through peak fitting. Further analysis (Fig. 2 (b)) shows that these characteristic peaks correspond to metal oxygen bonds (M-O, 530.8 eV), oxygen vacancies (VO, 531.6 eV), and hydroxyl groups (- OH, 532.5 eV), respectively. To investigate the interface control effect of ultra-thin MgO layer, Fig. 2 (c) shows the O 1s peak separation results of the bilayer thin film after 15 seconds of argon ion etching treatment. This

979-8-3315-2209-4/25 $31.00 © 2025 IEEE

method can effectively eliminate the detection interference of surface MgO. According to the quantitative analysis in Fig. 2 (f), the introduction of MgO layer has a significant regulatory effect on the ITMO back channel: the proportion of M-O bonds increased from 47% to 72%, while the concentration of oxygen vacancies decreased from 35% to 15%. XPS deep analysis reveals that Mg element (standard reduction potential -2.37 V) exhibits stronger oxygen affinity compared to In (+0.33 V) during annealing, and combines with active oxygen in the channel through diffusion to form a stable metal oxygen bond network.

To demonstrate the improvement in device stability enhanced by the MgO ultrathin layer, Table I show the PBS/NBS stability of the Initial and In-situ MgO devices. After applying ±20 V bias stress for 3600 seconds, the threshold voltage shifts of the Initial and In-situ MgO devices were -1.43/-0.29 V and -3.51/-1.15 V, respectively. The results from the bias stress stability tests further support the conclusion that the MgO ultrathin layer helps to reduce channel oxygen vacancies, improve film quality, and enhance device stability. The significant enhancement in the PBS stability of the In-situ MgO device further demonstrates that the MgO ultrathin layer effectively protects the channel, reducing the impact of water molecules and oxygen from the air on the device.

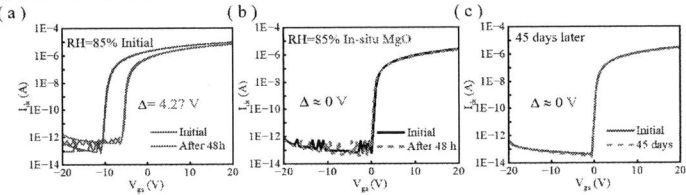

Fig. 3 (a) and (b) the transfer characteristic curves of the Initial and In-situ MgO devices before and after 48 hours of storage under 85% humidity, (c) the transfer characteristic curve of the In-situ MgO device before and after 45 days of atmospheric environment.

In addition, to confirm that the MgO ultra-thin layer can provide long term protection for the channel and maintain the electrical performance of the device in high humidity environments, the initial and in-situ MgO devices were stored in an 85% humidity environment for 48 hours. The transmission characteristic curves of the devices before and after storage are shown in Fig. 3 (a) and (b). As shown in Fig. 3 (c), the in-situ MgO device exhibits excellent environmental stability, with minimal changes in the transmission characteristic curve before and after humidity storage testing. In addition, after being stored in the atmospheric environment for 45 days, the performance of the in-situ MgO device remained almost unchanged, proving that the MgO ultra-thin layer can effectively protect the channel and enhance environmental stability for a long time.

REFERENCES

[1] M.Z. Hu, L. Xu, X.N. Zhang, H.Y. Hao, S. Zong, H.M. Chen, Z.C. Song, S.J. Luo, Z.H. Zhu, High mobility amorphous InSnO thin film transistors via low-temperature annealing, Appl. Phys. Lett. 122 (2023).

[2] K. Myny, The development of flexible integrated circuits based on thin-film transistors, Nat. Electron, 1 (2018) 30-39.

[3] M.T. Vijjapu, S.G. Surya, S. Yuvaraja, X.X. Zhang, H.N. Alshareef, K.N. Salama, Fully Integrated Indium Gallium Zinc Oxide NO2 Gas Detector, ACS Sensors, 5 (2020) 984-993.

[4] K. Nomura, H. Ohta, A. Takagi, T. Kamiya, M. Hirano, H. Hosono, Room-temperature fabrication of transparent flexible thin-film transistors using amorphous oxide semiconductors, Nature, 432 (2004) 488-492.

[5] K. Nomura, Recent progress of oxide-TFT-based inverter technology, J. Inform. Display, 22 (2021) 211-229.

[6] D. Geng, K. Wang, L. Li, K. Myny, A. Nathan, J. Jang, Y. Kuo, M. Liu, Thin-film transistors for large-area electronics, Nat. Electron, 6 (2023) 963-972. http://doi.org/10.1038/s41928-023-01095-8.

Voltage-dividing Capacitor Sharing Enabled High PPI Pixel Circuit with Extended Data Voltage Range for OLED on Silicon Micro-display

Yuanbo Sun[1], Lu Chang[1], Congwei Liao[1], Chen Chen[2], Xin Yuan[2], Zhiwei Ye[2], Xiufeng Zhou[2], Guangsheng Li[2] and Shengdong Zhang[1]*

1. School of Electronic and Computer Engineering, Peking University, Shenzhen, 518055, China

2. MianYang HKC Optoelectronics Technology Co., Ltd, MianYang, 621000, China

*Corresponding Author Email: zhangsd@pku.edu.cn

Abstract—**This paper proposes an OLED micro-display pixel circuit that employs two parallel voltage-dividing capacitors to extend the data voltage programming range, with one of the capacitors shared by multiple ajacent pixels. The proposed circuit not only expands the data voltage programming range but also effectively reduces the pixel area. Simulation results show that the programming voltage range is increased from 133 mV to 2.4 V, while the pixel size is minimized to 1.19 μm × 3.56 μm, achieving a PPI of 7134. Moreover, the pixel circuit renders non-uniformity suppressed to below 6% across the entire grayscale range thanks to the threshold voltage compensation function.**

Keywords—**organic light-emitting diode on silicon, shared capacitor, data voltage range expansion, high pixel per inch, threshold voltage compensation**

I. INTRODUCTION

Organic Light-Emitting Diodes on silicon substrates (OLEDoS) becomes the dominant micro-display technology for virtual reality (VR) and augmented reality (AR) applications [1]-[2]. Compared with other counterparts, OLEDoS micro-display exhibits high contrast ratios, fast response times, and low power consumption [3]. With the increase of pixel density (pixel per inch, PPI), the OLEDoS pixel current is continuously decreasing within the range of pA to nA. Consequently, the driving transistor of OLEDoS pixel works in the sub-threshold region, and the gate voltage range is compressed to several hundred millivolts correspondingly. On the other hand, the data voltage range should be over 2V in order to accurately control the display grayscale over 8 bit [4]. Furthermore, the high sensitivity of the devices to the process, voltage and temperature (PVT) variations becomes a major challenge since the drain current is exponentially dependent on the threshold voltage [5]. In order to address these variation issues, the pixel circuit of OLEDoS must have good compensation function.

Previously, data voltage scaling is used to expand data voltage range, avoiding the driver transistor working in the sub-threshold region. Wang et al. proposed connecting OLED and a controlled current source in parallel, then the OLED current is determined by the difference between the two current sources operating in the saturation region [6]. However, this approach brings additional power consumption and large mismatch. In addition, Shin et al. proposed a feedback structure where a resistor or a diode-connected MOS transistor is placed at the source electrode of the driving transistor, reducing the proportionality between the driving transistor's V_{GS} and the data voltage [7]. However, the introduction of source resistor or MOS transistor will increase the pixel layout area and inversely influence the pixel uniformity.

Moreover, K. Kimura et al. proposed a series capacitor structure, where the ratio of capacitance is directly linked to the expansion multiple of the data voltage range [8]. One advantage of this topology is that in CMOS process, the relative manufacturing accuracy of metal capacitor can reach several thousandths, so this technology can realize accurate control of display grayscale. Moreover, the metal-insulator-metal (MIM) capacitor and metal-oxide-metal (MOM) capacitor can be implemented using top metal layers to avoid occupying additional active area. However, integrating multiple capacitors within a single pixel remains a significant challenge, especially for ultra-high-density pixel circuits.

This work introduces an OLED micro-display pixel circuit with high resolution and high uniformity. The data voltage range is expanded and the threshold voltage dispersion is compensated by capacitor parallel structure and diode connection structure respectively. In addition, by sharing the voltage-dividing capacitor with 20 pixels, the average area of each pixel is 9.5% of the original.

II. CIRCUIT STRUCTURE AND OPERATION

Fig.1 (a) shows the proposed pixel circuit schematic, which consists of eight PMOS transistors (P1–P8) and two capacitors (Cc and Cs). Here, P1–P6 are switching transistors, P7 is the driving transistor, and P8 acts as the emission enabling transistor. For 20 pixels of the same column line, P4, P5 and Cc are shared. Fig.1 (b) shows the operation of the proposed pixel circuit, which consists of five consecutive phases: (1) Initialization, (2) V_{TH} extraction, (3) Charge transfer, (4) Voltage boosting, and (5) Emission. The proposed pixel circuit operates as follows.

979-8-3315-2209-4/25 $31.00 © 2025 IEEE

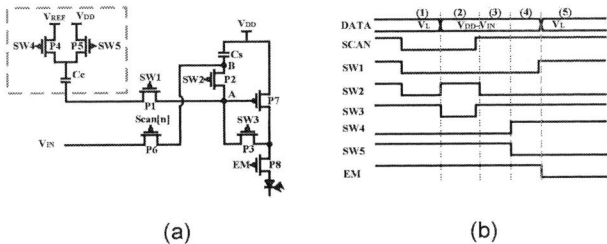

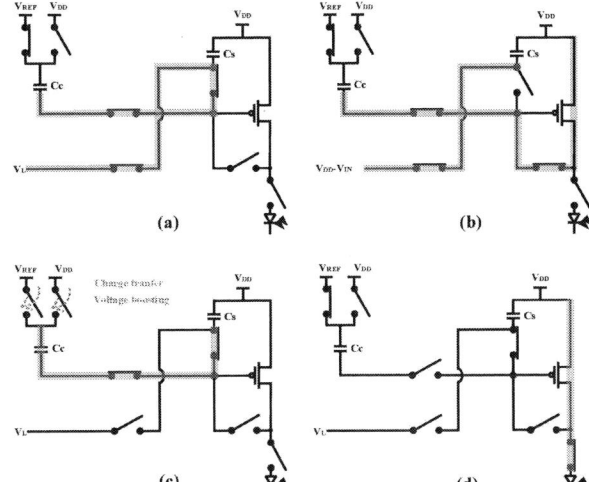

(a)

(b)

Fig. 1 (a) Schematic and (b)Timing diagram of the proposed pixel circuit.

A. Initialization phase

The level of Scan[n], SW1, SW2 and SW4 signal lines are all pulled down, thus P1, P2, P4 and P6 are turned on. The V_{IN} line provides the reference voltage V_L, which is simultaneously applied to the upper plate of capacitor Cs and Cc, as shown in Fig.2 (a). In this phase, there is no current flowing through the OLED.

B. V_{TH} extraction phase

The gate voltage of transistor P2 changes from GND to VDD, and the gate voltage of P3 drops from VDD to GND, while the gate voltages of other transistors remain unchanged. Consequently, the saturation current of the driving transistor P7 flows through P3 and P1 to charge the lower electrode of Cc until the voltage level reaches V_{DD} - $|V_{THP}|$, as shown in Fig.2 (b). Meanwhile, the lower plate of Cs is charged to V_{DD}-V_{IN} by the V_{IN} line. Then, charges stored in capacitors Cc and Cs, namely Q1 and Q2, can be expressed by:

$$Q1 = Cc*(V_{REF} - V_{DD} + |V_{THP}|) \tag{1}$$

$$Q2 = Cs*V_{IN} \tag{2}$$

C. Charge transfer phase

The switching transistors P4, P1 and P2 are turned on, and the charges on the lower electrode of Cc and Cs are redistributed. Here, the voltage at node A is named as VA', then according to the conservation of charge, the charge sum can be calculated as:

$$Cc*(V_{REF} - V_A') + Cs*(V_{DD} - V_A') = Q1 + Q2 \tag{3}$$

Combining equations (1), (2) and (3), VA' can be derived as:

$$V_A' = \frac{Cc}{Cc+Cs}*(V_{DD} - |V_{THP}|) + \frac{Cs}{Cc+Cs}*(V_{DD} - V_{IN}) \tag{4}$$

D. Voltage boosting phase

After charge transfer phase, the source-to-gate voltage of the driving transistor P7 is too large, in other words, P7 works in the saturation region. To scale down V_{SG} of P7 and avoid saturation operating, voltage boosting method through capacitor coupling is used. As shown in Fig.2 (c), P4 is turned off and P5 is turned on. As the voltage level of the upper electrode of Cc rises from V_{REF} to V_{DD}, the voltage of node A also rises due to the coupling effect, which can be

expressed by:

$$V_A = V_A' + \frac{Cc}{Cc+Cs}*(V_{DD} - V_{REF}) \tag{5}$$

Thus, the V_{SG} of P7 is :

$$V_{SG} = V_{DD} - V_A = \frac{Cs}{Cc+Cs}*V_{IN} + \frac{Cc}{Cc+Cs}*(V_{DD} + |V_{THP}| - V_{REF}) \tag{6}$$

E. Emission phase

The switching transistor P2, the driving transistor P7 and the enabling transistor P8 are turned on while the other transistors are turned off. The current flowing through the OLED can be expressed as:

$$
\begin{aligned}
I_{OLED} &= \frac{W}{L}I_0 \exp\left(\frac{V_{SG} - |V_{THP}|}{n*V_T}\right) \\
&= \frac{W}{L}I_0 \exp\left(\frac{\frac{Cs}{Cs+Cc}*(V_{IN} - |V_{THP}|) + \frac{Cc}{Cs+Cc}*(V_{REF} - V_{DD})}{n*V_T}\right)
\end{aligned} \tag{7}
$$

where W/L, n, V_{THP}, and V_T are the aspect ratio, subthreshold slope factor, threshold voltage of the driving transistor P7, and thermal voltage ($V_T = kT/q$), respectively. I_0 is the drain current at threshold normalized for W/L. In the proposed circuit, Cc is 20 times that of Cs, so the equation (7) can be expressed as

$$I_{OLED} = \frac{W}{L}I_0 \exp\left(\frac{\frac{1}{21}*(V_{IN} - |V_{THP}|) + \frac{20}{21}*(V_{REF} - V_{DD})}{n*V_T}\right) \tag{8}$$

From equations (6) and (8), it is clear that $\Delta V_{IN}/\Delta V_{SG}$=21, indicating that the data voltage range has been expanded by 21 times. And the influence of threshold voltage on V_{SG} is greatly weakened.

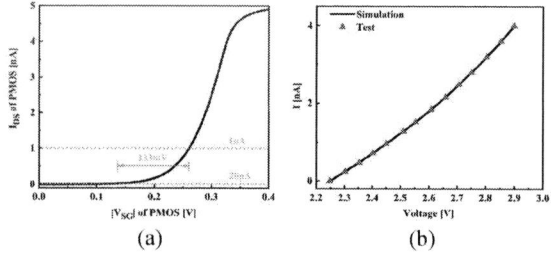

Fig. 3 (a) Current-voltage characteristic of PMOS. (b) Current-voltage characteristics for the OLED devices based on the experimental data.

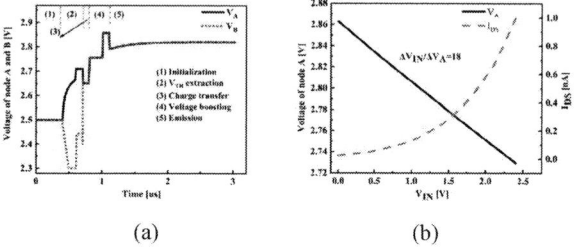

Fig. 4 (a) Voltage of node A and B during pixel programming. (b) The change of voltage of node A and current of P7 with the change of V_{IN}.

III. RESULTS AND DISCUSSIONS

The proposed OLEDoS pixel circuit is verified through simulations. The current-voltage characteristic of PMOS transistor is shown in Fig. 3(a). Considering the brightness requirements of OLED, the operating current range is set to 20 pA to 1 nA, which is typical for OLEDoS micro-display [7]. It can be observed that the gate voltage programming range of the driving transistor is only 133 mV. Fig. 3(b) compares the measured and simulated current-voltage characteristics of the OLED. As the OLED characteristic is well represented by the Verilog-A model, the developed OLED model is reliable for the SPICE simulations.

Simulation results for the timing diagrams of the proposed pixel circuit are shown in Fig.4 (a). Fig.4 (b) demonstrates that during the stable emission phase, the gate voltage of the driving transistor is almost linearly related to the data voltage V_{IN}. The data voltage range (2.4 V) is 18 times that of point A (133 mV), which is slightly smaller than the capacitor ratio (Cs+Cc)/Cs due to parasitic capacitances in the layout. The current of P7 under different data voltages is also shown in Fig.4 (b), which is consistent with the planned operating current range of OLED.

The threshold voltage compensation capability of the proposed pixel circuit was validated using Monte Carlo simulations with five thousand points. As shown in Fig.5, for the 3T1C pixel circuit without threshold voltage compensation, the ratio of the current standard deviation to the mean current exceeds 60% at the minimum grayscale level, consistent with previously reported studies [7]. For the

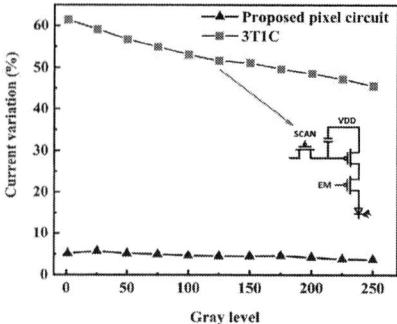

Fig. 5 Current variation in Monte Carlo simulations of the proposed pixel circuit and the pixel circuit without threshold voltage compensation function shown in the illustration.

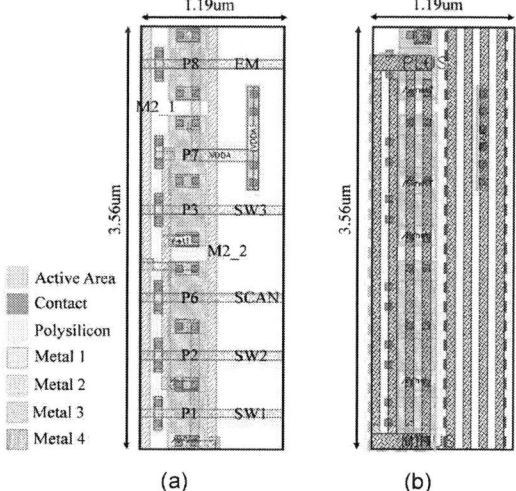

Fig. 6 Layout of pixel circuit: (a) Transistor. (b) MOM capacitor.

proposed shared-capacitor pixel circuit, this ratio is 5.2% at the minimum grayscale level, demonstrating a significant improvement in threshold voltage compensation. Fig.5 also illustrates the current variations across different grayscale levels under Monte Carlo simulations, showing that in general the relative current variation decreases as the grayscale level increases.

The pixel array is realized by 40nm CMOS process. Considering the voltage tolerance of the device, the transistor with the minimum channel length of 270 nm is selected from the PDK. Fig. 6(a) illustrates the primary layout components of a single pixel, showing the active area, polysilicon layer, substrate contacts, and source/drain contacts. Six transistors are arranged in a row to facilitate routing, and the layout area is optimized by source-drain multiplexing. Five horizontal Metal 1 lines, from top to bottom, are designated for the shared control signals EM, SW3, SCAN, SW2, and SW1 across each row of pixels in the display panel. To the left, a vertical Metal 2 line (M2_1) serves as the global V_{IN} LINE. M2_2 acts as a local metal

979-8-3315-2209-4/25 $31.00 © 2025 IEEE

TABLE I

COMPARISONS BETWEEN THE PROPOSED PIXEL CIRCUIT AND THE STATE-OF-THE-ARTS

Parameters	[7]	[8]	[9]	[10]	This work
CMOS Process	0.13 um	0.18 um	0.18 um	110 nm	40 nm
Pixel Per Inch	4670	3270	2125	4410	7134
Emission Current Deviation Error	17%~19%	2%	2.1%	1.16%	3%~6%
Data Voltage Expansion Ratio	3.4x	2x	6.2x	8.2x	18x

line connecting the drains of 20 adjacent transistors in the same column, which ultimately is linked to the lower plate of capacitor Cc. Capacitors Cc and Cs are implemented using MOM capacitors formed by Metal 3 and Metal 4 layers as shown in Fig.6 (b). The capacitor Cs is placed in the orange dotted box. Cc, placed between two red dotted lines, has the same width as Cs but extends to 20 times the height and one twentieth of the capacitor Cc is placed to the right of Cs.

A comprehensive comparison between the proposed pixel circuit and other prior pixel circuits are listed in Table I. Among various performance parameters, the proposed pixel circuit has obvious advantages in pixels per inch. The voltage amplification factor is also far beyond the existing literature research.

IV. CONCLUSION

A new OLEDoS micro-display pixel circuit that provides wide data voltage range and high uniformity is proposed. The data voltage range is expanded by 18 times by using capacitor parallel connection, and the area of each pixel is reduced to 1.19 μm × 3.56 μm by sharing capacitor Cc. Moreover, the proposed pixel circuit employs the diode-connect structure to reduce the influence of threshold voltage variation on the driving transistor. Therefore, the proposed pixel circuit is promising to achieve high-performance OLEDoS micro-displays with high resolution and high uniformity.

REFERENCRS

1. F. Yang, C. Wang, H. A floating high-voltage level shifter used in a pre-charge circuit for large-size AMOLED displays.

EDSSC. 2016

2. X. Guan, Y. Huang. A Pixel Circuit with Compensation for Threshold Voltage and Current-Resistance Voltage Drop. EDSSC. 2019

3. J. Wen et al. Design of a Peripheral-Circuit-Compensation Adjustable-Gamma-Voltage Driving Chip for OLED-on-Silicon Microdisplay. EDSSC. 2019

4. Lim G-W, Kang GG, Ma H, Jeong M, Kim HS. An area-efficient 10-bit source-driver IC with LSB-stacked LV-to-HV-amplify DAC for mobile OLED displays. IEEE J Solid-State Circuits.2023;58(11):3164–3178.

5. D. Judy and V. s. k. Bhaaskaran, Review and Analysis of the Impacts and Effects on Low Power VLSI Circuits Operating in Subthreshold Regime, International Journal of Engineering and Technology. 2013;5(5): 3870–3883.

6. Wang XH, Wang WB, Du H, Han ZS. New pixel driving circuit for active-matrix OLED-on-silicon. Electron Devices [Internet]. 2007;(05):1745-1748.

7. Shin H-J, Kim Y-D, Choi B-D. 4670-PPI OLEDoS pixel circuit design for wide data voltage range in a 5 V 0.13 μm CMOS process. J Soc Inf Display. 2024;32(5):165–173

8. K. Kimura, Y. Onoyama, T. Tanaka, N. Toyomura, and H. Kitagawa, New pixel driving circuit using self-discharging compensation method for high-resolution OLED micro displays on a silicon backplane, SID Symp. Dig., 2017;25(3): 167-176

9. X. Huo, C. Liao, M. Zhang, H. Jiao and S. Zhang, A Pixel Circuit With Wide Data Voltage Range for OLEDoS Microdisplays With High Uniformity. IEEE Transactions on Electron Devices, 2019;66(11): 4798-4804.

10. Na J-S, Hong S-K, Kwon O-K. A 4410-ppi resolution pixel circuit for high luminance uniformity of OLEDoS microdisplays. IEEE J Electron Devices Soc. 2019;7:1026–1032

Stress-Induced Band Structure Modulation in h-BN/HfSe$_2$ Heterostructures

Yunhao Liu

Abstract—**The h-BN/HfSe2 heterostructure holds great potential in the fields of optoelectronic and quantum devices[1]. Based on first-principles calculations under applied biaxial strain, this paper investigates the modulation of the optoelectronic properties of the h-BN/HfSe2 heterostructure. The results show that the h-BN/HfSe2 heterostructure exhibits a type-I direct bandgap of 0.718 eV. Within the selected strain range, the band structure of the heterostructure can be transformed between direct and indirect bandgaps, while also achieving fine-tuning of the bandgap. This broadens the application of this type of heterostructure in the field of light-emitting devices.**

I. Introduction

Two-dimensional hybrid heterostructures have attracted extensive interest in exploring the photoelectric effect and property modulation due to their excellent photoelectric performance and energy conversion efficiency[2]. Hexagonal boron nitride (h-BN) is one of the most concerned two-dimensional materials because of its outstanding properties. It possesses high thermal conductivity, a wide bandgap (5.86 eV), excellent electrical insulation, high mechanical strength, and strong chemical inertness. These characteristics enable h-BN to be widely applied in the fields of electronics, optoelectronics, and thermal management devices[6]. Applying strain to the heterostructure composed of h-BN and HfSe$_2$ can achieve the modulation of the properties of the h-BN/HfSe$_2$ heterojunction and provide a certain understanding of its applications. In this paper, a van der Waals (vdW) heterostructure composed of h-BN and monolayer HfSe$_2$ is constructed. By changing the lattice constant of the h-BN/HfSe$_2$ heterostructure, the regulation from -6% compressive strain to +6% tensile strain is realized. Based on density functional theory, this paper investigates the stability and photoelectric properties of the h-BN and HfSe$_2$ heterostructure system, including band structure, density of states, and band alignment[3]. Moreover, the band offset of h-BN/HfSe$_2$ is modulated by biaxial strain, and the electronic and optical properties of different systems are compared. It

Yunhao Liu is with *National Center for International Joint Research of Electronic Materials and Systems, International Joint-Laboratory of Electronic Materials and Systems of Henan Province, School of Electrical and Information Engineering, Zhengzhou University, Zhengzhou, Henan 450001, P. R. China*, Zhengzhou, Henan China. This work is supported by the [National Natural Science Foundation of China] under Grant [No.62174148]; the [National Key Research and Development Program] under Grant [No.2022YFE0112000 and 2016YFE0118400]; the [Key Program for International Joint Research of Henan Province] under Grant [No.231111520300]; the [Ningbo Major Project of 'Science, Technology and Innovation 2025'] under Grant [No.2019B10129]; and the [Zhengzhou 1125 Innovation Project] under Grant [No.ZZ2018-45]. Corresponding author: (Fang Wang, email: iefwang@zzu.edu.cn; Yuhuai Liu, email: ieyhliu@zzu.edu.cn)

can be found that the h-BN/HfSe$_2$ heterostructure with tunable characteristics has great application potential in the design and fabrication of optoelectronic devices.

II. Calculation Methods

All the calculations were performed using the plane-wave based PWmat code. The generalized gradient approximation Perdew-Burke-Ernzerhof (GGA-PBE) functional and the optimized norm-conserving Vanderbilt pseudopotentials were employed for structural optimization, self-consistent calculations, and electronic state analysis. Periodic boundary conditions were applied in the x and y directions for all structures, and a vacuum layer of 20 Å was placed in the z direction to avoid interactions between the top and bottom surfaces of the heterostructure. Additionally, the DFT-D3 method was introduced for van der Waals (vdW) correction to accurately describe the interactions between molecules. The plane-wave cutoff energy was set to 50 Ryd (680.3 eV), and the energy convergence precision and the force on a single atom during the structural relaxation process were set to 1×105 eV·Å3 and 0.02 eV/Å, respectively. For the interactions in the Brillouin zone, we used the 4 x 4 x 1 and 7 x 7 x 1 Monkhorst-Pack special k-point grids for structural optimization and self-consistent calculations. The band structures of the h-BN and HfSe2 unit cells were calculated using the Heyd-Scuseria-Ernzerhof (HSE 06) hybrid functional, which has minimal electron self-interaction error and produces accurate electronic energy levels. To save computational effort, the band structure, partial density of states, and band offset were calculated using the PBE functional. The optimized structures were visualized using the Vesta software. The binding energy (EBE) was used to describe the stability of the HfSe2/h-BN heterostructure. The expression is as follow:

$$E_{BE} = E_{h-BN/HfSe_2} - E_{h-BN} - E_{HfSe_2} \qquad (1)$$

Here, E_{h-BN}, E_{HfSe2} and $E_{h-BN/HfSe2}$ represent the total energies of the h-BN monolayer, the HfSe$_2$ monolayer and the heterostructure h-BN / HfSe$_2$, respectively.

III. Results and Discussions

A. Geometric structure

Firstly, in this work, we investigated the monolayers of h-BN and HfSe2, performing lattice optimization and property calculations for them[7]. The optimally obtained lattice constants for h-BN and HfSe$_2$ were a0 = 2.504 and a1 =

979-8-3315-2209-4/25 $31.00 © 2025 IEEE

3.763, respectively, which are close to the experimental data of 2.504 [5] and 3.75 [6]. The optimized B-N and Hf-Se bond lengths were 1.45 Å and 2.67 Å, respectively, consistent with previous research results. By combining h-BN (3x3x1) with HfSe$_2$ (2x2x1), as shown in Figure 1(e), the Eh-BN / HfSe$_2$ heterostructure was formed, with lattice constants of 7.54 Å and 7.52 Å for each, respectively. The lattice mismatch rate is less than 0.3%, and the negative binding energy indicates that the structure can exist stably.

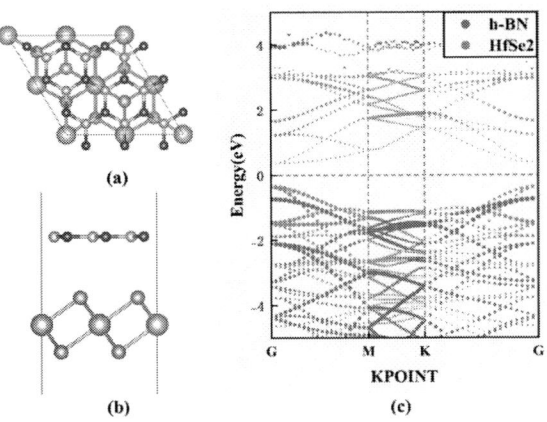

(a)

(b)   **(c)**

Fig. 1. (a) is the side view (lower layer) of the h-BN/HfSe2 heterostructure. (b) is the top view (upper layer) of the h-BN/HfSe2 heterostructure. (c) is the projected density of states of the h-BN/HfSe2 heterostructure.

B. Projected Band Structure and Density of States

The band structure of the Eh-BN / HfSe$_2$ heterostructure is not a simple superposition of the band structures of h-BN and HfSe$_2$. Their band structures are influenced by the van der Waals (vdW) interactions at the h-BN/HfSe$_2$ interface. Moreover, the intrinsic properties of h-BN and HfSe$_2$ are well preserved. We calculated the projected band structures and densities of states (DOS) of the heterojunction under seven different strains. When a -6% strain is applied, the heterojunction exhibits a bandgap of approximately 0.15 eV, tending towards metallic characteristics. Under -4%, -2%, 0%, +2%, +4%, and +6% strains, the heterojunctions all show distinct semiconductor characteristics, with the Fermi level moving to 0 eV. Different strains not only alter the electronic structure of the heterojunction but also change its electronic properties. The original h-BN/HfSe$_2$ system is shown in Figure d. Before any strain is applied, the heterostructure exhibits a direct bandgap, with both the conduction band minimum (CBM) and valence band maximum (VBM) located at the point, and a bandgap of 0.71 eV. Different strains not only change the electronic structure of the heterojunction but also its electronic properties. As can be seen from Figures a, b, c, d, e, f, and g, as the strain increases (-6%, -4%, -2%, 0%, +2%, +4%, +6%), the bandgap also increases (0.152 eV, 0.441 eV, 0.711 eV, 0.714 eV, 0.941 eV, 1.306 eV, 1.465 eV).

The CBM is always provided by HfSe$_2$ and increases with increasing strain (0.081 eV, 0.226 eV, 0.361 eV, 0.363 eV, 0.479 eV, 0.600 eV, 0.619 eV). For the VBM, the VBM provided by HfSe$_2$ gradually decreases (-0.071 eV, -0.216 eV, -0.350 eV, -0.351 eV, -0.462 eV, -0.706 eV, -0.776 eV). When the strain increases to +6%, the position of the valence band maximum changes, transitioning to an indirect bandgap[4].

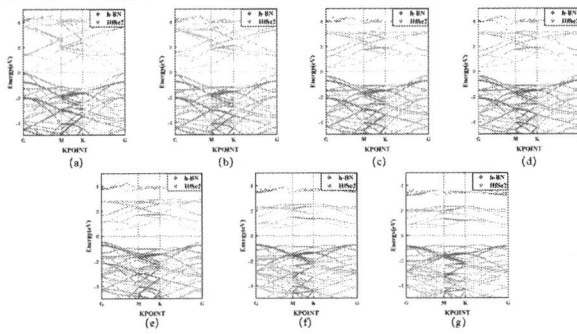

Fig. 2. (a) to (g) is Band Structure of the h-BN/HfSe2 heterostructures with appliedd -6% to +6% stresses,respectively.

The PDOS analysis of each heterojunction system is shown in Figure 2 shown, the original h-BN/HfSe$_2$ heterostructure, CBM is always contributed mainly by Hf atoms, at stresses of (-6%, -4%, -2%, 0%, +2%) VBM is mainly contributed by Se atoms, and at stresses of (+4%, +6%) VBM is contributed by N atoms, Se atoms together. The applied strain produces a change in the heterostructure.

The CBM of the h-BN/HfSe$_2$ heterostructure is always provided by HfSe$_2$ regardless of the applied stress (-6% to +6%), and the VBM of the h-BN/HfSe$_2$ heterostructure is mainly supplied by HfSe$_2$ at the stresses of -6%, -4%, -2%, 0%, and +2%, and is jointly supplied by h-BN, and HfSe$_2$ at the stresses of +4% and +6%. In addition to the electronic

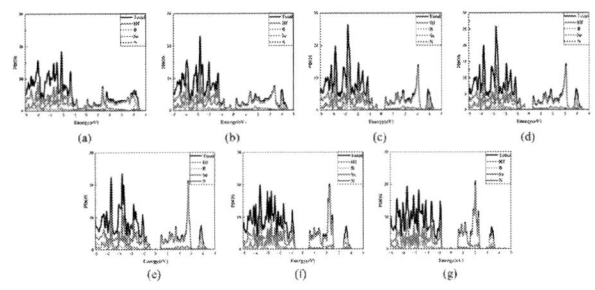

Fig. 3. (a) to (g) are the projected density of states of the h-BN/HfSe2 heterostructures with appliedd -6% to +6% stresses,respectively.

structure, the density of states also has a more significant effect on the energy band structure for heterojunctions. The localized density of states (LDOS) method was used to calculate the energy band offsets for each heterojunction system, and the conduction band offset (CBO) and valence band offset (VBO) were calculated as CBO = Eg 1 -Eg 2-VBO. In this work, Eg

1 and Eg 2 were set as the band gaps of h-BN and HfSe2, which are 5.9 eV and 1.363 eV, respectively. as shown in Fig. 8, the original h-BN/HfSe$_2$ heterojunction system, strained h-BN/HfSe$_2$, is shown in Fig. 8, and the original h-BN/HfSe$_2$ heterojunction system, strained h-BN/HfSe$_2$, is shown in Fig. 8, respectively. , the energy band alignments of the pristine h-BN/HfSe$_2$ heterojunction system, strained h-BN/HfSe$_2$ (-6%, -4%, -2%, 0%, +2%, +4%, +6%) system are shown, respectively.The individual heterostructures are all in type I energy band arrangement when -6%, -4%, -2%, 0%, +2%, +4%, +6% stress is applied, which increases the chance of electron-hole complexation.

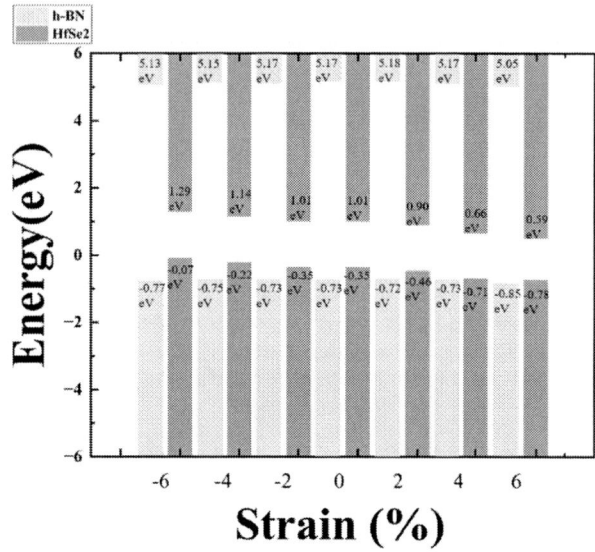

Fig. 4. The band edge positions of the h-BN/HfSe2 heterostructure with appliedd -6% to +6% stresses,respectively.

IV. CONCLUSIONS

In summary, the modulation of the optoelectronic properties of the h-BN/HfSe$_2$ heterostructures by biaxial strain has been investigated using a DFT-based first-principles approach. Observed from the perspective of projected energy bands, the bandgap of the h-BN/HfSe$_2$ heterostructure shows a positive correlation with the magnitude of the applied stress to the heterostructure; during the gradual increase of the stress from -6% to +6%, the CBM is always supplied by HfSe$_2$ and gradually rises, and the VBM supplied by HfSe$_2$ gradually decreases, which results in the gradual enlargement of the bandgap of the h-BN/HfSe$_2$ heterostructure. Meanwhile, the position of the valence band top changes to an indirect band gap when the stress increases to +6%.The individual heterostructures at applied -6%, -4%, -2%, 0%, +2%, +4%, and +6% stresses are all arranged in type I energy bands[5]. With the application of different stresses, the energy bands of the heterostructures can be interconverted from the direct bandgap to the indirect bandgap, and at the same time, the fine-tuning of the bandgap is realized, which broadens the application of the heterostructures in the field of light-emitting devices.

REFERENCES

[1] Zhang K, Feng Y, Wang F, Yang Z, Wang J. Two dimensional hexagonal boron nitride (2D-hBN): synthesis, properties and applications[J]. Journal of Materials Chemistry C, 2017, 5(46): 11992-12022.

[2] Niu T, Li A. From two-dimensional materials to heterostructures[J]. Progress in Surface Science, 2015, 90(1): 21-45.

[3] Jariwala D, Marks T J, Hersam M C. Mixed-dimensional van der Waals heterostructures[J]. Nature Materials, 2017, 16(2): 170-181.

[4] Wu Z, Jie W, Yang Z, Hao J. Hybrid heterostructures and devices based on two-dimensional layers and wide bandgap materials[J]. Materials Today Nano, 2020, 12: 100092.

[5] Ju W, Zhang Y, Li T, Wang D, Zhao E, Hu G, Xu Y, Li H. A type-II WSe2/HfSe2 van der Waals heterostructure with adjustable electronic and optical properties[J]. Results in Physics, 2021, 25: 104250.

[6] Lee S, Song M K, Zhang Y, Suh J M, Ryu J E, Kim J. Mixed-dimensional integration of 3D-on-2D heterostructures for advanced electronics[J]. Nano Letters, 2024, 24(30): 9117-9128.

[7] Barton A T, Yue R, Anwar S, Zhu H, Peng X, McDonnell S, Lu N, et al. Transition metal dichalcogenide and hexagonal boron nitride heterostructures grown by molecular beam epitaxy[J]. Microelectronic Engineering, 2015, 147: 306-309.

Band Structure Modulation of h-BN/HfSe2 Heterostructures Under an External Electric Field*

1st Shipei Ji 2nd Hang Xu 3rd Jiping Hu 4th Xiaotian Yang 5th Jun Zhang 6th Fang Wang 7th Juin J. Liou 8th Yuhuai Liu

Abstract—The h-BN/HfSe2 heterostructure shows great potential for applications in optoelectronic and power electronic devices. In this study, first-principles calculations under an applied vertical electrostatic field were conducted to investigate the band structure characteristics of the h-BN/HfSe2 heterostructure. The results indicate that the h-BN/HfSe2 heterostructure exhibits a type-I direct band gap of 0.718 eV. Within the selected range of applied electrostatic fields, the band edge positions of the h-BN/HfSe2 heterostructure can be modulated, while the band gap size, band gap type, and band alignment type remain stable. When an electric field of +0.2 V/Å is applied, both the valence band maximum (VBM) and conduction band minimum (CBM) of the heterostructure are above the Fermi level, suggesting the formation of Ohmic contact at the interface. These findings demonstrate that the properties of the h-BN/HfSe2 heterostructure can be effectively tuned by an external electrostatic field, providing valuable guidance for the research and fabrication of electronic devices based on the h-BN/HfSe2 heterostructure.

Index Terms—External electric field, heterostructure, 2D materials, first-principles calculations ,band modulation

I. INTRODUCTION

The unique dimensional characteristics of two-dimensional (2D) materials endow them with extraordinary physical and chemical properties, making them highly promising for applications in a wide range of fields. HfSe2 (hafnium diselenide) is a type of transition metal dichalcogenide (TMDC) with a distinctive layered crystal structure and an indirect band

Jun Zhang, Hang Xu, Jiping Hu, Xiaotian Yang and Shipei Ji are with *National Center for International Joint Research of Electronic Materials and Systems, International Joint-Laboratory of Electronic Materials and Systems of Henan Province, School of Electrical and Information Engineering, Zhengzhou University, Zhengzhou, Henan 450001, P. R. China,*Fang Wang and Yuhuai Liu are with *National Center for International Joint Research of Electronic Materials and Systems, International Joint-Laboratory of Electronic Materials and Systems of Henan Province, School of Electrical and Information Engineering, Zhengzhou University, Zhengzhou, Henan 450001, P. R. China, International Joint Laboratory for Integrated Circuits Design and Application, Ministry of Education, School of Physics, Zhengzhou University, Zhengzhou 450052, China*Institute of Intelligence Sensing, Research Institute of Industrial Technology Co. Ltd., Zhengzhou University, Zhengzhou, Henan 450001, P. R. China, Zhengzhou Way Do Electronics Co. Ltd., Zhengzhou, Henan 450001, P. R. China). Juin J. Liou is with *School of Electrical and Information Engineering, North Minzu University, Yinchuan, Ningxia 750001, P. R. China.* This work is supported by the [National Natural Science Foundation of China] under Grant [No.62174148]; the [National Key Research and Development Program] under Grant [No.2022YFE0112000 and 2016YFE0118400]; the [Key Program for International Joint Research of Henan Province] under Grant [No.231111520300]; the [Ningbo Major Project of 'Science, Technology and Innovation 2025'] under Grant [No.2019B10129]; and the [Zhengzhou 1125 Innovation Project] under Grant [No.ZZ2018-45]. Corresponding author: (Fang Wang, email:iefwang@zzu.edu.cn;Yuhuai Liu, email:ieyhliu@zzu.edu.cn)

gap semiconductor nature. [1] [2]Its excellent electron mobility, wide band gap range, and tunable optical properties make it an important member of the 2D materials family, particularly demonstrating great potential in applications such as field-effect transistors (FETs), photodetectors, and energy conversion devices. Meanwhile, h-BN (hexagonal boron nitride), as a wide-band gap 2D insulator, is widely used as a substrate material or dielectric layer in heterostructures due to its excellent chemical stability, atomically flat surface, and high thermal conductivity. Due to the weak van der Waals interactions between h-BN and other 2D materials, it effectively reduces the impact of interfacial defect scattering and surface roughness on electron transport, thereby enhancing the performance of heterostructure-based devices. [3] The heterostructure formed by combining HfSe2 and h-BN integrates the excellent semiconductor performance of HfSe2 with the insulating properties of h-BN, making it an ideal candidate system for constructing high-performance 2D electronic and optoelectronic devices. Through rational design of such heterostructures, precise modulation of electronic bandgaps, interfacial charge transfer, and optical absorption can be achieved, demonstrating promising applications in low-power electronics,photodetector, and flexible electronics. However, research on the interfacial physics and optical behavior of h-BN/HfSe2 heterostructures is still in its infancy, and further exploration is required to understand their electronic structure, interfacial interactions, and device performance potential. This paper aims to investigate the electronic modulation mechanisms of the h-BN/HfSe2 heterostructure under an applied vertical electric field. By varying the strength of the external electric field applied to the h-BN/HfSe2 heterostructure, modulation from -0.3 V/Å to 0.3 V/Å was achieved. Based on density functional theory (DFT), the stability and electronic properties of the h-BN/HfSe2 heterostructure were studied, including the energy band structure, the density of states, and the alignment of the band. Furthermore, it was found that the band offset of the h-BN / HfSe2 heterostructure was tuned by the external electric field. The electronic properties of different systems were compared, providing theoretical guidance and design ideas for the development of high-performance electronic devices based on 2D materials.

II. CALCULATION METHODS

All calculations were performed using the plane-wave-based PWmat code. Structural optimization, self-consistent calculations, and electronic state analysis were conducted using the generalized gradient approximation Perdew-Burke-Ernzerhof (GGA-PBE) functional and the optimized norm-conserving Vanderbilt pseudopotential. Periodic boundary conditions were applied in the x and y directions for all structures, while a 20 Å vacuum layer was introduced in the z-direction to avoid interactions between the top and bottom surfaces of the heterostructure. The DFT-D3 method was employed to include van der Waals (vdW) corrections for accurately describing intermolecular interactions. The plane-wave cutoff energy was set to 50 Ryd (680.3 eV), with energy convergence accuracy and single-atom forces during the structural relaxation process set to 1×10^{-3} eV·Å and 0.02 eV/Å, respectively. For Brillouin zone sampling, Monkhorst-Pack special k-point grids of 4 x 4 x 1 and 7 x 7 x 1 were used for structural optimization and self-consistent calculations. The Hseyd-Scuseria-Ernzerhof (HSE06) hybrid functional was employed to compute the band structures of the h-BN and HfSe2 unit cells. This functional introduces minimal electron self-interaction errors and produces accurate electronic energy levels. To reduce computational costs, the PBE functional was used to compute band structures, partial density of states (PDOS), and band alignment. The optimized structures were visualized using the VESTA software. The stability of the HfSe2/h-BN heterostructure was described using the binding energy (EBE)expressed as follows:

$$E_{BE} = E_{h-BN/HfSe2} - E_{h-BN} - E_{HfSe2} \quad (1)$$

E_{h-BN}, E_{HfSe2} and $E_{h-BN/HfSe2}$ represent the total energies of the h-BN monolayer, the HfSe2 monolayer, and the h-BN/HfSe2 heterostructure, respectively.

III. RESULT AND DISCUSSION

A. Geometric structure and properties

Firstly, in this work, we studied monolayer h-BN and monolayer HfSe2, performing lattice optimization and property calculations. The optimized lattice constants for h-BN and HfSe2 are a0 = 2.504 Å and a1 = 3.763 Å, respectively, which are close to the experimental data of 2.504 Å [5] and 3.75 Å [6]. The optimized B-N and Hf-Se bond lengths are 1.45 Å and 2.67 Å, respectively, as shown in Figure 1a and b, consistent with previous research results. As depicted in Figure.1.c and d, the band gaps of monolayer h-BN and diamond were calculated using the HSE method, with the Fermi level Ef adjusted. The band structures of monolayer h-BN and HfSe2 were calculated using HSE06, with the Fermi level Ef shifted to 0 eV. Both monolayer h-BN and HfSe2 exhibit indirect band gaps, measuring 5.60 eV and 1.36 eV, respectively, which are close to the experimental values. Combining h-BN (3x3x1) with HfSe2 (2x2x1) forms the h-BN/HfSe2 heterostructure, as shown in Figure.1e. The lattice constants for h-BN and HfSe2 are 7.54 Å and 7.52 Å, respectively, with a lattice mismatch of less than 0.3%. The interlayer distance d0 of the HfSe2/h-BN heterostructure is 3.58 Å. Additionally, the negative binding energy indicates that the structure can exist stably. graphicx

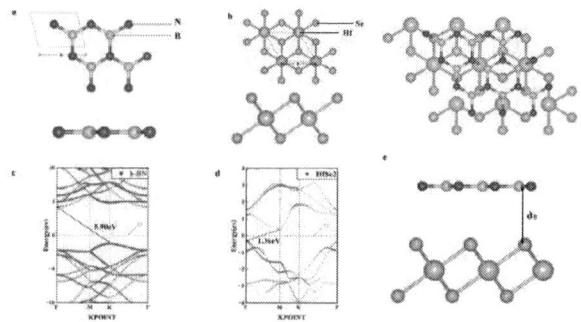

Fig. 1. a and b. Structural diagrams of monolayer h-BN and monolayer HfSe2. c and d. Band structure diagrams of monolayer h-BN and HfSe2. e. Structural diagram of the h-BN/HfSe2 heterostructure.

B. The influence of an external electric field

The band structure of the h-BN/HfSe2 heterostructure is not a simple superposition of the band structures of h-BN and HfSe2. Their band structures are influenced by the van der Waals (vdW) interactions at the h-BN/HfSe2 interface, while the intrinsic properties of h-BN and HfSe2 are well preserved. First, we calculated the projected band structure and density of states of the original h-BN/HfSe2 heterostructure. Due to the interlayer vdW interaction, compared to the monolayers, as shown in Figure 2a, the h-BN/HfSe2 heterostructure exhibits the characteristics of a direct bandgap semiconductor, with both the conduction band minimum (CBM) and valence band maximum (VBM) at the point, and a bandgap of 0.718 eV. It can be seen that the bandgap is of type I. Subsequently, we applied an external vertical electric field to the h-BN/HfSe2 heterostructure, where the direction of E from h-BN to HfSe2 is considered positive. To understand the bandgap transition under a vertical electric field, Figure 2 shows the band structure of the h-BN/HfSe2 heterostructure under selected external vertical electric fields. It can be observed that under electric field modulation from -0.3 to +0.3 V/Å, the bandgap value of the heterostructure is almost unaffected by the electric field and remains around 0.718 eV. This may be because changes in the bandgap of vdW heterostructures under an external electric field are mainly due to the relative shift of quasi-Fermi levels in the isolated monolayers caused by charge transfer between them, while their intrinsic band structures are individually preserved, resulting in minimal influence by the electric field [7]. Figure 3 shows the projected density of states of the h-BN/HfSe2 heterostructure under selected external vertical electric fields. It can be seen that under an external electric field from -0.3 to +0.3 V/Å, the VBM in the h-BN/HfSe2 heterostructure is mainly contributed by Se-4P states, while the CBM is mainly contributed by Hf-4P states. When the external electric field is in the same direction as the internal field of the heterostructure, the VBM and CBM of the h-BN/HfSe2 heterostructure shift upwards. When the external field is in the opposite direction, the VBM and CBM shift downwards, with increased deviation as the field strength increases. The shift of the band edges significantly affects the charge transfer behavior [8], consistent with the projected band structure of the heterostructure.

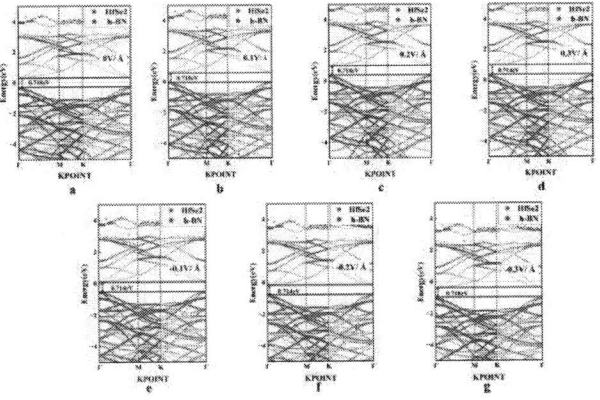

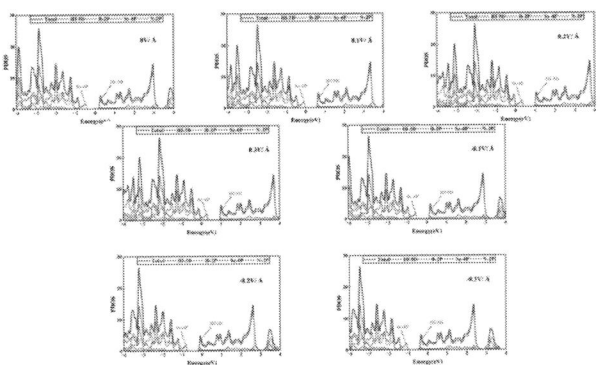

Fig. 2. a. Band structure of the original h-BN/HfSe2 heterostructure. b g. Band structures of the heterostructure under an applied uniform vertical electric field ranging from -0.3 to 0.3 V/Å.

Fig. 3. Density of states diagram of the original h-BN/HfSe2 heterostructure. b g. Density of states diagrams of the heterostructure under an applied uniform vertical electric field ranging from -0.3 to 0.3 V/Å.

C. Band alignment

The applied electric field also has a significant impact on the band structure. We used the local density of states (LDOS) method to calculate the band offsets of each heterojunction system. The formulas used for calculating the conduction band offset (CBO) and valence band offset (VBO) are: CBO = Eg1 - Eg2 - VBO. In this work, Eg1 and Eg2 are set as the band gaps of h-BN and HfSe2, which are 5.90 eV and 1.36 eV, respectively. Figure 4 shows the band alignment of the h-BN/HfSe2 heterostructure under an external vertical electric field ranging from -0.3 to 0.3 V/Å. Under the electric fields within the selected range, the heterostructure maintains a type I band gap, and the band gap remains almost unchanged, consistent with the projected band structure. Under an applied electric field from 0 to +0.3 V/Å, the VBM and CBM of both materials shift linearly upwards, and when the applied electric field is 0.2 V/Å, the VBM of both materials in the heterostructure are higher than the Fermi level, forming an Ohmic contact at the h-BN/HfSe2 interface. Under an

applied electric field from -0.3 to 0 V/Å, the VBM and CBM of both materials shift linearly downwards. Within the applied electric field range of -0.3 to +0.3 V/Å, the band gap of the h-BN/HfSe2 heterostructure is stable, and the CBO and VBO remain stable. Since the VBO of the h-BN/HfSe2 heterostructure is less than 1 eV, there is a risk of leakage, and an additional insulator is needed to ensure the stability of MEMS devices. [9] [10]

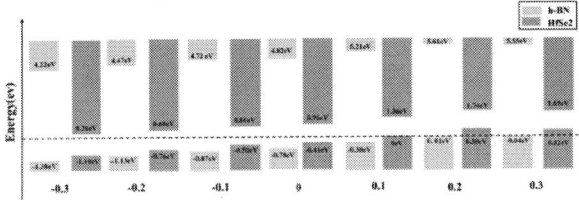

Fig. 4. Band alignment of the heterostructure under an applied uniform vertical electric field ranging from -0.3 to 0.3 V/Å.

IV. CONCLUSIONS

In summary, a DFT-based first-principles method was used to study the modulation of the properties of the h-BN/HfSe2 heterostructure under an applied vertical static electric field. From the perspective of the band structure, the h-BN/HfSe2 heterostructure is a direct bandgap semiconductor. Within the selected range of applied static electric fields, the band edge positions of the h-BN/HfSe2 heterostructure can be affected, but the bandgap size, bandgap type, and band alignment type remain stable. When the applied electric field is +0.2 V/Å, both the VBM and CBM are above the Fermi level, indicating the formation of an Ohmic contact at the interface within the heterostructure. The VBO of the h-BN/HfSe2 heterostructure is less than 1 eV, making it susceptible to leakage current. This study demonstrates that external electric fields within the selected range can modulate the band characteristics of the h-BN/HfSe2 heterostructure and provides guidance for the exploration and application of the h-BN/HfSe2 heterostructure in optoelectronic devices.

V. ACKNOWLEDGEMENT

This work was Supported by National Nature Science Foundation of China (Grant No. 62174148), National Key Research and Development Program (NKRDP Grant No. 2022YFE0112000), Key Program for International Joint Research of Henan Province (Grant No. 231111520300).

REFERENCES

[1] Huynh, Thi My Duyen, et al. "Geometric and electronic properties of monolayer HfX2 (X= S, Se, or Te): a first-principles calculation." Frontiers in Materials 7 (2021): 569756.

[2] Song, Hong-Yue, Jing-Jing Sun, and Meng Li. "Enhancement of monolayer HfSe2 thermoelectric performance by strain engineering: A DFT calculation." Chemical Physics Letters 784 (2021): 139109.

[3] Liu, Lei, Y. P. Feng, and Z. X. Shen. "Structural and electronic properties of h-BN." Physical Review B 68.10 (2003): 104102.

[4] Perdew, John P., Kieron Burke, and Matthias Ernzerhof. "Generalized gradient approximation made simple." Physical review letters 77.18 (1996): 3865.

[5] Morscher, M., et al. "Formation of single layer h-BN on Pd (1 1 1)." Surface science 600.16 (2006): 3280-3284.

[6] Mahajan, Vivek, et al. "Effect of biaxial strain on electronic and optical properties of vertically stacked HfS2/HfSe2 heterostructures." Physica Scripta 99.4 (2024): 045925.

[7] Peng, Zhirong, et al. "A heterostructure of C3N/h-BN with effectively regulated electronic properties by E-field and strain." Chemical Physics Letters 770 (2021): 138461.

[8] Ma, Yaqiang, et al. "Band structure engineering in a MoS 2/PbI 2 van der Waals heterostructure via an external electric field." Physical Chemistry Chemical Physics 18.41 (2016): 28466-28473.

[9] Gao, Xu, et al. "Graphene/GeTe van der Waals heterostructure: functional Schottky device with modulated Schottky barriers via external strain and electric field." Computational Materials Science 170 (2019): 109200.

[10] Gui, Qingzhong, et al. "Theoretical insights into the interface properties of hydrogen-terminated and oxidized silicon-terminated diamond field-effect transistors with h-BeO gate dielectric." IEEE Transactions on Electron Devices (2023).

Electronic properties and energy band alignments of h-BN/TiS$_2$ heterostructure

Cheng Li   Yunhao Liu   Fang Wang   Juin J. Liou   Yuhuai Liu

Abstract—In recent years, regulating of the photoelectric properties of 2D/2D van der Waals (vdW) heterostructures through different means has become a research focus. In this work, the stability, electronic properties and energy band arrangement of the h-BN/TiS$_2$ heterostructure are investigated based on the density-functional theory using first-principle calculations. The results indicate that the heterostructure is stable. Its band structure features an indirect band gap characteristic with a value of 0.876 eV. Moreover, the heterostructure demonstrates strong light absorption properties in the visible and ultraviolet regions. Compared with TiS$_2$, the h-BN/TiS$_2$ heterostructure is able to optimize the energy band structure, resulting in a significant reduction of the band gap. These properties provide a theoretical basis for the application of h-BN/TiS$_2$ heterostructure in the field of optoelectronic devices.

Index Terms—h-BN, electronic properties, VdW heterostructure, first-principles calculations

I. INTRODUCTION

2D materials have shown promising applications in many fields due to their unique physical and chemical properties [1] [2]. Among them, h-BN, as a wide bandgap insulating material with high thermal conductivity and chemical stability, can withstand extreme environmental conditions and has great potential for applications [3]. TiS$_2$, as a member of transition metal disulfides (TMDs), has a bandgap that is adjustable within a certain range, which provides flexibility for designing electronic devices with specific functions. Vertically stacked van der Waals (vdW) heterostructures not only preserve the

---

Cheng Li and Yunhao Liu are with *National Center for International Joint Research of Electronic Materials and Systems, International Joint-Laboratory of Electronic Materials and Systems of Henan Province, School of Electrical and Information Engineering, Zhengzhou University, Zhengzhou, Henan 450001, P. R. China*, Fang Wang and Yuhuai Liu are with *National Center for International Joint Research of Electronic Materials and Systems, International Joint-Laboratory of Electronic Materials and Systems of Henan Province, School of Electrical and Information Engineering, Zhengzhou University, Zhengzhou, Henan 450001, P. R. China, International Joint Laboratory for Integrated Circuits Design and Application, Ministry of Education, School of Physics, Zhengzhou University, Zhengzhou 450052, China,Institute of Intelligence Sensing, Research Institute of Industrial Technology Co. Ltd., Zhengzhou University, Zhengzhou, Henan 450001, P. R. China, Zhengzhou Way Do Electronics Co. Ltd., Zhengzhou, Henan 450001, P. R. China*. Juin J. Liou is with *School of Electrical and Information Engineering, North Minzu University, Yinchuan, Ningxia 750001, P. R. China*. This work is supported by the [National Natural Science Foundation of China] under Grant [No.62174148]; the [National Key Research and Development Program] under Grant [No.2022YFE0112000 and 2016YFE0118400]; the [Key Program for International Joint Research of Henan Province] under Grant [No.231111520300]; the [Ningbo Major Project of 'Science, Technology and Innovation 2025'] under Grant [No.2019B10129]; and the [Zhengzhou 1125 Innovation Project] under Grant [No.ZZ2018-45]. Corresponding author: (Fang Wang, email:iefwang@zzu.edu.cn;Yuhuai Liu, email:ieyhliu@zzu.edu.cn)

intrinsic properties of the component materials, but also enable new functions and performance optimization through the combination of different materials, which greatly broadens the application scope of 2D materials [4].

In recent years, heterostructures consisting of h-BN and TMDs have attracted much attention in the field of 2D materials [5] [6] [7]. However, the electronic properties and energy band arrangement of h-BN/TiS$_2$ heterostructure have yet to be studied in depth.

In this work, we constructed h-BN//TiS$_2$ heterostructure and systematically investigated their geometrical structures, electronic properties and optical properties using first-principle calculations to provide theoretical support for their applications in optoelectronic devices.

II. CALCULATION METHODS

In this work, all first-principles calculations are performed in the PWmat software. The calculations were performed using the SG15 pseudopotential for all atoms. The generalized gradient approximation (GGA)-Perdew-Burke-Ernzerhof (PBE) was used for structural optimization. The Heyd-Scuseria-Ernzerhof (HSE06) was used to calculate the energy bands. Periodic boundary conditions are applied on the structural boundaries along the x- and y-axis directions. A vacuum layer with a thickness of 18 Å is added along the z-axis direction to avoid interactions of neighboring layers. The interlayer van der Waals interactions are described using the DFT-D3 method and dipole corrections are taken into account in the calculation of the electrostatic potential. The plane-wave cutoff energy is set to 50 Ry, and the energy and force convergence criteria for structural optimization are 10^{-5} eV and 0.02 eV/Å. The k-points are generated using the Monkhorst-Pack method to ensure symmetry, and the k-point grids of 3 × 3 × 1 are used in the structural optimization and performance calculations, respectively. The constructed h-BN/TiS$_2$ heterostructure modeling maps and charge density difference maps were obtained with VESTA software.

III. RESULT AND DISCUSSION

A. Geometric structure

Firstly, the electronic structures and properties of the h-BN and TiS$_2$ monolayers were calculated. The lattice constants of h-BN were 2.5124 Å and that of TiS$_2$ was 3.34311 Å. The h-BN/TiS$_2$ heterostructure was constructed by vertically stacking 4 × 4 × 1 h-BN supercells and 3 × 3 × 1 TiS$_2$ supercells, and its structure is schematically shown in Fig. 1. The lattice mismatch is minimal, approximately 0.2%. The binding energy (E_b) is used to assess the structural stability, and its calculation formula is given as:

$$E_b = (E_{\text{h-BN/TiS}_2} - E_{\text{h-BN}} - E_{\text{TiS}_2})/n \qquad (1)$$

where $E_{\text{h-BN/TiS}_2}$ represents the energy of the entire h-BN/TiS$_2$ heterostructure, $E_{\text{h-BN}}$ represents the energy of the h-BN monolayer, E_{TiS_2} represents the energy of the TiS$_2$ monolayer, and n represents the total number of atoms in the heterostructure. The E_b of the h-BN/TiS$_2$ heterostructure is -21.919 meV, proving that the h-BN/TiS$_2$ heterostructure can exist stably. Currently, molecular dynamics is commonly used to observe the structural and thermal stability of two-dimensional materials. Therefore, we conducted 1000 fs AIMD simulations at 300 K on the TiS$_2$ monolayer and the h-BN/TiS$_2$ heterostructure to study their thermal stability, as shown in Fig. 2. The results indicate that the h-BN/TiS$_2$ heterostructure exhibits good thermodynamic stability, with minimal changes in free energy, less than 2.1 eV.

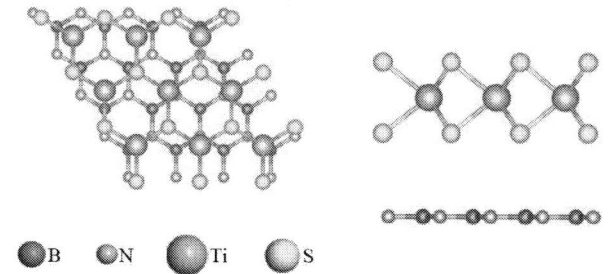

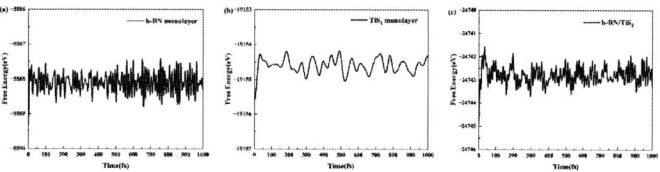

Fig. 1. The top (left) and side (right) views for h-BN//TiS$_2$ heterostructure.

Fig. 2. The curves showing the change in free energy over time during AIMD simulations at 300 K for different structures.

B. Electric properties

Next, we investigated the optoelectronic properties of the h-BN/TiS$_2$ heterostructure. Fig. 3 illustrates the energy band structures of the monolayers and the h-BN/TiS$_2$ heterostructure, the heterostructure exhibits an indirect band gap with the conduction band bottom (CBM) located between the Γ point and the M point, and the valence band top (VBM) located at the K point. The band gap is 0.876 eV. The band gap of the TiS$_2$ monolayer is 1.635 eV, and the introduction of h-BN significantly modulated the band gap of the heterostructure.

The projected density of states (PDOS) of the monolayers and the h-BN/TiS$_2$ is shown in Fig. 4. The CBM is mainly contributed by the Ti-3d states and the VBM is mainly contributed by the N-2p states.

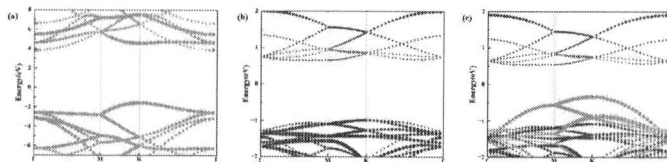

Fig. 3. The energy band structures of the h-BN monolayer (a), the TiS$_2$ monolayer (b) and the h-BN/TiS$_2$ heterostructure (c).

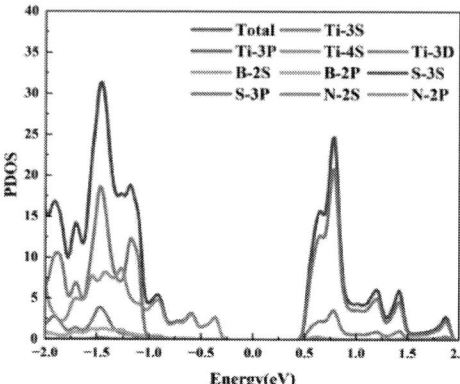

Fig. 4. The top (left) and side (right) views for h-BN//TiS$_2$ heterostructure.

In the subsequent analysis, we examined the charge transfer dynamics inside the h-BN/TiS$_2$ heterostructure. In order to visualize the charge transfer process, we analyzed the difference in charge density along the z-axis of the h-BN/TiS$_2$ heterostructure. The formula for calculating the differential charge density is:

$$\Delta \rho = \rho_{\text{h-BN/TiS}_2} - \rho_{\text{h-BN}} - \rho_{\text{TiS}_2} \qquad (2)$$

Where $\rho_{\text{h-BN/TiS}_2}$, $\rho_{\text{h-BN}}$, and ρ_{TiS_2} represent the charge densities of h-BN/TiS$_2$ heterostructure, h-BN monolayer, and TiS$_2$ monolayer, respectively. The differential charge density diagram is shown in Fig. 5, where the yellow areas represent the accumulation of electrons and the blue areas represent the depletion of electrons. The results show that electrons accumulate near the TiS$_2$ monolayer and are depleted near the h-BN monolayer, with electron transferring from the h-BN monolayer to the TiS$_2$ monolayer. TiS$_2$ forms n-type doping and h-BN forms p-type doping.

Fig. 5. The average charge density difference along the z-axis for h-BN/TiS$_2$ heterostructure. (isosurface: 8.96984×10^{-4} e/Å^3)

Optical properties are one of the most fundamental and important properties of optoelectronic devices, where the optical absorption coefficient $\alpha(\omega)$ can be used to quantitatively

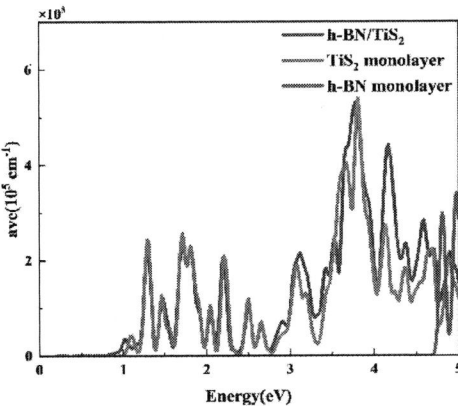

Fig. 6. The absorption spectra of the h-BN monolayer , the TiS_2 monolayer and the h-BN/TiS_2 heterostructure.

[7] Han, T., Liu, H., Wang, S., Chen, S., Yang, K. and Li, Z., 2021. Synthesis and Spectral Characteristics Investigation of the 2D-2D vdWs Heterostructure Materials. International Journal of Molecular Sciences, 22(3), p.1246.

evaluate the ability of a material to absorb light at different wavelengths, as shown in Fig. 6. The profile of the h-BN/TiS_2 heterostructure is very similar to that of the TiS_2 monolayer, showing multiple absorption peaks in the energy range from about 1 eV to 5 eV, with a broad light absorption range. The absorption is stronger in the visible region, followed by the UV region and weaker in the IR region. The shapes and intensities of these absorption peaks are not identical. This suggests that while the TiS_2 composition contributes to the absorption properties in this region, the presence of h-BN in the heterostructure alters the overall electronic structure and thus the light absorption properties.

IV. CONCLUSIONS

In conclusion, through first principles calculations, the geometrical structure and electronic properties of the h-BN/TiS_2 heterostructure were studied.The h-BN/TiS_2 heterostructure has a stable structure, an indirect band gap of 0.876 eV. The heterostructure exhibits strong light absorption in the visible and ultraviolet regions.These results provide a theoretical basis for its application in the fields of photovoltaic devices and solar cells.

REFERENCES

[1] Zhou, J., Wang, C., Zhang, X., Jiang, L. and Wu, R., 2024. Advances in two-dimensional layered materials for gas sensing. Materials Science and Engineering: R: Reports, 161, p.100872.
[2] Jaiswal, H.N., 2022. Exploring 2D Materials for Energy-Efficient Nanoelectronic Devices (Doctoral dissertation, State University of New York at Buffalo).
[3] An, L., Yu, Y., Cai, Q., Mateti, S., Li, L.H. and Chen, Y.I., 2023. Hexagonal boron nitride nanosheets: preparation, heat transport property and application as thermally conductive fillers. Progress in Materials Science, 138, p.101154.
[4] Qiu, H., Yu, Z., Zhao, T., Zhang, Q., Xu, M., Li, P., Li, T., Bao, W., Chai, Y., Chen, S. and Chen, Y., 2024. Two-dimensional materials for future information technology: status and prospects. Science China Information Sciences, 67(6), pp.1-147.
[5] Ma, L., Wang, Y. and Liu, Y., 2024. van der Waals Contact for Two-Dimensional Transition Metal Dichalcogenides. Chemical Reviews, 124(5), pp.2583-2616.
[6] Han, T., Liu, H., Chen, S., Chen, Y., Wang, S. and Li, Z., 2020. Fabrication and Characterization of MoS2/h-BN and WS2/h-BN Heterostructures. Micromachines, 11(12), p.1114.

1200V SOI N-channel LDMOS Adopting Partial Separation by Implantation of Oxygen

Mingxin Sun   Chengwu Pan   Shipeng Chang   Siyang Liu   Weifeng Sun   Long Zhang*

Abstract—The conventional silicon on insulator (SOI) based lateral diffused metal oxide semiconductor (LDMOS) is difficult to achieve operation voltage of 1200V because of the limitation of its vertical breakdown voltage (*BV*). In this work, a deep N-well (DN) in the substrate is used to establish the vertical voltage, and the BOX is obtained by partial separation by implantation of oxygen (SIMOX), which prevents the top silicon and the BOX from vertical voltage. The off-state *BV* of the proposed SOI N-channel LDMOS is 1600V and the on-state *BV* at V_{GS}=15V is 1440V. Moreover, the thickness of BOX is 0.4μm.

Index Terms—SOI, LDMOS, SIMOX, breakdown voltage

I. INTRODUCTION

High-voltage integrated circuits (HVICs) are extensively employed as gate drivers of insulated gate bipolar transistors (IGBTs) and SiC metal-oxide-semiconductor field effect transistors (MOSFETs)[1]-[4]. With the development of SiC MOSFETs, HVICs require enhanced robustness. Compared to bulk silicon devices, silicon on insulator (SOI) devices demonstrate superiority in enhanced latch-up immunity, suppressed inter-device crosstalk, and improved integration capability, making them well-suited for driving high-voltage SiC power devices. However, the buried oxide (BOX) disrupts the electrical connection between the top silicon and the substrate, preventing mutual depletion. Consequently, the substrate fails to establish vertical voltage, resulting in conventional SOI lateral diffused metal oxide semiconductor (LDMOS) exhibiting limited operation voltage around 600V, which proves insufficient to realize the 1200V level shift function. To solve this, Toyota developed a 2000V SOI LDMOS in [5] by adopting a 12.2μm BOX. However, the thick BOX layer introduced challenges in heat dissipation and manufacturing process complexity. Other methods based on the enhanced dielectric layer field (ENDIF) theory in [6]-[8] enhanced the *BV* but currently no LDMOS with a 1200V operation voltage has been developed.

In this paper, the BOX is formed by partial separation by implantation of oxygen (SIMOX), enabling the deep N-well (DN) in the substrate to establish vertical voltage. Moreover, the deep trench isolation (DTI), combined with the BOX, is formed to achieve complete isolation. This results in a higher *BV* compared to conventional SOI LDMOS and superior isolation characteristics compared to bulk silicon LDMOS.

Mingxin Sun, Chengwu Pan, Shipeng Chang, Siyang Liu and Weifeng Sun are with *National ASIC System Engineering Research Center, Southeast University*, Nanjing, China. This work is supported by the National Key R&D Program of China (2023YFB4403700) and the Natural Science Foundation of Jiangsu Province (BK20231150, BK20232006, BK20241292). Corresponding author: (Long Zhang, email: longzh@seu.edu.cn)

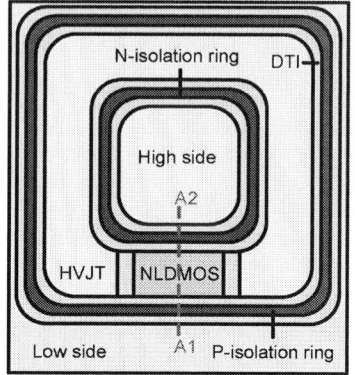

(a)

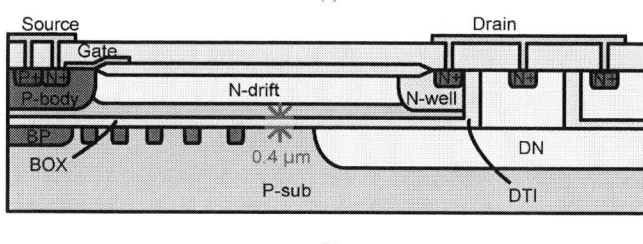

(b)

Fig. 1. (a) Schematic top view of the isolation structure of the HVIC with the proposed SOI NLDMOS. (b) Cross-sectional schematic diagram of the proposed SOI NLDMOS along A1-A2 indicated in Fig. 1(a).

II. STRUCTURE AND PROCESS

Fig. 1(a) shows the schematic top view of the isolation structure of the HVIC with the proposed 1200V SOI NLDMOS. The isolation structure is composed of an N-isolation ring, DTI, a high-voltage junction termination (HVJT), an NLDMOS and a P-isolation ring. The source of the NLDMOS is connected to the low-side region, while the drain is connected to the high-side region, serving as a level-up shifter.

Fig. 1(b) presents a cross-sectional schematic diagram of the proposed 1200V SOI NLDMOS along the A1-A2 line (indicated by the dashed line in Fig. 1(a)). The substrate with a DN and a P-type buried layer (BP) is used. The thickness of the BOX is 0.4μm and the proposed structure features an interrupted oxide at the drain side, allowing the depletion region to spread into the substrate through the silicon window left open. The DTI on either side of the silicon window achieves isolation by connecting to the BOX.

Fig. 2(a) shows the schematic process of partial SIMOX.

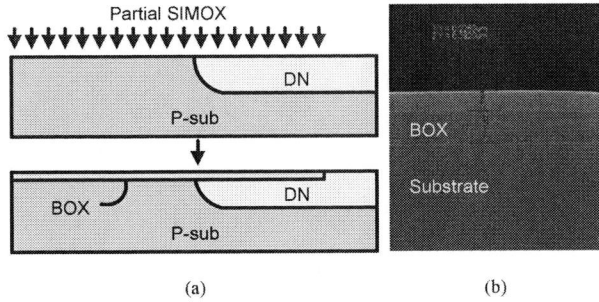

(a)   (b)

Fig. 2. (a) Schematic process of partial SIMOX of the proposed SOI NLDMOS. (b) SEM photo of the BOX in the proposed SOI NLDMOS.

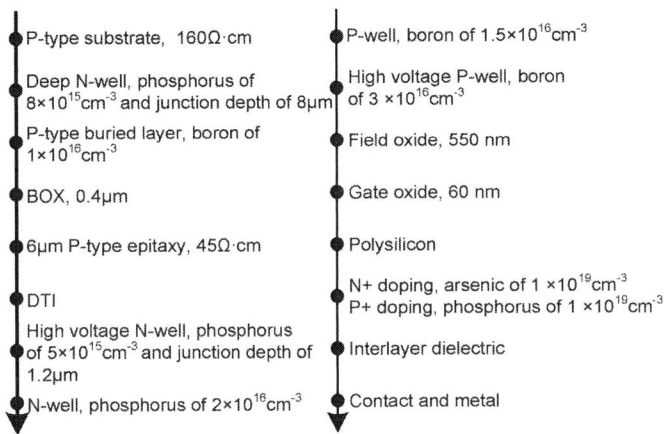

Fig. 3. Key processes and parameters of the proposed SOI NLDMOS.

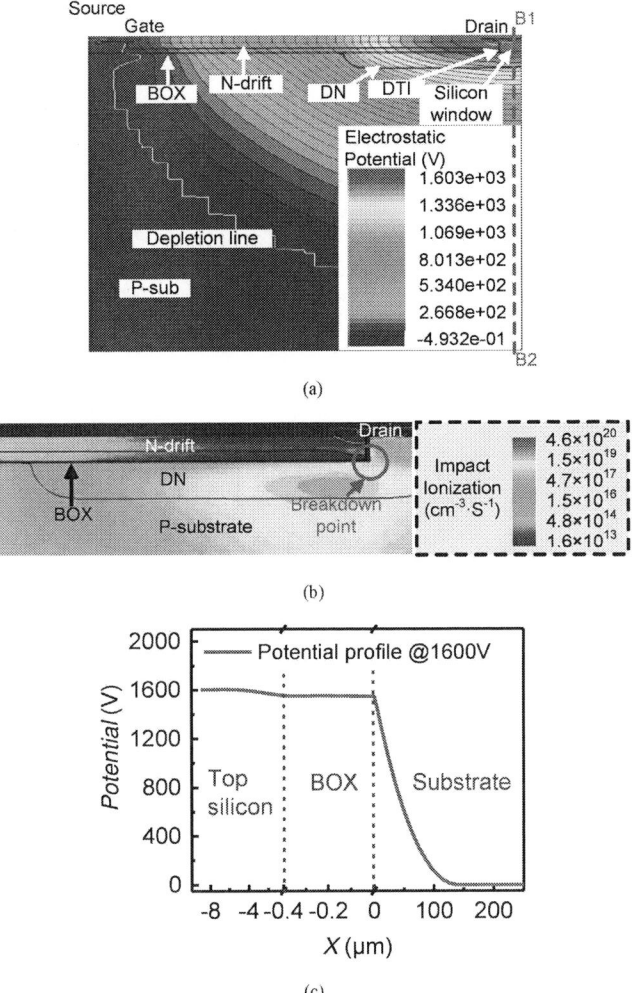

(a)

(b)

(c)

Fig. 4. Simulated off-state (a) equipotential contours, (b) impact ionization rate distribution and (c) electrostatic potential profiles for the proposed SOI NLDMOS along the line B1-B2 indicated in Fig. 4(a) near breakdown.

The BP and DN are formed in the substrate through ion implantation, and then in the specific region of the substrate, the BOX is formed by partial SIMOX. Fig. 2(b) shows the SEM photo of the BOX formed by partial SIMOX. The thickness of the BOX is 0.4µm and 0.2µm-thick silicon is left on the surface of the substrate. After that, a 6µm-thick P-type epitaxy is grown on the substrate.

Fig. 3. shows the key processes and parameters of the proposed 1200V SOI NLDMOS. The DTI is formed after the epitaxial growth to provide electrical isolation.

III. RESULTS AND DISCUSSION

Fig. 4. shows the potential and impact ionization distributions of the proposed 1200V SOI NLDMOS near breakdown at off state. Fig. 4(a) shows the off-state equipotential contours for the proposed structure. It can be clearly observed that they are almost uniform across the entire structure, including the substrate. And the depletion line (the white line in the picture) is located deep within the substrate, which implies that the silicon window formed by partial SIMOX enables the substrate to establish the vertical voltage. In this case, the crowding of the potential lines in the top silicon at the drain side is avoided, resulting in a near-ideal breakdown of the device. The off-state impact ionization rate distribution depicted in Fig. 4(b) shows that after full depletion, the device breaks down along the DTI

sidewall at its corner and inside the DN. Fig. 4(c) illustrates the off-state potential distribution along the line B1-B2 indicated in Fig. 4(a). As depicted, the vertical voltage of the proposed structure is primarily withstood by the substrate while there is almost no potential drop in the top silicon and the BOX along B1-B2.

The I-V characteristic curves of the proposed structure I-V are shown in Fig. 5. Since the DN is connected to the drain of the proposed SOI NLDMOS, both the current of the drain and the substrate should be focused in consideration of the BV. The breakdown point of the proposed structure is located in the substrate after optimization. The off-state BV of the proposed SOI NLDMOS is 1600V and the on-state BV at V_{GS}=5V, 10V, 15V and 20V is 1588V, 1525V, 1440V and 1322V respectively.

IV. CONCLUSION

In this work, a new 1200V SOI NLDMOS is proposed by partial SIMOX. The BV of the proposed SOI NLDMOS is

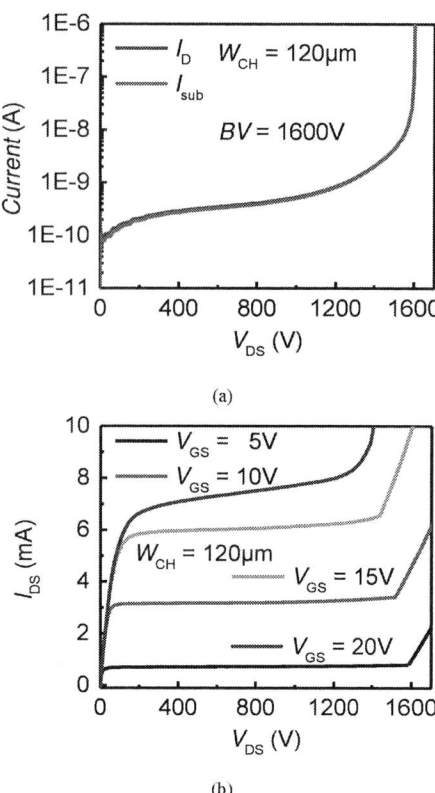

(a)

(b)

Fig. 5. Simulated (a) off-state *I-V* characteristic curve graph of the proposed SOI NLDMOS. (b) *I-V* characteristic curve graph of the proposed SOI NLDMOS at different V_{GS}.

1600V at off state and 1440V at V_{GS}=15V. Moreover, the thickness the BOX is only 0.4μm as it withstands no voltage, favorable for heat dissipation.

REFERENCES

[1] J. Zhu et al., "Noise Immunity and its Temperature Characteristics Study of the Capacitive-Loaded Level Shift Circuit for High Voltage Gate Drive IC," in IEEE Trans. Ind. Electron., vol. 65, no. 4, pp. 3027–3034, April 2018.

[2] L. Zhang et al., "An Isolation Structure Applying Potential Control Technique for 1200 V HVICs," in IEEE Trans. Electron Devices, vol. 71, no. 1, pp. 927–930, Jan. 2024.

[3] A. E. Khorasani and M. Griswold, "A Self-Protected 800V JFET with 6kV HBM robustness in 0.25 um BCD," 2020 32nd International Symposium on Power Semiconductor Devices and ICs (ISPSD), Vienna, Austria, 2020, pp. 412–414.

[4] Z. Yuan et al., "A Bootstrap Diode Emulator Integration to 600 V N-Type Epitaxial Platform for High Voltage Gate Driver IC," in IEEE Electron Device Letters, vol. 43, no. 11, pp. 1941–1944, Nov. 2022.

[5] T. Okawa, H. Eguchi, M. Taki and K. Hamada, "2000 V SOI LDMOS with new drift structure for HVICs," 2016 28th International Symposium on Power Semiconductor Devices and ICs (ISPSD), Prague, Czech Republic, 2016, pp. 435–438.

[6] X. Luo et al., "A High-Voltage LDMOS Compatible With High-Voltage Integrated Circuits on p-Type SOI Layer," in IEEE Electron Device Letters, vol. 30, no. 10, pp. 1093–1095, Oct. 2009.

[7] W. Zhang et al., "Novel Superjunction LDMOS (¿950 V) With a Thin Layer SOI," in IEEE Electron Device Letters, vol. 38, no. 11, pp. 1555–1558, Nov. 2017.

[8] W. Zhang et al., "A Novel High Voltage Ultra-Thin SOI-LDMOS With Sectional Linearly Doped Drift Region," in IEEE Electron Device Letters, vol. 40, no. 7, pp. 1151–1154, July 2019.

979-8-3315-2209-4/25 $31.00 © 2025 IEEE

Influence of the Charge-Imbalance Condition on the Short-Circuit Characteristics of the 4H-SiC Superjunction MOSFET

Huan Ning   Da Wang   Zhi Lin

Abstract—This paper conducts a study on the dynamic short-circuit characteristics of Silicon Carbide (SiC) Superjunction Metal-Oxide-Semiconductor Field Effect Transistors (SJ MOS-FETs) under charge imbalance conditions. By using Sentaurus TCAD simulation to compare the 1200 V traditional SiC MOSFET with the SJ device, it is found that the SJ structure significantly reduces the specific on-resistance ($R_{on,sp}$), while maintaining a breakdown voltage of 2000 V by increasing the epitaxial layer concentration and introducing the N/P pillar charge balance design. The simulation shows the key influence of the charge imbalance on the withstand time of the short circuit (T_{SC}). Specifically, the charge imbalance changes from -50% to 50%, and the difference in T_{SC} is 1.2 µs.

Index Terms—sic superjunction, charge imbalance, short circuit withstand time (T_{SC})

I. Introduction

With the increasing demand for high-efficiency and high-frequency operation in power electronic systems, Silicon Carbide, as a wide bandgap semiconductor material, has emerged as an ideal choice due to its exceptional physical and chemical properties [1]. Among these, the SiC SJ MOSFET has shown significant application potential in high frequency, high power converters, owing to its notable advantages such as low on-resistance and high breakdown voltage [2].

Some studies have explored the excellent short-circuit characteristics of SJ MOSFETs under charge-balanced conditions [3], as well as the static characteristics under imbalanced conditions [4]. However, there is relatively little discussion on the dynamic short-circuit characteristics under imbalanced conditions. In the manufacturing process, it is difficult to control the activation efficiency of p-type dopants, resulting in charge imbalance [5]. When the SiC SJ MOSFET is in a charge-imbalanced condition, the short-circuit current is likely to increase abnormally. This may cause a sharp rise in local temperature, thereby triggering the thermal failure of the device.

Therefore, this work explores the influence of the charge-imbalance condition of the superjunction structure, on its short-circuit characteristics and the magnitude of T_{SC}.

Huan Ning, Da Wang and Zhi Lin are with *dept. School of Microelectronics and Communication Engineering , Chongqing University*, Chongqing, China. This work is supported by the Natural Science Foundation of Chongqing, China, under Grant CSTB2022NSCQ-MSX1532. Corresponding author: (Zhi Lin, email: linzhi@cqu.edu.cn)

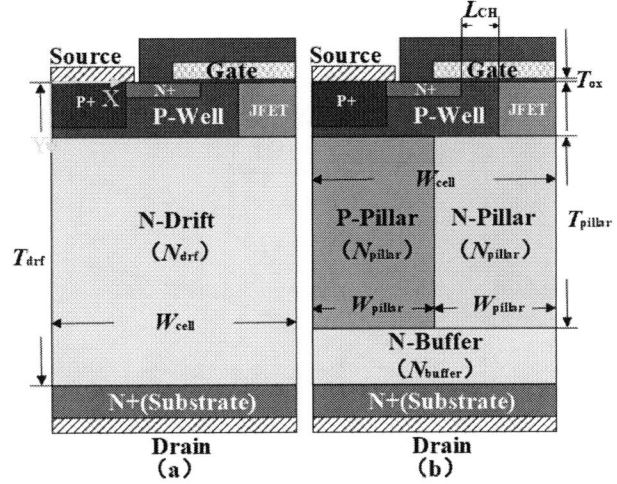

Fig. 1. Schematic cross-section of the half of (a) SiC VDMOS and (b) SiC SJ VDMOS.

TABLE I
MAJOR STRUCTURAL PARAMETERS OF TWO MOSFETs.

Parameters	Value	Unit
Half cell pitch (W_{cell})	2.4	µm
Half N/P-pillar width (W_{pillar})	1.2	µm
N-drift depth (T_{drf})	11	µm
N-drift concentration (N_{drf})	1×10^{16}	cm^{-3}
N-buffer doping (N_{buffer})	1.6×10^{16}	cm^{-3}
Gate oxide thickness (T_{ox})	42	nm
Channel length (L_{CH})	0.6	µm
N/P-pillar concentration (N_{pillar})	3.1×10^{16}	cm^{-3}
N/P-pillar depth (T_{pillar})	9	µm

II. Structure And Principle Of Operation

We conducted simulations of the 1200 V SiC MOSFET and SiC SJ MOSFET structures using Sentaurus TCAD. The cross-sectional views of the devices are shown in Fig. 1. The major parameters of the MOSFETs are shown in Table. I.

Both devices have a cell pitch of 4.8 µm and a gate oxide thickness of 42 nm. Furthermore, a JFET region with

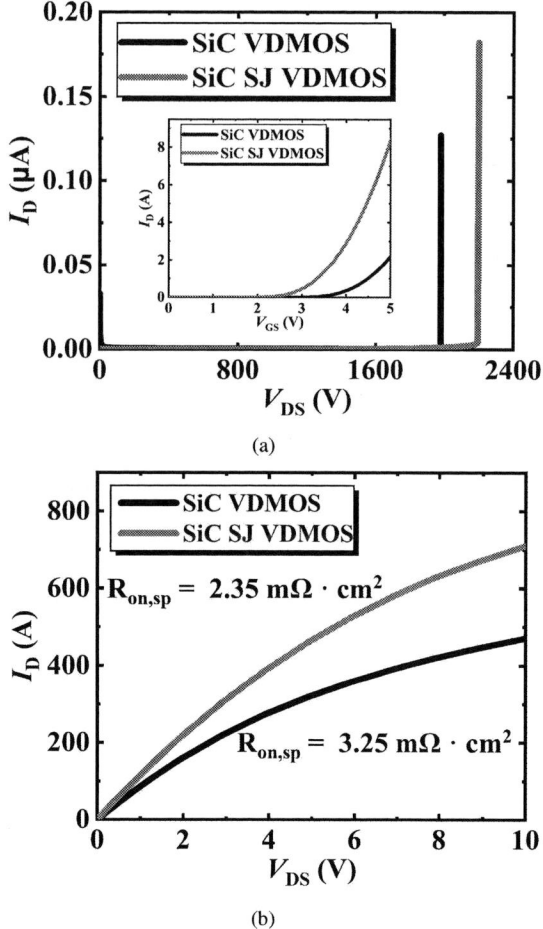

(a)

(b)

Fig. 2. (a) The blocking and transfer characteristics of the two MOSFETs. (b) The output characteristic of the two MOSFETs.

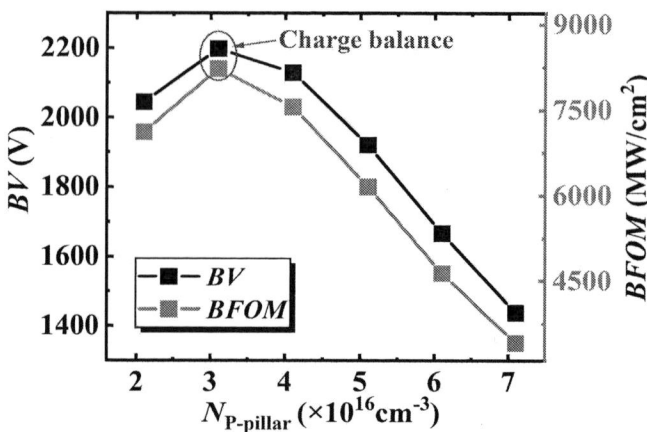

Fig. 3. Comparisons of BV and BFOM for SJ MOSFETs with p - pillar doping concentrations of 2.1×10^{16} cm^{-3}, 3.1×10^{16} cm^{-3}, 4.1×10^{16} cm^{-3}, 5.1×10^{16} cm^{-3}, 6.1×10^{16} cm^{-3} and 7.1×10^{16} cm^{-3}

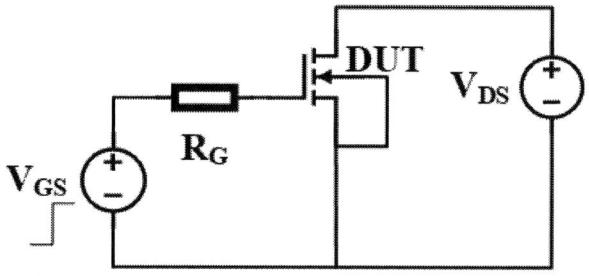

Fig. 4. Simulation circuit layout of the short-circuit safe operating area.

a doping concentration of 8×10^{16} cm^{-3} was incorporated into both MOSFET structures to optimize the forward conduction characteristics. The n-type epitaxial layer of the conventional SiC MOSFET has a doping concentration of 1×10^{16} cm^{-3} and a thickness of 11 μm. The doping concentration of the N-buffer layer in the SJ MOSFET can be increased to 1.6×10^{16} cm^{-3}. N and P pillars are introduced in the drift region, with a balanced doping concentration of 3.1×10^{16} cm^{-3}. Due to the complete depletion of the drift region, the breakdown voltage of the device is significantly enhanced, and the breakdown voltage is increased to over 2000 V, as shown in Fig. 2(a). Meanwhile, the introduced superjunction structure increases the concentration of voltage-withstand layer and reduces the $R_{on,sp}$, as depicted in Fig. 2(b).

Fig. 3 shows that when the charge is balanced, the BV of the SJ MOSFET and the Baliga's figure-of-merit (BFOM) reach their peak states. Both curves show a trend of first rising and then decreasing. When the doping concentration increases from 2.1×10^{16} cm^{-3} to 7.1×10^{16} cm^{-3}, the variation gradient of the breakdown voltage is approximately 200 V / 10^{16} cm^{-3}. The trend of variation in the power figure of merit curve is basically consistent with that of the breakdown voltage curve,

which is determined by the calculation formula of the power figure of merit [6].

III. OPTIMIZATION AND SIMULATED RESULTS

The test circuit is shown in Fig. 4. In order to analyze the short-circuit capability of the devices, a temperature of T=300 K, a DC bus voltage of V_{DS}=800 V, a gate voltage of V_{GS}= +18 V / 0 V and a gate resistance of R_G=1 Ω are applied.

Fig. 5 shows that at 1 μs, V_{GS} rises from 0 V to 18 V, the channel is formed, the MOSFET turns on, and the drain current continues to increase in the initial short period. Due to self-heating during short circuit, the temperature of the SiC MOSFET also keeps rising. As the junction temperature continues to increase, other scattering mechanisms (such as phonon scattering, etc.) intensify [7]. The mobility begins to decrease, causing the current to start decreasing after reaching its peak value. It can be seen that the short-circuit saturation current of the SiC superjunction MOSFET is not much different from that of the traditional SiC MOSFET. However, the short-circuit time when lattice temperature reaches the same value increases.

Fig. 6 shows the internal temperatures of different devices and temperature curves at X=2.4 μm. In the Fig. 6(a)-Fig. 6(d), Under the condition of charge imbalance, the more of the concentration of the P pillar, the farther the hot spot is from the surface and the channel. In the Fig. 6(e), On the one hand, the

979-8-3315-2209-4/25 $31.00 © 2025 IEEE

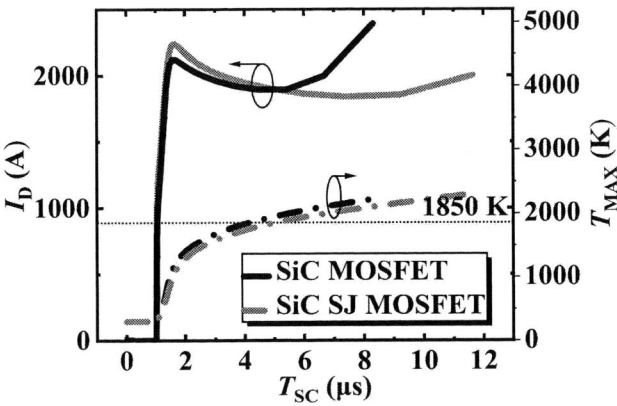

Fig. 5. Drain currents of SiC MOSFET and balanced SiC SJ MOSFET at V_{DS}=800 V, V_{GS}=18 V.

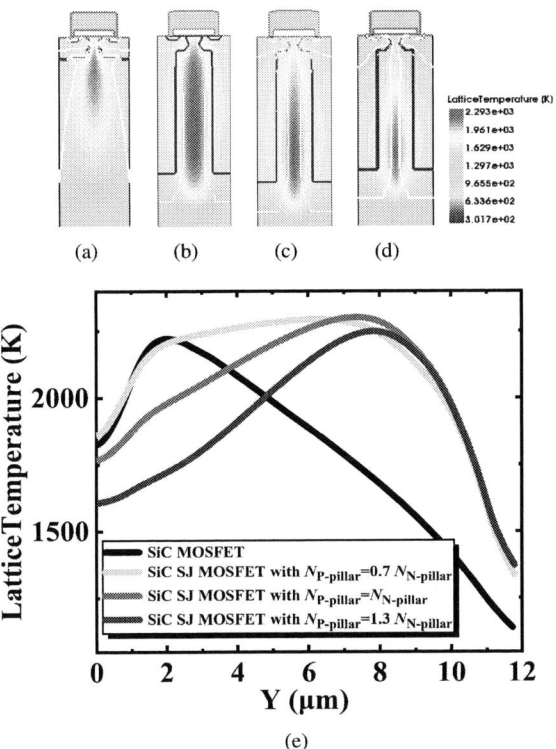

(a)   (b)   (c)   (d)

(e)

Fig. 6. Temperature distribution of (a) SiC MOSFET and (b) SiC SJ MOSFET with the $N_{P\text{-}pillar}$=0.7$N_{N\text{-}pillar}$ (c) $N_{P\text{-}pillar}$=$N_{N\text{-}pillar}$ (d) $N_{P\text{-}pillar}$=1.3 $N_{N\text{-}pillar}$ (e)Temperature distribution of SiC MOSFET and SiC SJ MOSFETs at X=2.4 μm under the break criteria lattice temperature at 2200 K.

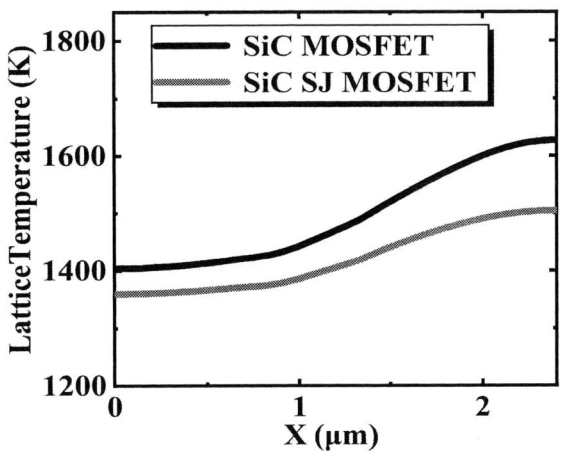

Fig. 7. Temperature distribution of SiC MOSFET and SiC SJ MOSFET at Y=0 μm under the break criteria lattice temperature at 1850 K.

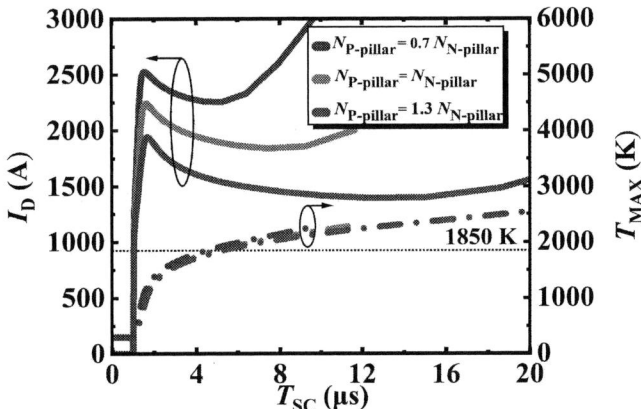

Fig. 8. The drain current and the T_{MAX} under different P-pillar concentrations and the break criteria lattice temperature at 2200 K.

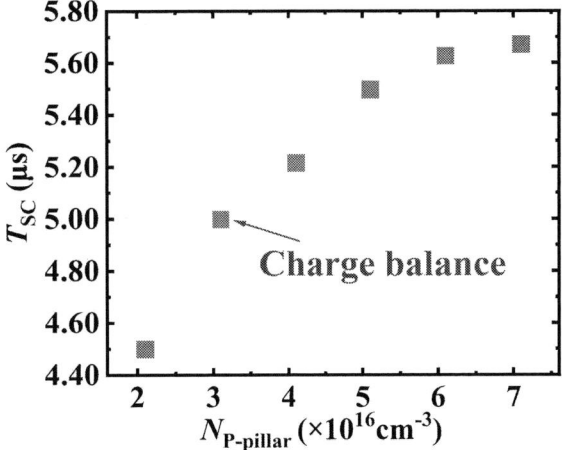

Fig. 9. The T_{SC} of different P-pillar concentrations under the break criteria lattice temperature at 1850 K.

position of the maximum temperature of the SiC SJ MOSFET is farther away from the surface, which can effectively avoid the melting of the metal material of the contact electrode and thermal failure caused by excessively, on the other hand, SiC SJ has a more uniform temperature distribution.

Typically, the near-surface region of SiC MOSFET is difficult to withstand an instantaneous high temperature above 1300-1500 K [8], [9]. During the simulation, we find that when the threshold temperature reaches 1850 K, the maximum temperature near the surface and channel of the charge-balanced SiC SJ MOSFET was 1500 K, as shown in the Fig. 7.

As shown in Fig. 8, compared to the charge-balanced state, when the concentration of the P-pillar increases, the temperature reaches 1850 K more slowly than in the balanced state. In contrast, when the concentration of the P pillar is decreased, the situation is just the opposite and T_{SC} becomes shorter. In these two charge imbalance conditions, the device with lower P-pillar concentration is more prone to overheating failure and a sharp increase in current.

Lastly, we extract the T_{SC} of different concentrations of P pillars under the lattice temperature of the break criterion at 1850 K. Fig. 9 shows that when the concentration of P-Pillar is continuously increased, T_{SC} rises simultaneously, but the upward trend is gradually slowing down. Specifically, when the concentration of P-Pillar increases from 3.1×10^{16} cm^{-3} to 7.1×10^{16} cm^{-3}, T_{SC} increases by 0.68 μs, when the concentration decreases from 3.1×10^{16} cm^{-3} to 2.1×10^{16} cm^{-3}, T_{SC} decreases by 0.5 μs. The charge imbalance changes from -50% to 50%, and the difference in T_{SC} is 1.2 μs.

IV. CONCLUSIONS

In this paper, Sentaurus TCAD simulation is used to compare the structures and short-circuit characteristics of traditional SiC MOSFETs and SiC superjunction MOSFETs. The result shows that, by making the epitaxial layer more concentrated and adding an N/P pillar charge balance, the SJ MOSFET design lowers resistance a lot while keeping a high breakdown voltage of 2000 V. The simulation also shows that the charge imbalance has a very significant impact on T_{SC}. When the concentration of the P-pillar is increased, the hot spot moves away from the surface and T_{SC} becomes longer. While reducing the P-pillar concentration accelerates the process of the temperature reaching the failure threshold. As the concentration of the P pillar decreases, T_{SC} becomes shorter. This means that adjusting the P-pillar concentration can make SiC SJ MOSFETs better at handling short circuits and more reliable. This research gives useful ideas for improving SiC SJ MOSFETs in high-frequency, high-power devices, helping to build stronger power systems.

V. ACKNOWLEDGMENT

This work was supported by the Natural Science Foundation of Chongqing, China, under Grant CSTB2022NSCQ-MSX1532.

REFERENCES

[1] Q. Wang, H. Hua, L. Zheng, J. Feng, C. Zhang, M. Gao, K Qiu, J Luo, X. H. Cheng, "Simultaneous improvement of high-frequency and Baliga figures of merit of 1.7 kV 4H-SiC MOSFET with retrograded JFET doping," Journal of Physics D: Applied Physics, vol. 57, no. 25, pp. 255105, 2024.

[2] S. H. Liu, M. M. Huang, M. J. Wang, M. Zhang, J. Wei, "Considerations for SiC Super Junction MOSFET: On-Resistance, Gate Structure, and Oxide Shield," Microelectronics Journal, vol. 137, pp. 105823, 2023.

[3] M. Okada, S. Kyogoku, T. Kumazawa, J. Saito, T. Morimoto, M. Takei, and S. Harada, "Superior Short-Circuit Performance of SiC Superjunction MOSFET," in 2020 32nd International Symposium on Power Semiconductor Devices and ICs (ISPSD), Vienna, Austria, 2020, pp. 70-73.

[4] Y. C. Lee, K. H. Lin, and K. Y. Lee, "The impact of the P-pillar structure design on Breakdown Voltage for 1.2kV 4H-SiC Superjunction DMOSFET," in 2021 IEEE International Future Energy Electronics Conference (IFEEC), Taipei, Taiwan, 16-19 Nov. 2021, pp. 1-4.

[5] K. Akshay and S. Karmalkar, "Quick Design of a Superjunction Considering Charge Imbalance Due to Process Variations," IEEE Transactions on Electron Devices, vol. 67, no. 8, pp. 3024-3029, 2020.

[6] B. J. Baliga, "Silicon Carbide Power Devices: Progress and Future Outlook," IEEE Journal of Emerging and Selected Topics in Power Electronics, vol. 11, no. 3, pp. 2400-2411, 2023

[7] M. Okawa, R. Aiba, T. Kanamori, Y. Kobayashi, S. Harada, H. Yang, N. Iwamuro, "First Demonstration of Short-Circuit Capability for a 1.2 kV SiC SWITCH-MOS," IEEE Journal of the Electron Devices Society, vol. 7, pp. 613-620, 2019.

[8] C. H. Yu, Y. Wang, M. T. Bao, X. J. Li, J. Q. Yang, and Z. H. Tang, "Simulation Study on Single-Event Burnout in Rated 1.2-kV 4H-SiC Super-Junction VDMOS," IEEE Transactions on Electron Devices, vol. 68, no. 10, pp. 5034-5040, 2021.

[9] M. Okawa, R. Aiba, T. Kanamori, Y. Kobayashi, S. Harada, H. Yang, N. Iwamuro, "First Demonstration of Short-Circuit Capability for a 1.2 kV SiC SWITCH-MOS," IEEE Journal of the Electron Devices Society, vol. 7, pp. 613-620, 2019.

Grounded Isolation Trenches in GaN-on-Si Power Integrated Circuits: An Electromagnetic Study for Trench Filling Considerations

Rui (Ray) Yao   Zijin Jiang   Miao Cui   Zhao Wang   Sang Lam   Stephen Taylor

Abstract—We report further work on isolation trenches for suppressing substrate coupling in GaN-on-Si power integrated circuits (ICs). Our computational electromagnetic (EM) investigation reveals that a grounded trench filled with highly-doped polysilicon or low-conductivity aluminium by low-temperature deposition is also effective for signal isolation (with $|S_{21}|$ close to -53 dB at 100 MHz for a 700-μm lateral separation distance). The results imply fabrication flexibility in the choices of filling materials and processing conditions for isolation trenches that are placed in the middle of the lateral distance.

Index Terms—*isolation trench, substrate coupling, trench filling material, GaN-on-Si technology, computational EM*

I. INTRODUCTION

Gallium nitride (GaN) has emerged as a promising semiconductor for making high-performance power electronic devices and circuits. In particular, GaN-on-Si technology is attractive for the development of power integrated circuits (ICs). This is due to various advantages associated with the silicon (Si) wafer base: the favourable cost of large wafer sizes, good thermal conductivity of Si, potential integration of optoelectronic devices etc. However, substrate coupling poses one big challenge in GaN-on-Si power ICs [1]-[5] that need reliable performance, operating especially at high frequencies beyond multi-megahertz (MHz). To suppress substrate coupling, trench isolation structures [2]-[4] are the common approach as investigated by researchers in the field.

Previously, we have reported our computational electromagnetic (EM) investigation into the substrate coupling mechanism in GaN-on-Si technology [6]-[7] and some varied designs of isolation trenches [8]. In this work, we continue the investigation into the isolation trenches with the manufacturing considerations of feasible trench-filling materials. This is based on our previous findings that the grounded trenches filled with aluminium (Al) help suppress more the substrate coupling. Apart from using Al metal with very high conductivity, polysilicon (polySi) with high doping as well as Al material formed by low-temperature deposition [9] can be alternative conducting materials for the grounded trenches. EM fields are computed in the grounded trenches filled with these conducting materials and S-parameter results are obtained to throw light on the suitable choice of filling materials. This would help save extensive resources for fabricating samples of different trench materials.

II. GROUNDED ISOLATION TRENCH IN GAN-ON-SI TECHNOLOGY

Fig. 1 shows the device structure constructed for solving the EM fields and then the S-parameter results to find out the substrate coupling. The semiconductor structure is comprised of four layers: a metallic base at the bottom for connecting the IC chip to the ground, and a 100-μm thick *p*-type silicon (Si) layer and a GaN buffer layer are sequentially built above, following with 100-nm thick *n*-doped-AlGaN layer at the top. An isolation trench, which is filled with a conducting material, is positioned in the middle between two protruded blocks at both ends. Each block has the same layout of 2 μm by 20 μm.

In fabrication, the trench can be formed first by etching through the GaN buffer layer (2 μm in this case), as a quite straightforward processing step with the dry etch rate of up to 1.3 μm/min [10]. Then the trench can be filled with polySi by chemical vapour deposition (CVD), or with low-conductivity (σ) Al material by low-temperature deposition [9]. Evaporation can be used for depositing normal Al metal with high conductivity. In the case of filling the trench with polySi, it needs an additional process step of doping by such as ion-implantation. With a doping concentration of 10^{19} /cm³, the polySi would become conducting with a reasonable electrical conductivity σ ≈ 1000 S/m [11]. With any of the three conducting materials, a grounded trench would be formed simply by electrical connection to ground.

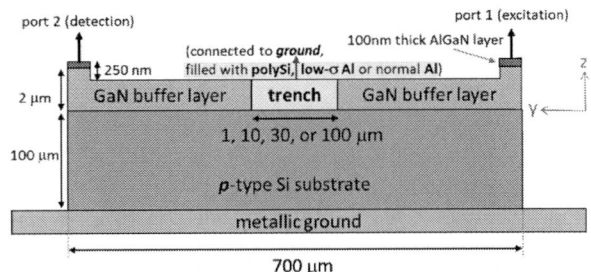

Fig. 1. A schematic cross-sectional diagram showing the GaN-on-Si device structure for computational EM investigation into the isolation trench for suppressing substrate coupling in GaN-on-Si power ICs, with the features not drawn in exact scale for better visualisation.

III. 3D EM SIMULATION & S-PARAMETER RESULTS

We keep using the commercial full-wave EM simulation software program, Ansys HFSS in our computational EM investigation. The electrical properties of the materials were manually set up in the EM simulation of the device structure (Fig. 1), as summarised in the following: ε_r = 8.9 and σ = 200 S/m for the GaN buffer layer; ε_r = 9.2 and σ = 5 × 10^3 S/m for the very thin *n*-doped-AlGaN layer; ε_r = 11.9 and σ = 2 × 10^3 S/m for the *p*-type Si substrate. As for the trench-filling materials, ε_r = 11.9 and σ = 1000 S/m for polySi; ε_r = 1.0 and σ = 1.25 × 10^5 S/m for Al material formed by low-temperature deposition [9]; ε_r = 1.0 and σ = 3.5 × 10^7 S/m for the normal Al metal. Other settings are the same as in our previous work [8] where more details can be found.

R.R. Yao, Z. Jiang, M. Cui, Z. Wang, S. Lam are with *School of Advanced Technology, Xi'an Jiaotong-Liverpool University*, Suzhou, China. S. Taylor, as well as R.R. Yao, is with *Department of Electrical Engineering and Electronics, The University of Liverpool*, Liverpool, UK. This work is supported by PGRS funding of XJTLU. Corresponding author: (S. Lam, email: s.lam.cn@ieee.org)

Fig. 2 shows the electric field intensity distribution (in the *y-z* plane) in colour scale obtained from three-dimensional (3D) EM simulations. Based on the colour scale, it can be seen about the electric field intensity being significantly weaker beyond the grounded trench on the lateral side of port 2 (for detection). The grounded trench filled with normal Al metal has the weakest electric field among the three trenches filled with different conducting materials. However, the electric field intensity in the Si substrate remains moderate (≈ 10 V/m) even close to the grounded trench. This implies that all three grounded trenches are equally effective to subdue electric field through the GaN buffer layer, but to different degree in suppressing the electric field in the Si substrate.

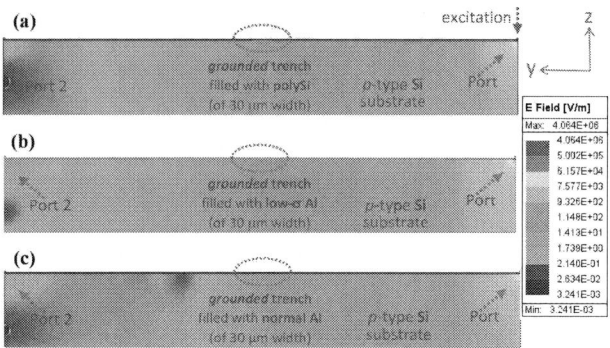

Fig. 2. Electric field distribution of the GaN-on-Si device structure with a grounded isolation trench, placed in the GaN layer, with (a) polysilicon, (b) low-σ Al material, and (c) normal Al metal as trench-filling material.

Fig. 3 to Fig. 5 show the electric field magnitude profiles of GaN-on-Si structures with the grounded trenches (of different width) filled with highly doped polySi, low-σ Al material and normal Al metal respectively. In all three grounded trenches, the electric field weakens sharply within a very short lateral distance (along y-axis) from the excitation port, then decreases considerably slower after. Getting close to the detection port, the electric field magnitude is about 0.1 V/m in all three grounded trenches filled with different conducting materials, for the three different trench widths. It means that all three different conducting materials and different trench widths have basically the same effectiveness in weakening the electric field at the far end.

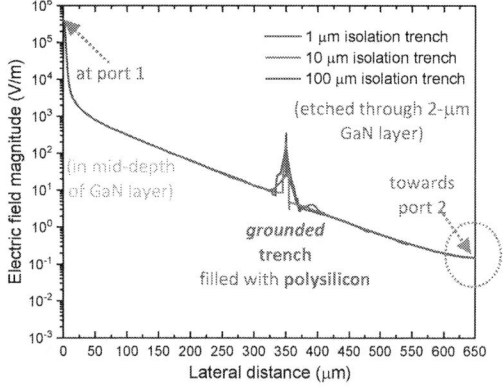

Fig. 3. Profile of the electric field (magnitude) in the middle-depth of the GaN buffer layer as a function of the lateral distance, showing the increased electric field strength in the grounded trench filled with polySi, but the same beyond the trench for three different widths (≈ 0.1 V/m at the detection end).

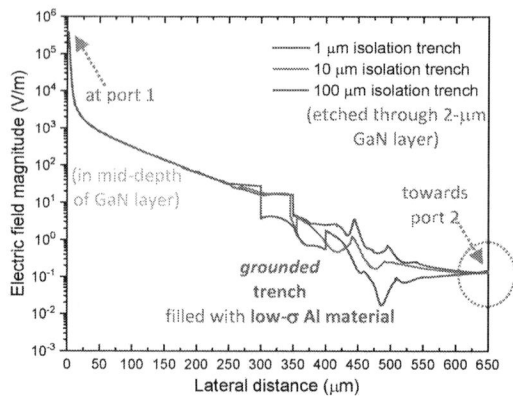

Fig. 4. Profile of the electric field in the GaN layer, showing the decreased electric field strength in the grounded trench filled with low-σ Al material, but the same (≈ 0.1 V/m) at the detection end for different trench widths.

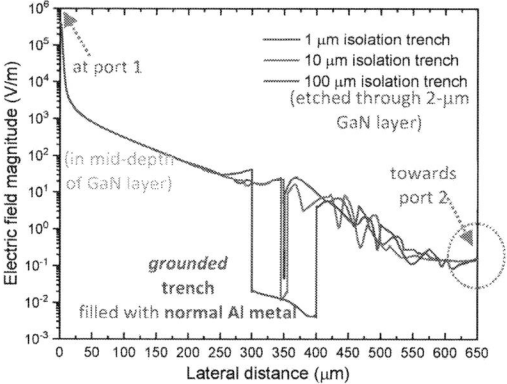

Fig. 5. Profile of the electric field in the GaN buffer layer, showing the sharply weakened electric field (by three orders of magnitude) in the grounded trench filled with normal Al metal, but about the same at the detection end (≈ 0.1 V/m) for three different trench widths.

While all three grounded trenches do not alter the overall trend of the electric field in the GaN buffer layer, the electric field does change within the local regions of the trenches to a different degree. In Fig. 3, the electric field becomes stronger in the grounded trench filled with polySi. It implies that the polySi behaves more as a dielectric material [8]. In both Fig. 4 and Fig. 5, the electric field has an obvious decrease in magnitude in the trench region; the decrease is much more significant in the grounded trench with normal Al metal (which has 100 times higher σ). However, the electric field seems to be unaltered beyond the trench region, with the trench widths being much smaller than the lateral distance.

Fig. 6 to Fig. 8 shows the electric field magnitude profile as a function of the vertical depth from the trench top through the 100-μm thick Si substrate. While the electric field within the trench region varies considerably in three different trenches, the finite electric field in the Si substrate is within the same range (around 10 V/m) in all three cases. Even in the grounded trench filled with normal Al metal, the electric field magnitude drops sharply within the trench region but the electric field in the Si substrate is not weakened by the same degree. These results imply that the grounded trenches cannot effectively weaken the electric field in the Si substrate, especially when the conductivity σ of the filling material is not high enough (in polySi and Al material by low-temperature deposition). It is similar for the effectiveness of the trench width to weaken the electric field in the Si substrate, when the trench width (maximum 100 μm in our EM study) is much smaller than the lateral separation distance (700 μm). These results imply again the same effectiveness of the grounded trenches (relatively narrow) filled with different conducting materials, when the isolation trench is placed in the middle position between the excitation and detection ports. The grounded trenches might have better effectiveness if placed at a critical position.

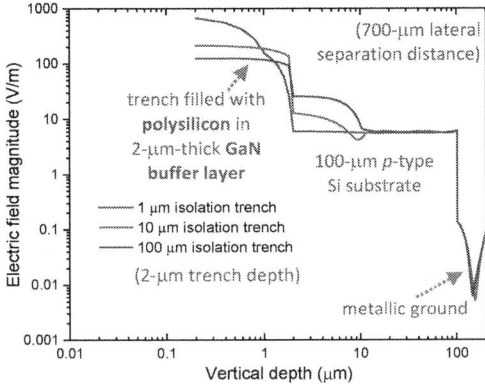

Fig. 6. Profile of electric field (magnitude) as a function of the vertical depth down through the GaN-on-Si device structure, at the mid-point of the trench, showing the intensified electric field in the polySi-filled trench and more or less the same electric field (≈ 4 V/m) deep down in the Si substrate.

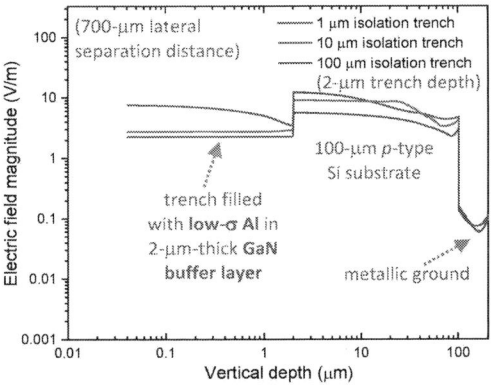

Fig. 7. Profile of electric field (magnitude) as a function of the vertical depth down through the GaN-on-Si device structure, at the mid-point of the trench, showing considerably weaker electric field in the grounded trench filled with low-σ Al material, with the electric field strength varies between 4 and 10 V/m in the Si substrate for different trench widths.

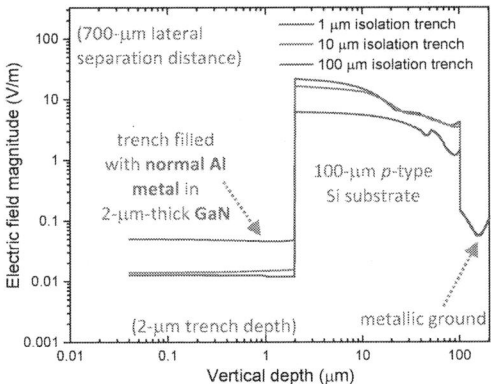

Fig. 8. Profile of electric field as a function of the vertical depth in the GaN-on-Si device structure, showing sharply weakened electric field (< 0.1 V/m) in the grounded trench filled with normal Al metal, with the field strength varies between 1 and 20 V/m in the Si substrate for different trench widths.

Fig. 9 shows the S-parameter results, specifically $|S_{21}|$ which can reveal the EM coupling in the GaN-on-Si structure. The results confirm the implications from the electric field profiles: the middle-positioned grounded trenches have almost the same effectiveness in suppressing the substrate coupling (about 8 dB improvement at high frequencies), regardless of the filling materials that have much varied electrical conductivity σ. As shown in Fig. 9(b), a wider grounded trench (up to 100 μm wide) has little help in the suppression of substrate coupling when the trench width is still small relative to the separation distance (700 μm).

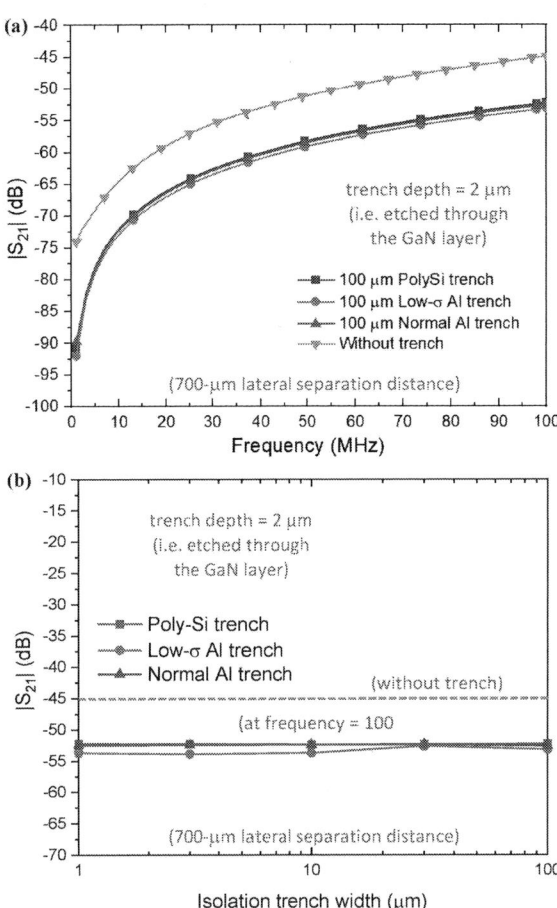

Fig. 9. $|S_{21}|$ as a measure of substrate coupling in the GaN-on-Si structure, showing almost the same signal isolation effectiveness of the grounded trenches, regardless of the trench filling materials (shown in (a)), and the trench width (shown in (b)) which is much smaller than the lateral distance.

When looking at the $|S_{21}|$ results closely, it might be somewhat counterintuitive to see that the low-σ Al material in the grounded trenches gives slightly better suppression (≈ 1 dB) in the substrate coupling than that of normal Al metal. The slight difference is likely due to the electric field paths in the Si substrate as illustrated conceptually in Fig. 10. The grounded trenches filled with the normal Al metal is far more effective weakening the electric field in the trench region (as shown in Fig. 8). As a result, the electric field path from the excitation port (as the interference source) to the detection port would be forced to have a larger curvature (path 3 in Fig. 10). The curvature is less in the case of the grounded trenches filled with the low-σ Al material, which is less effective in weakening the electric field in the trench region (as shown in

Fig. 7). A larger curvature of the electric field path would give at the detection port a larger electric field component along the z-axis (Fig. 10). It is mostly the z-component of the electric field (E_z) near the detection port to produce the coupling signal (namely the displacement current $J_{disp} = j\omega\varepsilon_{GaN}E_z$). As for the case of the grounded trench filled polysilicon (with even lower σ), the curvature of the electric field path is even less. However, it would mean relatively stronger electric field penetrating in the GaN buffer layer through the polySi-filled trench. Then, the coupling between the ports is partly through the Si substrate and partly through the GaN buffer layer, with the latter being the dominant coupling path [7] if without the isolation trench. More results including patterns of the electric field paths will be presented at the conference.

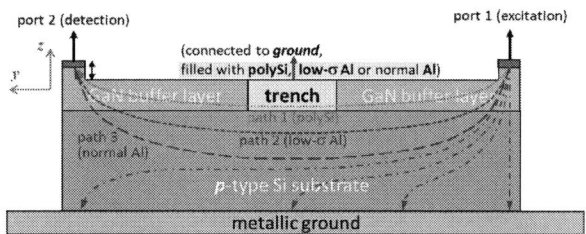

Fig. 10. A schematic cross-sectional diagram showing the GaN-on-Si device structure for computational EM investigation into the isolation trench for suppressing substrate coupling in GaN-on-Si power ICs, with the features not drawn in exact scale for better visualisation.

IV. CONCLUSION

We have reported computational EM investigation into the effectiveness of grounded trenches in suppressing substrate coupling in GaN-on-Si technology. Both the computed electric field profiles and S-parameter results confirmed the same improvement (≈ 8 dB in $|S_{21}|$) in the suppression of substrate coupling, regardless of the choice of trench filling materials (low or high σ). With the grounded trench symmetrically positioned in the middle position of the lateral separation distance, the improvement in signal isolation is similar by using different trench widths, especially when the trench width is much smaller than the lateral separation distance. These results imply helpful flexibility for the fabrication and monolithic integration of GaN-based power devices and circuits, paving also the way for future investigation into the grounded trenches placed unsymmetrically at critical positions for further improved suppression of substrate coupling.

ACKNOWLEDGMENT

The authors acknowledge the support from the Departments of Electrical and Electronic Engineering (EEE) and of Communications and Networking (CAN), School of Advanced Technology (SAT) of XJTLU as well as Department of Electrical Engineering and Electronics, The University of Liverpool. This work is supported in part by PGRS funding (FOSA2406036) of XJTLU. Both R. Yao and S. Lam sincerely acknowledge the technical and administrative support by Mr. Yubin Gu and other colleagues of Academic Enhancement Team in MITS for the arrangements of floating licenses of Ansys HFSS within the campus network of XJTLU.

REFERENCES

[1] Qimeng Jiang, Zhikai Tang, Chunhua Zhou, Shu Yang, and Kevin J. Chen "Substrate-coupled cross-talk effects on an AlGaN/GaN-on-Si smart power IC platform," *IEEE Transactions on Electron Devices*, vol. 61, no. 11, pp. 3808-3813, November 2014.

[2] Xiangdong Li, Marleen Van Hove, Ming Zhao, Karen Geens, Vesa-Pekka Lempinen, Jaakko Sorm, Guido Groeseneken, and Stefaan Decoutere, "200 V enhancement-mode p-GaN HEMTs fabricated on 200 mm GaN-on-SOI with trench isolation for monolithic integration," *IEEE Electron Device Letter*, vol. 38, no. 7, pp. 918–921, July 2017.

[3] Gang Lyu, Jin Wei, Tao Chen, Jie Zhang, and Kevin J. Chen, "Substrate and trench design for GaN-on-EBUS power IC platform," *IEEE Transactions on Electron Devices*, vol. 69, no. 7, pp. 3641-3647, July 2022.

[4] Gang Lyu, Jin Wei, Wenjie Song, Zheyang Zheng, Li Zhang, Jie Zhang, Sirui Feng, and Kevin J. Chen. "GaN on engineered bulk Si (GaN-on-EBUS) substrate for monolithic integration of high-/low-side switches in bridge circuits," *IEEE Transactions on Electron Devices*, vol. 69, no. 8, pp. 4162-4169, August 2022.

[5] Miao Cui and Sang Lam, "Use of DC probes for multi-mhz measurements of crosstalk and substrate coupling in gallium nitride power integrated circuits," 2024 *IEEE 36th International Conference on Microelectronic Test Structures* (ICMTS), Edinburgh, UK, 2024, pp. 1-5.

[6] Zijin Jiang, Rui (Ray) Yao, Miao Cui, Zhao Wang, Sang Lam and Stephen Taylor, "Impact of the resistive silicon base wafer on substrate coupling in power integrated circuits in GaN-on-Si technology," 2024 *IEEE 17th International Conference on Solid-State & Integrated Circuit Technology (ICSICT)*, Zhuhai, China, 2024, pp. 1-3.

[7] Rui (Ray) Yao, M. Cui, Zhao Wang, Sang Lam, and Stephen Taylor, "Electromagnetic investigation of substrate coupling in power integrated circuits on GaN-on-Si technology," 2024 *IEEE 11th Workshop on Wide Bandgap Power Devices & Applications (WiPDA)*, Dayton, OH, USA, 2024, pp. 1-5.

[8] Zijin Jiang, Rui (Ray) Yao, Miao Cui, Zhao Wang, Sang Lam, and Stephen Taylor, "On the design and effectiveness of isolation trenches to suppress substrate coupling in power integrated circuits in GaN-on-Si technology," 2024 *IEEE Workshop on Wide Bandgap Power Devices and Applications in Europe (WiPDA Europe)*, Cardiff, United Kingdom, 2024, pp. 1-5.

[9] Samuel P. Douglas and Caroline E. Knapp, "Low-temperature deposition of highly conductive aluminum metal films on flexible substrates using liquid alane MOD precursors," *ACS Applied Materials & Interfaces*, 12(23), 26193-26199, May 2020.

[10] S. J. Pearton, R. J. Shul, and Fan Ren, "A review of dry etching of GaN and related materials," *Materials Research Society Internet Journal of Nitride Semiconductor Research*, vol. 5, no. 1, pp. 1-38, 2000.

[11] Martin Peisl and Armin W. Wieder, "Conductivity in polycrystalline silicon—physics and rigorous numerical treatment," *IEEE Transactions on Electron Devices*, vol. 30, no. 12, pp. 1792-1797, December 1983.

Theoretical and Numerical Investigation on the On-state Characteristics of a 1200 V Multi-Gate 4H-SiC LDMOS

Da Wang   Huan Ning   Zhi Lin

Abstract—In this work, we investigate on the on-state characteristics of a 1200 V multi-gate Silicon Carbide Laterally Diffused Metal Oxide Semiconductor field effect transistor (4H-SiC LDMOS) with planar gate. Its specific on-resistance ($R_{on,sp}$) is studied by simulation and theoretical methods. First, we investigate the performance of a dual-gate SiC LDMOS (DG-LDMOS) using TCAD numerical simulation, achieving a breakdown voltage of 1306 V. And the specific on-resistance decreases from 32.4 mΩ·cm² to 17.4 mΩ·cm². Then the relationship between $R_{on,sp}$ and the number of parallel cells (N_{cell}) is theoretically derived. This theoretical result is consistent with the simulation result, which is $R_{on,sp}$ decreases and then increases with the increase of N_{cell}. When the cell pitch L_{cell} is 5 μm and three cells are connected in parallel, the device achieves the minimum on-resistance of 14.8 mΩ·cm². This enhancement does not increase the complexity of the process. Finally, the effect of the L_{cell} on $R_{on,sp}$ is explored. The result shows that wider cell requires fewer parallel cells to achieve minimum $R_{on,sp}$.

Index Terms—SiC LDMOS, specific on-resistance, multi-gate LDMOS

I. INTRODUCTION

Silicon carbide (SiC) have been widely used in power devices owing to its superior properties. Currently, the research direction of SiC power devices mainly includes SiC Vertical Double-diffused Metal Oxide Semiconductor field effect transistor (VDMOS) and SiC Lateral Double-diffused Metal Oxide Semiconductor field effect transistor (LDMOS). SiC VDMOS has the advantages of low specific on-resistance ($R_{on,sp}$) and small area. However, the structure of VDMOS determines that it is not easy to integrate with other devices [1]. Compared with VDMOS, LDMOS has the advantage of easy to integrate with other devices. Therefore, it finds widespread application in power integrated circuits [2]. Nevertheless, higher breakdown voltage (BV) usually leads to higher specific on-resistance ($R_{on,sp}$). Consequently, how to balance the relationship between the two is a hot research topic for LDMOS.

In recent years, trench gate LDMOS, reduced surface field (RESURF) technology and other technologies have effectively optimized the performance of LDMOS [3]–[10]. However, high electric field at the corners of the trench gate usually lead

Da Wang, Huan Ning and Zhi Lin are with *School of Microelectronics and Communication Engineering of Chongqing University*, Chongqing, China. This work is supported by the Natural Science Foundation of Chongqing, China, under Grant CSTB2022NSCQ-MSX1532. Corresponding author: (Zhi Lin, email: linzhi@cqu.edu.cn)

TABLE I
KEY STRUCTURAL PARAMETERS OF THE DUAL-GATE SiC LDMOS

Parameters	Value	Unit
N-drift length (L_{drift})	29	μm
N-drift depth (T_D)	12	μm
N-drift concentration (N_{drift})	7×10^{15}	cm⁻³
Channel length (L_{CH})	0.5	μm
Gate oxide thickness (T_{ox})	50	nm
Cell pitch (L_{cell})	5	μm
P-top width (L_{ptop})	15	μm
P-top depth (T_{ptop})	2	μm

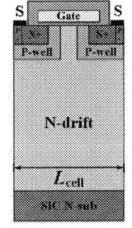 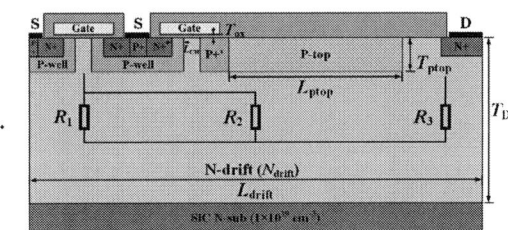

Fig. 1. Schematic cross-section of the SiC LDMOS.

to reliability problems and these technologies usually increase the complexity of the process.

In this article, we study a 1200 V multi-gate SiC LDMOS. This device adopts a planar gate structure, which simplifies the fabrication process. By paralleling multiple cells, $R_{on,sp}$ of the SiC LDMOS can be reduced. First, we explore the relationship between the number of cells (N_{cell}) and $R_{on,sp}$ by theoretical and numerical simulation methods, respectively. Additionally, we discuss the effect of the cell pitch (L_{cell}) on the relationship between N_{cell} and $R_{on,sp}$.

II. DEVICE STRUCTURE

Fig. 1 illustrates the cross-sectional schematic of the multi-gate LDMOS (MG-LDMOS). The major structure parameters are listed in Table I. The P-top RESURF region is introduced to minimize the electric field on the SiC surface. The length, depth and concentration of the P-top region play a very important role in the breakdown voltage of the device. If the concentration of the P-top region is too low or the length is too short, it is not enough to deplete the drift region, while if the concentration of the P-top region is too high or the length is

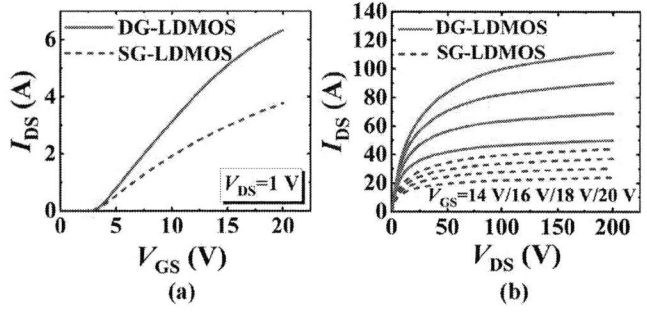

Fig. 2. (a) The transfer characteristic of the dual-gate LDMOS (DG-LDMOS) and single-gate LDMOS (SG-LDMOS). (b) The output characteristic of the DG-LDMOS and SG-LDMOS.

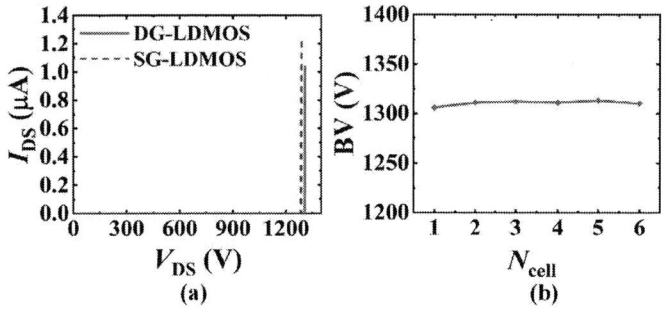

Fig. 3. (a) The blocking characteristic of the dual-gate LDMOS (DG-LDMOS) and single-gate LDMOS (SG-LDMOS). (b) Effect of N_{cell} on breakdown voltage of the multi-gate LDMOS.

too long, it will introduce a new electric field peak at the edge of the P-top region. Therefore, we optimize these parameters by simulation, the final parameters of the P-top region are shown in Table I.

III. RESULTS AND DISCUSSION

In this part, we first study the characteristics of the dual-gate LDMOS (DG-LDMOS), then more cells are paralleled and we explore the impact of the number of cells (N_{cell}) on breakdown voltage (BV) and $R_{on,sp}$ of the device. Finally, $R_{on,sp}$ of the multi-gate SiC LDMOS is modeled and we derive the relationship between N_{cell} and $R_{on,sp}$ theoretically, and compare the results of simulation and theoretical results.

Fig. 2(a) compares the transfer characteristic and Fig. 2(b) compares the output characteristic of the DG-LDMOS and SG-LDMOS. The active areas are both 10 mm². We can find that the threshold voltage (V_{th}) of the DG-LDMOS is about 2.78 V, which is just 0.03 V higher than that of the SG-LDMOS (2.75 V). $R_{on,sp}$ of the DG-LDMOS is 17.4 mΩ·cm² at (V_{GS}=20 V, which is reduced by 46% compared with SG-LDMOS (32.4 mΩ·cm²). Fig. 3(a) compares the blocking characteristic of the DG-LDMOS and SG-LDMOS, Fig. 3(b) demonstrates the effect of N_{cell} on breakdown voltage (BV) of the MG-LDMOS. By optimizing the length, depth and concentration of the P-top region, the device can achieve a breakdown voltage of 1306 V, while that of the SG-LDMOS is 1284 V. Moreover,

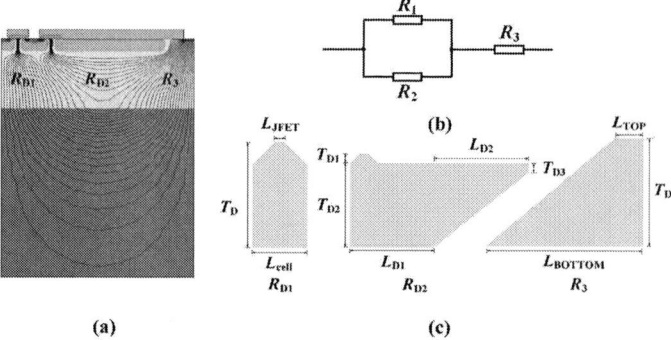

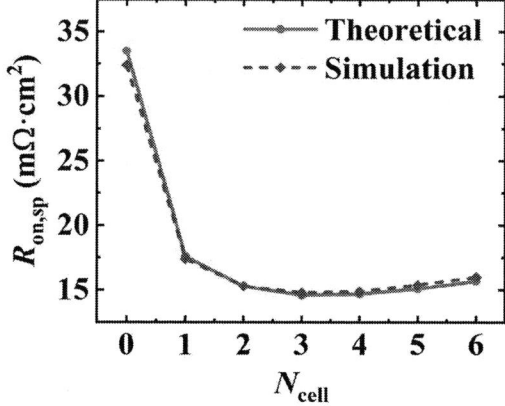

Fig. 4. (a) On-state current flowline of the dual-gate LDMOS. (b) Modeling of the R_{on} of the dual-gate LDMOS. (c) Resistance Modeling of the drift region of the dual-gate LDMOS.

Fig. 5. Comparison of theoretical and simulation results of the relationship between $R_{on,sp}$ and N_{cell}.

it is evident from the result that there is little influence on BV as N_{cell} increases.

Fig. 4(a) displays on-state current flowlines of the DG-LDMOS. According to the current flowlines, we can model R_{on} as shown in Fig. 4(b), where R_1 is the resistance of the vertical cell, R_2 is the resistance of the lateral cell, and R_3 is the series resistance at the drain side. Consequently, R_{on} can be calculated as:

$$R_{on} = R_1 \parallel R_2 + R_3$$
$$R_1 = \frac{R_{CH}}{2} + R_{JFET} + R_{D1} \qquad (1)$$
$$R_2 = R_{CH} + R_{JFET} + R_{D2}$$

where R_{CH} is the channel resistance, R_{JFET} is the JFET resistance, R_{D1} is the drift region resistance of the cell, and R_{D2} is the drift region resistance of the LDMOS. Due to the high doping concentration of the substrate, the substrate resistance can be ignored.

The expressions of R_{CH} and R_{JFET} are:

$$R_{CH} = \frac{L_{CH}}{W\mu_{CH}C_{OX}(V_G - V_{TH})}$$
$$R_{JFET} = \frac{\rho_{JFET}x_{JP}}{WL_{JFET}} \qquad (2)$$

979-8-3315-2209-4/25 $31.00 © 2025 IEEE

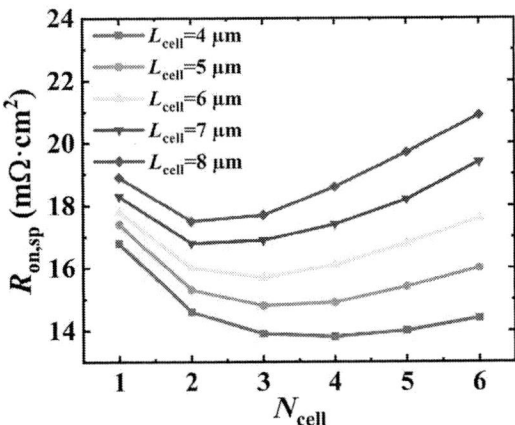

Fig. 6. The impact of different L_{cell} on the relationship between $R_{on,sp}$ and N_{cell}.

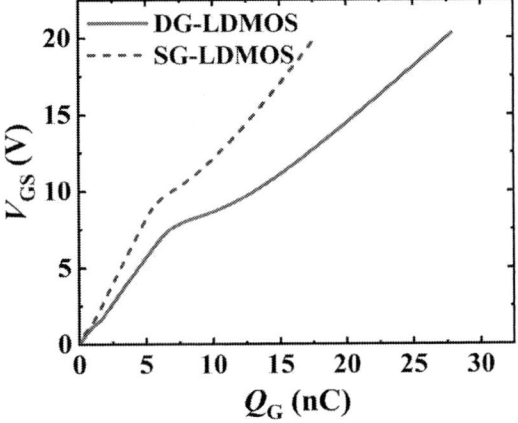

Fig. 7. The gate charge (Q_g) of the DG-LDMOS and SG-LDMOS.

where C_{OX} is the capacitance of the gate oxide, x_{JP} is the junction depth of P-well, L_{JFET} is the effective width of JFET region and W is the width of the channel. Furthermore, R_{D1}, R_{D2}, and R_3 in (1) can be modeled as shown in Fig. 4(c). Thus they can be expressed as follows:

$$R_{D1} = \frac{\rho_D}{2W} \times \ln \frac{L_{cell}}{L_{JFET}} + \frac{\rho_D}{WL_{cell}} \times (T_D - x_{JP} - \frac{L_{cell} - L_{JFET}}{2})$$

$$R_{D2} = \frac{\rho_D}{2W} \times \ln \frac{L_{JFET} + 2T_{D1}}{L_{JFET}} + \frac{\rho_D L_{D1}}{WT_{D2}} + \frac{\rho_D L_{D2}}{W(T_{D2} - T_{D3})} \times \ln \frac{T_{D2}}{T_{D3}} \quad (3)$$

$$R_3 = \frac{\rho_D \times T_{D4}}{W(L_{BOTTOM} - L_{TOP})} \times \ln \frac{L_{BOTTOM}}{L_{TOP}}$$

For multi-gate LDMOS, $R_{on,sp}$ is:

$$R_{on,sp} = (\frac{R_1}{N_{cell}} \parallel R_2 + R_3)(L_{LDMOS} + W_{cell}N_{cell}) \quad (4)$$

where L_{LDMOS} is the width of the single LDMOS (SG-LDMOS). Based on the above derivation, the comparison between theoretical and simulation results is shown in Fig. 5. It can be seen that the trend of theoretical result is consistent with simulation result, which is as N_{cell} increases, $R_{on,sp}$ first decreases and then increases. When N_{cell} is 3, $R_{on,sp}$ reaches its minimum. The theoretical minimum value of $R_{on,sp}$ is 14.6

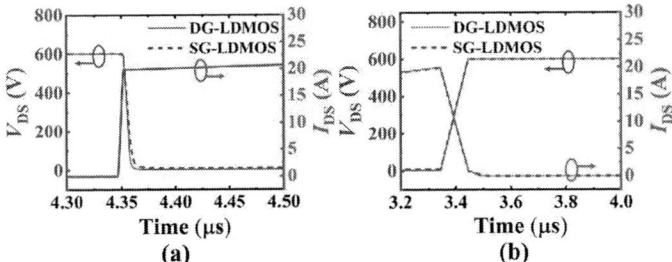

Fig. 8. (a) The turning-on characteristic of the DG-LDMOS and SG-LDMOS. (b) The turning-off characteristic of the DG-LDMOS and SG-LDMOS.

TABLE II
COMPARISION OF KEY ELECTRICAL PARAMETERS OF THE SG-LDMOS AND DG-LDMOS.

Parameters	SG-LDMOS	DG-LDMOS
V_{th} (V)	2.75	2.78
$R_{on,sp}$ (mΩ·cm²)	32.4	17.4
BV (V)	1284	1306
Q_g (nC)	17.7	27.5
t_{on} (ns)	8.4	5.3
t_{off} (ns)	81.4	80.3
E_{on} (μJ)	80.5	35.9
E_{off} (μJ)	37.8	34.6

mΩ·cm² and the simulation result is 14.8 mΩ·cm².

Fig. 6 reveals the impact of different L_{cell} on the relationship between $R_{on,sp}$ and N_{cell}, L_{cell} is from 4 μm to 8 μm. We can find that at different L_{cell}, $R_{on,sp}$ all tends to decrease and then increase as N_{cell} increases, but N_{cell} is not the same when $R_{on,sp}$ reaches the minimum. At $L_{cell} = 4$ μm, $R_{on,sp}$ is minimum when N_{cell} is 4, at $L_{cell} = 5$ μm and $L_{cell} = 6$ μm, $R_{on,sp}$ is minimum when N_{cell} is 3, at $L_{cell} = 7$ μm and $L_{cell} = 8$ μm, $R_{on,sp}$ is minimum when N_{cell} is 2. This indicates that the larger the cell pitch, the fewer cells are in parallel when $R_{on,sp}$ is minimum.

Fig. 7 shows the gate charge (Q_g) of the DG-LDMOS and SG-LDMOS. We can find that Q_g of the DG-LDMOS is 27.5 nC, which is 9.8 nC higher than that of the SG-LDMOS (17.7 nC). This is because C_{gs} and C_{gd} of the SG-LDMOS increases after paralleling the cells. Fig. 8(a) shows the turning-on characteristic of the DG-LDMOS and SG-LDMOS, Fig. 8(b) shows the turning-off characteristic of the DG-LDMOS and SG-LDMOS. It can be seen that the turn-on time (t_{on}) of the DG-LDMOS is 3.1 ns faster than that of the SG-LDMOS, and the fall time (t_{off}) is 1.1 ns faster.

TABLE II compares key electrical parameters of the SG-LDMOS and DG-LDMOS. It reveals that $R_{on,sp}$ of the DG-LDMOS is only half of that of the SG-LDMOS and BV is 22 V higher than that of the SG-LDMOS, but V_{th} is almost the same. Also, the switching time of the DG-LDMOS is shorter than that of the SG-LDMOS and the switching loss is lower than that of the SG-LDMOS.

IV. CONCLUSION

This work study a multi-gate SiC LDMOS with P-top RESURF structure, and $R_{on,sp}$ is studied by simulation and theory. We first investigate the DG-LDMOS, the device achieves a breakdown voltage of 1306 V by adjusting the parameters such as the length and depth of the P-top region. As more cells are connected in parallel, $R_{on,sp}$ of the device first decreases and then increases. When three cells are connected in parallel, $R_{on,sp}$ is minimum, which is 14.8 mΩ·cm². Besides, we theoretically analyze the relationship between $R_{on,sp}$ and N_{cell}, the result shows that the trend is consistent with the simulation result. Addtionally, we find that as the cell pitch (L_{cell}) increases, N_{cell} becomes smaller when $R_{on,sp}$ reaches the minimum.

REFERENCES

[1] M. Okamoto, A. Yao, H. Sato, and S. Harada, "First Demonstration of a Monolithic SiC Power IC Integrating a Vertical MOSFET with a CMOS Gate Buffer," in 33rd Int. Symp. Power Semiconductor Devices ICs (ISPSD), Nagoya, Japan, 2021, pp. 71-74.

[2] S. B. Isukapati, A. J. Morgan, and W. Sung, "Edge Termination and Peripheral Designs for SiC High-Voltage (HV) Lateral MOSFETs for Power IC Technology," in IEEE 34th Int. Symp. Power Semiconductor Devices ICs (ISPSD), Vancouver, BC, Canada, 2022, pp. 213-216.

[3] M. Kong, Z. Hu, J. Gao, Z. Chen, B. Zhang, and H. Yang, "A 1200-V-Class Ultra-Low Specific On-Resistance SiC Lateral MOSFET With Double Trench Gate and VLD Technique," IEEE J. Electron Devices Soc., vol. 10, pp. 83-88, 2022.

[4] H. Wang, B. Wang, L. Kong, L. Liu, H. Chen, T. Long, F. Udrea, and K. Sheng, "4H-SiC Trench Gate Lateral MOSFET With Dual Source Trenches for Improved Performance and Reliability," IEEE Trans. Device Mater. Reliab., vol. 23, no. 1, pp. 2-8, 2023.

[5] L. Liu, Q. Guo, J. Wang, M. Bai, J. Li, N. Ren, and K. Sheng, "Design and Experimental Demonstration of 4H-SiC Lateral High-Voltage MOSFETs With Double-RESURFs Technology for Power ICs," IEEE Trans. Electron Devices, vol. 71, no. 3, pp. 1572-1579, 2024.

[6] J. W. Hu, J. Y. Jiang, W. C. Chen, C. F. Huang, T. L. Wu, K. Y. Lee, and B. Y. Tsui, "1100 V, 22.9 mΩcm² 4H-SiC RESURF Lateral Double-Implanted MOSFET With Trench Isolation," IEEE Trans. Electron Devices, vol. 68, no. 10, pp. 5009-5013, 2021.

[7] H. Yu, J. Wang, L. Liu, and K. Sheng, "A Novel SiC LDMOS with Electric Field Optimization by Step Doping Technology," in 17th China International Forum on Solid State Lighting & International Forum on Wide Bandgap Semiconductors China (SSLCHINA: IFWS), Shenzhen, China, 2020, pp. 23-26.

[8] L. Zhang, L. Huang, P. Hu, C. Zhang, C. Wang, H. Wang, and K. Sheng, "Ultra-Low On-Resistance SiC LDMOS With Separated-Protected Trench Gates and Trench RESURF Technology," in 20th China International Forum on Solid State Lighting & 9th International Forum on Wide Bandgap Semiconductors (SSLCHINA: IFWS), Xiamen, China, 2023, pp. 96-99.

[9] M. Kong, Z. Cheng, N. Yu, R. Jin, J. Guo, and H. Yang, "An Ultra-low Specific On-resistance SiC LDMOS Using Double RESURF and Field Plate Techniques," in IEEE 15th International Conference on ASIC (ASICON), Nanjing, China, 2023, pp. 1-4.

[10] L. Liu, J. Wang, Z. Wang, M. Bai, J. Li, Z. Zhu, H. Xu, N. Ren, Q. Guo, and K. Sheng, "Electrical Characterization and Analysis of 4H-SiC Lateral MOSFET (LMOS) for High-Voltage Power Integrated Circuits," in 35th Int. Symp. Power Semiconductor Devices ICs (ISPSD), Hong Kong, 2023, pp. 366-369.

A Novel Hybrid Boost Converter With Continuous Output Current and RHP Zero Elimination

Zuyue Pang, Chenguang Lv, Ziyi Cui, Chi Zhang, Shuhai Chen, Shaowei Zhen

Abstract—A novel hybrid boost converter (HBC) with continuous output current and right-half-plane (RHP) zero elimination is proposed. The proposed converter consists of five power switches, a flying capacitor, an output capacitor, and an inductor. It operates within an input voltage range of 2.7 to 4.2 V and provides a 5 V output, making it suitable for portable electronic devices. Compared with previous hybrid boost converters and conventional boost converter (CBC), the proposed HBC features continuous output current, which equals inductor current, as Buck converter. Thus the RHP zero in CBC is eliminated and the transient response is significantly improved. Simulation results show that when the output current steps between 0 and 500 mA, the recovery time is merely 8.2μs / 7.5μs, with undershoot / overshoot voltage of 62 mV / 54 mV.

Index Terms—boost converter, right-half-plane (RHP) zero, low inductor current, fast load transient response.

I. INTRODUCTION

In recent years, the rapid growth of the portable electronics market has accelerated the development of DC-DC converters, especially boost converters. Boost converters step up the 2.7-4.2 V voltage from lithium-ion batteries to a stable 5 V, powering the internal components of portable devices. However, conventional boost converter (CBC) suffers from several drawbacks. Firstly, the power stage transfer function of CBC contains a right-half-plane (RHP) zero, severely limiting the converter's bandwidth and degrading transient response. Secondly, the output current I_O of CBCs is discontinuous, resulting in large output voltage V_O ripple. Additionally, the inductor current I_L in CBC is given by $I_O/(1-D)$, where D is the duty cycle. As D increases, the large I_L flowing through the inductor's direct current resistance (DCR) leads to higher losses, thereby reducing the converter's efficiency.

To address these issues, several novel hybrid boost converters (HBCs) and new control methods have been proposed [1–6]. However, [2] fails to eliminate the RHP zero. Although [1], [3], [5] and [6] eliminate the RHP zero, the I_L in these designs remains excessively high, which hinders further loss reduction. [4] is unsuitable for single-output applications. In light of these issues, this paper presents a novel hybrid boost converter with continuous output current and right-half-plane (RHP) zero elimination. The proposed HBC maintains an average inductor current equal to the output current, achieving both reduced losses and superior transient performance. The structure of this paper is as follows. Section II introduces the

The authors are with the *School of Integrated Circuit Science and Engineering, University of Electronic Science and Technology of China*, Chengdu, China. Corresponding author: (Shaowei Zhen, email: swzhen@uestc.edu.cn)

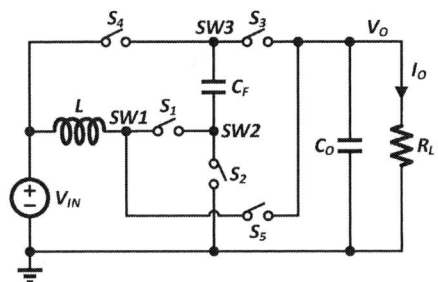

Fig. 1. Schematic of the proposed boost converter.

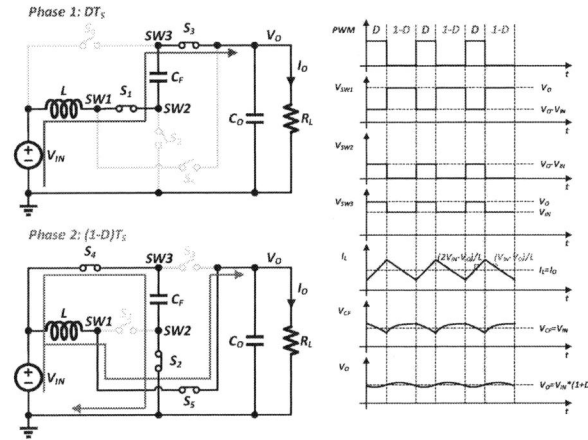

Fig. 2. Operation principle and waveforms of the proposed boost converter.

DC and AC operational principles and characteristics of the proposed converter, along with a discussion on I_L reduction and the RHP zero issue. Section III describes the circuit implementation and provides a small-signal model of the overall control loop. Section IV presents simulation results, followed by the conclusions drawn in Section V.

II. PROPOSED HYBRID BOOST CONVERTER

A. Operation Principle

The schematic of the proposed hybrid boost converter is illustrated in Fig. 1, which consists of five power switches (S_1–S_5), an inductor (L), an output capacitor (C_O), and a flying capacitor (C_F). Fig. 2 illustrates the operational principles and steady-state waveforms of the converter in continuous conduction mode (CCM). Here, V_{IN} denotes the input voltage, V_O the output voltage, I_O the output current, and V_{CF} is the

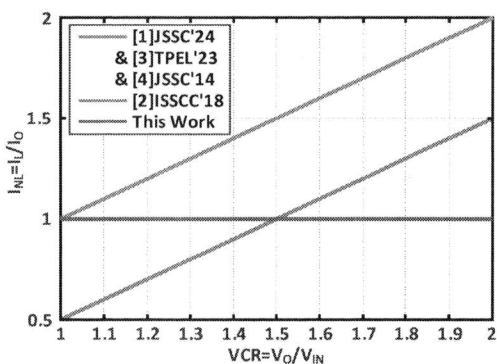

Fig. 3. Normalized I_L comparisons.

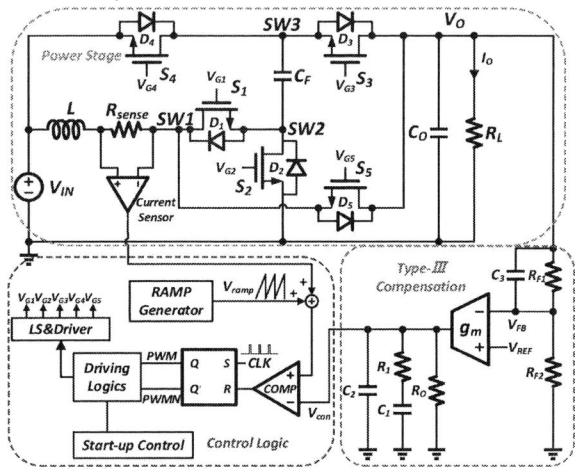

Fig. 4. Block diagram of the system.

TABLE I
SYSTEM PARAMETERS

Parameters	Value
Switching Frequency	1 MHz
L	2.2 μH
C_O/C_F	10 μF/10 μF
R_L/R_{sense}	10 Ω/30 mΩ
Current Sensor Gain	10
R_{F1}/R_{F2}	45 kΩ/15 kΩ
R_1/R_O	1.8 kΩ/10 MΩ
$C_1/C_2/C_3$	680 pF/33 pF/120 pF

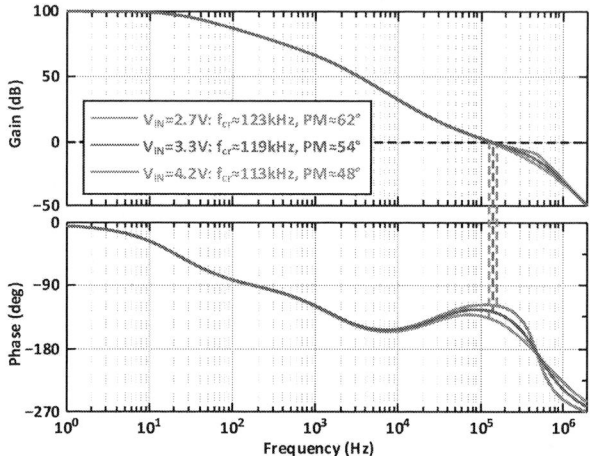

Fig. 5. Bode plot of the proposed boost converter.

voltage across C_F. V_{SW1} is the voltage at the switching node SW1 (right side of the L), V_{SW2} is the voltage at the switching node SW2 (bottom plate of C_F), V_{SW3} is the voltage at the switching node SW3 (top plate of C_F), I_L is the inductor current, D is the duty cycle of phase 1 (L charging phase), and T_S is the switching period.

During phase 1, S_1 and S_3 are turned on, while S_2, S_4, and S_5 are turned off. L is charged, and the flying C_F discharges. L and C_F are connected in series, supplying energy to the output capacitor C_O and the load R_L. The voltage across L is $(2V_{IN} - V_O)$, causing I_L to increase with a slope of $(2V_{IN} - V_O)/L$.

During phase 2, S_2, S_4, and S_5 are turned on, while S_1 and S_3 are turned off. C_F is charged by input source, while L discharges to supply energy to the load R_L. Neglecting the on-resistances of all switches and parasitic resistances, C_F is charged to V_{IN} since it is connected in parallel with the input source. The voltage across L is $(V_{IN} - V_O)$. Since $V_{IN} < V_O$, I_L decreases with a slope of $(V_{IN} - V_O)/L$.

Based on the topology of the proposed converter, it is evident that its small-signal model does not contain a RHP zero. When the load R_L suddenly decreases, causing the load current I_O to increase, the energy required by R_L over the

entire switching cycle increases. Consequently, D increases, enabling L to store more energy and meet the load demand. Since all the inductor current flows to the output during the $D \times T_s$, interval, increasing D directly transfers more energy from L to C_O and R_L, allowing V_O to recover to its steady-state value more rapidly. Additionally, the continuity between I_L and I_O results in a significantly reduced V_O ripple.

According to volt-second balance of L:

$$[V_{IN} + V_{CF} - V_O] DT_s + (V_{IN} - V_O)(1 - D) T_s = 0 \quad (1)$$

Neglecting on-resistances of all switches and parasitic resistances, it is assumed that $V_{CF} = V_{IN}$. Therefore, the voltage conversion ratio (VCR) of the converter is:

$$VCR = \frac{V_O}{V_{IN}} = 1 + D \quad (2)$$

The average inductor current is:

$$I_L = I_O \quad (3)$$

The average inductor current I_L of the proposed converter is always equal to I_O, regardless of D variations. As a result, the reduced I_L helps lower DCR losses and improves the converter's efficiency. The comparison of the normalized inductor current $I_{NL} = (I_L/I_O)$ of the proposed converter with those of the designs in [1–4] is shown in Fig. 3. In [1], [3], and

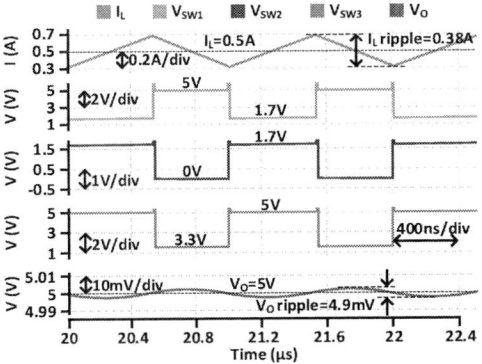

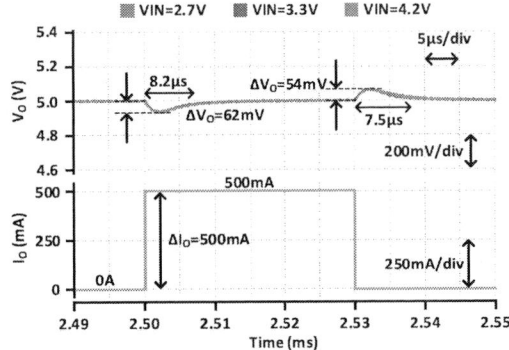

Fig. 6. Simulated steady-state waveforms (V_{IN}=3.3V,V_O=5V).

Fig. 8. Simulated load transient responses.

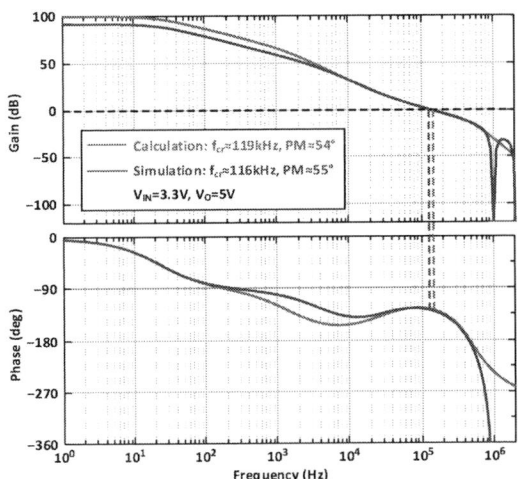

Fig. 7. Comparison of simulated and calculated loop bode plots.

[4], I_{NL} = VCR, whereas in [2], I_{NL} = VCR − 0.5. In contrast, the proposed design maintains I_{NL} = 1 at all times, contributing to higher converter efficiency, particularly under high VCR conditions.

B. Analysis of RHP Zero

This design employs fixed-frequency peak current mode (PCM) control , and thus the power stage transfer function of interest is the transfer function from the inductor current I_L to the output voltage V_O, denoted as $G_{vi}(s)$. For this converter, $G_{vi}(s)$ is equivalent to the impedance of the parallel network formed by C_O (including its equivalent series resistance r_c) and R_L, since the sum of the current flowing into C_O and the current flowing into R_L always equals the inductor current I_L. The expression of $G_{vi}(s)$ is:

$$G_{vi}(s) = \frac{\widehat{v}_O(s)}{\widehat{i}_L(s)} = R_L \frac{1 + sr_C C_O}{1 + s(R_L + r_C) C_O} \quad (4)$$

From eq4, it can be observed that, similar to a conventional buck converter, the power stage transfer function of the proposed converter contains only one pole and one ESR zero. The

impact of the ESR zero on the control loop can be neglected since it is located at a high frequency. Therefore, the proposed converter can be modeled as a single-pole system. Furthermore, unlike the designs in [1–5], where the power stage poles and zeros vary significantly with changes in the duty cycle (making the compensation circuit design challenging), the pole location of the proposed converter remains fixed. This greatly simplifies the compensation circuit design and ensures nearly identical loop characteristics across the entire operating range.

Equation (4) also confirms that the power stage of the proposed converter does not contain an RHP zero, resulting in superior transient performance.

III. CIRCUIT IMPLEMENTATION

A. System Design Scheme

The overall system architecture of the proposed boost converter employing PCM control and type-III compensation to ensure loop stability is depicted in Fig. 4. Switches S_1–S_5 are implemented with NMOSFETs. The voltage stress on S_1 is V_O, while S_2 and S_5 experience a voltage stress of V_{IN}, and S_3 and S_4 experience a voltage stress of $(V_O - V_{IN})$. Since the maximum voltage stress on all switches is V_O, the design does not require high-voltage devices under the conditions of V_{IN} = 2.7–4.2 V, and V_O = 5 V, which helps reduce conduction losses and further improves converter efficiency.

The system design parameters are summarized in TABLE I.

B. Loop Small-Signal Model

Based on the small-signal model for PCM control proposed by Ridley [7] and neglecting the feedforward gains of V_{IN} and V_O, a small-signal model of the control loop for the proposed converter is developed. As shown in Fig. 5, the bode plots of the proposed converter's loop are very similar when V_O = 5 V and V_{IN} is 2.7 V, 3.3 V, and 4.2 V, respectively. As V_{IN} varies from 2.7 V to 4.2 V, the loop crossover frequency f_{cr} ranges from 113 kHz to 123 kHz, and the phase margin (PM) ranges from 48° to 62°, ensuring consistent loop characteristics across the entire operating range.

TABLE II
PERFORMANCE COMPARISON WITH THE STATE-OF-THE-ART BOOST CONVERTERS

Work	ISSCC'18 [2]	ICSICT'20 [6]	TPEL'23 [3]	JSSC'24 [1]	This-Work
Topology	DPUC	SIDO	Time-based Boost	No RHP Zero Boost	No RHP Zero Boost
V_{IN}	2-4.2 V	2.7-4 V	2.5-4.5 V	2.7-4.2 V	2.7-4.2 V
V_O	3-5 V	4-5 V	5 V	5-6 V	5 V
L	4.7 μH	4.7 μH	2.2 μH	3.3 μH	2.2 μH
C_F/C_O	10 μF/10 μF	4.7 μF/4.7 μF	N.A./44 μF	20 μF/4.7 μF	10 μF/10 μF
Continuous Output Current	Yes	Yes	No	No	Yes
Switching Frequency f_s	1 MHz	1 MHz	1.5 MHz	2.2 MHz	1 MHz
RHP Zero Free	No	Yes	Yes	Yes	Yes
$I_{NL}=I_L/I_O$	VCR-0.5	1	VCR	VCR	1
Peak-to-Peak V_O ripple (V_O)	<15 mV (3-5 V)	N.A.	N.A.	45 mV (6 V)	4.9 mV (5 V)
Maximum undershoot/overshoot	190 mV	N.A.	N.A.	115 mV	62 mV
Settling Time: Up/Down (ΔI_O)	>80μs/>80μs (400 mA)	10μs/10μs (100 mA)	20-25μs/20-25μs (300 mA)	2.8μs/2.7μs (450 mA)	8.2μs/7.5μs (500 mA)

IV. SIMULATION RESULTS

Using the system parameters listed in TABLE I. and assuming a DCR of 60 mΩ for L and an ESR of 3 mΩ for C_F and C_O, the simulated waveforms of steady-state with $V_{IN} = 3.3$ V and $V_O = 5$ V are shown in Fig. 6. The average inductor current I_L is 0.5 A, which is equal to the output current I_O. The V_O peak-to-peak ripple is very small, measuring just 4.9 mV. As depicted in Fig. 7, the simulation results show that $f_{cr} = 116$ kHz and PM = 55°, which align closely with the calculated model, thus validating the accuracy of the model up to half the switching frequency. The discrepancy at low frequencies is due to the neglect of the V_{IN} and V_O feedforward gains in the model. The output voltage waveforms corresponding to a step change in I_O from 0 A to 500 mA are shown in Fig. 8. The response waveforms are nearly identical for different input voltages, with a recovery time of only 8.2μs / 7.5μs and undershoot / overshoot values of 62 mV / 54 mV. TABLE II. summarizes the performance of the proposed converter and compares it with that of state-of-the-art designs.

V. CONCLUSION

This paper proposes a novel hybrid boost converter with continuous output current and RHP zero elimination. The average inductor current is equal to the output current, which reduces DCR losses. Under PCM control, the power stage transfer function of the converter includes only a single fixed pole and a negligible high-frequency ESR zero, which fundamentally enables this converter with higher bandwidth, simpler compensation, and faster transient response across the entire operating range. Simulation results show that the proposed converter achieves a low average inductor current and small V_O ripple of 4.9 mV. When the output current steps between 0 and 500 mA, the recovery time is merely 8.2μs / 7.5μs, with undershoot / overshoot voltage of 62 mV / 54 mV, remaining within 1.25% of V_O.

REFERENCES

[1] J. Ruan et al., "A 2.8μs Response Time 95.1% Efficiency Hybrid Boost Converter With RHP Zero Elimination for Fast-Transient Applications," *IEEE J. Solid-State Circuits*, vol. 59, no. 9, pp. 2960–2970, Sep. 2024.

[2] S.-U. Shin et al., "A 95.2% efficiency dual-path DC-DC step-up converter with continuous output current delivery and low voltage ripple," in *IEEE Int. Solid-State Circuits Conf. (ISSCC) Dig. Tech. Papers*, Feb. 2018, pp. 430–432.

[3] M. Leoncini, A. Dago, A. Bertolini, A. Gasparini, S. Levantino, and M. Ghioni, "A Compact High-Efficiency Boost Converter With Time-Based Control, RHP Zero-Elimination, and Tracking Error Compensation," *IEEE Trans. Power Electron.*, vol. 38, no. 3, pp. 3100–3113, Mar. 2023.

[4] T.-H. Kong, S.-W. Hong, and G.-H. Cho, "A 0.791 mm^2 On-Chip Self-Aligned Comparator Controller for Boost DC-DC Converter Using Switching Noise Robust Charge-Pump," *IEEE J. Solid-State Circuits*, vol. 49, no. 2, pp. 502–512, Feb. 2014.

[5] Y.-K. Luo, Y.-P. Su, Y.-P. Huang, Y.-H. Lee, K.-H. Chen, and W.-C. Hsu, "Time-Multiplexing Current Balance Interleaved Current-Mode Boost DC-DC Converter for Alleviating the Effects of Right-half-plane Zero," *IEEE Trans. Power Electron.*, vol. 27, no. 9, pp. 4098–4112, Sep. 2012.

[6] Z. Tong, P. Cao, X. Zhang, and Z. Hong, "A Right-Half-Plane Zero-Free Single-Inductor Dual-Output Boost Converter with 92.44% Peak efficiency and Fast Transient Response," in *Proc. IEEE 15th Int. Conf. Solid-State Integr. Circuit Technol.*, Nov. 2020, pp. 1–3.

[7] R. B. Ridley, "A new, continuous-time model for current-mode control (power convertors)," *IEEE Trans. Power Electron.*, vol. 6, no. 2, pp. 271–280, Apr. 1991.

979-8-3315-2209-4/25 $31.00 © 2025 IEEE

A Novel Low Loss SOI-LIGBT with Carrier Stored Layer and P-drift

Haoru Wang, Yifan Shu, Jun Zhang, Jialei Tan, Kaiwei Dai, Pei Guo, Jiayuan Wang*, and Jie Wei*

Abstract—A novel lateral insulated gate bipolar transistor (LIGBT) on silicon on insulator (SOI) is studied by simulation, which is characterized by carrier stored layer and P-drift (CSP LIGBT). This structure replaces the N-drift of conventional LIGBT with P-drift, and introduces a N-type carrier stored (N-CS) layer at the bottom of the cathode P-well. In the on-state, the NMOS at the cathode is turned on, and the parasitic thyristor composed of the anode P+ / N-buffer / P-drift / N-CS is self-adaptively turned on. Thus, it increases the carrier concentration on the cathode side to enhance the conductivity modulation effect, which reduces the on-state voltage drop (V_{on}). During turning off period, the NMOS at the cathode is turned off, while the parasitic thyristor remains on-state in the initial stage. It not only suppresses the expansion of the depletion region to delay the rise of anode voltage (V_A), but also extracts excess carriers at low V_A, both of which help reduce the turn-off loss (E_{off}). Compared with conventional LIGBT and LIGBT with N-CS, the proposed CSP LIGBT decreases the E_{off} / V_{on} by 66.5%/9.8% and 59.0%/8.2% at the same V_{on} / E_{off}, respectively.

Keywords—*turnoff loss (E_{off}), self-adaptive, SOI LIGBT, P-drift, thyristors.*

I. INTRODUCTION

Insulated Gate Bipolar Transistor (IGBT) is the core device for energy conversion and transmission, and is the key to improve power efficiency and quality [1-2]. It owns the advantages of large current capability, high input impedance, simple control circuit, and high operating frequency. It is widely used in industrial control, electric vehicles, military aerospace and other fields [3-4]. With the development of Smart Power Integrated Circuit, lateral insulated gate bipolar transistor (LIGBT) has gradually emerged in response to the requirements of CMOS process compatibility and device integrability [5]. However, the low on-state voltage drop (V_{on}) of LIGBT always leads to a large turn-off loss (E_{off}), because of the intrinsic tail current. Common technologies to improve V_{on}-E_{off} trade-off relationship include trench gates technology [6], carrier storage technology [7], superjunction [8], and integrating MOS structure at the cathode side of conventional LIGBT [9-10]. Meanwhile, different shorted-anode structures are proposed to decrease E_{off} and suppress the impact of snapback [11-12].

This work proposes a novel low loss SOI LIGBT, which is studied by Sentaurus TCAD. This new structure reduces the V_{on} and E_{off}, and then optimizes the trade-off relationship between V_{on} and E_{off}.

II. STRUCTURE AND MECHANISM

Fig. 1(a) shows the cross-sectional schematic and equivalent circuit of the proposed LIGBT. It features a P-drift

---

Haoru Wang, Yifan Shu, Jun Zhang, Jialei Tan, Kaiwei Dai, Pei Guo and Jie Wei are with the State Key Laboratory of Electronic Thin Films and Integrated Devices, University of Electronic Science and Technology of China, Chengdu 610054, China (weijieuestc@uestc.edu.cn).

Jiayuan Wang is with Gingko College of Hospitality Management, Chengdu 611743, China (jiayuan.wang@gingkoc.edu.cn).

This work was supported in part by Fundamental Research Funds for the Central Universities under Grant ZYGX2024J028 and Xiaomi Young Scholars Fund. (Corresponding authors are Jie Wei and Jiayuan Wang)

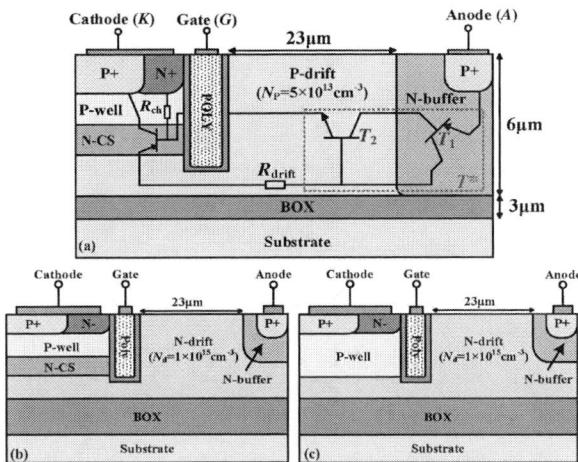

Fig. 1. Schematic cross section view of (a) CSP LIGBT, (b) CS LIGBT, and (c) Con. LIGBT

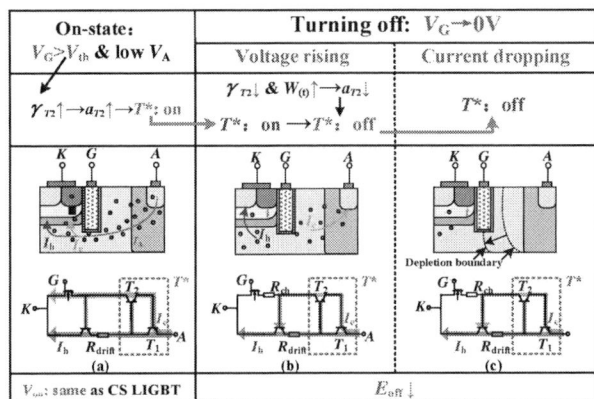

Fig. 2. Current schematic diagram and equivalent circuit of CSP LIGBT: (a) on-state. (b) The voltage rising and (c) current dropping during the turning off.

and an N-type carrier stored layer (N-CS) below the cathode P-well, named as CSP LIGBT. The P+ anode / N-buffer / P-drift form a PNP transistor (T_1), and the N-buffer / P-drift / N-CS layer form a PNP transistor (T_2). Both of them form parasitic thyristors (T^*). R_{drift} is the resistance of P-drift, and R_{ch} is the channel resistance. N_P is the P-drift doping concentration in CSP LIGBT and N_d is the N-drift doping concentration in other structures. Fig. 1(b) and 1(c) show the schematic view of CS LIGBT and conventional (Con.) LIGBT. The N-buffer doping of all devices is 3×10^{16} cm^{-3}, and the N-CS doping is 8×10^{15} cm^{-3}.

Fig. 2 shows the current flow and equivalent circuit diagram of CSP LIGBT in different states. Fig. 3 compares the carriers concentration distribution of CSP LIGBT and CS LIGBT at different operating state.

In the conductive on-state with $V_G > V_{th}$ in Fig. 2(a), an electron inversion layer is formed along the sidewall of the trench gate. Then the N-CS layer is shorted to the cathode N+, which improves the emitter junction injection efficiency of the

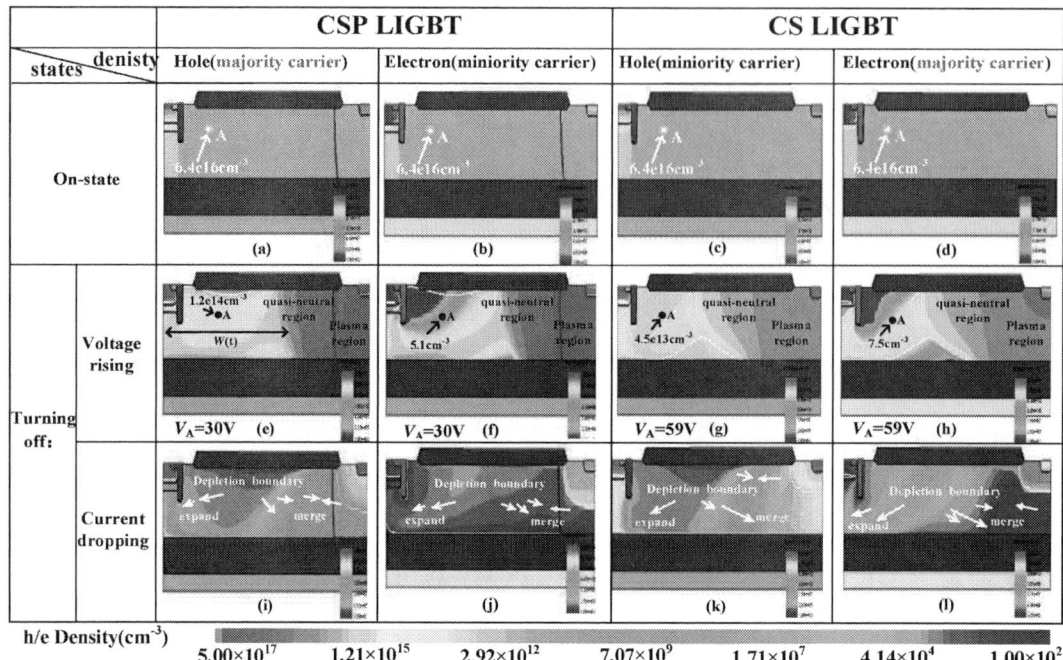

Fig. 3. The two-dimensional hole and electron concentration distribution of CSP LIGBT and CS LIGBT in different operating states: (a)-(d) for the conduction state, (e)-(h)for the voltage rising, (i)-(l) for the current dropping.

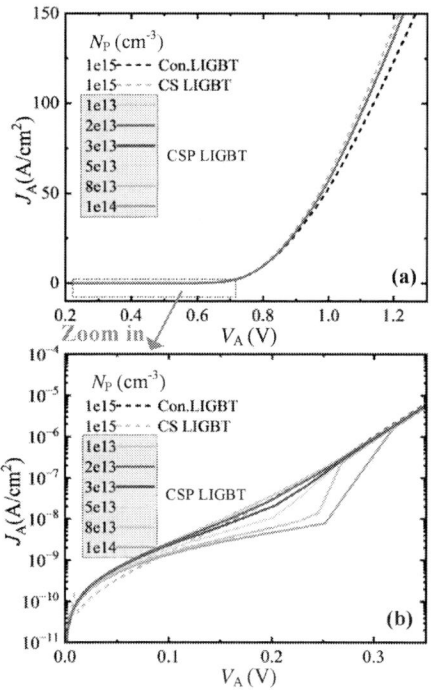

Fig. 4. (a) Forward I-V characteristics of different LIGBTs and (b) the effect of N_P on device snapback.

T_2 (γ_{T2}), and thereby increases the current gain of T_2 (α_{T2}). At this point, the parasitic thyristor T^* easily satisfies the conduction condition $\alpha_{T1} + \alpha_{T2} = 1$ to be turned on. Therefore, the P-drift of CSP LIGBT has the same hole and electron density as in Fig. 3(a)-(b), and the N-drift of CS LIGBT also has the same hole and electron density as in Fig. 3(c)-(d).

During turning off with voltage rising stage in Fig. 2(b). As V_{GS} rapidly decreases to 0 V, the electron inversion layer

along the trench gate sidewall disappears, and both γ_{T2} and electrons injected from the cathode N+ decrease sharply. Then the CSP LIGBT transforms from the forward conduction state to the voltage rising stage. However, the self-adaptive T^* still remains on-state, which makes the cathode side P-drift in a quasi-neutral state as shown in Fig. 3(e)-(f) without forming a depletion region. The anode voltage (V_A) is maintained by the quasi-neutral region with a low hole density gradient. Therefore, the CSP LIGBT not only suppresses the expansion of the depletion region, but also prolongs the voltage rising time with low V_A and extracts excess carriers during the voltage rising period. Both are beneficial for reducing E_{off}. On the contrary, the CS LIGBT forms a depletion region from the cathode side P-well/N-CS junction to maintain high V_A as shown in Fig. 3(g)-(h). Note that the width of the quasi-neutral P drift region with lower hole density is the equivalent base width $W(t)$ of T_2. As V_A continues to increase, $W(t)$ increases and then α_{T2} decreases. Finally, due to $\alpha_{T1} + \alpha_{T2} < 1$, T^* will be turned off and then the depletion region will quickly establish from the P-drift/N-buffer junction, and then the V_A will quickly reach the bus voltage (V_{bus}).

For the current dropping period of CSP LIGBT in Fig. 2(c) and Fig. 3(i)-(j), the depletion region rapidly expands, because most excess carriers in the P-drift are almost extracted by the on-state T^* during the voltage rising period, However, the CS LIGBT has a large amount of excess charge carriers to be recombined near the anode side as shown in Fig. 3(k)-(l). Therefore, the CSP LIGBT achieves a faster current dropping speed than CS LIGBT. Overall, the CSP LIGBT achieves a lower E_{off} compared to the CS LIGBT.

III. RESULTS AND DISCUSSION

Fig. 4 compares the forward I-V characteristics of different LIGBTs. It is reported that thyristor has the same ON-state voltage as a p-i-n diode at the same doping concentration. Therefore, the conductivity of CSP LIGBT is the same as that of CS LIGBT. As shown in Fig. 4(a), the CSP LIGBT exhibits

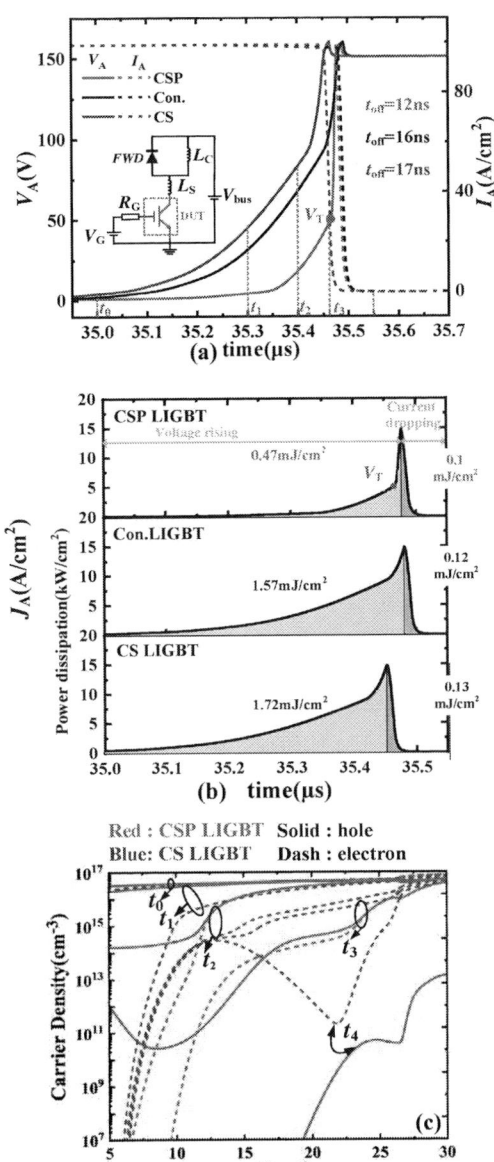

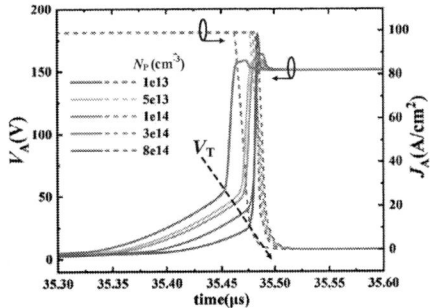

Fig. 6. Turn-off characteristics of CSP LIGBT with different N_P.

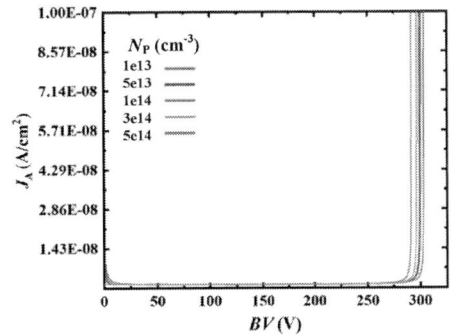

Fig. 7. Breakdown characteristics of CSP LIGBT under different N_P.

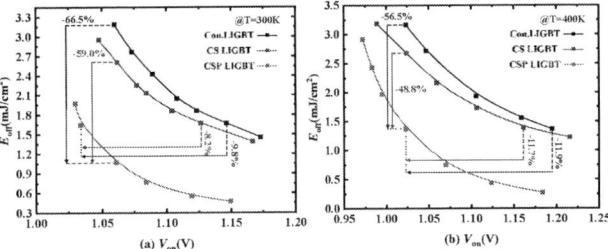

Fig. 8. Trade-off relationship between V_{on}-E_{off} of different LIGBTs at J_A=100A/cm^2 (a) at T=300 K and (b) at T=400K.

Fig. 5. (a) Different LIGBT turn-off characteristic curves. The bus voltage V_{bus}, gate resistance R_G, load inductance L_C, and stray inductance L_S are 150V, 10 Ω, 1 μH, and 1nH, respectively. (b) The power consumption of different LIGBTs changes over time during the turn-off. (c) Carrier concentration distribution of the two devices in (a) at different times.

the same V_{on}=1.12 V as CS LIGBT at J_A = 100 A/cm^2. As T^* switches from the forward blocking state ($\alpha_{T1} + \alpha_{T2} < 1$) to the forward conducting state ($\alpha_{T1} + \alpha_{T2} = 1$), it may cause a weak snapback phenomenon. Low N_P would increase the α_{T2} and then allow the T^* to turn on at a lower V_A , suppressing the snapback phenomenon. Fig. 4(b) shows that the CSP LIGBT completely eliminates the snapback phenomenon as $N_P < 3 \times 10^{13}$ cm^{-3}.

Fig. 5(a) shows the turn-off characteristics at the same V_{on} = 1.12 V for Con. LIGBT and CS LIGBT N_d =1 × 10^{15} cm^{-3}, and CSP LIGBT with N_P = 5 × 10^{13} cm^{-3}. Fig. 5(b) compares their power dissipation during turn-off. During the voltage rise phase, the V_A of CSP LIGBT remains at a low value until the turn-off voltage of parasitic thyristor (V_T), and then rapidly rises to V_{bus}. Thus, the CSP LIGBT achieved the lowest voltage rise period energy loss of 0.47 mJ/cm^2 among the three

LIGBTs. Fig. 5(c) shows that CSP LIGBT and CS LIGBT extract excess carriers at almost the same speed during $t_0 \sim t_1$. From t_2, CS LIGBT enters a current drop period to remove excess carriers through recombination, while CSP LIGBT still extracts excess carriers through the on-state T^* due to its extended voltage rising time. Therefore, the CSP LIGBT extracts excess carriers faster than CS LIGBT after t_2. Due to the majority of excess charge carriers in P-drift being extracted by the on-state T^* during the voltage rising period, the CSP LIGBT achieves the shortest turn-off time (t_{off}) of 12 ns. The t_{off} of CS LIGBT and Con. LIGBT are 16 ns and 17 ns, respectively. Thereby, CSP LIGBT reduces the energy loss during the current drop phase from 0.12 mJ/cm^2 in CS LIGBT to 0.1 mJ/cm^2. In addition, the Con. LIGBT requires a higher concentration of P+ anode doping to obtain the same V_{on}, thus it exhibits the longest t_{off}.

As shown in Fig. 6, the turn-off characteristics of CSP LIGBT with different N_d values are presented. Higher N_d values make it more difficult for the depletion region to expand during the initial turn-off period. Thus, it delays the voltage rising period and reduces the V_T to turning off the T^*, which helps to reduce E_{off}.

Fig. 7 shows the breakdown characteristics of CSP LIGBT. Its breakdown voltage is about 300 V as $N_P \le$5 × 10^{13} cm^{-3}. Fig. 8 shows the trade-off relationship between V_{on} and E_{off} in different LIGBTs. The CSP LIGBT achieves a better V_{on}-E_{off} relationship. Compared with CS and Con. LIGBT, the CSP LIGBT decreased E_{off} by 59.0% and 66.5% at the same V_{on}, and V_{on} decreased by 8.2% and 9.8% under the same E_{off}, respectively. Even under high-temperature conditions of 400 K, the CSP LIGBT maintains an excellent tradeoff between V_{on}-E_{off}.

IV. CONCLUSION

A novel low loss SOI LIGBT with carrier stored layer and P-drift (CSP LIGBT) is proposed. The CSP LIGBT forms a parasitic thyristor T^*. In the on-state, the parasitic thyristor T^* self-adaptively conducts to enhance the conductivity modulation effect, so as to realize a low V_{on}. During the turning off, the thyristor T^* suppresses the expansion of the depletion region at low V_A and extracts excess carriers quickly at the initial stage, so the CSP LIGBT achieves ultralow t_{off} of 12 ns and E_{off} of 0.57mJ/cm^2. The E_{off}/V_{on} of CSP LIGBT is reduced by 59.0% / 8.2% compared with the CS LIGBT at the same V_{on}/E_{off}.

REFERENCES

[1] N. Iwamuro and T. Laska, "IGBT History, State-of-the-Art, and Future Prospects, " IEEE Transactions on Electron Devices, vol. 64, no. 3, pp. 741-752, March 2017.

[2] D. Disney, T. Letavic, T. Trajkovic, T. Terashima and A. Nakagawa, "High-Voltage Integrated Circuits: History, State of the Art, and Future Prospects," IEEE Transactions on Electron Devices, vol. 64, no. 3, pp. 659-673, March 2017.

[3] Y. Gu, J. Ma, L. Zhang, J. Wei, S. Li and S. Liu, "Silicon-on-Insulator Lateral Insulated Gate Bipolar Transistor: Current Technologies and Prospects," IEEE Transactions on Electron Devices, vol. 71, no. 1, pp. 381-392, Jan. 2024.

[4] J. Tan, J. Wei, J. Lu, X. Liu, G. Deng, W. Song, P. Guo, B. Zhang and X. Luo, "High Short-Circuit Capability and Low-Loss SOI-LIGBT with Double-Integrated NMOS," 2024 IEEE 17th International Conference on Solid-State & Integrated Circuit Technology (ICSICT), Zhuhai, China, 2024, pp. 1-3.

[5] J. Cheng and X. Chen, "A novel low-side structure for OPTVLD-SPIC technologically compatible with BiCMOS," 2013 25th International Symposium on Power Semiconductor Devices & IC's (ISPSD), Kanazawa, Japan, 2013, pp. 123-126.

[6] M. Harada, T. Minato, H. Takahashi, H. Nishihara, K. Inoue and I. Takata, "600 V trench IGBT in comparison with planar IGBT-an evaluation of the limit of IGBT performance, " Proceedings of the 6th International Symposium on Power Semiconductor Devices and ICs, Davos, Switzerland, 1994, pp. 411-416.

[7] Shigeki, Akio, Youichi, Satoshi and Norihito, "Carrier-storage effect and extraction-enhanced lateral IGBT (E2LIGBT): A super-high speed and low on-state voltage LIGBT superior to LDMOSFET," 2012 24th International Symposium on Power Semiconductor Devices and ICs, Bruges, Belgium, 2012, pp. 393-396.

[8] W. Zhang, B. Zhang, M. Qiao, Z. Li, X. Luo and Z. Li, "The R_{ON}, min of Balanced Symmetric Vertical Super Junction Based on R-Well Model," IEEE Transactions on Electron Devices, vol. 64, no. 1, pp. 224-230, Jan. 2017.

[9] J. Wei, J. Lu, J. Tan, R. liu, P. Zhu, H. Li, B. Zhang and X. Luo, " Novel Integrated Double pMOS SOI-LIGBT With Low Loss and High Short-Circuit Capability," IEEE Transactions on Electron Devices, vol. 71, no. 11, pp. 7199-7203, Nov. 2024.

[10] X. Luo, J. Wang, K. Yang, J. Wei, K. Dai and P. Zhu, "Novel Ultralow Loss SOI LIGBT With a Self-Adaptive pMOS and Double Floating Ohmic Contacts," IEEE Transactions on Electron Devices, vol. 70, no. 10, pp. 5196-5202, Oct. 2023.

[11] J.-H. Chul, D.-S. Byeon, J.-K. Oh, M.-K. Han, and Y.-I. Choi, "A fastswitching SOI SA-LIGBT cc NDR region," in Proc. 12th Int.Symp. Power Semiconductor Devices ICs,Aug. 2000, pp. 149–152.

[12] L. Zhang, J. Zhu, W. Sun, Y. Du, H. Yu, K. Huang and L. Shi, "A high current density SOI-LIGBT with Segmented Trenches in the Anode region for suppressing negative differential resistance regime," 2015 IEEE 27th International Symposium on Power Semiconductor Devices & IC's (ISPSD), Hong Kong, China, 2015, pp. 49-52

Study on Mechanisms of Heavy Ion Induced Leakage Current and Single Event Burnout of FRD

Ailin Qiu[1], Huan Li[2], Xiaoping Dong[1], Mingmin Huang[1*], Yao Ma[1*], Qiang Yu[3]

1 College of Physics, Sichuan University, Chengdu, China
2 China ZhenHua Group YongGuang Electronics Company, Guiyang, China
3 Sichuan Suining Lippxin Microelectronics Co., Ltd, Suining, China
* Corresponding author's E-mail: mmhuang@scu.edu.cn, mayao@scu.edu.cn

Abstract—This paper studies irradiation effects of the fast recovery diode (FRD) by heavy ion irradiation, where 1332 MeV ^{181}Ta ions are used in experiments. It is revealed for the first time that heavy ion can cause damages along the ion track to reduce the carrier lifetimes around the ion track, which increases the reverse leakage current but hardly influences the forward conduction current. With zero bias and applied bias voltage, the leakage current increases by 9 times and 13 times before and after irradiation, respectively. Moreover, the experimental result shows that the single event burnout voltage (V_{SEB}) of FRD is 655 V, which can reach 50% of the rated reverse breakdown voltage. It is analyzed by simulations that the reduction in the carrier lifetimes around the ion track may have a positive role in increasing V_{SEB}.

Keywords—heavy ion irradiation, FRD, single event leakage current, single event burnout, carrier lifetime

I. INTRODUCTION

Fast recovery diode (FRD) is an important power semiconductor device with a short reverse recovery time and low power consumption [1][2]. With the increased demand for aerospace and deep space exploration, the application of power devices has been expanded to the aerospace field, which puts higher requirements on the radiation resistance of power devices. In the aerospace field, FRD can be applied in the solar cell power system and secondary electric power supply, etc. [3].

In recent years, many studies mainly focus on heavy ion induced single event effects (SEE) of power devices, e.g., Si power MOSFET [4], SiC MOSFET [5], SiC Schottky diode [6], etc. The SEE can be categorized into several types, i.e., single event burnout (SEB), single event gate rupture (SEGR), single event leakage current (SELC), etc. It is widely acknowledged that the power MOSFET suffers from SEB and SEGR [7]. The FRD also exhibits SEB failure [8]. Although the SiC power devices have a significant advantage in the voltage rate and power loss compared to the Si power devices, they are more sensitive to SEE [9]. For the SEB of Si FRD, most studies are based on simulations without experimental validation [10][11] or only describing the phenomenon in experiments without explaining the mechanism in depth [12].

In this paper, SEB mechanisms of FRD are studied in depth by combining the 181Ta heavy ion experiment and TCAD simulation. Section II presents heavy ion experiment conditions and results. Section III gives out TCAD simulation results and analysis. Section IV presents the conclusions.

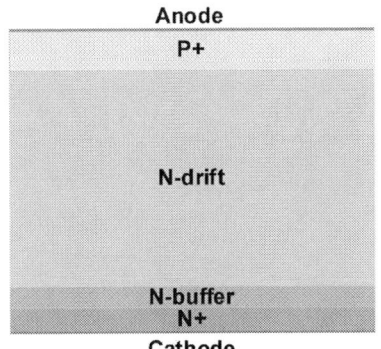

Fig. 1 Schematic diagram of FRD.

TABLE I PARAMETERS OF HEAVY ION

Ion	Energy (MeV)	Range (μm)	LET (MeV·cm²·mg⁻¹)
^{181}Ta	1332	79.69	82.1

II. EXPERIMENT CONDITIONS AND RESULTS

Fig. 1 shows the schematic diagram of FRD. Two commercial 1.2 kV FRD samples (FRD #1, FRD #2) were selected for the experiment. The two samples were decapped before the irradiation experiment. The heavy ion (^{181}Ta) irradiation experiment was conducted at the Harbin Institute of Technology Space Environment Simulation and Research Infrastructure (HITSESRI). Table I presents the parameters of the ^{181}Ta ion incident into the Si material simulated by the SRIM software.

During irradiation, the cathode bias voltage (V_R) was applied, and the cathode current (I_R) was monitored by the Keithley 2470 source meter. The V_R of FRD #1 was linearly scanned from 0 V until the SEB occurred (I_R increasing uncontrollably). The V_R of FRD #2 was set to be 0 V, aiming to study the effects of heavy ion incidence on the sample itself. The period of ion pulses was 12 s, and the averaged fluence rate was 5.13×10^3 cm⁻²/s. The total fluence (Φ) of FRD #1 was 4.9×10^5 cm⁻², and the Φ of FRD #2 was 1.45×10^6 cm⁻².

Fig. 2 shows waveforms of V_R and I_R of FRD #1 during irradiation. It shows that there is a current pulse when heavy ions are incident into the device. It can be found that the I_R is suddenly uncontrollably increased at V_R = 655 V. Thus, it can be considered that the SEB voltage of FRD is about 655 V.

979-8-3315-2209-4/25 $31.00 © 2025 IEEE

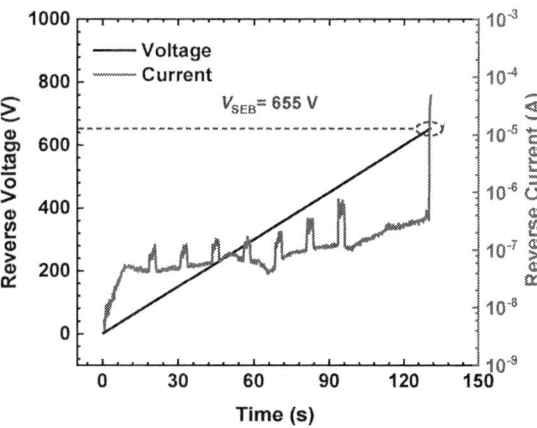

Fig. 2 Cathode leakage current and voltage waveforms of FRD#1 during irradiation.

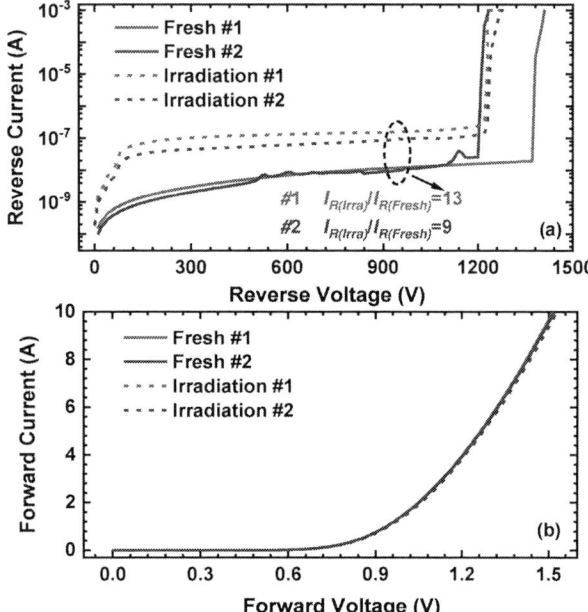

Fig. 3 Electrical characteristics of samples before and after irradiation. (a) Reverse I-V curves. (b) Forward I-V curves.

Fig. 3 shows the reverse I-V curves and forward I-V curves of the two samples. It can be found that the reverse leakage currents (I_L) of the two samples are increased after irradiation, while the forward currents (I_F) of them are almost unchanged. Even if the total fluence of the FRD #1 is only about 1/3 of that of the FRD #2, the increase in I_L of the FRD #1 is larger than the FRD #2, where the former is increased by 13 times and the latter is increased by 9 times. After irradiation, the breakdown voltage (BV) of the FRD #1 is decreased, while the BV of the FRD #2 remains almost unchanged. Note that the difference in original values of BV of the two samples is caused by decapping. It can be inferred from Fig. 3(a) that defects can be induced in the device by heavy ions themselves. This mechanism may have an effect on SEB of FRD. Besides, the applied voltage during ion irradiation can enhance defects in the device. Moreover, it seems strange that the forward I-V curves do not change after irradiation, which will also be explained in Section III.

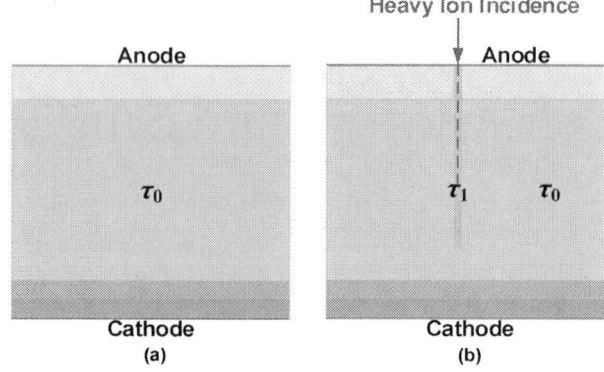

Fig. 4 Carrier lifetime distributions before and after irradiation. (a) before irradiation. (b) after irradiation. (τ_1 is the carrier lifetime of quite small damaged area, τ_0 is the carrier lifetime of undamaged area.)

III. TCAD SIMULATION AND ANALYSIS

According to results in Fig. 3(a), defects can be simultaneously induced in all probability when the heavy ion is incident into the device. In previous works, the effect of simultaneously induced defects by heavy ions on SEB of power semiconductor devices has been hardly considered. In fact, the simultaneously induced defects should largely reduce the carrier lifetimes along the ion track at the same time, which will play an important role in influencing the SEB. This section will study this mechanism in detail by TCAD simulations.

In simulations, the device length is 100 μm, the width is 1 μm, the N-drift doping concentration and thickness are 1.35×10^{14} cm^{-3} and 105 μm respectively, the N-buffer doping concentration and thickness are 2×10^{16} cm^{-3} and 2.5 μm respectively, the P+ depth is 5 μm, and the N+ depth is 0.5 μm. LET_F is set to be 0.85 pC/μm and r is set to 0.08 μm, where 0.85 pC/μm is equivalent to be 82.1 MeV·cm^2/mg (1 pC/μm (Si) = 96.53 MeV·cm^2/mg), and r is the damaged radius of one heavy ion simulated by the Geant4 software. The cathode thermal resistance is set to 1.49×10^{-6} cm^2·K/W, which is equal to 1.49 K/W. The heavy ion is vertically incident into the anode surface at the middle of the device. The definition of SEB in the simulation is that the lattice temperature reaches the melting point of Si, 1687 K.

Fig. 4 shows the carrier lifetime distributions before and after irradiation, which is used to explain the phenomena that I_L is obviously increased and I_F is almost unchanged in Fig. 3. When a heavy ion with high energy is incident into silicon, most energy of the heavy ion is converted to the energy for ionizing electron-hole pairs, and the left energy is lost due to hitting the silicon atoms to induce displacement damages (i.e., defects). Therefore, the carrier lifetimes along the ion track can be decreased from the original value (τ_0) to a much lower value (τ_1). At the reverse blocking state, I_L is mainly dependent on the damaged area (along heavy ion tracks) with low carrier lifetimes. Thus, I_L increases significantly. At the forward on-state, since the damaged area is very small, the modulation of the N-drift is hardly weakened overall. Thus, I_F is nearly unchanged. It is interesting to note that the simultaneously induced defects by heavy ions may have a positive role in improving the resistance to SEB, which will be discussed below.

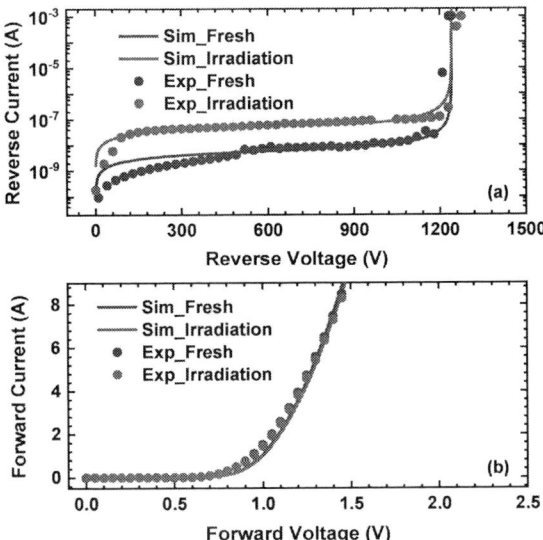

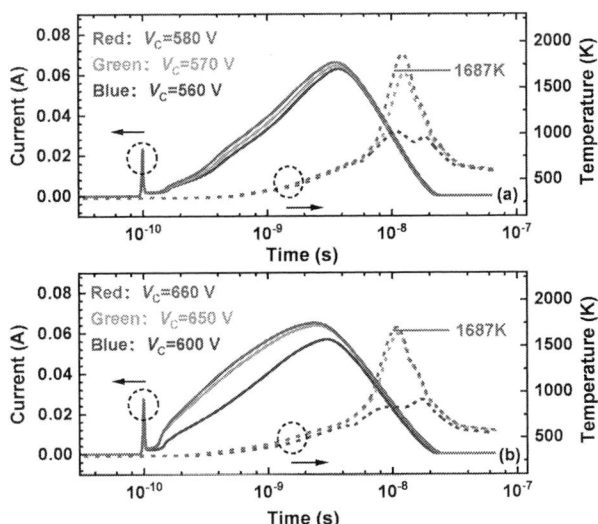

Fig. 5 Simulated and measured I-V curves of FRD #2 before and after heavy ion irradiation. (a) Reverse I-V curves. (b) Forward I-V curves.

Fig. 6 Cathode current and peak temperature waveforms. (a) $\tau_0 = \tau_1 = 1\times10^{-4}$ s at V_R=560 V, 570 V and 580 V. (b) $\tau_0 = 1\times10^{-4}$ s and $\tau_1 = 6\times10^{-10}$ s at V_R=600 V, 650 V and 660 V.

The I_L in the experiment can be used to calculate the carrier lifetimes in the damaged area and in the undamaged area by the equations below,

$$I_L = A \cdot J_R = A \cdot [(1-m)J_{R_0} + mJ_{R_1}] \qquad (1)$$

where A is the total active area of the sample (4×10^{-2} cm^2), J_R is the reverse current density, m is the ratio of the damaged active area to the total active area, J_{R0} is the reverse current density before irradiation, and J_{R1} is the reverse current density in the damaged area. The expression of m can be given by,

$$m = (\Phi \cdot A) \cdot \pi r^2 / A = \Phi \cdot \pi r^2 \qquad (2)$$

where Φ is the total fluence (1.45×10^6 cm^{-2}) and r is the damaged radius of one heavy ion (0.08 μm). By using (2), m is 0.03%. The generated current in the depleted region is the main component of the leakage current. Then, the expression of J_{R0} and J_{R1} can be given by,

$$J_{R_0} = \frac{qW_n n_i}{\tau_0} \qquad (3)$$

$$J_{R_1} = \frac{qD_{Irra}n_i}{\tau_1} + \frac{q(W_n - D_{Irra})n_i}{\tau_0} \approx \frac{qD_{Irra}n_i}{\tau_1} \qquad (4)$$

where n_i is the intrinsic carrier concentration, W_n is the width of N-drift region (105 μm, i.e., width of the depleted region at a high enough voltage), D_{Irra} is the depth of damaged region (79.69 μm), τ_0 ($= \tau_{n0} + \tau_{p0}$) is sum of electron and hole lifetimes of the region without irradiation, and τ_1 ($= \tau_{n1} + \tau_{p1}$) is sum of electron and hole lifetimes of the damaged region. If it is supposed that $\tau_{n0} = \tau_{p0}$ and $\tau_{n1} = \tau_{p1}$ and $\tau_1 = \tau_{n1} + \tau_{p1}$, it can be calculated by (1)-(4) that $\tau_0 = 1\times10^{-4}$ s and $\tau_1 = 3\times10^{-9}$ s.

By using values of τ_0, τ_1, m, r and D_{Irra}, I-V curves of FRD #2 are simulated. Fig. 5 shows simulated and measured reverse I-V curves and forward I-V curves of FRD #2 before and after heavy ion irradiation, where the simulation results are in good agreement with experiment results, which indicates that the calculated values of τ_0 and τ_1 are reasonable and credible.

Furthermore, the response of the FRD during the incidence of a heavy ion is simulated, where both the case without simultaneous damages ($\tau_1 = \tau_0$) and the case with simultaneous damages ($\tau_1 \neq \tau_0$) are considered. Fig. 6(a) shows the current and temperature waveforms of the FRD irradiated under $V_R = 560$ V, 570 V and 580 V, where the simultaneous damages are neglected ($\tau_1 = \tau_0 = 1\times10^{-4}$ s). It can be seen that the V_{SEB} is between 570 V and 580 V, which is quite smaller than the 655 V in the experiment. Fig. 6(b) shows the current and temperature waveforms of the FRD irradiated under $V_R = 600$ V, 650 V and 660 V, where simultaneous damages are considered and the value of τ_1 is adjusted to a lower value ($\tau_1 = 6\times10^{-10}$ s) than the calculated value to meet the experiment results. It can be found that V_{SEB} is between 650 V and 660 V in this case of simulation, which is consistent with the experiment. The reason why the τ_1 used in simulation needs to be lower than the calculation value is that the τ_1 at the moment of heavy ion incidence may be quite small, however, some defects generated by heavy ions may be eliminated because of the annealing caused by increased temperature, and τ_1 after irradiation can be increased compared to τ_1 during irradiation. Combining Fig. 5 and Fig. 6, it can be inferred that there should be reduction of local carrier lifetime caused by heavy ion incidence, and this mechanism may be beneficial to improve the resistance to SEB.

In order to analyze the positive effect of simultaneous damages on improving the resistance to SEB, the reason for SEB should be studied. In the following, the case of $V_R = 660$ V in Fig. 6(b) is used to explain the mechanism of SEB.

Fig. 7 shows carrier distributions at the moment of heavy ion incidence. It can be found that when heavy ions are incident, heavy ions generate numerous carriers along the ion track. The electrons will flow to the cathode and the holes will flow to the anode because of the electric field. Accumulation of electrons at the N-drift/N-buffer and holes at the P+/N-drift will cause high electric field at these locations.

979-8-3315-2209-4/25 $31.00 © 2025 IEEE

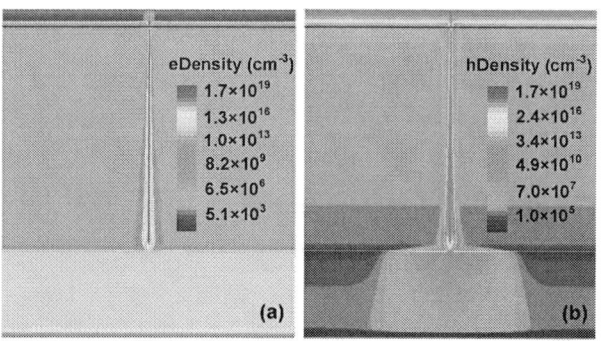

Fig. 7 Carrier distributions at the moment of heavy ion incidence (V_R = 660 V). (a) Electrons distribution. (b) Holes distribution.

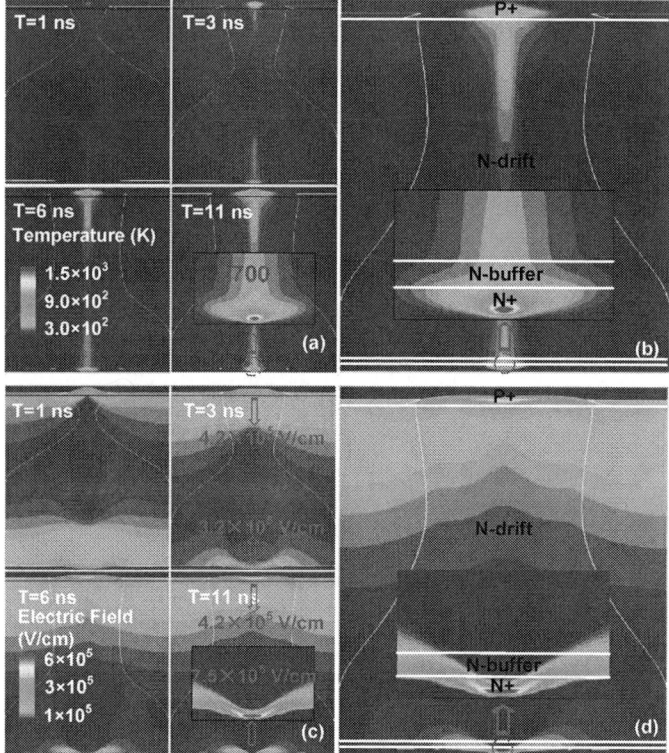

Fig. 8 Spatial distributions at different times (V_R=660 V). (a) Lattice temperature distributions. (b) Magnified temperature distribution at T = 11 ns. (c) Electric field distributions. (d) Magnified electric field distribution at T = 11 ns.

Fig. 8 shows lattice temperature and electric field distributions during heavy ion irradiation. When the electric field exceeds the critical breakdown electric field of Si (about 2×10^5 V/cm), avalanche multiplication occurs, and the impact ionization at these regions generates electron-hole pairs to form a huge current. As the current increases, power consumption and temperature also rise, creating positive feedback between current and temperature. If the temperature continues to rise and reaches the melting point of Si, SEB occurs. However, when defects generated by heavy ions capture the electrons and holes, it may reduce the accumulation of electrons and holes, reducing the electric field and current, which has a positive role in improving the resistance to SEB and makes FRD have a high V_{SEB} of 50% of breakdown voltage.

IV. CONCLUSIONS

The novel phenomenon of SEB in FRD is introduced by combining experiments with simulations. The reason why reverse leakage current increases but forward conduction current remains almost unchanged is that heavy ions cause damages in FRD to reduce carrier lifetimes in local but the damage area is too small to affect the conductivity modulation in the N-drift region, which is verified by TCAD simulations. It is worth mentioning that although the reduction of local carrier lifetime leads to an increase in leakage current, it also may increase the V_{SEB} of the FRD, which makes the FRD have positive resistance to heavy ion irradiation.

ACKNOWLEDGMENT

This work was supported by Sichuan University-Suining Strategic Cooperation Project under Grant No.2023CDSN-13.

REFERENCES

[1] W. C. Hung et al. "Leakage Current in Fast Recovery Diode Suppressed by Low Temperature Supercritical Fluid Treatment Process," in IEEE Electron Device Letters, vol. 41, no. 10, pp. 1540-1543, 2020.

[2] F. He et al., "Double Local Lifetime Control Enabling Lower Energy Loss of Fast Recovery Diode," 2020 IEEE 2nd International Conference on Civil Aviation Safety and Information Technology (ICCASIT), Weihai, China, 2020, pp. 252-255.

[3] Y. Hagiwara, "Pinned Buried PIN Photodiode Type Solar Cell," 2021 International Conference on Electrical, Computer and Energy Technologies (ICECET), Cape Town, South Africa, 2021, pp. 1-6.

[4] A. E. Waskiewicz et al., "Burnout of Power MOS Transistors with Heavy Ions of Californium-252," in IEEE Transactions on Nuclear Science, vol. 33, no. 6, pp. 1710-1713, 1986.

[5] E. Mizuta et al., "Investigation of Single-Event Damages on Silicon Carbide (SiC) Power MOSFETs," in IEEE Transactions on Nuclear Science, vol. 61, no. 4, pp. 1924-1928, 2014.

[6] A. F. Witulski et al., "Single-Event Burnout of SiC Junction Barrier Schottky Diode High-Voltage Power Devices," in IEEE Transactions on Nuclear Science, vol. 65, no. 1, pp. 256-261, 2018.

[7] M. Allenspach et al., "SEGR and SEB in n-channel power MOSFETs," in IEEE Transactions on Nuclear Science, vol. 43, no. 6, pp. 2927-2931, 1996.

[8] G. Soelkner et al., "Charge carrier avalanche multiplication in high-voltage diodes triggered by ionizing radiation," in IEEE Transactions on Nuclear Science, vol. 47, no. 6, pp. 2365-2372, 2000.

[9] J. M. Lauenstein. "Wide-Bandgap Semiconductors in Space: Appreciating the Benefits but Understanding the Risks." Conference on Radiation Effects on Components and Systems (RADECS 2018). No. 2018-561-NEPP, 2018.

[10] C. An et al., "Simulation Study On 4.5kV FRD Failure Induced By Neutron Irradiation," 2024 IEEE 2nd International Conference on Power Science and Technology (ICPST), Dali, China, 2024, pp. 396-400.

[11] X. Liao et al., "Simulation Aided Hardening of Power Diodes to Prevent Single Event Burnout," in IEEE Transactions on Electron Devices, vol. 69, no. 9, pp. 5088-5095, 2022.

[12] M. Mauguet et al., "Analysis of Heavy Ion Irradiation Test Results on Power Diodes," 2018 18th European Conference on Radiation and Its Effects on Components and Systems (RADECS), Goteborg, Sweden, 2018, pp. 1-4.

979-8-3315-2209-4/25 $31.00 © 2025 IEEE

Effect of Electron Irradiation and Post-Irradiation Annealing on Reverse-Conducting IGBT

Pengwei Chen[1], Huan Li[2], Xiaoping Dong[1], Mingmin Huang[1]*, Yao Ma[1]*, Chang Chen[1], Qiang Yu[3]

1 College of Physcs, Sichuan University, Chengdu, China

2 China ZhenHua Group YongGuang Electronics Company, Guiyang, China

3 Sichuan Suining Lippxin Microelectronics Co., Ltd, Suining, China

*Corresponding author's E-mail: mmhuang@scu.edu.cn, mayao@scu.edu.cn

Abstract—This article investigates the effect of electron irradiation and post-irradiation high-temperature annealing on static electrical characteristics of trench RC-IGBTs (1350 V/20 A) through experiments. The irradiation experimental results show that the threshold voltage, breakdown voltage, and conduction current (including saturation current) decrease and the collector leakage current increases observably. In the annealing processes mentioned in this article, the threshold voltage of the irradiated device recovers by 89.5%. Meanwhile, the blocking capability and conduction capability recover to pre-irradiation levels. The physical mechanisms of electrical characteristics variations by electron irradiation and post-irradiation annealing are explained from the view of device physics.

Keywords—RC-IGBTs, electron irradiation, post-irradiation annealing, static characteristics, fixed oxide charge, carrier lifetime

I. INTRODUCTION

The reverse-conducting insulated gate bipolar transistor (RC-IGBT) plays an important role in power electronics, because of its high voltage and power density [1]. Nevertheless, previous research has shown that silicon-based IGBTs have great challenges ahead when used in the power system of spacecrafts. In the space environment, radiation such as γ-ray, neutrons, protons, and electrons, will cause total ionizing dose (TID) effects on IGBT, affecting its electrical characteristics and hindering the long-term effective operation of IGBT [2-4]. Many researches in recent years have focused on the impact of γ-ray [5-7]. However, there are few studies on the TID effects induced by the electron irradiation.

To recover the electrical performance of semiconductor devices degenerated by the TID effects, high-temperature annealing is approached normally, because of its simplicity and high effectiveness. Previous studies have investigated TID and post-irradiation annealing effects on bipolar transistors and 0.18μm bulk n-channel MOSFETs exposed to different total doses of gamma irradiation [8-9]. The device performance has recovered to varying degrees after annealing. However, there are few studies on the post-irradiation effect of IGBT.

In this work, the TID and post-irradiation high-temperature annealing effects on electrical characteristics of RC-IGBTs were investigated through experiments. The effective annealing condition is studied, and variations in electrical characteristics by annealing and the mechanisms are studied.

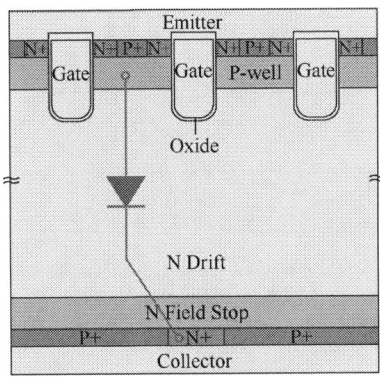

Fig. 1 Schematic cross section of the trench RC-IGBT.

II. SAMPLES AND EXPERIMENTAL SETUPS

Fig. 1 shows the schematic cross section of the trench reverse conducting IGBT (RC-IGBT), which integrates a PiN diode and an IGBT. The PiN diode is anti-parallel with the IGBT. Compared to the pair of discrete IGBT and PiN diode, the RC-IGBT increases power density and optimizes switching characteristics [10]. In the RC-IGBT, the n-type field stop layer is moderately doped to stop the electric field. The RC-IGBT turns on when the gate-emitter voltage (V_{GE}) exceeds the threshold voltage. When the collector-emitter voltage (V_{CE}) is low, the RC-IGBT operates in the MOS-mode. Furthermore, when V_{CE} is high enough, the electron current flows laterally in the n-type field stop layer, inducing a voltage drop higher than 0.7 V to turn on the pn junction of P+ collector and n-type field stop layer. Then, the RC-IGBT operates in the IGBT-mode.

Table I lists the electron irradiation conditions. The devices under test (DUTs, 1350V/20A RC-IGBTs), were irradiated by 1.7 MeV electron irradiation. The dose rate was around 5×10^{12} cm^{-2}·s^{-1}, and the total dose was 1×10^{15} cm^{-2}. During irradiation, the DUTs were at a gate bias of -6 V and the collector and emitter electrodes were grounded. The I-V curves of the DUTs are measured by Agilent B1505A. The threshold voltage (V_{TH}) was defined as the voltage at the collector current (I_C) of 10 mA in transfer I-V curves. To offset the effect of the V_{TH} shift (ΔV_{TH}) on the breakdown characteristics and the output characteristics after electron irradiation and annealing, a consistent value of $V_{GE} - V_{TH}$ (= 3.52 V) was used when testing.

979-8-3315-2209-4/25 $31.00 © 2025 IEEE

TABLE I   ELECTRON IRRADIATION CONDITIONS

Experiment Conditions	Values
energy	1.7 MeV
total dose	1×10^{15} cm^{-2}
dose rate	5×10^{12} cm$^{-2}\cdot$s^{-1}
voltage bias	$V_G = -6$ V, $V_C = V_E = 0$ V

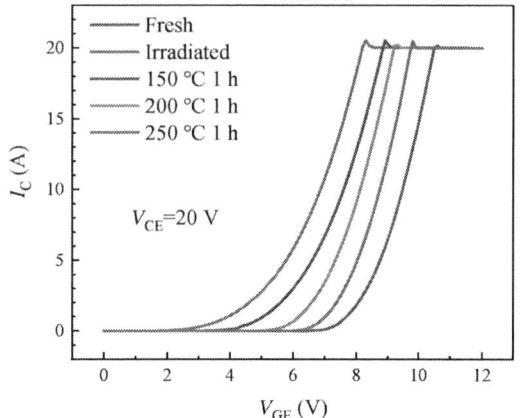

Fig. 2 Transfer characteristics of the DUT before and after irradiation, and after annealing for 1 h at 150 °C, 1 h at 200 °C, and 1 h at 250 °C.

The post-irradiation high-temperature annealing was completed by high-temperature furnace NBT T1500. Several DUTs were annealed at different temperatures to investigate the safe temperature range first. It was found that the DUTs can fully withstand the thermal stress as high as 250 °C. Then, one unannealed DUT was chosen to be annealed for 1 hour at 150 °C, 1 hour at 200 °C, and 1 hour at 250 °C consecutively. After each annealing, the electrical characteristics were measured once the DUT had been cooled to room temperature.

In addition, it was found that V_{TH} after annealing for 1 h was nearly the same as that annealing for 2 h at the same annealing temperature, which indicates that the recovery of electrical characteristics of the DUT is mainly depended on the annealing temperature but not the annealing time. Hence, the experimental results of the cases annealing for 1 h is taken to study the post-irradiation annealing effect.

III. EXPERIMENT RESULTS AND ANALYSES

Fig. 2 shows the transfer characteristics of the DUT before and after irradiation, and after annealing. The original V_{TH} is 6.48 V, and V_{TH} of DUT is decreased by about 5 V after electron irradiation. There are 1.66 V, 1.76 V, and 0.98 V increments in V_{TH} after each annealing at 150 °C, 200 °C, and 250 °C, respectively. The V_{TH} after annealing at 250 °C only differs by 0.68 V from the pre-irradiation value.

Fig. 3 shows the breakdown characteristics of the DUT before and after irradiation, and after annealing. The original breakdown voltage (BV) is 1478 V (at $I_C = 250$ µA) and it decreases by approximately 80 V after electron irradiation. There are 55 V, 23 V, and 2 V increments in BV after each annealing are at 150 °C, 200 °C, and 250 °C, respectively. Annealing above 200 °C is able to fully recover the BV.

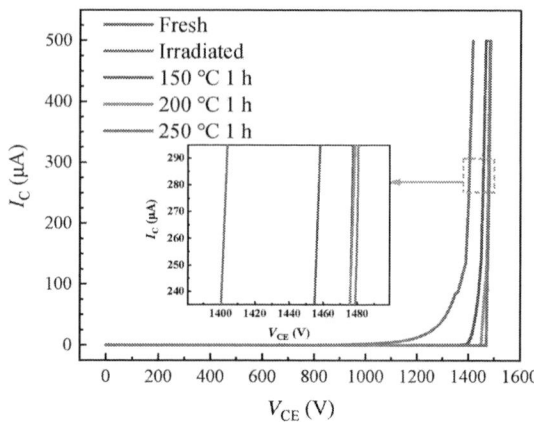

Fig. 3 Breakdown characteristics of the DUT before and after irradiation, and after annealing for 1 h at 150 °C, 1 h at 200 °C, and 1 h at 250 °C.

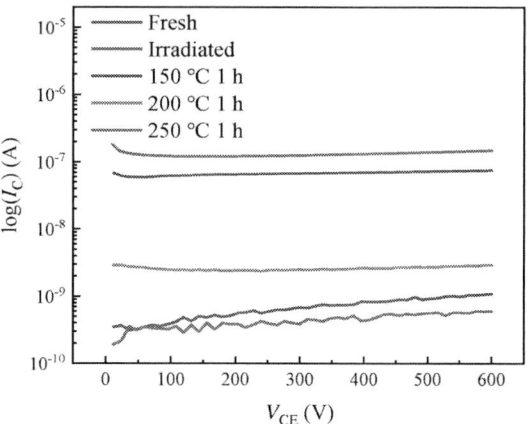

Fig. 4 Collector leakage current of the DUT before and after irradiation and after annealing for 1 h at 150 °C, 200 °C, and 250 °C.

Fig. 4 shows the collector leakage current (I_L) of the DUT before and after irradiation, and after annealing. The original I_L is 1 nA (at V_{CE}=300 V) and the I_L is increased by more than 100 times after irradiation. The I_L is significantly recovered after annealing at 200 °C. After annealing at 250 °C, I_L is fully recovered and even lightly lower than that before irradiation.

Fig. 5 shows the output characteristics of the DUT before and after irradiation, and after annealing. The saturation current is decreased by more than 10 A after electron irradiation, which means the conduction capability decreases severely. The saturation current is significantly recovered after annealing at 200 °C, and becomes slightly higher than that before irradiation after annealing at 250 °C. In the MOS-mode, the on-resistance consists of resistances of the n-drift region and the channel, where the resistance of the n-drift region should be higher than the channel resistance when $V_{GE} - V_{TH}$ is high enough. The on-resistance in the MOS-mode is nearly unchanged before and after irradiation. Therefore, the electron mobility and doping concentration in drift region should be unchanged.

Fig. 6 shows the RC curves of the DUT before irradiation and after irradiation and annealing at room temperature (RT) for 150 days. The reverse conduction voltage of DUT is unchanged. The forward leakage current is increased after irradiation and annealing at room temperature for 150 days.

979-8-3315-2209-4/25 $31.00 © 2025 IEEE

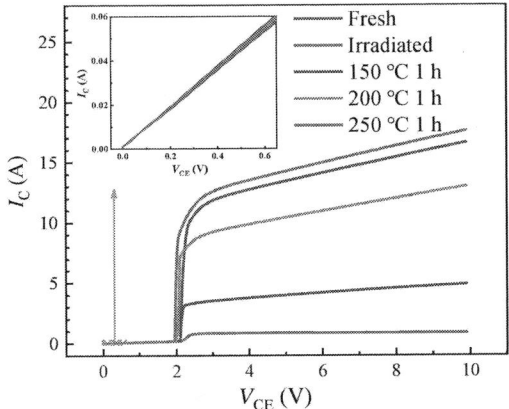

Fig. 5 Output characteristics of the DUT before and after irradiation and after annealing for 1 h at 150 °C, 1 h at 200 °C, and 1 h at 250 °C.

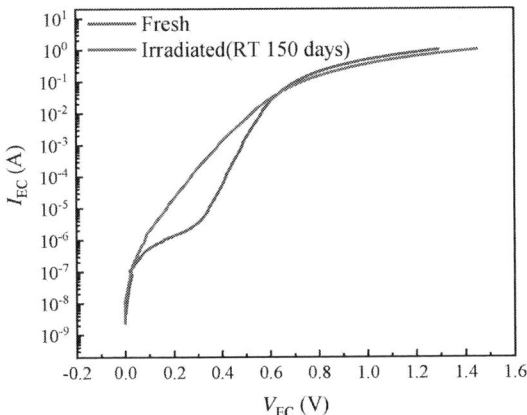

Fig. 6 The RC curves of the DUT before irradiation and after irradiation and annealing at room temperature for 150 days.

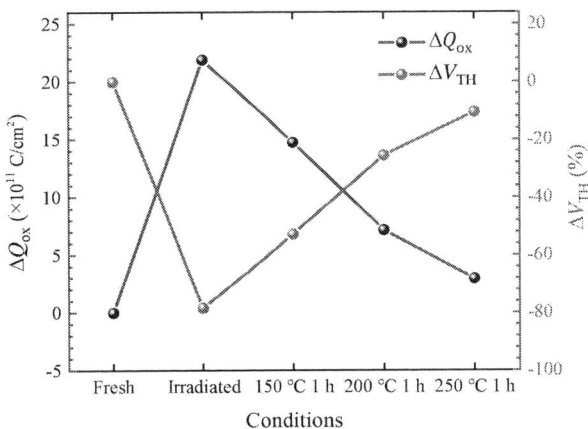

Fig. 7 The ΔQ_{ox} and ΔV_{TH} of the DUT before and after irradiation and after annealing for 1 h at 150 °C, 200 °C, and 250 °C.

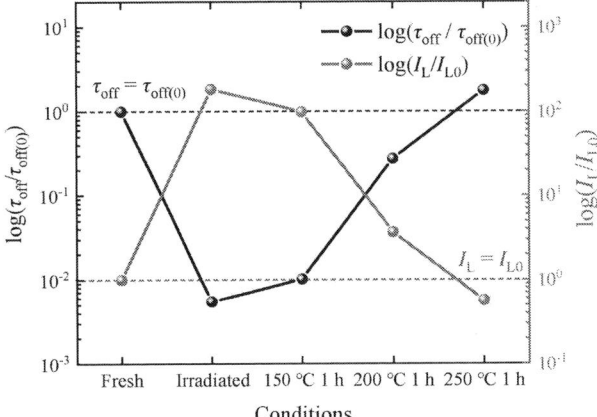

Fig. 8 The normalized τ_{off} and I_L of the DUT before and after irradiation and after annealing for 1 h at 150 °C, 200 °C, and 250 °C.

It is worthy to study reasons of variations of V_{TH}, BV, I_L, and conduction capability after irradiation and annealing. The reasons will be analyzed according to device physics as follows.

The V_{TH} shifts (ΔV_{TH}) can be equivalently attributed to the change of fixed oxide charge (ΔQ_{ox}) during irradiation and annealing [11]. The ΔQ_{ox} is given by,

$$\Delta Q_{ox} = \Delta V_{TH} \times C_{ox} = \Delta V_{TH} \varepsilon_{SiO2} / T_{ox} \quad (1)$$

where C_{ox} is the gate oxide capacitance per unit area, T_{ox} is the thickness of gate oxide, and ε_{SiO2} is the permittivity of oxide.

Fig. 7 shows ΔQ_{ox} and ΔV_{TH} of the DUT after irradiation, and annealing. During irradiation, the radiation-induced holes trapping leads to the increase of ΔQ_{ox}. The ΔQ_{ox} induces an extra electric field adding to gate oxide and the inversion layer forms more easily, leading to a decrease in V_{TH} after irradiation. After irradiation, V_{TH} is decreased by 78.4% and ΔQ_{ox} can be increased by 2.2×10^{12} cm^{-2}. After annealing at 250 °C, ΔQ_{ox} is only 3×10^{11} cm^{-2} and V_{TH} differs by 10.5% from the original value. The ΔQ_{ox} is mainly contributed by radiation-induced trapped holes. During high-temperature annealing, the trapped holes obtain enough energy to jump from the trap energy levels into the valence band to recombine with free electrons. Then, ΔQ_{ox} is reduced and the V_{TH} is recovered.

The variation of BV after irradiation and annealing may also be related to the fixed oxide charges in the field oxide at the termination of the DUT. With a positive value of ΔQ_{ox} induced by irradiation in the field oxide, the electric field at the termination may become more concentrated and lead to the reduction of BV [12]. After high-temperature annealing, ΔQ_{ox} in field oxide is also reduced and BV is then also recovered.

The reason for the variation of I_L after irradiation and annealing is analyzed below. At the off-state, the pn junction of P+ collector and n-type field stop layer is not turned on, and the RC-IGBT operates in the MOS-mode. Hence, the main component of I_L is the generated current in depleted n-drift region. Then, I_L can be given by,

$$I_L = A J_{SC} \quad (2)$$

where A is the active area, and J_{SC} is the space-charge generating current density. The J_{SC} can be defined as,

$$J_{SC} = n_i / \tau_{off} \sqrt{2q\varepsilon_s V_C / N_D} \quad (3)$$

where n_i is the intrinsic carrier concentration, τ_{off} is the carrier lifetime in the depleted drift region at the off-state, ε_s is the permittivity of silicon, V_C is the collector voltage, and N_D is the doping concentration in the drift region. According to (2)-(3),

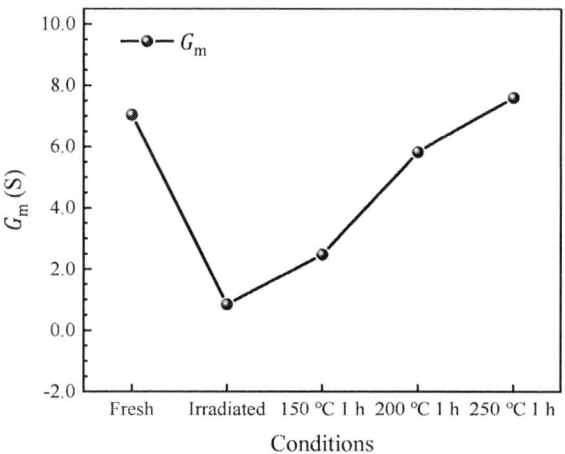

Fig. 9 The transconductance at saturation region of the DUTs before and after irradiation and after annealing for 1 h at 150 °C, 200 °C, and 250 °C.

the relation between I_L and τ_{off} can be given by,

$$I_L \propto \tau_{off}^{-1}. \tag{4}$$

With constant values of A, n_i, N_D, and V_C (=300 V), I_L is in fact only affected by the τ_{off}. Fig. 8 shows the normalized τ_{off} extracted from the I-V curves in Fig. 4 by using (4), where $\tau_{off(0)}$ is the original carrier lifetime in the depleted n-drift region. The τ_{off} decreases more than 100 times after electron irradiation. Previous research shows that the deep energy levels will be induced by electron irradiation [13]. The deep energy levels reduce the τ_{off} causing an increase of the I_L. After high-temperature annealing processes, the deep energy levels are gradually eliminated. After annealing at 250 °C, the τ_{off} of RC-IGBT become about 2 times of that before irradiation.

The trend of transconductance at the saturation region (G_m) of the DUT can explain the variations of conduction capability after irradiation and annealing. The G_m can be calculated by,

$$G_m = \partial I_C / \partial V_{GE} = \Delta I_C / \Delta V_{GE}$$
$$= \mu_n C_{ox} \left(1 - \alpha_{on}\right)^{-1} \left(V_{GE} - V_{TH}\right) Z / L \tag{5}$$

where μ_n is the electron mobility, α_{on} is the gain of the PNP transistor when the IGBT is in the on-state, and Z/L is the ratio of channel width to channel length.

The values of G_m can be calculated by (5) with two output I-V curves at V_{GE} - V_{TH}= 3.52 V and 3.02 V. Fig. 9 shows values of G_m at V_{CE} = 3 V (in current saturation region). After electron irradiation, the G_m is decreased from 7.04 S to 0.84 S, i.e., decreased by 88%. The increments in G_m by each annealing are 1.64 S, 3.34 S and 1.78 S. After annealing at 250 °C, G_m of RC-IGBT can be 8% higher than that before irradiation. It can be learned that the value of G_m is influenced by μ_n and α_{on}, where α_{on} is positively correlated with the carrier lifetime at the on-state (τ_{on}) (certainly positively correlated with τ_{off}). It can be found by comparing Fig. 8 and Fig. 9 that the variation trend of G_m is almost consistent with the variation trend of τ_{off}. Therefore, it can be inferred that the variation of G_m is mainly caused by inducing deep energy levels due to irradiation and removing of deep energy levels due to high-temperature annealing.

IV. CONCLUSIONS

This paper studies the effect of electron irradiation and post-irradiation high-temperature annealing on electrical characteristics of the RC-IGBT. The electrical characteristics of the RC-IGBT can be significantly degraded by electron irradiation with a dose of 1×10^{15} cm^{-2}. High-temperature annealing at 250 °C is able to recovery most of irradiation induced degradation in electrical characteristics, where BV, I_L and conduction capability can be fully recovered, and V_{TH} is recovered by 89.5%. This indicates that all irradiation induced defects in silicon has been removed, and most of fixed oxide charges induced by irradiation has been removed. In addition, it is found that the annealing temperature is much more effective to recover the degradation than the annealing time.

ACKNOWLEDGMENT

This work was supported by Sichuan University-Suining Strategic Cooperation Project under Grant No. 2023CDSN-13.

REFERENCES

[1] Findlay, Emma M., and Florin Udrea. "Reverse-conducting insulated gate bipolar transistor: A review of current technologies." *IEEE Transactions on Electron Devices* 66.1 (2018): 219-231.

[2] Yazdi, M. Baghaie, et al. "A concise study of neutron irradiation effects on power MOSFETs and IGBTs." *Microelectronics reliability* 62 (2016): 74-78.

[3] Baek, Hani, et al. "Effects of gamma irradiation on the electrical characteristics of trench-gate non-punch-through insulated gate bipolar transistor." *Semiconductor Science and Technology* 34.6 (2019): 065022.

[4] Liu, Zhenhua, et al. "Impact of proton-induced total ionizing dose effects on electrical characteristics and safe operating area of trench field-stop IGBT devices." *Microelectronics Reliability* 154 (2024): 115326.

[5] Marceau, M., C. Brisset, and M. Da Costa. "Study of dose effects on IGBT-type devices subjected to gamma irradiation." *IEEE Transactions on Nuclear Science* 46.6 (1999): 1680-1685.

[6] Tala-Ighil, Boubekeur, et al. "Experimental and comparative study of gamma radiation effects on Si-IGBT and SiC-JFET." *Microelectronics Reliability* 55.9-10 (2015): 1512-1516.

[7] Tala-Ighil, Boubekeur, et al. "Analysis of Commercial Punch-Through IGBTs Behavior Under ^{60}Co Irradiation: Turn-Off Switching Performances Evolution." *IEEE Transactions on Nuclear Science* 59.6 (2012): 3235-3243.

[8] Mo, Rigen, et al. "Study on annealing effect of bipolar transistors at different temperatures after total dose irradiation." *Microelectronics Reliability* 150 (2023): 115125.

[9] Aditya, Kritika, et al. "Effect of post radiation annealing on the TID response of 0.18 μm bulk NFETs." 2019 *Electron Devices Technology and Manufacturing Conference (EDTM)*. IEEE, 2019.

[10] Zhu, Liheng, et al. "Advanced High Voltage Reverse Conducting RC-IGBT Technology with Low Losses and Robust Switching Performance." 2020 32nd *International Symposium on Power Semiconductor Devices and ICs (ISPSD)*. IEEE, 2020.

[11] Oldham, Timothy R., and F. B. McLean. "Total ionizing dose effects in MOS oxides and devices." *IEEE transactions on nuclear science* 50.3 (2003): 483-499.

[12] Shu, Lei, et al. "Numerical and experimental investigation of TID radiation effects on the breakdown voltage of 400-V SOI NLDMOSFETs." *IEEE Transactions on Nuclear Science* 66.4 (2019): 710-715.

[13] Rai-Choudhury, P., J. O. H. N. Bartko, and JOSEPH E. Johnson. "Electron irradiation induced recombination centers in silicon-minority carrier lifetime control." *IEEE Transactions on Electron Devices* 23.8 (1976): 814-818.

1500V High-Voltage GaN HEMT Device with Multiple Field Plates for High Power Applications

Moufu Kong, Yaowen Zhang, Yingzhi Luo, Kangxiang Zhao, Bingke Zhang, Bo Yi, Hongqiang Yang

Abstract—In this paper, an optimized field plate structure is proposed for the high voltage application of gallium nitride (GaN) high electron mobility transistors (HEMTs). The proposed GaN HEMT (pro-HEMT) introduces a multi-layer field plate to reduce the peak electric field near the gate electrode, consequently enhancing the breakdown voltage of the device. The pro-HEMT was fabricated on a 6-inch sapphire substrate, experimental results demonstrate that the GaN HEMT device achieves a breakdown voltage of over 1500V with a gate-to-drain distance (L_{GD}) of ~33μm which is more than twice the breakdown voltage level of 650 V compared with the conventional GaN HEMT (conv-HEMT). In comparison to the conv-HEMT, the maximum electric field within the device is reduced from 4.75 MV/cm to 3.12 MV/cm with a decrease of approximately 34%. The threshold voltage (V_{th}) and the specific on-resistance ($R_{on,sp}$) of the proposed HEMT are -10.7V and 5.61 mΩ·cm², respectively. The results indicate that the pro-HEMT can be applied to most of the current high-voltage scenarios.

Index Terms—GaN HEMTs, field plate, breakdown voltage, on-resistance,

I. INTRODUCTION

Currently, the third-generation semiconductor power devices have demonstrated significant potential in the power electronics market. Gallium Nitride (GaN) has gained considerable attention due to its wider bandgap, higher electron saturation velocity, better high-temperature characteristics and higher breakdown electric field compared with silicon material. Leveraging these attributes, GaN high electron mobility transistors (HEMTs) are proposed, which utilize the two-dimensional electron gas (2DEG) generated by the AlGaN/GaN heterojunction to conduct current. The 2DEG has high channel electron concentration and high electron mobility, which help GaN HEMTs become a kind of devices with low on-resistance ($R_{DS,on}$) and high breakdown voltage (BV), and high switching speed [1]. However, GaN HEMT devices currently face limitations in terms of device structure

and process technology, making them less competitive for high voltage (e.g. >1200V) applications. The major challenge is that the GaN HEMT is a lateral device, the tradeoff relationship between BV and $R_{on,sp}$ is still severe.

The BV of GaN HEMTs depends on the maximum electric field inside the device when the high voltage is applied. The device will breakdown when the maximum electric field exceeds the critical breakdown electric field of the material. Therefore, how to reduce the maximum electric field in the structure has become the most important issue to improve the breakdown voltage. In the conventional structure (conv-HEMT) [2], [3], [4], since the electric field is generally concentrated at the sharp corner of the gate near the drain, a field plate is designed on the side of the gate near the drain to balance the electric field and decrease the maximum electric field. In high voltage applications, a single gate field plate is not enough to decrease the maximum electric field. Therefore, other methods are needed to balance the electric field and improve the breakdown voltage.

Based on the conv-HEMT with gate field plate [5], the proposed structure (pro-HEMT) is fabricated on a 6-inch sapphire substrate in this study. The parameters of the device structure are optimized under the premise of considering the characteristics of the device such as breakdown voltage, on-resistance and threshold voltage (V_{th}). Finally, the breakdown voltage of the pro-HEMT can be increased to higher than 1500 V, which is more than twice that of the conv-HEMT. On this basis, the structure layout of the device is optimized by the bonding pad over active-region (BPOA) [6], so as to reduce the area loss under the same current capacity. Through experimental analysis, the cell layout reduces the on-resistance under the premise that the GaN HEMTs meet the high voltage conditions, and realizes depletion-mode (D-mode) GaN HEMTs with high voltage and low on-resistance.

II. DEVICE STRUCTURE AND MECHANISM

Fig. 1(a) illustrates the cross-sectional architecture of a conventional GaN HEMT device, which only introduces a gate field plate to reduce the peak electric field near the gate electrode. Fig.1 (b) shows the cross-section of the proposed multi-field plates GaN HEMT device, in which the bottom-up structure is 1mm thick sapphire substrate, 2um GaN Buffer layer, 200nm GaN Channel layer, 0.5nm AlN layer, 15nm AlGaN layer and 20nm gate dielectric layer. And the

---

The authors Moufu Kong, Yaowen Zhang, Yingzhi Luo, Kangxiang Zhao , Bo Yi and Hongqiang Yang are with the State Key Laboratory of Electronic Thin Films and Integrated Devices of China, University of Electronic Science and Technology of China, Chengdu, China 611731; The author Bingke Zhang is with the Leshan Share Electronic Co., Ltd, No. 3, Nanshin East Road, High-Tech Zone, Leshan, Sichuan, 614099, China. This work was supported in part by the Central Guiding Local Science and Technology Development Special Project of Sichuan (2024ZYD0310) and the Key R & D project of science and technology plan of Sichuan province (Grant 2023YFG0005) (Corresponding author: Moufu Kong: kmf@uestc.edu.cn).

979-8-3315-2209-4/25 $31.00 © 2025 IEEE

pro-HEMT additionally adds a source field plate and a drain field plate serve as a terminal structure [7], [8], [9]. The lengths of the gate field plate, source field plate, and drain field plate are 5 μm, 9 μm, and 17 μm, respectively. The length of the source field plate extends beyond that of the gate field plate, helping to reduce the peak electric field at the edge of the gate. However, this structure will cause the peak electric field to concentrate near the drain side, so a drain metal field plate is also employed in the proposed device. Consequently, the maximum electric field within the entire device diminishes due to the combined influence of multiple field plates.

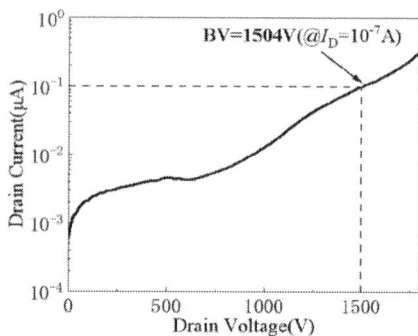

Fig. 3. The test results of the breakdown curve of the pro-HEMT.

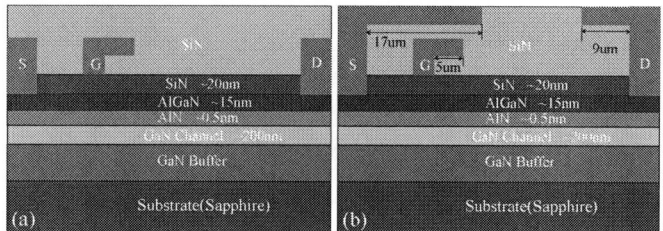

Fig. 1. (a) Schematics of the conv-HEMT and (b) the pro-HEMT.

On the basis of the design, we fabricated the pro-HEMT on a 6-inch sapphire substrate as shown in Fig.2, and designed two different pad position distributions, one is as shown in Fig.2(a), the pads are designed outside the active region. And the other is shown in Fig.2(b), the pads are designed above the active region, named device A and device B respectively.

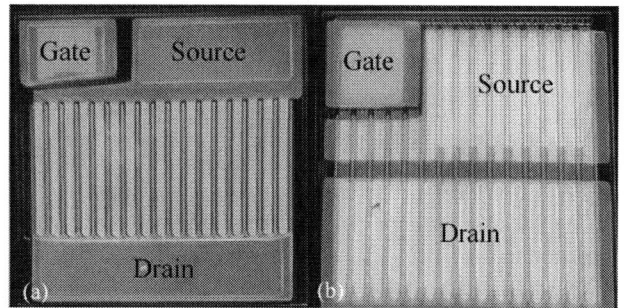

Fig. 2. The layout design of the proposed (a) device A: pads outside the active region and (b) device B: pads above the active region.

III. RESULTS AND DISCUSSION

In order to verify the proposed GaN HEMT device, the experimental test and TCAD simulation are both conducted. Fig. 3 illustrates the test result of the breakdown characteristic of the pro-HEMT. With a gate to source voltage (V_{GS}) of -20V and using a drain leakage current of I_{DS} = 0.1 μA as the device breakdown criterion, the device exhibits a breakdown voltage of higher than 1500 V, making it well-suited for 1200V-class high-voltage and high-power applications. And the leakage current in the blocking state, when V_{DS} is less than 1000V, remains at a low level, with a value of less than 10nA. Low leakage current reduces unnecessary power loss during the blocking state, leading to higher overall system efficiency. This is particularly important in power electronics applications where energy efficiency is a critical factor, such as in power supplies and motor drives.

When a blocking voltage V_{DS} of 1400 V is applied to the GaN HEMTs device, the peak electric field distributions of the two structures are shown in Fig.4 (a) and (b), respectively. The peak electric field of the conv-HEMT (Fig. 4(a)) exists only below the gate field plate. On the contrary, in the pro-HEMT, the peak electric field is transferred from the gate field plate to the source field plate as shown in Fig. 4(b). In Fig. 4 (c), the electric field distribution curves in GaN channel region between the gate and drain are shown. The electric field in the conv-HEMT is 4.75 MV/cm, which exceeds the critical breakdown electric field of GaN material (3.3 MV/cm) [10]. In the pro-HEMT, the source field plate shares part of the electric field and mitigates the peak electric field, which reduces the peak electric field of the channel from 4.75 MV/cm to 3.12 MV/cm, a decrease of 34%. In addition, with the addition of the multi-field plate, the electric field in the passivation layer of the device is also reduced from 8.1 MV/cm to 6.7 MV/cm, a decrease of approximately 17%, which greatly improves the reliability of the device.

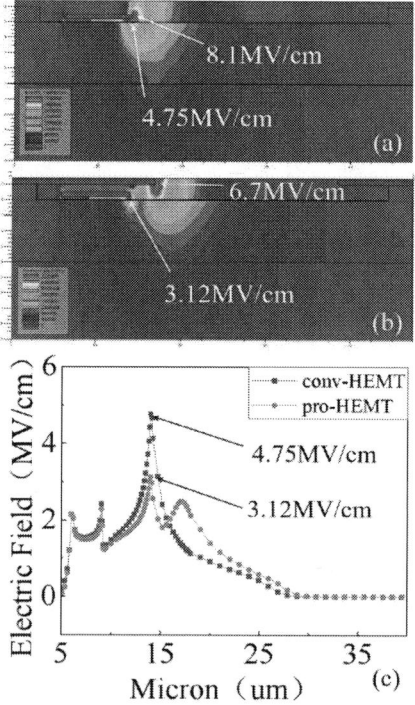

Fig. 4. (a) Electric field distribution of the conv-HEMT; (b) electric field distribution of the pro-HEMT; (c) electric field distribution in the channel of the conv-HEMT and the pro-HEMT.

Fig. 5 shows the test result of transfer characteristic of the pro-HEMT. It can be clearly seen that the V_{th} of the experimental results is approximately -10.7V, which indicates that the pro-HEMT is a depletion mode device.

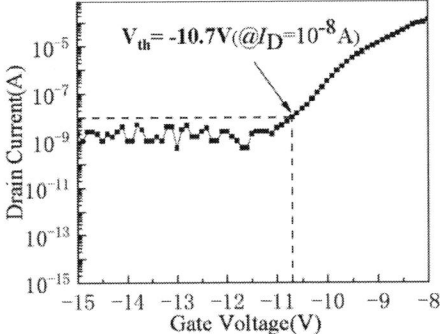

Fig. 5. The test result of the transfer characteristic curve of the pro-HEMT.

As shown in Fig.6, the output characteristic curves of simulation and experimental results are compared. In addition, we also compare the output characteristic curves of two different layouts in Fig.6. Comparing the simulation and test results of device A and device B, it can be proved that the simulation is in good agreement with the experiment. According to the test results, the $R_{DS,on}$ of device A and B are 127 mΩ and 95 mΩ, respectively, with the specific on-resistance ($R_{on,sp}$) of both devices are approximately 5.61 mΩ·cm². However, the test drain current of device B is about 27% higher than that of device A. This suggests that the pad placed above the active area increases the active area to enhance current density to reduce the $R_{DS,on}$ of the individual chip, maintaining a constant specific on-resistance. Furthermore, the pro-HEMT also achieves a low on-resistance at a high voltage of more than 1500 V.

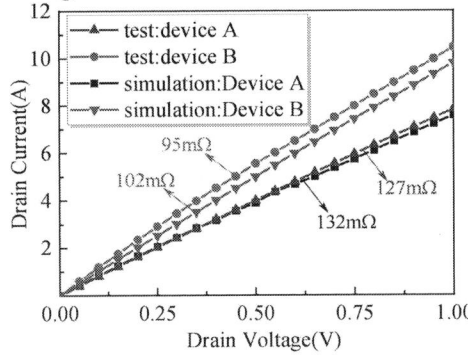

Fig. 6. The output characteristic curve of device A and device B

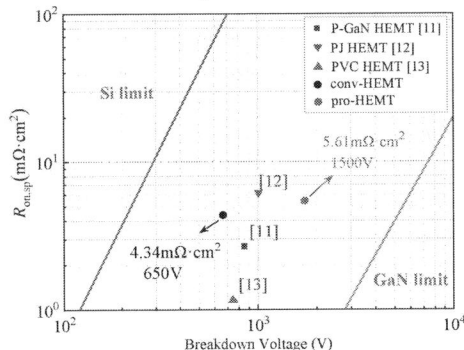

Fig. 7. The relationship between BV and $R_{on,sp}$ of different GaN HEMTs.

As shown in Fig.7, the relationship between BV and $R_{on,sp}$ between different GaN HEMT devices is depicted. It can be clearly seen that most of the GaN HEMT devices are only suitable for medium and low voltage applications [11], [12], [13]. but the pro-HEMT can increase BV to 1500V and its $R_{on,sp}$ only increases from 4.34 mΩ·cm² to 5.61 mΩ·cm² compared with the conv-HEMT, showing extraordinary potential in future high voltage applications..

IV. CONCLUSION

In this study, a novel high voltage GaN HEMTs structure with multiple field plates is proposed and fabricated. Compared with the conventional single gate field plate structure, it has better electric filed distribution and higher breakdown voltage, thus higher reliability. In addition, a novel method for positioning the pads above the active area has been used. The experimental results demonstrate that this approach increases the active area while maintaining a constant specific on-resistance, thereby enhancing the current capability. According to the test results of breakdown voltage, it shows that the proposed structure increases the breakdown voltage of GaN HEMTs to more than 1500 V, which is suitable for most future high-voltage scenarios.

REFERENCES

[1] L. Heuken, et al., "Analysis of an AlGaN/AlN Super-Lattice Buffer Concept for 650-V Low-Dispersion and High-Reliability GaN HEMTs," IEEE Transactions on Electron Devices, 2020, vol. 67, no. 3, pp. 1113-1119.

[2] U. K. Mishra, AlGaN/GaN transistors for power electronics, IEEE Electron Devices Meeting, 2010, pp. 13.2.1-13.2.4.

[3] O. Ambacher, et al., "Twodimensional electron gases induced by spontaneous and piezoelectric polarization in undoped". Journal of Applied Physics, 2000, vol.85, no.6, pp.3222-3233.

[4] G. Xie, B. Zhang, E. Xu, et al. "An AlGaN/GaN HEMT with a reduced surface electric field and an improved breakdown voltage". Chinese Physics B, 2012, vol.21, no.8, pp.086105.

[5] N. Q. Zhang, et al. "High Breakdown GaN HEMT with Overlappin g Gate Structure". IEEE Electron Device Letters, 2000, vol.21, no.9, pp.421-423

[6] S. K. Oh, et.al, "Bonding Pad Over Active Structure for Chip Shrinkage of High-Power AlGaN/GaN HFETs," IEEE Transactions on Electron Devices, 2016, vol. 63, no. 2, pp. 620-624.

[7] W. Saito, et al. "Design and demonstration of high breakdown voltage GaN high electron mobility transistor (HEMT) using field plate structure for power electronics applications". Japanese Journal of Applied Physics, 2004, pp.43,no.4,pp.2239-2242 .

[8] M. Zhang,et al., "Effect of field plate length on DC characteristics of high breakdown voltage GaN HEMTs for power switching application", in Proc.10th IEEE International Conference on Solid-State and Integrated Circuit Technology, 2010,pp.1356-1358.

[9] B. Liao, et al. "Simulation of AlGaN/GaN HEMTs' Breakdown Voltage Enhancement Using Gate Field-Plate, Source Field-Plate and Drain Field Plate". Electronics, 2019, vol.8, no.4, pp.406.

[10] T. P. Chow. "High-voltage SiC and GaN power devices". Microelectronic Engineering, 2005, vol.83, no.1, pp.112-122.

[11] X. Wei,et al., "Improvement of Breakdown Voltage and ON-Resistance in Normally-OFF AlGaN/GaN HEMTs Using Etching-Free p-GaN Stripe Array Gate," IEEE Transactions on Electron Devices, 2021, vol. 68, no. 10, pp. 5041-5047.

[12] A. Nakajima, et al. "GaN-Based Super Heterojunction Field Effect Transistors Using the Polarization Junction Concept," IEEE Electron Device Letters, 2011, vol. 32, no. 4, pp. 542-544.

[13] C. Yang, J. Xiong, J. Wei, et al. "Analytical model and new structure of the enhancement-mode polarization-junction HEMT with vertical conduction channel". Superlattices and Microstructures, 2016, 92: pp. 92-99.

A Novel SiC Trench MOSFET With Integrated Junction Barrier Schottky for Improved Performances

Bo Yi
School of Integrated Circuit Science and Engineering, University of Electronic Science and Technology of China
Chengdu, China
yb@uestc.edu.cn

JunFeng Duan
School of Integrated Circuit Science and Engineering, University of Electronic Science and Technology of China
Chengdu, China
junfeng.duan@outlook.com

XinYi Wu
School of Integrated Circuit Science and Engineering, University of Electronic Science and Technology of China
Chengdu, China
seeywo@163.com

Tao Zhu
School of Integrated Circuit Science and Engineering, University of Electronic Science and Technology of China, Chengdu, China, State Key Laboratory of Advanced Power Transmission Technology, Beijing Institute Of Smart Energy
Beijing, China
zhutao540611@126.com

JunJi Cheng
School of Integrated Circuit Science and Engineering, University of Electronic Science and Technology of China
Chengdu, China
chengjunji2005@126.com

HongQiang Yang*
School of Integrated Circuit Science and Engineering, University of Electronic Science and Technology of China
Chengdu, China
hqyang@uestc.edu.cn

Abstract—In this paper, a novel SiC MOSFET with an integrated Junction Barrier Schottky diode (JBS) on the surface (named JBS-MOS) is proposed, which enhances the reverse conduction capability and reduces the switching losses, offering superior performances in power converter applications. Simulation results show that the specific on-resistance ($R_{on,sp}$) and breakdown voltage (BV) of the JBS-MOS are 1.90 m$\Omega \cdot$cm^2 and 1604 V, respectively. Meanwhile, JBS-MOS achieves a reverse on-state voltage (V_{R_ON} @J_{DS} = -300 A/cm^2) of 1.91 V, reduced by 1.59 V compared to Conventional MOSFET (C-MOS). Additionally, its reverse recovery charge (Q_{rr}) is reduced by 78.7% at 450 K, resulting in a 53.7% reduction in turn-on loss at 450 K, indicating improved switching efficiency. These simulation results indicate that the JBS-MOS has potential for high-power and high-temperature applications.

Keywords—junction barrier Schottky (JBS), reverse on-state voltage drop, SiC trench MOSFET, reverse recovery charge.

I. INTRODUCTION

SiC trench MOSFET has emerged as one of the most attractive wide bandgap power device due to its high thermal conductivity, high critical electric field of ~3 MV/cm, and ultra-low specific on-resistance ($R_{on,sp}$) [1], which has been widely used in electric vehicles, and motor drives, etc [2]. However, experiments have proven that the electric field in the gate oxide can easily reach 4 MV/cm without protection [3], inducing severe reliability issues. Therefore, Rohm [4] and Infineon [3] have proposed Double-Trench and CoolSiC MOSFET to solve this challenge. Considering the dynamic resistance during switching, their p-shield layer should be well grounded [5]. Another significant issue is the reverse conduction performance of the MOSFET, including reverse on-state voltage (V_{R_ON}) and reverse recovery charge (Q_{rr}). An intrinsic body diode can offer economical choices, while the bipolar degradation [2] and high V_{R_ON} present challenges to reliability over time. To address these issues, Yen et al. [6] fabricated a planar SiC MOSFET with

integrated junction barrier Schottky (JBS). R. Aiba et al. [7] and Yi et al. [8] proposed a MOSFET with Schottky barrier diode (SBD) integrated at the sidewall of the trench gate.

In this paper, a MOSFET with an integrated junction barrier Schottky diode (JBS) is proposed, named JBS-MOS. The JBS structure is integrated on the surface of the MOSFET, consisting of a p-shield region, a current spreading layer (CSL) channel and a Schottky contact metal directly connected to the CSL channel. This structure enhances the reverse conduction capability and reduces the leakage current of the Schottky contact without affecting other characteristics. Moreover, the JBS structure offers unipolar operation and reduces switching losses, especially under high temperature.

II. DEVICE STRUCTURE AND MECHANISM

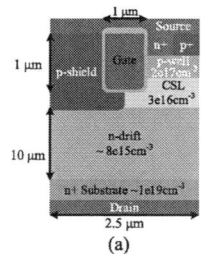 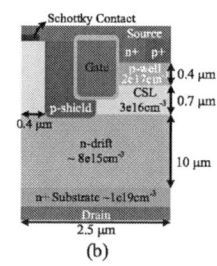

(a)   (b)

Fig. 1. Schematic cross-sectional view of (a) C-MOS and (b) JBS-MOS.

Fig. 1 illustrates the structure of the conventional MOSFET (C-MOS) and proposed SiC MOSFET with integrated junction barrier Schottky diode (JBS-MOS), both of which maintain the same cell pitch. The only difference between C-MOS and JBS-MOS is that a JBS is integrated near the p-shield. The same CSL is used under the Schottky contact to reduce the JFET resistance. For JBS-MOS, when it operates in blocking state, the heavily-doped deep p-shield provides a shielding effect for the Schottky contact, resulting in minimal leakage current across the JBS. In

979-8-3315-2209-4/25 $31.00 © 2025 IEEE

the reverse conducting state, the JBS effectively inactivates the conduction of the body diode, which effectively reduces the V_{R_ON} and eliminates conductivity modulation and bipolar degradation effects. The TCAD simulation models include Fermi, Auger, EffectiveIntrinsicDensity, Mobility (DopingDep (Masetti), HighFieldSaturation, Enormal (Lombardi, InterfaceCharge), IncompleteIonization). The mobility model is based on reference [8].

III. RESULTS AND DISCUSSION

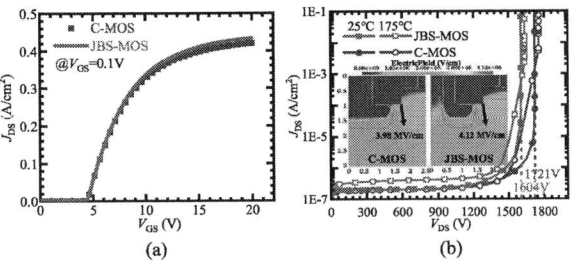

Fig. 2. Comparison of static characteristics of JBS-MOS and C-MOS. (a) Transfer characteristics. (b) Leakage current. The inserted graphs are the electric field distribution at V_{DS} = 1200 V.

Fig. 2(a) illustrates that the V_{TH} of C-MOS and JBS-MOS is both ~4.3 V. Compared to C-MOS, the JBS-MOS features a smaller p-shield area, and $R_{on,sp}$ slightly decreases from 1.96 $m\Omega\cdot cm^2$ to 1.90 $m\Omega\cdot cm^2$. However, this also results in a higher electric field at the corner of the p-shield for JBS-MOS. Its BV in 300 K decreases from 1721 V to 1604 V as shown in Fig. 2(b). Due to the shielding effect of p-shield, leakage current across the Schottky contact is significantly suppressed and almost the same as those of C-MOS under different temperatures.

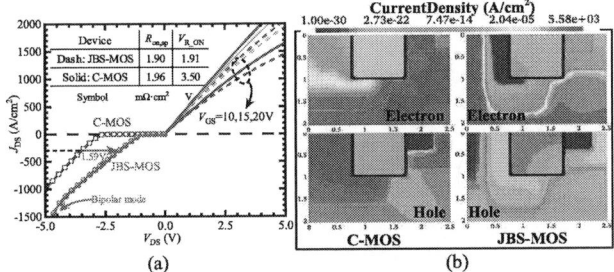

Fig. 3. Output and reverse performances. (a) Comparison of output and reverse characteristics for the C-MOS and JBS-MOS. (b) The electron and hole current density distribution of the C-MOS and JBS-MOS at J_{DS} = −300 A/cm².

Fig. 3(a) depicts the performances of the first and third quadrants. The difference in forward output characteristics confirms the reduction of $R_{on,sp}$ for the JBS-MOS. Compared with C-MOS, the V_{R_ON} of JBS-MOS is only 1.91 V, which is reduced by 1.59 V due to the introduction of JBS. Fig. 3(b) shows the electron and hole current density distribution of the two devices at J_{DS} = −300 A/cm². For JBS-MOS, JBS is activated without bipolar conduction. Thus, only electron current flows in the device. While, both electron and hole current exist in the C-MOS due to conductivity modulation. When the reverse bias voltage is close to -5 V, the device switches to bipolar operation mode [8], consistent with the behavior of C-MOS.

Fig. 4(a) depicts the turn-on curve at T = 300 K and T = 450 K. JBS-MOS exhibits a 19.5% lower Q_{rr} compared to C-MOS at 300 K. At T = 450 K, Q_{rr} of C-MOS reaches 2253 nC due to the bipolar degradation effect. While Q_{rr} of JBS-MOS remains at ~480 nC, reduced by 78.7%. In terms of switching loss, the turn-on loss of JBS-MOS at T = 450 K is reduced by 53.7% at a cost of 4.2% increase in turn-off loss as shown in Fig. 4(b).

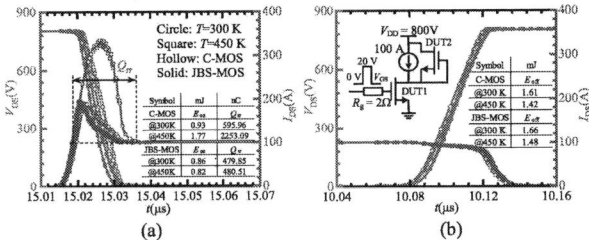

Fig. 4. Switch characteristics depends on temperature. (a) Turn-on curve. (b) Turn-off curve. The inset graph was the test circuit under current load where DUT1 and DUT2 are the same device. The rated current density is 300 A/cm².

IV. CONCLUSION

In this paper, a novel SiC MOSFET with an integrated Junction Barrier Schottky diode is proposed to improve the reverse conduction performance and reduce switching loss. Simulation results demonstrate that the JBS-MOS reduces V_{R_ON} by 1.59 V compared to C-MOS, while offering a 78.7% reduction in Q_{rr} at 450 K and a 53.7% reduction in turn-on loss at 450 K. These improvements make the JBS-MOS an excellent candidate for high-power and high-temperature systems.

REFERENCES

[1] M. Buffolo et al., "Review and Outlook on GaN and SiC Power Devices: Industrial State-of-the-Art, Applications, and Perspectives," IEEE Trans. Electron Devices, vol. 71, no. 3, pp. 1344–1355, Mar. 2024, doi: 10.1109/TED.2023.3346369.

[2] C. Langpoklakpam et al., "Review of Silicon Carbide Processing for Power MOSFET," Crystals, vol. 12, no. 2, p. 245, Feb. 2022, doi: 10.3390/cryst12020245.

[3] D. Peters et al., "The new CoolSiC™ trench MOSFET technology for low gate oxide stress and high performance," in Proc. Int. Exhib. Conf. Power Electron., Intell. Motion, Renew. Energy Energy Manage., May 2017, pp. 1–7.

[4] Y. Nakano, R. Nakamura, H. Sakairi, S. Mitani, and T. Nakamura, "690V, 1.00 mΩ·cm² 4H-SiC Double-Trench MOSFETs," MSF, vol. 717–720, pp. 1069–1072, May 2012, doi: 10.4028/www.scientific.net/MSF.717-720.1069.

[5] J. Wei, M. Zhang, H. Jiang, H. Wang, and K. J. Chen, "Dynamic degradation in SiC trench MOSFET with a floating p-shield revealed with numerical simulations," IEEE Trans. Electron Devices, vol. 64, no. 6, pp. 2592–2598, Jun. 2017, doi: 10.1109/TED.2017.2697763.

[6] C.-T. Yen et al., "1700V/30A 4H-SiC MOSFET with low cut-in voltage embedded diode and room temperature boron implanted termination," in 2015 IEEE 27th International Symposium on Power Semiconductor Devices & IC's (ISPSD), Hong Kong, China: IEEE, May 2015, pp. 265–268. doi: 10.1109/ISPSD.2015.7123440.

[7] R. Aiba et al., "Experimental Demonstration on Superior Switching Characteristics of 1.2 kV SiC SWITCH-MOS," in 2019 31st International Symposium on Power Semiconductor Devices and ICs (ISPSD), Shanghai, China: IEEE, May 2019, pp. 23–26. doi: 10.1109/ISPSD.2019.8757628.

[8] B. Yi et al., "Fabricating and TCAD Optimization for a SiC Trench MOSFET With Tilted P-Shielding Implantation and Integrated TJBS," IEEE Trans. Electron Devices, vol. 71, no. 3, pp. 1618–1625, Mar. 2024, doi: 10.1109/TED.2024.3361405.

979-8-3315-2209-4/25 $31.00 © 2025 IEEE

A Novel SiC Trench MOSFET with Integrated N⁺- PolySi/SiC Heterojunction Diode

Bo Yi
School of Integrated Circuit Science and Engineering, University of Electronic Science and Technology of China
Chengdu, China
yb@uestc.edu.cn

XinYi Wu
School of Integrated Circuit Science and Engineering, University of Electronic Science and Technology of China
Chengdu, China
seeywo@163.com

JunFeng Duan
School of Integrated Circuit Science and Engineering, University of Electronic Science and Technology of China
Chengdu, China
junfeng.duan@outlook.com

Tao Zhu
School of Integrated Circuit Science and Engineering, University of Electronic Science and Technology of China, Chengdu, China, State Key Laboratory of Advanced Power Transmission Technology, Beijing Institute Of Smart Energy Beijing, China
zhutao540611@126.com

JunJi Cheng
School of Integrated Circuit Science and Engineering, University of Electronic Science and Technology of China
Chengdu, China
chengjunji2005@126.com

HongQiang Yang*
School of Integrated Circuit Science and Engineering, University of Electronic Science and Technology of China
Chengdu, China
hqyang@uestc.edu.cn

Abstract—In this paper, we propose and investigate a novel 4H-SiC MOSFET with integrated N⁺-PolySi/N-SiC heterojunction diode (named as nHJD-MOS) by calibrated TCAD simulation. An N⁺ polysilicon region is set on the top of the drift region and connected to the source electrode to form a N⁺-PolySi/N-SiC heterojunction diode (nHJD), which is used to inactivate the body diode. Thus, the third quadrant performance and switching performances are improved. The numerical simulation results show that the specific on-resistance ($R_{on,sp}$) and breakdown voltage (BV) of the nHJD-MOS are 1.97 mΩ·cm² and 1565 V, respectively. At a current density of 300 A/cm², the reverse on-state voltage drop (V_{R_on}) is 1.42 V, which is 2.08 V lower than that of the Conventional MOS (Con-MOS). Furthermore, reverse recovery charge (Q_{rr}) and turning-on power loss (E_{on}) are also reduced by 17.8% and 6.3%, respectively. In addition, under high temperatures, E_{on} of the Con-MOS increases significantly due to increased Q_{rr}. While, E_{on} of the nHJD-MOS almost remains unchanged due to unipolar conduction of the HJD.

Keywords—SiC trench MOSFET, N⁺-PolySi/4H-SiC heterojunction, reverse on-state voltage, switching power loss

I. INTRODUCTION

SiC is a promising material for power electronics applications for its wide band gap, high critical electric field and excellent thermal properties [1]. However, the high intrinsic build-in voltage and degradation effect of the SiC PN diode limit the high efficiency applications [2]. In order to overcome these issues, SiC MOSFET integrated with Schottky diodes [3]-[4] and PolySi/4H-SiC heterojunction diode have been extensively studied [5]-[6]. For the PolySi/4H-SiC heterojunction with heavily doped silicon, the barrier height for electrons (Φ_{bn}) in the P-PolySi/N-SiC heterojunction is approximately 1.5 to 2.0 V, while for the N-PolySi/N-SiC heterojunction, it is around 0.7 V [7]-[8], which is benefit for low reverse on-state voltage drop (V_{R_on}). However, the low Φ_{bn} may also leads to large leakage current.

In this paper, a novel SiC trench MOSFET with integrated N⁺-PolySi/N-SiC heterojunction diode (named as nHJD-MOS) is proposed and investigated. Low V_{R_on} and low leakage current are simultaneously obtained for the nHJD-MOS.

II. DEVICE STRUCTURE AND CHARACTERISTICS

The cross-sectional view of Conventional SiC MOSFET (Con-MOS) and the proposed nHJD-MOS are shown in Fig. 1(a). The current spreading layer (CSL) concentration is varied between 1×10^{16}-3×10^{16} cm⁻³ to reduce the specific on-resistance ($R_{on,sp}$) and obtain a better trade-off with BV. In the nHJD-MOS, the doping concentration of N⁺ polysilicon is 1×10^{19} cm⁻³ and the thickness is 0.5 μm. Fig. 2 shows the band energy diagram of the N⁺-PolySi/N-SiC heterojunction with $N_{CSL} = 2\times10^{16}$ cm⁻³. The energy gaps of the conduction and valence bands are 0.79 and 1.43 eV, respectively. The barrier height of the electron is about 0.64 eV, which ensures a low V_{R_on} for third quadrant characteristics.

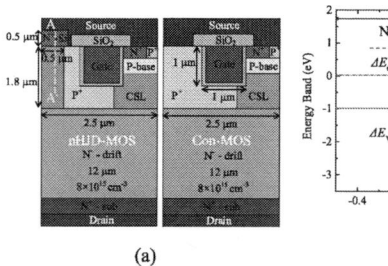
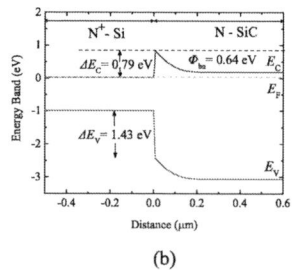

(a)   (b)

Fig.1. (a) Cross-sectional view of the nHJD-MOS and Con-MOS. (b) The energy-band diagram of the heterojunction along cutline A-A' at $N_{CSL} = 2\times10^{16}$ cm⁻³.

III. SIMULATION RESULTS AND DISCUSSION

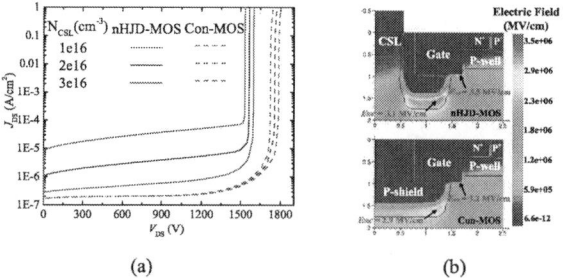

(a)   (b)

Fig. 2. Comparison of (a) Leakage current under different N_{CSL}. (b) Electric field distribution at $V_{DS} = 1500$ V with $N_{CSL} = 2\times10^{16}$ cm⁻³.

Fig. 2(a) shows that BV of the nHJD-MOS and Con-MOS are above 1500 V under different N_{CSL}. BV of nHJD-MOS is a little lower than that of the Con-MOS due to that the narrow P-shield region enhances the electric field crowding effect as demonstrated in Fig. 2(b). Under the same V_{DS}, the maximum electric field at the corner of the P-shield in nHJD-MOS reaches 3.1 MV/cm, which is larger than 2.9 MV/cm in the Con-MOS. As N_{CSL} increases from 1×10^{16} cm^{-3} to 3×10^{16} cm^{-3}, the leakage current for the nHJD-MOS increases. Fortunately, owing to the strong shielding effect of P-shield layer, the leakage current is still an ultra-low value of 1 µA/cm^2. Also, the electric field in gate oxide are suppressed within a reliable value of 4 MV/cm [9].

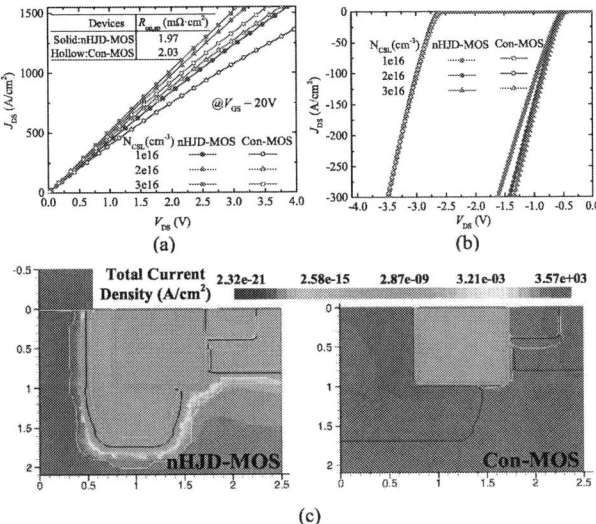

Fig.3. (a) Output curves at V_{GS} = 20 V for nHJD-MOS and Con-MOS. (b) Third quadrant characteristics at V_{GS} = -5 V. (c) Total current density distributions (at J_{DS} = -300A/cm^2) for nHJD-MOS and Con-MOS with N_{CSL} = 2×10^{16} cm^{-3}.

Fig. 3(a) shows that with the increase of N_{CSL}, $R_{on,sp}$ decreases for the two devices. When N_{CSL}=2×10^{16} cm^{-3}, $R_{on,sp}$ for the nHJD-MOS and Con-MOS are 1.97 mΩ·cm^2 and 2.03 mΩ·cm^2, respectively. Fig. 3(b) illustrates that under a typical current density of J_{DS} = -300 A/cm^2, as N_{CSL} increases from 1×10^{16} cm^{-3} to 3×10^{16} cm^{-3}, V_{R_on} of the nHJD-MOS decreases from 1.62 V to 1.34 V. In contrast, the Con-MOS maintains a stable V_{R_on} of approximately 3.50 V. The HJD not only significantly reduces V_{R_on}, but also inactivates the PN-body diode to prevent bipolar degradation problem. As demonstrated in Fig. 3(c), only electron current flows across the HJD in the proposal, while holes are injected from P region into N-drift region to form hole current in the Con-MOS.

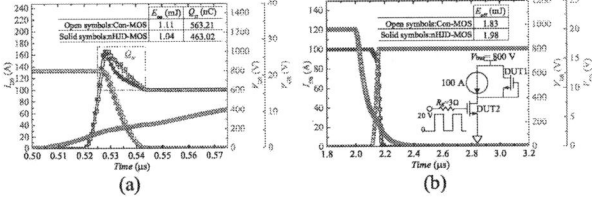

Fig.4. Transient process of DUT2 (device under test). The rated current density is 300 A/cm^2. Inserted is the test circuit. DUT1 and DUT2 are the same devices (nHJD-MOS or Con-MOS). (a) Turning-on process. (b) Turning-off process.

Fig. 4 illustrates that the switching speed of nHJD-MOS is almost the same as that of Con-MOS. However, due to

unipolar conduction of HJD in nHJD-MOS, its reverse recovery charge (Q_{rr}) is 17.8% lower than that of the Con-MOS. As a result, the turning-on power loss (E_{on}) of DUT2 from nHJD-MOS was reduced by 6.3%. The turning-off curves are almost coincident.

Fig. 5(a) shows that as the temperature increases, Q_{rr} of nHJD-MOS remains at ~ 463 nC due to unipolar characteristic, and E_{on} almost remains unchanged at ~1.04 mJ. However, for Con-MOS, the Q_{rr} increases from 563.21 nC to 2051.61 nC. Correspondingly, E_{on} increases from 1.11 mJ to 1.92 mJ. The turning-off process under different temperatures are almost the same.

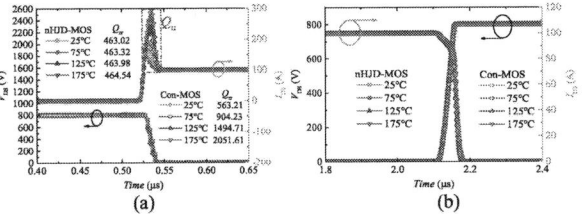

Fig.5. Impact of temperature on transient processes. (a) Turning-on process. (b) Turning-off process.

IV. CONCLUSIONS

In this paper, a SiC trench MOSFET with integrated N$^+$-PolySi/SiC heterojunction diode is proposed and investigated by TCAD simulation. Simulation results confirm that BV of the proposed nJHD-MOS is above 1565 V with $R_{on,sp}$ being only 1.97 mΩ·cm^2. The integrated nHJD operates in unipolar conduction and reduces V_{R_on} and Q_{rr} significantly. Moreover, the switching power loss of the nHJD-MOS significantly reduces and remains almost unchanged under high operating temperatures. While, turning-on power loss of the Con-MOS increases dramatically with the increase of temperature.

REFERENCES

[1] R. Wang et al., "A high-temperature sic three-phase ac-dc converter design for> 100/spl deg/c ambient temperature," IEEE Transactions on Power Electronics, vol. 28, no. 1, pp. 555-572, 2012.

[2] X. Jiang et al., "Investigation on degradation of SiC MOSFET under surge current stress of body diode," IEEE Journal of Emerging and Selected Topics in Power Electronics, vol. 8, no. 1, pp. 77-89, 2019.

[3] C.-T. Yen et al., "1700V/30A 4H-SiC MOSFET with low cut-in voltage embedded diode and room temperature boron implanted termination," in 2015 IEEE 27th International Symposium on Power Semiconductor Devices & IC's (ISPSD), 2015, pp. 265-268: IEEE.

[4] B. Yi et al., "Fabricating and TCAD Optimization for a SiC Trench MOSFET With Tilted P-Shielding Implantation and Integrated TJBS," IEEE Transactions on Electron Devices, 2024.

[5] W. Ni et al., "SiC trench MOSFET with an integrated low von unipolar heterojunction diode," in Materials Science Forum, 2014, vol. 778, pp. 923-926: Trans Tech Publ.

[6] H. Yu, S. Liang, H. Liu, J. Wang, and Z. J. Shen, "Numerical study of SiC MOSFET with integrated n-/n-type poly-Si/SiC heterojunction freewheeling diode," IEEE Transactions on Electron Devices, vol. 68, no. 9, pp. 4571-4576, 2021.

[7] Y. Wang, H.-Y. Wang, F. Cao, and H.-Y. Wang, "High performance of polysilicon/4H-SiC dual-heterojunction trench diode," IEEE Transactions on Electron Devices, vol. 64, no. 4, pp. 1653-1659, 2017.

[8] F. Triendl, G. Pfusterschmied, G. Pobegen, J. Konrath, and U. Schmid, "Theoretical and experimental investigations of barrier height inhomogeneities in poly-Si/4H-SiC heterojunction diodes," Semiconductor Science and Technology, vol. 35, no. 11, p. 115011, 2020.

[9] A. Saha and J. A. Cooper, "A 1-kV 4H-SiC power DMOSFET optimized for low on-resistance," IEEE Transactions on Electron Devices, vol. 54, no. 10, pp. 2786-2791, 2007.

Self-Heating Switching for Recovering Threshold Voltage of SiC MOSFETs after Electron Irradiation

Huanshi Guo[1], Huan Li[2], Mingmin Huang[1]*, Yao Ma, Nuoya Yang[1], Chang Chen[1], Yun Li[1], Qiang Yu[3]

1 College of Physics, Sichuan University, Chengdu, China

2 China ZhenHua Group YongGuang Electronics Company, Guiyang, China

3 Sichuan Suining Lippxin Microelectronics Co., Ltd, Suining, China

*Corresponding author's E-mail: mmhuang@scu.edu.cn, mayao@scu.edu.cn

Abstract—This paper provides a self-heating switching strategy to recover the threshold voltage of SiC MOSFET after electron irradiation, where the half-bridge topology is adopted. This strategy rapidly increases the junction temperature of the SiC MOSFET by using a high gate resistance (R_G) to increase the switching losses. After electron irradiation, the threshold voltage of 650 V SiC MOSFET sample is shifted from 3.9 V to 2.6 V, which is decreased by 1.3 V. An appropriate switching setting to recovery the threshold voltage is found by a combination of SPICE simulation and experiments. With R_G = 200 Ω, bus voltage of 400 V and inductive load of 185 μH, 27 switching pulses with a period of 3 μs and a duty cycle of 0.5 are performed. As a result, the threshold voltage of the irradiated sample (3.5 V) is recovered by 0.9 V, which recovers 70% of the threshold voltage shift by electron irradiation.

Keywords—electron irradiation, annealing, SiC MOSFET, threshold voltage, gate resistance, switching, self-heating

I. INTRODUCTION

In aerospace applications, power MOSFETs are typically used as switching devices in power electronic converter circuits [1][2]. The SiC MOSFET is an advanced commercial power device with advantages of high breakdown voltage, low on-resistance, and low switching loss [3]. Compared to silicon power MOSFETs, the SiC MOSFET is able to increase power density in power electronic converters, and also offers advantages of higher efficiency, faster switching speed, higher operating temperature, and longer lifetime [4], which is very attractive for aerospace applications. Research works also suggest that the SiC MOSFET appears to be more radiation-tolerant than Si MOSFETs due to its higher radiation resistance, making it of great potential in aerospace applications [5][6][7].

The space environment contains a wide variety of radioactive particles, e.g., gamma rays, X-rays, neutrons, electrons, heavy ion rays, etc [8]. Electron irradiation is able to create oxide traps in the oxide layer, leading to a reduction in the threshold voltage of MOSFETs. With the decrease of the threshold voltage, the risk of turn-off failure can be increased, which limits the life of the power MOSFET applied in the spacecraft [9][10]. Thermal annealing in a high-temperature furnace can remove trap charges in the oxide layer so as to recover the threshold voltage reduction by electron irradiation [11][12][13]. However, it is impractical to

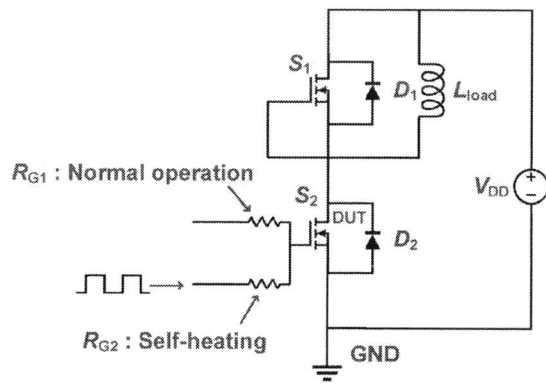

Fig. 1 Circuit schematic used for this strategy.

use high temperature annealing furnaces for thermal annealing in a space environment. Therefore, a strategy is needed to achieve thermal annealing of the irradiated power devices in the space environment. In 2023, a way to self-heat the irradiated SiC MOSFET by the weak short-circuit operation we propose, which recovers up to 58% of the threshold voltage reduction by the gama radiation. However, the weak short circuit operation may cause some unpredictable reliability problems.

This paper provides a new way to self-heat the irradiated SiC MOSFET by increasing gate resistance (R_G) to increase the switching losses, where a half-bridge topology is adopted to verify the effectiveness of the proposed new way. Section II explains the principle of the new way and studies the experiment conditions by SPICE simulations. Section III verifies the new way by experiments.

II. PRINCIPLE AND SIMULATIONS

Fig. 1 shows the schematic diagram of the half-bridge circuit based on SiC MOSFETs (S_1 and S_2). L_{load} is the load inductance, and V_{DD} is the DC voltage source. R_{G1} is used for normal switching operation of half-bridge circuits, R_{G2} is used for switching operation during self-heating annealing. During normal operation of the half-bridge circuit, R_{G2} is disconnected and R_{G1} is used as the gate resistance. When self-heating annealing is required, R_{G1} is disconnected and R_{G2} is used as the gate resistance. During self-heating, S_1 is shorted, S_2 is controlled by the gate signal. When S_2 is in the on state, S_2 and the inductor form a loop. When S_2 is in the off

979-8-3315-2209-4/25 $31.00 © 2025 IEEE

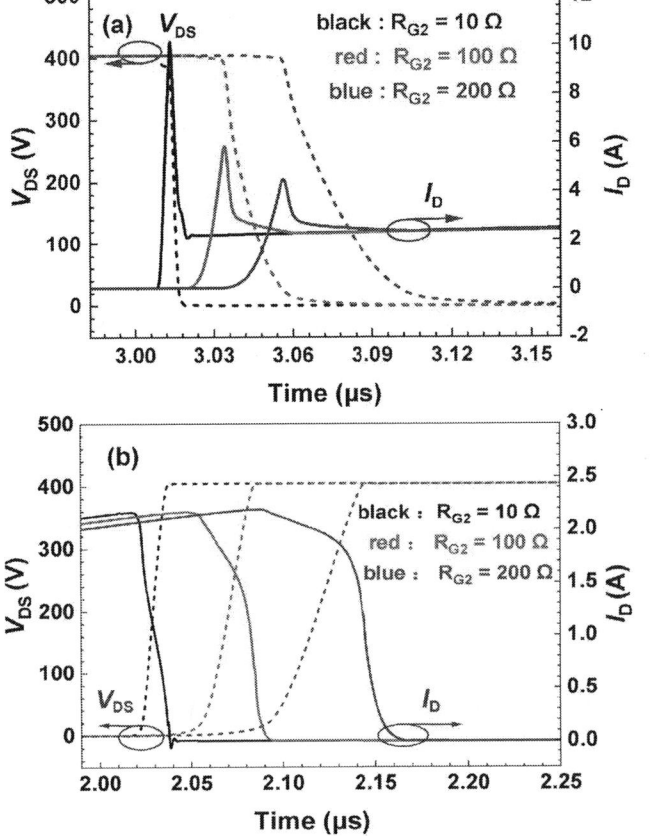

Fig. 2 (a) SiC MOSFET turn-on process at different gate resistances. (b) SiC MOSFET turn-off process at different gate resistances.

TABLE I SWITCHING CHANGES WITH GATE RESISTANCE.

R_{G2} (Ω)	E_{on} (μJ)	E_{off} (μJ)
10	12.73	2.96
100	27.94	16.10
200	54.85	25.20

state, the inductor current flows through the body diode of S_1, forming a freewheeling loop.

A 650 V commercial SiC MOSFET (C3M012065D) from Wolfspeed is utilized for both simulation and experiment. The threshold voltage V_{th} (V_{GS} @ 10 mA) of this sample is 3.9 V. Following electron irradiation with a dose of 1×10^{15} cm^{-2}, the threshold voltage of the sample decreases to 2.6 V.

Fig. 2 shows the switching processes of this sample at different .gate resistances (R_{G2} = 10 Ω, 100 Ω, and 200 Ω). where V_{DD} = 400 V and the on-state current of I_D = 2.5 A are used. The switching time significantly increases as R_{G2} increases. Table I shows the turn-on loss (E_{on}) and turn-off loss (E_{off}) for three different R_{G2} (R_{G2} = 10 Ω, 100 Ω, and 200 Ω). The E_{on} with a gate resistor of 200 Ω is 4.3 times higher than that with a gate resistor of 10 Ω, and the E_{off} is 7.5 times higher than that with a gate resistor of 10 Ω. Compare the E_{on} and E_{off} of R_{G2} = 100 Ω, E_{on} and E_{off} of R_{G2} = 200 Ω are also approximately doubled. When R_{G2} = 200 Ω, the switching loss of the device can be significantly increased, and the rate

of

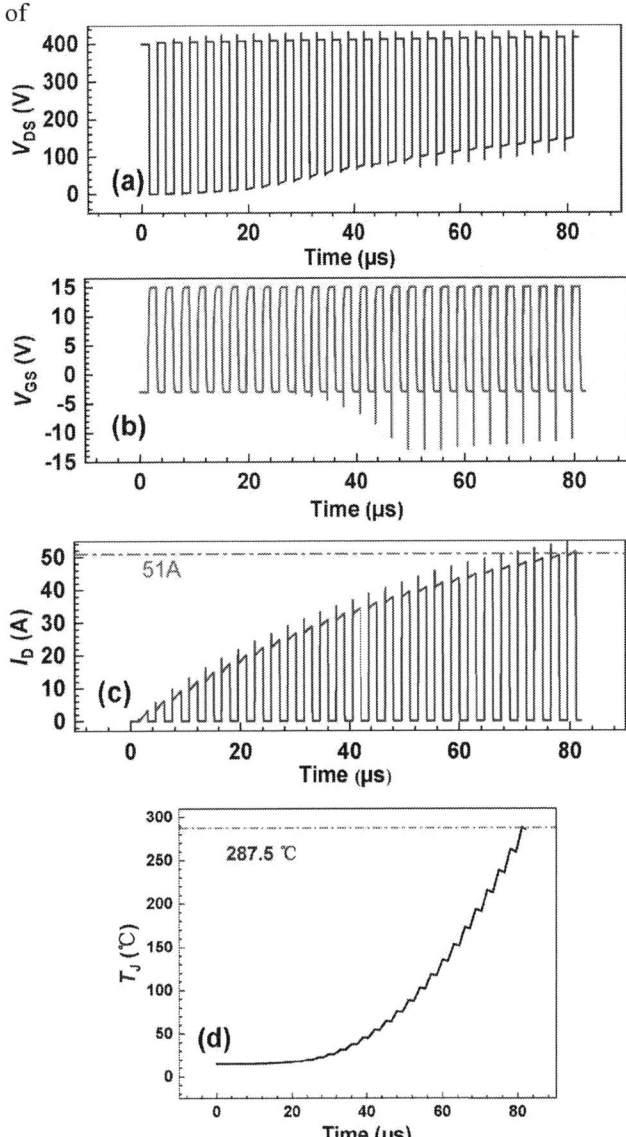

Fig. 3 (a) simulation waveforms of V_{DS}. (b) simulation waveforms of V_{GS}. (c) simulation waveforms of I_D. (d) simulation waveforms of junction temperature.

junction temperature rise is positively correlated with power dissipation. In this paper, R_{G2} = 200 Ω is selected to achieve the purpose of rapidly elevating the junction temperature.

Before the self-heating experiment, The circuit illustrated in Fig. 1 is simulated utilizing LTspice software to evaluate the waveforms of junction temperature and conduction current under continuously switching. In this simulation, a pulse waveform with a period of 3 μs and a duty cycle of 0.5 is applied, the gate-source voltage (V_{GS}) ranges from -3 V to 15 V, the supply voltage (V_{DD}) is set to 400 V, the load inductance (L_{load}) is 185 μH, and the gate resistance (R_{G2}) is 200 Ω. The simulation uses the SPICE model including the component of thermal model to determine the number of switching cycles needed for the self-heating experiment and to predict the final junction temperature, enabling the evaluation of the annealing effect.

979-8-3315-2209-4/25 $31.00 © 2025 IEEE

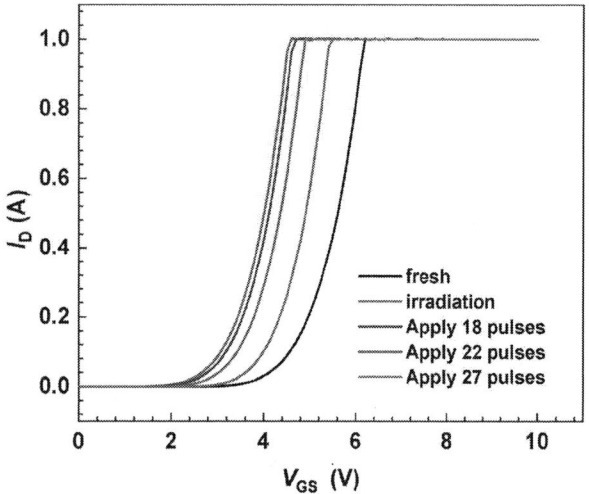

Fig. 4 Recovery of threshold voltage of SiC MOSFET.

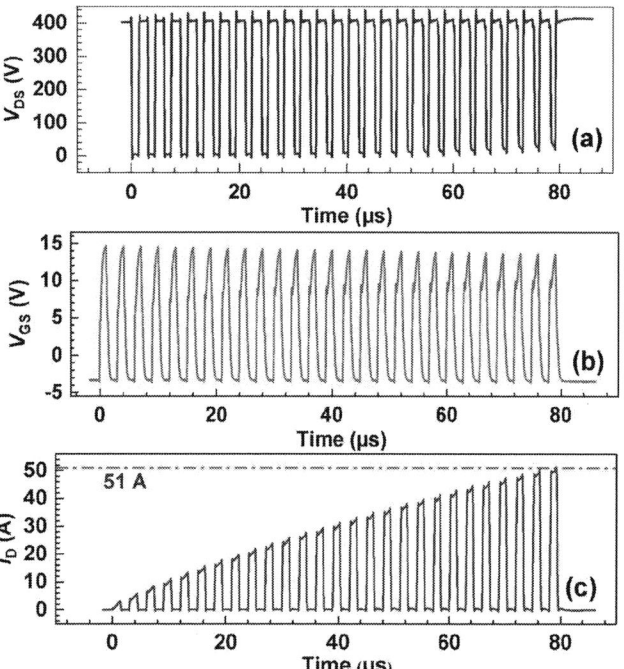

Fig. 5 (a) experiment waveforms of V_{DS}. (b) experiment waveforms of V_{GS}. (c) experiment waveforms of I_D.

Fig. 3 shows the simulation waveforms for V_{DS}, V_{GS}, I_D and junction temperature, respectively. As the temperature increases, the on-resistance of the device also increases, leading to a rise in V_{DS} during the on-state. By the 27th pulse, the drain current (I_D) reaches the maximum pulsed drain current specified in the sample's datasheet. The junction temperature ultimately reaches approximately 287.5 °C.

In previous works, the thermal annealing temperature for SiC MOSFETs after electron irradiation is usually between 160 °C and 360 °C [11][12][13][14][15]. The maximum operating junction temperature of commercial SiC MOSFETs is 175 °C. Considering that the self-heating in this strategy for SiC MOSFETs do not last for a long time; the junction temperature during self-heating may be kept in the range of 160 °C - 300 °C.

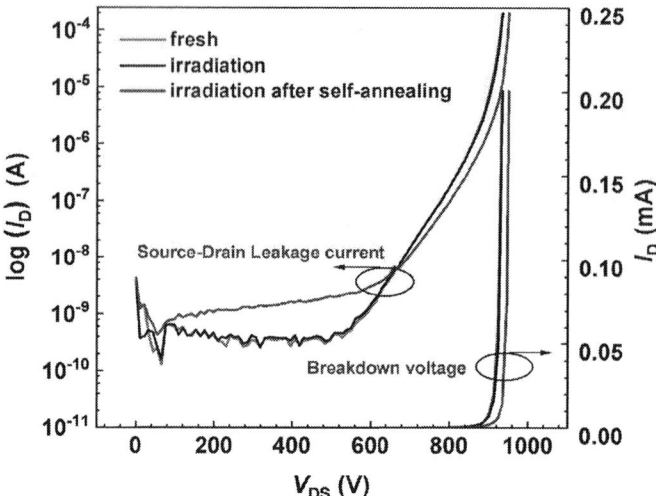

Fig. 6 Breakdown voltage and source-drain leakage current variation of SiC MOSFET.

III. EXPERIMENT AND RESULTS

In order to consider the effect of self-heating annealing junction temperature on the threshold voltage recovery, the self-heating annealing experiments used one sample, and were started with a small number of pulses. The threshold voltage was tested after each experiment when the junction temperature of this sample was cooled down to room temperature. Each subsequent experiment applied two additional pulses compared to the previous one. The experimental conditions for self-heating annealing are the same as the simulation.

Fig. 4 shows the recovery of the threshold voltage of the SiC MOSFET at various pulse counts. When 17 continous switching pulses were applied, there was no significant change in the threshold voltage of this sample. When increasing to 18 pulses, the threshold voltage begins to recover, the recovery is a little. It can be learned from the simulation results that the peak junction temperature in this case is roughly 125 °C. Based on the simulation results, with 22 pulses applied, the junction temperature reaches approximately 175 °C (the maximum junction temperature for normal operation of commercial devices), and the threshold voltage recovers by 0.3 V. With 27 pulses applied, the threshold voltage recovers to 3.5 V. Compare to the threshold voltage of this sample after electron irradiation, the threshold voltage recovers 0.9 V. Approximately 70% of the threshold voltage shift was recovered compare to the fresh sample. This sample is successfully self-heating annealed using this strategy. After the sample was cooled to room temperature and a sequence of 27 pulses was applied repeatedly. The threshold voltage did not recover further. It indicates that the recovery of V_{th} is mainly depended on the annealing temperature corresponding to the number of continuous switching pulses but is not depended on the annealing time corresponding to the times of repeating the treatment with the same switching pulses.

Fig. 5 shows the waveforms of V_{DS}, V_{GS}, and I_D captured by the oscilloscope during self-heating. The waveform data and variations closely align with the simulation results, indicating the reliability of the simulation for junction temperature prediction. This outcome is consistent with the previous findings [16]. From Fig. 5, the experimental maximum current flowing through the device reaches 51 A, which closely matches the maximum pulsed drain current specified in the datasheet. This ensures the sample operates in a safe self-heating environment.

Fig. 6 shows the variation in breakdown voltage and leakage current between the drain and source for the fresh sample, the irradiated sample, and the sample after self-heating annealing. After electron irradiation, it is interesting to find that the breakdown voltage and leakage current between the drain and source of this sample do not change significantly. After self-heating annealing, the breakdown voltage increases slightly, and the leakage current also increases. The increase in breakdown voltage after self-annealing may be attributed to the reduction of field oxide interface charge at the termination. The increase in leakage current between the source and drain may be due to the fact that, the charges in the gate oxide layer caused by irradiation are partially removed, but high temperatures might lead to the introduction of a limited number of defects in the bulk region.[14][15].

IV. CONCLUSIONS

This paper presents a novel strategy for recovering the threshold voltage of SiC MOSFETs after electron irradiation by utilizing circuit self-heating. The principle and strategy of safe self-heating for SiC MOSFETs are discussed, and the approach is validated through both simulation and experiment. The experiment successfully demonstrates the safe recovery of the threshold voltage of SiC MOSFETs following electron irradiation. This method can also be extended to other power switching devices under various irradiation conditions, such as gamma rays, heavy ions, and neutrons.

ACKNOWLEDGMENTS

This work was supported by Sichuan University-Suining Strategic Cooperation Project under Grant No. 2023 CDSN-13.

REFERENCES

[1] S. Hossain, N. K. et al. "MOSFET-based Three-Phase Inverter using Arduino - Applicable in Microgrid Systems," 2021 6th International Conference for Convergence in Technology (I2CT), pp. 1-5.Maharashtra, India, 2021.

[2] S. O'Donnell, J. et al. "Silicon carbide MOSFETs in more electric aircraft power converters: The performance and reliability benefits over silicon IGBTs for a specified flight mission profile," 2016 18th European Conference on Power Electronics and Applications (EPE'16 ECCE Europe), pp. 1-10, Karlsruhe, Germany, 2016.

[3] F. H. Ruddy, L. et al. "Silicon Carbide Neutron Detectors for Harsh Nuclear Environments: A Review of the State of the Art," in IEEE Transactions on Nuclear Science, vol. 69, no. 4, pp. 792-803, April 2022.

[4] D. -P. Sadik, J. et al. "Introduction of SiC MOSFETs in converters based on Si IGBTs," 2017 IEEE 3rd International Future Energy Electronics Conference and ECCE Asia (IFEEC 2017 - ECCE Asia), pp. 1680-1685, Kaohsiung, China Taiwan, 2017.

[5] T. Sakai et al. "Effects of gamma-ray irradiation on thin-gate-oxide VDMOSFET characteristics," in IEEE Transactions on Electron Devices, vol. 38, no. 6, pp. 1510-1515, June 1991.

[6] Q. Zeng, et al. "Research Progress on Radiation Damage Mechanism of SiC MOSFETs Under Various Irradiation Conditions," in IEEE Transactions on Electron Devices, vol. 71, no. 3, pp. 1718-1727, March 2024.

[7] J. Chen, S. et al. "Total Ionizing Dose Radiation Effect on Avalanche Robustness of 1200V Trench-type SiC Power MOSFETs," 2023 20th China International Forum on Solid State Lighting & 2023 9th International Forum on Wide Bandgap Semiconductors (SSLCHINA: IFWS), pp. 80-83. Xiamen, China, 2023.

[8] K. C. Mandal, et al. "High-Resolution Alpha Spectrometry Using 4H-SiC Detectors: A Review of the State-of-the-Art," in IEEE Transactions on Nuclear Science, vol. 70, no. 5, pp. 823-830. May 2023.

[9] Tamana Baba et al.,"Radiation-induced degradation of silicon carbide MOSFETs − A review," Materials Science and Engineering: B, Volume 300,2024,117096.

[10] S. Bonaldo et al., "Radiation-Induced Effects in SiC Vertical Power MOSFETs Irradiated at Ultrahigh Doses," in IEEE Transactions on Nuclear Science, vol. 71, no. 4, pp. 418-426, April 2024.

[11] J. M. Rafi et al., "Low-Temperature Annealing of Electron, Neutron, and Proton Irradiation Effects on SiC Radiation Detectors," in IEEE Transactions on Nuclear Science, vol. 70, no. 10, pp. 2285-2296, Oct. 2023.

[12] T. Liu et al., "Comparative Investigation on Ionizing Irradiation-Induced Threshold Voltage Degradation for 1200-V DT SiC MOSFET by Experiment and Simulation," in IEEE Transactions on Nuclear Science, vol. 71, no. 11, pp. 2386-2392, Nov. 2024.

[13] D. Hu et al. "Impact of Different Gate Biases on Irradiation and Annealing Responses of SiC MOSFETs," in IEEE Transactions on Electron Devices, vol. 65, no. 9, pp. 3719-3724, Sept. 2018.

[14] D. Hu et al. "Radiation and annealing effects of SiC MOSFETs at high voltage gate bias," 2018 20th European Conference on Power Electronics and Applications (EPE'18 ECCE Europe), pp. P.1-P.5, Riga, Latvia, 2018.

[15] Mu He et al. "Electron irradiation effects and room-temperature annealing mechanisms for SiC MOSFETs," Results in Physics, Volume 60, 2024.

[16] W. Luo et al., "A New Strategy to Recover Threshold Voltage of SiC MOSFET After Gamma Radiation by Self-Heating in Circuit," 2023 8th International Conference on Integrated Circuits and Microsystems (ICICM) pp. 205-208, Nanjing, China, 2023.

Study on Single Event Effects in High-Voltage Silicon Power Semiconductor Devices

Mingmin Huang[1*], Huan Li[2], Ailin Qiu[1], Yichu Qin[1], Xiaoping Dong[1], Dongyang Wang[1], Yao Ma[1*], Qiang Yu[3]

1 College of Physics, Sichuan University, Chengdu, China
2 China ZhenHua Group YongGuang Electronics Company, Guiyang, China
3 Sichuan Suining Lippxin Microelectronics Co., Ltd, Suining, China
*Corresponding author's E-mail: mmhuang@scu.edu.cn, mayao@scu.edu.cn

Abstract—This paper studies single event effects in high-voltage silicon power semiconductor devices, including fast recovery diode (FRD), Superjunction (SJ) MOSFET, IGBT, and reverse conducting (RC) IGBT. In experiments, 1332 MeV Ta+ ions with a linear energy transfer (LET) of 82.1 MeV·cm²/mg are used. The SJ MOSFET is sensitive to single event gate rupture (SEGR), and the SEGR voltage is about 125 V. The SEB in RC IGBT is close to that in the IGBT, where the SEB voltages of them are both about 250 V. The SEB voltage of FRD is about 640 V, which is much higher than the other devices. Simulation results show that, turning on of the parasitic npn transistor hardly aggravates SEB and SEGR in the SJ MOSFET but aggravates SEB in the IGBTs, and the p-collector in the IGBT does not aggravate the SEB.

Keywords—single event effect, superjunction MOSFET, IGBT, reverse conducting IGBT, fast recovery diode

I. INTRODUCTION

In the aerospace application, power semiconductor devices are the core component of the power supply system and electric propulsion system in the spacecrafts. However, there are many cosmic rays, which may cause degradation and failure of power semiconductor devices. Two single event effects caused by heavy ions, i.e., single event burnout (SEB) and single event gate rupture (SEGR), are the main concerns for power semiconductor devices [1]. With the bias voltage increasing, the SEB and SEGR become more likely to occur due to the fact that dynamic avalanches and dielectric breakdown become more serious. Consequently, there are few power switching devices above 600 V rated successfully applied in the aerospace application.

Recently, many works have focused on the SEB and SEGR of high-voltage power semiconductor devices above 600 V rated, e.g., superjunction (SJ) MOSFET [2], SiC junction barrier Schottky (JBS) diode [3], SiC MOSFET [4], and GaN high electron mobility transistor (HEMT) [5]. Studies on SEB and SEGR of high-voltage Si devices, e.g., IGBT, reverse conducting (RC) IGBT, and fast recovery diode (FRD), are very few. Since SiC and GaN have natural advantages of high critical electric field and high bandgap, SiC and GaN devices are expected to be more radiation-resistant and efficient than Si devices. Although SiC and GaN devices are found to be much more insensitive to the total ionizing dose (TID) effect than Si devices, the former two are sensitive to SEB and SEGR [6], where the SEB usually occurs at a bias voltage of 20% ~ 50% of breakdown voltage [7]. Therefore, the high-voltage Si power semiconductor devices seem to still be a candidate for aerospace application. More studies on SEB and SEGR of high-voltage Si power semiconductor devices are essential.

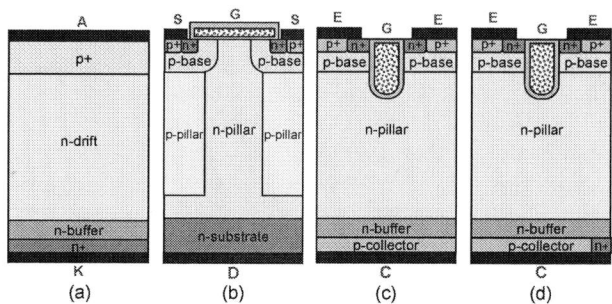

Fig. 1 Schematic diagrams of power semiconductor devices studied in this paper. (a) FRD. (b) SJ MOSFET. (c) IGBT. (d) RC IGBT.

In this paper, SEB and SEGR of high-voltage silicon power semiconductor devices (FRD, SJ MOSFET, IGBT, and RC IGBT) are studied by experiments and simulations. Section II presents experiment conditions and results. Section II gives out simulation results and analyses.

II. EXPERIMENT CONDITIONS AND RESULTS

Fig. 1 shows schematic diagrams of FRD, SJ MOSFET, IGBT, and RC IGBT, where the SJ MOSFET has planar gates, and the two IGBTs have trench gates. Different from the IGBT, an n+ region is introduced at the backside and shorted to the p-collector region in the RC IGBT, which acts as a MOSFET at a low collector (C) current (I_C) and acts as an IGBT when I_C is high enough to turn on the p-collector/n-buffer junction.

In experiments, 1.2 kV FRD, 650 V SJ MOSFET, 1.2 kV IGBT, and 1.35 kV RC IGBT were used. Pulsed 1332 MeV Ta+ ions were vertically incident into the decapped devices, where the linear energy transfer (LET) is 82.1 MeV·cm²/mg. The period of ion pulses is 12 s and the averaged fluence rate is 1×10^4 cm⁻²/s. During irradiation, the K/D/C bias voltages ($V_{KA}/V_{DS}/V_{CE}$) were applied and the K/D/C currents ($I_K/I_D/I_C$) were monitored by the Keithley 2470 source meter, the G current (I_G) was monitored by the Keithley 2450 source meter, and appropriate values of limited current were set. The SEB and SEGR are considered as cases with $I_K/I_D/I_C$ and I_G reaching the limited values, respectively.

Fig. 2 shows waveforms of I_D and I_G of SJ MOSFET during irradiation, where V_{DS} = 50 V, 100 V, 125 V, 150 V, and 200 V are used. At V_{DS} = 50 V, there is nearly no single event leakage current (SELC), i.e., permanent leakage current degradation. At V_{DS} = 100 V, the SELC of I_G occurs. It shows that $I_G = I_D$ at the time with no ion incident into the device at V_{DS} = 100V, which can be inferred that leakage current paths are formed in the gate oxide above the n-pillar.

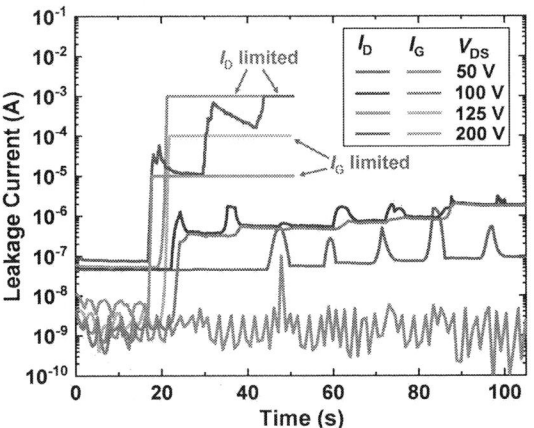

Fig.2 Waveforms of I_D and I_G of SJ-MOSFET during irradiation, where V_{DS} = 50V, 100 V, 125 V, 150 V, and 200 V are used.

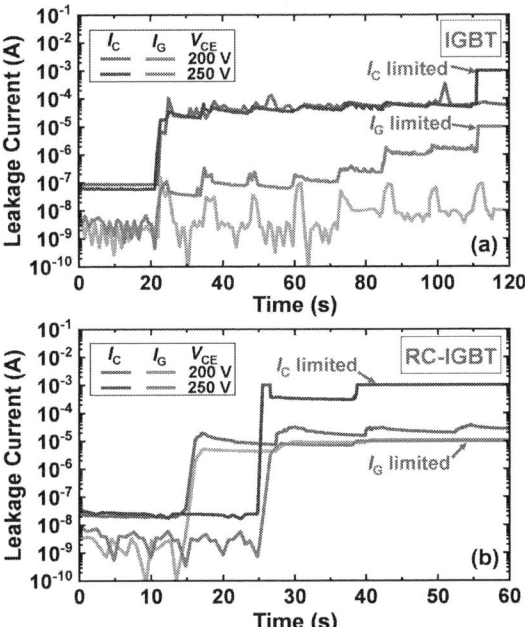

Fig.3 Waveforms of I_C and I_G of IGBT and RC IGBT during irradiation, where V_{CE} = 200V and 250 V are used. (a) IGBT. (b) RC IGBT.

At V_{DS} = 125 V, the SEB and SEGR occur together at the first pulse of ions, where the limited values of I_D and I_G are set by 1 mA and 100 μA, respectively. At V_{DS} = 200 V, a much lower limited value of I_G is set (10 μA). It is found that SEB does not occur at the first pulse and occurs at the third pulse. By comparing waveforms of cases at V_{DS} = 125 V and V_{DS} = 200 V, it can be inferred that the SJ MOSFET is more sensitive to SEGR than SEB, and the SEGR may be able to play a role in aggravating SEB. The SEGR voltage of SJ MOSFET is about 125 V, and the SEB voltage of SJ MOSFET is about 200 V.

Fig. 3 shows waveforms of I_C and I_G of IGBT and RC IGBT during irradiation, where V_{CE} = 200 V and 250 V are used. The limited values of I_C and I_G are set by 1 mA and 10 μA, respectively. Since the RC IGBT is at the MOS-mode when I_C is low enough, it is natural to predict that the RC IGBT may be more insensitive to SEB and SEGR than the IGBT. However, it is interesting to find that the RC IGBT is

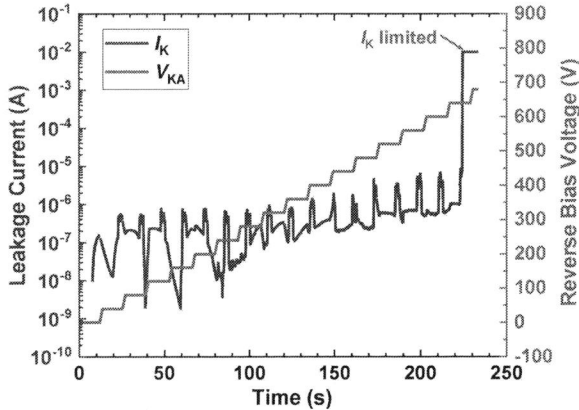

Fig.4 Waveforms of I_K of FRD during irradiation, where stepped increased V_{KA} is used.

not more insensitive but is a little more sensitive to SEB and SEGR than the IGBT. The SEB of the IGBT and RC IGBT both occur at V_{CE} = 250 V, where the SEB of the latter occurs at the first ion pulse and the SEB of the former occurs at the eighth ion pulse. The SEGR of the IGBT and RC IGBT occur at V_{CE} = 250 V and V_{CE} = 200 V, respectively. Experiment results indicate that the pnp transistor (p-collector/n-drift/p-base) may not aggravate SEB and SEGR. Besides, the SELC of I_C is more significant than that of I_G, where the permanent degradation of I_C by an ion pulse is in the order of 10 μA.

Fig. 4 shows waveforms of I_K of FRD during irradiation, where stepped increased V_{KA} is used. The value of V_{KA} is increased by 40 V in every 14 s. The SEB of FRD occurs at V_{KA} = 640 V, which is much higher than that of the SJ MOSFET, IGBT, and RC IGBT. In addition, the SELC of I_K is almost negligible in the FRD. Although performances of SJ MOSFET, IGBT and RC IGBT suffering heavy ions are not unsatisfactory, performance of the FRD (not hardened) suffering such high LET ions is inspiring. A noticeable improvement in resistance to SEEs is probably feasible for high-voltage Si power semiconductor devices.

SIMULTATION RESULTS AND ANALYSIS

In order to study mechanisms of high-sensitive SEB and SEGR in the SJ MOSFET, IGBT and RC IGBT, TCAD simulations are carried out. The LET value and incidence depth of 1332 MeV Ta$^+$ ions incident into Si are 82.1 MeV·cm²/mg and 79.7 μm. The n-pillar thickness and p-pillar thickness in the SJ MOSFET are 44 μm and 36 μm respectively, and the n-drift thickness in IGBT and RC IGBT is 94 μm. In simulations, the ion is considered to strike through the devices, the n-pillar and p-pillar widths are both 5 μm, the n-pillar and p-pillar doping concentrations are both 4×10^{15} cm^{-3}, the n-drift doping concentration is 6×10^{13} cm^{-3}, and the lattice heating model is used. The simulations have been calibrated according to experiment results.

Fig. 5 shows simulated waveforms of I_D and peak temperature (T_{peak}) for the SJ-MOSFET and the SJ-MOSFET with no n+ sources irradiated by Ta$^+$ ions at V_{DS} = 190 V and 200 V. Taking the melting point of Si (1685 K) as a criterion, the SEB voltage of the SJ-MOSFET is 200 V, which is in line with the experimental results in Fig. 2. It is interesting to find that the SEB voltage of the SJ-MOSFET with no n+

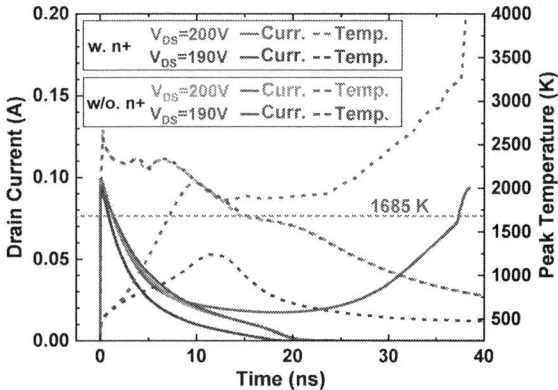

Fig.5 Simulated waveforms of I_D and T_{peak} for the SJ-MOSFET and the SJ-MOSFET with no n+ sources irradiated by the Ta$^+$ ion at V_{DS} = 190 V and 200 V.

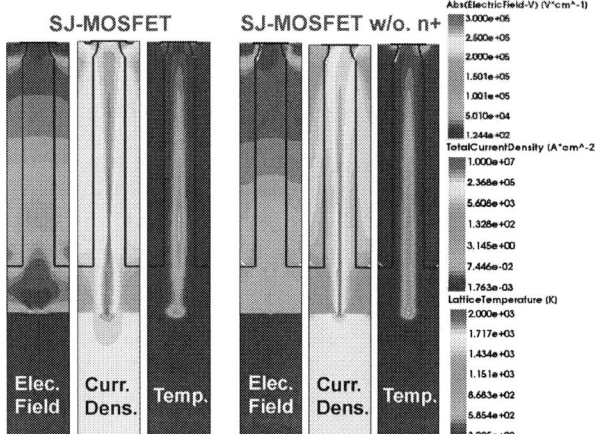

Fig.6 Distributions of electric field, temperature, and electron current density for the case at 10 ns and V_{DS} = 200 V in Fig. 5.

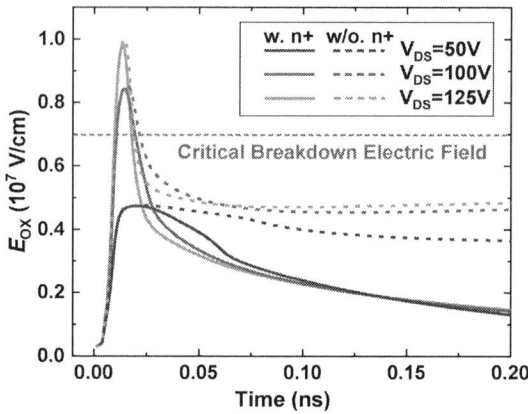

Fig.7 Peak electric field in the gate oxide for the SJ-MOSFET and SJ-MOSFET without n+ source irradiated by the Ta$^+$ ion at V_{DS} = 50 V, 100 V, and 125 V.

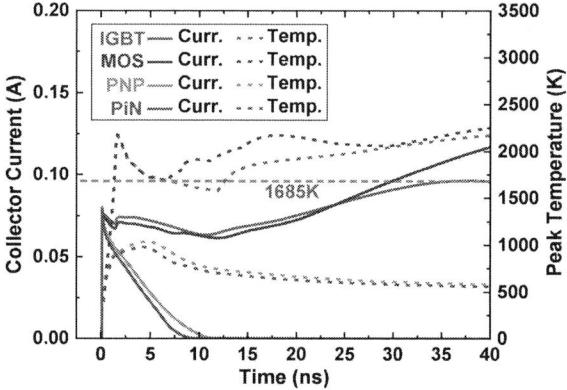

Fig.8 Simulated waveforms of I_C and T_{peak} for the IGBT, IGBT without p-collector (MOS), IGBT without n+ emitter (PNP), and IGBT without both p-collector and n+ emitter (PiN) irradiated by the Ta$^+$ ion at V_{DS} = 250 V.

sources (n+ sources replaced by p+ regions) is even lower than 190 V, although the current of it finally falls down. It seems that the positions with T_{peak} are different in the SJ-MOSFET and SJ-MOSFET with no n+ sources, and the turn-on of the parasitic npn transistor (n+/p-base/n-pillar) may be not the main cause of SEB.

In order to study the reason of the phenomena in Fig. 5, Fig. 6 shows distributions of electric field, current density, and lattice temperature for the case at 10 ns and V_{DS} = 200 V in Fig. 5. When the n+ sources are omitted, the electric field at the bottom is decreased but the electric field at the p-base/n-pillar is enhanced, since the electron charges at the bottom are decreased and the net positive charges (p-n) are increased. Moreover, current is more concentrated in the SJ-MOSFET with no n+ sources than that in the SJ-MOSFET, which leads to a higher peak lattice temperature at a different position (at the bottom). This may be due to that electrons injected from the n+ sources spread into the n-pillar and p-pillar to decentralize holes in the n-pillar and p-pillar to some extent. From this perspective, the turn-on of the parasitic npn transistor may not always play a negative role.

Fig. 7 shows the peak electric fields in the gate oxide (E_{ox}) for the SJ-MOSFET and SJ-MOSFET with no n+ sources irradiated by the Ta$^+$ ion at V_{DS} = 50 V, 100 V, and 125 V. It

can be found that the n+ sources have nearly no effect on SEGR. At V_{DS} = 50 V, the maximum E_{ox} is about 5×10^6 V/cm, which is lower than the critical electric field of gate oxide by measurement, i.e., 7×10^6 V/cm. Hence, there is nearly no permanent damages in the gate oxide at V_{DS} = 50 V as shown in Fig. 2. At V_{DS} = 100 V and 125 V, the maximum values of E_{ox} reach 8.4×10^6 V/cm and 9.8×10^6 V/cm, respectively. Such a high E_{ox} is able to form permanent or catastrophic damages in the gate oxide, which is in agreement with the experimental results in Fig. 2.

Fig. 8 shows simulated waveforms of I_C and T_{peak} for the IGBT, IGBT without p-collector (i.e., MOS), IGBT without n+ emitter (i.e., PNP), and IGBT without both p-collector and n+ emitter (i.e., PiN) irradiated by the Ta$^+$ ion at V_{DS} = 250 V, where the heavy ion in simulation is incident into the devices along the middle of p-base region. It can be found that waveforms of I_C and T_{peak} for the IGBT and IGBT without p-collector (MOS) are very close (T_{peak} reaches 1685 K within 1.5 ns) and are both uncontrollable, and waveforms of I_C and T_{peak} for the IGBT without n+ emitter (PNP) and IGBT without both p-collector and n+ emitter (PiN) are very close and both finally fall down. This indicates that the p-collector does not aggravate SEB but the n+ emitter aggravates SEB, which is in agreement with the experiment results, i.e., the SEB voltage of RC IGBT being not higher than that of IGBT and being much lower than that of FRD.

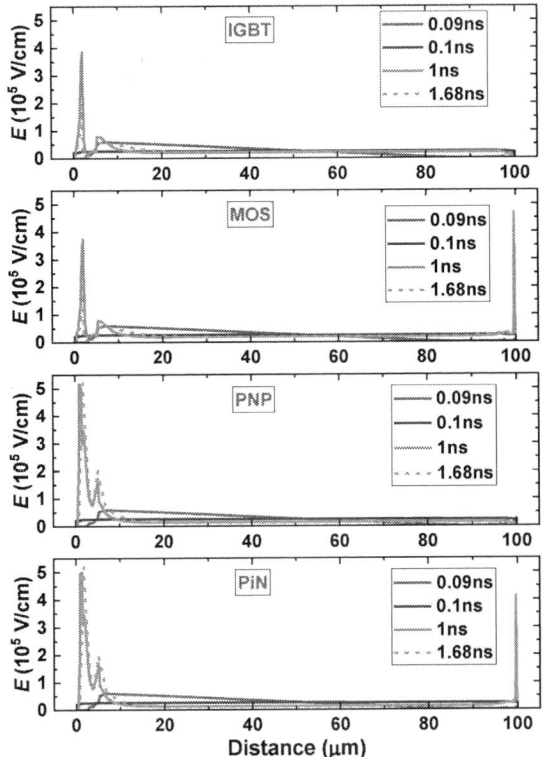

Fig.9 Electric field distributions along the ion track in the four structures in Fig. 8 at 0.09 ns, 0.1 ns, 1 ns, and 1.68 ns.

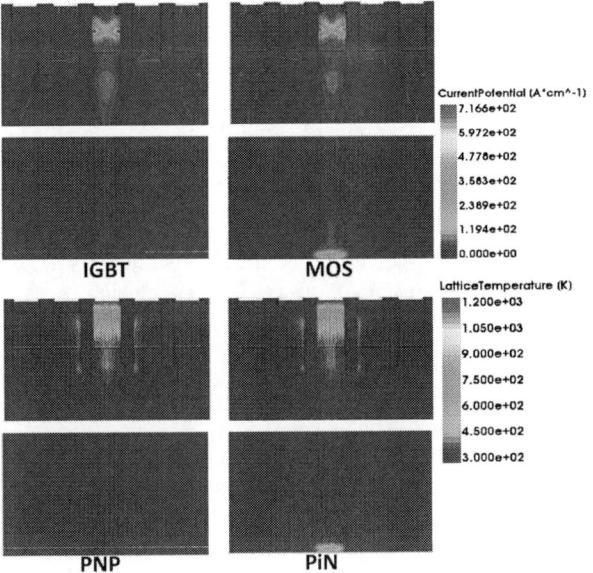

Fig. 10 Temperature distributions and current flowlines in the four structures in Fig. 8 at 1 ns.

In order to study the reason of the phenomena discussed in Fig. 8, Fig. 9 shows electric field distributions along the ion track in the four structures in Fig. 8 at 0.09 ns, 0.1 ns, 1 ns, and 1.68 ns. Evolutions of electric field in the IGBT and MOS structures are similar, and evolutions of electric field in the PNP and PiN structures are similar. A high electric field is formed at the bottom of the MOS and PiN structures, due to high-density electrons there. The IGBT and PNP

structures are able to inject holes into the n-drift, which offsets the electron charges to avoid the high electric fields at the bottom. Besides, it is found that the electric field at the top (in the p-base) of the PNP and PiN structures can be higher than that of the IGBT and MOS structures. Because the parasitic npn transistors in the IGBT and MOS structures are turned on to inject electrons to offset part of hole charges in the p-base. From the intuitive view, the peak temperature in the PNP and PiN structures should be higher than the other two structures. However, it should be noted that the joule heat is proportional to both electric field and current density.

Fig. 10 shows temperature distributions and current flowlines in the four structures in Fig. 8 at 1 ns. There are two high temperature portions (one at the top and the other at the bottom) in the MOS and PiN structures and one high temperature portion in the IGBT and PNP structures, which is consistent with electric field distributions in Fig. 9. Due to the turn-on of the parasitic npn transistors, the maximum current density in the p-base regions of the IGBT and MOS structures can be higher than that of the PiN and PNP structures. Therefore, T_{peak} in the IGBT and MOS structure are higher than that of the PiN and PNP structures. This may be used as an explanation for the phenomenon that the SEB voltages of IGBT and RC IGBT are very close and are much lower than that of the FRD.

III. CONCLUSIONS

In this paper, single event effects in FRD, SJ MOSFET, IGBT, and RC IGBT are studied by experiments and simulations, where 1332 MeV Ta$^+$ ions are used. The SJ MOSFET is more sensitive to SEGR than SEB. According to simulations, the parasitic npn transistor in the SJ MOSFET seems have very limited effect on the SEGR and SEB, since the peak temperature is not located around the n+/p-base junction but around the bottom of the n-pillar. The SEB voltages of IGBT and RC IGBT are both 250 V and are much lower than that of the FRD (640 V). The reason may be that the peak temperature is located around the n+/p-base junction due to both high electric field and high current density there.

REFERENCES

[1] J. L. Titus. "An updated perspective of single event gate rupture and single event burnout in power MOSFETs." IEEE Transactions on nuclear science, vol. 60, no. 3, pp. 1912–1928, 2013.

[2] K. Muthuseenu, et al. "Analysis of SEGR in silicon planar gate super-junction power MOSFETs." IEEE Transactions on Nuclear Science, vol. 68, no.5, pp. 611–616, 2021.

[3] X. Dong, et al. "Mechanism and Physical Model of the Single-Event Leakage Current for SiC JBS Diodes" IEEE Transactions on Nuclear Science, vol. 71, no. 10, pp. 2252–2259, 2024.

[4] D. R. Ball, , et al. "Effects of breakdown voltage on single-event burnout tolerance of high-voltage SiC power MOSFETs." IEEE Transactions on Nuclear Science, vol. 68, no. 7, pp. 1430–1435, 2021.

[5] R. D. Harris, et al. "Radiation Effects in AlGaN/GaN HEMTs." IEEE Transactions on Nuclear Science, vol. 69, no.5, pp. 1105–1119, 2022.

[6] J. M. Lauenstein. "Wide-Bandgap Semiconductors in Space: Appreciating the Benefits but Understanding the Risks." Conference on Radiation Effects on Components and Systems (RADECS 2018). No. 2018-561-NEPP, 2018.

[7] J. M. Lauenstein. "Wide-bandgap-power-SiC and GaN-radiation reliability." IEEE Nucl. Space Radiat. Effects Short Course III-1, 2020.

Electron Irradiation Effects on *I-V* Characteristics in p-GaN Gate HEMTs

Dongyang Wang [1], Yifei Huang [2,3], Shuting Wang [1], Mingmin Huang [1*], Qimeng Jiang [2,3*], Yao Ma [1]

[1] College of Physics, Sichuan University, Chengdu, 610065, China
[2] Institute of Microelectronics, Chinese Academy of Sciences, Beijing, 100029, China
[3] University of Chinese Academy of Sciences, Beijing, 100049, China

* Corresponding author's E-mail: mmhuang@scu.edu.cn, jiangqimeng@ime.ac.cn

Abstract—**Commercial p-GaN gate power HEMTs were irradiated with high-energy (1.7 MeV), high-fluence (1×10^{15}, 5×10^{15} and 1×10^{16} cm^{-2}) electrons. Significant degradation of the devices occurred only at the fluence of 1×10^{16} cm^{-2}. The drain-source saturation current was increased and the threshold voltage was slightly negatively shifted. Moreover, the leakage currents of gate and drain were increased significantly. The C-DLTS spectra show apparent changes in traps located in gate region. By four-terminal testing, it was found that the drain-substrate leakage current is the dominant component of the drain leakage current. The semi-insulating property of the buffer layer disappeared and it became a large resistor. After 7 days of annealing at room temperature, the gate and drain leakage currents were both recovered almost to pre-irradiation levels.**

Keywords—GaN HEMT, electron irradiation, p-GaN gate, leakage current, traps, total ionizing dose, DLTS

I. INTRODUCTION

GaN high electron mobility transistors (HEMTs) are widely used in power conversion devices because of their high critical breakdown electric field (about 3 MV/cm), low on-resistance, high switching frequency and small size. These advantages are highly consistent with the requirements for aerospace equipment. Therefore, there have been many studies using different kinds of particles to irradiate GaN HEMTs to evaluate their radiation tolerance [1-8].

The main space radiation effects include total ionizing dose (TID), single event effect (SEE) and displacement damage (DD). High-energy electrons are one of the main particles in Van Allen radiation belts, which pose a great threat to devices working in space. There are two forms of energy loss when electrons strike into devices, ionization energy loss (IEL) and non-ionization energy loss (NIEL), corresponding to the TID and DD, respectively. Several studies have reported the effects of electron irradiation on different GaN HEMTs. John *et al.* [5] found that the drain and gate current of AlGaN/GaN HFET increased after 0.45 MeV electron irradiation. They explained that the positive charge increased the concentration of 2-DEG (two-dimensional electron gas), and the enhancement of the trap-assisted-tunneling (TAT) effect increased the gate leakage current. Chen and Liu [6] observed an increase in saturation drain current but no change in threshold voltage or gate leakage current. The results were due to the IEL generating positive charges in AlGaN layer, particularly the NIEL generating N and Ga vacancies. Feng *et al.* [7] used time constant spectra (TCS) and differential amplitude spectra (DAS) to detect traps. They found that

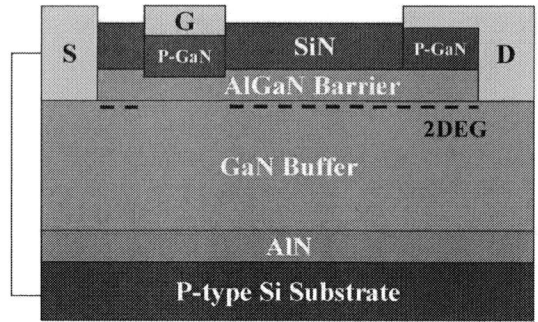

Fig. 1. Schematic cross-section of the HD-GIT structure.

the trap densities decreased near the drain and those increased near the gate after irradiation. Although most reports on GaN microwave transistors have shown little issue with the above radiation effects, there is little reason to assume that the high-voltage power devices will have the same radiation tolerance [4]. Moreover, there are few reports on changes in blocking characteristic, especially for high-voltage power GaN HEMTs.

In this article, commercial 600 V p-GaN gate HEMTs were used to investigate the effects of electron irradiation. Section II introduces the detailed information of samples and irradiation experiments. The results and analysis are presented in Section III, including the variations of *I-V* characteristics and C-DLTS (Capacitance-Deep Level Transient Spectroscope) spectra. The conclusion is given in Section IV.

II. SAMPLES AND EXPERIMENTAL SETUPS

The devices under test are HD-GITs (Hybrid-Drain-embedded Gate Injection Transistors), which are normally-off GaN HEMTs (IGLD60R190D1) produced by Infineon Corporation. Fig. 1 shows the schematic cross-section of the HD-GIT structure. All the devices were zero biased for irradiation with the fluences of 1×10^{15}, 5×10^{15} and 1×10^{16} cm^{-2}, respectively. The electron energy was 1.7 MeV, and the flux was stable at 5×10^{12} cm^{-2}/s. All samples were first tested in 12-24 hours after irradiation. The second test was conducted after 7 days of annealing at room temperature. DC measurements were performed using the Keysight B1505A power device analyzer.

III. RESULTS AND ANALYSIS

All devices irradiated at the same fluence exhibited the same degradation level. In addition, the devices irradiated at low fluences (1×10^{15} and 5×10^{15} cm^{-2}) had no apparent degradation after irradiation. Therefore, only

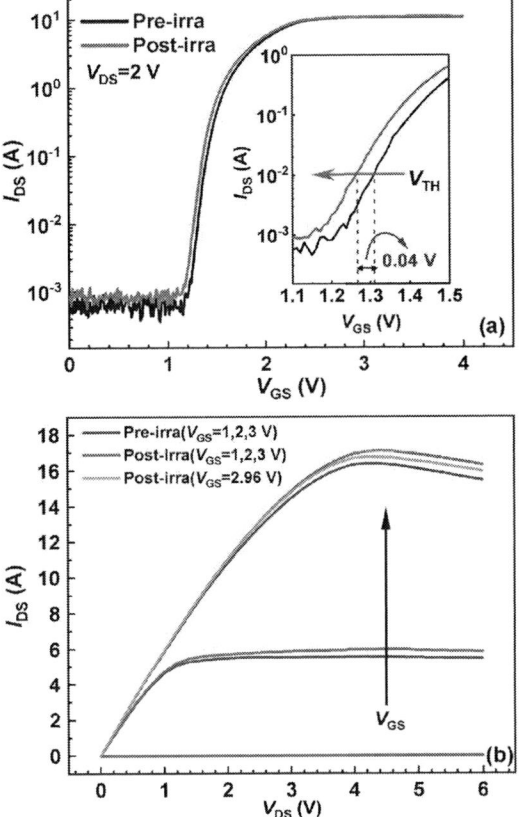

Fig. 2. Changes in (a) transfer curves and (b) output curves of the device at the fluence of 1×10^{16} cm^{-2}. V_{GS} = 2.96 V was supplied to eliminate the influence of the negative shifted V_{TH} after irradiation.

the devices at the highest fluence (1×10^{16} cm^{-2}) will be discussed in this article.

Fig. 2(a)-(b) show the transfer and output curves before and after irradiation. As shown in Fig. 2(a), the threshold voltage (V_{TH}) was slightly shifted toward the negative direction, which was not as severe as MOSFETs because of the absence of the gate oxide [9]. Due to the limitation of the test module, the minimum accuracy of the current is mA. Fig. 2(b) shows that the saturation current between drain and source ($I_{DS,Sat}$) increased significantly after irradiation. After testing, it was verified that the increase in $I_{DS,Sat}$ at high V_{GS} was not totally dominated by the negative shift of V_{TH} as the green line in Fig. 2(b) shows. The above phenomena are consistent with the results of other reports, which are related to the traps in AlGaN barrier layer [5-8]. The IEL produces a large number of electron-hole pairs, while the NIEL causes displacement damage introducing more immobile point defects that can capture ionized holes to become fixed positive charges [6]. The increase of positive charges in AlGaN barrier layer deepened the quantum well and thus increased the density of 2-DEG. Typically, the negative shift of V_{TH} and the increase in $I_{DS,Sat}$ occur simultaneously, both as a result of the increased density of 2-DEG. In addition, Pan et al. observed stress relaxation after irradiation and believed that the improvement in mobility and 2-DEG density was related to it [8].

The blocking characteristics of devices were measured at V_{GS} = 0 V and V_{DS} ranging from 0 to 600 V. Fig. 3(a)-(b)

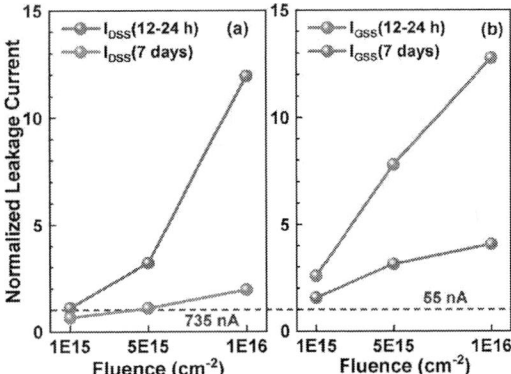

Fig. 3. Post-irradiation average leakage current at (a) drain and (b) gate normalized by pre-irradiation leakage current. The average I_{DSS} and I_{GSS} (V_{GS}=0 V, V_{DS}=600 V) before irradiation were 735 nA and 55 nA, respectively.

show the normalized leakage currents of drain (I_{DSS}) and gate (I_{GSS}). The leakage current was measured at V_{GS} = 0 V, V_{DS} = 600 V. Both the I_{DSS} and I_{GSS} increased at each fluence in the first measurement after irradiation. Obviously, the leakage current increased significantly as the irradiation fluence increased. The room temperature annealing effect was evident because of the substantial reduction in leakage current after 7 days. However, the recovery degree of I_{DSS} and I_{GSS} was not the same, probably because of the different leakage mechanisms. The drain-gate leakage mechanism is based on electron hopping along surface states [10-12]. As the voltage between the drain and gate gets higher, more electrons will tunnel into surface states from the gate edge where the electric field is highest [10]. To further investigate the causes of gate leakage current degradation by electron irradiation, C-DLTS was used to detect changes in gate area traps.

As shown in Fig. 1, the p-GaN/n-AlGaN heterojunction is below the gate. The thicknesses of the p-GaN and AlGaN barrier layer are about 100 nm and 20 nm, respectively. The depletion region at different gate biases may be distributed in both p-GaN and AlGaN layer, even in the GaN buffer layer. Fig. 4 shows the C-V characteristic of the heterojunction between gate and source. The total depletion region width (W_{dep}) is calculated roughly by the following equation

$$W_{dep} = \frac{A\varepsilon_S}{C_{dep}} \quad (1)$$

where A is the gate area (about 1.8×10^{-3} cm^2), ε_S is the permittivity of GaN and AlGaN (considered as the same value of $12\varepsilon_0$), and C_{dep} is the capacitance of the depletion region. As shown in the dashed frame in Fig. 4, within the voltage range of 1 - 1.5 V, the depletion region width increases rapidly from 50 nm to 130 nm as the voltage decreases. Moreover, it corresponds to the recovery or depletion of the 2-DEG, which means that the depletion region only distributed in p-GaN and AlGaN layer. Due to the extremely high density of 2-DEG and the thin thickness of the AlGaN layer, the depletion region within this voltage range is likely to be mainly distributed in p-GaN. Typically, the C-V characteristics are not analyzed when the pn junction is forward biased because of the influence of diffusion capacitance. As shown in the inset in

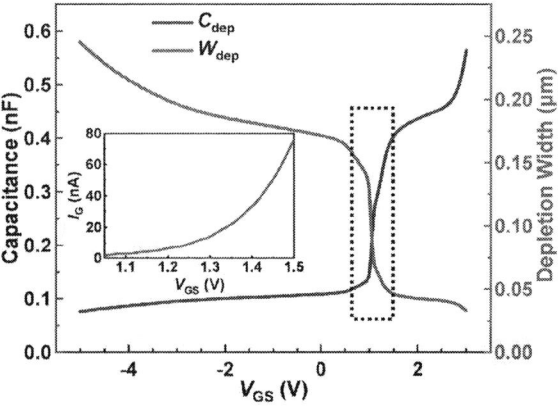

Fig. 4. *C-V* characteristic of p-GaN/n-AlGaN heterojunction between gate and source.

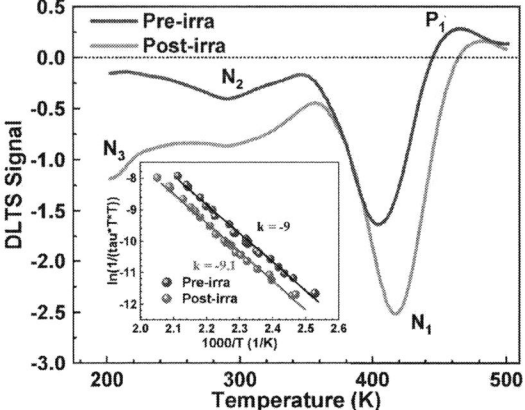

Fig. 5. DLTS spectra of p-GaN/n-AlGaN heterojunction between gate and source. $U_P = 1.8$ V, $U_R = 1$ V, $T_W = 1$ s, $T_P = 0.1$ s. The voltage was forward biased to the junction. (Inset) Arrhenius plots of the negative peak N_1.

Fig. 4, the gate current is only 80 nA at $V_{GS} = 1.5$ V, which means that the heterojunction is still in the off state. Therefore, the diffusion capacitance could be neglected in above analysis.

Fig. 5 shows the DLTS spectra changes of the heterojunction after irradiation. The intensity of negative signal peaks N_1 and N_2 increased significantly after irradiation, which implies a substantial increase in trap concentration. However, the density of the traps corresponding to the positive signal peak P_1 decreased. In addition, it seems that a new negative peak N_3 appeared after irradiation which was out of temperature measurement range. The Arrhenius plots of the negative signal peak N_1 are shown in the inset in Fig. 5. The activation energy of this electron trap is 0.78 - 0.8 eV, which almost did not change after irradiation. Moreover, it's close to the report that found an electron trap located in p-GaN with the energy level of 0.85 eV below the conduction band [13]. The nature of the trap is likely to be nitrogen interstitial or gallium interstitial [14]. The electrons collide with Ga/N atoms, knocking them out of their original position and thus increasing the concentration of this trap. The TAT effect becomes more significant as the electron trap concentration increases. And the surface state density may also increase. As a result, more electrons in p-GaN gate can cross the energy barrier and tunnel into surface states, which increases the gate leakage current.

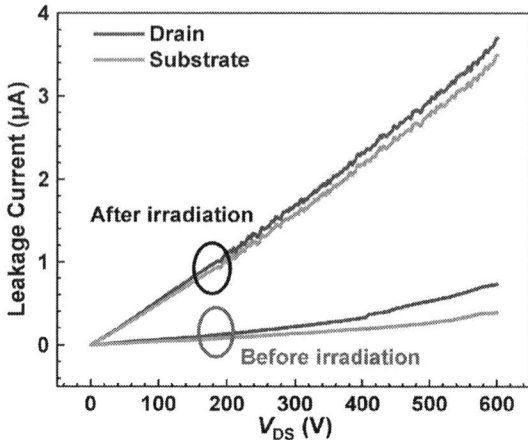

Fig. 6. Leakage current of drain and substrate in four-terminal measurements.

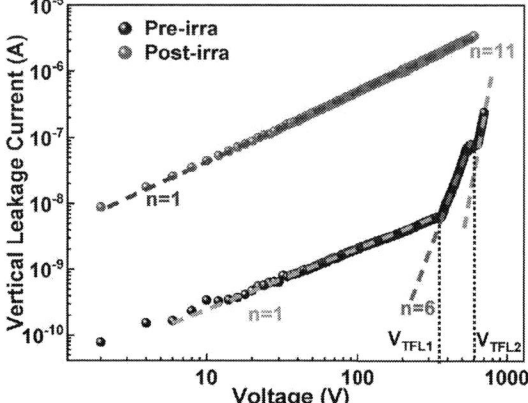

Fig. 7. Vertical blocking characteristic of the device before and after irradiation. Voltage is applied to the drain and substrate with the source and gate floating.

The positive signal peak P_1 probably corresponds to the electron traps located in AlGaN layer, which decreased in concentration after irradiation [7]. Unfortunately, detailed information on other traps could not be obtained accurately due to the broader or weaker signal peaks. However, it is clear that electron irradiation caused significant changes in trap concentration and even introduced new defects in gate region.

Although the gate leakage current increased significantly after irradiation, it was still less than 10% of the drain leakage current. Generally, for GaN HEMTs based on Si substrate, the substrate is internally shorted to the source to eliminate the substrate-coupled cross-talk effect [15]. Therefore, the leakage path of drain includes drain-gate, drain-source and drain-substrate. However, the drain-source leakage current is quite insignificant for E-mode power HEMTs with long gate length [11]. It seems that the drain-substrate leakage current is the main part of the drain leakage current. To verify this speculation, the connection between substrate and source was disconnected to perform four-terminal measurements.

Fig. 6 plots the leakage currents of the drain and substrate before and after irradiation in four-terminal measurements. The substrate current is almost equal to the drain current, which implies that the blocking ability in vertical direction was severely affected by irradiation.

979-8-3315-2209-4/25 $31.00 © 2025 IEEE   236

GaN buffer layer is usually doped with the high concentration of carbon to compensate unintentional doping. The deep acceptor traps introduced by carbon pin the Fermi level at $E_V + 0.9$ eV to achieve a semi-insulating state of the buffer layer [16]. It was found that the vertical breakdown of HEMTs based on Si is not dominated by impact ionization, but rather by space-charge-limited current (SCLC) conduction [17]. Fig. 7 shows the vertical blocking characteristics of the device before and after irradiation. The vertical I-V characteristic of device before irradiation shows a segmented relation of $I \propto V^n$. The logarithmic slope ($\log I$ vs. $\log V$) exhibits abrupt increases at $V_{TFL1} = 350$ V and $V_{TFL2} = 620$ V, from 1 to 6 and 6 to 11, respectively. The V_{TFL1} and V_{TFL2} represent the trap-filled-limit voltage of the acceptor and donor traps, respectively [18]. However, the slope stays at 1 after irradiation, which implies that the buffer layer is no longer in a semi-insulating state, but rather behaves as a large resistor. It was reported that the concentration of traps in GaN buffer reduced after electron irradiation, which might raise the Fermi level to enhance the conductivity of buffer layer [7]. In addition, the increased defects in AlN layer can form more leakage paths for electrons from the substrate to the buffer [19]. The holes generated by irradiation can be captured by deep donor traps, which decrease the negative space charge in the buffer [20], thus reducing the barrier for the injection of electrons from the substrate to the buffer. Therefore, the leakage current from drain to substrate increased significantly after irradiation.

IV. CONCLUSION

In this article, the high fluence electron irradiation effects on commercial p-GaN gate power HEMTs are reported. The threshold voltage of the devices negative shifted slightly and the drain-source saturation current increased after irradiation, which is associated with the changes in traps located in the AlGaN layer. In addition, the leakage current of gate and substrate increased significantly after irradiation. The enhancement of the TAT effect and surface state conductance are responsible for the increase of gate leakage current. The weakened semi-insulating characteristic of GaN buffer after irradiation made the vertical leakage current increase significantly, which is relevant to the buffer traps and defects in AlN layers. However, the leakage current recovered after several days of annealing at room temperature. Therefore, GaN HEMTs exhibited strong tolerance to the TID effect, giving them great potential for aerospace applications.

REFERENCES

[1] X. Zhou et al., "Total-Ionizing-Dose Radiation Effect on Dynamic Threshold Voltage in p-GaN Gate HEMTs," in IEEE Transactions on Electron Devices, vol. 70, no. 8, pp. 4081-4086, Aug. 2023.

[2] D. Wölk, S. K. Höffgen, E. Paschkowski, M. Steffens, C. Cazzaniga and C. Frost, "Single Event Effects by atmospheric neutrons in commercial (COTS) normally-off GaN HEMT," 2019 19th European Conference on Radiation and Its Effects on Components and Systems (RADECS), Montpellier, France, 2019.

[3] G. Sonia et al., "Proton and Heavy Ion Irradiation Effects on AlGaN/GaN HFET Devices," in IEEE Transactions on Nuclear Science, vol. 53, no. 6, pp. 3661-3666, Dec. 2006.

[4] M. J. Martinez et al., "Radiation Response of AlGaN-Channel HEMTs," in IEEE Transactions on Nuclear Science, vol. 66, no. 1, pp. 344-351, Jan. 2019.

[5] J. W. McClory, J. C. Petrosky, J. M. Sattler and T. A. Jarzen, "An Analysis of the Effects of Low-Energy Electron Irradiation of AlGaN/GaN HFETs," in IEEE Transactions on Nuclear Science, vol. 54, no. 6, pp. 1946-1952, Dec. 2007.

[6] Chen, Chao, and Xing Zhao Liu, "Effects of Low-Energy Electron Irradiation on Enhancement-Mode AlGaN/GaN High-Electron-Mobility Transistors," Advanced Materials Research, vol. 774–776, Trans Tech Publications, Ltd., Sept. 2013, pp. 876–880.

[7] Z. Feng et al., "Investigation of Electrical Characteristics and Trapping Effects in p-GaN Gate HEMTs Under Electron Irradiation," in IEEE Transactions on Electron Devices, vol. 71, no. 8, pp. 4543-4548, Aug. 2024.

[8] S. Pan et al., "Analysis of the Effects of High-Energy Electron Irradiation of GaN High-Electron-Mobility Transistors Using the Voltage-Transient Method," in IEEE Transactions on Electron Devices, vol. 68, no. 8, pp. 3968-3973, Aug. 2021.

[9] S. F. O. Abubakkar, N. F. Hasbullah, N. F. Zabah and Y. Abdullah, "3MeV-Electron Beam Induced Threshold Voltage Shifts and Drain Current Degradation on ZVN3320FTA & ZVP3310FTA Commercial MOSFETs," 2014 International Conference on Computer and Communication Engineering, Kuala Lumpur, Malaysia, 2014, pp. 273-276.

[10] W. S. Tan, P. A. Houston, P. J. Parbrook, D. A. Wood, G. Hill, C. R. Whitehouse, "Gate leakage effects and breakdown voltage in metalorganic vapor phase epitaxy AlGaN/GaN heterostructure field-effect transistors," Appl. Phys. Lett. 29 April 2002; 80 (17): 3207–3209.

[11] Gaudenzio Meneghesso et al., "Breakdown mechanisms in AlGaN/GaN HEMTs: An overview," 2014 Jpn. J. Appl. Phys. 53 100211.

[12] A. Stockman et al., "On the origin of the leakage current in p-gate AlGaN/GaN HEMTs," 2018 IEEE International Reliability Physics Symposium (IRPS), Burlingame, CA, USA, 2018, pp. 4B.5-1-4B.5-4.

[13] S. Yang et al., "Identification of Trap States in p-GaN Layer of a p-GaN/AlGaN/GaN Power HEMT Structure by Deep-Level Transient Spectroscopy," in IEEE Electron Device Letters, vol. 41, no. 5, pp. 685-688, May 2020.

[14] Fang, ZQ., Look, D.C., Kim, W. et al., "Characteristics of Deep Centers Observed in n-GaN Grown by Reactive Molecular Beam Epitaxy," MRS Internet Journal of Nitride Semiconductor Research 5 (Suppl 1), 943–949 (2000).

[15] Q. Jiang, Z. Tang, C. Zhou, S. Yang and K. J. Chen, "Substrate-Coupled Cross-Talk Effects on an AlGaN/GaN-on-Si Smart Power IC Platform," in IEEE Transactions on Electron Devices, vol. 61, no. 11, pp. 3808-3813, Nov. 2014.

[16] J. L. Lyons, A. Janotti, C. G. Van de Walle, "Carbon impurities and the yellow luminescence in GaN," Appl. Phys. Lett. 11 October 2010; 97 (15): 152108.

[17] C. Zhou, Q. Jiang, S. Huang and K. J. Chen, "Vertical Leakage/Breakdown Mechanisms in AlGaN/GaN-on-Si Devices," in IEEE Electron Device Letters, vol. 33, no. 8, pp. 1132-1134, Aug. 2012.

[18] M. A. Lampert, "Simplified theory of space-charge-limited currents in an insulator with traps," Phys. Rev., vol. 103, no. 6, pp. 1648–1656, Sep. 1956.

[19] J. J. Freedsman, A. Watanabe, Y. Yamaoka, T. Kubo and T. Egawa, "Influence of AlN nucleation layer on vertical breakdown characteristics for GaN-on-Si," Phys. Status Solidi A, 213: 424-428, Feb. 2016.

[20] H. Hanawa, H. Onodera, A. Nakajima and K. Horio, "Numerical Analysis of Breakdown Voltage Enhancement in AlGaN/GaN HEMTs with a High-k Passivation Layer," in IEEE Transactions on Electron Devices, vol. 61, no. 3, pp. 769-775, March 2014.

A Wide Input Range With A Small Area Bandgap Reference Voltage Source With Self-biased Cascode

Sini Wu, Bharatha Kumar Thangarasu, Nagarajan Mahalingam

Kaixue Ma, Fanyi Meng, Zhenghao Lu, Kiat Seng Yeo

Abstract—**A bandgap voltage reference (BGR) with self-biased folded cascode with wide input range and small area is proposed. The BGR using the self-biasing folded cascode can achieve a wide input range while ensuring a small area, and the effect of temperature on the output voltage is reduced by using the bipolar junction transistor (BJT) characteristics. The proposed BGR supply voltage range is 1.7-5V, and the output voltage is 1.1V. The simulation results of standard 130 nm BiCMOS process show that the proposed BGR line voltage is about 0.56mV/V in the VDD range of 1.7-5V. In the temperature range of -40 °C to 125 °C, The temperature coefficient (TC) is 35ppm/°C and power supply rejection ration (PSRR) is -32dB at DC. The effective region of the proposed BGR is $0.0071mm^2$.**

Index Terms—**bandgap reference, cascode, self-biasing, wide input range**

I. INTRODUCTION

Most of the existing bandgap voltage reference(BGR) structures are designed according to the characteristics of bipolar junction transistors. The main principle is to achieve a voltage proportional to the absolute temperature (PTAT) and complementary to the absolute temperature (CTAT), and scale and sum the two voltages to obtain a reference voltage independent of temperature [1], [2]. However, BGR usually deduces the bandgap voltage of Si based on the base-emitter voltage (V_{BE}) of BJTs [3]. When BJT is biased in the positive active region, its temperature dependence can be expressed as [4], [5] :

$$V_{BE} = V_{g0} - \frac{T}{T_0}[V_{g0} - V_{BE}(T_0)] - (\beta - \alpha)V_T \ln(\frac{T}{T_0}) \quad (1)$$

In Eq. (1), V_{g0} represents the temperature-independent bandgap voltage of Si at the reference temperature, in the temperature range of 150 K (-123°C) to 400 K (127°C, V_{g0} ≈ 1.17885 V) [6]; T_0 is the reference temperature, 300 K; β represents mobility, which is a constant, 3.54; α is the exponential temperature coefficient of the collector current, 0 or 1 [7]; $V_T = kT/q$ is the thermal voltage, k is the Boltzmann constant, and q is the electron charge. We introduces a bgr using self-biased folded cascode, which can achieve a wide input range while maintaining a small area.

Sini Wu, Bharatha Kumar Thangarasu, Nagarajan Mahalingam, Kaixue Ma, Fanyi Meng and Kiat Seng Yeo are with *the School of Microelectronics, Tianjin University*, Tianjin, China. Zhenghao Lu is with *the School of Electronic and Information Engineering, Soochow University*, Suzhou, China. Kiat Seng Yeo is also with *Engineering Product Development, Singapore University of Technology and Design,*, Singapore. Corresponding author: (Sini Wu, email: siniwu@qq.com)

II. A BGR USING SELF-BIASED FOLDED CASCODE

A. V_{bg} voltage generation

The proposed low temperaturen coefficient(TC) BGR is to offset the effect of temperature by adding CTAT and PTAT currents. For better presentation, Fig.1 shows the simplified circuit topology of the proposed BGR.

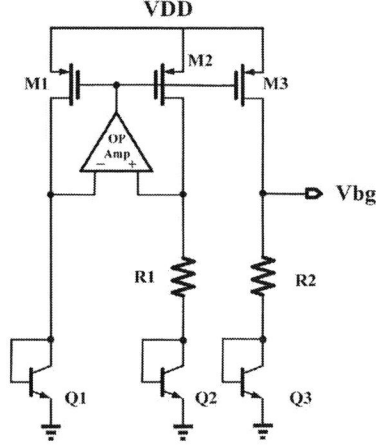

Fig. 1. Topology of the proposed BGR

$$I_{Q2} = \frac{V_{BE2} - V_{BE1}}{R_1} = \frac{V_T \ln N}{R_1} \quad (2)$$

where (2), as a constant, is the ratio of the saturation currents of Q1 and Q2 expressed as $N = \frac{I_{S1}}{I_{S2}}$; It can be seen from the formula that I_{Q2} will increase with the increase of temperature, which is PTAT current, so V_{bg} can be calculated:

$$V_{bg} = R_2 I_{Q3} + V_{BE3} \quad (3)$$

B. Stability analysis

We use a self-biased folded cascode as an op amp to make the circuit work properly over a wide input voltage range. The complete circuit is shown in Fig.2.

Assuming the op amp is not affected by VDD then:

$$Z_1 = \frac{1}{g_{mN}} \quad (4)$$

$$Z_2 = \frac{1}{g_{mN}} + R_1 \quad (5)$$

$$Z_3 = \frac{1}{g_{mN}} + R_2 \quad (6)$$

979-8-3315-2209-4/25 $31.00 © 2025 IEEE

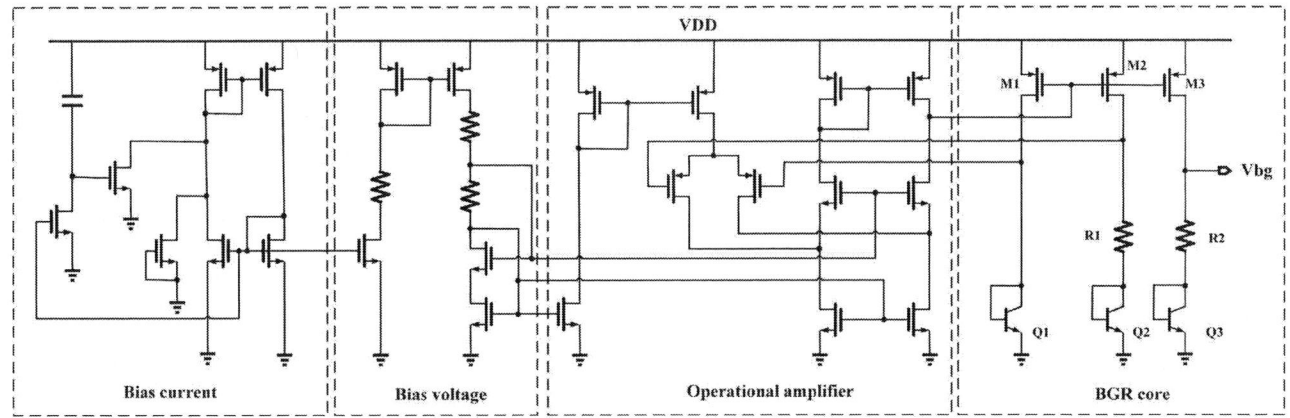

Fig. 2. Schematic of the proposed BGR circuit.

$$v_{bg1} = \Delta V \frac{Z_3}{r_o + Z_3} \frac{(1 + g_m r_o)(ro + Z1)(r_o + Z_2) - V_{BE1}}{(r_o + Z_1)(r_o + Z_2) + Ag_m r_o{}^2} \tag{7}$$

Here, r_o is the small-signal resistance between the source-drain of M1,M2,M3. g_m is the transconductance of M1,M2,M3. A is the gain of the op amp. ΔV is the fluctuation interference of the input voltage.

Assuming that only the OP Amp is affected by vdd then:

$$v_{bg2} = -g_m \beta \Delta V \frac{Z_3}{r_o + Z_3}$$
$$\frac{(r_o + Z_1)(r_o + Z_2)}{(r_o + Z_1)(r_o + Z_2) + Ag_m r_o{}^2 (Z_2 - Z_1)} \tag{8}$$

It can be obtained by the principle of superposition:

$$v_{bg} = v_{bg1} + v_{bg2}$$
$$= \Delta V \frac{Z_3}{r_o + Z_3} \frac{(r_o + Z_1)(r_o + Z_2)(1 + g_m r_o - g_m \beta r_o)}{(r_o + Z_1)(r_o + Z_2) + Ag_m r_o{}^2 R_1}$$
$$\approx \frac{R_2}{R_1 A} \left(\frac{1}{g_m r_o} + 1 - \beta \right) \Delta V \tag{9}$$

It can be seen that the use of self-biased folded cascode with large gain can reduce the influence of input voltage fluctuations, can ensure that the output voltage of the circuit is stable in a wide input voltage range, and has a good PSRR performance.

III. SIMULATION RESULTS AND DISCUSSION

The curve of reference voltage V_{bg} changing with temperature is shown in Fig.3. Simulation results show that TC can be as low as 35ppm/$^\circ C$ over a wide temperature range of -45$^\circ C$ to 125$^\circ C$.

Fig.4 shows the curve of V_{bg} changing with the input voltage. It can be seen that the output voltage works stably in a wide input range of 1.7v to 5v. When the VDD changes from 1.7V to 5V, the V_{bg} changes by 1.8 mV, and the linear adjustment rate is as low as 0.56mV/V in this range. This verifies that BGR can maintain consistent good performance in the case of large VDD fluctuations.

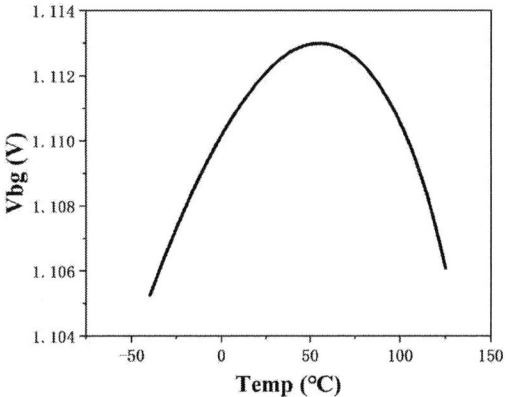

Fig. 3. Simulation results of V_{bg} as a function of T.

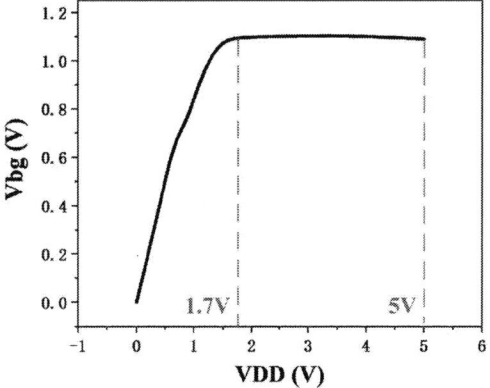

Fig. 4. Simulation results of V_{bg} as a function of VDD.

Fig.5 shows the PARR simulation results of V_{bg}. The proposed BGR has a PSRR of -32 dB at low frequency (¡ 1000 kHz). The PSRR decreases at frequencies above 1000kHz. As mentioned earlier, by using a self-biased folding cascode, it is possible to make the circuit achieve a good PSRR over a wide input range.

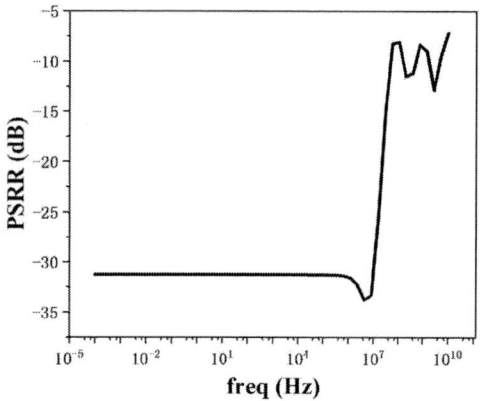

Fig. 5. PSRR of the V_{bg} simulation results of the proposed BGR.

Fig.6 shows the statistical distribution after 500 rounds of Monte Carlo simulation, and it can be seen that the mean value of V_{bg} is about 1.1V with a standard deviation of 42mV.

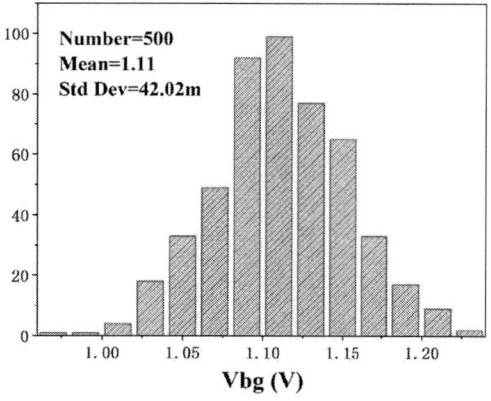

Fig. 6. Results of 500 Monte Carlo simulations of the V_{bg}.

Fig.7 shows the layout of our proposed voltage reference with an active area of $0.0071mm^2$ (83μm × 86μm), implemented in the 130nm BiCMOS process. Table 1 summarizes the performance of the proposed BGR circuit and compares it with several published BGRs. Although the BGR proposed in [8] also has a large input range, its area is large. The BGR proposed in [9] also has a small area, but its input range is limited. The comparison results show that the BGR structure proposed in this paper achieves a small area under the condition of a wide input range, and at the same time, its performance in terms of TC, PSRR, and linearity adjustment rate is quite good. However, as expected, the drawback of our design is accompanied by a relatively high TC.

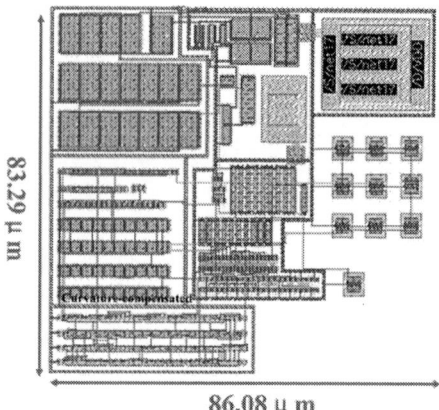

Fig. 7. Layout of the proposed BGR.

TABLE I
COMPARISON RESULTS OF THIS WORK WITH OTHER LITERATURES

Parameters	[8]	[9]	This work
Technology	130nm	130nm	130nm
Reference Voltage (V)	0.85	0.59	1.1
Temperature range (°C)	-40 to 125	-20 to 90	-40 to 125
PSRR (dB@DC)	110	87	32
TC (ppm/°C)	10	78	35
Line Regulation(mV/V)	-	-	0.56
Supply voltage range (V)	1.2-4	1.1-1.8	1.7-5
Area(mm^2)	0.62	0.008	0.0071

IV. CONCLUSION

Based on the traditional BGR, a wide input range and small area BGR with self-bias folding cascode is proposed. In the temperature range of $-40°C$-$125°C$, the TC is 35ppm/$°C$, and the PSRR is -32 dB at low frequencies. The BGR was achieved using a 130nm BiCMOS process and maintained at 1.1V. The results show that the proposed BGR is an ideal choice for circuit applications requiring high input voltage range and small area. The effective area of the proposed BGR is $0.0071mm^2$. In the supply voltage range of 1.7-5V, the BGR has the best linear regulation of 0.56mV/V. Through theoretical analysis and simulation verification, the results show that the proposed design method can not only effectively improve the accuracy of the band-gap reference voltage, but also has the advantages of wide input voltage range, simple structure and small layout area. The simulation results show the potential application of this method in power line system.

REFERENCES

[1] G. Rincon-Mora and P.E. Allen. "A 1.1-v current-mode and piecewise-linear curvature-corrected bandgap reference", IEEE Journal of Solid-State Circuits, 33(10):1551–1554, 1998.

[2] R. Nagulapalli, Rakesh Kumar Palani, and Srikar Bhagavatula. "A 24.4ppm/°c voltage mode bandgap reference with a 1.05v supply", IEEE Transactions on Circuits and Systems II: Express Briefs, 68(4):1088–1092, 2021.

[3] Guangqian Zhu, Yintang Yang, and Qidong Zhang. "A 4.6-ppm/°chigh-order curvature compensated bandgap reference for bmic". IEEE Transactions on Circuits and Systems II: Express Briefs, 66(9):1492–1496, 2019.

[4] Y.P. Tsividis. "Accurate analysis of temperature effects in i/sub c/v/sub be/ characteristics with application to bandgap reference sources". IEEE Journal ofSolid-State Circuits, 15(6):1076–1084, 1980.

[5] Xifeng Liu, Shiwei Yang, Jinfei Wang, and Zhenbang Xu. "A high-order curvature compensated voltage reference based on lateral bjt".AEU - International Journal of Electronics and Communications, 124:153325, 2020.

[6] H. Banba, H. Shiga, A. Umezawa, T. Miyaba, T. Tanzawa, S. Atsumi, and K. Sakui. "A cmos bandgap reference circuit with sub-1-voperation". IEEE Journal of Solid-State Circuits, 34(5):670–674,1999.

[7] Ximing Fu, Dalton Martini Colombo, Yadong Yin, and Kamal El-Sankary. "Low noise, high psrr, high-order piecewise curvature compensated cmos bandgap reference". IEEE Access, 10:110970–110982, 2022.

[8] A. I. Kamel, A. Saad and L. S. Siong, "A high wide band PSRR and fast start-up current mode bandgap reference in 130nm CMOS technology," 2016 IEEE International Symposium on Circuits and Systems (ISCAS), Montreal, QC, Canada, 2016, pp. 506-509.

[9] R. Papi, F. H. Noshahr and B. Gosselin, "A New Current-Mode Subthreshold, High-PSRR MOSFET-Only Bandgap Voltage Reference," 2023 21st IEEE Interregional NEWCAS Conference (NEWCAS), Edinburgh, United Kingdom, 2023, pp. 1-5.

A Novel High-Voltage Lateral GaN Diode with Hybrid p-GaN Anode

Kaijun Ding, Yimeng Tang, Yaowen Zhang, Yingzhi Luo, Bingke Zhang, Moufu Kong*

Abstract—This paper introduces a novel lateral AlGaN/GaN diode on a sapphire substrate, which features a hybrid p-type GaN (p-GaN) anode that allows for precise control of the on-state voltages while achieving high breakdown voltages. The proposed device achieves a highly adjustable and low forward turn-on voltage (knee voltage) with the change of the doping concentration of p-GaN region in the anode side. Under the conditions of an anode-cathode spacing (L_{AC}) of 8 μm and a p-GaN doping concentration of 5×10^{16} cm^{-3}, the proposed GaN diode demonstrates an impressively low knee voltage (V_{th}) of approximately 0.2 V and a high breakdown voltage nearing 1700 V. This combination of ultra-low knee voltage and exceptional high breakdown voltage underscores the advanced performance of the proposed GaN diode.

Index Terms—GaN diodes, p-GaN, high-voltage, knee voltage, breakdown voltage

I. INTRODUCTION

The third-generation semiconductors represented by GaN, possess several advantages over traditional silicon material, including wide bandgap, high-temperature tolerance, high critical breakdown electric field, and high electron saturation velocity. These properties make them ideal for applications requiring high frequency, high power and high temperature [1], [2].

Among various lateral devices, a notable advantage of GaN-based devices is the formation of a two-dimensional electron gas (2DEG) at the AlGaN/GaN heterojunction surface. Due to spontaneous and piezoelectric polarization effects, a high-density 2DEG is generated at the AlGaN/GaN interface, which can only move along the heterojunction boundary. The transport properties of this electron gas result in exceptionally high electron mobility. Leveraging the wide bandgap characteristics, electronic devices based on AlGaN/GaN heterojunctions offer high switching speed, low on-state voltage, high breakdown voltage, and high-temperature operation, making them the preferred choice for power electronics applications.

Based on the advantages of the 2DEG, the GaN high electron mobility transistor (HEMT) devices have been significantly developed and widely used in many low power and medium power electronics systems [3], [4]. GaN diodes offer advantages such as high operating frequency and excellent reverse recovery performance, which give them promising prospects for applications. Compared to GaN HEMTs, GaN diode devices have been rarely studied and have not been commercialized. In 2008, Wong et al. mentioned an AlGaN/GaN HEMT-compatible lateral field effect rectifier (L-FER) [5]. In 2010, Wang et al. proposed the hybrid-anode AlGaN/GaN field-effect rectifier (HA-FER) which significantly reduced the knee voltage of GaN L-FERs [6]. However, these lateral Schottky GaN diodes exhibit high on-state resistances, uneven electric field distribution, and the knee voltages are also difficult to control.

This paper presents a p-GaN hybrid anode GaN diode structure, in which Mg-doped p-GaN is introduced into the anode region. By adjusting the doping concentration of p-GaN, the turn-on voltages (knee voltages) can be easily controlled while achieving low on-resistance and maintaining high breakdown voltage.

II. DEVICE STRUCTURE AND MECHANISM

The device we designed is based on a sapphire substrate, and the structure is shown in Fig.1. The device process TCAD simulation is also conducted. Firstly, a 50-nm-thick AlN layer is grown on the sapphire substrate to improve the quality of subsequent epitaxial layers. On top of the AlN layer, a 2-μm-thick GaN buffer layer doped with carbon is set to reduce defects, suppress leakage current and enhance device performance. The carbon concentration is chosen as 1×10^{19} cm^{-3} because when it exceeds this value, the improvement of breakdown voltage is observed to be limited due to carbon self-compensation [7]. Subsequently, a 200-nm-thick GaN channel layer is grown, where 2DEG distributes intensively. Above the GaN channel layer, a 20-nm-thick Al$_{0.25}$Ga$_{0.75}$N barrier layer (AlGaN- Barrier) is deposited, forming a heterojunction with the GaN channel layer. This heterojunction generates two-dimensional electron gas (2DEG) at the interface, which serves as the primary conductive channel of the device. On top of the AlGaN barrier layer, a 100-nm-thick p-type GaN cap layer (p-GaN) is set to optimize the electric field distribution and conduction characteristics of the device. Following this, ohmic contacts for the cathode and anode metal layers are formed through metal deposition, and a

---

The authors Kaijun Ding, Yimeng Tang, Yaowen Zhang, Yingzhi Luo, Moufu Kong are with the State Key Laboratory of Electronic Thin Films and Integrated Devices of China, University of Electronic Science and Technology of China, Chengdu, China 611731; The author Bingke Zhang is with the Leshan Share Electronic Co., Ltd, No. 3, Nanshin East Road, High-Tech Zone, Leshan, Sichuan, 614099, China. This work was supported in part by the Central Guiding Local Science and Technology Development Special Project of Sichuan (2024ZYD0310).

(*Corresponding author: Moufu Kong: kmf@uestc.edu.cn).

Si₃N₄ passivation layer is deposited. Additionally, a 100-nm-thick cathode field plate and a second-order anode field plate are formed, along with the Si₃N₄ passivation layer, to further optimize the distribution of electric field and reduce the impact of surface states of the device.

The proposed GaN diode operates by utilizing the p-GaN anode to control the turn-on voltages. Under forward bias, since the p-GaN region is connected to the anode with a positive voltage with respect to the cathode, which reduces the depletion at the AlGaN/GaN heterojunction (under the p-GaN region) and allowing current to flow through the 2DEG channel. The different doping concentration of p-GaN region corresponds different knee voltages of the diode. Under reverse bias, the p-GaN has a negative voltage with respect to the cathode, the potential energy barrier between p-GaN and AlGaN rises and fully depletes the channel 2DEG under the p-GaN region, which effectively blocks electrons flow through the device. Additionally, the field plate and Si₃N₄ passivation layer optimizes the electric field distribution, enhancing breakdown voltage and reliability.

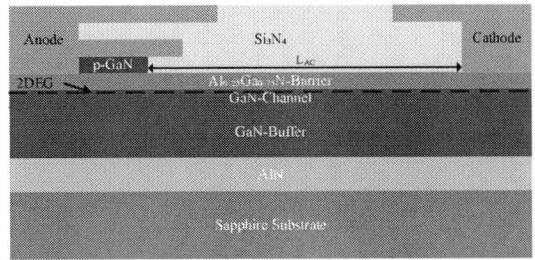

Fig. 1. Schematic cross-sectional view of the proposed GaN diode.

III. FORWARD AND REVERSE CHARACTERISTICS

The electrical performance of the proposed GaN diode is also investigated through TCAD device simulations. The current-voltage (I-V) and breakdown voltage characteristics are analyzed under various device structural parameters. We also gave emphasis on studying how different Mg concentrations in the p-GaN affect the forward characteristics of the proposed GaN diode.

Fig. 2 (a) shows the forward I-V characteristics of both the conventional Schottky contact structure and the proposed p-GaN hybrid anode structure under identical conditions with L_{AC} = 10 μm. The specific on-resistance is calculated using the voltage V_{AK} with a range of 2 V to 3 V, where the curve is closer to linear. The specific on-resistance of the conventional structure is 14.8 mΩ·cm², while the p-GaN hybrid anode structure exhibits a significantly lower specific on-resistance ($R_{on,sp}$) of 8.5 mΩ·cm² (both at L_{AC} = 10 μm). This highlights a substantial reduction by 42.57% in specific on-resistance for the proposed structure, compared to the traditional diode. Moreover, the new structure allows flexible adjustment of the turn-on voltage by modifying the doping profile of p-GaN region, which is more convenient and efficient than controlling the turn-on/knee voltage by adjusting the Schottky contact metal work functions of conventional structure.

Fig. 2 (b) illustrates how variations in L_{AC} affect the forward characteristics of the proposed diode. Notably, the

knee voltage of the proposed device remains almost unchanged with L_{AC}, maintaining a stable value of approximately 0.2 V when the forward current reaches 1 × 10⁻³ mA/mm. The I-V curve under a logarithmic scale for L_{AC} = 8 μm is also shown in the insert. As L_{AC} increases, the specific on-resistance of the device gradually rises.

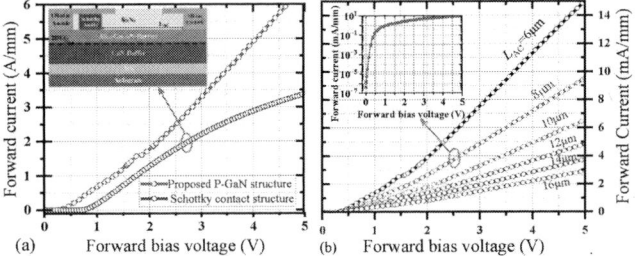

Fig. 2. The forward I-V curves (a) the conventional hybrid anode structure versus the p-GaN hybrid anode structure with L_{AC} of 10 μm, (b) forward I-V curves of the p-GaN hybrid anode structure with different L_{AC}.

From the forward bias curve, we can calculate the $R_{on,sp}$ of the proposed GaN diode. The $R_{on,sp}$ increases with the increasing of L_{AC}, showing an almost linear relationship, which is determined by the definition of $R_{on,sp}$, as shown in Fig. 3 (a). At the same time, a larger L_{AC} increases the distance over which electrons travel and their probability of scattering. As a result, the number of electrons actively participating in conduction decreases under the same voltage, leading to an increase in the $R_{on,sp}$.

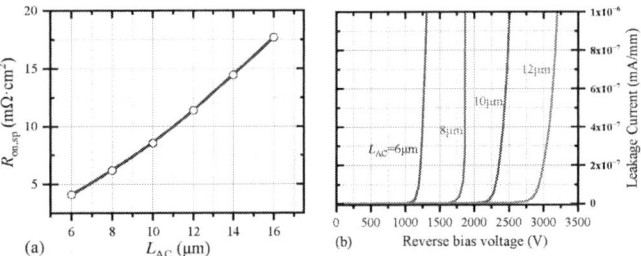

Fig. 3. (a) Specific on-resistance ($R_{on,sp}$) for different L_{AC}. (b) Breakdown Curves under different L_{AC} values.

When a reverse bias is applied to the diode, the breakdown voltage (V_{BR}) is defined as the cathode voltage corresponding to a leakage current of 1×10^{-8} mA/mm. Fig. 3 (b) shows that as L_{AC} increases from 6 μm to 12 μm, V_{BR} exhibits a significant linear growth trend. As L_{AC} becomes larger, V_{BR} increases further. This is based on the definition of current characterization for unit channel length, where a larger channel length requires a higher voltage to achieve the same unit current per channel length. However, as L_{AC} becomes larger, V_{BR} increases because when L_{AC} reaches a certain size, it helps to better distribute the electric field across the device, thereby achieving a higher breakdown voltage.

Fig. 4 shows that the electric field is predominantly distributed near the edge of the cathode and the p-GaN region. The electric field intensity near the cathode region arises from marginal discharge, while the breakdown near the p-GaN region is governed by the principle of P-N junction breakdown.

$$\nabla \cdot \vec{E} = \frac{q}{\epsilon_s}(p - n + N_D - N_A) \qquad (1)$$

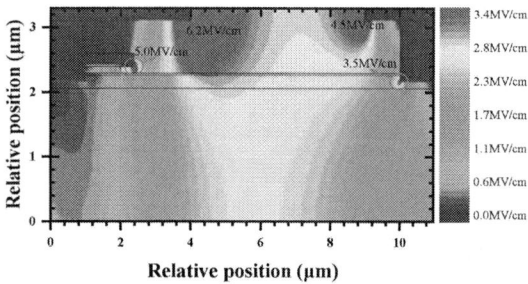

Fig. 4. The electric field distribution under the reverse bias of 1800 V when $L_{AC} = 8$ μm.

According to the Poisson equation (1), the built-in electric field can be derived. When the maximum electric field intensity reaches the critical breakdown field $E_C = 3.4 \times 10^6 \, V/cm$ of GaN material the device breaks down.

IV. KNEE VOLTAGE AND P-GAN DOPING DISCUSSION

In practical simulations, since AlGaN exhibits optimal electrical properties when the Si doping concentration is 3×10^{18} cm^{-3} [8], we set the AlGaN doping concentration to 1×10^{18} cm^{-3} and varied the Mg doping concentration in the p-GaN to investigate its impact on the knee voltages. Fig. 5 (a) demonstrates that adjusting the Mg doping concentration in p-GaN can significantly alter the knee voltages (V_{th}). As the Mg doping concentration in p-GaN increases, the knee voltage of the device rises notably. It is observed that when the Si concentration in AlGaN is 1×10^{18} cm^{-3}, the knee voltage is particularly sensitive to the doping concentration within the range of 1×10^{17} cm^{-3} to 5×10^{17} cm^{-3} in p-GaN. The slope of the I-V curve in Fig. 5 (a), which changes only slightly with Mg doping concentration, indicates that the doping concentration in the p-GaN region has a minor impact on the specific on-resistance. The reason for achieving a low knee voltage with low-concentration p-GaN region lies in the relatively lower barrier height induced by the lower doping concentration, which facilitates easier conduction under the same bias, thereby reducing V_{th}.

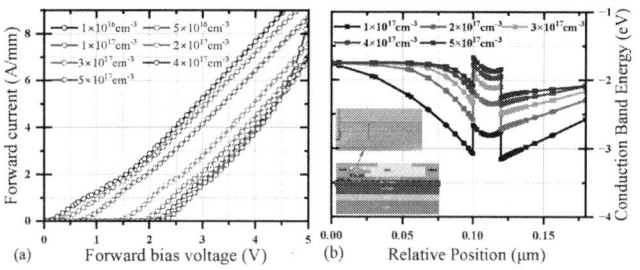

Fig. 5. (a) The I-V characteristic curves under different Mg concentrations. (b) The conduction band energy curves under different Mg concentrations.

Fig. 5 (b) illustrates the conduction band energy profiles along a 0.2 μm cut line at the interface between the p-GaN and AlGaN layers, under different Mg doping concentrations of p-GaN. The conduction band energy shows a significant increase as the doping concentration rises from 1×10^{17} cm^{-3} to 5×10^{17} cm^{-3}. Within this range, the turn-on voltage also increases notably with higher doping concentrations. This phenomenon occurs because, as the Mg doping concentration

in p-GaN increases and approaches the Si doping concentration in AlGaN, the barrier region gradually shifts from the interior of the p-GaN layer to the interface between p-GaN and AlGaN. This shift results in a higher barrier in the channel region, making it more difficult for the GaN device to turn on.

V. CONCLUSION

In this paper, a novel lateral diode structure incorporating p-GaN region has been designed, and the effects of L_{AC} and p-GaN doping concentration on knee voltage, specific on-resistance, and breakdown voltage have been thoroughly investigated. Compared to conventional Schottky contact structure, this design allows for the regulation of the turn-on voltage through doping control of p-GaN region and achieves a lower specific on-resistance than conventional structures. The turn-on voltage remains almost unaffected by L_{AC}, while the specific on-resistance is directly proportional to L_{AC}. Additionally, the breakdown voltage shows a positive correlation with L_{AC}. The Mg concentration in the p-GaN layer influences the knee voltage by altering the barrier height. Higher doping concentrations lead to a higher knee voltage. This device is compatible with the manufacturing process of p-GaN HEMTs, offering significant potential for integration and scalability in power electronics applications.

REFERENCES

[1] Y. Dora, A. Chakraborty, L. Mccarthy, S. Keller, S. P. Denbaars and U. K. Mishra, "High Breakdown Voltage Achieved on AlGaN/GaN HEMTs with Integrated Slant Field Plates," in IEEE Electron Device Letters, vol. 27, no. 9, pp. 713-715, Sept. 2006.

[2] W. Saito et al., "High breakdown voltage AlGaN-GaN power-HEMT design and high current density switching behavior," in IEEE Transactions on Electron Devices, vol. 50, no. 12, pp. 2528-2531, Dec. 2003.

[3] Islam N, Mohamed M F P, Khan M F A J, et al. Reliability, applications and challenges of GaN HEMT technology for modern power devices: A review[J]. Crystals, 2022, 12(11): 1581.

[4] Buffolo M, Favero D, Marcuzzi A, et al. Review and outlook on GaN and SiC power devices: industrial state-of-the-art, applications, and perspectives[J]. IEEE Transactions on Electron Devices, 2024.

[5] K. -Y. Wong, W. Chen, W. Huang and K. J. Chen, "Temperature dependence of AlGaN/GaN HEMT-compatible lateral field effect rectifier," 2008 IEEE International Conference on Electron Devices and Solid-State Circuits, Hong Kong, China, 2008, pp. 1-4, doi: 10.1109/EDSSC.2008.4760656.

[6] Z. Wang, B. Zhang, W. Chen and Z. Li, "A novel hybrid-anode AlGaN/GaN field-effect rectifier with low operation voltage," 2010 10th IEEE International Conference on Solid-State and Integrated Circuit Technology, Shanghai, China, 2010, pp. 1889-1891.

[7] Shuiming Li, Yu Zhou, Hongwei Gao, Shujun Dai, Guohao Yu, Qian Sun, Yong Cai, Baoshun Zhang, Sheng Liu, Hui Yang; Off-state electrical breakdown of AlGaN/GaN/Ga(Al)N HEMT heterostructure grown on Si(111). AIP Advances 1 March 2016; 6 (3): 035308.

[8] Min Xu, Fengpo Yuan, Guoying Chen, Dongmei Li, Jiayun Yin and Zhihong Feng. "Effect of Modulation-Doping on the Electrical Properties of AlGaN/GaN HEMT Materials," Semiconductor Technology, vol. 32, pp. 230-233, March 2007.

979-8-3315-2209-4/25 $31.00 © 2025 IEEE

A Robust P-Type Bubble+Shield-Gate Terminal Ring for A 750V IGBT

Kui Xiao[1], Shen Xu[1], Zheng Bian[2], Wei Yao[2]
[1]Southeast University, Nanjing, P. R. China;
[2]Technology Development Department, CSMC Technologies Corporation, Wuxi, P. R. China;

Abstract—This paper presents a novel high-robust 750V IGBT terminal ring structure, which reduces the surface electric field strength by moving the maximum electrical field region from the silicon surface to the body. Thus the influence of charges at each dielectric surface of device on the blocking voltage can be effectively reduced, when these surface charges are inevitably introduced by the manufacturing and packaging process of device. Resultly, such structure can significantly improve the HTRB, H3TBR, and HAST reliability of IGBT device. Additionally, the p-type bubble with the shield gate is connected to the emitter. Therefore, large number of hot carriers generated can be timely conducted through this channel when the device undergoes avalanche breakdown, avoiding the burnout problem of the device. The 750V IGBT terminal ring structure adopts a trench structure, and the bottom of trench is formed by p-type diffusion bubble. Then, the sidewalls are oxidized, etched, and filled with polycrystalline silicon (POLY) electrodes. By optimizing the process, the POLY electrodes can be perfectly fused, inducing good contact with the interface of p-type single crystal silicon, which results in a P-type withstand voltage bubble connected to POLY electrode structure. Thus, the problem of test degradation could be sovled.

Keywords—P-Type Bubble+Shield-Gate Terminal Ring, surface charge, breakdown voltage

I. INTRODUCTION

The IGBT is a representative product of the third technological revolution in power semiconductor devices, possessing excellent characteristics, such as high voltage, high current, and low on-resistance[1-2]. Due to these properties, IGBT is widely known as the "CPU" of industrial AC devices and find extensive applications in induction cookers, welding machines, white goods, photovoltaic wind power systems, and high-power energy management fields[3-4]. However, IGBT's reliability is infulenced inevitably by natural environment, which is strongly related to termination design[5-8]. For a long time, termination technology has been improved to be more effective[9-10]. Recently, with the significant growth of electric vehicles, photovoltaic energy storage systems, and industrial automation technologies, the requirements for IGBT's reliability have been increasing steadily. Therefore, it's necessary to take more optimization of termination technology for better ruggedness[11-12].

This paper focuses on studying a 750V IGBT with a p-type bubble+shield-gate terminal ring, which has a matchhead-like structure. The proposed terminal ring structure effectively transfers the maximum electric field into the silicon bulk，and reduces the impact of the surface charge introduced during inter level dielectric (ILD) and passivation or packaging processes on breakdown voltage performance. Additionally, by connecting the p-type bubble and shield polysilicon (POLY) in the terminal ring transition area to ground through emmitter, it allows for hot carriers to be imported into emitter from shield POLY to prevent burnout during IGBT breakdown test. Therefore, this design can be applied in scenarios with stringent reliability requirements.

II. STRUCTURE AND MECHANISM

The 750V IGBT terminal ring with a highly robust p-type bubble + shielded gate POLY electrode shaped like a matchhead structure is shown in Fig. 1 (a). This structure is different from common terminal ring structures such as field limit ring (FLR) shown in Fig. 1 (b). Firstly, it directly etched on the surface of light n-type epitaxial silicon to form trench of a certain depth. The bottom of trench is covered by p-type bubbles formed by injection and diffusion to form the main junction of IGBT device to withstand high voltage. The shielding gate POLY electrode is connected with p-type bubble, and the transition area of terminal ring is connected with emitter through contact and metal. The terminal ring is formed by floating half shield gate POLY electrode and oxide layer top filling.

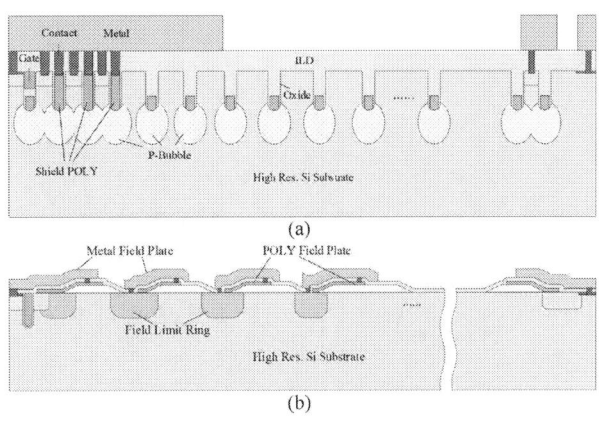

(a)

(b)

Fig. 1. The schematic diagram of (a) the 750V IGBT Shield POLY + p-bubble FLR structure and (b) normal FLR structure.

979-8-3315-2209-4/25 $31.00 © 2025 IEEE   245

Through the simulation of presented and normal terminal ring structures, their electric field distributions are shown as Fig. 2 (a) and (b), respectively. It can be seen that the electric field distribution of p-type bubble + shield gate POLY structure is uniform, and the maximum electric field is distributed at the bottom of p-type bubble near the transition area between terminal ring and cell. The maximum electric field of normal FLR structure is distributed at bottom of the last two rings.

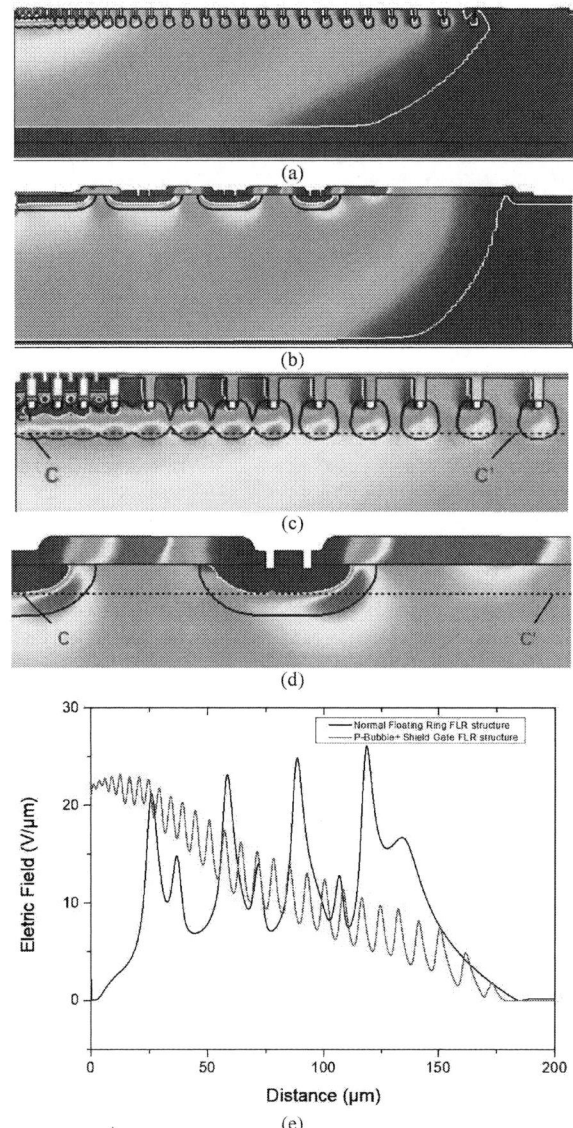

Fig. 2. The distribution of electric field of (a) 750V IGBT shield poly + p-bubble and (b) normal FLR structure; zoom of maximum electric field position of (c) P-bubble+ shield gate and (d) normal FLR structure; (e) cutline in silicon of dashed line as marked in (c) and (d).

Detailed zooms of the maximum electric field point of presented and normal terminal ring structure are illustrated in Fig. 2 (c) and (d), respectively. The position of maximum of electric field in presented structure is about 5.9um below the silicon surface, while the position of maximum electric field in normal FLR structure is about 3.3um below the silicon surface. Meanwhile, the corresponding cutline at the position of maximum electric

field point parallel to the silicon surface is shown in Fig. 2 (e), which compares the electric field distribution along the direction from the primitive cell region to the last terminal ring of two structure. The electric field peak of presented structure is below 22V/um, while the maximum electric field of normal FLR structure is more than 25V/um. It can be seen that within the same length of terminations, the proposed structure has a more uniform electric field distribution and a lower electric field peak, and the peak is located deeper in silicon then that in normal FLR structure. Thus, it is able to reduce the impact of surface charge in dielectric layer, passivation layer or package on the breakdown of device. These advantages can better improve the reliability of IGBT devices.

In addition, the location of electric field peak of p-type bubble + shield gate POLY structure is at the initial position of terminal ring, which is also the transition area between cell and terminal ring. When the device voltage triggers avalanche breakdown, the avalanche breakdown firstly occurs at the position of electric field peak, because the p-type bubble of the structure is connected with POLY electrode and emitter. The hot carrier generated by the avalanche breakdown can be guided into emitter effectively and quickly, because the p-type bubble + shield gate POLY structure is connect with emitter. Therefore, the burnout of device can be avoided when the IGBT device exceeds the voltage avalanche breakdown, which improves the robustness of device.

III. PROCESS FLOW AND EXPERIMENTS

Based on the process flow described in Section III, the p-bubble + shield gate IGBT with half shield POLY+p-bubble FLR device are fabricated.

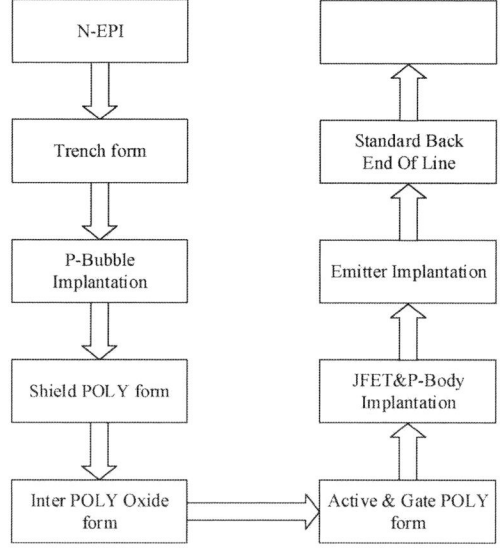

performed at the bottom of trench, and p-type bubble main junction is formed by thermal drive. After the shielded POLY electrode is deposited, part of the POLY electrode is etched by photolithography, while the other part of poly electrode is retained. Subsequently, the oxide layer is deposited and the source region is formed by photolithography, and the gate POLY is deposited and etched to form the gate electrode. Nextly, JFET and p-body lithography injection and drive-in were performed to form the channel region in cell. Meanwhile, standard IGBT backside processes, such as backside thinning and

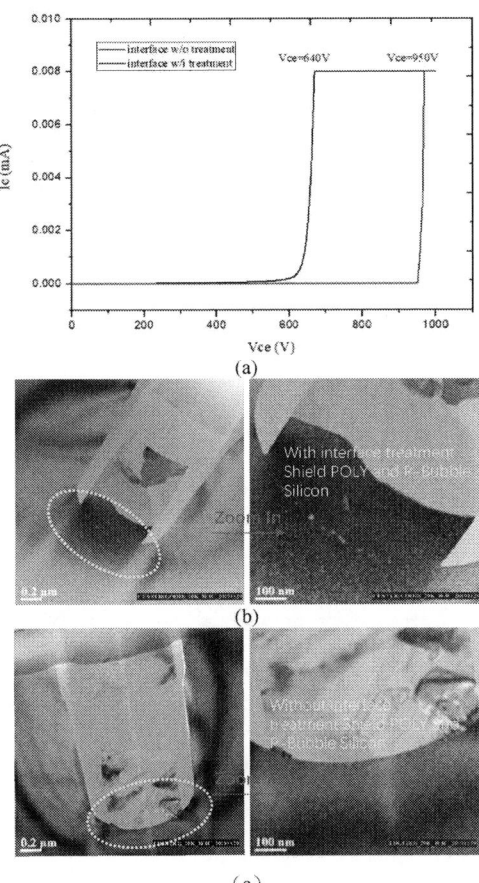

(a)

(b)

(c)

Fig.4. Measured breakdown and TEM FA. (a) Measured breakdown cure of the interface with and without treatment comparison.The Shiled POLY and P-Bubble Silicon interface TEM. (b) with porperly interface treatment the extension of single crystal Silicon to shield POLY forms a better quasi-single crystal structure (c) withou well tratment the Shield POLY and p-bubble Silicon boundary is obvious,not forming the single crystal sturcture.

injection metal sputter, are carried out. The final formation of cell region is a p-type bubble connected with a half shielded POLY electrode, followed by an intermediate oxide layer and the gate electrode structure, which is similar to a split gate DMOS. The transition region between cell and terminal ring is a p-type bubble connected with full shielded POLY electrode connected to emitter through the contact and metal. The main terminal ring is a p-type bubble connected with a half shielded POLY electrode and top filled with oxide layer floating ring structure.

During the process of forming the p-bubble+ shield POLY structure, the doping in shield POLY should be prevented to diffuse into p-bubble. Thus, undoped POLY is adopt as shield POLY, followed by Boron implant and annealing to form the structure connect between shield POLY and p-bubble. At the beginning, the breakdown voltage of the shield POLY and p-bubble structure is not enough and there is also a problem of large leakage. The breakdown curve is shown as black line in Fig. 4 (a). Through failure analysis (FA) and experimental verification, it is found that the treatment of interface between shield POLY and p-bubble is very critical, which directly affects the breakdown voltage and leakage of the

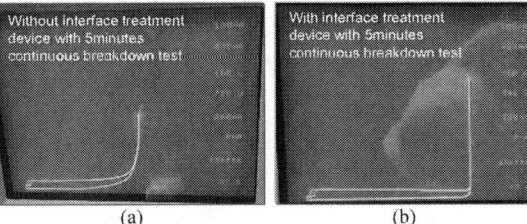

(a)   (b)

Fig. 5. The 750V IGBT device 5minutes breakdown voltage stress test with 370B equipment . (a) Withou interface treatment device after stress test, the BV shift from ~950V to ~720V; (b) With interface treatment device after stress test, the BV stable at ~950V without any shift.

device. If the interface was not treated porperly, there would be some inevitable SiO_2 and lattice defects, as shown in Fig. 4 (c) . Such defects would lead to poor breakdown voltage and leakage. With porper treatment, the interface could form single crystal liked silicon at p-bubble and shield POLY after the subsequent thermal processes. The extension of single crystal silicon to shield POLY forms a better quasi-single crystal structure, eliminating SiO_2 and lattice defects, which is shown as TEM picture in Fig. 4 (b). The decrease of such defects would improve breakdown voltage and leakage, and the improved breakdown curve is shown as blue line in Fig. 4 (a).

Fig. 5 (a) and (b) show the break down voltage curve comparison of device between without and with interface treatment, respectively. As a result, it could be seen that after 370B manual test equipment 5minutes continuous breakdown stress test, the breakdown voltage of the device without interface treatment walks in from 940V to 722V, while the breakdown voltage of device with inerface treatment is very stable at ~940V without any degradation.

IV. CONCLUSION

This paper proposed and experimentally demonstrated the a shield POLY + p-bubble FLR structure for 750V IGBT. The soft breakdown and Ice leakage problems are solved by optimizing and improving the contact interface between shield POLY and p-bubble silicon. This FLR structure maintains good robustness under continuous breakdown test on 370B test machine, and there is no breakdown voltage shift and device burn problem, which can be used for high reliability applications.

REFERENCE

[1] Zhigang Wang, et al . "A Novel Concept of Electron–Hole Enhancement for Superjunction Reverse-Conducting Insulated Gate

Bipolar Transistor with Electron-Blocking Layer." *Micromachines* 14(2023).

[2] Ruifen Nie, et al. "N-buffer design and optimization in 4500V class IGBT." *Journal of Physics: Conference Series* 2370 (2022)

[3] Yaolong Hong, et al. "Enhanced dynamic voltage clamping capability of clustered igbt at turn-off period." *Electronics and Energetics* 29(2016).

[4] Ge Zhao, et al. "A design of IGBT with the hole current bypass structure." *3rd International Conference on Mechatronics and Information Technology*，2016.

[5] C. Papadopoulos, et al. "Humidity Robustness of IGBT Guard Ring Termination." *International Exhibition and Conference for Power Electronics, Intelligent Motion, Renewable Energy and Energy Management*, 2019.

[6] U. Grossner, et al. "Passivation in High-Power Si Devices - An Overview." *ECS Transactions* 50(2013).

[7] Zhiliang Xu, et al. "Humidity related failure mechanism of IGBTs considering dynamic avalanche." *Microelectronics and reliability* Dec.(2023):151.

[8] C. Papadopoulos, et al. "The influence of humidity on the high voltage blocking reliability of power IGBT modules and means of protection." *Microelectronics Reliability* 88-90.SEP.(2018):470-475.

[9] P. Mirone, et al. "An area-effective termination technique for PT-Trench IGBTs." *IEEE International Conference on Microelectronics IEEE*, 2014.

[10] M. Antoniou, et al. "Deep p-ring trench termination: An innovative and cost-effective way to reduce silicon area." *IEEE Electron Device Letters* PP.2(2019):1-1.

[11] Xiaoli Tian, et al. "An Improved Edge Termination Structure to Optimize 3.3kV IGBTs Ruggedness." *14th IEEE International Conference on Solid-State and Integrated Circuit Technology (ICSICT)* (2018).

[12] Jing Zhu, et al. "Investigation on the Breakdown Failure in Stripe Trench-Gate Field-Stop Insulated Bipolar Transistor With Low-Saturation Voltage." *IEEE Transactions on Device and Materials Reliability* 16.3(2016):1-1.

Moore's Law at 60 – Still in Good Shape or Already Ailing?

Frank Schwierz

Department of Micro- and Nanoelectronic Systems, Technische Universität Ilmenau, Germany

Abstract—**April 19, 2025 marks the 60th anniversary of the publication a legendary paper by Gordon Moore in the trade journal *Electronics*. Even if the term Moore's Law was not coined until several years later, this paper marked the birth of a law that shaped the semiconductor industry for decades and that is still relevant today. In the present paper, we take a look at the history of Moore's Law, examine relevant trends related to this law, and discuss different views on the current status and future of Moore's Law.**

Keywords—*Moore's Law, MOSFET scaling, transistor count, energy efficiency, computer, FinFET, nanosheet transistor*

I. INTRODUCTION

Almost exactly 60 years before engineers and scientists meet at EDSSC 2025 to discuss their latest findings in semiconductor research, Gordon Moore published a landmark paper entitled *Cramming More Components onto Integrated Circuits* in the trade journal *Electronics* [1]. At the time, Moore was Director of the research and development laboratories at Fairchild Semiconductor, so he was a man who knew exactly what he was talking about.

In the above paper, Moore delivered the three following key messages that were to prove groundbreaking for the further evolution of semiconductor electronics. (i) Already in the first sentence, Moore stated that *the future of integrated electronics is the future of electronics itself.* (ii) Furthermore, he expressed the expectation that the number of devices integrated on a semiconductor die will double in regular intervals. (iii) Finally, Moore provided the motivation for increasing the number of components per chip. It is simply economics, or more precisely, the fact that increasing the number of devices per die in accordance to the achieved progress in processing technology leads to a falling price per device.

The statements (ii) and (iii) were later dubbed Moore's Law and not only proved to be correct, but also had a major impact on the evolution of the entire semiconductor industry for decades. In the following, we take the reader through the 60-year history of Moore's Law, discuss important trends in the evolution of ICs (integrated circuit) and expound on different opinions regarding the current status and future of Moore's Law.

II. THE EARLY YEARS OF MOORE'S LAW

During the 1950s and early 1960s the bipolar junction transistor was the dominating semiconductor device. In his 1965 paper, Moore considered five bipolar chip generations, starting with discrete transistors from 1959 up to ICs with 64 devices from 1965 fabricated at Fairchild. He plotted the logarithm of the device count for each generation versus the year of production and recognized that these five data points (one for each generation) were very close to a straight line. This means that the device count per die doubled every year, i.e., that it increased exponentially. This plot and his profound knowledge in semiconductor processing technology led Moore to the conclusion that the device count per die would continue to increase at the same rate for 10 more years until 1975.

In a paper presented at the International Electron Device Meeting (IEDM) in December 1975, Moore revisited his expectation and could show that his prediction was correct [2]. Moreover, he argued that the trend of doubling device count per die in regular intervals would continue, albeit at a lower rate (doubling every two years).

Sometime in the mid-1970s, the trend of the doubling component count per chip in regular intervals began to be referred to as Moore's Law. Despite intensive research, it has not been possible to determine who exactly coined the term Moore's Law [3]. What is certain, however, is that Caltech professor Carver Mead made a significant contribution to spreading the term Moore's Law and making it known in the community.

A major milestone in the evolution of semiconductor electronics was the invention of the microprocessor, frequently synonymously called CPU (central processing unit), in 1971. For many years, CPUs were undisputedly the most advanced integrated circuits and are still among the most complex ICs today. Therefore, large part of the following discussions focuses on CPUs, most notably on CMOS-based CPUs, although we are well aware that GPUs (graphics processing unit) are currently another very popular and successful type of CMOS-based logic ICs.

III. THE EVOLUTION OF CPU TRANSISTOR COUNT

Fig. 1 shows the evolution of the transistor count of MOSFET-based CPUs. The first CPU was the Intel 4004 introduced in 1971 consisting of around 2300 Si pMOSFETs, soon followed by the Intel 8008 with already 3500 pMOSFETs. Starting from 1974, a series of nMOSFET CPUs was released with the Intel 80286 from 1982 being the last one included in Fig. 1, which had a transistor count of 134,000. The motivation to use nMOSFETs instead of pMOSFET was the higher switching speed of n-channel MOSFETs, which is achieved because of the higher mobility of electrons compared to holes.

In 1985, Intel finally switched to CMOS CPUs, and CMOS became the dominant technology for digital logic due to its better energy efficiency compared to pMOS and nMOS. Note that Intel was and is not the only CPU manufacturer.

979-8-3315-2209-4/25 $31.00 © 2025 IEEE

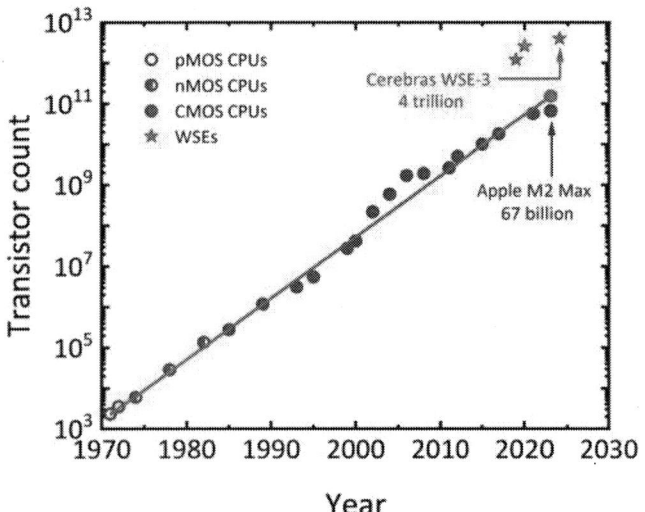

Fig. 1. Evolution of CPU transistor count. After [4], updated.

There were several chipmakers very active in the early days of the CPU business that later shut down their engagement, e.g., Hitachi or Motorola. In the recent past, collaborations between fabless companies designing complex logic ICs (e.g., Apple, AMD, or Qualcomm) and foundries like TSMC, Samsung, or GlobalFoundries fabricating these logic chips made significant inroads into the CPU market.

Over the years, CPUs have become increasingly complex and the transistor count has grown to 1 million in the late 1980s, to 1 billion around 2005, and exceeded 10 billion in 2015. This remarkable progress was achieved by a combination of three measures discussed by Moore in [2]: (i) Shrinking the lateral device dimensions, i.e., MOSFET scaling; (ii) Increasing the die size; (iii) Device and circuit cleverness. Transistor dimensions were reduced essentially following Robert Dennard's scaling rules [5-6] by shrinking the gate length, the decisive lateral dimension of a MOSFET, and then reducing all other dimensions and increasing the doping accordingly. In 1980, 3 µm was a typical gate length of MOSFETs in CPUs compared to around 30 nm in 2010.

Transistor scaling was combined with increasing die area, i.e., measure (ii). In 1980, CPU die areas were typically of the order of several tens of mm², while in 2010 the die area was increased up to several hundreds of mm². Quad-core Itanium dies, to name just one example, had a die area of almost 700 mm². In the right upper corner of Fig. 1, three red stars are shown. They belong to the three hitherto introduced generations of Cerebras wafer scale engines called WSE-1, WSE-2, and WSE-3. The WSEs represent a rather exotic example of complex ICs since a WSE die has an area of 46,225 mm², i.e., from a whole 400 mm wafer one gets one single WSE die. The 3rd generation WSEs are fabricated using the 5-nm technology at TSMC, contain four trillion transistors, and are designed for demanding AI applications [7].

Regarding measure (iii), an example for what Moore meant with device cleverness was the introduction of high-κ dielectrics and metal gates in 2007-2008. Another and particularly crucial innovation, which saved the survival of Moore's Law for more than a decade was the introduction of the FinFET architecture. In the early 2000s further shrinking the size of the traditional single-top-gate MOSFET became more and more challenging. As consequence of this, the increase of transistor count slowed down, see the rather flat evolution of transistor count between 2005 and 2011 in Fig. 1. Only after the introduction of the FinFET in 2011/2012 [8], the exponential increase of CPU transistor count could be restored.

The magenta full circle in Fig 1 shows the transistor count that would have been reached in 2023 if the number of transistors on CPU dies had doubled every two years since 1971. It can be seen that this circle is close to the transistor count of the most complex CPU on the market in 2023, the Apple M2 Max with 67 billion transistors. On the other hand, we see that the M2 Max data point is slightly below the exponential trend shown by the magenta line. It is difficult to say at the moment whether this is a sign of the beginning of the end of Moore's Law or if it is just a lean period (like the one between 2005 and 2011 mentioned above) that will soon be overcome.

IV   SOME MORE EXPONENTIALS

A. Price Per Transistor

We have mentioned in Sec. I that, provided the number of devices per die is not increased arbitrarily but in line with the progress in semiconductor technology, one will end up at falling cost per transistor with progressing time. In his 2003 plenary paper at the International Solid-State Circuit Conference (ISSCC) Moore presented a plot showing the average price per transistor versus year over the 35-year period from 1968 to 2002 [9]. This plot revealed a halving of the price about every 2 years, i.e., an exponential decrease of the price per transistor.

To find out whether the exponential fall in prices continues to the present day, we have collected the launching prices and transistor counts of CPUs introduced between 1971 and early 2025 from platforms such as [10-11] and calculated the price per transistor for these CPUs.

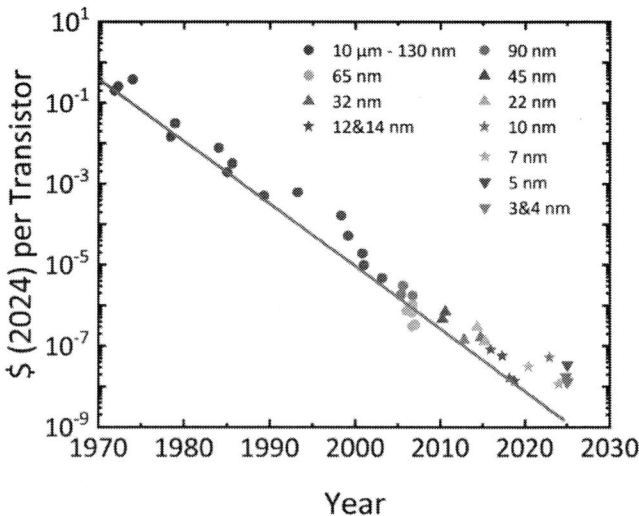

Fig. 2.   Evolution of the price per transistors given in constant dollars (based on the value of the dollar in 2024). The symbols indicate the technology nodes for the respective CPUs.

The result is depicted in Fig. 2, which shows the price per transistor in constant dollars (2024) versus time and the different symbols indicate the corresponding technology nodes. The magenta line beginning in 1971 with the Intel 4004 (launching price $60, which corresponds to around $465 in 2024) shows the price evolution that would have been expected if the price had halved every two years. Clearly the line describes the price descent over nearly five decades from the 10-μm technology down to the 12/14 nm node nicely.

There have been serious concerns that from the 28-nm node onwards not only would the exponential decrease in prices stop, but that prices per transistor would no longer decrease further at all [12-13]. As Fig. 2 clearly indicates, these concerns did not materialize. However, between 2020 and 2025, the picture for the 7 nm and below nodes becomes quite confused and no clear trend can be identified.

B. Energy Consumption per Logic Operation

Over the years, transistors became faster and the performance of CPUs and computers improved continuously. In addition, the energy needed to perform a given number of computations could be reduced year by year, i.e., the energy efficiency improved. At first glance, this may seem hard to believe given that the Cray 1, the world's fastest computer in 1976, had a power consumption of 115 kW, the BlueGene/L, number 1 in the Top500 supercomputer list [14] in November 2005 consumed 1.4 MW, and El Capitan (number 1 in the Top500 in November 2024) needed almost 30 MW. However, the computing performance increased much faster than the power consumption, so that the amount of energy needed to perform a given number of computations decreased dramatically over the years.

In 2011 Koomey et al. published a study on the energy efficiency of computing [15-16], and presented plots showing the evolution of computer performance (in terms of computations per second) and of computer energy efficiency (in terms of computations per kWh).

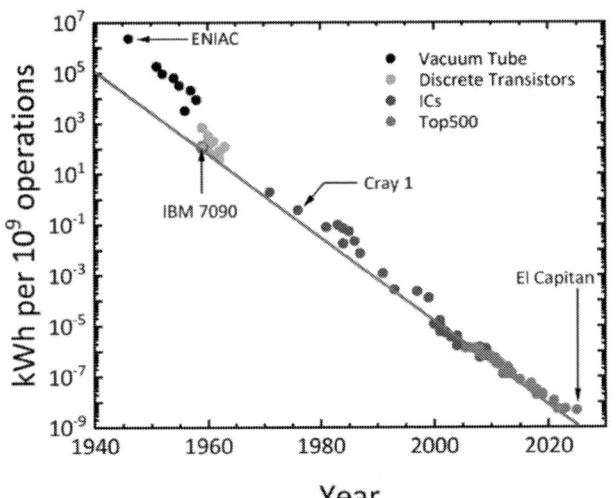

Fig. 3. Evolution of the energy consumption of computers when performing 10^9 computing operations versus time. Data taken from [14-16].

The latter inspired us to elaborate a plot showing the amount of energy needed to perform a given number of operations versus time. The resulting plot is shown in Fig. 3. Note that the data from [15-16] are related to CPS (computation per second) while those from [14] relate to Flops (full precision floating point operation per second). The plot in Fig. 3 reveals several interesting details.

First, it contains data for computers made of discrete, i.e., non-integrated, transistors (green symbols) and for computers using ICs (blue and red data points), and interestingly both types of computers follow the same trend. This is all the more remarkable since all transistorized computers (green symbols) and some of the early supercomputers like the Cray 1 used bipolar transistors, while the fast majority of the IC-based computers (blue data points) and all supercomputers from the top500 list (red symbols) rely on MOSFETs.

Second, the magenta line shows the evolution assuming a halving of the energy consumption per 10^9 operations every two years. As reference point for this line we took the data for the IBM 7090 mainframe computer from 1959 given in [16], which was based on discrete transistors. Note that we intentionally choose a slightly different symbol (green full circle with olive rim) for it than for the other computers using discrete transistors (full green circles) to make it easier to recognize in the dense cloud of green symbols.

Third, the plot also includes data points of old computers based on vacuum tubes from the 1940s and 1950s. Even if their energy consumption is by trend higher than that of transistorized computers, by and large they also follow the tendency of a halving of the energy consumption per operation every two years.

V. NO EXPONENTIAL IS FOREVER …

The question of whether Moore's Law still applies, and if so, for how much longer, has divided opinion for many years. For example, a study published in January 2014 [12] concluded that the decline of the cost per transistor from generation to generation observed for many years and discussed in Sec. IV A would stop at the 28 nm node and that at the following nodes the cost per transistor would even rise. Only a few months later, in August 2014, Mark Bohr introduced Intel's new 14 nm node technology and stated that the cost per transistor at Intel has continuously decreased exponentially from 130 nm to 14 nm [17]. A more recent example for contradicting opinions on the status of Moore's Law are the statements of Jensen Huang, CEO of Nvidia, the fabless company famous for its powerful GPUs, and Pat Gelsinger, former CEO of Intel. In September 2024, Huang said *Moore's Law is dead* [18], and only a few days later Gelsinger responded *Moore's Law is still alive and well* [19].

It is an indisputable fact that no exponential trend in technology can survive forever. However, by smart engineering solutions the life span of an exponential trend can be extended. Gordon Moore paraphrased this nicely in the title of his 2003 ISSCC paper [9] with *No exponential is forever: But "forever" can be delayed!*

We have mentioned in Sec. III that, at the time when the conventional Si MOSFET was approaching its scaling limits, the FinFET was the transistor architecture that helped to delay "forever". Today the potential of FinFET scaling is almost exhausted too and another innovation is needed to delay "forever" again. Fortunately, a replacement for the FinFET, the SNC (stacked-nanosheet-channel) MOSFET, is already in the starting blocks. The three leading-edge chipmakers TSMC, Samsung, and Intel and the recently founded Japanese company Rapidus in partnership with IBM are very active in research on SNC MOSFETs, see, e.g., [20-23] and either already started the mass production of SNC MOSFET logic chips or plan to do that in the near future. While currently the main interest of the chipmakers is focused on SNC transistors with Si nanosheet channels, research on introducing ultimately thin 2D (two-dimensional) transition metal dichalcogenide nanosheet channels (e.g., MoS_2, WSe_2) is underway, see, e.g. [24-25].

The SNC MOSFET, be it with Si or TMDC channels, shows promise to deliver superior electrical performance and better scalability compared to the FinFET. Whether the SNC MOSFET will actually be successful and whether it can establish itself as the successor to the FinFET will depend heavily on whether highly complex logic ICs can be manufactured cost-effectively on its basis.

VI. CONCLUSION

We have shown that important exponential trends in semiconductor electronics related to Moore's Law, namely the exponential increase of transistor count of CPUs over time as well as the exponential decay of the price per transistor and of the energy needed to perform a given number of computational operations could be maintained over decades. On the other hand, following Moore's Law was never easy and definitely not a self-runner. A closer look at the trends shown in Figs. 1-3 indicates that there were repeatedly lean periods that had to be overcome through innovations and smart engineering solutions. Examples for such periods were are the almost stagnating transistor count between 2006 and 2011 as well the temporarily only slightly decreasing price per transistor after 1989.

The progress recently made in developing SNC MOSFETs makes us optimistic that again an innovation is in place at the right time to further "delay forever" and to keep Moore's Law alive for another couple of technology generations. In the longer term, however, we will have to adjust to the fact that Moore's Law, at least in the form in which we know it, will come to an end. But this will definitely not mean the end of progress in electronics and computer technology.

For some time now, intensive research has been conducted into alternative approaches to information processing, e.g. neuromorphic computing or quantum computing, as well as alternative device concepts such as memristors or spin devices. These and other efforts will sooner or later enable new hardware options that complement our traditional digital CMOS and lead to improvements in information processing in terms of energy efficiency and speed. Furthermore, these alternative hardware options will enable completely new applications and likely lead to Moore's-Law-like exponentials – true to the title of the front cover of the April 2015 issue of IEEE Spectrum [26]: "Moore's Law is dead - Long live Moore's Law."

REFERENCES

[1] G. E. Moore, "Cramming more components onto integrated circuits", Electronics, vol. 38, no. 8, pp. 114-117, Apr. 19, 1965.

[2] G. E. Moore, "Progress in digital integrated electronics", Tech. Dig. IEDM, pp. 11-13, 1975.

[3] R. Courtland, "The Murky Origins of "Moore's Law", https://spectrum.ieee.org/the-murky-origins-of-moores-law.

[4] F. Schwierz and M. Ziegler, "Status and future prospects of CMOS scaling and Moore's Law – a personal perspective", Proc. IEEE LAEDC, 2020.

[5] R. H. Dennard et al., "Design of ion-implanted MOSFET'S with very small physical dimensions", IEEE J. Solid-State Circuits, vol. 9, pp. 256-268, 1974.

[6] D. J. Frank et al., "Device scaling limits of Si MOSFETs and their application dependencies", Proc. IEEE, vol. 89, pp. 259-288, 2001.

[7] S. K. Moore, "Cerebras unveils its next waferscale AI chip", see at https://spectrum.ieee.org/cerebras-chip-cs3.

[8] S. Damaraju et al., "A 22nm IA multi-CPU and GPU system-on-chip", Dig. ISSCC, pp. 56-57, 2012.

[9] G. E. Moore, "No exponential is forever: But "forever" can be delayed!", Dig. ISSCC, paper 1.1, 2003.

[10] CPU Specs Database, see at https://www.techpowerup.com/cpu-specs/?sort=name.

[11] DOS days. Old PC computing resource CPUs (Microprocessors), see at https://dosdays.co.uk/topics/cpus.php#i80286.

[12] H. Jones, "Why migration to 20 nm bulk CMOS and 16/14nm FinFET is not best approach for semiconductor industry", see at https://caxapa.ru/thumbs/598000/WP_handel-jones.pdf.

[13] D. O'Laughlin, "The rising tide of semiconductor cost", 2021, see at https://www.fabricatedknowledge.com/p/the-rising-tide-of-semiconductor.

[14] Top500. The List, see at https://top500.org/.

[15] J. G. Koomey, S. Berard, M. Sanchez, and H. Wong, "Implications of historical trends in the electrical efficiency of computing", IEEE Ann. Hist. Comput., vol. 33, iss. 3, pp. 46-54, 2011.

[16] J. G. Koomey, S. Berard, M. Sanchez, and H. Wong, "Assessing trends in the electrical efficiency of computation over time, see at https://www.researchgate.net/publication/229000231_Assessing_trends_in_the_electrical_efficiency_of_computation_over_time

[17] M. Bohr, 14 nm technology announcement, see at https://www.intel.com/content/dam/www/public/us/en/documents/presentation/advancing-moores-law-in-2014-presentation.pdf.

[18] https://www.marketwatch.com/story/moores-laws-dead-nvidia-ceo-jensen-says-in-justifying-gaming-card-price-hike-11663798618.

[19] https://www.cnbc.com/2022/09/27/intel-says-moores-law-is-still-alive-nvidia-says-its-ended.html.

[20] S. Liao et al., "First demonstration of monolthic CFET inverter at 48nm gate pitch toward future logic technology scaling", Tech. Dig. IEDM, pp. 1-4, 2024.

[21] J. Jeong et al., " World's first GAA 3nm foundry platform technology (SF3) with novel multi-bridge-channel-FET (MBCFETTM) process", Dig. VLSI Technol., pp. 1-2, 2023.

[22] A. Agrawal et al., "Silicon RibbonFET CMOS at 6nm gate length", Tech. Dig. IEDM, pp. 1-4, 2024.

[23] R. Bao et al., "Advanced Multi-Vt enabled by selective layer reductions for 2nm nanosheet technology and beyond", Tech. Dig. IEDM, pp. 1-4, 2024.

[24] F. Schwierz, M. Ziegler, and J. J. Liou, "MOSFETs with stacked 2D nanosheet channels – An auspicious option to delay "forever", Proc. SBMicro, pp. 1-4, 2023.

[25] C. Kim et al., "Transfer free 2D CMOS multi bridge channel FET", Tech. Dig. IEDM, pp. 1-4, 2024.

[26] Moore's Law is dead. Long live Moore's Law, see at https://ieeexplore.ieee.org/stamp/stamp.jsp?tp=&arnumber=7065391.

979-8-3315-2209-4/25 $31.00 © 2025 IEEE

A Hybrid Clause Deletion Framework for Enhanced SMT Solving in Hardware Formal Verification

Wenda Leng   Meihua Liu[*]   Yufeng Jin[*]

Abstract—Hardware formal verification transforms circuit designs from register transfer level descriptions into SMT problems which are then solved. As circuit complexity grows, conventional SMT solvers encounter efficiency bottlenecks, especially in managing learned clauses. Existing clause management strategies often rely solely on dynamic information or static characteristics, making it difficult to fully assess the impact of the clause and leading to inconsistent performance in complex verification scenarios. To address this concern, a hybrid clause deletion framework is introduced, which combines literal block distance with clause activity metrics to capture the underlying contribution of clauses. To demonstrate effectiveness and extensibility, two classical strategies—relevance deletion and BerkMin deletion—are integrated into this framework, producing two solver variants based on the Yices2 solver. Experimental results indicate that these variants achieve average speedups of $1.35\times$ and $1.37\times$ on SMT-LIB benchmarks, SMT-COMP benchmarks, and formal verification tasks derived from real-world circuits.

Index Terms—formal verification, SMT, clause deletion

I. INTRODUCTION

Formal verification is widely employed in hardware settings, where circuit properties are transformed into a Satisfiability Modulo Theories (SMT) formulation through systematic modeling and then verified by an SMT solver. As circuit complexity escalates, solving SMT problems becomes increasingly difficult, making solver efficiency ever more critical [1].

Classic improvement strategies for SMT solvers include restart strategies [2], [3], activity-based branching heuristics [4], and clause learning [5]. Verification tasks of large-scale circuits lead to a dramatic increase in the number of clauses, which underscores the importance of clause learning. Conflict-Driven Clause Learning (CDCL) is the most efficient algorithm, and many state-of-the-art solvers adopt it [6], [7]. CDCL generates a clause that captures the root cause of a conflict and includes it in the conjunctive normal form to prune the search space. However, repeated conflicts accumulate a large number of clauses, resulting in substantial memory consumption and reduced the efficiency of Boolean Constraint Propagation (BCP). Thus, clauses that contribute minimally to the solving process require removal.

Existing clause deletion strategies frequently rely on certain metrics, and once conflict thresholds are reached, clauses

regarded as less important are discarded. These metrics usually focus on dynamic information or static structural characteristics, making it difficult to comprehensively capture clause utility and leading to performance fluctuations in complex verification tasks. For instance, clauses with low Literal Block Distance (LBD) [8] values may lose relevance over time, whereas those with high activity [9] might become lengthy and intricate. Building on these observations, a hybrid clause deletion framework named H-LAD is introduced, which combines LBD with clause activity metrics, and is compatible with additional strategies to enhance clause management. The contributions of this paper are summarized as follows:

- A hybrid clause deletion framework that integrates LBD and clause activity, unifying dynamic and structural characteristics of learned clauses for more accurate deletion;
- The relevance deletion strategy and the BerkMin deletion strategy are respectively integrated into the proposed framework, demonstrating flexibility and extensibility;
- Experimental results validate the effectiveness and efficiency of two solver variants that incorporate H-LAD on various hardware formal verification benchmarks, as well as on benchmarks derived from actual circuits, compared with the state-of-the-art Yices2 solver.

II. BACKGROUND

A. Literal Block Distance

LBD is a widely adopted metric for evaluating the quality of learned clauses in state-of-the-art SMT solvers [8]. It is based on the assumption that fewer decision layers reflect stronger variable correlations and a higher likelihood of facilitating conflict resolution. Concretely, LBD is defined as the number of distinct decision levels in a learned clause. For instance, consider the learned clause C in (1), where V_1 is in layer 1, V_2 and V_3 are in layer 2, and V_4, V_5, and V_6 are in layer 4. The LBD value of C is therefore 3.

$$C = \neg V_1 \vee V_2 \vee V_3 \vee \neg V_4 \vee \neg V_5 \vee V_6 \tag{1}$$

Clauses with an LBD value of 2 are often referred to as Glue clauses [10]. Moreover, LBD is used to optimize other aspects of the solving process, such as guiding the restart strategy in the Glucose solver [11].

B. Activity

Activity is a metric introduced in MiniSAT and adopted by advanced solvers to measure the importance of learned clauses [9], [12], [13]. Each clause receives an initial activity

---

Wenda Leng and Yufeng Jin are with *School of Electronic and Computer Engineering (ECE)*, *Peking University Shenzhen Graduate School*, Shenzhen, China. Meihua Liu is with *Shenzhen GWX Technology Co., Ltd.*, Shenzhen, China. This work is supported by Shenzhen Science and Technology Program under Grant No. KJZD20230923115005009. It is also supported by the project under Grant No. KQTD20200820113105004. Corresponding author: Yufeng Jin (email: yfjin@pku.edu.cn); Meihua Liu (amo_jane@outlook.com).

979-8-3315-2209-4/25 $31.00 © 2025 IEEE   253

value, and this value increases whenever the clause participates in conflict analysis. To avoid retaining clauses that initially exhibit high activity but eventually lose relevance, a periodic decay mechanism is applied by multiplying activity values by a factor below 1 after a fixed number of conflicts. The primary rationale for employing activity-based deletion is to preserve relevant conflict information for solving SMT problems while maintaining the number of learned clauses within manageable limits. Specifically, once conflicts exceed a given threshold, learned clauses are sorted by their activity, and those with lower activity are removed.

III. METHODOLOGY

A. Hybrid LBD-Activity Clause Deletion Framework

The number and timing of learned clause deletions are two critical factors in developed deletion strategies [14]. Addressing these factors presents significant challenges, chiefly because of the NP-hard nature of the satisfiability problem, which introduces uncertainty into the solution process. Although the aforementioned metrics improve solution performance, they remain simplistic by focusing on a single type of characteristic and do not fully exploit conflicting information. To overcome these issues, the Hybrid LBD-Activity Clause Deletion framework (H-LAD) is introduced, combining structural and dynamic characteristics for fine-grained management of learned clauses.

At early stages of conflict, clause activity is prone to fluctuation, and relying on it as a primary guide may result in improper deletions. Hence, H-LAD initially prioritizes LBD for clause deletion. Once a certain conflict threshold is reached, clauses are sorted by activity; those with lower activity are eliminated based on considerations such as clause length, clause age, and relevance. Algorithm 1 outlines the workflow of the H-LAD framework. After each conflict analysis, LBD values are computed for learned clauses, and those whose LBD values exceed 6 are removed, since such clauses rarely contribute to conflicts [10]. When the conflict count reaches a specified threshold, the effectiveness of LBD-based deletion is evaluated to ensure that at least 10% of the total learned clauses are deleted. If this proportion is not achieved, clauses are ranked by activity, and additional strategies are applied to the half with lower activity.

B. Integration of Relevance and BerkMin Strategies in H-LAD

To validate the effectiveness and extensibility of the H-LAD framework, the additional strategy component within H-LAD is assigned to the simpler relevance deletion strategy, referred to as LAR, and the relatively complex BerkMin deletion strategy, referred to as LBM. The relevance evaluation criterion was originally proposed in the Chaff solver [15]. It assesses learned clauses by counting the number of unassigned literals present during subsequent conflict analysis. Clauses containing many unassigned literals often possess a lower probability of being involved in future conflicts, signaling reduced relevance.

Algorithm 1 Deletion Strategy Algorithm

Input:
    Set of learned clauses C
    Base conflict threshold T (e.g., 2000)
    Adjustment factor K
    Number of clause N
Output: Updated clause database C'

1: **if** $N < T + T \times K$ **then**
2:     **for int** $i = 0;\ i < n;\ i++$ **do**
3:         $\text{LBD}(c_i) = \text{LBDCalculation}(c_i),\quad c_i \in C$
4:         **if** $LBD(c_i) > 6$ **then**
5:             $\text{Remove}(C,\ c_i)$ ▷ Delete c_i
6:             LBDDeleteCount = LBDDeleteCount + 1
7:         **end if**
8:     **end for**
9: **else**
10:     **if** LBDDeleteCount $< 0.1 \cdot N$ **then**
11:         $\text{Order}(C,\ c_i)$ ▷ Reorder based on activity
12:     **end if**
13:     **for int** $j = N//2;\ j < N;\ j++$ **do**
14:         $\text{AdditionalStrategy}(c_j)$
                ▷ Delete by additional strategy
15:     **end for**
16: **end if**
17: **End**
18: **return** C'

The BerkMin deletion strategy takes into account clause length, clause age, and the activity metric [16]. Clause length is derived from the size of the clause, and shorter clauses are generally viewed as higher quality because they can be assigned more swiftly, thus accelerating BPC. Clause age represents the order in which clauses are generated, where older clauses are those created first and younger clauses are generated later. Specifically, the recently created 15/16 of the learned clauses form the new set, while the earliest 1/16 comprise the old set. In the new set, clauses longer than 43 or those with an activity below 7 are removed. In the old set, clauses longer than 9 or those with activity under a particular threshold are deleted.

IV. IMPLEMENTATION AND EXPERIMENTAL EVALUATION

A. Experimental Setup

LAR and LBM were integrated into Yices2, which is a state-of-the-art solver for hardware formal verification, resulting in the solver variants Yices2-LAR and Yices2-LBM. Meanwhile, the default strategies of Z3 and Yices2 serve as baselines.

For a comprehensive validation, benchmarks were collected from the SMT-LIB and SMT-COMP repositories, focusing on hardware formal verification tasks, which are regarded as both extensive and credible [17], [18]. Additional formal verification tasks derived from real-world circuits were also included, bringing the total to 1,122 benchmarks.

During the experiments, the standard rules of the international SMT competition were followed, with each bench-

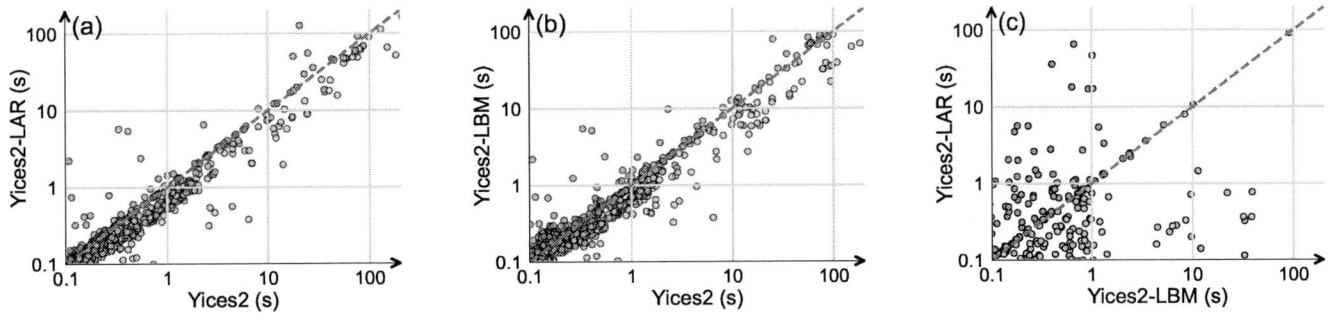

Fig. 1. Comparison of solving time : (a) Yices2-LAR vs. Yices2, (b) Yices2-LBM vs. Yices2, (c) Yices2-LAR vs. Yices2-LBM.

mark limited to 1,200 seconds of runtime. To prevent CPU resource allocation discrepancies from influencing the final solving time, only one solver was executed for each test. All experiments were carried out on a Linux system with an Intel® Core™ i5-1135G7 @ 2.40GHz processor.

B. Experimental Results

Figure 1 depicts the solving times of Yices2, Yices2-LAR, and Yices2-LBM across SMT-LIB benchmarks. Each point represents one benchmark, with logarithmic scales on both axes. They indicate solving times for two different solvers, and a diagonal line is included for reference. Red and green points respectively show cases where the solver on the x-axis is faster or slower. Figures 1(a) and 1(b) demonstrate that Yices2-LAR is faster than Yices2 on 78.3% of the benchmarks, and Yices2-LBM outperforms Yices2 on 62.9%. In the first half of Figures 1(a) and 1(b), Yices2, Yices2-LAR, and Yices2-LBM complete these tasks very quickly, and Yices2 is slightly faster without using the proposed framework. This marginal advantage arises from overhead produced by computing evaluation criteria—including LBD value, activity, clause length, and relevance—to guide deletion of learned clauses, which is less efficient for limited-size test scenarios. As problem complexity grows, H-LAD framework achieves significant speedups, showing that multiple deletion strategies can be effectively integrated to enhance performance. Furthermore, analyzing Figure 1(c) shows that Yices2-LAR outperforms Yices2-LBM in 60.6% of the benchmarks, indicating different adaptability profiles.

To further examine this observation, 622 cases in the SMT-COMP benchmarks set *Sosy_Lab* were studied, and Table I presents the results. Yices2-LAR and Yices2-LBM exceed the original Yices2 in terms of average solving time, decisions, and restarts, confirming the effectiveness and efficiency of the framework. On this particular set, Yices2-LBM surpasses Yices2-LAR by a small margin, reflecting that distinct deletion strategies adapt variably to different scenarios.

To comprehensively validate the effectiveness of the proposed framework, we conducted experiments using four solvers on both the SMT-LIB benchmarks and SMT-COMP benchmarks, as illustrated in Figure 2. The evaluation employed two critical metrics: the number of satisfiable in-

TABLE I
PERFORMANCE COMPARISON OF SOLVERS ON SOSY_LAB

Solver	#Solved	Time (s)	#Decisions	#Restarts
Yices2	622	37.268	1.224×10^7	13.188
Yices2-LAR	622	23.680	6.824×10^6	8.934
Yices2-LBM	622	21.047	6.306×10^6	8.209

stances solved (#SAT, depicted by red bars) and the total computation time in thousands of seconds (represented by teal bars). Experimental results reveal that while the baseline solvers Z3 and Yices2 successfully processed 737 and 735 SAT instances respectively, they required 24,327 and 20,178 seconds of total computation time. In contrast, our enhanced implementations—Yices2-LBM and Yices2-LAR—achieved equivalent instance coverage (737 instances, matching Z3's performance) while substantially reducing the computational overhead to 14,910 and 14,754 seconds, respectively. This empirical evidence demonstrates that our framework preserves the solving capabilities of the original solvers while significantly improving their computational efficiency.

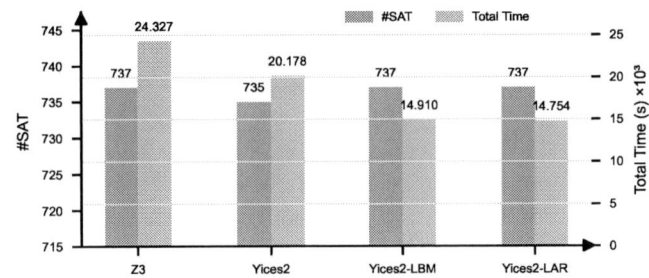

Fig. 2. Performance Comparison of Z3, Yices2, Yices2-LBM, and Yices2-LAR on Both the SMT-LIB Benchmarks and SMT-COMP Benchmarks.

The performance gains are particularly noteworthy, with Yices2-LBM and Yices2-LAR reducing total solving time by approximately 26% and 27% compared to baseline Yices2, and by a remarkable 39% compared to Z3. These substantial improvements in execution time, without compromising solution quality, conclusively validate the effectiveness of our proposed framework for enhancing SMT solver performance.

Performance in circuit-property verification tasks was also evaluated by running connectivity checks on seven circuits that vary in size. The process involved adding assertions into register transfer level files for open-source test circuits, converting them into comma-separated values format, and subsequently parsing, synthesizing, and transforming them via Yosys tools [19]. Figure 3 illustrates the outcomes. The *src_add*, *yhud580*, and *wz_al* circuits are smaller, each requiring less than 1 second overall. The average time savings amount to 14.7% for Yices2-LAR and 13.7% for Yices2-LBM. The *ac91_ale*, *ac92_ale*, *add_bx_ale*, and *add_tree_sv* circuits are more complex, where Yices2-LAR achieves an average saving of 37.0% and Yices2-LBM attains 36.8%. These results reveal that the proposed framework accelerates real-world circuit-property verification for all test circuits, with a more pronounced effect when problem complexity is higher.

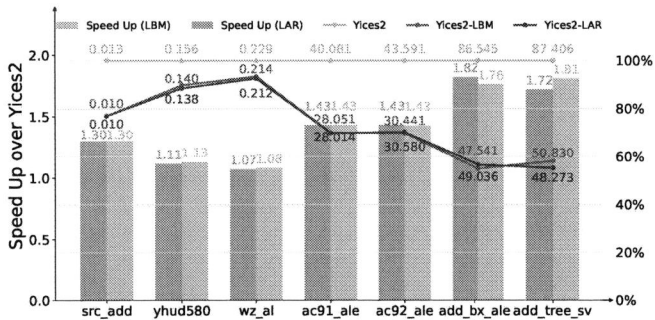

Fig. 3. Performance Improvement of Yices2-LAR and Yices2-LBM in Real-World Circuit Verification.

V. CONCLUSIONS

This paper introduces a hybrid clause deletion strategy framework that combines LBD values and clause activity, accounting for both the structural characteristics of learned clauses and the frequency of conflict. Subsequently, the relevance deletion strategy and the BerkMin deletion strategy are incorporated into this framework, achieving multidimensional evaluation and refined management of learned clauses. Experimental results indicate that solver variants based on the H-LAD framework significantly enhance the performance of Yices2, achieving substantial speedups of 1.35× and 1.37× on hardware formal verification benchmarks derived from SMT-LIB and SMT-COMP. The proposed framework also demonstrates effectiveness on real-world circuit formal verification, indicating its practical applicability. Furthermore, this framework can be extended across various verification domains by integrating a broader range of metrics, refining clause management, and adopting additional advanced strategies to further improve the capabilities of SMT solving.

REFERENCES

[1] C. Shim, J. Bae, and B. Kim, "30.3 vip-sat: A boolean satisfiability solver featuring 5× 12 variable in-memory processing elements with 98% solvability for 50-variables 218-clauses 3-sat problems," in *2024 IEEE International Solid-State Circuits Conference (ISSCC)*, vol. 67. IEEE, 2024, pp. 486–488.

[2] H. Kautz, E. Horvitz, Y. Ruan, C. Gomes, and B. Selman, "Dynamic restart policies," *Aaai/iaai*, vol. 97, pp. 674–681, 2002.

[3] C. P. Gomes, B. Selman, H. Kautz *et al.*, "Boosting combinatorial search through randomization," *AAAI/IAAI*, vol. 98, no. 1998, pp. 431–437, 1998.

[4] J. H. Liang, V. Ganesh, E. Zulkoski, A. Zaman, and K. Czarnecki, "Understanding vsids branching heuristics in conflict-driven clause-learning sat solvers," in *Hardware and Software: Verification and Testing: 11th International Haifa Verification Conference, HVC 2015, Haifa, Israel, November 17-19, 2015, Proceedings 11*. Springer, 2015, pp. 225–241.

[5] J. M. Silva and K. A. Sakallah, "Grasp-a new search algorithm for satisfiability," in *Proceedings of International Conference on Computer Aided Design*. IEEE, 1996, pp. 220–227.

[6] C. Oh, "Between sat and unsat: the fundamental difference in cdcl sat," in *International Conference on Theory and Applications of Satisfiability Testing*. Springer, 2015, pp. 307–323.

[7] L. Zhang, C. F. Madigan, M. H. Moskewicz, and S. Malik, "Efficient conflict driven learning in a boolean satisfiability solver," in *IEEE/ACM International Conference on Computer Aided Design. ICCAD 2001. IEEE/ACM Digest of Technical Papers (Cat. No. 01CH37281)*. IEEE, 2001, pp. 279–285.

[8] G. Audemard and L. Simon, "Refining restarts strategies for sat and unsat," in *Principles and Practice of Constraint Programming: 18th International Conference, CP 2012, Québec City, QC, Canada, October 8-12, 2012. Proceedings*. Springer, 2012, pp. 118–126.

[9] N. Eén and N. Sörensson, "An extensible sat-solver," in *International conference on theory and applications of satisfiability testing*. Springer, 2003, pp. 502–518.

[10] M. S. Chowdhury, M. Müller, and J.-H. You, "Exploiting glue clauses to design effective cdcl branching heuristics," in *Principles and Practice of Constraint Programming: 25th International Conference, CP 2019, Stamford, CT, USA, September 30–October 4, 2019, Proceedings 25*. Springer, 2019, pp. 126–143.

[11] G. Audemard and L. Simon, "On the glucose sat solver," *International Journal on Artificial Intelligence Tools*, vol. 27, no. 01, p. 1840001, 2018.

[12] L. De Moura and N. Bjørner, "Z3: An efficient smt solver," in *International conference on Tools and Algorithms for the Construction and Analysis of Systems*. Springer, 2008, pp. 337–340.

[13] B. Dutertre, "Yices 2.2," in *International Conference on Computer Aided Verification*. Springer, 2014, pp. 737–744.

[14] S. Jabbour, J. Lonlac, L. Sais, and Y. Salhi, "Revisiting the learned clauses database reduction strategies," *International Journal on Artificial Intelligence Tools*, vol. 27, no. 08, p. 1850033, 2018.

[15] M. W. Moskewicz, C. F. Madigan, Y. Zhao, L. Zhang, and S. Malik, "Chaff: Engineering an efficient sat solver," in *Proceedings of the 38th annual Design Automation Conference*, 2001, pp. 530–535.

[16] E. Goldberg and Y. Novikov, "Berkmin: A fast and robust sat-solver," *Discrete Applied Mathematics*, vol. 155, no. 12, pp. 1549–1561, 2007.

[17] C. Barrett, A. Stump, C. Tinelli *et al.*, "The smt-lib standard: Version 2.0," in *Proceedings of the 8th international workshop on satisfiability modulo theories (Edinburgh, UK)*, vol. 13, 2010, p. 14.

[18] L. Hadarean, A. Hyvärinen, A. Niemetz, and G. Reger, "Smt-comp 2019," *Int. Satisfiability Modulo Theories (SMT) Competition, Tech. Rep*, 2019.

[19] C. Wolf, "Yosys manual," *Retrieved January*, vol. 16, p. 2021, 2021.

A SOT-MRAM Based In-Memory Computing Macro with Enhanced Read-Current Stability and Multi-Bit MAC Operations

Saiya Wang[1,2], Junzhan Liu[1,2], Guangyao Wang[1,2], Wang Kang[1,2], Senior Member, IEEE

[1]School of Integrated Circuit Science and Engineering, Beihang University, Beijing, China

[2]National Key Laboratory of Spintronics, Hangzhou International Innovation Institute, Beihang University, Hangzhou, China

Email: wang.kang@buaa.edu.cn

Abstract—**This paper proposes an improved spin-orbit torque magnetic random access memory (SOT-MRAM) based in-memory computing (IMC) macro with multi-bit multiply-and-accumulation (MAC) operations and enhanced read-current stability. The proposed design integrates a 2T1MTJ cell with an optimized charge-reservoir-integrate-counter output (CRICO) unit, reducing the area and energy overhead associated with DAC and ADC circuits, while mitigating computational accuracy degradation caused by subtle result variations. The SOT-MRAM IMC macro utilizes the validated electrical performance of the SOT-MRAM array from a pre-fabricated 200 nm wafer. We implement a 32×32 bit-cell SOT-MRAM IMC macro using a 40 nm process, which enables 4-bit input through a 4-bit pulse width sequence, utilizes current ratio modulation between adjacent 4-column memory cells to achieve 4-bit weights, and generates a 4-bit output after accumulation. The simulation results show that this macro can achieve energy efficiency of 8.26 Tops/W at 4-bit input and output precision, while in comparison with conventional schemes, the design enhances the read current variation rate by four orders of magnitude, significantly improving the stability of the read current, which provides a highly reliable solution for IMC in next-generation intelligent computing systems.**

Index Terms—**SOT-MRAM, in-memory computing, read current stability, charge-reservoir-integrate-counter**

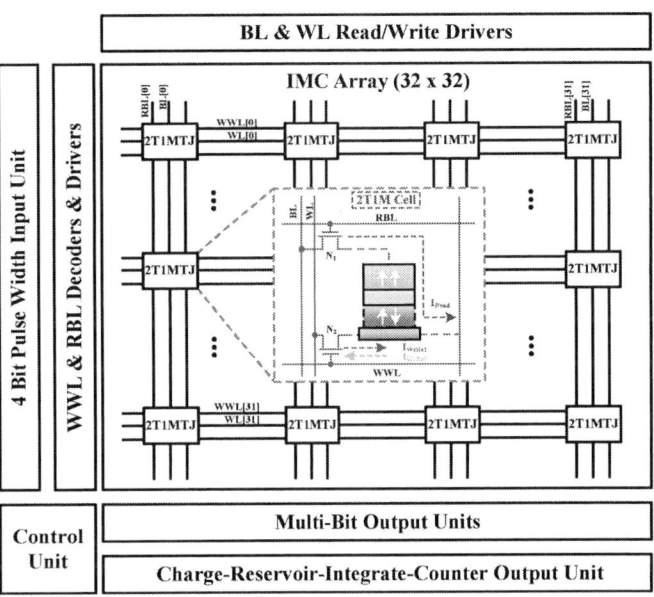

Fig. 1. The proposed 2T-MRAM IMC macro architecture.

I. INTRODUCTION

With the increasing demand for energy-efficient solutions, in-memory computing (IMC) has garnered significant attention due to its capability to perform data processing directly within memory cells, thereby reducing both power consumption and latency [1]–[3]. As a crucial subset of IMC, charge-domain IMC is enabled by the accumulation and distribution of charges, thereby facilitating low-power and high-speed processing. Among the promising candidates for implementing charge-domain IMC, spin-orbit torque magnetic random access memory (SOT-MRAM) is distinguished by its non-volatility, high density, rapid read/write capabilities, extended endurance, and strong compatibility with CMOS processes [4]. However, existing solutions face key challenges, including the degradation of computational accuracy caused by fluctuations in the reading current, the increased system complexity due to the reliance on external analog-to-digital converters, and the

limitations on array scalability caused by device parameter drift due to process variations. These issues severely constrain the potential breakthroughs in energy efficiency of the IMC architecture [5], [6]. To address this issue, an IMC macro is proposed based on SOT-MRAM. Experimental results indicate that under a 40 nm process, a read current stability enhancement of approximately 10^4 compared to conventional designs is achieved by the proposed scheme, which provides a high-energy-efficiency and high-reliability IMC solution for the next-generation intelligent computing systems.

II. SOT-MRAM COMPUTING: PROPOSED SCHEME AND READ STABILITY

Fig. 1 illustrates the overall architecture of the proposed macro, comprising a 32×32 bit-cell IMC array, multi-bit output units, a charge-reservoir-integrate-counter output (CRICO) unit, and a control unit [7]. The following sections detail the functionality of each module.

This work was supported by the Research Funding of Hangzhou International Innovation Institute of Beihang University (Grant No.2024KQ157) and Natural Science Foundation of China (62274008).

979-8-3315-2209-4/25 $31.00 © 2025 IEEE

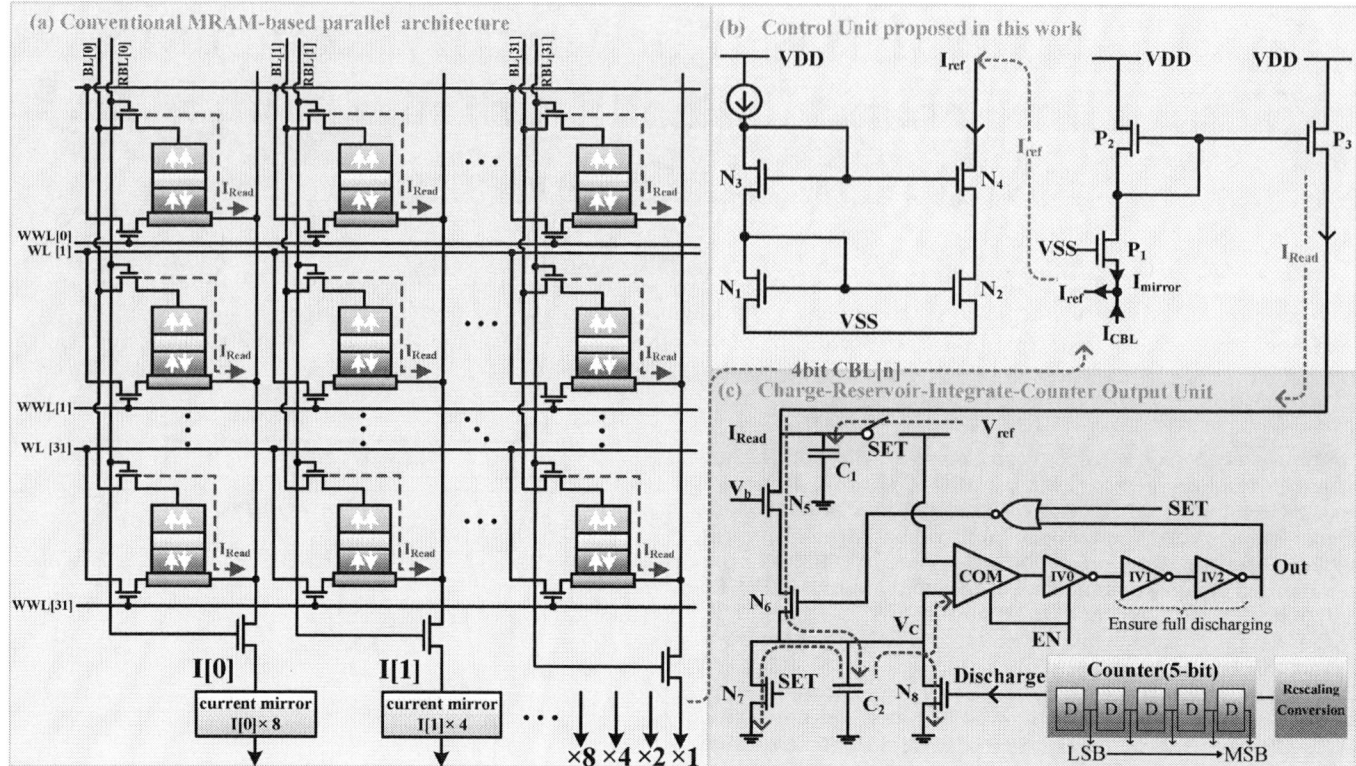

Fig. 2. Circuit diagram of the scheme proposed in this work. (a) Weight cell, (b) Control unit, (c) Charge-reservoir-integrate-counter output unit.

A. 2T1M Binary MRAM Based 4-Bit Synapse Array

To simplify circuit design and improve performance, a two-transistor, one-magnetic tunnel junction (2T1MTJ) cell is employed, as shown in Fig. 1. In this architecture, two transistors are integrated: a read transistor (N_1) and a write transistor (N_2). The drain of N_1 is connected to the bit line (BL), while its gate is linked to the read bit line (RBL). Meanwhile, the drain of N_2 is tied to the word line (WL), its gate to the write word line (WWL), and its source to a metal layer forming the write current path. During the write operation, N_1 is deactivated, N_2 is activated (WWL high, RBL low), allowing the write current to pass through the SOT track and switch the MTJ's free-layer magnetization via the SOT effect [8]–[10]. Conversely, during the read operation, N_1 is activated (RBL high, WWL low) while N_2 is deactivated, and the stored data are read by measuring the read current (I_{Read}).

B. Multi-Bit Implementation

As illustrated in Fig. 2(a), 4-bit synaptic weight values are stored in four adjacent binary MRAM cells, rather than in a single analog device. The output current from each column is multiplied by relevant factors through current mirrors. A 4-bit weight is formed by four adjacent binary MRAM cells, with column currents are multiplied by factors of 8, 4, 2, and 1, respectively. The terminals of each four-column group are collectively connected to the charge bit-line (CBL), then flows into the control unit. Finally, a 5-bit value resulting from the MAC operation is transmitted to the post-neuron.

C. An Optimized Reading Scheme

In IMC architectures, the intrinsic current of memory cells is often very small. The variation of column current is recorded as I_{CBL}, induced by cell-state switching, is similarly minute, thus making accurate measurement highly challenging. To mitigate this issue, a control unit circuit is proposed based on the current mirror principle, as shown in Fig. 2(b). The primary advantage of this circuit lies in the use of a current mirror to isolate the memory cell side from the read circuit side, while converting the subtle current variation I_{CBL} into a current change I_{mirror}.

According to Kirchhoff's current law (KCL), the drain current of P_1 can be expressed as

$$I_{\text{mirror}} = I_{\text{ref}} - I_{\text{CBL}}, \tag{1}$$

where I_{CBL} is the column output current of the memory cell, and I_{ref} is the reference current. Based on current-mirror principles, the drain current of P_3 can be expressed as

$$I_{\text{Read}} = k I_{\text{mirror}}, \tag{2}$$

where k represents proportionality constant defined by the mirror ratio. Consequently, small variations in $I[n]$ will be proportionally reflected in I_{Read} via I_{mirror}.

D. The CRICO Unit

As shown in Fig. 2(c), the CRICO unit collects charges from the CBL and generates an MAC output pulse sequence, which is subsequently processed by a counter composed of D flip-flops to produce the MAC result. The operation proceeds in two stages. During the "Set" stage, the SET signal is enabled, turning off NMOS N_6 and turning on N_7, thereby charging the reservoir capacitor C_1 to a reference voltage V_{ref} and discharging C_2 to ground via N_7. In the "Count" stage, the SET signal is disabled, N_6 is activated, and N_7 is deactivated. BL and C_1 then begin to charge C_2, while the comparator compares the voltage of C_2 (V_c) with V_{ref}. When $V_c > V_{\text{ref}}$, the comparator generates a high output signal ("Out"), which is stored in the counter. Simultaneously, "Out" activates N_8 to discharge C_2 and disables N_6 (via a NOR gate), thereby blocking the charging current from the BL to C_2. Conversely, when $V_c < V_{\text{ref}}$, "Out" transitions low, N_6 is activated, and N_8 is deactivated, permitting the BL to recharge C_2.

In conventional designs, C_2 may begin charging or discharging at different benchmark voltages, which can lead to computational errors. In contrast, in our design, N_2 (a clamp transistor) remains off at the start of the CBL discharging process, allowing the charges from the CBL to first be buffered into C_1. Once this occurs, N_2 turns on, and the charges stored in C_1 are transferred into C_2, facilitating smooth and stable charging and discharging of C_2. By utilizing C_1 as a "reservoir" capacitor to buffer the charge from the CBL, we ensure a consistent reference voltage (V_{ref}) and stable charging and discharging operations. According to (1) and (2), I_{CBL} can be expressed as

$$I_{\text{CBL}} = I_{\text{ref}} - \frac{1}{k} I_{\text{Read}}. \tag{3}$$

Therefore, the output of the counter needs to be processed by an additional rescaling conversion to represent I_{Read} as I_{CBL}.

III. MEASUREMENT RESULTS

The SOT-MRAM IMC macro is designed and verified for 40 nm nodes, utilizing the validated electrical performance of the SOT-MRAM array from a pre-fabricated 200 nm wafer [11]. Fig. 3(a) shows a magnified image of the SOT-MTJ. Fig. 3(b) illustrates a cross-sectional transmission electron microscope (TEM) image of a selected SOT-MTJ, obtained from the fully integrated 200 nm CMOS wafer. The design and simulation validations of this work are based on these verified experimental data. Table I provides a summary of the simulation parameters. By employing these high-resistance devices, overall power consumption can be significantly reduced, thereby increasing the feasibility of parallel operations. Based on these parameters, the read operations of both the conventional and proposed circuits are simulated, with particular emphasis on read-current stability.

Fig. 4(a) compares the read current (I_{Read}) of the conventional and proposed circuits between 60 ns and 80 ns, under conditions where all memory cells remain in the high-resistance state. In the conventional circuit, I_{Read} undergoes significant fluctuations, especially from 60 ns to 67 ns, and gradually stabilizes beyond 67 ns. By contrast, the proposed circuit maintains a stable I_{Read} throughout the entire observation interval. Although the stable read current in the proposed

TABLE I
SIMULATION PARAMETERS OF THE PROPOSED MACRO

Paraments	Value
Technology	40 nm
Bit-cell structure	2T1M
Supply voltage	1.1 V
Input precision	4-bit
Weight precision	4-bit
Output precision	4-bit

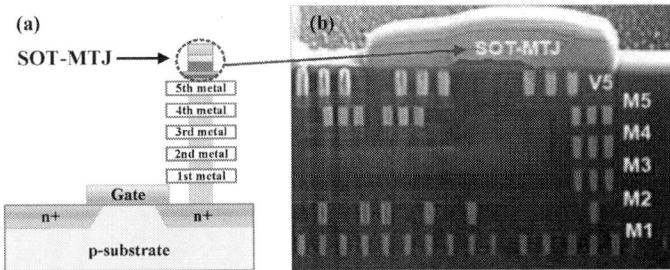

Fig. 3. (a) SOT-MTJ, (b) The cross-section TEM of a selected MTJ in the array using common 180 nm CMOS process technology node [11].

circuit (Fig. 4(a)) is promising, a more detailed evaluation of $\Delta I / \Delta t$ is required to quantify the improvement in stability.

To objectively evaluate the performance of the proposed circuit, further analysis is conducted on $\Delta I / \Delta t$ at five specific time points, 61 ns, 63 ns, 65 ns, 67 ns, and 69 ns, as shown in Fig. 4(b). In the conventional circuit, a maximum $\Delta I / \Delta t$ of $1010 \, \mu\text{A} \cdot \text{ns}^{-1}$ is exhibited at around 61 ns, whereas a minimum of $3.1 \, \mu\text{A} \cdot \text{ns}^{-1}$ is observed near 69 ns. By contrast, the proposed circuit consistently maintains a low $\Delta I / \Delta t$, with a maximum of only $2.6 \, \mu\text{A} \cdot \text{ns}^{-1}$ and a minimum that approaches $0 \, \mu\text{A} \cdot \text{ns}^{-1}$ (below 10^{-6} and therefore negligible). Notably, the proposed circuit's maximum $\Delta I / \Delta t$ is approximately 388 times lower than that of the conventional circuit.

To further assess the stability of the proposed circuit under various resistance-state distributions, scenarios are simulated in which the proportion of memory cells in the low-resistance ("1") state is set to 0%, 25%, 50%, 75%, and 100%. Since the read current stability improves over time (as shown in Fig. 4(b)), the midpoint of the read cycle (70 ns) is chosen as the basis for comparison. Fig. 4(c) illustrates the $\Delta I / \Delta t$ values of both circuits at 70 ns under these five resistance-state distributions. Overall, as the proportion of "1" states increases (i.e., as the overall resistance decreases), the $\Delta I / \Delta t$ in both circuits tends to decrease, indicating improved stability at higher read currents. Specifically, at 70 ns, the proposed circuit demonstrates superior stability under all five resistance-state distributions, exhibiting a $\Delta I / \Delta t$ approximately 10^4 times lower than that of the conventional circuit. Fig. 4(d) presents the operational waveforms of the circuit in a practical array configuration, thereby further validating the effectiveness of the proposed circuit. Table II compares our work with the previous works. Our approach achieves an energy efficiency of 8.26 Tops/W with 4-bit input and 4-bit weight precision.

979-8-3315-2209-4/25 $31.00 © 2025 IEEE   259

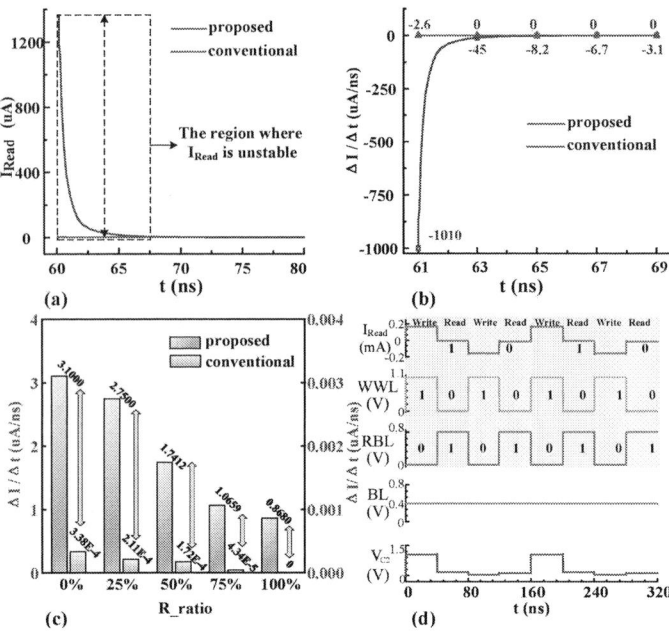

Fig. 4. Read Current and Stability Analysis. (a) Read current comparison, (b) Comparison of current fluctuation rates, (c) Read stability under various resistance states, (d) Circuit operating waveforms.

TABLE II
COMPARISON WITH PREVIOUS WORKS

	This work	**[12]**	**[13]**	**[14]**
Technology	40 nm	28 nm	28 nm	28 nm
Supply Voltage (V)	1.1	-	0.9	0.6-1.1
Weight Precision (bit)	4	8	7	8
Bit-Cell Type	2T	1T	-	8T
Efficiency (Tops/W)	8.26	2.24	1.02	5.27

IV. CONCLUSIONS

This paper presents a comprehensive investigation of SOT-MRAM-based IMC architecture, with a specific focus on charge-domain IMC scheme. By employing a 2T1MRAM cell and a high-sensitivity current-sensing circuit, read-current instability issues are effectively addressed, which are especially critical while integrating capacitors in IMC arrays. Specifically, the indirect monitoring of subtle current variations makes robust detection of the memory-cell state possible while preserving data accuracy. Furthermore, the CRICO unit demonstrates efficient MAC functionality through charge accumulation and pulse generation, thereby reducing the overall design complexity and power overhead.

Simulation results under a 40 nm process indicate that a read-current stability up to 10^4 times higher than conventional approaches is achieved, thereby enhancing overall reliability. In addition, a broad range of resistance state distributions is readily accommodated by the design without sacrificing performance. Taken together, these findings suggest a viable pathway toward high-precision, high-reliability IMC archi-

tectures based on SOT-MRAM, thereby providing a solid theoretical foundation and experimental validation for next-generation low-power, high-speed computing systems.

REFERENCES

[1] L. Liu, "Computing infrastructure for big data processing," *Frontiers of Computer Science (print)*, vol. 7, no. 2, pp. 165–170, Apr. 2013.

[2] G. Wang, Y. Lv, Y. Tian, J. Zhang, C. Guo, T. Bai, D. Wang, and W. Kang, "A 40nm 5-16tops/W@INT8 eFlash In-Memory computing SoC chip with noise suppression and compensation techniques to improve the accuracy," in *2023 IEEE International Conference on Integrated Circuits, Technologies and Applications (ICTA)*, Hefei, China, Oct. 2023, pp. 128–129.

[3] B. Pan, G. Wang, H. Zhang, W. Kang, and W. Zhao, "A mini tutorial of processing in memory: From principles, devices to prototypes," *IEEE Transactions on Circuits and Systems II: Express Briefs*, vol. 69, no. 7, pp. 3044–3050, Jul. 2022.

[4] M. Wang, W. Cai, D. Zhu, Z. Wang, J. Kan, Z. Zhao, K. Cao, Z. Wang, Y. Zhang, T. Zhang *et al.*, "Field-free switching of a perpendicular magnetic tunnel junction through the interplay of spin–orbit and spin-transfer torques," *Nature electronics*, vol. 1, no. 11, pp. 582–588, Nov. 2018.

[5] S. Peng, J. Lu, W. Li, L. Wang, H. Zhang, X. Li, K. Wang, and W. Zhao, "Field-free switching of perpendicular magnetization through voltage-gated spin-orbit torque," in *2019 IEEE International Electron Devices Meeting (IEDM)*. San Francisco, CA, USA: IEEE, Dec. 2019, pp. 28–6.

[6] I. M. Miron, K. Garello, G. Gaudin, P.-J. Zermatten, M. V. Costache, S. Auffret, S. Bandiera, B. Rodmacq, A. Schuhl, and P. Gambardella, "Perpendicular switching of a single ferromagnetic layer induced by in-plane current injection," *Nature*, vol. 476, no. 7359, pp. 189–193, Aug. 2011.

[7] H. Zhang, J. Liu, W. Kang, Y. Fan, S. Fu, J. Bai, B. Pan, Y. Liu, and W. Zhao, "A 40nm 33.6 tops/w 8t-sram computing-in-memory macro with dac-less spike-pulse-truncation input and adc-less charge-reservoir-integrate-counter output," in *2021 IEEE International Conference on Integrated Circuits, Technologies and Applications (ICTA)*. Zhuhai, China: IEEE, Nov. 2021, pp. 123–124.

[8] E. Grimaldi, V. Krizakova, G. Sala, F. Yasin, and P. Gambardella, "Single-shot dynamics of spin-orbit torque and spin transfer torque switching in three-terminal magnetic tunnel junctions," *Nature Nanotechnology*, vol. 15, no. 2, Jan. 2020.

[9] G. Hu, J. J. Nowak, M. G. Gottwald, S. L. Brown, B. Doris, C. P. D'Emic, P. Hashemi, D. Houssameddine, Q. He, D. Kim, J. Kim, C. Kothandaraman, G. Lauer, H. K. Lee, N. Marchack, M. Reuter, R. P. Robertazzi, J. Z. Sun, T. Suwannasiri, P. L. Trouilloud, S. Woo, and D. C. Worledge, "Spin-transfer torque MRAM with reliable 2 ns writing for last level cache applications," in *2019 IEEE International Electron Devices Meeting (IEDM)*, Dec. 2019, pp. 2.6.1–2.6.4.

[10] D. Apalkov, B. Dieny, and J. M. Slaughter, "Magnetoresistive random access memory," *Proceedings of the IEEE*, vol. 104, no. 10, pp. 1796–1830, Aug. 2016.

[11] C. Jiang, J. Li, H. Zhang, S. Lu, P. Li, C. Wang, Z. Zhong, Z. Hou, X. Liu, J. Feng, H. Zhang, H. Jin, G. Wang, H. Liu, K. Cao, Z. Wang, and W. Zhao, "Demonstration of a manufacturable SOT-MRAM multi-plexer array towards industrial applications," *Journal of Semiconductors*, vol. 44, no. 12, pp. 88–96, Dec. 2023.

[12] S. R. Kulkarni, D. V. Kadetotad, S. Yin, J.-S. Seo, and B. Rajendran, "Neuromorphic hardware accelerator for SNN inference based on STT-RAM crossbar arrays," in *2019 26th IEEE International Conference on Electronics, Circuits and Systems (ICECS)*, Nov 2019, pp. 438–441.

[13] S. Yin, S. K. Venkataramanaiah, G. K. Chen, R. Krishnamurthy, Y. Cao, C. Chakrabarti, and J.-s. Seo, "Algorithm and hardware design of discrete-time spiking neural networks based on back propagation with binary activations," in *2017 IEEE Biomedical Circuits and Systems Conference (BioCAS)*, Oct. 2017, pp. 1–5.

[14] J. Wang, X. Wang, C. Eckert, A. Subramaniyan, R. Das, D. Blaauw, and D. Sylvester, "A 28-nm compute SRAM with bit-serial logic/arithmetic operations for programmable in-memory vector computing," *IEEE Journal of Solid-State Circuits*, vol. 55, no. 1, pp. 76–86, Jan 2020.

Design of Low-power Spike-Time Dependent Plasticity Synaptic and Neuron Circuits on a SOI process

Qiaoyi Fu and Zhuojun Chen*
Hunan University
*Email: zjchen@hnu.edu.cn

Abstract—**Spiking Neural Networks (SNNs), as a type of brain-inspired computational neural network, exhibit powerful spatiotemporal information processing capabilities and low energy consumption characteristics. This paper presents a synaptic and neuronal circuit design based on a 0.18 μm SOI process, operating at a supply voltage range of 0.4 V to 0.8 V with transistors working in the subthreshold region. The synaptic circuit implements Spike-Timing-Dependent Plasticity (STDP), whose response time and weight variation range can be adjusted via four bias voltages, thereby extending the frequency adaptability. The neuronal circuit adopts the Leaky Integrate-and-Fire (LIF) model, featuring adjustable refractory periods, frequency adaptability, and tunable frequency, with a frequency tuning range of 4.76 kHz to 595.24 kHz at a 0.8 V supply voltage. At a 0.4 V supply voltage and 8 kHz frequency, the neuronal circuit consumes only 26.84 fJ per spike, and the total power consumption of the co-simulation circuit is 589 pW.**

Keywords—*Spiking Neural Network(SNN), Synaptic circuit, spike-time dependent plasticity (STDP)*

I. INTRODUCTION

In recent years, Artificial Neural Networks (ANNs) [1] have garnered significant attention in academia. However, ANNs rely on data-driven learning, which consumes substantial computational resources and struggles to accurately simulate the spatiotemporal correlation mechanisms of biological brains. In contrast, Spiking Neural Networks (SNNs), as a new generation of brain-inspired neural networks, offer greater biological plausibility and energy efficiency advantages [2]. Research on SNNs primarily focuses on two directions: software algorithms and hardware circuits, with hardware implementations being particularly favored due to their simple structure, high-speed operation, low power consumption, and large-scale integration capabilities.

Figure 1 illustrates the architecture of the SNN chip. Information is routed through Router R to axons, which then flow to neurons controlled by synapses. During each time step, if the synaptic value of a specific axon-neuron pair is non-zero and the axon is active, the neuron updates its state through synaptic weights, applies a leak mechanism, and triggers a spike. A Pseudo-Random Number Generator (PRNG) introduces noise to the spike threshold, while stochastic gate synapses and leak update probability calculation modules work in tandem. A buffer stores delayed input spikes, enabling the network to achieve information transmission between neurons and dynamic updates of synaptic weights.

Synaptic and neuronal circuits play a pivotal role in the hardware design of Spiking Neural Networks (SNNs), responsible for spike generation, transmission, reception, and integration. They enable inter-neuronal connections and

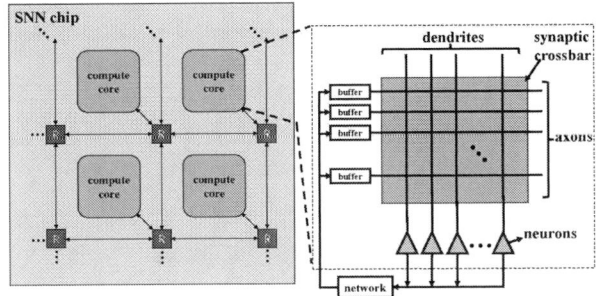

Fig.1 SNN chip architecture.

enhance network performance by optimizing energy consumption and latency. The design of these circuits must consider factors such as network scale, connectivity, and computational requirements to meet the demands of SNNs.

Hossein Eslahi et al. proposed a frequency-tunable neuronal circuit based on 22 nm FDSOI technology [3, 4]. This circuit leverages back-gate biasing of FDSOI for frequency tuning, offering low energy consumption and compact size. However, it lacks a refractory period function and does not demonstrate inter-neuronal information transfer. Arianna Rubino et al. implemented an efficient (low-power and slow-dynamic) analog neuronal circuit using 22 nm FDSOI technology [5, 6], optimizing both synaptic and neuronal circuits. Their mixed-signal analog/digital circuit mimics the dynamic properties of biological neurons and is optimized for large-scale SNN implementations in neuromorphic processors, albeit with a relatively large circuit footprint.

This paper presents a low-power Spike-Timing-Dependent Plasticity (STDP) synaptic circuit and a neuronal circuit with tunable refractory periods and frequency, designed using 0.18 μm SOI technology. The neuronal circuit exhibits complete biological functionality, and simulations demonstrate inter-neuronal information transfer, showcasing the ability of synaptic and neuronal circuits to emulate biological behaviors and transmit information via synaptic weights. Operating at a supply voltage of 0.4~0.8 V with transistors in the subthreshold region, the circuit achieves ultra-low energy consumption. At 0.8 V, the frequency tuning range is 4.76 kHz~595.24 kHz, while at 0.4 V and 8 kHz frequency, the energy consumption per spike is 26.84 fJ.

II. DESIGN OF SYNAPTIC AND NEURONAL CIRCUIT

A. Synaptic Circuit Design

The synaptic circuit converts the membrane voltage from the pre-neuron into current spike of corresponding intensity based on the weight magnitude. Its structure is shown in

979-8-3315-2209-4/25 $31.00 © 2025 IEEE

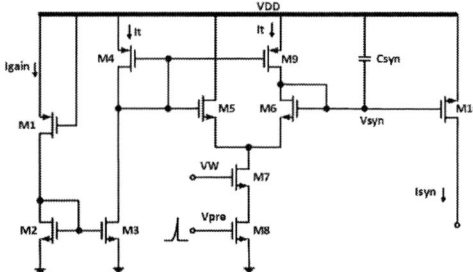

Fig.2 The proposed synaptic circuit.

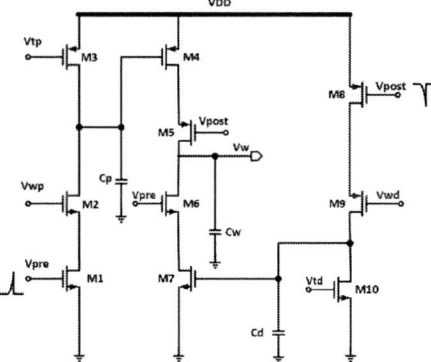

Fig.3 The proposed STDP circuit.

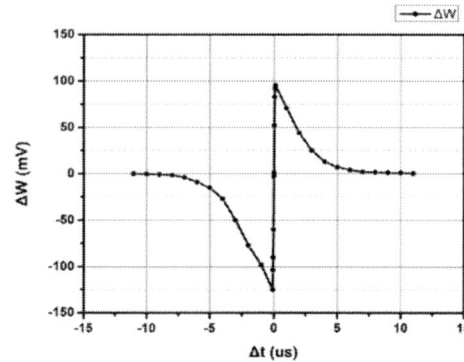

Fig.4 Numerical curves of Vw with the time difference between the front and rear neuron spikes.

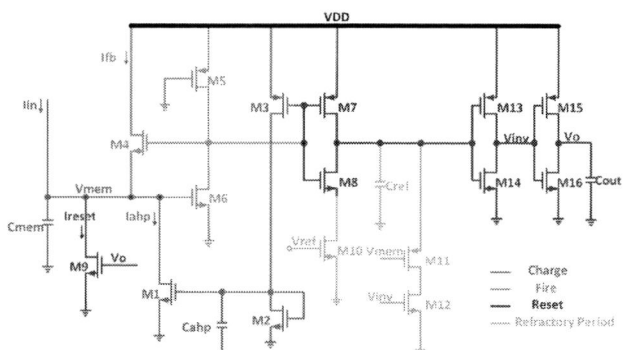

Fig.5 The proposed Neuron circuit

Figure 2. The reference current I_{gain} (approximately 1 pA) is amplified through a multi-stage current mirror, making I_t proportional to I_{gain}. Meanwhile, M3 and M4 provide an appropriate bias for M5 by setting suitable width-to-length ratios. When the pre-neuron spike signal V_{pre} is input, the capacitor C_{syn} begins to discharge, and M10 turns on, generating the synaptic current I_{syn}. When V_{pre} returns to a low voltage, I_t charges C_{syn} back to a high voltage, and M10 turns off. The weight voltage V_w regulates the discharge rate of the capacitor and the peak synaptic current, while the size of C_{syn} determines the charging time constant.

B. STDP Circuit Design

Spike-Timing-Dependent Plasticity (STDP) is crucial in artificial neural networks, as it mimics the memory and learning mechanisms of the human brain. The schematic of the STDP circuit is shown in Figure 3. V_w is adjusted by modulating the impedance of the transistor pairs above and below its node. When no spike signal is input, the circuit remains inactive.

The circuit features four tunable bias voltages, V_{wp}, V_{wd}, V_{tp}, and V_{td}, which control the charging and discharging processes of capacitors C_p and C_d. When the pre-neuron signal is input, C_p discharges and gradually recharges to a high voltage. If the post-neuron signal arrives while M4 is on, C_w charges, and V_w increases. If the post-neuron signal arrives first, C_w discharges, and V_w decreases. Figure 4 illustrates the variation curve of the synaptic weight V_w as a function of the time difference between pre- and post-neuron spikes.

C. Neuronal Circuit Design

The neuronal circuit proposed in this paper is designed using a 0.18 μm SOI process and adopts the Leaky Integrate-and-Fire (LIF) model to accommodate the current transmission characteristics of synapses. To minimize energy consumption, transistors operate in the subthreshold region. However, nanoscale CMOS devices exhibit significant leakage currents in the subthreshold region, while SOI technology significantly improves leakage and power consumption performance by introducing a buried oxide layer on the substrate, enhancing electrical isolation between the channel and substrate, thereby effectively reducing source-drain leakage, latch-up effects, parasitic capacitance, and substrate noise coupling. Additionally, forward or reverse biasing of the substrate terminal enhances channel and capacitance control, improving transistor drive performance or adjusting threshold voltage, providing unique design flexibility for back-gate modulation of device current. The circuit structure is shown in Figure 5, with the supply voltage V_{DD} set to 0.8 V to ensure MOS transistors operate in the subthreshold region.

Each spike consists of three main transient phases: charging, firing, and reset. The charging time is determined by the membrane capacitance C_{mem} and the difference between the synaptic current I_{in} (i.e., I_{syn}) and the current provided by the Afterhyperpolarization (AHP) mechanism, I_{ahp}. The AHP mechanism, composed of transistors M1-M3 and capacitor C_{ahp}, also regulates spike frequency. C_{mem} integrates the difference between I_{syn} and I_{ahp}, generating the membrane voltage V_{mem}. When $I_{in} > I_{ahp}$, V_{mem} rises from zero until it reaches the spike threshold V_{ths}.

When V_{mem} exceeds the threshold voltage, transistor M6 turns on, subsequently activating M4. The firing current I_{fb}

979-8-3315-2209-4/25 $31.00 © 2025 IEEE   262

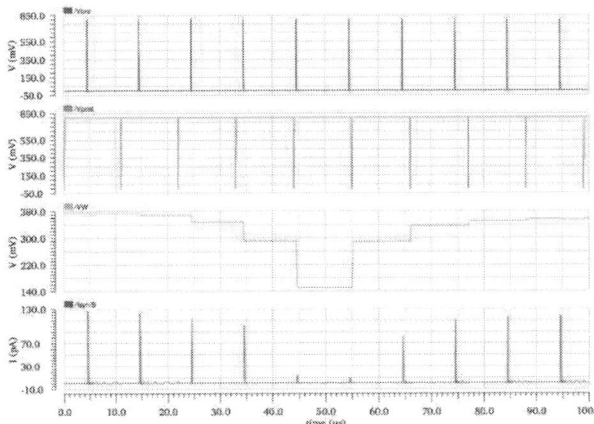

Fig.6 Simulation of the synaptic circuit with the STDP circuit.

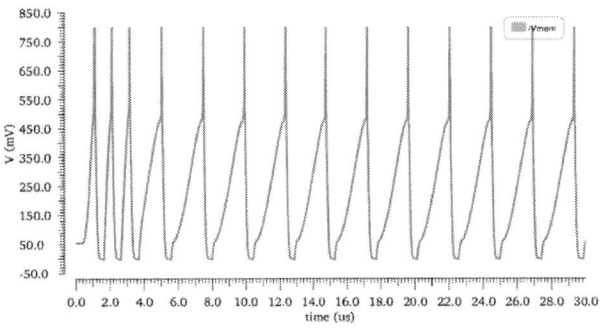

Fig.7 Simulation of the neuron circuit.

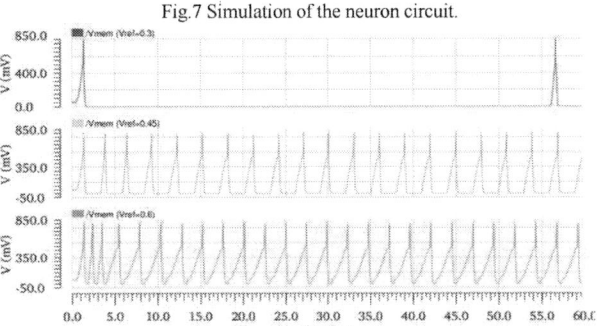

Fig.8 Simulation results of refractory period (V_{ref}=0.3 V (red), V_{ref}=0.45 V (yellow), V_{ref}=0.6 V (green)).

flows through M4, injecting additional charge into C_{mem} and pulling V_{mem} up to the supply voltage V_{DD}, forming a spike. Subsequently, after passing through three inverters (M7-M8, M13-M14, M15-M16), the output V_o drops to a low level, and the output capacitor C_{out} begins to charge. C_{out} must be larger than C_{mem} to ensure proper firing and reset operations. When C_{out} charges to a voltage sufficient to turn on M9, the reset process is completed.

The refractory period function is achieved by controlling the charging and discharging speed of capacitor C_{ref}, which regulates the output potential V_o and determines the conduction state of transistor M9. This mechanism adjusts the duration of the reset signal, thereby implementing the refractory period function. Additionally, by modulating the back-gate bias voltages V_{BN1} and V_{BN2} of M1 and M2, the threshold voltage of the transistors can be adjusted, altering the magnitude of the afterhyperpolarization current I_{ahp}. This further controls the charging time and modulates the spike frequency.

III. SIMULATION RESULTS

In this work, the STDP circuit was integrated into the basic synaptic circuit, and the time constant of the synaptic output current was adjusted to allow the synaptic current to decay to zero within a short period. By simulating pre- and post-neuronal spikes with varying time differences, the changes in the STDP circuit weight and synaptic current were observed, as shown in Figure 6. The experimental results demonstrate that when pre-neuronal spikes consistently precede post-neuronal spikes, the weight V_w gradually decreases, and the peak synaptic current also diminishes. Conversely, when post-neuronal spikes consistently precede pre-neuronal spikes, the weight V_w gradually increases, and the peak synaptic current rises accordingly.

The waveform of the neuron circuit was simulated with V_{DD} set to 0.8 V, an input current of 10 nA, initial values of V_{BN1} and V_{BN2} at 0 V, and V_{ref} at 0.8 V. The simulation results are shown in Figure 7. The results indicate that the neuronal waveform primarily consists of three transient phases: charging, firing, and resetting. When the input current exceeds the threshold voltage V_{ths} (approximately 450 mV), the neuron circuit generates a spike signal. During the charging phase, the membrane capacitor C_{mem} is gradually charged to V_{ths}, a process that lasts for a relatively long time. In the firing phase,

with the assistance of the feedback current I_{fb}, C_{mem} is rapidly charged, producing a spike signal. During the reset phase, the reset signal triggers M9 to conduct, causing C_{mem} to discharge quickly and the neuron circuit to return to its initial state.

The refractory period functionality of the neuron circuit was simulated with a V_{DD} of 0.8 V and an input signal of 10 nA current. The initial values of V_{BN1} and V_{BN2} were set to 0 V, and V_{ref} was set to 0.3 V, 0.45 V, and 0.6 V, respectively. The results are shown in Figure 8.

From the simulation waveform, it can be observed that when V_{ref} is 0.3 V and 0.45 V, the V_{mem} voltage remains at 0 V after a neural spike is generated, indicating that the refractory period functionality is active. As V_{ref} increases, the discharge rate of C_{ref} gradually accelerates, leading to a gradual shortening of the refractory period. When V_{ref} reaches a certain value (0.6 V in this case), the conduction level of transistor M10 essentially reaches its maximum, and the discharge rate of C_{ref} also approaches its maximum. At this point, the refractory period becomes nearly zero, and V_{mem} transitions into continuous spikes.

The synaptic circuit, STDP circuit, and neuronal circuit have been designed and analyzed in previous sections, demonstrating their biologically plausible characteristics. This section presents the co-simulation of these three circuits, with the structure shown in Figure 9.

Two neuronal circuits are connected via a synaptic circuit with an STDP module to observe the effect of spike timing differences on synaptic weights. Since the gate capacitance generates high-frequency currents under high-frequency V_{pre} inputs, causing significant fluctuations in bias voltages, four differential negative feedback operational amplifiers are added to the bias circuit for stabilization.

To better observe circuit characteristics, initial simulations used a 0.8 V supply voltage with all back-gate voltages set to

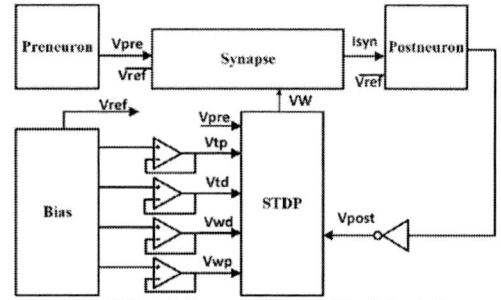

Fig.9 Circuit diagram of the combined simulation.

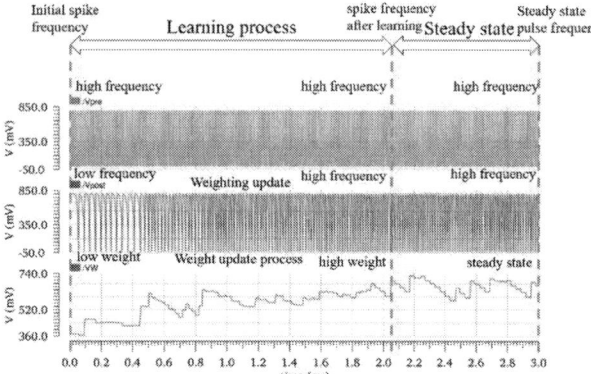

Fig.10 Comprehensive simulation results.

0 V. To reduce energy consumption, the supply voltage V_{DD} was adjusted to 0.4 V in the co-simulation, with appropriate back-gate voltages tuned to ensure proper functionality. The simulation results are presented in Figure 10. The pre-neuron generates high-frequency spikes, while the post-neuron produces low-frequency spikes after receiving the integrated current from the synaptic circuit. The spike timing difference between the pre- and post-neurons causes synaptic weight changes, which in turn modulate the synaptic current and adjust the post-neuron's spike frequency, gradually aligning it with the desired value. The simulation results show that the spike frequency of the post-neuron converges toward the one of the pre-neuron, with synaptic weights fluctuating within a reasonable range. This successfully mimics the learning behavior of neurons receiving signals from pre-neurons and validates the information transmission process of neuron-synapse-neuron.

The STDP circuit proposed in this study features adjustable response time range and weight change range. By tuning the parameters V_{td}, V_{wd}, V_{tp}, and V_{wp}, the frequency applicability can be extended, making it more suitable for synaptic signal transmission between different neurons. As shown in Table 1, compared to existing neuron circuits in the literature, the proposed design offers more comprehensive functionality. The adjustable refractory period, a fundamental physiological feature of neurons, allows for more accurate simulation of refractory characteristics in neurons with different functions. Additionally, the adaptive frequency modulation enables the neuron to automatically adjust its firing rate based on task requirements or the strength of presynaptic signals. When the weight is low, the input signal current decreases, leading to a reduction in the neuron's adaptive frequency, thereby achieving low-power operation that more closely aligns with the biological characteristics of neurons.

TABLE I. PROPOSED NEURON CIRCUIT COMPARED WITH OTHER WORKS

	[3]	[5]	[6]	This work
Technology	28nm FDSOI	22nm FDSOI	22nm FDSOI	0.18um SOI
Vdd	1.0V	0.3V	0.8V	0.8V
Refractory period	not adjustable	unsupportable	adjustable	adjustable
Adaptive frequency	not adjustable	adjustable	adjustable	adjustable

IV. CONCLUSION

This study presents a synaptic circuit with STDP functionality and a neuron circuit featuring adjustable refractory period and adaptive frequency modulation, designed based on a 0.18 μm SOI process. Through simulation experiments, the synaptic weight modulation capability of the synaptic circuit, as well as the refractory period adjustment and adaptive frequency modulation of the neuron circuit, were validated. Furthermore, co-simulation of the neuron and synaptic circuits was conducted to verify the information transmission process in the neuron-synapse-neuron pathway, demonstrating the learning behavior of the neuron upon receiving signals from a presynaptic neuron. Compared to existing works, the proposed design exhibits significant advantages in power efficiency. The simulation results effectively demonstrate the functional validity and superior low-power characteristics of the proposed circuits.

REFERENCES

[1] Hopfield J J. Neural networks and physical systems with emergent collective computational abilities[J]. Proceedings of the national academy of sciences, 1982, 79(8): 2554-2558.

[2] Yu Q, Tang H, Hu J, et al. Rapid feedforward computation by temporal encoding and learning with spiking neurons[J]. Neuromorphic Cognitive Systems: A Learning and Memory Centered Approach, 2017: 19-41.

[3] Qiao N, Indiveri G. Scaling mixed-signal neuromorphic processors to 28 nm FD-SOI technologies[C]//2016 IEEE Biomedical Circuits and Systems Conference (BioCAS). IEEE, 2016: 552-555.

[4] Eslahi H, Hamilton T J, Khandelwal S. Compact and energy efficient neuron with tunable spiking frequency in 22-nm FDSOI[J]. IEEE Transactions on Nanotechnology, 2022, 21: 189-195.

[5] Rubino A, Payvand M, Indiveri G. Ultra-low power silicon neuron circuit for extreme-edge neuromorphic intelligence[C]//2019 26th IEEE International Conference on Electronics, Circuits and Systems (ICECS). IEEE, 2019: 458-461.

[6] Rubino A, Livanelioglu C, Qiao N, et al. Ultra-low-power FDSOI neural circuits for extreme-edge neuromorphic intelligence[J]. IEEE Transactions on Circuits and Systems I: Regular Papers, 2020, 68(1): 45-56.

A Digital LDO Designed to Enhance the Side Channel Security with Random Noise Injection Technique

Xiaodan Gu, Jiaji He, Yu Long, Mao Ye*

School of Microelectronics, Tianjin University, Tianjin, 300072, China

*Email: mao_ye@tju.edu.cn

Abstract—To address security issues arising from side-channel attack (SCA), this paper proposes a dedicated digital low dropout regulator (DLDO) to protect the encryption engine. Taking advantage of the discrete nature of the DLDO, a parallel random noise injector circuit is integrated to randomize the current drawn from the external supply. The switched capacitor based injector is controlled with the on-chip random generator. The proposed architecture is implemented in a 55-nm CMOS process and operates at 40 MHz. With a 5 mA/0.1 ns load step and 50 pF capacitor, the DLDO has an undershoot voltage of 77.4 mV and a peak current efficiency of 99.9%. The power overhead of the protected 128b advanced encryption standard (AES) is 18.07%. Compared to an unprotected AES module for 100K-trace of correlation power analysis (CPA), the preliminary simulation result indicated that the protected AES can maintain security and the peak t-statistic in test vector leakage assessment (TVLA) reduces by 8.11×.

Keywords—Low dropout regulator (LDO), advanced encryption standard (AES), side-channel attack (SCA), correlation power analysis (CPA).

I. INTRODUCTION

With the rapid development of information technology, encryption engines are widely used in Internet of Things (IoT) devices due to their key management mechanisms and encryption algorithms. However, SCAs exploit physical signals such as power consumption and electromagnetic radiation leaked during the operation of encryption engines to steal secret information, posing a serious threat to device security.

In recent years, some protection schemes have enhanced the resistance of encryption engines to SCAs by using voltage regulators with protection functions. Previous works [1], [2] leveraged a current equalizer to balance the current drawn from the external supply, thus protecting circuit timing information and resisting CPA attacks. However, this method involves discharging capacitors to a predefined voltage, leading to increased power overhead and making it unsuitable for resource-constrained IoT devices. Moreover, multiple switching capacitors may cause unstable line regulation. [3], [4] adopted noise injection and randomization of reference voltage and resistance to confuse the current signature, improving the SCA resistance of AES chips. However, the noise current injected into the ground increases the power overhead.

In this paper, we propose a low-power DLDO with side-channel security, which includes a regulator core and a random

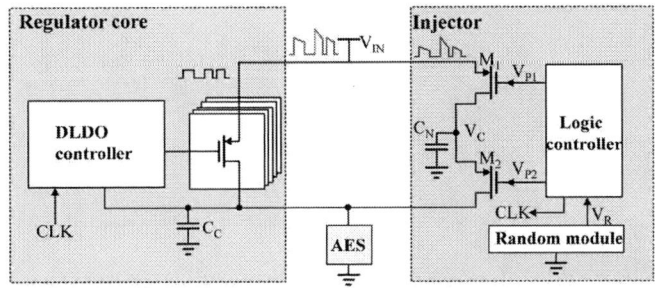

Fig. 1. Block diagram of the DLDO with Random Noise Injection Technique.

noise injector. This design not only enhances the defense against SCAs by injecting random noise without modifying the existing computational architecture and algorithms of the encryption engine, but also reuses the charge stored in the switching capacitors to reduce power overhead.

The structure of this paper is organized as follows : Section II introduces the system architecture and circuit implementation of the DLDO with random noise injection technique. Section III presents simulation results. Conclusions are drawn in Section IV.

II. SYSTEM ARCHITECTURE AND CIRCUIT IMPLEMENTATION

The overall DLDO with random noise injection (RNI) technology includes a regulator core and an injector.

A. Random Noise Injector

During CPA, the current signature of the encryption engine propagates through the PMOS transistor array of the DLDO to the measurement node, leaking circuit information. In order to resolve this issue, we propose an injector connected in parallel with the regulator core. Inject a random noise current during the operation of the encryption engine, preventing attackers from directly obtaining the operational information of the encryption engine from the external power supply terminal.

The injector consists of a 200 pF capacitor, two switches, and a logic controller, as shown in the right part of Fig. 1. When the on-chip random signal generator produces a pulse, switch M_1 closes and switch M_2 opens, thereby commencing the noise injection process. A portion of the input current

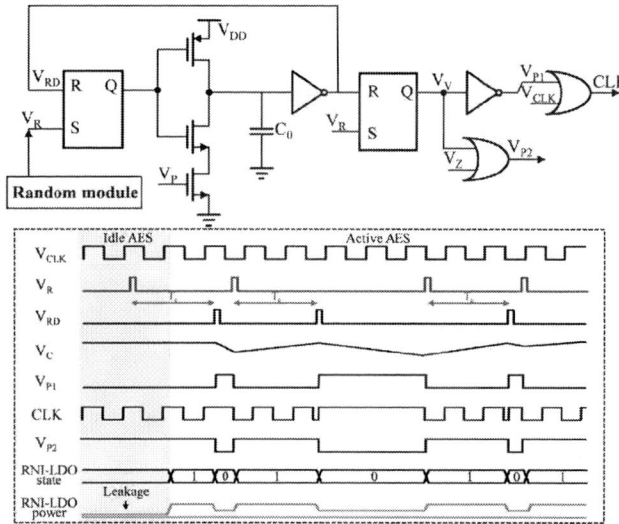

Fig. 2. Schematic and timing diagram of the logic controller operation.

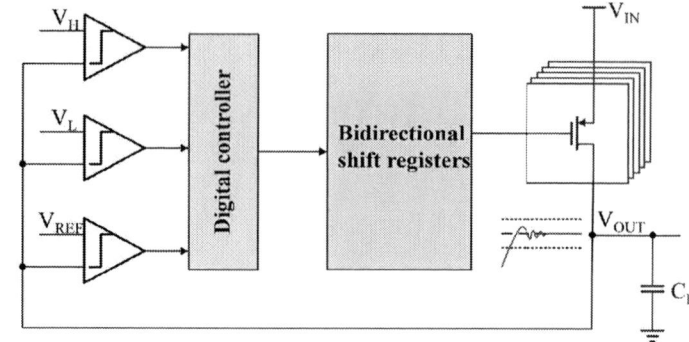

Fig. 3. Block diagram of the proposed regulator core.

provided by the external power supply is used to power the encryption engine through the regulator chip, while another portion serves as the noise current to charge the capacitor in the injector. After the charging period ends, the logic controller disconnects the injector from the external power supply. At this moment, the encryption engine simultaneously receives current from both the regulator core and the previously charged capacitor in the injector. The entire process blurs the current signature by injecting noise, effectively reducing the correlation between the externally provided current and the current demand of the encryption engine. Furthermore, compared to previous methods of injecting noise and current into the ground, it is more effective in reducing additional power overhead.

B. Logic Control Section

As shown in Fig. 2, when the random generator produces a pulse that activates the output voltage V_v of the trigger, the control voltage V_{p1} of the injector goes low, while V_{p2} goes high. The clock control of the regulator follows the external clock, independently supplying current to the encryption engine. At this time, the input current of the power supply includes the current required by the encryption engine and a noise current. After time T_c, V_v goes low, causing the controller to open switch M_1 and close switch M_2. The regulator clock CLK is latched high, meaning that the internal controller of the regulator continues to maintain the output before locking. This operation effectively masks the load current signatures, reducing information leakage.

C. Regulator Core

Low dropout regulator (LDO) is commonly used for power management in system-on-chip (SoC) due to its compact size, high current efficiency, and low output ripple. Compared to analog LDOs, digital LDOs with discrete characteristics

can operate at sub-threshold voltages and are becoming increasingly popular due to process scalability. Integrating a voltage regulator with the encryption engine helps provide faster transitions during significant current fluctuations and isolates the load from the measurement node.

Fig. 3 illustrates the block diagram of the voltage regulator, which can be divided into coarse and fine tuning parts. The coarse tuning part consists of a 20-bit shift register and an array of PMOS power transistors, while the fine tuning part consists of an 8-bit shift register and a PMOS array. C_L is 50 pF. Moreover, one comparator adjusts the direction of bidirectional shift registers, while the other two define the voltage range for trimming. When the output voltage exceeds the specified range, the controller initiates coarse tuning, providing the system with a faster transient response and mitigating overshoot and undershoot. Once the output voltage falls within the specified range, the controller enables fine tuning and generates a more precise output voltage, thus improving the power stability of the encryption engine.

III. SIMULATION RESULTS

In a 55-nm CMOS process, the proposed more secure DLDO structure powered by 1.2 V, provides a 1.1 V operating voltage to the encryption engine. Fig. 4(a) shows the transient waveforms of the regulator core. When the step current increases from 7 to 12 mA within 0.1 ns, the regulator exhibits an undershoot voltage of 77.4 mV with a stabilization time of 145 ns. Fig. 4(b) illustrates that the regulator core achieves a peak current efficiency of 99.9%.

Using Primetime PX to perform power analysis on 20 MHz AES, 100 K power traces under different plaintexts are collected. After data processing, these traces are used as the AES load for the DLDO. Based on circuit simulation results, we conduct a new CPA of the DLDO power curve. Fig. 4(d) shows that for the baseline AES, fewer than 200 power traces are sufficient to recover the correct encryption key. In contrast, after applying protection, the correlation coefficient decreases, making the key more difficult to crack, as shown in Fig. 4(e). Fig. 4(c) presents that the peak t statistic in TVLA for protected AES is reduced by 8.11× compared to unprotected AES. The preliminary measurement in Fig. 4(f)

979-8-3315-2209-4/25 $31.00 © 2025 IEEE

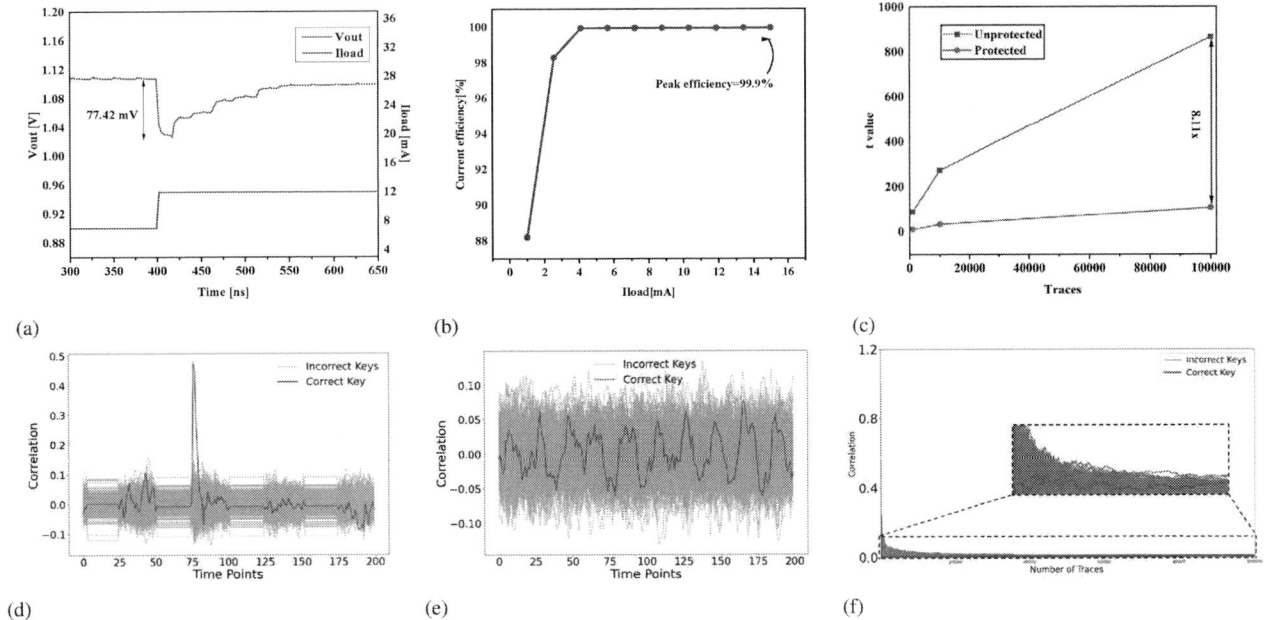

Fig. 4. Simulation results. (a) The transient waveforms of the regulator core when the load current changes from 7 to 12 mA. (b) The current efficiency of regulator core. (c) TVLA for unprotected AES and protected AES. (d) Correlation power analysis of the baseline AES. (e) Correlation power analysis of the protected AES. (f) MTD for CPA against the protected AES.

TABLE I
PERFORMANCE SUMMARY AND COMPARISON

	This work	JSSC 2024 [2]	JSSC 2020 [3]	ISSCC 2020 [4]
Process	55 nm	65 nm	130 nm	65 nm
Type	Randomizer	Equalizer	Randomizer	Quantizer
Attack Mode	TVLA,CPA	TVLA,CPA	TVLA,CPA	TVLA,CPA
VDROOP @I_L/T_{edge}	77.4 mV @ 5mA/0.1ns	46.4 mV @ 386uA/<0.1ns	235mV @ 40mA/<100ps	101 mV @ 20mA/506ns
Current Efficiency	99.9%	99.9%	N/A	99.4%
Power Overhead	18.07%	27.6%	32%	19.4%
TVLA Improvement	8.11×	13.36×	25×	18.42×

shows that the correct encryption key remains undisclosed even after analyzing 100 K power traces, achieving a more than 500× improvement in minimum-traces-to-disclose (MTD) and confirming its reliability as a security safeguard. Table 1 compares this work with other related works.

IV. CONCLUSION

The article presents a DLDO with a random noise injection technique, significantly enhancing the resistance of AES to SCAs. Under a load step condition of 5 mA/0.1 ns, the droop voltage of the regulator core is 77.4 mV, and the peak current efficiency is 99.9%. CPA demonstrates that the protected 128-bit AES achieves over 500× improvement in MTD. Moreover, we show that TVLA leakage is reduced by 8.11×, while the power overhead is controlled at 18.07%.

ACKNOWLEDGMENT

This work is supported by the National Key R&D Program of China under Grant 2023YFB4402800.

REFERENCES

[1] C. Tokunaga and D. Blaauw., "Secure AES engine with a local switched-capacitor current equalizer," in *IEEE International Solid-State Circuits Conference - Digest of Technical Papers*, pp. 64-65, Feb. 2019.

[2] M. Li et al., "EQZ-LDO: A Secure Digital Low Dropout Regulator Armed With Detection-Driven Protection Against Correlation Power Analysis," in *IEEE Journal of Solid-State Circuits*, vol. 59, no. 11, pp. 3806-3815, Nov. 2024.

[3] A. Singh et al., "Enhanced Power and Electromagnetic SCA Resistance of Encryption Engines via a Security-Aware Integrated All-Digital LDO," in *IEEE Journal of Solid-State Circuits*, vol. 55, no. 2, pp. 478-493, Feb. 2020.

[4] Y. He and K. Yang, "25.3 A 65nm Edge-Chasing Quantizer-Based Digital LDO Featuring 4.58ps-FoM and Side-Channel-Attack Resistance," in *IEEE International Solid-State Circuits Conference*, pp. 384-386, Feb. 2020.

Sparse Diffusion Accelerator with Pattern Pruning, Dynamic Detection and Quantization

Boran Cao   Sheng Zhang   Chen Tang   Xinyuan Lin   Leran Huang   Wentao Zhao   Yongpan Liu

Abstract—**Diffusion models, as the most advanced models in the fields of Artificial Intelligence Generated Content (AIGC), have excellent generative performance. However, diffusion models are characterized by large parameters and high computational costs, requiring large-scale storage and high-performance computing, which limits their application on resource-constrained edge devices. In this paper, we design a diffusion accelerator using a hardware-software collaborative approach to address this issue. We propose pattern pruning to achieve unified sparsity across multiple operators, apply sparsity in the Attention layer by selecting the Top4 elements, and implement quantization for diffusion models. The accelerator is in 28nm CMOS technology and achieves an energy efficiency of 12.88 TOPS/W and a peak performance of 1.61 TOPS. Compared with prior works, the accelerator achieves speedups of 1.4×-7.3× while improving energy and area efficiency by 1.1×-6.5× and 3.7×-26.5×, respectively.**

Index Terms—**diffusion model, pattern pruning, quantization-aware-training, dynamic detection, accelerator**

I. INTRODUCTION

Currently, diffusion models [1-3] have become state-of-the-art (SOTA) models for generative tasks, exhibiting outstanding performance in fields such as text-to-image and text-to-video generation. These domains demand low latency and high precision, which traditional hardware accelerators, such as CPUs and GPUs, fail to meet. In this paper, we propose an Application-Specific Integrated Circuit (ASIC) tailored to accelerate the DDPM-IP [4] model. The contributions of this work are as follows.

We propose a pattern pruning method to achieve a sparsity of 89% in the diffusion model. Additionally, we employ quantization-aware training (QAT) to quantize the model's data across all layers from FLOAT32 and FLOAT16 to INT10 while implementing Top4 dynamic detection in the Attention layer. As a result, the final model demonstrates an FID improvement of less than 0.6, ensuring the preservation of high-quality generative performance. Furthermore, we design an efficient hardware accelerator tailored to this algorithm.

The rest of this paper is organized as follows: Section II is a brief introduction to the architecture of diffusion models, weight pruning and data quantization. Section III describes the architecture of the whole accelerator, and the specific design details of its modules. In Section IV, the evaluation results and some conclusions are presented.

Author 1, Author 2 and Author 5 are with *Shenzhen International Graduate School, Tsinghua University*, Shenzhen, China. Author 3, Author 4, Author 6 and Author 7 are with *Department of Electronic Engineering, Tsinghua University*, Beijing, China. Corresponding author: (Author 7, email: ypliu@tsinghua.edu.cn)

II. DIFFUSION MODEL COMPRESSION

A. Diffusion Model Architecture

We have selected the DDPM-IP model, the structure of which is shown in Fig. 1. This diffusion model takes random noise as input and generates the target image through a denoising process at each step. The denoising process is a Markov chain process completed by a U-Net structure, which consists of ResNet and Attention blocks. It can be seen that the operators Conv2d, Conv1d, and Linear occupy the majority. Therefore, we aim to compress the parameters of these three operators.

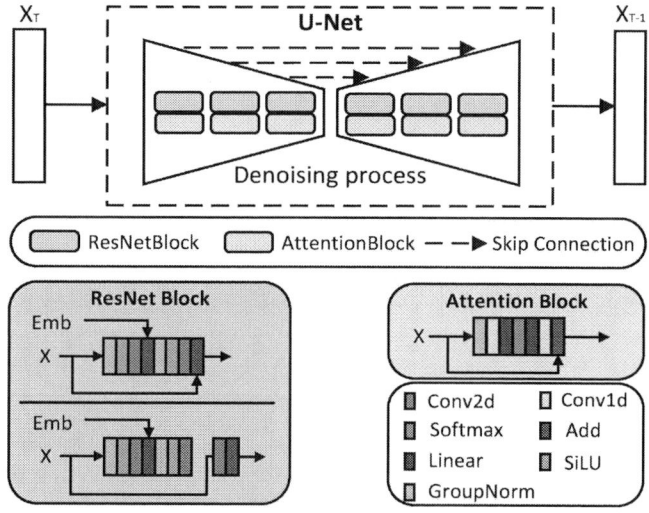

Fig. 1. Architecture of diffusion model

B. Pattern Pruning

For model compression, common pruning methods include structured pruning, unstructured pruning, and pattern pruning [5]. Compared to unstructured pruning, pattern pruning is easier to deploy and consumes fewer hardware resources; compared to structured pruning, it offers finer granularity and is easier to preserve accuracy. To implement the computation of three types of operators on the same hardware architecture while maintaining algorithmic accuracy, we choose pattern pruning. In this method, we group every 16 data points in the input channel dimension of the three operators for pattern pruning, as shown in Fig. 2.

During the training process, we select the 10 most frequently occurring patterns as the pattern set. Then, we use

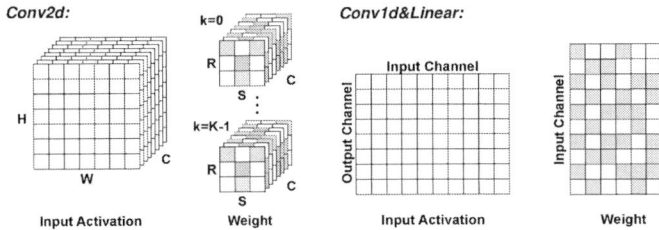

Fig. 2. Pattern pruning of operators

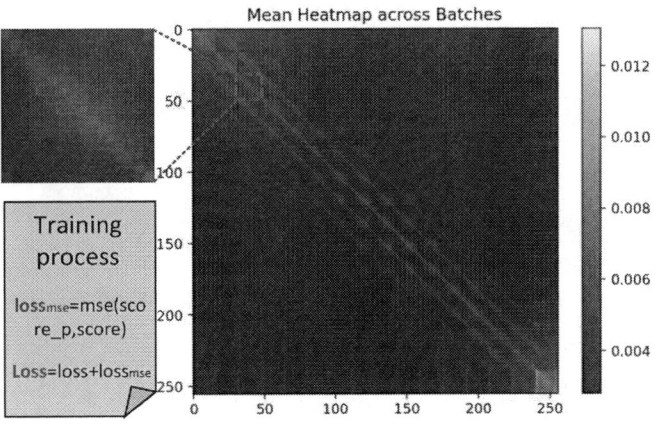

Fig. 3. Attention score heatmap

(1) to calculate the distance between other patterns and the 10 patterns in the pattern set, selecting the pattern with the smallest distance for matching. This helps to fix the number of patterns, reduce the model error caused by pruning, and aids the model in converging during the training process. Where k is original pattern vector and p is chosen pattern vector.

$$D_c(k, p) = 1 - \frac{k \cdot (k \otimes p)}{\|k\|_2 \cdot \|k \otimes p\|_2} \quad (1)$$

Additionally, we found that the sparsity level of different layers in the diffusion model has varying impacts on the final generation effect. The skip connection layers, which are significantly affected, are set to a lower sparsity level of 50%, while other layers are set to a sparsity level of 93.75%. Ultimately, the overall model's weight parameter sparsity reaches as high as 89%, with an increase in the FID value by 0.26, as shown in Table I.

TABLE I
ACCURACY WITH DIFFERENT METHODS

Model Process	Weight Precision	Activation Precision	Training Steps	FID ↓
Baseline	FLOAT32& FLOAT16	FLOAT32& FLOAT16	12w	3.25
Pruning(89%)	FLOAT32& FLOAT16	FLOAT32& FLOAT16	19w	3.51
Pruning+QAT	INT10	INT10	10w	3.60
Pruning+QAT +Top4	INT10	INT10	12w	3.76

C. Top 4 in Attention Block

In the Attention Block, the product of the Query and Key undergoes softmax to obtain a 256x256 Attention score matrix, as shown in the heatmap in Fig. 3. It can be observed that these data are automatically divided into 16x16 small matrix blocks. We extract the top 4 largest values from each row of the 16 data points, retaining them, and setting the remaining values to 0. Since the Attention score matrix is generated online, dynamic monitoring is required in real time. We add self.mse_loss = F.mse_loss(weight_partial, weight) to the total loss function for training. This approach reduces 75% of the parameters in the Attention score, and after training, the model's accuracy is not significantly impacted. This differs from offline weight parameter pruning.

D. Model Quantization

The two common quantization methods are post-training quantization (PTQ) and quantization-aware training (QAT). Although QAT requires retraining, it ensures higher model accuracy. Since our work focuses on designing inference chips, we use QAT to maintain the model's generation performance. In this approach, we need to separately quantize the model's weights and activations, and incorporate the quantization-dequantization algorithm into the model. During training, the model parameters learn the informational features brought by quantization errors, which makes the quantized model more stable in terms of performance. We quantize the model data from FLOAT32 and FLOAT16 precision to INT10 precision, which reduces the complexity of hardware design. The experimental results after training are shown in Table I.

III. ACCELERATOR IMPLEMENTATION

Fig. 4 shows the overall architecture of the proposed accelerator. It consists of a Memory buffer, Top-controller, PE Array, Quantization module, and Top4 module. The size of the Memory buffer is 2.23MB, which stores the weight parameters, mode information, and activation values required by one layer of the operator, which will reduce off-chip memory access. The details of each submodule are described below.

A. Unified Operator Module

Conv2d, Conv1d, and Linear operators have different computation sequences and accumulation methods. In the algorithm, we have already performed pattern pruning on the input channel dimension for these three operators, obtaining the data flow for the operators' computations. Therefore, to improve the chip's computational efficiency, we propose a method where, based on the data flow of these three operators, offline weight data, pattern information, and input activation values are stored in the corresponding memory buffers according to the sequence after the data flow split through a software script. This eliminates the need to design complex hardware logic to split unordered data and reorder it according to the operators'

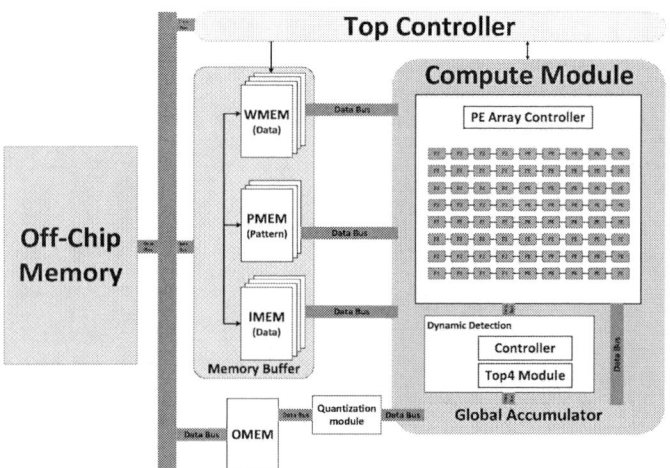

Fig. 4. Overall architecture of accelerator

computation sequence. This approach reduces the complexity of hardware control, avoids consuming a large amount of hardware resources, and improves computational efficiency.

Fig. 5 shows the hardware module architecture used for computing Conv2d, Conv1d, and Linear operators. The Controller is designed according to the characteristics of the data flows of these three operators, and it can configure the system accordingly to complete the computation tasks via the PE array. The PE array consists of 8x16 MACs, offering good spatial parallelism, enabling 256 multiply-accumulate operations to be performed simultaneously.

In the computation process, the pattern information marks which activation values need to be involved in the calculation, thereby avoiding unnecessary computations on all the data and saving computational resources. The effective activation values and weight values are multiplied and accumulated in the specified order to obtain the partial sum. The computation results are passed through the Quantization module and then output to the OMEM to complete the computation.

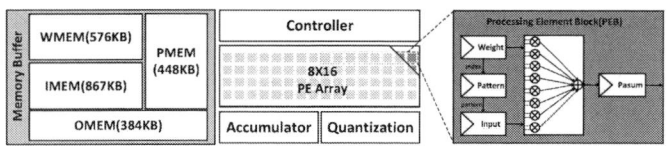

Fig. 5. Architecture of unified operator module

B. Dynamic Detection and Quantization Module

In the Attention Block, the product of the Query and Key is generated dynamically. Fig. 6 shows the working principle of dynamic detection, where the 4 largest values and their corresponding positions are selected from every set of 16 numbers. The specific process is as follows: First, the input 10-bit score data is combined with its corresponding 4-bit position information to form a new 14-bit value, denoted as Comb. Next, the Comb value is compared one by one with the other 15 values in the same group. If the current Comb

value is greater than another value, the corresponding Sort flag is incremented by 1; if it is less than or equal to another value, the Sort flag is incremented by 0. By comparing all 16 values, the sorting order of the 16 input values can be accurately determined. Based on this sorting, we select the top 4 largest values, each of which contains not only the numerical data but also its corresponding position information.

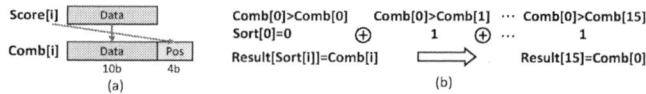

Fig. 6. Dynamic detection processing flow

In Fig. 7, the controller controls the data flow of the Attention deployment to the PE array to complete the computation. For the quantization module, the precision is defined as INT10. By multiplying the given scaling factor with the input INT32 data and then performing a shift operation, the resulting value is in INT10 precision. The scaling factor is calculated during the preprocessing stage through statistical analysis of hundreds of sample outputs to determine the average of the maximum values of each layer's output, thereby deriving the scaling factor for that layer. The shift value is determined based on the most significant bit of the scale and the representational range of the INT10 precision. During the quantization process, the input data is multiplied by the scaling factor to obtain the quantized data. Then, the quantized data is processed through a decision selection module to ensure its value is within the representable range of INT10, which is between the maximum value of 1023 and the minimum value of -1024. If the quantized data exceeds this range, truncation is performed to prevent overflow or underflow, ensuring the accuracy and stability of the computation results.

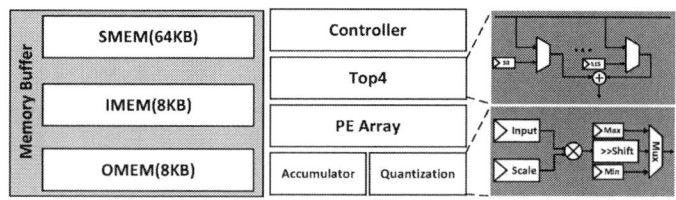

Fig. 7. Architecture of attention module

IV. EVALUATION

The simulation results of this diffusion model accelerator are based on a 28nm CMOS technology, with an area of 4.14 mm^2. It operates at 0.9V supply voltage under 50-630 MHz clock frequency. It achieves 1.61 TOPS peak performance, 12.88 TOPS/W energy efficiency, and 0.389 TOPS/mm^2 area efficiency with INT10×INT10 operation for diffusion on the CIFAR-10 dataset. Table II shows the comparison to prior works. We normalize peak performance, energy efficiency, and area efficiency based on the different CMOS technology nodes used in the prior works. Our work shows an 1.4×-7.3× speed-up, 1.1×-6.5× higher energy efficiency, and 3.7×-26.5× higher area efficiency.These results indicate that our designed accelerator has significant optimization potential in terms of computational performance, energy efficiency, and area utilization, providing strong hardware support for future diffusion model acceleration.

Our work primarily conducted experimental validation on the basic DDPM and DDPM-IP models. However, with the rapid development of artificial intelligence algorithms, more efficient generative models may emerge in the future, which could further improve training and inference speeds at the algorithmic level. Therefore, there is still significant room for improvement in optimizing diffusion model algorithms, and future exploration of more advanced algorithms could enhance the performance of accelerators.

TABLE II
COMPARISON WITH PRIOR WORKS

	HPCA'21[6]	JSSC'22[7]	ISSCC'23[8]	This Work
Technology	40nm	16nm	12nm	**28nm**
Supported Networks	Transformer	Attention,RNN, Linear	Transformer	**Diffusion Model, Transformer, Attention**
Data Precision	INT9	FP8	FP4 / FP8	**INT10**
Die Area	2.08 mm^2	8.84 mm^2	4.6 mm^2	**4.14 mm^2**
Core Frequency	1000MHz	130- 573 MHz	77 - 717 MHz	**50- 630 MHz**
Peak Performance [T(FL)OPS]	0.22	1.17	0.367	**1.61**
Energy Efficiency[a] [T(FL)OPS/W]	1.99	4.46	12.06 (FP8)	**12.88**
Area Efficiency[a] [T(FL)OPS/mm^2]	0.105	0.0432	0.0147	**0.389**

[a]Normalized to 28nm technology node: Energy efficiency∝ (Technology / 28), Area efficiency ∝(Technology / 28)^2

REFERENCES

[1] R. Rombach, A. Blattmann, D. Lorenz, P. Esser and B. Ommer, "High-resolution image synthesis with latent diffusion models," Proceedings of the IEEE/CVF conference on computer vision and pattern recognition (CVPR), 2022.

[2] J. Ho, et al. "Video diffusion models," Advances in Neural Information Processing Systems 35 (NeurIPS), 2022.

[3] C. Saharia, et al. "Photorealistic Text-to-Image Diffusion Models with Deep Language Understanding," Advances in neural information processing systems (NeurIPS), 2022.

[4] M. Ning, E. Sangineto, A. Porrello, S. Calderara and R. Cucchiara, "Input perturbation reduces exposure bias in diffusion models," In Proceedings of the 40th International Conference on Machine Learning(ICML), 2023.

[5] Wang J, Yu S, Yue J, et al. High pe utilization CNN accelerator with channel fusion supporting pattern-compressed sparse neural networks[C]//2020 57th ACM/IEEE Design Automation Conference (DAC). IEEE, 2020: 1-6.

[6] Ham T J, Jung S J, Kim S, et al. A^3: Accelerating attention mechanisms in neural networks with approximation[C]//2020 IEEE International Symposium on High Performance Computer Architecture (HPCA). IEEE, 2020: 328-341.

[7] T. Tambe, et al. "A 16-nm soc for noise-robust speech and nlp edge ai inference with bayesian sound source separation and attention-based dnns," IEEE Journal of Solid-State Circuits (JSSC), vol. 58, pp. 569-581, 2022.

[8] T. Tambe, et al. "22.9 A 12nm 18.1 TFLOPs/W sparse transformer processor with entropy-based early exit, mixed-precision predication and fine-grained power management," 2023 IEEE International Solid-State Circuits Conference (ISSCC), 2023.

40 nm Core, MV, HV Devices Integration Database for Future AI Development

Yi-Chuen Eng[1*], Daiki Qin[1*], Levi Chen[1*], Wenhao Wu[1*], Peter Liu[1*], Cloud Wang[1*], Chia-Yen Li[2], Nelson Yang[1], Junun Zhu[1], RR Zhu[1], XF Guan[1], Yuri Ma[1], Le Li[1], Hayato Wang[1], Leo Kuai[1], Carl Ma[1], Bo Yang[1], Jihong Yin[1], Kyle Xu[1], and Rumeng Qiu[1]

[1]Advanced Process Integration Division, Nexchip Semiconductor Corporation, Hefei 230012, China
[2]Research and Development Vice President Office, Nexchip Semiconductor Corporation, Hefei 230012, China
*Equal contribution. {engyichuen, daikiqin, levichen, wenhaowu, peterliu, cloudwang}@nexchip.com.cn; zjliuu1012@gmail.com; {eyc03m, qinxuwei0, 13155499570, wenhaowu123}@163.com

Abstract—**In this work, we meticulously collected and summarized raw data from 40 nm Core (including LV and 6T–SRAM), MV, and HV devices for AI analysis and prediction. Prior to developing AI systems to assist engineers in foundry operations, it is crucial to provide robust data sources and create a comprehensive technical knowledge library for AI. Therefore, we demonstrate the performance of 40 nm FET–based n–channel devices, presenting a real–world silicon dataset as the foundational step in AI development.**

Keywords—40 nm, Core, LV, 6T–SRAM, MV, HV, AI, performance metrics, FET.

I. INTRODUCTION

In the foundry work environment, data collection is a critical initial step for developing predictive models and conducting sensitivity analyses. It is essential to gain a comprehensive understanding of device performance by leveraging various databases (yield, tools, queue time, etc.). Artificial intelligence (AI) plays a crucial role in this task. Before AI can provide valuable assistance in the foundry, it is necessary to ingest and summarize raw data for analysis [1]–[6].

In this paper, we demonstrate the device performance of 40 nm Core, medium–voltage (MV), and high–voltage (HV) devices, as these architectures are based on field–effect transistors (FETs). The six–transistor–cell static random access memory (6T–SRAM) is classified as a Core device because the power supply voltage, V_{dd}, of 1.1 V is used for both low–voltage (LV) and SRAM cell. Although the 40 nm process node is a mature technology, our objective is to present real–world silicon (Si) data with high accuracy to support the development of Nexchip's device (DEV) AI functions.

II. EXPERIMENT

The 40 nm HV platform with devices of varying threshold voltages (V_{t}s) was utilized in this study. LV devices include low–V_t (LVT), standard–V_t (SVT), and high–V_t (HVT). The 6T–SRAM cell incorporates pull–down (PD), pass–gate (PG), and pull–up (PU) devices. But, this work we focus on n–channel devices, so we only present PD and PG. MV and HV devices were included to provide a comprehensive overview of transistor performance characteristics. The technology features gate lengths and oxide thicknesses of: 36–40 nm (2.05–2.11 a.u.) for LV, 49–53 nm (2.05–2.11 a.u.) for PD/PG, ≈ 0.9 μm (17.6–18 a.u.) for MV, and ≈ 2.5 μm (106–110 a.u.) for HV devices. Owing to confidentiality obligations, the transistor oxide thickness is reported in arbitrary units (a.u.).

The 40 nm Si measurement data were obtained from wafer acceptance testing (WAT) performed at the end of the fabrication process. Threshold voltage extraction uses the constant–current method [6], [11]. Linear threshold voltage, $V_{t,lin}$, is measured at a very low drain–source voltage, V_{ds}, of 100 mV (or $V_{ds,low}$), while saturation threshold voltage, $V_{t,sat}$, is measured at $V_{ds} = V_{dd}$. Drain–induced barrier lowering (DIBL) is calculated as $(V_{t,lin} - V_{t,sat})/(V_{dd} - V_{ds,low})$. On–state current is defined at $V_{gs} = V_{ds} = V_{dd}$, and off–state current at $V_{gs} = 0$ V and $V_{ds} = V_{dd}$. The short–channel effect (SCE) figure of merit [5], [6], $V_{SCE} = \Delta V_{DIBL}/(I_{on}/I_{off})$, quantifies device performance (lower values preferred). Units: $V_{t,lin}$, $V_{t,sat}$, DIBL, and V_{SCE} are expressed in mV; I_{on}/I_{off} is unitless and typically displayed on a logarithmic scale.

III. RESULTS AND DISCUSSION

A. MOSFET Performance: ΔV_{DIBL}, I_{on}/I_{off}, and $V_{t,sat}$

This paper investigates three key electrical parameters. The first is ΔV_{DIBL}, which accounts for DIBL. ΔV_{DIBL} (= $V_{t,lin} - V_{t,sat}$) is a quick measurement used to monitor short–channel performance, and excluding the measurement of subthreshold swing (SS) can save time and reduce costs [5]. The second parameter is I_{on}/I_{off}, which is the on/off current ratio. The final parameter is $\Delta V_{DIBL}/(I_{on}/I_{off})$, representing the SCE voltage, V_{SCE}, an established figure of merit used to evaluate SCE [5], [6]. By plotting the $V_{t,sat}$ on the X–axis, we can obtain an overall view of device performance.

Fig. 1 illustrates ΔV_{DIBL} versus $V_{t,sat}$. Core devices (including SRAM) and MV devices exhibit similar ΔV_{DIBL} values. However, it is important to note that SCE may lead to varying slope values (e.g., LV has a poorer ΔV_{DIBL} compared to PD and PG due to the shrinking of gate length). HV devices have distinct DIBL values due to a larger V_{dd} of 32 V.

I_{on}/I_{off} versus $V_{t,sat}$ is shown in Fig. 2. If Fig. 1 illustrates the short–channel characteristics of devices, Fig. 2 depicts the "performance" of devices affected by SCE (i.e., the gate's

979-8-3315-2209-4/25 $31.00 © 2025 IEEE

control over the channel region to turn on and off when V_{dd} is applied). Among the four device families, HV devices exhibit the highest I_{on}/I_{off} ratio, primarily due to their dominating I_{off}. A critical consideration arises: should the focus be on I_{on}/I_{off} or on I_{on} alone? The more important factor is I_{on}. For example, when LVT is designed to have the highest I_{on}, it inevitably also has the highest I_{off} among all devices, given that LVT has the smallest $V_{t,sat}$. Thus, it is anticipated that LVT will exhibit the lowest I_{on}/I_{off}.

B. Preliminary AI Framework Analysis

This study focuses on n–channel devices (Core, MV, and HV) to maintain scope consistency. Initial AI model development utilizes an 80/20 training–validation split with 2,000 wafer dies (measured under baseline process conditions) to establish the correlation framework. Future work will expand to p–channel devices and incorporate > 20,000 wafer dies from a test vehicle for broader analysis.

Fig. 3 presents a plot of V_{SCE} as a function of $V_{t,sat}$. It can be noted that each type of devices has different levels of ΔV_{DIBL} and I_{on}/I_{off} and thus different V_{SCE} values depending on the $V_{t,sat}$. This plot effectively summarizes and evaluates SCE and can serve as a foundation for future AI development, as it accounts for the DIBL and I_{on}/I_{off} of a device. All FET–based devices are represented in this plot, demonstrating the potential for comprehensive AI training.

But, is it possible for an inexperienced AI to accurately distinguish the differences and provide useful results? The outcome may be pessimistic, as the right input leads to the right output. Providing adequate "tagging" to our developing AI is essential to avoid miscomputations. We must ensure AI is informed of crucial parameters such as gate length, gate width, power supply, gate oxide thickness, and device titles. Without this information, AI may incorrectly assume that HV devices are the most important due to their highest I_{on}/I_{off}.

We need to instruct AI that an LVT is designed to achieve the highest I_{on} among all devices, provided no other devices have a smaller $V_{t,sat}$ than the LVT. In other words, HV devices have their own functions within a chip, just like other devices such as Core, SRAM, and MV devices. We need AI to summarize all devices in a plot and identify the direction for performance improvement.

Our Model employs a deep neural network (DNN)–based framework to predict WAT parameters through multivariate analysis of $V_{t,sat}$ correlations with other parameters (e.g., I_{on}/I_{off} ratio). This approach reduces required WAT measurements. Specifically, the model predicts $V_{t,sat}$ from three key inputs ($V_{t,lin}$, I_{on}, and I_{off}), establishing a foundation for device–level AI development. The model architecture of the DEV AI models, including Model 1 (M1) and Model 2 (M2), is illustrated in Fig. 4. Initially, raw data (all parameters) can be categorized into Group 1, comprising $V_{t,lin}$, I_{on}, I_{off}, and Group 2, encompassing other parameters. We utilize Group 1 and Group 2 as inputs to train our M1. Once both the Mean Absolute Percentage Error (MAPE) and the coefficient of determination (R^2) achieve our targets (MAPE < 5% and R^2 >

0.9) in both training and testing phases, we can consider this AI model to be robust. Subsequently, we proceed with further training of M1. Our ultimate objective is to use fewer WAT items (e.g., $V_{t,lin}$, I_{on}, and I_{off} in this work) to predict the remaining device parameters (phase one).

Preliminary results are shown in Fig. 5. We observe that the testing results from M1 closely align with the Si data. It is premature to celebrate, as we have only used three WAT items ($V_{t,lin}$, I_{on}, and I_{off}) to predict one WAT item ($V_{t,sat}$). Random sampling during the training/test set split resulted in underrepresentation of HV data in the test set, leaving only one data point. Nevertheless, this marks our initial step in DEV AI development. Our next step involves inputting extensive silicon data (> 20,000 wafer dies) for training and testing.

C. Future AI Implementation Directions

Turning our focus back to AI, we utilize all FEoL inline data and WAT data as inputs for AI training and testing, referring to this as M2. After successfully training and testing the AI model, M2 can also be considered robust. We then compare the performance of Models 1 and 2. Our ultimate goal is to use fewer WAT items to derive inline parameters in reverse (phase two). This allows us to collaborate with our modules for optimized recipe development aimed at improving device performance.

The AI framework reduces WAT prediction time to minutes, compared to 2–3 months for split experiments or hours/days required for TCAD simulations. While trained on 40 nm node data, its generalizability to advanced nodes (e.g., 28 nm) requires additional training data to avoid extrapolation errors. Future work will expand the dataset to ensure cross–node applicability.

The transformer model, a type of deep learning architecture, was first introduced in a 2017 paper (Fig. 6) [7]. Transformers operate efficiently by processing input sequences in parallel, optimizing both training and inference. [7], [8]. Transformers outperform convolutional neural networks (CNNs) and recurrent neural networks (RNNs) in long–range dependency modeling [7] and computational parallelization [7], making them particularly suitable for wafer metrology data analysis [10]. Our company is leveraging the self–attention mechanism and its extension, multi–head attention, to enhance AI model accuracy in device WAT predictions. Our objective is to plot all device WAT items and understand their interrelationships, a complex engineering task. With the support of the AI transformer model, we can expedite the study of device characteristics and swiftly establish the relationship between inline and device WAT. We are progressively moving towards our DEV AI with the aid of transformer models.

IV. CONCLUSION

We have demonstrated the performance of 40 nm HV platform FET–based devices. Visualizing all devices on a single plot enables effective comparison of transistor values.

This visualization serves as an initial input for AI development. More importantly, AI requires us to provide comprehensive Si data to enhance its problem–solving capabilities in real–world scenarios. Our next step is to employ transformer model to analyze and summarize Si data.

ACKNOWLEDGEMENT

The authors would like to thank IKAS Industries Co., Ltd. for supporting this work.

REFERENCES

[1] S. M. Sze, *Semiconductor Devices, Physics and Technology*, 2nd ed. New York, NY, USA: Wiley, 2001.

[2] D. A. Neamen, *An Introduction to Semiconductor Devices*. New York, NY, USA: McGraw-Hill, 2005.

[3] Jatmiko E. Suseno et al., "Artificial Intelligence Techniques for SPICE Optimization of MOSFET Modeling," in *CITISIA*, Jul. 2009, pp. 76–80, doi: 10.1109/CITISIA.2009.5224238.

[4] Amir Hossein Abdollahi Nohoji et al., "Performance Comparison of Artificial Intelligence Networks in Nanoscale MOSFET Modeling," in *ICNC*, Jul. 2011, pp. 807–810, doi: 10.1109/ICNC.2011.6022244.

[5] Y.-C. Eng et al., "Monitoring of FinFET Characteristics Using $\Delta V_{DIBLSS}/(I_{on}/I_{off})$ and $\Delta V_{DIBL}/(I_{on}/I_{off})$," *IEEE J. Electron Devices Soc.*, vol. 7, pp. 344–350, 2019, doi: 10.1109/JEDS.2019.2898697.

[6] Y.-C. Eng et al., "Significance of Overdrive Voltage in the Analysis of Short-Channel Behaviors of n-FinFET Devices," *IEEE J. Electron Devices Soc.*, vol. 10, pp. 281–288, 2022, doi: 10.1109/JEDS.2022.3160881.

[7] A. Vaswani et al., "Attention Is All You Need," in Advances in *Neural Information Processing Systems (NeurIPS)*, vol. 30, pp. 5998–6008, 2017, doi: 10.48550/arXiv.1706.03762.

[8] IBM. *What is a transformer model?* [Online]. Available: https://www.ibm.com/think/topics/transformer-model

[9] Z. Dai et al., "Transformer-XL: Attentive Language Models Beyond a Fixed-Length Context," in *IEEE Transactions on Pattern Analysis and Machine Intelligence (TPAMI)*, vol. 42, no. 1, pp. 187–200, Jan. 2020, doi: 10.1109/TPAMI.2020.3024209.

[10] J. Park et al., "Vision Transformer for Semiconductor Defect Detection and Classification," in *IEEE Transactions on Semiconductor Manufacturing*, vol. 35, no. 3, pp. 347-355, Aug. 2022, doi: 10.1109/TSM.2022.3176785.

[11] H.-G. Lee, S.-Y. Oh, and G. Fuller, "A simple and accurate method to measure the threshold voltage of an enhancement-mode MOSFET," *IEEE Trans. Electron Devices*, vol. TED-29, no. 2, pp. 346–348, Feb. 1982, doi: 10.1109/T-ED.1982.20707.

[12] Y. LeCun et al., "Gradient-Based Learning Applied to Document Recognition," in *Proceedings of the IEEE*, vol. 86, no. 11, pp. 2278-2324, Nov. 1998, doi: 10.1109/5.726791.

[13] K. Cho et al., "Learning Phrase Representations Using RNN Encoder-Decoder for Statistical Machine Translation," in Proceedings of the 2014 *IEEE Conference on Empirical Methods in Natural Language Processing (EMNLP)*, pp. 1724-1734, doi: 10.3115/v1/D14-1179.

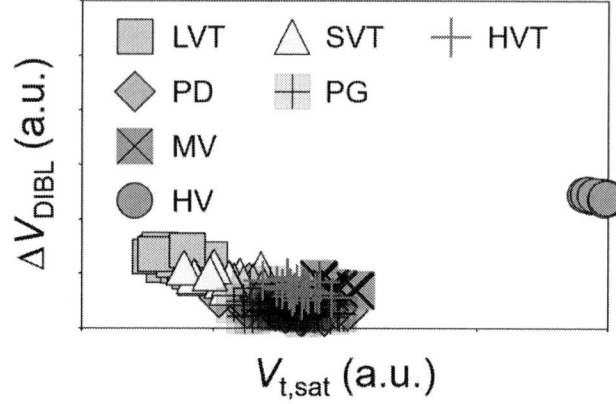

Fig. 1. ΔV_{DIBL} versus $V_{t,sat}$. LV (LVT, SVT, HVT): $V_{dd} = 1.1$ V. SRAM (PD, PG): $V_{dd} = 1.1$ V. MV: $V_{dd} = 8$ V. HV: $V_{dd} = 32$ V. LV, gate length ranges from 36 to 40 nm. PD and PG, gate length ranges from 49 to 53 nm. MV, gate length is approximately 0.9 µm. HV, gate length is around 2.5 µm.

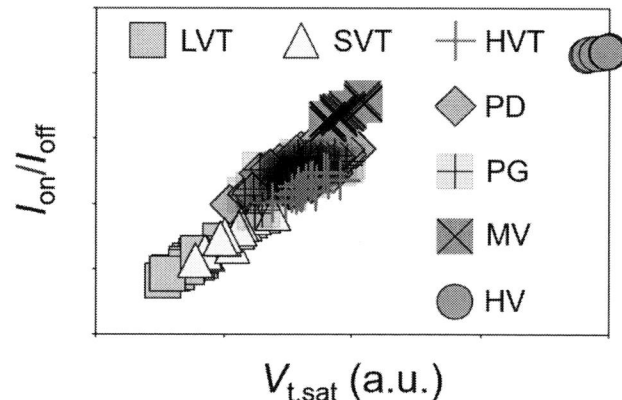

Fig. 2. I_{on}/I_{off} versus $V_{t,sat}$. LV (LVT, SVT, HVT): $V_{dd} = 1.1$ V. SRAM (PD, PG): $V_{dd} = 1.1$ V. MV: $V_{dd} = 8$ V. HV: $V_{dd} = 32$ V. LV, gate length ranges from 36 to 40 nm. PD and PG, gate length ranges from 49 to 53 nm. MV, gate length is approximately 0.9 µm. HV, gate length is around 2.5 µm.

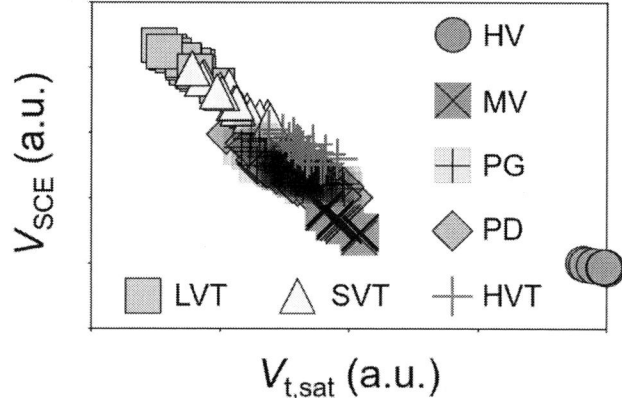

Fig. 3. V_{SCE} versus $V_{t,sat}$. LV (LVT, SVT, HVT): $V_{dd} = 1.1$ V. SRAM (PD, PG): $V_{dd} = 1.1$ V. MV: $V_{dd} = 8$ V. HV: $V_{dd} = 32$ V. LV, gate length ranges from 36 to 40 nm. PD and PG, gate length ranges from 49 to 53 nm. MV, gate length is approximately 0.9 µm. HV, gate length is around 2.5 µm.

979-8-3315-2209-4/25 $31.00 © 2025 IEEE

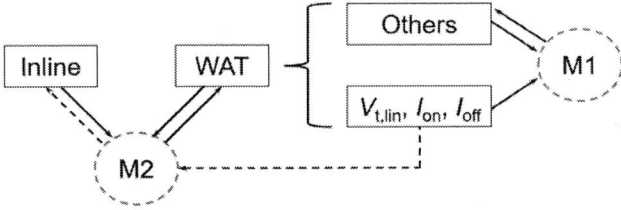

Fig. 4. Proposed DEV AI workflow: Model 1 (M1) uses Group 1 parameters ($V_{t,lin}$, I_{on}, I_{off}) and Group 2 parameters as inputs. Model validation requires MAPE < 5% and R^2 > 0.9. Phase one targets WAT item reduction; Phase two (M2) correlates inline and WAT data for reverse parameter derivation.

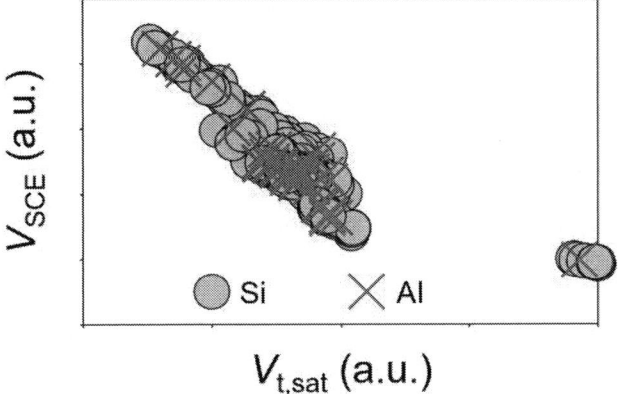

Fig. 5. The V_{SCE} vs. $V_{t,sat}$ correlation in the redrawn Fig. 3 demonstrates agreement between M1 model predictions (cross markers) and silicon measurement data (circle markers). Preliminary results demonstrate prediction capability using three WAT parameters to estimate $V_{t,sat}$ (MAPE < 5%, R^2 > 0.9).

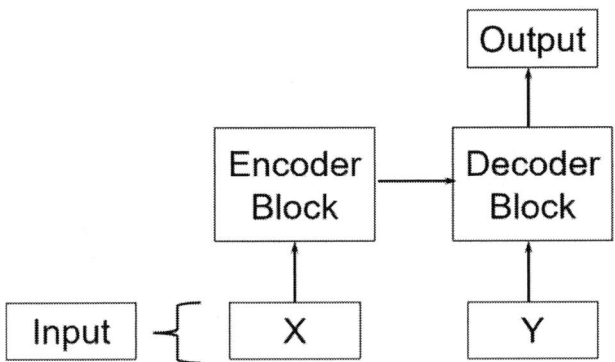

Fig. 6. Transformer model architecture (adapted from [7]) for DEV AI implementation. Multi–head attention mechanisms enable parallelized processing of FEoL/WAT parameter relationships, accelerating device characterization.

979-8-3315-2209-4/25 $31.00 © 2025 IEEE

Utilizing AI to Address the Poor Correlation of BEoL 1XDD VIA R_c and Its Inline Data at the 40 nm Node

Peter Liu[1], Daiki Qin[1], Levi Chen[1], Yi-Chuen Eng[1], Wenhao Wu[1], Cloud Wang[1], Chia-Yen Li[2], Nelson Yang[1], Carl Ma[1], Bo Yang[1], Jihong Yin[1], Kyle Xu[1], Rumeng Qiu[1], Leo Kuai[1], Junun Zhu[1], RR Zhu[1], XF Guan[1], Yuri Ma[1], Le Li[1], and Hayato Wang[1]

[1]Advanced Process Integration Division, Nexchip Semiconductor Corporation, Hefei 230012, China
[2]Research and Development Vice President Office, Nexchip Semiconductor Corporation, Hefei 230012, China
{peterliu, daikiqin, levichen, engyichuen, wenhaowu, cloudwang}@nexchip.com.cn; zjliuu1012@gmail.com; {qinxuwei0, 13155499570, eyc03m, wenhaowu123}@163.com

Abstract—This paper introduces an advanced AI model designed to address the poor correlation between BEoL 1XDD VIA R_c and its inline data at the 40 nm node. While engineers can identify WAT items with weak correlations, resolving persistent issues across multiple Si data readouts remains challenging. Utilizing AI neural networks, we successfully resolve these issues. The AI model weights (numerical values) help to clarify the problem, demonstrating that VIA R_c is highly dependent on its thickness and the top/bottom metal profile. Additionally, we employ SmartAPC for inline control in the BEoL process. Moreover, training data improves the accuracy and performance of neural networks. The insights gained from the AI model enhance accuracy and provide valuable hints for further improvement. Consequently, the implementation of powerful AI drives growth and efficiency.

Keywords—*AI, BEoL 1XDD VIA Rc, inline, 40 nm node, neural networks, weights, SmartAPC.*

I. INTRODUCTION

Artificial intelligence (AI) plays a crucial role in managing and solving complex problems. While humans find it challenging to handle enormous quantities of data, AI can process and analyze this information with ease. Essentially, AI can make predictions within seconds, a task that is significantly more time–consuming for humans. However, for AI to effectively process large datasets, it requires correctly labeled data; otherwise, it cannot discern which characteristics are important. Unlabeled data are almost challenging for AI training unless we "teach" or label these data. A well–trained AI model, when set up correctly, can quickly classify data and assist in making accurate predictions [1]–[5].

In this paper, we collaborate with AI to tackle complex tasks, specifically addressing the back–end–of–line (BEoL) 1X dual damascene (DD) via (VIA) contact resistance, R_c, and its poor correlation with inline data, using silicon (Si) data from the 40 nm technology node. By leveraging AI, we resolve our problem and develop a smart advanced process control (APC) (or SmartAPC) system to swiftly classify data and enhance inline control through neural networks.

II. ARTIFICIAL NEURAL NETWORK

A. VIA AI Model: A Deep Neural Network Backbone

The deep neural network architecture used in this study is depicted in Fig. 1. This architecture includes input layers, augmentation layers, hidden layers, and an output layer. The input layers are where the artificial neural network receives its data. The augmentation layers generate the connection between inputs through data augmentation, which aids in enhancing the machine learning models [6]. The hidden layers process the input data from augmentation layers, serving as the core computational units of the AI. The backbone employs a fully connected deep neural network (DNN) with a self–attention mechanism integrated between hidden layers to enhance feature interaction modeling. Finally, the output layer produces the final results of the AI model.

B. Data Collection & Cleaning

This study employs a 40 nm test vehicle dataset comprising > 3500 individual die measurements. To isolate baseline via resistance characteristics during model training, we intentionally excluded BEoL process variations in this initial phase. The dataset was rigorously filtered to remove: (1) systematic out–of–specification (OOS) cases (reserved for Phase II development), and (2) random outliers from defect mechanisms or transient process excursions. This curated dataset facilitates focused AI model development under nominal process conditions, enabling multidimensional analysis of critical parameters including top/bottom metal thicknesses (THK), CD, and via profiles.

C. AI vs. Traditional Methods

Traditional models, limited to the resistivity (R)–critical dimension (CD) relationship, fail to address modern production challenges [7]. Increasing process complexity and material variations further expose these shortcomings. Our AI model overcomes these limitations by integrating multiple inline parameters, enabling holistic process optimization, WAT prediction, and real–time adaptive control. Additionally, its dynamic learning capability enhances performance in complex processes and facilitates real–time feedback and intervention.

979-8-3315-2209-4/25 $31.00 © 2025 IEEE

III. RESULTS AND DISCUSSION

We evaluated forecast accuracy using Mean Absolute Percentage Error (MAPE), which measures the average absolute percentage deviation between predicted and actual values. For semiconductor metrology, a MAPE < 5% reflects high precision in Wafer Acceptance Test (WAT) parameter prediction [9]. To quantify model explanatory power, we used R–squared (R^2), where $R^2 > 0.9$ indicates robust predictive capability and validates parameter relationship capture [10]. Figs. 2–4 display the AI model testing results, which will be discussed in detail.

Fig. 2 shows the AI model's MAPE and R^2 after testing. Our AI training process is outlined below. We initially inputted all raw data, including inline and WAT items, and initiated our first collaboration with AI. We hoped that the AI could provide insights that are difficult for humans to uncover. Unfortunately, the initial results were not satisfactory, possibly due to human biases or empirical perceptions, leading us to dismiss these results. In the initial AI model testing, both MAPE and R^2 were about 6.09% and 0.68, respectively.

We continued to work diligently and study the outcomes. Subsequently, we achieved slightly improved results, but we were still not satisfied. Testing a is the first of the last six tests. After Testing a, we observed better MAPE and R^2 values in Testing b. However, challenges arose as we input more data (> 3500 wafer dies). The R^2 value decreased to around 0.3, and MAPE worsened, exceeding 4% because of extensive data sources from numerous wafers.

This experience highlighted the complexity of the AI model path. However, this is our journey toward achieving production efficiency in foundry manufacturing, or wafer per hour (WPH), which reflects the number of Si wafers processed within an hour. We continued training our BE AI model by labeling almost all information and "teaching" our BE AI model about the BEoL process. Ultimately, we achieved a MAPE of about 3.95% and an R^2 of around 0.91 (see Fig. 2, Testing f).

The highlight of the experiment is the BE AI model weights, as depicted in Fig. 3. These results are derived from Testing f, which is our final testing phase in this study. Partial VIA thickness (PV THK) is the thickness measured after all–in–one (metal and via) trench etching (TR ET). This thickness is crucial as it determines the VIA volume in the VIA R_c (Fig. 4). The primary influencing factor is the THK, while other factors are not disclosed here due to company confidentiality (not shown in the figure). By understanding the percentage contributions of each parameter, we can develop a SmartAPC system to efficiently monitor and adjust corresponding inline values to meet VIA R_c requirements. Currently, we are in the process of developing the optimized AI model that aims to predict the VIA R_c based on fewer inline items (key layers and compensation model).

The 40 nm node serves as our foundational platform for AI model development. We are currently collecting 28 nm process data to enable cross–node validation in future work. Generalization to other nodes (e.g., 28 nm) will require either complete process data for the target node or validated scaling methodologies to adapt the 40 nm model—a key focus of our ongoing research. As discussed in [8], post–process SPC and historical data form the basis of traditional APC feedback control. However, such methods primarily optimize individual process steps, often resulting in delayed observable effects on final WAT parameters. Our SmartAPC overcomes these limitations by combining deep learning with traditional APC. It establishes a nonlinear mapping between inline data and WAT results, enabling real–time WAT prediction and automatic parameter adjustments. This significantly improves both control scope and responsiveness.

Collaborating with AI allows us to "learn" from extensive Si data, enabling us to train machines to generate predictions and ultimately reduce costs. This represents the initial step toward achieving significant cost savings.

IV. CONCLUSION

The challenge of managing extensive data sources from numerous wafers has been significantly mitigated by AI's neural network capabilities. In this study, we utilized Si data from the 40 nm process node to develop an advanced AI model designed to improve the correlation between BEoL 1XDD VIA R_c and its inline data. Model weights determine that VIA R_c is significantly influenced by its thickness and the top/bottom metal profile.

The proposed AI model predicts BEoL VIA R_c values from key inline monitoring parameters, accounting for process interdependencies. Integration of this framework with APC (yielding SmartAPC) resolves the poor correlation between BEoL 1XDD VIA R_c and inline data at the 40 nm node in R&D. This method enables optimization of adjustable parameters for WAT outcome prediction.

In summary, the integration of robust AI technologies has been instrumental in driving growth and efficiency in the BEoL process. The proficiency of AI in analyzing vast databases has not only drastically reduced data mining time but also underscored its pivotal role in enhancing efficiency and productivity within semiconductor manufacturing. This achievement highlights the indispensable contribution of AI in modern foundry engineering.

ACKNOWLEDGEMENT

The authors would like to thank IKAS Industries Co., Ltd. for supporting this work.

REFERENCES

[1] Larry Hardesty. *Explained: Neural networks.* Accessed: Apr. 14, 2017. [Online]. Available: https://news.mit.edu/2017/explained-neural-networks-deep-learning-0414

[2] IBM Data and AI Team. *AI vs. machine learning vs. deep learning vs. neural networks: What's the difference?* Accessed: Jul. 6, 2023. [Online]. Available: https://www.ibm.com/think/topics/ai-vs-machine-learning-vs-deep-learning-vs-neural-networks

[3] Cole Stryker & Eda Kavlakoglu. *What is artificial intelligence (AI)?* Accessed: Aug. 9, 2024. [Online]. Available: https://www.ibm.com/think/topics/artificial-intelligence

[4] Daisuke Kobayashi et al., "Application of Natural Language Processing in Semiconductor Manufacturing," in *Int. Symp. Semiconductor Manufacturing*, Dec. 2022, pp. 1–3, doi: 10.1109/ISSM55802.2022.10026893.

[5] Hao-Chiang Shao et al., "LithoHoD: A Litho Simulator-Powered Framework for IC Layout Hotspot Detection," *IEEE Trans. Comput.-Aided Des. Integr. Circuits Syst.*, Early Access Article, doi: 10.1109/TCAD.2024.3463539.

[6] Vivienne Sze et al., "Efficient Processing of Deep Neural Networks: A Tutorial and Survey," *Proc. IEEE*, vol. 105, no. 12, pp. 2295–2329, Dec. 2017, doi: 10.1109/JPROC.2017.2761740.

[7] G. S. May and C. J. Spanos, *Fundamentals of Semiconductor Manufacturing and Process Control*. Hoboken, NJ: Wiley-IEEE Press, 2006.

[8] C. J. Spanos et al., "Statistical Process Control in Semiconductor Manufacturing," *Proc. IEEE*, vol. 88, no. 3, pp. 406–423, Mar. 2000, doi: 10.1109/5.837001.

[9] T. Nguyen et al., "Forecasting Semiconductor Yield Using Machine Learning: A MAPE-Centric Approach," *IEEE Transactions on Semiconductor Manufacturing*, vol. 35, no. 1, pp. 78–85, Feb. 2022, doi: 10.1109/TSM.2022.3141590.

[10] S. Patel and M. Zhang, "Model Validation for Wafer Test Parameters via R^2 and Cross-Industry Benchmarks," *IEEE Access*, vol. 10, pp. 45672–45683, 2022, doi: 10.1109/ACCESS.2022.3174567.

[11] J. Doe, A. Smith, and B. Johnson, "Optimization of AI Model Weights for Semiconductor Process Control," *IEEE Transactions on Semiconductor Manufacturing*, vol. 35, no. 2, pp. 123-130, May 2022, doi: 10.1109/TSM.2022.1234567.

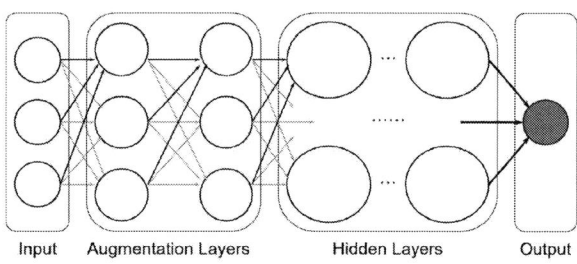

Fig. 1. Architecture of the artificial neural network used for VIA R_c prediction, showing input layers, augmentation layers, hidden layers, and output layer (R_c prediction).

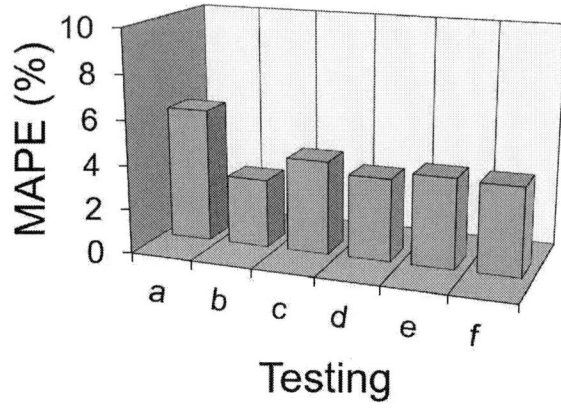

(a)

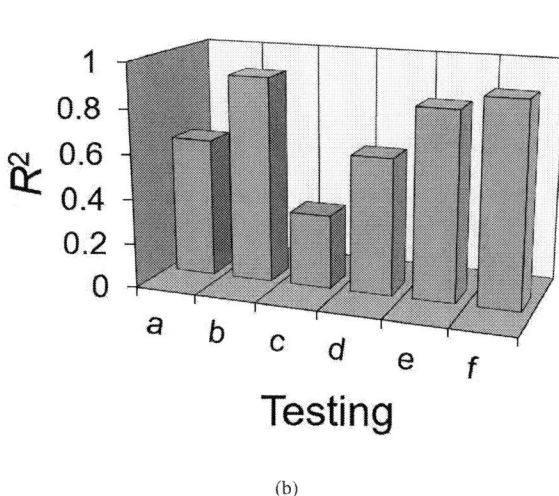

(b)

Fig. 2. Model performance metrics: (a) Mean Absolute Percentage Error (MAPE) and (b) R–squared (R^2) values across six testing phases (a–f), with final optimized performance of 3.95% MAPE and 0.91 R^2 achieved in Testing f.

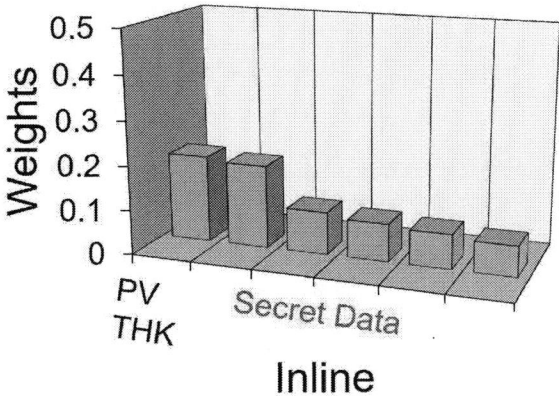

Fig. 3. Parameter weight distribution from the optimized AI model (Testing f), showing partial VIA thickness (PV THK) as the dominant factor in VIA R_c prediction. Other parameters are aggregated due to proprietary constraints.

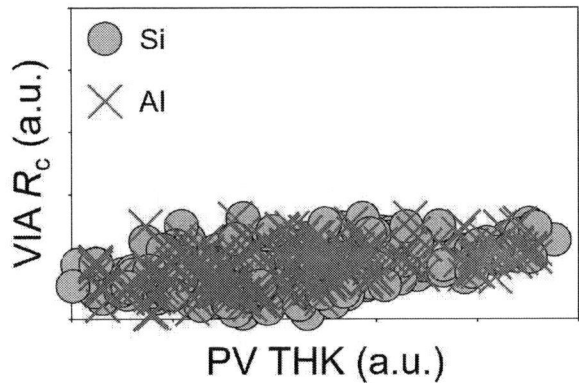

Fig. 4. Correlation between partial VIA thickness (PV THK) and final VIA R_c, demonstrating strong linear relationship critical for SmartAPC implementation.

A Sleep Posture Detection Method for Pregnant Women based on AI Algorithms and a Flexible Sleep Monitoring Belt with a MEMS IMU

Chunhua He   Shangle Ye   Jian Zhan   Jing Lin   Heng Wu   Maojin Liang*   Songqing Deng*

Abstract—**This paper proposes a new method for real-time detection of pregnant women's sleep postures, namely a sleep posture detection system using a flexible sleep monitoring belt equipped with a MEMS IMU (inertial measurement unit) and a machine learning algorithm. The system provides non-invasive, continuous, and high-precision monitoring to address key challenges in maternal health. By analyzing the changes in acceleration and angular velocity data captured by the IMU, we propose a multi-stage processing flow, including wavelet packet transform for denoising, feature extraction using short-time energy analysis, and classification through an integrated machine learning model. Experimental results show that the KDTREE model achieves a 100% accuracy, outperforming other models such as CNN and BP networks. This research helps improve sleep quality monitoring during pregnancy through low-cost, non-wearable solutions.**

Index Terms—**Sleep posture detection, pregnancy monitoring, machine learning, flexible sleep detection belt.**

I. INTRODUCTION

Real-time monitoring of pregnant women's sleep postures is valuable for fetal health assessment and perinatal clinical decision-making. Existing detection systems fall into two categories: wearable and non-wearable. Although wearable devices (e.g., chest strap pressure sensors, wrist acceleration sensors, etc.) have been structurally optimized, wearable deviations are prone to baseline drift of physiological signals, and improper operation may cause data confidence attenuation and sleep disruption [3-4]. In non-wearable solutions, although flexible conductive fabrics and fiber optic sensors can be integrated into beddings for non-sensing monitoring, they are

limited by the cost of device loss due to material fatigue [5-6]. Optical monitoring carries the risk of privacy leakage, and radar techniques (e.g., ultra-wideband / millimeter-wave) have the advantage of being non-invasive. However, their detection accuracies are susceptible to environmental multipath effects and high hardware costs [7]. In this study, we propose an intelligent system that integrates flexible sensor networks and deep learning algorithms to break through the bottleneck of existing techniques through multimodal data fusion and dynamic calibration mechanisms.

II. SYSTEM DESIGN AND METHODOLOGY

The distance between the heart and the MEMS inertial measurement unit (IMU, including a triaxial gyroscope and a triaxial accelerometer) varies with the sleep posture, thereby the three accelerations (namely ax, ay, az) and three angular velocities (namely $\omega x, \omega y, \omega z$) measured by the inertial sensor can be used for sleep posture recognition. The proposed sleep posture detection device is a flexible sleep monitoring belt consisting of a flexible sensor film and a digital signal processor (DSP) circuit, as shown in Fig. 1(a). The flexible polyethylene terephthalate (PET) film inside the belt consists of a MEMS IMU and pressure sensors. The structure and working principle of the pressure sensor unit are depicted in Fig. 1(b). The top and bottom layers are made of PET material, and the isolation layer is double-sided adhesive tape. The hardware system architecture is illustrated in Fig. 1(c), which consists of a power sub-system, a processor sub-system and a sensor sub-system. The sensor subsystem including the MEMS IMU and pressure sensors is the core of the belt.

The experimental platform is established for data acquisition of pregnant women. as depicted in Fig. 2(a). Then the inertial signals of different sleep postures can be acquired and filtered by wavelet package transform (WPT) to remove the noise. In addition, short-time energies of a few frames of the signals are applied as the feature matrices for sleep posture recognition, as shown in Fig. 2(b). Finally, four neural network models (including CNN, RNN, CNN&RNN and BP) and four other machine learning models (including KDTREE, DESCSION-TREE-LOG-LOSS, DESCSION-TREE-ENTROPY, and DESCSION-TREE-GINI) are adopted and optimized for smart sleep posture recognition. Taking KDTREE model as an example, its structure and recognition flow are depicted in Fig. 2(c).

This work was partially supported by the National Natural Science Foundation of China (Grant No. U22A2012), Guangdong Basic and Applied Basic Research Foundation (Grant Nos. 2024A1515220127 and 2024A1515011168), and Guangzhou Key Research and Development Program (Grant No. 2024B03J1251), and Smart Medical Innovation Technology Center, GDUT (Grant No. ZYZX24-037).

Chunhua He, Shangle Ye and Jian Zhan are with the School of Computer, Jing Lin is with the Classroom Management Center, Heng Wu is with the School of Automation, Guangdong University of Technology, Guangzhou 510000, China. (e-mail: hechunhua@pku.edu.cn, a852403278@163.com, 2112405121@mail2.gdut.edu.cn, lj@gdut.edu.cn, heng.wu@foxmail.com). Maojin Liang is with the Department of Otolaryngology, Sun Yat-Sen Memorial Hospital of Sun Yat-Sen University, Guangzhou 510120, China. (e-mail: liangmj3@mail.sysu.edu.cn). Songqing Deng is with the Department of Obstetrics and Gynecology, and Guangxi Hospital Division, the First Affiliated Hospital of Sun Yat-sen University, Guangzhou 510080, China. (e-mail:dengsq@mail.sysu.edu.cn). Corresponding author: (Maojin Liang, email: liangmj3@mail.sysu.edu.cn and Songqing Deng, email: dengsq@mail.sysu.edu.cn)

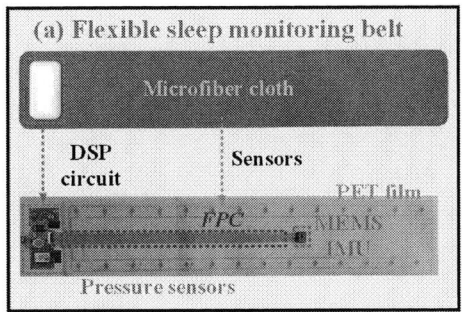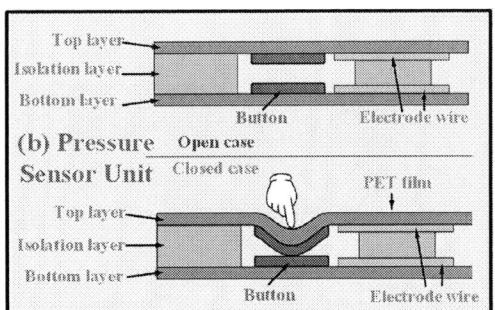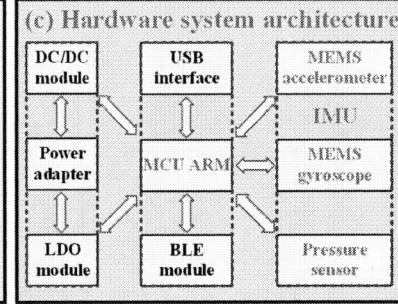

Fig. 1. (a) Structure of a flexible sleep monitoring belt with a MEMS IMU and pressure sensors; (b) Structure of a pressure sensor unit; (c) Hardware system architecture

III. EXPERIMENTAL RESULTS

A. Data Collection

This was a prospective study to include 27 healthy singleton pregnant women at 20-35 weeks of gestation as observation subjects. The researchers used a self-developed flexible sleep monitoring belt to collect nocturnal multimodal physiological data for seven consecutive nights for the subjects. The lying posture data were classified and labelled by standardized algorithms, which mainly included three postures: left latericumbent, supine and right latericumbent, and a total of 165 data files were acquired. The entire study process was performed in strict compliance with the study protocol approved by the Ethics Review Board (IRB) to ensure compliance with the ethical norms of medical research. Prior to enrollment, the researchers explained in detail the purpose of the study, experimental protocol, potential risks and social benefits through a structured informed consent process, and all participants were required to sign a written informed consent form before they could be enrolled in the study cohort.

B. Results Comparison

After experiments, four neural network models and four other machine learning models are adopted for sleep posture recognition, and the performance indices of accuracy, precision, recall, and F1 score are used for evaluation. The performance comparison of different recognition models is illustrated in Fig. 3. It figures out that the performance indices of BP model are the worst, just about 70%. However, those of DESCSION-TREE-LOG-LOSS, DESCSION-TREE-ENTROPY, and DESCSION-TREE-GINI models are improved to 93% ∼ 94%. Besides, those of CNN, RNN, and CNN&RNN models are better, as high as about 96%. The performance indices of KDTREE model are proved best, all reach 100%. Overall, the proposed sleep posture detection method is effective, which is beneficial for in-vitro medical applications.

IV. CONCLUSION

In this study, we propose a method to monitor the sleep postures of pregnant women, utilizing a flexible sleep monitoring belt equipped with a MEMS IMU and artificial intelligence algorithms, combined with short-term energy analysis and a lightweight KDTREE model, to achieve a 100% accuracy and low-cost real-time monitoring. This non-invasive solution promotes the intelligence of non-wearable devices. However, it should be noted that the current study has the following limitations: firstly, the sample size is still very small, and the complex postural patterns in the current study are insufficient; secondly, the generalization ability of the KDTREE model still needs to be further validated by more samples. Therefore, these will be our future work.

REFERENCES

[1] Lin B S, Peng C W, Lee I J, et al. System based on artificial intelligence edge computing for detecting bedside falls and sleep posture[J]. IEEE Journal of Biomedical and Health Informatics, 2023, 27(7): 3549-3558.

[2] Hu D, Gao W, Ang K K, et al. STConvSleepNet: A Spatiotemporal Convolutional Network for Sleep Posture Detection[C]//2024 46th Annual International Conference of the IEEE Engineering in Medicine and Biology Society (EMBC). IEEE, 2024: 1-5.

[3] Elnaggar O, Coenen F, Hopkinson A, et al. Sleep posture one-shot learning framework based on extremity joint kinematics: In-silico and in-vivo case studies[J]. Information Fusion, 2023, 95: 215-236.

[4] Liu J, Zhao W, Li J, et al. Multimodal and flexible hydrogel-based sensors for respiratory monitoring and posture recognition[J]. Biosensors and Bioelectronics, 2024, 243: 115773.

[5] Li T, Pei Q, Qin X, et al. Fusing the wireless technique optical fiber force sensor for remote monitoring of sleeping posture[J]. IEEE/ASME Transactions on Mechatronics, 2023, 28(5): 2703-2715.

[6] Diao H, Chen C, Yuan W, et al. Deep residual networks for sleep posture recognition with unobtrusive miniature scale smart mat system[J]. IEEE Transactions on Biomedical Circuits and Systems, 2021, 15(1): 111-121.

[7] Liu X, Jiang W, Chen S, et al. PosMonitor: Fine-grained sleep posture recognition with mmWave radar[J]. IEEE Internet of Things Journal, 2023, 11(7): 11175-11189.

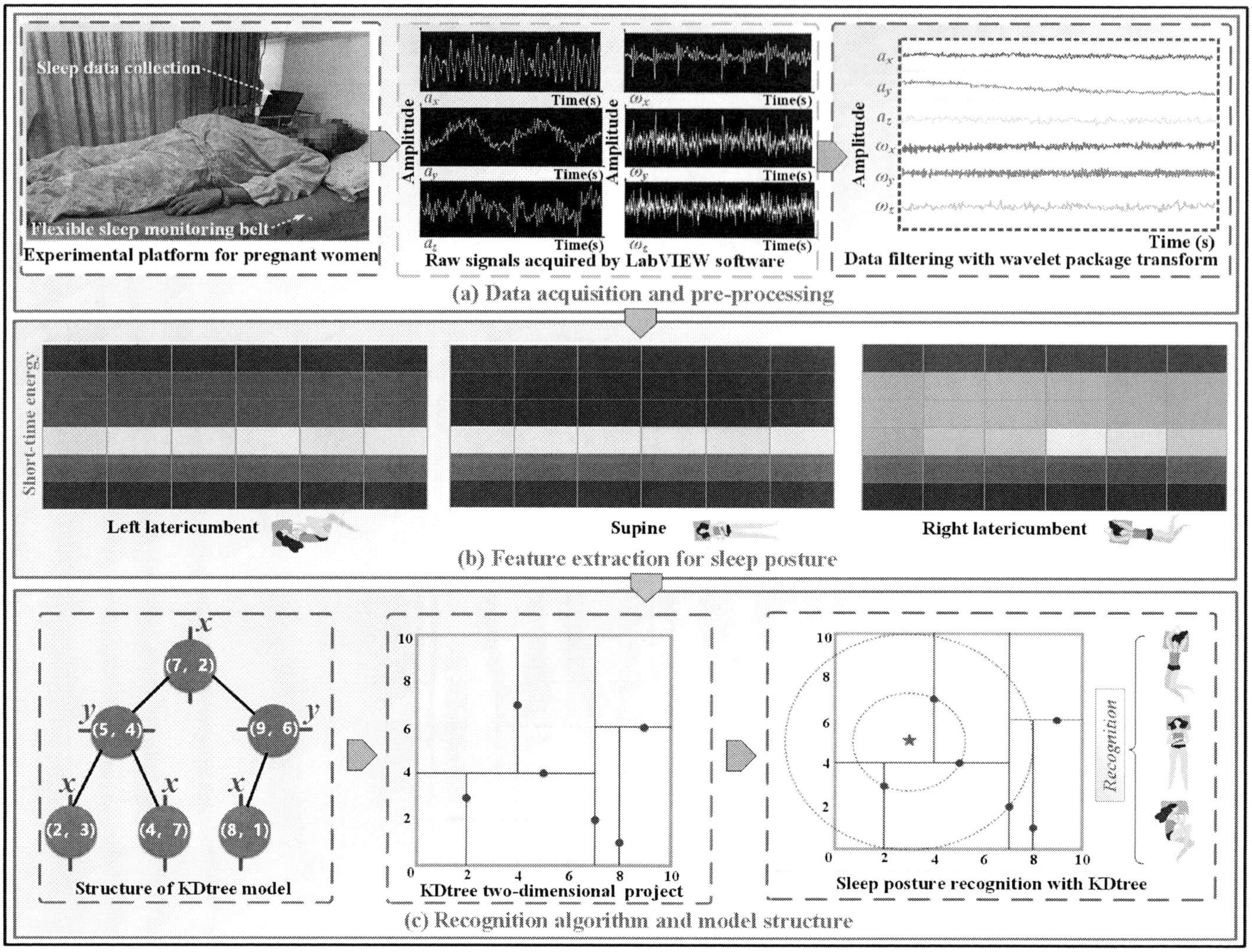

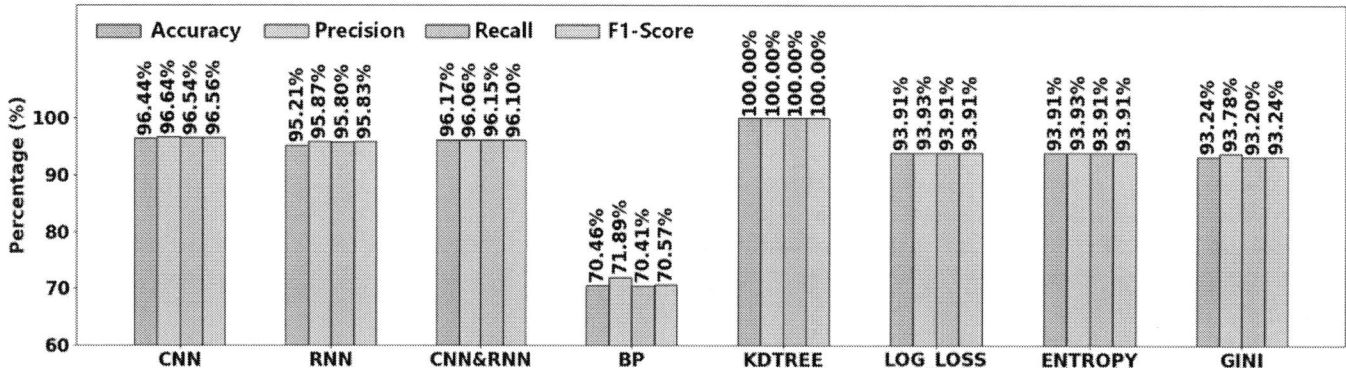

Fig. 2. (a) Data acquisition and pre-processing; (b) Feature matrices of three different sleep postures; (c) KDTREE model structure and recognition flow for sleep postures.

Fig. 3. Performance comparison of different recognition models for sleep postures detection of pregnant women.

Underwater Acoustic Target Localization Method Based on Frequency-Domain Feature Enhancement

Daoguang Zhang   Xianyou Zeng   Kai Yuan   Jing Lin   Heng Wu   Qinwen Huang*   Chunhua He*

Abstract—With the rapid development of artificial intelligence technology, deep learning has become a key method in underwater acoustic target localization. However, due to the inherent complex time-varying characteristics of the ocean environment, the influence of environmental noise, and the lack of experimental data, existing models still have room for improvement in localization performance. This paper proposes an underwater acoustic target localization method based on frequency-domain feature enhancement to address these challenges. The method first converts the time-domain sound pressure signal into a frequency-domain signal using the fast Fourier transform, analyzes the intensity and distribution characteristics of different frequency bands, then extracts the energy features of these frequency bands, and uses this feature to construct the sample covariance matrix as the input of the model. This paper validates the effectiveness of the proposed method using sea trial datasets. Experimental results show that the proposed feature enhancement method achieves significant advantages over traditional data augmentation methods in underwater passive localization tasks, with an absolute distance error of 0.12 km and an absolute depth error of 0.43 m, effectively improving the model's accuracy and noise robustness.

Index Terms—Underwater acoustic target localization, feature enhancement, convolutional neural network, hydrophone.

I. INTRODUCTION

With the development of society and the advancement of science and technology, the strategic position of the ocean has attracted widespread attention, and the importance of underwater detection technologies has become increasingly prominent. Underwater acoustic feature enhancement technique has gradually become a key research focus in underwater passive localization. Its core aim is to efficiently and accurately extract and enhance useful acoustic features in complex underwater environments while mitigating the interference

---

This work was partially supported by the National Natural Science Foundation of China (Grant No. U22A2012), Guangdong Basic and Applied Basic Research Foundation (Grant Nos. 2024A1515220127 and 2024A1515011168), and Guangzhou Key Research and Development Program (Grant No. 2024B03J1251), and Smart Medical Innovation Technology Center, GDUT (Grant No. ZYZX24-037).

Daoguang Zhang, Xianyou Zeng, Kai Yuan, and Chunhua He are with the School of Computer, Jing Lin is with the Classroom Management Center, and Heng Wu is with the School of Automation, Guangdong University of Technology, Guangzhou 510000, China. (e-mail: 2112305283@mail2.gdut.edu.cn, 2112305330@mail2.gdut.edu.cn, 2112305265@mail2.gdut.edu.cn, hechunhua@pku.edu.cn, lj@gdut.edu.cn, heng.wu@foxmail.com). Qinwen Huang is with the Science and Technology on Reliability Physics and Application of Electronic Component Laboratory, No.5 Electronics Research Institute of the Ministry of Industry and Information Technology, Guangzhou, 510000, P. R. China. (e-mail: 971230012@163.com). Corresponding author: (Qinwen Huang, e-mail: 971230012@163.com and Chunhua He, email: hechunhua@pku.edu.cn)

and errors caused by multipath propagation, environmental noise, and nonlinear effects [1], [2]. In deep-sea environments, sound propagation paths are often influenced by water temperature, salinity, and depth variations, leading to signal attenuation and distortion, increasing the difficulty of stable feature extraction. Therefore, to improve the accuracy and robustness of acoustic signal processing [3], [4], researchers have proposed various innovative methods to optimize feature extraction and enhancement. These methods synthesize signal processing theory, statistical tools, and machine learning algorithms [5]–[7], effectively overcoming the limitations of traditional techniques in complex environments and further advancing the application and development of underwater acoustic localization technique.

Modal signal decomposition and feature extraction techniques based on sample covariance have been widely applied in underwater acoustics research. Niu et al. [8] proposed a deep learning method based on big data for broadband acoustic source localization with an uncertain seabed environment. They employed sample covariance feature extraction techniques, which convert the original time-domain data into covariance matrices. The covariance matrices of multiple frequency bands are then combined as the model's input to enhance the feature representation ability of broadband signals. Liu et al. [9] proposed a convolutional neural network that integrates classification and regression for shallow-water acoustic source localization. After normalization using sample covariance based on broadband signals received by hydrophones, the data were used as input features for the model. This method constructs a feature that integrates both frequency-domain and spatial-domain information via sample covariance, effectively improving the accuracy of acoustic source localization in complex sound fields. Furthermore, Li et al. [10] proposed a time-domain signal enhancement method based on underwater acoustic channel modeling to address common noise interference issues in deep-sea environments. This method enhances the feature representation ability of underwater acoustic signals by generating simulation data and adding noise. By combining frequency-domain responses with time-domain signal generation, the global characteristics of the signal are preserved while enhancing the expression of local details, thus further optimizing feature extraction performance.

To reduce the impact of environmental noise on localization performance and enhance the localization accuracy of underwater acoustic target localization models, this study

---

979-8-3315-2209-4/25 $31.00 © 2025 IEEE

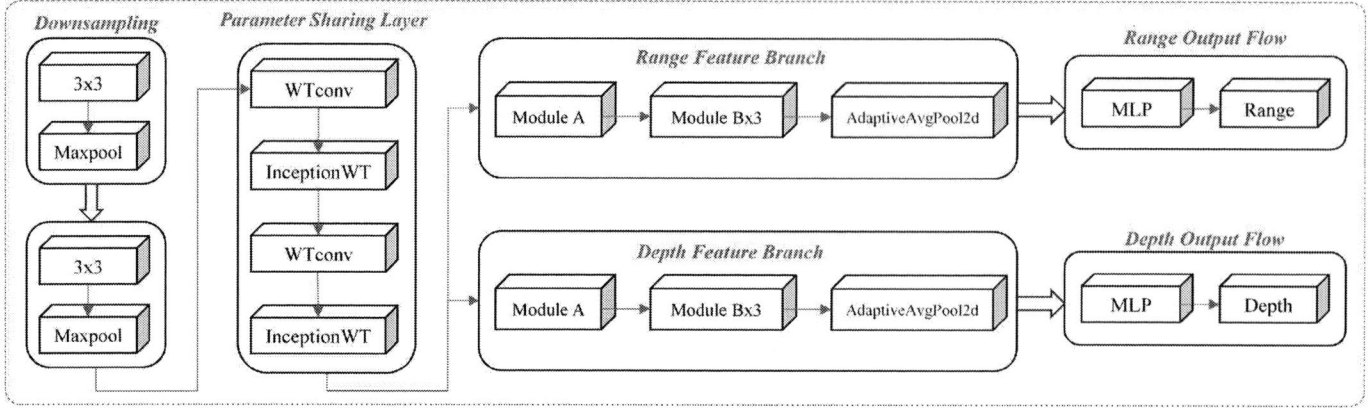

Fig. 1. The framework diagram of the InceptionWTnet

proposes a feature enhancement method based on the frequency domain. The process uses the fast Fourier transform to convert time-domain acoustic signals into frequency-domain signals, followed by feature extraction via energy spectrum analysis. The obtained features are then used to construct a sample covariance matrix, which is input into the model. The model primarily employs a multi-scale wavelet convolution module, which combines multi-scale convolution and wavelet transform. The wavelet transform decomposes the data across different scales, significantly extending the receptive field of the model. This enables the model to effectively extract low-frequency information from acoustic data and accurately capture frequency features closely related to target localization, thereby greatly enhancing the model's adaptability to the complexity and dynamic variations of the marine environment.

II. Feature enhancement

Since the target signals received by each hydrophone in the array are highly correlated, leveraging this characteristic can significantly improve the distinction of the signals, enhancing the accuracy of underwater acoustic target localization. First, the time-domain sound pressure signals collected from sea trial experiments are converted into frequency-domain signals using the fast Fourier transform. This conversion process helps extract the signal's frequency spectral characteristics, allowing for extracting more relevant features for target localization to insight analysis. During feature extraction, the energy of each frequency band is computed and combined to form initial features, which effectively compresses the information and highlights the valuable signal features. Subsequently, considering the correlations between the data, the obtained feature data is used to construct a normalized sample covariance matrix. The framework of the data preprocessing is shown in Fig. 2 (a), and the specific implementation steps are as follows:

First, the original time-domain signals are processed using the Fourier transform, with a sampling resolution set of N = 1024. The formula is as follows:

$$p[k] = \sum_{n=0}^{N-1} s[n]e^{-j\frac{2\pi}{N}kn}, \quad k = 0, 1, \cdots, \frac{N}{2} \tag{1}$$

Where $p[k]$ represents the data at frequency k, $S = [s_1, s_2, \cdots, s_t]$ represents the original time-domain signal and s_t represents the sound pressure signal at time t. Assuming the number of hydrophones in the vertical linear array is L. The complex sound pressure data from different frequency bands are extracted, as shown below:

$$p(f) = \begin{pmatrix} p_{1f_1} & \cdots & p_{1f_n} \\ \vdots & \ddots & \vdots \\ p_{Lf_1} & \cdots & p_{Lf_n} \end{pmatrix} \tag{2}$$

Where p_{Lf_n} represents the frequency-domain data at frequency n in channel L. Then, to reduce the excessive impact of random noise on the amplitude, the normalization operation is performed during data preprocessing. The L2 norm is used to normalize the complex sound pressure data, as shown below:

$$\widehat{p}(f) = \frac{p(f)}{\|p(f)\|_2} = \frac{[p_1(f), p_2(f), \ldots, p_L(f)]^T}{\sqrt{\sum_{i=1}^{L} |p(f)|^2}} \tag{3}$$

Where $\widehat{p}(f)$ represents the frequency-domain data at frequency f after normalization. The magnitude of the energy is then calculated, and the sample covariance matrix C(f) is computed from the vector obtained by the above equation. The spatiotemporal correlation features of the original signals are extracted, providing input features for the model:

$$E_l(f) = |\widehat{p}(f)|^2 \tag{4}$$

$$C(f) = E(f) \cdot E(f)^H \tag{5}$$

Where $E(f)$ is the energy magnitude at frequency f, $C(f)$ is the $L \times L$ sample covariance matrix (SCM), H is the conjugate transpose. The SCM of different frequencies will be stacked along the third dimension of the matrix, so the shape of the input feature data is $L \times L \times F$, where F is the number of frequencies. collaboration between functions within the multi-task learning framework.

979-8-3315-2209-4/25 $31.00 © 2025 IEEE

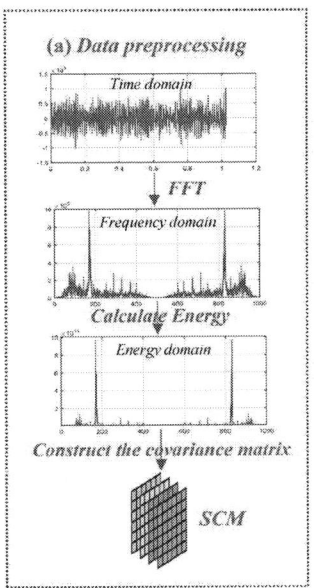
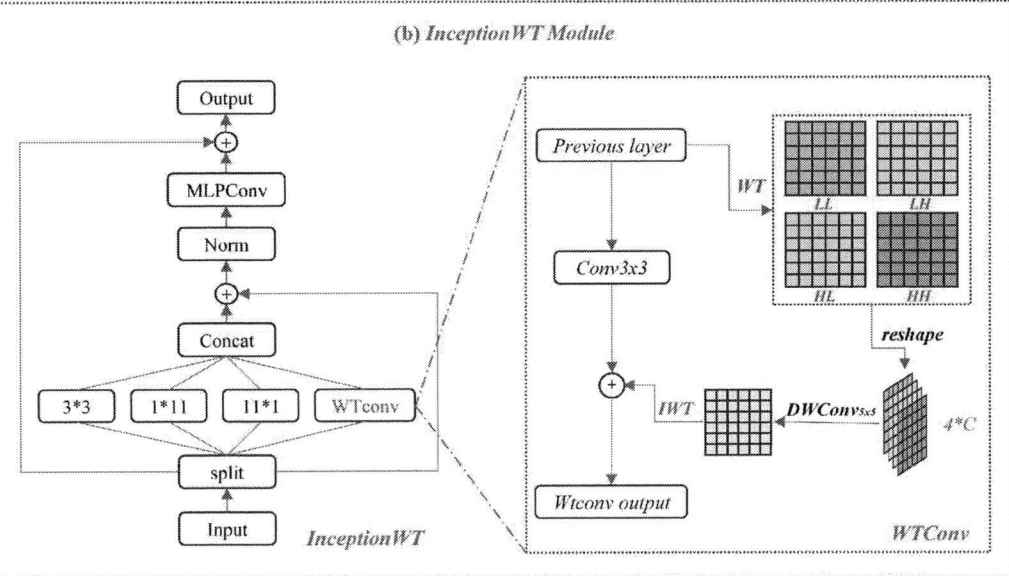

Fig. 2. (a) The framework of the Data preprocessing. (b) The framework of the InceptionWT module

III. LOCALIZATION ALGORITHM

A. Muti-task learning

To accurately localize the target source, this paper adopts a multi-task learning framework that allows the model to predict the distance and depth of the target source simultaneously. As shown in Fig. 1, the model first employs two downsampling modules to extract shallow features. The parameter-sharing layer is introduced to reduce the number of parameters, primarily consisting of a multi-scale wavelet convolution module to capture complex features. Two task-specific branches after the parameter-sharing layer enable simultaneous predictions of distance and depth: one for distance prediction and the other for depth prediction. Each branch extracts more prominent features for specific tasks, ensuring that the model maintains high prediction accuracy while fostering

B. InceptionWT module

To address the problem of underwater acoustic target localization, this paper proposes a multi-scale wavelet convolution module, as shown in Fig. 2 (b), which primarily consists of a combination of wavelet convolution and multi-scale convolution. The InceptionWT module is based on a multi-branch design, integrating depthwise separable convolutions and wavelet transforms to achieve multi-scale feature extraction. The module has four input branches, with three being depthwise separable convolutions and one being wavelet convolution. The features from the four branches are then fused using a feature fusion mechanism. This method endows the module with powerful multi-scale representation capabilities. The wavelet convolution module performs multi-resolution analysis of features through wavelet hierarchical decomposition, and the parameter-sharing mechanism between the depthwise separable convolutions and wavelet transform

effectively reduces the number of model parameters, decreasing the computational complexity of the model.

IV. EXPERIMENTAL RESULTS AND ANALYSIS

A. The dataset of the experiments

All experiments in this paper utilize vertical array data collected in a shallow-water region during an offshore experiment conducted by the SACLANT Centre in October 1993 in the Mediterranean Sea. The experimental site is located in a shallow-water area to the north of Elba Island, off the western coast of Italy, with the primary objective of validating geological acoustics and geometric parameter estimation methods based on sound field observation inversion. The data collection includes sound field data, equipment navigation data, environmental parameters, and other relevant information. During the deployment, an auxiliary vessel was used as a mobile sound source, and a vertical array consisting of 48 hydrophones was deployed for data collection. The hydrophones were spaced 2 meters apart, with a total aperture of approximately 94 meters, covering a depth range from 18.7 meters to 112.7 meters. To ensure the generalization capability of the model and the reliability of the validation results, the collected data were split into training and test sets in an 8:2 ratio.

B. The results of experiments

This section conducts a series of comparison experiments to validate the effectiveness of different feature extraction methods, with the results shown in Table.I. Various feature extraction methods were employed to train the model. Since the emission frequency of the sound source is 170Hz, data from the 160Hz-180Hz frequency band were selected as the Freq-specified feature. Freq-sum refers to the feature obtained by summing the data of several sub-bands within the 0-500Hz frequency range after dividing the range into multiple sub-bands. Freq-mean is the feature extracted by calculating the average value of the data within each sub-band of the 0-500Hz frequency range. Freq-max represents the maximum value from each sub-band within the 0-500Hz range, and Freq-energy refers to the maximum energy value from each sub-band within the same frequency range. After preprocessing, the extracted features were used as input features for model training. The experimental results show that, compared to other feature extraction methods, the Freq-energy feature significantly outperforms others regarding underwater acoustic target localization, with an average absolute error of 0.12 km and an average depth error of 0.43 m.

TABLE I
THE RESULTS OF COMPARATIVE EXPERIMENTS

Feature Extraction	Algorithm	MAE_r (km)	MAE_d (m)
Freq-specified	InceptionWTnet	0.26	0.67
Freq-sum	InceptionWTnet	0.51	1.04
Freq-mean	InceptionWTnet	0.47	0.81
Freq-max	InceptionWTnet	0.22	0.60
Freq-energy	InceptionWTnet	**0.12**	**0.43**

V. CONCLUSION

To enhance the accuracy of underwater acoustic target localization, this paper proposes an underwater acoustic target localization method based on frequency-domain feature enhancement. This method efficiently and accurately extracts and enhances useful acoustic features in complex underwater environments, effectively mitigating environmental noise's impact on localization performance. After feature enhancement, the extracted data is used to construct a sample covariance matrix, which is the input feature of the model. Additionally, this paper introduces a multi-scale wavelet convolution module that combines multi-scale convolution with wavelet transform. The model employs a multi-task learning architecture to address the underwater acoustic target localization task and is validated using data from sea trial experiments. Experimental results demonstrate that the proposed method performs better underwater passive localization tasks, with an average absolute error of 0.12 km and an average depth error of 0.43 m.

REFERENCES

[1] J. H. Schmidt, I. Kochańska, and A. M. Schmidt, "Performance of the Direct Sequence Spread Spectrum Underwater Acoustic Communication System with Differential Detection in Strong Multipath Propagation Conditions," *Archives of Acoustics*, pp. 129–140, Mar. 2024, doi: 10.24425/aoa.2024.148771.

[2] M. S. Mahmood and Y. Y. Al-Aboosi, "Effects of Multipath Propagation Channel in Tigris River," vol. 27, no. 02, 2023.

[3] W. Zhang, S. Jin, P. Zhuang, Z. Liang, and C. Li, "Underwater Image Enhancement via Piecewise Color Correction and Dual Prior Optimized Contrast Enhancement," *IEEE Signal Process. Lett.*, vol. 30, pp. 229–233, 2023, doi: 10.1109/LSP.2023.3255005.

[4] Q. Niu, Q. Zhang, and W. Shi, "Waveform design and signal processing method for integrated underwater detection and communication system," *IET Radar Sonar & Navig.*, vol. 17, no. 4, pp. 617–627, Apr. 2023, doi: 10.1049/rsn2.12365.

[5] J. Llor, "Wireless sensor networks; underwater acoustic communications; acoustic propagation; statistical modeling; network planning," 2013.

[6] A. C. Sing, J. K. Nelson, and S. S. Kozat, "Signal processing for underwater acoustic communications," *IEEE Commun. Mag.*, vol. 47, no. 1, pp. 90–96, Jan. 2009, doi: 10.1109/MCOM.2009.4752683.

[7] P. McDowell, "Environmental and Statistical Performance Mapping Model for Underwater Acoustic Detection Systems."

[8] H. Niu, Z. Gong, E. Ozanich, P. Gerstoft, H. Wang, and Z. Li, "Deep-learning source localization using multi-frequency magnitude-only data," *The Journal of the Acoustical Society of America*, vol. 146, no. 1, pp. 211–222, Jul. 2019, doi: 10.1121/1.5116016.

[9] M. Liu et al., "A Convolutional Neural Network Combining Classification and Regression for Source Localization in Shallow Water," *J. Phys.: Conf. Ser.*, vol. 2486, no. 1, p. 012068, May 2023, doi: 10.1088/1742-6596/2486/1/012068.

[10] D. Li, F. Liu, T. Shen, L. Chen, and D. Zhao, "Data augmentation method for underwater acoustic target recognition based on underwater acoustic channel modeling and transfer learning," *Applied Acoustics*, vol. 208, p. 109344, Jun. 2023, doi: 10.1016/j.apacoust.2023.109344.

Moire-based Overlay Metrology Enhanced By Deep Learning

Rui Liu*, Zhaokai Qiu*, Shaolin Zhou*‡, and Xugang Ma†

*School of Microelectronics, South China University of Technology, Guangzhou, China
†China Digital Industry Operation Group Co., Limited, Hong Kong, China
‡Email:eeslzhou@scut.edu.cn

Abstract—**Overlay (OVL) is one of the three critical metrics in Photolithography. As technology advances and critical sizes shrink, sub-nanoscale overlay metrology has become a major challenge. Diffraction superposition (DBO) technology is widely used for overlay metrology, with increasing attention on the circular moiré-based DBO method due to the symmetry, isotropy, and non-periodic nature of circular gratings. However, traditional analytical methods for circular moiré fringes suffer from low precision, limited to the tens of nanometers range. In this paper, we propose a hybrid deep learning architecture that combines wavelet convolution, octave convolution, and the Swin Transformer block for enhanced overlay accuracy. Our approach achieves sub-nanoscale overlay by using the overlay marks of circular grating marks in both x and y directions. Furthermore, this strategy demonstrates exceptional robustness to noise induced by the etching and chemical mechanical polishing processes.**

Index Terms—**Overlay Metrology, DBO, Deep Learning**

I. INTRODUCTION

Overlay (OVL) metrology is a critical factor affecting yield in advanced semiconductor manufacturing. For 7nm technology nodes, the International Roadmap for Devices and Systems specifies an overlay accuracy below 2.3nm [1]. The shrinking size of target devices has reduced the overlay error budget to the sub-nanometer scale .Although various diffraction-based overlay (DBO) techniques have been developed [2]–[4] , achieving sub-nanometer OVL metrology remains a significant challenge.

In photolithophic alignment, Wang Nan et al. [5]demonstrated that applying deep learning to a moiré fringe based alignment technique effectively improves accuracy and system robustness during the alignment process. This approach leverages convolutional neural networks (CNNs) to achieve nanometer accuracy using the circular alignment marks. Inspired by such a methodology, we have intended to apply the deep learning-based moiré fringe techniques to the OVL process to further improve the overlay measurement accuracy, with enhanced defect and noise robustness.

Recently, Transformer-based architectures have gained widespread use in computer vision due to their ability to capture global context and long-range dependencies through self-attention, overcoming the limitations of traditional convolutional neural networks. However, combining local and global features remains a challenge in practical applications. In order to apply it to the deep learning based Moiré fringe OVL metrology, we propose a hybrid deep learning architecture

WOATNet that combines wavelet convolution layer(WT Conv) [6], 2d-octave convolution model(OCM) [7], and Swin Transformer block(SWB) [8]. This hybrid approach effectively combines local detail preservation with global context utilization, improving the accuracy and robustness of moiré fringe etching error detection, which can achieves sub-nanoscale accuracy and it also exhibits exceptional robustness to noise caused by the etching and chemical mechanical polishing processes.

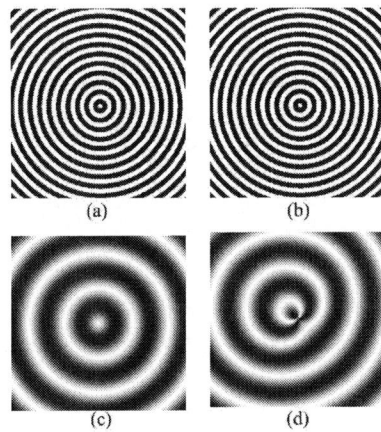

Fig. 1. Circular grating marks and circular moiré fringe. (a) the wafer mask. (b) the mask wafer. (c) simulated moire fringe for the zero displacement. (d) small displacement along x and y direction.

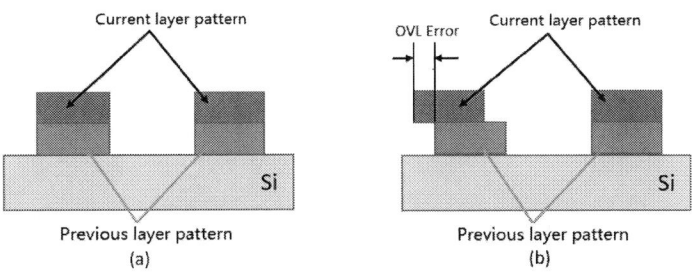

Fig. 2. The illustration of overlay (OVL) process for (a) ideal alignment with no offset between the current layer pattern (red) and the previous layer pattern (blue), and the (b) misalignment with none-zero displacement or overlay error.

979-8-3315-2209-4/25 $31.00 © 2025 IEEE

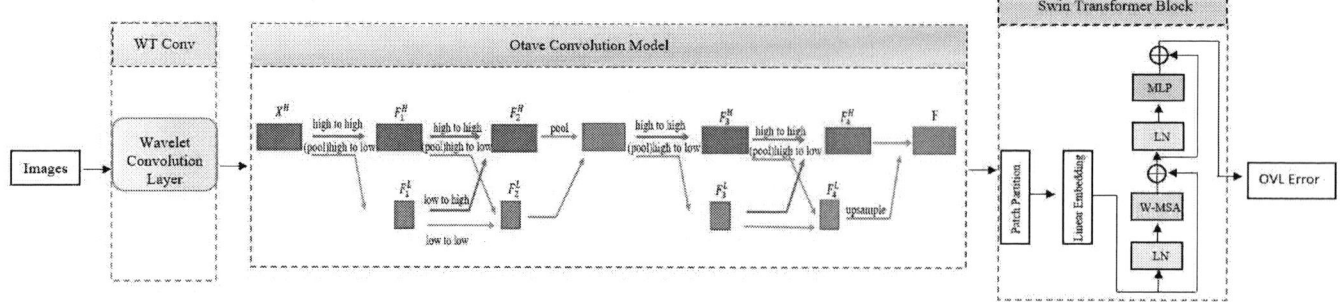

Fig. 3. The flowchart of our WOATNet for overlay offset analysis.

II. PROPOSED METHOD

A. Moiré fringe for OVL

A plane wave is incident perpendicularly on two circular gratings with similar periods $T1$ and $T2$, as shown in Fig1(a) and (b), resulting in multiple diffraction patterns. The diffracted harmonics overlap, creating interference and ultimately forming a multi-level moiré fringe field $E(x, y)$. When the position of one circular grating is fixed and the other circular grating is displaced relative to the fixed one, the spatial phase distribution of the moiré fringe field will change. Therefore, the two layers of circular grating markers are placed on the current layer and the previous layer, respectively. When the two grating markers are perfectly aligned, as shown in Fig. 2(a), their overlay generates an ideal moiré fringe distribution, as illustrated in Fig. 1(c).

However, when a relative displacement occurs between the two grating markers due to process-induced deviations (i.e., overlay error, as depicted in Fig. 2(b)), the resulting moiré fringe distribution changes accordingly, as shown in Fig. 1(d). This variation directly reflects the impact of overlay error on the spatial phase distribution of the moiré fringes and can be used to accurately characterize the alignment error between the two layers. As shown in Fig1(c) and (d), this change is primarily determined by the relative displacement. Among the moiré fringes, the (1,-1) order moiré fringes are easily distinguishable, When the wafer grating undergoes certain relative displacements Δx and Δy along the x and y directions, respectively, the complex amplitude distribution of the (-1, 1) order moiré fringes is given by:

$$E_{(-1,1)}(x,y) = \sum_{n=-\infty}^{+\infty} A_n B_{-n} \exp\left\{ i2\pi n \left[f_1 \times \right. \right. \tag{1}$$
$$\left. \left. \sqrt{(x+\Delta x)^2 + (y+\Delta y)^2} - f_2\sqrt{x^2+y^2} \right] \right\}$$

where A_n and B_{-n} are the Fourier transform coefficients, $f_1 = 1/T_1$ and $f_2 = 1/T_2$ are the reference frequency of two gratings. And by analyzing its phase information, the relative displacement between the mask and the silicon wafer can be obtained.

However, due to the complex phase information of circular moiré fringes, traditional analysis methods such as wavelet transform and windowed Fourier transform can only achieve accuracy on the order of tens of nanometers, which does not meet the current OVL process requirements. As a result, these methods can only be used as a coarse measurement solution.

In this paper, we introduce a WOATNet hybrid architecture to learn the mathematical mapping relation between the circular moiré fringes and the OVL error, enabling sub-nanometer level OVL error measurement.

B. WOATNET

The proposed framework of WOATNet is illustrated in Fig.3, which consists of the WT conv, OCM, and the Swin Transformer block. Firstly, wavelet convolution decomposes the input image, capturing features at different frequencies and spatial scales. This approach effectively reduces noise interference and focuses on learning the subtle effects of relative displacement on Moiré patterns. Secondly, in order to effectively fuse the multi-scale information obtained, an Octave Convolution model is proposed, which consists of four Octave Convolution blocks. Specifically, suppose the input and output data of the Oct-block are $X = \{X^H, X^L\}$ and $F = \{F^H, F^L\}$, where the superscripts H and L represent high-frequency and low-frequency, respectively. The Octave Convolution model defines $F^H = F^{H \to H} + F^{L \to H}$ and $F^L = F^{H \to L} + F^{L \to L}$, where $F^{H \to H}$ and $F^{L \to L}$ represent within-frequency transformations, while $F^{H \to L}$ and $F^{L \to H}$ represent inter-frequency updates. To complete the information update and interaction mentioned earlier, the weight of the Oct-Conv block W should also be divided into two parts, $[W^H, W^L]$. Furthermore, each element can be partitioned into within-band and inter-band components, for example, $W^H = [W^{H \to H}, W^{L \to H}]$ and $W^L = [W^{H \to L}, W^{L \to L}]$. Therefore, F^H and F^L can be computed using the following equations:

$$\begin{aligned} F^H &= F^{H \to H} + F^{L \to H} \\ &= \sum \left(W^H\right)^T X \\ &= \sum (W^{H \to H})^T X^H + \text{upsample}\left(\sum W^{L \to H^T} X^L\right) \end{aligned} \tag{2}$$

$$F^L = F^{H \to L} + F^{L \to L}$$
$$= \sum \left(W^L \right)^T X$$
$$= \sum (W^{H \to L})^T \text{pool}(X^H) + \sum (W^{L \to L})^T X^L \quad (3)$$

where T represents the transposition of the weights, and upsample and pool stand for the upsampling and average pooling operation, respectively. This model aims to separately fuse and learn low-frequency features (such as the overall structure of Moiré patterns) and high-frequency components (such as details, edges, and local features).

Finally, the Swin Transformer block is further introduced into the model. Through its hierarchical structure and local window-based self-attention mechanism, it effectively captures global image features and seamlessly integrates them with local features, enabling more precise and efficient image processing results.

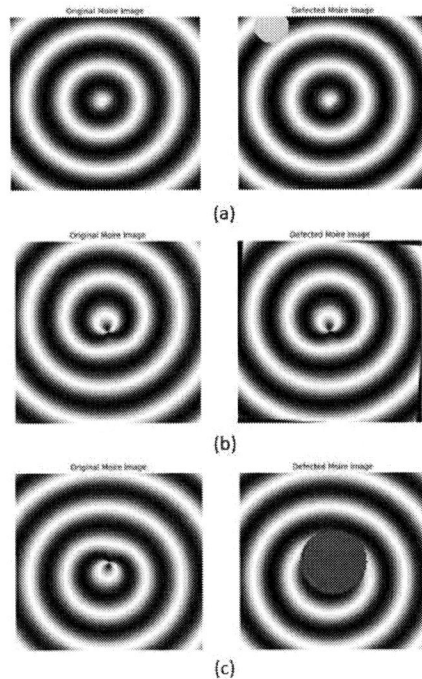

(a)

(b)

(c)

Fig. 4. The original ideal moiré fringes and the defected moiré fringes. (a) top circle corners defect. (b) sidewall corners defect. (c) the center of the circle defect.

III. EXPERIMENTS

A. Experiment Dataset

Unlike photolithophic alignment, which is performed prior to chip fabrication and requires large-area absolute precision to ensure the entire wafer pattern aligns with a reference, OVL detection is conducted after chip fabrication. Since the alignment has already been established, overlay detection focuses on measuring relative errors between layers, requiring error detection only within a small scale. Accordingly, we selected

two sets of circular gratings with periods of $1\,\mu$m and $1.1\,\mu$m, respectively, and simulated their interaction to generate experimental image dataset. These moiré fringe span displacement values in the x- and y-directions from $(-0.5\,\mu m, -0.5\,\mu m)$ to $(0.5\,\mu m, 0.5\,\mu m)$ with a step size of 1 nm, resulting in a total of one million image dataset. Each moiré fringe corresponds to a ground truth value $(\Delta x, \Delta y)$. The dataset was then randomly split into training (80%), validation (10%), and testing (10%) subsets. Additionally, to simulate the noise introduced by the etching and chemical mechanical polishing processes, defects are randomly added to the ideal dataset at the sidewall corners, top circle corners, or the center of the circle, as shown in Fig.4 following a normal distribution, within the 0.8 million training images. Gaussian noise with a mean of 0 and a variance of 0.1 is also applied.

B. Experimental Settings and Assessment Criteria

In this study, we select the Adam optimizer to train the proposed network. During training, the learning rate is dynamically adjusted through a warm-up phase followed by cosine annealing, and the warm-up period consists of 20 iterations. Through parameter tuning experiments, the minimum learning rate is set to 1×10^{-5}, and the total number of training epochs is set to 160, with a batch size of 64.

To assess the effectiveness of our models, we employ the Mean Absolute Error (MAE) between the ground truth and the predicted values as a key criterion to evaluate the performance of WOATNET. The MAE function is defined as:

$$\text{MAE} = \frac{1}{N} \sum_{i=1}^{N} \left(\left| \Delta x_i - \hat{\Delta x_i} \right| + \left| \Delta y_i - \hat{\Delta y_i} \right| \right) \quad (4)$$

Apparently, the smaller the MAE, the better the model's performance. Theoretically, when the MAE approaches zero, the predicted values are identical to the true values.

C. Experimental Results

Similarly, defects are randomly added to the dataset at the sidewall corners, top corners, or the center of the circle in the 100,000 Moiré fringe test dataset, following a normal distribution, also gaussian noise with a mean of 0 and a variance of 0.1 is also applied. The trained network was tested by the defected test dataset, whose results are shown in Table I. The experimental results indicate that the proposed hybrid architecture, WOATNet, exhibits robustness even under noisy and defective conditions, achieving sub-nanometer accuracy. Furthermore, the small standard deviation indicates that the model's predictions are consistent and reliable.

TABLE I
EXPERIMENTAL MEAN ABSOLUTE ERROR AND STANDARD DEVIATION FOR THE EVALUATION OF PREDICTED VALUES ACHIEVED WITH THE WOATNET IN THE DEFECTED TEST DATASET.

Direction	Mean Absolute Errors (nm)	Standard Deviation (nm)
X direction	0.146	0.173
Y direction	0.198	0.191

D. Ablation Study

The WOATNet comprises three components: WT Conv, OCM, and STB. To assess each submodel's contribution, we construct three networks for OVL metrology. Additionally, a fourth network, built entirely on the Swin Transformer, is designed to demonstrate that a Transformer-based approach alone may fall short compared to the proposed hybrid model. The four networks are as follows:

1) *NET1*: WT Conv + OCM.
2) *NET2*: WT Conv + STB.
3) *NET3*: OCM + STB
4) *NET4*: Swin Transformer.

Their MAE values are summarized in TableII. The results demonstrate that all three modules of WOATNet contribute positively to the performance of OVL metrology. Among the tested combinations, OCM and STB (NET3) achieve the best MAE values in the X and Y directions ($0.147nm$ and $0.272\ nm$), highlighting their ability to effectively integrate local and global features. In contrast, the Swin Transformer-only network (NET4) performs significantly worse ($41.628\ nm$ and $28.590\ nm$), underscoring the limitations of a pure Transformer architecture in lithography error metrology tasks. The ablation experiments validate the effectiveness of WOAT-Net's hybrid design, further highlighting the advantages of integrating convolutional and Transformer modules for this task."

TABLE II
MEAN ABSOLUTE ERROR FOR DIFFERENT SUBMODELS

Direction	NET1	NET2	NET3	NET4
X direction MAE (nm)	2.498	0.586	0.147	41.628
Y direction MAE (nm)	2.225	1.154	0.272	28.590

IV. CONCLUSION

In this paper, we propose a hybrid deep learning architecture, WOATNet, which seamlessly integrates Wavelet Convolution layers, a 2D Octave Convolution Model (OCM), and Swin Transformer blocks. Compared to traditional methods such as Windowed Fourier Transform(WFT) or Wavelet Transform(WT) that typically achieve tens of nanometers accuracy, WOATNet leverages the unique strengths of each component: WT Conv captures features across various frequencies and spatial scales, OCM efficiently integrates multi-scale information, and Swin Transformer blocks extracts global image features while fusing them with local details. Experimental results demonstrate that WOATNet achieves sub-nanometer accuracy in both the X and Y directions, even under simulated defect and noise conditions. Furthermore, ablation studies confirm the effectiveness of WOATNet's hybrid design, highlighting the advantages of combining convolutional and Transformer-based modules for this task.

REFERENCES

[1] B. Hoefflinger, "Irds—international roadmap for devices and systems, rebooting computing, s3s," *NANO-CHIPS 2030: On-Chip AI for an Efficient Data-Driven World*, pp. 9–17, 2020.

[2] B. Bringoltz, T. Marciano, T. Yaziv, Y. DeLeeuw, D. Klein, Y. Feler, I. Adam, E. Gurevich, N. Sella, Z. Lindenfeld *et al.*, "Accuracy in optical overlay metrology," in *Metrology, Inspection, and Process Control for Microlithography XXX*, vol. 9778. SPIE, 2016, pp. 483–501.

[3] M. Adel, D. Kandel, V. Levinski, J. Seligson, and A. Kuniavsky, "Diffraction order control in overlay metrology: a review of the roadmap options," *Metrology, Inspection, and Process Control for Microlithography XXII*, vol. 6922, pp. 23–41, 2008.

[4] Y.-S. Nam, S. Kim, J. H. Shin, Y. S. Choi, S. H. Yun, Y. H. Kim, S. W. Shin, J. H. Kong, Y. S. Kang, and H. H. Ha, "Overlay improvement methods with diffraction based overlay and integrated metrology," in *Optical Microlithography XXVIII*, vol. 9426. SPIE, 2015, pp. 256–263.

[5] N. Wang, W. Jiang, and Y. Zhang, "Deep learning–based moiré-fringe alignment with circular gratings for lithography," *Optics Letters*, vol. 46, no. 5, pp. 1113–1116, 2021.

[6] S. Finder, R. Amoyal, E. Treister, and O. Freifeld, "Wavelet convolutions for large receptive fields. arxiv 2024," *arXiv preprint arXiv:2407.05848*.

[7] X. Tang, F. Meng, X. Zhang, Y.-M. Cheung, J. Ma, F. Liu, and L. Jiao, "Hyperspectral image classification based on 3-d octave convolution with spatial–spectral attention network," *IEEE Transactions on Geoscience and Remote Sensing*, vol. 59, no. 3, pp. 2430–2447, 2020.

[8] Z. Liu, Y. Lin, Y. Cao, H. Hu, Y. Wei, Z. Zhang, S. Lin, and B. Guo, "Swin transformer: Hierarchical vision transformer using shifted windows," in *Proceedings of the IEEE/CVF international conference on computer vision*, 2021, pp. 10012–10022.

A Hybrid Algorithm for Automated Placement of CMOS Standard Cells

Weiteng Hu, Junlang Yu, Bin Li*, Zhaohui Wu

School of Microelectronics, South China University of Technology, Guangzhou 510641, China

*phlibin@scut.edu.cn

Abstract—As the complexity of standard cell library design increases, automated placement of CMOS standard cells becomes essential for shortening design cycles and reducing costs. This paper presents a hybrid algorithm that combines graph theory and heuristic methods to automatically generate and optimize the placement of CMOS standard cells. A custom graph theory algorithm is developed to find Euler paths for logic gates, while the simulated annealing algorithm is used to handle the placement of asymmetric standard cells. By strategically selecting algorithms based on the characteristics of the standard cells, the proposed approach efficiently optimizes area, net complexity, and pin density. Experimental results show that the proposed approach can achieve optimal layout quality and significantly improves the efficiency standard cell library layout development.

Index Terms—automated cell placement, graph theory, combinatorial optimization, simulated annealing

I. INTRODUCTION

CMOS standard cell design has become a fundamental part of integrated circuit design. The process of designing standard cell layouts plays a crucial role in determining the performance, power consumption, and area of the chip. However, the geometric scaling and the growing number of standard cells have made human-driven design procedure time-consuming and error-prone. As a result, automating the layout process is increasingly important, leading to an extensive research in Design Technology Co-Optimization (DTCO) [1].

Although some research has been conducted on automated layout techniques, the factors that need to be considered for advanced technology nodes are increasingly complex. Past research on automated layout algorithms typically focused on width minimization, whereas current research pay more attention on other design constraints such as symmetry, pin accessibility, and Design Rule Check (DRC) [2]. However, the design and optimization techniques for complex logic and sequential cells are still rarely discussed.

To address these challenges, this paper proposes a hybrid algorithm that combines graph theory and simulated annealing to automate the transistor placement for CMOS standard cells. By reducing routing complexity and optimizing pin accessibility, the algorithm minimizes routing space, which lead to lower standard cell height and more compact layout, and achieves a smaller layout area. Additionally, the consideration of design rules enhances the feasibility of the automatically generated standard cells.

It is demonstrated that the proposed hybrid algorithm can effectively balance demands of area optimization, pin density,

and symmetry, generating transistor placement that are both efficient and feasible for production. The rest of this paper is organized as follows: Section II introduces the proposed hybrid algorithm, including the graph theory and simulated annealing algorithm. Section III presents experimental results, and Section IV concludes the paper.

II. ALGORITHM IMPLEMENTATION

Stochastic algorithms, such as simulated annealing and genetic algorithm, are commonly used due to their flexibility in considering multiple factors [3], but it also took more runtime [4]. Alternatively, deterministic approaches, including graph theory algorithm [5], integer linear programming model, and Boolean satisfiability formulation, are more efficient but not applicable in all cases. By combining these two kinds of algorithms, the proposed approach can balance applicability and efficiency. The detailed workflow of the proposed algorithm is presented below.

Algorithm 1 The hybrid algorithmic workflow

Input: *CDL netlist, Design rules*
Output: *Transistor placement file*

1: Perform transistor folding to meet design rules.
2: **if** Standard cell exhibits duality characteristics **then**
3:     Construct graphs for PMOS and NMOS.
4:     Find Euler paths using Depth-First Search with backtracking.
5:     Evaluate legality and quality, output the optimal result.
6: **else**
7:     Transistor pairing and initialize parameters.
8:     **while** $T >$ TerminationTemperature **do**
9:         Generate new layout by modifying transistor pairs or chains.
10:         Assess layout quality based on width, net complexity, and pin density.
11:         Accept/reject layout based on Metropolis criterion.
12:         Update the placement and reduce T.
13:     **end while**
14:     Output the best placement.
15: **end if**

A. Graph Theory

To comply with design rules, large transistors are folded into smaller segments according to the specified standard cell

---

979-8-3315-2209-4/25 $31.00 © 2025 IEEE

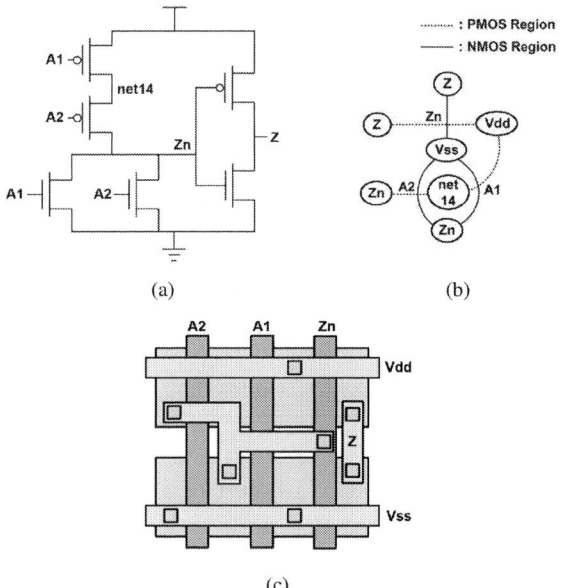

(a)   (b)

(c)

Fig. 1. Transistor placement using graph theory: (a) Netlist of the cell. (b) Graph construction from (a). (c) Layout result of the optimal solution.

height, ensuring their widths remain within acceptable ranges and avoiding violations of design rule constraints. During the graph construction, the source and drain nets of transistors are represented as vertices, while the transistors form the edges that connect these vertices and gate nets are used as labels for the edges [6], which will transform Fig. 1(a) to a pair of graphs representing the PMOS and NMOS regions respectively, as shown in Fig. 1(b).

Using the handshake theorem, graphs with more than two vertices of odd degree are immediately excluded from further processing, as they do not satisfy the necessary conditions for forming Euler paths, thereby reducing computational overhead. Euler paths in the remaining graphs are identified using a depth-first search (DFS) algorithm, which explores each branch of the graph as deeply as possible before backtracking. To ensure that all possible Euler paths are found, backtracking and pruning techniques are incorporated into the DFS process [7]. When the DFS reaches a dead-end, backtracking allows the algorithm to revert to a previous state and explore alternate path. Pruning further enhances efficiency by eliminating branches that cannot yield a valid Euler path, reducing the search space. For graphs with high complexity, independent short cycles are first recorded and temporarily removed before searching for Euler paths. When all the paths are identified, these short cycles are then reinserted into the path list to reconstruct the complete set of Euler paths corresponding to the original graph.

Under the one-dimensional layout condition, the Euler paths in both the PMOS and NMOS regions should have matching gate net orders to ensure valid pairing [8]. A hashing mechanism is employed to efficiently compare and match these paths. In this context, the hashing mechanism converts each Euler path's sequence of gate net orders into a

unique hash value, which is then stored in a hash table. This allows for constant-time lookups during the pairing process, significantly reducing the computational overhead when comparing numerous or complex paths [9]. By leveraging hashing, the comparison process is significantly accelerated, even when dealing with larger graphs. The successfully paired layouts achieve the minimum width for the standard cell. These layouts are then evaluated based on a combination of criteria, including net complexity, pin density, and design rule check. The optimal solution is selected based on this comprehensive evaluation, and it results in the layout shown in Fig. 1(c) after routing.

B. Simulated Annealing Algorithm

For standard cell layouts that cannot be processed by the graph theory, the transistor placement task is reformulated as a combinatorial optimization problem, which entails selecting the optimal solution from a finite but exponentially large set of possibilities, and is addressed using the simulated annealing algorithm. This algorithm emulates the physical cooling process of metals to iteratively refine placement configurations, gradually moving toward a globally optimal solution by exploring the solution space while avoiding being trapped in local optima.

In this approach, a transistor pair consisting of one PMOS and one NMOS is treated as a basic operational unit in the optimization process. The pairing of transistors follows the rule of matching transistors with the same gate net. By making small random modifications to the initial layout, new placement is generated. These modifications are realized through six different operations to explore diverse placement possibilities: moving transistor pairs, swapping transistor pairs, flipping a transistor pair or a single transistor, moving a transistor chain, flipping a transistor chain, and changing transistor pairings.

Each new placement aims to maximize active region sharing, forming transistor chains to make the placement as compact as possible while ensuring that it satisfies design rule. The new placement is then evaluated based on a combination of criteria, including total width, net complexity, pin density. The acceptance of the new placement is based on the Metropolis criterion, which determines the probability of accepting a new solution based on the following formula:

$$P = \begin{cases} 1, & \text{if } \Delta \leq 0 \\ \exp\left(-\frac{\Delta}{T}\right), & \text{if } \Delta > 0 \end{cases} \quad (1)$$

where Δ represents the change in the overall evaluation score between the new and original placements. At higher temperatures, the algorithm is more likely to accept suboptimal placements, allowing it to escape local optima and explore a broader solution space. As the temperature coefficient decreases during the annealing process, the probability of acceptance will decline and the algorithm converges to the placement with the optimal combined evaluation score.

Compared to the traditional simulated annealing algorithm, the developed algorithm can automatically adjust the runtime

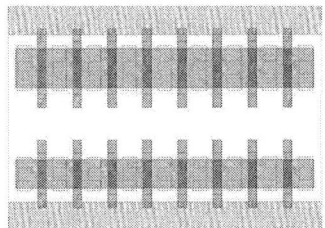

Fig. 2. Placement result using graph theory.

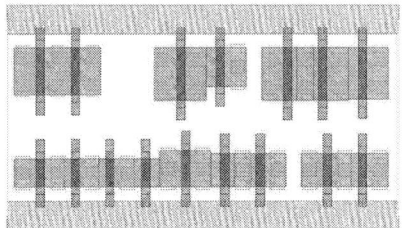

Fig. 4. Placement result using simulated annealing.

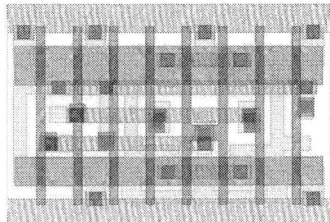

Fig. 3. Layout result using graph theory.

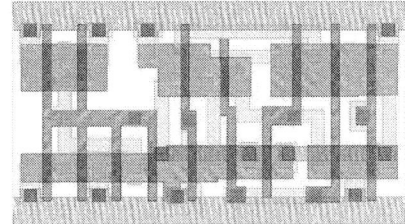

Fig. 5. Layout result using simulated annealing.

based on the size of the standard cell, ensuring that the computational effort is proportional to the problem's complexity. To reduce randomness and mitigate the risk of getting trapped in local optima, the simulated annealing processes are parallelized. The best layout is then selected from multiple independent runs, increasing the likelihood of finding the global optimum while reducing runtime for large-scale standard cell libraries.

III. EXPERIMENT RESULTS

The proposed algorithm was validated using a digital library containing 340 standard cells for transistor placement, with corresponding design rule constraints provided as inputs. During testing, the majority of logic gate circuits were processed using the graph theory algorithm. An example placement result is shown in the Fig. 2. It is demonstrated that the gates of all transistors are matched, and the amount of active area sharing is maximized to achieve the minimal cell width. No design rule violations were observed, while net complexity and pin density were simultaneously optimized. Furthermore, the execution time was approximately 0.05 seconds.

The placements generated by the proposed algorithm were further validated by manual routing to examine their compatibility. Fig.3 is the routed layout corresponding to the automated placement result presented in Fig.2. It is seen that the high-quality automated placement ensures smooth routing paths and optimized pin accessibility, thereby accelerating the overall design flow, which demonstrates the effectiveness of the proposed placement algorithm.

Standard cells without duality characteristics, as shown in the Fig. 4, were processed by the simulated annealing algorithm, demonstrating the approach's applicability. It is verified that the proposed algorithm can effectively balance

multiple metrics, including width, net complexity, and pin density, to generate optimal layouts while ensuring strict compliance with design rule constraints. For cells with fewer than 50 transistors, the program consistently generates optimal layouts within 5 minutes.

Similarly, the manually routed layout corresponding to the simulated annealing-based placement is shown in Fig. 5. For asymmetric layouts, the placement generated by the simulated annealing algorithm still enables efficient routing and pin access.

Table I summarizes the placement quality evaluation of several standard cells processed by the proposed method. Both the placement width and net complexity are measured per transistor, with net complexity defined as the sum of all net lengths. A lower net complexity value indicates easier routing. Pin accessibility is quantified by the standard deviation of inter-pin distances, and smaller value denotes a more uniform pin distribution, which conserves routing resources.

TABLE I: Placement Quality of Some Test Standard Cell

Cell Name	No. of tran-sistors	Width	Net Complexity	Pin Accessibility	Runtime (s)
AN2D2	8	4	3.5	0.06	0.05
IND3D2	14	7	20.5	0.16	0.05
AOI211D2	16	8	24	0.12	0.11
AOI33D2	24	12	38	0.1	31.22
AD1V4	28	16	62	0.08	46.33
LHCNDD2	27	19	49	0.07	87.23
DFCND2	38	29	117	0.2	237.84

Compared with [10], the proposed approach offers a significant speed advantage. Moreover, the simulated

979-8-3315-2209-4/25 $31.00 © 2025 IEEE

annealing algorithm is capable of handling complex logic and sequential cells that have not been addressed in most previous studies, thereby demonstrating its robustness and broad applicability.

IV. CONCLUSION

This paper presents a hybrid algorithm for the automated placement of standard cells, optimizing the layout performance by combining graph theory and simulated annealing algorithms. This approach significantly improves layout development efficiency and is validated on a standard cell library. Compared to previous work, this approach can automatically select the appropriate algorithm based on the characteristics of the standard cells, offering broad applicability. The algorithm not only minimizes area but also optimizes layout by considering net complexity, pin density and design rules, resulting in a well-balanced solution.

ACKNOWLEDGMENT

This work was supported by the Guangdong S&T programme, China [Grant number 2022B0101180001].

REFERENCES

[1] X. Xu, N. Shah, A. Evans, S. Sinha, B. Cline and G. Yeric, "Standard cell library design and optimization methodology for ASAP7 PDK: (Invited paper)," 2017 IEEE/ACM International Conference on Computer-Aided Design (ICCAD), Irvine, CA, USA, 2017, pp. 999-1004.

[2] H. Cho, H. Seo, S. Chung, K. -M. Choi and T. Kim, "Standard Cell Layout Generator Amenable to Design Technology Co-Optimization in Advanced Process Nodes," 2024 Design, Automation & Test in Europe Conference & Exhibition (DATE), Valencia, Spain, 2024, pp. 1-6.

[3] H. C. Prashanth, P. Jonna and M. Rao, "CellFlow: Automated Standard Cell Design Flow," 2023 IEEE Computer Society Annual Symposium on VLSI (ISVLSI), Foz do Iguacu, Brazil, 2023, pp. 1-5.

[4] A. Lu et al., "Simultaneous transistor pairing and placement for CMOS standard cells," in Design, Automation & Test in Europe Conference & Exhibition (DATE), Grenoble, France, 2015, pp. 1647-1652.

[5] J. Yoon and H. Park, "Design-Technology Co-Optimization with Standard Cell Layout Generator for Pin Configurations," 2024 25th International Symposium on Quality Electronic Design (ISQED), San Francisco, CA, USA, 2024, pp. 1-7.

[6] J. Schneider, "Transistor-level layout of integrated circuits," Ph.D. dissertation, Universitäts-und Landesbibliothek Bonn, 2014.

[7] Y. LEI, C. MA and B. YAN, "Automatic Standard Cell Layout Generator Integrated with Design Expertise," 2024 2nd International Symposium of Electronics Design Automation (ISEDA), Xi'an, China, 2024, pp. 59-63.

[8] K. Jo, S. Ahn, J. Do, T. Song, T. Kim and K. Choi, "Design rule evaluation framework using automatic cell layout generator for design technology co-optimization," in IEEE Transactions on Very Large Scale Integration (VLSI) Systems, vol. 27, no. 8, pp. 1933-1946, Aug. 2019.

[9] Z. Li, D. Zhang, X. Yuan and J. Zheng, "Deep Hash Model for Similarity Text Retrieval," 2022 5th International Conference on Artificial Intelligence and Big Data (ICAIBD), Chengdu, China, 2022, pp. 638-643.

[10] P. Van Cleeff, S. Hougardy, J. Silvanus and T. Werner, "BonnCell: Automatic Cell Layout in the 7-nm Era," in IEEE Transactions on Computer-Aided Design of Integrated Circuits and Systems, vol. 39, no. 10, pp. 2872-2885, Oct. 2020.

Design of Ecological Environment Monitoring System Based on RT-Thread and Cloud Platform

1st Jingyang Wen   2nd Lin Li   3rd Liangjing Bai   4rd Xiyuan Wu   5rd Wei Zhang   6th Hongxing Ma

Abstract—**A real-time, accurate, and cost-effective ecological environment monitoring system based on RT-Thread and cloud platform is proposed to address the issues of ecological environment monitoring. The system adopts a layered architecture: the acquisition layer collects high-frequency data through multiple sensors (temperature and humidity, PM2.5, soil nitrogen, phosphorus, potassium, etc.) and STM32. The combination of LTE Cat.1 and Ethernet dual channel redundant communication at the transport layer ensures low latency and high reliability. The application layer relies on Alibaba Cloud to achieve data storage and deep learning analysis, and develops a mobile client that supports second level recognition of plant images. Actual deployment has shown a 40% increase in system task scheduling efficiency, with communication latency as low as 120ms (LTE Cat.1 network environment) and a 41.7% reduction in memory usage, providing an efficient solution for ecological monitoring.**

Keywords—**Ecological environment monitoring; RT-ThreadCloud platform; Image recognition; Deep learning**

I. Introduction

Ecological environment monitoring is the core foundation of ecological protection and restoration [1], and its data quality directly affects the scientificity of decision-making [2]. Traditional manual monitoring methods have drawbacks such as high cost and poor real-time performance [3], while existing automation systems are often limited by the closed source, high cost, or insufficient real-time performance of embedded operating systems [4]. For example, although VxWorks is powerful, it is expensive, and RT-Linux has the problem of excessive resource consumption in real-time optimization [5]. In contrast, the domestically produced RT-Thread system provides an ideal solution for the design of distributed monitoring nodes due to its open-source, low resource consumption, and efficient task scheduling capabilities [6]. This article proposes a full stack monitoring system based on RT Thread, and its innovation lies in:

1. The hardware layer improves collection efficiency and resource utilization through RT-Thread task scheduling optimization (priority management, memory pooling mechanism).

2. The combination of transmission layer and dual channel redundancy design ensures communication reliability.

3. The application layer integrates cloud based deep learning and mobile image recognition, breaking through the bottleneck of insufficient interactivity in traditional systems.

II. Overall System Architecture

In order to improve the maintainability and scalability of the ecological environment monitoring system, the system architecture will be built according to the collection layer, transmission layer, and application layer. The collection layer integrates temperature and humidity sensors, PM2.5 sensors, soil nitrogen, phosphorus, and potassium sensors, and achieves high-frequency collection of multi-source environmental data through a microcontroller processor (such as STM32). The transport layer adopts LTE Cat.1 and Ethernet dual channel redundant transmission, supporting real-time data reporting and control command issuance, ensuring low latency and high reliability of the communication link. The application layer builds an ecological environment monitoring cloud service based on the Alibaba Cloud platform, providing data interfaces through the Flask framework to achieve data storage, deep learning analysis, and visualization, and developing a mobile client to support plant recognition in seconds. The architecture of the ecological environment monitoring system is shown in Figure 1.

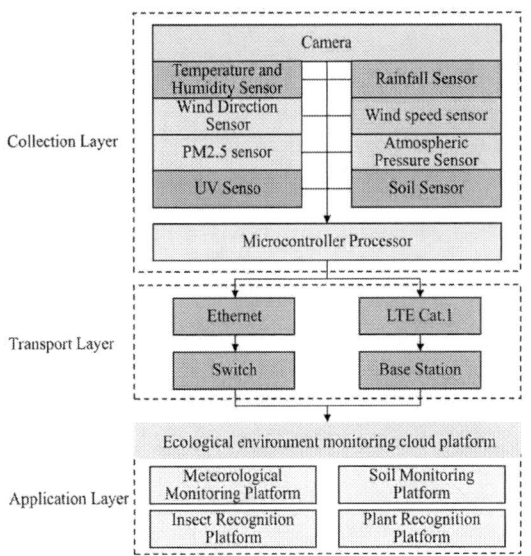

Fig. 1. Architecture of Ecological Environment Monitoring System

---

This work was supported by the Natural Science Foundation of Ningxia (2024AAC03154,2022AAC03246) , the High Level Construction Project of Ordinary Universities in Ningxia Province (bjg2024028), and the Teaching Project of North Minzu University (YKSZ202307, YJZT202431). Corresponding author: Hongxing Ma, E-mail: mhx@nmu.edu.cn.

979-8-3315-2209-4/25 $31.00 © 2025 IEEE

III. IMPLEMENTATION OF ECOLOGICAL ENVIRONMENT MONITORING SYSTEM

A. Multi Source Sensors and RT-Thread Task Scheduling Optimization

The hardware layer uses STM32 as the core controller and integrates more than ten types of sensors (with a sampling frequency of 1Hz). Through RT-Thread task scheduling optimization:

1. Set the sensor data collection task as the highest priority (priority 5) to ensure that there is no loss of sudden data.

2. Pre allocated memory pool reduces fragmentation, reducing system memory usage from 60MB in traditional Linux solutions to 35MB.

3. The hard real-time interrupt mechanism reduces data read latency and meets high-frequency acquisition requirements.

The hardware platform integrates a camera module that supports real-time collection of plant images (as shown in Figure 2). In actual deployment, the observation platform can synchronously monitor parameters such as light and rainfall, and track plant growth status, providing high-quality data sources for cloud analysis.

Fig. 2. Physical designated ecological information observation platform

B. Cloud Platform Service Deep Learning Integration

The cloud service is built on Alibaba Cloud and implements data interface services based on the Flask framework, supporting data storage and deep learning analysis. As of 2024, the system has accumulated over 7 million pieces of grassland ecological data (as shown in Figure 3), providing a rich foundation for ecological modeling.

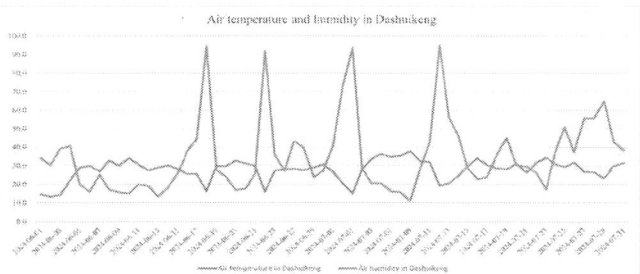

Fig. 3. Temperature and humidity statistics of Dashuikeng in 2024

The fixed-point ecological information observation platform has significant errors when using monocular cameras for plant recognition and height monitoring (as shown in Figure 4). Optimizing the ruler contour detection algorithm significantly improves measurement accuracy.

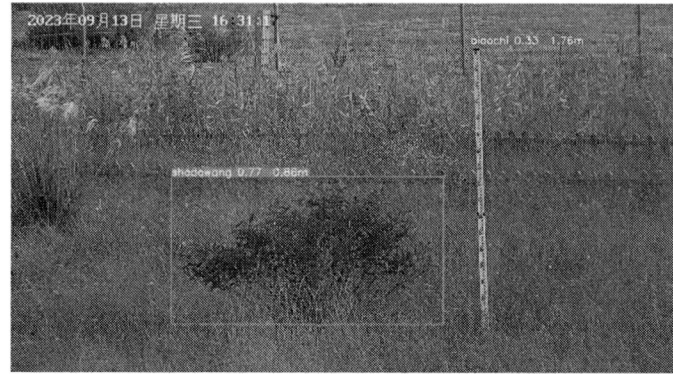

Fig. 4. Plant identification and height monitoring

IV. CONCLUSIONS

This system achieves efficient collection and stable transmission of multi-source data through RT-Thread task scheduling optimization (priority management, memory pool pre allocation) and dual channel redundant transmission design. Cloud based integrated deep learning models support second level recognition of plant images (accuracy¿92%) and dynamic assessment of ecological health. The actual deployment has verified the reliability and practicability of the system. In the future, the real-time can be optimized and the monitoring dimension can be expanded in combination with edge computing.

REFERENCES

[1] K. Jiangze, Y. Xueqing, L. Maoying, " Weave a dense monitoring network to care for green mountains and clear waters," People's Daily, p. 4, Jan. 09, 2024.

[2] Z. Yang, L. Jian, H. Minxin, "Problems and countermeasures in the quality management of ecological environment monitoring," Chemical Engineering Management, vol. 36, pp. 63-64, 2020.

[3] S. Yingjie, B. Jingdong, W. Yitao, et al, "Design of wireless sensor network system based on RT-Thread and ESP-NOW," Instrumentation Technology and Sensor, vol. 7, pp. 88-92+98, 2023.

[4] Y. Lei, W. Bangji, W. Bo, et al., "Design and implementation of real-time control system based on RT-Thread for Zynq-7000," Instrumentation Technology and Sensor, vol. 6, pp. 71-75, 2023.

[5] Liu J, Zhong Y, Wang Q. A Survey on Real-Time Modification Techniques for Embedded Linux. Aerospace Control, vol. 36, pp. 93-97, 2018.

[6] Kozlowski T, Noran O, Trevathan J. Designing an Evaluation Framework for IoT Environmental Monitoring Systems. Procedia Computer Science, vol. 219, pp. 220-227, 2023.

[7] Y. Zhang, T. Yan, Y. Wang and Q. Li, "Marine Environment Dynamic Monitoring System under Remote Sensing Ecological Index," 2024 International Conference on Integrated Circuits and Communication Systems (ICICACS), Raichur, India, pp. 1-5, 2024.

[8] Y. Zhang, S. Tan, X. Li and D. Xu, "Marine Ecological Environment Monitoring and Management System Based on Sensor Technology," 2024 International Conference on Integrated Circuits and Communication Systems (ICICACS), Raichur, India, pp. 1-6, 2024.

Content Caching-oriented Popularity Forecast Algorithm Design

Qi Chen, Member, IEEE,   Wenze Gao,   Takayuki Nakachi, Member, IEEE,

Yitu Wang, Member, IEEE,   Juinjei Liou, Fellow, IEEE

Abstract—To enable proactive content caching, file popularity forecast becomes an indispensable technique. Conventionally, the objective of information forecast lies in maximizing the accuracy, while neglecting other important metrics, such as forecast confidence and model complexity, which are crucial for the design of content caching. In this paper, we tailor Gaussian Process (GP)-based forecast algorithm for content caching so as to further improve the caching performance. Specifically, we analytically derive the influence of forecast confidence on caching performance, and propose the idea of Controlled Linear Model of Coregionalization (CLMC) to achieve a desired trade-off between forecast confidence and model complexity in terms of minimizing cache fetching loss. The performance improvement is verified by simulation.

I. Introduction

The long distance between user devices and cloud servers poses significant challenge on providing massive content delivery with latency constraint. To address this issue, cache-enabled network becomes a promising solution, which caches contents in close proximity to users. The commonly adopted caching criteria is frequency-based caching, where several the most popular contents are cached locally so as to maximize the hit ratio of content fetching [1], [2], [3], [4]. In this regard, it is essential to perceive and predict content popularity for achieving proactive caching, as popularity of contents is temporally evolving. In the literature, content popularity is directly predicted based on historical records by adopting time series prediction techniques, such as Auto Regression Integreate Moving Average (ARIMA) and Long Short Term Memory (LSTM) [5], while neglecting the interaction between popularity forecast and content caching. On top of

---

This work is supported in part by Key R&D projects in Ningxia Hui Autonomous Region (2024BEH04021), National Natural Science Foundation of China (No. 62301007,52065002), AI-Driven Research Paradigm Reform and Discipline Advancement Initiative Project of Shanghai Municipal Education Commission (Edge Intelligent Network Cooperation Benefit Allocation Supported by Privacy Rights under the 'Digital Intelligence-Legal' Framework), Humanities and Social Sciences Project of the Ministry of Education under Grant (24YJCZH023), JSPS Grant-in-Aid for Scientific Research (22K04089), and NingXia Natural Science Foundation for Young Elite Scientists Sponsorship Program. Corresponding author: Yitu Wang (yituwang@nmu.edu.cn).

Qi Chen is with the School of Artificial Intelligence and Law, Shanghai University of Political Science and Law, Shanghai 201701, China.

Yitu Wang, Wenze Gao, and Juinjei Liou are with the School of Electrical and Information Engineering, and Information Engineering and the Intelligent Equipment and Precision Measurement Technology Research and Development Group, North Minzu University, China.

Takayuki Nakachi is with the Information Technology Center, University of the Ryukyus, Nishihara-cho, Okinawa, Japan.

forecast accuracy, other metrics, especially forecast confidence and model complexity, also pose significant influence on the performance of content caching, which necessitates tailoring the conventional prediction techniques to satisfy special requirements of cache-enabled networks.

In this paper, we present some preliminary results of designing content caching-oriented popularity forecast algorithm. Especially, we consider a light-weight and non-parametric data-driven technique, i.e., GP, for popularity forecast [6], as a posterior distribution on the forecast could be derived, which poses vital influence on the performance of content caching. To analytically depict such influence, we first propose the caching problem as minimizing the expected loss of cache fetching, which is reformulated through considering four non-overlapping cases. Then, by analyzing the influence of forecast error on the expected loss, the probability of inaccurate forecast can be derived. Finally, the influence of forecast mean and variance on cache fetching can be analytically obtained, based on which we propose the idea of GP-CLMC to enhance the forecast performance w.r.t. caching performance.

II. System Model

The network of interest is with a cloud server, a cache-enabled edge server and N users, denoted as set $\mathcal{N}$. There exists F different files with the same size, denoted as set $\mathcal{F}$ [7]. Time is slotted, and the duration of a time slot is one unit of time. The popularity $c_i^k(t)$ is defined as hit counts of file k of user i in time slot t, and thus, the accumulated popularity of k in $\mathcal{N}$ can be calculated as

$$C^k(t) = \sum_{i \in \mathcal{N}} c_i^k(t). \tag{1}$$

Denote L as the constraint on cache space, i.e., the maximum number of files could be cached at the edge. Thus, the most popular L contents will be cached, denoted as set $\mathcal{L}(t)$. Based on $\mathcal{L}(t)$, the cache hit ratio is

$$s(t) = \frac{\sum_{k \in \mathcal{L}(t)} C^k(t)}{\sum_{k \in \mathcal{F}} C^k(t)}. \tag{2}$$

Our goal is to design a popularity forecast algorithm for maximizing the expected cache hit ratio. However, $C^k(t)$ cannot be explicitly known by the edge server at the beginning of t due to the causality constraint. To this end, popularity is casted into a time series $\{C^k(t-M), C^k(t-M+1), \cdots, C^k(t)\}$, and we predict $C^k(t)$ based on the historical records. We

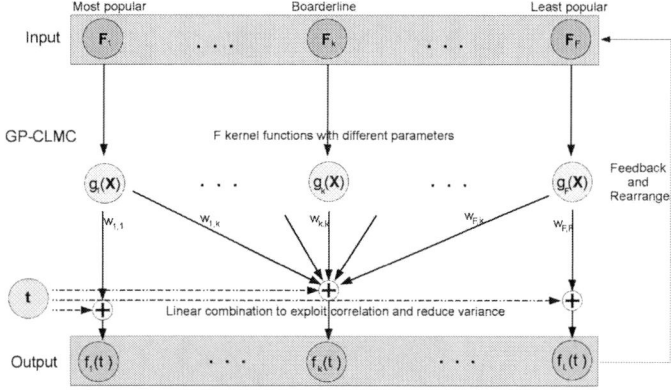

Fig. 1.   GP-CLMC Framework

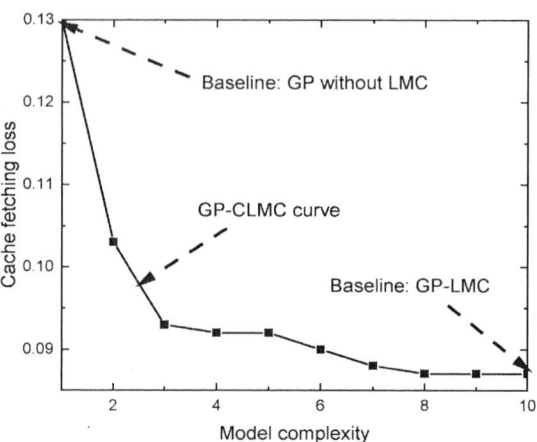

Fig. 2.   Cache fetching loss v.s. model complexity

formally formulate the optimization problem as follows,

$$\max_{h(\cdot)} \mathbb{E}_t\{s(t)\},$$

$$s.t. C^k(t) = h(C^k(t-M), C^k(t-M+1), \cdots, C^k(t-1)), \tag{3}$$

where $h(\cdot)$ denotes the forecast function.

III. PROPOSED FRAMEWORK

Since it is inviable to directly optimize Eq. (3), we first consider the following four non-overlapping circumstances,

1) Files that should be cached, are not cached due to inaccurate forecast. Denote the set as $\mathcal{F}_1(t)$, and the probability as $p_1^k(t), \forall k \in \mathcal{F}_1(t)$.
2) Files that should be cached, are cached. Denote the set as $\mathcal{F}_2(t)$, and the probability as $p_2^k(t), \forall k \in \mathcal{F}_2(t)$.
3) Files that should not be cached, are cached. Denote the set as $\mathcal{F}_3(t)$, and the probability as $p_3^k(t), \forall k \in \mathcal{F}_3(t)$.
4) Files that should not be cached, are cached due to inaccurate forecast. Denote the set as $\mathcal{F}_4(t)$, and the probability as $p_4^k(t), \forall k \in \mathcal{F}_4(t)$.

Clearly, circumstances 2 and 3 will not cause any deviation from optimal. Due to inaccurate forecast, files with higher actual popularity are replaced with those with lower actual popularity, which brings significant performance loss. Specifically, files in $\mathcal{F}_1(t)$ cause cache fetching loss, while files in $\mathcal{F}_4(t)$ compensate the fetching loss to some extent. Based on the above discussion, we can safely transform Eq. (3) into the following problem,

$$\min_{h(\cdot)} \mathbb{E}_t\left\{ \sum_{k \in \mathcal{F}_1(t)} p_1^k(t) C^k(t) - \sum_{k \in \mathcal{F}_4(t)} p_4^k(t) C^k(t) \right\}. \tag{4}$$

Given the training dataset $(\mathbf{X}, \mathbf{Y})$, the testing index x^* and kernel function $K(\cdot)$, GP models the corresponding testing output as a Gaussian distribution with mean $\hat{f}(x^*)$ and variance $\epsilon^2(x^*)$

$$\hat{f}(x^*) = \mathbf{K}_*^T(\mathbf{K}(\mathbf{X},\mathbf{X}) + \sigma^2\mathbf{I})^{-1}\mathbf{Y},$$
$$\epsilon^2(x^*) = K(x^*, x^*) - \mathbf{K}_*^T(\mathbf{K}(\mathbf{X},\mathbf{X}) + \sigma^2\mathbf{I})^{-1}\mathbf{K}_*, \tag{5}$$

where the parameters in $\mathbf{K}$ and ϵ are adjusted through minimizing the negative log marginal likelihood. Then, given the

cache threshold $\gamma(t)$ determined by cache size L, probabilities $p_1^k(t)$ and $p_4^k(t)$ can be calculated as [8],

$$p_1^k(t) = \frac{1}{2}\sqrt{\epsilon^k(t)}\left(1 - erf\left(\frac{\tau^k(t)}{\sqrt{2}\epsilon^k(t)}\right)\right),$$
$$p_4^k(t) = \frac{1}{2}\sqrt{\epsilon^k(t)}\left(1 - erf\left(\frac{-\tau^k(t)}{\sqrt{2}\epsilon^k(t)}\right)\right), \tag{6}$$

where $\tau^k(t) = \gamma(t) - C^k(t)$, and $erf(\cdot)$ represents the error function. It is observed that as for these "borderline files" in $\mathcal{F}_1(t) \cup \mathcal{F}_4(t)$, the smaller the associated variances are, the smaller cache fetching loss we suffer. Since the computation resource deployed on user devices is strictly limited, we propose the GP-CLMC framework for popularity forecast. As shown in Fig. 1, as for files in $\mathcal{F}_2(t) \cup \mathcal{F}_3(t)$, we generate the forecast using only one kernel function to reduce model complexity; As for files in $\mathcal{F}_1(t) \cup \mathcal{F}_4(t)$, we generate the forecast through linearly combining F kernel functions to reduce forecast variance in a best-effort way. In this sense, it becomes possible to achieve a desired trade-off between forecast confidence and model complexity in terms of minimizing cache fetching loss.

IV. SIMULATION

In this simulation, we evaluate the performance of the proposed forecast framework using a publicly available dataset LastFM [9]. The number of users is set to 40, the number of files is around 17000. Since the number of popular files is small, the cache size is set to 30. Denote model complexity as the number of nonzero weights in CLMC framework, which is proportion to F, i.e., the larger model complexity becomes, the larger number of borderline files we consider. It is demonstrated in Fig. 2 that the cache fetching loss becomes smaller when we increase the model complexity, while a too large complexity does not account for further improvement, which verifies the effectiveness of manipulating model structure in terms of minimizing cache fetching loss.

References

[1] M. C. Lee, A. F. Molisch, N. Sastry, and A. Raman, "Individual preference probability modeling and parameterization for video content in wireless caching networks," *IEEE/ACM Trans. Netw.*, vol. 27, no. 2, pp. 676-690, Apr. 2019.

[2] Y. Wang, M. Kong, G. Zhang, W. Wang, T. Nakachi and J. Liou, "Adaptive task offloading for mobile edge computing with forecast information," *IEEE Trans. Veh. Technol.*, vol. 74, no. 3, pp. 4132-4147, Mar. 2025.

[3] D. Wen, P. Liu, G. Zhu, Y. Shi, J. Xu, Y. C. Eldar, and S. Cui, "Task-oriented sensing, computation, and communication integration for multi-device edge AI," *IEEE Trans. Wireless Commun.*, vol. 23, no. 3, pp. 2486-2502, Mar. 2024.

[4] Y. Wang, W. Wang, V. K. Lau, T. Nakachi, and Z. Zhang, "Stochastic resource allocation and delay analysis for mobile edge computing systems," *IEEE Trans. Commun.*, vol. 71, no. 7, pp. 4018-4033, Jul. 2023.

[5] K. N. Doan, T. Van Nguyen, T. Q. S. Quek, and H. Shin, "Content-aware proactive caching for backhaul offloading in cellular network," *IEEE Trans. Wireless Commun.*, vol. 17, no. 5, pp. 3128-3140, May 2018.

[6] Y. Wang, T. Nakachi, and W. Wang, "Pattern Discovery and Multi-slot-ahead Forecast of Network Traffic: A Revisiting to Gaussian Process," *IEEE Trans. Netw. Service Manag.*. vol. 20, no. 2, pp. 1691-1706, Jun. 2023.

[7] S. E. Hajri and M. Assaad, "Caching improvement using adaptive user clustering," *Proc. of IEEE SPAWC 2016*, pp. 1-5, Jul. 2016.

[8] Q. Chen, W. Wang, F. R. Yu, M. Tao, and Z. Zhang, "Content caching oriented popularity prediction: A weighted clustering approach," *IEEE Trans. Wireless Commun.*, vol. 20, no. 1, pp. 623-636, Jan. 2021.

[9] G. Guo, J. Zhang, D. Thalmann, and N. Yorke-Smith, "ETAF: An extended trust antecedents framework for trust prediction," *Proc. of IEEE/ACM ASONAM 2014*, pp. 540-547 Aug. 2014.

Energy Efficient Scheduling with Forecast Information

Xuying Zhou, Member, IEEE,   Bijiao Yang,   Takayuki Nakachi, Member, IEEE,
Yitu Wang, Member, IEEE,   Juinjei Liou, Fellow, IEEE

Abstract—**To support massive data exchange for mobile devices with strictly limited battery, conventionally, water-filling-based power allocation is performed across different channels so as to maximize the utility of power consumption. With the proliferation of Machine Learning (ML) and Artificial Intelligence (AI), information forecast brings a new dimension to further enhance the performance. In this paper, we propose an analytical framework for water-filling-based power allocation in the temporal domain. Specifically, with L-step-ahead channel forecast, given sum data rate, we minimize power consumption based on Karush-Kuhn-Tucker (KKT) conditions to obtain closed-form power allocation. Following this, the performance improvements and the influence of inaccurate forecast information are evaluated through simulation.**

I. INTRODUCTION

With the advancement of Internet of Things (IoT), communication intensive applications are being proliferated, such as digital twin, smart factory and automated driving [1], [2], [3], [4], which bring unprecedented difficulties to the wireless communication systems due to the limited resource deployed on mobile devices. To address the challenge on communication capability, instead of allocating a designated channel to a user, some researches exploit channel diversity in Orthogonal Frequency Division Multiplexing Access (OFDMA) systems [5], some researches utilize techniques, such as spectrum aggregation, to allocate more channels to a user to support massive data transmission [6]. To address the constraint on power consumption, given channel allocation, the optimality of power consumption can be achieved through water-filling approach [7]. Due to the performance guarantee in terms of optimality, such solution is further extended to numerous wireless communication systems, including Intelligent Reflecting Surface (IRS) systems [8], energy harvesting scenarios [9], and Unmanned Aerial Vehicle (UAV) systems [10].

This work is supported in part by Key R&D projects in Ningxia Hui Autonomous Region (2024BEH04021), National Natural Science Foundation of China (No. 62301007, 62201539, 52065002), JSPS Grant-in-Aid for Scientific Research (22K04089), and NingXia Natural Science Foundation for Young Elite Scientists Sponsorship Program. Corresponding author: Yitu Wang (yituwang@nmu.edu.cn).

Xuying Zhou is with the College of Information Engineering, China Jiliang University, Hangzhou 310018, China.

Yitu Wang, Bijiao Yang, and Juinjei Liou are with the School of Electrical and Information Engineering, and Information Engineering and the Intelligent Equipment and Precision Measurement Technology Research and Development Group, North Minzu University, China.

Takayuki Nakachi is with the Information Technology Center, University of the Ryukyus, Nishihara-cho, Okinawa, Japan.

Besides the above discussions, ML and AI are finding application in wireless resource management. Conventionally, channel quality is obtained through channel estimation [11]. Recently, data-driven predictors have established themselves as strong competitors due to high prediction accuracy. For instance, it becomes possible to make real-time multi-step-ahead channel prediction based on Deep Neural Network (DNN) [12]. With such forecast information, we can extend water-filling to the temporal domain, i.e., manipulate power allocation in the temporal domain, to further improve the performance. However, limited prediction accuracy will negatively affect the utility of the forecast information in terms of deviation from the optimality of power allocation. In this paper, we present some preliminary results of energy efficient scheduling with forecast information. Specifically, the optimal scheduling is obtained in closed form based on convex optimization with KKT conditions. Then, the performance improvements and the influence of predict accuracy are demonstrated by simulation results.

II. SYSTEM MODEL

Consider a general wireless communication scenario, which includes one tranceiver pair with a designated wireless channel [1]. Time is slotted and the duration of a time slot is one unit of time. Let $x(t)$ denote the information symbol, the corresponding received signal is

$$y(t) = h(t)\frac{p(t)}{x}(t) + n(t), \tag{1}$$

where $h(t)$ denotes the time-varying channel quality, $p(t)$ is the transmit power, which is controlled by the scheduler, and $n(t)$ represents the i.i.d. complex Gaussian white noise with averaged power N_0. Following this, the received signal-to-noise ratio can be obtained as

$$\gamma(t) = \frac{|h(t)|^2}{N_0}p(t). \tag{2}$$

Denote $s(t) = \frac{|h(t)|^2}{N_0}$ as channel state, which remains constant within a time slot, and could vary across time slots in an i.i.d. manner. At the beginning of t, the scheduler determines $p(t)$, then the scheduled data rate can be represented as

$$r(t) = W \log_2(1 + p(t)s(t)), \tag{3}$$

where W denotes the bandwidth.

---

[1] The results of this paper can be easily extended to scenarios with multiple transceivers and multiple channels

In this paper, based on $L+1$-step-ahead channel forecast, generated by data-driven channel forecast algorithms, such as [12], [13], we try to minimize total power consumption with sum rate constraint that controlling the delay performance of the system, which can be formally formulated as

$$\min_{p(0),\cdots,p(t)} \sum_{i=0}^{L} p(t+i),$$

$$s.t. \sum_{i=0}^{L} W \log_2(1+s(t+i)p(t+i)) \geq (L+1)\lambda, \quad (4)$$

where λ denotes the averaged data arrival rate.

III. PROPOSED FRAMEWORK

In this section, we propose the energy efficient scheduling algorithm with forecast information in closed form.

To solve the optimization problem in Eq. (4), we establish the Lagrangian function as

$$G = \sum_{i=0}^{L} p(t+i) + \kappa\left((L+1)\lambda - \sum_{i=0}^{L} W \log_2(1+s(t+i)p(t+i))\right), \quad (5)$$

where κ is the Lagrange multiplier.

Theorem 1 (Optimal Power Allocation with Forecast Information). *For the optimality of Eq. (5), the optimal power allocation is*

$$p(t+i) = \left[\frac{2^{(L+1)\lambda}W}{\prod_{j=0}^{L} s^{1/(L+1)}(t+j)} - \frac{1}{s(t+i)}\right]^{+}. \quad (6)$$

Proof: Eq. (6) can be obtained by Lagrangian method with KKT conditions, i.e., letting the partial derivative of G w.r.t. $p(t+i)$ equals to zero. ∎

Remark 1. *Theorem 1 allocates power in a temporally water-filling manner. Specifically, as shown in Fig. 1, the water-filling levels of different time slot are balanced across the predicting horizon, i.e., $L+1$ time slots, which leads to enhancing the utility of transmit power. In addition, the power allocation deviates from optimal if forecast information is inaccurate, i.e., from time slot 6 to time slot 10, we demonstrate the case when the channel quality is over-estimated, resulting in smaller allocated power.*

IV. SIMULATION

In this simulation, channel state obeys Rayleigh distribution with the fading coefficient 4.5 and is i.i.d. over time slots. For performance comparison, we adopt a baseline scheme with $L = 0$, i.e., power allocation without channel forecast. The metrics of performance improvement is defined as $\gamma = \sum_{j=0}^{L} r(t+j)/\sum_{j=0}^{L} \hat{r}(t+j)$, where $\hat{r}(t+j)$ represents the rate allocation of the baseline algorithm.

It is demonstrated in Fig. 2 that with accurate forecast, it is always satisfied that $\gamma \geq 1$ despite of channel states, which confirms the performance improvement brought by incorporating forecast information; With inaccurate forecast, where the averaged deviation of predicted channel quality

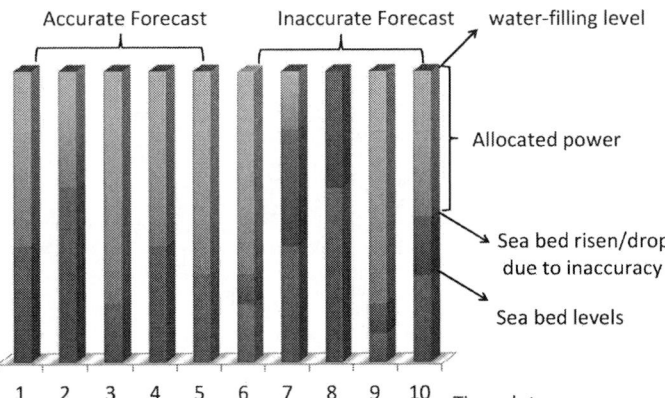

Fig. 1. Balancing the water-filling levels temporally

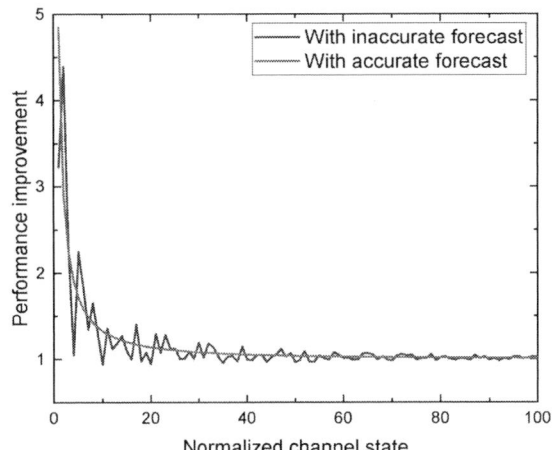

Fig. 2. Performance evaluation

from ground truth is approximately 10%, it is generally satisfied that $\gamma \geq 1$, while the performance improvement could be diminished for a few extreme cases, which necessitates the design of better handling prediction error for wireless communication systems.

REFERENCES

[1] L. Zhang, H. Wang, H. Xue, H. Zhang, Q. Liu, D. Niyato, and Z. Han, "Digital twin-assisted edge computation offloading in industrial Internet of Things with NOMA," *IEEE Trans. Veh. Technol.*, pp. 1-15, Apr. 2023.

[2] Y. Wang, M. Kong, G. Zhang, W. Wang, T. Nakachi and J. Liou, "Adaptive task offloading for mobile edge computing with forecast information," *IEEE Trans. Veh. Technol.*, vol. 74, no. 3, pp. 4132-4147, Mar. 2025.

[3] D. Wen, P. Liu, G. Zhu, Y. Shi, J. Xu, Y. C. Eldar, and S. Cui, "Task-oriented sensing, computation, and communication integration for multi-device edge AI," *IEEE Trans. Wireless Commun.*, vol. 23, no. 3, pp. 2486-2502, Mar. 2024.

[4] Y. Wang, W. Wang, V. K. Lau, T. Nakachi, and Z. Zhang, "Stochastic resource allocation and delay analysis for mobile edge computing systems," *IEEE Trans. Commun.*, vol. 71, no. 7, pp. 4018-4033, Jul. 2023.

[5] J. Huang, V. G. Subramanian, R. Agrawal, and R. A. Berry, "Downlink scheduling and resource allocation for OFDM systems," *IEEE Trans. Wireless Commun.*, vol. 8, no. 1, pp. 288-296, Jan. 2009.

[6] Y. Wang, W. Wang, L. Chen, P. Zhou, and Z. Zhang, "Energy efficient scheduling for delay-constrained spectrum aggregation: A differentiated water-filling approach," *IEEE Trans. Green Commun. Netw.*, vol. 1, no. 4, pp. 395-408, Dec. 2017.

[7] W. Yu, W. Rhee, S. Boyd, and J. M. Cioffi, "Iterative water-filling for Gaussian vector multiple-access channels," *IEEE Trans. Inf. Theory*, vol. 50, no. 1, pp. 145-152. Jan. 2004.

[8] O. Ozdogan, E. Bjornson, and E. G. Larsson, "Using intelligent reflecting surfaces for rank improvement in MIMO communications," *Proc. of IEEE ICASSP 2020*, pp. 9160-9164, May 2020.

[9] C. Qiu, Y. Hu, Y. Chen, and B. Zeng, "Deep deterministic policy gradient (DDPG)-based energy harvesting wireless communications," *IEEE Internet Things J.*, vol. 6, no. 5, pp. 8577-8588, Oct. 2019.

[10] J. Zhao, J. Liu, J. Jiang, and F. Gao, "Efficient deployment with geometric analysis for mmWave UAV communications," *IEEE Wireless Commun. Lett.*, vol. 9, no. 7, pp. 1115-1119, Jul. 2020.

[11] L. Wei, C. Huang, G. C. Alexandropoulos, C. Yuen, Z. Zhang, and M. Debbah, "Channel estimation for RIS-empowered multi-user MISO wireless communications," *IEEE Trans. Commun.*, vol. 69, no. 6, pp. 4144-4157, Jun. 2021.

[12] M. K. Shehzad, L. Rose, M. F. Azam, and M. Assaad, "Real-time massive MIMO channel prediction: A combination of deep learning and neuralprophet," *Proc. of IEEE GLOBECOM 2022*, pp. 1423-1428, Dec. 2022.

[13] Y. Wang, T. Nakachi, T. Inoue, T. Mano and R. Kudo, "Correlation discovery and channel prediction in mobile networks: A revisiting to Gaussian process," *Proc. of IEEE GLOBECOM 2021*, pp. 1-6, Dec. 2021.

Delay and Power Trade-off with Dynamic Lyapunov Function

Qi Chen, Member, IEEE,   Yang Yu,   Takayuki Nakachi, Member, IEEE,
Yitu Wang, Member, IEEE,   Juinjei Liou, Fellow, IEEE

Abstract—To balance delay and power in wireless communication systems, Lyapunov optimization is a commonly adopted technique. Through jointly considering queue backlog and power consumption, Lyapunov optimization in quadratic form achieves $O(V) - O(1/V)$ trade-off with unit power price V. However, as the distributions of channel quality and user transmission requests could be evolving over time, it remains an interesting yet important question *Should the form of the associated Lyapunov function adapt to such dynamic environment to achieve the best performance?* In this paper, we propose a scheduling framework to balance delay and power based on Lyapunov optimization, where the form of the associated Lyapunov function is dynamically determined according to the distance between the current average data arrival rate and the boundary of capacity region. The performance improvement is verified by simulation.

I. INTRODUCTION

Future wireless systems pose exigent requirements on ultra-high data rate and ultra-low delay to support numerous fancy applications, such as digital twin, smart factory and automated driving [1], [2], [3], [4]. However, the limited battery deployed on mobile devices necessitates the trade-off between delay and power consumption for practical implementation. For this purpose, there exists several techniques,

1) Large deviation transforms delay constraint into rate constraint, while it cannot adapt to instantaneous queue backlog [5].
2) Stochastic majorization minimizes delay with symmetric arrivals [6].
3) Markov Decision Process (MDP) minimizes delay performance with system state distributions [7].
4) Lyapunov optimization trade-off queue backlog and power consumption dynamically [8].

This work is supported in part by Key R&D projects in Ningxia Hui Autonomous Region (2024BEH04021), National Natural Science Foundation of China (No. 62301007,52065002), AI-Driven Research Paradigm Reform and Discipline Advancement Initiative Project of Shanghai Municipal Education Commission (Edge Intelligent Network Cooperation Benefit Allocation Supported by Privacy Rights under the 'Digital Intelligence-Legal' Framework), Humanities and Social Sciences Project of the Ministry of Education under Grant (24YJCZH023), JSPS Grant-in-Aid for Scientific Research (22K04089), and NingXia Natural Science Foundation for Young Elite Scientists Sponsorship Program. Corresponding author: Yitu Wang (yituwang@nmu.edu.cn).

Qi Chen is with the School of Artificial Intelligence and Law, Shanghai University of Political Science and Law, Shanghai 201701, China.

Yitu Wang, Yang Yu and Juinjei Liou are with the School of Electrical and Information Engineering, and Information Engineering and the Intelligent Equipment and Precision Measurement Technology Research and Development Group, North Minzu University, China.

Takayuki Nakachi is with the Information Technology Center, University of the Ryukyus, Nishihara-cho, Okinawa, Japan.

Among which Lyapunov optimization is advocated for its versatility and adaptability. In the literature, Lyapunov optimization with different Lyapunov functions are discussed. In [8], Lyapunov function in quadratic form is adopted to achieve $O(V) - O(1/V)$ trade-off. In [9], Lyapunov function in exponential form is adopted to achieve $O(\sqrt{V}) - O(1/V)$ trade-off. Most existing works adopt a fixed Lyapunov function for a designated problem. However, the performance of Lyapunov optimization is closely related to the distance between current rate arrival vector and the boundary of capacity region, where the distribution of the former could be time-evolving, and the latter is dependent on both the wireless channel quality and the scheduling algorithm. Therefore, it naturally raises a question whether the system performance can be improved with dynamic Lyapunov functions, the form of which is determined according to the current system state. In this paper, we present some preliminary results of designing delay and power trade-off scheduling algorithm with dynamic Lyapunov function. Especially, we adopt exponential Lyapunov function when rate vector is far inside the capacity region, while utilize quadratic Lyapunov function when rate vector becomes near the boundary of the capacity region. Through simulation, performance improvements in terms of delay and jitter of delay are verified.

II. SYSTEM MODEL

The wireless communication system of interest is consisted of a transceiver with a designated wireless channel. Time is slotted and the duration of a time slot is one unit of time. Denote $h(t)$ as the time-varying channel quality, $p(t)$ as the transmit power, B as the bandwidth, and $n(t)$ as the i.i.d. complex Gaussian white noise with average power N_0, we have the scheduled data rate as

$$r(t) = B \log_2(1 + p(t)s(t)), \qquad (1)$$

where $s(t) = \frac{|h(t)|^2}{N_0}$ represents channel state. Each transmitter maintains a packet queue with backlog $Q(t)$. Let $A(t)$ be the random packet arrival. To characterize the temporal evolution of the distribution of $A(t)$, we group K time slots into a frame, within which the average arrival rate is $\lambda_k, \forall k \in \mathbb{Z}^+$. The queue dynamics of $Q(t)$ is

$$Q(t + 1) = \max\{Q(t) - r(t), 0\} + A(t). \qquad (2)$$

In this paper, we are intended to find an appropriate Lyapunov function based on λ_k for each time frame, which

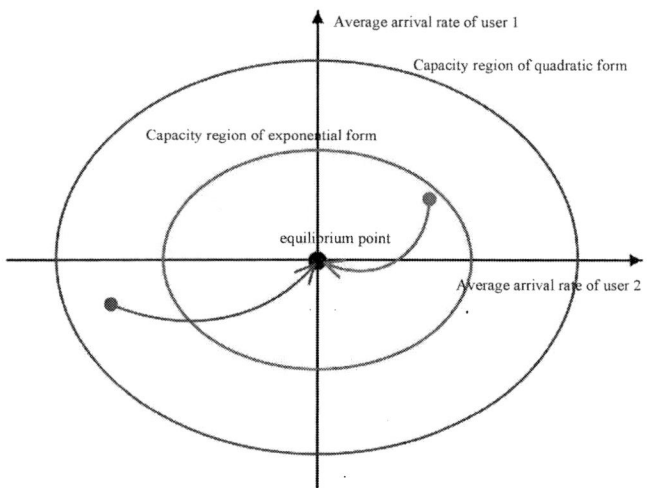

Fig. 1.   Illustration of stability region

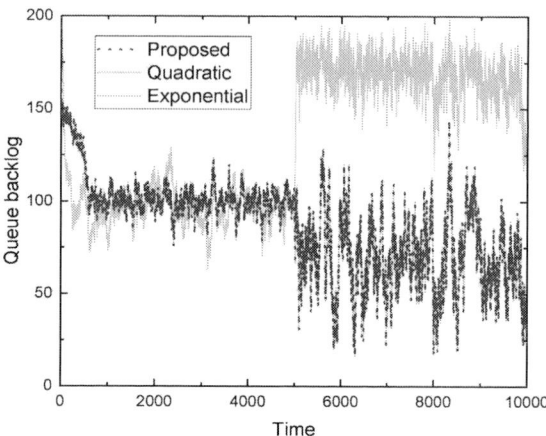

Fig. 2.   Performance comparison

can be formally formulated as

$$Q(t) \text{ is strongly stable,}$$
$$s.t. \quad \mathbf{P}(t) \in \Pi, \tag{3}$$

where Π denotes the power constraint, and strong stability is satisfied when

$$\lim_{T \to \infty} \frac{1}{T} \left(\sum_{t=0}^{T} \mathbb{E}[Q(t)] \right) < \infty. \tag{4}$$

III. Proposed Framework

We consider the following two Lyapunov functions [8][9],

$$L_1\big(Q(t)\big) = \frac{1}{2}\big(Q(t)\big)^2. \tag{5}$$

$$L_2\big(\mathbf{Q}(t)\big) = \big(e^{\omega(Q(t)-N)} + e^{\omega(N-Q(t))} - 2\big) + \frac{1}{2}\big(X(t)\big)^2, \tag{6}$$

where N denotes the objective average queue backlog, and $W(t)$ is a virtual queue whose dynamics follows

$$X(t) = \max\big\{X(t) - \big(r(t) + \epsilon 1_{Q(t)<N}(t)\big), 0\big\} \\ + A(t) + \epsilon 1_{Q(t) \geq N}(t), \tag{7}$$

where ϵ is a positive coefficient affecting the rate of exponential increase, and $1_{\cdot}(t)$ is an indicator function.

According to the Lyapunov drift minimization algorithm, the scheduling results are respectively as

$$p_1 = \frac{Q(t)B}{V \ln 2} - \frac{1}{s(t)}, \tag{8}$$

$$p_2 = \frac{W(t)B}{V \ln 2} - \frac{1}{s(t)}, \tag{9}$$

where

$$W(t) = 1_{Q(t) \geq N}(t)\left(\omega e^{\omega(Q(t)-N)} + 2X(t)\right) \\ + 1_{Q(t)<N}(t)\left(-\omega e^{\omega(N-Q(t))} + 2X(t)\right). \tag{10}$$

Since the above equations can be obtained through convex optimization, the proof is omitted for simplicity.

We switch the power scheduling policy for each time frame based on the distance between λ_k and the boundary of the capacity region, as shown in Fig. 1, when λ_k is far interior the capacity region of exponential form, p_2 amount of power is scheduled for transmission, whereas p_1 amount of power is scheduled when λ_k is close to and outside the boundary of the capacity region of exponential form.

IV. Simulation

In this simulation, channel state obeys Rayleigh distribution with the fading coefficient 6.5 and is i.i.d. over time slots. The average data arrival rate follows uniform distribution, where at the first 5000 time slots, the average data arrival rate is set to 5 packets per time slot, while jumps to 15 packets per time slot after time slot 5001. $N = 100$ throughout the 10000 time slots. According to Fig. 2, it is observed that

- When λ_k is far interior both the capacity regions, exponential Lyapunov function achieves superior performance in terms of delay jitter (the variance of exponential form is 124.71, while that of quadratic form is 166.25), as it provides a large Lyapunov drift penalty when $Q(t)$ deviates from N.
- When λ_k is near the capacity region of exponential form, even though N is set to 100, there is not enough power to maintain such short-term delay requirement, and quadratic Lyapunov function achieves superior performance due to the large distance between λ_k and its capacity region.

The above observations necessitate adopting dynamic Lyapunov function for temporally evolving wireless systems to better trade-off power and delay.

References

[1] L. Zhang, H. Wang, H. Xue, H. Zhang, Q. Liu, D. Niyato, and Z. Han, "Digital twin-assisted edge computation offloading in industrial Internet of Things with NOMA," *IEEE Trans. Veh. Technol.*, pp. 1-15, Apr. 2023.
[2] Y. Wang, M. Kong, G. Zhang, W. Wang, T. Nakachi and J. Liou, "Adaptive task offloading for mobile edge computing with forecast information," *IEEE Trans. Veh. Technol.*, vol. 74, no. 3, pp. 4132-4147, Mar. 2025.

979-8-3315-2209-4/25 $31.00 © 2025 IEEE

[3] D. Wen, P. Liu, G. Zhu, Y. Shi, J. Xu, Y. C. Eldar, and S. Cui, "Task-oriented sensing, computation, and communication integration for multi-device edge AI," *IEEE Trans. Wireless Commun.*, vol. 23, no. 3, pp. 2486-2502, Mar. 2024.

[4] Y. Wang, W. Wang, V. K. Lau, T. Nakachi, and Z. Zhang, "Stochastic resource allocation and delay analysis for mobile edge computing systems," *IEEE Trans. Commun.*, vol. 71, no. 7, pp. 4018-4033, Jul. 2023.

[5] Y. Cui, V. K. N. Lau, R. Wang, H. Huang and S. Zhang, "A survey on delay-aware resource control for wireless systems: large derivation theory, stochastic Lyapunov drift and distributed stochastic learning," *IEEE Trans. Inf. Theory*, vol. 58, no. 3, pp. 1677-1700, Mar. 2012.

[6] E. M. Yeh, *Multiaccess and fading in communication networks*, Ph.D. dissertation, MIT, Sept. 2001.

[7] D. P. Bertsekas, *Dynamic programming and optimal control*, 3rd Ed. Athena Scientific, 2007.

[8] M. J. Neely, "Stochastic network optimization with application to communication and queueing systems," *Synthesis Lect. Commun. Netw.*, vol. 3, no. 1, pp. 1-211, May 2010.

[9] Y. Wang, W. Wang, L. Chen, P. Zhou, and Z. Zhang, "Energy efficient scheduling for delay-constrained spectrum aggregation: A differentiated water-filling approach," *IEEE Trans. Green Commun. Netw.*, vol. 1, no. 4, pp. 395-408, Dec. 2017.

Codebook based SCMA with Message Passing for Multi-User Detection

Qixian Zheng   Hang Li   Lizhe Liu   Yashan Pang   Yangyang Guan   Zhiqun Cheng   Qinghua Guo

Abstract—**In this paper, a multiuser detection (MUD) scheme with optimized sparse code multiple access (SCMA) is proposed for high user density communication systems. By integrating a multi-dimensional sparse codebook design with a factor graph-optimized message passing algorithm (MPA), the proposed scheme improves the multiuser detection accuracy. Simulation results show that the proposed one demonstrates significant superiority in terms of active user identification error rate (AER) over the conventional MUD schemes with low-density signature (LDS).**

Index Terms—**SCMA, LDS, MPA, MUD, factor graph**

I. INTRODUCTION

Non-orthogonal multiple access (NOMA) has emerged as a promising technique to enhance spectral efficiency by allowing multiple users to share the same resource simultaneously. NOMA can be commonly categorized into two types: power-domain NOMA and code-domain NOMA [1]. Power-domain NOMA controls the transmission power of each user sharing the same resource, while code-domain NOMA employs user-specific spreading sequences characterized by low density or low correlation to mitigate interference on each chip. Multiple access techniques in code-domain NOMA include low-density signature code division multiple access (LDS-CDMA) [2], low-density signature orthogonal frequency division multiplexing (LDS-OFDM) [3] and sparse code multiple access (SCMA) [4]. In LDS-CDMA, each user's symbol is mapped to a sequence of low-density spreading chips. LDS-OFDM extends this concept by integrating LDS-CDMA with OFDM, where spread symbols are transmitted over multiple subcarriers.

SCMA advances the concept of LDS by directly mapping input bits to multidimensional constellations, thereby enabling each user to transmit distinct symbols across different resource elements (REs). This method removes the redundancy caused by symbol repetition in LDS, allowing SCMA to achieve superior spectral efficiency and enhanced interference suppression [5]. The sparse structure of SCMA also makes it well-suited for iterative detection using message passing algorithm (MPA), where resource nodes and user nodes exchange messages iteratively to update posterior probabilities of transmitted codewords [6]. Compared to LDS-CDMA, SCMA leverages its sparse codeword structure to exchange messages exclusively with non-zero nodes instead of interacting with all nodes, thereby significantly reducing computational complexity. However, most of the existing work [6], [7] focuses on the improvement of spectral efficiency of SCMA, and lacks the analysis of active user identification error rate (AER) in dynamic scenarios.

In this paper, we present a multiuser detection (MUD) scheme with optimized SCMA for high user density communication systems. Extensive simulation results reveal that the proposed one with MPA consistently outperforms the conventional MUD schemes under overloading conditions. This is attributed to the optimized sparse codebook structure in SCMA with MPA, which intrinsically mitigates multi-user interference, enabling high detection accuracy even in the overloading case.

II. SYSTEM MODEL

SCMA is a sparse codeword-based multi-access method that maps user bit streams directly to multiple REs. Fig.1 illustrates the transmitter block diagram of SCMA-OFDM systems. The user information bits are coded by forward error correction (FEC) coding, and then the coded bits are mapped to multi-dimensional non-orthogonal resource units by SCMA sparse codebook. Finally, the sparse symbols are loaded to the orthogonal subcarriers with OFDM modulation to complete the multi-access transmission. Due to the orthogonality of OFDM subcarriers, the codebook design of SCMA can satisfy the sparsity constraint and realize overload transmission through multi-dimensional constellation expansion.

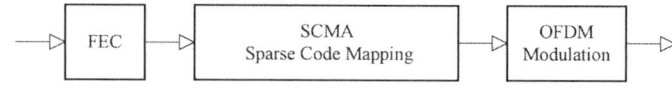

Fig. 1. Transmitter block diagram of SCMA-OFDM systems.

We follow the code design principle of point-to-point communication on fast-fading channel to design a SCMA codebook. In the case of rotating constellation, different operations (e.g,. phase rotation, complex conjugation, power offset, or

---

Qixian Zheng, Hang Li, Yangyang Guan and Zhiqun Cheng are with *School of Electronics and Information, Hangzhou Dianzi University*, Hangzhou, Zhejiang, China. Hang Li, Lizhe Liu and Yashan Pang are with *Key Laboratory of Advanced Communication Networks, Academy of Network and Communications of CETC*, Shijiazhuang, Hebei, China. Qinghua Guo is with *School of Electrical, Computer and Telecommunications Engineering, University of Wollongong*, Wollongong, NSW, Australia. This work is supported in part by the National Natural Science Foundation under Grant 62071163; in part by the Zhejiang Provincial Natural Science Foundation under Grant LY22F010003 and in part by the Project of Ministry of Science and Technology under Grant D20011. Corresponding author: (Hang Li, email: hangli@hdu.edu.cn)

979-8-3315-2209-4/25 $31.00 © 2025 IEEE

dimensional alignment), are applied to the constellation to build multiple sparse codebooks for many layers of SCMA.

The first step in designing the codebook is to initialize the mother constellation. Define a one-dimensional vector $\mathbf{m} = [m_1, m_2, \ldots, m_N]^T$, where $m_i \in \mathbb{C}$ denotes a complex constellation point, and N is the length (or dimension) of the constellation. These complex elements form the basis of the mother constellation. Subsequently, to enhance the distinguishability of constellation points, a phase rotation matrix $\boldsymbol{\Phi}$ is often introduced to expand $\mathbf{m}$ into a higher dimension. The phase rotation of the base constellation generates multiple different codebooks with different rotation angles available for each user layer, enhancing the differentiation of the constellation.

Secondly, the even dimensions of the mother constellation are interleaved (reordered) to optimize system performance. This interleaving reduces the correlation between constellation points and effectively lowers the peak-to-average power ratio (PAPR). At the receiver, the received signal can be expressed as

$$\mathbf{y} = \sum_{j=1}^{J} \text{diag}(\mathbf{h}_j)\mathbf{x}_j + \mathbf{n}, \qquad (1)$$

where J and K denote the number of users and REs, respectively. $\mathbf{x}_j = (x_{1,j}, \ldots, x_{K,j})^T$ is the SCMA codeword of user j, $\mathbf{h}_j = (h_{1,j}, \ldots, h_{K,j})^T$ is the channel vector of user j, and $\mathbf{n} \sim \mathcal{CN}(\mathbf{0}, \sigma^2\mathbf{I})$ denotes the Gaussian noise vector. Finally, through the above steps, sparse codebooks are created for all users. This design ensures the sparsity of user codewords and their distribution across specific REs, enabling efficient user separation and interference mitigation.

III. MPA DECONGING

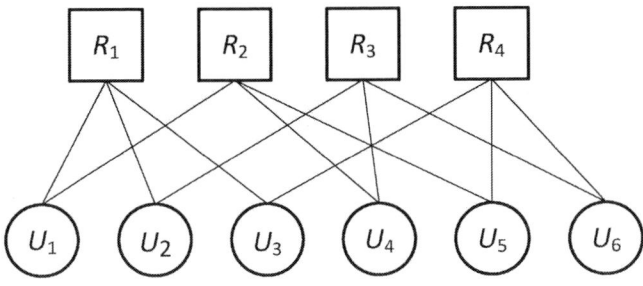

Fig. 2. Factor graph representation of a SCMA system.

An example of a bipartite factor graph designed for the SCMA system is shown in Fig. 2, where R_k denotes RE k ($k \in [1, K]$) and U_j denotes user j ($j \in [1, J]$), achieving a overload ratio of ($\lambda = \frac{J}{K}$) = 150%. Note that the number of users superimposing over one RE is 3, i.e., $d_f = 3$ and the number of non-zero dimension is 2, i.e., $N = 2$. Any RE aggregates signals superimposed from three overlapping users to form a deterministic sparse collision pattern: each RE receives contributions from three unique user nodes and every

user codeword spans two non-orthogonal REs. This sparsity of the structure is conducive to the superposition of multi-dimensional constellations, achieves better message passing detection. At the receiver, SCMA employs the MPA to achieve multi-user detection.

Message pass from RE R_k to user U_j is calculated as

$$m_{k \to j}(x_j) \propto \sum_{\substack{\{x_{j'}\} \in \mathcal{X}^{|\partial_k|-1} \\ j' \in \partial_k \setminus j}} p(y_k \mid \{x_{j'}\}) \prod_{\substack{j' \in \partial_k \\ j' \neq j}} m_{l \to k}(x_{j'}), \qquad (1)$$

where $\mathcal{X}^{|\partial_k|}$ denoted a set that the joint space of other user symbols on RE R_k, $\partial_k \setminus j$ denoted a set that excluding user U_j from the set of users connected to RE R_k. Then the user U_j computes the feedback message from user U_j to RE R_k is calculated as

$$m_{j \to k}(x_j) \propto \prod_{\substack{k' \in \partial_j \\ k' \neq k}} m_{k' \to j}(x_j), \qquad (3)$$

where $k' \in \partial_j$ denoted a set that excluding RE R_k from the set of REs connected to user U_j. After many iterations, the symbolic posterior probability of user U_j is

$$P(x_j = a) \propto \prod_{k \in \partial_j} m_{k \to j}(a), \quad \forall a \in \mathcal{X}, \qquad (4)$$

where $k \in \partial_j$ denoted a set that producting of connected REs' messages. $\hat{x}_j = argmaxP(x_j = a)$ denoted that user U_j employs the SCMA coodbook, so user U_j is identified.

The MPA iteratively updates messages through exchanges between resource nodes and user nodes in the factor graph. Resource nodes calculate posterior probabilities based on received signals and user prior information, then transmit updated messages to user nodes. User nodes refine symbol probability distributions using received information and their codebook characteristics, and pass the results back to resource nodes. As iterations proceed, posterior probabilities gradually converge, ultimately enabling user codeword detection.

IV. SIMULATION RESULTS

This section presents a comparative performance evaluation for different multi-user detection schemes. The parameters used in simulations are as follows. The number of REs K = 5, the number of user terminals $J = 10$, the number of codebook dimensions $N = 2$, and sparsity degree $d_f = 4$. QPSK modulation is employed in the codewords, and the results are obtained by averaging over 10^4 Monte Carlo trials. To examine the performance of user activity detection, we define the AER as

$$\text{AER} =$$

$$\frac{\#\text{of active users} - \#\text{of active users identified successfully}}{\#\text{of active users}}$$

Fig. 3 compares the active user detection performance for different multi-user detection methods versus signal-to-noise ratio (SNR) ranging from 0 dB to 20 dB with an

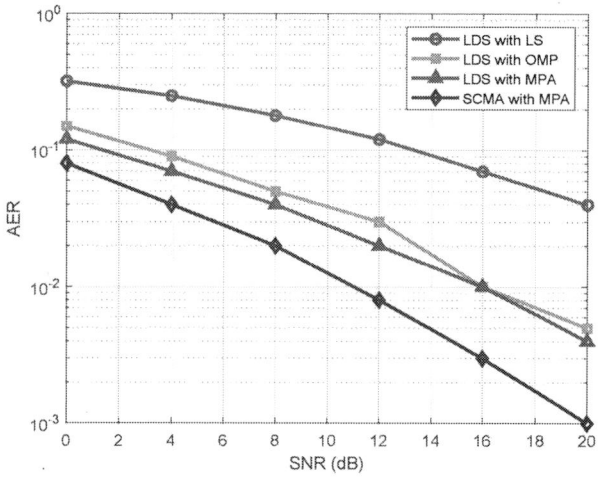

Fig. 3. AER performance comparison.

overloading factor of 2, which means the ratio of the number of actual users to the number of available resources. It is seen that the SCMA with MPA achieves superior detection performance through its multidimensional sparse codebook design, which effectively leverages the structural sparsity of signals. Although the LDS with MPA employs sparse spreading to mitigate multi-user interference, the codeword of low density spread spectrum sequence belongs to a single dimensional real number field. The randomness of spread spectrum sequence leads to a high collision probability, where collision probability is the probability that multiple users' spread spectrum sequences overlap non-zero elements on the same resource block between users, resulting in comparatively inferior performance relative to the SCMA with MPA. AER performance of the LDS with least squares (LS) is low in overloaded scenarios due to its inability to account for inter-user interference. The LDS with orthogonal matching pursuit (OMP) also incurs poor performance because it is highly sensitive to noise, which may cause the algorithm not to select the base vector of the inactive user, introducing false alarms or missing detection.

V. Conclusion

This paper conducts a comparative analysis of AER performance between SCMA and LDS multi-user detection systems. Through sparse-driven codebook design that suppresses interference via optimized resource sharing, the proposed scheme has superior performance in MUD. Simulation results show that the SCMA with MPA exhibits significant AER enhancement over the LDS with MPA under high-overload conditions. This is beneficial for massive machine-type communications in time-varying channels with dynamic user activity patterns.

References

[1] A. Benjebbour, Y. Saito, Y. Kishiyama, A. Li, A. Harada, and T. Nakamura, "Concept and practical considerations of non-orthogonal multiple access (NOMA) for future radio access," in *Proc. IEEE Int. Symp. Intell. Signal Process. Commun. Syst.*, 2013, pp. 770–774.

[2] R. Hoshyar, F. P. Wathan, and R. Tafazolli, "Novel Low-Density Signature for Synchronous CDMA Systems Over AWGN Channel," *IEEE Transactions on Signal Processing*, vol. 56, no. 4, pp. 1616–1626, 2008.

[3] R. Hoshyar, R. Razavi, and M. Al-Imari, "LDS-OFDM an Efficient Multiple Access Technique," in *2010 IEEE 71st Vehicular Technology Conference*, 2010, pp. 1–5.

[4] H. Nikopour and H. Baligh, "Sparse code multiple access," in *2013 IEEE 24th Annual International Symposium on Personal, Indoor, and Mobile Radio Communications (PIMRC)*, 2013, pp. 332–336.

[5] J. van de Beek and B. M. Popovic, "Multiple Access with Low-Density Signatures," in *GLOBECOM 2009 - 2009 IEEE Global Telecommunications Conference*, 2009, pp. 1–6.

[6] L. Yang, Y. Liu, and Y. Siu, "Low Complexity Message Passing Algorithm for SCMA System," *IEEE Communications Letters*, vol. 20, no. 12, pp. 2466–2469, 2016.

[7] S. Chaturvedi, Z. Liu, V. A. Bohara, A. Srivastava, and P. Xiao, "A Tutorial on Decoding Techniques of Sparse Code Multiple Access," *IEEE Access*, vol. 10, pp. 58 503–58 524, 2022.

A 12-bit 320MS/s Current-steering DAC With 70dBc SFDR in 22nm CMOS

Tinghua Chen   Qiji Huang   Xinpeng Xing   Haigang Feng

Abstract—A 12-bit 320MS/s current-steering digital-to-analog converter (DAC) is designed in a 22-nm CMOS process.A two-stage cascode current source structure was employed to enhance the finite output impedance, and two constant branch currents were added to further improve the linearity of the output. In this design, a common-centroid four-quadrant layout was adopted.Post-simulation results in 22-nm CMOS show that for an input signal up to 138MHz, the SFDR and the SNR of the presented 320MS/s DAC achieve 70dB and 71dB respectively, with a maximum output current of 2mA.The DAC consumes 3.293mW of power with supply voltages of 0.9V and 2.5V, and the chip area is 0.074mm².

Index Terms—current-steering (CS), digital-to-analog converter (DAC), high-speed,high-resolution,mismatch, cascode current source,SFDR.

I. INTRODUCTION

In communication systems, the Digital-to-Analog Converter (DAC) is essential for converting digital signals into analog signals for transmission over analog channels. The DAC's performance, particularly its resolution and sampling rate, critically influences communication quality by minimizing distortion and noise, thus improving signal reliability [1]. To meet the demands of modern communication systems, DACs must achieve high bandwidth, high linearity, low power consumption, and compact design [2]. Among various architectures, the current-steering DAC is the most suitable for these requirements. However, it faces challenges such as finite output impedance and current source mismatch, which are key non-ideal factors limiting its performance [3].

This paper presents a 12-bit, 320MS/s current-steering DAC for WIFI/BLE transceivers, featuring a two-stage decoder, a dual-stage cascode current source, and a common-centroid four-quadrant layout. The DAC achieves a spurious-free dynamic range (SFDR) exceeding 69 dB across the Nyquist frequency range. The paper is organized as follows: Section II outlines the architecture of the proposed DAC. Section III elaborates on the circuit and layout implementation, covering the core circuit, bit swapping, bias circuit, and floor plan. Post-layout simulation results are discussed in Section IV, followed by conclusions in Section V.

Author 1 is with *Shenzhen Graduate International School, Tsinghua University*, Shenzhen, China. Author 2 is with *Shenzhen Graduate International School, Tsinghua University*, Shenzhen, China. Author 3 is with *School of Integrated Circuits, Sun Yat-Sen University*, Shenzhen, China. Author 4 is with *Shenzhen Graduate International School, Tsinghua University*, Shenzhen, China.Corresponding author: (Author 3, email: xingxp@mail.sysu.edu.cn)

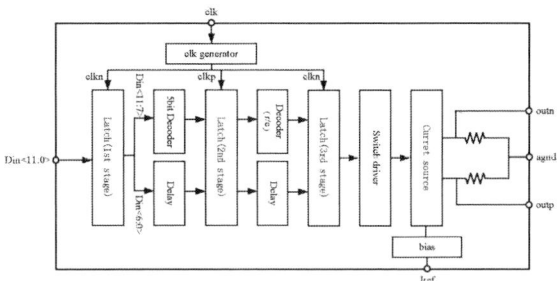

Fig. 1: Architecture of the presented DAC

II. DAC ARCHITECTURE

Fig. 1 illustrates the overall architecture of the 12-bit, 320 MS/s current-steering DAC designed in this paper. Integrated into a WIFI/BLE dual-mode transceiver, the DAC omits peripheral circuits such as data and clock interfaces.In this design, the least significant 7 bits(LSBs) are implemented in binary format, whereas the most significant 5 bits(MSBs) are transformed into a 31-segment thermometer code through a decoder. The digital decoder employs a two-stage row-column structure.To ensure error-free decoding and synchronized switch control signals, three-stage latches are implemented. An on-chip clock generation circuit converts the input clk into differential clk_n and clk_p, alternately driving the latches. By optimizing the signal swing and edge characteristics of the switch control signals, a properly aligned crossover point is achieved, minimizing switch-induced non-linearity.

Given that the DAC is integrated into a WIFI/BLE transceiver, it exhibits significant sensitivity to power dissipation. The system specification requires the DAC circuit to operate with a power consumption of less than 2.5mW. Additionally, the output common-mode voltage is restricted to 0.4V to ensure the optimal performance of the subsequent low-pass filter, thereby imposing design constraints on the current source array. To address these specifications, the full-scale output current of the DAC is configured at 2mA.

III. DAC CORE CIRCUIT AND LAYOUT DESIGN

A. Two-stage Row-Column Decoder

The two-stage row-column decoder adopted in this design is shown in Fig. 2. Here, a 5-bit binary input code 10011 is taken as an example for illustration. The shaded part represents the current sources selected to be turned on, while the white part represents the current sources not selected to be turned

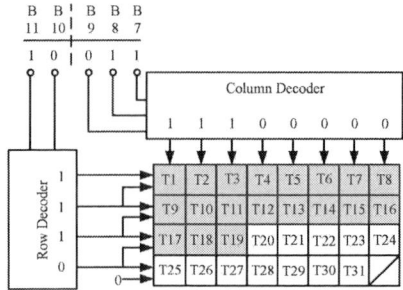

Fig. 2: 5-bit Two-Stage Row-Column Decoder

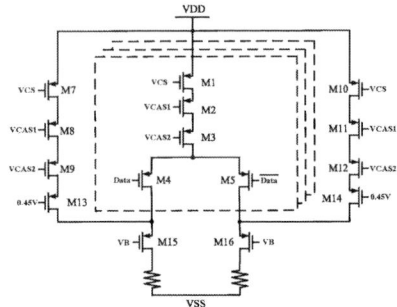

Fig. 3: Core circuit of the presented DAC

on. The second stage is the matrix decoder, and the result for each unit is:

$$T_{ij} = R_i C_j + C_{i+1} \qquad (1)$$

Where R_i is the output of the ith row decoder, C_j is the output of the jth column decoder.

B. Current Source Unit Circuit

The current source array is implemented as illustrated in Fig. 3. To achieve high linearity in the DAC output, the current-steering DAC utilizes current sources with high output impedance. This design employs a two-stage cascode current source to meet this requirement. Transistors M15 and M16, functioning as shared common-gate devices across all current sources, form a cascode switch structure. This architecture isolates the output voltage swing, minimizing its impact on the operating conditions of the switching transistors [4]. Two sets of constant current sources, composed of transistors M7–M9 and M10–M12, supply fixed currents to M15 and M16. Additionally, dummy switches M13 and M14 are incorporated to ensure branch current matching with the main DAC current sources. By introducing these small branch currents, M13 and M14 maintain a code-independent constant current, effectively reducing transient nonlinearities and code-dependent switching behavior, thereby enhancing the DAC's linearity.

Transistor M1 acts as the DAC's current source, whose MOS mismatch level critically influences the DAC's perfor-

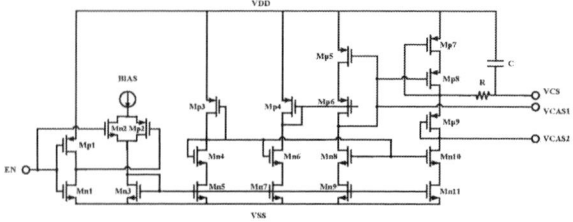

Fig. 4: Bias circuit of the presented DAC

mance, necessitating meticulous design. The transistor's final normalized mismatch current expression is [5]:

$$\left(\frac{\sigma(\Delta I)}{I}\right)^2 = \frac{1}{2WL}\left[A_\beta^2 + \frac{4A_{V_{TH}}^2}{(V_{GS} - V_{TH})^2}\right] \qquad (2)$$

Among this, A_β denotes the current gain factor, and $A_{V_{TH}}$ represents the threshold voltage mismatch coefficient, both of which are process-dependent constants. The parameters W and L correspond to the gate width and gate length of the transistor, respectively, while $V_{GS} - V_{TH}$ signifies the overdrive voltage of the transistor.

Herein, the DAC architecture employs a segmented current-source array with 5-bit thermometer-coded upper bits and 7-bit binary-weighted lower bits, totaling 31 unary current sources and 7 binary-scaled sources. To ensure precise current matching and uniform overdrive voltage across all bit weights while enabling shared biasing, the design implements a unit transistor approach. The B2 binary-weighted transistor serves as the fundamental unit cell, with all other current sources constructed through optimal series/parallel combinations of this identical unit device.

C. Bias Circuit

The biasing of a two-stage cascode current source is more complex than that of a single-stage cascode current source. The bias circuit designed in this work is shown in Fig. 4. The inverter formed by M_{p1} and M_{n1} delivers the bias circuit enable signal EN to the transmission gate comprising M_{p2} and M_{n2}. With M_{n3} as the current source, multiple current branches are generated via current mirror replication. To ensure current replication accuracy and minimize source-drain voltage mismatch effects, four cascode current mirror pairs are implemented: M_{n4}-M_{n5}, M_{n6}-M_{n7}, M_{n8}-M_{n9}, and M_{n10}-M_{n11}. Furthermore, the gate bias voltage noise of the current sources significantly contributes to the current-steering DAC's noise. To suppress gate voltage noise in the current source transistors, an RC filter is utilized to attenuate in-band noise. The filter's cutoff frequency is expressed as [6]:

$$\omega = \frac{1}{(1/g_{m,p7} + R)C} \qquad (3)$$

D. DAC layout design

This design employs a four-quadrant common-centroid layout, as depicted in Figure 5. For instance, the layout of 31 thermometer-code current sources is shown, where numbers

979-8-3315-2209-4/25 $31.00 © 2025 IEEE   310

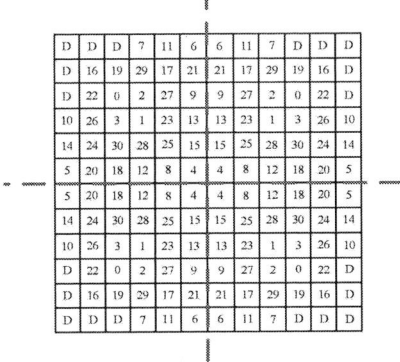

Fig. 5: Floorplan of the DAC current sources

0 to 30 represent the current sources and "D" indicates dummy transistors. The layout is symmetrically divided into four quadrants, each containing 36 units: 31 sub-thermometer-code current sources and 5 dummy current sources. Each sub-thermometer-code current source delivers one-fourth of the full-scale current. This common-centroid configuration effectively reduces gradient-induced mismatch.

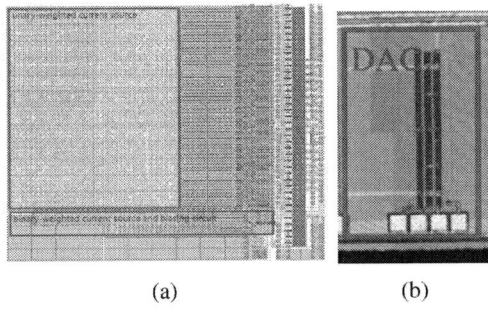

(a)   (b)

Fig. 6: (a)Layout of the presented DAC (b)Die of the presented DAC

The DAC's complete layout is depicted in Fig. 6(a). The layouts of both the thermometer-coded current sources and binary-weighted current sources are explicitly indicated in the schematic diagram.Integrated within a transceiver chip, the design excludes ESD protection and seal ring structures. Necessary PADs are connected at the top metal layer, with the DAC occupying a total area of 0.074mm². A die micrograph of the fabricated chip is shown in Fig. 6(b), where the DAC circuit is marked by a red box.

IV. SIMULATION RESULTS

The proposed current-steering DAC has been meticulously designed and extensively simulated in a 22-nm CMOS process. Fig. 7 illustrates its post-layout simulated output spectrum. Operating at a sampling rate of 320 MS/s, the DAC achieves a SFDR of 70.08 dB and a SNR of 71.87 dB for an input signal frequency of 138 MHz, demonstrating its robust performance in high-speed applications.

Simulations were performed on the circuit, including parasitic parameters, across a range of input frequencies. The

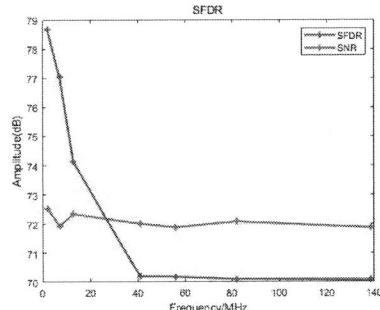

Fig. 7: DAC post-simulated performance as function of input frequency

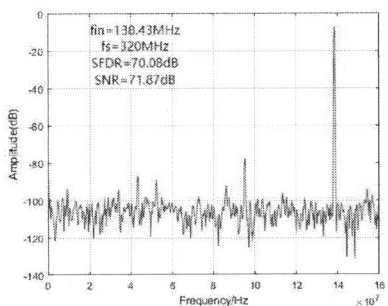

Fig. 8: DAC spectrum with 138MHz signal

post-layout simulation results, illustrated in Fig. 8, demonstrate that the DAC achieves excellent linearity under low-frequency input conditions. While the SFDR exhibits a slight degradation as the input frequency increases, the linearity remains well-maintained, underscoring the DAC's strong dynamic performance. The DAC consumes a total power of 3.293mW with supply voltages of 0.9V and 2.5V, and occupies a compact chip area of 0.074mm².

The performance of the CS DAC is summarized in Table 1, alongside a comparison with state-of-the-art designs.

TABLE I: CS DAC Performance Summary and State-of-the-Art Comparison

	This work	CICC'23 [1]	JSSC'11 [2]	VLSI'15 [3]
Process(nm)	22	28	140	20
Resolution, N	12	12	14	14
F_{clk} (Hz)	320M	1G	200M	750M
Supply(V)	0.9/2.5	0.9/1.5	1/1.8	1.0/1.8
I_L(mA)	2	8	20	2
P_{Total}(mW)	3.29	33	270	21.1
SFDR(dBc)	70	81	78	83

V. CONCLUSIONS

This work proposes a 12-bit 320MS/s current-steering DAC fabricated in a 22-nm CMOS technology. The design incorporates a two-stage row-column decoder to streamline the decoding logic, complemented by a three-stage latch chain to ensure precise timing synchronization. The current source unit is implemented using a two-stage cascode architecture, with two additional constant branch currents integrated to further improve linearity. In the layout design, a common-centroid quad-quadrant arrangement is employed to minimize both random and systematic mismatches in the DAC cells.Post-layout simulation results reveal that, operating at a sampling rate of 320MS/s, the DAC achieves a SFDR of 70.08dB and a SNR of 71.87dB for an input signal frequency of 138MHz, demonstrating its superior performance in high-speed applications.

REFERENCES

[1] C. -U. Park, J. -H. Chung and S. -T. Ryu, "A 12-bit 1GS/s Current-Steering DAC with Paired Current Source Switching Background Mismatch Calibration," 2023 IEEE Custom Integrated Circuits Conference (CICC), San Antonio, TX, USA, 2023, pp. 1-2.

[2] Y. Tang et al., "A 14 bit 200 MS/s DAC With SFDR >78 dBc, IM3 <-83 dBc and NSD <-163 dBm/Hz Across the Whole Nyquist Band Enabled by Dynamic-Mismatch Mapping," in IEEE Journal of Solid-State Circuits, vol. 46, no. 6, pp. 1371-1381, June 2011.

[3] S. M. Lee et al., "A 14b 750MS/s DAC in 20nm CMOS with <-168dBm/Hz noise floor beyond Nyquist and 79dBc SFDR utilizing a low glitch-noise hybrid R-2R architecture," 2015 Symposium on VLSI Circuits (VLSI Circuits), Kyoto, Japan, 2015, pp. C164-C165.

[4] C. -H. Lin et al., "A 12 bit 2.9 GS/s DAC With IM3 $\ll$- 60 dBc Beyond 1 GHz in 65 nm CMOS," in IEEE Journal of Solid-State Circuits, vol. 44, no. 12, pp. 3285-3293, Dec. 2009.

[5] E. Bechthum, G. I. Radulov, J. Briaire, G. J. G. M. Geelen and A. H. M. van Roermund, "A Wideband RF Mixing-DAC Achieving IMD <-82 dBc Up to 1.9 GHz," in IEEE Journal of Solid-State Circuits, vol. 51, no. 6, pp. 1374-1384, June 2016.

[6] B. Razavi, Design of Analog CMOS Integrated Circuits, 2nd ed., New York, NY: McGraw-Hill Education, 2017, pp. 228-231.

Robust Hybrid Holographic Beamforming Scheme Under the RHS Based UAV IoT System

Mingcheng Shen[1], Keer Chen[1], Jichong Guo[1*], Chunxia Su[1], and Chang Ding[2]

1. Suzhou University of Science and Technology, Suzhou 215000, China
2. Shaanxi Key Laboratory of Artificially Structured Functional Materials and Devices, Air Force Engineering University, Xi'an, China
*E-mail: guojichong@usts.edu.cn

Abstract—**The reconfigurable holographic surface (RHS) based unmanned aerial vehicle (UAV) internet of things (IoT) system is significantly impacted by complicated wind disturbance. Previous research on UAV attitude adjustment brings inevitable adjustment errors, indicating the necessity of mitigating these inaccuracies. In this context, we turn to explore a robust holographic hybrid beamforming scheme under the RHS based UAV IoT system to effectively deal with the adverse effect of non-ideal UAV attitude adjustment. The objective function is modeled as the sum mean square error (MSE) with the UAV attitude adjustment error. Based on it, a heuristic algorithm is proposed, where the digital part is designed on the minimum ergodic sum MSE subproblem, while the analog part is designed on the max-min signal to interference plus noise ratio (SINR) subproblem. Furthermore, simulation results show that our scheme outperforms existing benchmark schemes in sum-rate and bit error rate (BER).**

Index Terms—**IoT communications, holographic beamforming, UAV jitter**

I. INTRODUCTION

future sixth-generation (6G) network is expected to provide revolutionary mobile connectivity and high-throughput data services through low-cost communication devices. Benefiting from the programmability and tunability of metamaterials, reconfigurable holographic surfaces (RHS), embedded with numerous metamaterial radiating elements, exhibit great potential in realizing this bold vision of 6G. RHS acts as an ultra-thin and lightweight antenna integrated with transceivers, generating desired directional beams with low hardware costs and power consumption [1]. Existing research on RHS can be broadly categorized into hardware component design and radiation pattern control [2].

Because of the advantages and importance of unmanned aerial vehicles (UAVs), UAV-assisted internet of things (IoT) systems have received widespread attention. In [3], the author discusses the effective deployment and mobility of multiple UAVs functioning as aerial base stations for collecting data from ground-based IoT devices. The main challenge addressed here is the optimal deployment, mobility, and energy-efficient

---

The work of Chunxia Su and Jichong Guo was supported in part by the National Natural Science Foundation of China under Grant 62271085; in part by the Natural Science Foundation of the Higher Education Institutions of Jiangsu Province, China, under Grant 23KJB520035 and Grant 23KJB510031; in part by the Suzhou Science and Technology Plan (Basic Research) Project, under Grant SJC2023002; and in part by the Natural Science Foundation of Shandong Province, China, under Grant No. ZR2024MF035, ZR2021QF004. (*Corresponding author*: Jichong Guo)

usage of UAVs. However, few studies have considered the impact of jitter induced by external interference on UAV attitude during operation, which inevitably affects wireless communication. In [4], our team studied the issue of UAV jitter mitigation when equipped with a reconfigurable intelligent surface (RIS). Unlike RHS, RISs are widely used as passive relays due to their reflection characteristics [5]. Limited research has been conducted on the jitter problem of UAV-assisted IoT systems equipped with RHS, so implementing beamforming techniques to combat wind disturbances in UAV-assisted IoT systems based on RHS is necessary. Therefore, in this paper, we propose a solution to address the jitter issues caused by wind during the operation of UAV-assisted IoT systems equipped with RHS.

II. SYSTEM MODEL

In the RHS-based UAV IoT system, a single UAV and K user terminals (UTs) are considered. The UAV is equipped with an RHS, where it has L feeds and $N_x \times N_y$ radiation units. $\mathbf{M} \in \mathbb{C}^{N_x N_y \times L}$ denotes the holographic beamforming matrix, and $\mathbf{W} \in \mathbb{C}^{L \times K}$ represents the digital beamforming matrix, and $\mathbf{s} \in \mathbb{C}^{K \times 1}$ denotes the modulated signals of all the UTs.

Considering block fading, the whole channel matrix is $\mathbf{H} \in \mathbb{C}^{N_x N_y \times K}$. Consequently, the received signals are

$$\mathbf{y} = \mathbf{H}^{\mathrm{T}} \mathbf{M} \mathbf{W} \mathbf{s} + \mathbf{z}, \qquad (1)$$

where the Gaussian white noise vector $\mathbf{z} \sim \mathcal{CN}(0, \sigma^2 \mathbf{I})$, and σ^2 is the variance of noise.

The element of the holographic beamforming matrix is

$$\mathbf{M}_{n_x,n_y}^l = \sqrt{\eta} \widetilde{m}_{n_x,n_y} \cdot e^{-\alpha |\mathbf{r}_{n_x,n_y}^l|} \cdot e^{-j \mathbf{k}_s \mathbf{r}_{n_x,n_y}^l}, \quad (2)$$

where $\widetilde{m}_{n_x,n_y}$ is the normalized radiation amplitude of each RHS unit considering the effect of UAV jitter, and

$$\widetilde{m}_{n_x,n_y} = \sum_{k=1}^{K} \sum_{l=1}^{L} a_{k,l} m(\mathbf{r}_{n_x,n_y}^l, \theta_k + \widetilde{\Delta\theta}, \phi_k + \widetilde{\Delta\phi}), \quad (3)$$

where $a_{k,l}$ satisfies $\sum_{k=1}^{K} \sum_{l=1}^{L} a_{k,l} = 1$, ensuring that $\widetilde{m}_{n_x,n_y}$ lies in the interval $[0, 1]$. (θ_k, ϕ_k) is the three-dimensional direction angle designed for the k-th user. Specially, for the situation without UAV jitter, $\widetilde{\Delta\theta} = 0$ and $\widetilde{\Delta\phi} = 0$, at this moment, $\mathbf{M}$ is denoted as $\mathbf{M}_{00}$. The definitions of other variables in Eqns. (2) and (3) can be found in [6].

Therefore, the holographic beamforming matrix affected by wind disturbance includes two types of parameters, which are $a_{k,l}$ and (θ_k, ϕ_k), respectively. The weighted values $a_{k,l}$ and $\mathbf{W}$ consist of the hybrid holographic beamforming variables, while (θ_k, ϕ_k) constitutes the angle directing variables. Since the UAV attitude is adjusted in real time, the primary focus is to correct the direction angle variation of holographic beamforming caused by wind disturbances. These errors are subsequently mitigated through hybrid holographic beamforming techniques. Considering the particularity of the UAV IoT system, line of sight (LoS) component exists with a high probability. Thus, we have $\theta_k = \theta_k^{\text{LoS}}$ and $\phi_k = \phi_k^{\text{LoS}}$. Denote by $\Delta\theta$ and $\Delta\phi$ the random angle variables to describe the processing errors of UAV attitude adjustment. Then Eqn.(3) is updated by replacing $\theta_k + \widetilde{\Delta\theta}$ and $\phi_k + \widetilde{\Delta\phi}$ with $\theta_k^{\text{LoS}} + \Delta\theta$ and $\phi_k^{\text{LoS}} + \Delta\phi$.

III. DESIGN OF ROBUST HYBRID HOLOGRAPHIC BEAMFORMING SCHEME

In this study, we choose the sum mean square error (MSE) of all UTs during the block service time as the design criterion. Considering the sum MSE with UAV attitude adjustment error, the design problem is modeled as

$$
\min_{\{a_{k,l}\}, \mathbf{W}} \text{tr}\left[\left(\mathbf{H}^{\text{T}}\mathbf{M}\mathbf{W} - \mathbf{I}\right)\left(\mathbf{H}^{\text{T}}\mathbf{M}\mathbf{W} - \mathbf{I}\right)^{\text{H}}\right] + \sigma^2 K,
$$
$$
\text{s.t. } C1: \text{tr}[\mathbf{M}_{00}\mathbf{W}\left(\mathbf{M}_{00}\mathbf{W}\right)^{\text{H}}] \leq P,
$$
$$
C2: \gamma_k \geq \gamma_0, \ \forall k, \tag{4}
$$
$$
C3: \sum_{k=1}^{K}\sum_{l=1}^{L} a_{k,l} = 1, \ a_{k,l} \geq 0, \ \forall k, l,
$$

where γ_k is the signal to interference plus noise ratio (SINR) at the k-th UT and $\text{tr}[\cdot]$ is the trace operation.

Then the focus of work lies in decoupling hybrid holographic beamforming ($\{a_{k,l}\}$ and $\mathbf{W}$). The digital part ($\mathbf{W}$) is taken to decrease the processing error caused by the UAV attitude adjustment, while the analog part ($\{a_{k,l}\}$) is designed to improve the system performance.

We transform the objective function from sum MSE to ergodic sum MSE. Building on this, the design problem for the digital part ($\mathbf{W}$) is written as

$$
\min_{\mathbf{W}} \text{tr}\left[\mathbf{W}^{\text{T}}\mathbf{H}^{\text{T}}\mathbf{E}\left[\mathbf{M}\mathbf{M}^{\text{H}}\right]\mathbf{H}^*\mathbf{W}^*\right] + \sigma^2 K
$$
$$
- 2\text{Re}\left[\text{tr}\left[\mathbf{W}^{\text{T}}\mathbf{H}^{\text{T}}\mathbf{E}\left[\mathbf{M}\right]\right]\right], \tag{5}
$$
$$
\text{s.t. } C1: \text{tr}[\mathbf{M}_{00}\mathbf{W}\left(\mathbf{M}_{00}\mathbf{W}\right)^{\text{H}}] \leq P.
$$

Considering the power constraint $C1$, using Lagrange multiplier method, $\mathbf{W}$ is designed as

$$
\mathbf{W} = \rho\mathbf{E}\left[\mathbf{M}^{\text{H}}\right]\mathbf{H}^*\left(\mathbf{H}^{\text{T}}\mathbf{E}\left[\mathbf{M}\mathbf{M}^{\text{H}}\right]\mathbf{H}^* + \frac{\sigma^2 K}{P}\mathbf{M}_{00}^{\text{T}}\mathbf{M}_{00}^*\right)^{-1},
$$
$$
\tag{6}
$$

where $\rho = \sqrt{P / \text{tr}\left[\left(\mathbf{H}^{\text{T}}\mathbf{E}\left[\mathbf{M}\mathbf{M}^{\text{H}}\right]\mathbf{H}^* + \frac{\sigma^2 K}{P}\mathbf{M}_{00}^{\text{T}}\mathbf{M}_{00}^*\right)^{-2} \times (\mathbf{M}\mathbf{E}\left[\mathbf{M}^{\text{H}}\right]\mathbf{H}^*)(\mathbf{M}\mathbf{E}\left[\mathbf{M}^{\text{H}}\right]\mathbf{H}^*)^{\text{H}}\right]}$. Without consideration of

the holographic beam disturbance, it is easy to achieve the normal digital part based on Eqn.(6).

The design problem for the analog part becomes

$$
\max_{\{a_{k,l}\}} \min_k \ \gamma_k / \gamma_0,
$$
$$
\text{s.t. } C3: \sum_{k=1}^{K}\sum_{l=1}^{L} a_{k,l} = 1, \ a_{k,l} \geq 0, \ \forall k, l. \tag{7}
$$

Fortunately, the similar problem has been researched in [6].

IV. SIMULATION RESULTS

For an effective comparison, this section compares two contrastive schemes. One scheme is achieved by solving Eqn.(4) under the perfect UAV attitude adjustment. This scheme is referred to as 'Ideal Adjustment'. The other scheme is the holographic beamforming scheme in [6], which is recorded briefly as 'Holographic Beamforming'.

As is shown in Fig.1, 'Ideal Adjustment' performances the best, followed by our scheme and 'Holographic Beamforming'. The main reason is that non-ideal adjustment on UAV causes a mismatch in holographic beam angle correction. Compared to Holographic Beamforming, our scheme has a higher probability of covering the jitter direction, leading to improved bit error rate (BER) performance.

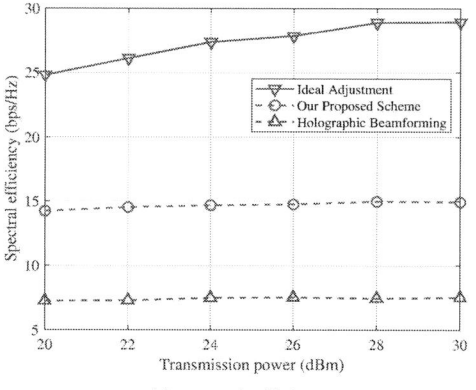

(a) spectral efficiency

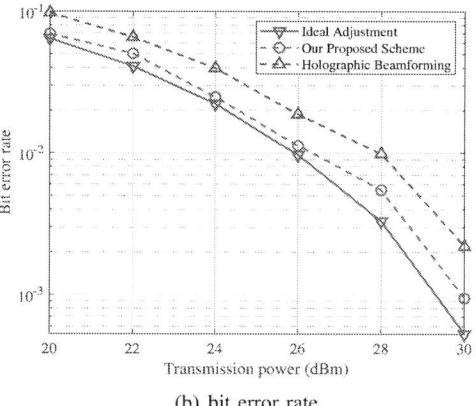

(b) bit error rate

Fig. 1. Performance comparison with 2° jitter deviation between the proposed scheme and other existing schemes. The number of RHS elements is 50×50.

979-8-3315-2209-4/25 $31.00 © 2025 IEEE

V. Conclusions

In this paper, we investigate a robust hybrid holographic beamforming scheme under the RHS-based UAV IoT system, which decouples the digital and analog parts of hybrid holographic beamforming. The digital beamforming is achieved by Lagrange multiplier method of minimizing ergodic sum MSE, and the analog holographic beamforming is obtained via quadratic cone programming of maximizing the minimum SINR. Simulation results validate the effectiveness of the proposed scheme.

References

[1] R. Deng, B. Di, H. Zhang, Y. Tan, and L. Song, "Reconfigurable holographic surface-enabled multi-user wireless communications: Amplitude-controlled holographic beamforming," IEEE Transactions on Wireless Communications, vol. 21, no. 8, pp. 6003–6017, 28 January 2022.

[2] ——, "Reconfigurable holographic surface: Holographic beamforming for metasurface-aided wireless communications," IEEE Transactions on Vehicular Technology, vol. 70, no. 6, pp. 6255–6259, 14 May 2021.

[3] M. Mozaffari, W. Saad, M. Bennis, and M. Debbah, "Mobile unmanned aerial vehicles (uavs) for energy-efficient internet of things communications," IEEE Transactions on Wireless Communications, vol. 16, no. 11, pp. 7574–7589, 15 September 2017.

[4] J. Hu, M. Shen, C. Su, Y. Xu, and J. Guo, "Rotator-aided ris beam stabilization method in uav communications," in Wireless and Satellite Systems, Springer Nature Switzerland, pp. 269–278, 27 March 2025, Harbin, China.

[5] R. Deng, B. Di, H. Zhang, D. Niyato, Z. Han, H. V. Poor, and L. Song, "Reconfigurable holographic surfaces for future wireless communications," IEEE Wireless Communications, vol. 28, no. 6, pp. 126–131, 21 January 2022.

[6] R. Deng, B. Di, H. Zhang, Y. Tan, and L. Song, "Reconfigurable holographic surface-enabled multi-user wireless communications: Amplitude-controlled holographic beamforming," IEEE Transactions on Wireless Communications, vol. 21, no. 8, pp. 6003–6017, 28 January 2022 .

LS and MMSE Channel Estimation in Dual Pulse Shaping Systems

Moting Deng Hang Li Lizhe Liu Yashan Pang Yangyang Guan Zhiqun Cheng Qixian Zheng Qinghua Guo

Abstract—The dual pulse shaping (DPS) technology is an innovative broadband wireless communication solution that enables full-rate transmission with low-performance and low-cost data converters. This is achieved by splitting the data stream into two half-symbol rate streams, and shaping them by Nyquist pulses and complementary Nyquist pulses. In this paper, we investigate linear channel estimation methods, e.g., least squares (LS) and minimum mean square error (MMSE) methods for DPS systems. Simulation results show that the MMSE channel estimation outperforms the LS one in terms of mean square error of channel estimation and bit error rate (BER) performance. The MMSE one exhibits about 2 dB and 5 dB difference at the BER of 20 dB compared to the case that the channel state information is perfectly known at the receiver, when the length of training sequences is set to 32 and 64, respectively.

Index Terms—channel estimation, least square (LS), minimum mean square error (MMSE), dual pulse shaping (DPS)

I. INTRODUCTION

With the continuous advancement of wireless communication technologies, the demand for signal transmission with larger bandwidth and more efficient bandwidth utilization has become increasingly urgent. However, this demand imposes stringent requirements on the sampling rate of data converters [1], [2]. To address this challenge, the dual pulse shaping (DPS) system has emerged as an innovative high-speed data transmission technique designed for millimeter-wave and terahertz communication systems [3], [4]. The DPS system operates by splitting the transmitted data into two parallel streams and shaping them using Nyquist and complementary Nyquist pulses [5]. This enables full-rate transmission using analog-to-digital (A/D) and digital-to analog (D/A) converters with half of the symbol rate, offering a cost-effective solution for large-bandwidth systems. Although the two data streams exhibit spectral overlap in the frequency domain, the pulse shaping ensures inter-symbol interference (ISI) free and the cross-symbol interference (CSI) free transmission theoretically.

Multipath fading in wireless channels inevitably introduces ISI and CSI, leading to degraded system performance. In

---

Moting Deng, Hang Li, Yangyang Guan, Zhiqun Cheng and Qixian Zheng are with *School of Electronics and Information, Hangzhou Dianzi University*, Hangzhou, Zhejiang, China. Hang Li, Lizhe Liu and Yashan Pang are with *Key Laboratory of Advanced Communication Networks, Academy of Network and Communications of CETC*, Shijiazhuang, Hebei, China. Qinghua Guo is with *School of Electrical, Computer and Telecommunications Engineering, University of Wollongong*, Wollongong, NSW, Australia. This work is supported in part by the National Natural Science Foundation under Grant 62071163; in part by the Zhejiang Provincial Natural Science Foundation under Grant LY22F010003 and in part by the Project of Ministry of Science and Technology under Grant D20011. Corresponding author: (Hang Li, email: hangli@hdu.edu.cn)

[6], the authors developed two UAMP-based low-complexity equalizers for DPS transmission, which achieved significant performance gain over conventional DPS equalizer. Although channel state information is assumed to be known at the receiver [6], accurate channel state information generally needs to be estimated through channel estimation methods, which is essential for effective equalization [7], [8]. Channel estimation for DPS systems has not yet been investigated.

In this paper, we introduce a channel estimation framework for the DPS system and investigate linear channel estimation methods, including least squares (LS) and minimum mean square error (MMSE) [9], [10]. Simulation results show that the MMSE estimator exhibits better channel estimation performance at low SNRs than the LS one, and has higher channel estimation and bit error rate (BER) performance as the length of training sequences increases.

Notations: Bold lower and upper case letters denote column vectors and matrices, respectively. The superscripts $(\cdot)^T$ and $(\cdot)^H$ denote the transpose and conjugate transpose. $\mathbf{a}^{<n>}$ denotes the column vector $\mathbf{a}$ with down circular shift by n position. $\mathbf{I}$ stands for the identity matrix. $\mathbb{E}[\mathbf{A}]$ represents the mathematical expectation of matrix $\mathbf{A}$.

II. SYSTEM MODEL

Fig. 1 illustrates the block diagram of the DPS transmission system, where an uncoded system is considered. At the transmitter, data symbols with a symbol duration of T_s are first generated through a mapper, and a sequence of training symbols is inserted before each data block to facilitate channel estimation. The cascaded symbols are then converted into two half-rate parallel data streams via a serial-to-parallel (S/P) converter and subsequently transformed into analog signals using D/A converters with a sampling rate of $\frac{1}{2T_s}$. These signals are individually processed through two pulse shaping filters, generating two independent spectrally shaped signals. Finally, the two signals are combined and filtered by a transmitter filter to form the transmitted signals.

At the receiver, the received baseband signal corrupted by additive white Gaussian noise (AWGN), is first processed through a receiver filter. The filtered signal is then sampled by two A/D converters at a rate of $\frac{1}{2T_s}$, producing two symbol streams. Note that, the second A/D converter operates with a time offset of T_s relative to the first. These streams are upsampled by a factor of two and then combined. The output of the received training sequence is processed by channel estimator to obtain the channel estimates for equalization.

979-8-3315-2209-4/25 $31.00 © 2025 IEEE

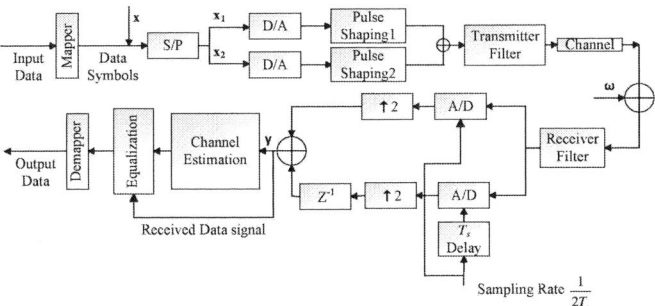

Fig. 1. Block diagram of the DPS transmission system for channel estimation

In a single-carrier DPS system, let $\mathbf{x} = [x_0, x_1, \cdots, x_{N-1}]^T$ denotes a training sequence within a transmission data block, where N represents the length of the training sequence. The training sequence is converted into two parallel streams via a S/P converter, which are defined as $\mathbf{x}_1 = [x_0, 0, \ldots, x_{N-2}, 0]^T$, and $\mathbf{x}_2 = [0, x_1, \ldots, 0, x_{N-1}]^T$. At the receiver, the received signal vector $\mathbf{y} \in \mathbb{C}^{M \times 1}$ for the DPS system is given by

$$\mathbf{y} = \sum_{i=1}^{2} \mathbf{X}_i \mathbf{H}_{pi} \cdot \mathbf{c} + \boldsymbol{\omega} = \mathbf{A} \cdot \mathbf{c} + \boldsymbol{\omega}, \quad (1)$$

where $\mathbf{X}_i \in \mathbb{C}^{N \times N}$ denotes the training sequence matrix, $\mathbf{c} \in \mathbb{C}^{M \times 1}$ denotes the channel channel impulse response, $\mathbf{H}_{pi} \in \mathbb{C}^{N \times M}$ denotes DPS filter response matrix, and $\boldsymbol{\omega}$ denotes an AWGN vector with zero mean and covariance matrix $\sigma^2 \mathbf{I}$. We assume that the cyclic prefix is used in this system, which leads to circulant matrices $\mathbf{X}_i = [\mathbf{x}_i^{<0>}, \mathbf{x}_i^{<1>}, \ldots, \mathbf{x}_i^{<M-1>}]$, $\mathbf{H}_{p1} = [\mathbf{h}_{p1}^{<0>}, \mathbf{h}_{p1}^{<1>}, \ldots \mathbf{h}_{p1}^{<M-1>}]$, and $\mathbf{H}_{p2} = [\mathbf{h}_{p2}^{<N-1>}, \mathbf{h}_{p2}^{<0>}, \ldots, \mathbf{h}_{p2}^{<M-2>}]$, where $\mathbf{h}_{pi} = [\mathbf{h}_{ri}^T, 0^T]^T$, $\mathbf{h}_{ri}$ represents the filter impulse response with a length of L_{PS}.

III. CHANNEL ESTIMATION FOR DPS

A. LS Channel Estimation

LS estimation is a classical linear estimation, and widely used in channel estimation due to its simplicity and computational efficiency. The primary goal of LS estimation is to estimate the channel response by minimizing the sum of squared errors between the received signal and its expected value. The objective function of LS channel estimation for the DPS system can be formulated as

$$\hat{\mathbf{c}} = \arg\min_{\mathbf{c}} \|\mathbf{y} - \mathbf{A}\mathbf{c}\|_2^2. \quad (2)$$

Taking the derivative with respect to $\mathbf{c}$ and setting it to zero gives the closed-form LS solution

$$\hat{\mathbf{c}} = (\mathbf{A}^H \mathbf{A})^{-1} \mathbf{A}^H \mathbf{y}. \quad (3)$$

B. MMSE Channel Estimation

The primary objective of MMSE estimation is to minimize the mean square error (MSE) of the estimation, thereby providing high-quality channel estimates even in the presence of noise and interference. For the DPS system, the cost function for MMSE channel estimation is defined as

$$J_{MMSE} = \mathbb{E}\left[\|\mathbf{c} - \hat{\mathbf{c}}\|^2\right] = \mathbb{E}\left[\|\mathbf{c} - \mathbf{W}\mathbf{y}\|^2\right], \quad (4)$$

where we assume that $\hat{\mathbf{c}} = \mathbf{W}\mathbf{y}$. $\mathbf{W}$ denotes filter matrix. Taking the derivative of J_{MMSE} with respect to $\mathbf{W}$ and setting it to zero yields

$$\frac{\partial J_{MMSE}}{\partial \mathbf{W}} = -2\mathbb{E}[\mathbf{c}\mathbf{y}^H] + 2\mathbf{W}\mathbb{E}[\mathbf{y}\mathbf{y}^H] = 0. \quad (5)$$

Solving (5) gives the MMSE filter matrix

$$\mathbf{W} = \mathbb{E}[\mathbf{c}\mathbf{y}^H]\left(\mathbb{E}[\mathbf{y}\mathbf{y}^H]\right)^{-1}. \quad (6)$$

Defining the cross-covariance and autocorrelation matrices as $\mathbf{R}_{\mathbf{cy}} = \mathbb{E}[\mathbf{c}\mathbf{y}^H]$ and $\mathbf{R}_{\mathbf{y}} = \mathbb{E}[\mathbf{y}\mathbf{y}^H]$ respectively. The estimated channel by MMSE, $\hat{\mathbf{h}}_{MMSE}$, can be expressed as

$$\hat{\mathbf{h}}_{MMSE} = \mathbf{W}\mathbf{y} = \mathbf{R}_{\mathbf{c}}\mathbf{A}^H(\mathbf{A}\mathbf{R}_{\mathbf{c}}\mathbf{A}^H + \sigma^2\mathbf{I})^{-1}\mathbf{y}, \quad (7)$$

where $\mathbf{R}_{\mathbf{c}} = \mathbb{E}[\mathbf{c}\mathbf{c}^H]$ represents the autocorrelation matrix of $\mathbf{c}$, and it might be updated continuously or periodically to track changes in the channel conditions due to mobility or environmental changes in practical systems.

IV. SIMULATION RESULT

We consider DPS transmission systems with a bandwidth of 5 GHz, and employ low-cost commercial data conversion devices with a sampling rate of 2.5 Gsps. The number of data symbols in one block is set to 128 QPSK modulation is employed. We assume that the coefficients $\{c_i\}$ of a multipath fading channel follow a circularly symmetric complex Gaussian distribution with zero mean and variance $\sigma_i^2 = \exp(-\alpha i)/\sum_i \exp(-\alpha i)$, with $\alpha = 0.1$ and a maximum delay spread of $4ns$. The channel impulse response length M is set to 21, with non-zero elements at $c_0, c_5, c_{10}, c_{15}, c_{20}$. The pulse shaping filter length L_{PS} is fixed to be 8. We assume that the channel statistical characteristics are unknown at the receiver, but M is known. Therefore, in simulations, we let $\mathbf{R}_{\mathbf{c}}$ be a diagonal matrix with the first 21 diagonal elements of $\frac{1}{21}$ and the remaining elements of zeros, where it is assumed that the power of each path is uniformly distributed. Both the transmitter and receiver filters are 10th order Butterworth low-pass filters. To compare the BER performance for different channel estimation methods, we adopt a linear MMSE equalizer [11] for signal detection.

Fig. 2 illustrates the BER performance using LS and MMSE channel estimation methods, demonstrating that the BER performance of MMSE channel estimation is higher than that of LS channel estimation. This difference arises primarily because MMSE estimation leverages channel statistical information, allowing for more accurate channel estimation. At the BER of 10^{-4}, the MMSE channel estimation exhibits a performance degradation of approximately 2 dB and 5 dB

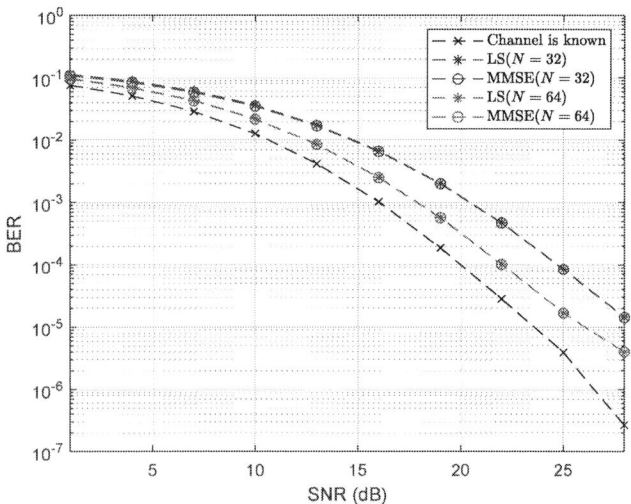

Fig. 2. BER comparison between MMSE and LS

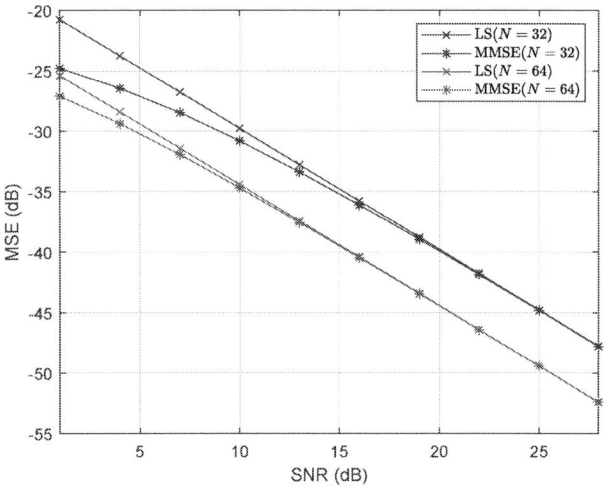

Fig. 3. MSE comparison between MMSE and LS

compared to the case assuming known channel, when N is set to 32 and 64. The BER performance of the DPS system enhances when the length of the training sequence N increases from 32 to 64. Fig. 3 shows that the MMSE channel estimation achieves lower MSE than LS estimation at low SNR. At the SNR of 10 dB, the MMSE method outperforms by up of 0.2 dB and 1.5 dB than the LS one, when N is set to 32 and 64 respectively. The accuracy of the estimated channel improves as the length of the training sequence increases.

V. CONCLUSION

In this paper, we investigate the LS and MMSE channel estimation performance in the DPS system. Simulation results show that the MMSE estimator outperforms the LS one in channel estimation at low SNRs and achieves improved channel estimation and BER performance with longer training sequences.

REFERENCES

[1] P. Rodríguez-Vázquez, J. Grzyb, N. Sarmah, B. Heinemann, and U. R. Pfeiffer, "Towards 100 Gbps: A Fully Electronic 90 Gbps One Meter Wireless Link at 230 GHz," in *2018 48th European Microwave Conference (EuMC)*, 2018, pp. 1389–1392.

[2] C.-X. Wang, F. Haider, X. Gao, X.-H. You, Y. Yang, D. Yuan, H. M. Aggoune, H. Haas, S. Fletcher, and E. Hepsaydir, "Cellular Architecture and Key Technologies for 5G Wireless Communication Networks," *IEEE Communications Magazine*, vol. 52, no. 2, pp. 122–130, 2014.

[3] H. Li, X. Huang, J. A. Zhang, H. Zhang, and Z. Cheng, "Dual Pulse Shaping Transmission with Sinc-function based Complementary Nyquist Pulses," *IET Communications*, vol. 16, no. 17, pp. 2091–2104, 2022.

[4] H. Zhang, X. Huang, J. A. Zhang, and Y. J. Guo, "Dual Pulse Shaping Transmission and Equalization for High-Speed Wideband Wireless Communication Systems," *IEEE Transactions on Circuits and Systems I: Regular Papers*, vol. 67, no. 7, pp. 2372–2382, 2020.

[5] A. Kumar and M. Magarini, "Improved Nyquist Pulse Shaping Filters for Generalized Frequency Division Multiplexing," in *2016 8th IEEE Latin-American Conference on Communications (LATINCOM)*, 2016, pp. 1–7.

[6] P. Cai, H. Li, Q. Guo, and X. Huang, "UAMP-Based Equalization for Dual Pulse Shaping Transmission Systems," *IEEE Wireless Communications Letters*, vol. 12, no. 7, pp. 1164–1168, 2023.

[7] J. Chen, Y.-C. Liang, H. V. Cheng, and W. Yu, "Channel Estimation for Reconfigurable Intelligent Surface Aided Multi-User mmWave MIMO Systems," *IEEE Transactions on Wireless Communications*, vol. 22, no. 10, pp. 6853–6869, 2023.

[8] H. Liu, J. Zhang, Q. Wu, H. Xiao, and B. Ai, "ADMM Based Channel Estimation for RISs Aided Millimeter Wave Communications," *IEEE Communications Letters*, vol. 25, no. 9, pp. 2894–2898, 2021.

[9] X. Wang, X. Shen, F. Hua, and Z. Jiang, "On Low-Complexity MMSE Channel Estimation for OCDM Systems," *IEEE Wireless Communications Letters*, vol. 10, no. 8, pp. 1697–1701, 2021.

[10] C. Wei, H. Liu, Z. Zhang, J. Dang, and L. Wu, "Near-Optimum Sparse Channel Estimation Based on Least Squares and Approximate Message Passing," *IEEE Wireless Communications Letters*, vol. 6, no. 6, pp. 754–757, 2017.

[11] M. Tuchler, A. Singer, and R. Koetter, "Minimum Mean Squared Error Equalization using a Priori Information," *IEEE Transactions on Signal Processing*, vol. 50, no. 3, pp. 673–683, 2002.

A Background Calibration in Pipelined-SAR ADCs using Capacitor Flipping and Gain Verification

Pengfei Ye   Xinpeng Xing   Haigang Feng

Abstract—**This paper presents a background inter-stage gain calibration technique for pipelined successive-approximation-register (SAR) analog-to-digital converters (ADCs). The gain verification module employs two capacitor-flipping to define the desired inter-stage gain, and compared it to the actual one in the circuit, residual gain error is then minimized by adaptively adjusting the feedback capacitance of the closed-loop amplifier. The proposed analog-domain calibration operates entirely in the background, with very limited additional hardware overhead and only 10K samples for convergence. Simulation results of a 14-bit 100-MS/s pipeline-SAR ADC in 28-nm CMOS show that the proposed calibration improves the ADC spurious-free dynamic range (SFDR) from 68 dB to 95.4 dB, and its signal-to-noise-and-distortion ratio (SNDR) from 55.5 dB to 83.3 dB.**

Index Terms—**pipelined SAR ADC ,analog calibration technique ,inter-stage gain error , gain verification**

I. INTRODUCTION

The pipelined SAR ADC, which integrates the merits of both Pipeline ADCs and successive-approximation ADCs, has found extensive applications due to its versatile performance. A pivotal operation in this ADC architecture is residue amplification [1], which guarantees that each stage processes a full-scale signal, thereby significantly alleviating the precision demands on the quantizer. Nevertheless, the conversion accuracy of the pipelined ADC is critically contingent upon the residue amplifier's ability to deliver precise inter-stage gain. Any deviation in the inter-stage gain can result in the leakage of quantization noise from the preceding sub-ADC stages into the overall digital output code. This leakage can manifest as decision threshold inaccuracies or missing codes, ultimately compromising the ADC's overall precision.

To mitigate this challenge, several techniques have been developed to minimize inter-stage gain errors. Prominent among these are correlated double sampling (CDS) [2], correlated level shifting (CLS) [3], and digital amplifiers [4]. While both CDS and CLS alleviate the stringent requirements for operational amplifiers, CDS introduces additional noise, and CLS increases the output load. Moreover, the necessity for extra clock phases in both methods reduces the available settling time. Digital amplifiers present another viable solution; however, they necessitate a high-precision comparator to detect the voltage difference at the operational amplifier's differential input, thereby escalating design complexity.

This paper presents a background inter-stage gain verification technique with rapid convergence, which enables the comparison between the actual gain and the ideal gain with nearly negligible additional analog overhead. Leveraging this comparison, the closed-loop residue amplifier dynamically corrects the actual gain by utilizing a feedback capacitor array.

II. THE PROPOSED CALIBRATION

A. Capacitor Flipping And Gain Verification

The ADC used to validate the proposed calibration technique comprises three stages: the first two stages are 5-bit with 1-bit redundancy each, while the final stage is 6-bit. The inter-stage residue amplifier utilizes a closed-loop configuration, with its gain determined by the ratio of the total capacitance of the preceding CDAC to the feedback capacitance. For simplicity, Fig. 1 consolidates the second and third stages into a single backend ADC. The ideal gain, as designed, is denoted as G_{ideal}, while the actual gain realized in the circuit is denoted as G_{real}. The first-stage SAR ADC performs a coarse quantization on the input signal V_{in}, producing the first-stage digital output code D_1. The residue signal V_{res} is generated by subtracting D_1 from V_{in}. This residue signal V_{res} is then amplified by the inter-stage residue amplifier with a gain of G_{real}, resulting in the amplified signal V_2. The backend SAR ADC subsequently performs a fine quantization on V_2, generating the backend digital output code D_2. Finally, the overall ADC output D_{out} is computed by combining D_2 and D_2/G_{ideal}.

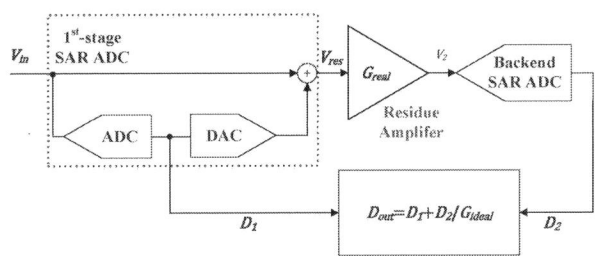

Fig. 1: The ADC block diagram.

Fig. 2 illustrates the timing diagram of the proposed calibration scheme, which operates in the background through three sequential steps. In the first two steps, two additional comparison cycles are inserted to validate the magnitude relationship between the actual inter-stage gain and the ideal design value

---

Author 1 and Author 3 are with *Shenzhen Graduate International School, Tsinghua University*, Shenzhen, China. Author 2 is with *School of Integrated Circuits, Sun Yat-Sen University*, Shenzhen, China. Corresponding author: (Author 2, email: xingxp@mail.sysu.edu.cn)

979-8-3315-2209-4/25 $31.00 © 2025 IEEE

via capacitor-flipping and comparisons. Subsequently, the third step triggers gain error correction by adaptively adjusting the feedback capacitance of the residue amplifier during its reset state, thereby ensuring seamless integration with the normal conversion cycle without introducing latency or interference.

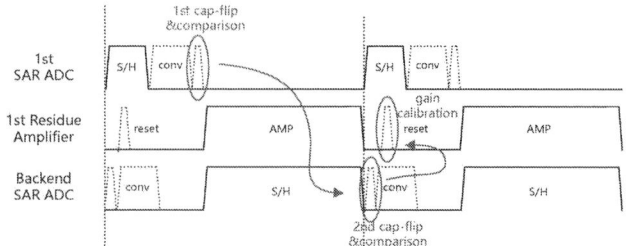

Fig. 2: The calibration timing diagram.

The initial step in gain error verification is to determine the range of the residue voltage. After the quantization process of the first-stage sub-SAR ADC is completed, the sign of the residue signal to be amplified on the first-stage CDAC can be determined based on the result of the last bit of quantization. Assuming the last bit of the first-stage sub-ADC quantization result is 0, once this bit is set, the residue voltage V_{res} remaining on the CDAC satisfies $0 < V_{res} < V_{LSB1}$. By flipping the capacitor $C_{1,d1}$ in the opposite direction, whose capacitance is half of the dummy capacitor as shown in Fig. 3, a new residue voltage V_1 is generated.which can be expressed as:

$$V_1 = V_{res} - V_{LSB1}/2 \qquad (1)$$

At this point, the comparator performs another comparison. If the comparison result is 1, it indicates that $V_1 > 0$, meaning $V_{LSB1}/2 < V_{res} < V_{LSB1}$. Conversely, if the comparison result is 0, it implies $0 < V_{res} < V_{LSB1}/2$. After determining the range of the residue voltage, the capacitor used for bit-setting must be restored to its original position. This ensures that the residue signal to be amplified returns to V_res, enabling the ADC's residue amplifier to accurately amplify the correct residue signal.

The residue signal from the first stage is sampled by the second-stage sub-SAR ADC after being processed by the residue amplifier, forming the signal to be quantized V_2 on the second-stage CDAC, which can be expressed as:

$$V_2 = G_{real} \cdot V_{res} \qquad (2)$$

In the second-stage sub-SAR ADC, identify the capacitor corresponding to $V_{LSB1}/2$. Given the 1-bit redundancy in the inter-stage design, this capacitor is the third-bit capacitor of the second-stage sub-ADC. Flipping this capacitor introduces a voltage change of ΔV. Flip the capacitor in the same direction as the flipping of $C_{1,d1}$ in the first stage, generating a new signal V_3 for comparison. ΔV and V_3 can be expressed as:

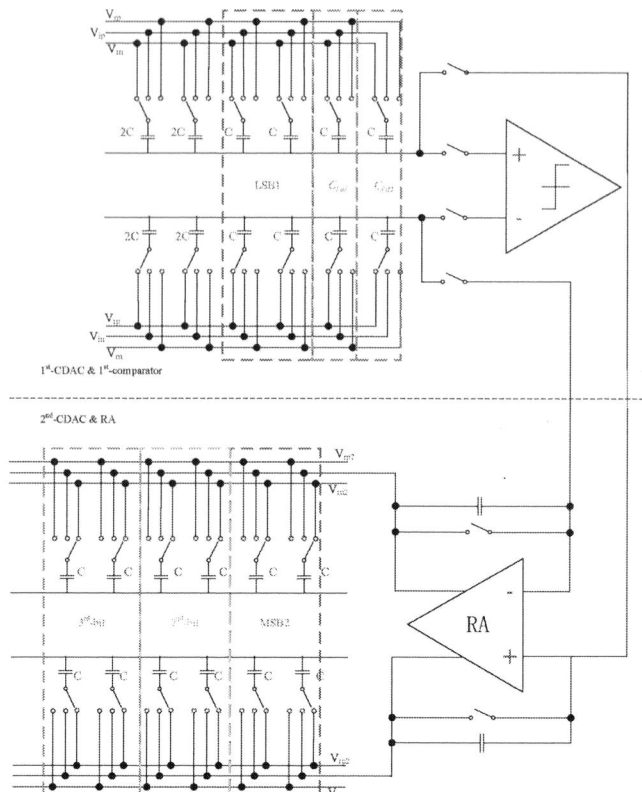

Fig. 3: The split dummy capacitor and corresponding capacitors.

$$\Delta V = G_{ideal} \cdot V_{LSB1}/2 \qquad (3)$$

$$V_3 = V_2 - \Delta V = G_{real} \cdot V_{res} - G_{ideal} \cdot V_{LSB1}/2 \qquad (4)$$

Before the quantization process begins, the comparator should perform a comparison on V_3. According to (2) and (4), discrepancy between the comparison results of V_1 and V_3 indicates that the actual gain deviates from the ideal gain. Specifically, if the comparison result of V_1 is 1 while that of V_3 is 0, it indicates that the actual gain G_{real} is lower than the ideal gain G_{ideal}. Conversely, if the comparison result of G_1 is 0 and that of G_3 is 1, it implies that G_{real} exceeds G_{ideal}.After determining the relationship between G_{real} and G_{ideal}, it is necessary to restore the capacitors in the second stage to their original configuration. This step ensures that the ADC can return to its normal operating state.

Similarly, if the last quantization bit of the first-stage sub-ADC is 1, flipping $C_{1,d2}$ in the opposite direction allows us to determine whether $-V_{LSB1}/2 < V_{res} < 0$ or $-V_{LSB1} < V_{res} < -V_{LSB1}/2$. In the second stage, the corresponding capacitor is flipped in the same direction as $C_{1,d2}$ in the first stage. After comparison, the relationship between G_{real} and G_{ideal} can also be determined.

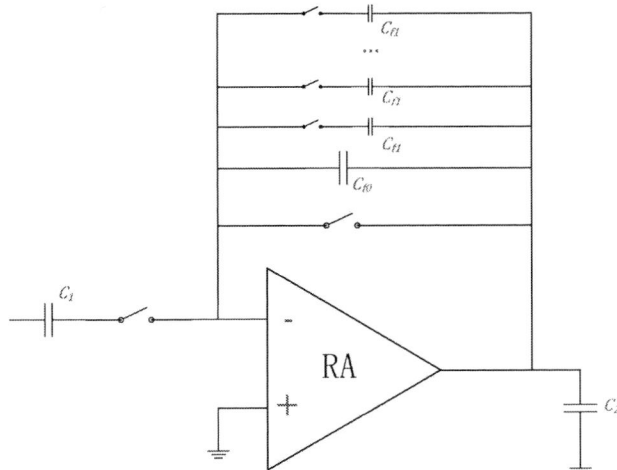

Fig. 4: The calibration capacitor array.

B. Gain Calibration

For a high-gain amplifier operating in negative feedback mode, its closed-loop gain depends on the ratio of the capacitors. The gain G can be expressed as:

$$G = \frac{G_0}{1 + \beta G_0} \approx \frac{1}{\beta} = \frac{C_1}{C_f} \tag{5}$$

where G_0 is the open-loop gain of the amplifier, C_1 is the total capacitance of the CDAC in the first-stage sub-ADC, and C_f is the feedback capacitance of the amplifier.Therefore, by altering the value of C_f, the gain of the amplifier can be adjusted. As shown in Fig. 3, the feedback capacitance C_f consists of a nominal capacitance C_{f0}, which is slightly smaller than the ideal value, and several calibration unit capacitors C_{f1}. Each calibration unit capacitor is connected in series with a switch and then paralleled to C_{f0}. By turning these switches on or off, the total capacitance of the feedback capacitor C_f can be adjusted, thereby modifying the closed-loop gain of the residue amplifier.

The calibration technique necessitates an additional comparison cycle for the sub-ADCs preceding and succeeding the residue amplifier. Since the main timing constraint in a pipelined SAR ADC is the residue amplification time, the extra comparison cycle is tolerable and does not significantly impact overall performance.

In this configuration, the switch positioned at the amplifier's input terminal maintains both terminals at near-identical common-mode potentials. Additionally, the residual voltage amplitude from the pre-stage is very small (<14 mV), making the nonlinear effects introduced by the switch's parasitic resistance negligible.

From a theoretical standpoint, the precision of this calibration method is fundamentally constrained by capacitor mismatch. In the proposed ADC architecture, the dummy capacitors in the first and second stages are designed to be $64fF$ and $16fF$, respectively. Monte Carlo simulations reveal that the standard deviations of mismatch for these capacitors

are 0.0838% and 0.1637%, respectively. The resulting gain errors are so minimal that they have no discernible impact on the overall performance of the ADC.

III. SIMULATION RESULTS

The 14-bit 100 MS/s pipelined SAR ADC designed to validate the proposed calibration technique is implemented in 28nm CMOS process without capacitor mismatch. In this design, both stages of the residue amplifiers undergo calibration. The ADC operates at a sampling rate of 100 MS/s. The input signal is a sinusoidal waveform with a frequency of $(487/1024) \times F_s$ and an amplitude of 90% of the full-scale range. Fig.5 illustrates the FFT spectra before and after calibration. Without calibration, the precision of the ADC is predominantly constrained by inter-stage gain errors. The spurious-free dynamic range (SFDR) is only 72.8 dB, and the signal-to-noise and distortion ratio (SNDR) is only 60.2 dB. With calibration, the SFDR is improved to 94.2 dB, and the SNDR is improved to 83.4 dB.

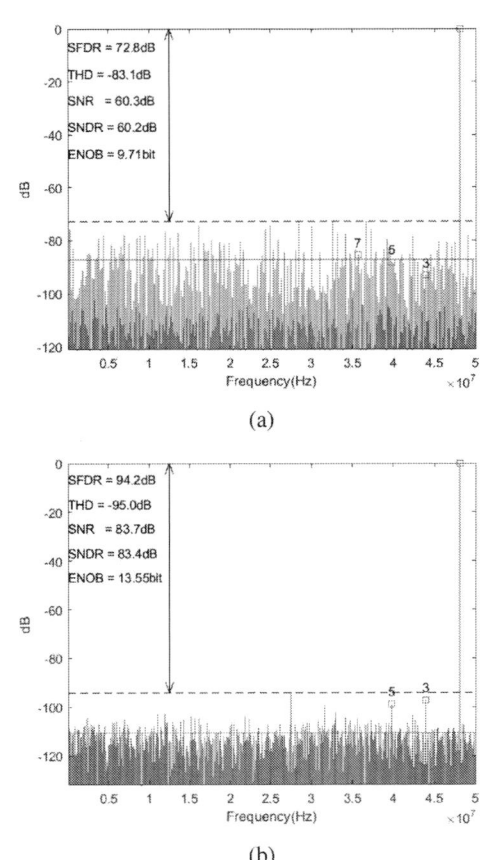

Fig. 5: FFT spectrum (a)Before calibration (b)After calibration.

Fig. 6 illustrates the convergence curve in background calibration, we can see the convergence at around 10k samples. During the initial calibration period, the gain error correction exhibits quasi-linear behavior. The convergence rate gradually

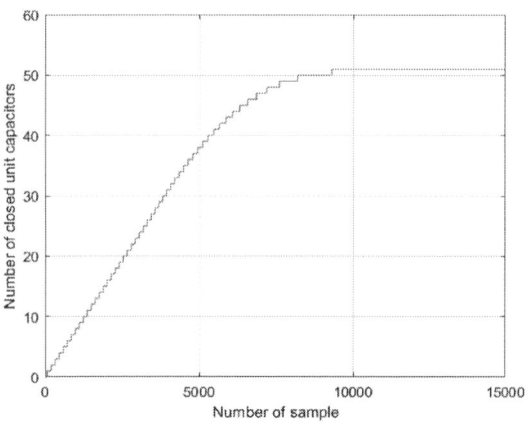

Fig. 6: Converge curve.

decelerates as the discrepancy between the actual gain and the ideal design value narrows.

Table 1 presents a comparison between the proposed calibration technique and other existing calibration methods. Compared to [5], this work eliminates the need for metastability detection, which is highly sensitive to clock timing. Compared to [6] and [7], the proposed technique demonstrates significantly faster convergence speed, as the error correction process is nearly linear. Moreover, the error verification module introduces almost no additional hardware overhead, utilizing only a unit capacitor array and a few switches for error minimization module. The proposed calibration technique eliminates the need for complex digital hardware, thereby minimizing additional overhead and simplifying the design.

TABLE I: Comparison With ADCs Using Calibration

	[5][a]	[6][b]	[7][a]	**This work** [a]
Process(nm)	40	/	40	**28**
Resolution	14	14	12	**14**
Fs(MS/s)	100	100	100	**100**
SFDR (wo. calib)	64.5	70.8	54.5	**72.8**
SFDR (wi. calib)	97.1	108.5	94	**94.2**
SNDR (wo. calib)	58.2	63.9	49	**60.2**
SNDR (wi. calib)	79.1	85.7	68.9	**83.4**
Convergence (Sample)	100k	30M	2M	**10k**
Analog Hardware Complexity	Dither Cap, Metastability Detector	Dither Cap, Logic Switch	Dither Cap, Comparator, Analog MUX	**Unit Cap Array, Logic Switch**

[a]Simulation Results. [b]Matlab Results.

IV. CONCLUSION

This paper presents an analog inter-stage calibration for pipeline-SAR ADC, using capacitor flipping and gain verification. The proposed analog calibration operates in the background, necessitating only a unit capacitor array and

several switches. Simulation results of a 28nm pipeline-SAR ADC indicate that after calibration, its SFDR and SNDR are improved by 21.4 dB and 23.2 dB, respectively.

REFERENCES

[1] J. Hao, Y. Shen, J. Zhang, Y. Zhang, S. Liu and Z. Zhu, "A 14b 180MS/s Pipeline-SAR ADC With Adaptive-Region-Selection Technique and Gain Error Calibration," in IEEE Transactions on Circuits and Systems II: Express Briefs, vol. 71, no. 1, pp. 16-20, Jan. 2024.

[2] O. A. Hafiz, X. Wang, P. J. Hurst and S. H. Lewis, "Immediate Calibration of Operational Amplifier Gain Error in Pipelined ADCs Using Extended Correlated Double Sampling," in IEEE Journal of Solid-State Circuits, vol. 48, no. 3, pp. 749-759, March 2013.

[3] J. -C. Wang, T. -C. Hung and T. -H. Kuo, "A Calibration-Free 14-b 0.7-mW 100-MS/s Pipelined-SAR ADC Using a Weighted- Averaging Correlated Level Shifting Technique," in IEEE Journal of Solid-State Circuits, vol. 55, no. 12, pp. 3271-3280, Dec. 2020.

[4] Dong-Young Chang, Jipeng Li and Un-Ku Moon, "Radix-based digital calibration techniques for multi-stage recycling pipelined ADCs," in IEEE Transactions on Circuits and Systems I: Regular Papers, vol. 51, no. 11, pp. 2133-2140, Nov. 2004.

[5] L. Zhang and J. Wu, "Background Calibration in Pipelined SAR ADCs Exploiting PVT-Tracking Metastability Detector," 2021 IEEE International Symposium on Circuits and Systems (ISCAS), Daegu, Korea, 2021, pp. 1-5.

[6] Y. Zhu, J. Sun and W. Liu, "A Hybrid Calibration Technique for Bit-Weight Errors in Pipelined-SAR ADCs," 2023 5th International Conference on Circuits and Systems (ICCS), Huzhou, China, 2023, pp. 165-169.

[7] J. Sun, M. Zhang, L. Qiu, J. Wu and W. Liu, "Background Calibration of Bit Weights in Pipelined-SAR ADCs Using Paired Comparators," in IEEE Transactions on Very Large Scale Integration (VLSI) Systems, vol. 28, no. 4, pp. 1074-1078, April 2020.

An 18-bit 1-MS/s SAR ADC With a Continuous-Time SAR-Based Pre-Quantization

Weifeng Qiao   Zengqing Liang   Haigang Feng   Xinpeng Xing

Abstract—This paper presents design of a power-efficient 18bit successive approximation register (SAR) analog-to-digital converter (ADC). A pre-quantization based on continuous-time (CT) SAR ADC is proposed to enhance its sampling frequency to 1 MS/s and improve its power efficiency. Furthermore, several techniques are applied to improve design performance, including a foreground calibration to mitigate CDAC weight errors, LSB repeating and residue estimation techniques to alleviate noise challenges. The proposed SAR ADC is designed in 180-nm CMOS, and the simulation results show that it achieves 102.9-dB SNDR and 112.4dB SFDR with Nyquist input, consumes 3.1 mW, corresponding to a state-of-the-art FoM of 185.1dB.

Index Terms—Analog-to-digital converter (ADC), continuous-time SAR (CT-SAR), high-precision, noise reduction, power efficiency, successive approximation register (SAR)

I. INTRODUCTION

High-precision, medium-speed analog-to-digital converter (ADC) is in high demand in fields such as medical imaging, instrumentation, and industrial control [1]. Compared to other ADC architectures, successive approximation register (SAR) ADC is better suited for process scaling, making it widely used in low-power applications [2]. However, as the resolution increases, comparator noise becomes the dominant noise source in the system, making the design of high-efficiency, low-noise systems a challenging task.

This paper presents a precision SAR ADC in 180-nm CMOS that employs a multiple-decision technique to reduce the system's sensitivity to noise. Additionally, it proposes a CT SAR ADC architecture based on a four-input comparator, enabling multi-bit pre-quantization with minimal additional power and time consumption. Simulations demonstrate that the pre-quantization stage consumes only 0.07 mW of power while achieving 5-bit fully zeroing delay pre-quantization. The overall system achieves an SNDR of 103.9 dB with a total power consumption of 3.11 mW.

II. ARCHITECTURAL DESIGN

A. ADC Top Level block

Fig. 1(a) shows the proposed ADC top-level block diagram. This 18-bit asynchronous SAR ADC incorporates a CT SAR ADC to quantify the MSBs. By significantly reducing the DAC output, the CT SAR ADC improves system reliability

---

Author 1 and Author 3 are with *Shenzhen Graduate International School, Tsinghua University,* Shenzhen, China. Author 2 and Author 4 are with *School of Integrated Circuits, Sun Yat-Sen University,* Shenzhen, China. Corresponding author: (Author 4, email: xingxp@mail.sysu.edu.cn)

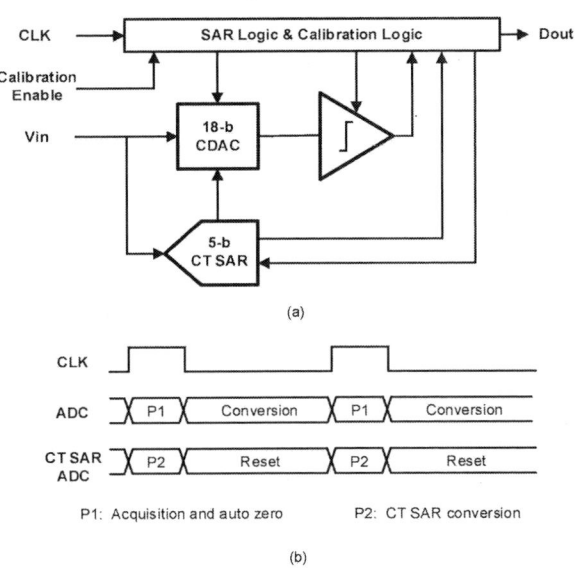

(a)

(b)

Fig. 1. (a) SAR ADC block diagram. (b) Conversion timings.

and speed while maintaining exceptionally low power consumption. Furthermore, the pre-quantization process does not consume any time during the conversion phase. Fig. 1(b) presents the corresponding operational phase diagram, illustrating that the first five bits are fully quantized during the sampling phase. The system's sampling time is determined by the CT SAR ADC. This requires additional calibration, which can be implemented simply and efficiently. During simpling phase, the CT SAR ADC performs quantization while the main ADC comparator undergoes auto-zeroing. Once the CT SAR ADC conversion is complete, the system transitions into the quantization phase to process the remaining bits.

The system is implemented using a 180nm CMOS process. The DAC bottom plate operates in a 5V voltage domain to ensure a sufficient input swing, making it highly suitable for a wide range of industrial measurement applications. Leveraging pre-quantization, the remaining circuits, including the top-plate switches, comparator, and digital engine, are designed within a 1.8V voltage domain.

B. Pre-Quantization Base on CT SAR ADC

In this design, the decision process for the five most significant bits is handled by the pre-quantizer, effectively reducing the voltage swing at the top plate to 1/32 of the con-

---

979-8-3315-2209-4/25 $31.00 © 2025 IEEE   323

ventional swing. Consequently, even with a full-scale bottom-plate sampling swing of ±5V, the top-plate sampling switches and comparator can be implemented using 1.8V transistors. For a comparator designed using the same gm/ID parameter, operating at a 1.8V supply voltage achieves equivalent noise performance while consuming only one-third of the power compared to a 5V supply voltage.

Unlike [2], our proposed method uses a CT SAR ADC as pre-quantizer to achieve concurrent quantization during the sampling phase of the main ADC. This effectively eliminates apparent pre-quantization delays. Compared to conventional ADC architectures employing Flash-based pre-quantization, our approach enables the quantization of additional bits without introducing timing constraints while maintaining superior power efficiency.

C. Calibration With LSB Capacitors

Reference [2] proposes a method for calibrating high-bit capacitors utilizing LSB capacitors. The calibration process involves quantizing the positive and negative residual voltages generated by the corresponding capacitors using a Z-ADC, which consists of LSB capacitors and a comparator. The resulting digital weights, representing the actual proportional values, are then determined in the digital domain and can be expressed by the following equation:

$$W_{i,p} = \sum_{j=0}^{i-1} D_{j,p} W_j + os \qquad (1)$$

$$W_{i,n} = \sum_{j=0}^{i-1} D_{j,n} W_j + os \qquad (2)$$

Offset elimination through differential subtraction. Following this approach, the calibration is extended to higher bits sequentially.

$$W_{i+1} = \left(\sum_{j=0}^{i} D_{j,p} W_j - \sum_{j=0}^{i} D_{j,n} W_j \right)/2 \qquad (3)$$

To minimize the influence of noise during the calibration process, each bit undergoes multiple conversion cycles, with the calibration results subsequently averaged. Unlike the calibration method in [2], the proposed approach leverages the stochastic nature of thermal noise, allowing multiple calibration iterations to estimate weight values that exceed the maximum resolution of the ADC. The proposed ADC retains the results of multiple calibration iterations in a single-bit floating-point format, thereby enhancing calibration accuracy.

D. LSB repeats and Residual estimation

In a SAR ADC, the primary sources of noise stem from sampling noise and conversion noise. To mitigate the system's sensitivity to conversion noise, this design incorporates two techniques: LSB repeat and residual estimation.

Unlike [4], which employs repeated bit decisions followed by a voting mechanism. Following each repeated decision, a corresponding redundant unit capacitor is set. This approach

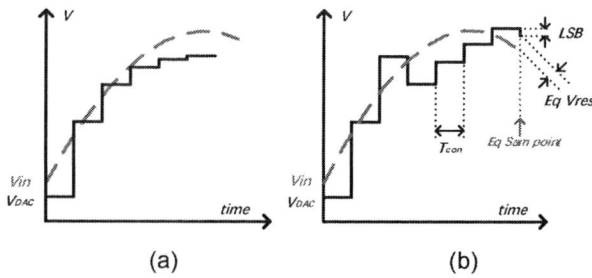

Fig. 2. The approximation process of CT SAR ADC.

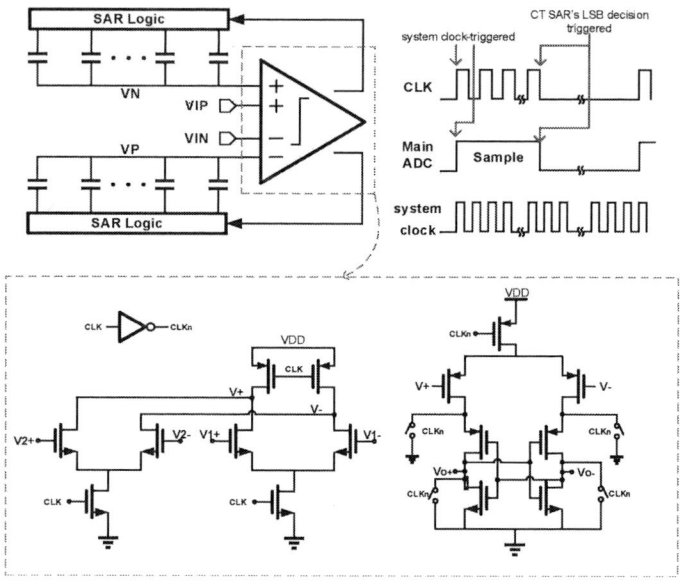

Fig. 3. Design of CT SAR ADC based on a four-input comparator.

offers the advantage of mitigating potential errors in higher-bit decisions while gradually converging the top-plate residual voltage toward VCM, thereby eliminating the need for additional metastability detection circuitry.

In [5], the residual voltage after conversion undergoes multiple repeated evaluations, with the distribution of 1/0 in the decision results utilized to estimate the residual voltage. In this design, the residual voltage is assessed over 9 consecutive iterations. The result mapping register is configurable to optimize residual estimation across varying noise levels.

III. CIRCUIT DESIGN

A. CT SAR

Owing to CT SAR ADC's continuous-time nature, the final decision point can be effectively treated as the sampling point, enabling synchronous conversion during the sampling phase without consuming the main ADC's conversion time. Compared to conventional Flash ADCs, this approach facilitates higher-bit pre-quantization without incurring additional power consumption or latency. In contrast to traditional discrete-time (DT) SAR ADCs, the bit conversion time in a CT SAR ADC is inherently correlated with its resolution.

979-8-3315-2209-4/25 $31.00 © 2025 IEEE

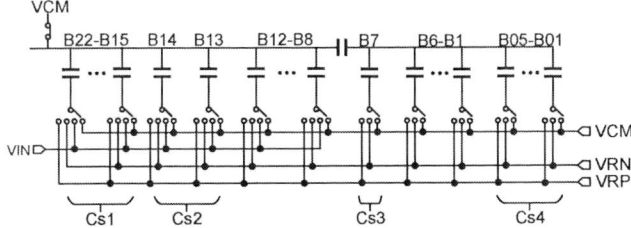

Cs1: MSBs determined by CT SAR ADC
Cs2: Redundant capacitors for improved robustness
Cs3: Redundant capacitors for calibration
Cs4: 5 LSB repeats

Fig. 4. DAC structure.

In addition to the non-ideal factors inherent in traditional DT SAR ADC, the CT SAR ADC exhibits its own distinct sources of error. Redundancy is essential for CT SAR ADC. Fig. 2 compares the approximation process of the CT SAR ADC, highlighting the difference between the processes with and without redundancy. For a differential ADC, the least significant bit (LSB) is typically half the magnitude of LSB+1. Since the input signal continues to evolve after the LSB+1 decision is finalized, there is a potential risk that the residual signal may exceed the quantization range of the LSB. This relationship can be mathematically expressed as follows:

$$V_{res,max} = \left| \frac{dV_{in}}{dt} \right|_{max} \times T_{con} - LSB \qquad (4)$$

where T_{con} represents a single SAR cycle time, V_{in} and V_{res} are the input signal voltage and equivalent residual voltage, respectively. In this design, the value of T_{con}/T_{Vin} is greater than 100, and for a 5-bit, 500 kHz bandwidth system, it is almost unaffected by the aforementioned errors.

Fig. 3 presents the fundamental structure and timing sequence of the proposed CT SAR ADC, along with the design of the four-input comparator. The CDAC of the CT SAR ADC is reset during the conversion phase of the main ADC. The system clock concurrently triggers both the conversion mode of the CT SAR ADC and the sampling phase of the main ADC. Upon completion of the predefined number of conversions by the CT SAR, the main ADC simultaneously concludes its sampling phase.

In [6], the CT SAR ADC employs capacitive coupling, which results in the loss of DC information, thus limiting its applicability to bandpass scenarios. The proposed CT SAR ADC architecture, which leverages a four-input comparator-based coupling scheme, addresses this limitation.

B. CDAC

For a four-input comparator, the comparison result can be represented as follows:

$$D = sign[(V_{1+} + V_{2+}) - (V_{1-} + V_{2-})] \qquad (5)$$

Where the sign function is used to determine the positive or negative sign. To achieve a stable summing process, the VCM of the two input pairs needs to be stable. Therefore, the

CDAC is designed in a differential form. In this design, the comparator output can be expressed as:

$$D = sign[(V_{IP} + V_N) - (V_{IN} + V_P)] \qquad (6)$$

Where V_P and V_N represent the voltages at the differential CDAC top plates, and the common-mode voltage is always maintained at VCM, which is equal to the common-mode voltage of the input signal. The SAR logic is the same as that of the DT SAR ADC.

Compared to conventional current-mode DAC coupling and capacitive coupling techniques, this approach effectively preserves DC information while incurring only a minimal increase in power consumption.

Fig. 4 depicts the CDAC architecture adopted in this work, where the weighting of the 8 MSB capacitors is designed to align with that of the CT SAR ADC. In addition to the pre-quantization capacitors, an additional 1/8 redundant capacitor is incorporated to mitigate potential output errors from the CT SAR ADC and compensate for gain mismatches between the two ADCs. To enhance matching accuracy, the bridge capacitor is implemented using two unit capacitors, while redundancy is strategically introduced in the LSB segment. This design approach simplifies the implementation of calibration algorithms for capacitor mismatch and bridge capacitor parasitics, ensuring robust operation without exceeding the dynamic range. Five unit capacitors with equal weighting are employed to mitigate decision errors in higher-order bits.

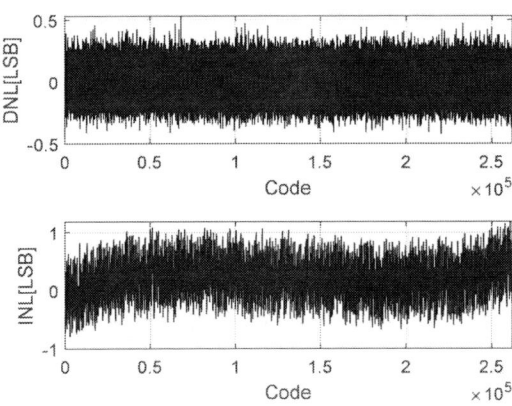

Fig. 5. Simulated DNL and INL.

IV. SIMULATION RESULTS

The system is designed in a 180-nm CMOS process. The ADC input interface operates and CT SAR ADC at 5 V while all other circuits use 1.8 V. Fig. 5 shows the differential non-linearity (DNL) and integral non-linearity (INL), which are +0.53/-0.45 differential LSB and +1.18/-0.79 differential LSB, respectively. Fig. 6 shows the simulated spectrum with a near-Nyquist-rate input signal. With a 0-dBFS near-Nyquist-rate input of 467 kHz, the measured SNDR and SFDR are 102.9 and 112.4 dB, respectively. As discussed in Section II, the foreground calibration is performed for capacitor mismatches

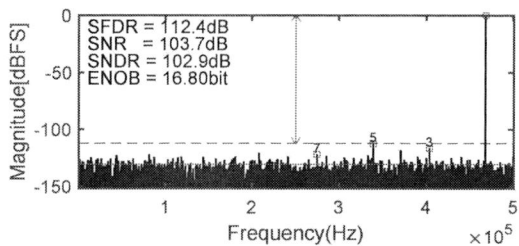

Fig. 6. Typical ac spectrum.

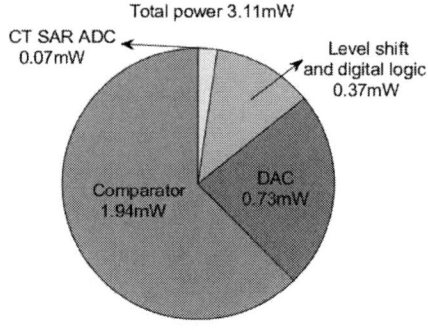

Fig. 7. Power consumption breakdown.

and bridge capacitor parasitics, without which the SNDR and SFDR would be limited to 80.8 and 88.7 dB, respectively.

Running at 1 MS/s, the prototype ADC consumes 3.11 mW. The power breakdown is plotted in Fig. 7. It shows that the comparator occupies 62% of the total power, which is achieved by LSB repeat and residual estimation to relax design constraints. The CT SAR ADC achieves 5-bit pre-quantization with an additional power consumption of only 0.07mW.

The proposed ADC is compared with a selection of benchmark high-resolution SAR ADCs in Table I. Leveraging the low power consumption, high resolution, and zero-latency characteristics of the CT SAR ADC, the system's timing constraints are significantly relaxed, enabling the implementation of redundant comparisons to further optimize power efficiency. The Schreier figure-of-merits (FoMs) with the Nyquist frequency input is 185.1 dB, and is in line with the state-of-thearts.

V. CONCLUSION

A 18-bit 1 MS/s SAR ADC in 180-nm CMOS with a 5-bit CT SAR ADC pre-quantizer has been presented in this paper. To improves both ADC speed and energy efficiency, a digital foreground CDAC calibration, an LSB repeats and a residual estimation technique are implemented with minimal hardware overhead. Based on these, a SNDR of 102.9-dB and a SFDR of 112.4dB are realized for a Nyquist input signal, the 185.1dB FoM makes it very suitable for high-accuracy power-efficient applications.

REFERENCES

[1] B. Gönen, F. Sebastiano, R. Quan, R. van Veldhoven and K. A. A. Makinwa, "A Dynamic Zoom ADC With 109-dB DR for Audio Applications," in IEEE Journal of Solid-State Circuits, vol. 52, no. 6, pp. 1542-1550, June 2017.

[2] J. Shen et al., "A 16-bit 16-MS/s SAR ADC With On-Chip Calibration in 55-nm CMOS," in IEEE Journal of Solid-State Circuits, vol. 53, no. 4, pp. 1149-1160, April 2018.

[3] P. Harpe, E. Cantatore and A. van Roermund, "A 10b/12b 40 kS/s SAR ADC With Data-Driven Noise Reduction Achieving up to 10.1b ENOB at 2.2 fJ/Conversion-Step," in IEEE Journal of Solid-State Circuits, vol. 48, no. 12, pp. 3011-3018, Dec. 2013.

[4] Q. Zhao, Q. Huang, Y. Chen, Y. Fan, S. Huang and J. Yuan, "A 16-bit 1-MS/s SAR ADC With Asynchronous LSB Averaging Achieving 95.1-dB SNDR and 98.1-dB DR," in IEEE Transactions on Circuits and Systems I: Regular Papers, vol. 71, no. 12, pp. 6447-6458, Dec. 2024.

[5] L. Chen, X. Tang, A. Sanyal, Y. Yoon, J. Cong and N. Sun, "A 0.7-V 0.6- μW 100-kS/s Low-Power SAR ADC With Statistical Estimation-Based Noise Reduction," in IEEE Journal of Solid-State Circuits, vol. 52, no. 5, pp. 1388-1398, May 2017.

[6] L. Shen et al., "A Two-Step ADC With a Continuous-Time SAR-Based First Stage," in IEEE Journal of Solid-State Circuits, vol. 54, no. 12, pp. 3375-3385, Dec. 2019.

[7] C. P. Hurrell, C. Lyden, D. Laing, D. Hummerston and M. Vickery, "An 18 b 12.5 MS/s ADC With 93 dB SNR," in IEEE Journal of Solid-State Circuits, vol. 45, no. 12, pp. 2647-2654, Dec. 2010.

[8] S. Konno, Y. Miyahara, K. Sobue and K. Hamashita, "A 16b 1.62MS/s Calibration-free SAR ADC with 86.6dB SNDR utilizing DAC Mismatch Cancellation Based on Symmetry," 2020 IEEE Asian Solid-State Circuits Conference (A-SSCC), Hiroshima, Japan, 2020.

[9] Z. Gao, Y. Zhong, L. Jie and N. Sun, "A 92.9dB-SNDR 178.2dB-FoMs SAR ADC with kT/C noise cancellation and FIA," 2024 IEEE International Conference on Integrated Circuits, Technologies and Applications (ICTA), Hangzhou, China, 2024.

TABLE I
COMPARISON WITH RECENT HIGH-RESOLUTION SAR ADCS

Specifications	This work*	[2]	[7]	[8]	[9]*
Architecture	SAR	SAR	Pipe SAR	SAR	SAR
Technology	180nm	55nm	250nm	350nm	28nm
Resolution[bit]	18	16	18	16	17
Speed[MS/s]	1	16	12.5	1.62	0.7
Power[mW]	3.11	16.3	105	8.1	1.04
SFDR[dB]	112	98	82	87	108
SNDR[dB]	103	78	80	100	93
INL[LSB]	-0.8/1.2	-1.9/2.3	-2.5/2.5	NA	NA
FoMs[dB]	185.1	165	157.7	168.4	178.2

* is simulation result.

Study on the properties of composite double-layer films of anthocyanin agar and different concentrated TiO$_2$

Yuanzhan Sheng, Shiyu Zuo, Yan Liu*, Pao-Hsun Huang*

Abstract—This work primarily investigates the optical and physical aspects of double composite films derived from agar, including anthocyanin and TiO$_2$ nanoparticles. To address the existing health and environmental issues associated with food packaging films, the solution casting method was refined to produce degradable multifunctional nanocomposite films, while further investigating the enhancement of various properties of these films with varying concentrations of TiO$_2$ as a protective layer. Films composed of 25% natural dye bilberry anthocyanins served as pH-responsive color indicators, while TiO$_2$ functioned as a light-blocking and antibacterial agent. The swelling ratio, drying rate, and spectrum of the film were analyzed in further detail. The experimental results indicate that the incorporation of TiO$_2$ into the double-layer composite film significantly enhances its swelling ratio. As the concentration of TiO$_2$ increases, the swelling ratio progressively diminishes, reaching its minimum at a concentration of 5 g/L. Optical spectrum study demonstrates that the incorporation of TiO$_2$ efficiently mitigates UV radiation and certain visible light wavelengths, hence safeguarding the stability of anthocyanin in the bottom layer under sunshine. The efficacy of these nanocomposite films was demonstrated by observing the color change resulting from the deterioration of chicken breast.

Index Terms——*TiO$_2$, Anthocyanin, Monitoring pH, Anti-ultraviolet, Photocatalysis*

I. INTRODUCTION

Titanium dioxide (TiO$_2$), as a common semiconductor material, not only has the advantages of photocatalytic performance and good anti-UV performance, but also has low price, abundant reserves, non-toxic and good electrochemical stability, so it is a very promising food packaging material. Anthocyanin (ATH), a natural water-soluble pigment, is sensitive to pH value, which shows different colors in different acidic and alkaline environments, so it can be used to monitor food spoilage and sense whether gases affecting the pH value of the environment are evolved. However, anthocyanins have a series of disadvantages, such as poor stability and sensitivity to light. Therefore, how to protect them using TiO$_2$ is a particularly suitable research option. As shown in Figure 1.1, this paper combines semiconductor materials and food engineering across domains to prepare TiO$_2$ films through the solution casting method in the semiconductor process, so as to improve the properties of food packaging films and explore a healthier and more environmentally friendly food packaging film.

II. CIRCUIT STRUCTURE DESIGN and ANALYSIS

Food packaging plays a vital role in the storage, distribution and retail chain of food products with its protective, portable and barrier roles[1]. The important role of food packaging is the protection and preservation of food during processing, transportation and storage. If packaging is not suitable or mechanically damaged, food products may change their original properties at biological, chemical and physical levels. In addition, packaging is an important tool for marketing and communicating with consumers, and the ability to better present food to consumers is an important responsibility of packaging. Today, the large amount of traditional petroleum-based

979-8-3315-2209-4/25 $31.00 © 2025 IEEE

packaging places a huge burden on the ecological environment, and there is a serious global problem of environmental pollution from plastics. In addition, packaging systems contain harmful compounds that carry the risk of serious health disorders, so improvements in food packaging films are worth exploring in depth, and finding more environmentally friendly alternatives is critical. Edible and functional food packaging films may be a suitable option to avoid the use of non-biodegradable plastics. Food quality is determined by consumer acceptability, which is largely determined by the taste, appearance and smell of the food[2]. Extending the shelf life of food is the primary goal of food packaging. Different packaging materials are used to protect the contents, to maintain the quality and safety of the food before it is consumed, and to prevent adverse reactions. Packaging materials must have good closure and strong resistance to mechanical damage, prevent microorganisms and spoilage of the food, and act as oxygen and water vapor barriers. The most widely used materials in the food packaging industry are glass, plastic, metal and paper, etc. Most flexible and rigid synthetic packaging materials are preferred because of the importance of material mechanical properties for food protection. For example, polyethylene terephthalate (PET), polyethylene (PE), polypropylene (PP), polyvinyl chloride (PVC), polystyrene (Polystyrene, PS) and other such synthetic petrochemical plastics are the most popular packaging materials today[3]. Petrochemical plastics are cost-effective and have excellent barriers to oxygen and aromatic compounds, as well as good tensile strength and elongation at break ratios, and are also soft, lightweight, and transparent. Despite the advantages of such packaging materials, they are a huge source of the world's generated waste. The non-recyclable, non-renewable and non-degradable nature of petrochemical plastics is a serious environmental hazard. Their accumulation of non-degradable plastics during manufacture and use and the formation of hazardous chemicals

pollute the environment and endanger human health. The packaging industry accounts for 47% of global bioplastic production and is expected to reach 2.87 million tons by 2025[4], indicating the inevitable trend of natural, environmentally friendly and biodegradable films in the future of commerce. The high priority given to food safety and quality has promoted innovation in the food packaging industry. Natural colorants with environmentally friendly and multifunctional properties are widely used in packaging films. Natural colorant films not only protect food from the surrounding environment, but also function as a promising new food packaging ingredient for active use. Thus, in addition to having a similar function of sensing the deterioration of the food in the package, natural colorants are also able to maintain the quality of the package due to their excellent active function. Therefore, natural colorants play an important role in the development of the packaging industry, broadening the application prospects of food packaging. Due to the poor stability and light sensitivity of anthocyanins, the search for materials to improve their performance as new packaging materials with more functions has attracted a lot of attention in recent years.

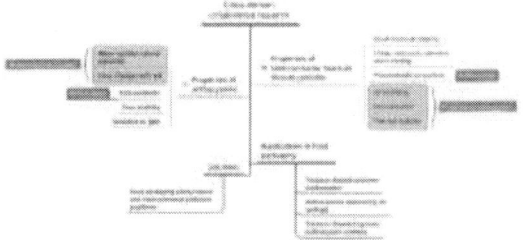

Fig.1 Cross-domain collaboration research content overview map.

III. RSULT AND DISCUSSION

Each film sample solution was successfully prepared and cast to form a gel, and individual films in the final dried products produced curling and shrinking, which was in accordance with the preparation results of the solution casting method of this experimental method, and the composite bilayer films of agar anthocyanin and TiO₂ with different concentrations were successfully prepared and the relevant data information was tested and completed.

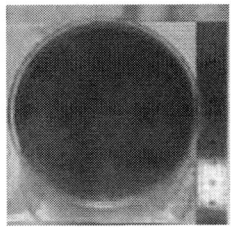

Fig. 2 Top and side views of monolayer agar-anthocyanin gel

IV. CONCLUSION

This work investigates food packaging films using agar, anthocyanins, and TiO₂. Anthocyanin was enhanced by monitoring the film for pH changes from food spoiling, instability, and optical penetration. Solution casting produced agar-anthocyanin monolayer and bilayer films with varied TiO₂ concentrations. The film swelling and drying rates were assessed for each variable. Anthocyanin films generated by this material lost some anthocyanin when dissolved in water due to its water solubility. TiO₂ may greatly increase water absorption in bilayer composite films. The film swelling coefficient decreases with TiO₂ concentration, reaching its lowest at 5 g/L. The drying rate of each sample film was approximately the same, hence TiO₂ would not alter anthocyanin drying. The optical properties of each sample film were also tested, and the results showed that TiO₂ added to the composite bilayer film at 5 g/L concentration could resist UV light and part of the visible light band, keeping the lower anthocyanin layer stable in sunlight. The film's performance for monitoring meat spoilage corresponding to its color change is effective. The thickness of the film affects the color change's differentiation, with thinner films having a more obvious color change. TiO₂ layer can improve color change differentiation.

Due to their eco-friendliness, versatility, and wide range of sources, natural colorant films are intriguing alternatives to petroleum-based food packaging, and research is booming. Some gel-like films curled up after drying, which may be related to their thickness, drying temperature, and time, but the experimental measurements showed that the film is effective and feasible for application. Solution casting is inexpensive and safe, but it takes a long time to prepare, has an unstable drying form, and is laborious, making it difficult to adapt to large-scale manufacturing.

At temperatures above 60 °C, anthocyanin degrades and deactivates. We can pick a low-temperature approach or create agar-anthocyanin film by solution casting, then prepare TiO₂ layer by spin coating, and mix the two. Electrochemically printing labels using anthocyanin-TiO₂ composite films is another promising use. Industrially viable smart film fabrication processes that meet reduced production costs and better commercial viability will promote healthy and environmentally friendly smart film packaging to the public and be extensively used.

REFERENCE

[1] Huang Jiayin, Hu Zhiheng, Li Gaoshang, et al. Make your packaging colorful and multifunctional: The molecular interaction and properties characterization of natural colorant-based films and their applications in food industry[J]. Trends in Food Science & Technology, 2022, 124.

[2] Lai WingFu. Design of Polymeric Films for Antioxidant Active Food Packaging[J]. International Journal of Molecular Sciences, 2021, 23(1).

[3] Bayram Banu, Ozkan Gulay, Kostka Tina, et al. Valorization and Application of Fruit and Vegetable Wastes and By-Products for Food Packaging Materials[J]. Molecules, 2021, 26(13).

[4] Milad Hadidi, Shima Jafarzadeh, Mehrdad Forough, et al. Plant protein-based food packaging films; recent advances in fabrication, characterization, and applications[J]. Trends in Food Science & Technology, 2022, Volume 120, Pages 154-173.

A Negative Voltage Convertor Using Switch Capacitor Structure

Shumin You, Zhipeng Liu, Ronglin Yang, Xixian Wang, Pao-Hsun Huang, Yan Liu*

Abstract —This paper presents a high-efficiency switched-capacitor DC-DC converter implementing a negative unity voltage conversion ratio. Based on a charge pump architecture, the proposed design achieves superior performance with 5V input supply. Simulation shows that it delivers regulated output voltage from -4.95V to -4.53V while the load current range of 1mA to 50mA, maintains an output voltage ripple below 39mV and has a peak conversion efficiency of 92.1%. This implementation offers a competitive solution for applications requiring efficient negative voltage generation with strict area constraints.

Index Terms——switched capacitor convertor , level shift , bootstrap driver

I. INTRODUCTION

With the rapid development of portable electronic devices such as smartphones, wearable devices, etc., the demand for efficient, small-volume power management circuits is growing. Switched capacitor chips have become an ideal choice for low- and medium-power power management modules due to their high efficiency and easy integration.

The evolution of switched capacitor DC-DC converter technology has progressed from early architectures like the Dickson[1][2] and Cockcroft-Walton multipliers to more advanced topologies, including Series-Parallel, Ladder[3],, Fibonacci, and Divider configurations. These developments have been driven by the need to meet diverse voltage conversion ratio (VCR) requirements[4]. Each topology offers distinct trade-offs in terms of efficiency, area, and complexity, making them suitable for specific applications. In recent years, the

introduction of novel topologies and control methodologies has signified the maturation of switched capacitor DC-DC converter technology[5][6]. Study[7] proposes a continuous approximation-based capacitor DC-DC converter capable of precise output voltage regulation. Another research[8] designs a circuit that utilizes clock-driven sampling and holding processes to generate both positive and negative voltages from a single charge pump.

Based on the traditional voltage double structure, this paper designs a switching capacitor circuit that can achieve a negative transformer ratio. Efficient negative voltage power supply is achieved through optimized control circuits and switching strategies, providing a new solution for power management of portable devices.

II. CIRCUIT STRUCTURE DESIGN and ANALYSIS

The architecture of the proposed negative-voltage switched-capacitor (SC) converter is illustrated in Figure 1, comprising two key functional blocks: switched-capacitor power stage and control and driving circuitry. The switched capacitor circuit implements charge transfer and voltage conversion through two-phase operation: in phase 1, switches M1 and M3 are activated (M2/M4 OFF), transferring charge from the input source to the flying capacitors and in phase 2, switches M2 and M4 are activated (M1/M3 OFF), delivering charge to the output node to generate the negative output voltage. The control and driving circuitry is critical designed featuring a high-frequency oscillator for timing generation, a dead-time control module to prevent cross-conduction[9]. and dual gate-driver circuits for robust switch control.

This work was funded by The Scientific Research Project of Jimei University, grant nos. Q2010030

979-8-3315-2209-4/25 $31.00 © 2025 IEEE

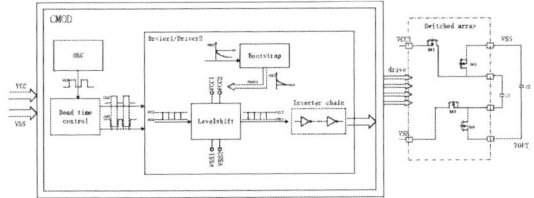

Fig.1 Architecture of the proposed Chip

Due to the varying topology configurations of the switched-capacitor array during different operating phases, the switching control nodes require distinct voltage levels. A level shift circuit is designed for converting logic-level inputs to appropriate voltage standards for different power switches. And a bootstrap power supply circuit is applied to provide dynamic voltage sourcing for the level-shifter. To improve drive signal propagation speed and enhance switching responsiveness an implemented inverter chains The architecture maintains switching efficiency across all operational states while meeting the stringent timing requirements of high-frequency switched-capacitor conversion.

III. PARAMETER DESIGN

To achieve optimal circuit parameter configuration and maximize device size utilization efficiency, this work develops an equivalent output voltage model for the negative-voltage switched-capacitor converter. This model enables systematic determination of critical design parameters including: clock frequency, switch(MOS) size, and external capacitor value. The model establishes the fundamental relationship between output characteristics and design variables, particularly the dependence of output voltage ripple (U_{o_ripple}), load current (I_{load}), operating frequency (f) and output capacitor (C_2). The quantitative relationship is expressed as:

$$U_{o\_ripple} = I_{load}/(2 * f * C_2) \quad (1)$$

The output voltage U_o is:

$$U_o = U_i - \frac{I_{load}}{2*f*C_1} - U_{SW1} - U_{SW2} \quad (2)$$

Where U_i is the input voltage, U_{SW1} and U_{SW2} are the conduction voltage of the MOS at the end of phase 1 and 2 respectively.

In terms of switching transistor parameter design, the equivalent conduction voltages of M1 and M3 are calculated by:

$$U_{SW1} = (U_i - U_o) * \exp(-\frac{1}{2*f*R_{SW1}*C_1}) \quad (3)$$

Where R_{SW1} is in phase 1, the sum of the equivalent on-resistances of the M1 and M3 . The equivalent on voltage U_{SW2} of M2 and M4 is calculated by the deep transistor resistance formula :

$$R = \frac{U_{DS}}{I_{DS}} = \frac{1}{\mu*C_{OX}*\frac{W}{L}*|U_{GS}-U_{TH}|} \quad (4)$$

And the current flowing through M2 and M4[10] is expressed as:

$$I_{DS} = \frac{C_1}{C_1+C_2} * I_{load} \quad (5)$$

The model enables co-optimization of all key parameters while considering ripple performance, silicon area efficiency and power conversion efficiency. This analytical framework significantly enhances the design efficiency and enables first-pass success in practical implementations..

IV. CIRCUIT VERIFICATION

Figure 2 presents the key performance characteristics of the proposed negative-voltage power management IC. The simulation results demonstrate that it maintains stable output ranging from -4.95V to -4.53V while supporting a wide load current range from 5mA to 50mA. And the maximum output ripple is 39mV. It has fast start-up time of 166μs. In addition, the system maintains >82% efficiency across entire load range and a peak efficiency at 92.1% at the load current of 20mA. Detailed efficiency metrics provided in Table 1.

Max. Output Ripple	39mV	54mV	N/R
Max. Load Current	50mA	0.3mA	100uA
Peak Efficiency	92%	72%	92%

V. CONCLUSION

This work presents a switched-capacitor-based power management IC that implements a negative voltage conversion with unity transformation ratio, specifically designed for low-voltage supply applications in medium-to-low power systems and an analytical model for parameter optimization. The design demonstrates an effective solution for negative voltage generation in portable electronics.

REFERENCE

[1] A. Singh, T. Singh, I. Pindoo, A. Choudhary, R. Kumar and P. Bhullar, "Transient response and dynamic power dissipation comparison of various Dickson charge pump configurations based on charge transfer switches," 2015 6th International Conference on Computing, Communication and Networking Technologies (ICCCNT), Dallas-Fortworth, TX, USA, 2015, pp. 1-6, doi: 10.1109/ICCCNT.2015.7395219.

[2] M. El Alaoui et al., "Analysis and design of dickson charge pump for EEPROM in 180nm CMOS technology," 2018 International Conference on Intelligent Systems and Computer Vision (ISCV), Fez, Morocco, 2018, pp. 1-5, doi: 10.1109/ISACV.2018.8354067.

[3] S. Hasani and R. Beiranvand, "A Transformerless Switched-Capacitor Converter Applicable for Photovoltaic Systems," 2022 13th Power Electronics, Drive Systems, and Technologies Conference (PEDSTC), Tehran, Iran, Islamic Republic of, 2022, pp. 196-201, doi: 10.1109/PEDSTC53976.2022.9767351.

[4] M. D. Seeman and S. R. Sanders, "Analysis and Optimization of Switched-Capacitor DC-DC Converters," IEEE Transactions on Power Electronics, vol. 23, no. 2, pp. 841–851, Mar. 2008.W.

[5] H. Ki, F. Su, and C.-Y. Tsui, "Charge redistribution loss consideration in optimal charge pump design," in IEEE International Symposium on Circuits and Systems (ISCAS), May 2005, pp. 1895–1898, Vol. 2.

[6] Huang mo,Chen ZhongJun,etc. "Switched Capacitor

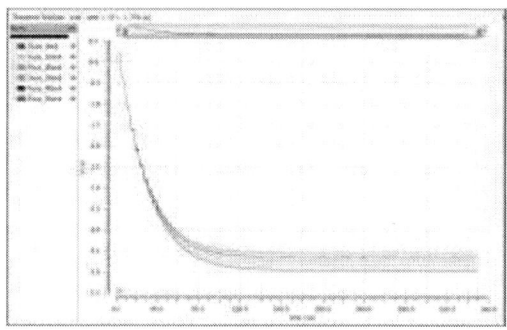

(a) Output voltage

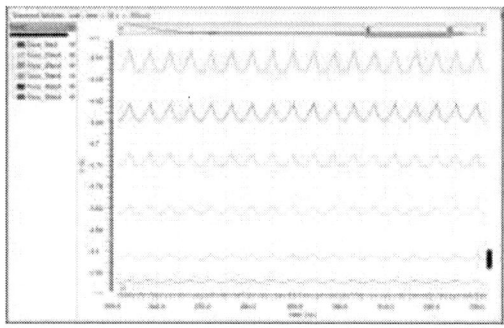

(b) Output ripple

Fig 2: Simulation result of output voltage and output ripple

TABLE I: System conversion efficiency under different load current (U_i=5V)

I_{load}	Efficiency
5mA	82.6%
10mA	89.3%
20mA	92.1%
30mA	91.8%
40mA	90.7%
50mA	89.4%

The comparison with state-of-the-art designs is shown in Table 2.

TABLE II: Comparison of different circuit performances(N/R=Not reported)

Metric	This work	[7]	[8]
Input	5V	4V	2.5V
Maximum Output	-4.95V	1.5V	-9.15V
Technology(nm)	180	180	110
Clock Frequency	223kHz	80k-1.7MHz	50MHz

DC-DC Converters: Evolution from Transformer Model to Circuit", Journal of Electronics & Information Technology, 2024. DOI: 10.11999/JEIT231216.

[7] S. Bang, A. Wang, B. Giridhar, D. Blaauw and D. Sylvester, "A fully integrated successive-approximation switched-capacitor DC-DC converter with 31mV output voltage resolution," 2013 IEEE International Solid-State Circuits Conference Digest of Technical Papers, San Francisco, CA, USA, 2013, pp. 370-371, doi: 10.1109/ISSCC.2013.6487774.

[8] V. Rana and A. Mittal, "Switched Capacitor based High Positive and Negative Voltage Charge-pump using Sample and Hold technique," 2018 IEEE Asia Pacific Conference on Circuits and Systems (APCCAS), Chengdu, China, 2018, pp. 78-81, doi: 10.1109/APCCAS.2018.8605637.

[9] CHENG Han, YE Yidie, PAN Chunbiao, XI Zhenghui. An Adaptive Dead Time Control Circuit for GaN Gate Drive [J]. Microelectronics,2009,39(01):77-80

[10] ZHENG Xinyi, LUO Ping, WANG Hao. A Static Model for Switched Capacitor Power Converters Based on Loss Voltage Analysis [J]Microelectronics, 2022,52(05):758-763. DOI:10.13911/j.cnki.1004-3365.220345.

A Method of Locusts Detection

1st Hongxing Ma   2nd Yingfei Wang   3rd Xuan Liu   4th Jintai Chi   5th Fuyuan Wang   6th Junjie Liu

Abstract—**Agricultural pest identification and monitoring constitute the fundamental basis for effective pest control. Addressing the technical challenge of insufficient feature extraction in agricultural pest identification that leads to difficulties in small target recognition, this paper proposes the YOLOv8-TL method. The core innovation involves the design of Trefoil Dilated Residual Convolution (TC) to compensate for detail loss during feature extraction of small targets. Through hierarchical feature extraction via Region Residualization, Semantic Residualization, and multiple residual operations, the method constructs more detailed feature representations, thereby enhancing recognition accuracy for small-scale insects. Additionally, we introduce the Large Separable Kernel with Attention (LSKA) mechanism to improve model sensitivity towards target features by capturing interdependencies between spatial and channel domains. Experimental validation on a self-built locust dataset demonstrates that the improved YOLOv8-TL model achieves a mean average precision (mAP) of 88.6%, representing a 4.4% improvement over the baseline YOLOv8 model.**

Index Terms—**pest identification, object detection, YOLOv8, CNN**

I. INTRODUCTION

Global climate change and agricultural intensification threaten agroecosystems through multidimensional environmental stress. Insect population dynamics, as key ecological indicators, critically impact crop health and sustainable production. Climate-driven extreme weather (e.g., droughts, heatwaves) alters insect migration and reproduction by disrupting microhabitats [1], while monoculture practices and pesticide overuse accelerate habitat fragmentation, fostering heterogeneous insect communities [2].

Traditional insect identification techniques relying on manual morphological classification and laboratory microscopic analysis exhibit inherent limitations including poor real-time performance, high labor costs, and limited spatial coverage. These deficiencies become particularly pronounced in complex agricultural scenarios involving polyculture systems, multi-layered canopy structures, and dynamically changing illumination conditions. Consequently, deep learning-based visual detection technologies have emerged as a research focus. Zhou [3] reconstructed the YOLOv4 backbone network CSPDarkNet53 into a GhostNet architecture combined with

The first four authors are with *School of Electrical and Information Engineering, North Minzu University,*, Yinchuan City, China. This work was supported by the High Level Construction Project of Ordinary Universities in Ningxia Province (bjg2024028), and the Teaching Project of North Minzu University (YKSZ202307, YJZT202431), and the Key Teaching Project of North Minzu University (2021JY005), and the Ministry of Education's Supply and Demand Coordination Employment and the Education Project (2024011087854). Corresponding author: (Yingfei Wang, email: wang.yingfei@foxmail.com)

transfer learning strategies, achieving a 79.38% mean average precision (mAP) for rice disease detection. To address inter-class confusion caused by high morphological-color similarity in Solanaceae crop pests, Zhang [4] integrated CBAM attention modules and ASFF feature fusion mechanisms into the YOLOv5 framework, effectively enhancing model specificity through spatial-channel dual-dimensional feature recalibration and cross-scale information interaction. Tian [5] introduced DenseNet connectivity and adaptive attention modules (AAM) to construct a multi-level feature enhancement network, attaining 86.2% mAP under complex background interference.

Current models, however, inadequately capture tiny insect details and exhibit limited receptive fields. We propose YOLOv8-TL with two enhancements: 1) Trefoil Convolution (TC) and Large Separable Kernel with Attention (LSKA) employs large-kernel separable convolutions for long-range dependencies and cross-attention for adaptive feature enhancement.

II. MATERIALS AND METHODS

A. Trefoil Convolution

The Trefoil Convolution(TC) module employs a multi-stage residual architecture (Fig. 1.(a)) that achieves efficient multi-scale context extraction and fusion through three functionally distinct processing stages shown in Fig. 2.

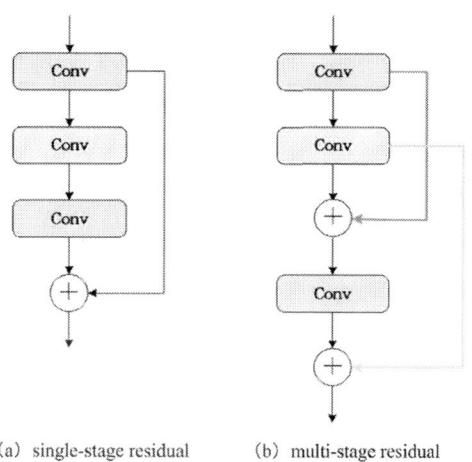

(a) single-stage residual   (b) multi-stage residual

Fig. 1. Residual structures

- Region Residualization: A 3×3 separable convolution combined with batch normalization (BN) and SiLU activation generates region residual features. Compared to conventional CBS (Conv-BN-SiLU) modules, TC adopts

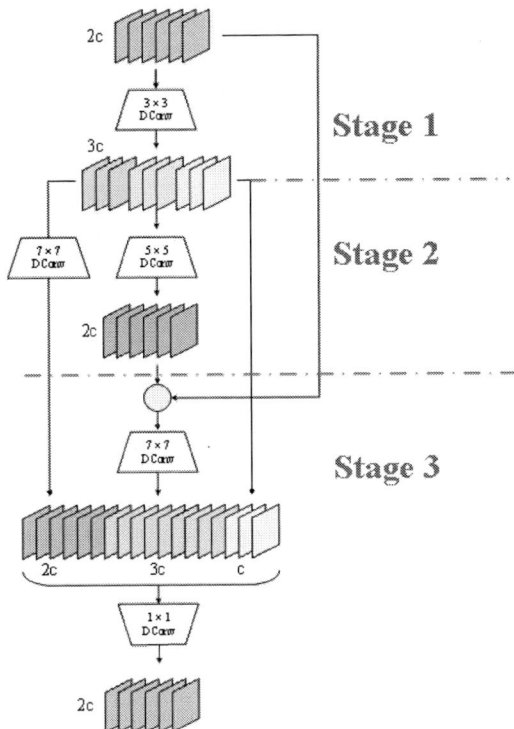

Fig. 2. the structure of the TC module

separable convolutions to reduce parameters and control model complexity. This stage produces structurally coherent feature maps with well-defined spatial organization, serving as standardized inputs for subsequent morphological filtering.

- Semantic Residualizatio: Multi-scale dilated depthwise separable convolutions perform morphological filtering on regional features. The normalized features from Stage 1 enable precise capture of scale-specific semantics through convolutional kernels with varying dilation rates (3×3, 5×5, 7×7), eliminating interference from redundant receptive fields. Grouped convolution ensures independent learning of optimal scale features across channels, enhancing multi-scale extraction efficiency by 22%.

- Multi-Residual Fusion: The dual residual connection mechanism enhances feature representation through coordinated spatial-channel optimization and hierarchical feature integration. Initially, spatial compensation is achieved by performing element-wise addition between 5×5convolution outputs and original inputs, followed by 7×7 dilated convolution to synergistically fuse spatial and channel information, thereby reinforcing long-range dependency modeling. Concurrently, feature refinement is implemented through systematic fusion of concise spatial features from initial processing stages with high-level semantic features derived from deep network layers, constructing comprehensive feature representations.

This architecture progressively supplements spatial details while stabilizing gradient propagation through residual path-

ways, effectively preventing gradient explosion during model training while maintaining feature discriminability.

Given an input feature map $\mathbf{X} \in \mathbb{R}^{H \times W \times C}$ where H, W and C denote height, width, and channel dimensions respectively, the TC module executes hierarchical feature transformation through sequential operations.

First, undergoes region residualization via a 3×3 depthwise separable convolution ($*_{\mathrm{DW3}}$) to generate spatially structured features:

$$F_r = \mathrm{BN}(\mathrm{SiLU}(X *_{\mathrm{DW3}})) \tag{1}$$

Subsequently, multi-scale semantic extraction is performed by applying 5×5 and 7×7 depthwise separable convolutions ($*_{\mathrm{DW5}}$, $*_{\mathrm{DW7}}$) to F_r , yielding scale-specific features:

$$F_{D5} = F_r *_{\mathrm{DW5}} \tag{2}$$

$$F_{D7} = F_r *_{\mathrm{DW7}} \tag{3}$$

Spatial-channel integration is then achieved through residual fusion: Intermediate features F_{D5} are element-wise summed with , followed by a 7×7 dilated convolution ($*_{\mathrm{DW7-d}}$) to produce enhanced spatial features:

$$F_X = (F_{D5} \oplus X) *_{\mathrm{DW7-d}} \tag{4}$$

Final feature aggregation combines F_{D7}, F_X and F_r through summation and 1×1 convolution ($*_{\mathrm{Conv1}}$), formulated as

$$Y = (F_{D7} \oplus F_X \oplus F_r) *_{\mathrm{Conv1}} \tag{5}$$

This cascaded design ensures progressive refinement of multi-scale contextual representations while preserving structural coherence.

B. Large Separable Kernel with Attention

Large Kernel Attention (LKA) encounters challenges when handling extremely large kernel sizes, as it results in heightened computational complexity and memory usage. To tackle these problems, an improved approach named Large Separable Kernel Attention (LSKA) has been put forward by lau [6], which further boosts the efficiency and scalability of the attention mechanism.As defined in Equation (6), this method first applies a vertical $(2d - 1) \times 1$ depthwise convolution and then a horizontal $1 \times (2d - 1)$ depthwise convolution to the input feature map $F_C \in \mathbb{R}^{C \times H \times W}$, where C represents the number of channels and $H \times W$ denotes the spatial dimensions. The convolution operator ($*$) in this scenario preserves the spatial resolution and establishes long - range dependencies through expanded receptive fields.

$$Z_C = X_{H,W}^{W_C}(2d - 1) \times 1 * \left(X_{H,W}^{W_C} 1 \times (2d - 1) * F_C \right) \tag{6}$$

This hierarchical decomposition effectively reduces the computational complexity from $O(k^2)$ (quadratic) to $O(k)$ (linear) with respect to the kernel size k. This is accomplished by factorizing the 2D kernel into orthogonal 1D components that jointly maintain equivalent contextual coverage.For extremely large kernels, LSKA further alleviates the computational burden caused by large kernels via recursive hierarchical decomposition. Here, $\lfloor k/d \rfloor$ indicates the step - by - step reduction of the kernel size, and $\overline{Z_C}$ represents the output of the previous stage.

$$Z_c = X_{H,W}^{W_C} \left\lfloor \frac{k}{d} \right\rfloor \times 1 * \left(X_{H,W}^{W_C} 1 \times \left\lfloor \frac{k}{d} \right\rfloor * \overline{Z_C} \right) \quad (7)$$

To generate spatially - aware attention weights while maintaining dimensional consistency, the proposed method utilizes a 1×1 convolutional layer ($W^{1\times 1}$) to conduct channel - wise fusion on the intermediate features Z_c.

$$A_c = W^{1\times 1} * Z_c \quad (8)$$

This operation preserves the original spatial resolution ($H \times W$) and generates an attention map A_c that encodes the region - specific importance across channels. Subsequently, feature enhancement is achieved through element - wise multiplication ($\otimes$) between the attention weights A_c and the input features F_c.

$$\overline{F_c} = A_c \otimes F_c \quad (9)$$

This modulation mechanism selectively amplifies the feature responses in salient regions while suppressing irrelevant patterns, effectively guiding the network to focus on discriminative spatial contexts without introducing additional dimensional transformations.

III. RESULTS AND DISCUSSION

A. Experimental Environment Construction

In this experiment, the server is configured with PyTorch as the deep learning environment. Considering the stability of the system, the server operating system is Linux, and the version of CUDA is 12.4. The experimental graphics card is NVIDIA RTX 4090, and the size of the video memory is 12G.The parameters of the YOLOv8 training scripts are shown in TABLE I.

TABLE I
TABLE OF TRAINING PARAMETERS

Model Parameter	Configure
Image size	640px
Optimizer	AdamW
Batch size	128
Epoch	200

B. Dataset

The experimental use of self-build locusts dataset has a total of 4082 images, before the experiment using Python script to divide the dataset, the division of the ratio is 7: 2: 1. The main goal of this experiment is to detect the target is the locusts, there are six classes of objects to be detected: Oedaleus infernalis, Oedaleus asiaticus, Brachytrupes portentosus, Pseudoxya rufipes, Chorthippus albomarginatus and Tenodera sinensis.

C. Experimental Results and Analysis

This study systematically integrated the custom-designed TC operator and Large Separable Kernel Attention (LSKA) module into the YOLOv8 architecture to develop the enhanced YOLOv8-TL detection network. Under identical hyperparameter configurations, comparative experiments were conducted among YOLOv3 [7], YOLOv5, YOLOv8, and the proposed model, with mean Average Precision (mAP) comparisons shown in Figure 3.. Experimental results demonstrate that after 200 training epochs, YOLOv8-TL achieved superior performance with an mAP of 88.6%, representing a 4.4% absolute improvement over the baseline YOLOv8. Convergence analysis revealed stable optimization trajectories for both YOLOv8-TL and YOLOv8 after 150 epochs, while YOLOv3 exhibited oscillatory convergence behavior and YOLOv5 failed to converge with severe gradient oscillations, attaining only 54.5% mAP.

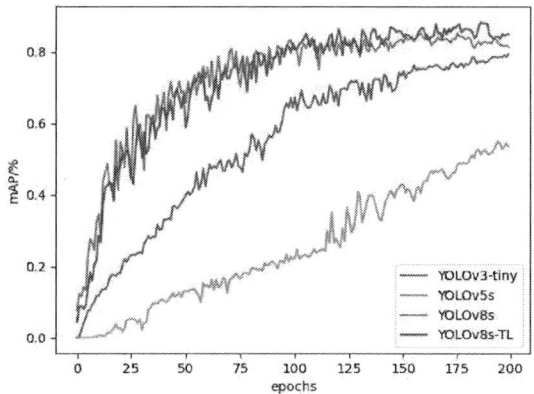

Fig. 3. Performance comparison of mAP across YOLOv3, YOLOv5, YOLOv8, and YOLOv8-TL

D. Abbreviations and Acronyms

In order to verify the reasonableness of the YOLOv8-TL algorithm, this paper conducts comparative experiments on the models as a way to verify the effect of the TC and LSKA moudle acting alone. A total of four experiments were conducted in this paper, as shown in Table II, where "" indicates that the method was used, "TC" denotes the Trefoil Dilated Residual Convolution and "LSKA" is the Large Separable Kernel with Attention. Experiment 1 is the original YOLOv8 network and

Experiment 4 is the YOLOv8-TL algorithm. From Table II, it can be seen that the improved network improves the accuracy by 4.4% over the original network.

TABLE II
COMPARATIVE EXPERIMENTS ON INFLUENCING FACTORS

Experimental	TC	LSKA	mAP50
1			0.842
2	√		0.854
3		√	0.860
4	√	√	**0.886**

E. Visualization

In this paper, the trained model is used to reason about the locusts, as shown in Figure 4. It can be observed from the picture that the model in this paper is excellent for pests detection, and the model has good generalization ability in different environments, which can be applied to different usage scenarios.

Fig. 4. Visualization of Experimental Effects

IV. CONCLUSION

To address the challenges of low detection accuracy for agricultural pests in complex field environments and insufficient feature extraction capability in conventional deep learning networks, this study proposes an enhanced algorithm incorporating two key innovations: 1) A Trefoil Convolution (TC)

module designed to overcome feature extraction limitations through multi-scale hierarchical residual learning, significantly improving model discriminative power; 2) Integration of Large Separable Kernel Attention (LSKA) mechanism to enhance spatial-channel feature representation. Experimental validation demonstrates the proposed YOLOv8-TL model achieves 88.6% mAP on pest datasets, representing a 4.4% improvement over baseline YOLOv8. The optimized architecture has been successfully deployed at Ningxia Grassland Station for real-time pest monitoring, demonstrating practical agricultural applicability. While achieving superior accuracy, current implementation exhibits suboptimal detection accuracy for overlapping pest instances, particularly in scenarios involving dense pest clusters or partial occlusions, and future work will be devoted to further enhance the generalization ability of the model.

REFERENCES

[1] Jörg Müller, Torsten Hothorn, Ye Yuan, Sebastian Seibold, Oliver Mitesser, Julia Rothacher, Julia Freund, Clara Wild, Marina Wolz, and Annette Menzel. Weather explains the decline and rise of insect biomass over 34 years. *Nature*, 628(8007):349–354, 2024.

[2] Daijiang Li, Michael Belitz, Lindsay Campbell, and Robert Guralnick. Extreme weather events have strong but different impacts on plant and insect phenology. *Nature Climate Change*, pages 1–8, 2025.

[3] Zhao Bai, Zhan Tang, Lei Diao, Shuhan Lu, Xuchao Guo, Han Zhou, Chengqi Liu, and Lin Li. Video target detection of east asian migratory locust based on the mog2-yolov4 network. *International Journal of Tropical Insect Science*, pages 1–14, 2022.

[4] Pan Zhang and Daoliang Li. Cbam+ asff-yoloxs: An improved yoloxs for guiding agronomic operation based on the identification of key growth stages of lettuce. *Computers and Electronics in Agriculture*, 203:107491, 2022.

[5] Yunong Tian, Shihui Wang, En Li, Guodong Yang, Zize Liang, and Min Tan. Md-yolo: Multi-scale dense yolo for small target pest detection. *Computers and Electronics in Agriculture*, 213:108233, 2023.

[6] Kin Wai Lau, Lai-Man Po, and Yasar Abbas Ur Rehman. Large separable kernel attention: Rethinking the large kernel attention design in cnn. *Expert Systems with Applications*, 236:121352, 2024.

[7] Joseph Redmon and Ali Farhadi. Yolov3: An incremental improvement. *arXiv preprint arXiv:1804.02767*, 2018.

Detection, Tracking, and Activity Monitoring of Small and Micro Insects Based on Deep Learning

1st Haobo Jia   2nd Jintai Chi   3rd Yun Ma   4th Xuan Liu   5th Xiaobin Ren   6th Hongxing Ma

Abstract—Tetranychus urticae is a significant pest in agricultural ecosystems, and its activity patterns and level of infestation have a substantial impact on crop growth. Traditional monitoring methods face numerous challenges due to the mite's small size and rapid movements. This study leverages deep learning techniques, employing an enhanced YOLOv8 object detection model for precise identification and localization of T. urticae. Additionally, by integrating the DeepSORT multi-object tracking algorithm, we successfully achieve real-time and continuous tracking of individual mites, accurately recording their movement trajectories. By extracting motion feature parameters, we constructed an activity level assessment model. Experimental results demonstrate that this approach can track T. urticae movements in real-time with high accuracy and reliably assess fluctuations in their activity levels.

Index Terms—Deep Learning, YOLOv8, Tetranychus urticae, Object Tracking, Activity Monitoring

I. INTRODUCTION

In agricultural ecosystems, pest monitoring and control are critical for ensuring crop yield and quality. Tetranychus urticae, a widely distributed and significant agricultural pest, can severely damage various crops. Its outbreaks may even lead to total crop failure, causing substantial economic losses. Traditional monitoring methods primarily rely on manual field surveys, which are labor- and resource-intensive, inefficient, and unable to support large-scale, real-time monitoring.

The activity level of Tetranychus urticae is closely related to its reproduction, spread, and damage severity. Monitoring its activity can provide valuable insights into population dynamics, helping to develop effective pest control strategies. However, traditional insect behavior study methods have numerous limitations, highlighting the urgent need for more precise and efficient technologies to enable real-time, long-term, and continuous monitoring of mite behavior.

Shen Yi (2024) [1]explored deep learning-based algorithms for the identification and tracking of Tetranychus urticae. He proposed an efficient monitoring method using the YOLO series algorithms, emphasizing the potential of deep learning in detecting small pests, particularly its advantages in real-time monitoring and accuracy. This technology is expected to significantly improve the efficiency of agricultural pest management. Liu Siqi (2021) [2] focused on the application of video object tracking algorithms for assessing the activity levels of storage pests. She proposed a comprehensive method based on video monitoring, analyzing pest behavior characteristics to evaluate their activity. The study demonstrated that combining video analysis with deep learning techniques enhances the monitoring capability of storage pests, providing crucial data support for food security. Liu Siqi (2021) [3] further explored the video object tracking algorithm for determining the activity levels of storage pests, emphasizing the importance of real-time video monitoring for obtaining pest behavior data. The integration of deep learning-based monitoring technology allows for more accurate assessments of pest activity, providing a scientific basis for pest control strategy development. Zhou Yizhe (2019) [4] introduced a video-based algorithm for evaluating pest mortality in stored grain, examining how visual data analysis can accurately assess pest death. Traditional methods were found to be inadequate in terms of efficiency and accuracy, while video monitoring combined with deep learning offers a more reliable evaluation approach, with broad application prospects.

This study proposes a method that combines the improved YOLOv8 object detection model with the DeepSORT multi-object tracking algorithm, aiming to achieve precise identification, localization, and activity monitoring of Tetranychus urticae.

II. DEEPSORT MULTI-OBJECT TRACKING TECHNOLOGY

DeepSORT is a deep learning-based feature representation method that tracks objects by extracting key point features. It utilizes deep convolutional networks to learn object feature representations, enhancing the ability to distinguish between targets and maintaining consistent tracking of both identity and position across multiple frames.

The DeepSORT algorithm is an improved version of the SORT (Simple Online and Real-Time Tracking) algorithm. Unlike SORT, DeepSORT incorporates Faster R-CNN for object detection, combined with the Kalman filter and Hungarian algorithm. This enhancement significantly increases the speed of multi-object tracking while achieving state-of-the-art (SOTA) accuracy. In the SORT algorithm workflow, Faster R-CNN is used as the object detection method, and the Intersection over Union (IoU) of different bounding boxes serves as the metric for measuring the relationship between objects across frames—this forms the association cost matrix.

All the authors are from *Northern University for Nationalities* in Ningxia Province, China.Except Xiaobin Ren is Institute of Desertification Control, Ningxia Academy of Agriculture and Forestry Sciences Ningxia province, China. This work was supported by Ningxia Hui Autonomous Region Agricultural Science and Technology Independent Innovation Fund Project(NGSB-2021-15-01), the High Level Construction Project of Ordinary Universities in Ningxia Province (bjg2024028) and the Teaching Project of North Minzu University (YKSZ202307, YJZT202431). Corresponding author: (Hongxing Ma, email: mhx@nmu.edu.cn)

979-8-3315-2209-4/25 $31.00 © 2025 IEEE

The Kalman filter is then used to predict the object's position in the current frame, and the Hungarian algorithm is applied to associate detection boxes with specific targets.

III. TETRANYCHUS URTICAE DATASET COLLECTION AND PREPROCESSING

Tetranychus urticae samples were collected by the College of Plant Protection at Southwest University. The samples were captured under a laboratory microscope using cowpea seedling leaves. A total of 274 images of Tetranychus urticae were collected, each with a resolution of 9024 × 12032 pixels. Tetranychus urticae can be classified into four stages based on its morphological characteristics: eggs, larvae, nymphs, and adults, with each stage exhibiting distinct features.

The collected 274 images underwent data augmentation, including horizontal flipping, vertical flipping, isotropic scaling, non-isotropic scaling, random translation, and HSV transformation , resulting in a total of 1911 images of Tetranychus urticae.

To more effectively assess the model's generalization ability during dataset construction, the 1911 images were divided into training and test sets at an 80:20 ratio. Additionally, to validate the model's performance, 105 additional images of Tetranychus urticae were captured for the validation set. The details of Tetranychus urticae dataset are shown in Table I.

TABLE I
TETRANYCHUS URTICAE MORPHOLOGICAL FEATURE DATASET (NUMBER OF INDIVIDUALS)

Category	Training Set	Test Set	Total Sample
Eggs	68130	13681	81811
Larvae	8762	1543	10305
Nymphs	7582	1834	9416
Adults	19493	4212	23705

IV. YOLOv8-BASED TETRANYCHUS URTICAE OBJECT DETECTION MODEL IMPROVEMENT AND TRAINING

A. Model Improvement Strategy

To address the challenges posed by the small size and complex morphological features of Tetranychus urticae, improvements are proposed for the YOLOv8 model. The modified network model is named YOLOv8s-FP2.

The YOLOv8s-FP2 model consists of four main components. The input section receives the raw image, adjusts its dimensions to the required width and height, and passes it on to the backbone network for effective training and inference. The backbone network is responsible for extracting features from the input image. The neck network, situated between the backbone and detection head, further processes and combines features from the backbone. The detection head generates the prediction results for object detection, including the bounding box coordinates, confidence, and class probabilities.

YOLOv8s-FP2 introduces the Focus module in the backbone network, which converts the low-resolution feature map into a higher-resolution feature map while retaining the original feature information. This enhances the model's ability to perceive finer details and local information, improving detection accuracy. Additionally, a multi-scale detection head is added to detect very small objects, as these often occupy fewer pixels and have more dispersed and subtle features in images, requiring more refined detection methods. Given that the training data primarily consists of very small targets (Tetranychus urticaes), retaining the P5 detection head could lead to the model focusing too much on larger objects, potentially causing overfitting. Therefore, the P5 detection head is removed to prevent this issue, allowing the model to better focus on learning from very small targets.

B. Model Training

In the experiment, input images were resized to 640 × 640 pixels, and Mosaic data augmentation was applied to enhance the model's robustness and generalization ability. During the training process, the initial learning rate (Learning Rate, LR) was set to 0.01, and the Adam optimizer was used to adjust the learning rate. Additionally, the batch size was set to 64, and the number of epochs was set to 600. In the performance testing phase, the IoU threshold was set to 0.5, and Non-maximum Suppression (NMS) was applied to refine the prediction results. In NMS, the IoU threshold was also set to 0.5 to ensure that the final predictions were accurate and reliable.

C. Model Evaluation

To further validate the performance of the YOLOv8s-FP2 model, its performance was compared with the original YOLOv3s, YOLOv4s, YOLOv5s, and YOLOv8s models. After 600 iterations on Tetranychus urticae dataset, the improved YOLOv8s-FP2 model achieved a precision (P) of 83.3%, recall (R) of 69.8%, and mean average precision (mAP) of 74.7%. As shown in Figure 1, the comparison of average precision values demonstrates a significant improvement in the performance of the improved model over the original models.

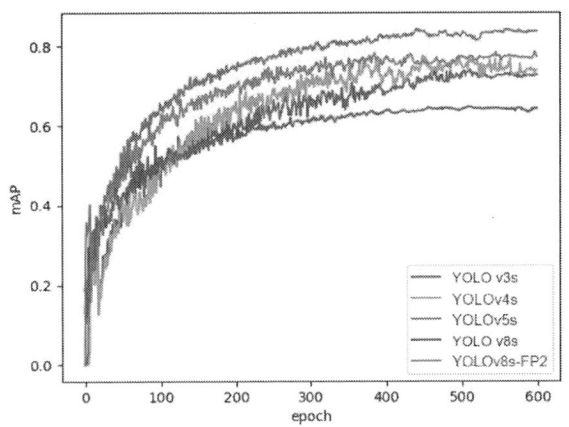

Fig. 1. Comparison of Average Precision Values

Through the experimental comparison and analysis, it can be concluded that the improved YOLOv8s-FP2 model demonstrates the best performance in detecting the morphological features of Tetranychus urticae.

V. Tetranychus urticae Tracking Algorithm Design and Implementation

A. Selection of Tracking Algorithm

To achieve dynamic tracking of Tetranychus urticae, this study effectively combines the YOLOv8 model with the DeepSORT algorithm. DeepSORT, based on Kalman filtering, enables target state prediction and enhances the representation of target appearance features through deep learning. This combination not only improves the algorithm's ability to recognize targets but also maintains tracking continuity in complex situations such as target occlusion, significantly increasing the accuracy and reliability of monitoring.

The DeepSORT algorithm builds upon the SORT (Simple Online and Real-Time Tracking) algorithm by incorporating appearance information, which is used in conjunction with motion information for better target association. The matching degree between detection results and tracking results is calculated using a fusion metric, which is a cascaded matching of appearance and motion information.

By integrating the Mahalanobis distance for motion information and the cosine similarity for appearance features, the DeepSORT algorithm merges both types of information. On one hand, the Mahalanobis distance, based on motion information, provides possible target location estimates, which are effective for short-term prediction. On the other hand, cosine similarity, based on appearance features, is more effective for recovering target IDs after long-term occlusion, as motion information alone may not be distinguishable in such cases. In other words, the cost matrix of the DeepSORT algorithm is derived from both the Mahalanobis distance and Euclidean distance.

1) Association of Motion Information: The DeepSORT algorithm uses the Mahalanobis distance between detection boxes and predicted tracker boxes to describe the association of motion information.

$$d^{(1)}(i,j) = (\boldsymbol{d}_j - \boldsymbol{y}_i)^{\mathrm{T}} \boldsymbol{S}_i^{-1} (\boldsymbol{d}_j - \boldsymbol{y}_i) \qquad (1)$$

The term d_j represents the position of the j-th detection box, y_i denotes the predicted position of the target by the i-th tracker, and S_i refers to the covariance matrix between the detection position and the average tracking position. In other words, the Mahalanobis distance accounts for the uncertainty of state measurements by calculating the standard deviation between the detection position and the average tracking position. Additionally, DeepSORT thresholds the Mahalanobis distance using the 95% confidence interval derived from the inverse chi-squared distribution.

$$b_{i,j}^{(2)} = 1 \left[d^{(2)}(i,j) \le t^{(2)} \right] \qquad (2)$$

If the Mahalanobis distance of a particular association is smaller than the specified threshold t_1, the association of the motion state is considered successful.

2) Association of Target Appearance Information: Only when the uncertainty of motion is low, the Mahalanobis distance-based motion measurement method is accurate. When the camera undergoes significant displacement, camera movement introduces rapid shifts in the image plane, which makes the Mahalanobis distance measure highly inaccurate in occlusion scenarios, causing the association method to fail and leading to ID switch phenomena. Therefore, a second association measurement method is introduced: for each tracked target, an appearance feature vector is extracted using a neural network. The association is determined by calculating the minimum cosine distance between the detection of the current frame and the feature vector of the historically tracked target, which is used for appearance-based information association. The cost matrix for the appearance association of targets is calculated as follows:

$$d^{(2)}(i,j) = \min \left\{ 1 - r_j^{\mathrm{T}} r_k^{(i)} \mid r_k^{(i)} \in \mathcal{R}_i \right\} \qquad (3)$$

Finally, a linear weighting of the motion and appearance metrics is applied to form the final metric. The final cost matrix for association, computed using the Hungarian algorithm, is as follows:

$$c_{i,j} = \lambda d^{(1)}(i,j) + (1 - \lambda)d^{(2)}(i,j) \qquad (4)$$

If the distance calculated by Equation (2) is smaller than the specified threshold, the association is considered successful. Only when both appearance and motion metrics meet their respective threshold conditions will the fusion in Equation (2) be applied. The distance metric works well for short-term prediction and matching, but for long-term occlusion scenarios, the use of appearance features for measurement is more effective.

B. Tracking System Implementation

The OpenCV library is used for video capture and processing, including reading video frames, displaying detection results, and drawing target trajectories. A single-stage detection framework quickly identifies the bounding boxes (in xywh format) and class IDs (cls_id) of targets in each video frame for real-time processing. The target tracking module integrates YOLO's built-in tracking functionality (using the model.track method with the persist=True parameter) and employs a defaultdict data structure to record the location information of each target, making it easier to store and update the trajectories of individual targets. The defaultdict automatically initializes data structures, simplifying code complexity. By tracking the detected targets, the code manages the ID of each target and records its positional changes throughout the video. The target IDs (track_id) are associated across frames, and the track_history dictionary stores the sequence of trajectory points. The Numpy library is used to process and visualize the target trajectory points, while OpenCV's cv2.polylines function dynamically draws trajectory lines, enabling the visualization of the motion path.

979-8-3315-2209-4/25 $31.00 © 2025 IEEE   340

In summary, the target detection module in the code relies on the YOLO model for real-time detection, while the target tracking module achieves continuous tracking by recording and visualizing the target's historical trajectory. The integration of these technologies enables efficient detection and tracking of Tetranychus urticae in the video. The movement trajectories of Tetranychus urticae in the video are shown in Figures 2.

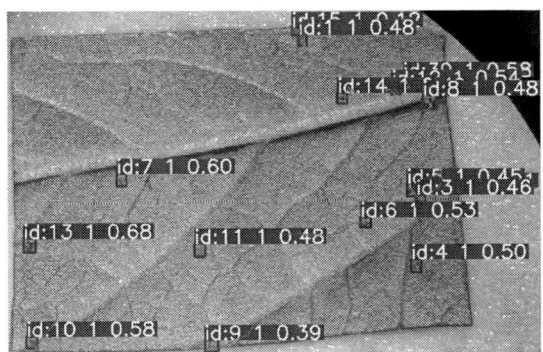

Fig. 2. Movement Trajectory of Tetranychus urticae at 6 Seconds in the Video

C. Tracking Performance Evaluation

The video contains a total of 15 Tetranychus urticaes. The recognition and tracking system successfully identifies all of them with 100% accuracy, assigning unique IDs to each mite, labeled as ID 1 through 15. The system also accurately displays the movement trajectories of Tetranychus urticaes in the video.

VI. CONSTRUCTION AND ANALYSIS OF TETRANYCHUS URTICAE ACTIVITY INDEX

Based on the movement speed of the mites, three activity levels for Tetranychus urticae were defined: Death, Low Activity, and Active. The "Death" state corresponds to an average speed of 0 cm/s, indicating complete stillness. The "Low Activity" state is defined by a speed range of 0 cm/s to 0.1 cm/s, where the mite occasionally shows small movements. The "Active" state corresponds to speeds above 0.1 cm/s, indicating frequent movement and normal physiological activities (such as foraging and mating).

By monitoring the movement speed of the insects in real-time and using the Kalman filter-based tracking algorithm (DeepSORT), the active state of each insect can be effectively determined. Using this classification method, the activity levels of Tetranychus urticaes in the video segments were evaluated, with the results shown in Table II: Activity Level Evaluation Results.

By defining different activity levels and combining video data with speed metrics, a method for quantifying insect activity has been proposed. This approach significantly enhances the scientific rigor and effectiveness of behavior analysis for Tetranychus urticae, providing new insights for optimizing pest control strategies for small and micro pests.

TABLE II
ACTIVITY LEVELS OF TETRANYCHUS URTICAES

Mite ID	Average Speed (cm/s)	Activity Level
1	0.01	Low Activity
6	0.22	Active
10	0.07	Low Activity
11	0.00	Dead
13	0.28	Active
15	0.16	Active

VII. CONCLUSION

This study, based on deep learning techniques—specifically the YOLOv8 algorithm—successfully implements detection, tracking, and activity monitoring of Tetranychus urticac. Using the improved YOLOv8 model, we have significantly increased detection accuracy and recall rates in complex environments, overcoming many challenges associated with small pest identification in traditional monitoring methods. Additionally, by integrating the DeepSORT algorithm, continuous tracking of individual Tetranychus urticaes is achieved, effectively recording their movement trajectories and reducing identity misalignment, thus providing a reliable tool for further behavioral studies.In terms of activity monitoring, activity classification criteria were established. Real-time monitoring of insect movement speed categorizes their activity into three levels: Death, Low Activity, and Active. This classification method offers a quantifiable indicator for practical monitoring and enhances the scientific and effective analysis of Tetranychus urticae behavior.Through an in-depth analysis of experimental data, the system's performance under various conditions has been validated, emphasizing the effectiveness of the proposed approach in improving detection accuracy, tracking stability, and activity monitoring precision.

REFERENCES

[1] Z. Wen, P. Zhang, Q. Zhu and A. Pang, "Target Detection and Tracking Techniques for Vehicles in Real-Time Scenarios Based on YOLOv8," 2023 International Conference on Industrial IoT, Big Data and Supply Chain (IIoTBDSC), Wuhan, China, 2023, pp. 423-427.

[2] Y. Wang and H. Zhang, "Research on machine vision technology based detection and tracking of objects on video image," 2022 International Conference on Image Processing, Computer Vision and Machine Learning (ICICML), Xi'an, China, pp. 267-270, 2022.

[3] Z. Sun, J. Chen, L. Chao, W. Ruan and M. Mukherjee, "A Survey of Multiple Pedestrian Tracking Based on Tracking-by-Detection Framework," in IEEE Transactions on Circuits and Systems for Video Technology, vol. 31, no. 5, pp. 1819-1833, 2021.

[4] T. Meng et al., "Localization-Guided Track: A Deep Association Multiobject Tracking Framework Based on Localization Confidence of Camera Detections," in IEEE Sensors Journal, vol. 25, no. 3, pp. 5282-5293, 1 Feb.1, 2025.

[5] S. Liu et al., "Embedded Online Fish Detection and Tracking System via YOLOv3 and Parallel Correlation Filter," OCEANS 2018 MTS/IEEE Charleston, Charleston, SC, USA, 2018, pp. 1-6.

Locust Object Detection in Complex Backgrounds Using SPD-Conv and PPA Fusion in YOLOv11

1st Xuan Liu   2nd Kaiwen Chen   3rd Lin Li   4th Haobo Jia   5th Wei Sun   6th Hongxing Ma

Abstract—To address the issues of misidentification and missed detection of small targets, as well as the difficulty in distinguishing between targets and background during the locust object detection process in complex backgrounds, we propose a locust detection algorithm based on the integration of SPD-Conv and PPA into YOLOv11. SPD-Conv is utilized to optimize the object detection network model of YOLOv11; it retains all information when downsampling feature maps, avoiding the loss of fine-grained information caused by traditional stride convolutions and pooling operations, thus showing better performance in handling small objects and low-resolution images. On this basis, the PPA module is added, which, by improving the traditional convolution operation in C2k2, can better preserve critical information of small targets. Experimental results show that this algorithm achieves an mAP50-95 of 85.8% on a self-built locust dataset, representing an improvement of 2.2% over the original model, while also meeting real-time inference speed requirements.

Index Terms—YOLOv11, SPD-Conv, PPA, Complex Background, Locust target detection

I. INTRODUCTION

The destruction caused by locust plagues to agricultural ecosystems is unquestionable, and there is an urgent need for efficient real-time detection methods [1]. Traditional macro-monitoring approaches mainly rely on technologies such as remote sensing and geographic information systems. However, high costs and resolution limitations make these methods less effective for monitoring small targets, particularly lacking in predictive capability for large-scale pest infestations [2].

In recent years, object detection and classification based on deep learning models have become mainstream. Li Y et al. [3] introduced a scale-aware module into the network to construct a parallel multi-branch architecture, where each branch shares identical transformation parameters but possesses different receptive fields. Xu C et al. [4] developed a Multi-Scale Convolution-Capsule Network (MSCCN) for crop pest identification, comprising a multi-scale convolution module, a Capsule Network (CapsNet) module, and a SoftMax classification module. This approach leverages multi-scale convolutions to extract discriminative features at various scales

and utilizes CapsNet to encode hierarchical structures of pests with size variations in crop images, employing Softmax for pest recognition. Hu J et al. [5] proposed the Squeeze-and-Excitation (SE) attention mechanism, which, when integrated into networks as an SE module, focuses on interdependencies between feature channels, enabling the model to automatically learn the importance of different channel features and improve performance through weighted summation of channel features. Yang X et al. [6] utilized attention mechanisms and multi-task learning to design a novel network model SCRDet, where the attention mechanism emphasizes significant regions and multi-task learning allows the model to share features across different tasks, thereby enhancing the model's generalization capability. However, challenges remain in the identification of small targets in real-world environments, including the loss, misidentification, and missed detection of small targets, as well as difficulties in distinguishing targets from the background. Therefore, there is an urgent need to improve existing object detection methods to achieve stable and accurate identification of pest quantities and locust disaster situations.

In this paper, based on the YOLOv11n model, we utilize SPD-Conv (Space-to-Depth Layer Followed by a Non-strided Convolution Layer) to optimize the object detection network model of YOLOv11 and further enhance it by incorporating the PPA (Parallelized Patch-Aware Attention) module. This approach improves the distinction between targets and background while preserving critical information of small targets. The innovations in the model algorithm improvements are reflected in:

1. At the network architecture level: Introducing SPD-Conv to replace traditional downsampling operations effectively reduces the loss of feature map resolution, retains more details of small targets, and enhances the model's detection capability for small targets.

2. At the feature extraction level: Adding the PPA module, which combines multi-scale pooling with attention mechanisms, enhances the model's ability to distinguish between targets and background. This is particularly effective for object detection tasks in complex backgrounds.

3. At the performance optimization level: While maintaining the efficient inference speed of YOLOv11, the collaborative optimization of SPD-Conv and PPA modules significantly improves the model's detection accuracy and robustness in scenarios with dense targets and small targets.

The first four authors and the sixth author are with *School of Electrical and Information Engineering, North Minzu University*, Yinchuan City, China. The fifth author Wei Sun is Institute of Desertification Control, Ningxia Academy of Agriculture and Forestry Sciences Ningxia province, China. This work is supported by Key Research and Development Program of Ningxia Hui Autonomous Region (2024BF0101302), Graduate Innovation Project of Northern University for Nationalities (YCX24319), Graduate Education Quality Improvement Project of Northern Nationalities University (YJZT202431), Collaborative Education Project of the Ministry of Education (220603879075555). Corresponding author: (Hongxing Ma, email: mhx@nmu.edu.cn)

979-8-3315-2209-4/25 $31.00 © 2025 IEEE   342

II. METHOD

A. Algorithm Improvement

1) Improvement of SPD-Conv Module: SPD-Conv consists of a Spatial-to-Depth (SPD) layer and a non-strided convolutional layer (i.e., a convolutional layer with a stride of 1). The core idea is to downsample the feature maps through the SPD layer while retaining all information across channel dimensions, thus avoiding information loss [7]. Subsequently, the non-strided convolutional layer further processes the feature maps by reducing the number of channels and extracting effective feature representations.

The core operation of the SPD (Spatial-to-Depth) layer is to decompose the input feature map X into multiple sub-feature maps and concatenate these sub-feature maps along the channel dimension, thereby achieving downsampling. Suppose the size of the input feature map X is:

$$S \times S \times C_1 \tag{1}$$

The SPD layer decomposes it into scale×scale sub-feature maps. Each sub-feature map is obtained by sampling the original feature map at intervals of scale. For example, when scale=2, the feature map X is divided into 4 sub-feature maps, and the size of each sub-feature map is:

$$\frac{S}{2} \times \frac{S}{2} \times C_1 \tag{2}$$

The sub-feature maps obtained from the decomposition are concatenated along the channel dimension to generate a new feature map X', with its spatial dimensions reduced to:

$$\frac{\text{scale}}{S} \times \frac{\text{scale}}{S} \tag{3}$$

And the channel dimension increases to

$$\text{scale}^2 \times C_1 \tag{4}$$

Through this approach, the SPD layer achieves downsampling while retaining all original feature information.

After the SPD layer, the SPD-Conv module introduces a non-strided convolutional layer to further process the concatenated feature maps. The purpose of this convolutional layer is to reduce the number of channels through learnable parameters while retaining as much discriminative feature information as possible. Unlike strided convolutions, non-strided convolutions do not result in asymmetric sampling or loss of information in the feature maps, thereby ensuring effective feature extraction for low-resolution images and small object detection tasks. YOLO is a series of highly popular object detection models, and in this paper, the latest YOLOv11 is chosen for demonstration. By applying the method described in this section to YOLOv11, an improved model called YOLOv11-SPDConv can be obtained by simply replacing the original modules at the 7 positions shown in Figure 1 and changing the stride of YOLOv11 to 1.

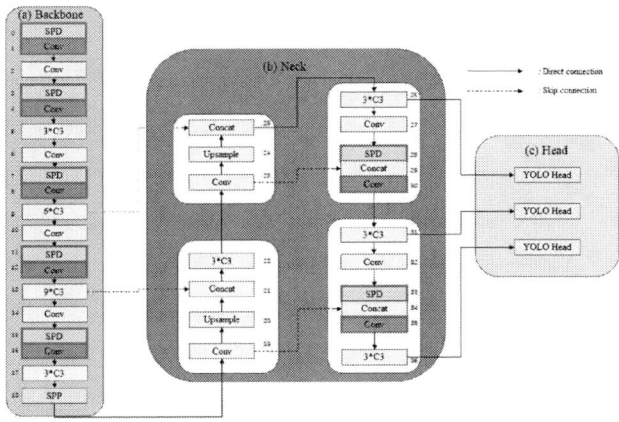

Fig. 1. YOLOv11-SPDConv model, the red rectangular box is the replaced part.

2) Improvement of PPA Module: The core of the PPA (Parallelized Perception Block Attention) module lies in its multi-branch feature extraction strategy, which enhances the accuracy of small object detection by processing feature information at different scales and levels in parallel [8]. The input feature map $X \in R^{H' \times W' \times C}$ is first adjusted through a pointwise convolution to obtain $X' \in R^{H' \times W' \times C'}$.

The PPA module consists of three parallel branches: a local branch, a global branch, and a serial convolution branch. Each branch is responsible for extracting features at different scales. The local branch extracts features by controlling the block size parameter p, dividing the input feature map into multiple non-overlapping local blocks and performing feature extraction on these blocks; this branch focuses on local detail information, making it suitable for capturing the detailed features of small objects. The global branch also controls the block size parameter p, but instead extracts global features, focusing on overall contextual information to capture the relationship between small objects and the background. The serial convolution branch extracts features through multiple 3×3 convolution layers, replacing traditional 7×7, 5×5, and 3×3 convolution layers, and by stacking multiple convolution layers, it extracts multi-level feature information. The features extracted by the three branches, X_{local}, X_{global}, and X_{conv}, are fused through summation to obtain the fused feature map $\tilde{X} \in R^{H' \times W' \times C'}$.

After feature extraction, the PPA module enhances the features through an attention mechanism, further improving the discriminative power of the features, and ultimately outputs the enhanced feature map(As shown in Fig. 2).

B. Dataset Construction

The dataset originates from a self-built locust dataset by the laboratory. The images were captured in the wild and the dataset was augmented using the CycleGAN network and the Python library ImgAug. For dataset annotation, the X-AnyLabeling tool was used for semi-automatic labeling. The specific process involved manually annotating 100 images per

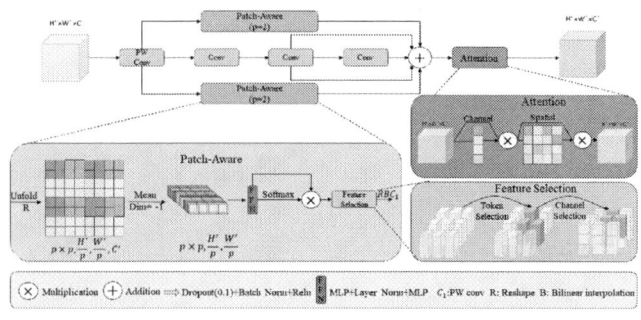

Fig. 2. The detailed structure of the parallelized perceptual block attention module is described. This module is primarily composed of two components: multi-branch fusion and an attention mechanism. The multi-branch fusion component includes perceptual block convolutions and concatenated convolutions. In the perceptual block, the 'p' parameter is set to 2 and 4, representing the local branch and the global branch, respectively.

class first, then training a YOLOv8n model for 300 epochs. The trained weights were exported in ONNX format. Following the official instructions, the YAML configuration file was set up, and the weights were imported into the X-AnyLabeling software to complete the labeling of the remaining images, followed by manual corrections.

(a) Oedaleus decorus asiaticus (b) Oedaleus infernalis saussure

Fig. 3. Partial display of self built locust dataset.

III. EXPERIMENTS AND RESULTS

A. Experimental Environment Parameters

The experimental environment for this study is a PC running Windows 11 as the operating system. The CPU and GPU configurations are i9 14900kf + 4090D with 24GB of video memory. The development language used is Python 3.9, and the deep learning framework employed is PyTorch 2.5.1. The unified training parameters include an input resolution of 640×640, a batch size of 128, and 300 training epochs.

B. Experimental Results and Analysis

The experimental results (Table 1) show that after 300 epochs of training, the improved model achieved an accuracy of 97.8%, a recall rate of 96.1%, an F1 score of 96.94, and mAP50-95 of 85.8%, with an inference speed of 91 frames per second. Compared to the original models YOLOv5n, YOLOv8n, and YOLOv11n, the YOLOv11n-SPDConv-PPA model demonstrates superior inference accuracy. In terms of

response speed, it is only slightly slower than YOLOv5n. This indicates that the improved model not only responds very quickly in terms of processing speed but also more accurately locates small target objects.

TABLE I
COMPARISON OF TYPICAL MODEL DETECTION RESULTS

Module	P/%	R/%	F1	mAP50-95%	FPS/s
YOLOv5n	71.6	83.3	77.01	80.8	93
YOLOv8n	89.5	90.6	90.05	82.7	82
YOLOv11n	92.4	90.5	91.44	83.6	87
YOLOv11n-SPDConv-PPA	97.8	96.1	96.94	85.8	91

Table 2 shows the results of the ablation study. From the data in the table, it is clear that the detection speed improved after adopting non-strided convolutional layers, with an increase of 8% compared to the original model. Although the addition of the SPD and PPA modules slightly increased the model complexity, the detection speed still improved by 4.5%, accuracy increased by 5.4%, recall improved by 5.6%, the F1 score rose by 5.5, and the mAP50-95 increased by 2.2% compared to the original model.

TABLE II
COMPARISON OF ABLATION EXPERIMENT RESULTS

Module	+SPD	+Conv(stride=1)	+PPA	P/%	R/%	F1	mAP50-95%	FPS/s
YOLOv11n				92.4	90.5	91.44	83.6	87
	✓			94.1	91.3	92.68	84.1	85
		✓		92.6	91.5	92.05	83.7	94
			✓	94.7	93.9	94.30	84.5	86
	✓	✓		94.3	93.7	94.00	84.2	92
		✓	✓	95.5	95.0	95.25	84.5	93
	✓		✓	97.2	95.8	96.49	85.6	90
	✓	✓	✓	97.8	96.1	96.94	85.8	91

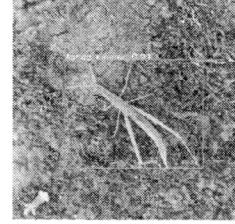
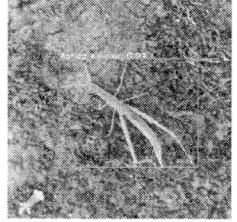
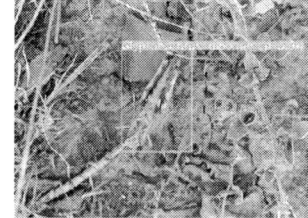

(a) Recognition results of Acrida cinerea (b) Recognition results of Glyptobothrus albonennus

Fig. 4. Display of recognition results.

IV. CONCLUSION

This study addresses the challenges of detecting small targets in real-world environments, such as the tendency for small targets to be lost, misidentified, or missed, and difficulties in distinguishing targets from the background. We propose an improved YOLOv11n detection method by replacing traditional downsampling operations with SPD-Conv, which effectively reduces feature map resolution loss, retains more details of small targets, and enhances both the model's detection capability and inference speed for small targets. Building on this, we introduce the PPA module, combining multi-scale pooling with attention mechanisms to strengthen the model's ability to distinguish targets from the background. Compared to classic models and through ablation studies, YOLOv11n-SPDConv-PPA shows a 4.5% increase in Frames Per Second (FPS), a 5.4% improvement in accuracy, a 5.6% increase in recall rate, an F1 score increase of 5.5, and a 2.2% improvement in mAP50-95 compared to the original model. Therefore, when detecting small locust targets against complex backgrounds, this method not only offers superior detection performance but also faster speeds, meeting the precision requirements and real-time response needs during use.In the future, we will continue to explore lightweight technologies such as model pruning and quantization for easier deployment on edge devices,introduce multi-source data fusion technology, combining multi-modal data like visible light images, videos, and audio to enhance detection performance and expand the use cases for other pest detection scenarios. The ultimate goal is to build a complete real-time monitoring and early warning system for small insects, achieving a full closed-loop from data collection, processing to warning, providing strong technical support for smart agriculture and ecological monitoring.

REFERENCES

[1] H. Zhao, B. Huang, H. Wang, and Y. Yue, "Pest identification method in complex farmland environment based on improved YOLO v7," Nongye Jixie Xuebao/Transactions of the Chinese Society for Agricultural Machinery, vol. 54, no. 10, pp. 246–254, 2023.

[2] X. Guo, X. Hao, X. Yao, and L. Li, "Joint intent detection and slot filling of knowledge question answering for agricultural diseases and pests," Nongye Jixie Xuebao/Transactions of the Chinese Society for Agricultural Machinery, vol. 54, no. 1, pp. 205–215, 2023.

[3] Y. Li, Y. Chen, N. Wang, et al., "Scale-aware trident networks for object detection," in Proceedings of the IEEE/CVF International Conference on Computer Vision, 2019, pp. 6054–6063.

[4] C. Xu, C. Yu, S. Zhang, et al., "Multi-scale convolution-capsule network for crop insect pest recognition," Electronics, vol. 11, no. 10, p. 1630, 2022.

[5] J. Hu, L. Shen, and G. Sun, "Squeeze-and-excitation networks," in Proceedings of the IEEE Conference on Computer Vision and Pattern Recognition, 2018, pp. 7132–7141.

[6] X. Yang, J. Yang, J. Yan, et al., "Scrdet: Towards more robust detection for small, cluttered and rotated objects," in Proceedings of the IEEE/CVF International Conference on Computer Vision, 2019, pp. 8232–8241.

[7] R. Sunkara and T. Luo, "No more strided convolutions or pooling: A new CNN building block for low-resolution images and small objects," in Machine Learning and Knowledge Discovery in Databases. ECML PKDD 2022, M. R. Amini, S. Canu, A. Fischer, T. Guns, P. Kralj Novak, and G. Tsoumakas, Eds., Cham: Springer, 2023, pp. 463–479, doi: 10.1007/978-3-031-26409-2_27.

[8] S. Xu, S. Zheng, W. Xu, et al., "HCF-Net: Hierarchical context fusion network for infrared small object detection," in 2024 IEEE International Conference on Multimedia and Expo (ICME), IEEE, 2024, pp. 1–6.

[9] J. Liu, X. Wang, and G. Liu, "Tomato pests recognition algorithm based on improved YOLOv4," Frontiers in Plant Science, p. 1894, 2022.

[10] S. M. S. U. Sourav and H. Wang, "Intelligent identification of jute pests based on transfer learning and deep convolutional neural networks," Neural Processing Letters, pp. 1–18, 2022.

[11] Y. Rao, W. Zhao, B. Liu, et al., "Dynamicvit: Efficient vision transformers with dynamic token sparsification," Advances in Neural Information Processing Systems, vol. 34, pp. 13937–13949, 2021.

[12] H. Sajjad, F. Dalvi, N. Durrani, et al., "On the effect of dropping layers of pre-trained transformer models," Computer Speech & Language, vol. 77, p. 101429, 2023.

Research and Application of Mushroom Growth Status Monitoring Technology

1st Yun Ma   2nd Fuyuan Wang   3rd Haobo Jia   4rd Kaiwen Chen   5th Xiaobin Ren   6rd Hongxing Ma

Abstract—Maintaining optimal growth conditions is a critical prerequisite for enhancing yield and profitability in modern mushroom cultivation facilities. However, existing cultivation environments are characterized by complexity, with challenges in the stability and accuracy of environmental data monitoring. Manual assessment of mushroom growth status is further hindered by inefficiency, high costs, and low precision. To address these challenges, this study conducts research on mushroom growth status monitoring technologies. A Mushroom Cultivation Room and Growth Status Monitoring Terminal was designed and developed to resolve data acquisition and processing issues for environmental parameters (e.g., temperature, humidity, CO_2 levels) and substrate biomass metrics. A lightweight GhostConv module was introduced to optimize the YOLO11s model, enabling dynamic monitoring and analysis of growth indicators such as cap dimensions and maturity levels. The optimized model was deployed on a Mushroom Identification App, facilitating long-term monitoring and predictive analytics for mushroom classification and growth status at the Ningxia Dawukou Mushroom Technology Hub (NDMTH). This system provides a scientific decision-making framework for mushroom cultivation.

Index Terms—Deep Learning, Mushroom Recognition, Environmental Monitoring, YOLO11s, Mushroom Cultivation Expert System

I. INTRODUCTION

The 2024 Central No. 1 Document, released in February, outlines a strategic "roadmap" for advancing rural revitalization. It emphasizes strengthening rural talent development by "promoting the Science and Technology Hub model and encouraging experts from research institutes and universities to serve agriculture and rural communities [1]."The establishment of Science and Technology Hubs is an important initiative for advancing the reform of graduate training models and supporting the revitalization of rural talent.The establishment of Science and Technology Hubs is an important initiative for advancing the reform of graduate training models and supporting the revitalization of rural talent.

To actively promote the Science and Technology Hub model, the Ningxia Dawukou Mushroom Technology Hub (NDMTH) has been established. Mushroom cultivation is one

---

Xiaobin Ren is Institute of Desertification Control,Ningxia Academy of Agriculture and Forestry Sciences Ningxia province,China.Other authors are from *Northern University for Nationalities* in Ningxia Province, China.This work was supported by Ningxia Hui Autonomous Region Agricultural Science and Technology Independent Innovation Fund Project(NGSB-2021-15-01), Graduate Innovation Project of North Minzu University (YCX24333), the High Level Construction Project of Ordinary Universities in Ningxia Province (bjg2024028) , and the Teaching Project of North Minzu University (YKSZ202307, YJZT202431). Corresponding author: (Hongxing Ma, email: mhx@nmu.edu.com)

of Ningxia's advantageous and distinctive industries, with continuous expansion in variety, significant improvements in quality, and growing cultivation scales. The application of smart greenhouse technologies has elevated Ningxia's agricultural production to a new level, serving as a critical driver for mushroom industry development [2]. Currently, the mushroom industry benefits from the widespread adoption of factory-based and mechanized production, modern information technologies, and advanced sensor systems. Automated cultivation and precision monitoring strategies have become key components in transitioning China's mushroom industry toward high-value-added economic products.

Current mainstream mushroom monitoring systems are primarily built on advancements in environmental sensing technologies and data transmission [3]. During mushroom growth, the focus of monitoring has been on external environmental factors such as temperature, humidity, and CO_2 concentration, with limited attention given to the mushrooms themselves as the primary subjects of monitoring. Therefore, developing a monitoring strategy that focuses on the mushrooms themselves and integrating it with traditional monitoring systems has become a critical research direction for automated mushroom production [4].

This study, grounded in the principles of precision agriculture and leveraging the Ningxia Dawukou Mushroom Technology Hub (NDMTH), investigates technologies for monitoring mushroom growth status in modern cultivation facilities. The research aims to achieve real-time data acquisition, automated recognition, decision support, and visualization of mushroom growth status, ultimately forming an expert cultivation system. This system provides scientific and personalized guidance and technical services for mushroom growers at NDMTH.

II. MATERIALS AND METHODS

This study leverages advanced electronic information technologies, including the Internet of Things (IoT), machine learning, and machine vision, to investigate and apply mushroom growth status monitoring techniques. The overall system architecture is illustrated in Fig 1.

The overall system architecture is divided into three components:

- Mushroom Cultivation Room and Growth Status Monitoring Terminal: Enables real-time data acquisition of environmental parameters (e.g., temperature, humidity, CO_2 concentration, light intensity) and substrate growth

---

979-8-3315-2209-4/25 $31.00 © 2025 IEEE   346

status metrics (e.g., temperature, humidity, pH, dissolved oxygen levels).

- Mushroom Growth Status Monitoring: Performs dynamic analysis and model prediction of growth indicators, including color changes, contour morphology, cap dimensions, and maturity levels.
- Mushroom Cultivation Expert System: Designs and develops a Mushroom Image Recognition Application to enable visualization, predictive analytics, and decision support for growth status monitoring.

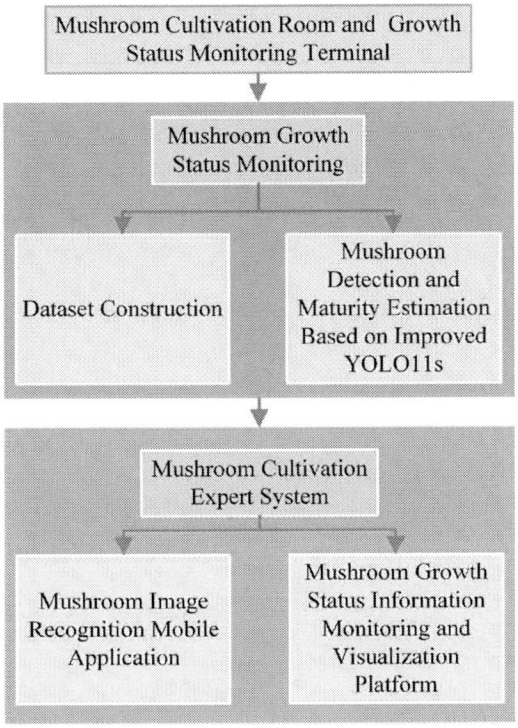

Fig. 1. Overall System Architecture.

III. MUSHROOM CULTIVATION ROOM AND GROWTH STATUS MONITORING TERMINAL

Currently, most modern mushroom cultivation rooms feature complex environments characterized by high humidity, temperature fluctuations, and other challenging conditions. Traditional environmental monitoring technologies are no longer sufficient to meet the growing demand for precise mushroom growth status monitoring. Therefore, this study leverages Internet of Things and machine learning technologies to design a Mushroom Cultivation Room and Growth Status Monitoring Terminal, which consists of two modules: an edge-side data acquisition module and a cloud-based processing module. The overall design block diagram is illustrated in Fig 2.

The core task of the edge-side data acquisition module is to collect real-time environmental data (e.g., temperature, humidity, CO_2 concentration, light intensity) from the monitored cultivation room, as well as growth-related metrics (e.g., temperature, humidity, digital dissolved oxygen content, TDS

coefficient) from the substrate blocks. These data are securely transmitted to the cloud server via a 4G communication module. The module connects various environmental sensors distributed across different locations to the core controller through a sensor expansion board, enabling direct communication with the main controller (STM32) without the need for relay devices. This design significantly enhances system flexibility and scalability. To ensure stable signal transmission and efficient communication, all sensors are mounted on the system's RS485 bus, and the Modbus-RTU protocol is used for communication between the sensors and the STM32 controller.

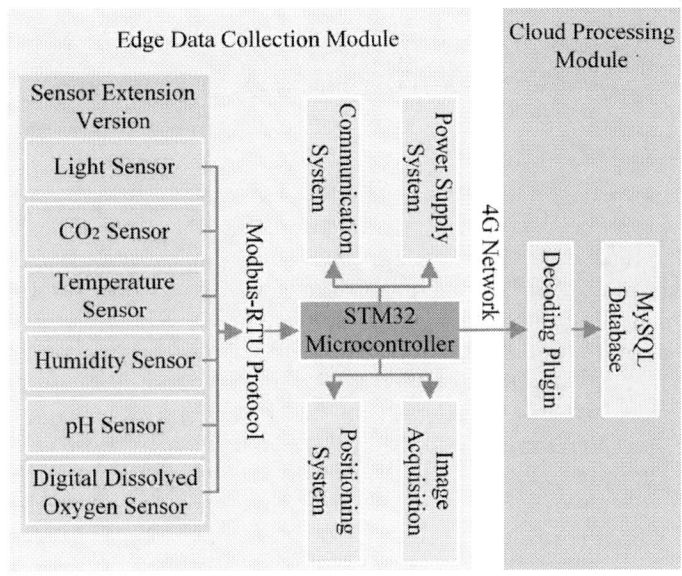

Fig. 2. Monitoring Terminal Design Block Diagram.

The physical implementation of the Mushroom Cultivation Room and Growth Status Monitoring Terminal is shown in Fig 3. This system is currently installed and operating stably at the shiitake mushroom cultivation base within the Ningxia Dawukou Mushroom Technology Hub (NDMTH).

Fig. 3. Monitoring Terminal Hardware.

IV. MUSHROOM GROWTH STATUS MONITORING

Mushroom growth status monitoring plays a critical role in identification, classification, and growth analysis [5]. However, its effectiveness largely depends on the availability of representative and diverse datasets. Challenges such as uneven surfaces of mushroom racks, dense distribution of substrate blocks, irregular cap shapes, and mutual occlusion between mushrooms significantly limit image recognition accuracy. To address these challenges and improve precision, robust deep learning algorithms are required for data processing and evaluation. For analyzing growth indicators such as color changes, contour morphology, cap dimensions, and maturity levels, mushroom growth status monitoring is divided into two components: dataset construction and mushroom detection and maturity estimation based on an improved YOLO11s model.

A. Dataset Construction

To enhance the model's ability to recognize and analyze mushrooms, images were captured from multiple perspectives and under varying lighting conditions, thereby increasing dataset diversity and improving the generalization capability of the mushroom recognition model. The images were primarily captured at a resolution of 3000 × 4000 pixels, with a total of 3,118 valid images collected. A subset of these images is shown in Fig 4, which includes two categories: Mushroom Cultivation Substrate Blocks (Fig4a) and Individual Mushroom Cultivation Substrate Block (Fig 4b). The dataset currently covers 6 mushroom species, with detailed statistics provided in Table 1.

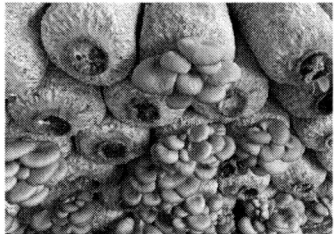 

a) **Mushroom Cultivation Substrate Blocks** b) **Individual Mushroom Cultivation Substrate Block**

Fig. 4. Sample Captured Images.

The dataset currently covers 6 mushroom species, with detailed statistics provided in TABLE I.

TABLE I
MUSHROOM IMAGE DATASET.

Mushroom Image Dataset	Dataset Size
Lentinula edodes	1028
Pleurotus ostreatus	573
Pleurotus geesteranus	435
Agaricus bisporus	331
Pleurotus eryngii	424
Volvariella volvacea	327

To ensure the diversity of the mushroom dataset and improve the accuracy and robustness of the mushroom growth status recognition model while mitigating potential overfitting issues, five data augmentation techniques were applied: random rotation, random sharpness, random color adjustment, Gaussian noise, and salt-and-pepper noise. These augmented images were then mixed to create a comprehensive dataset.

By expanding the dataset, the diversity of the images was enhanced, thereby improving the model's robustness and generalization capabilities. Finally, the LabelImg-1.8.6 tool was used to annotate the mushroom images, creating a VOC-format dataset for model training and validation purposes.

B. Mushroom Detection and Maturity Estimation Based on Improved YOLO11s

This study selects the YOLO11s model, which balances accuracy and model efficiency, for mushroom detection. The YOLO11s network architecture primarily consists of four modules: Input Layer, Backbone Network, Neck Network, and Head Network [6]. To address the issues of high computational cost, model complexity, and deployment challenges on edge devices associated with traditional Conv (convolution) operations, a lightweight GhostConv module is introduced to replace the Conv modules in the backbone network for feature extraction. The structure of the GhostConv module is illustrated in Fig 5.

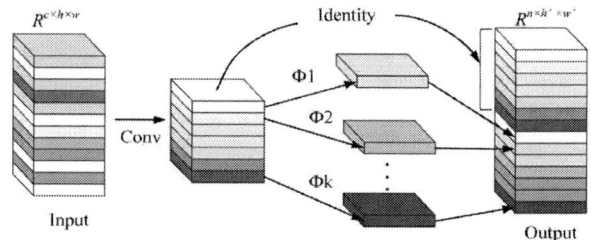

Fig. 5. GhostConv Network Structure.

The Ghost module generates feature maps through a unique three-step process.

- Base Feature Map Generation: A standard Conv operation generates the base feature map with m channels. Computational cost and parameter count are calculated using Equations (1) and (2).

$$F_2 = n' \times h' \times w' \times c \times k \times k \qquad (1)$$

$$P_2 = n' \times c \times k \times k \qquad (2)$$

- Ghost Feature Map Generation: The base feature map undergoes a cheap operation (e.g., depthwise convolution) to generate the Ghost feature map, which requires significantly less computation. Computational cost is calculated using Equation (3).

$$F_3 = n'' \times h' \times w' \times m \times k \times k \qquad (3)$$

- Feature Map Concatenation: The base feature map and Ghost feature map are concatenated to produce the final output feature map. The total computational cost is F_2+ F_2.

The improved YOLO11s network model is employed to achieve mushroom detection and maturity estimation. The input layer receives images containing substrate blocks and mushrooms, preprocesses them, and identifies mushrooms by cropping each anchor box in the captured images. It then outputs the corresponding adaptive anchor box information. The backbone network extracts features such as cap contour morphology and texture from the mushroom images. Using a size measurement method, the cap dimensions are calculated, enabling maturity estimation. This provides data support for model predictions in the mushroom cultivation expert system.

Using the improved YOLO11s model, the mushroom dataset was successfully recognized, with partial recognition results shown in Fig 6. Experimental results indicate that after 300 training epochs, the YOLO11s model achieves a mean average precision (mAP) of 90.7%The performance change curve is illustrated in Fig 7.

Fig. 6. Mushroom Image Recognition Results.

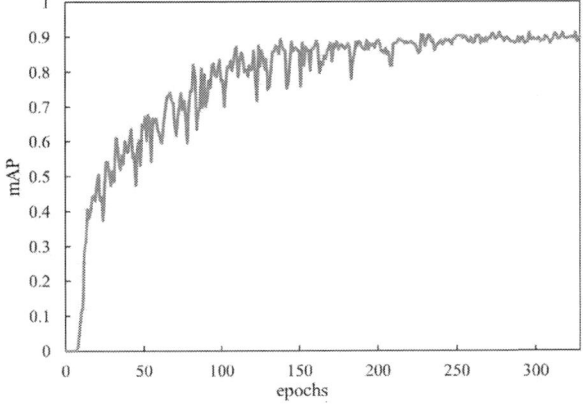

Fig. 7. Contrast curves of mAP50.

V. MUSHROOM CULTIVATION EXPERT SYSTEM

To support research on mushroom species and maturity levels at the Ningxia Dawukou Mushroom Technology Hub (NDMTH), an improved YOLO11s model was utilized to monitor mushroom growth status. A Mushroom Image Recognition Application was designed and developed, integrating the mushroom recognition model with a growth status monitoring and visualization platform to enable remote tracking. The user interface of each module in the mushroom image recognition app is shown in Fig 8.

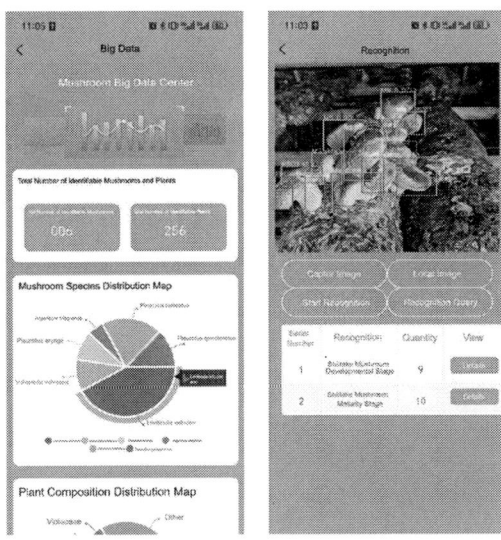

Fig. 8. Mushroom Recognition App Interface.

VI. CONCLUSION

This study addresses challenges such as poor stability and low accuracy in monitoring mushroom growth status in complex modern cultivation environments, as well as high complexity and computational costs in traditional deep learning models. Through extensive experiments and rigorous evaluation, the developed monitoring terminal has proven its practicality in accurately, stably, and rapidly detecting mushroom growth indicators in complex cultivation environments. The improved YOLO11s model demonstrates high precision and deployability, while its deployment at the Dawukou Shiitake Cultivation Base for real-time growth status monitoring highlights its agricultural applicability. This system provides visual reference data for cultivation decision-making and expert guidance for disease prevention and control.

REFERENCES

[1] Central Committee of the Communist Party of China, State Council. (2024). Opinions on learning and applying the experience of the 'Thousand Villages Demonstration, Ten Thousand Villages Improvement' project to effectively promote comprehensive rural revitalization. Issued on January 1, 2024.

[2] D. Garg and M. Alam, "Smart agriculture: a literature review," Journal of Management Analytics, vol. 10, no. 2, pp. 359–415, 2023.

[3] P. M. Szczypiński, A. Klepaczko, and P. Zapotoczny, "Identifying barley varieties by computer vision," Computers and Electronics in Agriculture, vol. 110, pp. 1-8, 2015.

[4] V. Marinoudi, C. G. Sørensen, S. Pearson, and D. Bochtis, "Robotics and labour in agriculture. A context consideration," Biosystems Engineering, vol. 184, pp. 111-121, 2019.

[5] Y. L., Z. Jia, Y. Xuling, et al., "Monitoring the growth status of corn crop from UAV images based on dense convolutional neural network," International Journal of Pattern Recognition and Artificial Intelligence, vol. 36, pp. 12, 2022.

[6] Y. Dai and X. Fang, "An armature defect self-adaptation quantitative assessment system based on improved YOLO11 and the segment anything model," Processes, vol. 13, no. 2, pp. 532, 2025.

RSMA Receiver Detection based on Neural Network Supervised Learning

Dou Pei
College of Electronic and Information
Hangzhou Dianzi University
Hangzhou, China
1609102659@qq.com

Zhigang Zhou
College of Electronic and Information
Hangzhou Dianzi University
Hangzhou, China
zgzhou@hdu.edu.cn

Jian-gong Ni
College of Electronic and Information
Hangzhou Dianzi University
Hangzhou, China
nijiangong@hdu.edu.cn

Yejiang Lin
College of Electronic and Information
Hangzhou Dianzi University
Hangzhou, China
1419785480@qq.com

Gang Chen
Shanghai Ketai Information Technology
limited company
Shanghai, China
chengang@shketai.com

Jian Zhou
Shanghai Institute of Microsystem
and Information Technology
Shanghai, China
zjian@mail.sim.ac.cn

Abstract—This paper proposes a deep neural network Rate-Splitting Multiple Access (RSMA) receiving detection algorithm based on a 2-user 1-layer RS transceiver structure. First, data is transmitted over a flat fading channel in the offline state to generate a training dataset, which is then used to train the neural network. Next, experimental simulations are performed using Matlab to detect the receiver signal's BER in three cases: no cyclic prefix (CP), clipping and ideal state, under multipath Rayleigh fading channel transmission. The experimental results are compared with the decoding performance using the Least Squares (LS) algorithm and the Minimum Mean Square Error (MMSE) algorithm, respectively. The experimental results show that the decoding performance of the regression neural network with Long Short-Term Memory (LSTM-RegNN) outperforms the LS algorithm and MMSE algorithm in the absence of CP, clipping, and ideal state. In addition, the decoding performance of the Fully Connected Classification Network (FCCN) is lower than that of the MMSE in the clipping experiments with high signal-to-noise ratio (SNR), while the decoding performance of FCCN is higher than that of both the LS algorithm and the MMSE algorithm in the clipping experiments with low SNR, as well as in the case of no CP and ideal state. The experimental results show that deep neural networks (DNNs) are a promising tool for RSMA signal detection.

Index Terms—rate-splitting multiple access, signal detection, neural network supervised learning.

I. INTRODUCTION

In the domain of Rate-Splitting Multiple Access (RSMA), Deep Neural Networks (DNNs) are frequently employed to optimize resource allocation, enhance performance metrics in transmission [1],verify the efficacy of RSMA [2], and balance the relationship between computational complexity and performance [3] etc.

The challenge of error transmission in RSMA transmission, caused by the use of Successive Interference Cancellation (SIC) for data recovery, is addressed by leveraging the high accuracy of deep neural networks. These networks are employed

as an alternative to SIC-based signal detection, significantly enhancing the quality of the transmitted information [4]. For instance, Anagha K. Kowshik et al. employed Long Short-Term Memory (LSTM) classification neural networks in the uplink and downlink of RSMA, respectively, to demonstrate the efficacy of their decoding process and examine the impact of neural network parameter changes on decoding performance [5]. However, these neural networks did not consider the performance characteristics of the neural network under non-ideal channel transmission conditions, nor did they use regression neural networks for performance evaluation at the decoding end.

In this paper, the regression neural network with LSTM (LSTM-RegNN) and the Fully Connected Classification Network (FCCN) of supervised learning are utilized to process the received information under different environmental states. The Bit Error Rate (BER) performance of the recovered data is then compared with LS and MMSE. The experimental results demonstrate that both the regression neural network and the classification neural network exhibit superior performance in terms of RSMA signal detection. These findings show that both regression and classification neural networks exhibit strong robustness in RSMA signal detection, demonstrating potential application opportunities.

II. SYSTEM MODEL

The fundamental concept of RSMA involves dividing a user's transmitted message into two parts at the transmitter side, referred to as the public part and the private part. Subsequently, the public parts from all users are combined into a unified data stream, which is transmitted concurrently and shares the same time and frequency resources as the private parts' data streams. At the receiver end, the public part is received and decoded by all users, while the private part is decoded only by the corresponding user.

Funding information:Shanghai Industrial Collaborative Innovation Project under grant HCXBCY-2024-051

A. RSMA Received Signal Detection Model

The transmission framework of 2-user 1-layer RS is adopted and the signal processing flow at the transmitter is as follows: firstly, the bitstreams of the two users are organized into the public and private streams; next, the sorted data streams undergo the modulation process; at the same time, the modulated signals are then precoded; and then, the data streams are converted from the frequency domain to the time domain by using the inverse Fast Fourier Transform (IFFT). A cyclic prefix (CP) is inserted to mitigate inter-code interference, resulting in the transmitted data information being $x(n)$.

Assuming the transmitted channel is a multipath Rayleigh fading channel with the channel variable is $h(n)$, the signal $y(n)$ received at the receiver can be expressed as

$$y(n) = h(n) * x(n) + z(n) \tag{1}$$

Where, $z(n)$ is the additive Gaussian white noise generated during the transmission of the signal. The frequency domain signal, after the prefix is removed at the receiving end and the execution of a Fast Fourier Transform (FFT) on the signal, is

$$Y(k) = H(k)X(k) + Z(k) \tag{2}$$

The signal detection approach proposed in this paper applies neural networks to the data recovery processing stage on the receiver side, without considering implicit operations like removing CP and performing FFTs. The specific flow is illustrated in Fig.1 [6].

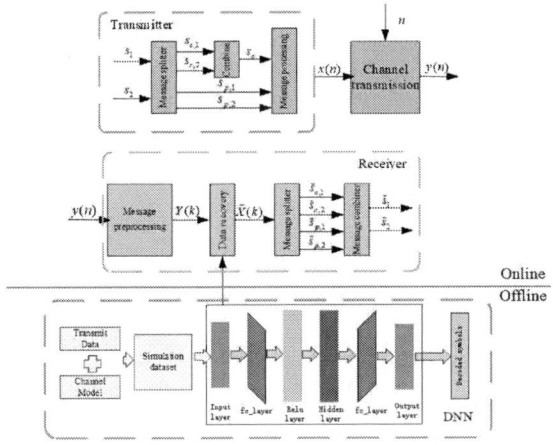

Fig. 1. Detection Model of Neural Network RSMA Received Signal.

B. Parameter Setting of Neural Network

- Loss Function: LSTM-RegNN uses the mean squared error function, also known as L2 Loss. On the other hand, FCCN employs the cross-entropy function, which aims to minimize the loss and bring the model's predictions closer to the correct category.
- Optimization Algorithm: Adam's algorithm, a first-order gradients-based optimization method, is used. It combines the concepts of momentum and RMSprop to adaptively adjust the learning rate of each parameter. This algorithm

is particularly effective for optimization problems that involve large-scale data and parameters. Therefore, Adam's algorithm is selected as the optimization method for both neural networks.

- Model Structure Layer: Both neural network structures are comprised of five layers, with the number of neurons in each layer being 512, 500, 250, 120, and 16, respectively. The input numbers correspond to the number of real and imaginary parts of the four OFDM blocks. In the FCCN framework, the final layer employs the Softmax function as the activation layer, while the majority of other layers utilize the ReLu function as the activation function.

III. PERFORMANCE EVALUATION

Based on the proposed neural networks supervised learning RSMA signal detection model, experiments are conducted using Matlab under different transmission conditions to evaluate the SNR required by each algorithm when 10^{-2} is set as the false bit performance threshold (as indicated by the green line in Fig.2.3.4).

A. Impact of Cyclic Prefix

In a multipath propagation environment, the signal reaches the receiver through multiple paths, causing delay spread and inter-symbol interference. The CP is added at the beginning of each transmitted signal as a repeated portion, preventing multipath delays from affecting the next symbol. However, adding the CP reduces spectral efficiency and increases system overhead. Therefore, the performance of LSTM-RegNN and FCCN is evaluated without CP. The resulting outcomes are presented in Fig.2.

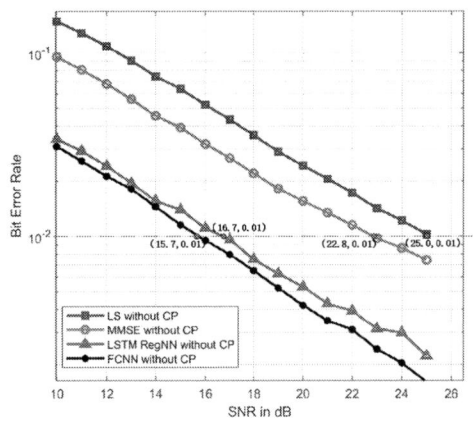

Fig. 2. Bit Error Rate without CP.

As illustrated in Fig.2, the absence of CP in LSTM-RegNN results in improvements of 6.1 dB and 8.3 dB compared to MMSE and LS, respectively. Similarly, the implementation of FCCN leads to improvements of 7.1 dB and 9.3 dB over MMSE and LS with 10^{-2} is designated as the false bit performance threshold.

979-8-3315-2209-4/25 $31.00 © 2025 IEEE

B. Impact of Clipping

In the context of wireless transmission, peak-to-average power ratio(PAPR) is a crucial metric for evaluating the ratio of peak to average power in a signal. This metric serves as a foundation for signal design and optimization. To reduce the PAPR of the signal, a common approach is the implementation of clipping [7]. The clipped signal becomes:

$$\tilde{x}(n) = \begin{cases} x(n), & \text{if } |x(n)| \leq A \\ Ae^{j\angle x(n)}, & \text{otherwise} \end{cases} \quad (3)$$

Due to LS's sensitivity to the nonlinear distortion induced by clipping, a comparison is made between LSTM-RegNN, FCCN, and MMSE. After normalizing the signal's energy, performance comparisons are conducted with A set to 2 and 4, and a peak value of 8.13. The results of this comparison are presented in Fig.3.

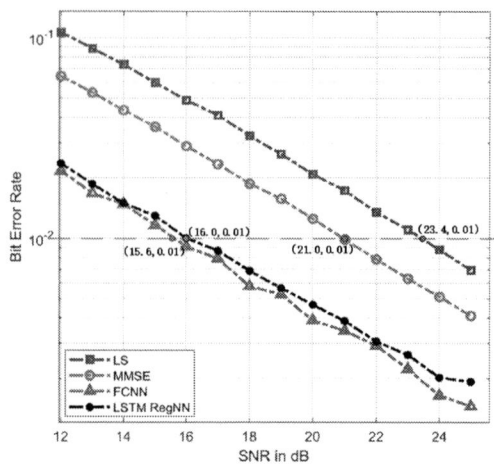

Fig. 4. Bit Error Rate in Ideal State.

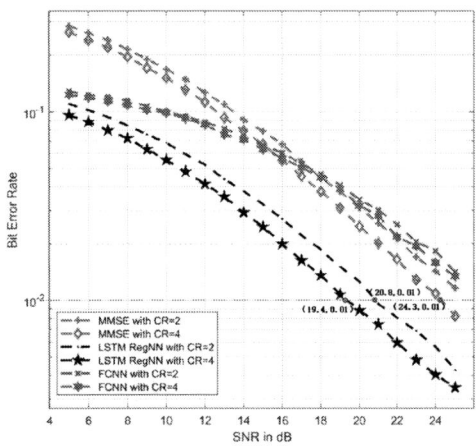

Fig. 3. Bit Error Rate in Clipping.

As illustrated in Fig.3, compared to MMSE, FCCN demonstrates superior performance in low SNRs scenarios but underperforms in high SNRs conditions. In contrast, LSTM-RegNN consistently exhibits significant advantages across all SNR ranges.

C. Ideal State

Experiments are conducted on LSTM-RegNN and FCCN under ideal conditions, and their performance is compared with LS and MMSE. The results are shown in Fig.4.

As shown in Fig.4, when operating under ideal conditions, LSTM-RegNN improves the signal by 5 dB compared to MMSE and by 7.4 dB compared to LS. Similarly, FCCN shows a 5.4 dB improvement over MMSE and a 7.8 dB improvement over LS with 10^{-2} is designated as the false bit performance threshold.

IV. CONCLUSION

In the 2-user 1-layer RS transceiver architecture of RSMA, a deep neural network-based detection algorithm is proposed to address the error transmission effects of the traditional SIC

algorithm. Experimental results under various transmission conditions show that, in without CP experiments, when the threshold is 10^{-2}, the performance of LSTM-RegNN is 8.3 dB and 6.1 dB better than the LS and MMSE algorithms, respectively. Similarly, FCCN improves by 9.3 dB and 7.1 dB, respectively. In clipping experiments with a peak value of 8.13, at truncation thresholds of 2 and 4, LSTM-RegNN performs significantly better than MMSE, while FCCN outperforms MMSE at low SNR but performs weaker at high SNR, demonstrating stronger robustness compared to MMSE. Under ideal conditions with a threshold of 10^{-2}, LSTM-RegNN improves by 7.4 dB over LS and 5 dB over MMSE, while FCCN improves by 7.8 dB and 5.4 dB, respectively. These results highlight the potential of deep neural networks for RSMA receiver signal detection.

REFERENCES

[1] Sonia Pala, Mayur Katwe, Keshav Singh, Theodoros A. Tsiftsis, and Chih-Peng Li, "Robust Transmission Design for RIS-Aided Full-Duplex-RSMA V2X Communications via Multi-Agent DRL," IEEE Transactions on Vehicular Technology. 2024, pp. 1-15

[2] Shimaa A. Naser, Abubakar Sani Ali, and Sami Muhaidat, "Deep Reinforcement Learning for RSMA-Based Multi-Functional Wireless Networks," In GLOBECOM 2023 - 2023 IEEE Global Communications Conference, 2023, pp. 2967–2972.

[3] Yiwen Wang, Yijie Mao, and Sijie Ji, "RS-BNN: A Deep Learning Framework for the Optimal Beamforming Design of Rate-Splitting Multiple Access," IEEE Transactions on Vehicular Technology, vol.73, no.11, 2024, pp. 17830-17835.

[4] Rafael Cerna Loli, Onur Dizdar, Bruno Clerckx, and Cong Ling, "Model-Based Deep Learning Receiver Design for Rate-Splitting Multiple Access," IEEE Transactions on Wireless Communications, vol.22, no.11, 2023, pp. 8352-8365.

[5] A. K. Kowshik, A. H. Raghavendra, S. Gurugopinath and S. Muhaidat, "Deep Learning-Based Signal Detection for Rate-Splitting Multiple Access Under Generalized Gaussian Noise," IEEE Open Journal of Vehicular Technology, vol. 4, 2023, pp.257-270.

[6] B. Clerckx, H. Joudeh, C. Hao, M. Dai and B. Rassouli,"Rate splitting for MIMO wireless networks: a promising PHY-layer strategy for LTE evolution," IEEE Communications Magazine, vol. 54, no. 5, pp. 98-105, May 2016.

[7] T. Lee and H. Ochiai, "Experimental Analysis of Clipping and Filtering Effects on OFDM Systems," 2010 IEEE International Conference on Communications, Cape Town, South Africa, 2010, pp. 1-5.

A Efficient Reed-Solomon Codes Recognizer Based on Galois Field Fourier Transform

Weiran Cao   Wei Zhang   Yihan Wang

Abstract—The blind recognition technology for Reed-Solomon (RS) codes has been widely applied in scenarios such as non-cooperative communications. However, the high latency and significant hardware resource consumption caused by numerous iterative operations limit the development of hardware implementations for the recognizer. This paper presents an half Galois field Fourier transform (H-GFFT) algorithm with a calculation complexity of (n-1)/2 compared to the traditional GFFT. Furthermore, a multi-path parallel circuit architecture suitable for H-GFFT is proposed, which effectively reduces calculation latency. Building on this, a hardware design for a recognizer that can identify the length and rate of RS codes has been completed, which has been implemented on the Xilinx KC705 FPGA development platform. Experimental results indicate that the recognition performance of the recognizer approaches the theoretical recognition limit, with a hardware acceleration ratio of 188.22.

Index Terms—Reed-Solomon codes, Blind recognition, Galois field Fourier transform, FPGA implementation

I. Introduction

In the field of non-cooperative communications, blind parameter recognition technology plays a crucial role in accurately interpreting the modulation and coding information of intercepted signals[1]. The current mainstream blind recognition method for Reed-Solomon (RS) codes is based on Galois field Fourier transforms (GFFT), which offer higher accuracy and robustness compared to rank-deficient blind recognition algorithms[2–4]. However, this method also has the drawbacks of requiring many iterations and consuming significant hardware resources[5].

This paper presents a hardware implementation of an RS codes blind recognizer based on half Galois field Fourier transforms (H-GFFT). H-GFFT utilizes the property of spectrum zero components of correct codewords appearing in pairs, reducing the hardware consumption of calculating the spectrum for each codeword by (n+1)/2. To address the issue of high iteration counts in traditional GFFT hardware implementations, a parallel architecture is employed to complete the H-GFFT hardware, reducing the calculation latency for a single codeword to 1/4 or 1/8. Additionally, the parameter estimation module leverages the special characteristics of the H-GFFT spectrum to improve recognition rates through channel partitioning.

The rest of this paper is organized as follows: Section II provides a brief introduction of RS codes and the Galois Field Fourier Transform. Section III describes the architecture and algorithm details of the recognizer, along with a mathematical proof of the proposed simplification. Section IV presents the experimental results. Section V concludes the paper.

II. Preliminaries

A. Basic Theory of RS Codes

The RS (n, k, t, p) codes studied in this paper are a class of non-binary linear block cyclic codes defined over the Galois field $GF(2^m)$. Each RS code has a length of n, composed of k information symbols and $2t$ parity symbols, with each symbol containing m bits. The parameter $t = (n - k)/2$ characterizes the error correction capability of the RS codes[6]. The variable p represents the integer form of the primitive polynomial $p(x)$, with its roots denoted as the primitive element α. The generator polynomial $g(x)$ of the RS codes is defined as the lowest degree polynomial that has the roots $\alpha, \alpha^2, \ldots, \alpha^{2t}$. Therefore, it can be expressed as follows:

$$g(x) = (x - \alpha)\left(x - \alpha^2\right)\left(x - \alpha^3\right)\cdots\left(x - \alpha^{2t}\right) \quad (1)$$

B. Galois Field Fourier Transform

Define the message polynomial as $m(x) = m_0 + m_1 x + m_2 x^2 + \ldots + m_{k-1} x^{k-1}$, and define the codeword polynomial as $r(x) = r_0 + r_1 x + r_2 x^2 + \ldots + r_{n-1} x^{n-1}$. The relationship between the two is $r(x) = g(x)m(x)$. Perform GFFT operation on the codeword $r(x)$ to obtain $R(z)$, as shown in equation (2), where $R_j = r(\alpha_j) = \sum_{i=0}^{n-1} r_i \alpha_j^i$, R_j represents the j-th spectral component of $R(z)$[7]. If the codeword polynomial $r(x)$ has α^j as root, then $R_j = 0$. Therefore, a correctly transmitted codeword certainly takes $\alpha, \alpha^2, \ldots, \alpha^{2t}$ as roots resulting in $R_1 = R_2 = \cdots = 0$.

$$R(z) = R_0 + R_1 z^1 + R_2 z^2 + \cdots + R_{n-1} z^{n-1} \quad (2)$$

III. Proposed Implementation

Fig. 1 illustrates the architecture of the recognizer. The storage, segmentation, transformation, and estimation of the received information are all implemented on FPGA. In the first stage, the codewords are received in a serial mode and enter the matrix reconstruction module, where they are divided into analysis matrices of different code lengths and initialized in the corresponding RAM. In the second stage, under the influence of a read enable signal, symbols are output in parallel to spectrum information statistics module, where the H-GFFT operation is performed on the codewords to compute the spectral components. In the third stage, the parameter

---

Weiran Cao, Wei Zhang, Yihan Wang are with *the School of Microelectronics, Tianjin University*, Tianjin, China. Corresponding author: (Weiran Cao, email: 2023232027@tju.edu.cn)

TABLE I
SIMULATION RESULTS OF TWO GFFT MODES UNDER DIFFERENT SNR

SNR(dB)	H-GFFT				GFFT			
	LCR=2	*LCR=4*	*LCR=6*	*LCR=8*	*LCR=2*	*LCR=4*	*LCR=6*	*LCR=8*
5	642	8	38602	1230	25	0	38645	29
4	1676	26	13865	441	34	0	13891	11
3	2640	72	2661	104	83	0	2675	2
2	2664	78	259	13	98	0	260	0
1	1435	51	14	2	93	0	15	0

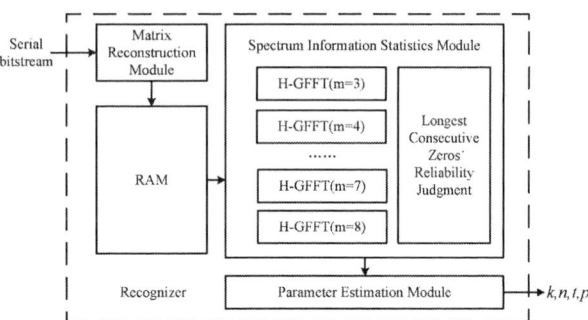

Fig. 1. Architecture diagram of the recognizer

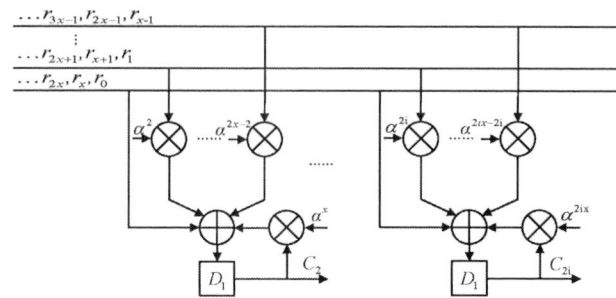

Fig. 2. Multi-path parallel H-GFFT architecture

estimation module analyzes the spectral information obtained from the H-GFFT and calculates the recognition results.

A. H-GFFT Hardware Design

The recognition method based on GFFT mainly utilizes a majority voting approach to select the parameters with the most frequent occurrences of consecutive zero spectra. When performing GFFT on a correct RS codeword, the length of the consecutive zero spectrum in its spectrum equals $2t$, meaning that the roots always appear in pairs. Existing research uses GFFT to obtain the spectrum, requiring $n(n-1)$ additions and n^2 multiplications. The proposed H-GFFT can reduce this to $(n-1)^2/2$ additions and $(n^2+n)/2$ multiplications.

Due to the paired occurrences of zeros in the frequency spectrum, the number of pairs matches the error correction capability t. H-GFFT simplifies the traditional method by only retaining the calculations for α^2, α^4 and so on, effectively halving the hardware resources while maintaining similar performance. This simplification method has a most direct impact on performance, which is reflected in a significant increase in the number of low consecutive roots(LCR) in the spectrum. Specifically, for erroneous codewords, the probability that both α and α^2 are zero is $1/2^{2m}$, while the probability that α^2 is zero is $1/2^m$. The former can be ignored, while the latter will result in erroneous codewords being judged as codewords with roots α and α^2. To evaluate the impact of simplification on performance, we simulated both traditional GFFT and H-GFFT on 100,000 bits of RS$(31,25)$, with the LCR information at different SNR is recorded in Table I.

Data indicates that the channel environment has a significant impact on the accuracy of the H-GFFT conversion results. To minimize the impact, a parameter selection method based on the mode of consecutive zeros in the frequency spectrum is proposed. For a single GFFT operation, if the occurrence

frequency of the mode u of consecutive zeros is significantly higher than $LCR = 2$, the channel is considered to be fine, and the parameter corresponding to the mode is selected as the reliable recognition result. If this condition is not met, the channel is considered poor, and the maximum value is chosen as the result of that GFFT operation.

B. Multi-path Parallel H-GFFT Architecture

As mentioned in Section II, completing a single GFFT spectral computation requires n multiplications and $n-1$ additions, which is inefficient in terms of both time and space. Horner's Rule is an efficient and concise approach to polynomial evaluation, however, the standard Horner's method significantly reduces time efficiency. By equivalently transforming the formula and applying it to the calculation of spectral components, the computation of C_{ij} can be transformed to equation(3) [8].

$$C_{ij} = \left(\left(r_{i,n-1}\alpha^{jx} + \cdots + r_{i,n-p-1} \right) \alpha^{jx} \cdots \right) + \cdots + r_{i,0} \tag{3}$$

Designing the above equation as a circuit results in a highly parallel GFFT computation architecture, as shown in the Fig.2.

The computational efficiency of the multi-path parallel GFFT spectrum calculation unit has increased by a factor of x, but the number of Galois field multipliers has increased by x. Using constant multipliers can reducing resource consumption. By constructing XOR gates, we completed the design of constant multipliers for different code lengths based on primitive elements and their higher powers. The resource consumption and comparisons are shown in Table II.

Under the same code length, the resource consumption of the GF-Multiplier is similar to that of the adders, both only use a few XOR gates. Compared to traditional multipliers, the hardware overhead is much smaller. Therefore, the highly

979-8-3315-2209-4/25 $31.00 © 2025 IEEE

TABLE II
COMPARISON OF HARDWARE RESOURCE CONSUMPTIONS

Bit Width	Multiplier	GF-Multiplier	GF-Adder
4	19	4	4
5	27	5	5
6	38	5	6
7	52	7	7
8	71	9	8

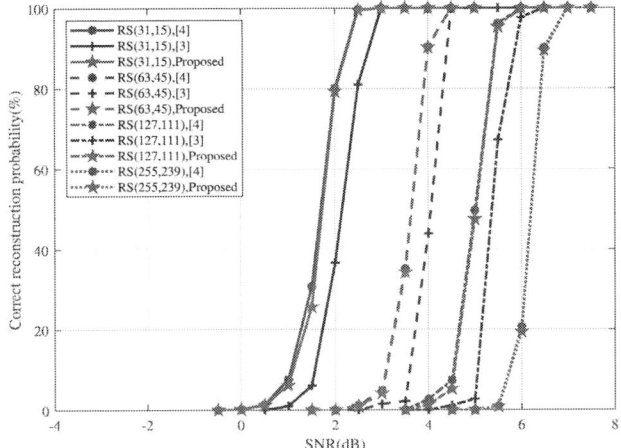

Fig. 3. Comparison of recognition performance for RS codes based on different method

parallel GFFT spectral computation unit still has very low hardware resource consumption.

In the context of fixed input bits, the calculation delay is proportional to m, with a delay ratio of the shortest to the longest code length being 2.3. The proposed recognizer categorizes the code lengths into short codes and long codes, using 8-path or 4-path parallel architecture, which reduces the delay ratio to 1.2.

IV. RESULTS AND COMPARISONS

A. Performance Comparison

This paper compared the recognition performance of the proposed recognizer with the method described in t he literature [3][4]. The algorithm proposed in [4] approaches or achieves the theoretical recognition limit. the probability of correct recognition is used to evaluate the performance of the recognizer. All simulations utilize binary phase shift keying modulation in an additive white Gaussian noise channel. The results are derived from 1,000 Monte Carlo experiments, with the number of received bits set to 100,000. The comparison results are shown in Fig. 3. Overall, the performance of the recognizer is superior to that in [3] and is close to the performance in [4]. For RS(63, 45), it maintains a 90% recognition rate at an SNR of 3.5, approaching the recognition limit while exceeding the recognition rate of 46.3% proposed in [3]. The recognition rate being close to the algorithm proposed in [4] indicates that the recognizer approaches the theoretical recognition limit even with half the GFFT hardware. This demonstrates the feasibility of the recognizer in identifying different code-length RS codes. The performance deficiencies of the LC-GFFT proposed in [3] can be attributed to its

oversight of the unreliability of majority voting under harsh channel conditions. The recognizer effectively addresses this issue through channel partitioning method.

B. Hardware Deployment

The proposed recognizer was synthesized and implemented on the Xilinx KC705 FPGA development platform, using the device model xc7k325tffg676-2. The recognizer occupies 29,581 LUTs, which is 14.51% of the total LUT resources; it utilizes 63,077 registers, accounting for 15.48% of the total register resources. The software algorithm running on the Intel Core i5-10500 platform takes 163 ms to execute, while the hardware computation takes 0.867 ms when the recognizer's operating clock is set to 120 MHz, resulting in a hardware acceleration ratio of 188.22.

V. CONCLUSIONS

In this paper, we propose an RS codes recognizer based on a multi-path parallel H-GFFT. The experimental results show that our recognizer achieves a high hardware acceleration ratio through a highly parallel design while ensuring recognition accuracy, and it further reduces hardware resource consumption by utilizing H-GFFT.

REFERENCES

[1] M. Song, J. Kim, and D. Shin, "Blind reconstruction of bch and rs codes using single-error correction," *IEEE Transactions on Signal Processing*, vol. 69, pp. 5120–5133, 2021.

[2] D. Jo, S. Kwon, and D. Shin, "Blind reconstruction of bch codes based on consecutive roots of generator polynomials," *IEEE Communications Letters*, vol. 22, no. 5, pp. 894–897, 2018.

[3] L. Shi, W. Zhang, Y. Chang, H. Wang, and Y. Liu, "Blind recognition of reed-solomon codes based on galois field fourier transform and reliability verification," *IEEE Communications Letters*, vol. 27, no. 8, pp. 2137–2141, 2023.

[4] Y. Wang, W. Zhang, L. Shi, Y. Chang, and Y. Liu, "Blind recognition of rs codes based on channel condition determination," *IEEE Communications Letters*, vol. 28, no. 5, pp. 1132–1136, 2024.

[5] Y. Chang, W. Zhang, D. Wei, H. Wang, and Y. Liu, "Reduced-complexity rs codes recognizer based on spectra update algorithm," *IEEE Communications Letters*, vol. 27, no. 7, pp. 1704–1708, 2023.

[6] S. Lin, "Error control coding second edition," *Upper Saddle River: Pearson Education*, pp. 255–262, 1983.

[7] C. Li, T. Zhang, and Y. Liu, "Blind recognition of rs codes based on galois field columns gaussian elimination," in *2014 7th International Congress on Image and Signal Processing*, pp. 836–841, 2014.

[8] R. Burden and J. Faires, "Numberical analysis," *Thomson Learning*, pp. 92–97, 2001.

Effects of 3d electron/hole doping on the static and high-frequency magnetic properties of c-oriented hcp-(CoIr) thin films with easy-plane magnetocrystalline anisotropy

1st Tianyong Ma   2nd Ao Han   3rd Qi Liu   4th Zisheng Li   5th Sha Zhang

Abstract—In this study, we investigated the effects of 3d electron/hole doping on both static and high-frequency magnetic properties of c-axis oriented hcp-CoIr soft magnetic thin films. As expected, both Ni and Cr doping enter into the Co site of the hcp-(CoIr) crystal structure. We found that electron from Ni doping increase magnetic moment, while hole from Cr doping decrease magnetic moment. And the coercivity also show the opposite trend with the increasing doping values. The high-frequency magnetic properties can be adjusted by 3d electron/hole doping due to the change of the intrinsic magnetocrystalline anisotropy and saturation magnetization. Moreover, similar to the coercivity, the damping factor increase with the increasing doping values for 3d electron doping while decrease for 3d hole doping. And an applied magnetic field has important effects on the damping factor of the Ni doping and Cr doping films. These results are expected to facilitate the soft magnetic properties of the thin films, which is benefical to meeting the requirements of specific applications in the future.

Index Terms—magnetic thin film, negative magnetocrystalline anisotropy, saturation magnetization, damping factor

I. INTRODUCTION

Textured soft magnetic films (SMFs) with easy-plane magnetocrystalline anisotropy have been attracted special attention due to their wide applications in high-frequency electromagnetic devices, such as film inductors[1, 2], micro-transformers[3] and microwave noise absorbers[4, 5]. In perpendicular magnetic recording, the suppression of spike noise and wide adjacent track erasure was achieved, as these films has higher probability to form the Néel type domain wall than traditional SMFs[6, 7]. In high-frequency devices, with the introduction of easy-plane magnetocrystalline anisotropy, the f_r could be significantly increased while the μ_i being unaffected[8]. And at micrometer thickness, these films still present in-plane domains and excellent high-frequency magnetic properties, hence could overcome the problem of insufficient magnetic flux signal for traditional SMFs[9] However, further adjustments in static and high-frequency magnetic properties are still needed to exceed their potential in various applications. The adding a third component into the Fe-, Co-, Ni-based SMFs is an effective method of adjusting and improving the magnetic properties, such as the anisotropy field of CoFe films can be effectively tuned by adding Cr element[10]; the improved soft magnetic properties of FeCoNi film by Zr[11]; the adjustable magnetocrystalline anisotropy of MnAl using Ni[12]. The essence of the above regulation lies in that these doped elements enter the lattice, causing changes or shifts in the band structure of the host metal. Cr and Ni as 3d transition metal element is adjacent to Co in the periodic table, and the number of 3d electrons of Ni is one more than that of Co, while that of Cr is two electrons less than that of Co. Doping the Ni or Cr into the c-textured hcp-CoIr SMFs with easy-plane magnetocrystalline anisotropy, both Ni and Cr dopants maybe replace Co in CoIr grains, and serve as donors and acceptors, respectively. Moreover, the 3d electron/hole doping may lead to the opposite movement of Fermi level and then significantly change the magnetocrystalline anisotropy of CoIr grains[13, 14]. Therefore, it is significant to systematically research the effects of 3d electron/hole doping on both the static and high-frequency properties of the hcp-CoIr SMFs. In this work, we presented our study of hcp-$(CoIr)_{100-x}Ni_x$ SMFs and compared the different effects of Ni and Cr doping. We have found that dopants atoms may replace Co atoms in CoIr phase. The saturation magnetization and the coercivity show the opposite trend with the increasing doping values for 3d electron and hole doping. The negative magnetocrystalline anisotropy can be changed in a much broader range from about -5×10^6 erg/cm^3 to almost zero. Moreover, the dynamic magnetic properties and the damping factor are easily controlled by adding the transition element Ni and Cr to meet different applications.

II. EXPERIENCE

All the hcp-CoIr SMFs with a layered structure of substrate/Ti/Au/$(CoIr)_{100-x}Ni_x$ were fabricated by DC magnetron sputtering technique. For the growth of the films, pure Ar was used as the sputtering working gas. Si wafer with (100) surface orientation was used as substrate, then an 3 nm amorphous layer of Ti was deposited on the substrate in 0.25 Pa Ar gas. The clean and flat surface of amorphous

All authors are with *The Key Laboratory of Physics and Photoelectric Information Functional Materials, North Minzu University*, Yinchuan 750021, China. This work is supported by the research start-up fund of China (Nos.2021KYQD22). Corresponding author: (Author 5, email: zhangsha_126@126.com)

979-8-3315-2209-4/25 $31.00 © 2025 IEEE

356

Ti can improve adhesiveness of the Au seed-layer. The 10 nm (111)-plane oriented Au seed layer was sputtered under pure Ar with 0.25 Pa pressure. The deposited seed layer of Ti(3 nm)/Au(10 nm) is beneficial to induce the c-axis oriented hcp-(CoIr)$_{100-x}$Ni$_x$ SMFs. Both the Co target with several Ir chips on top of it and the Ni target were used in preparation process of magnetic layer which was deposited at 0.3 Pa Ar gas. The sputtering power of the Ni target can be changed in order to adjust Ni doping content while maintaining the Co target power at 60 W unchanged. The chemical composition of our films was quantified using energy dispersive spectroscopy (EDS). Subsequent the grain morphology was obtained by transmission electron microscopy (TEM). Crystalline phase identification was performed using X-ray diffraction (XRD). Static magnetic properties and dynamic magnetic properties were measured with a vibrating sample magnetometer (VSM) and an electron spin resonance spectrometer (ESR), respectively. The high-frequency magnetic properties and damping factor for all our films were thoroughly investigated by vector network analyzer (VNA).

III. RESULTS AND DISCUSSION

Fig.1 (a) shows the XRD patterns of our hcp-(CoIr)$_{100-x}$Ni$_x$ films with x=0~10.6. The Au (111) peaks and hcp-(CoIr)$_{100-x}$Ni$_x$ (002) peaks are the only observed, which indicating the perfect orientation of the c-axis of the magnetic layer. As shown in Fig.1 (b) and Fig.1 (c), the changes of the lattice parameter c and FWHM are very small as a function of x, namely the changes are smaller than 0.3% and 6%, respectively. Considering the similar atomic size of Co and Ni, these results may suggest that the doped Ni actually enters into the Co site of the hcp-(CoIr) structure. This is

the high magnification image of the magnetic layer reveals a high orientation of the c-plane, which indicating good film crystallinity. As our expected, the doped Ni atoms replace Co atoms in CoIr phase. To have a better comparison, the lattice parameter c and FWHM of Cr doping films are shown in Fig.1 (b) and Fig.1 (c). These facts suggest that Cr also enter into the Co site of the hcp-(CoIr) crystal structure[15]. The similar doping dependencies of c and FWHM can be seen, which may be resulted from the similar atomic size of Cr and Ni. Interestingly, the bigger values of Ni doping films is due to Shannon radius for Ni larger than that for Cr and/or increasing lattice volume caused by donated electrons[16]. Fig.2 (a) shows the normalized in-plane magnetic hysteresis

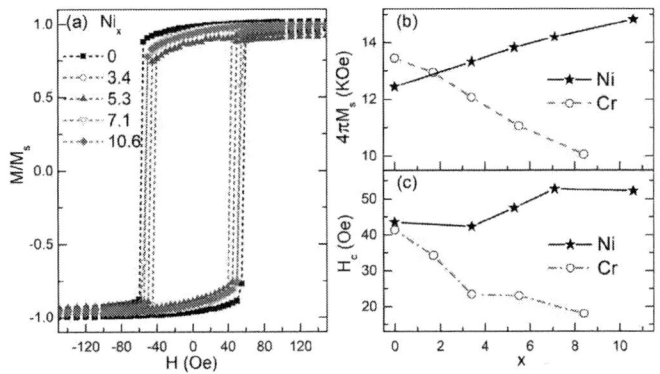

Fig. 2. (a) Normalized in-plane magnetic hysteresis loops of the oriented hcp-(CoIr)$_{100-x}$Ni$_x$ films with different x values. (b)-(c) x dependence of the determined and.

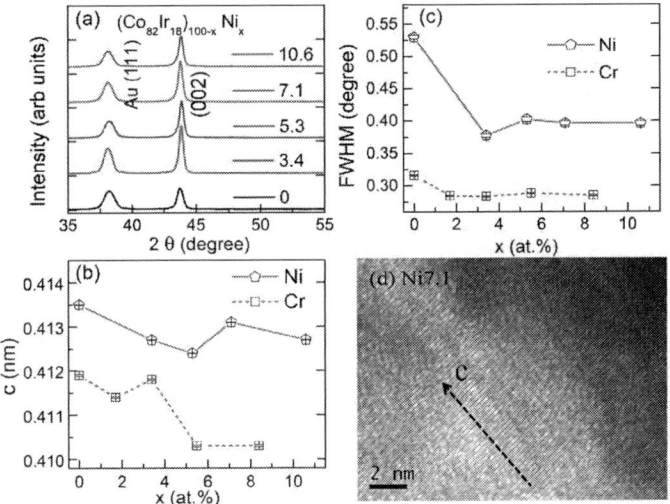

Fig. 1. (a)XRD patterns of the oriented hcp-(CoIr)$_{100-x}$Ni$_x$ films with indicated x values. (b-c) the c lattice parameter and FWHM plotted as a function of x. (d) show the high-resolution images of the Ni doped magnetic layers, the arrow indicate the c-axis direction.

consistent with the fact that the Ni$_x$ system still have well observed (002) peak at highest doping. As shown in Fig1. (d),

loops of our oriented hcp-(CoIr)$_{100-x}$Ni$_x$ films. The determined the $4\pi M_s$ and H_c are plotted in Fig.2 (b)-(c) as a function of Ni content x. With increasing x, $4\pi M_s$ exhibits a linear increasing behavior from 12.6 KOe to about 14.8 KOe at the highest doping level in the current work due to the magnetism of Ni doping. H_c has the smallest value at x=3.4 and the biggest value at x=7.1. Overall, however, it has a visible trend of increase with increasing x. We thought the increasing of H_c may caused by pinning effect of Ni dopants[17]. For Cr doping system, $4\pi M_s$ have an opposite trend with doping Ni system, which may be induced by nonmagnetic Cr dopants. The deeper causes are related to the introduction of 3d electron/hole which can cause changes in the electronic structure and the magnetic ordered structure[13]. Similarly to SrC and BaC system[18], electrons from Ni doping increase the magnetic moment, while holes from Cr doping decrease the magnetic moment. The smalle H_c can be caused by reduction of defects and/or internal stress resulted in Cr dopants.

Fig.3(a) presents the resonance field H_r determined from our ESR measurements for hcp-(CoIr)$_{100-x}$Ni$_x$ films as a function of φ_H. We have fitted the experimental data by the following equations[19]:

$$\left(\frac{\omega}{\gamma}\right)^2 = \left[H_\theta + H_u \cos^2 \varphi_0 + H_r \cos(\varphi_0 - \varphi_H)\right]$$
$$\cdot \left[H_u \cos(2\varphi_0) + H_r \cos(\varphi_0 - \varphi_H)\right], \quad (1)$$
$$2H_r \sin(\varphi_0 - \varphi_H) = H_u \sin(2\varphi_0)$$

, where φ_0 is the value of φ_M with the equilibrium positions, ω is the angular frequency about 18π, and γ is the gyromagnetic ratio. The determined in-plane and out-of-plane anisotropy field are plotted in Fig.3 (b)-(c) as a function of x. We can see from Fig.3 (b), H_u with minor level variation shows the highest value at x=3.4 and then decreases with further creasing x. As shown in Fig.3 (c), H_θ decreases monotonically with increasing x. According to the equation $H_\theta = 4\pi M_s - 2K_{grain}/M_s$, we may conclude that the decrease of H_θ is caused by the increase of the intrinsic magnetocrystalline anisotropy constant K_{grain} as shown in Fig.3 (d). Here, since the crystal quality does not change much with Ni doping, the observed change of K_{grain} suggests that the Ni doping might change the local crystal symmetry and/or spin-orbital coupling interaction that leads to the increase of the intrinsic magnetocrystalline anisotropy constant. As shown by our XRD results, the Cr doping shows some similarities to the Ni doping, so the similar doping dependencies of H_θ and K_{grain} can be seen in Fig.3 (c)-(d). Therefore, both

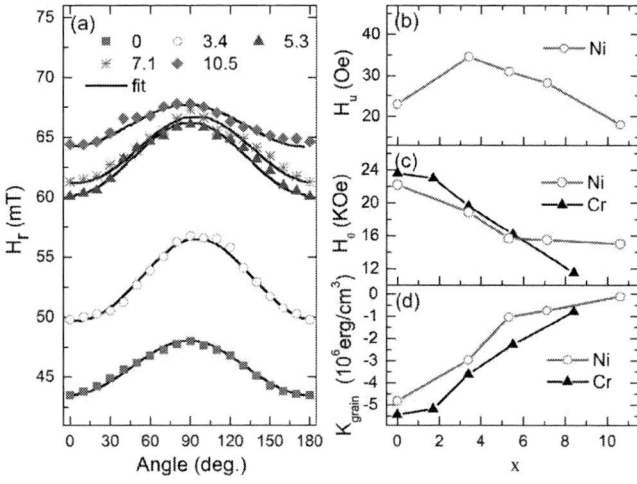

Fig. 3. (a) In-plane angle dependence of the resonance field for *hcp*-(CoIr)$_{100-x}$Ni$_x$ films with indicated x values. The black solid lines are theoretical fits to the experimental datum. (b) The determined effective in-plane anisotropy field, (c) the total effective out-of-plane anisotropy field and (d) the deduced as a function of doping content x.

electron doping and hole doping significantly may change the band structure of CoIr grains and thus regulate the intrinsic magnetocrystalline anisotropy.

In order to check their high frequency properties, the microwave permeability spectra of these films were measured by VNA using the shorted microstrip method with 160 Oe applied field, and are shown in Fig. 4 (a)-(e) as black lines.

The red solid lines are fitted by the Landau–Lifshitz–Gilbert (LLG) equation[20, 21]:

$$\mu = 1 + 4\pi M_s \gamma \frac{\omega_1 + i\alpha\omega}{\omega_1\omega_2 - \omega^2 + i\alpha\omega(\omega_1 + \omega_2)} \quad (2)$$

in which $\omega_1 = \gamma(4\pi M_s + H_{grain} + H_u) = \gamma(H_\theta + H_u)$, where ω is the angular frequency, α is the damping factor. As plotted in Fig. 4 (f) and (g), the and an be systematically changed as a function of doping Ni content. And the determined μ_i^{exp} and f_r^{exp} values match well with the values calculated by the equations $\mu_i = 1 + 4\pi M_s/H_u$ and $f_r = \frac{\gamma}{2\pi}\sqrt{H_u H_\theta}$[22]. The initial magnetic permeability increases with doping Ni content because of the increasing $4\pi M_s$ and nearly invariable H_u. The decreased resonance frequency from the maximum value of 5.7 GHz to 4.6 GHz is due to the reduced H_θ caused by the decreased K_{grain}. As shown in Fig.4(h), the product $(\mu_i - 1)f_r^2$ exhibits a first decreasing and then increasing trend for Ni doping films because of two competitive factors, namely, the decreasing K_{grain} and increasing $4\pi M_s$, while a decreasing trend for Cr doping films because of decreasing both K_{grain} and $4\pi M_s$. As shown in Fig. 4 (i), the damping factor α shows an upward trend for Ni doping films while decrease for Cr doping films, namely, $3d$ electron and hole have opposite effects on damping factor. It is well known that the damping factor consists of two parts contributions, namely, the intrinsic part and extrinsic part[23, 24]. Because the saturation magnetization is linearly increase as a function of x, the intrinsic part have an important effect on . For extrinsic part, the changed local crystal symmetry might cause internal stress which can result in the increase magnetic inhomogeneities.

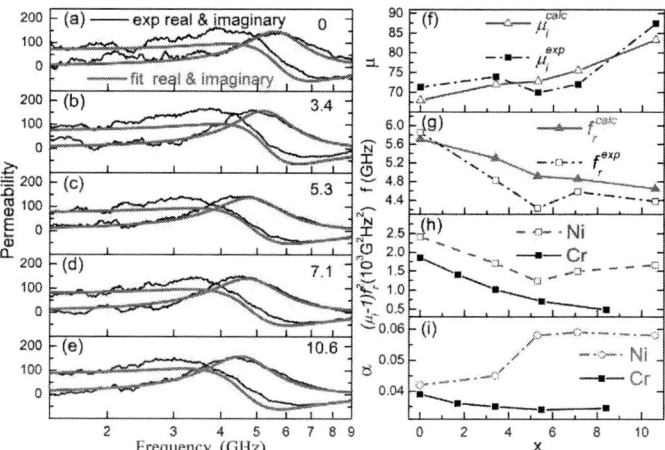

Fig. 4. (a)-(e)show the measured and fitted magnetic spectra of the oriented *hcp*-(CoIr)$_{100-x}$Ni$_x$ films samples with different compositions. (f)-(g) show the measured and calculated initial magnetic permeability and natural resonance frequency as a function of x. (h) In-plane anisotropy field, and (i) the damping factor as a function of doping content x.

In order to investigate the source of damping in depth, we have measured the permeability spectra with an applied field from 0 Oe to 160 Oe. The FWHM (Δf) were determined from the profile fit to the imaginary part of measured permeability spectra. The damping factor was calculated by the equation $\Delta f = (H_\theta + 2H_u + 2H_{app})\gamma \cdot 2\pi$[25, 26], and shown in Fig.5 (a)-(b). With the increasing applied field, the damping factor shows a decrease trend for $3d$ electron/hole doping films. This may be explained by the decrease magnetic inhomogeneity. Moreover, the decreasing trend is gradually diminishing, which suggests the proportion of extrinsic part decrease. Therefore, at the 160 Oe applied field, the intrinsic part plays a vital role in the increase damping factor with increasing x value. For Cr doping films, the damping factor α first increases and then decreases at the 160 Oe applied field. The discrepancy between Ni doping films and Cr doping films may be mainly caused by the fundamental properties, such as and the spin-orbit coupling effect, which is determined by the density states of $3d$ electron/hole near the Fermi surface[13, 14].

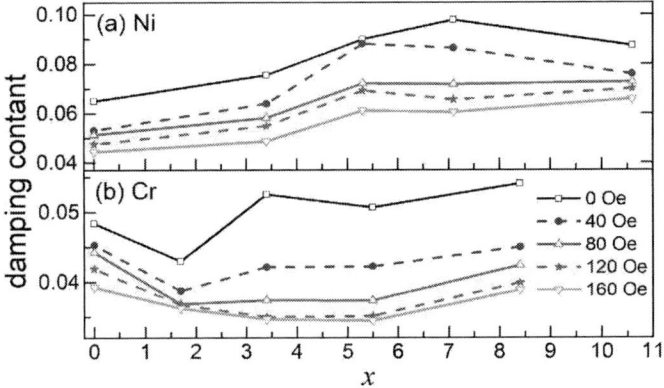

Fig. 5. the derived damping factor (a) for the Ni doping films and (b) for Cr doping films with applied magnetic field from 0 Oe to 160 Oe.

IV. CONCLUSION

We have investigated the magnetic properties of c-axis oriented hcp-$(CoIr)_{100-x}Ni_x$ SMFs. The third component of Ni doping can significantly change the saturation magnetization and the coercivity. The electrons from Ni doping increase the magnetic moment, while holes from Cr doping decrease the magnetic moment. And the change in coercivity is caused by pinning effect and defects/internal stress for Ni dopants and Cr dopants, respectively. The high frequency properties of resonance frequency and initial permeability are dependent on doping Ni content x. Moreover, we found that the damping factor increase with the increasing doping values for $3d$ electron doping while decrease for $3d$ hole doping, and an applied magnetic field has important effects on the damping factor of the Ni doping and Cr doping films. These results are expected to facilitate the soft magnetic properties of the thin films, enabling us to better meet the requirements of specific applications in the future.

REFERENCES

[1] H. Wu, S. Zhao, D. S. Gardner, H. Yu, "Improved high frequency response and quality factor of on-chip ferro-magnetic thin film inductors by laminating and patterning Co-Zr-Ta-B films." IEEE transactions on magnetics 49.7 (2013): 4176-4179.

[2] S. Itapu, D. G. Georgiev, V. Devabhaktuni, Itapu, Srikanth, Daniel G. Georgiev, and Vijay Devabhaktuni. "Improvement in inductance and Q-factor by laser microstructuring of ferromagnetic on-chip thin film inductors." Journal of Electromagnetic Waves and Applications 29.12 (2015): 1547-1556.

[3] A. Abdeldjebbar, A. Hamid, Y. Guettaf, R. Melati, "Design of micro-transformer in monolithic technology for high-frequency flyback-type converters." Electrical Engineering 100.4 (2018): 2589-2601.

[4] Phuoc, Nguyen N., Feng Xu, and C. K. Ong. "Ultra-wideband microwave noise filter: Hybrid antiferromagnet/ferromagnet exchange-coupled multilayers." Applied Physics Letters 94.9 (2009).

[5] J. Lu, H. Chi, X. Zhang, L. Shen, "Noise reduction using photonic microwave filter for radio over fiber system." Microwave and Optical Technology Letters 48.2 (2006): 305-307.

[6] Hashimoto, Atsushi, Shin Saito, and Migaku Takahashi. "A soft magnetic underlayer with negative uniaxial magnetocrystalline anisotropy for suppression of spike noise and wide adjacent track erasure in perpendicular recording media." Journal of applied physics 99.8 (2006).

[7] F. Xu, T. Wang, T. Ma, Y. Wang, S. Zhu, F. Li, "Enhanced film thickness for Néel wall in soft magnetic film by introducing strong magnetocrystalline anisotropy." Scientific reports 6.1 (2016): 20140.

[8] T. Wang, Y. Wang, G. Tan, F. Li, S. Ishio, "Microwave magnetic properties of the oriented CoIr soft magnetic film with negative magnetocrystalline anisotropy." Physica B: Condensed Matter 417 (2013): 24-27.

[9] T. Y. Ma, J. Y. Jiao, L. Qiao, T. Wang, F. S. Li, "Micrometer thick soft magnetic films with magnetic moments restricted strictly in plane by negative magnetocrystalline anisotropy." Journal of Magnetism and Magnetic Materials 444 (2017): 119-124.

[10] A. Devonport, A. Vishina, R. K. Singh, M. Edwards, K. Zheng, J. Domenico, "Magnetic properties of chromium-doped $Ni_{80}Fe_{20}$ thin films." Journal of Magnetism and Magnetic Materials 460 (2018): 193-202.

[11] Z. Li, F. Wang, C. Zhao, Y. Liao, M. Gao, H. Zhang, "Microstructural evolution and enhanced magnetic properties of $FeCoNiZr_x$ medium entropy alloy films." Journal of Alloys and Compounds 971 (2024): 172649.

[12] M. Choi, Y. K. Hong, H. Won, C. D. Yeo, N. M. Shah, B. C. Choi, "Tuning the magnetocrystalline anisotropy of rare-earth free $L1_0$-ordered $Mn_{1-x}TM_xAl$ magnetic alloy (TM= Fe, Co, or Ni) with transition elements." Journal of Magnetism and Magnetic Materials 589 (2024): 171513.

[13] M. Schoen, D. Thonig, M. Schneider, "Ultra-low magnetic damping of a metallic ferromagnet." Nature Physics 12.9 (2016): 839-842.

[14] S. J. Xu, J. Y. Shi, Y. S. Hou, "Tuning of the intrinsic

979-8-3315-2209-4/25 $31.00 © 2025 IEEE

magnetic damping parameter in epitaxial CoNi (001) films: Role of the band-filling effect." Physical Review B 100.2 (2019): 024403.

[15] T. Y. Ma, J. Y. Jiao, Z. W. Li, L. Qiao, T. Wang, F. S. Li, "Tuning the static and dynamic magnetic properties of c-axis oriented hcp-(CoIr) thin films by the addition of Cr." Applied Surface Science 457 (2018): 598-603.

[16] Shannon, Robert D. "Revised effective ionic radii and systematic studies of interatomic distances in halides and chalcogenides." Foundations of Crystallography 32.5 (1976): 751-767.

[17] Y. K. Takahashi, T. O. Seki, K. Hono, T. Shima, K. Takanashi, "Microstructure and magnetic properties of FePt and Fe/FePt polycrystalline films with high coercivity." Journal of applied physics 96.1 (2004): 475-481.

[18] S. J. Dong, First-principles study on several novel functional magnetic materials, Master's thesis, Tianjin Normal University.

[19] T. Ma, F. Xu, J. Jiao, Y. Wang, T. Wang, "Adjustable high-frequency magnetic properties of oriented $Co_{80}Ir_{20}$ soft magnetic thin films with strong negative magnetocrystalline anisotropy." Applied Physics A 122 (2016): 1-5.

[20] Phuoc, Nguyen N., Feng Xu, and C. K. Ong. "Tuning magnetization dynamic properties of Fe–SiO$_2$ multilayers by oblique deposition." Journal of Applied Physics 105.11 (2009).

[21] T. Wang, S. Zhang, F. Xu, X. Ma, J. Zhang, F. Li, "The improvement of high-frequency magnetic properties in oriented hcp-$Co_{78}Ir_{22}$ soft magnetic films fabricated at high substrate temperature." Journal of Magnetism and Magnetic Materials 406 (2016): 118-122.

[22] X. De-Sheng, L. Fa-Shen, F. Xiao-Long, W. Fu-Sheng, "Bianisotropy picture of higher permeability at higher frequencies." Chinese Physics Letters 25.11 (2008): 4120.

[23] X. Guo, L. Xi, Y. Li, X. Han, D. Li, Z. Wang, Y. Zuo, "Reduction of magnetic damping constant of FeCo films by rare-earth Gd doping." Applied Physics Letters 105.7 (2014).

[24] Seemann, K., H. Leiste, and Ch Klever. "On the relation between the effective ferromagnetic resonance linewidth feff and damping parameter eff in ferromagnetic Fe–Co–Hf–N nanocomposite films." Journal of magnetism and magnetic materials 321.19 (2009): 3149-3154.

[25] Y. Yu, Q. Zhan, J. Wei, J. Wang, G. Dai, Z. Zuo, "Static and high frequency magnetic properties of FeGa thin films deposited on convex flexible substrates." Applied physics letters 106.16 (2015).

[26] J. B. Youssef, N. Vukadinovic, D. Billet, M. Labrune, "Thickness-dependent magnetic excitations in Permalloy films with nonuniform magnetization." Physical Review B 69.17 (2004): 174402.

Investigation on the static and dynamic magnetic properties of hcp-$(CoIr)_{100-x}M_x$ (M=B, SiO$_2$) thin films

1st Tianyong Ma   2nd Qi Liu   3rd Ao Han   4th Zisheng Li   5th Sha Zhang

Abstract—**In this study, though B and SiO$_2$ doping, we present systematic investigations of both static and dynamic magnetic properties of c-axis oriented hcp-CoIr soft magnetic thin films with easy-plane magnetocrystalline anisotropy. It is found that the coercivity could be well optimized by the B and SiO$_2$ doping. The minimum value is about 46% and 73% of its original value for B$_x$ system and SiO$_x$ system, respectively. The high-frequency magnetic properties and the extracted damping constant can also be controlled by using the doping method. We have found that the optimal concentration is 3 and 5.2 for the B and SiO$_2$ dopants, respectively. These findings shall help to tune the soft magnetic properties of thin films to meet specific applications in the future.**

Index Terms—**magnetic thin film, easy-plane magnetocrystalline anisotropy, dynamic properties**

I. INTRODUCTION

Soft magnetic films (SMFs) have great applications in the high-frequency devices, such as film inductors[1, 2], microtransformers[3] and microwave noise filters[4, 5], due to their high initial permeability μ_i and natural resonance frequency f_r . Practically, textured SMFs with easy-plane magnetocrystalline anisotropy present significant advantages: 1) at micrometer thickness, still possess the in-plane domains and excellent high-frequency magnetic properties, which can overcome the problem of insufficient magnetic flux signal for traditional SMFs, such as micron oriented hcp-CoIr SMFs[6]; 2) can extend Acher limit between μ_i and f_r so that the product $(\mu_i - 1)f_r^2$ has been improved several times[7]; 3) own the higher critical thickness for the Néel type domain wall changing into Bloch type, which can reduce the signal-to-noise ratio of magnetic recording media[8]. However, these films still have coercivities H_c in the range of several militesla which are too large for their application in actual devices. Moreover, damping also needs to be regulated to achieve practical applications. For the traditional Fe-/Co-based films, magnetic properties (such as H_c, damping constant, μ_i and f_r) were frequently tuned by adding B[9, 10], SiO$_2$[11, 12] or Fe/Co-oxides[13] into the thin film which optimizes the exchange coupling interaction between different crystallite grains. For c-axis oriented hcp-(CoIr) SMFs, the effect of

All authors are with *The Key Laboratory of Physics and Photoelectric Information Functional Materials, North Minzu University*, Yinchuan 750021, China. This work is supported by the research start-up fund of China (Nos.2021KYQD22). Corresponding author: (Author 1, email: tianyongma@nmu.edu.cn)

B and SiO$_2$ dopants on the soft magnetic property and the damping have not been systematically studied. Moreover, the doped elements or oxides may reduce the easy-plane magnetocrystalline anisotropy, as it is possible for these dopants to enter into the lattice or induce amorphous at higher doping levels. Therefore, the concentration range of B and SiO$_2$ dopants and optimal doping concentration also need to be systematically discussed. In this work, by employing the B and SiO$_2$ dopants, we systematically investigate the soft and high-frenquency magnetic properties of the hcp-(CoIr)$_{100-x}$M$_x$ films. We have found that dopants at different positions have very different effects in changing the magnetic properties of our films. The minimum coercivity is about 46% and 73% of its original value for B$_x$ system and SiO$_x$ system, respectively. The negative effective magnetocrystalline anisotropy changes significantly from -5.46×10^6 erg/cm^3 to almost zero. Moreover, the damping constant can be tuned as a function of doping level. These results will have a benefit for tuning the the soft magnetic properties of our thin films to meet different applications.

II. EXPERIENCE

Prepared by DC magnetron sputtering technique, these SMFs have a layered structure of substrate/Ti/Au/(CoIr)$_{100-x}$M$_x$ (M=B, SiO$_2$). Si wafer with (100) surface orientation was used as substrate, then an amorphous layer of Ti(3 nm) was deposited to provide a flat surface which also improves the adhesiveness of the Au seed-layer. Above the Ti layer, the Au(10 nm) seed-layer was deposited with its [111]-direction being perpendicular to the film plane, which will induce the c-axis orientation of the hcp-(CoIr)$_{100-x}$M$_x$ soft magnetic layer. For the growth of the seed layer, pure Ar gas was used with a pressure of 0.25 Pa. For the magnetic layer, 0.3 Pa Ar gas was used as the sputtering atmosphere. The magnetic layer was deposited by using a Co target with several Ir chips on top of it together with a second target of B/SiO$_2$. The content of B/SiO$_2$ can be changed by changing the sputtering gun power on the B/SiO$_2$ target while maintaining the gun power on the Co target with a constant value of 60 W. Before any detailed discussion on the structural and magnetic properties, for simplicity, we would like to summarize all the investigated films in the following sections with their composition and naming method in Table1. The Chemical compostion have

979-8-3315-2209-4/25 $31.00 © 2025 IEEE

been characterized by using energy dispersive spectrometer (EDS). Microstructure and grain morphology of the films were measured using transmission electron microscope (TEM). Crystal structure of our samples were characterized by x-ray diffraction (XRD) with Curadiation K_{α_1}. Static and dynamic magnetic properties were measured with a vibrating sample magnetometer (VSM) and an electron spin resonance spectrometer (ESR), respectively. The microwave permeability measurements were performed by a vector network analyzer (VNA).

TABLE I

TABLE FOR hcp-(CoIr)$_{100-x}$M$_x$ SAMPLE SYSTEMS STUDIED IN THIS WORK. THE COIR COMPOSITION, THE X VALUES FOR M=B, SiO$_2$ AND THE NAMING METHOD USED IN THE CURRENT WORK FOR THESE SAMPLES.

CoIr	M/x	Naming method
Co$_{79}$Ir$_{21}$	B/x = 0, 3, 6.5, 8.7, 10.8, 13	B$_x$
Co$_{79}$Ir$_{21}$	SiO$_2$/x = 0, 1.7, 3.5, 5.2, 7.0, 8.7, 10.5	SiO$_x$

III. RESULTS AND DISCUSSION

Fig. 1(a) and Fig. 1(b) show the XRD patterns of the oriented hcp-(CoIr)$_{100-x}$B$_x$ and hcp-(CoIr)$_{100-x}$(SiO$_2$)$_x$ films, respectively. Similar to our earlier work[6], two diffraction peaks corresponding to Au (111)-plane and hcp-(CoIr)$_{100-x}$B$_x$ (002)-plane can be seen. The (002) peak of the magnetic layer first became stronger and sharper with increasing B content, and then the peak got broader and weaker with further increasing x and eventually vanishes at x=13. Similarly, the observed two peaks also are Au (111) peak and (CoIr)$_{100-x}$(SiO$_2$)$_x$ (002) peak, respectively. The (002) peak became less intense and more broader with increasing SiO$_2$ content x. The determined c lattice constant from the (002) peak and grain size are plotted with respect to x in Fig. 1(c) and Fig. 1(d), respectively. For B$_x$ system, with increasing x value, the c lattice parameter increases from 0.4135 nm to 0.4166 nm. Considering that the shannon radius of B is much smaller than that of Co and Ir[14], it is very likely that it will segregate at the grain boundaries at lower B doping levels and then it may reside at the interstitial site of the hcp-(CoIr) crystal structure.

This agrees well with the initial slow increase of the c lattice constant at lower doping levels and then faster increase at higher doping levels, which is also similar to the behavior of doping B into FeCo films[9]. These results suggest that B dopants can lead to a refined crystallite size as shown in Fig. 1(d). For SiO$_x$ system, below x=3.5, c decreases a little bit with increasing x. Above this doping level, c increases with increasing x monotonically. For grain size shown in Fig. 1(d), it first increases and then decreases again with increasing x, and have the biggest value around x=3.5~5.2. These results suggest that SiO$_2$ first resides at the grain boundaries. Then, with further increasing SiO$_2$, the Si/O atoms may enter into the interstitial site which causes the expansion of the crystal lattice and eventually destabilizes the crystal structure.

Fig. 2(a) and Fig. 2(b) show the high magnification image of the magnetic layer for the SiO$_{8.7}$ film and B$_{8.7}$ film,

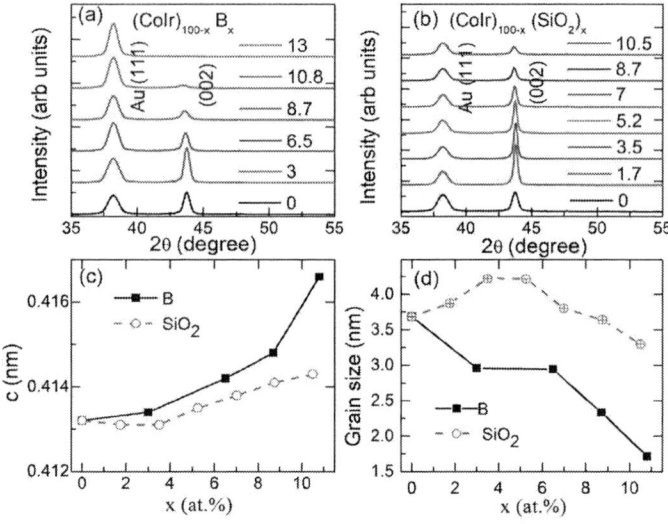

Fig. 1. (a-b) XRD patterns of the oriented hcp-(CoIr)$_{100-x}$B$_x$ films and hcp-(CoIr)$_{100-x}$(SiO$_2$)$_x$ films, respectively. (c-d) The c lattice parameter and grain size plotted as a function of x, respectively.

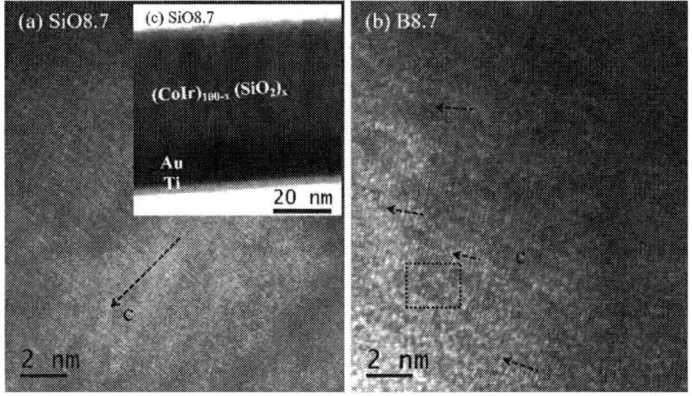

Fig. 2. (a-b) Images of the morphology of the SiO$_{8.7}$ film and B$_{8.7}$ film, respectively. (c) the layered structure of our hcp-SiO$_{8.7}$ film insetting in (a).

respectively. Inset in (a) shows the total cross section of the layered structure for our films. One can be seen that a high degree of crystallinity and orientation of the c-plane for the SiO$_{8.7}$ film, while the mixing of crystal and amorphous regions for B$_{8.7}$ film. Consistent with the above analysis, doped elements may resides at the grain boundaries, or/and enter into the interstitial site, or/and replace Co atoms in CoIr phase.

Fig. 3(a) and (b) show the normalized in-plane hysteresis loops for the oriented hcp-(CoIr)$_{100-x}$B$_x$ and hcp-(CoIr)$_{100-x}$(SiO$_2$)$_x$ thin films, respectively. The determined are plotted in Fig. 3(c) as a function of content x, and plotted in Fig. 3(d). For Bx system, with increasing x, the $4\pi M_s$ slowly decreases to about 88% of the original value at x=13. For H_c, it decreases much faster with increasing x, about 46% of its original value at a doping level of x=8.7. With further increasing x, starts to increase again. This first decrease behavior may be caused by the refined crystallite size[15], and then increase may be arise from both the defects and/or

979-8-3315-2209-4/25 $31.00 © 2025 IEEE

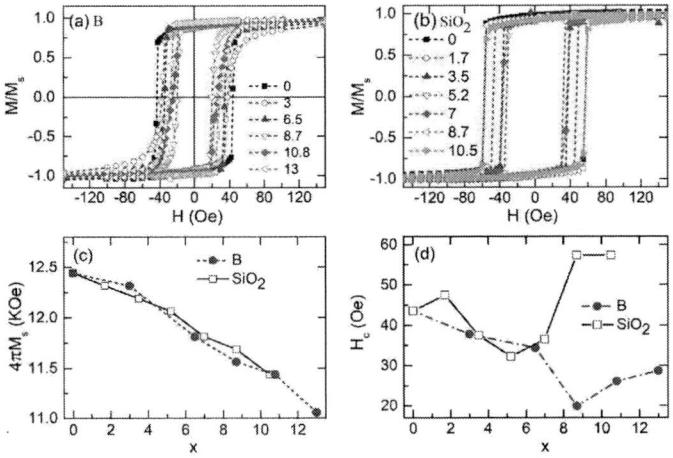

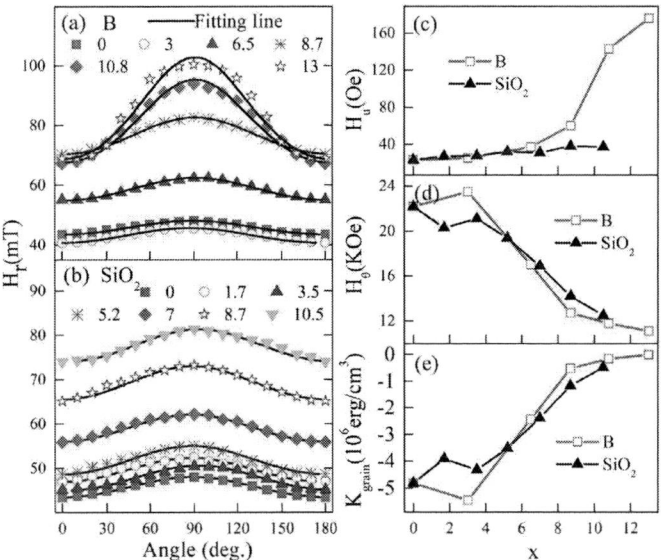

Fig. 3. (a) Normalized in-plane magnetic hysteresis loops of the oriented hcp-$(CoIr)_{100-x}B_x$ films with different x values. (b) Normalized in-plane magnetic hysteresis loops for the oriented hcp-$(CoIr)_{100-x}(SiO_2)_x$ films. (c) x dependence of the determined $4\pi M_s$. (d) x dependence of the determined H_c.

Fig. 4. (a-b) In-plane angle dependence of the resonance field H_r for hcp-$(CoIr)_{100-x}B_x$ films and hcp-$(CoIr)_{100-x}(SiO_2)_x$ films, respectively. The black solid lines are theoretical fits to the experimental data. (c) The determined effective in-plane anisotropy field H_u, (d) the total effective out-of-plane anisotropy field H_θ and (e) the deduced magnetocrystalline anisotropy constant K_{grain} as a function of doping content x.

internal stress and the miss alignment of small grains[16, 17]. For SiO_x system, one can see that $4\pi M_s$ decreases linearly with increasing SiO_2 content, which is similar to the case of the B_x system. With increasing SiO_2 content x, H_c first decreases and then increases again. Different from the B_x system, the decrease of H_c may be caused by the decreased defects and/or internal stress because the SiO_2 dopants of a small amount at the grain boundaries improves the film quality and/or optimizes the inter-gain coupling interaction[18]. With further increasing SiO_2, the increase may be caused by the increased defects and/or internal stress arising from the interstitial or/and replacing Si/O atoms.

Fig. 4(a-b) shows the angle dependence of resonance field from our ESR measurements. Theoretical fits were performed by following equation (1) as shown by the black solid curve[6, 19].

$$
\begin{aligned}
\left(\frac{\omega}{\gamma}\right)^2 &= \left[H_\theta + H_u \cos^2\varphi_0 + H_r \cos(\varphi_0 - \varphi_H)\right] \\
&\cdot \left[H_u \cos(2\varphi_0) + H_r \cos(\varphi_0 - \varphi_H)\right], \\
2H_r &\sin(\varphi_0 - \varphi_H) = H_u \sin(2\varphi_0)
\end{aligned}
\tag{1}
$$

where ω is the value of φ_M with the equilibrium positions, ω is the angular frequency about 18π, and γ is the gyromagnetic ratio. The determined in-plane and out-of-plane anisotropy field H_θ were plotted in Fig. 4(c) and (d) as a function of x, respectively. The calculated intrinsic $K_{grain} = \left(4\pi M_s - H_\theta\right)M_s/2$ is shown in Fig. 4(e). For Bx system, below x=8.7, the increase of H_u is small, while above x=8.7, the increase of H_u is very fast because of the very strong strain and the misalignment of small grain. As shown in Fig. 4(d), has the biggest value at x=3, then decreases with increasing x which can be understood by the reduced crystal quality. The very fast decrease of together with the slow decrease of the crystal quality suggest that higher B content may also cause

the breaking down of the c-plane orientation of our film. Correspondingly, the deduced magnetocrystalline anisotropy constant has the smallest value at x=3 and then increases very fast and saturate to the value of about zero in the amorphous state. For SiO_x system, as shown in Fig. 4(c)-(e), that is the increase of H_u with x can be understood by the stronger effect of the defects and/or internal stress induced by the interstitial site Si/O atoms. H_θ and K_{grain} are mostly determined by the film quality and the c-axis orientation of the film. Therefore, the decrease of H_θ and increase of K_{grain} suggest that the film quality and the degree of the c-axis orientation decrease with increasing doping level of the SiO_2 doped into our films. To verify the potential applicability, we measured the microwave permeability spectra of all films by VNA with 160 Oe applied field. The complex permeability for the B_x films and SiO_x films are shown in Fig. 5(a)-(f) and shown in Fig. 6(a)-(f), respectively. For Bx system, as plotted in Fig. 5(g) and Fig. 5(h), the determined and values from the measured spectra match well with the calculated values using the extended Kittel equation $\mu_i = 1 + 4\pi M_s/H_u$ and $f_r = \frac{\gamma}{2\pi}\sqrt{H_u H_\theta}$ [20]. The initial magnetic permeability decreases with the increase of B content, which is caused by the decrease and the increase . The resonance frequency decreases from the value of 5.71 GHz to 4.61 GHz , and then increases to 5.95 GHz. The decrease of f_r can be understood due to the reduced K_{grain} and the increase may be caused by the increased H_u . For SiO_x system, the determined μ_i^{exp} and f_r^{exp} values are plotted in Fig. 6(g) and Fig. 6(h), and match well with the calculated values. The similar doping dependency of μ_i for the B_x films and SiO_x films can be expected due to the

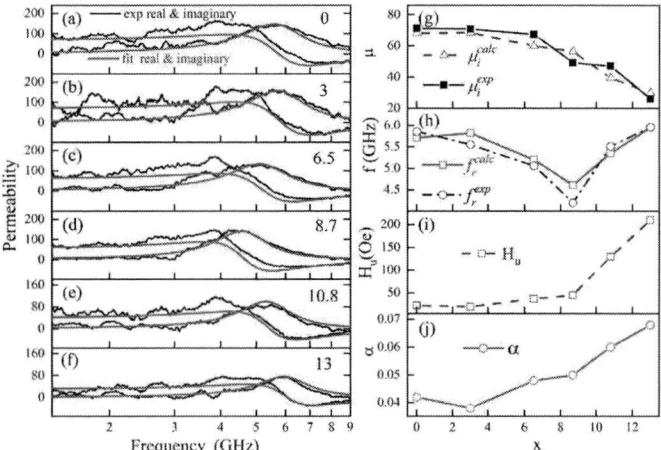

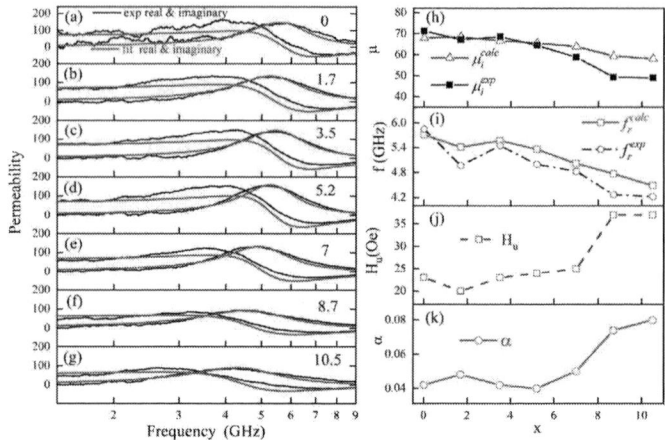

Fig. 5. (a)-(f) Permeability spectra of hcp-$(CoIr)_{100-x}B_x$ films with indicated x values. The red solid lines are theoretical fits to the experimental data. The initial permeability μ_i and natural resonance frequency f_r determined from the measured spectra are plotted as a function of x in panel (g) and (h), respectively. The calculated values using the determined $4\pi M_s$ and anisotropy constant from the VSM and ESR measurements are also shown for comparison. (i) The determined effective in-plane anisotropy field H_u, (j) The determined effective damping coefficient α .

Fig. 6. (a-g) Permeability spectra of hcp-$(CoIr)_{100-x}(SiO_2)_x$ films. (h)The initial permeability μ_i as a function of x, (i) natural resonance frequency f_r, (j) The determined effective in-plane anisotropy field H_u, (k) The determined effective damping coefficient .

same tendency of $4\pi M_s$ shown in Fig. 3(c) and H_u shown in Fig. 4(c). As the SiO_2 content increases, the resonance frequency shifts gradually towards lower values, which can be explained with the reduced K_{grain} . The fitting curve of microwave permeability spectra is represented by the red solid line in Fig. 5(a)-(f) and Fig. 6(a)-(f) according to the Landau–Lifshitz–Gilbert (LLG) equation[21, 22]:

$$\mu = 1 + 4\pi M_s \gamma \frac{\omega_1 + i\alpha\omega}{\omega_1\omega_2 - \omega^2 + i\alpha\omega(\omega_1 + \omega_2)} \quad (2)$$

in which $\omega_1 = \gamma(4\pi M_s + H_{grain} + H_u) = \gamma(H_\theta + H_u)$, $\omega_2 = \gamma(H_u)$, where ω is the angular frequency, α is the damping constant and H_u is the in-plane anisotropy field. One can see that the fitting curves are in high agreement with the experimental data. For B_x system, the extracted in-plane anisotropy field have the same tendency with shown in Fig. 4(c). As shown in Fig. 5(j), the damping coefficient α first decreases and then changes towards higher values, which is the result of two factors[23, 24]. One is the intrinsic part affected by saturation magnetization or spin orbit coupling, this part has largely influence on α because saturation magnetization is linearly decrease with doping level.

Another is the extrinsic part related to magnetic inhomogeneity caused by the defects and/or internal stress and misalignment of grains. For SiO_x system, the similar doping dependency of H_u can be seen as shown in Fig. 4(c). The damping coefficient plotted in Fig. 6(j) first fluctuates at low doping levels and then increases at high doping levels. Similar to the situation found for the B_x system, the damping constant can be caused either by the intrinsic contributions and/or extrinsic contributions.

IV. CONCLUSION

In summary, we have investigated both the static and dynamic magnetic properties of c-axis oriented hcp-$(CoIr)_{100-x}M_x$ (M=B, SiO_2) soft magnetic thin films. As expected, the B and SiO_2 dopants first reside at the grain boundaries and then, at higher doping levels, enter into the interstitial site of the hcp-$(CoIr)$ crystal structure. The change in coercivity is the result of multiple factors, mainly including decrease of the crystallite size, defects and/or internal stress caused by the interstitial dopants, and grain orientation. Especially, the intrinsic magnetocrystalline anisotropy constant can be changed from very large negative values of -5.46×10^6 erg/cm^3 to almost zero. Moreover, the high-frequency magnetic properties and the extracted damping constant can be controlled by using the doping method. We have found that the optimal concentration is 3 and 5.2 for the B and SiO_2 dopants, respectively. Interestingly, all these dopants can greatly change the magnetic properties, which is very important for potential applications of our SMFs.

REFERENCES

[1] J. Lou, D. Reed, M. Liu, N. X. Sun, "Electrostatically tunable magnetoelectric inductors with large inductance tunability." Applied Physics Letters 94.11 (2009).

[2] G. Liu, X. Cui, S. Dong, "A tunable ring-type magnetoelectric inductor." Journal of Applied Physics 108.9 (2010).

[3] H. Zhang, Y. Liu, Z. Zhong, "An improved microchip thin film transformer formed by vacuum evaporation and sputtering." Vacuum 62.1 (2001): 1-6.

[4] Phuoc, Nguyen N., Feng Xu, and C. K. Ong. "Ultrawideband microwave noise filter: Hybrid antiferromagnet/ferromagnet exchange-coupled multilayers." Applied Physics Letters 94.9 (2009).

[5] M. Yamaguchi, Y. Miyazawa, K. Kaminishi, H. Kikuchi, S. Yabukami, K. Arai, T. Suzuki, "Soft magnetic ap-

plications in the RF range." Journal of magnetism and magnetic materials 268.1-2 (2004): 170-177.

[6] T. Y. Ma, J. Y. Jiao, L. Qiao, T. Wang, F. S. Li, "Micrometer thick soft magnetic films with magnetic moments restricted strictly in plane by negative magnetocrystalline anisotropy." Journal of Magnetism and Magnetic Materials 444 (2017): 119-124.

[7] T. Ma, F. Xu, J. Jiao, Y. Wang, T. Wang, "Adjustable high-frequency magnetic properties of oriented $Co_{80}Ir_{20}$ soft magnetic thin films with strong negative magnetocrystalline anisotropy." Applied Physics A 122 (2016): 1-5.

[8] F. Xu, T. Wang, T. Ma, Y. Wang, S. Zhu, F. Li, "Enhanced film thickness for Néel wall in soft magnetic film by introducing strong magnetocrystalline anisotropy." Scientific reports 6.1 (2016): 20140.

[9] I. Kim, J. Kim, K. H. Kim, M. Yamaguchi, "Effects of boron contents on magnetic properties of Fe-Co-B thin films." IEEE transactions on magnetics 40.4 (2004): 2706-2708.

[10] Munakata, Makoto, Shin-Ichi Aoqui, and Masaaki Yagi. "B-concentration dependence on anisotropy field of CoFeB thin film for gigahertz frequency use." IEEE transactions on magnetics 41.10 (2005): 3262-3264.

[11] S. Liu, Y. Ma, L. Chang, G. Li, J. Wang, Q. Wang, "Effect of doping SiO_2 and applying high magnetic field during the film growth on structure and magnetic properties of evaporated Fe films." Thin Solid Films 651 (2018): 1-6.

[12] D. Yao, X. Zhou, H. Zuo, B. Zhang, "Fabrication, magnetism and high frequency application of exchange-coupled $Fe_{65}Co_{35\pm2} - SiO_{1.7\pm0.2}$ granular films." Applied surface science 254.8 (2008): 2556-2561.

[13] W. Wang, Y. Chen, G. H. Yue, K. Sumiyama, T. Hihara, D. L. Peng, "Magnetic softness and high-frequency characteristics of $Fe_{65}Co_{35}O$ alloy films." Journal of Applied Physics 106.1 (2009).

[14] Shannon, Robert D. "Revised effective ionic radii and systematic studies of interatomic distances in halides and chalcogenides." Foundations of Crystallography 32.5 (1976): 751-767.

[15] Herzer, G. "Grain size dependence of coercivity and permeability in nanocrystalline ferromagnets." IEEE Transactions on magnetics 26.5 (1990): 1397-1402.

[16] Morris. R. A, Inaba. Y, Harrell. J. W and Thompson. G. B, "Influence of underlayers on the c-axis distribution in $Co_{80}Pt_{20}$ thin films." Thin solid films 518.17 (2010): 4970-4976.

[17] Feng E X, Wang Z K, Du H W, Wei J W, Cao D R, Liu Q F and Wang J B, "Electrodeposition of FeCoCd films with in-plane uniaxial magnetic anisotropy for microwave applications." Journal of Applied Physics 115.17 (2014).

[18] C. Wang, Y. Zhang, P. Zhang, Y. Rong, and T. Y. Hsu, "Influence of annealing on microstructure and magnetic-transport of $FeCo–SiO_2$ nanogranular films." Journal of magnetism and magnetic materials 320.5 (2008): 683-690.

[19] Koji S. Nakayama, Tomoaki Chiba, Susumu Tsukimoto, Yoshihiko Yokoyama, Toshiyuki Shima, Shin Yabukami, "Ferromagnetic resonance in soft-magnetic metallic glass nanowire and microwire." Applied Physics Letters 105.20 (2014).

[20] Shujuan Yuan, Baojuan Kang, Liming Yu, Shixun Cao, Xinluo Zhao, "Increased ferromagnetic resonance linewidth and exchange anisotropy in NiFe/FeMn bilayers." Journal of Applied Physics 105.6 (2009).

[21] Phuoc, Nguyen N., Feng Xu, and C. K. Ong. "Tuning magnetization dynamic properties of Fe–SiO_2 multilayers by oblique deposition." Journal of Applied Physics 105.11 (2009).

[22] S. Ge, S. Yao, M. Yamaguchi, X. Yang, H. Zuo, T. Ishii, D. Zhou, F. Li, "Microstructure and magnetism of FeCo–SiO_2 nano-granular films for high frequency application." Journal of Physics D: Applied Physics 40.12 (2007): 3660.

[23] Kamberský, V. "On the Landau–Lifshitz relaxation in ferromagnetic metals." Canadian Journal of Physics 48.24 (1970): 2906-2911.

[24] Arias, Rodrigo, and D. L. Mills. "Extrinsic contributions to the ferromagnetic resonance response of ultrathin films." Physical review B 60.10 (1999): 7395.

High-performance Piezoresistive Pressure Sensor Based on MEMS Technology

Zhikang Lan   Weiping Li   Tongqing Liu   Yizhou Ye   Xiaodong Huang*

Abstract—**Piezoresistive pressure sensors based on MEMS technology are widely used in various applications. However, when it comes to practical applications, a major limitation of these sensors is their sensitivity to temperature variations, which causes significant temperature drift and impacts their measurement accuracy and reliability. To address this issue, this paper proposes a novel approach to mitigate the temperature drift of piezoresistive pressure sensors. This is achieved by integrating a temperature sensor directly onto the sensor chip, coupled with software-based compensation algorithms. The integrated system effectively compensates for temperature-induced variations in the piezoresistive material, resulting in a more stable and accurate pressure measurement. The designed sensor chip can be fabricated using standard MEMS processes, ensuring minimal impact on production complexity and cost. This solution provides a practical and scalable method to enhance the performance of MEMS-based piezoresistive pressure sensors, making it highly suitable for applications that demand high temperature stability and precision.**

Index Terms—**MEMS, piezoresistive sensor, pressure sensor, mechanical sensor**

I. INTRODUCTION

With the rapid progress of silicon-based MEMS microfabrication techniques, a wide range of MEMS sensors have been successfully commercialized [1]–[3]. Among these, MEMS pressure sensors stand out as one of the most prominent categories, thanks to their outstanding performance, affordability, and ease of mass production [4]–[8]. These sensors have become essential components in many fields, including automotive systems, industrial automation, environmental monitoring, and biomedical applications [9]. Specifically, piezoresistive pressure sensors are highly regarded within the MEMS pressure sensor family due to their straightforward structure, ease of manufacturing, and high sensitivity. These sensors function by detecting variations in the resistance of a piezoresistive material when exposed to pressure, offering a cost-effective and dependable solution for diverse applications [10], [11].

However, despite their many benefits, piezoresistive pressure sensors encounter certain limitations that can impact their performance in real-world scenarios. A major issue is

Zhikang Lan and Xiaodong Huang are with *Key Laboratory of MEMS of the Ministry of Education, Southeast University*, Nanjing, China. Weiping Li is with *Nanjing Gaohua Technology Co., Ltd*, Nanjing, China. Tongqing Liu is with *Wuxi Sencoch Technology Co., Ltd.*, Wuxi, China. Yizhou Ye is with *Key Laboratory for Optoelectronic Technology and Systems of Ministry of Education, Chongqing University*, Chongqing, China. Corresponding author: (Xiaodong Huang, email: xdhuang@seu.edu.cn)

their sensitivity to temperature changes, which can result in significant measurement errors. Fluctuations in temperature can alter the resistance of the piezoresistive material, causing inaccuracies in pressure readings and undermining the sensor's long-term stability and precision. While various temperature compensation methods have been suggested, such as algorithm-based adjustments or the inclusion of external temperature sensors, these solutions often introduce additional complexity and can reduce the real-time response efficiency of the sensors [12]–[14].

In response to this challenge, this paper presents a simple design for piezoresistive pressure sensors that enhances their resistance to temperature variations. A key feature of this design is the integration of a platinum resistor on the chip surface, which facilitates precise, real-time temperature monitoring. This enables dynamic compensation for temperature-induced effects, significantly improving both the reliability and long-term stability of the sensor.

II. PRINCIPLE AND DESIGN

As depicted in Fig. 1(a), the schematic diagram of the proposed MEMS piezoresistive pressure sensor highlights its fundamental design, which includes a suspended membrane integrated with four piezoresistors. These piezoresistors are arranged in a Wheatstone bridge configuration, as shown in Fig. 1(b), a common setup in pressure sensing applications for its high sensitivity and precision in detecting small resistance changes [15], [16]. In the absence of external pressure, the four piezoresistors within the sensor remain balanced, with equal resistance across each one. As a result, the Wheatstone bridge output voltage is zero or at its baseline value. When an external pressure is applied, it causes the suspended membrane

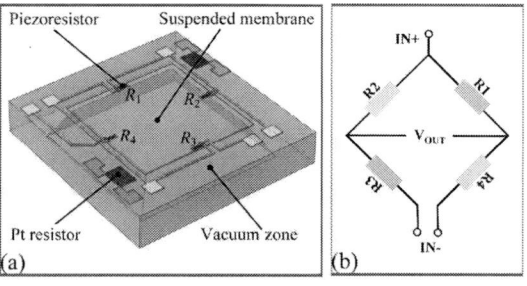

Fig. 1. (a) Perspective view and (b) the circuit diagram of the MEMS piezoresistive pressure sensor.

to deform, inducing mechanical strain on the piezoresistors. This strain alters the electrical resistance of the piezoresistors through the piezoresistive effect. Two of the piezoresistors are compressed, while the other two experience tension, leading to differential resistance variations. These differences disrupt the balance of the Wheatstone bridge, producing an output voltage that is proportional to the applied pressure. The output signal is then processed and calibrated to deliver precise pressure measurements.

To address the temperature drift problem commonly encountered by MEMS piezoresistive pressure sensors, this paper proposes an innovative solution. A platinum resistor, known for its excellent temperature coefficient of resistance (TCR) and stability, is directly integrated onto the sensor chip. This integration enables real-time monitoring of temperature fluctuations and dynamic compensation for temperature-induced changes. By implementing this approach, the temperature stability of the MEMS piezoresistive pressure sensor can be significantly enhanced.

III. FABRICATION AND PACKAGING

The proposed piezoresistive pressure sensor was fabricated using a standard MEMS process, which involves eight key steps, as shown in Fig. 2. These steps include: (a) boron ion implantation to create the piezoresistors, (b) formation of heavily doped regions, (c) oxidation and nitride deposition, (d) etching of contact holes, (e) aluminum metallization to form electrical leads, (f) platinum deposition to create the temperature sensor, (g) backside etching to form the cavity, and (h) backside bonding to establish the vacuum zone.

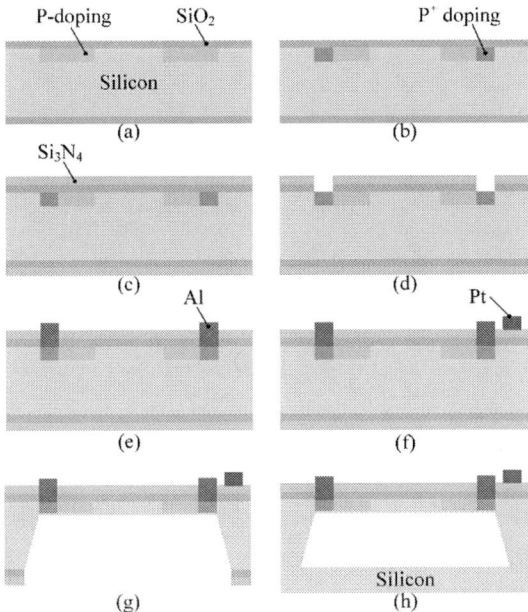

Fig. 2. Fabrication process flow of the MEMS piezoresistive pressure sensor. (a) Boron ion implantation. (b) Heavily doped region formation. (c) Oxidation and nitride deposition. (d) Etching contact holes. (e) Aluminum metallization. (f) platinum deposition for temperature sensing. (g) Backside etching for cavity formation. (h) Backside bonding.

Fig. 3. Photographs of the fabricated piezoresistive pressure sensor. (a) SEM of the front side of the chip. (b) SEM of the platinum temperature sensor. (c) Photograph of the packaged sensor.

The MEMS piezoresistive pressure sensor is produced through this fabrication process, and Fig. 3(a) illustrates the photographs of the fabricated device. After fabrication, the pressure sensor chip is packaged using a corrugated stainless steel diaphragm, which is commonly employed in the packaging of MEMS piezoresistive pressure sensors [17]. Fig. 3(c) provides a photograph of the packaged MEMS pressure sensor.

IV. EXPERIMENT AND DISSCUSSION

The packaged MEMS pressure sensor was installed in a vacuum chamber to evaluate its performance. The vacuum chamber, modified from an oven, allows temperature control from -45°C to 125°C, while an external pressure controller adjusts the pressure from 0 to 10 MPa.

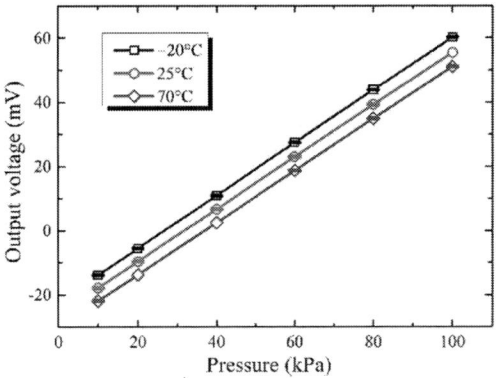

Fig. 4. The output response of the MEMS pressure sensor at three temperature conditions (-20°C, 25°C, and 70°C) during four repeated tests.

The sensor was first tested at temperatures of -20°C, 25°C, and 70°C. For each temperature, the vacuum level was varied from 10 kPa to 100 kPa, and four repeated tests were conducted at each condition. The results, shown in Fig. 4, reveal that the output voltage of the sensor increases nearly linearly with applied pressure at a constant temperature, indicating excellent linearity. Additionally, the sensor maintains a consistent linear relationship with pressure despite temperature variations, with the output characteristic curves at different temperatures showing nearly identical slopes.

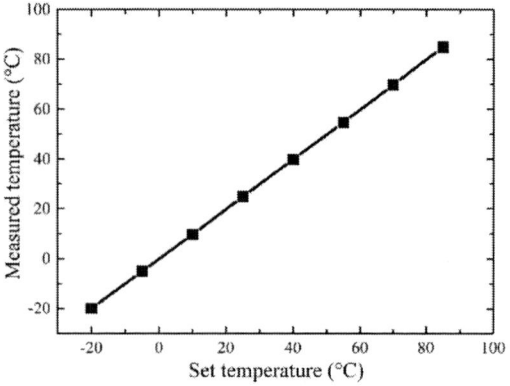

Fig. 5. The testing results of the platinum resistor for temperature measurement.

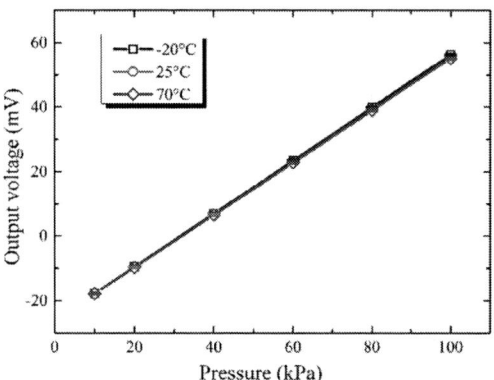

Fig. 6. The output characteristic curves of the MEMS pressure sensor at different temperatures after compensation.

In addition, we also characterized the temperature measurement performance of the fabricated platinum resistor on the sensor chip, and the results are shown in Fig. 5. As seen in the figure, the platinum resistor demonstrates accurate temperature measurement capability. Since temperature changes did not affect the slope of the pressure sensor's output curve, and the resistance of the platinum resistor varies linearly with temperature, temperature-induced drift in the pressure sensor can be effectively compensated using a linear interpolation algorithm. After compensation, the output characteristics of the MEMS pressure sensor at different temperatures, shown in Fig. 6, demonstrate that the temperature drift has been effectively mitigated, making the output almost independent of temperature.

V. CONCLUSIONS

In this paper, a MEMS piezoresistive pressure sensor with enhanced temperature drift suppression capability has been fabricated and characterized. The sensor features a suspended membrane integrated with four piezoresistors, along with a platinum resistor designed on the chip surface. Fabricated using a standard MEMS process, the sensor underwent experimental evaluation in a vacuum chamber. The results demonstrate that, with compensation from the fabricated platinum resistor, the MEMS pressure sensor maintains nearly identical output characteristic curves across a range of temperatures, highlighting its excellent temperature drift suppression performance. The temperature drift compensation method proposed in this study offers significant potential for the application of piezoresistive pressure sensors in fields such as automotive, industrial monitoring, and medical devices.

REFERENCES

[1] H. Yanazawa and K. Homma, "Growing market of MEMS and technology development in process and tools specialized to MEMS," in *IEEE EDTM*, 2017, pp. 143-144.

[2] S. Kaminaga, "Vision for commercialization of MEMS - A view from the industry," in *Proc. IEEE 29th Int. Conf. Micro Electro Mech. Syst. Conf. (MEMS)*, Shanghai, China, 2016, pp. 242-242.

[3] W. Fang, S. -S. Li and M. -H. Li, "Leveraging Semiconductor Ecosystems to MEMS," in *Proc. IEEE 36th Int. Conf. Micro Electro Mech. Syst. Conf. (MEMS)*, Munich, Germany, 2023, pp. 143-148.

[4] C. Cheng *et al.*, "A MEMS Resonant Differential Pressure Sensor With High Accuracy by Integrated Temperature Sensor and Static Pressure Sensor," in *IEEE Electron Device lett.*, vol. 43, no. 12, pp. 2157-2160, Dec. 2022.

[5] Min-Xin Zhou, Qing-An Huang, Ming Qin and Wei Zhou, "A novel capacitive pressure sensor based on sandwich structures," in *J. Microelectromech. Syst.*, vol. 14, no. 6, pp. 1272-1282, Dec. 2005.

[6] H. Gao, Y. Jiang, Y. Cui, L. Zhang, J. Jia and J. Hu, "Dual-Cavity Fabry–Perot Interferometric Sensors for the Simultaneous Measurement of High Temperature and High Pressure," in *IEEE Sensors J.*, vol. 18, no. 24, pp. 10028-10033, Dec. 2018.

[7] M. Basov and D. Prigodskiy, "Investigation of high-sensitivity piezoresistive pressure sensors at ultra-low differential pressures," in *IEEE Sensors J.*, vol. 20, no. 5, pp. 7646–7652, Jul. 2020.

[8] S. Peng *et al.*, "Recent Advances in 3-D Printed, Wearable Pressure Sensors for Plantar Pressure Monitoring: A Review," in *IEEE Sensors J.*, vol. 24, no. 21, pp. 33903-33921, Nov. 2024.

[9] Y. Zhao, L. Miao, Y. Xiao and P. Sun, "Research Progress of Flexible Piezoresistive Pressure Sensor: A Review," in *IEEE Sensors J.*, vol. 24, no. 20, pp. 31624-31644, Oct. 2024.

[10] S. S. Kumar and B. D. Pant, "Effect of piezoresistor configuration on output characteristics of piezoresistive pressure sensor: an experimental study," Microsyst Technol 22, 709–719, 2016.

[11] K. V. Meena, Ribu Mathew, Jyothi Leelavathi, A. Ravi Sankar, "Performance comparison of a single element piezoresistor with a half-active Wheatstone bridge for miniaturized pressure sensors," in *Measurement*, vol. 111, pp. 340-350, 2017.

[12] P. Nathan, R. Manning and D. M. Birch, "Dynamic Compensation of Ultra-Low-Range Pressure Sensors," in *IEEE Sensors J.*, vol. 21, no. 9, pp. 11094-11100, May 2021.

[13] C. Cheng, J. Yao, Y. Lu, C. Xiang, J. Chen, D. Chen, and J. Wang, "A bulk-micromachined resonant differential pressure microsensor insensitive to temperature and static pressure," in *Proc. IEEE 35th Int. Conf. Micro Electro Mech. Syst. Conf. (MEMS)*, Jan. 2022, pp. 656–659.

[14] E. G. Mohammed, M. Serigne, G. Chantal, L. Benjamin, H. Li, "A simple and effective method to compensate the thermal drift of implantable blood pressure sensors," in *Sens. Actuators A, Phys.*, vol. 376, 2024.

[15] J. W. Song, J. S. Lee, J. E. An, and C. G. Park, "Design of a MEMS piezoresistive differential pressure sensor with small thermal hysteresis for air data modules," in *Rev. Sci. Instrum.*, vol. 86, no. 6, pp. 433–436, Jun. 2015.

[16] L. M. Middelburg, H. W. V. Zeijl, S. Vollebregt, et al, "Toward a Self-Sensing Piezoresistive Pressure Sensor for All-SiC Monolithic Integration," in *IEEE Sensors J.*, pp(99):1-1, 2020.

[17] S. F. Moosavian, D. Borzuei, M. Farajollahi, "Stress, sensitivity, and frequency analysis of the corrugated diaphragm for different corrugation structures," in *Smart Structures and Systems*, 2021.

979-8-3315-2209-4/25 $31.00 © 2025 IEEE

Improved Plasmonic Scattering Imaging based on super-resolution algorithm for image reconstruction

Zhaochen Huo[1], Bing Chen[2], Yu Li[1], Ya Li[2], Xiaonan Yang[1]

1 School of Electrical and Information Engineering, Zhengzhou University, Zhengzhou 450001, China

2 Department of Gastroenterology, The First Affiliated Hospital of Zhengzhou University, Zhengzhou, 450052, China

E-mail addresses: iexnyang@zzu.edu.cn

Abstract-- Plasmon scattering imaging (PSI) has been shown to be effective in detecting tiny targets, approximately 100 nm. However, due to noise interference, the detection accuracy of small targets still needs to be improved. Here, we propose an image processing strategy based on a novel blind super-resolution deep learning neural network (ESRGAN-SE) to improve image resolution without increasing experimental complexity. To verify the effectiveness of the system, 100nm Au nanoparticles were imaged and analyzed. The experimental results of PSI images show that the proposed super-resolution detection method has strong generalization and robustness, obtaining the best evaluation result of 0.6129 in Structure Similarity Index Measure (SSIM). This method has the potential to greatly improve the accuracy and efficiency of targets analysis, thereby more accurately diagnosing cancer and potentially improving patient treatment outcomes.

Keyword--*Plasma Scattering Imaging , Blind Super-resolution, Deep Learning*

I. INTRODUCTION

Surface plasmon resonance imaging (SPRI), as a label-free, highly sensitive imaging platform, has become an effective tool for quantitative analysis of tiny targets, approximately 100 nm [1]. According to the different coupling devices, most SPRI systems can be divided into two categories: prism coupling and objective coupling. Prism-coupled SPRI systems have the advantage of providing a millimeter-level field of view (FoV), but are susceptible to interference from the primary lens and cannot obtain sufficient spatial resolution or sufficient signal-to-noise ratio (SNR) [2]. In addition, although SPRI systems with high numerical aperture objectives have greater magnification, the small imaging range and limited throughput of the system limit its application.

Subsequently, plasmon scattering imaging (PSI) was proposed based on SPRI technology. This technology has the same millimeter-level large field of view as prism-coupled SPRI, but with higher image resolution [3]. However, PSI technology uses light scattering from the sample for imaging, which is inevitably strongly interfered by background scattering. Although some techniques have been proposed to improve the signal-to-noise ratio and enhance imaging, such as label-free single-molecule pull-down (LFSMP) [4] and wide-field plasmon thermal microscopy (W-PTM) [5], these improvements are still limited in suppressing the noise scattered light and complicate the experiment. Therefore, it is still challenging to accurately and conveniently eliminate the scattering interference and obtain the real scattered light of the object from the PSI image.

Currently, deep learning technology with powerful feature extraction has been proven an efficient way to identify and analyze scattered light from complex biological systems [6]. Among them, super-resolution image reconstruction technology has shown remarkable capability in feature extraction and eliminating scattering interference. This strategy is highly attractive for the PSI system to increase the SNR and obtain the high resolution images without increase the complexity of imaging system [7]. However, most super-resolution image reconstruction technologies based on supervised networks, which required both the high-resolution (HR) and low-resolution (LR) images to form image pairs for the training process. Due to the applications of gold-coated chips in PSI system and the limitation in imaging process, it is still difficult to collect the HR-LR image pairs for the PSI system. The reliable super-resolution image reconstruction algorithm is still lacking for the PSI system.

Herein, we introduced super-resolution technology into PSI systems by combining the Real-ESRGAN with Squeeze-and-Excitation Network (SE-Net). Real-ESRGAN is a typical blind super-resolution network (BSRN) [8], which can solve the issue of image pairs in super-resolution models but fails in detecting the tiny target in the PSI system. Then we optimized the Real-ESRGAN with the SE-Net [9] to build the feature recalibration capability and enhance the feature extraction of tiny objects. We named the new network as ESRGAN-SE and verified it using PSI images of AuNPs. For measurement reproducibility and simpler operability, we trained this model with thousands of PSI images collected in a Kretschmann prism-coupled PSI system. The trained model can directly generate HR images from the original experimental images. Simultaneously, the reconstructed images of AuNPs were

979-8-3315-2209-4/25 $31.00 © 2025 IEEE

experimentally studied to evaluate the effectiveness of the ESRGAN-SE network. The structural similarity of the images reconstructed by different algorithms was compared with the images enlarged by bilinear interpolation. The proposed network obtained the best evaluation result of 0.6129. It is shown that the proposed super-resolution detection method has stronger generalization ability and robustness, providing a new strategy for PSI imaging analysis.

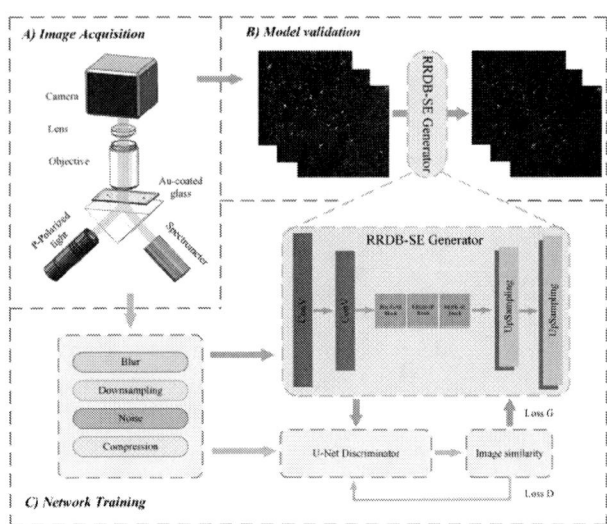

Fig. 1. A) PSI system based on prism. The original image of PSI is collected by camera 1. B) In the reconstruction stage, the preprocessed image is directly input to the generator network, which generates high-resolution PSI images at the output. C) In the training phase, the preprocessed images are input into the blind super-resolution GAN network. The network is trained to extract and recognize features from single images. The performance of the generator and discriminator is improved through iterating and optimizing continuously.

II. EXPERIMENTAL SECTION

A. Working Principle

Fig. 1 illustrates the schematic process of our blind super-resolution method for scattering light analysis of Au nanoparticles (AuNPs). The PSI images were obtained experimentally through a conventional Kretschmann prism-coupled system (Fig. 1A). After obtaining the PSI image of AuNPs, the differential image was obtained by difference calculation. Five images in each group were averaged. The first group was considered as background noise. The background noise was subtracted from each group to form a PSI difference image. Then, the training dataset was obtained through image preprocessing, which was used to train the ESRGAN-SE network. In the image preprocessing process, the number of pixels of the original PSI image is 2048×2048, which requires too much calculation. Therefore, the original PSI image is selected with a region of interest (512×512) to facilitate network training. In the

training stage, the images in the training dataset were served as HR images. And, these HR images were degraded into LR images by high-order degradation models to meet the image pair requirements of the network in the field of blind super-resolution. Subsequently, these LR images were reconstructed by the generator as new HR images. The discriminator distinguished the difference between the new HR images obtained by the generator and the original images, and generated the Loss. And, the generator is affected through backpropagation to achieve the effect of adversarial training (Fig. 1C). In the super-resolution reconstruction stage, the trained generator was extracted separately and considered as the reconstruction module. Different from the training stage, the images in the training dataset were regarded as LR images and directly input into the reconstruction module to obtain super-resolution images (Fig. 1B).

B. Experimental Setup

The experimental setup was constructed on a classic prism-coupled PSI system. A 100 mW laser diode (L63163DG, Thorlabs) with a center wavelength of 633 nm was installed on a laser diode mount (LDM9T, Thorlabs) as the light source, which was driven by a desktop laser diode driver controller (ITC4001, Thorlabs). The laser was first adjusted and collimated via a fiber attenuator (FVA-UV, Wyoptics) and a collimator (75-UV, Wyoptics) connected by optical fibers. Then the light was adjusted through a pair of lenses with a focal length of 50 mm (AC254-050-A, Thorlabs), and P-polarized light was acquired through a polarizer (LPVISC100-MP2, Thorlabs). P-polarized light was projected onto the prism surface. The incident angle was adjusted by a 5-axis translation stage (PY005-K1, Thorlabs) to generate surface plasmon resonance. Subse-quently, a top-mounted 40× objective (Nikon, ELWD ADM 40XC, NA = 0.60) collected the scattered light. Then, PSI images were collected by a camera (GC6600, Hama-matsu) through a lens with a focal length of 200 mm (AC254-200-A, Thorlabs). Likewise, the light reflected from the gold-coated glass slide was collected by the other camera (MER-132-43U3M-L, Hangzhou Hengyang Optics) with a polarizer and a lens with a focal length of 50 mm.2.3. Sample Preparation

C. Blind Super-resolution GAN Framework

A new blind super-resolution network named ESRGAN-SE was built by combining the features of the Real-ESRGAN and SE-Net models (Fig. 1C). Similar with the traditional BSRN, the proposed BSRN framework consists of three parts: High-order Degradation Model, Generator, and Discriminator. In generator part, a new structure was build to obtain the feature recalibration capability and enhance the feature extraction of tiny objects.

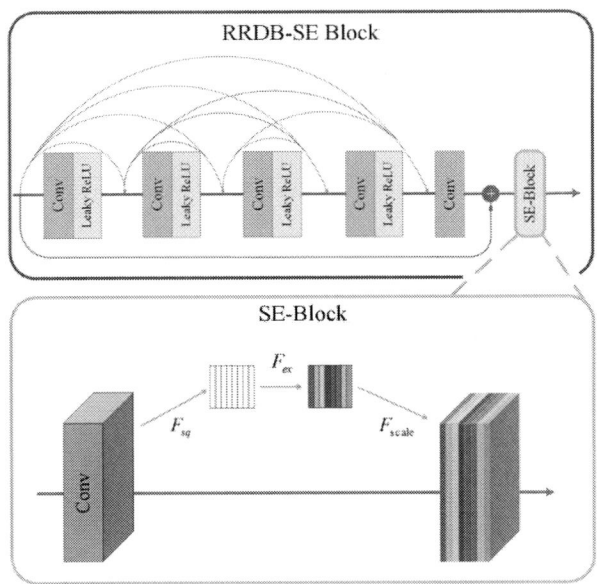

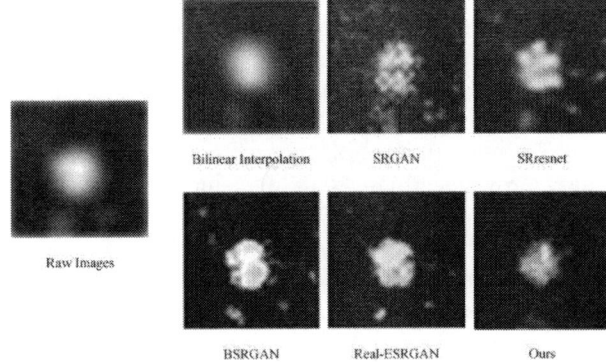

Fig. 3. The comparison diagram of the results from experimental methods. Single particle image reconstructed by each method, includes differential images, bilinear interpolation, SRGAN, SRresnet, BSRGAN, Real-ESRGAN, and ESRGAN-SE.

Fig. 2. Schematic diagram of the network structure of RRDB-SE Block and SE-Block.

After acquiring the LR images from the high-order degradation model, the generator reconstructs these LR images into super-resolution HR images. However, the original generator network has insufficient extraction capabilities for tiny targets and cannot adapt to a large number of small targets in PSI images. Therefore, the Residual-in-Residual Dense Blocks with SE-Net (RRDB-SE Block) is proposed (Fig. 2). SE-Net Block with a recalibration strategy can adaptively learn useful information in the convolutional feature maps of dense residual blocks and weight them accordingly. The newly assigned weights can highlight the characteristic texture of small targets, thereby improving the capabilities of the network to detect small targets.

III. RESULTS AND DISCUSSION

The PSI images of AuNPs were recorded in multiple experiments, and single particle images (10×10 pixels) were cropped from the differential images (Fig. 3). There were three pixels around the inside of the particle image that were brighter and the middle is darker, which also makes the center of the image darker in all network experiment results, such as traditional Bilinear Interpolation, SRGAN [10], SRresnet [10], BSRGAN [11], Real-ESRGAN [8] and ESRGAN-SE (The proposed).   .

Traditional bilinear interpolation only enlarges the original image, which is almost similar to the original image. SRGAN processes single particle points into several brighter scattered small dots. The network results are seriously inconsistent with the original image. The reason for this situation is that the convolution kernel and downsampling layer lose part of the feature information when the network is trained forward without residual connection. Compared with SRGAN, the resulting graph of SRresnet is more overall, and there is no separation of

bright spots. Since the weight of the original image is enhanced by the residual connection, the noise and targets are not over-sharpened, which makes this result closer to the original image. BSRGAN focuses more on using the values of surrounding pixels to calculate the center pixel value. After super-resolution processing, the target becomes a white circle because the difference between the pixels on the edge and the surrounding pixels is large. However, the difference between the central pixel and the surrounding pixels is not large, which makes the center darker. This network focused on highlighting the edge features of the target during convolution while ignoring the overall target. It is obvious that the whole AuNP was divided into upper and lower parts. Similarly, the same situation occurred in the results of Real-ESRGAN. The center brightness was darker and seemed to be divided into three parts. Although the super-resolution results of our network are darker than those of other networks in terms of brightness, they do not turn the surrounding background noise into bright spots. Compared with Real-ESRGAN, the experimental results of ESRGAN-SE are more like a whole, and the difference between the center of the target and the surrounding pixels is relatively small. The edge of the target is also brighter than the center, but there is no obvious sense of segmentation.

In addition, we compared the Structure Similarity Index Measure (SSIM) of these images. The structural similarity of the images reconstructed by different algorithms is compared with the images enlarged by bilinear interpolation. The proposed network obtained the best evaluation result of 0.6129. The average values of other reconstruction results were all lower than the optimal value, which were 0.3965, 0.5432, 0.5329, and 0.5741 respectively.

IV. CONCLUSIONS

In summary, we developed a novel analytical method for PSI images by using blind super-resolution technology. To verify the effectiveness of the system, we analyzed the PSI images of 100nm gold nanoparticles. The trained RRDB-SE generator can directly reconstruct the PSI image, which can effectively suppress the scattered light of the background and eliminate noise interference. Furthermore, this technology enabled the system to have high stability since no additional equipment was required. The experimental results show that super-resolution technology is extremely feasible in the field of PSI and has great potential in scattering measurement and imaging analysis.

REFERENCES

[1] Wu W, Yu X, Wu J, et al. Surface plasmon resonance imaging-based biosensor for multiplex and ultrasensitive detection of NSCLC-associated exosomal miRNAs using DNA programmed heterostructure of Au-on-Ag. Biosens Bioelectron. 2021;175:112835.

[2] Zhang P, Jiang J, Zhou X, et al. Label-free imaging and biomarker analysis of exosomes with plasmonic scattering microscopy. Chem Sci. 2022;13(43):12760-12768.

[3] Zhang P, Ma G, Dong W, Wan Z, Wang S, Tao N. Plas-monic scattering imaging of single proteins and binding kinetics. Nat Methods. 2020;17(10):1010-1017.

[4] Ma G, Zhang P, Zhou X, et al. Label-Free Single-Molecule Pulldown for the Detection of Released Cellular Protein Complexes. ACS central science, 2022, 8(9):1272-1281.

[5] Wang R, Jiang J, Zhou X, Wan Z, Zhang P, Wang S. Rapid Regulation of Local Temperature and Transient Receptor Potential Vanilloid 1 Ion Channels with Wide-Field Plasmonic Thermal Microscopy. Anal Chem. 2022;94(42):14503-14508.

[6] Xu Y, Wang X, Zhai C, et al. A Single-Shot Autofocus Approach for Surface Plasmon Resonance Microscopy. Anal Chem. 2021;93(4):2433-2439.

[7] Li W, Zhou K, Qi L, Lu L, Jiang N, Lu J, & Jia, J. Best-Buddy GANs for Highly Detailed Image Super-Resolution. AAAI Conference on Artificial Intelligence. 2021.

[8] Wang X, Xie L, Dong C, & Shan Y. Real-ESRGAN: Train-ing Real-World Blind Super-Resolution with Pure Synthetic Data. 2021 IEEE/CVF International Conference on Com-puter Vision Workshops (ICCVW). 2021:1905-1914.

[9] Hu J, Shen L, Albanie S, Sun G, Wu E. Squeeze-and-Excitation Networks. IEEE Trans Pattern Anal Mach Intell. 2020;42(8):2011-2023.

[10] Ledig C, Theis L, Huszár F, Caballero J, Aitken AP, Tejani A, Totz J, Wang Z, & Shi W. Photo-Realistic Single Image Super-Resolution Using a Generative Adversarial Net-work. 2017 IEEE Conference on Computer Vision and Pattern Recognition (CVPR), 105-114. 2017.

[11] Zhang K, Liang J, Gool LV, & Timofte R. Designing a Prac-tical Degradation Model for Deep Blind Image Super-Resolution. 2021 IEEE/CVF International Conference on Computer Vision (ICCV), 2021;4771-4780.

A Novel Active-Pixel Readout Circuit Based on Bias Current Cancellation for Dynamic X-ray Imaging

Jiangbo Hu[1], Yuhan Zhang[1], Congwei Liao[2] and Shengdong Zhang[1]*

[1]School of Electronic and Computer Engineering,

Peking University, Shenzhen, China

[2]College of Integrated Circuits and Optoelectronic Chips, Shenzhen Technology University

*Email: zhangsd@pku.edu.cn

Abstract—This paper presents a novel readout circuit technique, which addresses the issue that the integration capacitors are easily saturated for current-mode active pixels. The proposed circuit utilizes a column-shared current source with threshold voltage compensation to cancel the static operating current of the active pixel. Consequently, the current flowing into the readout circuit consists solely of the changing current caused by X-ray exposure, avoiding that unwanted charges are injected into the ROIC, and significantly mitigating the demand for the integration capacitor. Furthermore, the DC offset of the current sources among columns can be eliminated through correlated double sampling, thereby improving uniformity. Compared to the conventional active-pixel readout circuit, the proposed one features a significant increase of the charge gain from 20.9 to 52.8 and a remarkable decrease of the nonlinearity from 1.83% to 0.39% with the integration capacitor of 8 pF and the ROIC voltage swing of 1.8 V.

Keywords—*current-mode active pixels, bias current cancellation, integration capacitor, correlated double sampling, charge gain, nonlinearity*

I. Introduction

Active amplification within pixels is the increasingly preferred technology in large-area X-ray flat-panel detectors (FPD) [1]-[3]. Compared to passive pixels, active pixels of constant currents during readout enable a higher readout frame rate, and the in-pixel amplification results in a higher signal-to-noise ratio (SNR) for the active pixel sensor (APS), thereby improving the dynamic range. High SNR implies high charge gain. For medical applications like digital breast tomosynthesis, the APS must achieve a charge gain greater than 30 for the input signal [4]. Consequently, the output current of the single active pixel is typically higher than 10 μA, with an integration time on the order of tens of microseconds. Therefore, the output current of the APS is much larger than that of the passive pixel sensor. This necessitates a large integration capacitor for the external readout circuit.

However, since the readout IC (ROIC) is a large-scale parallel device, each integrated circuit can be connected to up to 256 lines, with each line capable of accepting a maximum capacitor of about 10 pF [3]. A limited voltage swing of 1.8 V (typical ROIC voltage) leads to the integration capacitance saturating in less than 1.8 μs of integration time.

Previous studies demonstrated methods to avoid the large integration capacitor by utilizing transimpedance amplifiers [3], [5]. The transimpedance amplifier can directly convert current to voltage by placing a resistor across the amplifier output and the negative input. However, compared to integration capacitors, resistors introduce additional noise. Moreover, the matching of resistors among different channels is not as good as that of capacitors.

This work proposes an active-pixel readout circuit based on bias current cancellation. This design eliminates the static operating current in the pixel output by using a column-shared bias current source so that only the changing current caused by X-ray exposure enters the ROIC for integration. This significantly reduces the need for external capacitors in the ROIC and ensures that the changing current occupies the full swing of the amplifier, improving the dynamic range and linearity. Additionally, the bias current source provides threshold voltage (V_T) compensation, enhancing the stability of the current source. Furthermore, the proposed readout circuit is compatible with correlated double sampling (CDS), which eliminates DC offset caused by different bias currents among channels. Therefore, the proposed active-pixel readout circuit is promising for dynamic X-ray medical imaging applications.

II. Proposed Active-Pixel Readout Circuit and Operation

The schematic of the proposed active-pixel readout circuit architecture is shown in Fig. 1. The pixel is a 3-T source degenerated active-pixel circuit [6]. T_{R1} implements the reset operation between the two subsampling periods. T_A is the amplifying transistor, and T_{S1} is the switching transistor. The bias current source consists of one resetting transistor (T_{R2}), one switching transistor (T_{S2}), and one dual-gate transistor (T_{BIAS}). T_{R2} and the storage capacitor C_{TH} at node C are used to compensate for threshold voltage shift (ΔV_T). The timing diagram is illustrated in Fig. 2.

A. Reset Phase

During the reset phase, Sense[n] and RST[n] are low to make that the output current of pixels is zero. RST_2 and RST_3 are both high to turn T_{R2} and T_{S2} on. Therefore, node C is charged to the reference voltage V_{REF1}. In the meantime, the drain and the gate at node C of T_{BIAS} are at a high voltage level.

979-8-3315-2209-4/25 $31.00 © 2025 IEEE

On the other hand, node A has been reset to V_{RST} during the previous readout phase.

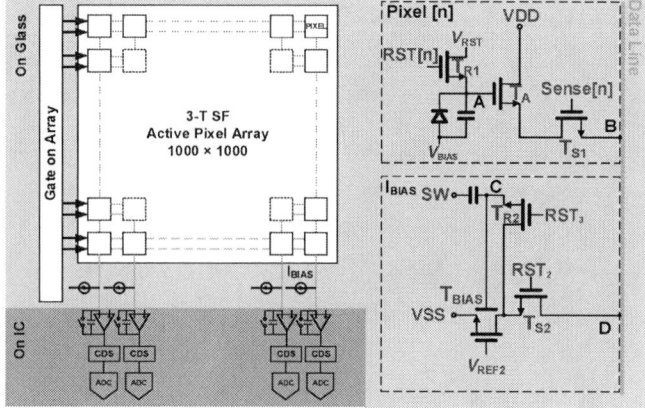

Fig. 1. The proposed active-pixel readout circuit architecture.

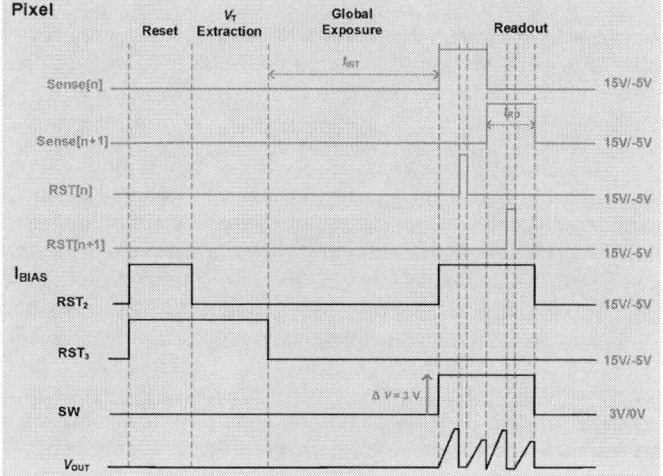

Fig. 2. The timing diagram.

B. Threshold Voltage Extraction Phase

In the V_T extraction phase, RST$_3$ is also high, and other control signals are low. Then, node C is discharged through the diode connection until the transient current of T$_{BIAS}$ approaches zero. Consequently, the V_T of T$_{BIAS}$ is always set to 0 V after the V_T extraction phase, no matter how it varies before. Since V_{REF2} is lower than VSS, the circuit can still compensate for the negative V_T of T$_{BIAS}$.

C. Global Exposure Phase

The photodiode (PD) is in a reverse-biased condition, and the movement of holes and electrons forms the PD photocurrent I_{PH} flowing from the PD cathode to the PD anode. Therefore, the charge in the storage capacitor at node A is reduced. RST[n] is low to turn T$_{R1}$ off to prevent the leakage of data charge in the parasitic capacitor of the PD.

D. Readout Phase

As shown in Fig. 2, the voltage level of SW is elevated by 3 V to make T$_{BIAS}$ into the saturation region. Each row is read out twice to achieve CDS for eliminating DC offset and low-

frequency noise, with a reset signal applied between the two subsamples. In the readout phase, Sense[n] is high to turn TS1 on during the first readout. Before the second subsampling, RST[n] is set to high voltage to reset the data voltage at node A. After resetting, the second subsampling is starting. During the two subsamples, SW is high all the time. The relationship between the output current I_{OUT1} and I_{OUT2} flowing into the ROIC and the bias current I_{BIAS} can be expressed as:

$$I_{OUT1} = I_{PIXEL1} - I_{BIAS} \tag{1}$$

$$I_{OUT2} = I_{PIXEL2} - I_{BIAS} \tag{2}$$

Here, I_{PIXEL1} and I_{PIXEL2} represent the output currents of the pixel during the first and second readout, respectively. I_{OUT1} and I_{OUT2} represent the currents flowing into the external readout circuit during the first and second readouts, respectively. After CDS, the equivalent current flowing into the external readout circuit is expressed as $I_{OUT1} - I_{OUT2}$, which is equal to $I_{PIXEL1} - I_{PIXEL2}$ and independent of I_{BIAS}. Therefore, CDS can eliminate DC offset caused by bias current mismatch among different columns.

I_{BIAS} can be expressed as:

$$I_{BIAS} = \frac{1}{2} \mu C_{OX} \frac{W}{L} (\Delta V)^2 \tag{3}$$

After V_T compensation, the bias current I_{BIAS} of T$_{BIAS}$ is directly proportional to the square of ΔV according to (2), which is independent of V_T thanks to the V_T extraction phase. The purpose of V_T compensation is to improve the stability of I_{BIAS}, which does not require high precision, as the static operating current in the output current of pixels with high charge gain is much higher than the changing current caused by X-ray exposure. Variation in I_{BIAS} due to ΔV_T can lead to significant changes in the current flowing into the ROIC, which may cause the integrator to saturate. V_T compensation addresses this issue, and CDS can further mitigate the problem of bias current mismatch among channels resulting from insufficient V_T compensation accuracy.

III. RESULTS AND DISCUSSIONS

The proposed bias current source of the active-pixel readout circuit is implemented using DG a-IGZO TFT technology, with a self-aligned structure to minimize parasitic capacitance [7], [8], which can increase reset speed and reduce charge coupling. Fig. 3(a) illustrates the cross-sectional view and equivalent symbol of the DG TFT.

The pixel adopts a source degenerated topology, which exhibits higher linearity compared to the common-source topology. To achieve the pixel size of $75 \times 75 \ \mu m^2$, the W/L ratio of T$_A$ is 50 μm/6 μm and that of T$_{S1}$ is 25 μm/6 μm, while the W/L ratio of T$_{R1}$ is 6 μm /6 μm and that of T$_{R2}$ is 20 μm /6 μm. The W/L ratio of T$_{BIAS}$ is 105 μm /6 μm, and the W/L ratio of T$_{S2}$ is 200 μm /6 μm. The parasitic capacitance of PD is 0.6 pF, and the storage capacitor at node C is 3 pF. The global reset time is set to 50 μs, the threshold voltage extraction time is 150 μs, and the global exposure time is 1 ms. Considering a typical array comprising of 1000 $\times$1000 pixels operating in

979-8-3315-2209-4/25 $31.00 © 2025 IEEE   374

real-time at 30 Hz, the total readout time per row is 32 μs, which includes a single readout time of 13 μs, a reset time of 2 μs, and a hold time of 4 μs.

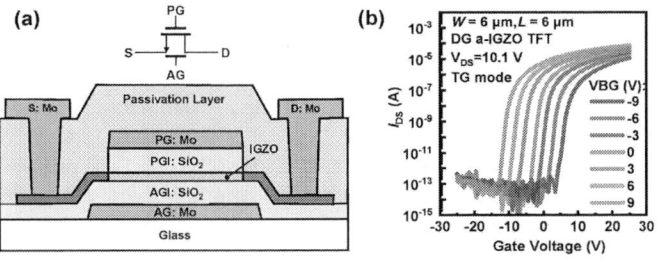

Fig. 3. (a) Cross-sectional schematic of DG a-IGZO TFT and its equivalent electronic symbol. (b) Measured transfer characteristic curves of the DG a-IGZO TFT.

A. Output Swing and Linearity

Fig. 4(a) and (b) respectively illustrate the transient output voltage of APS with and without current subtraction. Under the condition that the voltage swing of ROIC is 1.8 V and the integration capacitor is 8 pF, the APS with current subtraction can achieve an output swing of 1.72 V and a charge gain of 52.78. In contrast, the APS without current subtraction exhibits an output swing of 0.68 V, with a charge gain of only 20.87.

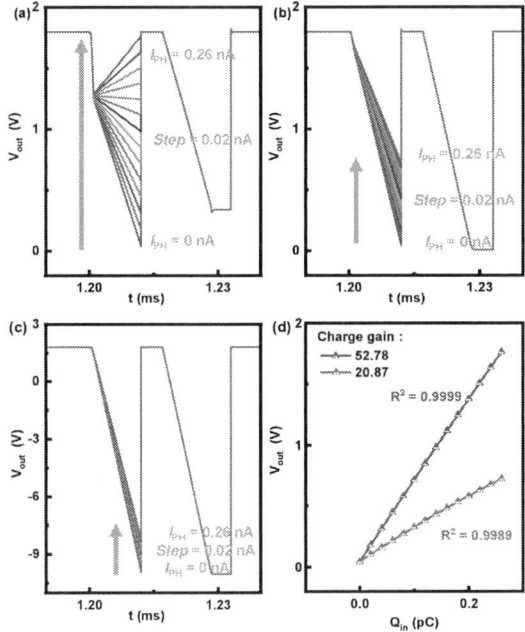

Fig. 4. (a) The transient response of V_{OUT} with current subtraction and (b) without current subtraction. (c) The relationship between the active pixel output current and the input charge with a charge gain of 52.78 and without current subtraction. (d) Relationship between input charge and output voltage after CDS with current subtraction (charge gain of 52.78) and without current subtraction (charge gain of 20.87) under the condition that the voltage swing of ROIC is 1.8 V and the integration capacitor is 8 pF.

After performing linear fitting on the input-output curves, the results are shown in Fig. 4(d). The APS with current subtraction exhibits a nonlinearity of 0.39%, while the APS without current subtraction shows a nonlinearity exceeding 1.8%.

B. V_T Compensation and CDS

Fig. 5 illustrates the compensation effect. The proposed bias current source effectively compensates for both positive and negative V_T shifts of T_{BIAS}. The output current error remains within 0.74% under ΔV_T of ± 1 V in the proposed circuit. The output current error rate of the bias current source without internal compensation is 93.35%. As shown in Fig. 4(c), without current subtraction, the bias current accounts for a significantly larger proportion of the current flowing into the ROIC compared to the changing current induced by X-ray exposure. Therefore, the purpose of compensation is to prevent excessive I_{BIAS} deviation, which could cause the current flowing into the external readout circuit to exceed the swing range of ROIC.

Although V_T compensation can enhance the stability of the bias current source, its precision still cannot exceed 10 bits. According to (1) and (2), after applying CDS technology, the output current becomes independent of I_{BIAS}. This effectively eliminates precision errors caused by the DC offset of the bias current source. The results after applying CDS are shown in Fig. 6. With a ΔV_T of ± 1 V, CDS reduces the DC offset of the bias current to within 0.001 V, meeting the precision requirement of 10 bits.

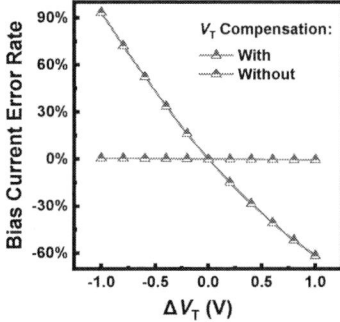

Fig. 5. Compensation effect for positive and negative threshold voltage shift. The bias current error does not exceed 0.74% under a threshold voltage shift of ± 1 V.

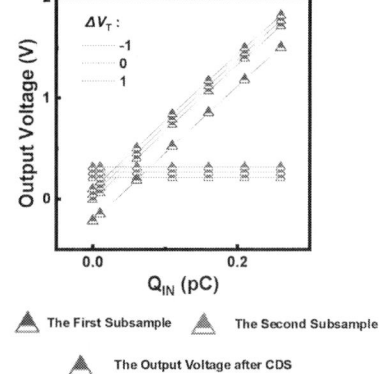

Fig. 6. The effectiveness of CDS in eliminating bias current source errors. Under a ΔV_T of ± 1 V, CDS can effectively reduce the DC offset of the bias current to meet the precision requirement of 10 bits.

C. Layout

Fig. 7 shows the test APS chip and the bias current source of the proposed active-pixel readout circuit. PD is widely used

vertically stacked on top of the TFT backplane. Therefore, the PD is not shown in the layout.

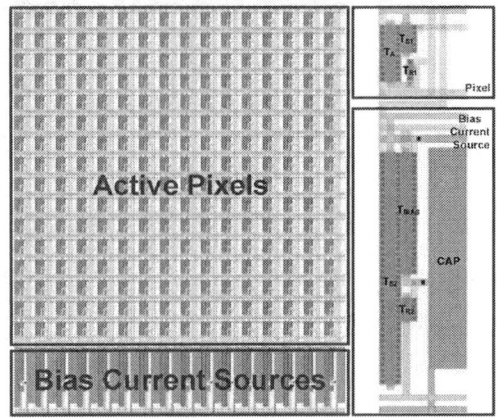

Fig. 7. Layout of the pixel array (left), the bias current source, and the pixel (right). The pixel area is $75 \times 75\ \mu m^2$, with a minimum metal line width of 5 μm and a minimum spacing of 5 μm.

Table I presents a comparison between the proposed pixel circuit and previous works.

TABLE I. COMPARISON BETWEEN THE PROPOSED READOUT CIRCUIT AND OTHER WORKS

	This Work	[9]	[4]	[5]
Technology	DG a-IGZO	a-IGZO	a-IGZO	a-IGZO
Critical dimension (μm)	6	4	5	5
Operating principle	Integrating	Integrating	Integrating	Trans-impedance
Pixel pitch (μm)	75	>135	<75	100
Integration capacitor (pF)	8	100	100	N/A
Output swing (V)	1.72	<1.5	<1.75	1.05
Full-scale range (pC)	0.26	46	2	1.86
Charge-to-current gain ($\mu A/pC$)	4.54	0.011	4.38	3.54
Nonlinearity	0.39%	0.95%	N/A	3.50%

*Numbers in italic are extracted by fitting the figures given in the referenced paper.

IV. CONCLUSION

This design eliminates the static operating current in the pixel output by using a column-shared bias current source, improving the dynamic range and linearity. Additionally, the bias current source provides V_T compensation, enhancing the stability of the current source. Furthermore, the proposed readout circuit is compatible with CDS, which eliminates the DC offset caused by different bias currents among channels. The proposed circuit with the 6 μm DG a-IGZO TFT process was demonstrated in terms of SPICE simulations, layout, and merits comparison. Research results show the CtC gain is 4.54 $\mu A/pC$, with a nonlinearity of 0.39%. The circuit achieves a pixel area of $75 \times 75\ \mu m^2$ at a frame rate of 30 Hz. Under a ΔV_T of ± 1 V, CDS can effectively reduce the DC offset of the bias current to meet the precision requirement of 10 bits. This work effectively addresses the limitations of external large capacitors in the ROIC, and is therefore promising for dynamic X-ray imaging applications.

ACKNOWLEGEMENT

This work was carried out at Guangdong Provincial Center for Oxide Semiconductor Devices and ICs, and supported financially by National Natural Science Foundation of China under Grant U24A20297, the Ministry of Science and Technology Key Research and Development Program under Grant 2022YFB3607200, and Shenzhen Municipal Scientific Program under Grant JCYJ20220818100808019.

REFERENCES

[1] Dandekar M, Myny K, Dehaene W. An Active-Pixel Readout Circuit Technique towards all LTPS-TFT-on-foil Large-Area Imagers with Inherent Nonlinearity Compensation[C]//2023 IEEE International Symposium on Circuits and Systems (ISCAS). IEEE, 2023: 1-5.

[2] Tai Y H, Lin C H, Yeh S, et al. LTPS active pixel circuit with threshold voltage compensation for X-ray imaging applications[J]. IEEE Transactions on Electron Devices, 2019, 66(10): 4216-4220.

[3] De Roose F, Myny K, Steudel S, et al. 16.5 A flexible thin-film pixel array with a charge-to-current gain of 59μA/pC and 0.33% nonlinearity and a cost effective readout circuit for large-area X-ray imaging[C]//2016 IEEE International Solid-State Circuits Conference (ISSCC). IEEE, 2016: 296-297.

[4] Zhao C, Kanicki J. Amorphous In‐Ga‐Zn‐O thin‐film transistor active pixel sensor x‐ray imager for digital breast tomosynthesis[J]. Medical physics, 2014, 41(9): 091902.

[5] Roose F D, Tedde S, Myny K, et al. A large-area a-igzo 256x256 imager using a current-mode transimpedance readout for mammography applications[C]//2019 International Image Sensor Workshop. image sensors; Snowbird Resort, Utah, USA, 2019.

[6] Karim K S, Nathan A. Readout circuit in active pixel sensors in amorphous silicon technology[J]. IEEE Electron Device Letters, 2001, 22(10): 469-471.

[7] Zhang Y, Li J, Zhang Y, et al. Deep sub-micron self-aligned bottom-gate amorphous InGaZnO thin-film transistors with low-resistance source/drain[J]. IEEE Electron Device Letters, 2023, 44(8): 1300-1303.

[8] Zhang Y, Chang L, Lu L, et al. Asymmetric Double-Gate (ADG) Oxide Thin-Film Transistor Technology for Medium-and Small-Sized AMOLED Displays[J]. IEEE Electron Device Letters, 2025.

[9] Zhang R, Bie L, Fung T C, et al. High performance amorphous metal-oxide semiconductors thin-film passive and active pixel sensors[C]//2013 IEEE International Electron Devices Meeting. IEEE, 2013: 27.3. 1-27.3. 4.

Design of 10MHz Isolated Current Sense Amplifier Based on FDDA and Frequency Modulation

Hongwei Shen[1], Xinghong Chen[2], Shaowei Zhen[2*], Jingying Sun[1], Zupei Gu[1], Yongwang Ma[1], Yidong Yuan[1], Bo Zhang[2]

[1]Beijing Smartchip Microelectronics Technology Co., Ltd., Beijing 100089, China
[2]University of Electronic Science and Technology of China, Chengdu 611731, China
*swzhen@uestc.edu.cn

Abstract—**A 10MHz high precision isolated current sense amplifier (ICSA) based on fully differential difference amplifier (FDDA) and frequency modulation is presented. The proposed ICSA generates high frequency pulse utilizing voltage controlled oscillator (VCO), modulated by sensed voltage. The frequency is stabilized by feedback loop consisted by frequency-to-voltage converter (FVC) and clamp amplifier. Then the modulated high frequency pulse is transmitted across the isolation barrier. The FDDA is employed to provide high sensing bandwidth while suppressing offset and noise. The proposed ICSA is implemented with 0.18μm BCD process and the bandwidth is high to 10 MHz with fixed gain of 4.**

Keywords—**Fully differential difference amplifier (FDDA), voltage controlled oscillator (VCO) , switch capacitor (SC), frequency-to-voltage converter (FVC)**

I. INTRODUCTION

Isolated current sense amplifiers (ICSA) are essential in power systems, where they detect current signals and transmit them across an isolation barrier to eliminate interference between high-voltage and low-voltage sides. Current sensing typically employs a small series resistor to detect signals, thereby minimizing the impact on the original circuit which results in extremely weak sensed signals. As a consequence, ICSA require substantial noise and offset rejection capabilities to ensure the accuracy of the sensed signals. The current-sensing block of the ICSA is shown in Fig. 1.

Besides, as the operating frequency of power systems increases, the demand for higher bandwidth (BW) in ICSA is becoming more pronounced. However, most commercially available ICSA products exhibit bandwidths in the kHz range, with very few achieving MHz-level performance [1].

In this article, a ICSA based on the fully differential difference amplifier (FDDA) and frequency modulation is presented. As will be shown, FDDA offers several benefits in terms of BW, noise and offset performance. In addition, The frequency modulation technique employing a voltage controlled oscillator (VCO) and frequency-to-voltage converter (FVC) provides a simple solution for high-frequency signal transmission across an isolation barrier.

This article is organized as follows. Section II explains the architecture of the proposed circuit, and Section III describes the measurement results of the ICSA.

II. ARCHITECTURE OF PROPOSED ICSA

The main blocks consisting of the proposed ICSA are FDDA, VCO and FVC, the block diagram is shown in Fig. 2. FDDA is used as the analog front end to sense the signal and suppress noise and offset. The VCO and FVC form a feedback network through an operational amplifier. This configuration ensures that the output voltage of the FVC remains clamped to the input voltage of the operational amplifier, even when affected by process, voltage, and temperature (PVT) variations. This significantly enhances the signal accuracy of the ICSA. The FVC at the receiver end is matched with the transmitter end, ensuring that the demodulated voltage after frequency demodulation equals the clamped voltage of the feedback loop. The signal is then transmitted through a single-ended-to-differential converter, enabling signal transmission across the isolation barrier.

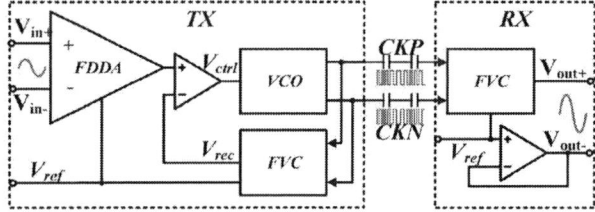

Fig. 2. Implemented ICSA based on FDDA with frequency modulation using VCO and FVC

A. Fully Differential Difference Amplifier

The FDDA is used as the primary module for achieving high bandwidth and suppressing noise and offset. The proposed circuit architecture comprises two main components: a differential difference amplifier (DDA) and a resistive feedback network [2]. The top-level structure of the circuit is depicted in Fig. 3.

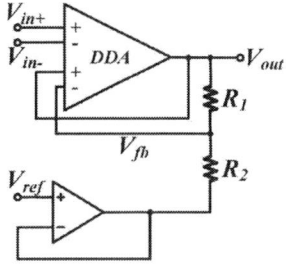

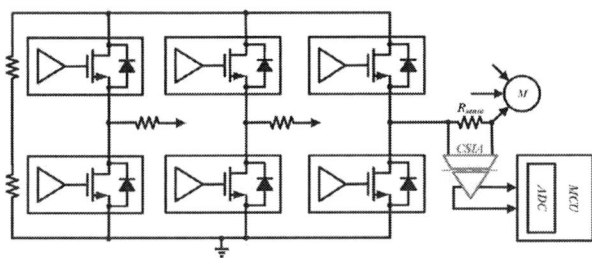

Fig. 1. Current-sensing block of the ICSA

Fig. 3. FDDA based on DDA with a resistive feedback network

979-8-3315-2209-4/25 $31.00 © 2025 IEEE

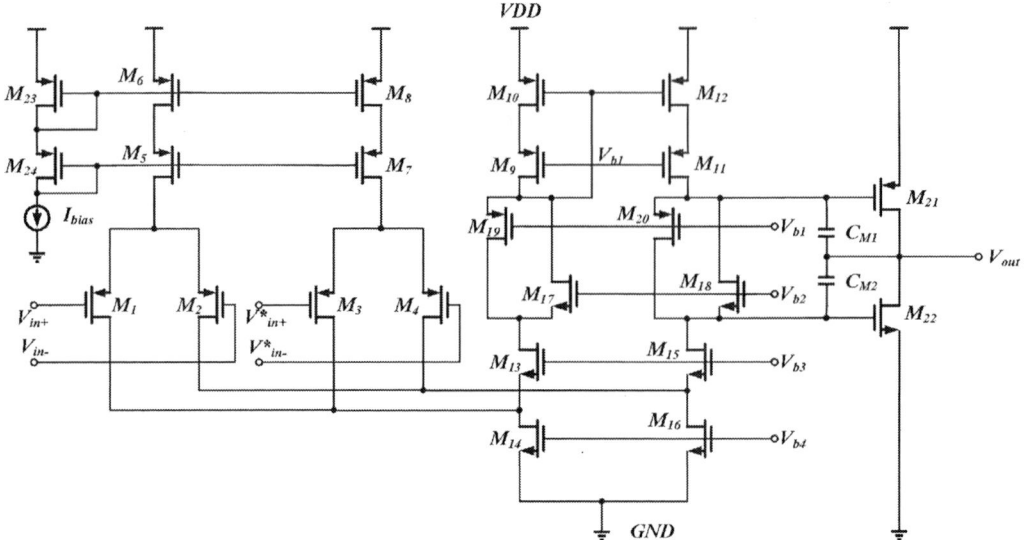

Fig. 4. Internal structure of DDA with folded cascode amplifiers and a floating class-AB output stage

The resistive feedback network is used to determine the closed-loop gain and consists of two identical feedback structures that symmetrically amplify the differential signals. The feedback resistors need to be sufficiently large to achieve a high loop gain (LG), ensuring gain accuracy. The closed-loop gain can be obtained by

$$(V_{out} - V_{ref}) = \frac{R_2 + R_1}{R_1}(V_{in+} - V_{in-}) \qquad (1)$$

The DDA is composed of two pairs of folded cascode amplifiers and a floating class-AB output stage. The internal structure of the DDA on transistor level without bias network is shown in Fig. 4. The input pairs consist of transistors M_1-M_4, which are designed with larger channel lengths to minimize mismatch. The floating class-AB output stage, consists of M_{21} and M_{22}, is driven by transistors M_{18} and M_{20}.

B. Voltage Controlled Oscillator

VCO is constructed from four cascaded fully differential inverters, as depicted in Fig. 5. The control voltage V_{ctrl} is converted into biasing voltages V_N and V_P through a voltage-to-current conversion circuit. This ensures that the subsequent fully differential pair has a current I_N that is twice the value of I_P. By adjusting the magnitudes of the currents I_N and I_P, the delay time T_D of each stage is altered, which in turn modifies the output frequency of the VCO. The frequency is determined by

$$f = \frac{\frac{I_N}{2} - I_P}{8C_L V_{DD}} \qquad (2)$$

Where C_L is the load capacitance, V_{DD} is the supply voltage, I_N is the tail current of the NMOS current source and I_P is the current of the PMOS current source.

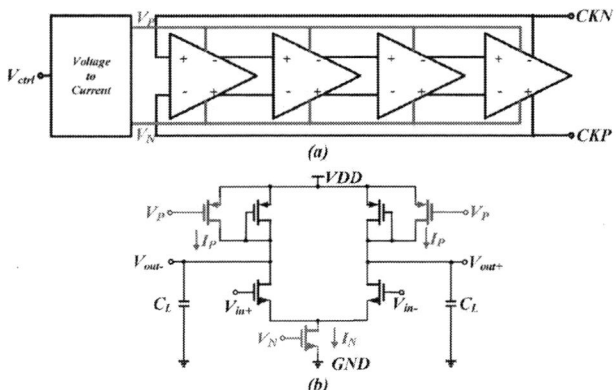

Fig. 5. (a) Block of VCO inverter cascade. (b) Internal structure of inverter.

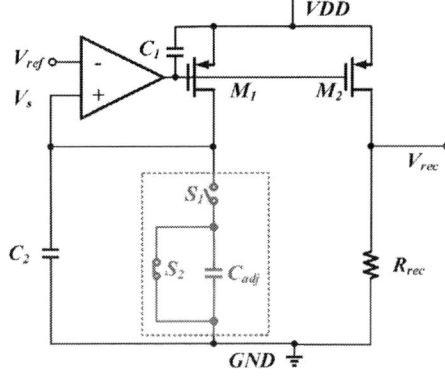

Fig. 6. Internal structure of FVC circuit

C. Frequency-to-Voltage Converter

The FVC is composed of a switch-capacitor feedback loop (SCFBL) and a voltage recovery circuit (VRC), as shown in Fig. 6. The SCFBL uses an amplifier to clamp the voltage V_S to a reference value. The switch-capacitor network forms a adjustable resistor, whose resistance is inversely proportional to the frequency.

979-8-3315-2209-4/25 $31.00 © 2025 IEEE 378

Capacitor C_1 is utilized to provide compensation for the feedback loop, thereby enhancing loop stability. The selection of C_1 is challenging because an excessively large value can lead to insufficient loop bandwidth, while a value that is too small may not provide adequate phase margin. Capacitor C_2 is employed to filter out switching ripples. Its value must also be chosen appropriately, a value that is too high can reduce the non-dominant pole frequency, degrading loop stability, whereas a value that is too low may be insufficient to eliminate switching ripples.

The VRC copies the current from SCFBL and converts it back into a voltage V_{rec} through resistor R_{rec}. The V_{rec} is given by

$$V_{rec} = V_{ref} C_{adj} R_{rec} f \qquad (3)$$

Where V_{ref} is the reference voltage of the clamp amplifier, C_{adj} is the switch-controlled capacitor.

III. SIMULATION RESULTS

The proposed ICSA is implemented with 0.18μm BCD process and operates at a supply voltage of 1.8 V. Fig. 7 shows the bandwidth simulation results of the ICSA, achieving a bandwidth of 10 MHz.

In addition, the output of the ICSA is verified by simulation using input signals with different amplitudes and frequencies. Fig. 8 shows the output waveform with a typical operating frequency of 1 MHz and a peak-to-peak amplitude of 100 mV for the triangular wave input. The simulation results indicate that the output waveform maintains a frequency of 1 MHz and amplifies the signal peak-to-peak amplitude to 414 mV, achieving a fixed gain of 4. The comparison of ICSAs performance is shown in TABLE I.

TABLE I.   COMPARISON OF ICASs PERFORMANCE

	This work	[1]	[3]	[4]	[5]	[6]
Tech	**0.18μm BCD**	0.18μm CMOS	-	-	-	-
Supply Voltage [V]	**1.8**	5	3 to 20	3.3/5	3.3/5	3.3/5
Bandwidth [kHz]	**10000**	35200	1000	310	220	310
Gain	**4**	-	20	8.2	8.2	8.2

IV. CONCLUSION

A high-bandwidth, high-precision ICSA realized with 0.18μm BCD process is presented. To achieve high bandwidth and precision, FDDA is employed. Frequency modulation is implemented using a VCO and a SC circuit. The ICSA achieves a bandwidth of 10 MHz, a fixed gain of 4.

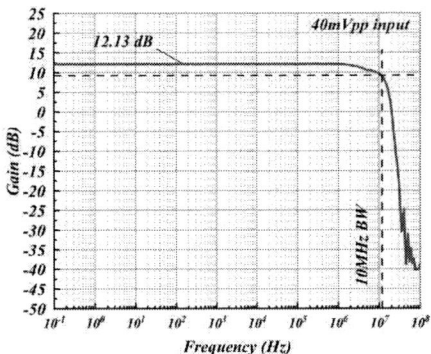

Fig. 7. Simulated bandwidth results of the ICSA

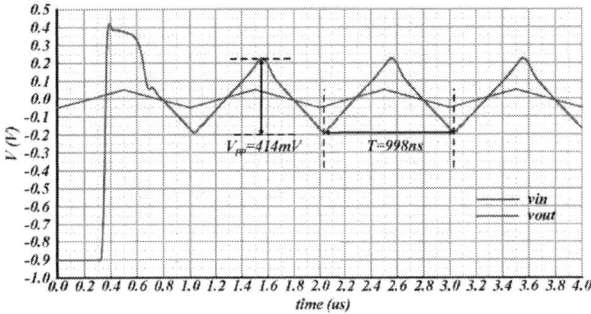

Fig. 8. Simulated results of the input voltage (vin) and output differential voltage (vout) for a triangular wave input with a frequency of 1 MHz and a peak-to-peak amplitude of 100 mV

ACKNOWLEDGMENT

This work is supported by CIE-Smartchip research fund No.2023-006.

REFERENCES

[1] S. Takaya, H. Ishihara and K. Onizuka, "18.7 A DC to 35MHz Fully Integrated Single-Power-Supply Isolation Amplifier for Current- and Voltage-Sensing Front-Ends of Power Electronics," 2020 IEEE International Solid-State Circuits Conference - (ISSCC), San Francisco, CA, USA, 2020, pp. 298-300.

[2] Y. Zhang, "A High-Gain, Low-Power Fully Differential Amplifier with Common-Mode Feedback for High-Speed and High-Precision Analog-to-Digital Converters," 2024 IEEE 7th International Conference on Automation, Electronics and Electrical Engineering (AUTEEE), Shenyang, China, 2024, pp. 979-985.

[3] "TPA158 Datasheet," 3PEAK, 2024. [Online]. Available: https://static.3peak.com/res/doc/ds/Datasheet_TPA158.pdf.

[4] "AMC1400 Datasheet," Texas Instruments, July 2022. [Online]. Available: https://www.ti.com/lit/ds/symlink/amc1400.pdf?ts=1714484324325& ref_url=https%253A%252F%252Fwww.ti.com%252Fproduct%252F de-de%252FAMC1400.

[5] "NSI1400D Datasheet," Novosense, Rev. 1.0, Oct. 2023. [Online]. Available: https://www.novosns.com/Public/Uploads/uploadfile/files/20231206/ NSI1400_Datasheet_Rev1.0_EN.pdf.

[6] "CA-IS1300x Datasheet," Chipanalog Inc., Version 1.05, June 2024. [Online]. Available: https://www.chipanalog.com/web/bocupload/2024/06/11/ca-is1300x_datasheet_cn_version1.05.pdf

Magnetic Field Measurement Using Resonance Frequency-Optimized Fluxgate Sensors

Wenbo Wang   Shanglin Yang   Hesen Su

Abstract—Fluxgate sensor based on the second harmonic method is the mainstream of fluxgate sensor. However, due to the complexity of post-processing circuit and power consumption, a fluxgate sensor measurement method based on resonance frequency optimization is proposed in this paper to improve the efficiency and accuracy of fluxgate magnetic field measurement. Finally, it is found that this method can measure the external magnetic field efficiently, which provides a new reference for improving the performance of fluxgate sensor.

Index Terms—fluxgate; Resonance, sensitivity

I. INTRODUCTION

Fluxgate sensors usually contain one or more coils. Fluxgate uses the characteristics of nonlinear magnetic material magnetic field changes to generate induced electromotive force on the conductor. When the external magnetic field passes through the primary coil, the magnetic field generated by the excitation current changes the magnetic flux in the secondary coil. The nonlinear magnetic saturation characteristics cause the induced voltage harmonic signal containing the external magnetic field information to be generated in the secondary coil. The magnitude of the external magnetic field can be obtained by measuring the induced voltage [1].

Resonant frequency is widely used in other measurement fields, among which the most widely used is parity time-symmetric LC resonant sensor, which is mostly used to measure humidity [2], pressure and so on. In this paper, resonance is applied to fluxgate sensor to realize a new method of measuring magnetic field, so as to improve the sensitivity of fluxgate detection magnetic field.

II. THEORETICAL BASIS

A. Working principle of fluxgate sensor

The structure of fluxgate is shown in Figure 1. The primary coil is reversely connected and the secondary coil is directly connected [3]. When the excitation current passes through the primary coil, a magnetic field of the same size and direction is generated in the two coils, which is combined with the external magnetic field in the magnetic core to become a new magnetic field that changes with the external magnetic field. The induction coil at this time due to the role of the internal magnetic field, the induction current will be generated in the coil, but because it is reversed, the magnetic field information generated by the excitation voltage can cancel each other, thus

Wenbo Wang, Shanglin Yang, Hesen Su are with *School of Electrical and Information Engineering, North Minzu University*, Ningxia 750021, China. Corresponding author: (Shanglin Yang, email: ysl029@163.com)

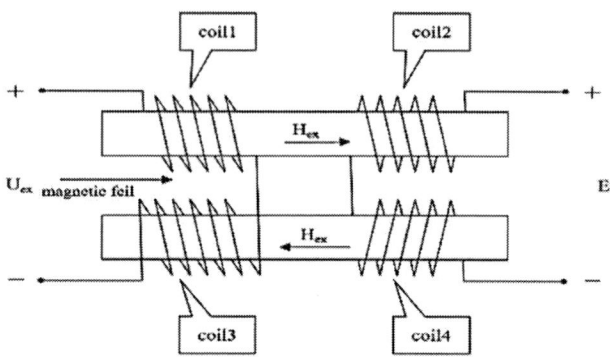

Fig. 1. Hysteresis loop of soft magnetic material

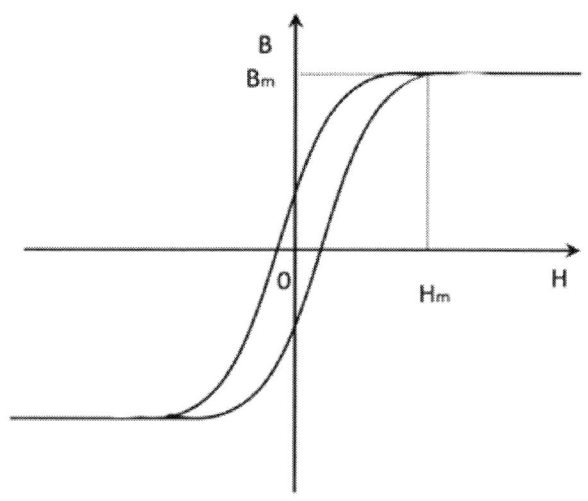

Fig. 2. fluxgate structure

retaining the information of the external magnetic field in the output current.

The traditional fluxgate uses the saturation characteristics of the magnetic core material to measure the magnetic field [4]. As shown in Fig. 2, under the changing external magnetic field, the magnetic field strength of soft magnetic materials will reach saturation with the change of magnetic induction intensity. Therefore, under the excitation current of the fluxgate, the soft magnetic core will complete cycle saturation.

The second harmonic method to analyze the amplitude of

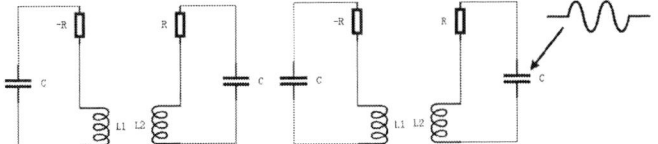

Fig. 3. Parity time-symmetric LC sensing system with no perturbations and perturbations

the second harmonic of the time-domain signal of the output voltage is used in fluxgate, and then obtains the external magnetic field as shown in formula 1 through the relationship between the second harmonic and the external magnetic field. The following formula is the sensitivity of the fluxgate, which is related to the structure of the fluxgate and the material of the magnetic core[5].

$$G = \frac{d(U_{2m})}{d(H_x)}, \qquad (1)$$

Where U_{2m} is the amplitude of the second harmonic of the output voltage and H_x is the external magnetic field.

III. PHYSICAL MECHANISM OF RESONANT FREQUENCY

For a simple LC loop, its resonant frequency $\omega = 1/(LC)^{1/2}$. A variety of sensor fields have used resonance to improve their sensitivity, the most mature is the use of parity time-symmetric LC resonance measurement [6], as shown in the figure, in a Hermitic system, the left and right parts of the LC circuit through inductors, capacitors or direct coupling, in the absence of external perturbation only one resonance peak. When a perturbation occurs on one side, such as the inductance change of ΔL, a frequency shift is generated, and the relationship between the external perturbation and the frequency shift is obtained by measuring the size of the frequency shift under different perturbations[7].

IV. METHOD AND DESIGN

A. System Design

In this paper, a dual-core magnetic structure fluxgate is designed, the magnetic core is Permalloy, and the hysteresis loop is fitted with the inverse tangent function, as shown in Fig. 4.

The dual-core fluxgate can be equivalent to a series and parallel LC circuit, where ΔL is a time-varying inductance, calculated by the following formula.

$$\Delta L = \frac{N^2 \cdot S}{l} \cdot \frac{a \cdot b}{1 + [b \cdot (\frac{N \cdot i}{l} + H_X) + c]^2} \qquad (2)$$

Where N is the number of coil turns, S is the cross-sectional area of magnetic flux, l is the coil length, and a, b, and c are constants in the hysteresis loop curve fitting function of the magnetic core

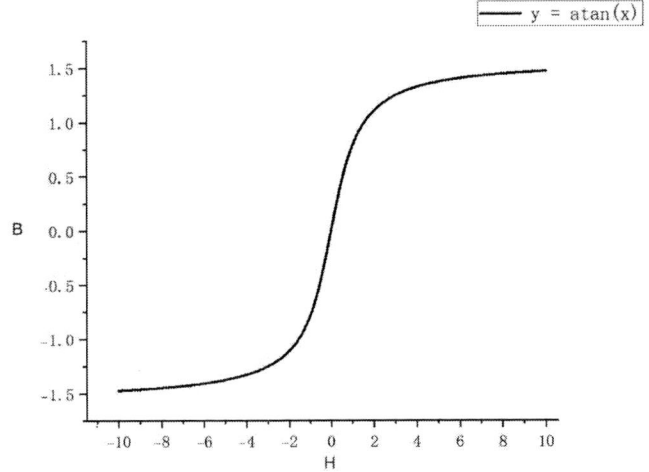

Fig. 4. Hysteresis loop fitted by inverse tangent function

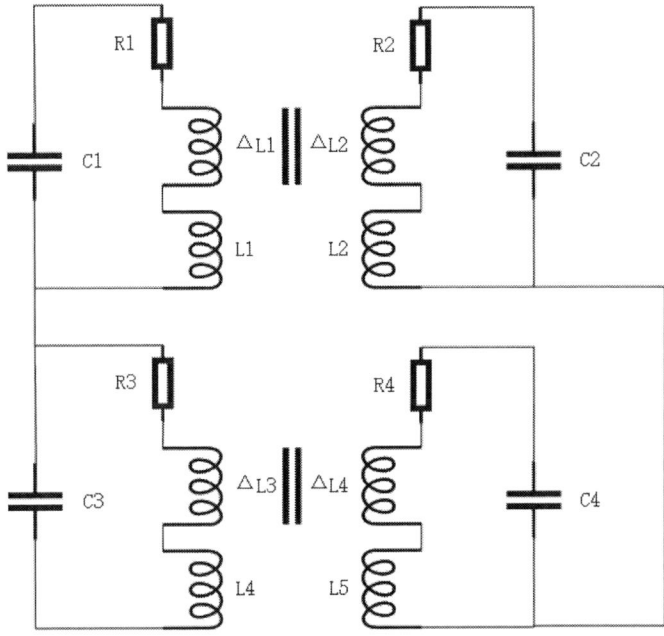

Fig. 5. Circuit model of fluxgate

B. Simulation design

In this paper, simulation software is used for simulation, the circuit with the same primary and secondary parameters is adopted, and the baud instrument is used for detection. Permalloy has a very low saturation magnetic field strength, in order to obtain higher sensitivity, it is selected as the magnetic core in this paper.

The values of capacitance, inductance and resistance of the equivalent circuit we choose to simulate are from the previous research of Yang, The specific parameters are shown in Table 1. In order to compare the measurement method proposed in this paper, we use the excitation current of $1KHz$, $160mA$ to obtain the results of the traditional me asurement method.

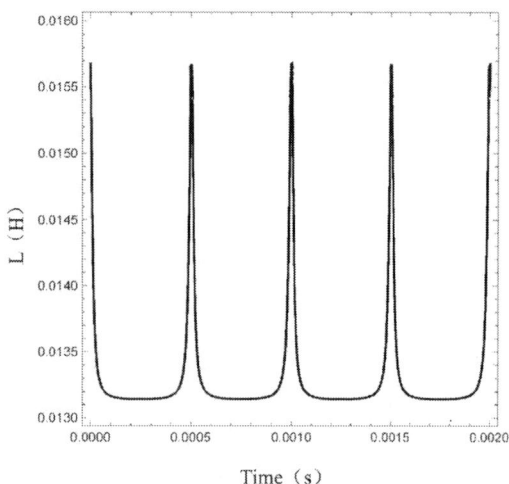

Fig. 6. The value of L changes over time

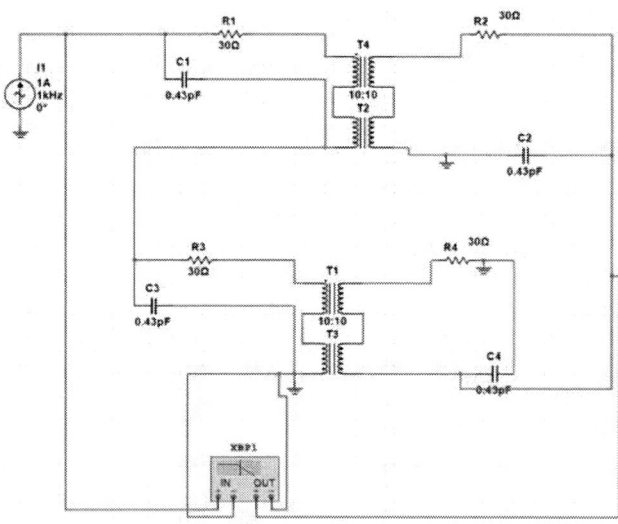

Fig. 7. Simulation circuit adopted in this paper

TABLE I
SPECIFIC PARAMETERS SIMULATED IN THIS PAPER

parameter	Unit	Value
L	mH	1.313
R	Ω	30
C	pF	0.43
Turns	-	500
Coupling coefficient	-	1

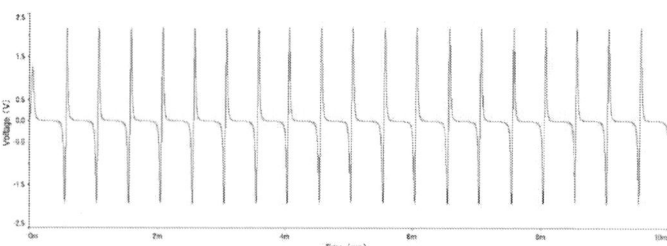

Fig. 8. Transient output voltage signal

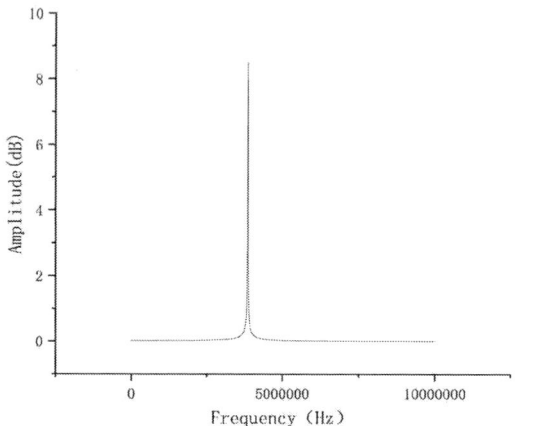

Fig. 9. Fluxgate output voltage Baud diagram

By measuring the magnetic field interval of different external magnetic fields from -100 to $100A/m$, the simulation results show that the simulated fluxgate conforms to the curve in Fig. 10.

According to Fig. 10, it can be found that there is a quadratic power relationship between the resonant frequency of fluxgate and the magnitude of the external magnetic field. The sensi-

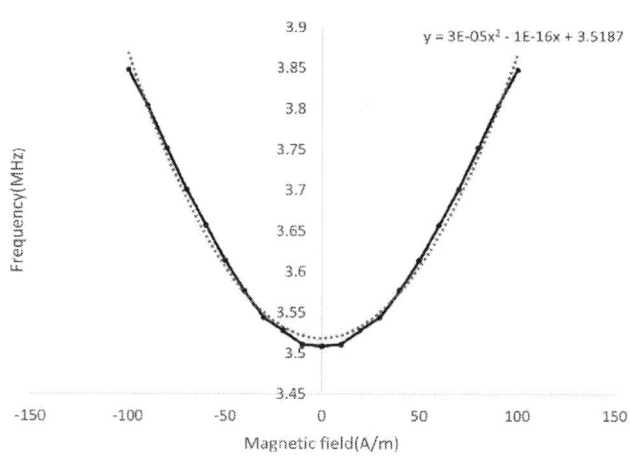

Fig. 10. Relation between resonant frequency and external magnetic field

979-8-3315-2209-4/25 $31.00 © 2025 IEEE

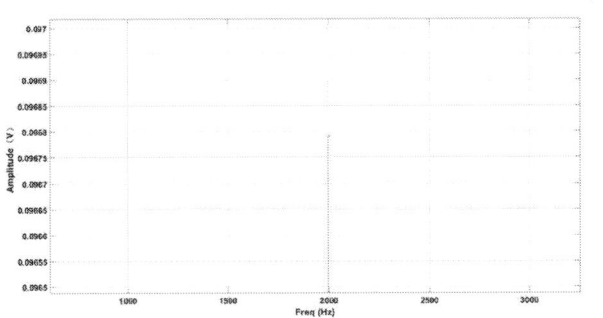

Fig. 11. Frequency domain diagram obtained by Fourier transform

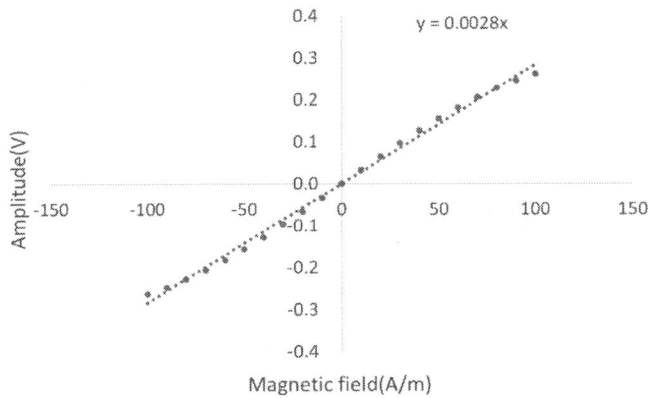

Fig. 12. Second harmonic amplitude of output voltage under different external magnetic fields

tivity is about $3KHz/(A/m)^2$. The traditional fluxgate measurement uses the second harmonic method. The relationship between the amplitude of the second harmonic and the external magnetic field is obtained by Fourier transform of the time-domain diagram of the output voltage, as shown in Fig.10. It can be seen from Fig.12 that the sensitivity of the traditional method to measure the magnetic field is $2.8mV/(A/m)$. Obviously, the method presented in this paper can effectively improve the sensitivity of fluxgate measurement of external magnetic field.

V. CONCLUSION AND FUTURE WORK

This paper presents a new measurement method based on fluxgate sensor, which can fit the curve well by using the relationship between harmonic frequency and measured magnetic field. This method can greatly improve the detection sensitivity. It provides a new method for magnetic field measurement. However, this study did not analyze the effectiveness of this method from the experiment. The next step is to apply this method to the actual fluxgate sensor to further improve its performance.

REFERENCES

[1] D. Hrakova, P. Ripka and T. Kmječ, "Enhancing Performance of Fluxgate Sensors Using Annealed Nanocrystalline Core," in IEEE Transactions on Magnetics, vol. 59, no. 9, pp. 1-9, Sept. 2023.

[2] D. Hrakova, P. Ripka and M. Butta, "Sensitivity and noise of parallel fluxgate sensor with amorphous wire cores," Journal of Magnetism and Magnetic Materials, Volume 563, 2022, 169981.

[3] P. l Ripka, M. Mirzaei and J. Blažek, "Magnetic position sensors," Meas. Sci. Technol. 2022, 33 022002.

[4] J. Maier, P. Ripka and P. Chen, "CMOS-based micro-fluxgate with racetrack core and solenoid coils," Sensors and Actuators A: Physical, Volume 379, 2024, 115886.

[5] P. Priftis, S. Angelopoulos, A. Ktena and E. Hristoforou, "Development of a high-sensitivity orthogonal fluxgate sensor, Journal of Magnetism and Magnetic Materials," Volume 590, 2024. 171646.

[6] L. Wang, S. Zhang and Q. Yuan, "Strain-Induced Frequency Splitting in PT Symmetric Coupled Silicon Resonators," Micromachines 2024, 15, 1278.

[7] C. Zhang, Y. Zhang, X. Wang and H. Meng, "Study of Nonlinear Excitation Circuits for Fluxgate Magnetometer," Sensors 2023, 23, 2618.

A Narrow Band Surface Acoustic Wave Filter Based On 112°XY-LiTaO3

Yusuo Wang   Shanglin Yang   Zhijuan Zhao   Shunjing Lei   Hong Zhang   Juin Jei Liou

Abstract—In this paper, the finite element method is used to design a narrow band filter with a center frequency of 110MHz, and the narrow band and stop band suppression is achieved by cascading two-port resonators. The surface acoustic wave (SAW) filter was simulated by finite element method, the sample was compared with simulation, and the parameters were adjusted according to the actual situation to achieve a narrow band filter with a bandwidth of about 0.29%.

Index Terms—Finite element, narrow band, SAW filter

I. INTRODUCTION

Surface acoustic waves are generated on the solid surface and propagate along the surface. Acoustic RF technology uses these sound waves to intercept and process signals, especially in the rapidly developing RF filter technology. Acoustic filters have been widely used in mobile communication and other fields because of their advantages of small size, low cost and stable performance [1]. Lithium tantalate (LiTaO3) substrate with X cut Y rotation 112° propagation [2], a relatively small frequency temperature coefficient (TCF) and large electrome-chanical factor (K2) are simultaneously achieved, In addition, due to the lack of beam steering and sufficient suppression of the spurious BAW response, is widely used as a substrate for Rayleigh-type SAWs [3].

Finite element analysis (FEA) is a numerical calculation method used to solve problems in structural mechanics, ther-modynamics, fluid mechanics and other fields [4]. Using COMSOL software to perform FEA of surface acoustic wave devices can predict the performance of devices in advance and reduce unnecessary material loss.

In this paper, a high quality SAWF design simulation model based on 112°XY-LiTaO3 is constructed by the two-stage cascade resonant SAWF design model using the FEA design method. Finally, a SAWF whose performance index exceeds the current level and meets the needs of high-end high-fidelity wireless voice system intermediate frequency filtering is de-veloped. The existing requirements are that the insertion loss of the device is greater than -5db, the stopband suppression is less than -30db, and the stopband suppression is less than -35db after the experiment.

Yusuo Wang, Shanglin Yang, Juin Jei Liou are with *School of Electrical and Information Engineering, North Minzu University*, Ningxia 750021, China. Zhijuan Zhao, Shunjing Lei, Hong Zhang were with *the Ningxia Sawtech co., Ltd.*, Ningxia 750011, China. This work is supported by the 2024 Autonomous Region Key Research and Development Program, and the Project ID:2024BEE03005. Corresponding author: (Shanglin Yang, email: ysl029@163.com)

II. METHODOLOGY

A two-port resonator, whose two reflective gates are located on either side of a conventional transversal filter [5], is shown in Figure 1.If the resonant frequency of the device on the reflection gate is the same as the resonant frequency of the interfinger transducer (IDT), the transmission admittance will become large when resonant, so that narrow band and low loss passband can be achieved [3]. The out-of-band suppression can be doubled by series cascade, and the insertion loss will be increased at the same time, so the two-stage cascade structure is used in this paper, as shown in Figure 2.

Fig. 1. Two-port resonator.

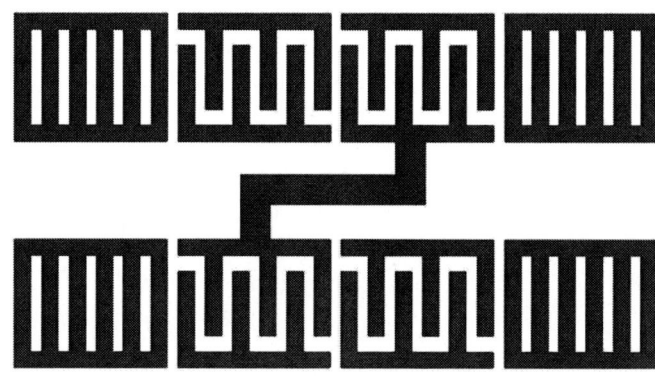

Fig. 2. Two-port resonator two-stage cascade.

COMSOL software is used to build the model. Due to the huge computational amount of the 3D model, as shown in Figure 3, the section 2D model can be used for simulation. The 2D model studies the section plane of the vertical cut in the aperture direction on the basis of the 3D model, and the length of the aperture is added in the simulation process, as shown in Figure 4. However, in the simulation process of 2D model, the effect of electrode edge effect, cascade acoustic radiation and shallow body wave cannot be considered, which will affect stopband suppression and parasitic response. Although 2D analytical models give very good speed performance for precise filter design, their application is limited to structures with well-managed 3D properties such that 2D approximation is fair enough [6].

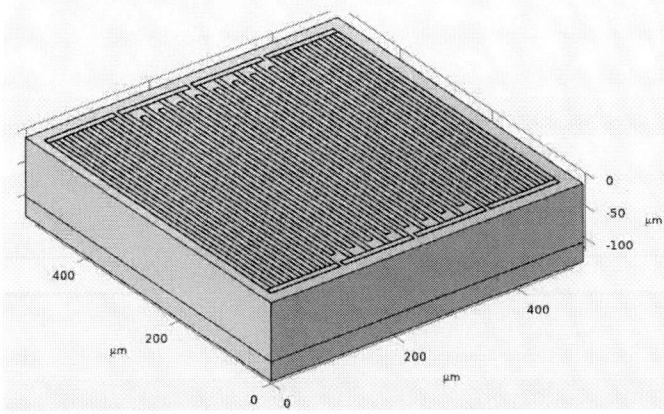

Fig. 3. 3D model of two-port resonator.

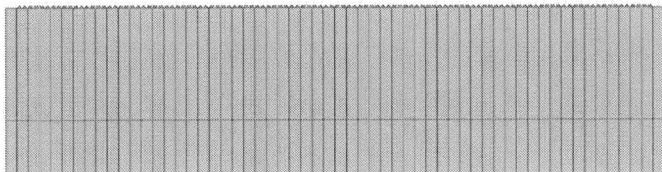

Fig. 4. 2D model of two-port resonator.

III. SIMULATED AND EXPERIMENTAL RESULTS

In this section, the design of a narrow band SAW filter with a center frequency of 110MHz is introduced. As Figure 5 shows, according to the working principle of SAW, the SAW stimulated by IDT can be regarded as the superposition of multiple pairs of IDT excitation signals. Then the central frequency [7] of IDT is:

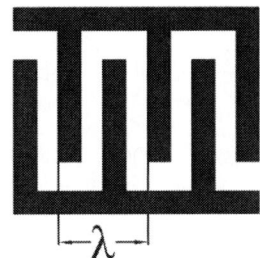

Fig. 5. The basic model of the IDT.

$$f_0 = \frac{v_s}{\lambda} \qquad (1)$$

where v_s is the SAW velocity and f_0 is the SAW central frequency and λ is the SAW wavelength. According to the

existing process, the propagation speed of SAW is 3270m/s, considering and the center frequency of 110MHz the acoustic wavelength is calculated to be 29.727μm.

The bandwidth is affected by the logarithm of the IDT and the electromechanical coupling coefficient of the substrate, which limits the maximum bandwidth. When the logarithm of the IDT is large enough, the double end pair can be roughly calculated by the following formula.

$$\frac{\Delta f_{-3dB}}{f_0} = 0.3184 \frac{2}{N} \qquad (2)$$

Where Δf_{-3dB} is -3db bandwidth and N is IDT refers to logarithm. According to the formula, the logarithm of the cross finger is 58, and the logarithm is finally 70 after experimental adjustment.

COMSOL was used to simulate levels 1 to 3 cascades in a 2D model, as shown in Figure 6. As can be seen from the figure, the insertion loss gradually increases with the increase of order, and the stopband suppression becomes better and better. Due to the actual size of the device, as well as insertion loss and stopband suppression, the two-stage cascade is selected.

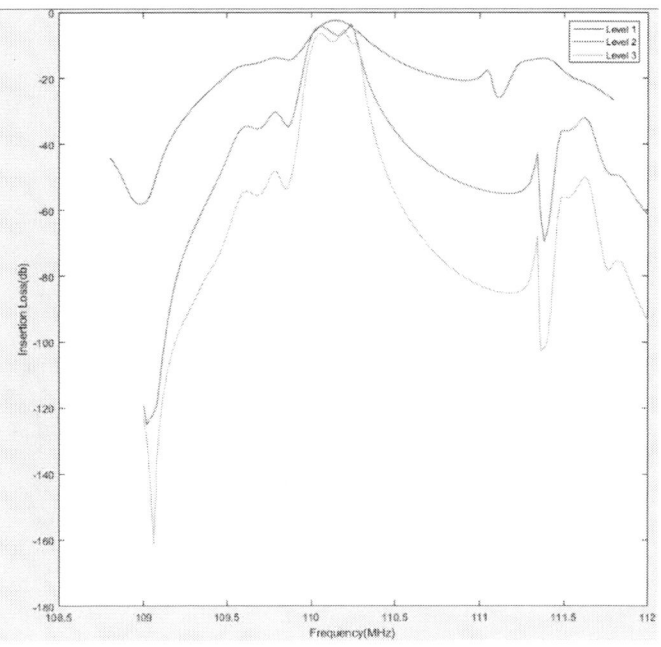

Fig. 6. Cascaded simulation of 1 to 3 levels in a 2D model.

In COMSOL software, the 2D model uses suspension potential for two-stage cascade connection, perfectly matching layer is set at the boundary [8], uniform fingertip transducer is set, and the film thickness ranges from 0.65μm to 1μm with an interval of 0.05μm. The simulation results are shown in Figure 7.

The simulation results show that as the film thickness increases and the center frequency shifts to the left, the passband defect becomes larger and larger. According to the

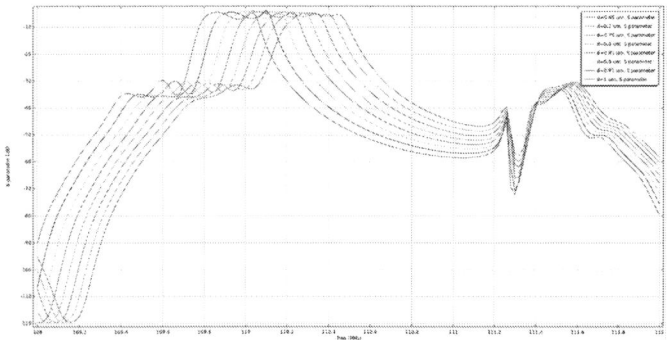

Fig. 7. Simulation film thickness parameters.

simulation results, samples were made to test different film thicknesses, as shown in Figure 8.

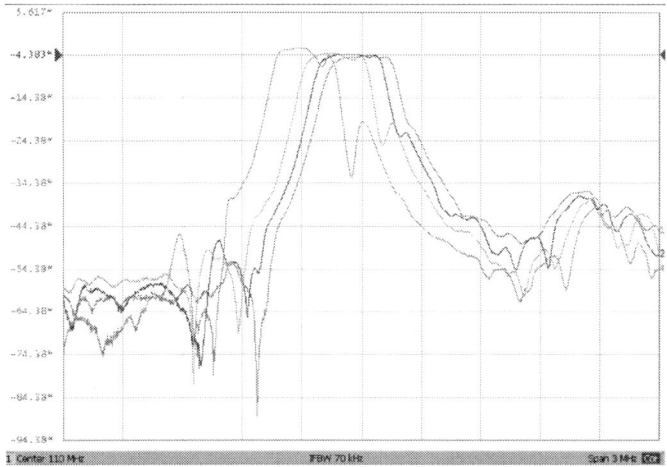

Fig. 8. Sample film thickness parameters.

The film thickness of the sample from left to right was 1μm, 0.9μm, 0.75μm and 0.7μm. The film thickness of 0.9μm was selected to compare the simulation results with the sample results as shown in Figure 9.

The comparison results show that dielectric loss, resistance caused by metal device connection and interaction between two sets of transducers are not considered in the software simulation process. But the trend is about the same. The film thickness of 0.7μm was obtained by adjusting parameters according to the actual sample conditions, as shown in Figure 10.

According to the experimental results of the sample, the center frequency is 110MHz, the bandwidth is 317.79KHz, and the insertion loss is 4.383db. The narrow band filter with a bandwidth of about 0.29% is achieved.

IV. CONCLUSION

In this paper, COMSOL software is used to design a narrow band filter with a center frequency of 110MHz, and 2D simulation is carried out according to the given parameters, and the real test results are compared with it. In the simulation

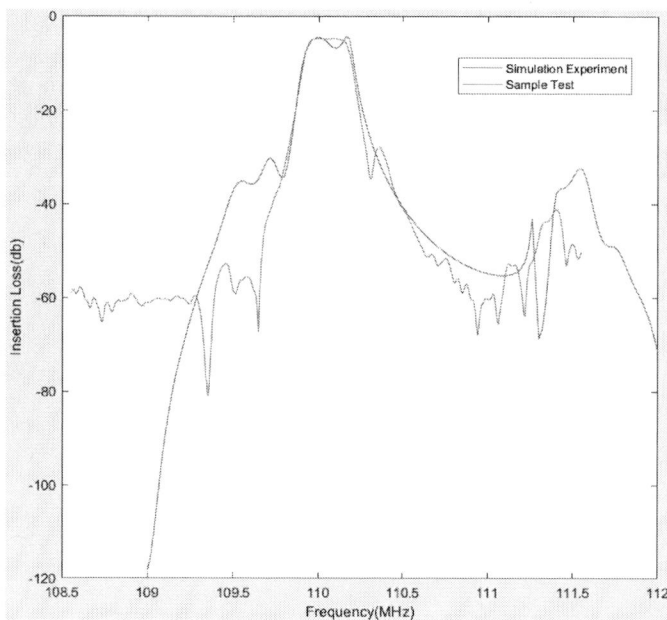

Fig. 9. The simulation results were compared with the experimental results..

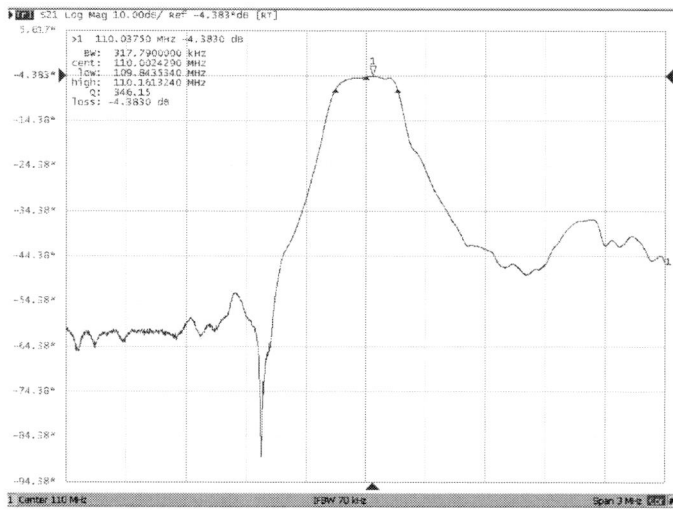

Fig. 10. Simulation of samples with film thickness of 0.7μm.

process, dielectric loss and resistance effect caused by metal connection are not considered, so parameters need to be adjusted according to the actual situation after making the real object. Finally, a narrow band filter with a bandwidth of about 0.29% is realized.

REFERENCES

[1] Chen Chong, Ma Ming-Yuan, Pan Feng, Song Cheng. Magneto-acoustic coupling: Physics, materials, and devices. Acta Phys. Sin., 2024, 73(5): 058502.

[2] D. Irzhak, K. Pundikov, D. Roshchupkin,Measuring of the surface acoustic wave amplitude in the X-112° Y-cut of a LiTaO3 crystal using X-ray diffraction at the laboratory X-ray source, Materials Letters, Volume 374, 2024, 137191, ISSN0167-577X.

[3] Hashimoto, Ken-ya. (2000). Surface acoustic wave devices in telecommunications. Modelling and simulation. 10.1007/978-3-662-04223-6.

[4] K. Dbich, T. Laroche, S. Ballandras, M. Mayer, X. Perois and K. Wagner, "An optimal 2D and 3D modelling of finite SAW and BAW devices based on Perfectly Matched Layer method," 2012 European Frequency and Time Forum, Gothenburg, Sweden, 2012, pp. 200-205.

[5] C. -M. Lin, Y. -Y. Chen, V. V. Felmetsge, D. G. Senesky and A. P. Pisano, "Two-port filters and resonators on AlN/3C-SiC plates utilizing high-order Lamb wave modes," 2013 IEEE 26th International Conference on Micro Electro Mechanical Systems (MEMS), Taipei, Taiwan, 2013, pp. 789-792.

[6] V. Yantchev, P. Turner and V. Plessky, "COMSOL modeling of SAW resonators," 2016 IEEE International Ultrasonics Symposium (IUS), Tours, France, 2016, pp. 1-4.

[7] Li Y, Shao M, Jiang B, et al. Surface acoustic wave pressure sensor and its matched antenna design[J]. Measurement and Control, 2019, 52(7-8): 947-954.

[8] Zhang H, Wang H. Investigation of Surface Acoustic Wave Propagation Characteristics in New Multilayer Structure: SiO2/IDT/LiNbO3/Diamond/Si. Micromachines. 2021; 12(11):1286.

A 1280×1024 DI ROIC with On-chip ADC and LVDS For Infrared Focal Plane Arrays

Jia Shen, Yan Dong, Yaoxu An, Yao Li, Mao Ye*

School of Microelectronics, Tianjin University, Tianjin, 300072, China
*Email: mao_ye@tju.edu.cn

Abstract—This paper proposes a readout integrated circuit (ROIC) for Germanium Infrared Focal Plane Arrays (IRFPA), which adopts a pixel-level compensation circuit to offset the integrated output voltage, thereby mitigating performance degradation caused by dark current. A programmable gain amplifier (PGA) is integrated into the column channel to amplify the reduced voltage swing, thereby enhancing the dynamic range. The output voltage of the column processing circuit is then digitized using 11.98-bits pipelined successive approximation (Pipeline SAR) analog-to-digital converter (ADC). Due to the multi-channel output of the ROIC, multiple ADCs also output digital codes in parallel and transmit them through the Low Voltage Differential Signaling (LVDS) interface. The LVDS eye diagram has an max eye height of 686 mV and a deterministic jitter of 59.6 ps. The clock signals required by the ADC and LVDS are provided by a phase-locked loop (PLL) circuit.

Keywords—ROIC, dark current cancellation, Pipelined SAR ADC, LVDS, PLL.

I. INTRODUCTION

Germanium-based infrared detectors exhibit a high infrared absorption coefficient in the wavelength range from 0.8 μm to 1.8 μm and hold immense potential for development in fields such as aerospace remote sensing, industrial inspection and military security. With the advancement of infrared technology, the demand for high-quality imaging using germanium detectors is increasing. The defects caused by lattice mismatch between germanium and silicon substrates, the narrow bandgap nature of germanium material itself, and the effects of surface and interface states all contribute to a significant dark current in germanium detectors. The existing dark current greatly reduces the voltage dynamic range, making it necessary to design high performance readout circuits to mitigate this issue with germanium-based detectors [1].

This paper proposes a 1280×1024 DI ROIC for germanium-based detector with a pixel pitch of 10 μm, which has the function of dark current compensation. It is equipped with an on-chip ADC and LVDS for further processing. Additionally, the sampling clock for the ADC and the clock signals for the LVDS interface are provided by a PLL. Simulation results show that the proposed ROIC can improve the readout efficiency by outputting voltage signals through 8 channels. The Pipeline SAR ADC used for converting the analog voltage signals output from the ROIC achieves an effective number of bits (ENOB) of 11.98 bit. The parallel output data of the ADC is converted into serial data through the LVDS interface and then output. Simulation results show that the LVDS eye

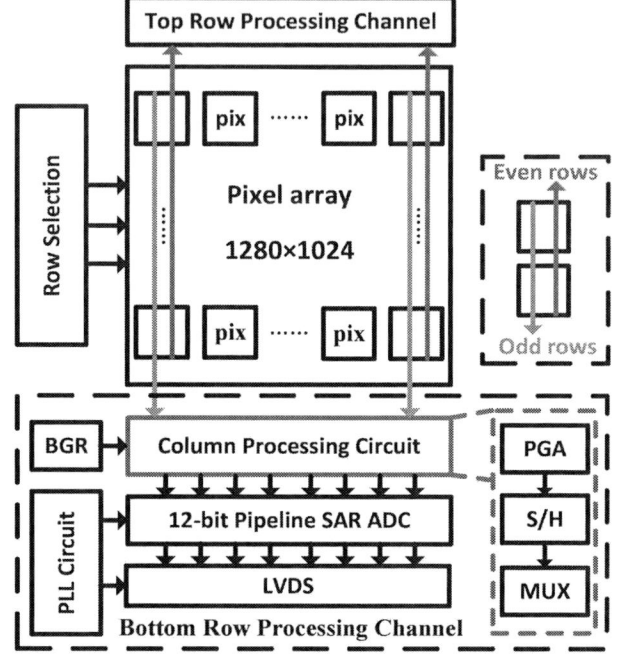

Fig. 1. System architecture of the proposed infrared focal plane array readout circuit.

diagram has an max eye height of 686 mV and a deterministic jitter of 59.6 ps. The input clock frequency of the PLL is 5 MHz, and simulation results show that the lock frequency is 320 MHz.

II. CIRCUIT DESIGN

A. ROIC Architecture

The overall architecture of the readout circuit is shown in Fig. 1, which mainly comprises a DI-based pixel array, column processing circuit, 12-bit Pipeline SAR ADC, and LVDS. In this architecture, the pixel circuit which has the function of dark current compensation, converts the photocurrent signal from the detector into a voltage signal and outputs it in the form of a source follower. The core circuits in the column channel are the PGA, the sample-hold circuit, and the MUX driver circuit. The output voltage from the pixel units is further amplified by the PGA, and after sampling, it is selected and driven for output via the MUX driver. In order to improve the readout rate, the readout circuit adopts multi-channel

979-8-3315-2209-4/25 $31.00 © 2025 IEEE
388

output mode, and the column output bus is segmented to achieve 8-channel outputs. The column readout channels are placed at both ends of the pixel array, separately reading out the odd and even rows. The multi-channel outputs from the column processing circuit are quantized and converted to digital signals by 12-bit Pipeline SAR ADC. The parallel output data from the ADC is serialized and then output through the LVDS interface. The sampling clock of the ADC and the clock signal of the LVDS are provided by a PLL circuit. The bandgap reference (BGR) circuit provides reference voltages and bias currents for the aforementioned circuit modules.

Through the row, column selection circuit and other digital control circuit, control the working sequence of the whole circuit. The readout circuit can be configured through SPI protocol, which supports integrate then read (ITR) / integrate while read (IWR) mode, global / rolling shutter mode, bias current adjustment and reference voltage trimming.

B. Pixel Circuit Design

The design of pixel circuits is crucial in the readout circuit, as it integrates the photocurrent signals generated by the detector and converts them into voltage signals, directly affecting the performance of the imaging system and the quality of the image.

The proposed direct injection pixel schematic is illustrated in Fig. 2. When the RST_INT switch is open, the integration process begins, and the photocurrent is integrated onto the integration capacitor Cint. The voltage Vint across Cint is proportional to both the integration time and the value of the photocurrent. Before the integration process ends, the RST_SH switch is closed to reset the sampled voltage from the previous frame, and the voltage on the sampling capacitor CSH is reset to GND. Subsequently, the SH switch is closed, causing the voltage on CSH to briefly follow the integration voltage of the current frame, and the output is ultimately controlled by the Row_en switch through a source follower. Finally, the RST_INT switch is closed to terminate the integration process and reset the integration capacitor Cint [2].

To mitigate the issue of dark current, a dark current compensation structure is implemented within the pixel unit. Through this structure, during the sampling phase, the switch rstb_vos connects the bottom plate of the sampling capacitor CSH to the supply voltage VDD, so that the sampled charge on the capacitor is (VDD-Vint)×CSH. During the readout phase, the voltage on the bottom plate of CSH is switched from VDD to Vos via the switch rst_vos. According to the principle of charge conservation, the voltage on the bottom plate becomes Vint-(VDD-Vos). Therefore, the output voltage of the pixel circuit can be adjusted by setting different Vos voltages to compensate for varying levels of dark current. The pixel output voltage reaches a maximum of 2 V, and the linearity exceeds 99.95%.

Fig. 2 illustrates the layout of the pixel cell circuit. In the design, MOS capacitors and MOM capacitors are employed. The DI structure itself occupies a compact area, enabling higher capacitance density within the limited space, thereby

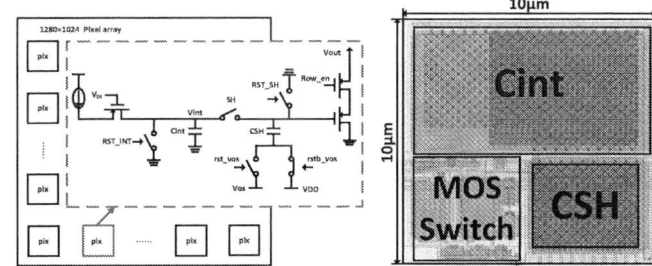

Fig. 2. Schematic and layout of the proposed pixel circuit with dark current compensation.

increasing the full-well capacity of the pixel cell and enhancing its ability to integrate light signals. The full-well charge capacity of the pixel cell can reach up to 17.5 Me-.

C. Column Channel Circuit Design

After the pixel circuit completes the sample-and-hold process, the pixel signals are transferred to the column processing channel under row control. Fig. 3 shows the circuit diagram of the column processing channel, which is mainly composed of PGA , interleaved sampling circuit and MUX driver.

The PGA is designed in the column processing circuit to compensate for the reduced output voltage swing caused by the dark current cancellation circuit in the pixels. The operation of the PGA is divided into two stages. During the sampling phase, switch CK_PGA is closed. The pixel voltage is sampled onto the sampling capacitor Cs, while the operational amplifier is reset to a unity feedback mode. The voltage at the operational amplifier output node is Vref. Subsequently, the switch CK_PGA opens, placing the PGA circuit in the amplification state. Meanwhile, the voltage at the operational amplifier output node becomes (Vin-Vref)×Cs/Cf+Vref. Ultimately, voltage amplification is achieved through the charge relationship between these two stages.

In the design of traditional readout circuits, the time during which the PGA drives a large sampling capacitor within each row cycle is very short, while most of the time is used for multiplexing voltages between columns. When the PGA is configured for high gain, the bandwidth requirements for the operational amplifier increase, leading to higher power consumption. This paper uses an interlaced sampling technique, which optimizes time allocation and reduces power consumption to address the aforementioned issues. As shown in Fig. 3, the switches CK_odd and CK_even perform interleaved sampling. When the switch CK_odd is closed and CK_even is open, Cs1 samples, while the voltage previously sampled by Cs2 is output to the subsequent MUX driver. The MUX driver selects and outputs the sampled voltage, and performs the reset operation through the switch CK_MUX. Interlaced sampling extends the settling time of the PGA to nearly a full row cycle. This approach reduces the bandwidth requirements of the PGA operational amplifier, decreases power consumption, and maintains high frame rate readout efficiency. The layout of the

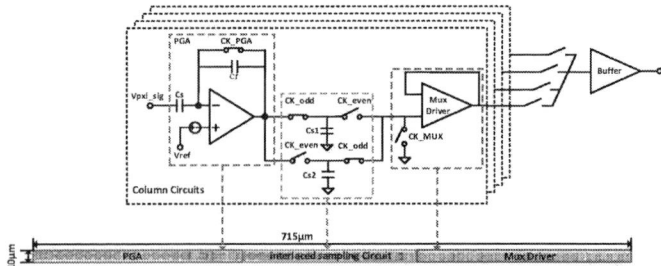

Fig. 3.   Schematic and layout of the proposed Column Channel.

column processing channel is shown in Fig. 3, corresponding to the circuit diagram.

D. ADC and LVDS Circuit Design

The multi-channel analog voltages by the ROIC are further processed, convert into 12-bit parallel digital data through a Pipeline SAR ADC, and finally output as high-speed serial digital data via an LVDS interface.

The analog voltage output from the ROIC is quantized and converted into a digital signal by the ADC. The Pipeline SAR ADC architecture used in this paper is shown in Fig. 4(a). The ADC consists of a cascade of a 6-bit and an 8-bit SAR ADC [3]. In the SAR ADC, the CDAC (capacitive digital-to-analog converter) converts digital signals into analog signals for comparison by the comparator. The SAR logic adjusts the state of the CDAC based on the output of the comparator to approach the input analog signal, and generates the final digital output signal. The 6-bit digital data from the first stage and the 8-bit digital data from the second stage are processed through delay calibration and digital correction logic to obtain a complete 12-bit digital code output. The sampling rate is 5 MHz, and the ENOB is 11.98 bits. The layout of the Pipelined-SAR ADC in this paper is shown in Fig. 4(a), with an area of 0.23 mm². The differential circuit of the CDAC is symmetrically laid out to reduce the impact of noise on circuit performance.

The parallel 12-bit data output from the ADC is further processed and then output via the LVDS interface. Fig. 4(b) illustrates the schematic and layout of the LVDS, occupying an area of 0.02 mm². The data is first converted into serial data through a parallel to serial circuit and then transmitted via the LVDS interface. In the LVDS_core, the transmitter drives the differential signal line with a 3.5mA constant current source [4]. The current flows through a 100Ω termination resistor to generate a voltage swing of 350mV. This design not only achieves low power consumption and low voltage swing but also improves interference resistance and signal integrity through the characteristics of differential signaling.

E. PLL Circuit Design

This paper designs a programmable PLL circuit with a frequency divider that uses binary coding to control the division ratio, providing clock signals for the ADC and LVDS, and achieving frequency multiplication of the PLL through

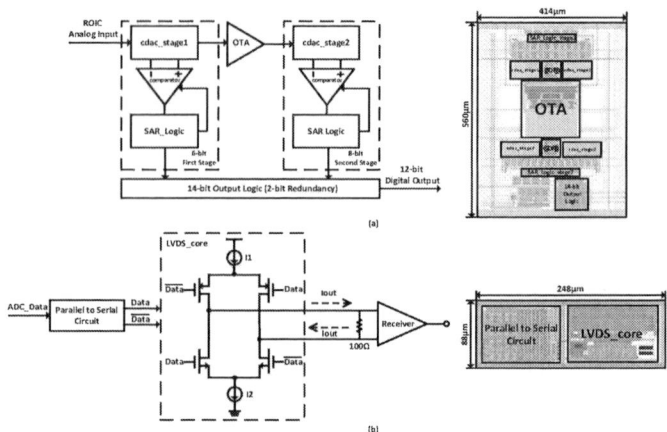

Fig. 4.   ROIC Output Processing Circuit. (a) 12-bit Pipeline SAR ADC architecture and layout; (b) LVDS main architecture and layout.

coding control [5]. Fig. 5 illustrates the overall structure of the proposed PLL, which is composed of a phase frequency detector (PFD), a charge pump (CP), a loop filter (LPF), a voltage controlled oscillator (VCO), and a frequency divider. The PFD compares the frequency and phase difference between the input clock CLKref and the feedback clock CLKfed. The CP charges and discharges the LPF based on the output signals from the PFD. The LPF filters the current signal output from the charge pump, generating a control voltage. The VCO adjusts its output frequency based on this control voltage. The output frequency is then divided by the divider to generate the feedback clock CLKfed. The VCO adjusts its output frequency until the phase difference reaches its minimum value, achieving frequency and phase synchronization.

The basic schematic of the programmable frequency divider is shown in Fig. 5. The main circuit consists of a series of D Flip-Flops (DFF) that can be set to 0 or 1, forming a sequential state machine. The division ratio is determined by setting the initial value of in<0-6>, which establishes the initial state of the Q values. When the input clock CLK_in arrives, the output of the DFF toggles, and the state machine begins to count down. With each input clock edge, the state value decreases by one. When the state value reaches the detection value set by the period detection circuit, an output reload clock is generated, completing one counting cycle. The reload clock is then utilized to transmit the reload signal and reset Q<0-6>.

By setting different division ratios, the entire PLL can be programmed to set the frequency multiplication ratio. In this paper, a 5 MHz square wave signal is used as the input to the PLL. Through programming, clock signals with frequencies of 80 MHz, 160 MHz, and 320 MHz are generated to serve as the clock signals required by the ADC and LVDS. The layout of the PLL is depicted in Fig. 5, where the digital modules are separated from the analog modules, with an area of 0.13 mm².

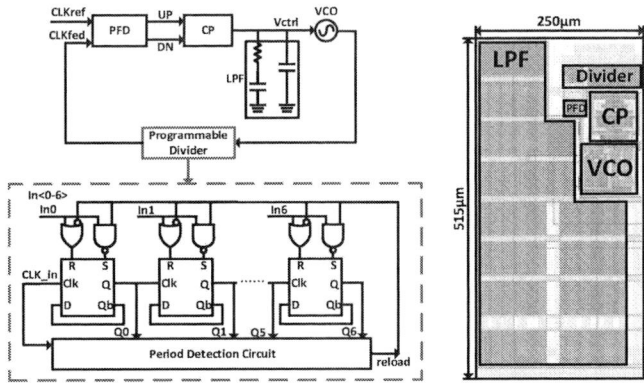

Fig. 5. Schematic and layout of Programmable Frequency Divider.

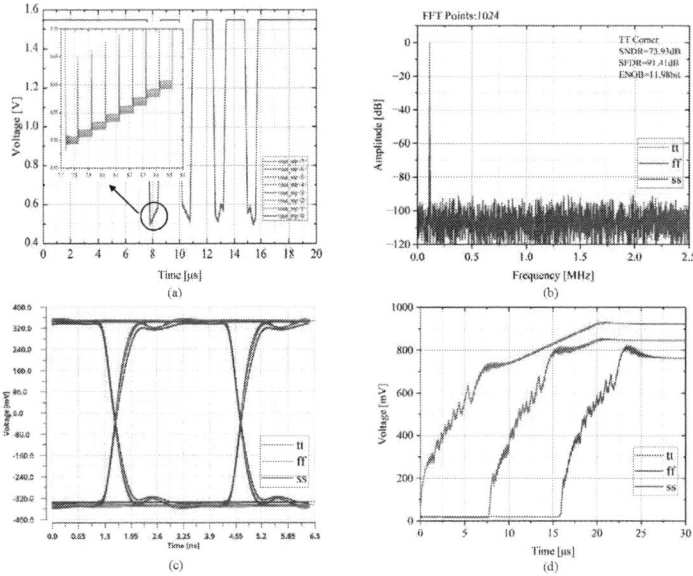

Fig. 6. The simulation results of the core main modules: (a) the simulation output voltage of an 8-channel ROIC for a frame; (b) ADC output spectrum under different process corners; (c) LVDS Eye Diagram Results under different process corners; (d) the control voltage Vctrl waveform of the PLL under different process corners.

III. SIMULATION RESULTS

Fig. 6 shows the simulation results of the main modules of the circuit. Due to the large scale of the entire array in the circuit, a 64×8 array is selected for the overall simulation of the infrared focal plane readout circuit. Fig. 6(a) shows the simulated output voltage waveforms of the top processing channel for one frame in eight-channel mode. Since the circuit reads out the odd and even rows separately, the figure displays the output voltage of the top processing channel for 4 rows. Due to the 8-channel output, the original 64 analog voltages in each row are divided into 8×8 voltage values through the 8 channels. As shown in the enlarged portion of Fig. 6(a), the voltage of one row is output through eight channels, vout_top<0:7>.

Fig. 6(b) illustrates the output spectrum of the Pipeline SAR ADC under different process corners at a sampling frequency of 5 MHz. Under the tt process corner, it achieves a signal-to-noise ratio (SNDR) of 73.93 dB and a spurious-free dynamic range (SFDR) of 91.41 dB at a sampling rate of 5 MS/s, with an ENOB of 11.98 bit.

Fig. 6(c) shows the eye diagram simulation results of LVDS with eight parallel 12-bit binary signal inputs under different process corners. Among them, under the tt process corner, the deterministic jitter of the LVDS is 59.6 ps, with an max eye height of 686 mV and an eye width of 2.99 ns. The simulation results indicate that the eye height is sufficiently high to correctly interpret the transmitted data.

In this paper, the PLL is designed with an input reference clock frequency of 5 MHz and a division ratio set to 64. A time-domain simulation of the entire PLL is conducted, resulting in the control voltage Vctrl waveform as shown in Fig. 6(d). Simulation results show that the PLL can lock under different process corners with a 1.8V power supply voltage at room temperature. In the simulation, the power supply voltage is set to rise to 1.8 V within 20 us. Under the tt process corner, the lock voltage stabilizes at 845 mV, and the lock time is less than 10 us. Subsequently, signals that meet the requirements of other modules are obtained by further processing the PLL output.

IV. CONCLUSION

In conclusion, we propose a 1280×1024 array infrared readout circuit with on-chip ADC and LVDS. A PGA is designed in the column processing circuit to expand the pixel output swing, which is reduced due to the dark current compensation function. Interlaced sampling techniques and 8-channel output are employed to enhance the readout rate. The output of the ROIC is processed on-chip through ADC and LVDS, with the clock signals required by the ADC and LVDS provided by a programmable frequency multiplier PLL.

REFERENCES

[1] L. Colace, G. M. A. Altieri and G. Assanto, "Waveguide photodetectors for the near-infrared in polycrystalline Germanium on silicon," in IEEE Photonics Technology Letters, vol. 18, no. 9, pp. 1094-1096, May 1, 2006.

[2] C. N. Kunnatharayil, S. Abbasi, O. Ceylan and Y. Gurbuz, "A Low-Noise 320×240 Digital ROIC for SiGe Microbolometers with a Fast Converging Offset Calibration Technique," 2021 IEEE International Symposium on Circuits and Systems (ISCAS), Daegu, Korea, 2021, pp. 1-5.

[3] Z. Xu, M. Miyahara and A. Matsuzawa, "A 3.6 GHz fractional-N digital PLL using SAR-ADC-based TDC with-110 dBc/Hz in-band phase noise," 2015 IEEE Asian Solid-State Circuits Conference (A-SSCC), Xiamen, China, 2015, pp. 1-4.

[4] R. A. Melo and B. Valinoti, "Serial QDR LVDS High-Speed ADCs on Xilinx Series 7 FPGAs," 2019 X Southern Conference on Programmable Logic (SPL), Buenos Aires, Argentina, 2019, pp. 25-30.

[5] J. Wadekar, B. Chattopadhyay, R. Mehta and G. Nayak, "A 0.5-4GHz Programmable-Bandwidth Fractional-N PLL for Multi-protocol SERDES in 28nm CMOS," 2016 29th International Conference on VLSI Design and 2016 15th International Conference on Embedded Systems (VLSID), Kolkata, India, 2016, pp. 236-239.

POCT system based on lens-less imaging for complete blood counting

Yu Li   Zhaochen Huo   Jianing Li   Bing Chen   Ya Li   Xiaonan Yang

Abstract—**As a basic and important test item in clinical medicine, complete blood count (CBC) has a wide range of application prospects. This study presents a portable blood cell detection method based on lens-free imaging technology and the YOLO model, enabling high-throughput classification and counting of blood cells over a large field of view. Experimental results demonstrate that the detection models for RBC, WBC, and PLT achieve precision rates of 95.94%, 98.83%, and 97.40%, respectively, on the test set. This work provides a novel approach for the miniaturization and intelligent automation of CBC technology.**

Index Terms—**lens-free, large field of view, CBC, high throughput, POCT**

I. Introduction

In recent years, with the rapid advancement of microfluidic technologies and biomedical detection systems, portable diagnostic devices have found widespread applications in medical diagnostics, health monitoring, and POCT (Point-of-Care Testing). The core of these technologies lies in the use of miniaturized systems and algorithms to efficiently detect and analyze target samples, bypassing the need for traditional, costly, and complex equipment. These devices, characterized by their low cost, portability, and ease of use, hold significant promise in resource-limited healthcare settings and primary care environments [1].

Complete Blood Count (CBC) is one of the most fundamental and critical diagnostic tests in clinical medicine, playing a vital role in disease diagnosis, treatment evaluation, and health monitoring. As a cornerstone of hematological assessments, CBC is widely employed in the diagnosis and management of conditions such as anemia, infections, and leukemia, and remains one of the most common laboratory tests [2]. However, current mainstream methods for CBC still have certain limitations in terms of equipment, operational complexity, and applicability.

At present, CBC is primarily performed using three methods: manual counting, automated counting, and flow cytometry. Manual counting involves direct cell enumeration from blood smears under a microscope, it is simple to operate but highly dependent on the experience of the technician prone to human error, and suffers from low efficiency and limited throughput. Automated blood cell counters, based

on the Coulter principle, employ high precision systems to achieve rapid and accurate cell counts [3]; flow cytometry, which uses fluorescent dyes to label cell surface antigens, offers high sensitivity and the ability to detect complex blood compositions [4],However, these devices are costly, require specialized operational environments, and are typically confined to professional laboratories, making them unsuitable for portable or widespread use.

The advent of lens-free imaging technology provides a promising solution for POCT-based CBC. In 2006, Changhuei Yang's research team introduced the lens-free imaging system [5], which discards traditional optical lenses and instead uses a digital photodetector array to directly capture the diffraction patterns of transmitted light. This innovative approach significantly simplifies the design of traditional imaging systems, substantially reducing the size and cost of the device, and offers considerable advantages for POCT applications. For instance, Carlos Buitrago Duque and colleagues developed a POCT device for blood smear observation using lens-free imaging [6], while Tongge Li and his team utilized lens-free systems for high throughput viral infection cell visualization [7]. These studies demonstrate that lens-free imaging, with its wide field of view, simple structure, and high throughput, is particularly well-suited for integration into POCT devices for rapid processing of large sample volumes.

Compared to traditional imaging systems, lens-free imaging technology offers several notable advantages, including simplicity, ease of integration, high throughput, and large field of view. By directly coupling optical sensors with the sample region, lens-free systems can cover a broader detection area while preserving the ability to resolve single cells, making them ideal for fast detection of densely packed samples. Moreover, since lens-free systems do not require complex optical alignment or precision mechanical structures, they can operate stably in resource-constrained environments, meeting the stringent portability and adaptability requirements of POCT devices.

Building on these technological advantages, we developed a lens-free imaging POCT system integrated with microfluidic technology. This system dynamically captures blood cell images through lens-free imaging and utilizes YOLOv11 for blood cell classification. Experimental results show that the system achieves detection accuracies of 95.94%, 98.83%, and 97.40% for RBC, WBC, and PLT, respectively.

Yu Li , Zhaochen Huo ,Jianing Li and Xiaonan Yang are with *School of Electrical and Information Engineering , Zhengzhou University*, Zhengzhou, China. Bing Chen and Ya Li are with *Department of Gastroenterology , The First Affiliated Hospital of Zhengzhou University* , Zhengzhou, China. Corresponding author: (Xiaonan Yang, email:iexnyang@zzu.edu.cn)

979-8-3315-2209-4/25 $31.00 © 2025 IEEE

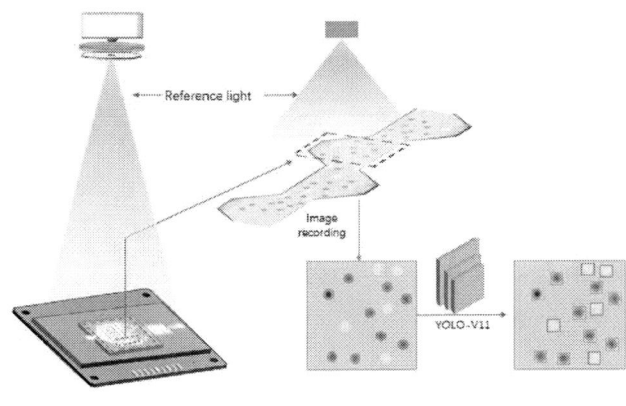

Fig. 1. System schematic diagram.

II. SYSTEM OVERVIEW AND PRINCIPLE

The schematic diagram of the lens-free imaging system is shown in Fig. 1. The system primarily consists of three modules: the image acquisition module, the light source module, and the sample introduction module. The image acquisition module uses a black-and-white CMOS image sensor. Unlike color sensors, which have a color filter array on each pixel, black-and-white sensors reduce light loss during the imaging process, improving image quality. The light source module includes a blue LED, a 455 nm optical filter, and a 50 μm aperture, meeting the temporal and spatial coherence requirements for holographic imaging. The light source module is integrated into a top light shield and positioned approximately 5 cm from the sample plane.

During operation, the microfluidic chip is placed directly on the CMOS image sensor, ensuring that the imaging area matches the effective area of the sensor. The image sensor used is the Hikvision MV-CB120-10UM-B black-and-white board camera, with an imaging area of 7.459 mm by 5.616 mm (pixel size = 1.85 μm × 1.85 μm), and a field of view (FOV) of 41.89 mm², approximately 50 times that of a conventional microscope. Due to the large field of view, eight sub-channels were designed within the microfluidic imaging region. The system captures video at a frame rate of 28 fps, and after video capture, the frames are transferred to a computer for processing. Once the video is decomposed into individual frames, crop each frame into 8 areas of 680 x 2700 according to the channel position. Subsequent processing steps include background removal, holographic image reconstruction, super-resolution enhancement, and object identification and counting.

A. Sample Preparation

To ensure high-quality representative data for training the YOLOv11 network, the samples of RBC, WBC, and PLT were separated and purified. WBC preparation utilized RBC lysing solution (Cat R1010) provided by Solabio. Fresh whole blood was diluted and mixed with the RBC lysing solution, incubated on ice for 15 minutes to lyse the red blood cells. The sample was then centrifuged at 450×g for 10 minutes to remove the supernatant, yielding WBC sediment. The WBC pellet was washed twice with phosphate-buffered saline (PBS) to remove any residual lysing solution and other impurities. The resulting WBC suspension was used directly for subsequent experiments. PLT preparation was based on low-speed centrifugation. Fresh whole blood was centrifuged at a low speed (500×g for 10 minutes) to separate the plasma layer, from which the PLT-containing supernatant was carefully extracted. For further purification, the plasma layer was centrifuged again at low speed (100×g for 20 minutes) to obtain a purified PLT sample. For RBC purification, the blood was first centrifuged at 1000×g for 15 min. Due to the different densities of the components in the blood, the RBCs will be clustered in the lowest layer, the upper plasma and leukocyte layers were discarded, and the RBCs were washed with PBS at a ratio of 1:1. Then the blood was centrifuged again at 1000×g for 10 min and after the second centrifugation, the supernatant was discarded and the remaining RBCs were resuspended to obtain the final RBC sample.

B. Dataset

The dataset was constructed by imaging pure samples of the three blood cell types to ensure the authenticity of each cell type's image. The dataset used for the RBC and WBC detection model contains 23,024 RBC, 42,331 WBC, 12,425 PLT, and 10,782 Overlap instances. The dataset was split into training, validation, and test sets at a ratio of 8:1:1, with no overlap between the sets.

Before training, each cell was labeled with the corresponding cell type using the "labelbee" annotation tool. This ensured that only relevant information from the labeled bounding boxes was extracted during training. The neural network was progressively fine-tuned using the training set to approximate the real-world ground truth. The model's performance was evaluated on the test set, and training was stopped if the accuracy on the validation set significantly decreased or plateaued. All images in the dataset were standardized to a size of 680 × 2700 pixels.

III. RESULTS AND DISCUSSION

A. Dataset

When the light source satisfies both temporal and spatial coherence conditions, the untransmitted light interferes with the transmitted light passing through the sample. This interference pattern is then recorded as both amplitude and phase on the image sensor plane. The prepared RBC, WBC, and PLT samples were introduced into the microfluidic device, and dynamic images were captured. The resulting blood cell images are shown in Fig. 2a.

Due to structural differences among the three types of cells, their holographic images exhibit variations. RBC, with their biconcave shape and relatively smooth surface, induce minimal phase disturbance to the light waves, resulting in fewer diffraction rings and simple, clear patterns in their

---

979-8-3315-2209-4/25 $31.00 © 2025 IEEE

holograms. WBC, with their more complex, irregular surfaces and typically larger diameter, as well as their nucleated structure, cause significant phase shifts in the light, leading to stronger amplitude and more complex interference patterns with additional diffraction rings. PLT, which are small (1.5 μm–4 μm) and irregularly shaped, induce weak scattering and phase disturbances, thus producing shallow diffraction fringes with fewer diffraction rings in their holographic images. Since the PLT imaging fringes are very faint, they are challenging to use in dataset creation and network training, necessitating individual enhancement. We employed the angular spectrum reconstruction algorithm and a blind super-resolution network to enhance the PLT images, as shown in Fig. 2b.

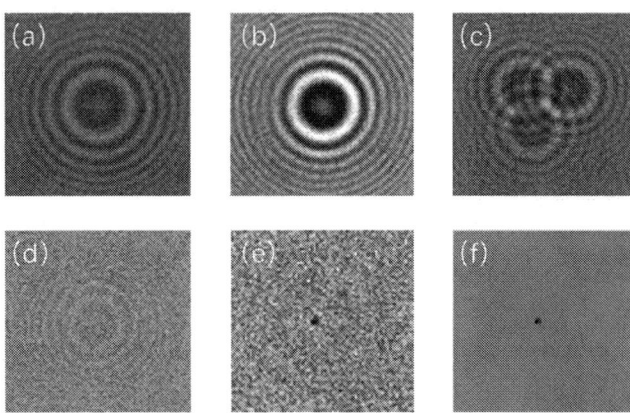

Fig. 2. (a) RBC holographic image (b) WBC holographic image (c) Image formed by interference patterns of overlap holographic fringes (d) PLT holographic image (e) PLT image reconstructed by angular spectrum method (f) PLT image after blind super-resolution processing.

During the flow of cells, the diffraction fringes of overlapping cells are inevitable, which may impact accurate cell counting. In our designed trapezoidal channel, cells move with irregular flow fields and change their relative positions with respect to other cells. To prevent double counting, overlapping cells are labeled as overlap in the dataset. These overlapping targets can be accurately detected, and the integrated Byte-Track algorithm of YOLOv11 is used for continuous tracking until the cells separate into distinct ones, ensuring accurate counting.

B. Model Training

The model was trained based on the YOLOv11 framework. Due to the significant differences between holographic images and traditional RGB images, we employed transfer learning to fine-tune the model. The YOLOv11n model, pre-trained on the COCO dataset, was used as the initial weight. Only the fully connected layers and some convolutional layers were trained, significantly reducing the training time and improving detection performance.

During the training process, we monitored the changes in the Average Train Loss and mean Average Precision (mAP50), as shown in the Fig.3. The validation loss dropped rapidly in the early stages of training and then gradually stabilized,

indicating that the model's loss had converged and further training had limited impact on the train loss. Meanwhile, the mAP50 curve approached 1 and stabilized, verifying that the model had reached an optimal state for object detection tasks.

C. Model Testing

After training, the model's performance was evaluated using an independent test set. The key metrics for the testing phase include Precision and Recall. Precision refers to the proportion of true positives among the predicted positives, while Recall represents the proportion of actual positives correctly predicted by the model.

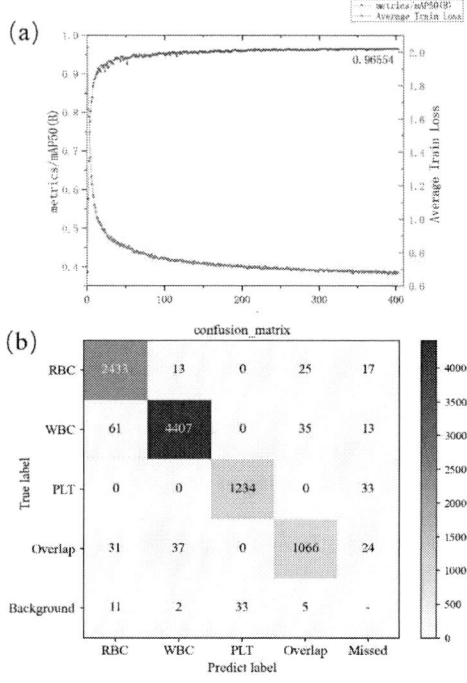

Fig. 3. (a) Average Train Loss and mAP curves of the training process (b) Confusion Matrix of the Model on the Test Set.

TABLE I
CONFUSION MATRIX AND PERFORMANCE EVALUATION

Ground Truth	Predicted					Performance Evaluation	
	RBC	WBC	PLT	Overlap	Missed	Precision	Recall
RBC	2433	12	0	25	17	95.94%	97.79%
WBC	11	903	21	35	13	98.83%	97.59%
PLT	0	0	1234	0	33	94.40%	97.40%
Overlap	31	37	0	1066	24	94.25%	92.06%

As shown in Fig. 3, the detection results of a single frame image validate the accuracy of our method. Table 1 summarizes the detection results for cells in each frame of the test set.

From Table 1, it can be seen that the model achieved detection accuracy rates of 95.94% for RBC, 98.83% for

WBC, and 97.40% for PLT, with corresponding recall rates of 97.79%, 97.59%, and 97.40%. Some misidentifications occurred due to overlapping sizes between WBC and PLT. In addition, during cell flow, the small size of PLT and the fact that they may not be at the same height lead to some missed detections. Meanwhile, cells overlap with each other during flow, and the model achieved 94.25% precision and 92.06% recall for the detection of such cell overlap region.

IV. CONCLUSION

By leveraging the characteristic differences of various cell types in holographic images, we have developed a stable, high-precision, and efficient cell detection system. Experimental results demonstrate that the system can effectively distinguish and count cells flowing through the microfluidic channels. Specifically, the detection accuracies for RBCs, WBCs, and PLTs reached 95.94%, 98.83%, and 97.40%, respectively, while the detection accuracy for overlaps was 94.25%. In the future, the system's throughput and reliability can be further improved by integrating more advanced imaging techniques and optimized detection algorithms.

REFERENCES

[1] Zeng L, et al. AI-Based Portable White Blood Cells Classification and Counting System in POCT. IEEE Sensors Journal, 2024.

[2] Seo I H, Lee Y J. Usefulness of complete blood count (CBC) to assess cardiovascular and metabolic diseases in clinical settings: a comprehensive literature review. Biomedicines, 2022, 10(11): 2697.

[3] Y Xu, X Xie, Y Duan, L Wang, Z Cheng, J Cheng. A review of impedance measurements of whole cells. Biosensors and Bioelectronics, 2016, 77: 824-836.

[4] Z Chao, Y Han, Z Jiao, Z You, J Zhao. Prism Design for Spectral Flow Cytometry. Micromachines, 2023, 14(2): 315.

[5] Heng X, et al. Optofluidic microscopy—a method for implementing a high resolution optical microscope on a chip. Lab on a Chip, 2006, 6(10): 1274-1276.

[6] Buitrago-Duque C, Patiño-Jurado B, Garcia-Sucerquia J. Robust and compact digital Lensless Holographic microscope for Label-Free blood smear imaging. HardwareX, 2023, 13: e00408.

[7] Li T,et al. Virus detection light diffraction fingerprints for biological applications. Science Advances, 2024, 10(11): eadl3466.

A 320×256 15μm-Pitch Readout Circuit with Column-Parallel 84dB-DR 167 KS/s Incremental Sigma-Delta ADCs for IRFPA

Chenxu Zhao   Yao Li   Mao Ye   Qiuwei Wang   Jun Du   Yiqiang Zhao

Abstract—To achieve high-precision, high-frame-rate infrared detectors for wearable devices, this paper proposes a digital readout circuit based on a column-parallel 16-bit incremental sigma-delta ADC for IRFPA. The ADC adopts a third-order feed-forward discrete-time modulator cascaded with a fully customized low-pass digital downsampling filter. By utilizing digital correlated double sampling (CDS) technology, the offset, low-frequency noise, and Fixed Pattern Noise (FPN) in the entire signal chain are reduced, achieving a larger dynamic range. At a speed of 167 KS/s, it achieves 13.2-bit ENOB and 84-dB DR, yielding a Walden Figure-of-Merit (FoM$_W$) of 0.25 pJ/conv, a Schreier Figure-of-Merit (FoM$_{SNDR}$) of 165 dB, and a Dynamic Range Figure-of-Merit (FoM$_{DR}$) of 167 dB. It is applied to a 320×256 15μm-pitch 60dB-DR Capacitive Transimpedance Amplifier (CTIA)-based pixel array designed in a 180 nm 1P4M SOI process. This design achieves a maximum frame rate of 600 Hz and a linearity of over 98%. By applying the digitally correlated double sampling (CDS) technique, the readout noise is significantly reduced by 61% in high gain mode, and 53% in low gain mode, and the maximum dynamic range is expanded to 70 dB, enabling the precise digital readout of weak current signals.

Index Terms—Column-Parallel Sigma-Delta ADC, Digital Correlated Double Sample (CDS), Infrared Focal Plane Array (IRFPA), Readout Integrated Circuit (ROIC), Capacitive Transimpedance Amplifier (CTIA)

I. INTRODUCTION

Infrared detectors used in wearable devices need to be miniaturized, high-precision, and high-frame-rate, requiring customized design to achieve the optimal solution [1]. On-chip digitization has become a trend in infrared detection technology, leading to the development of ADCs from off-chip to cellular, column-parallel, and in-pixel designs. Moving the ADC forward in the signal chain has many benefits, such as reducing the interference of analog signals, improving accuracy, and increasing frame rate. The column-wise ADC is currently a more effective solution [2]–[6]. With limited power consumption and area constraints, the 13-bit accuracy of various column-wise ADCs in imaging applications has become a bottleneck.

This paper proposes a column-parallel digitized readout channel that meets the 84-dB DR and 167 KS/s application

Chenxu Zhao, Qiuwei Wang, and Jun Du are with *the School of Microelectronics, Tianjin University*, Tianjin 300072, China. Yiqiang Zhao, Yao Li, and Mao Ye are with *the School of Microelectronics and Tianjin Key Laboratory of Imaging and Sensing Microelectronic Technology, Tianjin University*, Tianjin 300072, China. Corresponding authors: (Yiqiang Zhao, email: yq_zhao@tju.edu.cn; Yao Li, email: liyao@tju.edu.cn)

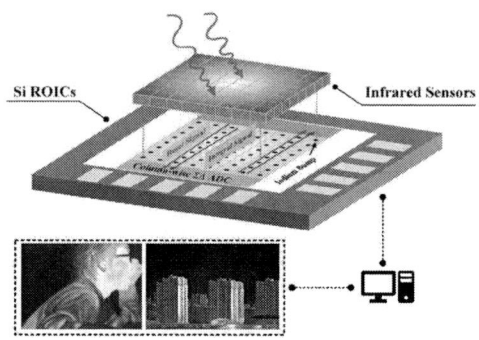

Fig. 1. Infrared focal plane array.

requirements to achieve low noise readout of infrared signals. It adopts a third-order feed-forward discrete-time modulator cascaded with a fully customized low-pass digital downsampling filter. It is applied to a 320×256 15μm-pitch 60dB-DR CTIA-based pixel array designed in a 180 nm 1P4M SOI process. By incorporating digital CDS technology, the design eliminates the readout noise in the entire signal chain by 61% in high gain mode and 53% in low gain mode. The maximum dynamic range is increased to 70 dB, achieving over 98% linearity and a maximum frame rate of 600 Hz. The proposed digitized readout channel proves an excellent candidate for achieving high-precision infrared readout circuits.

II. CIRCUIT IMPLEMENTATION

Fig. 1 illustrates the infrared focal plane array readout circuit of a 320×256 array with a pixel size of 15 μm. The pixel outputs the reset and integration signals parallel to the column-wise readout channel for quantization and digital CDS processing. The IRFPA ROIC employs a rolling shutter readout mechanism.

A. CTIA-Based Pixel

The proposed pixel shown in Fig. 2 adopts a CTIA structure with two gain modes to adapt to different illumination scenarios. C_s and C_r store the integral and reset signals, respectively. The two signals are fed parallel through the column-wise source followers to the readout channel.

B. Column-wise Readout Channel

As a time-for-accuracy structure, sigma-delta ADC has become a possible solution for improving imaging accuracy.

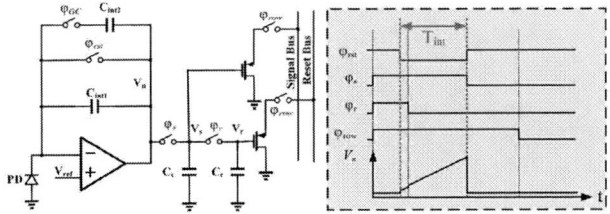

Fig. 2. The proposed CTIA pixel and timing diagram.

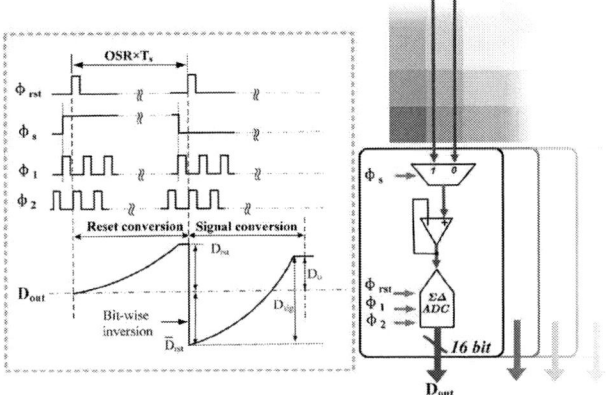

Fig. 3. Proposed column-wise readout channel with digital CDS.

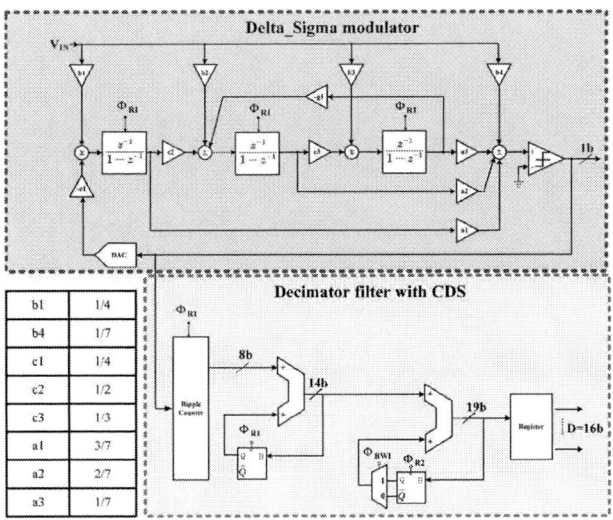

Fig. 4. Architecture of the proposed column-wise sigma-delta ADC.

However, there are still several challenges: (1) efficient topology, (2) conservation of timing resources, and (3) reasonable circuit design. In large-array imaging applications, the main challenge is maximizing the performance of sigma-delta ADCs while working within limited area and power consumption constraints. This requires finding a good balance between speed and accuracy to better align with the needs of the imaging system.

Fig. 3 shows the entire column-wise readout channel, consisting of a multiplexer, an output buffer, and a 16-bit sigma-delta ADC. As indicated by D_{out}, under the control of the Φ_s signal, the multiplexer first selects the pixel reset signal to be input into the ADC for quantization. Then, it takes the opposite number after the quantization is complete. Subsequently, the multiplexer inputs the pixel integration signal, accumulating continuously based on the complementary code, thereby achieving digital CDS functionality.

The column-wise sigma-delta ADC proposed in this paper is shown in Fig. 4. An incremental discrete-time third-order feed-forward single-bit quantized modulator cascaded with a three-stage digital decimation filter structure is adopted, with a digital CDS technology at the end. Discrete-time structures have more significant advantages for high-precision applications. Additionally, the feedforward structure can mitigate overload issues in integrators when the input signal has an extensive dynamic range, reducing the nonlinearity introduced by this and improving the SNDR. In column-wise imaging applications, the limited area makes it challenging to apply multi-bit quantizers, as the mismatch noise and nonlinear

harmonics caused by the mismatch in the feedback CDAC are significant. Therefore, a single-bit quantizer structure is adopted in this modulator.

Due to the infrared signal being a low-frequency quasi-static signal and the strict constraints on the area and power consumption of the column-wise ADC, the CIC (Cascaded Integrator-Comb) filter becomes the optimal choice. Compared to traditional structures, the incremental sigma-delta ADC simultaneously resets devices with memory functions, such as capacitors and registers, at the end of each sampling period. This enables its digital decimation filter to achieve movement without constructing a finite-length moving window through hardware by utilizing periodic reset operations. As a result, the comb filter is eliminated, leaving only the digital integrator and decimator. Moreover, compared to a complete CIC structure, there is no need to expand the bit width. Each register can be truncated according to requirements, further reducing power consumption and area. To prevent data overflow, the bit widths of the three stages are set to 8 bits, 14 bits, and 19 bits, respectively. The accumulators are composed of full adders and registers. To implement digital CDS in the final stage integrator, a data selector is introduced at the end. Finally, a 16-bit register is used to downsample and store the output results, converting the high-speed single-bit modulated bitstream into a Nyquist-rate codeword, thereby filtering out high-frequency noise.

Fig. 5 illustrates the circuit structure of the modulator. Due to area constraints, the entire loop adopts a single-ended structure, where the integrators are switched-capacitor integrators, the adder is the passive adder, and the comparator is a dynamic comparator. Each integrator is equipped with a reset switch to clear any residual charge from the previous quantization cycle at the beginning of each cycle. Two non-overlapping clocks control the integrators' sampling and integration operations. The Φ_1 clock controls the sampling process of all three integrators and the addition setup process of the adder, while

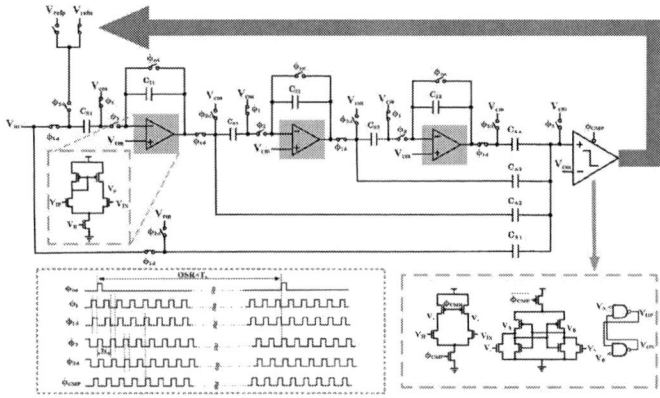

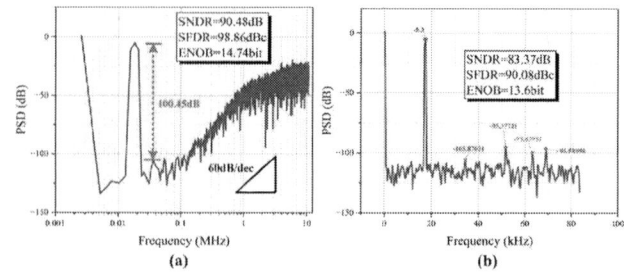

Fig. 7. Spectral analysis diagram with hanning window of (a) the proposed three-order modulator output for a -5.3 dBFS and 18.26 kHz input and (b) the proposed incremental ADC output after downsampling and filtering for a -5.3 dBFS and 17.28 kHz input (OSR=128).

Fig. 5. Proposed incremental sigma-delta modulator with timing diagram.

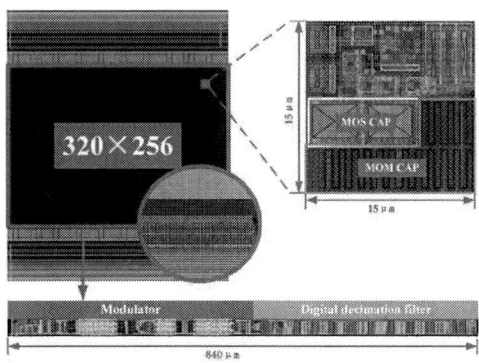

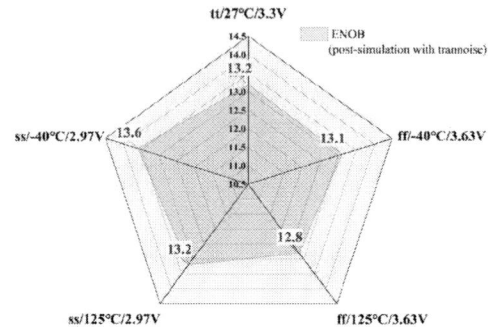

Fig. 8. Post-simulation with transient noise under various process corners for a -5 dBFS and 16.95 kHz input.

Fig. 6. Layout of the proposed IRFPA readout circuit.

the Φ_2 clock controls the integration process of the integrators and the reset process of the adder. To eliminate the interference caused by charge injection related to the input signal in the sampling switch channel, which can lead to nonlinear harmonics, the clock on the left side of the sampling capacitor lags slightly behind the clock on the right side. After the adder signal is established, the comparator feeds the comparison result back to the input DAC, subtracting it from the input signal to form a stable overall negative feedback structure.

III. LAYOUT AND POST-SIMULATION RESULTS

Fig. 6 displays the overall layout of the digitized infrared focal plane array readout circuit with a 320×256 array size and 15-µm pixel pitch. It is designed in a 180 nm 1P4M SOI process, where the column-wise readout channels are placed at both ends of the pixel array, separating odd and even columns. The upper half of the pixel layout consists of active devices, while the lower half comprises passive devices. The integration capacitor employs a higher-precision MOM capacitor, and the sampling capacitor uses a larger-capacitance MOS capacitor. The layout of the sigma-delta ADC is divided into an analog modulator section and a digital decimation filter section, with both sections having approximately the same area.

Fig. 7(a) shows the spectral analysis diagram of the proposed modulator circuit, which shows the effect of third-order noise shaping. Fig. 7(b) presents the FFT diagram of the

incremental ADC output after downsampling and filtering. It shows that high-frequency noise has been effectively filtered, and the ADC outputs at the Nyquist rate.

Rigorous simulations, including PVT (Process, Voltage, Temperature) corner simulation, Transient noise simulation, and post-simulation, are conducted on this ADC, with the results recorded in Fig. 8. These results prove that the ADC design exhibits good robustness and stability.

The variations in results across different process corners are due to several factors, including (1) coefficient deviations caused by capacitor mismatch, (2) variations in noise characteristics, and (3) variations in the driving capability of switched capacitor circuits. The system is observed to be more sensitive to clock frequency and levels under the SS process corner, exhibiting speed-sensitive characteristics. Under the FF process corner, the absolute values of capacitance decrease, leading to an increase in KTC noise. At high temperatures, the thermal noise of the circuit increases, and the signal's settling time becomes longer, resulting in a decrease in accuracy. In summary, reducing the operating speed and increasing the capacitance can enhance overall system performance.

The pixel has two gain modes, 12.4 µV/e⁻ in high gain mode and 3 µV/e⁻ in low gain mode. The single-pixel power consumption is 0.33 µW, and the full-well capacity is 93 ke⁻ in high-gain mode and 380 ke⁻ in low-gain mode. From the noise statistical distribution of the pixel output without CDS in Fig. 9(a), (b), it can be obtained that the readout noise is

979-8-3315-2209-4/25 $31.00 © 2025 IEEE

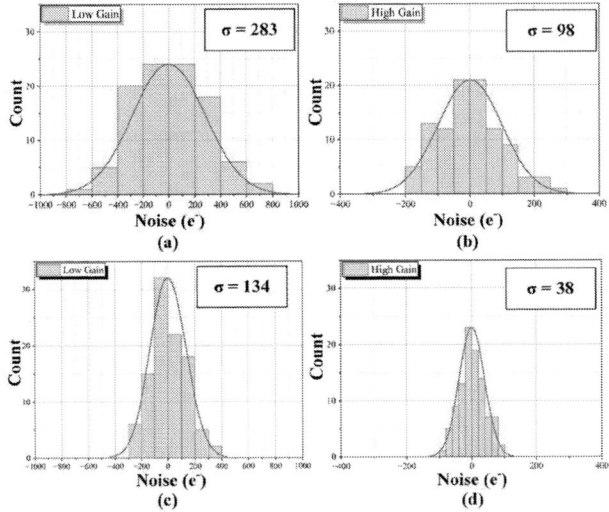

Fig. 9. The noise statistical distribution without CDS in (a) low-gain and (b) high-gain modes, as well as the noise statistical distribution with CDS in (c) low-gain and (d) high-gain modes.

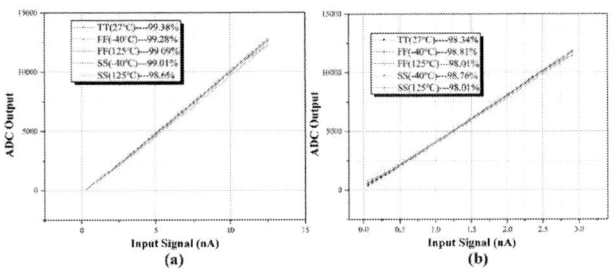

Fig. 10. Linearity fitting results of the ADC output (a) in low gain mode and (b) in high gain mode at different corners.

283 e⁻ in low gain mode and 98 e⁻ in high gain mode. The pixel's dynamic range is 63 dB in low gain mode and 60 dB in high gain mode. From the noise statistical distribution of the ADC output with CDS in Fig. 9(c), (d), it can be obtained that the readout noise is 134 e⁻ in low gain mode and 38 e⁻ in high gain mode. Compared to the readout noise without CDS, it has been reduced by 61% in high gain mode and 53% in low gain mode. The dynamic range has increased to 70 dB and 68 dB in two modes, demonstrating the crucial role of digital CDS.

Fig. 10 presents the post-simulation linearity of the entire signal chain from the photocurrent to the 16-bit codeword output of the ADC at different corners. The linearity is over 98% and demonstrates good PVT robustness and stability.

TABLE I presents the performance of this column-wise sigma-delta ADC and comparison.

IV. CONCLUSIONS

This paper proposes a column-wise 84dB-DR, 167 KS/s readout channel based on sigma-delta ADC. Applied to a 60dB-DR CTIA-based pixel array, the system achieves 98% linearity and a maximum frame rate of 600 Hz. After digital

TABLE I
PERFORMANCE METRICS OF THE COLUMN-WISE ADC AND COMPARISON

Parameters	This work	TCASI -15 [2]	JSSC -09 [3]	TCASII -18 [4]
Process	180nm SOI 1P4M	130nm CMOS 1P4M	350nm CMOS	180nm CMOS
ADC type	$\Delta\Sigma$	$\Delta\Sigma$	$\Delta\Sigma$	EC ($\Delta\Sigma$+SS)
Power supply (V)	3.3/1.8	2.8/1.2	1.2	1.8
Sample frequency (kS/s)	166.933	187.6	16	185
Bandwidth (kHz)	83.46688	93.8	8	92.5
Resolution (bit)	16	12	-	12
ENOB (bit)	13.2*	10.3	10.17	10.5
DR (dB)	84*	60.9	76	68.1
FoM$_W$ (pJ/conv)	0.25*	0.23	0.303	0.082
FoM$_{SNDR}$ (dB)	165*	155	154	160
FoM$_{DR}$ (dB)	167*	153	167	164

a FoM$_W$=Power/($2\times$BW$\times2^{(SNDR-1.76)/6.02}$).
b FoM$_{SNDR}$=SNDR+10log$_{10}$(BW/Power).
c FoM$_{DR}$=DR+10log$_{10}$(BW/Power).
* Post-simulated with transient noise

CDS processing, the readout noise is reduced by 61% in high gain mode and 53% in low gain mode, and the maximum dynamic range is improved to 70 dB, achieving a high-precision digital readout.

REFERENCES

[1] W. Lei, J. Antoszewski, and L. Faraone, "Progress, challenges, and opportunities for HgCdTe infrared materials and detectors," Appl. Phys. Rev., vol. 2, no. 4, 041303, 2015.

[2] Y. R. Jo, S. K. Hong, and O. K. Kwon, "A Multi-Bit Incremental ADC Based on Successive Approximation for Low Noise and High Resolution Column-Parallel Readout Circuits," IEEE Trans. Circuits Syst. I Regul. Pap., vol. 62, no. 9, pp. 2156–2166, 2015.

[3] Y. Chae and G. Han, "Low Voltage, Low Power, Inverter-Based Switched-Capacitor Delta-Sigma Modulator," IEEE J. Solid-State Circuit, vol. 44, no. 2, pp. 458–472, 2009.

[4] B. K. Jeon, S. K. Hong, and O. K. Kwon, "A Low-Power 12-Bit Extended Counting ADC Without Calibration for CMOS Image Sensors," IEEE Trans. Circuits Syst. II Express Briefs, vol. 65, no. 7, pp. 824–828, 2018.

[5] S. Lee, J. Jeong, T. Kim, C. Park, T. Kim, and Y. Chae, "A 5.2-Mpixel 88.4-dB DR 12-in CMOS X-Ray Detector With 16-bit Column-Parallel Continuous-Time Incremental $\Delta\Sigma$ ADCs," IEEE J. Solid-State Circuits, vol. 55, no. 11, pp. 2878–2888, 2020.

[6] I. Lee, B. Kim, and B. G. Lee, "A Low-Power Incremental Delta-Sigma ADC for CMOS Image Sensors," IEEE Trans. Circuits Syst. II Express Briefs, vol. 63, no. 4, pp. 371–375, 2016.

Two-Step Current Integration Enabled High Dynamic Range X-Ray Image Sensing

Yuezhong Duan, Tengyan Huang, Congwei Liao, and Shengdong Zhang, *Senior Member, IEEE*

Abstract—**This paper proposed a two-step current integration approach to achieve high dynamic range X-Ray image sensing. During the X-Ray exposing stage, the photo-current (I_{ph}) is collected and integrated for the first integration step. Following that, the dark current (I_{dark}) in the reverse direction is integrated for the second integration step. Thus, the sensing result is determined by the net value of the two integration steps, and even very weak optical signals can be detected for extended detection dynamic range. Amorphous-silicon photo thin film transistors (a-Si photo TFTs) operating in the on region is used for the photo-to-current conversion, which exhibit higher photoresponsivity. The proposed two-step current integration approach effectively eliminates I_{dark} for the exposing stage. Simulation of a $100\mu m \times 100\mu m$ pitch pixel shows that the two-step current integration can enable X-Ray sensing to achieve a dynamic range of 73.98dB, increased by 13.98dB compared to the conventional scheme.**

Keywords—**sensing circuit, a-Si photo TFTs, high dynamic range**

I. INTRODUCTION

AMORPHOUS silicon thin film transistors (a-Si TFTs) are widely used in static flat X-Ray medical imaging applications, including dental detection, chest X-Ray, digital radiography (DR), etc [1] [2] [3]. This is attributed to advantages of a-Si TFT technology in terms of good uniformity, mature fabrication process, and low production cost over large-area manufacturing [4] [5].

Conventionally, passive pixel sensing (PPS) detectors are based on a-Si TFTs and a-Si photodiodes (PDs). Due to the hybrid integration of TFTs and PDs, the traditional manufacturing process of X-ray flat-panel detectors is complex and expensive. In recent years, a-Si photo TFTs based detectors have garnered significant research interest, primarily because the merit of fully TFT-compatible process, and high photoresponsivity through proper voltage biasing [6] [7]. Currently, the photoresponsivity of Si p-i-n PDs is approximately 100 mA/W - 620 mA/W. Bablich et al. introduced an a-Si:H p-i-n PD with a photoresponsivity reaching 744 mA/W in 2023 [8], and Mohammad R. Esmaeili-Rad et al. introduced an a-Si photo TFT operating in the off region with a photoresponsivity of 92mA/W, which is nearly comparable to that of a-Si PDs [9]. As for a-Si photo TFTs in the on region and illuminated with 520 nm wavelength light, the photoresponsivity can reach up to 55.4 A/W, which is significantly higher than that of the traditional a-Si PDs. A higher photoresponsivity means that

under the same light intensity, a-Si photo TFTs can generate a larger I_{ph}, leading to efficient reduction of X-Ray radiation dose [10]. However, to date, X-ray detection based on a-Si photo TFTs is affected by significant I_{dark}, which limits the dynamic range and impacts image quality [11] [12] [13].

This paper proposed a two-step current integration approach to extend the dynamic range of X-Ray sensing based on a-Si photo TFTs. By sequentially integrating I_{ph} and reverse I_{dark} generated by the a-Si photo TFTs, it maintains the a-Si photo TFTs operating in the on region during the exposing stage to achieve higher photoresponsivity and eliminates the large dark current charges generated in the on region after the exposing stage. Hence, this approach effectively enlarge the dynamic range of X-Ray sensing.

II. THE PROPOSED SENSING METHOD

A. Two-step current integration approach

Fig.1 (a) and (b) illustrate the principle of the proposed two-step current integration for the exposing and eliminating stages, respectively. For the first step current integration, due to the high level of the $S[n]$ signal, the a-Si photo TFT is turned on and outputting I_{ph} with photoresponsivity larger than that of the a-Si PD. But the I_{ph} value comprises both illumination-related current and the dark current. Thus the dark current is also converted into the voltage signal by the C_{ST}. The dynamic range after the first integration is then expressed by [2]:

$$DR = 20 \times \log_{10} \frac{I_{max}}{I_{minon}} \qquad (1)$$

where I_{minon} is I_{dark} in the on region of the a-Si photo TFT.

For the second step current integration, $S[n]$ is kept with high level, while V_{BIAS} is pulled down, thus the source and drain electrodes of the a-Si photo TFT is swapped and a reverse I_{dark} in the on region is generated flowing from C_{ST} to V_{BIAS} signal line. For the a-Si photo TFT, the non-gate controlled region can be regarded as a linear resistance in the absence of light. Therefore, the same magnitude of current in the opposite direction will be generated with the opposite voltage biasing [14].

Then, the dynamic range after the second step current integration is changed to:

$$DR' = 20 \times \log_{10} \frac{I_{max} - I_{minon}}{I_{minoff}} \qquad (2)$$

where I_{minoff} is I_{dark} in the off region. As shown in Fig.1 (c), I_{minon} is three orders of magnitude smaller than I_{max} in the on region, while it dozens of times more than I_{minoff}.

All the authors are with the School of Electronic and Computer Engineering, Peking University, Shenzhen 518055, China. Shengdong Zhang is also with the School of Integrated Circuits, Peking University, Beijing 100871, China (e-mail: zhangsd@pku.edu.cn). Congwei Liao is also with the College of Integrated Circuits and Optoelectronic Chips, Shenzhen Technology University.

Therefore, it could be ignored when compared to I_{max} in Equation1, and the dynamic range of the sensing circuit after the proposed two-step current integration can be approximated as:

$$DR' = 20 \times \log_{10} \frac{I_{max}}{I_{minoff}} \qquad (3)$$

which means that the dynamic range is increased approximately by 30 dB.

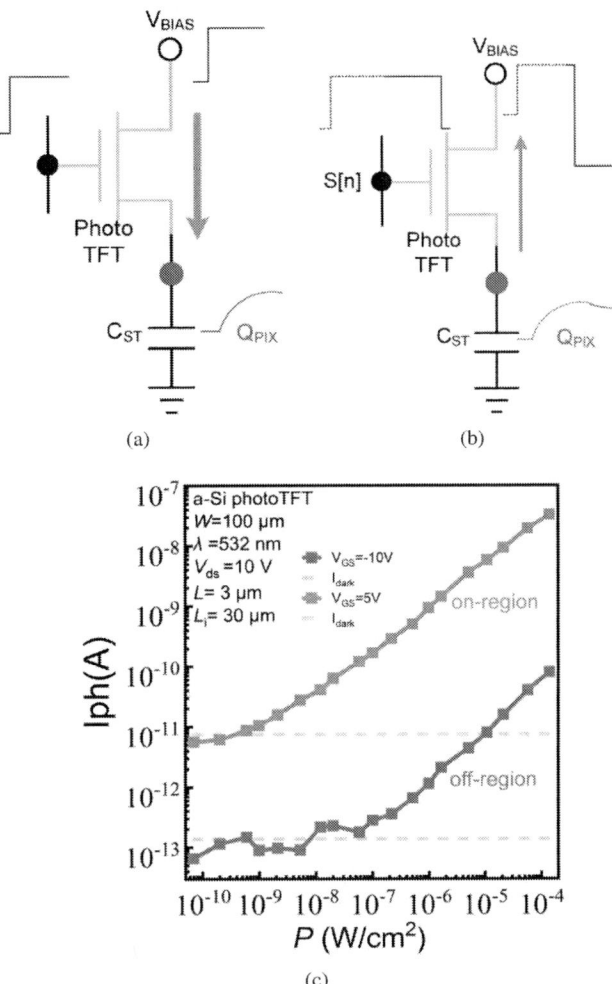

(a)

(b)

(c)

Fig. 1: Principle of the proposed two-step current integration approach, (a) the first step (exposing stage), (b) the second step (eliminating stage), (c) photoelectric characteristic of a-Si photo TFTs (V_{GS}=5V and V_{GS}=-10V).

B. The proposed two-step current integration pixel circuit

The proposed two-step current integration pixel circuit is composed of an a-Si photo TFT, an a-Si readout TFT (TR), and a storage capacitor (C_{ST}). Fig.2 (a) and (b) schematically show the pixel circuit and operating timing, $S[n]$ and $G[n]$ are gate controlling signals for a-Si photo TFTs and TR, respectively.

Instead of using a-Si PDs, which requires additional fabrication processes, the pixel circuit uses a-Si photo TFTs and traditional structure TFTs, which are completely compatible with the TFT process, to achieve photoelectric conversion, readout, reset functions. Furthermore, it uses the proposed two-step current integration approach to reduce the effect of the large I_{dark} generated by the a-Si photo TFTs in the on region.

It works in the following sequence:

(1) **T1 (resetting stage):** Before the first detection frame, SW and $G[n]$ are at a high level and TR is turned on line by line to reset the feedback capacitor and the restore capacitor. Thus, C_{ST} and C_{FB} are resetting, and remaining charges are cleared.

(2) **T2 (exposing stage, the 1-st step current integration):** During global exposure, $S[n]$ is kept at a high level to achieve high sensitivity, while $G[n]$ is at a low level, and V_{BIAS} is at a high level. The a-Si photo TFT is activated, and the different I_{ph} is generated according to the incident X-Ray. Thus, I_{ph} is integrated in C_{ST}, so that the Q_{PIX} in C_{ST} after the exposing stage is given by the equation:

$$Q_{PIX} = I_{ph} \times T_{exposing} \qquad (4)$$

(3) **T3 (eliminating stage, the 2-nd step current integration):** $S[n]$ and $G[n]$ are at a high level, and V_{BIAS} is set to negative voltage, so that the reverse I_{dark} is generated and integrated for the same duration of the exposing stage, leading to a portion of the stored signal charge of C_{ST} to flow through the a-Si photo TFT, which means the I_{dark} charges caused by the exposing stage can be eliminated. The Q'_{PIX} in C_{ST} after the eliminating stage is given by the equation:

$$Q'_{PIX} = (I_{ph} - I_{dark}) \times T_{exposing} \qquad (5)$$

Therefore, it shows that I_{dark} generated by the a-Si photo TFTs in the on region during the exposing stage has been eliminated.

(4) **T4 (readout stage):** $S[n]$ turns to low level, $G[n]$ turns to high level, thus TR is turned on and the a-Si photo TFT is operating in the off region. The optical signal charges stored by C_{ST} transfer to the external readout circuit so that the entire array is read out line by line.

C. On-region operating mode of a-Si photo TFTs

Fig.1 (c) shows the photoelectric characteristic of a-Si photo TFT with a size of W/L/L_i=100μm/3μm/30μm. The maximum I_{ph} (0.033μA) in the on region (V_G=5V) is 417.8 times that of (0.079nA) in the off region (V_G = -10V) at a maximum optical power of 139μW/cm^2. However, its dynamic range is 55.22dB in the off region, surpassing 53.51dB in the on region.

Therefore, I_{dark} in the on region limits the improvement of the dynamic range and the ideal operating mode is to utilize the larger I_{ph} of a-Si photo TFTs in the on region and eliminate the influence of I_{dark} during this stage, while keeping a-Si photo TFTs in the off region at non-exposing stages.

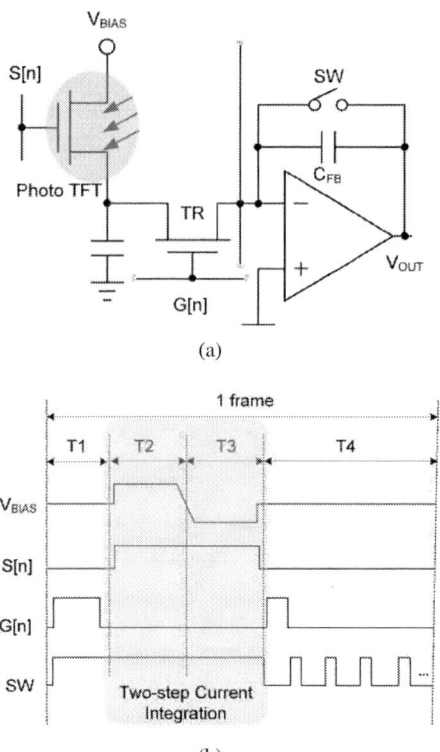

(a)

(b)

Fig. 2: Sensing circuit based on a-Si photo TFTs using two-step current integration approach, (a) pixel circuit, (b) the timing diagram.

III. RESULT AND DISCUSSION

Fig. 3 (a), (b) compares the transient response corresponding to different I_{ph} using the two-step current integration and the traditional one-step current integration methods. It can be seen that, due to the added second step current integration, the voltage on the C_{ST} in the circuit with the added second step current integration decreases slowly during the eliminating stage. This indicates that Q_{PIX} is decreasing to mitigate the effect of I_{dark} in the on region during the exposing stage.

Fig. 4 (a), (b), (c), and (d) show merits comparison of the sensing circuit using proposed two-step current integration approach versus the one-step current integrated circuit, in terms of Q_{PIX}, V_{OUT}, $\Delta Q/Q$, and DR, by simulation for a $100\mu m \times 100\mu m$ pitch pixel. In this pixel, the size of the a-Si photo TFT is W/L/L$_i$=77μm/3μm/27μm, and the size of TR is W/L=20μm/3μm. TABLE I shows the parameters of the simulation of the proposed two-step current integration pixel circuit.

It shows that, thanks to the two-step current integration approach, the stored charge Q_{PIX} on C_{ST} and the output voltage $|V_{OUT}|$ are reduced, and the amount of these changes also increases as the I_{ph} increases. This is because the larger the I_{ph} during the exposing stage, the higher the voltage at C_{ST}, leading to a larger reverse I_{dark} generated during the eliminating stage.

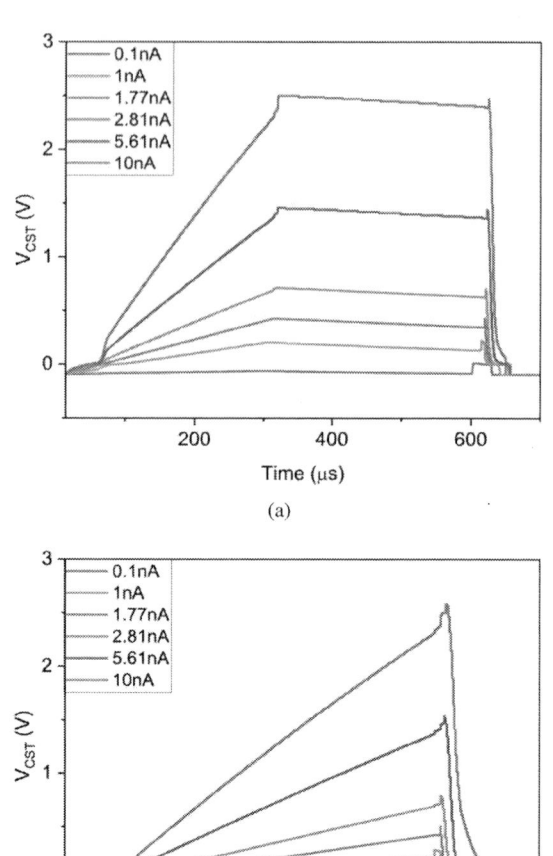

(a)

(b)

Fig. 3: Transient response of the voltage of the C_{ST} inside the pixel corresponding to different I_{ph}, (a) 2-step current integration, (b) 1-step current integration.

$\Delta Q/Q$ represents the degree of elimination of dark current noise charge, with up to 54% reduction. From Fig.4 (c), it can be seen that the smaller the I_{ph} during the exposing stage, the more significant the elimination effect after the eliminating stage. Fig.4 (d) shows that the dynamic range (DR) can be improved by approximately 22dB in 10μW/cm^2 compared to the one-step current integration circuit, and it will increase with increasing optical power ($Pmax$).

Compared to the circuit based in a-Si:H PDs Fig. [4], this work reduces the exposure time from 16ms to 300μs, which means lower X-Ray radiation dose, while increasing the dynamic range from 60dB to 73.98dB. The dynamic range of the indirect-conversion X-Ray detector with an embedded PD formed by a 3-D dual-gate TFT [15] is also less 13.98dB than that of this work. Compared to the circuit that the photo TFT operates in the off region and is reused as a readout TFT, without correlated double sampling to reduce noise signal

979-8-3315-2209-4/25 $31.00 © 2025 IEEE

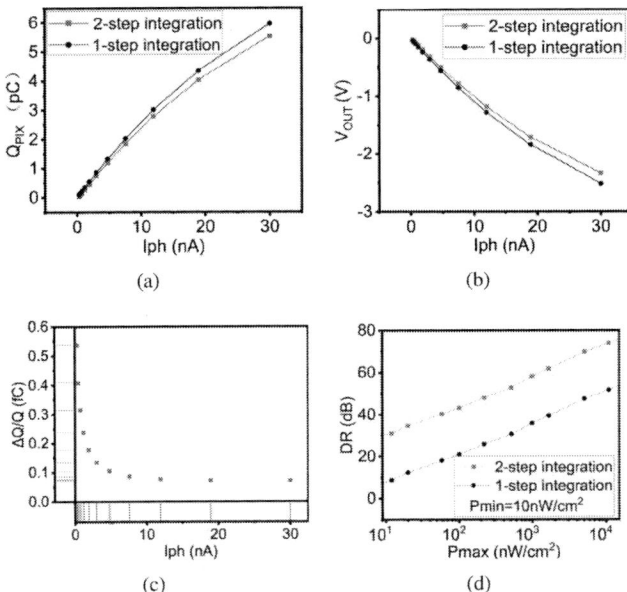

Fig. 4: Merits comparison of the sensing circuit using proposed two-step current integration approach versus the original circuit, in terms of (a) Q_{PIX}, (b) V_{OUT}, (c) $\Delta Q/Q$, and (d) DR.

effects, this work operates the photo TFT in the on region to obtain larger photocurrents, and uses two-step integration approach to eliminate dark current signals and has a separate readout transistor with shading treatment that can avoid the time consumed by the state switching of the photo TFT [9]. In addition, the dynamic range of the two-step current integration pixel circuit is increased by 22.28dB compared to that of the one-step current integration pixel circuit.

Hence, compared to conventional PPS circuits based on a-Si PDs and sensing circuits based on a-Si photo TFTs, the proposed schematic significantly improves the dynamic range, which can effectively enhance imaging quality.

TABLE I: Parameters of the proposed pixel circuit

Symbol	Interpretation	Values
$T_{exposure}$	time of exposing stage	$300\mu s$
$T_{eliminate}$	time of eliminating stage	$300\mu s$
W_{photo}	channel width of photo TFT	$77\mu m$
L_{photo}	channel length of photo TFT	$3\mu m$
L_i	non-gate controlled channel length	$27\mu m$
W_{read}	channel width of readout TFT	$20\mu m$
L_{read}	channel length of readout TFT	$3\mu m$
P_{max}	maximum optical power	$10\mu W/cm^2$
P_{min}	minimum optical power	$10nW/cm^2$
λ	light wavelength	$532nm$

IV. CONCLUSION

In conclusion, this paper demonstrated a two-step current integration approach to achieve high dynamic range X-Ray image sensing based on a-Si photo TFTs. The first step current integration occurs during the X-Ray exposing stage, the a-Si photo TFT operates in the on region to achieve higher photoresponsivity and the generated photocurrent is integrated into the restore capacitor of the pixel. Furthermore, to address the issue of dark current in the on region affecting the detection dynamic range, the second step current integration takes place during the eliminating stage to integrate the reverse dark current, which can significantly eliminate up to 54% of the dark current charge. Thus the proposed two-step current integration approach enables the X-Ray image sensing to achieve a dynamic range of 73.98dB, increased by 13.98dB compared to the conventional a-Si PDs based pixel sensing circuit under the same illumination.

REFERENCES

[1] K. Karim, A. Nathan, and J. Rowlands, "Amorphous silicon active pixel sensor readout circuit for digital imaging," *IEEE Transactions on Electron Devices*, vol. 50, no. 1, pp. 200–208, 2003.

[2] A. Nathan, P. Servati, and K. Karim, "TFT circuit integration in a-Si:H technology," in *2002 23rd International Conference on Microelectronics. Proceedings (Cat. No.02TH8595)*, vol. 1, pp. 115–124, 2002.

[3] M. H. Izadi, O. Tousignant, M. F. Mokam, and K. S. Karim, "An a-Si Active Pixel Sensor (APS) Array for Medical X-ray Imaging," *IEEE Transactions on Electron Devices*, vol. 57, no. 11, pp. 3020–3026, 2010.

[4] G. Yoo, T. ching Fung, D. Radtke, M. Stumpf, U. Zeitner, and J. Kanicki, "Hemispherical thin-film transistor passive pixel sensors," *Sensors and Actuators A: Physical*, vol. 158, no. 2, pp. 280–283, 2010.

[5] D. Geng, K. Wang, L. Li, K. Myny, A. Nathan, J. Jang, Y. Kuo, and M. Liu, "Thin-film transistors for large-area electronics," *Nature Electronics*, vol. 6, no. 12, pp. 963–972, Dec 2023.

[6] Y. Lee, I. Omkaram, J. Park, H.-S. Kim, K.-U. Kyung, W. Park, and S. Kim, "A α-Si:H Thin-Film Phototransistor for a Near-Infrared Touch Sensor," *IEEE Electron Device Letters*, vol. 36, no. 1, pp. 41–43, 2015.

[7] T. Huang, H. Liu, F. Tang, L. Lu, M. Zhang, and S. Zhang, "Physical Insight Into Multiple Gate-Voltage Dependencies of Off-State Photocurrent in Amorphous InZnO Thin-Film Transistors," *IEEE Transactions on Electron Devices*, vol. 71, no. 3, pp. 2243–2246, 2024.

[8] A. Bablich, M. Müller, R. Bornemann, A. Nachtigal, and P. Haring Bolívar, "High Responsivity and Ultra-Low Detection Limits in Nonlinear a-Si:H p-i-n Photodiodes Enabled by Photogating," *Photonic Sensors*, vol. 13, no. 4, pp. 230415, Jul 2023.

[9] M. R. Esmaeili-Rad, N. P.·Papadopoulos, M. Bauza, A. Nathan and W. S. Wong, "Blue-Light-Sensitive Phototransistor for Indirect X-Ray Image Sensors," *IEEE Electron Device Letters*, vol. 33, no. 4, pp. 567-569, 2012.

[10] H. Liu, X. Zhou, T. Huang, J. Chen, F. Tang, L. Lu, and S. Zhang, "P-1.9: Effect of wavelength on photoresponse characteristics of amorphous InZnO thin film transistors," *SID Symposium Digest of Technical Papers*, vol. 53, no. S1, pp. 596–599, 2022.

[11] Y.-H. Tai, L.-S. Chou, and H.-L. Chiu, "Gap-Type a-Si TFTs for Front Light Sensing Application," *Journal of Display Technology*, vol. 7, no. 12, pp. 679–683, 2011.

[12] C.-L. Lin, C.-L. Lee, C.-E. Wu, F.-H. Chen, W.-S. Liao, R. W. Chuang, and J.-S. Yu, "Optical Pixel Sensor Based on a-Si:H TFTs to Detect Combined Optical Signals for Multiuser Interactive Displays," *IEEE Transactions on Electron Devices*, vol. 67, no. 6, pp. 2425–2431, 2020.

[13] C.-L. Lin, C.-L. Lee, C.-H. Ke, P.-C. Lai, C.-T. Chiu, Y.-C. Chiu, and C.-W. Kuo, "Lifetime Optimization of Optical Sensing System With Highly Reliable a-Si:H TFT-Based Optical Sensor and Driver Circuit in Large AMLCDs," *IEEE Access*, vol. 12, pp. 78122–78131, 2024.

[14] H. Liu, X. Zhou, C. Fan, J. Chen, L. Lu, H. Zhou, and S. Zhang, "Thorough Elimination of Persistent Photoconduction in Amorphous InZnO Thin-Film Transistor via Dual-Gate Pulses," *IEEE Electron Device Letters*, vol. 43, no. 8, pp. 1247–1250, 2022.

[15] Y. Xu, Q. Zhou, J. Huang, W. Li, J. Chen, and K. Wang, "Highly-Sensitive Indirect-Conversion X-Ray Detector With an Embedded Photodiode Formed by a Three-Dimensional Dual-Gate Thin-Film Transistor," *Journal of Lightwave Technology*, vol. 38, no. 14, pp. 3775–3780, 2020.

A Low-Noise Charge-Balanced Readout Circuit for MEMS Accelerometer

Wenting Wang, Yao Li, Cheng Yuan, Mao Ye [*]

School of Microelectronics, Tianjin University, Tianjin, 300072, China

*Email: mao_ye@tju.edu.cn

Abstract—**This paper presents a low-noise and high linearity readout circuit for MEMS capacitive accelerometers. To meet the stringent linearity requirements for automotive applications, charge balanced architecture is adopted to balance the electrostatic force on the sensing capacitance imposed by the C/V readout circuit. A boxcar sampler with auto-zero(AZ) and correlated double sampling(CDS) techniques is integrated to further suppress overall noise. Additionally, the adoption of a floating inverter amplifier (FIA) significantly reduces power consumption. Designed using a 180nm BCD process, the proposed readout circuit achieves a noise floor of 63.1 $\mu g/\sqrt{Hz}$, a dynamic range of 80.5 dB, and a nonlinearity of 0.045%, with a signal bandwidth of 1 kHz, as verified through simulation results.**

Keywords—*MEMS accelerometer readout circuit, charge-balanced, boxcar sampler, switched-capacitor circuit, FIA*

I. INTRODUCTION

Unlike traditional open-loop and closed-loop readout circuits, the charge-balanced architecture, as a unique type of open-loop design, combines the advantages of both. Compared to open-loop architectures, it offers higher linearity, while its complexity is significantly lower than that of closed-loop architectures [1]- [2]. Among these circuit design techniques, the amplifier typically has a wide bandwidth, which allows more high-frequency components of thermal noise from the input to enter the amplifier's input stage. These noise components are aliased into the low-frequency signal band, increasing the noise power in the low-frequency range and thereby reducing the SNR. In [3], the boxcar sampler principle, equivalent to a sinc filter with its first notch at $1/T_{int}$,where T_{int} is the integration time, is used to effectively reduce noise aliasing.

In this work, the boxcar sampler, which combines a transconductance amplifier with an integrator, effectively reduces the switching noise sampled on the sensor capacitance and minimizes the impact of the parasitic capacitance between the common-mode electrode and the substrate. Meanwhile, the use of FIA in the integrator effectively reduces power consumption while improving area efficiency.

Section II describes the architecture and circuit details of the proposed readout circuit, while Section III presents the simulation results.

II. SYSTEM ARCHITECTURE AND CIRCUIT IMPLEMENTATION

A. Architecture

The accelerometer readout circuit is composed of a self-balancing bridge(SBB), a boxcar sampler, and a sample-and-

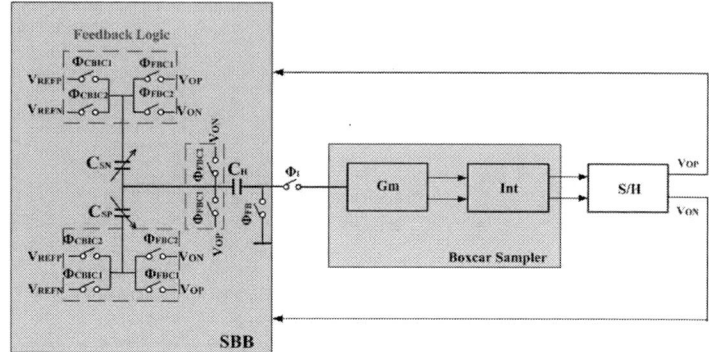

Fig. 1.   Block diagram of the readout circuit.

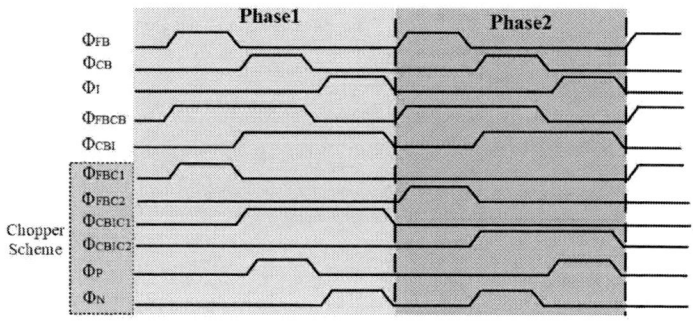

Fig. 2.   Clock diagram of the readout circuit.

hold circuit, as shown in Fig.1. In the self-balancing bridge, the sensing capacitance electrodes are first connected to the output voltage for charge balancing, followed by a connection to the reference voltage. The signal is then transmitted from the common-mode electrode of the sensing capacitance to the boxcar sampler. Within the boxcar sampler, the charge signal is first converted into a differential current signal by a transcon-

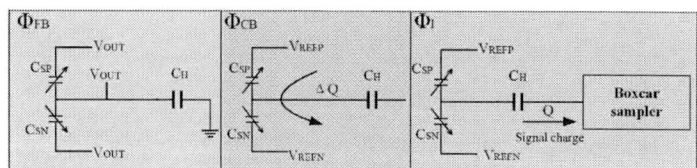

Fig. 3.   Circuit diagrams of three phases of SBB

979-8-3315-2209-4/25 $31.00 © 2025 IEEE   404

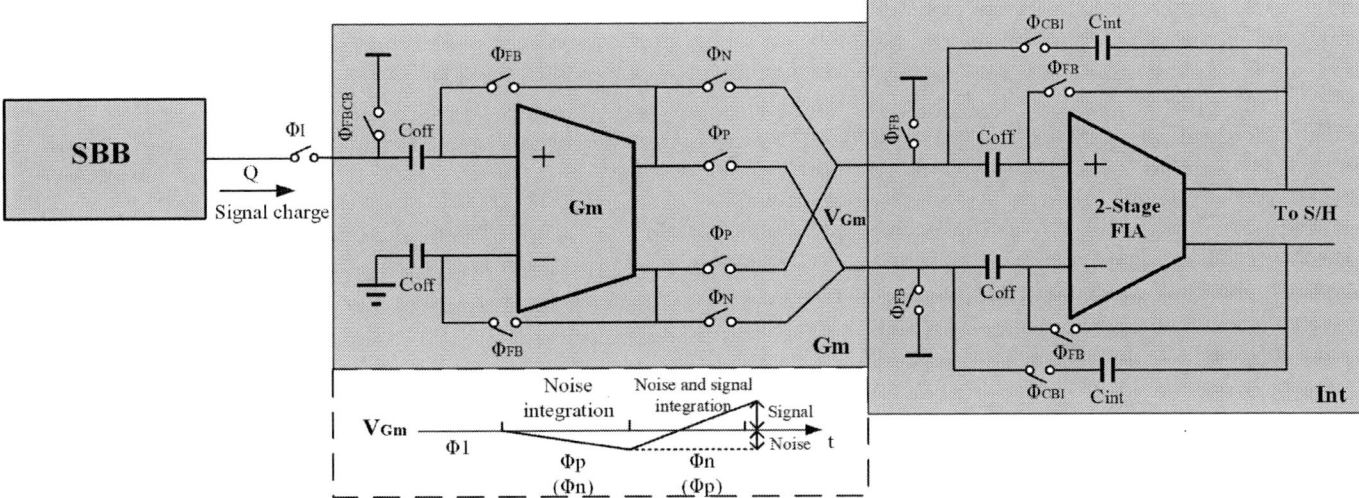

Fig. 4. Schematic of boxcar sampler.

ductance amplifier (Gm). The integrator then integrates this current signal into a voltage signal. The resulting voltage signal is sampled and held, providing an output for further ADC processing while simultaneously feeding the result to the sensing capacitance of the accelerometer to achieve charge balancing.

The readout circuit uses three non-overlapping phases Φ_{FB}, Φ_{CB} and Φ_I. The chopper is implemented at the output of the Gm and at the output of the sample-and-hold circuit. By combining with the chopper scheme, Φ_{FBC1}, Φ_{FBC2}, Φ_{CBIC1}, Φ_{CBIC2}, Φ_N and Φ_P are generated and applied to the logic control of the feedback switches and the switching control of the transconductance amplifier's output. The clock diagram is shown in Fig. 2.

B. Charge-Balanced Capacitive Sensing

The charge-balanced readout technique adopts a self-balancing bridge (SBB) structure, which controls the charge balance between the MEMS sensing capacitors to ensure that the output maintains a linear relationship with the displacement of the proof mass. This improves the accelerometer's linear range and dynamic range.

Φ_{FB} represents the feedback phase, during which the top and bottom plates of the capacitors C_{SP} and C_{SN}, as well as the intermediate holding capacitor C_H, are connected to the output voltage. This charges C_H to V_{OUT}, ensuring that feedback is achieved under the condition where the voltages on the top and bottom plates of the two sensing capacitors are equal. This ensures that no voltage difference exists, effectively eliminating the electrostatic force that could induce additional displacement in the accelerometer. Φ_{CB} represents the charge-balanced phase, during which the top and bottom plates of C_{SP} and C_{SN} are connected to the reference voltages V_{REFN} and V_{REFP}, respectively. The output voltage V_{OUT} on C_H is biased to the intermediate plate of the capacitors. Due to non-ideal factors such as parasitic capacitance C_P, charge leakage occurs

between the sensing capacitors. During charge balancing, charge transfer takes place between the two sensing capacitors, resulting in an error charge $\triangle Q$. Φ_I represents the charge transfer phase, during which the charge bridge is connected to the frontend circuit, transferring the corresponding charge signals for further processing. The circuit diagrams of the self-balancing charge bridge during its three phases are detailed in Fig.3.

After several cycles, the charge distribution is completed, and the error charge becomes zero ($\triangle Q=0$). The charge on the top and bottom capacitors is balanced, resulting in a stable output voltage:

$$V_{OUT} = \frac{C_{SP} - C_{SN}}{C_{SP} + C_{SN}} V_{ref} \qquad (1)$$

C. Boxcar Sampler

The transconductance amplifier and integrator are employed as a boxcar sampler, effectively acting as an anti-aliasing filter before the charge signal is processed in the frontend. It filters out signal and noise components above the Nyquist frequency, resulting in a significant reduction in noise. The gain of the boxcar sampler is defined by the ratio between the transconductance gm of the Gm, the integrating capacitance C_{int} and the integration time T_{int}:

$$GAIN = \frac{gm}{C_{int}} \cdot T_{int} \qquad (2)$$

This indicates that, compared to traditional integrators, the GM-C integrator decouples the relationship between the sampling capacitor and the integration capacitor. This allows for more flexible selection of the values for the sense capacitors and integration capacitors. A more detailed comparison between conventional integrators and the boxcar sampler is presented in Table 1.

A detailed circuit-level block diagram of the implemented boxcar sampler is given in Fig. 4. The Gm stage and the integrator perform offset cancellation during Φ_{FB}, with the Gm stage achieving it through auto-zeroing (AZ) and the integrator through correlated double sampling (CDS). At this point, the capacitor C_{off} stores the offset voltage and the input noise charge.

The output switches of the Gm are controlled by opposite signals, Φ_p and Φ_n. $\Phi_p(\Phi_n)$corresponds to noise transfer integration, during which the residual non-ideal noise charges not stored in the capacitor C_{off} are integrated. $\Phi_n(\Phi_p)$ corresponds to signal and noise transfer integration, where the input signal is connected, and both the signal and non-ideal noise charges are integrated. During this phase, the output current is integrated in the opposite direction to that of Φ_p (Φ_n), allowing the offset voltage and noise to cancel out through the integration processes of Φ_n and Φ_p.

The transconductance of the Gm and the integration capacitance C_{int} must be carefully balanced based on noise, stability, and area requirements. Increasing the transconductance of the Gm stage reduces the noise contribution of the overall readout circuit, as it enhances the overall gain. However, an increase in transconductance can lead to instability in the system. This instability needs to be compensated by increasing the integration capacitance C_{int}, which maintains the same gain but reduces area efficiency. Meanwhile, the chopping switches are placed at the input of the readout circuit and the output of the transconductance amplifier. The chopping frequency is set to half of the sampling frequency to effectively eliminate flicker noise.

TABLE I: Comparison between Integrator and Boxcar Sampler

	Integrator	Boxcar Sampler
Gain	$\dfrac{C_{int}}{C_S}$	$\dfrac{gm}{C_{int}} \cdot T_{int}$
Feedback factor	$\dfrac{C_{int}}{C_S + C_P + C_{int}}$	$\dfrac{C_{int}}{C_P + C_{int}}$
Transfer function	$\dfrac{C_{int}}{C_S} \cdot \dfrac{Z^{-1}}{1 - Z^{-1}}$	$\dfrac{gm}{C_{int}} \cdot \dfrac{1 - e^{-sT_{int}}}{s}$

D. Two-Stage FIA

The amplifier in the integrator adopts a two-stage Floating Inverter Amplifier (FIA), with the specific circuit shown in Fig. 5. Φ_{FB} represents the reset phase of the FIA. During this phase, the top and bottom plates of the reservoir capacitor C_{RES} are connected to the supply voltage V_{DD} and ground, respectively. The voltage across the capacitor is charged to V_{DD}, while the voltage on the external load capacitor is reset to the common-mode level V_{CM}. At this moment, both the NMOS and PMOS transistors are in the cutoff state, and no static current flows through the circuit. Φ_{CBI} represents the amplification phase of the FIA. The switches on the top and bottom plates of C_{RES} are connected to the source terminals of the PMOS and NMOS transistors, respectively. Subsequently, the charge stored on C_{RES} provides current to the differential inverter for amplifying the input voltage. As C_{RES} gradually discharges, the source voltage of the NMOS transistor increases while the source voltage of the PMOS transistor decreases. The

common-mode current flowing through the differential inverter decreases over time until the MOS transistors enter the cutoff region, completing the amplification process of the FIA.

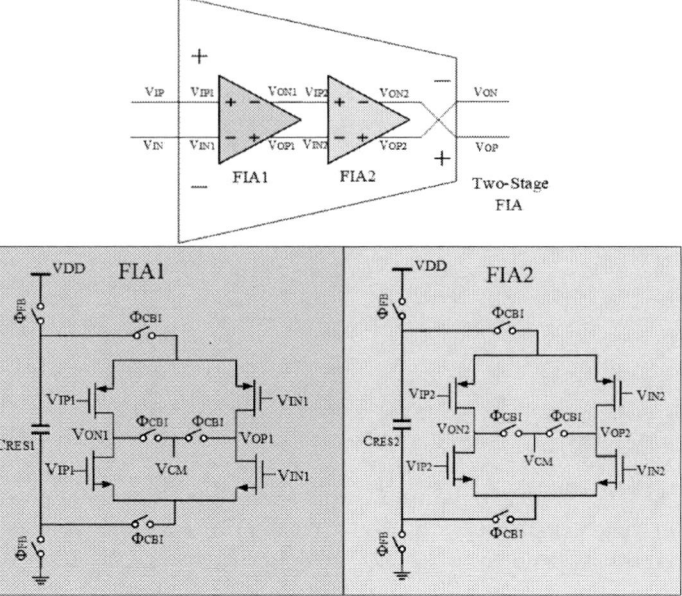

Fig. 5. The circuit diagrams of two-stage FIA.

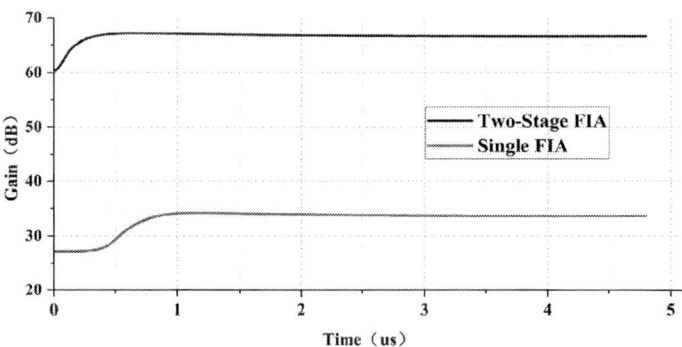

Fig. 6. The gain of FIA during amplification phase

The differential voltage established across the load capacitor represents the output voltage of the FIA. The output voltage gain is given by:

$$A(t) = \frac{2C_{RES} \cdot \Delta V_S(t)}{nV_T C_L} \tag{3}$$

$\triangle V_S(t)$ represents the change in the source voltages of the NMOS and PMOS transistors, n is the process parameter, and V_T is the thermal voltage.

The gain of a single-stage FIA is typically 30 dB. To meet the precision requirements of the readout circuit, two FIAs are cascaded to form a two-stage configuration, achieving a gain of over 60 dB, as illustrated in Fig. 6. Compared to the conventional folded cascode amplifier, the FIA has several advantages. The FIA utilizes the capacitor C_{RES} to provide current for the differential inverter-based amplifier.

979-8-3315-2209-4/25 $31.00 © 2025 IEEE

After each amplification cycle, the switches connected to the top and bottom plates of the capacitor are reset, and no static current flows through the circuit. Therefore, unlike the constant current consumption of the traditional operational amplifier, the FIA achieves significantly lower power consumption. By employing the FIA amplifier, the total power consumption of the readout circuit is reduced by approximately 30%. Additionally, the output common-mode voltage of the FIA remains at the constant common-mode level V_{CM} throughout the amplification phase, eliminating the need for a common-mode feedback circuit. This improves the area efficiency of the circuit.

III. SIMULATION RESULT

The proposed readout circuit is designed with 180nm BCD process. In the simulation, the bandwidth of the acceleration input signal is 1 kHz, and the sampling frequency is set to 100 kHz. The measurement results are converted into acceleration expressed in g by dividing the output voltage values by the respective gains in the signal path. Fig.7.(a) visualizes the measured static non-linearity of the readout circuit over an input range between 0g and 20g. The maximum non-linearity determined from the measurements over the complete input range is 0.045%. Fig.7.(b) shows the variation of the mean offset value with temperature in the range of -20°C to 140°C under different process corners. The maximum mean offset value is limited to within 12.8 mg. Fig. 8 shows the output FFT spectrum with the inclusion of noise and mismatch as non-ideal factors, the resulting noise density in the baseband is 63.1 $\mu g/\sqrt{Hz}$. Hence the equivalent dynamic range is 80.5dB. Table. II. compares the proposed readout circuit with previously published charge-balanced readout circuit.

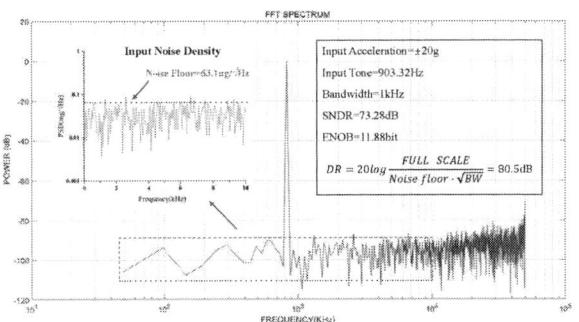

Fig. 8.   FFT spectrum and noise spectrum density of the readout circuit.

TABLE II: Performance Comparison of Readout Circuit

Reference	[4]	[5]	This work
Full scale (g)	±50g	±8	±20g
Bandwidth (Hz)	86	1K	1K
Sample rate (Hz)	1K	50K	100K
Noise floor ($\mu g/\sqrt{Hz}$)	84	250	63.1
Power (μW)	1400	62	419
Nonlinearity (%)	0.1	< 1	0.045
DR (dB)	95.5	-	80.5
Process (CMOS)	130nm	180nm BCD	180nm BCD

the circuit effectively reduces power consumption. The final circuit achieves a noise floor of 63.1 $\mu g/\sqrt{Hz}$, a dynamic range of 80.5 dB, and a nonlinearity of 0.045% with a signal bandwidth of 1 kHz.

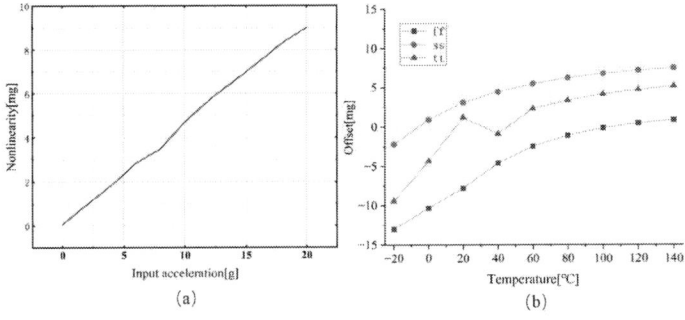

Fig. 7.   (a) Non-linearity of the readout circuit. (b) Mean offset of the readout circuit.

IV. CONCLUSION

A low-noise charge-balanced readout circuit for MEMS accelerometer was proposed in this paper. The charge-balanced technique is employed to improve linearity through negative feedback. By utilizing a boxcar sampler, the combination of the transconductance amplifier and integrator enables an improved noise performance. Meanwhile, the use of FIA in

REFERENCES

[1] M. Yucetas, M. Pulkkinen, A. Kalanti, J. Salomaa, L. Aaltonen and K. Halonen, "A High-Resolution Accelerometer With Electrostatic Damping and Improved Supply Sensitivity," in IEEE Journal of Solid-State Circuits, vol. 47, no. 7, pp. 1721-1730, July 2012.

[2] V. P. Petkov, G. K. Balachandran and J. Beintner, "A Fully Differential Charge-Balanced Accelerometer for Electronic Stability Control," in IEEE Journal of Solid-State Circuits, vol. 49, no. 1, pp. 262-270, Jan. 2014.

[3] A. Lanniel, T. Boeser, T. Alpert and M. Ortmanns, "Noise Analysis of Charge-Balanced Readout Circuits for MEMS Accelerometers," in IEEE Transactions on Circuits and Systems I: Regular Papers, vol. 68, no. 1, pp. 175-184, Jan. 2021.

[4] A. Lanniel, T. Boeser, T. Alpert and M. Ortmanns, "Low-Noise Readout Circuit for an Automotive MEMS Accelerometer," in IEEE Open Journal of the Solid-State Circuits Society, vol. 1, pp. 140-148, 2021.

[5] Lai X, Wang Y, Li Q, Habib K, "Reset Noise Sampling Feedforward Technique (RNSF) for Low Noise MEMS Capacitive Accelerometer, " Electronics. 2022; 11(17):2693.

A 18.3-ENOB 160.2-μW Fully Dynamic Discrete-Time Delta-Sigma ADC Using Floating Inverter Amplifiers

Shipeng Zhang, Jixiang Zhang, Yao Li, Mao Ye*
School of Microelectronics, Tianjin University, Tianjin, 300072, China
*Email: mao_ye@tju.edu.cn

Abstract—**This paper presents a fourth-order discrete-time delta-sigma modulator that employs a two-stage floating inverter amplifier (FIA) in the first integrator. The first integrator utilizes a two-stage FIA with a DC gain of up to 74 dB, while the subsequent three integrators employ simple FIAs with a DC gain of up to 40 dB. The modulator coefficients are meticulously optimized to ensure a lower output swing at each stage, making the design particularly suitable for modulators. The proposed design is fabricated using a 180 nm CMOS process with a supply voltage of 1.8 V. Tested under a sampling frequency of 2 MHz, an OSR of 128, and an input signal amplitude of -5 dBFS, the modulator achieves a power consumption of only 160.2 μW. Simulation results indicate that the modulator achieves a signal-to-noise-and-distortion ratio (SNDR) of 112.1 dB, a spurious-free dynamic range (SFDR) of 118.19 dB, and a Schreier figure of merit (FOM) based on SNDR of 188.98 dB.**

Index Terms—**Analog-to-digital converters, floating inverter amplifiers (FIAs), delta–sigma modulators (DSMs), dynamic**

I. INTRODUCTION

With the rapid development of smart sensors and Internet of Things (IoT) technologies, the demand for low-power, high-resolution ADCs has become increasingly urgent. Delta-sigma ADCs can achieve high precision through noise shaping and oversampling techniques. However, traditional OTAs consume a significant portion of the power budget. Compared to conventional OTAs, FIAs reduce power consumption through their fully dynamic operation [1]. In addition, the FIA is powered by a floating reservoir capacitor, which allows it to provide a stable output common-mode voltage. This feature enhances the scalability of delta-sigma ADCs. This benefit supersedes the previous use of dynamic amplifiers, whose unstable output common-mode voltage could compromise the accuracy of the modulator [2].

This paper proposes a low-power, high-resolution delta-sigma ADC tailored for IoT applications, employing fully dynamic FIAs as amplifiers in the integrators. According to [3], a simple FIA can achieve a DC gain of approximately 35 dB, while the cascoded FIA used in [3] achieves a gain greater than 50 dB. However, the limited DC gain of cascoded FIAs is insufficient for suppressing low-frequency noise effectively in the fourth-order CIFF structure of the modulator. To address this issue, a two-stage FIA is proposed. The two-stage FIA achieves a DC gain of up to 74 dB, satisfying the gain requirements for the first integrator of the modulator, while

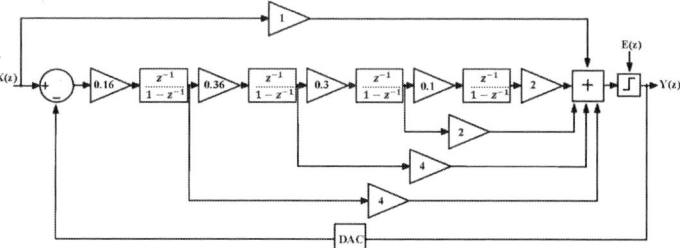

Fig. 1. Block diagram of the proposed DSM.

simple FIAs suffice for the subsequent integrators. To enable a fully dynamic modulator, a dynamic comparator is used as the 1-bit quantizer.

The structure of this paper is organized as follows: Section II introduces the overall architecture of the modulator. Section III discusses the circuit implementation of the two-stage FIA and the dynamic comparator. Section IV presents simulation results and comparisons with other designs. Conclusions are drawn in Section V.

II. SYSTEM ARCHITECTURE

This designed structure is a fourth-order CIFF modulator. The CIFF architecture offers the advantage of reducing the output swing requirements of the integrators. The complete block diagram of the modulator is shown in Fig. 1, with a comparator used as the 1-bit quantizer. The coefficients within the structure are meticulously optimized to ensure that the output swing of each stage remains within a lower range (approximately within ±0.5 V), thereby stabilizing the gain of the FIA. The noise transfer function (NTF) of this structure is given in (1). Analysis confirms that both the zeros and poles of the NTF lie within the unit circle, ensuring the stability of the structure.

$$NTF(z) = \frac{(z-1)^4}{z^4 - 3.36z^3 + 4.3104z^2 - 2.506z + 0.5593} \quad (1)$$

The complete circuit diagram of the designed modulator is illustrated in Fig. 2. This design operates at a sampling rate of 2 MHz with an oversampling ratio of 128, achieving a bandwidth of 7812.5 Hz. This circuit is fully differential, designed to suppress even-order harmonics and improve the

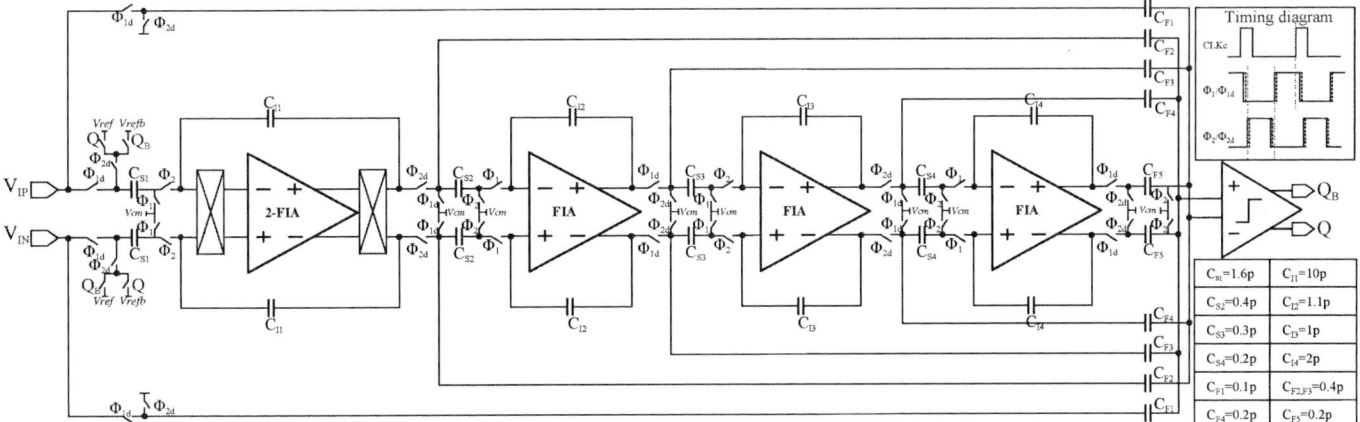

Fig. 2. Circuit diagram of the proposed DSM.

SNR. A two-phase, non-overlapping clock is utilized to prevent unwanted distortion in the integrators, and the timing diagram of this circuit is shown in Fig. 2, where CLKc is the clock of the comparator.

The first integrator is composed of a two-stage FIA and incorporates a chopper circuit to eliminate flicker noise and mitigate mismatch effects. The sampling capacitor of the first integrator is the largest to suppress the influence of thermal noise. The subsequent three integrators, which do not require high DC gain, are implemented using simple FIAs. During sampling, the FIA connects its output to V_{cm}, which causes periodic refreshing of the charge on the integration capacitor and prevents retention of the output result from the previous cycle. Therefore, the sampling period of each integrator must be synchronized with the integration period of the preceding integrator to ensure that the integration result is accurately and directly transferred to the next stage.

To minimize power consumption, a dynamic latch comparator is employed as the 1-bit quantizer. The quantizer outputs, Q and Q_B, are used for DAC feedback within the first integrator stage. Finally, a passive adder is implemented using a switched-capacitor circuit, which further reduces power consumption.

III. CIRCUIT IMPLEMENTATION

A. Proposed Two-stage FIA

The amplifier in the integrator of this design abandons traditional OTA and adopts a low-power, fully dynamic, closed-loop two-stage FIA as the amplifier, with the circuit structure shown in Fig. 3.

The simple FIA represented by FIA1 in Fig. 3 operates in two phases: charging and amplification. During the charging phase Φ_1, the floating reservoir capacitor C_{res} is charged to V_{DD}, and the output of the FIA is reset to the common-mode voltage V_{cm}. During the amplification phase Φ_2, the floating reservoir capacitor C_{res} powers the FIA, amplifying the input signal as current flows through the circuit. Once the charge on C_{res} is fully consumed, the FIA automatically shuts down, ensuring low power consumption and preventing unnecessary

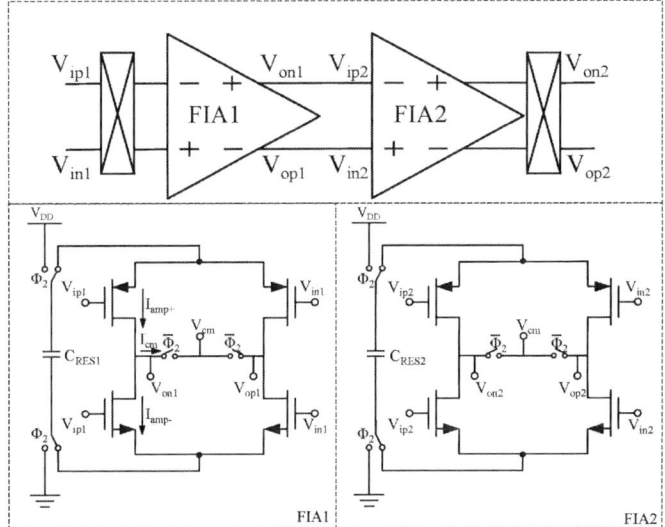

Fig. 3. Two-stage FIA circuit structure.

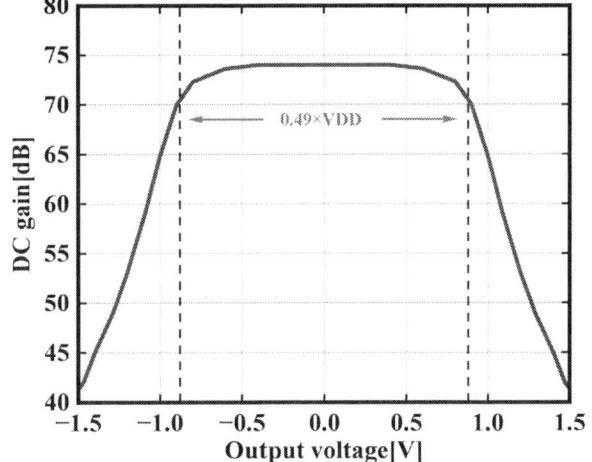

Fig. 4. DC gain versus output voltage of the proposed FIA.

979-8-3315-2209-4/25 $31.00 © 2025 IEEE   409

operation [4]. The FIA does not require a common-mode feedback (CMFB) circuit. When the FIA operates, the input current I_{amp+} at the top of C_{res} must equal the output current I_{amp-} at the bottom of C_{res}. As a result, the current I_{cm} flowing out of V_{on1} must be zero, generating a stable output common-mode voltage without requiring a CMFB circuit [1]. Additionally, due to its self-quenching mechanism, the open-loop gain of the FIA is insensitive to clock-related errors such as jitter.

Since the simple FIA achieves only a modest DC gain of approximately 35 dB, two FIAs are cascaded to form a two-stage FIA, as shown in Fig. 3. This configuration achieves a DC gain of approximately 74 dB, which satisfies the gain requirements of the first integrator in this architecture. Simulations of the two-stage FIA reveal the relationship between DC gain and output voltage, as shown in Fig. 4, where the output swing reaches $0.49 \times V_{DD}$-3dB. With the coefficient design discussed in the previous section, the necessary DC gain is attained across the output swing range.

The two-stage FIA operates as a closed-loop system, and its stability needs to be considered. According to the derivation in [4], the stability can be improved by ensuring that C_{res1} is two to three times larger than C_{res2}. This causes the dominant pole to shift inward more quickly during the discharge phase of the loop, improving the phase margin and enhancing overall stability. Therefore, this design does not require Miller compensation.

B. Proposed Dynamic Comparator Design

To achieve a fully dynamic structure for the modulator, this design employs a dynamic double-tail latch-type comparator. This structure eliminates the need for a static operating point, thereby avoiding static power consumption. The circuit implementation is shown in Fig. 5.

The detailed operation is as follows: when the comparator clock (CLKc) is low, the comparator resets, $M3$ and $M4$ conduct, pulling nodes AP and AN up to V_{DD}, while the outputs V_{op} and V_{on} reset to 0. When CLKc goes high, $M3$ and $M4$ are turned off, and the charges stored in the parasitic capacitance of $M3/M4$ and $M1/M2$ during the reset phase are discharged through the tail current transistor $M5$. The magnitude of the

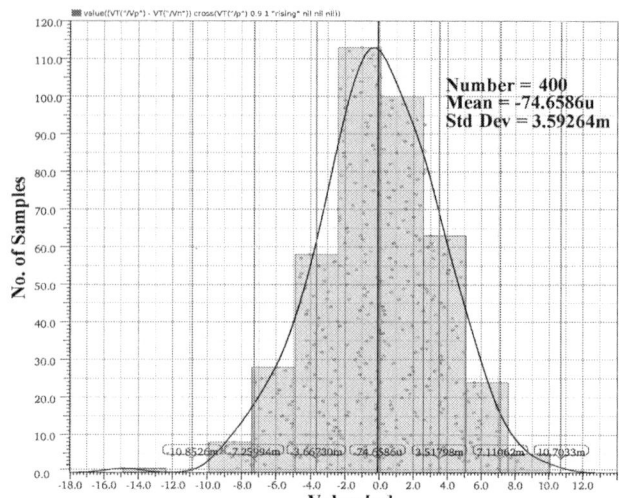

Fig. 6. Monte Carlo simulation results of dynamic comparator.

discharge current is determined by the input differential level, leading to different voltage drop rates at nodes AP and AN, which reflect the input voltage difference. The second-stage latch circuit amplifies this voltage difference, generating the comparison result.

A Monte Carlo simulation was conducted on this dynamic comparator. The simulation results are shown in Fig. 6, where the standard deviation of the offset is 3.6 mV, indicating that the input offset has a minimal impact on the performance of the modulator.

IV. SIMULATION RESULTS

The design presented in this paper was implemented using a 180 nm process, with a power consumption of 160.2 μW at a sampling frequency of 2 MHz. The first integrator consumes 66.95 μW, the remaining three integrators consume 44.75 μW, the clock consumes 40.5 μW, while other components collectively consume 7.9 μW. Fig. 7(a) illustrates the power consumption of the proposed modulator as a function of the sampling frequency, showing a linear relationship. Fig. 7(b) depicts the power consumption distribution across different parts of the modulator. Fig. 8 presents the output spectrum of the modulator under a 2 MHz sampling frequency, an

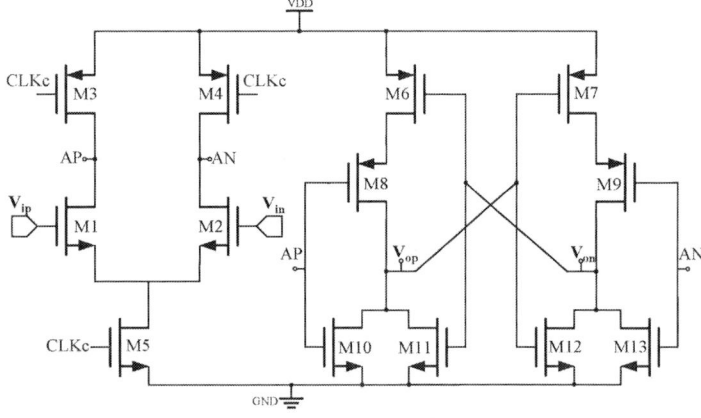

Fig. 5. Dynamic comparator circuit implementation.

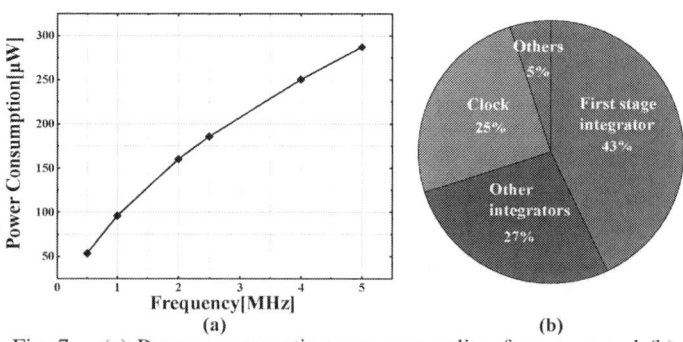

Fig. 7. (a) Power consumption versus sampling frequency and (b) measured power breakdown.

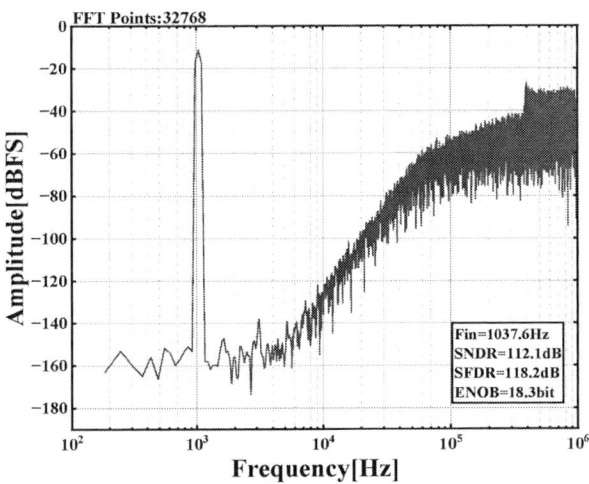

Fig. 8. Measured output spectrum at 2MHz sampling frequency.

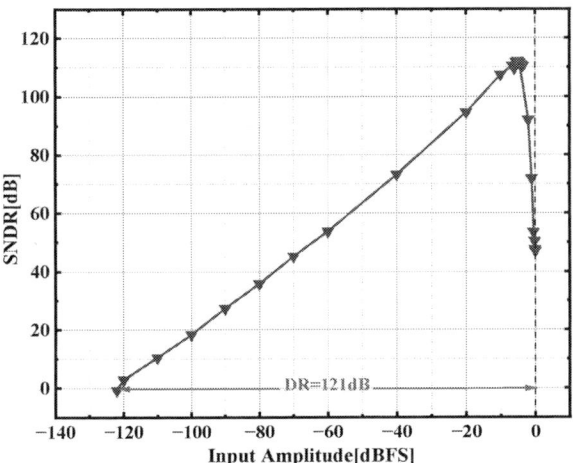

Fig. 9. Measured SNDR versus input amplitude.

TABLE I
PERFORMANCE SUMMARY AND COMPARISON

	This work	JSSC [3] 2022	TCAS-II [5] 2024	TCAS-II [2] 2022
Process(nm)	180	180	65	65
Type	DT Two-stage FIA	DT Cascoded FIA	DT Two-stage FIA	DT Two-stage Cascoded FIA
Supply(V)	1.8	1.8	1.5	1
BW(Hz)	7.8125k	24k	390	19.5k
Fs(Hz)	2M	3.072M	100k	10M
Power(μW)	160.2	340	3.5	43.5
SNDR(dB)	112.1	96.2	102.7	88.5
FOM$_{SNDR}$(dB)	188.98	174.7	183.1	175.01

REFERENCES

[1] X. Tang et al., "An Energy-Efficient Comparator With Dynamic Floating Inverter Amplifier," in *IEEE Journal of Solid-State Circuits*, vol. 55, no. 4, pp. 1011-1022, April 2020.

[2] A. Matsuoka, T. Nezuka and T. Iizuka, "Fully Dynamic Discrete-Time $\Delta\Sigma$ ADC Using Closed-Loop Two-Stage Cascoded Floating Inverter Amplifiers," in *IEEE Transactions on Circuits and Systems II: Express Briefs*, vol. 69, no. 3, pp. 944-948, March 2022.

[3] R. S. A. Kumar, N. Krishnapura and P. Banerjee, "Analysis and Design of a Discrete-Time Delta-Sigma Modulator Using a Cascoded Floating-Inverter-Based Dynamic Amplifier," in *IEEE Journal of Solid-State Circuits*, vol. 57, no. 11, pp. 3384-3395, Nov. 2022.

[4] X. Tang et al., "A 13.5-ENOB, 107-μW Noise-Shaping SAR ADC With PVT-Robust Closed-Loop Dynamic Amplifier," in *IEEE Journal of Solid-State Circuits*, vol. 55, no. 12, pp. 3248-3259, Dec. 2020.

[5] L. Liu, Z. Qin, J. Yin, X. Liao and Y. Tian, "A 16.8-ENOB, 3.5-μW Fourth-Order Discrete-Time Delta-Sigma ADC for Biosignal Acquisition Applications," in *IEEE Transactions on Circuits and Systems II: Express Briefs*, vol. 71, no. 4, pp. 1749-1753, April 2024.

oversampling ratio of 128, and an input signal amplitude of -5 dBFS. Fig. 9 shows the measured SNDR of the proposed modulator for different input signal amplitudes, achieving a dynamic range (DR) of 121 dB.

Table I summarizes the performance of the delta-sigma ADC presented in this paper and compares it with other advanced delta-sigma ADCs.

V. CONCLUSION

A forth-order fully dynamic discrete-time delta-sigma ADC using two-stage FIA is presented in this paper. This ADC addresses the high power consumption of conventional amplifiers by replacing them with a two-stage FIA. The two-stage FIA offers advantages such as fully dynamic operation, high efficiency, and the elimination of additional CMFB circuits. By integrating the two-stage FIA into a fully dynamic ADC, the proposed design achieves a simulated SNDR of 112.1 dB at a sampling frequency of 2 MHz while consuming only 160.2 μW. This makes it highly suitable for the growing demand for low-power, high-resolution ADCs in rapidly developing IoT applications and smart sensors.

979-8-3315-2209-4/25 $31.00 © 2025 IEEE

RCS-SAR: A 12-bit 20MS/s Secure ADC with Random Capacitor Switching Against Power Side-Channel Attacks

Jixiang Zhang, Shipeng Zhang, Jiaji He, Mao Ye*

School of Microelectronics, Tianjin University, Tianjin, 300072, China

*Email: mao_ye@tju.edu.cn

Abstract—**Analog-to-digital converters (ADCs) exhibit a hardware security vulnerability that allows attackers to steal sensitive output information by observing current waveforms from exposed power supply pins. This paper presents a successive approximation register (SAR) ADC that employs a mechanism of random capacitor switching to the V_{CM} reference voltage. This approach effectively encrypts the current waveform, providing protection against power side-channel attacks (PSCA). A 12-bit, 20 MS/s secure SAR ADC using 55-nm CMOS validates the proposed design. It encrypts the reference current waveform, reducing the amplitude of the current signal from -46 dB to -81 dB and decreasing the signal-to-noise ratio (SNDR) of current by 4 bits. In protected mode, the Nyquist spurious-free dynamic range (SFDR) and SNDR at 20 MS/s were simulated at 90.4 dB and 70.4 dB, resulting in a figure of merit (FoM) of 22.1 fJ/conv.-step.**

Index Terms—**successive approximation register, hardware security, power side-channel attack.**

I. INTRODUCTION

SAR ADCs are popular in sensor applications for their scalability, ease of implementation, and energy efficiency. As shown in Fig. 1, ensuring information security in ADC applications has become a pressing concern. The mechanism of successive approximation causes them to show different information in the time, electromagnetic, and power domains during quantization. Recent studies [1] [2] have explored these characteristics and employed various methodologies to execute side-channel attacks. In this context, power channel attacks and corresponding protective measures have been developed. Reference [2] proposes the use of a current equalizer to isolate the supply current from the digital output, while Reference [3] advocates for the independent control of the cell capacitance of the most significant bit (MSB) and the randomized modulation of power consumption. Furthermore, RaM-SAR and Sniff-SAR enhance secure data conversion by implementing stochastic mapping and integrating PSA and EMSA detectors, respectively [4] [5].

This paper proposes a RCS-SAR that randomly switches capacitors during quantization. After each comparison, the bottom plate of the random capacitor switches to V_{CM} and then to V_{REF} or G_{ND} based on the result. This method ensures accurate signal quantization and randomizes the current waveform with minimal hardware costs.

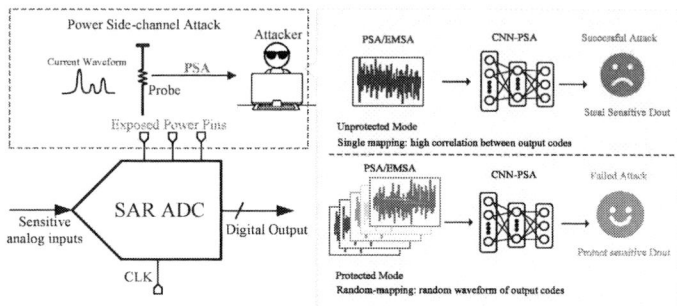

Fig. 1. Side-channel security challenges for ADCs and protection mechanisms for random current waveforms.

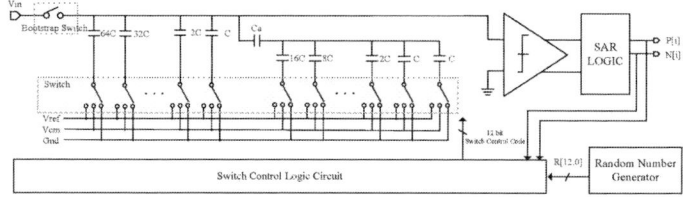

Fig. 2. Single-ended SAR ADC architecture for this design.

II. PROPOSED SIDE-CHANNEL PROTECTIONS

The design of SAR ADCs is crucial to understanding side-channel attacks. This paper presents the design of a 12-bit fully differential SAR ADC, with the single-ended architecture illustrated in Fig. 2. It comprises four primary modules: a sample-and-hold (S/H) circuit, a capacitor digital-to-analog converter (CDAC), a comparator, and a digital module that implements the sar logic. In this design, bootstrap switching circuits are employed as sampling switches for top-sampling technology. The CDAC use a binary segment-weighted capacitor structure. To achieve fast response and minimize power consumption, a two-stage dynamic comparator is used, as shown in Fig. 3. Digital logic circuitry includes dynamic flip-flops and the requisite logic circuits to generate the bottom plate switching control signals based on the comparison results P[i], N[i] and random numbers R[12:0].

In CDAC, higher bits require the use of larger parallel cell capacitors for simultaneous switching. This operation generates side-channel leakage that persists for one clock

979-8-3315-2209-4/25 $31.00 © 2025 IEEE

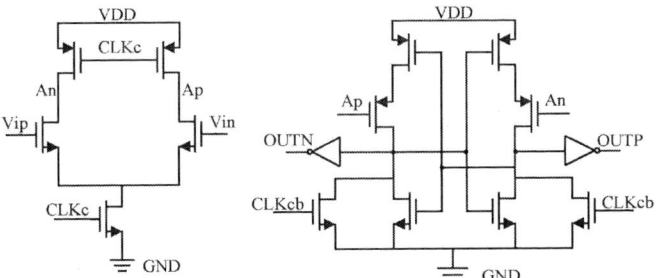

Fig. 3. Two-stage dynamic comparator.

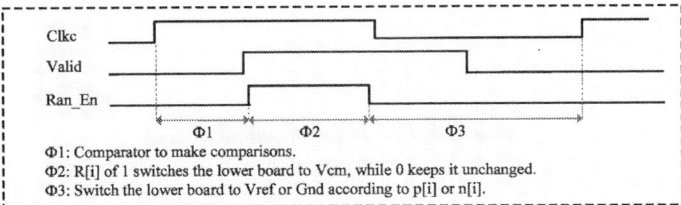

Φ1: Comparator to make comparisons.
Φ2: R[i] of 1 switches the lower board to Vcm, while 0 keeps it unchanged.
Φ3: Switch the lower board to Vref or Gnd according to p[i] or n[i].

Fig. 4. Timing of this design when operating in protected mode

cycle per transition, allowing the current waveform to signify whether the bit is '0' or '1'. The SAR ADC, without protective measures, generates a fixed V_{REF} current waveform when outputting a specific digital code. Attackers can exploit the correlation between the digital output code and the V_{REF} current waveform to steal the output code.

During the quantization process, the encryption of the V_{REF} current waveform can be accomplished by switching the V_{CM} with a randomly selected capacitor. This approach significantly reduces the potential for an attacker to reconstruct data from side-channel leakage. The timing of this design when operating in protected mode is shown in Fig. 4.

The comparison cycle can be divided into three phases. In the first phase, the comparator begins comparison during the high phase of CLK_C. Upon completion of the comparison, the results are latched into P[i] and N[i]. In the second phase, based on the random number R[12:0], the bottom plate switches of the capacitor with R[i] set to 1 are connected to V_{CM}, while the others remain in their original state. This action induces a random alteration in the voltage of the bottom plate capacitors. In the third phase, when CLK_C is low, the comparator is reset, and the comparator output is pushed to a high level. Subsequently, the lower plate switches are connected to the reference voltage V_{REF} and G_{ND} based on P[i] and N[i], thereby ensuring the accurate establishment of the successive approximation.

Through this process, each successive approximation consists of two steps: the top plate voltage is switched to a random voltage, and then it is adjusted to the correct quantization voltage. For a given analog input, the voltage of the top plate switches between the same two voltages to achieve the correct conversion. However, the path of quantization of the input signal changes significantly due to the random number R[12:

0]. This alteration results in different current waveforms during the same quantization process, effectively encrypting the reference current waveform and complicating the determination of whether the output bit is 0 or 1. In unprotected mode, after the comparator completes its comparison, the lower plate switches are connected to the reference source V_{REF} and G_{ND} based on P[i] and N[i], without encrypting the reference current waveform.

The random number generator (RNG) used in this design is generated by an off-chip pseudo-random number generator excitation source. The primary overheads associated with this architecture include the RNG, added switches, and logic circuits, which may lead to a modest increase in the area and power consumption.

III. SIMULATION RESULTS

A. Power side-channel

Primary side channel attacks against SAR ADC include template-based attacks [1], multilayer perceptron-based attacks (MLP-PSA), and convolutional neural network-based attacks (CNN-PSA) [2]. These attacks do not directly target the digital logic or data processing components of the ADC. Instead, attackers exploit the correlation between the current waveform of the reference voltage pin and the resulting digital output code, allowing them to steal sensitive information.

The 0.4V differential input voltage was quantized through multiple quantization processes to assess the randomness of the current waveform encryption. Fig. 5(a) illustrates a comparison of the quantized voltages on the top plate of the capacitor in both the unprotected and protected modes. It is apparent that the quantization path for the input signal undergoes significant alterations due to the random numbers R[12:0]. However, each successive approximation operates between two fixed voltages to achieve the correct ADC conversion. Importantly, as shown in Fig. 5(b), the variability in the quantization path leads to the generation of distinct reference current waveforms during the quantization of the identical signal, thereby completing the encryption of the reference current waveform and demonstrating a degree of randomization. As a result, if a side-channel attack is conducted on such ADCs, the correlation and prediction of the digital output become increasingly difficult, as the energy waveforms are no longer unique for each specific input voltage. Additionally, varying input voltages may produce similar energy dissipation values across each switching state.

To assess the degree of encryption of the reference current waveform, a sinusoidal signal with an input frequency of 2 MHz was analyzed. The initial successive approximation of the peak-to-peak value of the current was conducted to perform the Fast Fourier Transform (FFT), with the results shown in Fig. 6. In the unprotected mode, the amplitude response of the current signal exhibited considerable fluctuations across all frequencies, accompanied by distinct peaks. This observation indicates a strong correlation between the input signal and the current waveform. Conversely, in the protected mode, both the amplitude response peaks and the harmonics of the signal

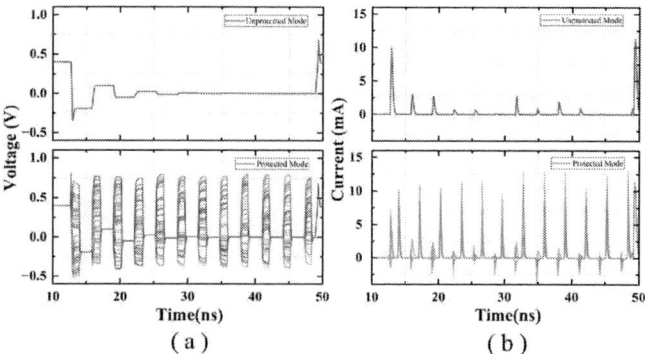

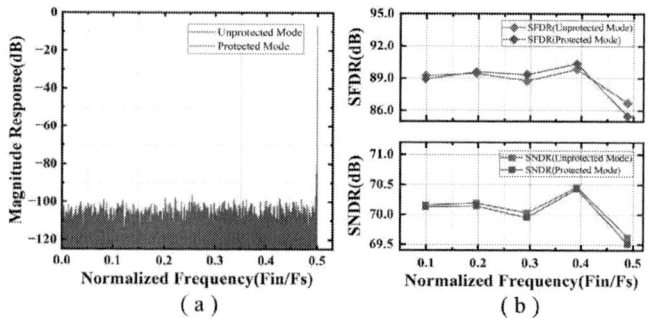

Fig. 7. (a) the FFT spectrum for fs/2. (b) overall performance.

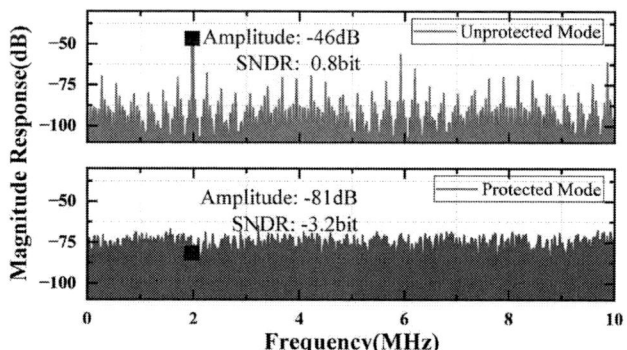

Fig. 5. (a) voltage waveforms of the capacitor's top plate during quantization. (b) reference current waveforms during quantization.

TABLE I
PERFORMANCE COMPARSION OF SECURE ADC

	This Work		JSSC'21[2]	CICC'22[3]	VLSI'22[4]	CICC'23[5]
Architecture	RS-SAR[1]		S2:ADC[2]	RS-SAR[2]	RaM-SAR[2]	Sniff-SAR[2]
Technology[nm]	65		65	65	65	65
Supply Voltage[V]	1.2		1.2	1.2	1.2	1.2
Resolution[bit]	12		12	8	12	12
Protect Mode	Unprotected	Protected	Current Equalizer	Random Switching	Random Mapping	Random Mapping
Sampling Rate[MS/s]	20	20	1.25	2	25	40
Power[μW]	450.2	1192.4[3]	158.5[3]	50.2	539.8	698
SNDR[dB]	70.5	70.4	69.2	48.1	67.2	66.6
SFDR[dB]	89.8	90.4	89.6	N/A	80.5	80.2
FoMw[fJ/conv.-step]	8.3	22.1	54.3	120.7	8.5	9.8

1.Simulated. 2. Silicon Results. 3. Doesn't Include RNG.

and is applicable to other differential ADCs. Ultimately, this technique effectively mitigates security vulnerabilities in SAR ADCs, thereby safeguarding sensitive information in sensor applications.

ACKNOWLEDGMENT

This work is supported by the National Key R&D Program of China under Grant 2023YFB4402800.

REFERENCES

[1] T. Miki, N. Miura, H. Sonoda, K. Mizuta and M. Nagata, "A Random Interrupt Dithering SAR Technique for Secure ADC against Reference-Charge Side-Channel Attack," *2020 IEEE International Symposium on Circuits and Systems (ISCAS)*, Seville, Spain, 2020, pp. 1-1

[2] T. Jeong, A. P. Chandrakasan and H. -S. Lee, "S2ADC: A 12-bit, 1.25-MS/s Secure SAR ADC With Power Side-Channel Attack Resistance," in *IEEE Journal of Solid-State Circuits*, vol. 56, no. 3, pp. 844-854, March 2021

[3] M. Ashok, E. V. Levine and A. P. Chandrakasan, "Randomized Switching SAR (RS-SAR) ADC Protections for Power and Electromagnetic Side Channel Security," *2022 IEEE Custom Integrated Circuits Conference (CICC)*, 2022, pp. 1-2.

[4] R. Chen, H. Wang, A. Chandrakasan, and H.-S. Lee, "Ram-sar: A low energy and area overhead, 11.3fj/conv.-step 12b 25ms/s secure randommapping sar adc with power and em side-channel attack resilience," in *2022 IEEE Symposium on VLSI Technology and Circuits (VLSI Technology and Circuits)*, 2022, pp. 94–95.

[5] R. Chen, A. Chandrakasan and H. -S. Lee, "Sniff-SAR: A 9.8fJ/c.-s 12b secure ADC with detectiondriven protection against power and EM side-channel attack," *2023 IEEE Custom Integrated Circuits Conference (CICC)*, San Antonio, TX, USA, 2023, pp. 1-2.

Fig. 6. FFT Spectrum of the peak-to-peak value of the first successive approximation of the current.

were significantly diminished, suggesting that the implemented protection measures effectively reduce the correlation between the input signal and the reference current waveform.

B. ADC Performance

The SAR ADC is well suited for sensor applications that require moderate resolution, bandwidth, and low power consumption. The dynamic performance of the ADC, without calibration, is illustrated in Fig. 7. Table I shows a comparison of the designed ADCs with the secure ADCs. Notably, the performance remains stable despite variations caused by time and random switching modifications. The dynamic performance in both unprotected and protected modes is comparable at low-frequency inputs, with an SNDR of 70.4 dB and an SFDR of 90.4 dB. Power consumption is simulated at 1.19 mW, resulting in a FoM of 22.1 fJ/conv.-step.

IV. CONCLUSION

This paper presents a comprehensive study of security issues in SAR ADCs. We have designed and analyzed a 12-bit SAR ADC to identify potential security vulnerabilities. We propose a novel protection technique using random switching during quantization, which encrypts the current waveform with minimal impact on power consumption and performance. Our architecture integrates seamlessly into existing designs

979-8-3315-2209-4/25 $31.00 © 2025 IEEE

A high integration and high stability design of analog driving ASIC for MEMS Gyroscopes

Hetian Sun

Mao Ye

Abstract—This paper presents a driving application-specific integrated circuit (ASIC) for micro-electromechanical system (MEMS) gyroscopes. Active MOS resistors are introduced in the trans-impedance amplifier (TIA) and capacitor scaler technique is adopted for the proportional-integral (PI) controller to improve the integration. Meanwhile, the improved TIA ensures sufficient loop gain, and the enhanced PI controller generates an appropriate control voltage in the driving loop, maintaining the amplitude and frequency stability of MEMS gyroscope vibrations. An optimized band-pass filter with high frequency selectivity is introduced to reduce the harmonic distortion in the driving signal, mitigating the risk of oscillation failure and reducing the set-up time of the driving loop. The driving circuits are implemented using the SMIC 0.18 μm BCD process, its layout area is only 482 × 223 μm². Post-layout simulation results demonstrate that the set-up time of the proposed driving loop is reduced to 80 ms, with an amplitude stability of 2.14ppm and a frequency stability of 7.16ppm. Compared with the traditional MEMS gyroscope driving circuits, this work achieves higher integration, amplitude stability and frequency stability.

Index Terms—MEMS gyroscope, TIA, PI controller, band-pass filter, integration, stability

I. INTRODUCTION

In practical applications, the driving circuits for most micro-electromechanical systems (MEMS) gyroscopes were typically implemented using PCB designs. However, the conventional PCB designs often fail to meet the practical requirements due to their large size and high power consumption, whereas ASICs can effectively meet the demands for high integration and low power consumption.

Recently, aiming to enhance the integration of the driving circuits, significant advancements have been made in driving loop ASICs based on the automatic gain control (AGC) technique [1], and the basic structure is illustrated in Fig. 1, which is composed of the trans-impedance amplifier (TIA), rectifier circuit, proportional-integral (PI) controller, variable gain amplifier (VGA) and other auxiliary circuits.

To satisfy the vibration requirement of gyroscopes, certain modules, such as TIA and PI controller, rely on multiple MΩ-level off-chip resistors and nF-level off-chip capacitors. Large resistor R_3 in the several hundreds of megohms range is required for the TIA to achieve high loop gain to ensure that the driving loop oscillates properly. In the PI controller, large capacitor C_1 in the nF range is required to achieve the

---

Hetian Sun is with *the School of Microelectronics, Tianjin University* , Tianjin 300072, China. Mao Ye is with *the School of Microelectronics and Tianjin Key Laboratory of Imaging and Sensing Microelectronic Technology, Tianjin University*, Tianjin 300072, China. Corresponding author: (Mao Ye, email: mao ye@tju.edu.cn)

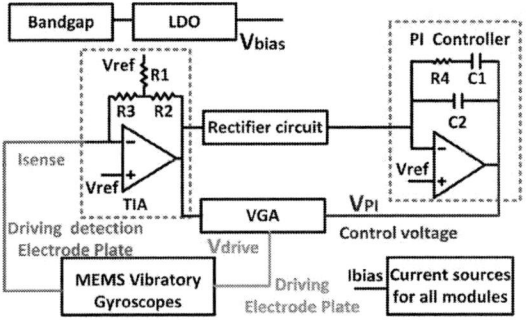

Fig. 1. Structure diagram of the driving loop circuits based on AGC system

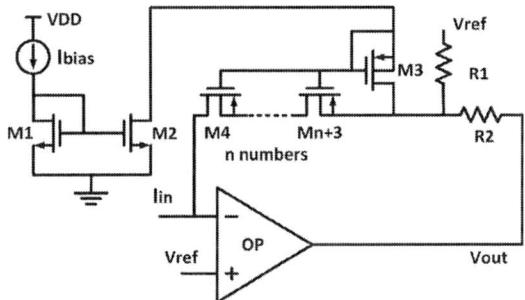

Fig. 2. Circuit diagram of the improved TIA

proper integral coefficient, and C_2 in the nF range is required to achieve the lower low-pass cutoff frequency. Thereby, the PI controller ensures the stability of the driving loop oscillation amplitude and frequency precisely. To further enhance the integration of the driving loop circuits, improved TIA and PI controller are proposed in this paper to reduce the use of off-chip resistors and capacitors.

Additionally, harmonic distortion issues are urgently required to be improved [2]. In the gyroscope sensors, parasitic capacitors between the driving electrode plate and the driving detection electrode plate lead to insufficient bandpass frequency selectivity, reducing the ability to filter harmonic signals. This deficiency may cause oscillation failure in the driving loop [3-4]. Meanwhile, circuit modules, such as the improved TIA, PI controllers, and VGA, exhibit nonlinear behavior and may generate unwanted harmonic signals. To address these problems, this work introduces an enhanced band-pass filter with high frequency selectivity. The improved filter minimizes harmonic distortion in the driving signal,

979-8-3315-2209-4/25 $31.00 © 2025 IEEE

mitigates the risk of oscillation failure, and reduces the set-up time of the driving circuits. By fine-tuning the parameters of the entire driving circuits after introducing these improved modules, the MEMS gyroscopes achieve more stable vibration in driving mode.

II. PROPOSED CIRCUITS

A. Transimpedance amplifier with active MOS resistors

The proposed TIA improves the integration of the driving circuits by replacing passive resistors with MOS active resistors. As shown in Fig. 1, the gain of the conventional TIA is given by $R_2+R_3+R_2R_3/R_1$, which is typically in the GΩ range to provide adequate loop gain for the driving loop. However, a large resistance ratio R_2/R_1 may result in severe DC offset in V_{out}. Therefore, smaller R_2 and large off-chip resistors R_3 which is in the range of several hundreds of megohms are commonly chosen for the conventional TIA.

The improved TIA is shown in Fig. 2, with an appropriately large W/L of M_3 and n extremely small W/L MOSFETs, labeled M_4 to M_{n+3}. A small bias current in the uA range drives $M_3 \sim M_{n+3}$ into the subthreshold region, ensuring that $M_4 \sim M_{n+3}$ behave like resistors with very large resistance and introduces very low leakage current, thereby eliminating the use off-chip resistors. Meanwhile, extremely small leakage currents in $M_4 \sim M_{n+3}$ do not affect the normal operation of TIA. Improved TIA can provide even higher loop gain with proper circuit parameterization, ensuring constant amplitude vibration of the MEMS gyroscopes. However, the linearity of the active MOS resistors is much worse than the passive resistors. The good filters are required to remove the harmonic signals resulting from the nonlinearity problem.

B. Proportional-Integral controller with capacitor scalers

The proposed PI controller is depicted in Fig. 3, which improves the integration by introducing two capacitor scalers. As shown in Fig. 1, in the conventional PI controller, C_1 and C_2 are off-chip capacitors with values in the nF range. In this work, two capacitor scalers are introduced to ensure that the improved PI controller operates correctly, even when C_1 and C_2 are only in the pF range. In the first capacitor scaler enclosed in the red box, R_6=nR_5, $V_n \approx V_p$. According to the I-V relationship, the equivalent impedance of the circuit branch in the red box is:

$$Z_{eq} = \frac{1}{n+1}(R_6 + \frac{1}{sC_1}) \tag{1}$$

Equation (1) demonstrates that C_1 is equivalently amplified by a factor of n and the equivalent resistance of R_6 is greatly reduced. Using R_6 in the hundreds of kilo-ohms range, R_5 in the kilo-ohm range, and C_1 in the pF range, the capacitor scaler achieves the equivalence of a resistor in the kilo-ohm range and a large capacitor in the nF range. Obviously, the area of on-chip resistor R_6 in the hundreds of kilo-ohms range is much smaller than the off-chip capacitor in the nF range, which is better for the integration of driving circuits.

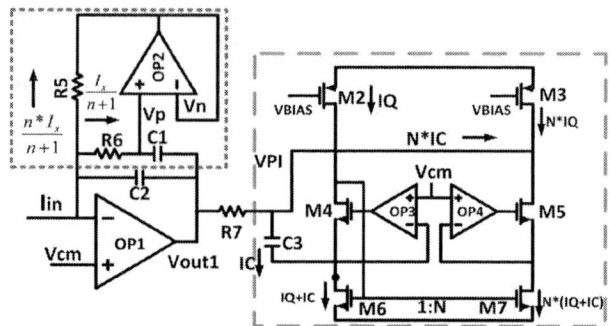

Fig. 3. Circuit diagram of the improved PI Controller

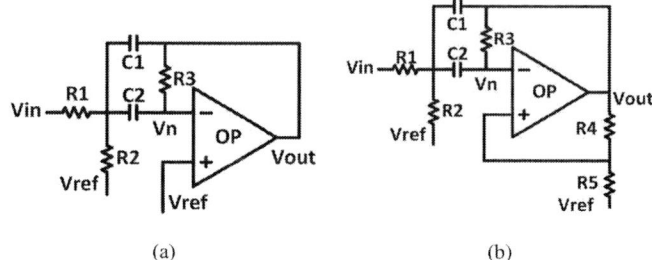

(a)   (b)

Fig. 4. (a) Circuit diagram of the conventional band-pass filter (b) Circuit diagram of the improved band-pass filter

In the conventional PI controller, the capacitor C_2 in the nF range plays the role of low-pass filtering. In order to achieve the same low cutoff frequency with a much smaller C_2, an extra passive filter composed of R_7 and C_3 is designed. The second capacitor scaler enclosed in the blue box is introduced to increase the equivalent capacitance of C_3. The W/L of the M_2 and M_3 is 1:N and the W/L of the M_6 and M_7 is also 1:N, according to the current relationship, the effective current flowing through C_3 is 1/(N+1) of the input current flowing through R_7, thereby the capacitor scaler increases the equivalent capacitance of C_3 by a factor of N. In the capacitor scaler, OP3 and OP4 are in the gain-boost topology to enhance the output impedance of the current mirror composed of M_4, M_5, M_6 and M_7 and stabilize the drain voltages of M_4 and M_5, to make the PI controller robust even after the introduction of the complex capacitor scalers.

Compared to the off-chip resistors and capacitors used in conventional TIA and PI controllers, as well as smaller size for improved integration, on-chip resistors and capacitors used in improved TIA and optimized PI controllers provide several advantages. These include better temperature coefficients, improved long-term stability, higher accuracy, and lower surface resistance values. Therefore, the optimized PI controller provides a much more appropriate integral coefficient and low-pass cutoff frequency for the driving loop, generating a proper control voltage for the driving loop and ensuring that MEMS gyroscopes achieve more stable vibration in driving mode.

C. High frequency selectivity Band-pass filter

In order to solve the harmonic distortion problems, this work introduces an improved band-pass filter between the VGA and

979-8-3315-2209-4/25 $31.00 © 2025 IEEE   416

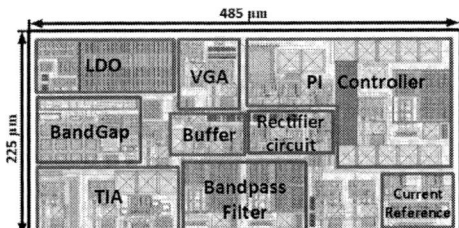

Fig. 5. MEMS gyroscope driving loop ASIC layout

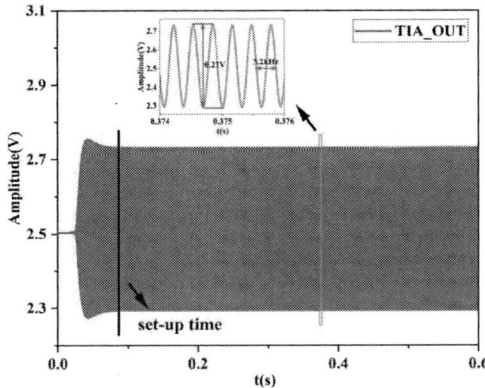

Fig. 6. Waveform diagram of the driving detection signal

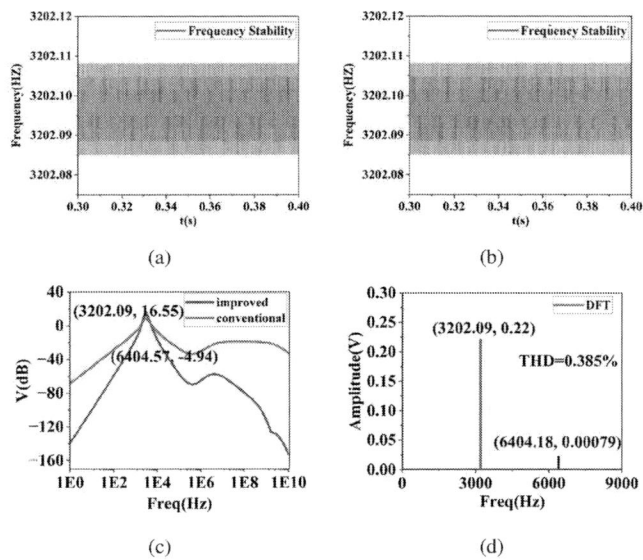

Fig. 7. (a) Amplitude fluctuation waveform diagram of the driving detection signal (b) Frequency fluctuation waveform diagram of the driving detection signal (c) Frequency response waveform diagram of the band-pass filters (d) Waveform diagram of the spectrum of the driving signal

the driving electrode plate of the MEMS gyroscopes. The conventional band-pass filter is illustrated in Fig. 4(a). The optimized band-pass filter improves the frequency selectivity by introducing a positive feedback composed of R_4 and R_5 branches, as is illustrated in Fig. 4(b). The transfer function of the conventional filter is given by:

$$\frac{V_{out}}{V_{in}}(s) = \frac{\frac{-1}{R_1 C_1}s}{s^2 + \frac{C_1+C_2}{C_1 C_2 R_3}s + \frac{R_1+R_2}{C_1 C_2 R_1 R_2 R_3}} \quad (2)$$

Equation (2) demonstrates that the center frequency $\omega_{con} = [(R_1+R_2)/(C_1 C_2 R_1 R_2 R_3)]^{1/2}$, and the quality factor Q_{con} of the conventional bandpass filter is derived as shown in:

$$Q_{con} = \frac{\sqrt{(R_1+R_2)C_1 C_2 R_3}}{\sqrt{R_1 R_2}(C_1+C_2)} \quad (3)$$

The transfer function of the improved filter is given by:

$$\frac{V_{out}}{V_{in}}(s) = \frac{-\frac{R_4+R_5}{C_1 R_1 R_4}s}{s^2 + \frac{R_1 R_2 R_4(C_1+C_2) - R_1 R_2 R_4(C_1+C_2)}{C_1 C_2 R_1 R_2 R_3 R_4}s + \frac{R_1+R_2}{C_1 C_2 R_1 R_2 R_3}} \quad (4)$$

Equation (4) demonstrates that the center frequency $\omega_{imp} = [(R_1+R_2)/(C_1 C_2 R_1 R_2 R_3)]^{1/2}$, and the quality factor Q_{imp} of the improved bandpass filter is derived as shown in:

$$Q_{imp} = \frac{\sqrt{(R_1+R_2)C_1 C_2 R_3}}{\sqrt{R_1 R_2}(C_1+C_2) - \frac{C_2 R_3 R_5(R_1+R_2)}{\sqrt{R_1 R_2 R_4}}} \quad (5)$$

Obviously, the center frequency ω_{imp} of the improved filter remains identical to ω_{con} of the conventional filter. To maintain the stability of the second-order system in (4) after introducing the positive feedback, it is essential to ensure that $R_1 R_2 R_4(C_1+C_2) - C_2 R_3 R_5(R_1+R_2) > 0$.

In this work, Q_{imp} in (5) improves by 446% compared to Q_{con} in (3) with positive feedback. The higher Q_{imp} indicates that the improved band-pass filter exhibits superior frequency selectivity and effectively reduces harmonic distortion in the driving signal, making the MEMS gyroscopes vibrate stably.

III. SIMULATION RESULTS

The layout of the driving loop circuits proposed in this paper is illustrated in Fig. 5. It is implemented using the SMIC 0.18 μm BCD process. The complete driving loop circuits underwent post-layout simulation, with the waveform of the drive detection signal depicted in Fig. 6. The results indicate that the set-up time of the driving loop is approximately 80 ms, the amplitude of the drive detection signal is around 0.22 V. As shown in Fig. 7(a), the amplitude fluctuation is approximately 4.7×10^{-7} V, consequently the amplitude stability of the drive detection signal is calculated to be approximately 2.14ppm. The frequency fluctuation of the drive detection signal is presented in Fig. 7(b), the simulation results show that the frequency fluctuation is approximately 0.02294 Hz, while the oscillating frequency is about 3202.096 kHz. Therefore, the frequency stability is determined to be about 7.16ppm.

The frequency response waveform diagram of the bandpass filter is illustrated in Fig. 7(c), which shows more intuitively that the improved band-pass filter has better frequency selectivity ,amplifying the fundamental frequency signal while

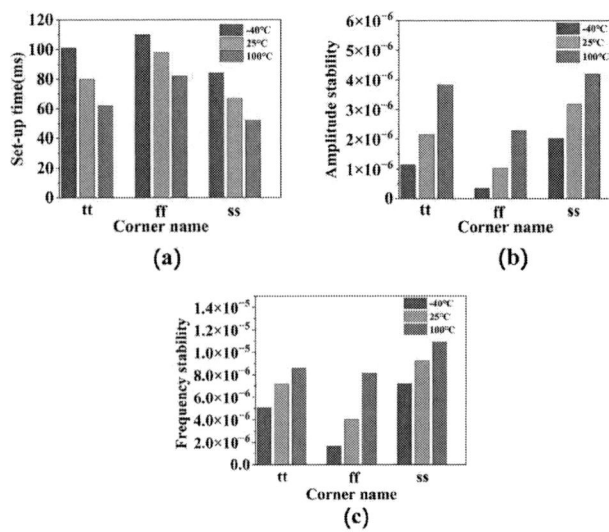

(a) ... **(b)**

(c)

Fig. 8. Simulated (a) set-up time, (b) amplitude stability, (c) frequency stability of the driving circuits at different working conditions

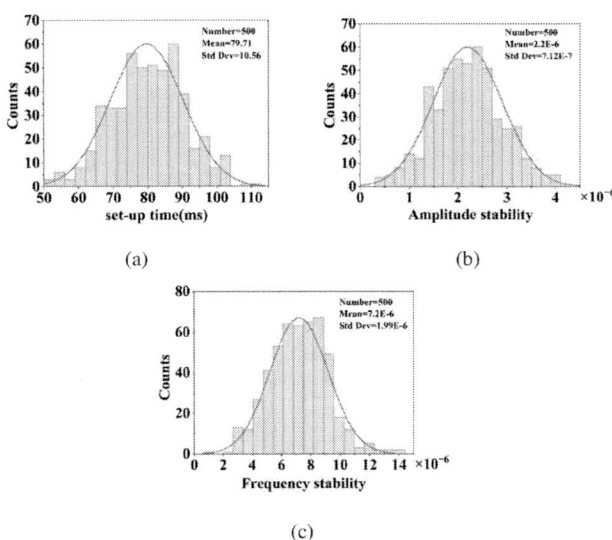

(a) ... (b)

(c)

Fig. 9. Monte Carlo simulation results of (a) set-up time (b) amplitude stability (c) frequency stability

greatly reducing the harmonic signal components. The frequency spectrum of the driving signal is illustrated in Fig. 7(d), it shows that the total harmonic distortion (THD) of the driving signal is calculated to be only 0.358%.

The simulation results above are based on the most general operating conditions of tt process corner, 5V power supply voltage, and 25°C. The performance evaluations about the process and temperature of the entire driving circuits are shown in Fig. 8. Under ss process corners or low-temperature operating conditions, since the gain of the improved TIA increases, the overall drive loop gain also increases, therefore the driving loop vibrates more quickly but becomes less stable.

TABLE I: Performance Comparison

Ref	[5]	[6]	[7]	this work
Frequency stability	25ppm	30ppm	0.93ppm	7.16ppm
Amplitude stability	50ppm	25ppm	—	2.14ppm
Set-up time (ms)	200	150	500	80
Power consumption (mW)	180	25	90	11.3

Conversely, under ff process corners or high-temperature operating conditions, the driving loop vibrates more slowly but exhibits greater stability. The simulation results demonstrate that the driving circuits exhibit excellent robustness.

Monte Carlo simulations are also performed on this driving circuits. The simulation results are shown in Fig. 9, where the standard deviation of the set-up time, amplitude stability and frequency stability are respectively 10.56ms, 0.712ppm and 1.99ppm, indicating that the mismatch has minimal impact on the performance of the driving circuits. The simulation results shown in Fig. 8 and Fig. 9 indicate that the performance of the circuits remain stable even the operating conditions changed.

IV. CONCLUSION

TABLE I presents performance comparisons between the proposed driving circuits and several other driving circuits.

The driving circuits presented in [5], [6], and [7] include various complex modules, such as charge amplifiers, phase shifter circuits, multipliers, and mixers. These circuits occupy significant layout area and consume substantial power. In contrast, this work eliminates the need for various complex modules, minimizes the use of numerous off-chip resistors and capacitors. These optimizations significantly reduce the layout area and simultaneously lower power consumption. In addition, the improved band-pass filter is introduced to this work to make the MEMS gyroscopes vibrate stably. The performance comparison results clearly demonstrate that this work exhibits significant advantages in terms of power consumption, set-up time, amplitude stability and frequency stability. As well, the core circuits layout occupies an area of only $482 \times 223 \ \mu m^2$.

REFERENCES

[1] Cui J , Chi X Z , Ding H T ,et al.Transient response and stability of the AGC-PI closed-loop controlled MEMS vibratory gyroscopes [J]. Journal of Micromechanics & Microengineering, 2009, 19(12): 125015.

[2] Zou P, Ma, K and Mou, S. A Low Phase Noise VCO Based on Substrate-Integrated Suspended Line Technology[J]. IEEE Microwave and Wireless Components Letters, 2017, 27(8):727-729.

[3] Hajimiri A, Lee T H. The Design of Low Noise Oscillators[M]. American:Kluwer Academic Publishers, 2002: 5-10.

[4] Hao S, Hu T, Gu Q J. A CMOS Phase Noise Filter With Passive Delay Line and PD/CP-Based Frequency Discriminator[J]. IEEE Transactions on Microwave Theory & Techniques, 2017, (99):1-11.

[5] Han T , Wang G , Dong C ,et al.A Self-Oscillating Driving Circuit for Low-Q MEMS Vibratory Gyroscopes[J].Micromachines, 2023, 14(5).

[6] Zhang H , Chen W , Yin L ,et al.An Interface ASIC Design of MEMS Gyroscope with Analog Closed Loop Driving[J].Sensors (14248220), 2023, 23(5).

[7] Li Q, Ding L, Liu X, et al. Research on a silicon gyroscope interface circuit based on closed-loop controlled drive loop[J]. Sensors, 2022, 22(3): 834.

Design of a 12-bit 200MS/s asynchronous SAR ADC in 12nm FinFET technology

Jingwen Li[1,2], Shaohui Pan[2], Yukun Fu[2], Huachao Xu[2], Bin Li[1*]

[1] School of Microelectronics, South China University of Technology, Guangzhou, China
[2] Guangzhou Anyka Microelectronics Co., Ltd., Guangzhou, China
*phlibin@scut.edu.cn

Abstract—**This paper presents a 12-bit 200MS/s successive approximation register analog-to-digital converter (SAR ADC) for low power applications. The ADC is designed and simulated in a 12nm FinFET process with a supply voltage of 0.9V. A fully-differential segmented capacitor structure is employed to shrink chip area. A FIA (floating inverter amplifier)-assisted dynamic latch comparator with cross-coupled devices is introduced to reduce the power consumption and the differential kickback noise. The simulated results show that the proposed SAR ADC achieves 72.3dB SNDR with a low consumption of 0.49mW at 200MS/s, translating to a 0.91fJ/conv.step FoM.**

Keywords—**SAR ADC, dynamic comparator, low power**

I. INTRODUCTION

SAR ADC stands as a low power solution in the domain of analog-digital conversion, particularly suited for medium to high-resolution applications [1]. This technology is celebrated for its balanced blend of speed, power efficiency and resolution, making it suitable for a diverse array of applications including portable battery-operated devices, sensor interfaces, and industrial control systems [2][3].

As the new wireless products requires large data throughput as well as low power consumption, FinFET technology has been embraced as a critical way to contribute to the advanced IC fabrication [4]. This paper demonstrates a 12-bit SAR ADC in 12nm FinFET achieving 200MS/s sampling rate for wireless communication. Section 2 presents the architecture of the proposed SAR ADC, including digital-to-analog (DAC) array, SAR logic circuit and dynamic comparator. Section 3 illustrates the simulation results and the comparison with prior works. Section 4 gives a summary.

II. PROPOSED CIRCUITS

A. Architecture of the proposed SAR ADC

As shown in Fig.1, the proposed SAR ADC architecture is composed of a fully differential binary weighted capacitive digital to analog converter (DAC) array, a dynamic latch comparator and an asynchronous SAR logic timing circuit. The reference voltage is the same as the supply voltage VDD. Top-plate sampling is employed to reduce charge injection of the switch and a bootstrapped switch is used to achieve high linearity sampling.

B. Asynchronous SAR logic

Compared to the synchronous SAR ADCs, asynchronous SAR ADCs can generate the clock by an internal logic circuit, which means it doesn't need an extra phase lock loop (PLL) circuit. The SAR logic block is controlled by a Valid signal, generating a CLK signal to pulse the comparator block. Reversely, the comparator is driven by the start/end signal of the proposed ADC to generate the Valid signal and send it to SAR logic block. Valid signal means that the comparison period is over.

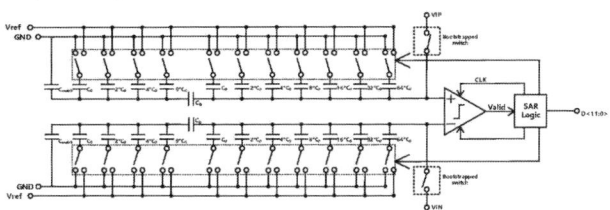

Fig.1 Architecture of the proposed SAR ADC

C. Segmented capacitor array

Here we segment the DAC capacitor array into MSB (4 bits) and LSB (7 bits) sub arrays and the differential structure is regarded as 1 bit. The fully differential segmented capacitor array is shown in Fig. 1. The unit capacitance C_0 of 5.76fF is used to balance the area and the kT/C noise, while the $C_b=2C_0$ and the $C_{match}=15C_0$ are used to linearize the 12-bit capacitor array. For better suppression of parasitic effects, a home-made capacitor unit is used here instead of the mom capacitor provided by the process library. Capacitors are completely surrounded by metal layers in order to be isolated from the substrate and VSS. The parasitic capacitances are combined to the top plate of the capacitor, which does not affect the linearity of the entire capacitor array.

A bootstrapped switch is proposed to improve the linearity of the sampling circuit. As shown in Fig.2, the input signal Vip is sampled by a bootstrapped switch at the top plates of capacitors. An identical dummy transistor M12 is added in the circuit to eliminate the coupling effect of the drain-source capacitor C_{ds} of the sampling transistor M11. M3 and M8 are always on to alleviate the transient rise of bootstrap capacitor C_b. When CLK is on, the input signal is transferred to the comparison block. When CLK is off, the SAR ADC turns into the cycling period.

D. FIA-assisted comparator with de-coupling units

In terms of the relatively low power consumption, fast speed and small area, purely dynamic comparator serves as a fundamental building block in this SAR ADC and FIA structure realizes power reuse and avoids the complete discharge of. The structure of comparator is as shown in Fig.3. When CLK is high, the reservoir capacitor CRES is

979-8-3315-2209-4/25 $31.00 © 2025 IEEE

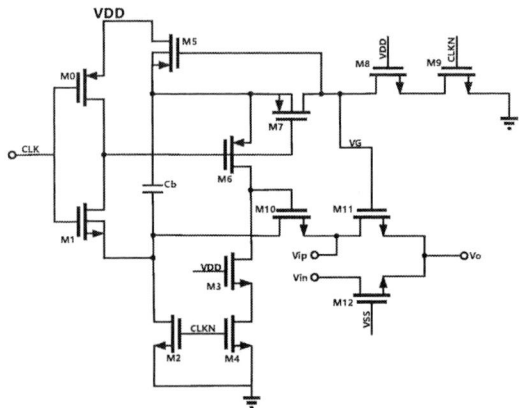

Fig. 2 Schematic of the proposed bootstrapped switch

pre-charged to VDD/GND. In this phase, M0 turns off to avoid static current and the output signals Von/Vop are reset to GND. When CLK is low, C_{RES} is discharged through the inverter pairs, resulting in the decrease of V_{SP} and the increase of V_{SN}. As the overdrive voltage (Vgs-Vth) of the inverter pairs also decrease, a boosted gm/Id is provided to realize high energy efficiency. During this period, M0 turns on, the source voltage V_{source} of the differential pairs is instantaneously raised to high voltage.

As parasitic capacitance and resistance grow dramatically in the FinFET technology, the variation of V_{source} will be coupled to the inputs of the differential pairs via gate-source capacitor C_{gs}. The coupling effect will influence the comparison result, so the addition of de-coupling units is necessary for high-precision comparator. M11/M12 are introduced as decoupling units to weaken kickback noise.

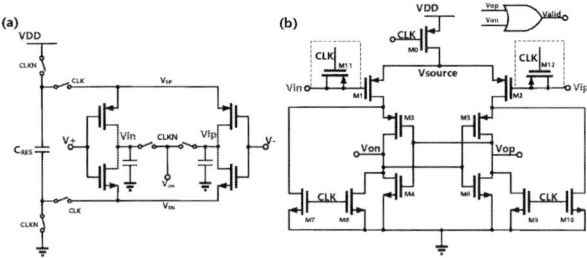

Fig. 3 Schematic of (a) FIA structure and (b) dynamic latch comparator

III. SIMULATION RESULTS

The proposed SAR ADC is implemented in a 1P9M 12nm FinFET process. Fig.4 shows the layout zoomed-in view of the ADC, which occupied an active area of 130μm×116μm. FFT spectrum is present in Fig 5(a). At an input signal frequency of 8.59 MHz, the designed SAR ADC can achieve 72.3dB SNDR and 11.4bits ENOB with a supply voltage of 0.9V. The inset in Fig 5(a) gives the performance of this SAR ADC at different input frequency.

Table 1 is the comparison of the figure of merit with prior works using FinFET process. Compared with prior works, this SAR ADC performs a high sampling rate of

200MS/s and a relatively low power consumption of 494μW, ultimately translating into 0.91fJ/conv.step FoM.

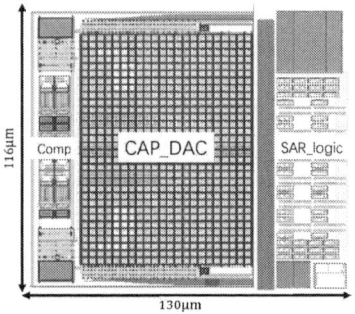

Fig. 4 Layout of the proposed SAR ADC

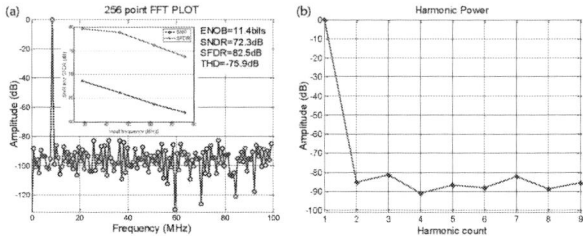

Fig. 5 (a) FFT spectrum and (b) harmonic power contribution of the FFT spectrum

TABLE I. PERFORMANCE COMPARISON

	This work[a]	REF[5][a]	REF[6][b]	REF[7][b]
Process(nm)	12	14	14	14
Resolution(bit)	12	10	12	12
Supply(V)	0.9	0.8	1.0	0.8
Sampling Rate(MHz)	200	50	70	100
Power(mW)	0.49	0.04	4.3	0.78
SNDR(dB)	72.3	59.59	68.1	61.29
Area(mm²)	0.015	-	0.019	0.112
ENOB(bits)	11.40	9.59	11.05	9.89
FoM(fJ/conv.step)	0.91	1.07	29.6	8.2

a Post-simulation results
b Chip-test results

REFERENCES

[1] X. Tang et al., "Low-Power SAR ADC Design: Overview and Survey of State-of-the-Art Techniques," in IEEE Transactions on Circuits and Systems I: Regular Papers, vol. 69, no. 6, pp. 2249-2262, June 2022.

[2] S. Lee, H. Kang and M. Lee, "9.9 A 2.72fJ/conv 13b 2MS/s SAR ADC Using Dynamic Capacitive Comparator with Wide Input Common Mode," 2024 IEEE International Solid-State Circuits Conference (ISSCC), San Francisco, CA, USA, 2024, pp. 184-186.

[3] C. Liu, S. Chang, G. Huang and Y. Lin, "A 10-bit 50-MS/s SAR ADC with a Monotonic Capacitor Switching Procedure," IEEE JSSC, vol. 45, no. 4, pp. 731-740, Apr. 2010.

[4] S. Lee et al., "A 1.2V High-Voltage-Tolerant Bootstrapped Analog Sampler in 12-bit SAR ADC Using 3nm GAA's 0.7V Thin-Gate-Oxide Transistor," 2024 IEEE International Solid-State Circuits Conference (ISSCC), San Francisco, CA, USA, 2024, pp. 70-72.

[5] A. Wang, C.-J.R. Shi, "A 10-bit 50-MS/s SAR ADC with 1 fJ/Conversion in 14 nm SOI FinFET CMOS", Integration, the VLSI Journal (2018), https://doi.org/10.1016/j.vlsi.2018.03.010.

[6] C. C. Lee, C. Lu, R. Narayanaswamy and J. B. Rizk, "A 12b 70MS/s SAR ADC with digital startup calibration in 14nm CMOS," 2015

Symposium on VLSI Circuits (VLSI Circuits), Kyoto, 2015, pp. C62- C63.

[7] Y. Zheng, J. Lan, F. Ye and J. Ren, "A 12-bit 100MS/s SAR ADC with Equivalent Split-Capacitor and LSB-Averaging in 14-nm CMOS FinFET," in IEEE Access, vol. 9, pp. 169107-169121, 2021.

A 100.3dB-SNDR 10.8mW Pipeline-SAR ADC with second-order gain error shaping and one-time foreground calibration

1st Haowen Wu   2nd Weihui Liu   3rd Haigang Feng   4th Xinpeng Xing

Abstract—**Pipeline-SAR is a competitive hybrid ADC structure in high-precision design, but resulting into inter-stage gain error. Furthermore, the nonlinearity caused by CDAC capacitance mismatch cannot be ignored in the coarse SAR sub-ADC. This work presents a second-order gain error shaping (GES) technique to mitigate inter-stage gain error, and an one-time foreground calibration to resolve CDAC nonlinearity, in an 18-bit 1MS/s pipeline-SAR ADC. The simulation results in 0.18μm BCD demonstrates that the ADC achieves an SNDR of 100.3dB with an OSR of 5, and consumes 10.8mW core power from 5V supply with an area of 3.125mm² , corresponding to a state-of-the-art FOM of 170.08 dB.**

Keywords—*high-precision design, Pipeline-SAR ADC, gain error shaping, calibration.*

I.   INTRODUCTION

To achieve high resolution (>16bit), hybrid SAR ADC is adopted like Pipeline-SAR ADC which combines concepts of SAR and pipeline ADC [1]. For a Pipeline-SAR ADC, the main accuracy requirement for SAR ADC is transferred to the residual amplifier (RA) and the difficulty of high-precision design mainly depends on the residual amplifier design.

A sufficient accuracy RA for overall Pipeline-SAR ADC is difficult to realize. Correlated level shifting (CLS) [2] can relax the requirement for the opamp by an additional level shift phase compared to the typical amplification timing, and its equivalent loop gain is squared. But the extra phase and capacitance is timing unfriendly. Noise shaping (NS) [3] provides a new way that the noise caused by gain error in the bandwidth is shaped, only a mild oversampling (4~8) can realize excellent high-pass shaping effect.

This paper offers an analysis and design of an 18bit Pipelined-SAR ADC. Several key techniques like gain error shaping (GES), One-time foreground calibration are proposed to resolve non-ideal factors in ADC.

II.   ANALYSIS AND TECHNIQUE THEORY

A. Analysis of ADC nonidealities

As is shown in Fig.1, because the RA has gain error *GE*,

Author 1 and Author 3 are with *Tsinghua Shenzhen International graduate school*, Shenzhen, China. Author 2 is with *Chipsea Technologies Co Ltd*, Shenzhen, China. Author 4 is with *Sun Yat-Sen University*, Shenzhen, China. Corresponding author: Xinpeng Xing, email: xingxp@mail.sysu.edu.cn)

---

the system's transfer function varies to

$$D_{out} = V_{in} - \frac{Q_2}{G} - GEQ_1 \tag{1}$$

D_{out} is overall ADC output code, V_{in} is input signal, Q_1 is residue voltage of the coarse SAR ADC, Q_2 is residue voltage of the fine SAR ADC, G is ideal gain, GE is gain error. For an 18bit ADC, the value of GEQ_1 is inevitably far greater than Q_2/G, resulting in a tremendous drop in overall ADC SNR.

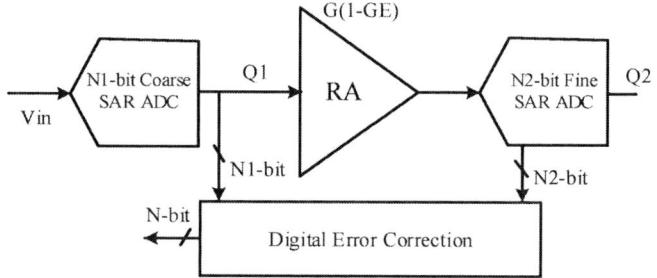

Fig. 1. Structure of two step Pipeline-SAR ADC

Additionally, the actual capacitance will have random mismatch which is associated with the input signal, thus resulting in nonlinearity in conversion. For a N-bit SAR ADC with the unit capacitance mismatch rate ΔC, the standard deviation of Differential Non-linearity (DNL) σ_{DNL} caused by mismatch is

$$\sigma_{DNL} = \frac{1}{2}\sqrt{2^N}\Delta C LSB \tag{2}$$

LSB is the minimum resolution voltage. In order for accuracy of the N_1-bit coarse SAR ADC to meet the overall N-bit ADC, it demands

$$3\sqrt{2^{N_1}}\Delta C \frac{V_{REF}}{2^{N_1}}G < \frac{V_{REF}}{2^{N-N_1}} \tag{3}$$

V_{REF} is the reference voltage. Because of large G and N_1, it is necessary to add calibration in high-resolution Pipeline-SAR ADC.

B. Gain error shaping

The proposed gain error shaping (GES) is a type of noise shaping with aim of shaping gain error noise out of the signal bandwidth. Formula (1) shows that GEQ_1 contributes noise component to the output signal spectrum. Instead of minimizing gain error GE, GES disposes the leakage Q_1. It predicts the previous period of Q_1, executing the subtraction $Q_1(n) - Q_1(n-1)$ at the input of RA and executing the addition with $Q_1(n-1)$ at the output of RA. $Q_1(n)$ is current period quantization noise from the coarse ADC and $Q_1(n-1)$ is last period's.

For a Pipeline-SAR ADC, $Q_1(n-1)$ can be easily predicted with output code of the fine SAR ADC. Fig. 2 shows the structure of Pipeline-SAR ADC equipped with GES. After the coarse SAR ADC finishes its conversion and leaves $Q_1(n)$ for RA, the last period fine SAR ADC output code $D_2(n-1)$ is added to the input of RA with scaling of $-G$. Then the RA begins to amplify $Q_1(n) - D_2(n-1)/G$. Meanwhile, the coarse SAR ADC samples the signal from the output of RA with $D_2(n-1)$ added to the output. In this way, the prediction of $Q_1(n-1)$ does not disturb the normal proceed of Pipeline-SAR ADC by bringing two opposite polarity feedback loop.

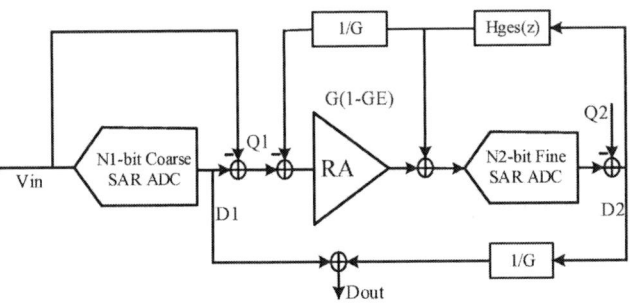

Fig. 2. Structure of two-stage Pipeline-SAR ADC with GES.

The system's transfer function can be acquired through mason's formula

$$D_{out} = V_{in} - \frac{Q_2}{G}\frac{1}{1-GEHges(z)} - GEQ_1\frac{1-Hges(z)}{1-GEHges(z)} \quad (4)$$

Hges (z) is the GES transfer function with value of z^{-1} for first order GES and $z^{-2} - 2z^{-1} + z$ for second order GES. Fig. 3 compares theoretical output spectrum result with and without GES, it realizes a high-pass shaping for GEQ_1, effectively weakening the in-band noise component.

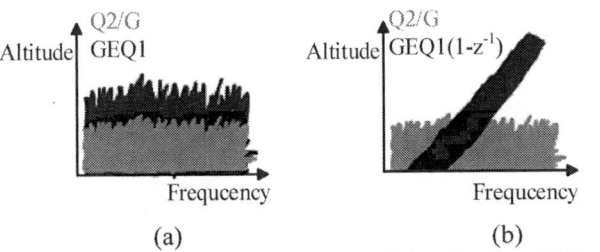

Fig. 3. Theoretical output spectrum of the Pipeline-SAR ADC (a) without GES circuit, (b) with first-order GES circuit.

C. One-time foreground calibration

One-time foreground calibration is implemented before the ADC normal operating sequence. The key thought is employing the fine SAR ADC to quantify the residue voltage caused by mismatch capacitance from the coarse SAR ADC when several specific program is executed.

Fig. 4 shows two steps to acquire the weight error from MSB capacitance. First step is to acquire difference error. In initial state, it connects the capacitance bottom plate reference voltage of the calibrated bit to V_{CM} as shown in Fig.4 (a), and the RA is in unit negative feedback state, the fine SAR ADC is in reset state. After this, the calibrated bit is connected to the calibration code programed in the calibration circuit as is shown in Fig.4 (b). The second step is to acquire cumulative error. Similarly, it is composed of two state as shown in Fig.4 (c) and Fig.4 (d). Considering the KT/C thermal noise, the calibration process should take the average multiple times. The weight error information of each capacitance is stored, and the final calibration code is supplemented in the digital field according to the output code of the coarse SAR ADC.

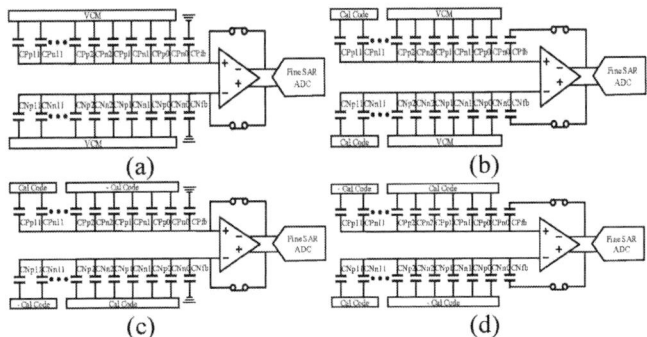

Fig. 4. The procedure of the calibration is composed of (a) initial state of acquiring differential error, (b) conversion state of acquiring differential error, (c) initial state of cumulative error, (d) conversion state of acquiring cumulative error.

III. CIRCUIT IMPLEMENTATION

A. System architecture

The proposed 18-bit Pipeline-SAR ADC is demonstrated in Fig. 5. Considering area, speed, noise, power and linearity, the coarse SAR ADC is set to be an 11-bit SAR ADC, and the fine SAR ADC is set to be a 9-bit SAR ADC. There is 2 bits inter-stage redundancy between the two SAR ADC to tolerate the offset from comparator and RA. Second order GES is implemented in the Pipeline-SAR ADC by bringing in two feedback loop to the input and output of RA. They do not affect the entire Pipeline-SAR ADC process. Due to bring an extra signal in the input of RA, it may saturate its output if we use 512× as gain. So a scaled reference voltage is used in the fine SAR ADC, V_{REF2} is a quarter of V_{REF1}, and in this case, the gain of RA is set to be 128×. Calibration circuits are all digital, and its calibration code will be calculated with ADC output code in digital domain to get a final output code.

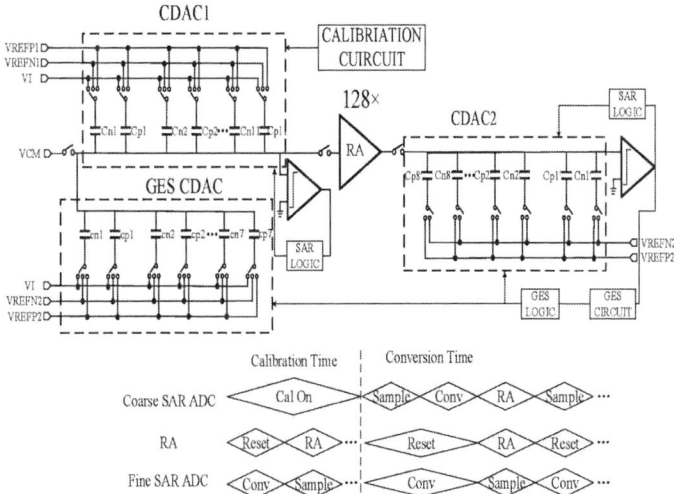

Fig. 5. The proposed 18-bit Pipeline-SAR ADC circuit and timing diagram.

B. Implementation of the gain error shaping

The Fig. 6 (a) – (d) shows detailed implementation of the GES circuit. To realize a 1/G scaled $D_2(n-1)$, a GES capacitance array is parallelly connected with the main CDAC array in coarse SAR ADC and the value of the main CDAC array C_{main} is G times larger than GES capacitance array C_{GES}. As for the compensation loop in fine SAR ADC, no extra capacitance array is demanded, because it can be realized by top-plate sampling with the main CDAC itself.

At beginning, the coarse SAR ADC samples the input signal, the RA is in reset state and the fine SAR ADC converts the sampled value of the last cycle as shown in Fig.8 (a). Secondly, the main CDAC is switched to convert the input signal and the GES CDAC is in reset stage as shown in Fig.8 (b). Thirdly, the coarse SAR ADC completes its conversion and triggers the feedback code value stored in register to be added to the bottom plate of GES CDAC. The fine SAR ADC finishes its conversion and stores the feedback code value for the next period in the register as shown in Fig.8 (c). Fourthly, when the fine SAR ADC samples the signal from RA on its top plate, its bottom plate is switched to $-D_2(n-1)$ as shown in Fig.8 (d). After sampling finishes and the bottom plate is switched to V_{CM}, $D_2(n-1)$ is added to the signal automaticly.

Compared with gain error shaping in [3], this work realizes the second order gain error shaping with an one-time feedback code generated from the digital domain, avoiding the complex process in second order feedback.

To realize a second order transfer function, a frequency bisector is used to divide the trigger clock signal and get two feedback code value. The adder array is used to calculate the feedback code of the current period which instinctively becomes a 10-bit code. For a $128 \times$ G, if feedback code is 9 bits, the unit capacitance of GES CDAC is C_{main} / 262144 which is process unrealizable. Therefore, the feedback code truncates the MSB 7 bits and the reference voltage of the GES CDAC becomes a quarter of the main CDAC, thus making the unit capacitance of GES CDAC to be C_{main} / 8192.

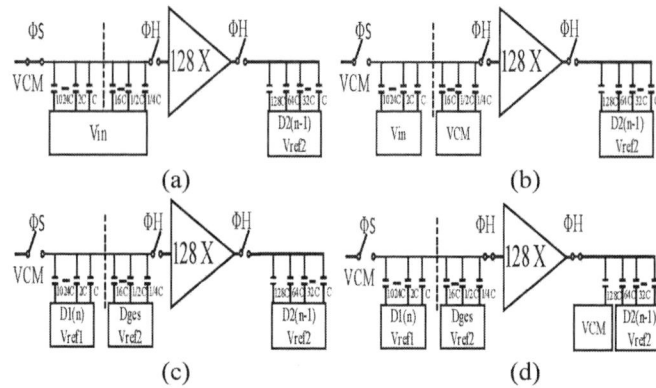

Fig. 6 (a)~(d). The detailed implementation of the GES.

IV. SIMULATION RESULTS.

The proposed 18-bit Pipeline-SAR ADC was implemented in 0.18μm CMOS technology and occupies 2.5 mm $\times$1.25mm as shown in Fig. 7.

Fig. 7. Layout of the proposed Pipeline-SAR ADC.

The ADC operates at 5V supply voltage, the conversion reference voltage for the coarse SAR ADC is 4.096V and 1.024V for the fine SAR ADC. With a 32MHz outside synchronous clock, the sampling rate is set as 1 MS/s.

Fig.8 shows the simulation result (ignoring the capacitance mismatch) with and without the second order GES when there is a gain error caused by design margin error and PVT variation in SS corner 27 ℃ whose gain error is about 7.8%. The input signal frequency is 5/4096 MHz and the output spectrum is based on 4096 FFT points with an OSR=5. It can be seen that without GES, the simulation SNDR and SFDR are 94.61 and 108.21 dB and with GES, SNDR and SFDR are improved to 98.58 and 113.43 dB. The noise caused by gain error is evidently high-pass shaping in the spectrum.

979-8-3315-2209-4/25 $31.00 © 2025 IEEE

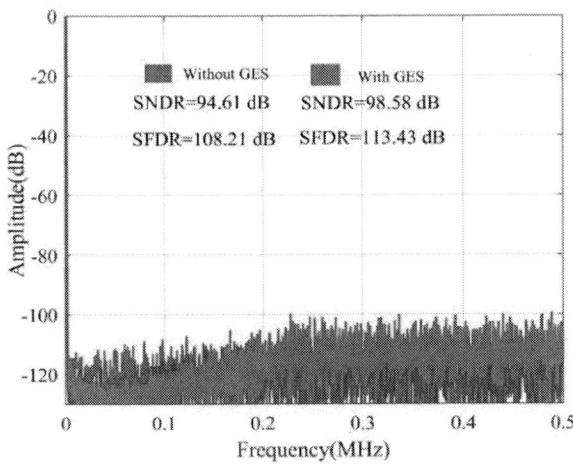

Fig.8. Simulation spectrum with GES and without GES.

Detailed output simulation results under various PVT corners are shown in Table I. It contains a mild OSR=5.

Table I ENOB(BITS) UNDER VARRIOUS PVT CORNERS

Process Temp ℃	TT	FF	SS	FS	SF
-40	16.40	16.51	16.23	16.29	16.49
27	16.25	16.37	16.07	16.13	16.35
105	16.10	16.21	15.87	15.97	16.18

Fig.9 shows the overall simulation results with foreground calibration on and off. Without calibration, the SFDR and SNDR is only 96.26 and 89.36 dB and with calibration, the SFDR and SNDR is improved to 110.40 and 100.38 dB. The harmonic component caused by capacitance mismatch is effectively decreased.

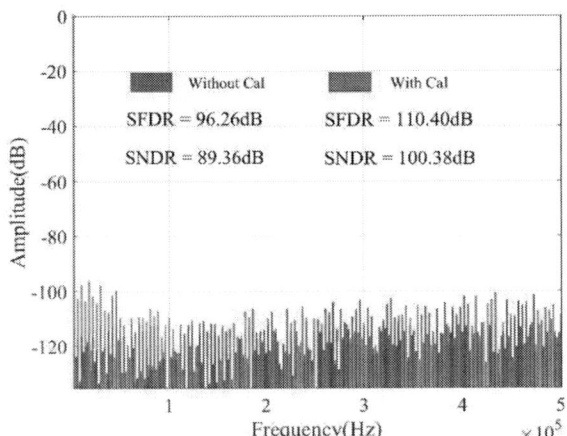

Fig.9. Simulation spectrum with and without calibration.

The overall ADC consumes a power of 10.9 mW at a supply of 5 V and the FOM of the ADC is 170.08 dB.

Table II lists the proposed work with some Pipeline-SAR ADC with resolution greater than 16 bits. The proposed work can realize a high SNDR with less power consumption and less area.

Table II. PERFORMANCE SUMMARY AND COMPARISON

	This work	[4]	[5]	[6]	[7]
Resolution（bits）	18	18	18	16	22
BW (MSPS)	0.1	6.3	2.5	7.5	6
SNDR(dB)	100.38	90	98.6	90.8	91.3
Power (mW)	10.8	105	30.52	9.82	9.4
FOM (dB)	170.08	167.8	178.0	182.7	179.3
Process (nm)	180	250	180	180	180
Area (mm²)	3.125	6	5.74	1.82	0.6

V. CONCLUSIONS

This work presents an 18-bit Pipeline-SAR ADC with second-order GES and one-time foreground calibration to resolve inter-stage gain error and capacitance mismatch, respectively. The detailed circuit implementations of these techniques are also presented, which can be integrated in any Pipeline-SAR ADC. The overall Pipeline-SAR ADC simulated in 0.18μm BCD achieves SNDRs above 100 dB and SFDRs higher than 110 dB. The state-of-the-art 170dB FoM, together with the small area and less power consumption makes the presented Pipeline-SAR ADC widely used in high-performance application.

REFERENCES

[1] C. C. Lee and M. P. Flynn, "A SAR-Assisted Two-Stage Pipeline ADC," in *IEEE Journal of Solid-State Circuits*, vol. 46, no. 4, pp. 859-869.

[2] B. R. Gregoire and U. -K. Moon, "An Over-60 dB True Rail-to-Rail Performance Using Correlated Level Shifting and an Opamp With Only 30 dB Loop Gain," in *IEEE Journal of Solid-State Circuits*, vol. 43, no. 12, pp. 2620-2630.

[3] C. -K. Hsu, T. R. Andeen and N. Sun, "A Pipeline SAR ADC With Second-Order Interstage Gain Error Shaping," in *IEEE Journal of Solid-State Circuits*, vol. 55, no. 4, pp. 1032-1042.

[4] C. P. Hurrell, C. Lyden, D. Laing, D. Hummerston and M. Vickery, "An 18 b 12.5 MS/s ADC With 93 dB SNR," in *IEEE Journal of Solid-State Circuits*, vol. 45, no. 12, pp. 2647-2654.

[5] A. Bannon, C. P. Hurrell, D. Hummerston and C. Lyden, "An 18 b 5 MS/s SAR ADC with 100.2 dB dynamic range," *2014 Symposium on VLSI Circuits Digest of Technical Papers*, Honolulu, HI, USA, 2014, pp. 1-2.

[6] A. ElShater *et al.*, "3.7 A 10mW 16b 15MS/s Two-Step SAR ADC with 95dB DR Using Dual-Deadzone Ring-Amplifier," *2019 IEEE International Solid-State Circuits Conference - (ISSCC)*, San Francisco, CA, USA, 2019, pp. 70-7.

[7] M. Li *et al.*, "10.4 A Rail-to-Rail 12MS 91.3dB SNDR 94.1dB DR Two-Step SAR ADC with Integrated Input Buffer Using Predictive Level-Shifting," *2023 IEEE International Solid-State Circuits Conference (ISSCC)*, San Francisco, CA, USA, 2023, pp. 1-3.

Electromagnetic Investigation of Substrate Coupling for Monolithic Microwave Integrated Circuits in GaN-on-Si Technology

Rui (Ray) Yao   Miao Cui   Zhao Wang   Sang Lam   Stephen Taylor

Abstract— In this work, we present electromagnetic (EM) investigation into the substrate coupling in monolithic microwave integrated circuits (MMICs) in GaN-on-Si technology for high signal frequency up to 25 GHz. By numerically solving Maxwell's equations based on finite element method (FEM), the electric field distribution and then the electromagnetic (EM) coupling are determined in the GaN-on-Si device structures. The S-parameter results disclose that $|S_{21}|$ increases sharply to about -10 dB for multi-gigahertz signals over a physical separation distance of 700 μm. Such poor signal isolation remains more or less the same extending to 25 GHz. The computed electric field intensity distribution reveals the EM coupling through the GaN buffer layer rather than the resistive silicon (Si) substrate. These results pose important implications for the implementation of GaN-on-Si MMICs to avoid serious crosstalk and noise coupling among nearby devices and circuits operating at gigahertz frequencies. Effective isolation trench structures or alike would be necessary for suppressing the substrate coupling in GaN-on-Si MMICs.

Index Terms—substrate coupling, gallium nitride (GaN) devices, GaN-on-Si technology, monolithic microwave integrated circuits, crosstalk, microwave signal isolation.

I. INTRODUCTION

Over the past decades, we have witnessed rapid development of gallium nitride (GaN) semiconductor devices, with their high breakdown voltage, high electron mobility and wide bandgap, making GaN a promising semiconductor candidate for next-generation integrated circuit (IC) technology, particularly for high performance power electronic devices and high frequency applications [1]. Compared to the traditional silicon carbide (SiC) substrate [2] for GaN-based ICs, the silicon (Si) substrate can be an attractive alternative due to various advantages associated with the Si wafer base: low-cost and abundant availability in large wafer sizes [3, 4], good thermal conductivity of Si wafers, potential integration of optoelectronic devices [5] etc.

There has been significant progress made in enhancing the performance of GaN-based power electronic devices fabricated on a Si substrate of high resistivity in recent years, leading to an increasing interest in the research on GaN-based monolithic microwave integrated circuits (MMICs). There has been some research on devices and circuits in GaN-on-Si technology operating at microwave frequencies [6, 7]. However, there are very few reports on the substrate effects [8] or signal isolation [9] in GaN-based MMICs, especially for devices operating at microwave frequencies. Despite the

persistent efforts made in advancing the power ICs in GaN-on-Si technology, crosstalk and substrate coupling remain one big challenge hindering the monolithic integration of high power electronic devices and logic circuits built on the same GaN chip [10]. For high level integration of GaN-based circuits on the same chip, signal integrity is crucially important. It is due to the fact that crosstalk and interference signal coupling through the substrate [11] can lethally damage the performance and stability of power ICs operating at higher frequencies beyond multiple-megahertz (multi-MHz).

Based on some experimental findings of multi-MHz crosstalk and substrate coupling in power electronic devices in GaN-on-Si technology [9, 12] as well as our previous works reporting the computational electromagnetic (EM) investigation into the substrate coupling mechanism in power ICs in GaN-on-Si technology [13], hence in this work, we continue our well-established EM investigation into the EM coupling in a typical GaN-on-Si IC device structure (Fig. 1). It is to find out the substrate effects in terms of signal isolation in the microwave regime (above 1 GHz) in order to provide further insights into substrate coupling in MMICs in GaN-on-Si technology. Then, there could be future improvement in the design and fabrication of GaN-on-Si ICs operating at mutli-GHz frequencies and beyond, with acceptable signal integrity.

II. GaN-ON-Si DEVICE STRUCTURE FOR EM COUPLING INVESTIGATION

To investigate the crosstalk and substrate coupling effect in GaN-based microwave devices and circuits, we adopted a device structure as illustrated in Fig. 1. The structure is devised to be as close as possible to the real structure for our computational EM studies. By numerically solving Maxwell's equations based on finite element method (FEM), the electric fields and S-parameter results are obtained and then to investigate the substrate coupling mechanism.

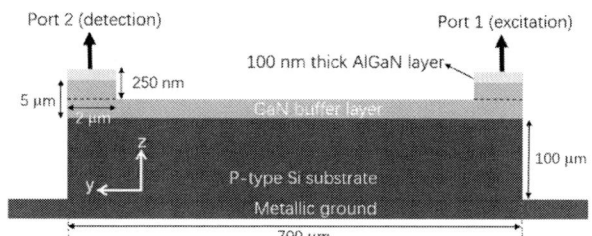

Fig. 1. A schematic cross-sectional diagram showing the GaN-on-Si device structure for computational EM investigation into the crosstalk and substrate coupling mechanism, with the features not drawn in exact scale for better visualisation.

It can be seen that the physical geometry of this GaN-on-Si device structure is comprised of four layers: a metallic base at the bottom-most layer for connecting the IC chip to the ground, and then a 100-μm thick *p*-type Si layer and a 5-μm

---

R.R. Yao, M. Cui, Z. Wang, S. Lam are with *School of Advanced Technology (SAT), Xi'an Jiaotong-Liverpool University (XJTLU)*, Suzhou, China. S. Taylor, as well as R.R. Yao, is with *Department of Electrical Engineering and Electronics, The University of Liverpool*, Liverpool, UK. M. Cui and S. Lam are with *Department of Electrical and Electronic Engineering*, Z. Wang with *Department of Communications and Networking, SAT, XJTLU*. This work is supported by PGRS funding of XJTLU. Corresponding authors: (R.R. Yao and S. Lam, email: Rui.Yao@liverpool.ac.uk; s.lam.cn@ieee.org)

thick GaN buffer layer built sequentially above, following with a 100-nm thick doped-AlGaN barrier layer at the topmost level. Note that in the actual device structure, the AlGaN barrier layer thickness is only 25 nm, which is however too thin compared to other layers' thickness for the FEM-based computation. For this reason, a layer thickness of 100 nm is adopted as it does not cause computational problems. Such a change to the barrier layer thickness is considered to be acceptable for producing basically the same computational EM results, because of the ohmic contact through the doped-AlGaN layer.

As shown in Fig. 1, the central region is slightly etched out (with the AlGaN barrier layer removed), leaving two protruded blocks at both ends. Each protruded block has the same layout area of 2 μm by 50 μm. The two protruded blocks are to model device structures and circuits built on the same IC chip but physically separated by certain lateral distance. Such a GaN-on-Si testing structure is more or less the same as those of experimental investigation in some reported works [9, 12, 14]. In practice, two aluminium (Al) strips are in contact with the 100-nm thick doped-AlGaN layer to represent the terminals (e.g. drain) of two separate HEMTs, which share the same substrate. The two HEMTs have certain lateral separation distance (700 μm for the device structure shown in this work) apart from each other. Since the doped-AlGaN barrier layer is very thin, ohmic contacts are formed between the Al strips and the GaN layer. The effect of this extremely thin AlGaN layer is negligible in our EM simulation as mentioned before. The two Al strips have equal area (2 μm by 50 μm) and serve as two 50-ohm signal ports: one with excitation voltage of 1.0 V and the other to detect the input signal in our EM coupling investigation. As for the computational EM investigation, we set the excitation port at one end (port 1) and then detecting the microwave signal at the other end (port 2). As a result, any electric fields computed at port 2 compared with that at port 1 can reveal the EM coupling in the device structure.

III. Full-Wave 3D EM Simulation & Substrate Coupling Results

The computational EM investigation was carried out using a commercial full-wave EM simulation software program, Ansys HFSS. Since the EM simulation was performed at microwave frequencies, we had to manually set up the appropriate electrical properties for the materials used in modeling the GaN-on-Si device structure (as shown in Fig. 1) during the EM simulation. They are summarized in the following: $\varepsilon_{rGaN} = 8.9$ and $\sigma_{GaN} = 200$ S/m are assigned to the GaN buffer layer; $\varepsilon_{rAlGaN} = 5$ and $\sigma_{AlGaN} = 5 \times 10^3$ S/m are assigned to the very thin doped-AlGaN layer, while $\varepsilon_{rSi} = 11.9$ and $\sigma_{Si} = 2 \times 10^3$ S/m are assigned to the p-type silicon substrate. In order to prevent the uncontrollable complex situations from occurring in the EM studies, the dielectric loss tangent has not been taken into account in our EM simulation. Other settings are similar to our previous work in which more details can be found [13].

During the EM simulation, Ansys HFSS employs the FEM to finely divide the structures into many smaller meshes and solve Maxwell's equations across inter-element boundaries between the adjacent finite elements, thus giving a solution for the EM fields at different spatial points in the structure being simulated. The computed EM fields can be used to calculate the voltage and current of the signal and then the S-parameters. Among the S-parameters, $|S_{21}|$ can reveal the EM coupling in the GaN-on-Si structure (shown in Fig. 1). Readers interested in the detailed mathematical calculations can refer to the relevant sections in our previous work [13, 15].

The frequency response ($|S_{21}|$) of the GaN-on-Si device structure spanning from 0.5 to 25 GHz is depicted in Fig. 2. As the value of $|S_{21}|$ indicates the level of crosstalk or EM coupling in the device structure, it can be seen that with a large lateral separation distance of 700 μm, the crosstalk or EM coupling is relatively weaker at lower signal operation frequencies (around -30 dB at 0.5 GHz); the EM coupling then becomes significantly stronger as the frequency increases, with $|S_{21}| \approx -11$ dB from 10 GHz to 25 GHz. It implies that there is apparently serious EM coupling at gigahertz frequencies. Therefore, the microwave signal isolation beyond multi-GHz would be poor among devices and circuits built on the same GaN-on-Si chip. The situation can be alleviated by shrinking the device areas (with smaller transistor sizes) at the expense of the transistor performance. Fig. 3 shows how the lateral separation distance of the GaN-on-Si device structure would weaken the EM coupling with the EM simulation carried out at various signal frequencies. As shown in the S-parameter results, even a considerably large separation distance (700 μm) seems not very helpful in improving the poor microwave signal isolation beyond about 3 GHz.

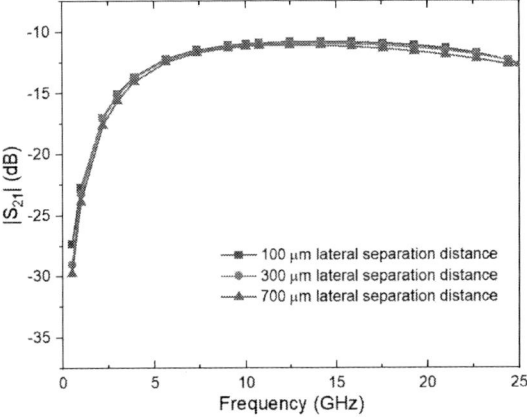

Fig. 2. $|S_{21}|$ of the GaN-on-Si device structure as a function of the frequency spanning from 0.5 to 25 GHz, showing inferior signal isolation at higher signal frequency for a lateral separation distance of 700 μm.

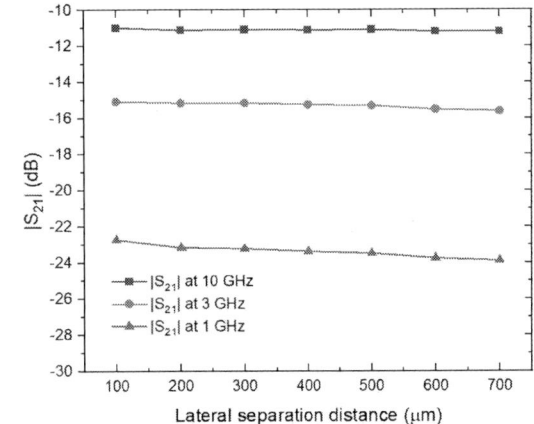

Fig. 3. $|S_{21}|$ of the GaN-on-Si device structure as a function of the lateral separation distance at signal frequencies of 1, 3, 10 GHz respectively.

Apart from the lateral separation distance of the GaN-on-Si device structure employed in our computational EM investigation, we also investigated the EM coupling effect with the possible dependence on the thickness of the GaN buffer layer. The $|S_{21}|$ results are shown in Fig. 4, in which the coloured curves represent the computed $|S_{21}|$ of the device structure with varied thickness of the GaN buffer layer (from 1 μm to 5 μm); the lateral separation distance of the device structure being studied was kept constant at 700 μm. It can be seen that, regardless of the GaN buffer layer thickness, the $|S_{21}|$ curves behave in a similar fashion (i.e. intensifying sharply over low-GHz signal frequency range and then gradually leveling off after 10 GHz). By adopting a thicker GaN buffer layer (5 μm), the EM coupling can be slightly suppressed by only about 4 dB, compared to the case of 1 μm thick GaN buffer layer.

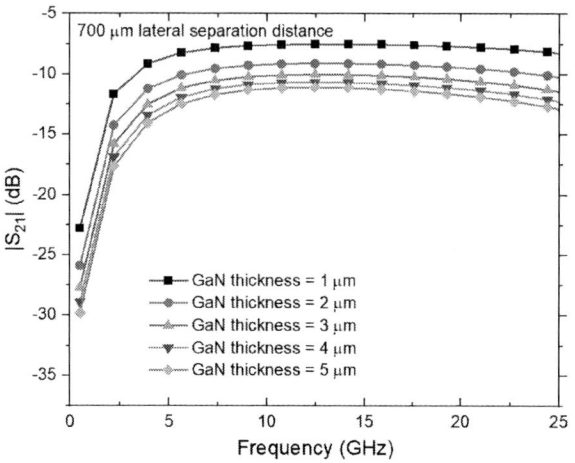

Fig. 4. $|S_{21}|$ of the GaN-on-Si device structure as a function of frequency ranging from 0.5 to 25 GHz with the thickness of the GaN buffer layer varying from 1 to 5 μm while the lateral length is fixed to 700 μm.

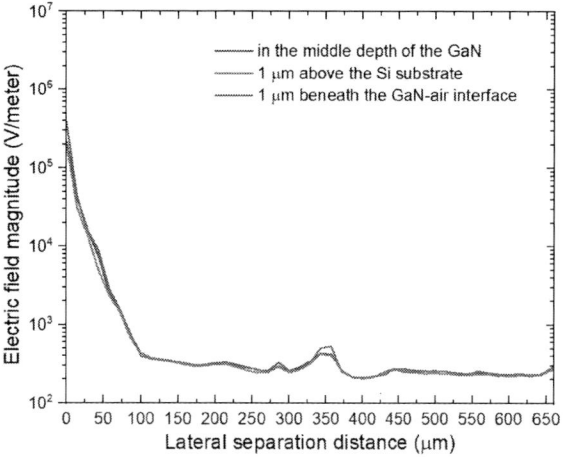

Fig. 5. Profiles of the electric field (magnitude) in different lateral positions of the vertical levels within the GaN buffer layer (at 10 GHz signal operation frequency), starting from the excitation port to detection port, along the GaN-on-Si device structure as a function of the lateral separation distance (with the y-axis in logarithmic scale for better visualisation).

The profiles of the electric field magnitude taken from the front surface of and within the GaN buffer layer are plotted in Fig. 5, with the electric field magnitude measured at three different vertical levels of the GaN buffer layer: right in the middle depth, 1 μm above the Si substrate and 1 μm beneath the interface between the GaN buffer layer and air respectively. The electric field profiles serve as additional evidence supporting the $|S_{21}|$ response for the EM coupling shown in Fig. 2 and Fig. 3.

To investigate deeper the EM coupling in the GaN-on-Si device structure, the colour map of the electric field intensity distribution (in the y-z plane) is exhibited in Fig. 6. In this case, only one excitation port is activated in the EM simulations to plot the electric field intensity distribution which extends to both ends of the modeling device structure rather than a two-port layout shown in Fig. 1. In Fig. 6, the electric field intensity distribution (in the y-z plane) is displayed in colour scale (with the associated scale bar indicating the electric field magnitude). While only the cross-sectional distribution is shown, the electric field was computed by three-dimensional (3D) EM simulations. According to this detailed plot, it can be clearly seen that the electric field intensity gradually attenuates, starting from the positions away from the excitation port (the reddish region). Meanwhile, the weakening electric field permeates to both end of the modeling device structure in a quasi-symmetrical manner. Furthermore, by examining how the electric field decays in the GaN-on-Si device structure shown in Fig. 6, the computed electric field intensity distribution can also reveal that the EM coupling primarily takes place through the GaN buffer layer rather than the resistive Si substrate.

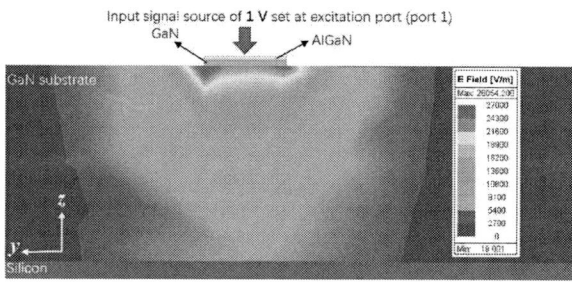

Fig. 6. Electric field intensity distribution (in the y-z plane) of the GaN-on-Si device structure, at the excitation port (with the signal frequency of 10 GHz and the electric field magnitude marked by the scale bar).

Having verified the EM coupling path predominantly through the GaN buffer layer, isolation trench structures by etching through the GaN buffer layer can be used to suppress the substrate coupling in the microwave regime. In particular, a trench filled with aluminium metal and connected to ground [15] is considered to be very effective in cutting the electric field penetrating from the excitation port to the detection port. In order to significantly improve the inferior signal isolation above multi-GHz frequencies, the grounded trench is to be placed at critical positions to effectively cut the electric field originating from excitation port. We have chosen to place a grounded trench at 1 μm away from the excitation port. Fig. 7 shows the $|S_{21}|$ results of the GaN-on-Si device structure with such a grounded trench of 1 μm width filled with Al metal. The layout of the grounded trench has the same length and orientation as the metal strip and protruded block of the excitation port. Compared with the $|S_{21}|$ results of that without a grounded trench (as plotted on the same graph in Fig. 7), the improvement in the signal isolation is ≈13 dB at 25 GHz, for the same lateral separation distance of 700 μm. At lower frequencies, the improvement can be as impressive as 32 dB, lowering $|S_{21}|$ from -23 dB to -55 dB at 1 GHz. Some further improvement in the microwave signal isolation is expected for optimised placement of the grounded trench filled with Al metal. More results will be presented at the conference.

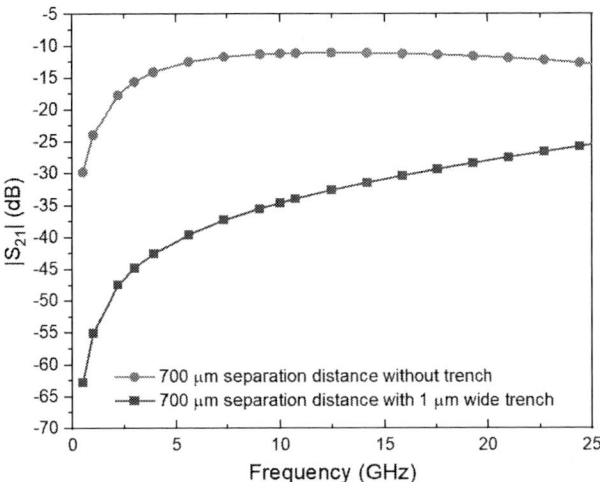

Fig. 7. $|S_{21}|$ of the GaN-on-Si device structure with a grounded trench filled with Al metal, showing an improvement in the signal isolation by about 13 dB at 25 GHz and up to 32 dB at lower frequencies from 0.5 GHz, for a lateral separation distance of 700 μm.

IV. CONCLUSION

We have, to date as the first one, performed FEM-based 3D EM simulations to investigate the crosstalk and substrate coupling in GaN-on-Si technology for monolithic microwave integrated circuits (MMICs) in the microwave frequency regime, ranging from 0.5 to 25 GHz. The S-parameter results along with the corresponding electric field magnitude profiles clearly indicate that the EM coupling can be considerably detrimental to the power electronic devices and circuits built on the same GaN chip especially above multi-GHz frequencies; at such high frequencies, adopting a thicker GaN buffer layer has little help in the suppression of EM coupling. In addition, the computed electric field intensity distribution reveals that the EM coupling takes place mainly through the GaN buffer layer rather than the resistive Si substrate even at microwave frequencies. The inferior microwave signal isolation would undermine the performance of GaN-based MMICs. To the best of our knowledgeable, isolation trench structures or alike (etching through the GaN buffer layer) would be needed to improve the suppression of crosstalk and substrate coupling in the microwave frequency regime, with $|S_{21}|$ lowered by about 13 dB or more.

ACKNOWLEDGMENT

The authors acknowledge the support from the Departments of Electrical and Electronic Engineering (EEE) and of Communications and Networking (CAN), School of Advanced Technology (SAT) of XJTLU as well as Department of Electrical Engineering and Electronics, The University of Liverpool. This work is supported in part by PGRS funding (FOSA2406036) of XJTLU. Both R. Yao and

S. Lam sincerely acknowledge the technical and administrative support by Mr. Yubin Gu and other colleagues of Academic Enhancement Team in MITS for the arrangements of floating licenses of Ansys HFSS within the campus network of XJTLU.

REFERENCES

[1] U. K. Mishra, P. Parikh, and Y.-F. Wu, "AlGaN/GaN HEMTs-an overview of device operation and applications," *Proceedings of the IEEE,* vol. 90, no. 6, pp. 1022-1031, 2002.

[2] J. Millan, P. Godignon, X. Perpiñà, A. Pérez-Tomás, and J. Rebollo, "A survey of wide bandgap power semiconductor devices," *IEEE Transactions on Power Electronics,* vol. 29, no. 5, pp. 2155-2163, 2013.

[3] N. Posthuma *et al.*, "An industry-ready 200 mm p-GaN E-mode GaN-on-Si power technology," *2018 IEEE 30th International Symposium on Power Semiconductor Devices and ICs (ISPSD)*, 2018: IEEE, pp. 284-287.

[4] H. Wui Then *et al.*, "Advances in Research on 300mm gallium nitride-on-Si(111) NMOS transistor and silicon CMOS integration," *2020 IEEE International Electron Devices Meeting (IEDM)*, 2020-12-12 2020: IEEE.

[5] L. Zhang *et al.*, "High Brightness GaN-on-Si Based Blue LEDs Grown on 150 mm Si Substrates Using Thin Buffer Layer Technology," *IEEE Journal of the Electron Devices Society,* vol. 3, no. 6, pp. 457-462, 2015.

[6] C. Florian, P. A. Traverso, and A. Santarelli, "A Ka-band MMIC LNA in GaN-on-Si 100-nm technology for high dynamic range radar receivers," *IEEE Microwave and Wireless Components Letters,* vol. 31, no. 2, pp. 161-164, 2021.

[7] H. W. Then *et al.*, "Enhancement-mode 300-mm GaN-on-Si (111) with integrated Si CMOS for future mm-wave RF applications," *IEEE Microwave and Wireless Technology Letters,* vol. 33, no. 6, pp. 835-838, 2023.

[8] J. Wei, M. Zhang, G. Lyu, and K. J. Chen, "Substrate effects in GaN-on-Si integrated bridge circuit and proposal of engineered bulk silicon substrate for GaN power ICs," in *2020 IEEE Workshop on Wide Bandgap Power Devices and Applications in Asia (Wipda Asia)*, 2020: IEEE, pp. 1-4.

[9] M. Cui and S. Lam, "Use of DC probes for multi-MHz measurements of crosstalk and substrate coupling in gallium nitride power integrated circuits," in *2024 IEEE 36th International Conference on Microelectronic Test Structures (ICMTS)*, 2024: IEEE, pp. 1-5.

[10] D. Yan and D. B. Ma, "A monolithic GaN power IC with on-chip gate driving, level shifting, and temperature sensing, achieving direct 48-V/1-V DC–DC conversion," *IEEE Journal of Solid-State Circuits,* vol. 57, no. 12, pp. 3865-3876, 2022.

[11] J. Wei *et al.*, "GaN power integration technology and its future prospects," *IEEE Transactions on Electron Devices,* vol. 71, no. 3, pp. 1365-1382, 2024-03-01 2024.

[12] M. Cui *et al.*, "Characterization of transient threshold voltage shifts in enhancement-and depletion-mode AlGaN/GaN metal-insulator-semiconductor (MIS)-HEMTs," in *2018 IEEE International Conference on Electron Devices and Solid State Circuits (EDSSC)*, 2018: IEEE, pp. 1-2.

[13] R. R. Yao, M. Cui, Z. Wang, S. Lam, and S. Taylor, "Electromagnetic Investigation of Substrate Coupling in Power Integrated Circuits in GaN-on-Si Technology," *2024 IEEE 11th Workshop on Wide Bandgap Power Devices & Applications (WiPDA)*, 2024: IEEE, pp. 1-5.

[14] M. Cui *et al.*, "Monolithic integration design of GaN-based power chip including gate driver for high-temperature DC–DC converters," *Japanese Journal of Applied Physics,* vol. 58, no. 5, p. 056505, 2019.

[15] Z. Jiang, R. Yao, M. Cui, Z. Wang, S. Lam, and S. Taylor, "On the design and effectiveness of isolation trenches to suppress substrate coupling in power integrated circuits in GaN-on-Si technology," *2024 IEEE Workshop on Wide Bandgap Power Devices and Applications in Europe (WiPDA Europe)*, 2024: IEEE, pp. 1-5.

A multi-parameter tunable bandstop filter adopting parallel RF switch

Xiang Li, Xingkun You, Kuangyong Gao, Huafei Cheng, Huanghao Ying, Zhen Shi, Haizhen Guo, and Zhonghai Zhang

Abstract—**A multi-parameter tunable bandstop filter was proposed in this letter. The tunable bandstop filter was a parallel structure and mainly constituted of coaxial transmission line, RF switches and tunable resonant • cavities. Unlike the traditional notch points switching based on series RF switch, the parallel RF switch was adopted to realize the number adjust of the notch points. This scheme allows multiple resonators to be connected in parallel on the same transmission line, with the advantage of low passband insertion loss. In order to verify the effectiveness of the method described in this paper, a dual cavity multi parameter tunable notch filter was designed and fabricated, and the test results validated the effectiveness of the proposed scheme.**

Index Terms—**tunable bandstop filter, coaxial transmission line, RF switches, multi-parameter**

I. INTRODUCTION

Tunable band stop filters can suppress high-level noise within the communication band effectively, and this function can improve the signal-to-noise ratio of the entire communication system[1-4]. A multi-parameter tunable band stop filter that can tune the frequency and number of notch points independently and this function can suppress frequency variable interference signals effectively. At the same time, if there was no high-level interference signal in the passband, the notch points inside the passband can be cut off to minimize the impact of insertion loss on the passband signal-to-noise ratio.

There were two general methods for multi parameter tunable notch filters design. The first method was to adjust the frequency and number of notch points adopting a special resonator structure[5-7]. This form of tunable notch filter was difficult to improve the notch depth by adding the number of resonators. For the second method, the number of notch points and frequency can be achieved by switching the series of tunable bandstop filter through an RF switch[8-10]. This method will result in significant passband insertion loss due to the insertion of the series tunable bandstop filter. And with the number increasing of notch points, the passband insertion loss will sharply increase.

In this letter, a multi-parameter tunable bandstop filter

Xiang Li, Xingkun You, Kuangyong Gao, Huafei Cheng, Huanghao Ying, and Zhonghai Zhang was School of Electronic Information, Hangzhou Dianzi University. Hangzhou, R.P. China.

Haizhen Guo and Zhen Shi was Zhejiang Post&Telecommunication Construction Co.,Ltd, No.99, Tai'an Road, Binjiang District, 310051, Hangzhou Zhejiang, China. (e-mail: 18958134369@189.cn).

This work was supported by the National Key Research and Development Project (No. 2022YFF1400101) .Corresponding author:Zhonghai Zhang, email: zhangzhonghai@hdu.edu.cn.

adopting RF switch with parallel structure. The tunable notch filter implemented in this structural allows multiple notch points to be connected in parallel on the same transmission line, and this function can achieve switchable notch points and maintain a small insertion loss within the passband at same time.

II. MULTI-PARAMETER TUNABLE BANDSTOP FILTER DESIGN

The structure of multi-parameter tunable bandstop filter based on parallel RF switch in this letter was shown in Fig.1.

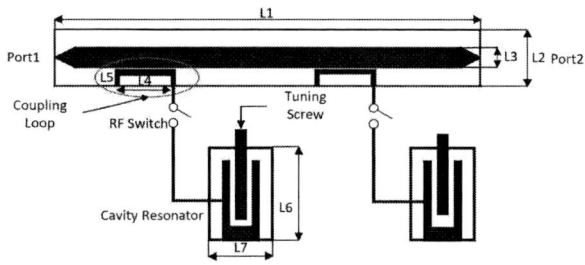

Fig. 1. The structure of the multi-parameter tunable bandstop filter.

In Fig.1, the notch point was achieved through a coupling loop between the main transmission line and the tunable resonant cavity. An RF switch was embedded in the middle of the coupling loop. When the switch was closed, a notch point was introduced in the passband, and when the switch was opened, the notch point disappears. This function can achieve a tunable number of notch points. At the same time, through adjusting the sleeve capacitance value of the tunable resonant cavity, the notch frequency of the entire notch filter can also be tuned.

III. THE FABRICATED AND MEASURED OF THE MULTI-PARAMETER TUNABLE BANDSTOP FILTER

To verify the effectiveness of the proposed tunable bandstop filter, a multi-parameter tunable notch filter with two tunable resonant cavities was designed and fabricated. The size parameters of the tunable notch filter described in this article were as follows(Unit:mm): L1=515, L2=42, L3=12, L4=40, L5=9, L6=133, L7=40. The fabricated tunable bandstop filter was shown in Fig.2.

Fig. 2. The fabricated multi-parameter tunable bandstop filter.

In Fig. 2, the main transmission line was a coaxial structure, with copper as the inner conductor and aluminum alloy as the outer conductor. The coupling ring was made of silver plated copper wire. The inner conductor of the tunable resonant cavity was also copper, and the outer conductor was a silver plated metal cavity. The mechanical RF switches were adopted to achieve the notch switching. The coupling ring was connected to the tunable resonant cavity through coaxial lines and RF switches.

A. With notch point

When the RF switch was closed, a tunable notch point appears in the passband, and the measured performance was shown in Fig. 3.

From Fig. 3, through tuning the tunable resonators, the notch point can change in the range of 263MHz-320MHz and the depth of the notch is about 10dB. With the frequency increasing, the 3dB bandwidth of the notch point will also widen.

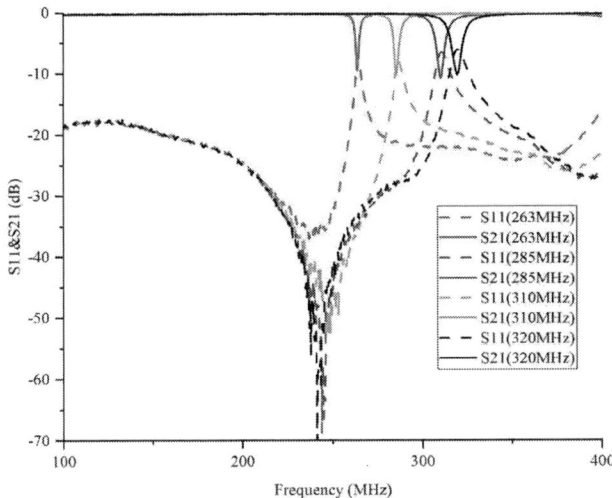

Fig. 3. The transmission performance of the multi-parameter tunable bandstop filter(with notch point).

B. Without notch point

When the RF switch was opened, the notch point disappears in the passband, and the measured performance was shown in Fig. 4.

From Fig. 4, there were no notch point in the passband. The insertion loss was less than 0.2dB and can maintain a good signal-to-noise ratio in the communication system.

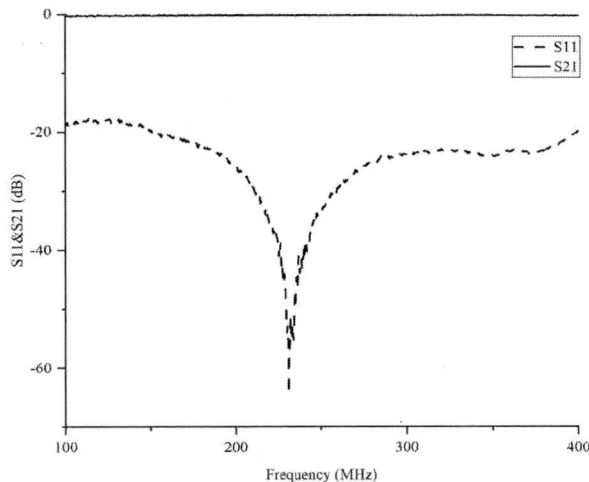

Fig. 4. The transmission performance of the multi-parameter tunable bandstop filter(without notch point).

The performance comparisons of the proposed tunable bandstop filter with other state-of-the-art designs are summarized in Table I. It shows that the proposed filter has the good performance in a wide tuning range.

TABLE I.   TABLE TYPE STYLES

Ref	Fractional tuning range	Insertion Loss(dB)	The notch point disappears
[11]	34%	1.9-4.3	No
[12]	11.9%	1.5-3	No
This work	20.5%	0.1-0.2	Yes

IV. CONCLUSION

A multi-parameter tunable notch filter based on a parallel RF switch was presented in this letter. The frequency and number of notch points can be adjusted independently as needed. This characteristic results in the tunable notch filter proposed in this article having a relatively low passband insertion loss performance, and the notch point can be turned on or off according to application. The tunable notch filter described in this article was suitable for the suppressing of high-level interference signals in broadband communication systems and maintaining good passband signal-to-noise ratio at the same time.

REFERENCES

[1] C. -W. Tang and Y. -H. Fan, "Design of A Tunable Dualband Microstrip Bandstop Filter," 2022 Asia-Pacific Microwave Conference (APMC), Yokohama, Japan, 2022, pp. 815-817,

[2] Q. Li, X. Chen, P. -L. Chi and T. Yang, "Tunable Bandstop Filter Using Distributed Coupling Microstrip Resonators With Capacitive Terminal," in IEEE Microwave and Wireless Components Letters, vol. 30, no. 1, pp. 35-38, Jan. 2020.

[3] H. B. Lou, S. Y. Chen, J. R. Chen and X. G. Huang, "Analysis of Tuning Range and Quality Factor Characteristics and Its Application to Tunable Bandstop Filter," 2021 13th International Symposium on Antennas, Propagation and EM Theory (ISAPE), Zhuhai, China, 2021, pp. 1-3.

[4] F. H. Almansour, G. H. Alyami, M. A. Alshammari and H. N. Shaman, "Ultra-Wideband (UWB) Microstrip Bandstop Filter with Transmission Zeros," 2023 SBMO/IEEE MTT-S International Microwave and Optoelectronics Conference (IMOC), Castelldefels, Spain, 2023, pp. 106-108.

[5] Y. -H. Cho, C. Park and S. -W. Yun, "4-Pole Tunable Absorptive Bandstop Filters Using Folded Coupled-Lines With an Inductor," in IEEE Access, vol. 10, pp. 120191-120199, 2022.

[6] Y. Ning, Z. Wei, P. -L. Chi and T. Yang, "A Novel Filter Architecture With Five Reconfigurable Filtering Functions," 2023 IEEE/MTT-S International Microwave Symposium - IMS 2023, San Diego, CA, USA, 2023, pp. 831-834.

[7] B. H. Ahmad, M. Z. Ramlan, M. K. Zahari, N. Hassan and A. H. Ab Rashid, "T-Shaped Matched Resonator Bandstop Filter for Microwave Applications," 2024 IEEE Asia-Pacific Conference on Applied Electromagnetics (APACE), Langkawi, Kedah, Malaysia, 2024, pp. 160-162.

[8] M. A. Sanchez-Soriano et al., "Wideband Reconfigurable Bandpass-to-Bandstop Filter Based on Embedded Switches on Silicon Technology," 2023 53rd European Microwave Conference (EuMC), Berlin, Germany, 2023, pp. 174-177.

[9] W. Chen, L. -F. Qiu and L. -S. Wu, "Reconfigurable Filter With Switchable Operating Frequency, Bandwidth, and Response Type Using RF Switches and Circulators," in IEEE Transactions on Microwave Theory and Techniques.

[10] A. K. Gorur and D. Psychogiou, "Single-/Dual-Band Bandpass-to-Bandstop Filters With Center Frequency Tunability," in IEEE Access, vol. 12, 2024,pp. 90697-90706.

[11] Y.-H. Cho and G. M. Rebeiz, "Two- and four-pole tunable 0.7 – 1.1-GHz bandpass-to-bandstop filters with bandwidth control," IEEE Trans. Microw. Theory Techn., vol. 62, no. 3, pp. 457 – 463, Mar. 2014.

[12] N. Kumar and Y. K. Singh, "RF-MEMS-based bandpass-to-bandstop switchable single- and dual-band filters with variable FBW and reconfigurable selectivity," IEEE Trans. Microw. Theory Techn., vol. 65, no. 10, pp. 3824 – 3837, Oct. 2017.

Design Method of Ultra-Wideband Power Amplifier Based on Optimization Approach

1st Xiang Chen   2nd Guohua Liu   3rd En Hong

Abstract—This paper presents an ultra-wideband power amplifier (PA) design method based on particle swarm optimization (PSO) algorithm. The initial PA is designed using the simplified real frequency technique (SRFT). Subsequently, the PSO algorithm is employed to optimize the drain efficiency (DE), output power (Pout), and gain of the PA. The proposed method achieves a comprehensive optimization of these parameters across the 0.7–3.1 GHz frequency band. The final design demonstrates an output power exceeding 40.7 dBm and a drain efficiency greater than 61% within the target bandwidth.

Index Terms—ultra-wideband, particle swarm optimization, power amplifier, output power, drain efficiency

I. INTRODUCTION

With the evolution of wireless communication technologies toward higher frequencies and broader bandwidths, 5G/6G systems have demonstrated significant potential in cellular high-speed communication, automotive radar, and industrial IoT due to their nanosecond-level pulse signals, low power consumption, and high anti-interference capabilities. However, as a vital module of 5G/6G systems, power amplifiers (PAs) must simultaneously achieve high output power, high drain efficiency (DE), and flat gain across ultra-wide bandwidths (typically spanning multiple octaves). These requirements nearly contradictory challenges for traditional design methodologies: the complexity of broadband impedance matching networks and the multi-objective trade-offs among efficiency, bandwidth, and gain. Existing broadband PA designs primarily rely on empirically driven parameter adjustments, such as load-pull based impedance matching or harmonic tuning techniques [1]. While these methods can optimize efficiency to over 70% in narrowband scenarios, they exhibit significant limitations in broadband applications. First, load-pull techniques require exhaustive impedance point traversals, leading to exponentially increasing computational costs with bandwidth. Second, harmonic tuning depends on frequency-specific harmonic control, making it difficult to ensure balanced performance across wide frequency bands. Additionally, traditional gradient-descent algorithms are prone to local optima, resulting in design outcomes that are highly sensitive to initial values and suffer from efficiency collapse at band edges. In recent years, intelligent optimization algorithms have provided new avenues for multi-parameter radio frequency

(RF) circuit design. Global search methods such as genetic algorithms (GA) [2] and simulated annealing (SA) [3] have been applied to PA matching network optimization. However, their slow convergence and complex parameter tuning hinder practical applications. In contrast, the particle swarm optimization (PSO) algorithm, leveraging swarm intelligence and low computational complexity, has emerged as a promising approach for multi-objective optimization in microwave circuits. Nevertheless, existing research predominantly focuses on narrowband PAs or single performance metrics. Integrating PSO with classical circuit synthesis methods to achieve multi-objective co-optimization for Ultra-Wideband PAs remains an unresolved challenge. This paper proposes a hybrid "SRFT-PSO" design framework to address these challenges. First, the simplified real frequency technique (SRFT) [4] is utilized to generate the initial topology of a broadband matching network, circumventing the numerical instability of conventional real-frequency methods. Subsequently, an improved PSO algorithm is introduced to dynamically adjust the matching network parameters across the 0.7–3.1 GHz band, with DE, P_{out}, and $Gain$ as joint optimization targets. The remainder of this paper is organized as follows: Section II introduces the SRFT principle, Section III details the PSO algorithm and multi-objective optimization framework, Section IV presents pre- and post-optimization PA performance, and Section V concludes the work.

II. SIMPLIFIED REAL FREQUENCY TECHNIQUE

The performance of a matching circuit is typically evaluated using the transducer power gain (TPG). TPG is defined as the ratio of the power delivered to the load to the power available from the source, characterizing the loss of the matching network. Figure 1 illustrates the scattering parameter representation of a matching network E.

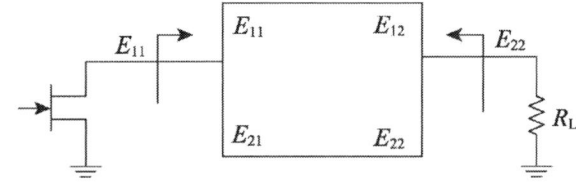

Fig. 1. Matching network of scattering parameters.

Any bounded-real reflection coefficient $E_{11}(s)$ can be realized by a lossless two-port network terminated with a pure resistor. Once $E_{11}(s)$ is determined, the parameters of the

---

Xiang Chen and Guohua Liu are with *the School of Electronics and Information(of Aff.)*, Hangzhou Dianzi University, *(of Aff.)*, Hangzhou 310018, China. En Hong is with *the School of Electronics and Information (of Aff.),Hangzhou Dianzi University, (of Aff.)*, Hangzhou 310018, China. Corresponding author: (Guohua Liu, email: ghliu@hdu.edu.cn)

network E can be derived. Thus, the real frequency method expresses the matching network E in terms of the S parameter:

$$E_{11} = \frac{h(s)}{g(s)} = \frac{h_0 + h_1 s + \dots + h_n s^n}{g_0 + g_1 s + \dots + g_n s^n} \tag{1}$$

$$f(s) = s^k \tag{2}$$

$$E_{21}(s) = E_{12}(s) = \pm\frac{f(s)}{g(s)} \tag{3}$$

$$E_{22}(s) = -(-1)^k \frac{h(-s)}{g(s)} \tag{4}$$

Where $h(s)$ and $g(s)$ are n-term Hurwitz polynomials. For a lossless network, the following condition must be satisfied:

$$g(s) \cdot g(-s) = h(s) \cdot h(-s) + f(s) \cdot f(-s) \tag{5}$$

The TPG expression is derived as:

$$TPG = \frac{(1 - |E_{11}|^2)(1 - |E_L|^2)}{|1 - E_{22}E_L|^2} \tag{6}$$

where E_L is the input reflection coefficient of the load network. An error function $Terr(s)$ is constructed to minimize deviations from the target $TPG(T_0)$ across the entire frequency band. This yields optimized h and g polynomials for constructing the broadband matching circuit:

$$Terr(s) = |T(s) - T_0| \tag{7}$$

III. Multi-Objective PSO Optimization Framework

A. Particle swarm optimization

The PSO algorithm mimics the foraging behavior of bird flocks. A swarm of particles is randomly distributed in the solution space to locate the global optimum. Each particle adjusts its position by learning from its historical best position (P_{best}) and the global best position (G_{best}). The velocity and position update equations in PSO are defined as:

$$V_{i,j}(t+1) = wV_{i,j}(t) + c_1 r_1(P_{best,i,j}(t) - X_{i,j}(t)) \tag{8}$$

$$X_{i,j}(t+1) = X_{i,j}(t) + V_{i,j}(t+1) \tag{9}$$

where $V_{i,j}$ and $X_{i,j}$ denote the velocity and position of particle i in dimension j at iteration t, w is the inertia weight balancing global and local exploration, c_1 and c_2 are acceleration coefficients, and r_1, r_2 are random numbers.

B. Multi-Objective Fitness Function Modeling

Broadband PA design requires simultaneous optimization of DE, P_{out}, and $Gain$, which exhibit complex nonlinear interdependency. For instance, increasing output power often necessitates higher transistor bias voltages, which may degrade DE. Meanwhile, gain flatness is directly influenced by the frequency response of the matching network. To balance these objectives, a multi-objective fitness function is formulated:

$$\begin{aligned} Fitness = {} & a \cdot abs(DE - DE_{goal}) + b \cdot abs(Pout \\ & - Pout_{goal}) + c \cdot abs(Gain - Gain_{goal}) \\ & + punishment \end{aligned} \tag{10}$$

where DE_{goal}, $Pout_{goal}$, and $Gain_{goal}$ represent optimization goals. Coefficients a, b, and c are weights satisfying, and *punishment* penalizes deviations from target values.

C. SRFT-PSO Co-Optimization Workflow

A multi-octave PA is designed using Wolfspeed's 10 W GaN HEMT device (CGH40010F). The drain and gate bias voltages are set to 28 V and -2.8 V, respectively. The source-pull impedance method is employed to acquire matching impedance values. The SRFT program generates a matching network comprising capacitors and inductors, which are converted into practical microstrip lines using stepped-impedance formulas. This yields the initial PA topology. The PA structure is fabricated on a Rogers 450b substrate ($\varepsilon_r = 3.66$, thickness = 0.762 mm). The PSO algorithm uses a population size of 50 and 100 iterations, totaling 5,000 evaluations. The fitness function weights (DE_{goal}, $Pout_{goal}$, $Gain_{goal}$) are set to 61, 40, and 10, respectively.

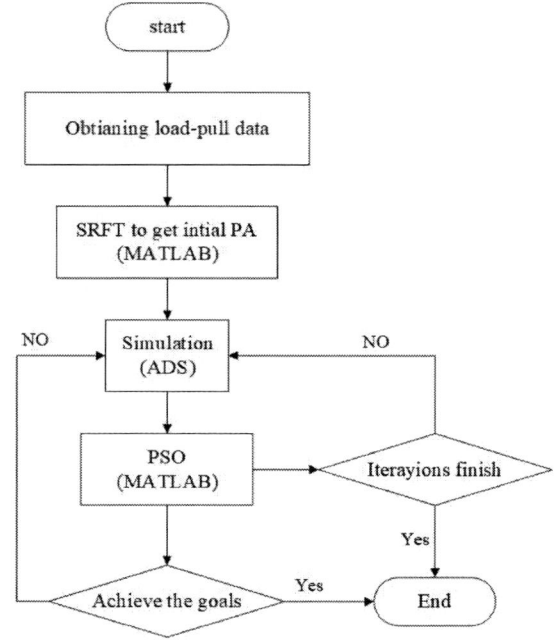

Fig. 2. The flow chart of ADS-MATLAB co-simulation.

Figure 2 illustrates the PSO optimization process via ADS-MATLAB co-simulation. Figure 4 illustrates the Schematic diagram of the designed PA and Figure 5 illustrates the Layout of the designed PA. Compared with the traditional design process of broadband PA, the proposed automatic design method in this article is much faster and more convenient. It eliminates the time-consuming and laborious manual design. Compared with other optimization methods [2] [3], this design adopts the SRFT algorithm to generate the initial PA, which is more convenient. Besides, the complexity of the proposed algorithm is also smaller, making it easier to implement. This method is an excellent approach for rapid design optimization of ultra-wideband PA.

IV. RESULTS

Figure 3 compares the performance of the multi-octave PA before and after optimization. The initial SRFT-based design exhibits acceptable performance at low frequencies but suffers from rapid degradation in P_{out}, $Gain$, and DE at high frequencies. After PSO optimization, DE exceeds 61%, P_{out} ranges from 40 to 42 dBm, and $Gain$ surpasses 10 dB across the entire 0.7–3.1 GHz band.

V. CONCLUSION

This paper proposes a hybrid SRFT-PSO method for Ultra-Wideband PA design. The SRFT generates the initial PA topology, which is then optimized using the PSO algorithm. The final design achieves an output power over 40.7 dBm and a drain efficiency over 61% across the 0.7–3.1 GHz frequency band.

TABLE I
PERFORMANCE COMPARISON WITH RECTIFIERS

Ref.	Freq(GHz)	DE(%)	Pout(dBm)	Gain(dB)
[2]	2-3	60-65	41.6-42	\
[5]	0.5-2.3	60-81	39.2-41.2	≥ 11.7
[6]	0.5-3	53.4-70	39.6-41.4	10.6-12.4
[7]	1.85-2.1	58.5-73	38.6-39	8.9-10.2
This work	0.7-3.1	61-70	40.8-41.4	10.8-11.4

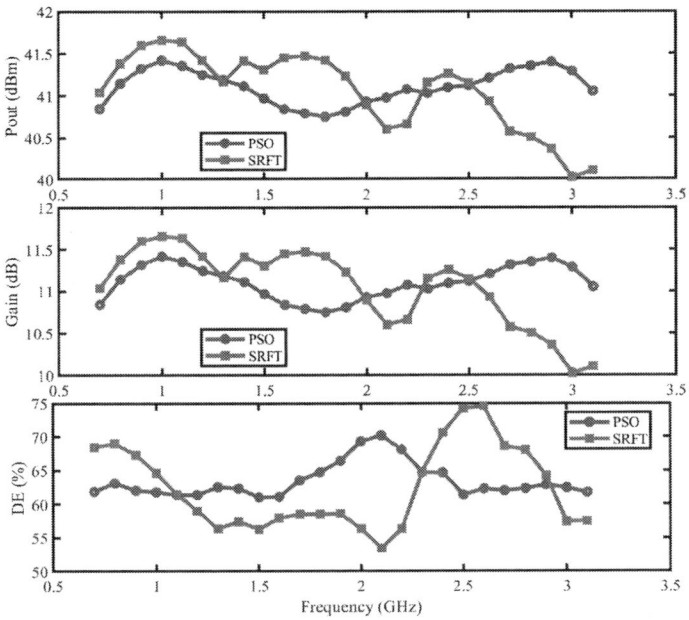

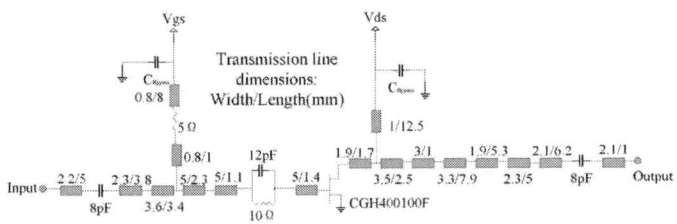

Fig. 3. Simulated results : Output power, Gain, DE.

Fig. 4. Schematic diagram of the designed PA.

REFERENCES

[1] S. Eskandari and A. B. Kouki, "The Second Source Harmonic Optimization in Continuous Class-GF Power Amplifiers," in IEEE Microwave and Wireless Components Letters, vol. 32, no. 4, pp. 316-319, April 2022, doi: 10.1109/LMWC.2021.3128497.

[2] B. Sun, "A Wide-band Power Amplifier based on Genetic Algorithm and Direct Layout Optimization," 2022 IEEE MTT-S International Microwave Workshop Series on Advanced Materials and Processes for RF and THz Applications (IMWS-AMP), Guangzhou, China, 2022, pp. 1-3, doi: 10.1109/IMWS-AMP54652.2022.10107103.

[3] C. Li, "Simulated Annealing Particle Swarm Optimization for High-Efficiency Power Amplifier Design," in IEEE Transactions on Microwave Theory and Techniques, vol. 69, no. 5, pp. 2494-2505, May 2021, doi: 10.1109/TMTT.2021.3061547.

[4] Yildiz, A. Aksen and S. B. Yarman, "Multiband filter design from bandpass prototype filter with using frequency transformation via SRFT," 2016 National Conference on Electrical, Electronics and Biomedical Engineering (ELECO), Bursa, Turkey, 2016, pp. 432-436.

[5] S. Y. Zheng, Z. W. Liu, X. Y. Zhang, X. Y. Zhou, and W. S. Chan,"Design of ultrawideband high-efficiency extended continuous class-F power amplifier," IEEE Trans. Ind. Electron., vol. 65, no. 6,pp. 4661–4669, Jun. 2018.

[6] S. Kilinc, B. S. Yarman, and S. Ozoguz, "Broadband power amplifier design via fictitious matching," IEEE Trans. Circuits Syst. II, Exp. Briefs,vol. 69, no. 12, pp. 4844–4848, Dec. 2022.

[7] S. Li, Y. Wu, Y. Yang, X. Chen, W. Wang, and Z. Chen, "Bandpass filtering power amplifier with wide stopband and high out-of-band rejection," IEEE Trans. Circuits Syst. II, Exp. Briefs, vol. 70, no. 3,pp. 969–973, Mar. 2023.

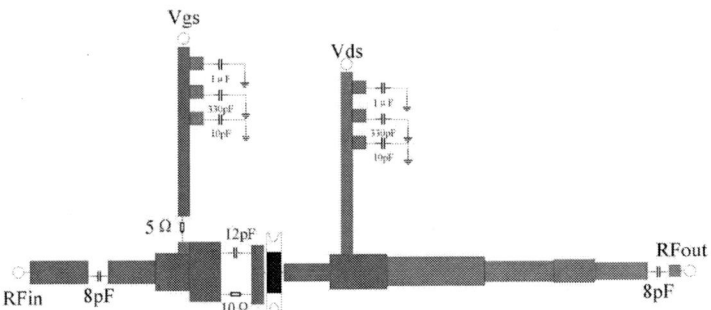

Fig. 5. Layout of the designed PA

979-8-3315-2209-4/25 $31.00 © 2025 IEEE

A Wide Band Reconfigurable PVT Robust high precision quarter phase detector for clock multiplier

Anqing Chen   Nagarajan Mahalingam   Bharatha Kumar Thangarasu   Kaixue Ma

Fanyi Meng   Zhenghao Lu   Kiat Seng Yeo

Abstract—**We present the design and simulation results of a differential charge pump quarter phase detector. The use of differential structure and dynamic comparator avoids the non-ideal factors caused by the charge pump, allowing us to achieve a PVT robust and simple design. Digitally adjustable current source allows the phase detector to operate over a wide frequency range. Simulation results show that the phase detector has the ability to detect a quarter phase deviation better than 0.1% cycle at various process corners from 1MHz to 1GHz.**

Index Terms—**quarter phase detector, PVT robust, wide frequency range**

I. INTRODUCTION

Compared with integer PLL, fractional-N PLL has been widely used in the communication field because it can provide finer frequency accuracy. But the Delta Sigma Modulator (DSM) of the fractional-N PLL introduces quantization noise, which makes the phase noise of the fractional-N PLL worse. Using a higher reference clock means a lower multiplication ratio, which can reduce the noise of the PFD/CP and shape the quantization noise of the DSM to a higher frequency [1].There are some ways to increase the reference clock frequency: Directly use a high-frequency crystal oscillator [2], but this will significantly increase the cost. Multi-level PLLs can be used [3], but this will increase power consumption and require a large area overhead. Another way is to use a clock multiplier [4].

A clock with a 50% duty cycle can have the highest utilization efficiency, and many double data rate (DDR) systems also require the clock signal duty cycle to be maintained at 50%. As PVT (Process Voltage Temperature) changes become more severe, circuit robustness becomes one of the most important characteristics in clock multiplier design.

Usually the reference clock multiplier has a self-calibration function, so that the waveform after multiplication is close to 50% duty cycle. For the XOR gate structure multiplier, there are usually two solutions which are shown in Fig. 1 to make the duty cycle of the multiplier output 50%. The first solution [5] [6] is to use only a simple and rough delay unit in the delay unit. Then directly shape the output of the multiplier. This means that the duty cycle correction (DCC) circuit works at the highest frequency, which will increase power consumption.

Anqing Chen 1, Nagarajan Mahalingam 2, Bharatha Kumar Thangarasu 3, Kaixue Ma 4, Fanyi Meng 5, Kiat Seng Yeo 7 are with *School of microelectronics, Tianjin University*, Tianjin, China. Zhenghao Lu 6 is with *School of Electronics and Information, Soochow University*, Suzhou, China. Corresponding author: (Anqing Chen 1, email: chenanqing@tju.edu.cn)

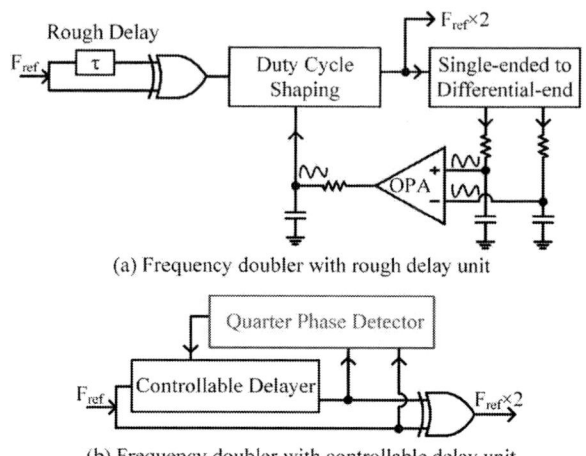

(a) Frequency doubler with rough delay unit

(b) Frequency doubler with controllable delay unit

Fig. 1. Structure diagrams of DCC with (a) rough delay unit and (b) controllable delayer.

The DCC structure needs to use an RC circuit for filtering, and an operational amplifier is used after the RC filter to obtain the phase difference information. This is necessary to ensure that the input fluctuation of the operational amplifier is as small as possible. Therefore, the area of the RC filter has to be relatively large, especially when the clock input frequency is very low. This makes the area overhead of this structure relatively large, and at the same time has certain requirements on the input frequency. This solution cannot achieve a wide enough bandwidth. The correction result of the DCC structure mainly depends on the differential voltage of the operational amplifier input. Therefore, the operational amplifier needs to have a sufficiently high gain. The situation where there are multiple poles in the loop will cause the circuit to oscillate easily. Therefore, this brings a lot of difficulties to the design. Another solution [7] is to use a controllable delay device at the input of the XOR gate to control the duty cycle of the frequency multiplier output. However, the delay device is very susceptible to PVT. Therefore, it is necessary to use a quarter phase detector for phase detection and delay correction. Therefore, the robustness of the phase detector plays a decisive role in the frequency multiplier output.

Typically a charge pump is used in phase detectors [8]. It has a simple structure. It uses a current source to charge and discharge the capacitor to obtain duty cycle or phase

---

979-8-3315-2209-4/25 $31.00 © 2025 IEEE

information. However, the charge pump structure has inherent defects, such as clock feedthrough and charge injection, that is, when the drain current of the current tube switches between on and off, it will extract charge from the output. In addition, the current source is not an ideal device, and there will be a mismatch between the charging and discharging current sources. Mismatch in the current mirror can cause errors in the charge pump's upper and lower currents, but this is not the main source of error. The most important source of error comes from the finite output resistance of the MOSFET. The channel modulation effect will cause current mismatch, that is, the charge pump output voltage will cause charge pump charging and discharging current mismatch, which will significantly affect the accuracy of the charge pump phase detector. Directly increasing the channel length can reduce the impact of the channel length modulation effect, but correspondingly, this will make the charge sharing effect more serious due to the increase in the parasitic capacitance of the current source. Therefore, research on charge pumps mainly focuses on how to reduce current mismatch [8], but this usually significantly increases the complexity of the circuit and cannot completely eliminate those effects.

In summary, the inherent defects of the charge pump have become the main obstacle to improving the performance of the phase detector based on the charge pump. This paper proposes a charge pump quarter phase detector with a differential structure, which can effectively avoid the effects of current mirror mismatch, clock feedthrough, charge sharing, etc. The quarter phase detector can achieve a phase detection accuracy of 0.1% cycle in the range of 1MHz ~ 1GHz at various process angles and temperatures.

II. QUARTER PHASE DETECTOR

Fig. 2 shows the core circuit of the quarter phase detector. A differential structure is used to compare the phases of the clock signal and the delayed clock signal. Usually, the mismatch between the charging current source and the discharging current source is more serious. Therefore, the charging current source and the discharging current source are marked in blue and red respectively. At the same time, the main parasitic effects are marked in the figure.

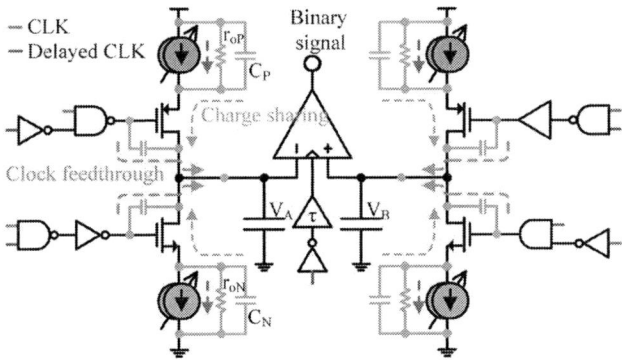

Fig. 2. Architecture of the proposed quarter phase detector.

The working timing diagram of the phase detector is shown in Fig. 3. Due to the limited linear output range of the charge pump, even if the amount of charge charged and discharged by the current source in each cycle is inconsistent at the initial. The voltage on the capacitor cannot be infinite or infinitely small. Therefore, after several cycles, due to the influence of parasitics, there must be a steady state so that the voltage on the capacitors V_A and V_B remains unchanged after each cycle.

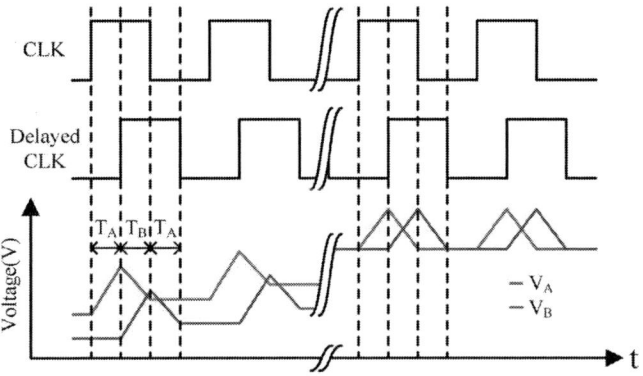

Fig. 3. Timing operation diagram.

The MOSFET used as a current source can be equivalent to a combination of an ideal current source and parasitic resistance and capacitance. Because the MOSFET used as a switch is turned on and off once per cycle, the clock feedthrough and charge injection effects introduced by the switching MOSFET can be ignored. If the duty cycle of the input clock is 50%, when the phase detector reaches a steady state. According to the law of conservation of charge. The phase detector system has the following equation:

$$\left(I_{up} + \frac{V_{DD} - V_A}{r_{oP}} \right) T_A - \left(I_{down} + \frac{V_A}{r_{oN}} \right) T_B \\ + (V_{DD} - V_A) C_P - V_A C_N = 0 \tag{1}$$

$$\left(I_{up} + \frac{V_{DD} - V_B}{r_{oP}} \right) T_B - \left(I_{down} + \frac{V_B}{r_{oN}} \right) T_A \\ + (V_{DD} - V_B) C_N - V_B C_N = 0 \tag{2}$$

If both sides have the same steady-state voltage, that is, $V_A = V_B$, then by combining the above formulas we can get:

$$\left(I_{up} - I_{down} + \frac{V_{DD} - V_A}{r_{oP}} - \frac{V_A}{r_{oN}} \right) (T_A - T_B) = 0 \tag{3}$$

Where $I_{up} - I_{down}$ is the mismatch current of charge and discharge, $\frac{V_{DD} - V_A}{r_{oP}}$ and $\frac{V_A}{r_{oN}}$ represents the influence of the limited output resistance of the current source. Then when $T_A - T_B = 0$, that is, when the input clock and the delayed clock differ by a quarter of the phase, the equation always holds true. By using a strong arm comparator to sample the voltage on the load capacitors V_A and V_B and a small delay τ is implemented to avoid the influence of voltage fluctuations

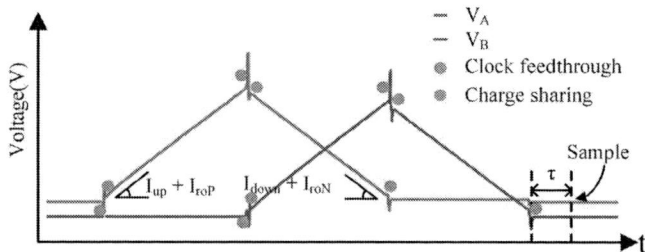

Fig. 4. Example of a cycle with non-ideal factors.

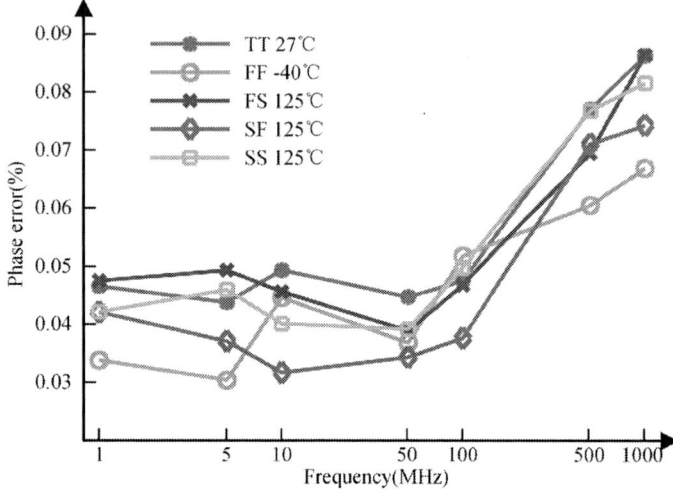

Fig. 5. The result of the quarter phase detector detects deviation from one quarter in different process corner and temperature.

on the load capacitor. How the appropriate comparator sampling time can avoid the influence of fluctuations on the load capacitance is shown in Fig. 4. The use of the comparator means that the information output by the phase detector is binary. Therefore, even if there is a mismatch between the charging current and the discharging current, the existence of parasitic effects such as clock feedthrough, charge sharing, and charge injection, it does not affect the function of the phase detector. This shows that the structure has good robustness.

The operating frequency and sensitivity of the phase detector are directly affected by the charge pump current and load capacitance. This is because it is necessary to ensure that the current source transistor operates in the saturation region as much as possible. Therefore, in order to achieve wideband, the charging current and the discharging current are digitally controllable. When the charge pump current is small, the circuit can operate at a lower frequency, and when the charge pump current is large, the phase detector can operate at a higher frequency.

III. SIMULATION RESULT

The implementation has been evaluated by using 0.13um BiCMOS process technology. To verify the influence of process corner and temperature, the simulation was conducted in FF -40°C, TT 27°C, SF 125°C, FS125°C and SS 125°C environments. A clock with 50% duty cycle and a delayed version of the clock were used as inputs. When the phase difference between the two clocks exceeds one quarter, the comparator will output a high level, otherwise it will output a low level. Fig. 5 shows the phase detector's ability to identify a quarter-phase shift. By controlling the size of the current source, the phase detector can operate over a wide range of frequencies. The results shows that the phase detector has a phase detection capability better than 0.1% cycle at each process corner and from 1MHz to 1GHz.

IV. CONCLUSION

We present the design and simulation results of a differential charge pump quarter phase detector. The use of differential structure and dynamic comparator allows us to achieve a robust design, avoiding the impact of various non-ideal factors of the charge pump. The structure is simple and the implementation complexity is low. The use of digitally controlled current sources allows the circuit to operate over a wider range of input frequency variations. This allows the design to have a wider range of uses. The simulation provides results from 1MHz to 1GHz, at various process corners and temperatures. It shows that the design has good PVT robustness. This broadband PVT robust quarter phase detector can be used in clock multipliers to adjust the duty cycle.

REFERENCES

[1] K. Lee, J. Jung, S. Kim, S. Oh, J. Lee and S. M. Park, "A 208-MHz, 0.75-mW Self-Calibrated Reference Frequency Quadrupler for a 2-GHz Fractional-N Ring-PLL in 4-nm FinFET CMOS," in IEEE Transactions on Circuits and Systems II: Express Briefs, vol. 70, no. 8, pp. 2719-2723, Aug. 2023.

[2] S. Ek, T. Påhlsson, C. Elgaard, A. Carlsson, A. Axholt, A. -K. Stenman et al., "A 28-nm FD-SOI 115-fs Jitter PLL-Based LO System for 24–30-GHz Sliding-IF 5G Transceivers," in IEEE Journal of Solid-State Circuits, vol. 53, no. 7, pp. 1988-2000, July 2018.

[3] Y. Liang, C. C. Boon and Q. Chen, "A 14.2 mW 29-39.3-GHz Two-Stage PLL with a Current-Reuse Coupled Mixer Phase Detector," 2023 IEEE Radio Frequency Integrated Circuits Symposium (RFIC), San Diego, CA, USA, 2023.

[4] F. Song, Y. Zhao, B. Wu, L. Tang, L. Lin and B. Razavi, "16.5 A Fractional-N Synthesizer with 110fsrms Jitter and a Reference Quadrupler for Wideband 802.11ax," 2019 IEEE International Solid-State Circuits Conference - (ISSCC), San Francisco, CA, USA, pp. 264-266, 2019.

[5] S. Jung, J. Jung, B. Han, S. Oh and J. Lee, "A 9.4MHz-to-2.4GHz Jitter-Power Reconfigurable Fractional-N Ring PLL for Multi-Standard Applications in 7nm FinFET CMOS Technology," 2019 IEEE Asian Solid-State Circuits Conference (A-SSCC), Macau, Macao, pp. 87-90, 2019.

[6] M. M. Ghahramani, Y. Rajavi, A. Khalili, A. Kavousian, B. Kim and M. P. Flynn, "A 192MHz differential XO based frequency quadrupler with sub-picosecond jitter in 28nm CMOS," 2015 IEEE Radio Frequency Integrated Circuits Symposium (RFIC), Phoenix, AZ, USA, pp. 59-62, 2015.

[7] Jung, Gunok, G. Park, U. Cho and J. Son. "Fully digital clock frequency doubler." IEICE Electron. Express 7, pp. 416-420, 2010.

[8] N. Liu, J. Todsen and D. Chen, "A Low-Power and Area-Efficient Analog Duty Cycle Corrector for ADC's External Clocks," 2020 IEEE International Symposium on Circuits and Systems (ISCAS), Seville, Spain, pp. 1-4, 2020.

An OTA-C bandpass filter for infrared receiver chip

1st Huilin Huang   2nd Hongyi Liu   3rd Shuxin Xu   4th Wanghui Zou

All authors are with School of Physics and Electronics, Changsha University of Science and Technology, Changsha, China. This work is partially supported by Research Project of Hunan Provincial Department of Education (23A0260). Corresponding author: (Author 4, email: zouwh@csust.edu.cn)

Abstract—**Different infrared (IR) communication devices may operate at varying modulation frequencies. In this work, an OTA-C bandpass filter with a tunable center frequency is designed for infrared receiver chips, enabling adaptability to different IR communication devices. The circuit and layout design were carried out using a 0.18 μm 3.3V CMOS process. Simulation results demonstrate that under various process corners and within the temperature range of −20 °C to 85 °C, the tuning frequency range effectively covers 36–56.7 kHz, and a passband gain of ~16 dB, a bandwidth of ~4 kHz, and a quality factor (Q) as high as ~15 are exhibited. The core layout area is only 336 × 190 μm².**

Index Terms—**IR receiver chip, bandpass filter, operational amplifier, OTA-C, linearity**

I. INTRODUCTION

Infrared (IR) communication technology is widely used in various remote control applications due to its advantages of high security, low power consumption, and low cost. Commonly used IR communication frequencies include 36 kHz, 38 kHz, 40 kHz, and 56.7 kHz. As one of the core components of an IR receiver chip, the bandpass filter (BPF) should possess tunability, enabling its center frequency to cover various communication frequencies. Additionally, a high quality factor (Q) is also required to enhance frequency selectivity.

The most widely used bandpass filters are OTA-C BPFs, which are usually composed of operational transconductance amplifiers (OTAs) and capacitors (C). An OTA-C BPF consisting of five OTAs, two grounded capacitors, and one resistor was proposed in [1], while a tunable OTA-C BPF architecture employing four OTAs and two grounded capacitors, with adjustable center frequency and quality factor, was introduced in [2]. Though these structures offer good performance, they also exhibit relatively high circuit complexity. The linearity of the OTA is a critical factor influencing the performance of OTA-C BPFs. Source degeneration is a regular technique used to improve linearity. In [3], an auxiliary differential pair was further introduced to drive the bulk terminals of the main pair, resulting in an enhanced linearity. In [4], a complementary differential pair was employed in combination with source degeneration to achieve additional improvements in linearity. The center frequency of the OTA-C BPF can be tuned by varying the transconductance, g_m, of the OTA. However, the tuning range is limited and significantly affected by process, voltage, and temperature (PVT) variations. In [5], an array of fully differential transconductance circuits was employed to control the filter parameters, and fine-tuning was obtained by using programmable capacitors. Although a tenfold frequency tuning range was achieved, the circuit complexity was considerably increased.

In this work, an OTA-C BPF consisting of three OTAs and two grounded capacitors was designed. Two OTA architectures with enhanced linearity were also developed. A digitally controlled tuning circuit was implemented, enabling adjustment of the OTA bias current and capacitance values. This approach allows the center frequency to be tuned to accommodate commonly used frequencies.

II. CIRCUIT DESIGN

A. Source-degenerated OTA

OTAs with source degeneration are widely used in practical applications due to their favorable linearity and relatively simple circuit implementation. A typical source-degenerated OTA structure is shown in Fig. 1. A pair of MOS transistors, M_{R1} and M_{R2}, are introduced at the sources of the input differential pair transistors M1 and M2. Operated in the linear region, these two transistors act as source degeneration resistors controlled by V_{IN+} and V_{IN-}, and improve the linearity of the OTA.

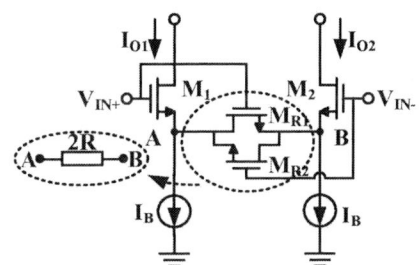

Fig. 1. A typical source-degenerated OTA

For a basic OTA, if the differential input, $V_{id}=V_{IN+}-V_{IN-}$, is sufficiently small, the transconductance, g_m, can be derived as follows:

$$g_m = \sqrt{2K_n \cdot I_B} \qquad (1)$$

here, $K_n =\mu_n C_{ox} (W/L)$, where μ_n is the carrier mobility, C_{ox} is the gate oxide capacitance per unit area, and W/L is the transistor width-to-length ratio. For the differential output current, $I_o=I_{o1}-I_{o2}$, there always exists third-order and higher-order odd harmonics. Among them, the third-order harmonic component has the most significant impact on the nonlinearity of the output current, and the third harmonic distortion coefficient can be expressed as follows [6]:

$$HD_3 = \frac{V_{id}^2}{16(V_{GS} - V_{TH})^2} \qquad (2)$$

979-8-3315-2209-4/25 $31.00 © 2025 IEEE

here, V_{GS} denotes the gate-to-source voltage, and V_{TH} is the threshold voltage. For a source-degenerated OTA, assuming a source degeneration resistance of $2R$, the equivalent transconductance, g'_m, and equivalent third harmonic distortion coefficient, HD'_3, will be determined as follows [7]:

$$g'_m = \frac{g_m}{1 + g_m R} \tag{3}$$

$$HD'_3 = \frac{HD_3}{(1 + g_m R)^2} \tag{4}$$

It can be observed that the source degeneration resistance reduces the effective transconductance of the OTA by a factor of $1 + g_m R$, and decreases the third harmonic distortion coefficient by a factor of $(1 + g_m R)^2$. The reduction in the distortion components in the output current extends the linear operating range, thereby allowing the transconductance to remain relatively constant over a wider input voltage range and improving the linearity of the OTA.

B. Linearity-improved OTAs

To further improve the linearity of the OTA and facilitate center frequency tuning, two OTA structures were designed in this work. The first structure, OTA1, as shown in Fig. 2, incorporates separated source degeneration resistors. M_1–M_4 and M_{11}–M_{14} form cascade current mirrors. M_5 and M_6 serve as the input differential pair, while M_7–M_{10} implement the source degeneration. Based on the basic OTA, separated MOS resistors, controlled by V_{IN+} and V_{IN-}, are introduced at the sources of M5 and M6, respectively. The separated source degeneration structure uses only one single tail current; however, the drain-source voltage of M7/M10 and M8/M9 consumes part of the voltage swing. Due to negative feedback effects, when the same device dimensions are used, it can be easily derived that the equivalent resistance $2R'$ of transistors M7/M10 and M8/M9 is larger than the equivalent resistance $2R$ of M_{R1}/M_{R2} in Fig. 1. Moreover, under differential input, the two source degeneration resistors in the separated structure are effectively in series, resulting in a differential equivalent resistance of $4R'$. Hence, the equivalent transconductance of this structure is reduced by a factor of $1 + g_m 2R'$, and the third harmonic distortion coefficient is suppressed by $(1 + g_m 2R')^2$, thereby achieving an improved linearity.

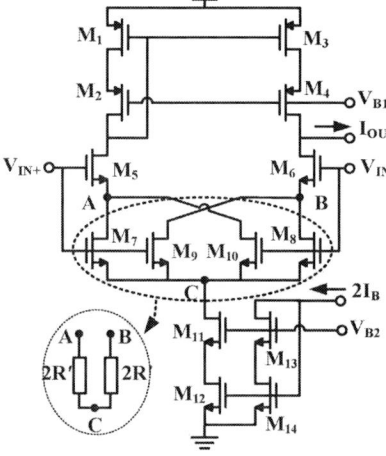

Fig. 2. Circuit structure of OTA1

The second OTA structure, OTA2, combines an auxiliary differential pair with source degeneration, as illustrated in Fig. 3. Its equivalent circuit is also shown in the figure. Transistors M5 and M6, together with M9 and M10, form two differential pairs with identical aspect ratios. Because of the auxiliary differential pair, the differential input voltage is divided, and a portion of the input voltage is applied to the auxiliary pair. According to [8], when two differential pairs are employed, the OTA's transconductance and third harmonic distortion are reduced by factors of 2 and 4, respectively. Assuming that OTA2 utilizes the same source degeneration resistors as in Fig. 2, and the equivalent resistance is denoted as R'', the effective transconductance is then reduced by a factor of $2(1 + g_m R'')$, and the third harmonic distortion coefficient is decreased by a factor of $4(1 + g_m R'')^2$, thereby further enhancing the linearity.

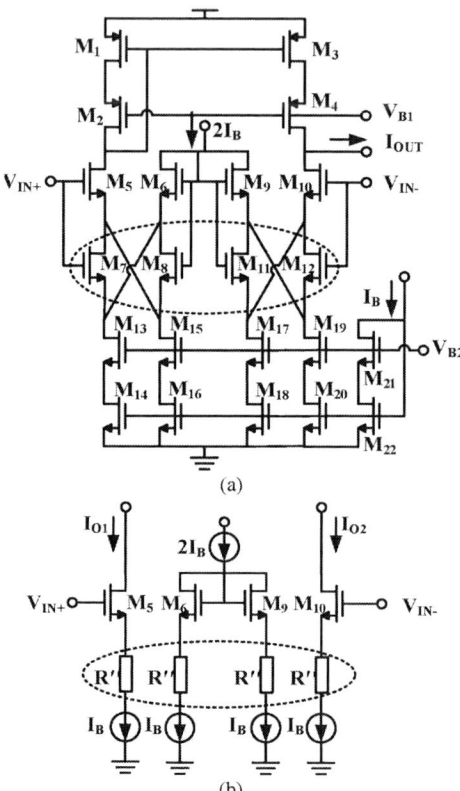

Fig. 3. (a) Circuit structure and (b) Equivalent circuit of OTA2

C. Structure of BPF

IR receiver chips impose stringent requirements on the selectivity and stability of BPF. Through precise circuit parameter design and tuning, the BPF can effectively pick up the target received signals. In this work, an OTA-C BPF was built using the two aforementioned OTAs. As shown in Fig. 4, the proposed OTA-C BPF consists of three OTAs, specifically two OTA1s and one OTA2, and two grounded capacitors, C_A and C_B, resulting in a relatively simple circuit structure compared to those in [1] and [2]. At node A, according to Kirchhoff's Current Law (KCL), the following relationship holds:

$$C_A s V_A(s) = -g_{m1} V_O(s) \tag{5}$$

And at node B, there has

979-8-3315-2209-4/25 $31.00 © 2025 IEEE

$$g_{m1}\left[V_A(s) - V_{IN}(s)\right] + g_{m2}\left[V_{IN}(s) - V_O(s)\right] = C_B s V_O(s) \quad (6)$$

here, g_{m1} and g_{m2} represent the transconductance of the OTA1 and OTA2, respectively. Based on Equ.5 and 6, the transfer function can be derived as follows:

$$H(s) = \frac{V_O(s)}{V_{IN}(s)} = \frac{C_A(g_{m2} - g_{m1})s}{C_A C_B s^2 + g_{m2} C_A s + g^2{}_{m1}} \quad (7)$$

Consequently, the expressions for the center frequency ω_0, quality factor Q, and passband gain H_0 can be obtained as follows:

$$\omega_0 = \frac{g_{m1}}{\sqrt{C_A C_B}}, \; Q = \frac{g_{m1}}{g_{m2}}\sqrt{\frac{C_B}{C_A}}, \; |H_0| = \left|1 - \frac{g_{m1}}{g_{m2}}\right| \quad (8)$$

According to Equ.8, by adjusting the values of g_{m1}, g_{m2}, C_A, and C_B, the center frequency can be tuned, while the quality factor and the passband gain can be stabilized and maximized. Specifically, by increasing g_{m1} or the product of C_A and C_B, while maintaining the ratios of g_{m1}/g_{m2} and C_A/C_B constant, and ensuring that g_{m1}/g_{m2} is greater than C_A/C_B, the circuit can simultaneously achieve a tunable center frequency, as well as a stable and high quality factor and passband gain. This configuration effectively meets the requirements for handling different communication frequencies. A simple first-order low-pass filter (LPF) is employed before BPF in order to reduce high-frequency interference.

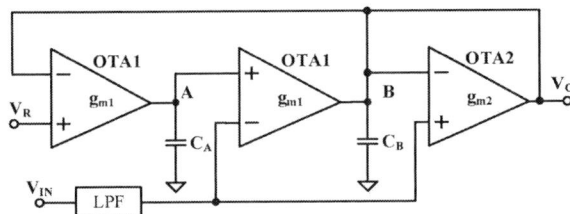

Fig. 4. Structure of the proposed OTA-C BFP

D. Frequency tuning

From Equ.1, the transconductance is proportional to the square root of the bias current. Considering the four target frequencies, 36 kHz, 38 kHz, 40 kHz, and 56.7 kHz, it is obvious that 36 kHz, 38 kHz, and 40 kHz are relatively close to each other, whereas 56.7 kHz differs to a certain extent. As a result, current-based tuning alone is adequate for the lower three frequencies. However, for 56.7 kHz, relying solely on bias current tuning would necessitate a large adjustment in OTA bias current to achieve the required center frequency shift. To address this, a combined scheme utilizing both bias current and capacitance tuning is employed for 56.7 kHz.

The tuning circuit is illustrated in Fig. 5. Inputs T_1–T_4 form a 4-bit digital control signal that corresponds to 16 possible control states. Of all 15 outputs, I_1–I_9 are assigned for bias current tuning, while C_1–C_6 are used for capacitance tuning, enabling coverage of the desired frequency range with sufficient tuning resolution. The current tuning unit is shown in Fig. 5(b). When the control signal YI_n is high, the current I_0 is bypassed to ground; when YI_n is low, I_0 contributes to the OTA's total bias current, thereby adjusting its transconductance. As depicted in Fig. 5(c), the capacitance tuning unit employs a switched MOS capacitor

M_C, operating in the deep linear region. When the control input YC_n transitions from high to low, the equivalent capacitance decreases accordingly.

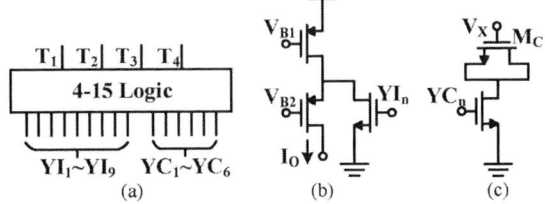

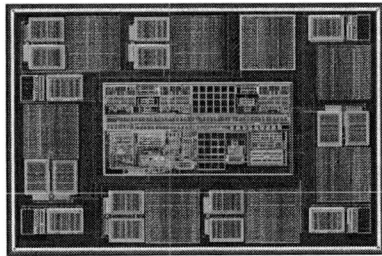

Fig. 5. Tuning circuit, (a) 4-15 logic; (b) Current tuning cell; (c) Capacitance tuning cell

III. SIMULATION RESULTS AND ANALYSIS

The OTA-C band-pass filter circuit and layout were designed based on a 0.18μm 3.3V 1P6M CMOS process. Fig. 6 shows the final chip layout. The total chip area including PADs and IOs is $632 \times 419\,\mu m^2$, and the core area is only $336 \times 190\,\mu m^2$.

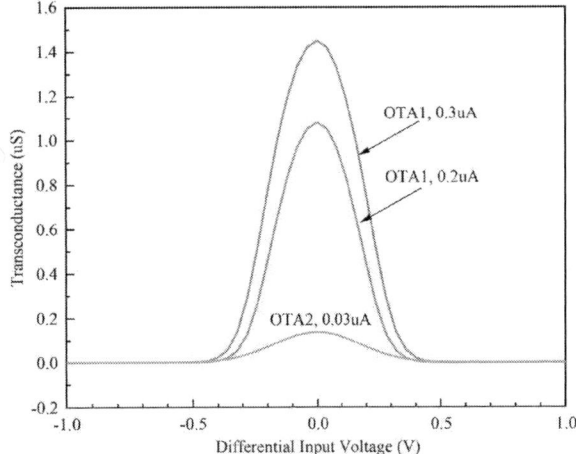

Fig. 6. Chip layout including PADs and IOs.

Based on the analysis in section II, the bias current of OTA2 is fixed at 0.03μA, and that of OTA1 is set to be adjustable from 0.2μA to 0.3μA. The size of C_A is fixed, and that of C_B is set to be adjustable from 10μm×10μm×24 to 10μm×10μm×42. The cutoff frequency of the low-pass filter is set at ~160 kHz. Based on the final chip layout, the post-layout simulations were conducted. Fig. 7 illustrates the simulated transconductance of OTA1 and OTA2 under TT process corner and 25°C temperature. For OTA1, when the bias current is between 0.2μA and 0.3μA, the corresponding peak transconductance ranges from 1.08μS to 1.45μS. For OTA2, the peak transconductance is fixed 0.14μS. It's obvious that the OTA2 has a much lower transconductance but a higher linearity that the OTA1.

Fig. 7. Tansconductance of OTA1 and OTA2

Fig. 8 shows the 16 frequency response curves at the TT process corner and 25°C temperature, corresponding to digital control input T_1-T_4 from 1111 to 0000. It can be observed that the center frequency ranges from 30.2 kHz to 67.5 kHz, the gain ranges from 17.3 dB to 15.0 dB, and the quality factor ranges from 20.1 to 10.4. The quality factor reflects bandwidth and selectivity. Compared with the low frequency end, the gain and quality factor at the high frequency end are lower. The above parameters are strongly affected by process and temperature changes. Considering a temperature range of -20°C to 80°C, the extreme cases occur at SS corner and 80°C, as well as FF corner and -20°C. At SS corner and 80°C, the center frequency decreases, while at FF corner and -20°C, the center frequency increases. Fig. 9 shows the changes under extreme conditions, and it can be seen that, at SS corner and 80°C, the center frequency ranges from 35.8 to 79.9 kHz, while at FF corner and -20°C, it's from 25.3 to 56.8 kHz. The results demonstrate that the proposed design fully meets the design requirements.

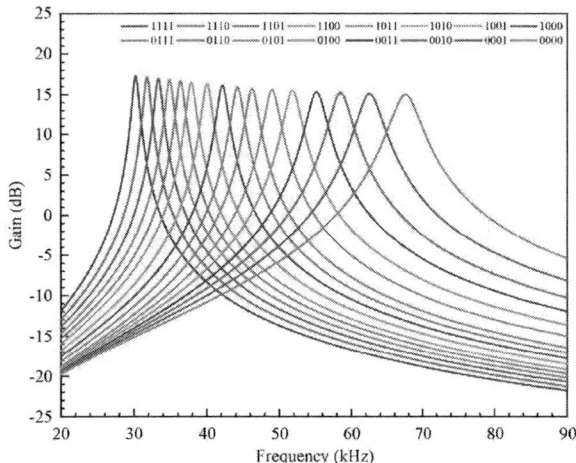

Fig. 8. Frequency response at TT process corner and 25°C temperature.

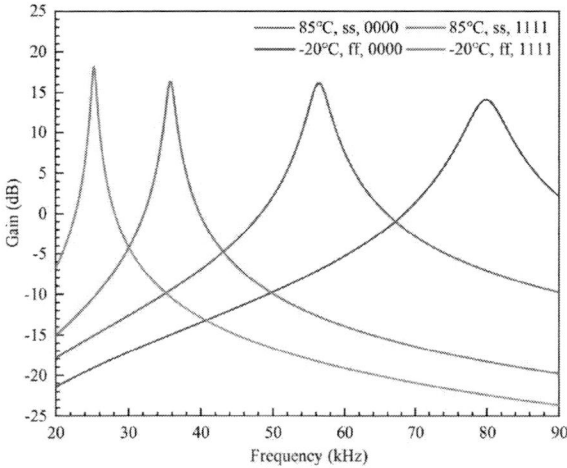

Fig. 9. Frequency response at extreme condtions and settings.

Table 1 presents a performance comparison between this work and other resent literatures. The results indicate that the presented work offers advantages such as wider frequency range, narrower bandwidth, and higher quality factor.

TABLE I.   COMPARISONS WITH RESENT PUBLICATIONS

Parameters	This work	[9]	[10]	[11]
Process (μm)	0.18	0.25	0.5	0.5
Supply voltage (V)	3.3	-	2.5~5.5	1.8~5.5
Frequency(KHz)	36~56.7	36	38	38
Gain (dB)	~16	23.8	18.3	16.5
Band width (kHz)	~4	5.9	7.5	8.2
Quality factor	~15	6.1	5.1	4.6

IV. CONCLUSION

This paper presents an OTA-C BPF for infrared receiver chips, consisting of 3 OTAs and 2 grounded capacitors, combined with a tuning circuit to achieve a tunable center frequency. The OTAs incorporate source degeneration and auxiliary differential pair structures to enhance linearity. The tuning circuit adjusts the center frequency by tuning the OTA bias current and the value of the grounded capacitors. The design was implemented in a 0.18μm 3.3V CMOS process. The simulation results indicate that, across different process corners and a temperature range of -20°C to 85°C, the presented OTA-C BPF can effectively cover the frequencies of 36kHz, 38kHz, 40kHz, and 56.7kHz, and obtain a bandwidth of ~4kHz and a quality factor of ~15, which demonstrates superior performance compared with other works.

REFERENCES

[1] M. Kumngern and K. Dejhan, "Capacitor-grounded electronically tunable voltage-mode OTA-C multifunction filter with three inputs and five outputs," 2012 10th IEEE International Conference on Semiconductor Electronics (ICSE), Kuala Lumpur, Malaysia, 2012.

[2] K. Baxevanaki and C. Psychalinos, "Second-Order Bandpass OTA-C Filter Designs for Extracting Waves from Electroencephalogram," 2019 8th International Conference on Modern Circuits and Systems Technologies (MOCAST), Thessaloniki, Greece, 2019.

[3] P. Monsurro, S. Pennisi, G. Scotti and A. Trifiletti, "Linearization Technique for Source-Degenerated CMOS Differential Transconductors," IEEE Transactions on Circuits and Systems II: Express Briefs, 54.10 (2007: 848-852.

[4] P. Pandey, J. Silva-Martinez and Xuemei Liu, "A CMOS 140-mW fourth-order continuous-time low-pass filter stabilized with a class AB common-mode feedback operating at 550 MHz," IEEE Transactions on Circuits and Systems I: Regular Papers, 53.4 (2006): 811-820..

[5] E. Lebel, A. Assi and M. Sawan, "Field programmable Gm-C array for wide frequency range bandpass filter applications," 2005 IEEE International Symposium on Circuits and Systems (ISCAS), Kobe, Japan, 2005.

[6] T. Kumar,, S. Kar and D. Boolchandani, "wide linear range CMOS OTA and its application in continuous-time filters," Analog Integrated Circuits and Signal Processing, 103. 2 (2020): 283–290.

[7] F. Krummenacher and N. Joehl, "A 4-MHz CMOS continuous-time filter with on-chip automatic tuning." IEEE Journal of Solid-State Circuits, 23.3 (1988): 750-758.

[8] R. Torrance, T. Viswanathan and J. Hanson, "CMOS voltage to current transducers," IEEE Transactions on Circuits and Systems, 32.11 (1985): 1097-1104.

[9] Y. Yang and Q. Feng, "Design of a Gm-C bandpass filter with Trimming circuit [in Chinese]," Electronic Devices and Materials, 38.11 (2019): 37-42.

[10] C. Gao, Noise Analysis and Performance Optimization on Infrared Receiver IC [in Chinese], M.S. thesis, Huaqiao University, Quanzhou, China, 2018.

[11] X. Xu, Design of a Low-Power Infrared Receiver Chip [in Chinese], M.S. thesis, Hunan University, Changsha, China, 2022.

Design of Dual Band Synchronous Rectifier With a Novel Matching Network

1st Guoqing Chen   2nd Guohua Liu   3rd Rui Zhang

Abstract—**This work proposes the use of a dual-path matching network to enhance the conversion efficiency of RF rectifiers in high-input-power wireless power transmission (WPT) systems. By combining L-type and T-type matching networks, the dual-path matching network allows the rectifier to achieve high efficiency in both frequency bands with a compact size and reduced design complexity. The results of the experimental layout testing demonstrate that, at input power of 40 dBm at 2.4 GHz and 5.8 GHz, the efficiency of RF-DC in the layout reaches 77.6% and 48.8%, respectively. This can drive the realization of remote wireless charging for Internet of Things (IoT) devices or communication equipment.**

Index Terms—**Dual band, GaN, synchronous rectifier, matching network.**

I. INTRODUCTION

With the development of wireless power transmission technology, a new and commercially valuable area of wireless charging has been extensively explored, including applications such as radio frequency identification (RFID), electric vehicles, and drone charging. Energy harvesting systems composed of diode rectifiers have been widely reported. However, the power capacity of diodes is limited and fails to meet the high-power requirements of applications like drone charging. Therefore, synchronous rectifiers based on GaN transistors have been proposed. These rectifiers can handle power levels above 10W, meeting the demands of current high-power applications. Wireless power transmission systems comprising transistor rectifiers and antennas could be used in space solar power stations or locations where wired power is not easily accessible. Such systems operate in the industrial, scientific, and medical (ISM) frequency bands to avoid interference with mobile communications. Multi-band rectifiers in ISM can support high-performance power transmission while reducing size and cost.

Although there have been many achievements in transistor-based multi-band synchronous rectifiers, there is still a lack of solutions for high-input-power, high-efficiency rectifiers using transistors. This paper proposes a novel dual-path matching network that merges circuits designed for different frequency bands, achieving efficient rectifier design under high-input-power conditions. This rectifier could significantly contribute to the advancement of wireless charging technology.

Guoqing Chen, Guohua Liu and Rui Zhang are with *School of Electronics and Information Engineering, Hangzhou Dianzi University*, Hangzhou, China. Corresponding author: (Guohua Liu, email: ghliu@hdu.edu.cn)

II. ANALYSIS OF THE PROPOSED RECTIFIER

According to the time-reversal duality principle, the input of the rectifier is the output of the power amplifier (PA), and the drain bias of the amplifier becomes the DC output of the rectifier. The proposed dual-band rectifier with a dual-band phase shifting network (DPSN) has been illustrated to control the gate-drain phase shift, ensuring high efficiency of the rectifier [1].

The schematic diagram of the proposed dual-path matching network is shown in Fig. 1, which includes two matching paths, making the phase-shift network that conducts at one frequency cut-off at another frequency, thereby achieving impedance matching at two different frequency points.

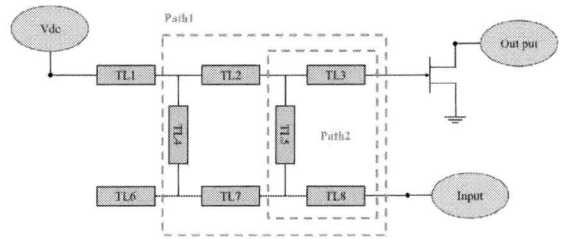

Fig. 1. Dual-path matching network.

Since the rectifier needs to maintain good rectification efficiency even at high input signal frequencies and power levels, after comparing multiple models of transistors currently available on the market, the CREE company's CG2H40010F transistor was ultimately chosen. To control the operating state of the transistor, a gate bias voltage of -5.3V was selected.

Typically, a simple L-type matching network with series microstrip lines and parallel open or short-circuited stubs can be utilized. For the series microstrip line, the input impedance can be described as:

$$Z_{in} = Z_0 \frac{Z_L + jZ_0 \tan\theta}{Z_0 + jZ_L \tan\theta} \quad (1)$$

where Z_L is the resistance and θ is the reactance.

The dual-path matching network can be viewed as combining an L-type matching network with a T-type matching network. This allows the design of the matching network to be transformed into the design of series and parallel microstrip lines, which can significantly improve the design efficiency and provide a standardized design approach.

979-8-3315-2209-4/25 $31.00 © 2025 IEEE   443

III. SIMULATION OF DUAL-BAND SYNCHRONOUS RECTIFIER

The layout of the rectifier is shown in Fig. 2. The simulation results from both the schematic and layout simulations are shown in Fig. 3.

In the schematic simulation results, at an input power of 40 dBm, the rectification efficiencies for the 2.4 GHz and 5.8 GHz input signals reach 80.9% and 64.4% when the load value is 50 ohms. In the layout simulation, when the load value is 50 ohms, at an input power of 40 dBm, the rectification efficiencies for the 2.4 GHz and 5.8 GHz input signals are 77.6% and 48.8% , respectively.

It can be observed that the 5.8 GHz signal shows significant attenuation in the layout simulation. This is because as the input signal frequency and power increase, energy is more likely to escape from the network. Therefore, there is considerable room for optimization in this circuit.

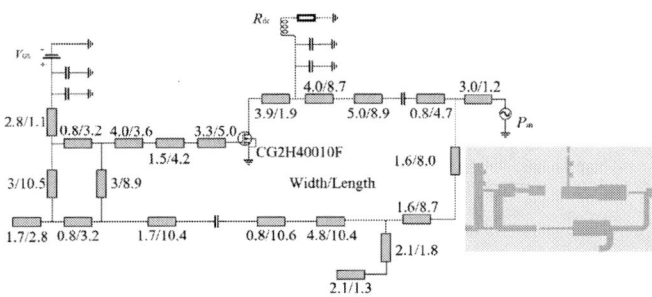

Fig. 2. Rectifier circuit schematic and layout.

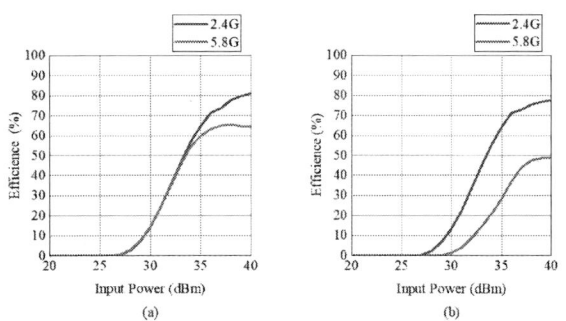

Fig. 3. Simulation results from (a) schematic and (b) layout at 2.4 GHz and 5.8 GHz.

Fig. 4 illustrates the relationship between the rectifier's rectification efficiency and the input signal frequency. The data in the figure indicate that the rectification efficiency peaks at 2.4 GHz and 5.8 GHz, demonstrating the dual-band rectification characteristics. This characteristic is advantageous for the rectifier to operate in work environments that require multi-band input.

Table I lists the performance comparison of some related rectifiers. We can see that the proposed rectifier exhibits a higher input power capacity compared to the rectifiers using diodes. Compared with a transistor-based rectifier, this work has the highest frequency ratio.

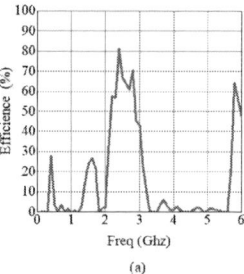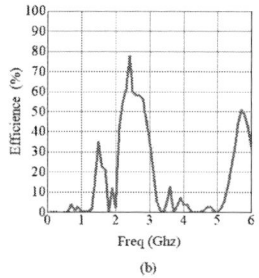

Fig. 4. Effect of frequency in (a) schematic and (b) layout.

TABLE I
PERFORMANCE COMPARISON WITH RECTIFIERS

Ref.	Device	Freq(GHz)	Pin(dBm)	Efficiency(%)
[2]	schottky diode	2.45/5.8	10	66.8/51.5
[3]	schottky diode	3.5/5.8	0	45.6/33
[4]	schottky diode	2.45/5.8	0	57.6/33.6
[5]	GaN HEMT	1.9/2.4	40	75/76
This work	GaN HEMT	2.4/5.8	40	77.6/48.8

IV. CONCLUSION

This paper proposes a dual-path matching network designed to control input signals within two operating bands, achieving good matching performance even at high input power levels, thus improving rectification efficiency. The test results show that, at frequencies of 2.4 GHz and 5.8 GHz, the maximum rectification efficiencies of the layout reach 77.6% and 48.8%, respectively. The proposed rectifier still has potential for optimization and provides ideas and references for subsequent dual-band rectifier designs.

REFERENCES

[1] M. A. Hoque, S. N. Ali, M. A. Mokri, S. Gopal, M. Chahardori and D. Heo, "A Highly Efficient Dual-band Harmonic-tuned GaN RF Synchronous Rectifier with Integrated Coupler and Phase Shifter," 2019 IEEE MTT-S International Microwave Symposium (IMS), Boston, MA, USA, 2019, pp. 1320-1323, doi: 10.1109

[2] D. Wang and R. Negra, "Design of a dual-band rectifier for wireless power transmission," 2013 IEEE Wireless Power Transfer (WPT), Perugia, Italy, 2013, pp. 127-130, doi: 10.1109

[3] D. Surender et al., "Analysis of Facet-Loaded Rectangular DR-Rectenna Designs for Multisource RF Energy-Harvesting Applications," in IEEE Transactions on Antennas and Propagation, vol. 71, no. 2, pp. 1273-1284, Feb. 2023, doi: 10.1109

[4] X. -B. Huang, J. -J. Wang, X. -Y. Wu and M. -X. Liu, "A dual-band rectifier for low-power Wireless Power Transmission system," 2015 Asia-Pacific Microwave Conference (APMC), Nanjing, China, 2015, pp. 1-3, doi: 10.1109

[5] Z. Zhang, V. Fusco, Z. Cheng, N. Buchanan, and C. Gu, "A transistor-based dual-band high-efficiency rectifier with dual-polarity modes,"IEEE Microw. Wireless Compon. Lett., vol. 32, no. 2, pp. 169–172,Feb. 2022.

An Optimization Design for Planar Microwave Sensors Based on Slow-Wave Transmission Line

Huayi Wu

Guohua Liu

Abstract—This paper proposes a novel optimization design for planar microwave sensors. The design applies slow-wave transmission lines to enhance the internal electric field strength, thus improving sensor performance. The optimization is validated through theoretical analysis, simulations, and experimental testing. In the simulations, the performance of a CSRR based sensor using slow-wave transmission lines is compared to that of conventional transmission lines. The results show a 6.07% increase in sensitivity, a 20.5% improvement in resonance depth, and a 19.2% rise in the quality factor (Q-factor). Finally, the optimized sensor prototype was tested, showing a max error of 1.1%.

Index Terms—Planar microwave sensor, slow-wave transmission line, optimization design, dielectric property measurement.

I. INTRODUCTION

Planar microwave sensors are widely used for measuring the dielectric properties[1-3]. The sensing region is typically a type of resonator, such as the complementary curved ring resonator (CCRR)[1], split ring resonator (SRR)[2], or complementary split ring resonator (CSRR)[3]. These resonators cannot be directly connected to the pre-stage or post-stage circuits. Therefore, microstrip lines are designed within the sensor to connect or couple the resonators. When materials under test (MUTs) with different dielectric properties are placed in the sensing region, the resonant frequency of the resonators within the region changes accordingly. This is the operating theory of planar microwave sensors.

Currently, researchers are optimizing the structure of resonators to strengthen the distributed electric field, thereby improving the performance of sensors, such as resonant depth and sensitivity. In [1], the researchers designed a CCRR structure in the center region of a quarter-mode substrate integrated waveguide, resulting in a high electric field distribution, which in turn achieved higher sensitivity. In [2], the researchers studied the effect of the number of rings in the SRR structure on the electric field distribution and selected the structure with the strongest electric field intensity. In this study, a new optimization scheme is proposed, where the microstrip line is optimized into a slow-wave transmission line.

Slow-wave transmission lines (SWTLs) feature a periodic structure, also referred to as slow-wave structures [4]. This periodicity alters the impedance characteristics of the transmission line. Among these effects, the phase velocity decreases due to the slow-wave structure, leading to a slower propagation of electromagnetic waves, which is known as the slow-wave effect. In planar printed circuits, SWTLs are widely used in power divider design [5-7]. Researchers primarily utilize the controllable impedance characteristics of SWTLs to achieve miniaturization and suppress higher-order harmonics in power dividers. Further, in this study, the electric field distribution characteristics of SWTLs are investigated. The differences between SWTLs and conventional transmission lines are used as the theoretical basis for sensor optimization.

In summary, an optimization design based on SWTL is proposed in this paper. And its theoretical basis is presented in Section II. Then, it's applied in a CSRR-based sensors. The comparison before and after optimization is in Section III. Finally, experimental validation is provided in Section IV.

II. THEORETICAL ANALYSIS

A. Theory of Sensors

The operating theory of the sensor is based on the perturbation method. When the MUT is loaded onto the resonator, the field distribution of the resonator changes, which in turn affects its resonant frequency. The relationship between the dielectric properties of the MUT and the resonant frequency is

$$\frac{\Delta f_r}{f_r} = \frac{\int_V \left(\Delta \varepsilon E_0 E_1 + \Delta \mu H_0 H_1 \right) dV}{\int_V \left(\varepsilon_0 |E_0|^2 + \mu_0 |H_0|^2 \right) dV} \quad (1)$$

where f_r denotes the resonant frequency of the unloaded resonator, while Δf_r represents its frequency shift after loading the MUT. E_0 and H_0 correspond to the electric and magnetic field distributions of the unloaded resonator, whereas E_1 and H_1 describe these distributions after the MUT is introduced. ε_0 and μ_0 represent the permittivity and permeability of air, respectively. $\Delta \varepsilon$ and $\Delta \mu$ denote the variations in the electromagnetic properties of the MUT relative to air.

Since the perturbation is small, E_1 and H_1 can be approximated as equal to E_0 and H_0, respectively. Thus, the simplified expression is

$$\frac{\Delta f_r}{f_r} = \frac{\int_V \left(\Delta \varepsilon |E_0|^2 + \Delta \mu |H_0|^2 \right) dV}{\int_V \left(\varepsilon_0 |E_0|^2 + \mu_0 |H_0|^2 \right) dV} \quad (2)$$

According to Eq. (2), optimizing the sensor's performance requires maximizing the intensity of the electromagnetic field

---

Huayi Wu and Guohua Liu are all with *the School of Electronics and Information Engineering, Hangzhou Dianzi University*, Hangzhou, China. The work was supported by National College Students' Innovative Entrepreneurial Training Plan Program (Grant Number: 202410336076). Corresponding author: (Guohua Liu, email: ghliu@hdu.edu.cn)

979-8-3315-2209-4/25 $31.00 © 2025 IEEE

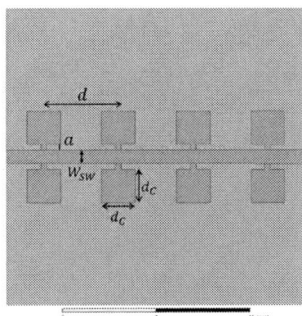

Fig. 1. Structure of SWTL.

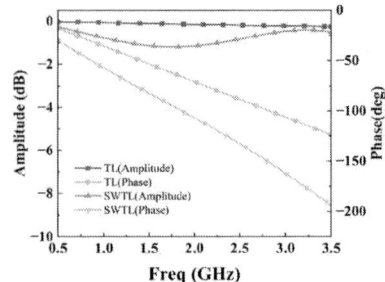

Fig. 2. SWTL and TL simulation results.

distribution and ensuring that the MUT is positioned within the resonant electromagnetic field.

B. Theory of SWTL

For transmission lines, if its equivalent inductance and capacitance increase proportionally, its equivalent impedance remains unchanged, while equivalent phase velocity decreases accordingly. This phenomenon is known as the slow-wave effect. To quantify the extent of the slow-wave effect, the slow-wave ratio (SWR) is defined as

$$SWR = \frac{v_p}{v_{p0}} \qquad (3)$$

where v_{p0} and v_p represent the phase velocities of the conventional transmission line and the SWTL, respectively.

Electromagnetic waves of the same frequency have different wavelengths in transmission lines with distinct SWR. Thus, the same energy is distributed over different ranges. Then

$$\frac{E_{SW}}{E_0} = \sqrt{\frac{\lambda_0}{\lambda_{SW}}} = \sqrt{\frac{v_p/f}{v_{p0}/f}} = \frac{1}{\sqrt{SWR}} \qquad (4)$$

In summary, for electromagnetic waves of the same frequency, a lower phase velocity in the transmission line leads to a shorter wavelength. As a result, the generated electromagnetic field becomes stronger. This, in turn, enhances the excitation of the resonant unit.

III. SIMULATION OF SWTL AND SENSOR

A. Design and Simulation of SWTL

A transmission line (TL) with an impedance of 50 Ω is fabricated on a dielectric substrate made of FR4 with a thickness of 0.8 mm, using a microstrip line with a width of 1.45 mm and a copper thickness of 0.035 mm. This structure is used as a reference. A SWTL with the same impedance (as shown in Fig.1) is constructed by periodically adding parallel plate capacitors to a thinner microstrip line. The width of the thinner microstrip line W_{SW} is 0.75 mm. Each parallel plate capacitor is a square with a width d_C of 1.8 mm. The distance between the capacitors, d, is 4 mm. And the distance from the capacitors to the microstrip line, a, is 0.3 mm.

Electromagnetic simulations of both TL and SWTL were performed using Ansys Electronics Desktop, yielding the amplitude and phase shift values of S_{21} for each, as shown

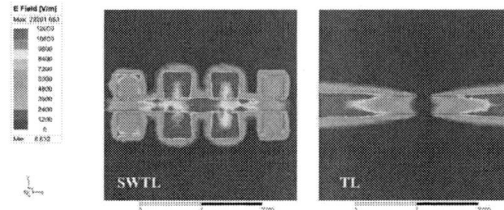

Fig. 3. Comparison of electric field distribution simulation between SWTL and TL.

TABLE I
SIZE PARAMETERS OF CSRR.

parameter	W_r	r_1	d_{S1}	r_2	d_{S2}
Val. (mm)	0.4	4.4	1.5	3	1

in Fig.2. Both lines have the same length of 16 mm, but their phase delays at the same frequency differ. This is one manifestation of the slow-wave effect. Based on the relationship between phase delay and phase velocity, the SWTL's SWR is 0.64.

From the perspective of the equivalent circuit, the equivalent inductance L_{d0} of the TL is 1.10 nH. Assuming that the inductance of the SWTL is only considered from the microstrip line, the equivalent inductance L_{ds} is 1.61 nH. Therefore, the SWTL's SWR is 0.68. It indicates that the theoretical calculation results are consistent with the simulation results.

The simulated electric field distributions of the two structures are shown in Fig. 3. This is consistent with the conclusion in Section II. The electric field distribution within the SWTL is more concentrated and stronger.

B. Simulation Verification of the Sensor's Optimization

The base layer of both transmission line structures is etched with a CSRR structure, as shown in Fig. 4. The etched CSRR structure consists of two concentric rings. Size parameters are shown in Table I. SWTL and CSRR form the SW-CSRR structure. A MUT with dimensions 1 mm × 1 mm × 0.762 mm is placed on the surface of the CSRR.

Simulations of CSRR and SW-CSRR with MUTs having different dielectric constants were performed, as shown in Fig. 5. Sensitivity, resonance depth, and quality factor (Q-factor) were used to evaluate the sensor's performance. Sensitivity reflects the degree to which the sensor changes in response to variations in the dielectric properties of the MUT. This sensor

979-8-3315-2209-4/25 $31.00 © 2025 IEEE

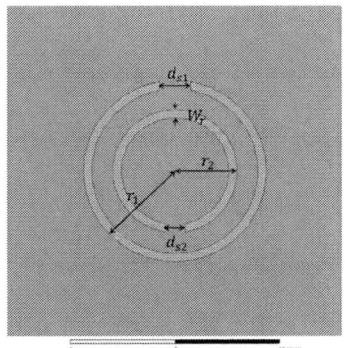

Fig. 4. Structure of CSRR.

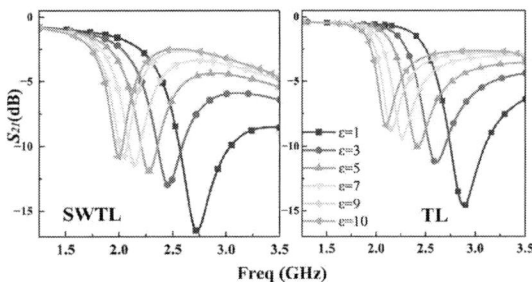

Fig. 5. Simulation comparison of SW-CSRR and CSRR.

detects the dielectric properties of the MUT through with the resonant frequency shifts. The sensitivity is defined as [1]

$$S = \frac{\Delta f_r}{f_r} \times \frac{1}{\Delta \varepsilon} \qquad (5)$$

A higher sensitivity indicates stronger resolution capability of the sensor. However, this leads to a broader operating bandwidth. The resonance depth is the magnitude of S_{21} when the circuit reaches resonance. The Q-factor is the ratio of the resonant frequency to the -3 dB bandwidth. The resonance depth and Q-factor determine how easily the resonance point can be detected. They also affect the sensor's noise resistance performance.

The sensors was optimized from CSRR to SW-CSRR. As a result, the average relative increases in sensitivity, resonance depth and Q-factor were 6.07%, 20.5%, 19.2%.

IV. EXPERIMENTAL VALIDATION

In order to verify the optimization scheme, the SW-CSRR was fabricated and tested, with the test photos shown in Fig. 6. The dielectric constants of the samples are as follows: Rogers AD255C (ε=2.55), Taconic RF35 (ε=3.5), Rogers 4350 (ε=3.66), and Taconic RF60A (ε=6.15). To reduce the occasional factor, two sensor prototypes were made before the test. After testing, the measurement results of the two sensors are consistent. This also proves the robustness of the design.

The measurement results are shown in Fig. 7. The max error between the resonant frequency f_r obtained from the physical test and the simulation results is 1.1%. The physical

TABLE II
COMPARISON OF PERFORMANCE ENHANCEMENTS AMONG DIFFERENT OPTIMIZATION

Item	Str.	Sens. Enhanc.	Depth Enhanc.	Q-factor Enhanc.
[8]	IDC-CSRR	26.9%	14.3%	not given
[9]	IDS-CSRR	50.1%	-7.29%	not given
This work	SW-CSRR	6.07%	20.5%	19.20%

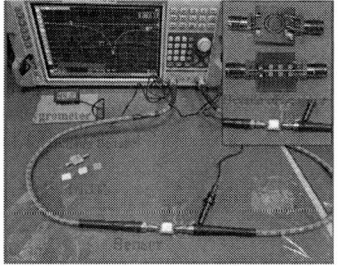

Fig. 6. SW-CSRR sensor experimental setup.

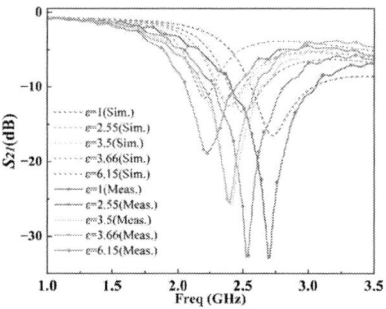

Fig. 7. Comparison of sensor measurement and simulation.

test verified the correctness of the simulation, indicating that the optimization scheme is feasible.

The previous optimization of CSRR is to combine with interdigital capacitor [8] [9]. This optimization improves the sensitivity significantly. However, the enhancement of resonance depth is lower than the proposed optimization, and even negative enhancement [9]. The proposed scheme is compared with related work in Table II. For the design of sensors, proposed scheme can be used together with existing schemes.

V. CONCLUSION

SW-CSRR is proposed in this paper with a 20.5% increase in resonant depth, a 19.2% increase in Q-factor and a 6.07% increase in sensitivity. The resonant frequency deviation between simulation and real object is less than 1.1%. This approach provides a new perspective for designing sensors with higher sensitivity and Q-factor. In applications such as material testing and evaluation, existing sensors suffer from insufficient Q-factor. And the findings of this study may contribute to developing an improved solution.

REFERENCES

[1] T. Qi, G. Liu, J. Yu and S. Jiang, "A Quarter-Mode Substrate Integrated Waveguide Microwave Sensor Loaded With CCRR for Solid Material Measurement," IEEE Sensors J., vol. 23, no. 18, pp. 21105-21112, 15 Sept.15, 2023.

[2] Y. Gong, G. Liu, S. Jiang, J. Yu and T. Qi, "A DGS-CPW Microwave Sensor Loaded With SRR for Solid Material Measurement," IEEE Trans. Instrum. Meas., vol. 73, pp. 1-8, 2024.

[3] A. Buragohain, A. T. T. Mostako and G. S. Das, "Low-Cost CSRR Based Sensor for Determination of Dielectric Constant of Liquid Samples," IEEE Sensors J., vol. 21, no. 24, pp. 27450-27457, 15 Dec.15, 2021.

[4] A. Ebrahimi et al., "Highly Sensitive Phase-Variation Dielectric Constant Sensor Based on a Capacitively-Loaded Slow-Wave Transmission Line," IEEE Trans. Circuits Syst. I, vol. 68, no. 7, pp. 2787-2799, July 2021.

[5] B. Wu, Z. Sun, X. Wang, Z. Ma and C.-P. Chen, "A Reconfigurable Wilkinson Power Divider With Flexible Tuning Range Configuration," IEEE Trans. Circuits Syst. II Exp. Briefs, vol. 67, no. 7, pp. 1219-1223, July 2020.

[6] K. Kim and C. Nguyen, "An Ultra-Wideband Low-Loss Millimeter-Wave Slow-Wave Wilkinson Power Divider on 0.18 μm SiGe BiCMOS Process," IEEE Micro. Wireless Compon. Lett., vol. 25, no. 5, pp. 331-333, May 2015.

[7] Y. Zhou et al., "Slow-Wave Half-Mode Substrate Integrated Waveguide 3-dB Wilkinson Power Divider/Combiner Incorporating Nonperiodic Patterning," IEEE Micro. Wireless Compon. Lett., vol. 28, no. 9, pp. 765-767, Sept. 2018.

[8] A. Paul, A. Mazumder and M. Kar, "Design of Optimized IDC Loaded CSRR Sensor for Detection of Adulteration in Edible Oils," in Proc. 2024 IEEE Silchar Subsect. Conf. (SILCON 2024), Agartala, India, 2024, pp. 1-5.

[9] S. A. Enche Ab Rahim, N. F. N. Zulkifli, M. F. Sapuri, N. E. Abd Rashid, Z. I. Khan and N. A. Zakaria, "Sensitivity Enhancement of CSRR Sensor Using Interdigital Structure in Detecting Ammoniacal Nitrogen for Water Quality Applications," in Proc. 2022 IEEE Symp. Wirel. Technol. Appl. (ISWTA), Kuala Lumpur, Malaysia, 2022, pp. 53-56

A High-Gain and Low Profile ESPAR Antenna Base on Alford Loop Antenna

Jiayuan Fan
Hangzhou Dianzi University
Hangzhou, China
State Key Laboratoy of Millimete
Waves
Nanjing, China
2076058311@qq.com

Chao Gu
Queen's University Belfast
Belfast, United Kingdom
chao.gu@qub.ac.uk

Zhiwei Zhang
Hangzhou Dianzi University
Hangzhou, China
State Key Laboratoy of Millimete
Waves
Nanjing, China
2361051379@qq.com

Dengfa Zhou
Hangzhou Dianzi University
Hangzhou, China
State Key Laboratoy of Millimete
Nanjing, China
zdf1372@qq.com

Abstract—This paper designs a low-profile, high-gain, electronically steerable passive array radiator antenna(ESPAR). The antenna is composed of a single dielectric substrate, with complementary Alford loop antennas serving as the radiating elements on both the top and bottom surfaces of the substrate. On the bottom surface, parasitic elements and PIN diode biasing circuits are placed around the radiating elements, and microstrip lines are located near the parasitic elements, with inductive elements loaded in the middle to provide a certain impedance. The design employs coaxial feeding technology, and the main lobe radiation direction is changed by the forward and reverse biasing of the PIN diodes. The antenna has seven radiation modes, enabling 360° beam switching. The overall height is 1.6mm, the center frequency at 2.472GHz, and the gain reaches 6.93dBi.

Keywords—Low-profile, High-gain, 360° beam scanning, ESPAR antenna

I. INTRODUCTION

In recent years, with the rapid development of wireless communication technology, the requirements for antenna performance have become increasingly high. First, the antenna needs higher gain to transfer enough energy; further, in the practical application of UAV and Internet of Things[1], the antenna should be flexible by concentrating the beam and switching the antenna.

Although traditional phased array antennas can achieve beam steering, their complex feeding structures and high costs limit their promotion in practical applications. ESPAR antennas, which are simple and easy to control, have become increasingly important in modern communication systems. They control the beam direction by loading adjustable reactance to influence the phase difference between parasitic elements and have beam scanning capabilities[2]. This feature makes them widely used in mobile communication systems such as those in vehicles and aircraft. ESPAR antennas not only have a simple structure and are easy to control but also can effectively reduce system complexity and costs, thus attracting extensive attention in both academia and industry.

The seven-element ESPAR antennas proposed in references [3] to [6] generate omnidirectional radiation by placing a monopole at the center as the radiating element and evenly distributing six monopoles around it as parasitic elements. They add varactor diodes or switch circuits to the parasitic elements to achieve beam scanning. However, due to the monopoles being approximately one-quarter of the working frequency wavelength in length, the size is too large, and a certain structure is needed to solve the problem of beam direction deviation from the horizontal direction. Patch antennas are favored by industry personnel due to their ease of processing and integration. The patch ESPAR antenna adopted in [7] achieves low profile and beam scanning effects.

In recent years, research on ESPAR antennas has been continuously deepened, and various new structures and optimization design methods have emerged. For example, by integrating the microstrip patch antenna structure with the ESPAR antenna structure, antennas with higher gain and larger scanning angles can be designed. By introducing slot coupling technology, the radiation efficiency and scanning performance of the antenna can be further improved. In addition, the application of ESPAR antennas in signal processing fields has also attracted much attention, such as using ESPAR antennas for direction-of-arrival (DOA) estimation[8,9].

This paper aims to deeply explore the principles, structures, optimization design methods, and applications of ESPAR antennas in wireless communication systems. Through a comprehensive analysis of existing literature and research results, this paper proposes a new ESPAR antenna structure and conducts performance tests on it.

II. STRUCTURE OF ANTENNA

The antenna structure proposed in this paper is shown in Fig. 1. The antenna adopts coaxial feeding technology and uses a Rogers 5880 dielectric substrate with a size of 140mm*140mm*1.575mm (relative permittivity of 2.2 and loss tangent of 0.0009). Alford loop antennas with complementary structures are placed on the upper and lower surfaces as the radiating elements. Around the radiating elements on the lower surface, a circular patch antenna composed of 12 arc-shaped patches is placed as the parasitic element. Each two arc-shaped patches form a group, and a PIN diode (BAR65-02V) is loaded in the middle. The antenna beam is switched by the forward and reverse bias of the PIN diode. Secondly, 12 microstrip lines are placed around the parasitic element and connected to the parasitic element by welding 100nh inductance to provide an

979-8-3315-2209-4/25 $31.00 © 2025 IEEE

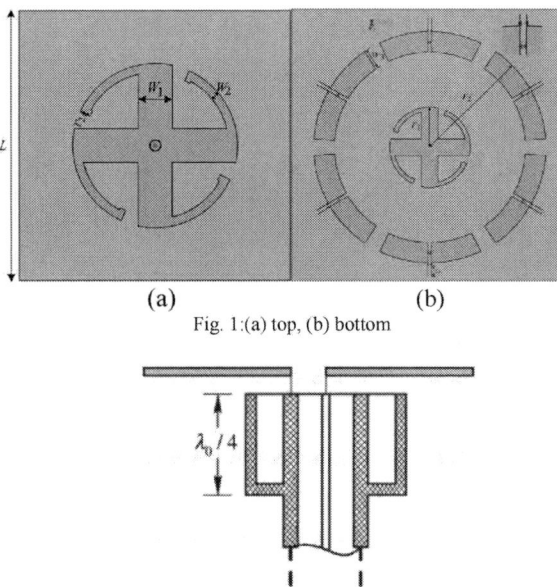

(a)   (b)

Fig. 1:(a) top, (b) bottom

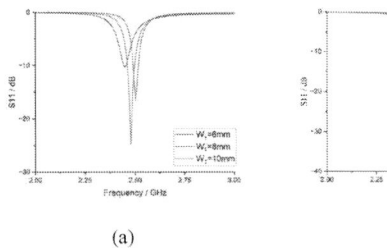

Fig. 2: The structure of Balun

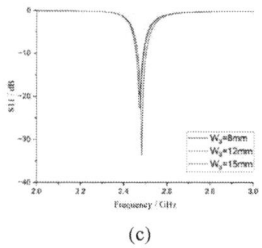

(c)

Fig. 3:(a) The relation of W_1 and S11, (b) The relation of W_2 and S11, (c)The relation of W_3 and S11

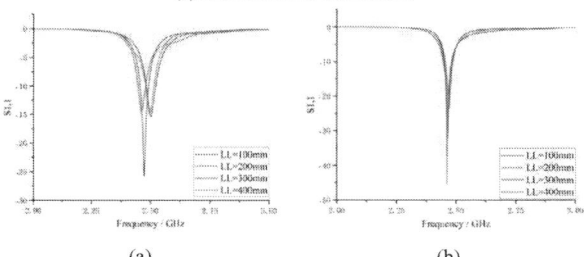

(a)   (b)

Fig. 4: The influence of Balun (a) without Balun, (b) with Balun

TABLE I.   THE PARAMETERS OF ANTENNA (UNITS: MM)

W_1	W_2	W_3	r_1	r_2
8	2	12	19.85	2.5
r_3	L	LL	W_4	
58.56	140	300	0.2	

impedance of $X_L \approx j1551\ \Omega$ at the operating frequency of 2.472GHz. As shown in Fig. 2, a copper tube balun structure with a length of approximately one quarter of the wavelength (29.2mm), an outer diameter of 2.5mm and an inner diameter of 2mm is desig ned in the coaxial feeding part to reduce the influence of the change in the length of the coaxial line on the antenna's S11 during the processing and measurement.

III. THE SIMULATION OF ANTENNA

In this paper, simulation software cst2022 is used to simulate the proposed antenna. It can be seen from Fig. 3(a) that the influence of the width W1 of the cruciform structure in the radiation unit on the antenna reflection coefficient. It can be seen from Fig. 3(b) that the influence of the width W_2 of the arc patch on the antenna reflection coefficient. It can be seen from Fig. 3(c) that the influence of the width W_3 of the parasitic element arc patch on the antenna radiation coefficient. The red line in the figure is the reflection coefficient of the antenna after the optimization of various parameters. It can be seen from Fig. 4 that the antenna can effectively reduce the influence of the change of coaxial line length LL on S11 when loading the Balun structure. Table I shows the final parameters of the antenna.

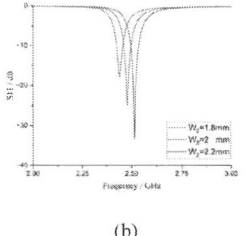

(a)   (b)

The positive and negative bias states of 6 pin diodes are expressed as "1" and "0". "1": the pin diode is positive bias, which is equivalent to a resistance of about 0.2 ohm; "0": The pin diode is negative bias, it is equivalent to a small capacitor at this time. The parasitic unit loaded with the capacitor will play the effect of a guide to achieve the deflection of the antenna main lobe radiation direction. We machined and tested the antenna gain and direction pattern when using a single pin diode forward bias and multiple pin diodes forward bias respectively. Fig. 5 shows the omnidirectional pattern of the antenna when the pin diode state is "000000". The maximum gain is 1.12dBi and S11 has a minimum of -13dB. Fig. 6 shows the six directional diagrams of the antenna when the single pin diode state of the antenna is "1", and the combinations are 100000, 010000, 001000, 000100, 000010 and 000001. The maximum gain is 4.9dBi and S11 has a minimum of -41dB.. Fig. 7 shows the six directional diagrams of the antenna when the two pin diodes of the antenna are in the state of "1", and the combinations are 110000, 011000, 001100, 000110, 000011 and 100001. The maximum gain is 6.93dBi and S11 has a minimum of -37dB. When controlling three or more pin diodes with a state of "1", excessive beam width and gain drop, which are unnecessary.

979-8-3315-2209-4/25 $31.00 © 2025 IEEE   450

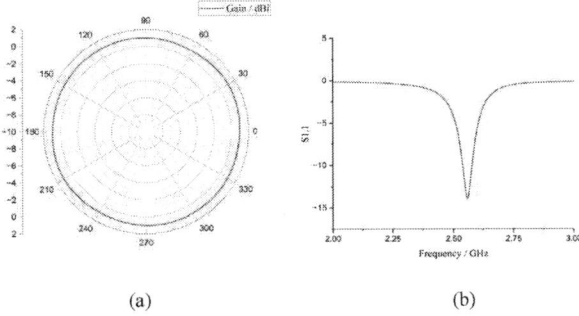

(a)   (b)

Fig. 5:Condition 1(a) Field direction diagram, (b) S11

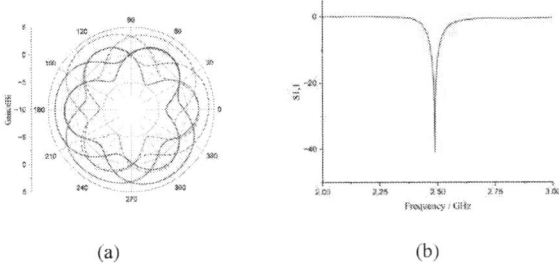

(a)   (b)

Fig. 6:Condition 2(a) Directional Beampattern, (b) S11

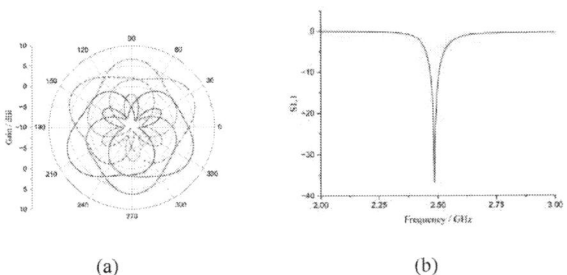

(a)   (b)

Fig. 7:Condition 3(a) Directional Beampattern, (b) S11

Comparing this work's antenna to those of the same type[4], [5], [7] and [9]. [5] and [7] exhibit a much wider fraction bandwidth(FBW), but their size are much larger than the design, and their gain are lower than the design. [4] has a gain of 7.63dBi and a 19% fraction bandwidth, but it uses 13 units along with larger sizes and higher profiles.[9] has a gain of 8.2dBi, while it can only scan for 75°.

TABLE II.   ANTENNA PERFORMANCE COMPARISON

Work	Num. Of radiating elements	Gain (dBi)	FBW (%)	Steer Range (deg)	Dimension (mm*mm *mm)
[4]	13	7.63	19	±180	Π*100²*28
[5]	7	5.1	NONE	±180	Π*30²*51
[7]	5	3.95	3.7	±180	200*200*16.6
[9]	9	8.2	1.8	75	50*70*1.858
This work	8	6.93	1.3	±180	140*140*1.6

IV. CONCLUSIONS

In this paper, a low profile, high gain ESPAR antenna can realize 360° beam scanning. The antenna adopts coaxial feeding technology to control the different characteristics of the pin diode on the parasitic unit, so that the current phase on the parasitic unit is different, and realize the deflection of the main lobe radiation direction of the antenna direction diagram. The actual test results show that the antenna achieves the best effect in controlling the positive bias of two pin diodes. At this time, the overall height of the antenna is 1.6mm, -10dB bandwidth is 32 MHz, 2.472GHz in the central frequency point, and the highest gain of 6.93 dBi in the Theta = 90 direction.

ACKNOWLEDGMENT

This work was partly supported by the Open Foundation of the State Key Laboratory of Millimeter-Waves (Grant K202540) and partly by Zhejiang Provincial Natural Science Foundation of China (Grant LMS25F010005).

REFERENCES

[1] P. Sotres, J. R. Santana, L. Sánchez, J. Lanza, and L. Muñoz,"Practical lessons from the deployment and management of a smart city Internet-of-Things infrastructure: The Smart Santander testbed case," IEEE Access, vol. 5, pp. 14309-14322, 2017.

[2] K. Gyoda and T. Ohira,"Design of electronically steerable passive array radiator (ESPAR) antennas," in Proc. IEEE Antennas Propag. Symp., Salt Lake City, UT, USA, vol. 2, Jul. 2000, pp. 922-925.

[3] H. Kawakami and T. Ohira, "Electrically steerable passive array radiator (ESPAR) antennas," in IEEE Antennas and Propagation Magazine, vol. 47, no. 2, pp. 43-50, April 2005.

[4] M. M. Rahman and H. -G. Ryu, "ESPAR Antenna with Double Ring Placement of Parasitic Elements," 2022 International Workshop on Antenna Technology (iWAT), Dublin, Ireland, 2022, pp. 216-219.

[5] Junwei Lu, D. Ireland and R. Schlub, "Dielectric embedded ESPAR (DE-ESPAR) antenna array for wireless communications," in IEEE Transactions on Antennas and Propagation, vol. 53, no. 8, pp. 2437-2443, Aug. 2005.

[6] T. Hassan, A. Kausar, H. Umair and M. A. Anis, "Gain optimization of a seven element ESPAR Antenna using Quasi-Newton method," 2011 IEEE International Conference on Microwave Technology & Computational Electromagnetics, Beijing, China, 2011, pp. 293-296.

[7] Positano F, Lizzi L, Staraj R. Design and on-field test of ESPAR antenna for UAV-based long-range IoT applications[J]. Frontiers in Antennas and Propagation, 2024, 2: 1429710.

[8] M. Groth, M. Rzymowski, K. Nyka and L. Kulas, "ESPAR Antenna-Based WSN Node With DoA Estimation Capability," in IEEE Access, vol. 8, pp. 91435-91447, 2020.

[9] R. De Marco, E. Arnieri, G. Amendola and L. Boccia, "Microstrip ESPAR Antenna With Conical Beam Scanning," in IEEE Antennas and Wireless Propagation Letters, vol. 23, no. 1, pp. 174-178, Jan. 2024.

[10] L. Kulas, "Simple 2-D Direction-of-Arrival Estimation Using an ESPAR Antenna," in IEEE Antennas and Wireless Propagation Letters, vol. 16, pp. 2513-2516, 2017.

[11] Kulas, Łukasz, "Direction-of-Arrival Estimation Using an ESPAR Antenna with Simplified Beam Steering," 47th European Microwave Conference 2017, pp. 1-4.

[12] L. Santamaria, F. Ferrero, R. Staraj and L. Lizzi, "Slot-Based Pattern Reconfigurable ESPAR Antenna for IoT Applications," in IEEE Transactions on Antennas and Propagation, vol. 69, no. 7, pp. 3635-3644, July 2021

A New Scalable De-embedding Method With a Distributed Pad Model and Lumped Compensation for Pad-Line Discontinuities in RF On-Wafer Characterization

Hongfei Su   Bharatha Kumar Thangarasu   Nagarajan Mahalingam   Kaixue Ma

Fanyi Meng   Zhenghao Lu   Kiat Seng Yeo

Abstract—This paper presents a scalable cascaded de-embedding technique for precise RF device characterization. The method utilizes the LINE and LINE-PAD-LINE structures to extract pad parasitics without the inaccuracies introduced by lumped assumptions. A T-model is then constructed by combining the extracted pad parasitic model with the LINE and 2LINE structures to compensate for the discontinuities between the interconnection and the pad. By combining transmission line theory with the cascade-configuration concept to characterize the line properties, the method can achieve scalability. The proposed de-embedding method has been validated on the RF parameters of inductor device up to 120 GHz, with an average S-parameter error below 1.4%. The results show that the proposed method achieves higher accur acy compared to traditional lumped-element-based approaches.

Index Terms—De-embedding, distributed, discontinuity, RF device

I. INTRODUCTION

The high-frequency characteristics of RF devices are becoming increasingly important with the expansion of millimeter-wave applications, such as 28-GHz 5G communications, 60-GHz wireless communication, 77-GHz automotive radar, and 94-GHz imaging systems. Accurate and reliable device models are essential for the design of RF/microwave circuits. As operating frequencies continue to rise, the impact of interconnect and pad parasitics on device's behavior becomes increasingly critical for on-wafer device characterization. Hence, over the past few decades, many de-embedding methods have been proposed to remove parasitics. The lumped element circuit model is one of the most traditional approaches, in which equivalent circuits are established by the parallel-series combination of parasitic elements, and these elements are sequentially de-embedded from the external to the internal one by one [1]- [3]. However, this type of method has limitations, it require more de-embedding structures to achieve higher accuracy [3], and is not recommended for millimeter-wave applications due to the distributed nature of

---

Hongfei Su 1, Bharatha Kumar Thangarasu 2, Nagarajan Mahalingam 3, Kaixue Ma 4, Fanyi Meng 5, Kiat Seng Yeo 7 are with *School of microelectronics, Tianjin University*, Tianjin, China. Zhenghao Lu 6 is with *School of Electronics and Information, Soochow University*, Suzhou, China. Corresponding author: (Hongfei Su 1, email: suhongfei@tju.edu.cn)

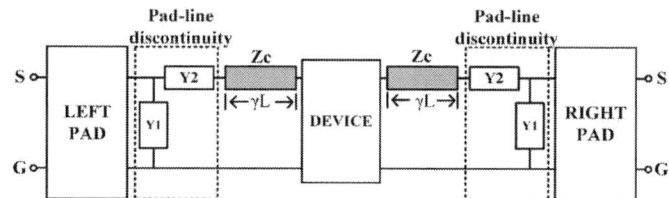

Fig. 1. Proposed equivalent circuit model of the DUT structure.

metal interconnect parasitics. Moreover, de-embedding errors due to imperfections in SHORT interconnections cannot be fully eliminated using these techniques which will cause over de-embeding. To address these issues mentioned above, several distributed cascade-based de-embedding methodologies [4]–[5] have been proposed, where interconnect lines are modeled as cascaded stages of RLGC elements based on transmission line theory. However, the distributed effects of pad parasitics are neglected in [4]– [7], as they are approximated by lumped elements, which limits the accuracy of the method. Further [8] proposed a fully distributed de-embedding method in which both the pad and line are modeled as distributed elements using two special test structures. Compared with lumped de-embedding method, this fully distributed method can achieve higher accuracy. However, the test structures used in this method can only be used for a fixed configuration and is difficult to apply to devices of different sizes and this lack of scalability will result in higher de-embedding costs. Additionally, the method does not fully address the discontinuities between the pad and interconnection, limiting its applicability in Higher frequency.

In this paper, a scalable distributed de-embedding technique is presented. It takes into account the distributed effects of the pad, overcoming the limitations of lumped pad models while achieving scalability in the de-embedding process. A lumped pad-line discontinuity model is also developed to further calibrate the accuracy of the model. The details of the proposed de-embedding technique are presented in the following section and validated on an inductor fabricated using 0.13 μm BiCMOS technology, up to 120 GHz.

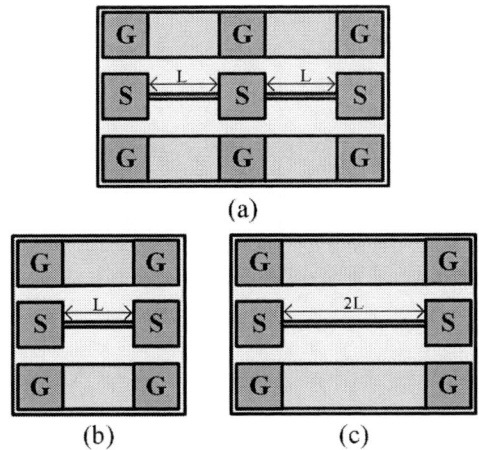

(a)

(b)   (c)

Fig. 2. De-embedding structures employed in the proposed method.(a) LINE-PAD-LINE. (b) LINE. (c) 2LINE.

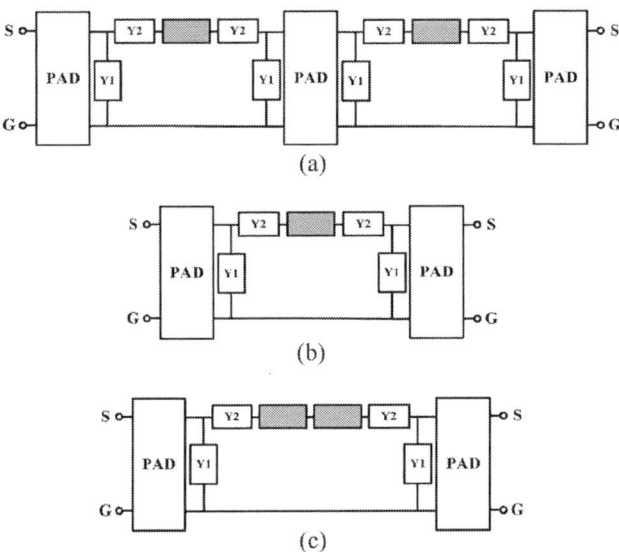

(a)

(b)

(c)

Fig. 3. Equivalent parasitic network models of de-embedding structures.(a) LINE-PAD-LINE. (b) LINE. (c) 2LINE.

II. PROPOSED DE-EMBEDDING THEORY

A. De-embedding Structures and Their Models

Fig. 1 illustrates the equivalent circuit model of the DUT with GSG pads and metal interconnects. The GSG pads are characterized by distributed parameter ABCD matrixes to represent the distributed parasitic effects of the pad. The characteristic impedance γ and propagation constant Z_C are calculated according to transmission line theory to characterize the distributed parasitic effects of the interconnects. To accurately represent the step discontinuity between the pad and the interconnect, lumped series and parallel conductances are modeled in a T-network configuration. Thus, the DUT model is represented by a cascade of ABCD matrices:

$$[A_{DUT}] = [P_L][D_L][TL][A_{INT}][TL][D_R][P_R] \quad (1)$$

Fig. 2 and Fig. 3 present the proposed de-embedding test structures and their corresponding equivalent circuit models, including LINE, 2LINE, and LINE-PAD-LINE. The three test structures are fully symmetrical to ensure the accuracy of the de-embedding process. The LINE and LINE-PAD-LINE structures are used to extract the distributed parasitics of the pad, while the introduction of the 2LINE structure provides the necessary information for extracting the discontinuity between the pad and the interconnects. The de-embedding process can be broadly divided into three stages, which will be described in detail in the following sections.

B. Extraction of Pad Parasitics

In order to accurately represent the parasitic effects of the pad, the pad model in this method does not make any lumped assumptions. Instead, it directly employs cascade-based ABCD matrices for extraction, without the need to compute the elements of a lumped model. The ABCD matrices of the LINE and LINE-PAD-LINE structures for de-embedding can be expressed as:

$$[A_{LINE}] = [P_L][D_L][TL][D_R][P_R] \quad (2)$$

$$[A_{L-PAD-L}] = [P_L][D_L][TL][D_R][P_M][[D_L]][TL][D_R][P_R] \quad (3)$$

The next step is to calculate P_L and P_R using (4)-(5). It is evident that during this calculation, the pad in the center of the LINE-PAD-LINE structure is affected by discontinuity effects at both ends, while in the LINE structure, the pad experiences discontinuity effects only at the end connected to the interconnect. This difference leads to the neglect of the discontinuity effects in the calculation process. Therefore, in this work, the discontinuity between the pad and the interconnect is modeled, as will be presented in the following.

$$[P_L] = [A_{LINE}][[A_{LINE}]^{-1}[A_{L-PAD-L}]]^{-1} \quad (4)$$

$$[P_R] = [[A_{L-PAD-L}][A_{LINE}]^{-1}]^{-1}[A_{LINE}] \quad (5)$$

C. Extraction of Discontinuities between Pad and Interconnect

To compensate for the discontinuity between the pad and the interconnect, a lumped T-network is constructed as shown in Fig. 1. The left and right pad in cascade can be represented by matrices formed using the LINE and 2LINE structures, with the corresponding equivalent circuit diagram shown in Fig. 4(a). It is evident that the discontinuity between the pad and the interconnect is preserved, the multiplication sum of ABCD matrix $[P_L][D_L]$ and $[D_R][P_R]$ can be extracted using the following [5]:

$$[P_L][D_L][D_R][P_R] = [A_{LINE}][A_{2LINE}]^{-1}[A_{LINE}]$$
$$= \begin{bmatrix} A_c & B_c \\ C_c & D_c \end{bmatrix} \quad (6)$$

Since the widths of the interconnects on the left and right sides are identical, the discontinuities between the pad and the interconnect are symmetric. Therefore, through matrix

979-8-3315-2209-4/25 $31.00 © 2025 IEEE   453

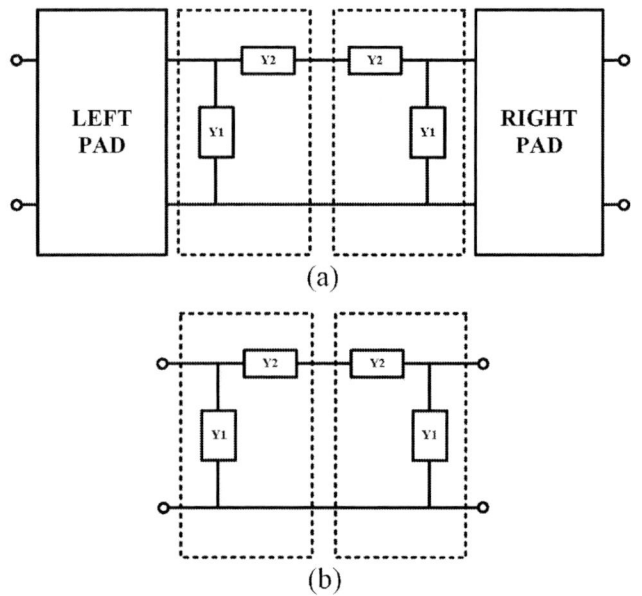

(a)

(b)

Fig. 4. (a) The multiplication sum of pad with discontinuity. (b) The multiplication sum of discontinuity.

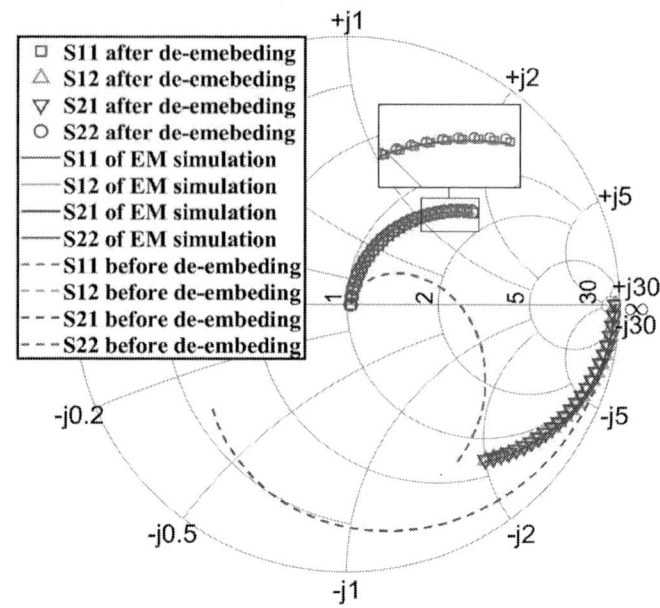

Fig. 5. Comparison of the results before and after de-embedding with the EM simulation results.

operations, $[D_L][D_R]$ as shown in Fig.4(b) can be calculated:

$$
\begin{aligned}
[D_L][D_R] &= [P_L]^{-1} \begin{bmatrix} A_c & B_c \\ C_c & D_c \end{bmatrix} [P_R]^{-1} \\
&= \begin{bmatrix} 1 & 1/Y_2 \\ Y_1 & 1+Y_1/Y_2 \end{bmatrix} \begin{bmatrix} 1+Y_1/Y_2 & 1/Y_2 \\ Y_1 & 1 \end{bmatrix} \\
&= \begin{bmatrix} A_D & B_D \\ C_D & D_D \end{bmatrix}
\end{aligned} \tag{7}
$$

The D_L and D_R can be obtained using (8) and (9).

$$
[D_\mathrm{L}] = \begin{bmatrix} 1 & B_\mathrm{D}/2 \\ \frac{C_\mathrm{D}}{1+2(A_D+D_D)} & 1+\frac{B_\mathrm{D} C_\mathrm{D}}{2(1+2(A_\mathrm{D}+D_\mathrm{D}))} \end{bmatrix} \tag{8}
$$

$$
[D_\mathrm{R}] = \begin{bmatrix} 1+\frac{B_\mathrm{D} C_\mathrm{D}}{2(1+2(A_\mathrm{D}+D_\mathrm{D}))} & B_\mathrm{D}/2 \\ \frac{C_\mathrm{D}}{1+2(A_\mathrm{D}+D_\mathrm{D})} & 1 \end{bmatrix} \tag{9}
$$

D. Extraction of Interconnect Lines

For the modeling of interconnects, the LINE structure is employed.Using the known Pad and discontinuity models, the LINE structure is able to be de-embedded.then by applying the ABCD matrix presented in (10) and utilizing the de-embedding results from the LINE structure, the characteristic impedance Z_c and propagation constant γ of the interconnect line are able to be calculated [9]. The scalability of the de-embedding structures is achieved by adjusting the interconnect line length l.

$$
[TL] = \begin{bmatrix} \cosh \gamma l & Z_c \sinh \gamma l \\ \frac{1}{Z_c} \sinh \gamma l & \cosh \gamma l \end{bmatrix} \tag{10}
$$

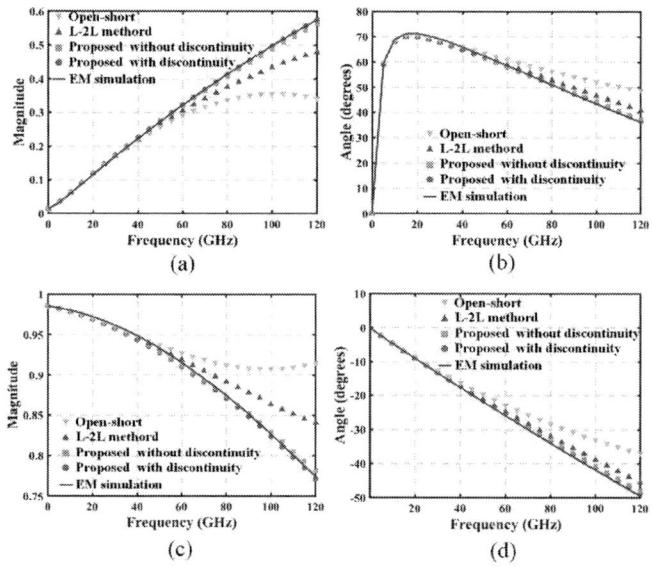

Fig. 6. Comparison of S-parameters after de-embedding using different methods.(a) Magnitude of S11. (b) Phase of S11. (c) Magnitude of S21. (d) Phase of S21.

III. SIMULATION RESULTS AND DISCUSSION

To validate the proposed de-embedding method and compare it with conventional approaches, a high-frequency spiral inductor along with its corresponding dummy structures were implemented using 0.13 μm BiCMOS technology. In the test structures, the wide M1 metal is used as the reference ground plane for all ports, with the top metal layers serving as the pad metal. Additionally, the G-pads of the two GSG pads are interconnected by the top metal layers to reduce transmission

979-8-3315-2209-4/25 $31.00 © 2025 IEEE

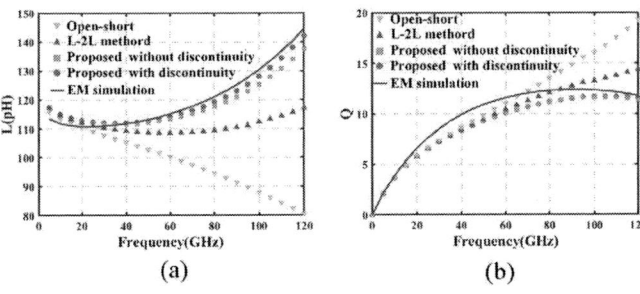

Fig. 7. Comparison between the proposed de-embedding method and conventional techniques for inductor. (a) L. (b) Q.

line losses. The interconnect length of the DUT structure is 50 μm, and the corresponding interconnect length of the LINE structure is 100 μm.

Fig. 5 illustrates the EM simulation results before and after de-embedding. Based on the comparison of these results, the maximum deviation between simulation and calculated S-parameters is less than 3% within 120 GHz , with the average deviation being less than 1.4%, which validates the feasibility of the proposed de-embedding method.

To assess the improvement of the proposed method over traditional approaches, Fig. 6 presents the S11 and S21 parameters obtained using different de-embedding methods. The Open-Short method and L-2L method, due to their lumped-element assumptions, result in average S-parameter errors of 12.5% and 5.5% within the 120 GHz frequency range, respectively. Additionally, to prove the validity of the discontinuity model, a comparison of the S-parameters, with and without considering discontinuities, is also presented as shown in Fig. 6, When discontinuities are not considered, the average S-parameter error is 1.9%, which is 0.5% higher than when discontinuities are considered.

Fig. 7 presents the extracted inductance (L) and Figure of merit (Q). It can be observed that the proposed method shows excellent agreement with the EM simulation results within the 120 GHz range. Compared to the de-embedding method with lumped-element assumptions for the pad, the proposed method demonstrates higher accuracy.

IV. CONCLUSION

This paper presents a new scalable cascaded de-embedding technique for precise RF device characterization. Using three test structures,including LINE,2LINE and LINE-PAD-LINE, the proposed method constructs a distributed model for the pad and a lumped model for pad-line discontinuity. Compared to other techniques, the proposed method enables accurate de-embedding of surrounding parasitics, effectively addressing the discontinuities and distributed effects of the pad and interconnects. The proposed de-embedding technique has been validated on a spiral inductor fabricated in 0.13-μm BiCMOS technology, over a frequency range up to 120 GHz, demonstrating excellent accuracy.

REFERENCES

[1] M. C. A. M. Koolen, J. A. M. Geelen and M. P. J. G. Versleijen, "An improved de-embedding technique for on-wafer high-frequency characterization," Proceedings of the 1991 Bipolar Circuits and Technology Meeting, Minneapolis, MN, USA, 1991, pp. 188-191.

[2] T. E. Kolding, "A four-step method for de-embedding gigahertz on-wafer CMOS measurements," in IEEE Transactions on Electron Devices, vol. 47, no. 4, pp. 734-740, April 2000.

[3] I. M. Kang et al., "Five-Step (Pad–Pad Short–Pad Open–Short–Open) De-Embedding Method and Its Verification," in IEEE Electron Device Letters, vol. 30, no. 4, pp. 398-400, April 2009.

[4] A. M. Mangan, S. P. Voinigescu, Ming-Ta Yang and M. Tazlauanu, "De-embedding transmission line measurements for accurate modeling of IC designs," in IEEE Transactions on Electron Devices, vol. 53, no. 2, pp. 235-241, Feb. 2006.

[5] H. -Y. Cho, J. -K. Huang, C. -W. Kuo, S. Liu and C. -Y. Wu, "A Novel Transmission-Line Deembedding Technique for RF Device Characterization," in IEEE Transactions on Electron Devices, vol. 56, no. 12, pp. 3160-3167, Dec. 2009.

[6] X. S. Loo, K. S. Yeo and K. W. J. Chew, "THRU-Based Cascade De-embedding Technique for On-Wafer Characterization of RF CMOS Devices," in IEEE Transactions on Electron Devices, vol. 60, no. 9, pp. 2892-2899, Sept. 2013.

[7] R. Wang, C. Li and Y. Wang, "An Improved Through-Only De-Embedding Method for 110-GHz On-Wafer RF Device Characterization," in IEEE Microwave and Wireless Components Letters, vol. 32, no. 10, pp. 1219-1222, Oct. 2022.

[8] X. S. Loo et al., "A New Millimeter-Wave Fixture Deembedding Method Based on Generalized Cascade Network Model," in IEEE Electron Device Letters, vol. 34, no. 3, pp. 447-449, March 2013.

[9] Cho, Ming-Hsiang et al. "A Cascade Open-Short-Thru (COST) De-Embedding Method for Microwave On-Wafer Characterization and Automatic Measurement." IEICE Trans. Electron. 88-C (2005): 845-850.

A Thru-Line De-Embedding Method With Double-Type Pad Model for On-Wafer Device Characterization

Yutong Wu   Bharatha Kumar Thangarasu   Nagarajan Mahalingam   Kaixue Ma

Fanyi Meng   Zhenghao Lu   Kiat Seng Yeo

Abstract—This paper presented a Thru-Line de-embedding method based on double-T and double-π pad models, suitable for the 110 GHz frequency range. The method is characterized by the use of two pad models that are more accurate at high frequencies, combined with an odd-even technique to simplify the calculations. It also takes into account the distribution effects of the metal interconnects between the pads and the intrinsic devices. Based on this, parasitic effects are eliminated using cascaded ABCD matrices. An on-chip inductors were fabricated using a 130 nm BiCMOS process to verify the accuracy of the proposed method in the 0-110 GHz frequency range. The de-embedded measurement results are in good agreement with the electromagnetic simulation results of the intrinsic devices, demonstrating that the proposed method outperforms traditional methods.

Index Terms—De-embedding methods, millimeter-wave, odd-even technique, inductor

I. INTRODUCTION

With the continuous advancement of BiCMOS technology, millimeter-wave (mm-wave) integrated circuits have become a key area of research, demanding precise measurement techniques for accurate RF-integrated circuit device modeling [1]. However, the intrinsic performance characteristics of on-wafer devices are often obscured by the embedded ground-signal-ground (GSG) pads and complex metal interconnections, making direct extraction from measurements challenging. As a result, the use of de-embedding techniques to eliminate parasitic effects from the measurement data is essential. The most widely used de-embedding method is the open-short technique [2]. However, its accuracy decreases as the frequency increases. To enhance de-embedding accuracy, several improvements have been proposed, primarily including lumped and cascaded de-embedding approaches. In lumped methods, additional virtual measurement structures are introduced into the traditional open-short technique, such as the open-short-load method and the pad-open-short method [3]. These three-step de-embedding methods provide more accurate characterization of the pad-contact structure in the measurement setup.

However, due to their reliance on lumped-element interconnect models, these methods cannot achieve high precision at millimeter-wave frequencies. As the frequency increases, parasitic electromagnetic effects become more complex. Consequently, four-step de-embedding methods, such as the open-short-open-short method [4], and five-step methods [5], have been proposed. These methods significantly increase model complexity and the number of required virtual structures, leading to higher design and testing costs. In contrast, cascaded de-embedding methods based on two-port and four-port networks have also been introduced. While four-port cascaded networks provide more comprehensive solutions, two-port networks offer simpler computations and are more widely applicable [6]. To further streamline the de-embedding process, methods utilizing through-only methods have been proposed [7]-[9]. However, the method in [8] is not suitable for on-wafer devices fabricated using BiCMOS technology. The approach in [7] neglects the transmission line length and models the through structure as a π-equivalent circuit, resulting in insufficient de-embedding accuracy at high frequencies.[9] is applicable only to fixed interconnect lengths. [10] proposed an L-2L de-embedding method for load-assisted structures, but only verified the results below 40GHz.

In reality, the pad response is not symmetrical but reciprocal. The accuracy of the traditional lumped-pad model based on two parameters significantly decreases at high frequencies. Therefore, double-T-type and double-π-type pad models with three variables were used in this paper. Additionally, to simplify the calculation process, the odd-even mode analysis method is used. To obtain a scalable interconnect model, a line test structure is employed to calculate the characteristic impedance Zc of the transmission line.

The structure of this paper is as follows. Section II introduces the thru-line de-embedding method. Section III presents experimental validation of the proposed method using two inductors. The conclusions are summarized in Section IV.

II. PROPOSED DE-EMBEDDING METHOD

A. An Equivalent Circuit Model of the Through

As shown in Fig. 1, the model of the measured Device Under Test (DUT) takes into account the probe pads and interconnect lines. Since the thru structure necessarily includes

---

Yutong Wu 1, Bharatha Kumar Thangarasu 2, Nagarajan Mahalingam 3, Kaixue Ma 4, Fanyi Meng 5, Kiat Seng Yeo 7 are with *School of microelectronics, Tianjin University*, Tianjin, China. Zhenghao Lu 6 is with *School of Electronics and Information, Soochow University*, Suzhou, China. Corresponding author: (Yutong Wu 1, email: wuyutong_202@tju.edu.cn)

979-8-3315-2209-4/25 $31.00 © 2025 IEEE

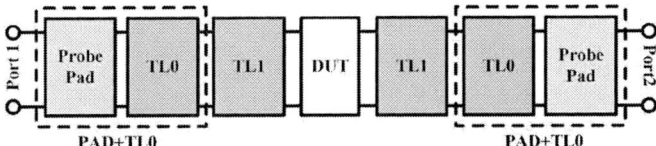

Fig. 1. Model of a measured DUT considering the probe pads and interconnects into account.

short transmission lines between the two pads, the portion of the interconnect with length L_0 in the DUT test structure is included in the pad model. The DUT can thus be represented by a cascade ABCD matrix as follows:

$$A_{\text{complete}} = A_{\text{PAD+TL0}} A_{\text{TL1}} A_{\text{DUT}} A_{\text{TL1}} A_{\text{PAD+TL0}} \quad (1)$$

The accuracy of de-embedding is primarily determined by

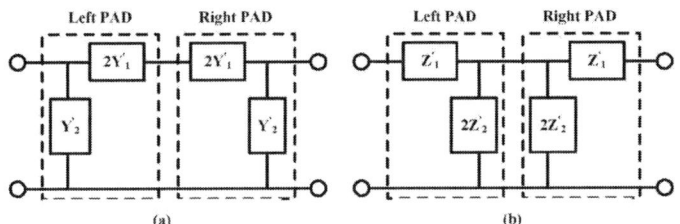

Fig. 2. Conventional pad models.(a) π-type pad model.(b) T-type pad model.

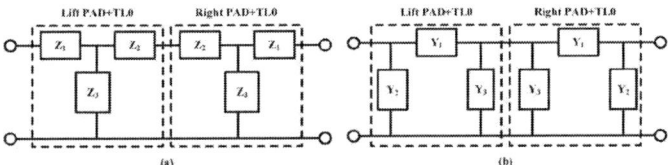

Fig. 3. Proposed double-type pad model.(a) double-T-type pad model.(b) double-π-type pad model.

the equivalent circuit models of the dummy patterns. The interconnect lines are modeled as transmission lines with length L_1, characteristic impedance Z_{c0}, and propagation constant γ_0. Due to the reciprocity of the pads, two more precise models with three parameters are used, based on the traditional two-pad models, as shown in Figure 3. In the double-T model, the discontinuity of the pad and interconnect is represented by Z_2. Since the transmission line between the two pads in the thru structure is relatively short, the series impedance is treated as a whole and included in Z_2. Similarly, in the double-π-type model, the parasitics are incorporated into Y_2.

B. Odd-Even Mode Theory

Based on the odd-even mode theory, the odd-mode and even-mode circuits of the pad equivalent circuit, as shown in Fig. 4, can be derived. The input admittances of the odd-mode and even-mode circuits are denoted as Y_o and Y_e, respectively. The relationship between the admittance of the symmetric

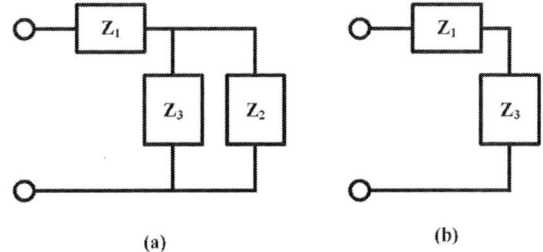

Fig. 4. The odd- and even-mode equivalent circuits of the double-T pad model.(a) Odd mode circuit.(b) Even mode circuit.

network and its odd-even mode admittances can be described by Equation (2).

$$\mathbf{Y} = \begin{bmatrix} Y_{11} & Y_{12} \\ Y_{21} & Y_{22} \end{bmatrix} = \frac{1}{2} \begin{bmatrix} Y_e + Y_o & Y_e - Y_o \\ Y_e - Y_o & Y_e + Y_o \end{bmatrix} \quad (2)$$

From Equation (2), the odd-mode and even-mode admittances of the symmetric network can be derived, which are expressed in Equations (3) and (4), respectively.

$$Y_o = Y_{11} - Y_{12} \quad (3)$$

$$Y_e = Y_{11} + Y_{12} \quad (4)$$

The parameters Y_{11} and Y_{12} are obtained from the measurement results of the thru structure. Taking the double-T-type model as an example, the odd-mode and even-mode impedances, Z_o and Z_e, can be determined based on the odd-mode and even-mode equivalent circuits as follows:

$$Z_o = Z_1 + Z_3 \parallel Z_2 = \frac{1}{Y_o} \quad (5)$$

$$Z_e = Z_1 + Z_3 = \frac{1}{Y_e} \quad (6)$$

These equations, labeled as Equations (5) and (6), are independent, but they involve three unknown parameters. To address this, it can be assumed that $Z_2 = kZ_1$, where k is a constant for a given pad structure. The value of k can be determined from electromagnetic (EM) simulations by evaluating the inductance at low frequencies, where it remains constant. In this study, k is found to be 0.96. The pad model, $A_{\text{PAD+TL0}}$, can be simplified and expressed as follows:

$$A_{\text{pad+TL0}} = \begin{bmatrix} \frac{z_1+z_3}{z_3} & \frac{(z_1+z_3)(z_2+z_3)-z_3^2}{z_3} \\ \frac{1}{z_2} & \frac{z_2+z_3}{z_3} \end{bmatrix} \quad (7)$$

C. Extraction of Transmission Line Parameters

In the line dummy structure shown in Fig. 1, the parasitics from the pad and TL_0 are removed, resulting in the transmission matrix of TL_2, denoted as A_{TL2}:

$$A_{\text{TL2}} = A_{\text{PAD+TL0}}^{-1} A_{\text{line}} A_{\text{PAD+TL0}}^{-1} \quad (8)$$

979-8-3315-2209-4/25 $31.00 © 2025 IEEE   457

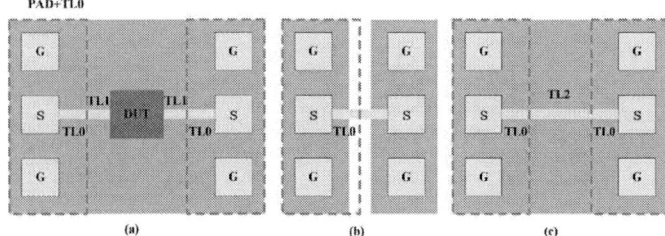

Fig. 5. Configurations of the DUT and the corresponding deembedding test structures for the proposed method. (a) DUT. (b) thru. (c) line.

The equivalent ABCD matrix of the transmission line TL_2 is expressed as:

$$A_{\text{TL2}} = \begin{bmatrix} \cosh(\gamma_0 l_2) & Z_{c0}\sinh(\gamma_0 l_2) \\ \dfrac{\sinh(\gamma_0 l_2)}{Z_{c0}} & \cosh(\gamma_0 l_2) \end{bmatrix} \quad (9)$$

By combining Equations (8) and (9), the parameters γ_0 and Z_{c0} can be determined. Consequently, the A_{TL1} matrix of the DUT test structure, which corresponds to the ABCD matrix of a transmission line with a length of l_1, can be obtained.

The characteristics of the intrinsic device A_{DUT} can then be de-embedded as:

$$A_{\text{DUT}} = A_{\text{PAD+TL0}}^{-1} A_{\text{TL1}}^{-1} A_{\text{complete}} A_{\text{TL1}}^{-1} A_{\text{PAD+TL0}}^{-1} \quad (10)$$

III. RESULT AND DISCUSSION

To verify the accuracy of the proposed method from 0 to 110 GHz, a comparison and validation were performed on an inductor fabricated in a 130-nm BiCMOS process. The proposed de-embedding technique was evaluated using three dummy structures, as illustrated in Fig.5. The topmost metal layer (MA) was used as the signal line, while the bottommost metal layer (M1) served as the ground. During the actual measurements, the following parameters were set: $L_0 = 10\,\mu\text{m}$, $L_1 = 30\,\mu\text{m}$, and $L_2 = 70\,\mu\text{m}$. The spacing between the two pads in the thru structure was $30\,\mu\text{m}$, designed to maximize the pad distance and minimize coupling effects between the pads. The experimental results confirmed that the coupling effects between the pads can be neglected when $L_0 = 15\,\mu\text{m}$.

Fig. 6 compares the measured S-parameters before and after de-embedding with electromagnetic (EM) simulation results, validating the effectiveness of the proposed method. To further evaluate its improvement, Table 1 summarizes the comparison between the de-embedded results obtained using different methods and the EM simulation results. Over the frequency range of 0–110 GHz, the proposed method using the double-T pad model achieved an error of 3.15%, while the double-Π pad model showed an error of 5.80%. Both methods outperformed the L-2L method, which had an error of 6.93%. Particularly at higher frequencies, the proposed method demonstrates significant improvement over the L-2L method. Furthermore, the data in Table 1 highlight that the double-T pad model provides more accurate modeling for uncertainties

and short transmission lines, resulting in smaller errors. The error is calculated as follows:

$$Error = \sum_{m=1}^{M} \frac{1/4 * \sum_{j=1}^{2}\sum_{i=1}^{2} abs\left[\dfrac{S_{ij-de}-S_{ij-EM}}{S_{ij-EM}}\right]}{M} \quad (11)$$

The term $S_{ij,\text{de}}$ denotes the de-embedded S-parameter, whereas $S_{ij,\text{EM}}$ represents the S-parameter obtained through electromagnetic (EM) simulation. Here, M signifies the number of test points.

Figs.7 and 8 illustrate that the inductance (L) and quality factor (Q) extracted using the proposed method exhibit better agreement with the intrinsic device's EM simulation results compared to the L-2L method across the frequency range of 0–110 GHz.

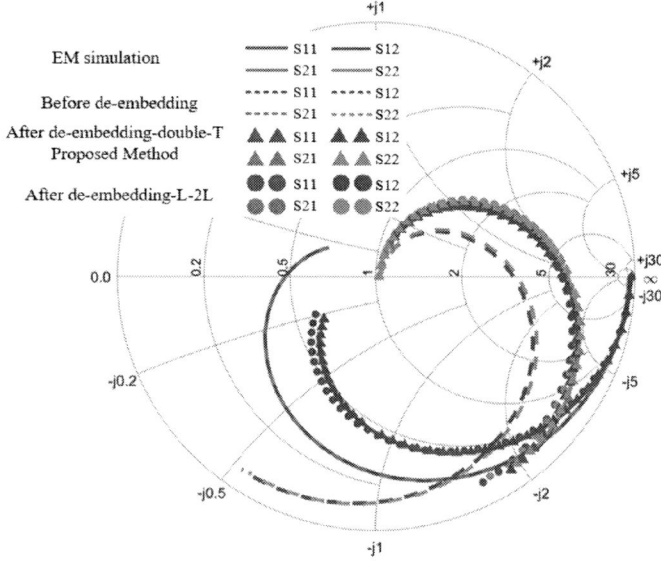

Fig. 6. S-parameter comparisons between de-embedding results and EM simulation results

TABLE I
COMPARISON OF DE-EMBEDDING ERRORS ACROSS DIFFERENT METHODS

Method	0–70 GHz Error	70–110 GHz Error	0–110 GHz Error
L-2L	0.0456	0.1898	0.0693
Double-T ($k = 0.96$)	0.0114	0.0672	0.0310
Double-Π ($k = 1$)	0.0499	0.0738	0.0483

IV. CONCLUSION

In this letter, an improved Thru-Line de-embedding method for on-wafer RF device characterization up to 110 GHz is proposed. Test structures were fabricated using a 130 nm BiCMOS process, and experimental data were measured at frequencies up to 110 GHz to validate the effectiveness of the proposed method. Compared to the conventional L-$2L$ de-embedding approach, the proposed method demonstrates

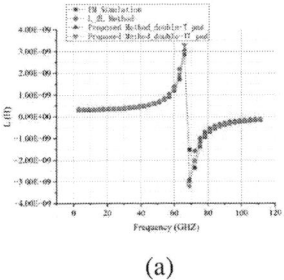 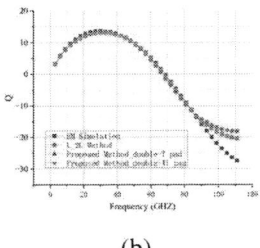

(a)   (b)

Fig. 7. Comparison of L and Q between de-embedding results and EM simulation results. (a) L. (b) Q.

improved accuracy, particularly at higher frequencies. Moreover, in comparison with the Thru-only method, this technique exhibits better scalability and accounts for the discontinuities between pads and interconnect lines.

REFERENCES

[1] S. Kawai, S. Sato, S. Maki, K. K. Tokgoz, K. Okada, and A. Matsuzawa, "Accurate transistor modeling by three-parameter pad model for millimeter-wave CMOS circuit design," IEEE Trans. Microw. Theory Techn., vol. 64, no. 6, pp. 1736–1744, Jun. 2016.

[2] M. C.A. M.Koolen,J.A. M.Geelen, and M. P. J. G. Versleijen, "An improved de-embedding technique for on-wafer high-frequency characterization," in Proc. Bipolar Circuits Technol. Meeting, 1991, pp. 188–191.

[3] L. F. Tiemeijer, R. J. Havens, A. B. M. Jansman, and Y. Bouttement, "Comparison of the 'pad-open-short' and 'open-short-load' deembedding techniques for accurate on-wafer RF characterization of high-quality passives," IEEE Trans. Microw. Theory Techn., vol. 53, no. 2, pp. 723–729, Feb. 2005.

[4] T. E. Kolding, "A four-step method for de-embedding gigahertz on-wafer CMOS measurements," IEEE Trans. Electron Devices, vol. 47, no. 4, pp. 734–740, Apr. 2000.

[5] K. Katayama, S. Amakawa, K. Takano, and M. Fujishima, "300-GHz MOSFET model extracted by an accurate cold-bias de-embedding technique," in IEEE MTT-S Int. Microw. Symp. Dig., Phoenix, AZ, USA, May 2015, pp. 1–4.

[6] C.-H. Chen and M. J. Deen, "A general noise and S-parameter deembedding procedure for on-wafer high-frequency noise measurement of MOSFETs," IEEE Trans. Microw. Theory Techn., vol. 49, no. 5 pp. 1004–1005, May 2001.

[7] H. Ito and K. Masuy, "A simple through-only de-embedding method for on-wafer S-parameter measurements up to 110 GHz," in IEEE MTT-S Int. Microw. Symp. Dig., Jun. 2008, pp. 383–386.

[8] X. Wang, W. Wang, Y. Liang, and R. Jin, "1-Thru deembedding method for one-port microwave device characterization," IEEE Microw. Wireless Compon. Lett., vol. 32, no. 4, pp. 355–358, Apr. 2022.

[9] R. Wang, C. Li and Y. Wang, "An Improved Through-Only De-Embedding Method for 110-GHz On-Wafer RF Device Characterization," in IEEE Microwave and Wireless Components Letters, vol. 32, no. 10, pp. 1219-1222, Oct. 2022.

[10] M. Lingmei, S. Xiaotian, C. Tan, Y. Wei, Q. Rong and W. Liang, "A High-Precision De-Embedding Method for GaAs IPD Wafer-Level Testing," 2024 2nd International Symposium of Electronics Design Automation (ISEDA), Xi'an, China, 2024, pp. 765-767.

979-8-3315-2209-4/25 $31.00 © 2025 IEEE

An Adaptive Digital Loop Filter For Fast-Locking TDC-Based ADPLL

Kaiqiang Qin   Nagarajan Mahalingam   Bharatha Kumar Thangarasu   Kaixue Ma

Fanyi Meng   Zhenghao Lu   Kiat Seng Yeo

Abstract—An adaptive digital loop filter (DLF) for fast-locking TDC-based all-digital phase-locked loop (ADPLL) is proposed. The proposed DLF utilizes a proportional-integral filter structure with dynamically adjustable integral coefficients, which is controlled by a finite state machine (FSM). The adaptive method optimizes the response speed and stability according to the magnitude of the phase error and its variation. An adaptive DLF based ADPLL is implemented in 130 nm SiGe BiCMOS technology. The simulation result shows that the proposed ADPLL with adaptive DLF can achieve a locking time of 2.46 μs at 3.5 GHz output frequency.

Index Terms—ADPLL, DLF, adaptive, fasting locking

I. INTRODUCTION

The All-Digital Phase-Locked Loop (ADPLL) is a digital implementation of the conventional analog phase-locked loop (PLL). The DLF uses standard digital cells instead of the large-area resistors and capacitors as passive components used in loop filters, which not only reduces layout area but also increases flexibility. Typically, the feedback loop composed of a frequency divider forms a frequency synthesizer. In addition to using time-to-digital converter (TDC) to quantify phase error, BB-PD also quantifies phase error but exhibits a nonlinear response that depends on the size of the phase error [1]–[3]. A novel DLF for BBPLL has been proposed in [1], where an adaptive locking monitor (ALM) is utilized to assist in predicting frequency errors and divides the locking process into three steps, effectively reducing locking time without compromising jitter performance. A reconfigurable digital loop filter for ADPLL has been proposed in [4], where the coefficients can be controlled through an external interface or internal options in the digital module, allowing the DCO to operate in a specific operating region. The architecture in [5] is based on a hybrid TDC-BB ADPLL that utilizes an adaptive proportional derivative digital loop filter that enhances loop stability and build-up time.

In this work, an adaptive digital loop filter is proposed for fast-locking TDC-based ADPLL, and the proposed DLF dynamically adjusts the integral coefficients based on phase error.

Kaiqiang Qin, Nagarajan Mahalingam, Bharatha Kumar Thangarasu, Kaixue Ma, Fanyi Meng and Kiat Seng Yeo are with *School of microelectronics, Tianjin University*, Tian jin, China. Zhenghao Lu is with *School of Electronics and Information, Soochow University*, Su Zhou, China. Corresponding author: (Kaiqiang Qin, email: qkq@tju.edu.cn)

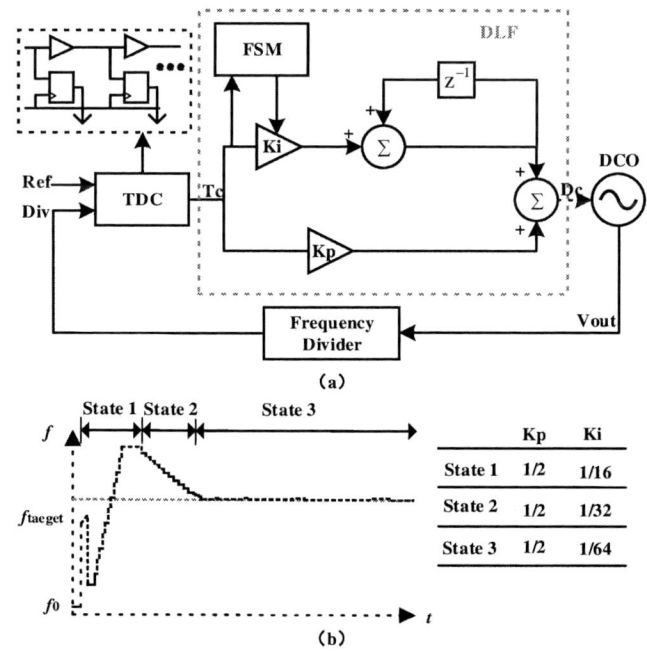

Fig. 1. (a) Proposed DLF implemented in ADPLL architecture (b) Locking timing diagram.

II. DLF ARCHITECTURE

The proposed DLF, along with the related modules within the ADPLL—namely, TDC, DCO, and Divider—are shown in Fig. 1(a). The TDC is modeled with a delay chain and flip-flops. The DLF comprises a Proportional-Integral (PI) filter and a finite state machine (FSM), all of which are implemented using standard logic cells.

The integral coefficient serves to accumulate phase error and is typically used to modulate the system's response speed and stability. Consequently, the integral coefficients can be dynamically adjusted to correspond with the various locking states within the loop. According to the DLF architecture shown in Fig. 1(a), the integral coefficients are adaptive and controlled by the FSM. The DCO controlling code can be given in the discrete-time domain as follows:

$$Dc[n] = Kp * Tc[n] + Ki * (Tc[n] + Tc[n-1] + ...Tc[1]) \tag{1}$$

Tc is the TDC output code, Dc is the DCO input controlling code, Kp is the proportional coefficient, and Ki is the integral coefficient.

According to the adjustment of Ki, the locking process of the ADPLL is divided into three distinct states, as shown in Fig. 1(b). In the beginning, the first two cycles maximize Tc so that the initial frequency of the DCO output frequency is close to the locked frequency. Then comes state 1, which is referred to as phase capture. In this state, as the phase error increases, the DCO frequency continually converges toward the target frequency until the phase error ceases to grow. In state 1, a broader bandwidth is necessary to promptly respond to phase error; hence, Kp equals 1/2 and Ki equals 1/16. As the Tc decreases, transitioning to state 2, the phase error begins to decrease and the DCO frequency decreases to near the target frequency; further adjustment of the integral coefficients is required to ensure the system's stability, so Kp keeps 1/2 and Ki changes to 1/32. In state 3, fine-tuning is required, so Kp keeps 1/2, Ki changes to 1/64, and the phase error gradually decreases to the resolution range of TDC. At this point, the Tc oscillates between 7'd0 and 7'd1, so the phase error is accumulated from the difference between the DCO frequency and the target frequency to adjust the output frequency, making this state the most time-consuming.

Fig. 2 shows the FSM in DLF, which controls the selection of the MUX according to the sign of Tc[n]-Tc[n-1] and the magnitude of Tc[n] to realize the adaptive integral coefficient Ki, ensuring that Ki has different values at different locking states.

In summary, the FSM enters an initial state in state 1, setting a larger value of Ki for a faster system response. In state 2, a shift in the direction of the slope of Tc is detected and MUX outputs a smaller Ki, and then in state 3, when the value of Tc is detected to be sufficiently small and MUX outputs an even smaller value of Ki. The final Kp and Ki are set as 1/2 and 1/64 respectively, because the steady state jitter performance is taken into account.

III. SIMULATION

Based on the proposed adaptive digital loop filter, it is implemented using 130 nm SiGe BiCMOS technology, while the other modules of ADPLL are implemented by Verilog-A behavioral level modeling without including circuit-induced imperfections. Fig. 3 shows the layout of the adaptive DLF with the FSM and PI filter positioned on either side of the layout. Adhering to the digital layout drawing principle, every other row is flipped along the horizontal axis to make the power and ground lines match to save space and metal line length.

In this design, the TDC output is 7 bits, the DCO input is 13 bits, and the DLF performs a 7-bit to 13-bit conversion. Closed loop transient simulations are performed with 100 MHz reference clock and 1.5 V supply voltage. Fig. 4 shows the closed loop transient simulation result of ADPLL for the 3.5 GHz target frequency. The TDC code's maximum value serves as the initial value of the ADPLL output frequency, and the output frequency varies with phase error until it achieves its maximum value when the phase error begins to decrease. The output frequency gradually approaches the target frequency, and the phase error reduces steadily. When the phase error is reduced to the TDC resolution, coarse phase locking is achieved. The frequency and phase locking are then achieved by accumulating phase error.

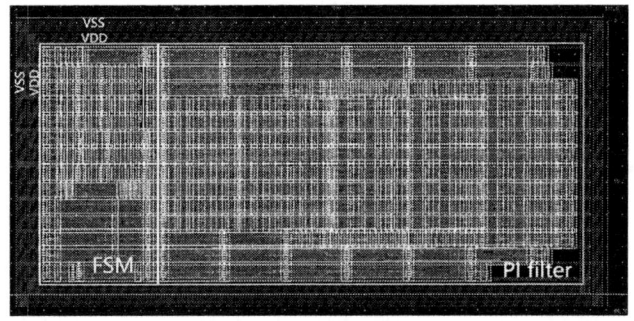

Fig. 3. Adaptive DLF layout.

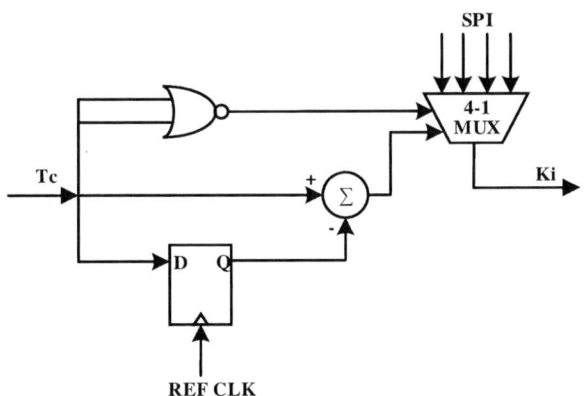

Fig. 2. Proposed FSM.

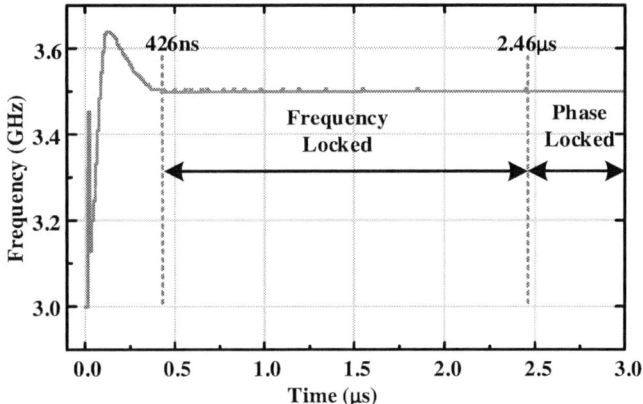

Fig. 4. Post layout closed loop transient simulation of ADPLL for 3.5GHz target frequency.

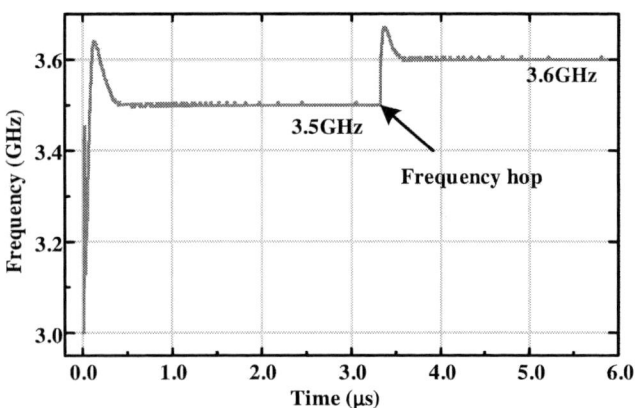

Fig. 5. Transient step response from 3.5GHz to 3.6GHz.

TABLE I
PERFORMANCE COMPARISON

	[2]	[6]	[7]	**This work**
Technology (nm)	65	40	180	**130**
Type	Bang-Bang	Bang-Bang	Linear TDC	**Linear TDC**
Frequency (GHz)	4	4.96	0.8	**3.5**
Locking time (μs)	17.4	3.8	7.52	**2.46**
Power (mW)	17.2 @4GHz	5.1 @4.96GHz	18.2 @0.8GHz	**0.32 @3.5GHz**
Reference Frequency (MHz)	500	40	-	**100**
Area (mm^2)	0.3	0.0106	0.225	**0.0165**

As Fig. 4 illustrates, it validates the design concepts shown in Fig. 1(b), which optimize locking time and loop stability. In this simulation, it takes 426 ns (43 clock cycles) to achieve the frequency locking and only 2.46 μs to realize the frequency and phase locking, which greatly reduces the locking time compared to the traditional BBPLL. The total area of the proposed DLF is 183×90 μm^2 and the power consumed by the DLF is 0.32 mW at an output frequency of 3.5 GHz. After multiple simulations for different output frequencies, it is confirmed that this DLF can be applied to ADPLL with frequency ranging from 2.4 GHz to 6 GHz. Moreover, the transient step response simulation shown in Fig. 5 verifies the proper step response by changing the crossover ratio after the output frequency is locked at 3.5 GHz to make the output frequency jump to 3.6 GHz. The simulation result shows that the hopping time is less than 0.2 μs.

The performance comparison of the ADPLL built based on the proposed DLF with other published works is shown in Tabel I, Only the DLF performance for this work is shown in this table. This work achieves fast locking with low power consumption as compared to other works.

IV. CONCLUSION

This work proposes an adaptive coefficient DLF in TDC-based ADPLL, where the FSM adjusts the integral coefficients in response to the magnitude or variation of the phase error to shorten the locking time while maintaining loop stability. The DLF is synthesized using a 130 nm SiGe BiCMOS technology with an overall estimated area of 183×90 μm^2, while the other modules, such as the TDC, are modeled using Verilog-A. Together, they form the ADPLL and are simulated to reach a lock time of 2.46 μs at 3.5 GHz output frequency while achieving less than 200 ns hopping time at 100 MHz frequency hopping.

REFERENCES

[1] L. Shi, W. Gai, L. Tang, and X. Xiang, "A novel digital loop filter with frequency error prediction for fast-locking bang-bang adpll," in *2017 international conference on electron devices and solid-state circuits (EDSSC)*. IEEE, 2017, pp. 1–2.

[2] J. A. Tierno, A. V. Rylyakov, and D. J. Friedman, "A wide power supply range, wide tuning range, all static cmos all digital pll in 65 nm soi," *IEEE Journal of Solid-State Circuits*, vol. 43, no. 1, pp. 42–51, 2008.

[3] S.-Y. Yang, W.-Z. Chen, and T.-Y. Lu, "A 7.1 mw, 10 ghz all digital frequency synthesizer with dynamically reconfigured digital loop filter in 90 nm cmos technology," *IEEE Journal of Solid-State Circuits*, vol. 45, no. 3, pp. 578–586, 2010.

[4] N. Ahmad and K. Y. Lee, "Design and implementation of a low-area reconfigurable and synthesizable digital loop filter for adpll," in *2023 fourteenth international conference on ubiquitous and future networks (ICUFN)*. IEEE, 2023, pp. 580–582.

[5] A. Lotfy, M. Ghoneima, and M. Abdel-Moneum, "A fast locking hybrid tdc-bb adpll utilizing proportional derivative digital loop filter and power gated dco," in *2016 IEEE international symposium on circuits and systems (ISCAS)*. IEEE, 2016, pp. 1646–1649.

[6] C.-C. Chang, Y.-T. Chin, H. A. Ibrahim, K. Y. Chang, and S.-J. Jou, "A low-jitter adpll with adaptive high-order loop filter and fine grain varactor based dco," in *2021 IEEE International Symposium on Circuits and Systems (ISCAS)*. IEEE, 2021, pp. 1–5.

[7] H.-C. Chu, Y.-H. Hua, and C.-C. Hung, "A fast-locking all-digital phased-locked loop with a 1 ps resolution time-to-digital converter using calibrated time amplifier and interpolation digitally-controlled-oscillator," in *2016 IEEE International Conference on Electron Devices and Solid-State Circuits (EDSSC)*. IEEE, 2016, pp. 375–378.

A 2.4 GHz Dual-Path Rectifier With Wide-Dynamic-Range for RF Energy Harvesting

Wansi Ge   Bharatha Kumar Thangarasu   Nagarajan Mahalingam   Kaixue Ma

Fanyi Meng   Zhenghao Lu   Kiat Seng Yeo

Abstract—This paper presents a dual-path rectifier operating at 2.4 GHz, designed to provide a wide input range with high power conversion efficiency (PCE), also known as the power dynamic range (PDR). The low-power path employs a cross-coupled differential-drive (CCDD) rectifier, while a differential Dickson full-wave rectifier is designed for a high-power path. A low-power adaptive control circuit is implemented to sense the output voltage of the rectifier, enabling automatic switching between two paths without external reference or power supply. Implemented in a 130-nm CMOS technology, the proposed dual-path rectifier achieves a sensitivity of -20.2 dBm and a PDR exceeding 25 dB for loads of 20 kΩ, 50 kΩ, and 80 kΩ. Notably, the dual-path rectifier achieves peak PCEs of 64.3% and 56.2% at input power of -11 dBm and 3.4 dBm with a load of 20 kΩ, respectively.

Index Terms—Cross-coupled differential-drive rectifier, Dickson rectifier, dual-path rectifier, RF energy harvesting

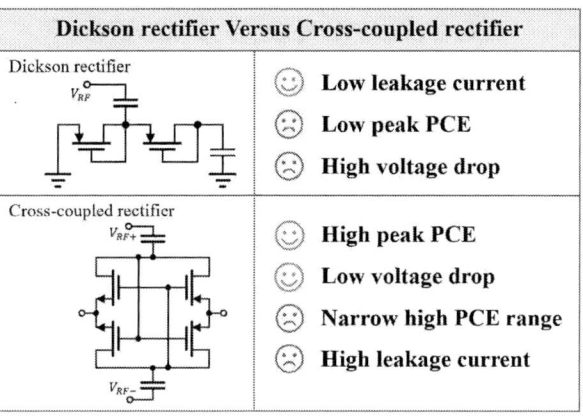

Fig. 1. Pros and Cons of Dickson rectifier and CCDD rectifier.

I. INTRODUCTION

Radio frequency (RF) energy harvesting has garnered significant attention as a reliable energy solution to replace batteries in various fields, including the Internet of Things (IoT), wireless sensor networks, healthcare, industry, and agriculture [1]. Moreover, the Industrial, Scientific, and Medical (ISM) bands, particularly the 2.4 GHz band, are of considerable interest within the RF bands due to their widespread applications and established standards [2].

In RFEH system, the rectifier is the critical block to transform the received RF power to DC power, often limiting the overall performance of the whole system [2]. Power conversion efficiency (PCE) is one of the key metrics for rectifier, as it directly impacts the distance of wireless power transmission [3]. Additionally, due to the variability of the RF energy density in real-world environments, the power dynamic range (PDR) is a vital indicator of rectifier to assess reliability in different RF conditions, which refers to the RF input power range where the PCE of rectifier remains above 20% [3]. Furthermore, the ability of rectifier to store and retain collected energy is another significant aspect, which is typically measured by leakage current [4].

Currently, mainstream CMOS rectifiers can be categorized into two types [5]: Dickson rectifiers and cross-coupled differential-drive (CCDD) rectifiers, as illustrated in Fig. 1, each type has distinct advantages and disadvantages. The Dickson rectifier performs well during high-power operation but suffers from low peak PCE due to high conduction voltage drop. In contrast, CCDD rectifier has higher peak PCE, but their narrow PDR and high leakage current affect reliability in RF environments with varying power densities.

Although several studies have reported various topologies [6] [7] designed to reduce reverse leakage current and extend dynamic range, a significant proportion of reverse leakage current remains when input power is high enough, severely limiting the PCE of rectifiers under high input power conditions. Furthermore, the dual-path rectifier proposed in [8] utilizes auxiliary circuits to provide reference voltage; however, these auxiliary circuits introduce additional losses, thereby impacting the overall efficiency of the system.

In this paper, we present a dual-path rectifier that effectively combines Dickson rectifier with CCDD rectifier and achieves adaptive path switching without the need for auxiliary circuits. The structure of this paper is organized as follows: Section II analyzes the operation principles of the dual-path rectifier; Section III introduces the simulation results and a conclusion is drawn in Section IV.

Wansi Ge, Bharatha Kumar Thangarasu, Nagarajan Mahalingam, Kaixue Ma, Fanyi Meng, and Kiat Seng Yeo are with the *School of Microelectronics, Tianjin University*, Tianjin, China. Zhenghao Lu is with the *School of Electronic and Information Engineering, Soochow University*, Suzhou, China. Kiat Seng Yeo is also with *Engineering Product Development, Singapore University of Technology and Design*, Singapore. Corresponding author: (Wansi Ge, email: gewansi@tju.edu.cn)

979-8-3315-2209-4/25 $31.00 © 2025 IEEE

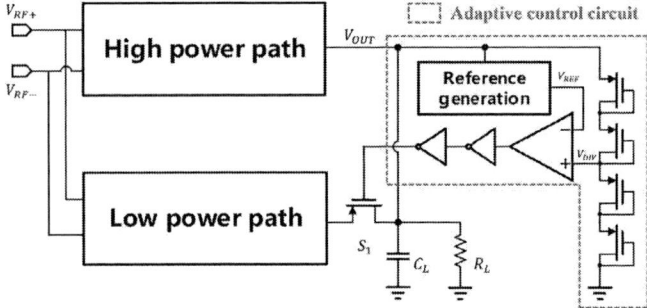

Fig. 2. The system architecture of the proposed dual-path rectifier.

II. CIRCUIT DESIGN

A. Operation Principle of Proposed System

Fig. 2 illustrates the block diagram of the proposed dual-path rectifier, which consists of a low-power path, a high-power path, an adaptive control circuit, and switch S_1. The low-power path employs a traditional CCDD rectifier to achieve better sensitivity and PCE at low input power. The high-power path utilizes a Dickson rectifier, extended into a differential full-wave rectifier to facilitate better integration with the CCDD rectifier, thereby achieving higher PCE at high input power. When the input power is low, the high input power path is disabled due to its high conduction voltage drop, while the low power path is activated. As the input power gradually increases, the rectifier output voltage V_{OUT} rises until the voltage output from the voltage divider circuit V_{DIV} exceeds the reference voltage V_{REF}, the comparator will output "1", then the low-power path is disabled and the high-power path starts to operate.

B. Rectifier Design

To enhance PDR, the proposed dual-path rectifier integrates the advantages of Dickson rectifier and CCDD rectifier. As shown in Fig. 3(a), the low power path employs a two-stage CCDD rectifier, where the cross-coupled differential structure significantly reduces the effective threshold voltage, thereby minimizing conduction losses caused by the on-resistance and improving both PCE and sensitivity [9]. However, it suffers from the poor performance at high input power due to the reverse current leakage. Therefore, when the input power is high, the low input path needs to be disconnected, otherwise, the low-power path will generate a large reverse leakage current that limits the V_{OUT}.

As shown in Fig. 3(b), the high input path utilizes an improved two-stage Dickson rectifier, which is extended into a differential full-wave rectifier. This design not only facilitates better integration with the CCDD rectifier for matching network optimization, but also significantly reduces reverse leakage current, enhances performance at high input power levels, and increases system reliability.

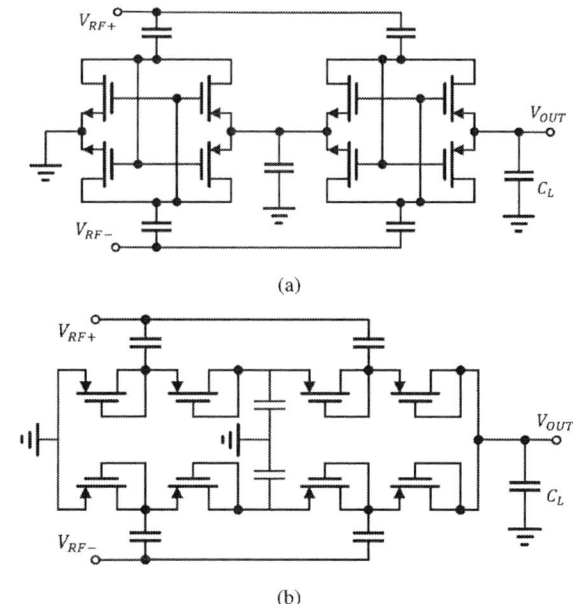

(a)

(b)

Fig. 3. (a) The two-stage CCDD rectifier for low-power path; (b) The two-stage full-wave rectifier for high-power path.

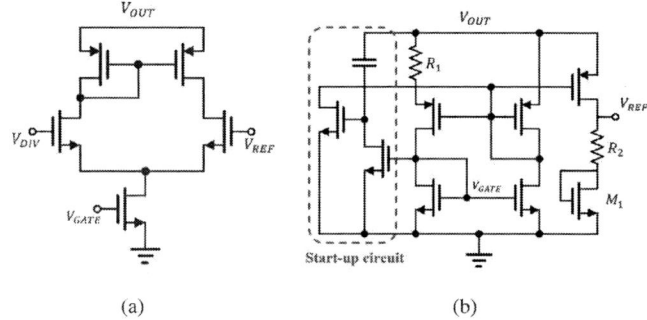

(a)　　　　　(b)

Fig. 4. Schematic of (a) the reference generation circuit and (b) comparator for adaptive control.

C. Adaptive Control Circuit

The proposed adaptive control circuit consists of a voltage divider network, a reference generation circuit, and a comparator. Due to the limited energy harvested by the rectifier, the adaptive control circuit employs long-channel high-voltage devices with relatively low current drive capability to meet low power requirements. As shown in Fig. 4(a), the comparator utilizes a conventional five-transistor OTA circuit, providing strong portability. Fig. 4(b) shows the reference generation circuit which simultaneously produces a reference voltage and a reference current, with the output stage employing NMOS transistor M_1 and resistor R_2 for temperature compensation, achieving a temperature coefficient of 36 ppm/°C. When the output voltage V_{OUT} is 1.5 V, the rectifier switches from the low-power path to the high-power path, with the total current consumption of the adaptive control circuit only 177 nA.

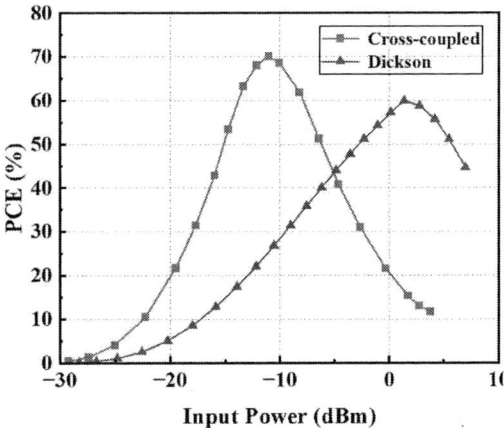

Fig. 5. Simulated PCE of the two-stage CCDD rectifier and the two-stage Dickson rectifier.

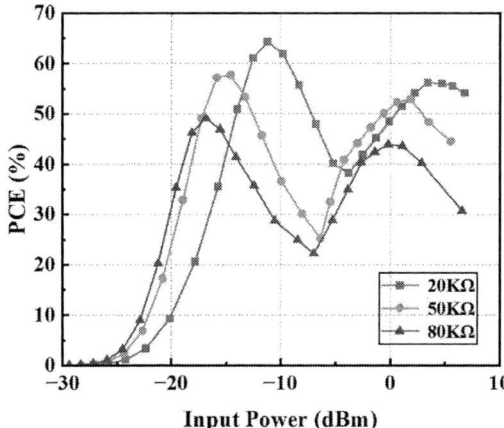

Fig. 7. Simulated PCE of the proposed dual-path rectifier with different load.

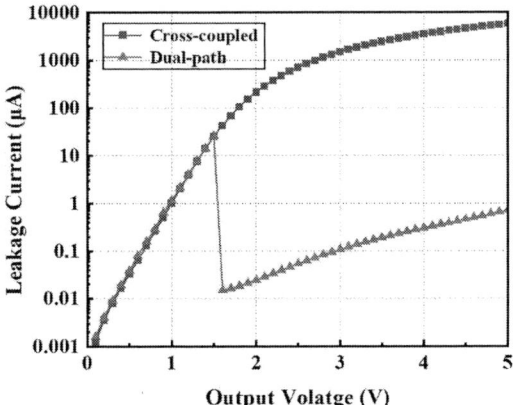

Fig. 6. Simulated leakage current of the two-stage CCDD rectifier and the two-stage Dickson rectifier.

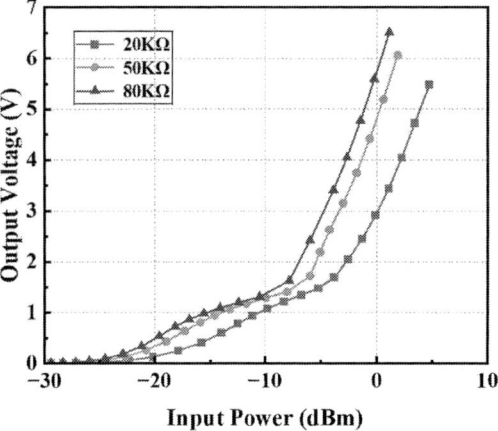

Fig. 8. Simulated output voltage of the proposed dual-path rectifier with different load.

III. SIMULATION RESULTS

The proposed dual-path rectifier is designed in a 0.13-um CMOS technology and the low-power path NMOS is used in the deep N-well process to eliminate the body effect. The definition of PCE is as follows [3]:

$$PCE_{rec} = \frac{P_{out}}{P_{in,rec}} \tag{1}$$

where P_{out} is the DC output power, $P_{in,rec}$ is the average received power of the rectifier.

The Fig. 5 shows the simulated PCE of the two-stage CCDD rectifier and the two-stage Dickson rectifier, it can be observed that the CCDD rectifier exhibits better PCE at low input power, while the Dickson rectifier demonstrates superior performance at high input power, attributed to its lower reverse leakage current. As shown in Fig. 6, with the continuous increase in output voltage, the low-power path is disconnected, and the high-power path starts to operate, resulting in a dramatic reduction in system leakage current.

The simulated PCEs of the proposed dual-path rectifier with adaptive control circuit under different loads are shown in Fig. 7. under a load of 20 kΩ, the PCE is higher than 20% with input power $P_{in,rec}$ ranging from -18 to 7 dBm, maintaining the PDR of 25 dB, and it achieves two PCEs of 64.3% at -11dBm and 56.2% at 3.4 dBm. Additionally, the PDR still exceeds 25 dB at loads of 50 kΩ and 80 kΩ.

Fig. 8 shows the output voltage of the dual-path rectifier versus $P_{in,rec}$ with different load. As can be seen, path switching is achieved at an output voltage of 1.5 V and there is a significant increase in the V_{OUT}. The minimum $P_{in,rec}$ required to charge up the dual-path rectifier with 1 V output is -20.2 dBm with pure capacitive load. From the above simulation results, it can be concluded that the proposed dual rectifier can have better performance at different input powers compared to the CCDD rectifier and Dickson rectifier with only single peak PCE.

Table I shows a performance summary and comparison with the published state-of-the-art works. This paper proposed a

TABLE I
PERFORMANCE COMPARISON WITH STATE-OF-THE ART WORKS

	This work*	TCAS-II [8]	VSLI [10]	TCAS-II [5]	MWCL [11]	TCAS-I [12]
Frequency	2.4 GHz	900 MHz	2.4 GHz	900 MHz	1 GHz	953 MHz
Matching network	Not included	off-chip	Not included	Not included	Not included	Not included
Technology	130 nm CMOS	65 nm CMOS	180 nm CMOS	65 nm CMOS	180 nm CMOS	130 nm CMOS
No.of stages	2	2	3	3	1	3
Sensitivity @V_{out}=1 V	-20.02 dBm for R_L=∞	-17.7 dBm for R_L=∞	N.A	-15.5 dBm for R_L=100 kΩ	-18 dBm for R_L=100 kΩ	N.A.
Peak PCE @$P_{in.rec}$	64.3% @-11 dBm 56.2% @3.4 dBm	32.5% @-14 dBm 36.5% @-10 dBm	41% @2.1 dBm 47% @8.9 dBm	79.77% @-17.5 dBm	65% @-20 dBm	69.5% @-1.5 dBm
PDR**	25 dB	11 dB	11 dB	21 dB	13 dB***	13 dB***

*Simulation result **PCE above 20%, ***Estimated form the figure.

dual-path rectifier to achieve a better and comparable PDR of 25 dB, in which the PCE can be maintained above 20%. Moreover, it exhibits a substantial peak PCE. The proposed rectifier can operate efficiently at a wide input RF power range, improving the robustness and reliability of RF energy harvesting over different transmission distances or in unstable environments.

IV. CONCLUSION

This paper presents a 2.4 GHz dual-path rectifier with adaptive control circuit to maintain high PCE for a wide input power range. The low-power path utilizes two-stage CCDD rectifier, while the two-stage different full-wave rectifier based on Dickson rectifier is designed for high-power path. The low-power adaptive control circuit can automatically switch paths without auxiliary circuits or external power supply. The simulation results show that the proposed rectifier achieves a PDR of 25dB from -18 to 7 dBm with a sensitivity of -20.2 dBm and two peak PCEs, which are 64.3% and 56.2% at input powers of 11 and 3.4 dBm, respectively.

REFERENCES

[1] D. Chen, R. Li, J. Xu, D. Li, C. Fei, and Y. Yang, "Recent progress and development of radio frequency energy harvesting devices and circuits," Nano Energy, p. 108845, 2023.

[2] T. Hu, M. Huang, Y. Lu, X. Y. Zhang, F. Maloberti, and R. P. Martins, "A 2.4-GHz CMOS differential class-DE rectifier with coupled inductors," IEEE Trans. Power Electron, vol. 36, no. 9, pp. 9864–9875, 2021.

[3] A. C. C. Chun, H. Ramiah, and S. Mekhilef, "Wide power dynamic range CMOS RF-DC rectifier for RF energy harvesting system: A review," IEEE Access, vol. 10, pp. 23 948–23 963, 2022.

[4] W. Ge, B. K. Thangarasu and K. S. Yeo, "A Wide Input Range RF Energy Harvester with Low Leakage Current for Far-Field Wireless Power Transfer in 28nm CMOS Technology," 2024 IEEE International Conference on IC Design and Technology (ICICDT), Singapore, Singapore, pp. 1-4, 2024

[5] A. Choo, Y. C. Lee, H. Ramiah, Y. Chen, P. -I. Mak and R. P. Martins, "A High-PCE Range-Extension CMOS Rectifier Employing Advanced Topology Amalgamation Technique for Ambient RF Energy Harvesting," in IEEE Transactions on Circuits and Systems II: Express Briefs, vol. 70, no. 10, pp. 3747-3751, Oct. 2023.

[6] A. S. Almansouri, M. H. Ouda, and K. N. Salama, "A CMOS RF-to-DC Power Converter With 86% Efficiency and -19.2-dBm Sensitivity," IEEE Transactions on Microwave Theory and Techniques, vol. 66, no. 5, pp. 2409–2415, 2018.

[7] X. Li, F. Mao, Y. Lu and R. P. Martins, "A VHF wide-input range CMOS passive rectifier with active bias tuning", IEEE J. Solid-State Circuits, vol. 55, no. 10, pp. 2629-2638, Oct. 2020.

[8] Y. Lu, H. Dai, M. Huang, M. Law, S. Sin, S. U, and R. P. Martins, "A Wide Input Range Dual-Path CMOS Rectifier for RF Energy Harvesting," IEEE Transactions on Circuits and Systems II: Express Briefs, vol. 64, no. 2, pp. 166–170, 2017.

[9] K. Kotani, A. Sasaki, and T. Ito, "High-efficiency differential-driven CMOS rectifier for UHF RFIDs," IEEE J. Solid-State Circuits, vol. 44, no. 11, pp. 3011–3018, 2009.

[10] C.-J. Li and T.-C. Lee, "2.4-GHz high-efficiency adaptive power," IEEE Trans. Very Large Scale Integr. (VLSI) Syst., vol. 22, no. 2, pp. 434–438, Feb. 2014.

[11] M. H. Ouda, W. Khalil and K. N. Salama, "Wide-Range Adaptive RF-to-DC Power Converter for UHF RFIDs," in IEEE Microwave and Wireless Components Letters, vol. 26, no. 8, pp. 634-636, Aug. 2016

[12] A. K. Moghaddam, J. H. Chuah, H. Ramiah, J. Ahmadian, P. -I. Mak and R. P. Martins, "A 73.9%-Efficiency CMOS Rectifier Using a Lower DC Feeding (LDCF) Self-Body-Biasing Technique for Far-Field RF Energy-Harvesting Systems," in IEEE Transactions on Circuits and Systems I: Regular Papers, vol. 64, no. 4, pp. 992-1002, April 2017

A 2 - 9 GHz SiGe Broadband High-Gain Low Noise Amplifier Using Resistive Feedback Technology

Xiaozheng Guo Bharatha Kumar Thangarasu Nagarajan Mahalingam Kaixue Ma Fanyi Meng Zhenghao Lu

Kiat Seng Yeo

Abstract—**This paper presents the design and implementation of a SiGe BiCMOS broadband high-gain low noise amplifier (LNA) operating in the 2 to 9 GHz frequency range. The LNA utilizes advanced SiGe heterojunction bipolar transistor (HBT) technology, which offers high cutoff frequencies and low noise performance, making it ideal for broadband applications. The resistive-feedback and inductive-peaking techniques are employed to enlarge the bandwidth. Moreover, the inductive-degeneration technique and L-matching network are utilized to realize a wideband noise and input impedance matching simultaneously.The LNA is fabricated in a 0.13-μm SiGe BiCMOS technology. It achieves a peak gain of 22.8 dB and a minimum noise figure (NF) of 2.7 dB at room temperature. The output-referred 1-dB compression point (OP1dB) is 3.58 dBm. The S11 is below −10 dB from 1.4 to 9.3 GHz and S22 is below −10 dB from 1.9 to 10.5 GHz.**

Index Terms—**Bandwidth (BW), BiCMOS, low-noise amplifier (LNA), noise figure (NF)**

I. INTRODUCTION

Currently, with the rapid expansion of the wireless market, numerous wireless standards have been proposed, and research into multi-standard operations has significantly intensified. The concept of designing a multi-standard communication system is undeniably intriguing. The primary objective of such systems is to integrate as many components as feasible across various standards, thereby minimizing both the cost and footprint of the systems. However, constructing a system capable of accommodating diverse frequency bands and requirements poses a considerable challenge. The designers employ two strategies to meet the requirements: one option involves utilizing tunable LNAs (Low Noise Amplifiers) with narrowband-based RF front ends. However, this approach entails significant drawbacks, including large chip area occupation, design complexity, and increased costs. Alternatively, they can adopt a broadband RF front end, which has the capability to fulfill the needs of any standard within a wide frequency range.

Broadband low-noise RF amplifiers (LNAs) are essential for multiband/multi-standard RF front-ends, UWB radios, and potentially other standards. Designing such wideband amplifiers poses challenges due to difficulties in achieving wideband impedance matching and the excessive voltage drop across resistive loads. To address these challenges, various topologies have been employed in designing broadband LNAs, including distributed amplifiers, multi-section LC matching networks, common-gate configurations, and shunt resistive feedback.

Nowadays, many approaches have been proved to be very effective in expanding the bandwidth of low noise amplifier (LNA), and the most commonly used approaches are distributed topologies, common base/gate (CB/CG) LNAs, shunt-resistive feedback and inductive peaking techniques.While distributed topology-based ultra-wideband (UWB) LNAs [1], [2] can achieve ultra-wide bandwidth, they typically suffer from disadvantages such as large area, high power consumption, and elevated noise figure (NF). To address wideband input matching, the common base/gate topology has been adopted [3]. However, this approach can lead to output current coupling directly to the input terminal, ultimately degrading the NF. To mitigate this issue, noise-canceling techniques [4], are often employed in common base/common gate (CB/CG) LNAs. Additionally, the shunt-resistive feedback technique has been reported in numerous studies [5], for effectively broadening the LNA's bandwidth. Another common method is the inductive peaking technique [6], which, while effective for bandwidth extension, tends to result in a larger chip area. Besides bandwidth, NF is a crucial performance metric for LNAs. A low and flat NF across the desired bandwidth is essential for UWB LNAs. .

II. CIRCUIT DESIGN AND ANALYSIS

A. *Topology of the proposed LNA*

Fig. 1 depicts the schematic of the proposed ultra-wideband LNA, which features a one-stage cascode structure. To achieve a high and flat gain across a wide frequency range, the series inductive-peaking technique is employed in both stages. In

---

Xiaozheng Guo, Bharatha Kumar Thangarasu, Nagarajan Mahalingam, Kaixue Ma, Fanyi Meng and Kiat Seng Yeo are with *the School of Microelectronics, Tianjin University*, Tianjin, China. Zhenghao Lu is with *the School of Electronic and Information Engineering, Soochow University*, Suzhou, China. Kiat Seng Yeo is also with *Engineering Product Development, Singapore University of Technology and Design,*, Singapore. Corresponding author: (Xiaozheng Guo, email: guoxiaozheng@tju.edu.cn)

979-8-3315-2209-4/25 $31.00 © 2025 IEEE

the first stage, resistive-feedback and inductive-degeneration techniques are utilized to compensate for the real part of the input impedance (Rin) and enhance stability, albeit at the slight cost of gain. Furthermore, a Second-order LC-matching network is incorporated at the LNA's input to simultaneously achieve wideband noise matching and input impedance matching. Although the feedback resistance is to improve the stability of the structure, the value of the feedback resistance will affect the overall gain to a certain extent.

In the first stage of the LNA, the series inductor L3, resistor R3 and the shunt parasitic capacitor Cp of the output of the first stage form the RLC resonant network, thereby extending the bandwidth and improving the gain flatness.

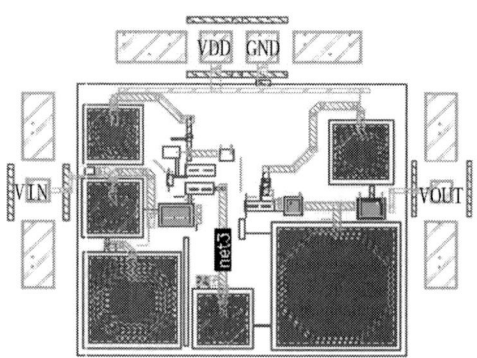

Fig. 2. Layout of the broadband LNA.

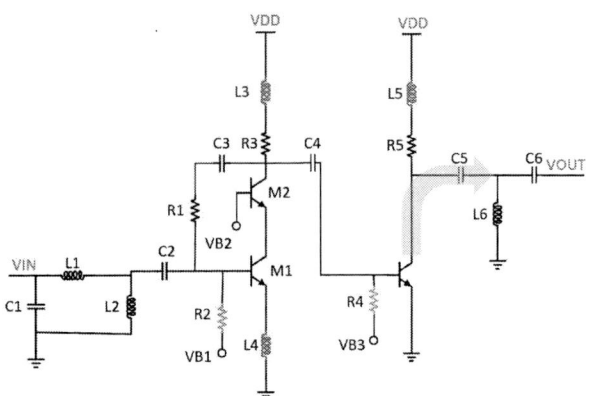

Fig. 1. Simplified schematic of the proposed broadband LNA.

B. Wideband Noise and Input Impedance Matching

The traditional method for achieving simultaneous noise and input impedance matching involves using emitter/source degeneration and base/gate inductors. However, this approach is limited to a single frequency, resulting in a narrow bandwidth. In the proposed Ultra-Wideband (UWB) Low-Noise Amplifier (LNA), the first stage employs a resistive-feedback technique along with a second-order LC-matching network to accomplish wideband simultaneous noise and input impedance matching.

III. SIMULATION RESULTS

The proposed broadband LNA was designed in 0.13-μm SiGe BiCMOS process in 2 V. Fig. 2 shows the LNA layout with a die size of 0.712 x 0.596 mm². And obtained the post simulation results. The simulated S21 reaches 22.8 dB as its peak value, 19.8 dB at 2.2 GHz, and 22.8 dB at 9 GHz, as being shown by Fig. 3 (b). Fig. 3 (a) and Fig. 3 (c) shows the S11 and S22 versus frequency characteristics of the broadband LNA. S11 is less than 10 dB in the entire 2 GHz to 9 GHz bandwidth, and S22 is less than 10 dB in the entire 2 GHz-9 GHz bandwidth. It demonstrates good input and output matching. Fig. 3 (d) shows the measured and simulated NF, which varies from 2.7 dB to 3.3dB within the bandwidth. As shown in Fig. 4 (a), it is found that the OP1dB is 3.58 dBm.

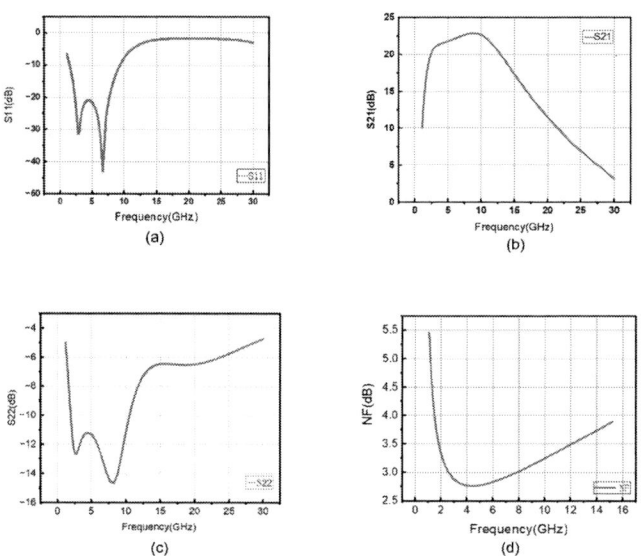

Fig. 3. Simulation results of (a) S11 (b)S21 (c)S22 and (d)NF .

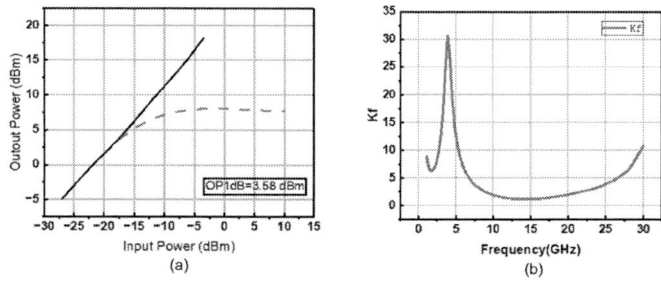

Fig. 4. Simulation results of (a) OP1dB and (b) stability factor, Kf.

979-8-3315-2209-4/25 $31.00 © 2025 IEEE   468

IV. CONCLUSION

In this brief, a two-stage UWB LNA with a 3-dB gain BW from 2 to 9 GHz is presented. Self-biased resistive feedback topology is employed. Inductance peak techniques are used to increase bandwidth and gain. To accomplish the input impedance matching, the LNA uses a second-order LC network. The LNA achieves a peak gain of 22.8 dB and a minimum NF of 2.7 dB at room temperature. The output-referred 1-dB compression point (OP1dB) is 3.58 dBm. The S11 is below –10 dB from 1.4 to 9.3 GHz and S22 is below –10 dB from 1.9 to 10.5 GHz. Table 1 summarizes the performance of the UWB LNA.

TABLE I
PERFORMANCE SUMMARY AND COMPARISON

Reference	[7]	[8]	[9]	This work
Technology(SiGe)	130nm	130 nm	130nm	130 nm
BW (GHz)	1-27	5.1-33.6	0.3-15	2-9
FBW(%)	185.7	147.2	192.1	127.2
Gain (dB)	27.0	14.3	37.3	22.8
NF (dB)	1.4	3.3	1.8	2.7
IIP3 (dBm)	-8.0	1.7	-27.3	-7.13
OP1dB (dBm)	7.1	-	-	3.58
Pdc (mW)	85.0	56.7	52.0	82.06

REFERENCES

[1] G. Nikandish and A. Medi, "A 40-GHz bandwidth tapered distributed LNA," IEEE Trans. Circuits Syst. II, Exp. Briefs, vol. 65, no. 11, pp. 1614–1618, Nov. 2018.

[2] J. X. Yan, H. Luo, J. Zhang, H. Zhang, and Y. Guo, "Design and analysis of a cascode distributed LNA with gain and noise improvement in 0.15-μm GaAs pHEMT technology," IEEE Trans. Circuits Syst. II, Exp. Briefs, vol. 69, no. 12, pp. 4659–4663, Dec. 2022.

[3] R.-M. Weng, C.-Y. Liu, and P.-C. Lin,"A low-power full-band low-noise amplifier for ultra-wideband receivers," IEEE Trans. Microw. Theory Techn., vol. 58, no. 8, pp. 2077–2083, Aug. 2010.

[4] Z. Liu and C. C. Boon, "A 0.092-mm2 2–12-GHz noise-cancelling low-noise amplifier with gain improvement and noise reduction," IEEE Trans. Circuits Syst. II, Exp. Briefs, vol. 69, no. 10, pp. 4013–4017, Oct. 2022.

[5] C. Feng, X. P. Yu, W. M. Lim, and K. S. Yeo, "A compact 2.1–39 GHz self-biased low-noise amplifier in 65 nm CMOS technology," IEEE Microw. Wireless Compon. Lett., vol. 23, no. 12, pp. 662–664, Dec. 2013.

[6] N. Li, W. Feng, and X. Li, "A CMOS 3–12-GHz ultrawideband low noise amplifier by dual-resonance network," IEEE Microw. Wireless Compon. Lett., vol. 27, no. 4, pp. 383–385, Apr. 2017.

[7] Z. Wang et al., "A 1–27 GHz SiGe Low Noise Amplifier With 27-dB Peak Gain and 2.85±1.45 dB NF," in IEEE Trans. on Circuits Syst. II: Exp. Briefs, vol. 71, no. 5, pp. 2629-2633, May 2024.

[8] C. Çaışkan, I. Kalyoncu, M. Yazici and Y. Gurbuz, "Sub-1-dB and Wide-band SiGe BiCMOS Low-Noise Amplifiers for X -Band Applications," in IEEE Trans. Circuits Syst. I, Reg. Papers, vol. 66, no. 4, pp. 1419-1430, April 2019.

[9] S. Zeinolabedinzadeh, A. Ç. Ulusoy, M. A. Oakley, N. E. Lourenco and J. D. Cressler, "A 0.3–15 GHz SiGe LNA With >1 THz Gain-Bandwidth Product," in IEEE Microw. and Wireless Compon. Lett., vol. 27, no. 4, pp. 380-382, April 2017.

979-8-3315-2209-4/25 $31.00 © 2025 IEEE

A superconducting tunable filter with fixed input/output position

Mingchao Li, Kuangyong Gao, Xingkun You, Xiang Li, Huanghao Ying, Huafei Cheng, Haizhen Guo, Zhen Shi, and Zhonghai Zhang

Abstract—A superconducting tunable filter with fixed input/output position was proposed in this letter. The superconducting tunable filter mainly consists of a folded open ring and two tunable capacitors loaded at the open end of the ring. In order to maintain the input/output ports in a fixed position, the input/output resonant cavity adopts a zigzag resonant structure to achieve the adjustment of superconducting input/output position in different frequency bands and ensure the fixed position. A superconducting tunable filter composing of six tunable resonators has been designed, fabricated and measured. The measured results demonstrate the effectiveness of the method described in this article.

Index Terms—superconducting tunable filter, rectangular tunable coupling rings, tunable capacitors.

I. INTRODUCTION

THE superconducting filters have been widely used due to their low insertion loss in the passband and steep out of band suppression performance. [1-10]. Tunable superconducting filters can adjust their operating frequency in real-time according to practical applications, which can improve spectrum utilization effectively.

The design methods of superconducting tunable filters were generally divided into two categories. The first type was implemented using resonators with special structures[1-2]. This form of superconducting tunable filter has a compact structure and excellent performance, but it was difficult to achieve steeper out of band suppression performance by expanding the number of resonators. In the second type, a resonant coupling model was adopted, people can improve the out of band suppression performance of the filter by adding cross coupling and increasing the number of resonators[3-10].

For resonators in the same form, the input/output feeding positions of superconducting filters with different frequencies were different. In order to reduce the design cost of metal shielding shells, it was necessary to have the input/output ports of the filter in the same position. At the same time, due to applications such as multiplexers, it was required that the length of the input/output transmission lines of superconducting filters be minimized. Therefore, it was necessary to improve the form of resonators while meeting the requirements of fixed positions of input/output ports and the shortest input/output transmission lines.

A superconducting tunable filter employing a zigzag resonant structure was proposed in this letter. By introducing the zigzag structure of the input-output resonator, the loaded Q values of the input-output can be adjusted without changing the position of the input-output structure, thereby maintaining the fixed position of the input-output structure while ensuring the performance of the filter.

II. THE DESIGN OF SUPERCONDUCTING TUNABLE FILTER

The structure of the superconducting tunable filter in this letter was shown in Fig.1.

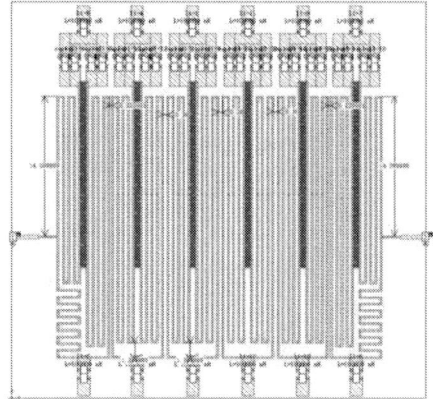

Fig.1. The structure of the superconducting tunable filter

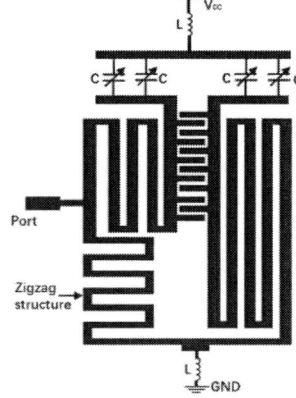

Fig.2. The Zigzag structure of the input resonator

For the superconducting tunable filter shown in Fig.1, the

Mingchao Li, Kuangyong Gao, Xingkun You, Xiang Li, Huanghao Ying, Huafei Cheng, and Zhonghai Zhang was School of Electronic Information, Hangzhou Dianzi University. Hangzhou, R.P. China. Hangzhou Dianzi Univ., Xiasha University Park, Street #2, 310018, Hangzhou Zhejiang, China. (e-mail: zhanghaidong6388@163.com). Haizhen Guo and Zhen Shi was Zhejiang

Post&Telecommunication Construction Co.,Ltd, No.99, Tai'an Road, Binjiang District, 310051, Hangzhou Zhejiang, China. (e-mail: 18958134369@189.cn). This work was supported by the National Key Research and Development Project (No.2022YFF1400101). Corresponding author: Zhonghai Zhang, email: zhangzhonghai@hdu.edu.cn.

979-8-3315-2209-4/25 $31.00 © 2025 IEEE

resonator was realized of U type structure. In order to achieve miniaturization of the filter, a folded U-shaped resonator was used, and an interdigital capacitor was added at the open end for further miniaturization.

The zigzag resonant structure on the input/output resonant cavity was shown in Fig.2.

From Fig.2, it can be seen that by adjusting the length of the zigzag structure, the loaded Q values of the input and output can be dynamically adjusted. According to reference 1, while straight-line resonators exhibit a resonance frequency comparable to their zigzag counterparts, their physical dimensions are substantially larger. However, zigzag resonators demonstrate significantly weaker coupling strength compared to straight-line configurations. This reduced coupling enables closer spatial arrangements between adjacent resonators, ultimately facilitating the implementation of narrowband filters with exceptionally compact layouts.

III. THE DESIGN OF THE FULLY TUNABLE FILTER

To verify the performance of the superconducting tunable filter proposed in this paper, a 6-cavity superconducting tunable filter operating at VHF band was designed and measured in this letter. The layout of the superconducting tunable filter was shown in Fig.3.

Fig.3. The layout of the superconducting tunable filter.

In Fig.3, the superconducting substrate was made of lanthanum aluminate material with a dielectric constant of 23.75 and a substrate thickness of 0.5mm. In Fig.3, frequency tuning was achieved using Mac46H202 varactor diode from Macom Corporation.This varactor diode has hyperabrupt junctions with constant gamma of 0.75 from 0 to 20 volts and very high quality factor approaching that of abrupt junction varactors. The equivalent circuit diagram of this varactor is shown in the Fig.4.

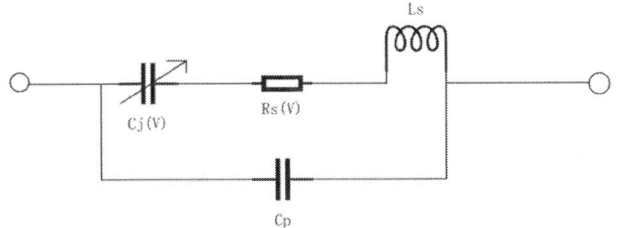

Fig.4. Tuning Varactor Equivalent Circuit

The measured results were shown in Fig.5 and Fig.6.

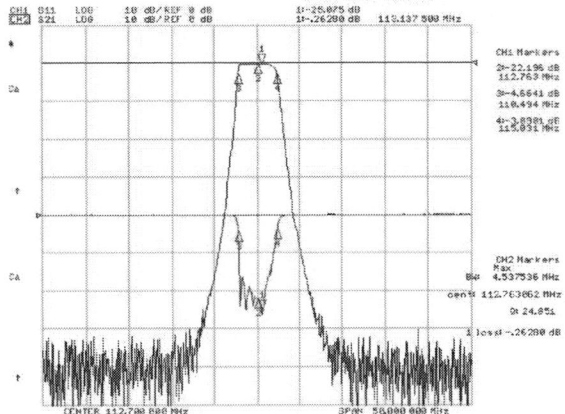

Fig.5. The measured transmission performance of the superconducting tunable filter(low frequency).

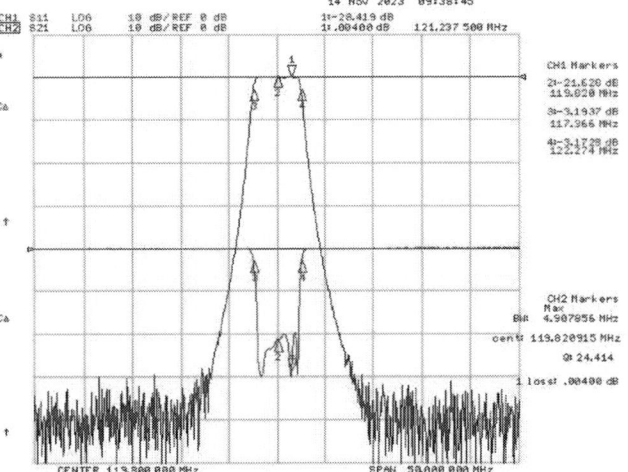

Fig.6. The measured transmission performance of the superconducting tunable filter(high frequency).

From Fig.5 and Fig.6, it can be seen that as the capacitance value of the varactor diode varies from 1pf to 5pf, this superconducting tunable filter can be tuned from 122MHz to 129MHz.

IV. CONCLUSION

A superconducting tunable filter with a fixed position of input/output structure that does not vary with the operating frequency band was presented in this letter. By using a zigzag structure in the input/output resonant cavity to dynamically adjust the loaded Q values of the input-output at different operating frequencies, the position of the input-output structure was ensured to be fixed. The structure described in this article can reduce the design cost of filter shielding boxes, ensure the shortest length of input and output transmission lines, and can be applied to the circuit design of complex functions such as superconducting multi-couplers.

REFERENCES

[1] X. Wang et al., "A Compact Four-Pole Tunable HTS Bandpass Filter at VHF Band," in IEEE Transactions on Applied Superconductivity, vol. 31, no. 6, pp. 1-7, Sept. 2021.

[2] W. Yang, H. Zhang and Z. Shang, "Wide-Stopband HTS Dual-Band Bandpass Filter Using a Folded Step Impedance Hairpin-Ring Resonator," in IEEE Transactions on Applied Superconductivity, vol. 35, no. 3, pp. 1-6, May 2025.

[3] H. Liu, S. Wang, J. Kuang, R. Wang and H. Tian, "Compact HTS Narrowband Bandpass Filter Based on Spiral D-CRLH Resonator," in IEEE Transactions on Circuits and Systems II: Express Briefs, vol. 71, no. 3, pp. 1007-1011, March 2024.

[4] H. Zhang and S. Hai, "Design of Double Passband HTS Filter Based on SIR," 2024 9th International Conference on Intelligent Informatics and Biomedical Sciences (ICIIBMS), Okinawa, Japan, 2024, pp. 798-802.

[5] L. Feng, Z. Li, C. Lu, X. Yu and J. Wang, "Differential HTS Bandpass Filters Based on Coplanar Stripline," 2023 Cross Strait Radio Science and Wireless Technology Conference (CSRSWTC), Guilin, China, 2023, pp. 1-3.

[6] N. Sekiya and T. Tsuruoka, "Improvement of Filter Properties of Independently Tunable Superconducting Dual-Band Bandpass Filter," in IEEE Transactions on Applied Superconductivity, vol. 29, no. 5, pp. 1-4, Aug. 2019.

[7] L. Zhou, H. Li, Z. Long, S. Cao, M. Jiang and T. Zhang, "Design of High-Temperature Superconducting Dual-Band Filter With Multiple Transmission Zeros," in IEEE Transactions on Applied Superconductivity, vol. 29, no. 6, pp. 1-12, Sept. 2019.

[8] H. Liu, Y. Wang, P. Wen and S. Zheng, "Novel Tri-Band High-Temperature Superconducting Bandpass Filters Using Asymmetric Shunted-Line Stepped-Impedance Resonator (SLSIR)," in IEEE Access, vol. 7, pp. 32504-32509, 2019.

[9] L. Tao, B. Wei, X. Guo and B. Cao, "Compact Ultra-Narrowband Superconducting Filter Using Asymmetric Twin-Spiral Resonators," 2018 Asia-Pacific Microwave Conference (APMC), Kyoto, Japan, 2018, pp. 1348-1350.

[10] B. Ren *et al.*, "Differential Dual-Band Superconducting Bandpass Filter Using Multimode Square Ring Loaded Resonators With Controllable Bandwidths," in *IEEE Transactions on Microwave Theory and Techniques*, vol. 67, no. 2, pp. 726-737, Feb. 2019.

A 5.5-6-GHz, 4.8mW, 5.7dBm IIP3 LNA in 22-nm CMOS Technology for Wi-Fi Application

Yingqi Liu, Kaiyun Deng, Haoyu Dong, Bozhi Qiu and Haigang Feng[*]
Shenzhen International Graduate School,Tsinghua University, Shenzhen, China
*feng.haigang@sz.tsinghua.edu.cn

Abstract—**This paper presents a low-noise amplifier (LNA) designed for Wi-Fi applications, maintaining high linearity while achieving low power consumption, superior gain, and a low noise figure (NF). The proposed LNA employs a cascode topology with a two-stage capacitive cross-coupling technique to enhance the effective transconductance and output impedance, thereby improving gain and reducing NF. The two-stage capacitive cross-coupling structure continuously suppresses the second harmonic, which effectively eliminates the generated third-order harmonic, leading to improved linearity. In the proposed design, the LNA combines a balun to achieve input matching and single-ended-to-differential conversion. Fabricated in a 22 nm CMOS process, the post-layout simulation results demonstrate a peak gain of 13.5 dB, a minimum noise figure of 2.43 dB and an input reflection coefficient (S11) better than -16 dB over the bandwidth of 5.5-6 GHz, while achieving an input third-order intercept point (IIP3) of 5.7 dBm at 5.85 GHz and a power consumption of 4.8 mW with 0.9 V supply voltage.**

Index Terms—**CMOS, LNA, capacitive cross-coupling, linearity improvement**

I. INTRODUCTION

To accommodate increasing application demands [1] and achieve superior communication quality, the Wi-Fi Alliance has consistently updated the 802.11 standards, now reaching Wi-Fi 7 (802.11be). Meanwhile, the designs of the transceiver chip also must be continuously improved. The 5 GHz frequency range, preferred over 2.4 GHz for its better interference immunity, stability, and application potential, faces challenges like increased parasitic effects at higher frequencies. Additionally, with spectrum resources becoming increasingly scarce and signal interference intensifying, RF front-ends must process signals without introducing excessive noise or distortion. The low-noise amplifier (LNA), as the first active stage in the receiver, is crucial for improving signal-to-noise ratio (SNR) and sensitivity [2]- [4].

To enhance the noise performance of the LNA, noise optimization techniques are essential. The mainstream noise optimization techniques include noise-canceling [5]- [8] and g_m-boost techniques [9]. However, noise-canceling techniques typically require auxiliary paths to cancel noise and is less advantageous for low-power applications. Active g_m-boost techniques introduce additional transistors as active boosting stages, inevitably adding extra noise and power consumption. In contrast, capacitive cross-coupling techniques can enhance the effective transconductance (G_m) without these drawbacks, thereby significantly improving noise performance.

Within this context, we propose a novel LNA topology based on capacitive cross-coupling techniques to support Wi-Fi 7 applications in the 5 GHz frequency range. Section II presents the overall architecture. A detailed analysis of the LNA's key performance metrics is provided in Section III. Post-layout simulation results are discussed in Section IV, and conclusions are drawn in Section V.

II. PROPOSED LNA DESIGN

The design of the proposed LNA is illustrated in Fig. 1. The LNA consists of a two-stage capacitive cross-coupling differential cascode amplifier, a balun and two buffers. The input signal is first processed by a step-up balun achieving single-ended-to-differential conversion. Meanwhile, the balun transforms the source resistance to match the input impedance of the differential CG stage. The drain of M1 and M2 is further cascoded with CG transistors M3 and M4, and the differential output at the drains of M3 and M4 is processed through a buffer to achieve output impedance matching.

In a traditional CG LNA, to achieve impedance matching, the g_m must satisfy $g_m = 1/R_s = 20mS$ [10]. While increasing g_m further reduces the noise figure, it simultaneously degrading the input matching. However, the g_m-boost achieved by capacitive cross-coupling enhances the degree of freedom in adjusting the g_m, effectively breaks the trade-off between input matching and noise performance. Meanwhile, the proposed LNA introduces two-stage capacitive

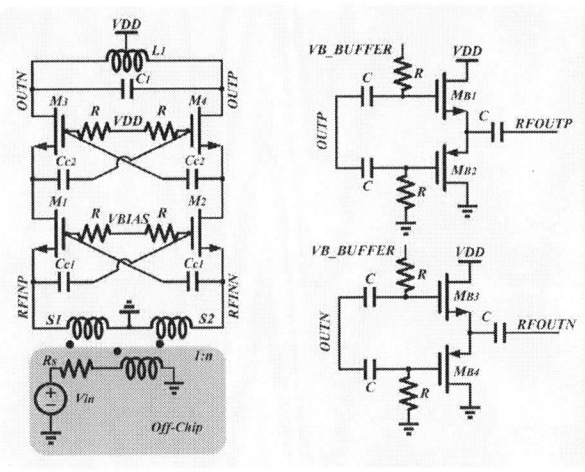

Fig. 1. Proposed capacitive cross-coupling cascode LNA.

979-8-3315-2209-4/25 $31.00 © 2025 IEEE

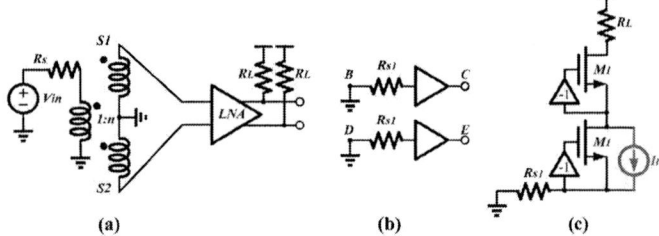

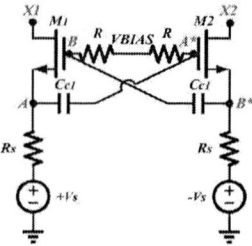

Fig. 2. (a) The cascaded structure of the balun and LNA, (b) the equivalent circuit of (a), and (c) noise analysis of (a).

Fig. 3. Fundamental capacitive cross-coupling structure.

cross-coupling. This topology continuously suppresses second-order distortion, thereby reducing third-order distortion and significantly improving linearity under high-gain condition, comprehensively optimizing the overall performance of the LNA.

III. DESIGN DETAILS

A. Input Matching

Consider the source impedance as R_s, due to the secondary impedances seen by S1 and S2 of the balun are equal, the single-ended input impedance of each half-circuit of the LNA needs to be equal to $(n^2 R_s)/2$ to achieve impedance matching. The capacitive cross-coupling result in $G_m = 2g_m(g_m = g_{m1} = g_{m2})$, consequently, the input impedance of each half-circuit can be simplified to $1/2g_m$. By properly adjusting the circuit parameters and the transformer turns ratio, impedance matching can be achieved in accordance with $1/2g_m = (n^2 R_s)/2 = R_{s1}$.

B. Noise Analysis

Under the condition of achieving input matching, the cascaded structure of the balun and LNA in Fig. 2(a) can be equivalently represented as shown in Fig. 2(b) for noise analysis, based on the impedance transformation relationship of the balun. Fig. 2(c) illustrates the specific circuit to analyze the noise of the input transistor. The noise voltage at point B is denoted as v_B. Given that the noise currents flowing through R_{s1} and R_L are equal, we obtain $v_B = -R_{s1}v_{outn}/R_1$. Consequently, let $A = 1$, the noise voltage across the gate and source of the input transistor is $-(A+1)v_A = -(A+1)(-R_{s1}v_{out}/R_1)$. Therefore, the output noise generated by M1 is:

$$I_{n1} + \frac{R_{s1}}{R_1}v_{n,M1out}(A+1)g_m = -\frac{v_{n,M1out}}{R_1} \quad (1)$$

$$\overline{v_{n,M1out}^2} = \left(\frac{R_L I_{n1}}{(A+1)g_m R_{s1}+1}\right)^2 = \frac{4kT\gamma g_m R_L^2}{(2g_m R_{s1}+1)^2} \quad (2)$$

Since the circuit achieves input matching, the gain from point B to point C is $1/2A_v = g_m R_L$ where R_L represents the loss of the load inductor. Thus, the output noise of the source impedance R_{S1} in the half-circuit can be expressed as

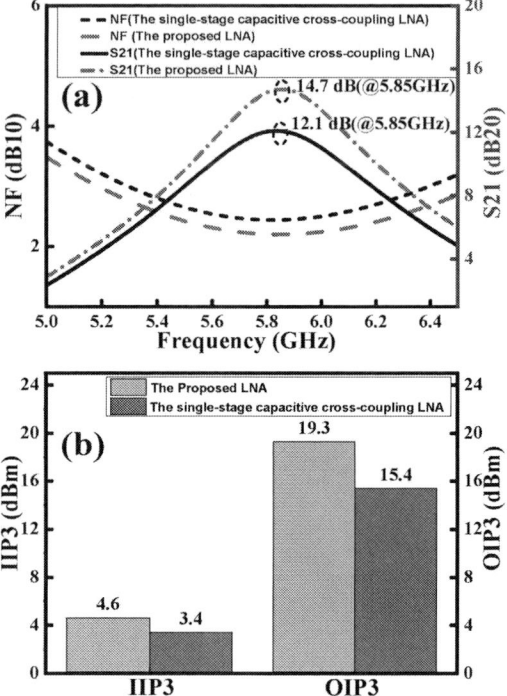

Fig. 4. Pre simulated (a) S21 and NF, (b) IIP3 and OIP3.

$4kTR_{S1} * (g_m R_L)^2$. The output noise of the load resistor is $4kTR_L$. At this point, the output noise of the half-circuit is:

$$\overline{v_{n,out}^2} = \frac{4kT\gamma g_m R_L^2}{(2g_m R_{s1}+1)^2} + 4kTR_{s1}*(g_m R_L)^2 + 4kTR_L \quad (3)$$

Using (3), the noise figure of the proposed balun-assisted capacitive cross-coupling LNA can be derived as (4). The noise figure of a traditional cascode circuit with a load R_L can be expressed as $NF = 1 + \gamma + \frac{4R_s}{R_L}$ [11]. Comparing the noise figure of the traditional cascode LNA and the proposed LNA, it is evident that the proposed LNA achieves input matching while enhancing the G_m, thereby reducing the noise figure compared to the traditional CG structure. This effectively breaks the trade-off between input matching and noise figure.

$$NF = \frac{\overline{v_{n,out}^2}}{\left(\frac{1}{2}A_\nu\right)^2 4kTR_{s1}} = 1 + \frac{\gamma}{2} + \frac{2n^2 R_s}{R_L} \quad (4)$$

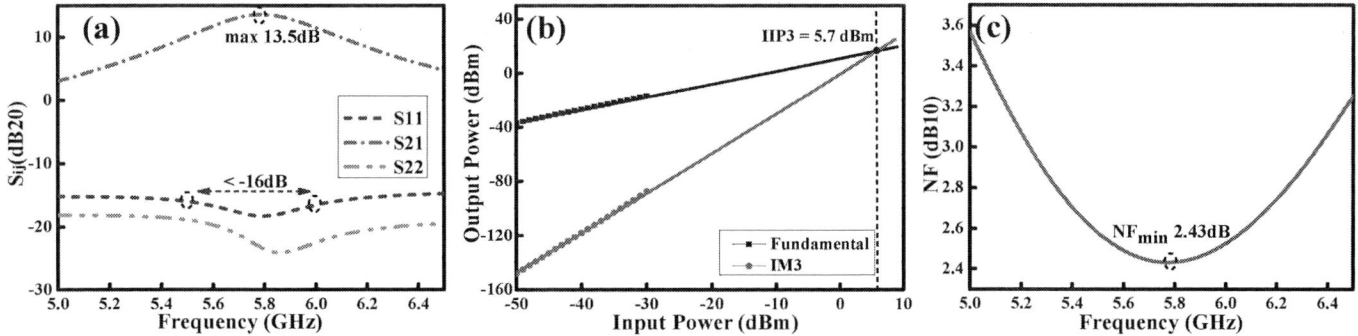

Fig. 5. Post-layout simulated (a) S-parameters, (b) IIP3, (c) NF.

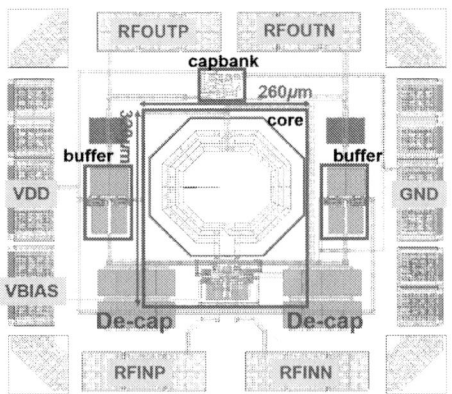

Fig. 6. Layout of the propoesd LNA.

TABLE I
PERFORMANCE COMPARISON

	RFIC'20 [12]	JSSC'21 [13]	TCASII'22 [14][a]	This work[a]
CMOS Tech (nm)	22	28	65	22
Freq (GHz)	5.4~6.1	0.02~4.5	0.47~3.3	5.5~6
Gain (dB)	28.3~31.1	15.2	19.45~22	10.5~13.5
S11(dB)	<-8	≤-10	≤-10	<-16
NF (dB)	1.9~3.3	2.7[b]	2.57~3.5	2.43~2.58
IIP3 (dBm)	-19.9 @5.7GHz	-4.62~-3.53	2.81	**5.7 @5.85GHz**
VDD (V)	1	1	1.5	**0.9**
Power (mW)	39	4.5	12.5	**4.8**
Active area (mm²)	0.061	0.03	0.0057	0.0832

[a] Post-layout simulation results; [b] Measured over 100MHz-6.5GHz;

C. Linearity

Considering the the fundamental capacitive cross-coupling structure in Fig. 3, gate-source voltage can be expressed as $v_{gs} = a_1 v_s + a_2 v_s^2 + a_3 v_s^3$ due to the input signal containing nonlinearity. Based on the expression for the output current $i_D = g_{m1} v_{gs} + g'_{m1} v_{gs}^2/2 + g''_{m1} v_{gs}^3/6$ and the definition of the third-order intercept point (IIP3), the expression for IIP3 can be derived as:

$$AIIP3 = \sqrt{\frac{4}{3}\left|\frac{2g_{m1}a_1}{2g_{m1}a_3 + 2g_{m1}'a_1 a_2 + \frac{g''_{m1}}{3}a_1^3}\right|} \quad (5)$$

At high frequencies, taking transistor M1 as an example, the first-order component of the gate-source voltage of M1 is equal to the difference between the first-order components at point B* and point A: $v_{gs,fund} = v_{B^*,fund} - v_{A,fund} = v_{B,fund} - v_{A,fund} = 2v_{B,fund}$. It can be observed that the fundamental component of the gate-source voltage of M1 is effectively doubled, while the second-order harmonic component is canceled i.e., $a_2 = 0$. Thus, the denominator of the IIP3 expression becomes smaller, the capacitive cross-coupling structure enhances linearity by canceling the second-order nonlinearity.

From the expression of the output current, it can be observed that the output nodes X1 and X2 still contain second-order nonlinear components due to the inherent nonlinear behavior of the circuit. However, by eliminating the second-order components in v_{gs}, the nonlinear second-order components in the output are reduced. When a cascode structure is introduced, the second harmonic generated by the first-stage transistor interacts with the second-order nonlinearity of the second-stage transistor, producing additional third harmonics at the output, which degrades the circuit's linearity. To address this, the proposed design incorporates capacitive cross-coupling in each stage of the cascode structure, the second-stage transistors M3 and M4 utilize capacitive cross-coupling to further eliminate the second harmonics at nodes X1 and X2, thereby suppressing the additional third harmonics and enhancing the overall linearity of the circuit. The derived expression for the overall IIP3 of the circuit is:

$$A_{IIP3} \approx \sqrt{\frac{4}{3}\left|\frac{4a_1\beta_1 g_{m1}g_{m2}}{8a_1^3 g_{m1}^3 \cdot \frac{g''_{m2}}{3}\beta_1^3}\right|}$$

$$a_1 \approx 2g_{m1}R_S + 1 \quad (6)$$

$$\beta_1 \approx 2\frac{C_{c2}}{C_{c2} + C_{gs2}}g_{m2}r_{o1} + 1$$

979-8-3315-2209-4/25 $31.00 © 2025 IEEE

Where β_1 represents the second-order nonlinear coefficient of the gate-source voltage of M3 and M4 with respect to nodes X1 and X2. We observe that the smaller the value of β_1, the higher the linearity, and β_1 is also determined by C_{c2}. Specifically, the smaller C_{c2}, the higher the linearity. As shown in Fig. 4, the comparison of pre-simulation performance parameters between the proposed LNA and the cascode LNA based on single-stage capacitive cross-coupling technology reveals that the proposed LNA achieves a 3.9 dBm improvement in Output Third-Order Intercept Point (OIP3) , thereby breaking the trade-off between gain and linearity.

IV. SIMULATION RESULTS

The proposed LNA is implemented using 22-nm CMOS technology. Fig. 5(a) presents the post-layout simulation results of the S-parameters, demonstrating excellent matching, gain, and isolation over the bandwidth of 5.5 GHz to 6 GHz. Fig. 5(b) shows that the NF within this frequency range is as low as 2.43 dB, while Fig. 5(c) indicates that the IIP3 at 5.85 GHz reaches 5.73 dBm. The layout of the proposed LNA is illustrated in Fig. 6 , with a core area of $260um \times 320um$. Table 1 provides a performance comparison with previously published LNAs, highlighting that the proposed LNA achieves enhanced linearity under a low supply voltage of 0.9 V and superior overall performance.

V. CONCLUSION

We propose a cascode LNA based on a two-stage capacitive cross-coupling technique, which optimizes linearity under low power consumption. In traditional LNA topologies, there exists a significant trade-off between input matching and noise figure, as well as between linearity and gain. The proposed LNA alleviates the transconductance requirement for input transistor matching through the use of a step-up balun and g_m-boost technique. Furthermore, the two-stage capacitive cross-coupling effectively suppresses nonlinearity, thereby breaking the trade-off relationship. This circuit demonstrates low NF and high linearity under low power consumption, exhibiting excellent overall performance.

REFERENCES

[1] Chen Q, Luo Z, Deng H, et al. A single chain 800M/1.8 G/2.4 GHz multistandard transceiver with multibranch transformer for low-cost IoT applications[J]. IEEE Transactions on Circuits and Systems II: Express Briefs, 2022, 69(12): 4684-4688.

[2] TR S K, Rajanna P K T, Deepika P N, et al. Design and Simulation of Low Noise Amplifier for Sub-6GHz 5G Application[C]//2023 International Conference on Smart Systems for applications in Electrical Sciences (ICSSES). IEEE, 2023: 1-5.

[3] Vardhan S H, Pathak D, Dutta A. A Gain Reconfigurable CMOS Wideband LNA for Sub-7GHz 5G NR Receiver[J]. IEEE Transactions on Circuits and Systems II: Express Briefs, 2023.

[4] Zheng Y, Li Q. A 2.4-5.25 GHz Active Feedback Balun-LNA With Variable Gain in 40nm CMOS Technology[C]//2023 8th International Conference on Integrated Circuits and Microsystems (ICICM). IEEE, 2023: 468-472.

[5] F. Bruccoleri, E. A. M. Klumperink, B. Nauta. Noise cancelling in wideband CMOS LNAs[C]. International Solid-state Circuits Conference, 2002: 406-407.

[6] F. Bruccoleri, E. A. M. Klumperink, B. Nauta. Wide-band CMOS low-noise amplifier exploiting thermal noise canceling[J]. IEEE Journal of Solid-state Circuits, 2004, 39(2): 275-282.

[7] Blaakmeer S C, Klumperink E A M, Nauta B, et al. Wideband CMOS receivers exploiting simultaneous output balancing and noise/distortion canceling[C]//2008 European Microwave Integrated Circuit Conference. IEEE, 2008: 163-166.

[8] W. Chen, G. Liu, B. Zdravko, A. M. Niknejad. A Highly Linear Broadband CMOS LNA Employing Noise and Distortion Cancellation. IEEE Journal of Solid-State Circuits, 2008, 43(5): 1164-1176.

[9] F. Belmas, F. Hameau, J. Fournier. A Low Power Inductorless LNA With Double Gm Enhancement in 130 nm CMOS[J]. IEEE Journal of Solid-State Circuits, 2012, 47(5): 1094-1103.

[10] Z. Luo et al., "A 0.4–6-GHz Blocker-Tolerant Receiver in 65-nm CMOS With Bandwidth-Extended Technologies for Future V2X Applications," IEEE Trans. Circuits Syst. II, Exp. Briefs, vol. 71, no. 5, pp. 2634-2638, May 2024.

[11] B. Razavi, "RF Microelectronics," in Prentice-Hall Communications Engineering and Emerging Technologies Series, 2nd ed. Upper Saddle River, NJ, USA: Prentice-Hall, 2011.

[12] Yeh Y S, Lee H J. A 5-6 GHz Low-Noise Amplifier with¿ 65-dB Variable-Gain Control in 22nm FinFET CMOS Technology[C]//2020 IEEE Radio Frequency Integrated Circuits Symposium (RFIC). IEEE, 2020: 371-374.

[13] Bozorg A, Staszewski R B. A 0.02–4.5-GHz LN (T) A in 28-nm CMOS for 5G exploiting noise reduction and current reuse[J]. IEEE Journal of Solid-State Circuits, 2021, 56(2): 404-415.

[14] Shirmohammadi B, Yavari M. A linear wideband CMOS balun-LNA with balanced loads[J]. IEEE Transactions on Circuits and Systems II: Express Briefs, 2022, 69(3): 754-758.

High-Sensitivity Displacement Sensor Based on Frequency Selective Surface

1st Bo Qi   2nd Mi Lin   3rd Guohua Liu

Abstract—In this work, a high-sensitivity displacement sensor based on frequency selective surface (FSS), capable of achieving precise displacement measurements in both horizontal and vertical directions is introduced. The optimized design enables a highly sensitive response within a displacement range of 0 to 7 mm. To further enhance sensitivity, a dual frequency collaborative analysis method is employed. By monitoring the variations in the difference between these two frequency points, the sensor can more accurately detect minute displacements. The sensor exhibits sensitivities of 160 MHz/mm (horizontal) and 145.7 MHz/mm (vertical), with corresponding relative sensitivities of 4.07% and 3.71%, respectively. This dual-frequency approach not only improves sensitivity but also increases the robustness and interference resistance of the sensor.

Index Terms—frequency selective surface, displacement sensor, wireless sensor, dual-frequency

I. INTRODUCTION

Structural health monitoring (SHM) is a critical approach for ensuring the safety of infrastructure such as buildings and bridges. Accurate measurement of displacement parameters can directly reflect potential risks such as structural deformation and settlement [1] [2]. Traditional displacement sensors often rely on wired connections or embedded circuits, which suffer from complex installation, high maintenance costs, and insufficient long-term stability, making them unsuitable for large-scale distributed monitoring. Therefore, the development of low-cost, high-sensitivity, and physically disconnected wireless displacement sensors has become an important research direction in this field.

In recent years, wireless sensing technology based on Frequency selective surface (FSS) has provided a novel solution to address these challenges. As a periodic array of resonant elements, FSS can achieve cable-free signal transmission through electromagnetic property modulation, significantly reducing the complexity of sensor hardware. Research has demonstrated the unique advantages of FSS structures in applications such as strain detection [3], temperature monitoring, and dielectric constant measurement of chemical liquids [4]. Their high sensitivity makes them particularly suitable for micro-displacement recognition. In this paper, a novel high-sensitivity wireless displacement sensor based on FSS is proposed. By analyzing the variation in resonant characteristics of FSS units under small displacements, a dual-direction displacement-sensitive structure is designed. Simula-

tion experiments demonstrate that the sensor exhibits excellent sensitivity in both horizontal and vertical directions, achieving sensitivities of 160 MHz/mm (horizontal) and 145.7 MHz/mm (vertical), with corresponding relative sensitivities of 4.07% and 3.71%, respectively. The proposed design offers a more practical and efficient wireless sensing solution for structural health monitoring.

II. SENSOR DESIGN AND ANALYSIS

The patch-type FSS exhibits reflective properties for incident waves at specific frequencies. When an electromagnetic wave strikes the FSS metal patch, its surface electrons are excited and oscillate, generating an induced current. At resonance, electromagnetic energy causes electron oscillations, and the energy from the incident wave becomes confined to the surface, reflecting back into the upper half space, thereby exhibiting a reflective filtering characteristic. When there is a relative displacement between the two FSS layers, the resonant structure of the sensor changes, altering the resonance conditions, which are primarily reflected in the frequency of the resonance point. This characteristic can be used to wirelessly measure the transmission or reflection properties of the FSS, observing the changes in the resonant frequency to enable wireless displacement sensing. The proposed displacement sensor consists of two layers of identical FSS, and the resonant structure is a combination of square and circular metal ring, designed to achieve higher sensitivity. Fig.1(a) shows a perspective view of the sensor unit along with the optimized structural parameters. Fig.1(b) demonstrates the sensor undergoing relative displacement changes in both horizontal and vertical directions from its initial stationary state.

III. SIMULATION RESULT

A. Displacement in Horizontal Direction

Software simulation and analysis of the relative displacement in the horizontal direction is performed, with displacement varying from 0 to 7 mm. The results are shown in Fig.2(a). Two resonant points experience significant frequency shifts. As the relative displacement Δx increases from 0 to 7 mm, the resonant frequency f_1 of the point 1 shifts from 4.55 GHz to a higher frequency of 5.23 GHz, while resonant frequency f_2 of the point 2 shifts from 3.31 GHz to a lower frequency of 2.87 GHz. The frequency difference between the

---

Bo Qi, Mi Lin and Guohua Liu are with *School of Electronic Information, Hangzhou Dianzi University*, Hangzhou, China. Corresponding author: (Mi Lin, email: linmi@hdu.edu.cn)

979-8-3315-2209-4/25 $31.00 © 2025 IEEE

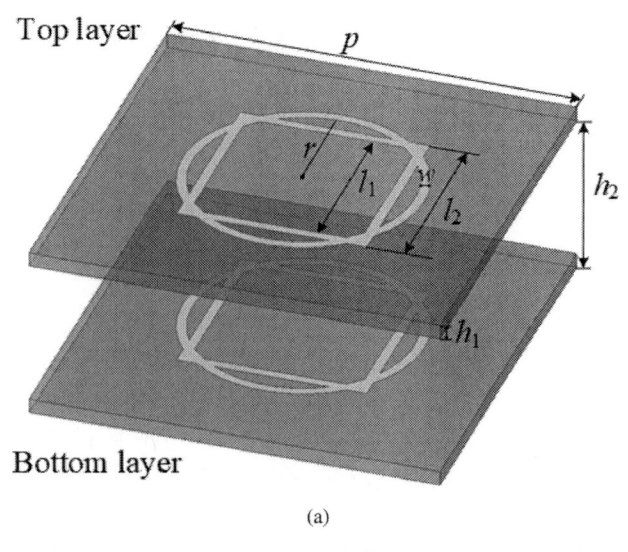

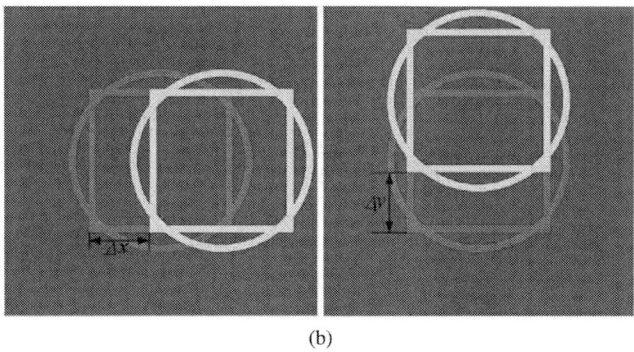

Fig. 1: The proposed FSS displacement sensor. (a) Structure of the unit cell; (b) Schematic diagram with displacement. Detailed dimensions: $l_1 = 17$, $l_2 = 19$, $w = 1$, $p = 2$, $h_1 = 0.762$, $h_2 = 1.5$, $r = 12.1$ (unit: mm).

two points increases monotonically as the horizontal displacement changes, as mathematically described by:

$$\Delta f = |f_1 - f_2| \tag{1}$$

Therefore, a cooperative analysis of these two points can not only achieve higher sensor sensitivity but also enhance the robustness and anti-interference capabilities of the sensor. The trend of the frequency difference between the two resonant points as the relative displacement changes is shown in Fig.2(b). The directional sensitivity characteristics of the FSS sensor can be quantified through the following constitutive relationships: S_d, which represents the sensitivity, quantifies how much the resonant frequency changes per unit displacement. Meanwhile, S_{d_res}, the relative sensitivity, is the sensitivity normalized by the initial resonant frequency, providing a dimensionless measure:

$$S_d = \frac{|\Delta f_h - \Delta f_l|}{\Delta d} \tag{2}$$

$$S_{d\_res} = \frac{S_d}{f_c} \times 100\% \tag{3}$$

where $d \in \{x, y\}$ denotes the displacement direction (x: horizontal, y: vertical) and Δd denotes corresponding di-

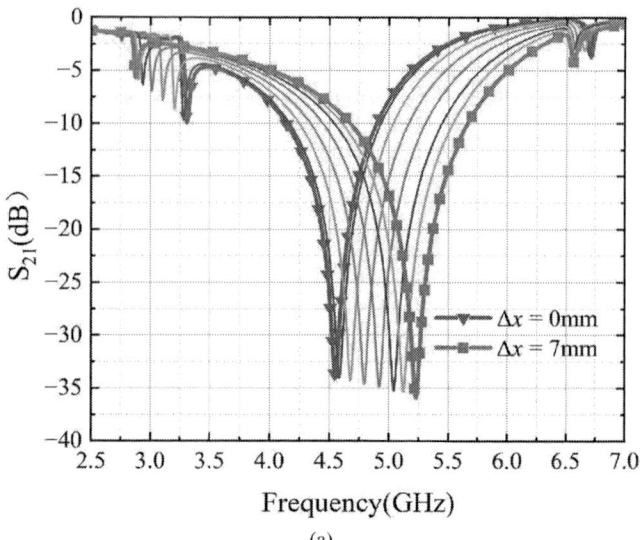

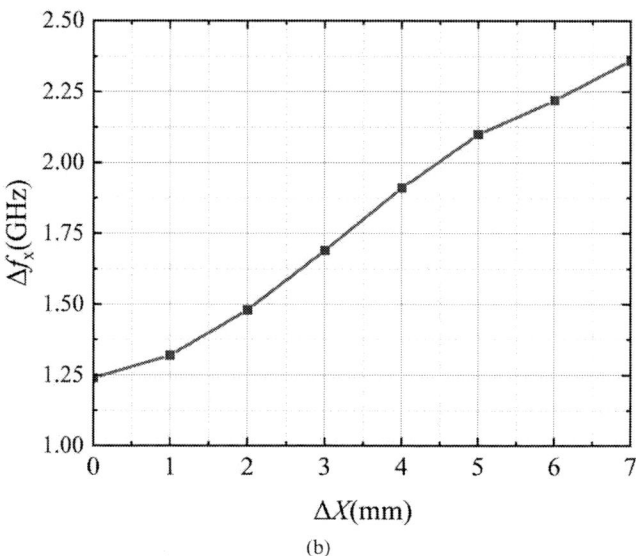

Fig. 2: The FSS displacement sensor with horizontal direction displacement: (a) Variation of S_{21} with Δx from 0 to 7 mm. (b) Variation of Δf_x with Δx from 0 to 7 mm.

rectional displacement changes, $\Delta f_h = \max(|f_{1,d} - f_{2,d}|)$ and $\Delta f_l = \min(|f_{1,d} - f_{2,d}|)$ represent the maximum and minimum differential frequencies between resonant modes, respectively. The characteristic frequency $f_c = (f_{1,0} + f_{2,0})/2$ corresponds to the mean resonance frequency at zero displacement, with $f_{n,0}$ indicating the initial resonance frequency of resonant point 1 and 2. The horizontal sensing sensitivity of the sensor is $160\,\mathrm{MHz/mm}$, with a relative sensitivity of 4.07%.

B. Displacement in Vertical Direction

When there is vertical displacement Δy between two layers, the software simulation results are shown in Fig.3(a). Similarly, both resonant points experience noticeable shifts. As the relative displacement increases from 0 to 7 mm, f_1 shifts from 4.55 GHz to a lower frequency of 4.15 GHz, while f_2

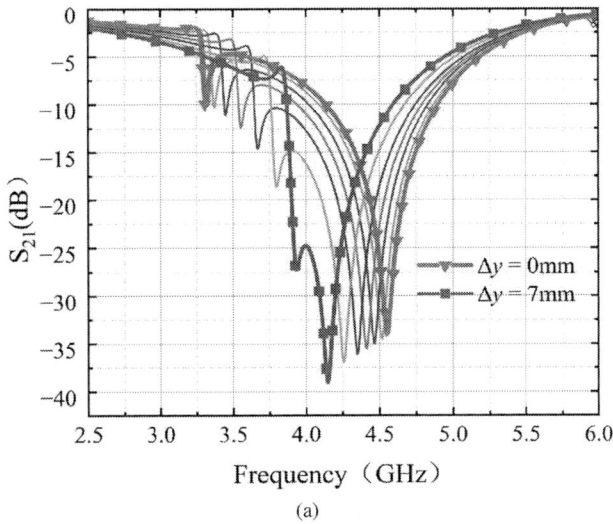

(a)

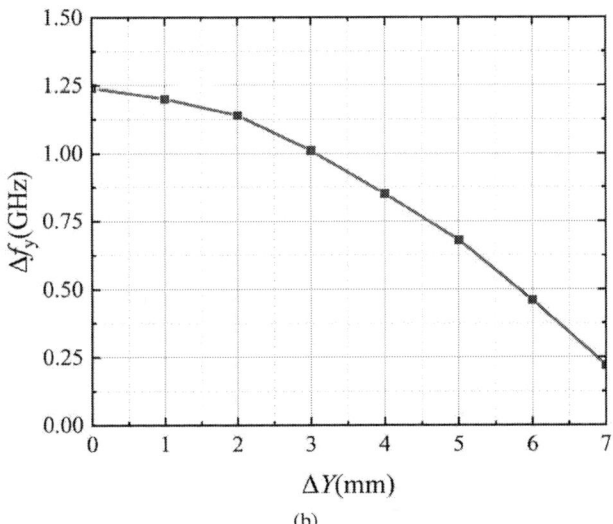

(b)

Fig. 3: The FSS displacement sensor with vertical direction displacement: (a) Variation of S_{21} with Δy from 0 to 7 mm.(b) Variation of Δf_y with Δy from 0 to 7 mm.

TABLE I: PERFORMANCE COMPARISON WITH PREVIOUS RESEARCH

Ref.	Dynamic Range	Sensitivity (/mm)	Wireless	Numbers of Directions	Relative Sensitivity
[2]	10mm	52MHz	Yes	1	2.6%
[5]	7mm	52MHz	No	2	2.62%
[6]	5mm	116MHz	Yes	1	-
[7]	-3-6mm	11.1MHz	Yes	2	0.69%
	0-3mm	42.3MHz			2.67%
This Work	7mm	160MHz	Yes	2	4.07%
		146MHz			3.71%

shifts from 3.31 GHz to a higher frequency of 3.93 GHz. The frequency difference between the two points decreases monotonically as the vertical displacement increases. The trend of the frequency difference between the two resonant points as the relative displacement changes is shown in Fig.3(b). The vertical sensing sensitivity of the sensor is 145.7 MHz/mm,

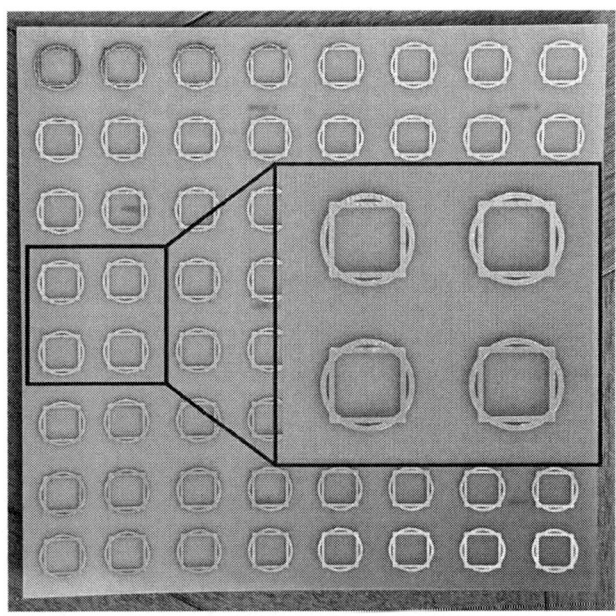

Fig. 4: Prototype of the proposed FSS

with a relative sensitivity of 3.71%.

IV. FABRICATION AND MEASUREMENT

To experimentally validate the performance of the FSS sensor, an 8×8 units prototype was fabricated on an FR-4 substrate using standard printed circuit board (PCB) photolithography. As shown in Fig.4, the fabricated device exhibits an effective sensing area of 328×328 mm^2 with dimensional tolerances maintained within ± 0.1 mm. The substrate consists of a 0.8 mm thick FR-4 laminate ($\varepsilon_r = 4.4 \pm 0.3$, $\tan \delta = 0.02$) with a 35 μm copper cladding ($\sigma = 5.8 \times 10^7$ S/m) patterned through wet chemical etching. This fabrication process achieved critical dimensions of 200 μm for both minimum conductive line width and inter-element spacing. The implemented manufacturing protocol ensures repeatable geometric accuracy while maintaining cost-effectiveness through conventional PCB processing techniques. The comparison between the proposed sensor with other similar works is shown in TABLE.I.

V. CONCLUSION

In this paper, a wireless two-dimensional displacement sensor based on FSS is proposed. It can be observed that the proposed FSS sensor exhibits superior performance in terms of both displacement measurement range and sensing sensitivity. Additionally, the capability for wireless measurement offers significant convenience for practical applications, making it highly suitable for structural health monitoring in buildings. Future research will focus on fabricating prototypes for testing in real-world scenarios.

REFERENCES

[1] J.-J. Zhang, B. Wu, Y.-T. Zhang, et al., "Two-Dimensional Highly Sensitive Wireless Displacement Sensor With Bilayer Graphene-Based Frequency Selective Surface," IEEE Sensors J., vol. 21, no. 21, pp. 23889–23897, Nov. 2021, doi: 10.1109/JSEN.2021.3116457.

[2] P. Njogu, B. Sanz-Izquierdo, Z. Chen, and E. A. Parker, "Three-Dimensional Frequency Selective Surface Displacement Sensor Using Complementary Dielectrics," IEEE Sensors J., vol. 23, no. 16, pp. 18692–18699, Aug. 2023, doi: 10.1109/JSEN.2023.3289777.

[3] S. Rodini, S. Genovesi, G. Manara, and F. Costa, "Wireless mm-Wave Chipless Pressure Sensor," IEEE Trans. Microwave Theory Techn., vol. 72, no. 7, pp. 4163–4173, Jul. 2024, doi: 10.1109/TMTT.2023.3347473.

[4] P. M. Njogu, B. Sanz-Izquierdo, and E. A. Parker, "A Liquid Sensor Based on Frequency Selective Surfaces," IEEE Trans. Antennas Propagat., vol. 71, no. 1, pp. 631–638, Jan. 2023, doi: 10.1109/TAP.2022.3219540.

[5] Z. Mehrjoo, A. Ebrahimi, and K. Ghorbani, "Microwave Resonance-Based Reflective Mode Displacement Sensor With Wide Dynamic Range," IEEE Trans. Instrum. Meas., vol. 71, pp. 1–9, 2022, doi: 10.1109/TIM.2021.3130669.

[6] M. Hernandez-Aguila, J.-L. Olvera-Cervantes, A.-E. Perez-Ramos, and A. Corona-Chavez, "WiFi Sensor Node With High Sensitivity and Linearity Based on a Quarter-Wavelength Resonator for Measuring Crack Width," IEEE Sensors J., vol. 23, no. 23, pp. 28883–28890, Dec. 2023, doi: 10.1109/JSEN.2023.3324929.

[7] J. Wang, L. Chen, and G. Wan, "A Multibranch Displacement Sensor With Bidirectional Detection for Structural Health Monitoring," IEEE Sensors J., vol. 24, no. 19, pp. 29831–29840, Oct. 2024, doi: 10.1109/JSEN.2024.3448210.

Design of a wireless charging system for airborne sensor of a tunnel boring machine

Zhenxuan Zhang   Xiaolong Wei   Zhen Huang   Yao Wang

Abstract—In the current application of power supply for sensors in shield machines, traditional wired power systems face safety hazards such as wire aging and the susceptibility of interfaces to damage. Additionally, maintenance is challenged by the difficulty of disassembly and assembly. To address the issues associated with wired power supply, this paper proposes a wireless power transfer (WPT) system as a replacement. In the sensor power supply scenario of shield machines, coils are installed within metal structures, which are influenced by the surrounding metallic environment. This paper optimizes the coil design by introducing suppression mechanisms and ferrite magnetic isolation sheets, followed by simulation and experimental validation. The experimental results demonstrate that the wireless power transfer system with the optimized coil structure exhibits higher transmission efficiency and power in a metallic environment.

Index Terms—Wireless power transmission; Magnetic coupling mechanism; Transmission efficiency; Tunnel boring machine

I. INTRODUCTION

At present, the intelligentization and informatization of shield and tunneling equipment have become the major development directions of the industry [1], and the number of sensors installed on the equipment has significantly increased. However, traditional wired power supply has problems such as the aging of wire ends, the easy damage of sockets, and the inability to meet the waterproof requirements in the harsh underground construction application environment. These problems lead to the decline in the reliability of sensors and the shortening of their service life. In contrast, Wireless Power Transfer (WPT) technology offers advantages such as higher reliability, greater flexibility and convenience, and better sealing performance compared to wired power supply. Thus, it provides an effective solution for powering sensors in shield machines [2].

However, in practical applications, the WPT system in tunnel boring machines is often subject to interference from the metal environment. During the energy transfer process in a WPT system, the presence of a metallic environment induces eddy current effects, which cause changes in the coil parameters. This leads to a shift in the system's resonant frequency, resulting in a decrease in both transmission efficiency and transmitted power [3]. There are various methods to improve the performance of WPT systems in metal environments, such

as laser cutting on metal surfaces to reduce eddy currents [4] and designing nonlinear WPT transmission systems that are insensitive to frequency variations [5]. Currently, the most commonly used approach is to optimize the coil structure, such as reducing the coupling between the coil and the metal to enhance the WPT system performance [6] or using ferrite materials to increase the coupling between coils [7].

Aiming at the problems of low transmission efficiency and power of the wireless power supply system in a metal environment. A small magnetic coupling WPT system for powering sensors in tunnel boring machines is designed and modeled. The coil is optimized by adding suppression mechanisms and magnetic shielding materials outside the coil to improve the transmission power and efficiency of the wireless power system in metal environments. Finally, an experimental platform being constructed to validate the performance of the designed wireless power transfer system in the metal environment.

II. CIRCUIT ANALYSIS OF WPT SYSTEM

Since the resonant current at the transmitter in the PS-type topology does not flow directly through the power supply, the power supply experiences lower stress, and the system achieves higher efficiency when delivering high output power. Therefore, this paper adopts the PS resonant topology. The equivalent circuit model of the PS-type magnetic coupling wireless power transfer system is shown in Fig. 1, U represents the equivalent AC voltage source ; ω is the frequency of the equivalent voltage source; I_1 and I_2 represent the current at the input and output ends, respectively; I_L is the current flowing through the transmitter coil; C_1 and C_2 are the resonance capacitors at the input and output ends; R_1 and R_2 are the internal resistances of the transmitter and receiver coils; R_L is the load resistance; L_1 and L_2 are the equivalent inductance of the transmitter and receiver coils.

In Fig. 1, the equivalent impedance of the transmitting and receiving sides can be derived as follows:

$$\begin{cases} Z_1 = \dfrac{1}{j\omega C_1} / (j\omega L_1 + R_1 + Z_{21}) \\ Z_2 = j\omega L_2 + R_2 + R_L + \dfrac{1}{j\omega C_2} \end{cases} \tag{1}$$

Where Z_{21} is the equivalent impedance mapped from the receiving side to the transmitting side, which can be defined as follows:

Zhenxuan Zhang and Yao Wang are with *School of Electrical and Information Engineering, Zhengzhou University*, Zhengzhou, China. Xiaolong Wei and Zhen Huang are with *China Railway Engineering Equipment Group*, Zhengzhou, China. Corresponding author: (Yao Wang, email: ieyaowang@zzu.edu.cn)

979-8-3315-2209-4/25 $31.00 © 2025 IEEE   481

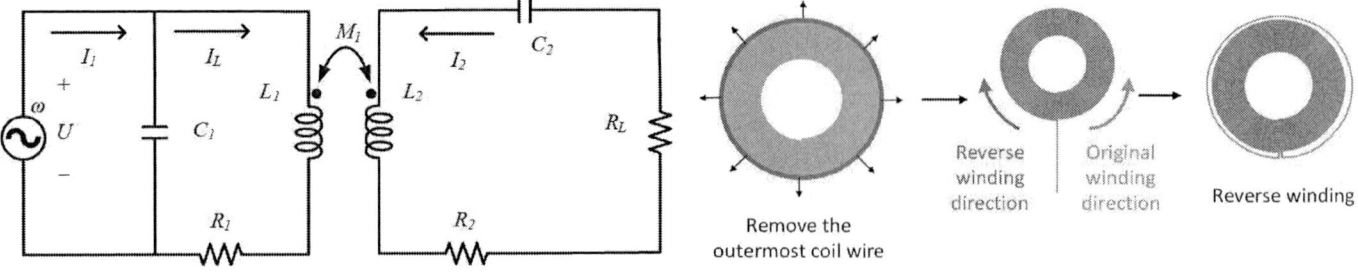

Fig. 1. Equivalent circuit of WPT system.

Fig. 2. The coil optimization process.

$$Z_{21} = \frac{\omega^2 M_1^2}{Z_2} \tag{2}$$

According to Kirchhoff's laws, the equivalent circuit of WPT system satisfies the following equations:

$$\begin{cases} I_1 = \dfrac{U}{Z_1} \\[2mm] I_L = \dfrac{U}{j\omega L_1 + R_1 + Z_{21}} \\[2mm] I_2 = \dfrac{j\omega M I_L}{R_2 + R_L} \end{cases} \tag{3}$$

According to formula (3), P_{out} and η can be solved by:

$$\begin{cases} P_{out} = I_2^2 \cdot R_L = (\dfrac{j\omega M I_L}{R_2 + R_L})^2 \cdot R_L \\[2mm] \eta = \dfrac{P_{out}}{U \cdot I_1} \end{cases} \tag{4}$$

Observing formula (4), it can be concluded that when the circuit is operating normally, the angular frequency of the power supply, the resonance capacitance, and the load resistance remain constant. The power transferred by the system and its efficiency are primarily influenced by the variations in coil resistance, self-inductance, and mutual inductance parameters.

III. DESIGN OF THE OPTIMIZED COIL

The coil optimization process is shown in Fig.2. Firstly, the outermost wire of the traditional circular coil is removed. Secondly, the removed wire is wound in the opposite direction. Finally, ferrite material is added to the back of the coil for compensation. The added outermost suppression mechanism can effectively control the magnetic field distribution at the edges of the coupled coils, thereby reducing the impact of the inner wall of the metal pipeline at the edges on the characteristics of the coupled coils.

Although the reduction in the number of turns and the addition of the suppression coil also lead to adverse changes in the coil parameters, the guiding effect of the added ferrite material with high magnetic permeability on the magnetic field can effectively compensate for the degradation of the coil parameters.

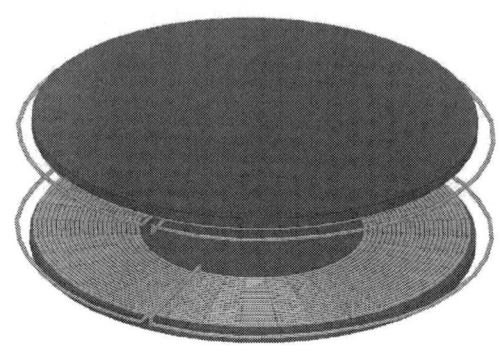

Fig. 3. The optimized coil.

IV. MODELING AND SIMULATION

The finite element simulation software MAXWELL was utilized for modeling and analysis, simulating the changes in coil parameters as well as the magnetic field distribution near the coil, to verify the performance of the proposed optimized coil. The distance between the two coils was set to 5 mm, the current was set to 2 A, and the current frequency was set to 120 kHz.

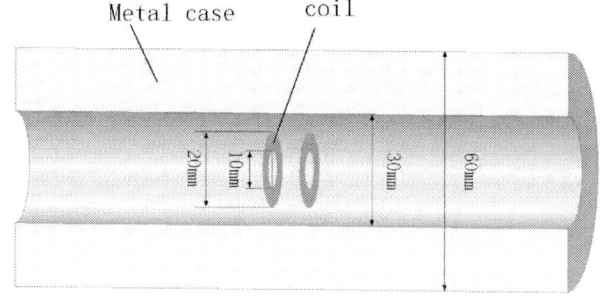

Fig. 4. The wireless power transfer system model.

As shown in Fig. 4, two coils are placed face-to-face at the center of a metal pipeline with 5 mm between them. The transmitter coil and the receiver coil have identical dimensions, with an inner diameter of 10 mm and an outer diameter of 20 mm. The metal pipeline has a diameter of 60 mm and an inner diameter of 30 mm. The coils are wound using 0.2 mm diameter Litz wire.

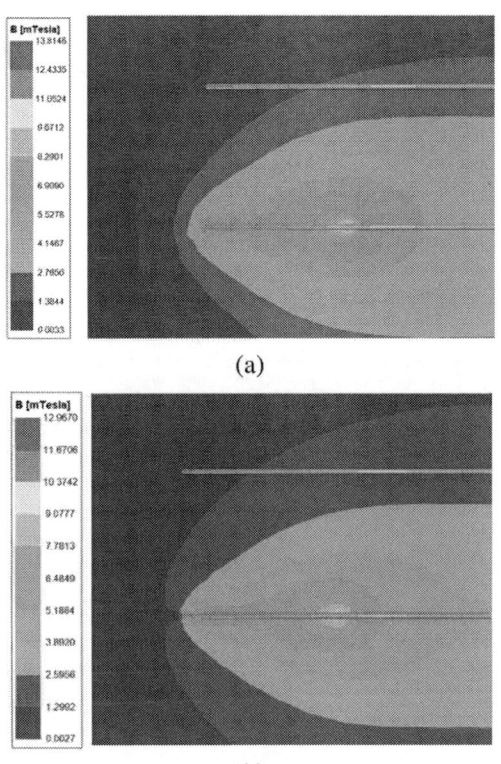

(a)

(b)

Fig. 5. (a)The magnetic flux density distribution of the original coil. (b)The magnetic flux density distribution of the optimized coil.

Fig. 5 shows the simulation results of the magnetic flux density distribution for different coils. It can be observed that the coil with the suppression structure significantly reduces the magnetic flux density on both sides compared to the original coil. The magnetic flux density drops from 6.9 mT to less than 1 mT, successfully suppressing the magnetic flux leakage towards the metal on both sides of the coil. This improvement in the design demonstrates the effectiveness of the suppression structure in controlling the magnetic field distribution and minimizing unwanted interactions with nearby metallic surfaces.

TABLE I
SIMULATION RESULTS OF COIL PARAMETERS

The type of coil	L(uH)	M(uH)	R(mΩ)
original coil(air)	10.75	3.3	754.61
original coil(metal)	9.87	2.51	844.91
optimized coil(air)	11.69	3.92	770.12
optimized coil(metal)	10.98	3.25	841.53

The materials of the metal pipeline were set as air and metal respectively to simulate the influence of the metal environment on the coil parameters. The simulation results are shown in Table 1. The self-inductance of the original coil decreased

from 10.75 uH to 9.87 uH, a reduction of 8.2%. The mutual inductance decreased from 3.3 uH to 2.51 uH, a reduction of 24%. The internal resistance increased from 754.61 mΩ to 844.91 mΩ, an increase of 11.9%. In contrast, the self-inductance of the optimized coil decreased from 11.69 uH to 10.98 uH, a reduction of 6%. The mutual inductance decreased from 3.92 uH to 3.25 uH, a reduction of 17.1%, and the internal resistance increased from 770.12 mΩ to 841.53 mΩ, an increase of 9.2%. The simulation results indicate that the optimized coil performs better in the metal environment.

V. MEASUREMENT RESULTS

To validate the actual performance of the wireless power transfer system, an experimental platform was established to test the transmission efficiency and power of the wireless power transfer system under different conditions. The specific parameters of the testing platform are as follows: power supply voltage of 12 V, resonant frequency of 120 kHz, the self-inductance of the unoptimized coil is 10.7 uH, the self-inductance of the optimized coil is 10.94 uH, the resonant capacitance C_1 is 100 nF, the resonant capacitance C_2 is 168 nF, and the load is 5 Ω.

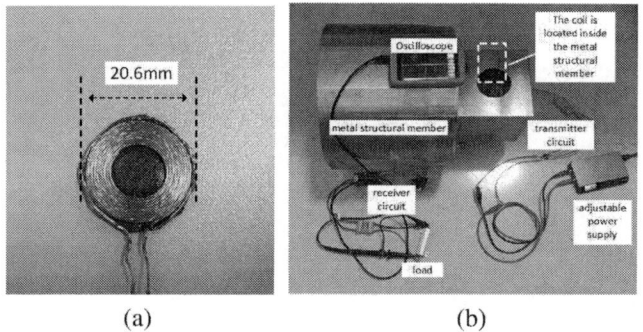

(a)   (b)

Fig. 6. (a) The optimized coil.(b)The experimental circuits.

The experiment tested the performance of two types of coils in a metallic environment. During the test, both the transmitting and receiving coils were of the same type. The received power at the load end was measured, and the output efficiency was calculated by comparing it with the output power at the power supply end. A comparison graph of power and efficiency as a function of coil distance was plotted based on the measurements. The parameters obtained from the experiment are shown in Fig. 7.

As shown in Fig.7(a), the load power of the optimized coil is consistently higher than that of the original coil within the coil distance range of 0-8 mm. As the coil distance increases, the load power of both coils decreases sharply. The load power of both coils gradually approaches each other with increasing distance, but the optimized coil's power remains superior to the unoptimized coil.

From Fig.7(b), it can be observed that the efficiency when using the optimized coil is significantly higher than that of the unoptimized coil, and the efficiency gap between the two changes little as the coil distance increases. After adopting

979-8-3315-2209-4/25 $31.00 © 2025 IEEE   483

the optimized coil, the system's maximum efficiency reaches 77.16%, and the maximum load power reaches 5.02 W. In the typical application scenario, where the coil distance is 4 mm, the efficiency of the optimized coil is 48.78%, and the load power is 1.8 W. In contrast, the efficiency of the unoptimized coil is 22.72%, and the load power is 1.35 W, resulting in a 114% improvement in efficiency and a 33.3% increase in power.

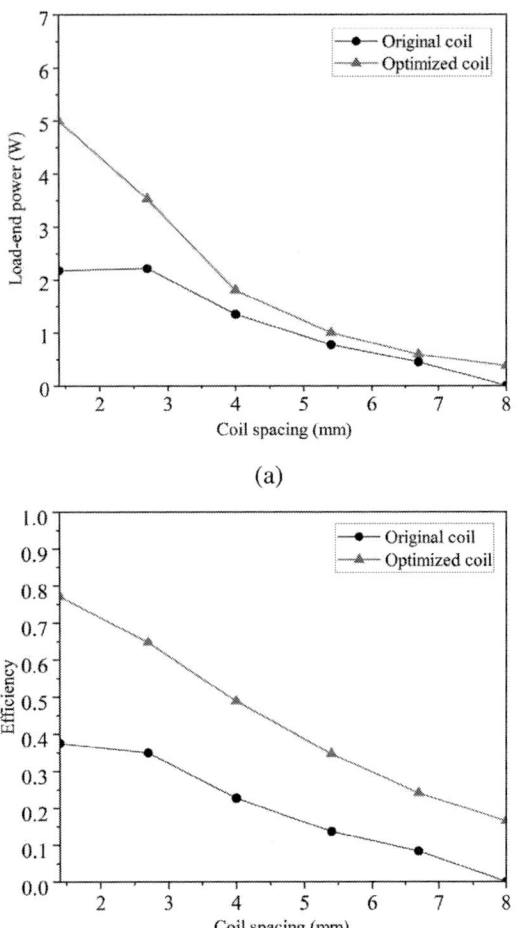

Fig. 7. (a) The transfer power. (b) The transfer efficiency.

The experimental results confirm that the optimized coil exhibits higher transmission efficiency and load-end received power under metal interference, and the designed wireless power transfer system performs significantly better than one using an unoptimized coil.

VI. CONCLUSION

In this paper, a wireless power transfer system using an optimized coil design is presented. Experimental results demonstrate that the optimized coil structure improves the transmission power by 33.3% and increases system efficiency by 114% compared to the original coil design in a metallic environment. This makes it more suitable for wirelessly powering sensors located within the metallic environment of the tunnel boring machine.

REFERENCES

[1] L. Chen, Z. Liu, W. Mao, H. Su and F. Lin, "Real-Time Prediction of TBM Driving Parameters Using Geological and Operation Data," IEEE/ASME Transactions on Mechatronics, vol. 27, no. 5, pp. 4165-4176, Oct. 2022.

[2] D. Newell and M. Duffy, "Review of power conversion and energy management for low-power, low-voltage energy harvesting powered wireless sensors," IEEE Trans. Power Electron, vol. 34, no. 10, pp. 9794–9805, Oct. 2019.

[3] Deng Yongsheng, Yang Yongmin, Luo Yanting, Dai Zhuoyue, Zhang Shigang, and Luo Xu, "Modeling and Analysis of Wireless Power Transfer System in Metal Environment," 2020 IEEE Wireless Power Transfer Conference (WPTC), Seoul, Korea, pp.15-19, November 2020.

[4] Y. Wang, J. Yu and S. Chen, "Analysis of Low-eddy Current WPT Applied in Underground Coal Mines," 2024 27th International Conference on Electrical Machines and Systems (ICEMS), Fukuoka, Japan, pp. 2546-2549, 2024.

[5] M. Wang, G. Song, Y. Shi and R. Yin, "A Nonferromagnetic Metal-Insensitive Robust Wireless Power Transfer System," IEEE Microwave and Wireless Technology Letters, vol. 33, no. 12, pp. 1670-1673, Dec. 2023.

[6] Wenjiang Yuan, Xian Zhang, Lin Sha and Zhixin Chen. "A Method of Metal Interference Suppression by Circular Internal Flux for Rotating Wireless Power Supply System," IEEE Transactions on Electromagnetic Compatibility, vol. 66, no. 3, pp. 809-820, June. 2024.

[7] D. Wang, X. Dai and P. Deng, "Design and Optimization of DD Coupling Mechanism of Inductively Coupled Power Transmission System in Metal Environment," 2020 IEEE 5th Information Technology and Mechatronics Engineering Conference (ITOEC), Chongqing, China, pp. 206-210, 2020.

An Ultra-Wideband Power Amplifier based on 130nm SiGe Process

1st Yu Hongshi 2nd Bharatha Kumar Thangarasu 3rd Nagarajan Mahalingam

4th Kaixue Ma 5th Fanyi Meng 6th Zhenghao Lu 7th Kiat Seng Yeo

Abstract—This paper presents a compact ultra-wideband differential RF power amplifier (PA) operating across the 1–9 GHz frequency range, tailored for next-generation communication and radar systemsbrackets. The proposed design, simulated using 130nm Global Foundries BiCMOS technology, employs a broadband impedance matching network and differential topology to achieve consistent gain and output power across the entire bandwidth while maintaining excellent linearity. By leveraging an optimized transistor biasing scheme and careful layout considerations, the PA achieves high efficiency and minimizes intermodulation distortion. Post-simulation results demonstrate a flat gain response of 3 dB with a 27dB maximum gain, an output power of 14 dBm at 1 dB compression, and a peak power-added efficiency (PAE) of 17% across the band. The maximum PAE achieves 21.9%. The compact size and robust performance make this PA a strong candidate for ultra-wideband applications requiring high linearity and efficiency.

Index Terms—Controllable gain, power added efficiency (PAE), power amplifier (PA), power down (PD) mode, ultra wideband.

I. INTRODUCTION

Power amplifiers (PAs) operating in the 1-9 GHz frequency range are critical components in the RF front-end of modern wireless communication systems. They amplify weak RF signals to levels suitable for transmission, enabling reliable communication across various applications such as 5G, Wi-Fi, IoT, and radar systems [1]. With the growing demand for multi-standard, high-frequency devices, ultra-wide bandwidth PAs have become essential for addressing the increasing complexity of communication networks. The RF power amplifier market is projected to exceed 6 billion dollars by 2026, highlighting the growing importance of advanced RF technologies [2].

In the 1-9 GHz frequency band, PAs must meet stringent requirements for linearity and bandwidth. Linearity ensures that the amplified signal remains free of distortion, preserving data integrity across wide frequency ranges. Meanwhile, ultra-wide bandwidth PAs enable support for multiple communication standards and protocols within the same hardware, reducing the need for multiple components and enhancing system flexibility.

Another key aspect of modern PA design is energy efficiency, particularly through the incorporation of power-down

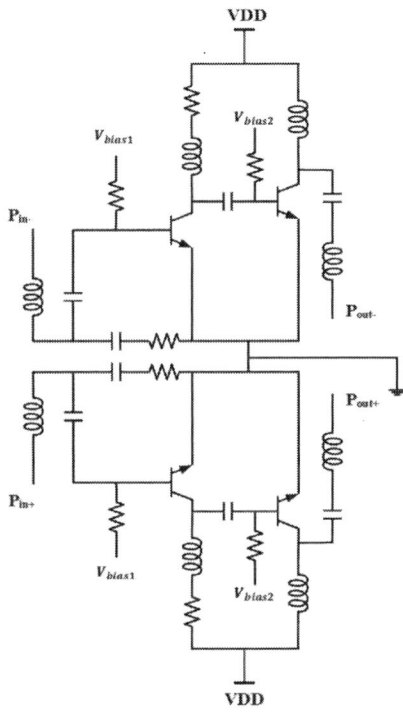

Fig. 1. The schematic of the PA.

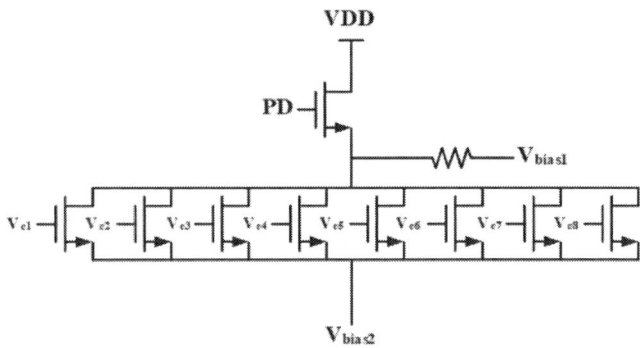

Fig. 2. The bias circuit of the PA.

Yu Hongshi, Bharatha Kumar Thangarasu, Nagarajan Mahalingam, Kaixue Ma, Fanyi Meng and Kiat Seng Yeo are with *the School of Microelectronics, Tianjin University*, Tianjin, China. Zhenghao Lu is with *the School of Electronic and Information Engineering, Soochow University*, Suzhou, China. Kiat Seng Yeo is also with *Engineering Product Development, Singapore University of Technology and Design,*, Singapore. Corresponding author: (Yu Hongshi, email: yuhongshi@tju.edu.cn)

979-8-3315-2209-4/25 $31.00 © 2025 IEEE

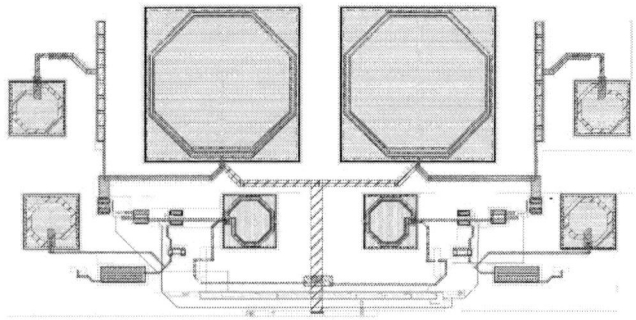

Fig. 3. The layout of the proposed PA.

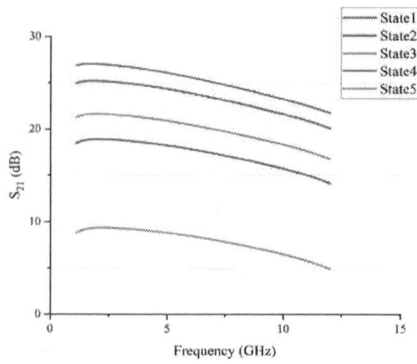

Fig. 4. The S_{21} parameters from the input and output ports.

(PD) modes. These modes allow the PA to minimize power consumption during idle or low-demand periods, significantly reducing energy usage. This not only extends battery life in portable devices but also helps lower the energy footprint of larger systems, such as base stations and satellite communication platforms. By leveraging PD capabilities, PAs contribute to greener communication networks and lower operational costs, aligning with global sustainability goals.

Designing a 1-9 GHz PA involves addressing the trade-offs between linearity, bandwidth, and power efficiency. Through advanced circuit design, it is possible to achieve a PA that delivers high signal quality while optimizing energy usage.

II. THE CIRCUIT DESIGN

A. Main Part of the PA

The proposed design of PA circuit is shown in Fig. 1. It is a two stages common emitter (CE) topology. The first stage is the driven stage and the second stage is the power stage which determines the total gain of the PA. For the first stage, the VDD is given by a resistor which helps to increase the stability and reduce the Q factor of the inductor. Therefore, we can get a stable and ultra-wideband PA. The concise matching network achieves a smaller core area.

B. Bias Circuit Design

The bias circuit schematic is shown in Fig. 2. Nine field effect transistors (FETs) are used to control the gain of the power amplifier (PA). As depicted in part A, the driving stage does not require precise bias current; therefore, a large-value resistor is used to provide a stable bias current. To enable controllable output power, eight FETs are connected in parallel, with finger widths increasing in a sequence of 1x, 2x, 4x, and so on. The bias current can be adjusted by applying a high level to each FET, allowing for 256 distinct gain states.

In PD mode, when the PD transistor is off, both bias 1 and bias 2 currents are zero, causing the PA to enter a sleep state. As a result, the PA will not work, thereby saving energy.

C. Layout Photo

As shown in Fig. 3, the core area of the differential PA is $1.265 \times 0.634 = 0.802mm^2$. The topology is totally symmetric which helps to get a good differential performance.

III. SIMULATION RESULTS

A. Gain

Because of the variable gain control model, the gain can be set to 256 different states. Some typical results are shown in Fig. 4. The maximum gain is 27 dB at 2GHz.

B. S Parameters

For a wideband matching network, the S_{11} is totally less than -8 dB and S_{22} is less than -5 dB from 1-9 GHz.

C. OP_{1dB} and PAE

From the Fig. 6, a OP_{1dB} of 14.1 dBm and saturation power of 16.4 dBm can be observed. And the PAE at OP_{1dB} point is 16%. It also achieve a 21.9% PAE which is shown in Fig. 7.

D. Bandwidth

From Fig. 4, each state provides a wide bandwidth ranging from 1 to 9 GHz and the gain difference is less than 3 dB.

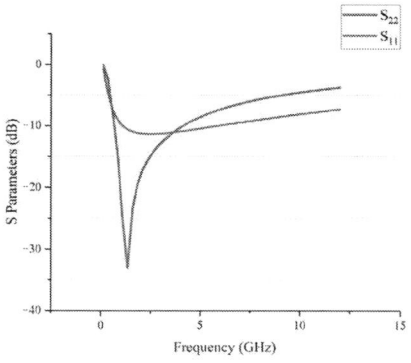

Fig. 5. The S_{11} and S_{22} parameters of the amplifier.

E. PD Mode

Fig. 8 illustrates the PD mode of the amplifier. When it works in the PD mode, the maximum DC current is less than 20 nA while the output power is less than -35 dBm.

979-8-3315-2209-4/25 $31.00 © 2025 IEEE

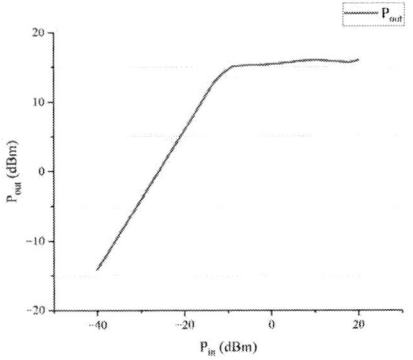

Fig. 6. The output power versus the input power.

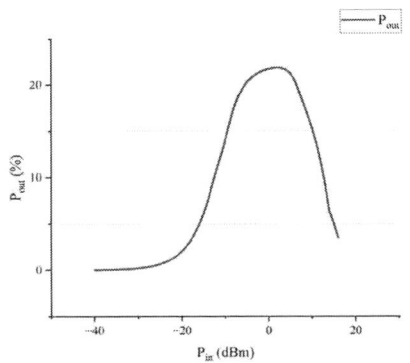

Fig. 7. The PAE of the proposed PA.

IV. COMPARISON

Table I provides a performance comparison with other reported CMOS power amplifiers (PAs). The proposed design demonstrates comparable wideband performance, making it suitable for applications requiring operation across a broad frequency range. In addition, it achieves the highest output power among the compared PAs, highlighting its superior capability to deliver strong signal amplification. A key advantage of this

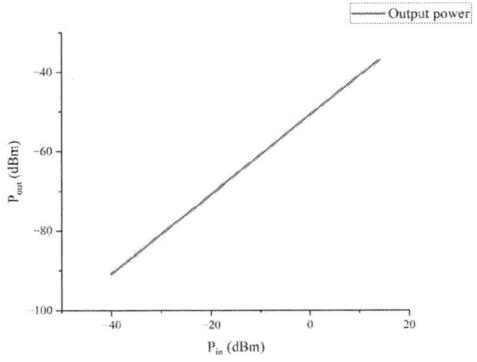

Fig. 8. The PD mode output power of the amplifier.

work is its ability to achieve different gain levels by digitally controlling the biasing, providing enhanced flexibility to adapt to various operating conditions and application requirements. Furthermore, the PA incorporates a PD mode, which significantly reduces power consumption during idle periods, improving overall energy efficiency. This combination of wideband operation, digitally tunable gain, high output power, and energy-saving features makes the proposed PA a highly competitive solution for next-generation wireless communication systems. The figure of merit of the PA is defined as

$$FoM\left(GHz/mm^2\right) = \frac{Gain \times Bandwidth(GHz) \times OP_{1dB}(mW)}{Size(mm^2) \times DC\ Power(mW)}$$

(1)

TABLE I
COMPARISON OF PAPERS

Paper	Gain (dB)	Bandwidth (GHz)	Chip Size (mm^2)	OP_{1dB} (dBm)	Power Consumption (mW)	FoM (GHz/mm^2)
[1]	15	3	0.85	7	24	23.3
[2]	15.8	2.8	1.65	11.4	25	35.6
[3]	13.7	1.2	0.61	17.4	158	16.2
[4]	13	4	0.88	1	15	7.6
[5]	21	5	2.86	30	1000	219.8
This work	27	8	0.8	14.3	150	903.6

V. CONCLUSION

This paper presents a design approach for ultra-wideband CMOS power amplifiers with variable gain capabilities. Post-simulation results confirm the feasibility and potential of the proposed PA structure for wideband applications. Future work will focus on optimizing the gain response and further enhancing the design for ultra-wideband performance across a broader frequency range.

REFERENCES

[1] S. A. Z. Murad, R. K. Pokharel, H. Kanaya, K. Yoshida and S. A. Z. Murad, "A 3.0–7.5 GHz CMOS UWB PA for group 1 3 MB-OFDM application using current-reused and shunt-shunt feedback," 2009 International Conference on Wireless Communications & Signal Processing, Nanjing, China, 2009, pp. 1-4.

[2] S. -K. Wong, S. Maisurah, M. N. Osman, F. Kung and J. -H. See, "High Efficiency CMOS Power Amplifier for 3 to 5 GHz Ultra-Wideband (UWB) Application," in IEEE Transactions on Consumer Electronics, vol. 55, no. 3, pp. 1546-1550, August 2009.

[3] D. Polge, A. Ghiotto, E. Kerhervé and P. Fabre, "3.4 to 4.8 GHz 65 nm CMOS power amplifier for ultra wideband location tracking application in emergency and disaster situations," 2016 11th European Microwave Integrated Circuits Conference (EuMIC), London, UK, 2016, pp. 269-272.

[4] S. A. Z. Murad, R. K. Pokharel, A. I. A. Galal, R. Sapawi, H. Kanaya and K. Yoshida, "An Excellent Gain Flatness 3.0–7.0 GHz CMOS PA for UWB Applications," in IEEE Microwave and Wireless Components Letters, vol. 20, no. 9, pp. 510-512, Sept. 2010.

[5] Q. Cai, H. Zhu, D. Zeng, Q. Xue and W. Che, "A Three-Stage Wideband GaN PA for 5G mm-Wave Applications," in IEEE Transactions on Circuits and Systems II: Express Briefs, vol. 69, no. 12, pp. 4724-4728, Dec. 2022.

A SAW Filter with 3-dB FBW of 15.5% on LiNbO$_3$/SiO$_2$/Quartz Substrate

Lijun Feng   Changjian Zhou   Xiuyin Zhang

Abstract—**The advancement of communication technologies has imposed higher demands on surface acoustic wave filters. In this work, we utilize a heterogeneous integrated substrate made of 42°YX-LiNbO$_3$/SiO$_2$/36°YX-Quartz (LNOQ). Through meticulous optimization, the optimal combination of layer thicknesses for resonators that resonate around 2 GHz was identified. In addition, this is achieved innovatively by designing a reflection grating period that differs from the electrode period, thereby suppressing a portion of the noise in the resonator, resulting in a smoother frequency response for the filter based on these resonators. The final fifth-order filter achieves a center frequency of 2156 MHz, with a 3-dB fractional bandwidth (FBW) of 15.5% and out-of-band suppression of more than 20 dB across the range of 0.5 to 4 GHz. Our investigations validate the potential of the LNOQ platform for fabricating wide fractional bandwidth filters.**

Index Terms—**piezoelectric resonators, piezoelectric filters, MEMS, LiNbO$_3$ on Quartz(LNOQ)**

I. INTRODUCTION

Radio-frequency acoustic devices such as filters play a crucial role in the front ends of consumer electronics and mobile phone systems. The advancement of communication systems necessitates high-performance filters that offer larger bandwidths, higher frequencies, lower insertion losses, and steeper roll-offs. To address these requirements, surface acoustic wave (SAW) resonators based on layered LiTaO$_3$ (LT) and LiNbO$_3$ (LN) on insulator substrates have been extensively studied. These resonators have recently demonstrated a high electromechanical coupling coefficient (k^2) and Bode-Q (Q_{max}) with shear-horizontal surface acoustic waves (SH-SAW) [1]–[3]. However, they have not been utilized in the design of filters.

In this work, after optimizing the thickness of the electrode and piezoelectric layers for the LN/SiO$_2$/Quartz structure, the resulting resonators achieved an electromechanical coupling coefficient of up to 27%. Subsequently, the period of the reflective grating was set to a value different from that of the resonator's electrode period, effectively eliminating the spurious modes on the right side of the anti-resonant frequency of the resonator, thereby avoiding spikes at the upper frequency edge of the filter. Finally, we designed a fifth-order filter with

a minimum insertion loss of 0.51 dB and a 3-dB bandwidth of 330 MHz.

II. STRUCTURE DESIGN AND SIMULATION

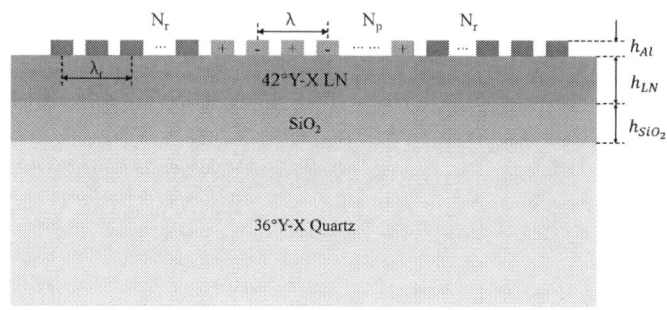

Fig. 1. Schematic of LNOQ sturcture.

Fig. 1 presents a cross-sectional view of the acoustic surface wave resonator utilized in this study. LT and LN are commonly employed piezoelectric films in acoustic surface wave resonators. In comparison to LT, LN facilitates a greater electromechanical coupling coefficient, which is a prerequisite for the design of wideband filters. Considering that LN is an anisotropic material, its electromechanical coupling coefficient varies with changes in cut angle; hence, a judicious design of the cut angle is essential. For the rotated YX-cut LN, when the rotation angle lies between 40° and 90°, the electromechanical coupling coefficient for SH-SAW exceeds 25% [3]. For this work, a 42° YX LN is chosen as the piezoelectric layer. The SiO$_2$ layer underlying the piezoelectric layer serves as a functional layer, enhancing the temperature stability of the resonators. The supporting substrate is selected as 36°YX quartz due to its superior stability and reliability; furthermore, its relatively low phase velocity mitigates the excitation of high-frequency spurious waves [4].

The material of the electrode is aluminum (Al). In Fig. 1, N_p denotes the number of electrode pairs within the resonator, while N_r indicates the number of reflective grating pairs. The period of the electrodes is represented as λ, and the period of the reflective gratings is denoted as λ_r, both referring to the distance between the centers of two fingers separated by a single interdigitated finger.

After qualitatively determining the substrate and electrode materials, a quantitative investigation of the electrode thickness, as well as the piezoelectric layer thickness (h_{Al} and h_{LN}),

---

Lijun Feng and Xiuyin Zhang are with *School of Electronic and Information Engineering, South China University of Technology,* Guangzhou, China. Changjian Zhou is with *School of Microelectronics, South China University of Technology,* Guangzhou, China. This work is supported by the National Natural Science Foundation of China (62293520, 62293524). Corresponding authors: zhoucj@scut.edu.cn (Changjian Zhou), eexyz@scut.edu.cn (Xiuyin Zhang)

979-8-3315-2209-4/25 $31.00 © 2025 IEEE

will be conducted to identify the optimal combination. The thickness of the SiO$_2$ layer has been established at 240 nm before the study. And the supporting substrate is configured to 4λ.

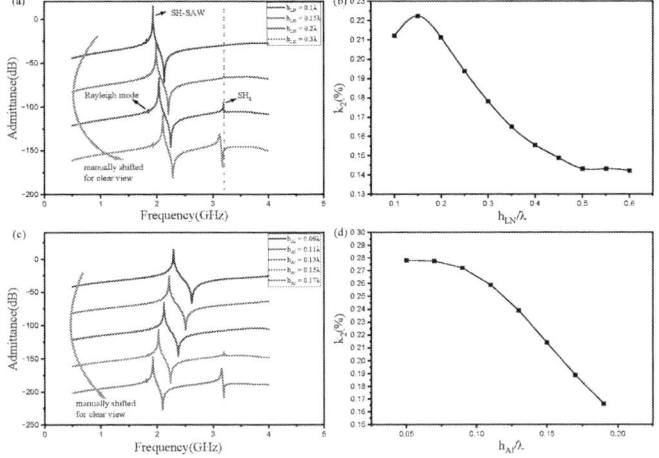

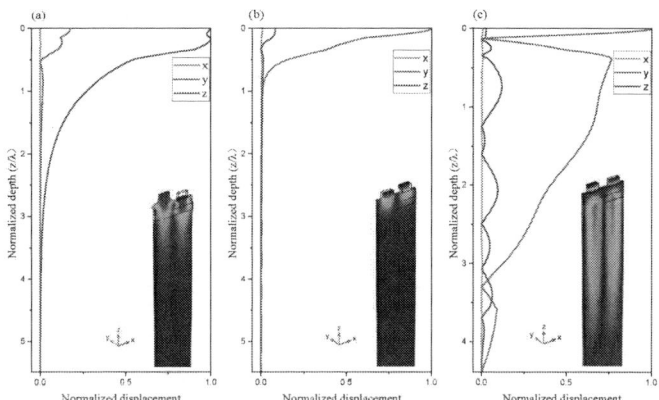

Fig. 2. (a) Resonator admittance curves of resonator at different h_{LN}. (b) Variation of k^2 with h_{LN}. (c) Resonator admittance curves of resonator at different h_{Al}. (d) Variation of k^2 with h_{Al}.

We established a FEM model in COMSOL. λ was set to 1.6 μm, and the value of h_{LN} was varied to simulate the frequency response of the resonator, as illustrated in Fig. 2(a). It is observed that at various h_{LN} values, a spurious response appears to the left of the resonance point. At this frequency, the distribution of acoustic displacement in the x, y, and z components was calculated in the depth direction, confirming the presence of Rayleigh waves, as depicted in Fig. 3(a). The primary mode employed by this resonator is the SH mode, with its acoustic displacement distribution shown in Fig. 3(b). The frequency separation between the resonance and anti-resonance points for SH mode approaches 200 MHz, with a admittance ratio of approximately 70 dB, making it highly suitable for the design of wideband and low-loss filters. When h_{LN} exceeds 0.2λ, a significant spurious response appears to the right of the anti-resonance point. Through the calculation of its acoustic displacement distribution, it is identified as the SH$_1$ mode, as shown in Fig. 3(c). Therefore, to achieve better out-of-band suppression in the filter design, it is preferable for h_{LN} to remain below 0.2λ.

The formula for calculating the electromechanical coupling coefficient in Fig. 2(b) is given by $k^2 = \pi/2 \cdot (f_a/f_r) \cdot \tan(\pi/2 \cdot (f_a/f_r))$. It can be seen that k^2 reaches its maximum value when h_{LN} is between 0.1 and 0.2λ. Considering the limitations of the substrate preparation process, the final thickness of the LN film was set at 300 nm(0.1875λ).

Similarly, when the electrode thickness exceeds 0.15λ, the SH$_1$ mode is excited, and the Rayleigh mode becomes increasingly prominent, as illustrated in Fig. 2(c). The electromechanical coupling coefficient for various electrode thicknesses was calculated and presented in Fig. 2(d). It is observed that with decreasing electrode thickness, the electromechanical coupling

coefficient increases. However, excessively small electrode thickness can lead to increased ohmic losses in the resonator, which in turn results in greater insertion loss for the filter. After careful consideration, the value of h_{Al} was chosen to be 180 nm(0.1125λ).

The resonators using this substrate can achieve a quality factor exceeding 600 at resonance points near 2 GHz.

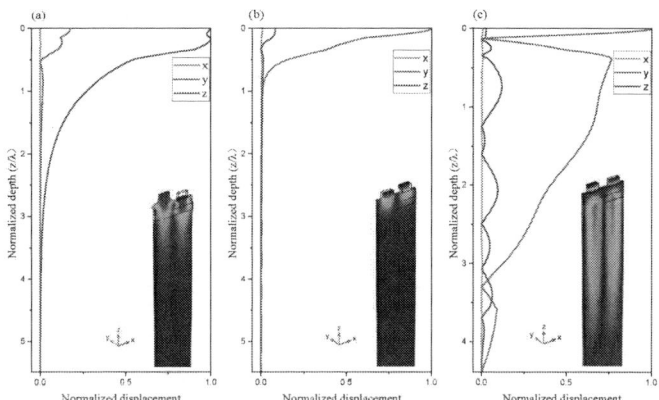

Fig. 3. Calculated relative displacement of acoustic waves in depth direction. (a) Rayleigh mode. (b) SH mode. (c) SH$_1$ mode.

III. SH-SAW-BASED LADDER-TYPE FILTERS

After determining the specific parameters of the substrate, we employed general synthesis methodology to design a fifth-order filter [5]. The filter topology comprises three series resonators and two parallel resonators. The three set of series resonators are abbreviated as S$_1$, S$_2$, and S$_3$, with S$_1$ being equal to S$_3$. The parallel resonators, P$_1$ and P$_2$, are also identical. Utilizing the synthesis method, we obtained the static capacitance(C$_0$), dynamic capacitance(C$_a$), and dynamic inductance (L$_a$) for each resonator.

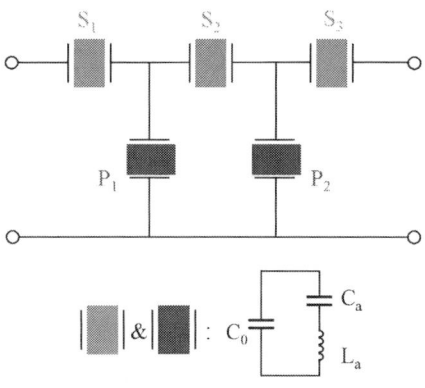

Fig. 4. Filter topology.

Based on the obtained data, a filter circuit was constructed in Advanced Design System, with its S-parameters shown in Fig. 5. The passband of the filter is 2000 MHz to 2330 MHz.

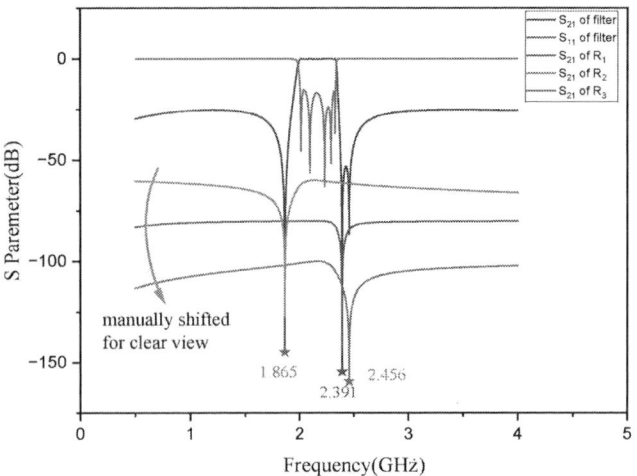

Fig. 5. S parameter of the filter and resonators.

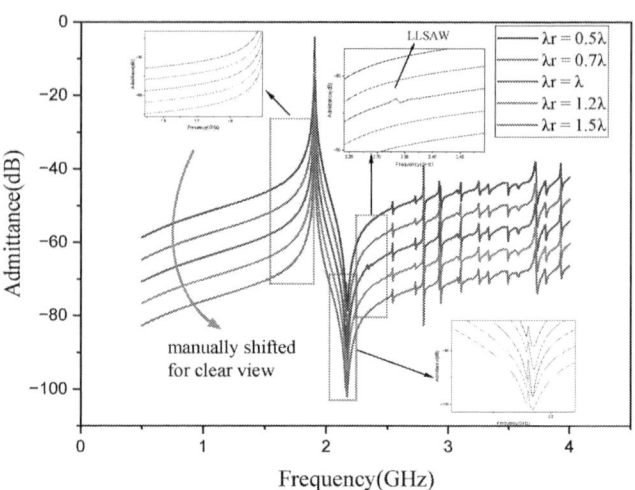

Fig. 6. Resonator admittance curves of resonator at different λ_r.

There are five poles within the passband, provided by five resonators. In the low frequency band, there is a zero located at 1.865 GHz, realized by resonators P_1 and P_2. In the high frequency band, there are two zeros located at 2.391 GHz and 2.456 GHz. The former is realized by S_1 and S_3, while the latter is achieved through S_2.

A. Reflective Grating Design

The simulation indicates that the spurious modes to the left of the resonant frequency are effectively controlled, while a significant amount of noise appears to the right of the anti-resonant frequency. Therefore, due to the wide bandwidth of the filter, higher demands are placed on the parallel resonator, necessitating further design improvements.

With λ_r set at 1.9 μm, the resonant frequency of the resonator is observed to be approximately 1.8 GHz. When λ_r is equal to λ, Fig. 6 illustrates that there is a noise component to the left of the anti-resonant frequency due to energy leakage. Concurrently, the LLSAW mode is excited near 2.33 GHz, which will cause irregularities in the transition band of the filter [6].

By adjusting the value of λ_r to be unequal to λ, the LLSAW can be effectively suppressed. This adjustment also provides a certain degree of suppression for the noise to the left of the anti-resonant point, although it comes at the cost of a slight reduction in the admittance ratio. In this study, a reflection grating period of $\lambda_r = 0.7\lambda$ for the parallel resonator is chosen, as it achieves the objective of noise suppression while maintaining a relatively stable admittance ratio.

B. Design and Characterization of SAW Filters

After obtaining the component values of each resonator through filter synthesis, the corresponding curves were fitted using COMSOL simulations. Through meticulous design adjustments, the final specific values of the physical parameters of the resonators are presented in Table I.

The S-parameters of the symmetric filter obtained from simulations are shown in Fig. 7(a). Within the frequency range of 0.5 to 4 GHz, the overall out-of-band suppression of the filter remains below 20 dB. Notably, at frequencies ranging from 1.68 GHz to 1.92 GHz and 2.36 GHz to 2.67 GHz, the out-of-band suppression can even reach below 30 dB. The center frequency of the filter is 2156 MHz, with a passband of 1996 to 2330 MHz, indicating a 3-dB bandwidth of 15.5%. Fig. 7(b) shows an enlarged view of the S21 parameter near the passband.

Table II summarizes the parameters of the reported SAW filters including f_c, IL and 3-dB FBW. Among these results, our LN/SiO$_2$/Quartz deveices demonstrate superior performance compared to peers by maintaining a large fractional bandwidth at high frequency.

TABLE I
RESONATOR PARAMETERS OF FILTER.

	S1 & S2	P1 & P2	S3
$\lambda(\mu m)$	1.682	1.94	1.624
N_p	50	100	50
Aperture(um)	109.33	77.6	24.36
$\lambda r(\mu m)$	1.682	1.358	1.624
N_r	15	11	15
Static Capcitance(pF)	2.397	2.941	0.543

IV. CONCLUSION

In summary, we have presented the design of a resonator that exhibits no significant spurious responses within 0.5 to 4 GHz. This is a critical advancement, as minimizing spurious responses is essential for achieving high performance and reliability in communication systems. Building on this robust resonator platform, we designed a filter that operates at a center frequency of 2156 MHz, featuring a relative bandwidth

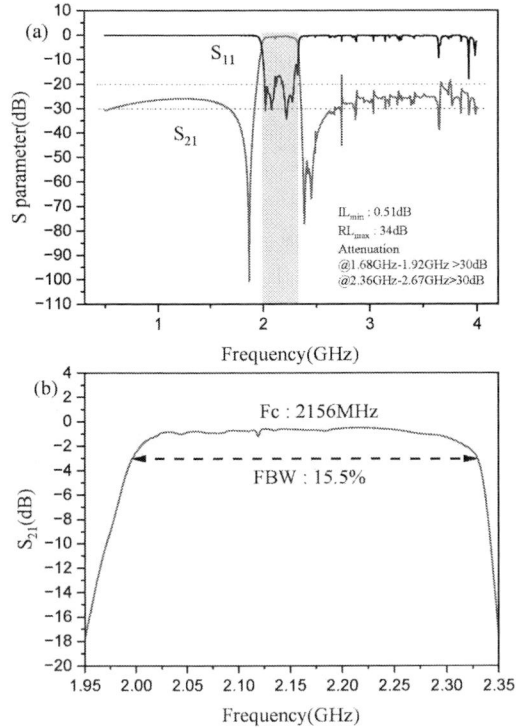

Fig. 7. (a) Simulated S-parameters of SAW filter. (b) Zoomed-in measured S_{21}.

TABLE II
COMPARISONS OF STATE-OF-ART SAW FILTERS.

Ref	Material	fc (GHz)	IL (dB)	3-dB FBW (%)
[7]	LT/SiO$_2$/Quartz	1.977	0.89	4.8
[8]	LT/SiO$_2$/polySi/Si	2.669	2	6
[9]	LN/SiO$_2$/Si	1.27	0.7	20.1
This work	LN/SiO$_2$/Quartz	2.156	0.51	15.5

of 15.5%. Our findings highlight the potential of our resonator and filter design to meet the increasing demands of modern communication technologies, paving the way for future developments in high-frequency devices.

REFERENCES

[1] J. Wu et al.,"SAW Resonators on 15°YX-LiNbO3/SiO2/Sapphire Substrate with Excellent Electromechanical Coupling," 2023 Joint Conference of the European Frequency and Time Forum and IEEE International Frequency Control Symposium (EFTF/IFCS), Toyama, Japan, 2023, pp. 1-4.

[2] T. -H. Hsu et al.,"Harnessing Acoustic Dispersions in YX-LN/SiO2/Si SH-SAW Resonators for Electromechanical Coupling Optimization and Rayleigh Mode Suppression," in IEEE Transactions on Ultrasonics, Ferroelectrics, and Frequency Control, vol. 70, no. 12, pp. 1786-1793, Dec. 2023.

[3] X. Ke et al.,"Heterogeneous Integration of 42°YX LiNbO3/SiO2/Quartz for Wideband and Spurious-Free SAW Resonators," in IEEE Transactions on Electron Devices, vol. 71, no. 10, pp. 6343-6349, Oct. 2024.

[4] B. Xiao et al.,"Inherent Suppression of Transverse Modes on LiTaO3/AT-Quartz SAW Devices," in IEEE Transactions on Electron Devices, vol. 71, no. 2, pp. 1266-1273, Feb. 2024.

[5] S. Amari and G. Macchiarella,"Synthesis of inline filters with arbitrarily placed attenuation poles by using nonresonating nodes," in IEEE Transactions on Microwave Theory and Techniques, vol. 53, no. 10, pp. 3075-3081, Oct. 2005.

[6] T. Sato and H. Abe,"Longitudinal leaky surface waves for high frequency SAW device applications," 1995 IEEE Ultrasonics Symposium. Proceedings. An International Symposium, Seattle, WA, USA, 1995, pp. 305-315 vol.1.

[7] B. Xiao et al.,"Spurious-Free SAW Filters With Inherent Suppression of Transverse Modes on LiTaO3/SiO2/Quartz Platform," 2024 IEEE MTT-S International Conference on Microwave Acoustics & Mechanics (IC-MAM), Chengdu, China, 2024, pp. 141-144.

[8] Y. Xiang et al.,"High-frequency temperature-stable SAW filters based on a heterogeneous integrated multilayered structure," 2024 IEEE Ultrasonics, Ferroelectrics, and Frequency Control Joint Symposium (UFFC-JS), Taipei, Taiwan, 2024, pp. 1-3.

[9] H. Xu et al.,"Large-Range Spurious Mode Elimination for Wideband SAW Filters on LiNbO3/SiO2/Si Platform by LiNbO3 Cut Angle Modulation," in IEEE Transactions on Ultrasonics, Ferroelectrics, and Frequency Control, vol. 69, no. 11, pp. 3117-3125, Nov. 2022.

A 130nm SiGe BiCMOS Ultra-Wideband High-Gain Power Amplifier With 86% Fractional Bandwidth and Digital Gain Control

Yuqing Liu   Bharatha Kumar Thangarasu   Nagarajan Mahalingam   Kaixue Ma

Fanyi Meng   Zhenghao Lu   Kiat Seng Yeo

Abstract—This paper introduces a compact, high-performance broadband power amplifier (PA) designed in 130nm SiGe BiC-MOS process, featuring a high-order RLC matching network and digital control for 8-step variable gain. The PA covers an ultra-wide bandwidth from 10.5 to 26.4 GHz, achieving a fractional bandwidth of 86%, and delivers a saturated output power (P_{sat}) of 15.4 dBm, an output 1-dB compression point (OP1dB) of 13.5 dBm, and a gain of 24.5 dB at 22 GHz, with a gain ripple of only 1.0 dB within the 11-22 GHz band, making it ideal for advanced communication systems.

Index Terms—Digital Control, High gain, High-order RLC matching, SiGe BiCMOS, Ultra-Wideband PA

I. INTRODUCTION

The escalating demand for high-speed data transmission and the proliferation of wireless communication technologies have underscored the significance of the Ku-band in modern communication systems [1]. This band is particularly crucial for applications such as satellite communications, radar systems, and 5G wireless networks, where the need for broad bandwidths allows for the employment of simple modulation schemes, thereby reducing the complexity of the front-end electronics and significantly shortening the time-to-market for new services.

A broadband PA is capable of operating over a wide frequency range, enabling it to support multiple communication bands and standards simultaneously. Several technical approaches have been employed to realize broadband Pas, such as magnetically coupled resonator (MCR), High-order matching network, varactor-based and Switched-capacitor-based tunable matching methods [2] [3] [4] [5]. The MCR network can effectively increase the gain-bandwidth product (GBW) of the PA by forming a high-order matching network with the area of one inductor. However, the transformer loss in the MCR, which becomes significant in the mm-wave band, has often been neglected or brought difficulty calculation in theoretical

derivations, leading to limitations in achieving a flat frequency response. Tunable matching methods have been explored to achieve optimal loads at different center frequencies. While these methods offer the potential for reconfigurability and adaptability, they introduce added system complexity and may suffer from linearity issues due to large voltage swings at the PA's output.

In this paper, a ultra-wideband PA with high-order RLC matching network and digital control for variable gain is proposed. By tuning the specific capacitor and inductor in the input and output matching network, the PA can realize the trade-off between high gain, good gainflatness and wide bandwidth. The paper is organized as follows: Section II discusses the design detail of the PA. Section III presents the overall PA topology and post-layout simulation results. Section IV gives the conclusion.

II. DESIGN DETAIL OF THE PA

A. PA Transistor Selection

The broadband power amplifier (PA) described in this work is meticulously crafted using a 130nm SiGe BiCMOS process, prioritizing high-f_T NPN transistors for their exceptional speed and power handling capabilities. These transistors, with a maximum width of $18\mu m$, are strategically chosen to optimize the balance between gain, output power, and efficiency. The entire PA is constructed in a fully differential configuration to enhance linearity and reduce even-order harmonics, ensuring signal integrity across the extensive operating frequency range.

B. Matching Network Selection

The input and output matching networks are engineered with high-order RLC components to achieve broadband impedance matching. This design is crucial for maximizing power transfer to the load while minimizing reflections and losses, ensuring a flat and broad frequency response. Despite their higher Q-factors, more compact areas, and direct power supply capabilities compared to single-ended inductors, differential inductors present layout challenges due to their elongated shape, which can lead to unnecessary area consumption. To address this, only two differential inductors are strategically placed in the layout, following the sequence of devices in the

---

Yuqing Liu, Bharatha Kumar Thangarasu, Nagarajan Mahalingam, Kaixue Ma, Fanyi Meng and Kiat Seng Yeo are with *the School of Microelectronics, Tianjin University*, Tianjin, China. Zhenghao Lu is with *the School of Electronic and Information Engineering, Soochow University*, Suzhou, China. Kiat Seng Yeo is also with *Engineering Product Development, Singapore University of Technology and Design*, Singapore. Corresponding author: (Yuqing Liu, email: 3020232024@tju.edu.cn)

979-8-3315-2209-4/25 $31.00 © 2025 IEEE

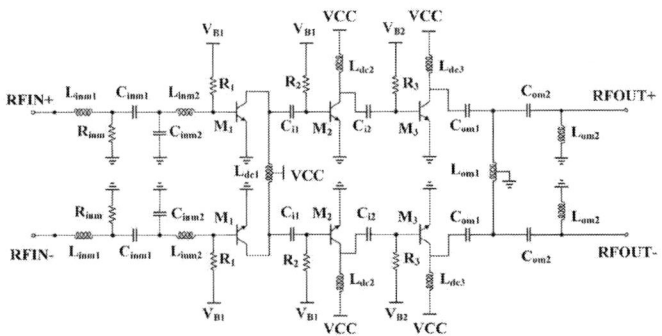

Fig. 1. The architecture of the proposed PA.

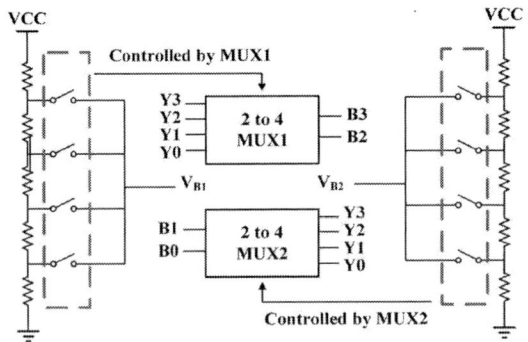

Fig. 2. The architecture of the proposed PA.

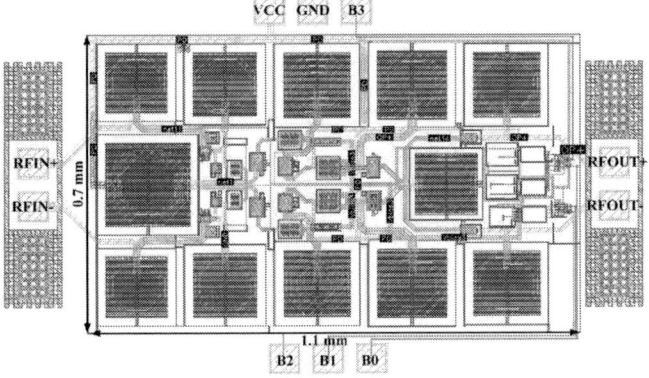

Fig. 3. Layout of the proposed PA (1462 μm x 881 μm).

schematic, to optimize space utilization while maintaining the benefits of differential topology.

III. PA DESIGN AND SIMULATION RESULTS

A. Overall PA design

The overall design of the PA, as depicted in Fig. 1, revolves around a compact and efficient architecture that incorporates a fully differential structure for improved performance. Comprising three stages operating in Class A, the PA ensures linear amplification across a broad frequency spectrum. The PA is powered at 1.5V, and the biasing network is designed with digital control using two 2-4 multiplexers (MUXs), as

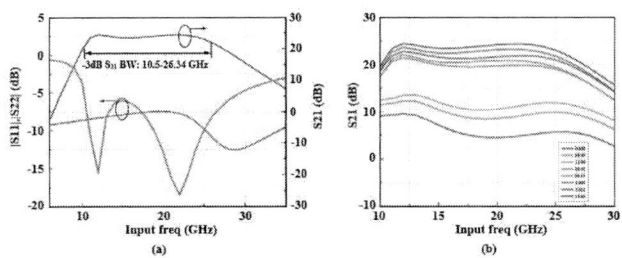

Fig. 4. Post-layout simulation result. (a) SP-performance. (b) Variable gain.

TABLE I
COMPARISON WITH SILICON-BASED BROADBAND PAs

Ref.	This work	[6]TCASII'23	[7]ISSCC'17	[2]TCASI'18
Process(um)	130nm SiGe	130nm	130nm SiGe	65nm
Supply(V)	1.5	1.2	1.5	1.0
BW$_{-3dB}$(GHz)	10.5-26.4	21.3-27.9	23.3-39.7	21.6-41.6
Gain(dB)	24.5	17.5	18.2	20.8
Variable Gain	YES	NO	NO	NO
P$_{sat}$(dBm)	15.6	15.7	16.8	15.3
OP1dB(dBm)	13.5	12.1	15.2	12.9
Area(mm^2)	0.78	0.11	1.76*	0.11

*Including pads.

depicted in Fig. 2, to adjust the voltage division and provide the necessary bias voltages for each stage. This digital control mechanism allows for fine-tuning of the PA's gain and output power characteristics over a 4-bit range with 8-step resolution, offering high flexibility and adaptability.

B. Post-layout simulation results

As shown in Fig. 3, the compact layout is achieved through careful integration of the matching networks, biasing circuits, and PA stages, ensuring that the PA occupies a minimal footprint while maintaining high performance. The layout design also incorporates guard rings and grounding strategies to minimize parasitic effects and ensure stability over temperature variations.

The PA's post-layout simulation results are shown in Fig. 4. Within the critical 11-22 GHz frequency band, a gain ripple of only 1.0 dB is achieved, ensuring a flat and consistent gain across the entire bandwidth. At 22 GHz, the PA delivers a P$_{sat}$ of 15.4 dBm, an OP1dB of 13.5 dBm, and a gain of 24.5 dB. With 4-bit control, an 8-step variable gain is achieved. The comparison with the state-of-the-arts work is listed in Table I. These performance metrics underscore the PA's ability to provide high output power and high gain while maintaining linearity within a wider bandwidth, which is vital for modern communication systems.

IV. CONCLUSION

This paper presents an ultra-broadband power amplifier (PA) designed in 130nm SiGe BiCMOS process, featuring a high-order RLC matching network and digital control for variable gain. The PA achieves an ultra-wide bandwidth of

10.5-26.4 GHz with an 86% fractional bandwidth with high gain of 24.5 dB at 22 GHz, while maintaining a low gain ripple of 1.0 dB within the target bandwidth of 11-22 GHz. The compact design demonstrates exceptional performance, making it suitable for Ku-band and a broad range of advanced communication system applications.

REFERENCES

[1] P. M. Asbeck, N. Rostomyan, M. Özen, B. Rabet and J. A. Jayamon, "Power Amplifiers for mm-Wave 5G Applications: Technology Comparisons and CMOS-SOI Demonstration Circuits," in IEEE Transactions on Microwave Theory and Techniques, vol. 67, no. 7, pp. 3099-3109, July 2019.

[2] H. Jia, C. C. Prawoto, B. Chi, Z. Wang and C. P. Yue, "A Full Ka-Band Power Amplifier With 32.9% PAE and 15.3-dBm Power in 65-nm CMOS," in IEEE Transactions on Circuits and Systems I: Regular Papers, vol. 65, no. 9, pp. 2657-2668, Sept. 2018.

[3] R. Wang, C. Li, K. Qiu, Y. Liu, S. Yin and Y. Wang, "A Ka-Band Broadband Power Amplifier With Transformer-Based Lossy Magnetically Coupled Resonator Network: Analysis and Design," in IEEE Transactions on Microwave Theory and Techniques, vol. 72, no. 12, pp. 6857-6870, Dec. 2024.

[4] Z. Bai, A. Azam and J. S. Walling, "A Frequency Tuneable Switched-Capacitor PA in 65nm CMOS," 2019 IEEE Radio Frequency Integrated Circuits Symposium (RFIC), Boston, MA, USA, 2019, pp. 295-298.

[5] M. Vigilante and P. Reynaert, "A 29-to-57GHz AM-PM compensated class-AB power amplifier for 5G phased arrays in 0.9V 28nm bulk CMOS," 2017 IEEE Radio Frequency Integrated Circuits Symposium (RFIC), Honolulu, HI, USA, 2017, pp. 116-119.

[6] J. Li et al., "A K-Band Broadband Power Amplifier With 15.7 dBm Power and 30.4% PAE in 0.13 um CMOS," in IEEE Transactions on Circuits and Systems II: Express Briefs, vol. 70, no. 4, pp. 1321-1325, April 2023.

[7] S. Hu, F. Wang and H. Wang, "A 28-/37-/39-GHz Linear Doherty Power Amplifier in Silicon for 5G Applications," in IEEE Journal of Solid-State Circuits, vol. 54, no. 6, pp. 1586-1599, June 2019.

Accurate Gummel Parameter Extraction for SiGe HBTs with Temperature-Dependent Saturation Current and Ideality Factor Models

Xudong Cai, Guofang Yu, Jie Cui, Yue Zhao, Jun Fu

School of Integrated Circuits, and Beijing National Research Center for Information Science and Technology (BNRist), Tsinghua University, Beijing 100084, China. (fujun@tinghua.edu.cn)

Abstract—**This paper presents the temperature-dependent DC characteristics of Silicon-Germanium Heterojunction Bipolar Transistors (SiGe HBTs) using a modified and extended Mextram model, covering 80 K to 400 K. An improved method for extracting Gummel characteristics, with a focus on the ideality factor and saturation current, is proposed. The method incorporates a temperature-dependent saturation current model and a correction factor, improving accuracy. Simulation results match experimental data with a maximum deviation of 15%. This approach enables accurate performance predictions for SiGe HBT circuit simulation and design.**

Keywords—*SiGe HBTs, forward-Gummel measurements, compact model, wide temperature parameter extraction.*

I. Introduction

Silicon-Germanium (SiGe) technology is widely recognized for its unparalleled analog and RF performance in extremely wide temperature environments [1], [2]. This is primarily attributed to the unique energy band structure and device architecture of SiGe HBTs, which exert a positive influence on the low-temperature operations of bipolar transistors, enabling reliable functionality across a broad temperature range from sub-1 K to over 400 K [3], [4], [5], [6]. Therefore, analyzing the characteristics of SiGe HBTs under wide or cryogenic temperature conditions is crucial, especially for the design of low-temperature or wide-temperature-range circuits [7].

This study conducts an in-depth investigation of the DC characteristics of SiGe heterojunction bipolar transistors (HBTs) over a temperature range of 80 K to 400 K. The temperature dependence of the device's quiescent operating point is analyzed. Parameters of the DC model, such as saturation current and ideality factor, are extracted across this wide temperature range, and the relationship between these model parameters and temperature is examined. Finally, simulations of the Gummel characteristics and output characteristics of the device are performed using the extracted wide-temperature-range DC model parameters.

II. Experimental Details and Model Description

A. Device fabrication and measurement configuration

The investigated NPN SiGe HBT devices were fabricated using 0.5 μm SiGe technology, adopting a collector-base-emitter-base-collector (CBEBC) layout configuration. The emitter window width (We) is 0.6 μm, and the emitter lengths (Le) are 8, 12, 16, and 20 μm, respectively. Based on the increasing sizes of Le, the four samples are labeled as HBT-x (x=1, 2, 3, 4). These devices are constructed with a lightly selectively implanted collector (SIC), heavily doped polysilicon emitters, and heavily boron-doped extrinsic polysilicon bases. Local oxidation of silicon (LOCOS) is employed for device isolation.

The DC characteristics of the devices across a wide temperature range (80 K to 400 K) were characterized using a Lakeshore Cryogenic probe station. The system comprises a high-low temperature test system (including a cryogenic probe station, vacuum system, temperature control system, cooling system, and microscope system) and a Keysight B1500 semiconductor parameter analyzer. The primary focus is on extracting and fitting the compact model parameters for HBT-x (x=1, 2, 3, 4) to analyze the carrier transport mechanisms and non-ideal current effects.

B. Description of wide temperature range compact model

The main current I_N is modeled as [8]:

$$I_N = \frac{I_{S,F}e^{\frac{V_{B_2E_1}}{N_FV_T}} - I_{S,R}e^{\frac{V^*_{B_2C_2}}{N_RV_T}}}{q_B} \quad (1.1)$$

where all the symbols have their usual meanings in Mextram model except for forward and reverse saturation current $I_{S,F}$ and $I_{S,R}$, forward and reverse ideality factor N_F and N_R, respectively. Under wide operating temperature conditions, the Mextram model cannot accurately describe the electrical characteristics of SiGe HBTs. Thus, the model requires extension. To describe the wide temperature characteristics, the primary current model is modified as follows:

$$I_N = \frac{I_S(T)}{q_B}\left[\exp\left(\frac{V_{B2E1}}{N_F(T)V_T}\right) - \exp\left(\frac{V_{B2C2}}{N_R(T)V_T}\right)\right] \quad (1.2)$$

Here q_B accounts for modulation of base charge (or effective charge for HBTs), $I_S(T)$ is the saturation current, which is temperature-dependent, and $N_F(T)$ is the forward ideality factor and $N_R(T)$ is reverse ideality factor. $V_T = kT/q$. The non-

979-8-3315-2209-4/25 $31.00 © 2025 IEEE

ideal ideality factors $N_F(T)$ and $N_R(T)$ are not equal to 1 but are described as temperature-dependent functions [9]:

$$N_F(T) = N_{F,nom}\left[1 - \frac{T - T_{nom}}{T_{nom}}\left(A_{NF}\frac{T_{nom}}{T}\right)^{X_{NF}}\right] \quad (2.1)$$

$$N_R(T) = N_{R,nom}\left[1 - \frac{T - T_{nom}}{T_{nom}}\left(A_{NR}\frac{T_{nom}}{T}\right)^{X_{NR}}\right] \quad (2.2)$$

where T_{nom} is the nominal temperature, typically set at 300 K, while $N_{F,nom}$ and $N_{R,nom}$ denote the ideality factors at the nominal temperature. A_{NF}、X_{NF}、A_{NR}, and X_{NR} are fitting parameters. Furthermore, the relationship between saturation current and temperature considers the non-linear variation of the bandgap with temperature, lower temperature ranges, and the effects of the ideality factors. The variation of saturation current with temperature is given by [9]:

$$I_S(T) = I_{S,nom}\left(\frac{T}{T_{nom}}\right)^{\frac{X_{IS}}{N_F(T)}} exp\left(\frac{-E_{a,t}\left(1 - \frac{T}{T_{nom}}\right)}{N_F(T)V_T}\right) \quad (3.1)$$

$$E_{a,t} = E_{a,nom} - \frac{\alpha\beta T_{nom}^2}{(T_{nom}+\beta)^2} + \frac{\alpha\beta T T_{nom}}{(T_{nom}+\beta)(T+\beta)} \quad (3.2)$$

where E_{g0} represents the bandgap width at 0 K, α=4.45e-4V/K and $\beta = 686K$ are material-related constants. $E_{a,t}$ denotes the activation energy with a non-linear temperature relationship, and $E_{a,nom}$ is the activation energy at T_{nom}. X_{IS} is a fitting parameter.

III. RESULTS AND DISCUSSION

A. Gummel characteristics

In Fig. 1, the forward Gummel characteristics of HBT-x (x = 1, 2, 3, 4) under conditions of 400K, 300K, 200K, and 80K demonstrate a consistent electrical behavior across devices of different sizes. As shown in Fig. 2, at 300K, the collector current of HBT-x (x = 1, 2, 3, 4) is measured under a fixed bias voltage of V_{BE}=0.615V. The horizontal axis represents the normalized emitter length (Le/8 μm), while the vertical axis denotes the corresponding collector current. The circles indicate measured data, and the straight line represents the fitted linear equation. The experimental and fitted results reveal a linear relationship between the electrical behavior and device dimensions. Leveraging this linearity can streamline the modeling process by reducing the need for multiple parameter extractions, allowing for scalable models across different device sizes. Consequently, parameter extraction for HBT-x (x=1, 2, 3, 4) is conducted for only one device size.

In Fig. 3, the forward-active mode Gummel characteristics of SiGe HBT-1 are illustrated for temperatures ranging from 400K to 80K (step: 20K). The measurements are conducted with the CB junction at zero bias, mitigating additional effects between the base and collector due to junction biasing. The experimental results indicate that as the temperature decreases, the non-ideal current mechanisms in the low-current region become increasingly prominent, particularly the non-ideal base current. This non-ideal base current, attributed to trap-assisted

tunneling (TAT) in the heavily doped base-emitter junction, is caused by the numerous traps in the base-emitter space charge region (BE SCR) [10]. This excessive base current in the low-current region leads to a reduced effective amplification range, which is a critical aspect to address in modeling. Furthermore, the output characteristics reveal good performance at various temperatures, with a high Early voltage, which is advantageous for designing high-performance low-temperature circuits [7].

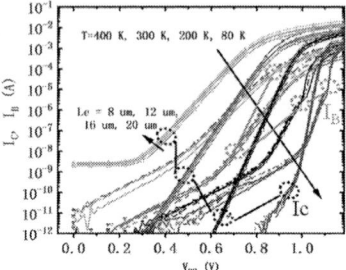

Fig. 1. Forward Gummel characteristics of HBT-x (x=1, 2, 3, and 4) at 400K, 300K, 200K, ,and 80K.

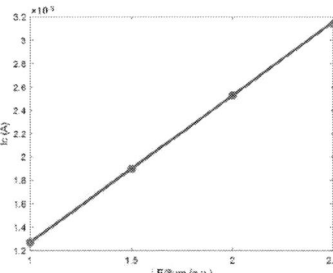

Fig. 2. Collector current of different HBT-x (x=1, 2, 3, and 4) under identical bias conditions at 300K.

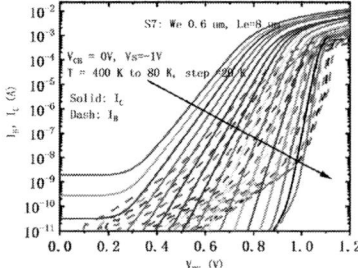

Fig. 3. Forward Gummel characteristics of HBT-1 for temperatures ranging from 400K to 80K

B. Extraction of Forward Gummel Parameters

For the extraction of forward Gummel parameters, the first step is to extract the parameters in the low injection region, including the extraction of saturation current and ideal factor. The range of this extraction refers primarily to the extraction of parameters in the ideal region. The method for extracting saturation current in a wide temperature range is consistent with the classical Mextram extraction method. Equation (1.1) shows the expression for the collector current. Under conventional testing conditions, due to the BC junction zero-bias or reverse bias, the contribution of the BC junction's second term can be considered negligible. Furthermore, the forward base current depends only on the internal voltage at the BE junction, V_{B2E1}. In the ideal region, the voltage drops across the variable base resistance can be ignored, yielding $V_{B2E1} = V_{BE}$. Thus, equation (1) simplifies as follows:

$$I_N = \frac{I_S(T)}{q_B} \exp\left(\frac{V_{BE}}{N_F(T)V_T}\right) \quad (4.1)$$

Here, I_N refers to the collector current I_C. In the ideal voltage range, q_B can be approximated as 1 to simplify the extraction of saturation current. Hence, the saturation current is given by:

$$I_S(T) = I_N \exp\left(-\frac{V_{BE}}{N_F(T)V_T}\right) \quad (4.2)$$

During the extraction process, equation (4.1) can be logged as follows:

$$\ln(I_N) = \ln(I_S(T)) + \frac{V_{BE}}{N_F(T)V_T} \quad (4.3)$$

V_{BE} can be calculated using external node voltages. The intercept of the above equation gives the logarithm of the saturation current $\ln(I_S(T))$, and the ideality factor $\frac{1}{N_F(T)V_T}$ can be obtained from the slope $N_F(T)$.

The Gummel characteristics directly reflect device performance, incorporating effects from current, capacitance, and resistance models. The primary parameters extracted from the Gummel curves include the ideality factor and saturation current. Fig. 4 illustrate the extracted parameters and fitted model calculation of the saturation current and ideality factor for collector current of HBT-1 for varying temperatures. The model calculated results align well with experimentally extracted data. However, further investigation reveals that, although the model appears to match the experimental data well under certain conditions, it exhibits significant discrepancies when simulating the Gummel characteristics over a wide temperature range. For instance, the saturation current at 80 K is on the order of 10^{-70}. This demonstrates that even small changes in the order of magnitude can result in substantial percentage errors. Even errors below 5% pose significant challenges in calculating the Gummel characteristics. As such, there is a pressing need to refine the model for more accurate representation of the wide-temperature-range behavior.

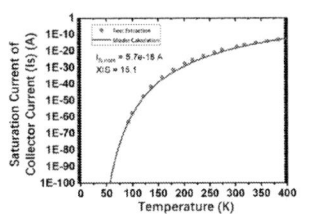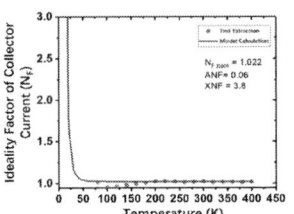

Fig. 4. Extraction and fitted model calculation of the saturation current and ideal factor of collector current of HBT-1 for varying temperatures.

C. Model and Parameter Extraction Improvement

To address these issues, we propose an improved model and parameter extraction strategy. The ideality factor is modeled using the approach detailed in (2.1). For the saturation current, we improve the model and apply a correction factor $f(T)$ to account for discrepancies. This approach yields accurate results across different device sizes (HBT-x, where x=1, 2, 3, and 4).

The temperature model for saturation current is refined as follows:

$$I_s{}'(T) = \exp(A_{IS}T^{X_{IS}} + C_{IS}) \quad (5.1)$$

Applying the correction factor $f(T)$ results in:

$$I_s(T) = f(T) * I_s{}'(T) \quad (5.2)$$

$$f(T) = a_0 + \sum_{n=1}^{5}[a_i \cos(n\omega T) + b_i \sin(n\omega T)] \quad (5.3)$$

The parameters a_0, a_i, b_i, and ω are obtained through extraction.

The specific solution process using this method is as follows:

1) Extract the saturation current and ideality factor in the ideal region of the Gummel plot using the extraction methods from Equations (4.1) to (4.3);

2) Simulate the ideality factor extracted in the first step using the ideality factor model (2.1);

3) Re-extract the saturation current based on the ideality factor obtained in the second step, using the method described in the first step;

4) Extract the three parameters in Equation (5.1) based on the saturation current obtained in the third step, and simulate the saturation current using this model;

5) Calculate the average relative error based on the results from the models in steps two and four.

$$E_{MS} = \frac{1}{N} \sum_{i=1}^{N} \frac{|Meas_i - Simu_i|}{Meas_i} \quad (6)$$

where N represents the number of data points.

6) If $E_{MS} > E_{aim}$, where E_{aim} is the target error, the following expression is used to tune the parameter $f(T)$:

$$E_f = 1 + \frac{1}{N} \sum_{i=1}^{N} \frac{|Meas_i - Simu_i|}{Meas_i} \quad (7)$$

7) Using the parameter tuning results obtained earlier, the saturation current is calculated using Equation (4.2).

TABLE I: PARAMETERS EXTRACTED USING THE IMPROVED MODEL

$I_s{}'(T)$	A_{IS}	X_{IS}	C_{IS}			
	-10540	-0.9353	10.7			
$f(T)$	a_0	a_1	a_2	a_3	a_4	a_5
	0.9916	-0.0719	0.01053	-0.1781	0.00531	0.03748
	b_1	b_2	b_3	b_4	b_5	w
	0.0248	0.1191	-0.00365	-0.06485	-0.03059	0.01606

Fig. 5 shows the relationship between the saturation current and temperature for HBT-1 collector current simulated using the improved model. A comparison with the experimental results indicates that the model provides a good simulation result. The extracted parameters are shown in Table I.

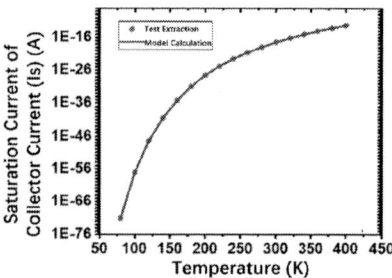

Fig. 5. Relationship between saturation current and temperature for HBT-1 collector current simulated using the improved model.

D. Simulation Analysis of the Improved Model

In above section, the main parameters of the model were extracted. Based on these extracted parameters, simulations were performed and compared with experimental results to calculate the error. As shown in Fig. 6 (a), the collector current of HBT-1's Gummel plot was simulated for temperatures ranging from 400K to 80K. Fig. 6 (b) presents the average relative error between the simulation and experimental results. It can be observed that the maximum error in the simulation of the collector current does not exceed 15%, indicating that the model can accurately simulate the device. At 100K and 80K, non-ideal current appears in the low current region due to the trap-assisted tunneling (TAT) current analyzed earlier.

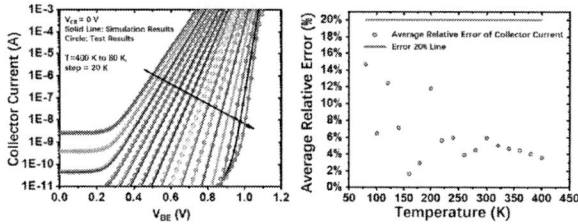

Fig. 6. Simulation of collector current Gummel curve for HBT-1 from 400 k to 80 k, (a) Gummel curve simulation, (b) calculation of average relative error between simulation and experiment.

As shown in Fig. 7, the output characteristics of device HBT-1 were simulated at different temperatures (400 K, 300 K, 200 K, and 80 K).

IV. Conclusion

This work introduces an improved extraction method for Gummel characteristics, achieving better accuracy in saturation current and ideality factor modeling. The proposed temperature-dependent model and correction factor reduce simulation errors, particularly at low temperatures. The method shows strong agreement with experimental data, ensuring accurate HBT device simulations for diverse temperature conditions. This approach enhances modeling, providing a solid foundation for future research in semiconductor device simulation.

Acknowledgment

This work was supported by Grant QYJS-2022-1700-B, and is partly supported by the National Natural Science Foundation of China (Grant number: 62474101 and 92064002).

References

[1] G. Yu, R. Liang, X. Wang, J. Xu, and T.-L. Ren, "Operation of silicon-germanium heterojunction bipolar transistors with different structures at deep cryogenic temperature," Sci. Bull., vol. 64, no. 7, pp. 469–477, Apr. 2019, doi: 10.1016/j.scib.2019.03.005.

[2] G. Yu, J. Cui, Y. Zhao, J. Fu, and T.-L. Ren, "Investigation of Base Transport Mechanism in Silicon-Germanium Heterojunction Bipolar Transistor Operating over Wide Temperature Range," J. Phys. Conf. Ser., p. 012013 (6 pp.)-012013 (6 pp.), 2021, doi: 10.1088/1742-6596/2065/1/012013.

[3] M. D. Ganeriwala, A. Singh, A. Dubey, R. Kaur, and N. R. Mohapatra, "A Bottom-Up Scalable Compact Model for Quantum Confined Nanosheet FETs," IEEE Trans. ELECTRON DEVICES, vol. 69, no. 1, pp. 380–387, Jan. 2022, doi: 10.1109/TED.2021.3130015.

[4] G. Yu, J. Cui, Y. Zhao, W. Cui, and J. Fu, "Effects of co-60 gamma-ray irradiation on the DC and RF characteristics of SiGe HBTs," Microelectron. Reliab., vol. 159, Aug. 2024, doi: 10.1016/j.microrel.2024.115443.

[5] K. Ishimaru, Future of Non-Volatile Memory -From Storage to Computing-. in 2019 IEEE International Electron Devices Meeting (IEDM). 2019, p. 1.3 (6 pp.). doi: 10.1109/IEDM19573.2019.8993609.

[6] N. Waldhoff, F. Danneville, G. Dambrine, B. Geynet, and P. Chevalier, Investigation of SiGe HBT potentialities under cryogenic temperature. in Proceedings of the 39th European Solid-State Device Research Conference. ESSDERC 2009. 2009, p. 4. doi: 10.1109/ESSDERC.2009.5331374.

[7] M. Varonen et al., "Cryogenic W-Band SiGe BiCMOS Low-Noise Amplifier," presented at the PROCEEDINGS OF THE 2020 IEEE/MTT-S INTERNATIONAL MICROWAVE SYMPOSIUM (IMS), 2020, pp. 185–188.

[8] L. Luo et al., "Wide temperature range compact modeling of SiGe HBTs for space applications," in 2011 IEEE 43rd Southeastern Symposium on System Theory, Auburn, AL, USA: IEEE, Mar. 2011, pp. 110–113. doi: 10.1109/SSST.2011.5753786.

Fig. 7. Output characteristic simulation at 400 K (a), 300 K (b), 200 K (c), and 80 K (d).

[9] Ziyan Xu, Xiaoyun Wei, Guofu Niu, Lan Luo, D. Thomas, and J. D. Cressler, "Modeling of temperature dependent IC-VBE characteristics of SiGe HBTs from 43–400K," in 2008 IEEE Bipolar/BiCMOS Circuits and Technology Meeting, Monterey, CA: IEEE, Oct. 2008, pp. 81–84. doi: 10.1109/BIPOL.2008.4662717 .

[10] G. A. M. Hurkx, D. B. M. Klaassen, M. P. G. Knuvers, and F. G. O'Hara, "A new recombination model describing heavy-doping effects and low-temperature behaviour," Int. Electron Devices Meet. 1989 Tech. Dig. Cat No89CH2637-7, pp. 307–10, 1989, doi: 10.1109/IEDM.1989.74285.

A Capacitorless Fully Synthesizable Digital-LDO Using scalable APR-friendly Power MOS Cell with 13.71A/mm2 current density

Yunxin Wang[1], Xuliang Wang[1], Boran Zhang[1], Renwei Chen[1], Xuchen Men[1], Xiaosen Liu[1*]

1.the School of Integrated Circuits, Tsinghua University, Beijing 100084, China

wangyunx22@mails.tsinghua.edu.cn

Abstract—**This paper introduces a capacitorless fully synthesizable Digital Low-Dropout Regulator (DLDO). The design features a voltage sensor based on a Time-to-Digital Converter (TDC) and a current estimation algorithm for control. A Digitally Controlled Ring Oscillator (DCRO) generates the system clock. The current estimation algorithm quickly calculates the required number of PMOSs and turns on those PMOSs in next cycle, enhancing transient response. Additionally, a scalable and APR-friendly PMOS cell is presented to increase power density. This design implemented in 65-nm CMOS with a core area of 0.0729 mm², simulations show the DLDO supports 100 μA to 1 A and achieves a 13.7 ns recovery time for a 500 mA change within 5.5ns.**

Keywords—*fully synthesizable, Digital low-dropout regulator(DLDO),current estimation algorithm, digital control ring oscillator, scalable APR-friendly Power MOS Cell, hierarchical auto place-and-route*

I. INTRODUCTION

The development of the IC industry drives SoC design iterations, spurring power management innovations. Modern SoCs feature multiple modules with varying supply voltage requirements, such as analog IP operating at $\geq$ 1 V and digital IP operating at $\leq$ 0.7 V for energy efficiency. While switching converters offer high efficiency, they consume excessive chip area. Low Dropout Regulators (LDOs) convert input voltage to suitable supplies for different IPs. Analog LDOs (ALDOs) provide high PSRR and low output ripple, but face challenges with advanced process nodes and low supply voltages. Digital LDOs (DLDOs) [1]–[2], can operate at low voltages and have attracted significant attention in recent years. Synthesizable DLDOs offer scalability, low design costs and can be easily distributed across VLSI chips. However, gate-based comparators struggle with large comparator offsets [3]–[5], and tristate logic in the power stage faces challenges with low current density [6].

In this paper, we present a fully synthesizable DLDO that utilizes a scalable APR-friendly Power MOS cell. A fully digitally controlled ring oscillator (DCRO) generates the necessary clock for the control system. Additionally, a TDC-based voltage sensor converts the output voltage into digital code, while a current estimation algorithm is employed as a control strategy to improve transient response.

II. PROPOSED FULLY SYNTHESIZABLE DLDO

Fig.1 shows the proposed fully synthesizable DLDO architecture, A fully digital ring oscillator generates the system clock used to drive the registers and the TDC. In the TDC-based voltage sensor, V_{OUT} functions as the supply voltage for the delay line. By utilizing the system clock, V_{OUT} is converted into a digital code. The current estimation algorithm, based on the number of PMOS transistors currently active, the difference between V_{IN} and V_{OUT}, as well as the difference

between V_{IN} and V_{REF}, calculates the appropriate number of PMOS transistors for the current load and activates these PMOSs in the next clock cycle. Additionally, the proposed fully synthesizable DLDO uses a scalable, APR-friendly Power MOS cell, which is well-suited for digital backend tools. The PMOS fully occupy the entire standard cell area, enhancing power density.

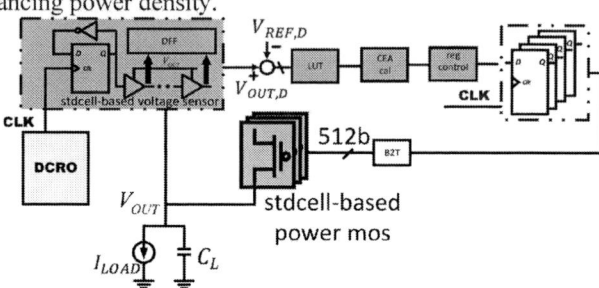

Fig. 1. Fully Synthesizable DLDO Architectures

A. CEA-BASED Control strategy

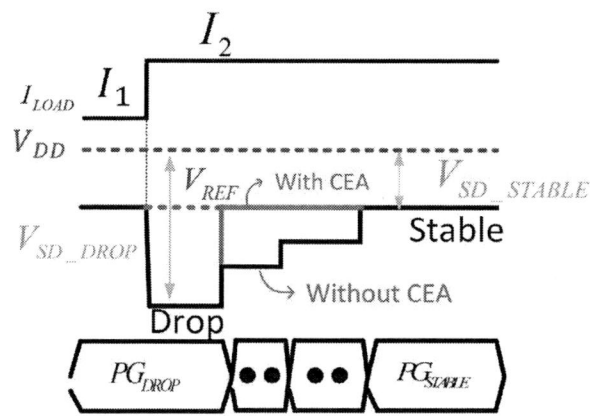

Fig. 2. Current-Estimation Algorithm

In the DLDO, PMOSs have V_{SG} equal to the supply voltage and a small V_{SD}, causing PMOSs to operate in the deep triode region with current linearly dependent on V_{SD}. The $PG_{code}[N-1:0]$ indicates the number of active PMOSs, determining the load current as shown Equations (1), where K includes V_{SG}, W, L, and other parameters. During a current step, PG_{drop} reflects the previous current, and insufficient PMOSs cause a voltage drop. The control system then adjusts V_{OUT} until it equals V_{REF}.. Eventually, PG_{stable} reflects active PMOS count which adequately supplies the load current, as shown in Fig.2 and Equations (2).

$$I_{LOAD} = K \times PG_{code} \times V_{SD} \qquad (1)$$

$$PG_{drop} \times V_{SD\_drop} = PG_{stable} \times V_{SD\_stable} \qquad (2)$$

$$PG_{stable} = PG_{drop} \times \frac{V_{DD}-V_{OUT}}{V_{DD}-V_{REF}} \qquad (3)$$

Control Equation (3) can be obtained from equation (2). The division in equation (3) can be replaced with a lookup table and a multiplication to reduce the latency of circuit.

B. Scalable APR-friendly Power MOS and Layout

Fig 3 illustrates the custom PMOS standard cell layout. The cell height matches one row, and the width is an integer multiple of the site for APR tool compatibility. The power rails are V_{IN} and V_{OUT}, respectively. The backend uses a hierarchical physical design flow. After completing the Power MOS Array APR, an additional APR is performed at the top level with other logic gates. The final layout is shown in Fig. 4

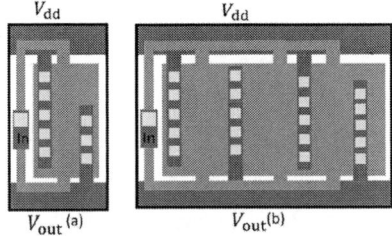

Fig. 3. Scalable APR-friendly Power MOS Cell

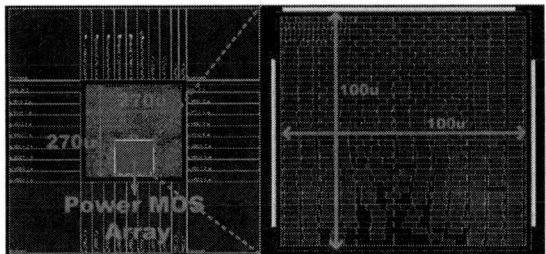

Fig. 4. Lay out of Fully Synthesizable DLDO

III. SIMULATION RESULT

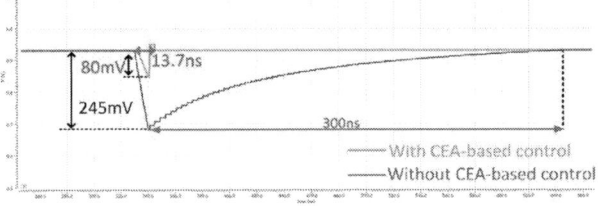

Fig. 5. Transient response Simulate result with 500mA-1A load step

This design is implemented and simulated in TSMC 65 nm technology, occupying a core area of 0.0729 mm². and the area of power Mos array is 0.01 mm². The simulation conditions are set as follows: V_{IN} = 1V, V_{OUT} = 0.93V, F_s = 250Mhz C_{total} = 1pF,(Parasitic capacitance of load) and the current step from 500mA to 1A in 5.5ns

Simulation results of transient response are shown in Fig.5 result compares the performance of the CEA-based control strategy with a non-CEA-based control strategy. In addition to the CEA-based control, the proposed DLDO also uses nonlinear compensation. When the system detects V_{OUT} drops below a certain threshold, it activates all transistors to charge the load, preventing excessive voltage drops. The drawback of nonlinear compensation is a minor voltage overshoot. For digital

loads, the negative impact of voltage drops is more severe than that of slight overshoots, making the small overshoot acceptable.

IV. CONCLUSION

This paper implemented and simulated a Capacitorless fully synthesizable DLDO using TSMC 65nm technology. The power MOS array uses customized cells to achieve higher power density and is compatible with APR tools. V_{out} is directly delivered through the power grid, resulting in reduced parasitic resistance and lower IR drop. All circuitry outside the power MOS array utilizes digital standard cells, with TDC-based voltage sensors and DCRO serving as gate-level netlists, while CEA control and other blocks are described at the RTL level. After APR, the GDS and netlist are imported into Virtuoso for post simulation. Simulation results show that when the current transitions from 500 mA to 1 A, the voltage drops by 80mV and recovers in 13.7 ns. The chip core area is 0.079 mm² and can support load currents ranging from 100 µA to 1 A.

TABLE I. PERFOEMANCE COMPARASION

	[2]	[3]	[4]	**This work**
Process[nm]	28	40	28	**65**
synthesizable	no	Yes	No	**Yes**
Active area[mm²]	0.0055	0.04	0.0152	**0.0729**
V_{IN}[V]	N/A	0.6-1.1	0.65-1.0	**0.7-1.1**
V_{OUT}[V]	N/A	0.55-1.05	0.6-0.95	**0.6-1.05**
C_{total}	Cap-free	Cap-free	Cap-free	**Cap-free**
V_{drop}@load step	137mV @400mA	126mV@ 120mA	122mV@ 30mA	**80mV@ 500mA**
Load range[mA]	N/A	20-200	1.4uA-80	**90-1000**
Fom[a][ps]	0.0048	N/A	5.09	**0.0032**

a. $FOM=\dfrac{C_{total}\times\Delta V_{OUT}\times I_Q}{\Delta I_{LOAD}^2}$

[1] Yasuyuki Okuma et al., "0.5-V input digital LDO with 98.7% current efficiency and 2.7-µA quiescent current in 65nm CMOS," IEEE Custom Integrated Circuits Conference 2010, San Jose, CA, 2010, pp. 1-4.

[2] D. -J. Min, J. -G. Lee, K. Cho and J. H. Shim, "An Output-Capacitor-Free Adaptive-Frequency Digital LDO with a 420-mA Load Current and a Fast Settling Time," 2023 IEEE International Symposium on Circuits and Systems (ISCAS), Monterey, CA, USA, 2023, pp. 1-5

[3] C. Cao, Y. Tang, X. Huang, Z. Zou and L. Zheng, "A Fully Synthesizable Capacitorless Digital LDO for Distributed Power Delivery Network," 2024 IEEE International Symposium on Circuits and Systems (ISCAS), Singapore, Singapore, 2024, pp. 1-5

[4] G. Koo, J. Kim, S. Lee, J. H. Shim and K. Cho, "An Ultrawide Load-Range Fast-Transient Output Capacitor-Less Digital LDO With Adaptive Gate Modulation and Droop Detection," in IEEE Solid-State Circuits Letters, vol. 7, pp. 199-202, 2024

[5] J. Oh, Y. -H. Hwang, J. -E. Park, M. Seok and D. -K. Jeong, "An Output-Capacitor-Free Synthesizable Digital LDO Using CMP-Triggered Oscillator and Droop Detector," in IEEE Journal of Solid-State Circuits, vol. 58, no. 6, pp. 1769-1781, June 2023

[6] N. Ojima, T. Nakura, T. Iizuka and K. Asada, "A Synthesizable Digital Low-Dropout Regulator Based on Voltage-to-Time Conversion," 2018 IFIP/IEEE International Conference on Very Large Scale Integration (VLSI-SoC), Verona, Italy, 2018, pp. 55-58

[7] A. Fahmy, J. Liu, P. Terdal, R. Madler, R. Bashirullah and N. Maghari, "A synthesizable time-based LDO using digital standard cells and analog pass transistor," ESSCIRC 2017 - 43rd IEEE European Solid State Circuits Conference, Leuven, Belgium, 2017, pp. 271-274

[8] C. Liu, C. Zhao and C. . -J. R. Shi, "A Fully-Synthesizable Fast-Response Digital LDO Using Automatic Offset Control and Reuse," 2021 IEEE International Symposium on Circuits and Systems (ISCAS), Daegu, Korea, 2021, pp. 1-5

A Novel LVDDSCR Structure for Improved Latch-Up Immunity and Reduced Transient Overshoot Voltage

Yilidana Mamuti, Hongyi Li, Zhiyuan He, Zihan Zheng, Jiyuan Huang, Yimu Yang, JiXiang Shang, JiZhi Liu, Ruibo Chen and Zhiwei Liu

Abstract—This paper introduces a novel low-voltage dual-directional silicon-controlled rectifier (LVDDSCR) structure designed to address the latch-up effect and excessive transient overvoltage observed in traditional devices. By incorporating an additional low-resistance trigger path, the conduction characteristics of the device are optimized. This modification results in a 20% increase in the holding voltage (Vh) and a 35% reduction in transient overshoot voltage. TCAD simulations and experimental results validate the improvements, making this new LVDDSCR structure particularly suitable for applications sensitive to transient voltage, such as high-speed communication interfaces and sensor I/O ports.

Keywords—Dual-directional silicon-controlled rectifiers, latch-up immunity, transient overvoltage, TCAD simulation, ESD protection, high-speed communication.

I. INTRODUCTION

Dual-directional silicon-controlled rectifiers (DDSCRs) are widely employed for electrostatic discharge (ESD) protection in high-voltage and harsh environmental conditions[1, 2]. These devices offer high reliability and are integral to applications such as high-voltage direct current (HVDC) transmission and power electronics converters[3, 4, 5, 6]. Unlike unidirectional SCRs, DDSCRs provide symmetrical voltage clamping in both forward and reverse directions, making them ideal for circuits with differential signals, such as high-speed communication interfaces[7, 8]. However, after triggering, the parasitic transistor structure within the device can form a positive feedback loop, leading to latch-up[9, 10], which results in sustained current flow, excessive power consumption, and potential device failure, compromising system reliability.

Several approaches have been proposed to mitigate latch-up, including dual-gate DDSCR structures that enhance ESD current release and capacitance coupling techniques that stabilize the voltage during triggering[11, 12, 13]. However, these methods do not fully address the issue of transient overshoot voltage[14, 15], which can damage sensitive components and degrade circuit performance, potentially causing false triggering and signal distortion. To further optimize DDSCR performance, especially during transient events, additional solutions have been explored, such as introducing fast-response diodes and MOSFETs to limit signal swing[16]. However, these solutions only partially address overshoot voltage and do not resolve the latch-up issue.

This paper proposes a new LVDDSCR structure that incorporates an additional bidirectional current discharge path, which enhances holding voltage and reduces transient overshoot voltage. Simulation results indicate a 20% improvement in holding voltage and a 35% reduction in overshoot voltage compared to traditional DDSCR designs. These improvements make the new structure suitable for fast-response, high-reliability ESD protection in applications like high-speed communication and sensor I/O ports.

II. DESIGN AND STRUCTURE

A. Traditional LVDDSCR Structure

The schematic of the traditional LVDDSCR structure is shown in Figure 1(a), which is an enhancement of DDSCR with the addition of two N+ regions, primarily aimed at improving the device's electrostatic discharge (ESD) resistance and the stability of its trigger voltage. Its equivalent circuit is depicted in Figure 2.

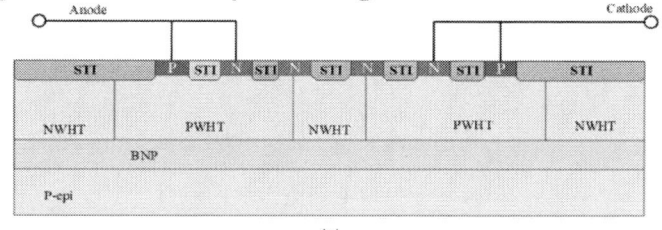

(a)

Yilidana Mamuti, Zhiyuan He, Zihan Zheng, Jiyuan Huang, Yimu Yang, Jixiang Shang, Jizhi Liu, Ruibo Chen and Zhiwei Liu are with the School of Integrated Circuit Science and Engineering (Demonstrative Microelectronics School), University of Electronic Science and Technology of China (UESTC), Chengdu, Sichuan, China. Hongyi Li is with the Glasgow College, University of Electronic Science and Technology of China (UESTC) Chengdu, Sichuan, China. This work is supported by the School of Integrated Circuit Science and Engineering, UESTC. Corresponding author: (Zhiwei Liu, email: ziv_liu@hotmail.com)

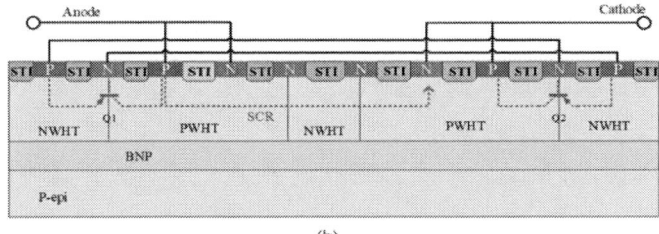

(b)

Fig. 1(a). The schematic of the traditional LVDDSCR structure. (b). New LVDDSCR structure with additional current paths.

This structure provides a bidirectional ESD discharge path but suffers from slow turn-on speeds and excessive overshoot voltage due to the long diffusion region in the right-side PWHT. These limitations restrict the application of DDSCRs in systems requiring fast transient response and stable voltage, such as high-speed communication systems and high-precision sensor interfaces.

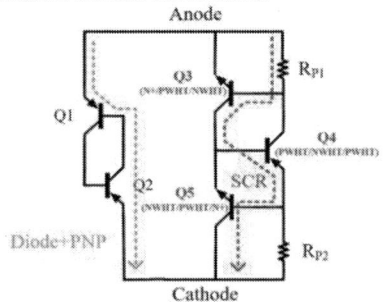

Fig. 2. The equivalent circuit of traditional LVDDSCR

B. *New LVDDSCR Structure*

To address these limitations, the new design introduces additional current paths in the NWHT regions on both sides of the device, as shown in Figure 1(b). This modification introduces two PNP transistors, Q1 and Q2, which manage the breakdown junction of the device. During a forward ESD event, the current flows through the base-emitter diode of Q1, reaching the emitter of Q2. As external voltage increases, the collector of Q2 undergoes breakdown, and the transistor turns on, facilitating current conduction and reducing the overshoot voltage. Additionally, the positive feedback of the SCR is shunted by the Q1 and Q2 transistors, increasing the holding voltage of the device.

III. TCAD SIMULATION AND EXPERIMENTAL RESULTS

A. *TCAD Simulation*

To simulate the behavior of the new LVDDSCR structure, a pulse with a 5ns width and a 300ps rise time was applied between the anode and cathode, with a pulse current of 2.5A. The current density distribution at different time points during the transient event is shown in Figure 3. At 100ps, transistor Q2 begins conducting, and by 300ps, the SCR path starts to modulate conductivity while the Q1 and Q2 path performs current shunting, enhancing device performance.

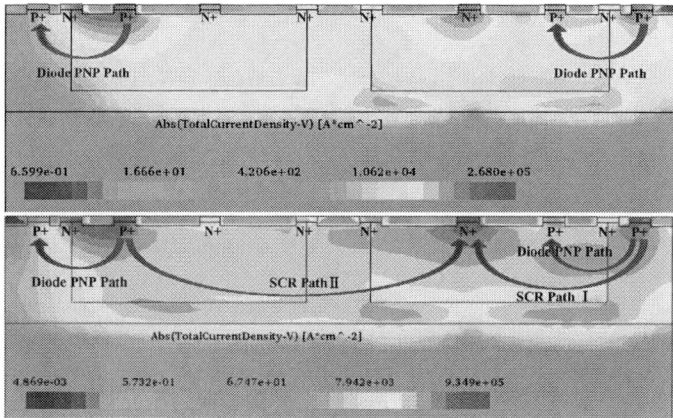

Fig. 3. Current density distribution in different time points.

B. *Experimental Results*

TCAD simulations were validated through Transmission Line Pulse (TLP) testing, as shown in Figure 4. The new LVDDSCR structure exhibited a holding voltage of 11.5V, a 20% improvement over the traditional LVDDSCR, which had a holding voltage of 9.5V. Further VFTLP testing, conducted with a rise time of 300ps and a pulse current of 2.5A, revealed a 35% reduction in overshoot voltage for the new LVDDSCR design, confirming its superior dynamic response in fast transient events, as shown in Figure 5.

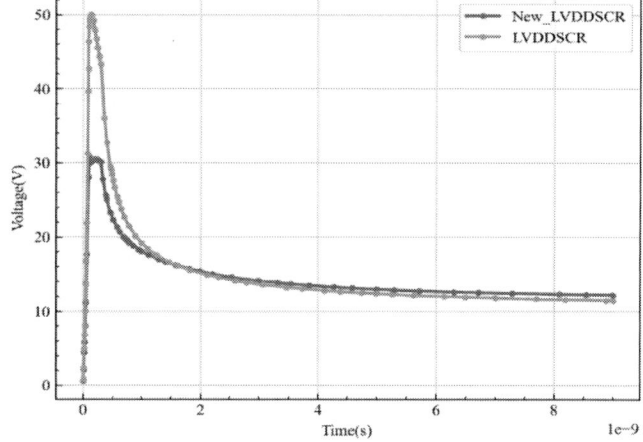

Fig. 4. TLP simulation results for New_LVDDSCR and LVDDSCR

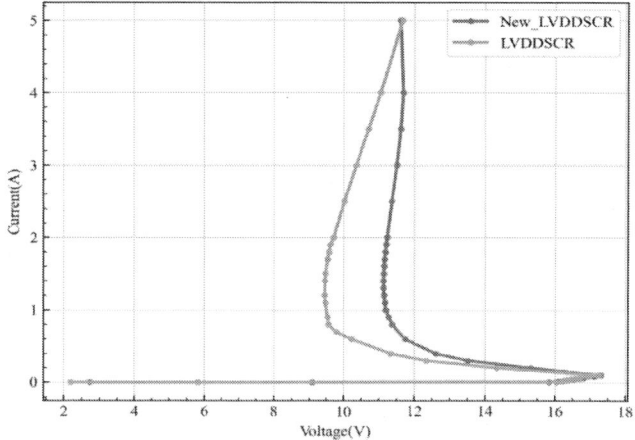

Fig. 5. VFTLP simulation results for New_LVDDSCR and DDSCR

979-8-3315-2209-4/25 $31.00 © 2025 IEEE 503

IV. CONCLUSION

This paper presents a novel DDSCR structure that addresses the critical issues of latch-up and excessive transient overshoot voltage. By introducing an additional low-resistance trigger path, the holding voltage of the device is increased by 20%, and the transient overshoot voltage is reduced by 35%. These enhancements improve the device's latch-up immunity and dynamic response, making it ideal for applications in high-speed communication interfaces and sensor I/O ports. Future work will focus on further optimizing the device structure to accommodate more complex application environments.

REFERENCES

[1] 2019 Index IEEE Transactions on Device and Materials Reliability Vol. 19, IEEE Transactions on Device and Materials Reliability, vol. 19, no. 4, pp. 797-819, 2019.

[2] Wang Y, Jin X, Yang L, et al., "Robust dual-direction SCR with low trigger voltage, tunable holding voltage for high-voltage ESD protection," Microelectronics Reliability, vol. 55, no. 3-4, pp. 520-526, 2015.

[3] Zhou Z J, Jin X L, Wang Y, et al., "New DDSCR structure with high holding voltage for robust ESD applications," Chinese Physics B, vol. 30, no. 3, art. 038501, 2021.

[4] Bao X, Wang Y, Liu Y, et al., "Design and manufacture of dual-gate DDSCR with high failure current and holding voltage," Chinese Journal of Electrical Engineering, vol. 10, no. 2, pp. 116-125, 2024.

[5] K. I. Do, Y. S. Koo, "A novel low dynamic resistance dual-directional SCR with high holding voltage for 12 V applications," IEEE Journal of the Electron Devices Society, vol. 8, pp. 635-639, 2020.

[6] J. Y. Li, Y. Wang, D. D. Jia, et al., "New embedded DDSCR structure with high holding voltage and high robustness for 12-V applications," Chinese Physics B, vol. 29, no. 10, art. 108501, 2020.

[7] "International Journal of Circuit Theory and Applications Reviewer Acknowledgement," International Journal of Circuit Theory and Applications, vol. 38, no. 1, pp. 108-110, 2010.

[8] K. I. Do, B. S. Lee, Y. S. Koo, "A new dual-direction SCR with high holding voltage and low dynamic resistance for 5 V application," IEEE Journal of the Electron Devices Society, vol. 7, pp. 601-605, 2019, doi: 10.1109/JEDS.2019.2916399.

[9] H. Xiaozong, L. Zhiwei, L. Fan, C. Hui, J. J. Liou, "High holding voltage SCR with shunt-transistors to avoid the latch-up effect," in 2016 IEEE International Nanoelectronics Conference (INEC), Chengdu, China, 2016, pp. 1-2, doi: 10.1109/INEC.2016.7589357.

[10] Z. Zhu, S. Wang, X. Fan, "A novel latch-up-immune DDSCR used for 12 V applications," in 2022 IEEE International Reliability Physics Symposium (IRPS), Dallas, TX, USA, 2022, pp. P15-1-P15-4, doi: 10.1109/IRPS48227.2022.9764527.

[11] X. Bao, Y. Wang, Y. Liu, X. Jin, "Design and manufacture of dual-gate DDSCR with high failure current and holding voltage," in Chinese Journal of Electrical Engineering, vol. 10, no. 2, pp. 116-125, June 2024, doi: 10.23919/CJEE.2024.000061.

[12] S. Dong, H. Jin, M. Miao, J. Wu, J. J. Liou, "Novel capacitance coupling complementary dual-direction SCR for high-voltage ESD," IEEE Electron Device Letters, vol. 33, no. 5, pp. 640-642, May 2012, doi: 10.1109/LED.2012.2188015.

[13] Z. Liu, J. J. Liou, S. Dong, Y. Han, "Silicon-controlled rectifier stacking structure for high-voltage ESD protection applications," IEEE Electron Device Letters, vol. 31, no. 8, pp. 845-847, Aug. 2010, doi: 10.1109/LED.2010.2050575.

[14] J. Liu, X. Fu, X. Cao, et al., "Research on mechanism and ESD protection characteristics of a novel DP_DDSCR with low trigger voltage," in 2024 3rd International Symposium on Semiconductor and Electronic Technology (ISSET), IEEE, pp. 370-375.

[15] Pan J, Li F, Wen L, et al., "A self-biased triggered dual-direction silicon-controlled rectifier device for low supply voltage pplication-specific integrated circuit electrostatic discharge protection," IEEE Electronics, vol. 13, no. 17, art. 3458, 2024.

[16] H. He, J. A. Salcedo, S. Parthasarathy, et al., "Compact and fast-response voltage clamp for bi-directional signal swing interface applications," IEEE Electron Device Letters, vol. 39, no. 12, pp. 1880-1883, 2018.

3D TCAD Optimization of Segmented SCRs for High-Voltage ESD Protection

Tianyi Zhang   Qiang Cui   Boris Dobrichkov   Dimitar Nikolov   Vladimir Garistov

Abstract—**The rapid advancement of automotive electronics has intensified the demand for robust electrostatic discharge (ESD) protection in high-voltage applications such as the in-vehicle network. Conventional silicon-controlled rectifier (SCR) devices suffer from latch-up risks because of their inherent low holding voltages. To address this challenge, segmented SCR structures have emerged as a promising solution. However, the conventional 2D TCAD simulation methodology fails to accurately characterize these three-dimensional topological variations; hence, it is unable to guide the design of this structure. This paper proposes a novel 3D TCAD-based methodology that can systematically investigate the electrical characteristics of segmented SCR devices. The simulation framework enables: (1) quantitative analysis of holding voltage enhancement through emitter segmentation ratio optimization. (2) comprehensive evaluation of robustness variations under different geometric configurations. This study presents the first TCAD-based investigation of segmented SCR structures' operational characteristics under varying temperature conditions, establishing a valuable framework for further high-voltage automotive ESD protection research.**

Index Terms—**ESD, Segmented SCR, 3D simulation, Robustness**

I. INTRODUCTION

The rapid development of automotive electronics has heightened requirements for ESD reliability in high-voltage applications. [1]With supply voltages escalating, conventional silicon-controlled rectifier devices face critical latch-up issues due to their inherent low holding voltages, which calls for optimized design of SCR with higher holding voltage. Recent studies [2] [3] [4] [5]have identified emitter segmentation as an effective approach to suppress latch-up by reducing parasitic bipolar gain through geometric optimization. Traditional 2D simulation fails to characterize this unique 3D structure, hence unable to optimize this structure. This simulation gap forces designers to rely on costly fabrication iterations without sufficient pre-silicon verification.

To address this challenge, this paper builds a 3D technology computer-aided design (TCAD) simulation framework for segmented SCR analysis. Our research conducts two experiments:

(1) we investigate the correlation between emitter segmentation ratio and holding voltage under vary temperatures. (2) we analyze the impact of geometric configurations on device robustness. This work provides valuable information for future detailed studies on high-voltage automotive electronics ESD protection.

II. SIMULATION SETUP

Accurate simulation of ESD phenomena requires comprehensive modeling of multi-physics interactions under high-current conditions. Conventional SPICE-based circuit simulators, which rely on compact behavioral models, exhibit limitations in capturing ESD events. To address this challenge, this study employs TCAD tools that numerically solve coupled carrier transport equations and Poisson's equation at the device level.

The key to this experiment is determining how to apply stimuli to the modeled devices in order to study their behavior under ESD stresses. The following sections explain how to model both TLP and HBM excitations.

For the emulation of TLP stress, a 50-Ω resistor is added to the Anode to emulate the TLP impedance, and TLP-like pulses with a 100-ns width and 10-ns rise time are used as the input,as illustrated in figure 1(a). Similar to real-world TLP measurements, the simulated quasistatic I-V curves are extracted by taking the average value over a time window 70%-90% in the voltage and current versus time waveforms. [6]

For the emulation of HBM ESD events, an RLC model, as illustrated in figure 1(b), is applied to yield the prescribed HBM waveform. The component values are individually set to Cesd=100pF, Resd=1.5kΩ, and Lesd=7500nH.

In the modeling process, as the Segmented SCR studied in this paper features a unique emitter segmentation optimization along the Z-axis, only a 3D model can accurately capture this characteristic in the modeling process. In the simulation process,while traditional 2D methods merely approximate lateral current spreading, the 3D framework explicitly resolves the 3D carrier transport dynamics essential for accurate holding voltage prediction. The capability to model structural dependencies makes this approach indispensable for modern SCR design, particularly when evaluating geometric optimization strategies prior to costly fabrication cycles.

Author 1 and Author 2 are with *College of Integrated Circuits, Zhejiang University*, Hang Zhou, China. Author 3, Author 4 and Author 5 are with *Department of Electronics, Faculty of Electronic Engineering and Technologies, TechnicalUniversity of Sofia*, Sofia, Bulgaria. This work is supported by the National Natural Science Foundation of China (Grant No.92373106) and Zhejiang ICsprout Semiconductor Co.,LTD. Corresponding author: (Author 2, email: qiang_cui@zju.edu.cn)

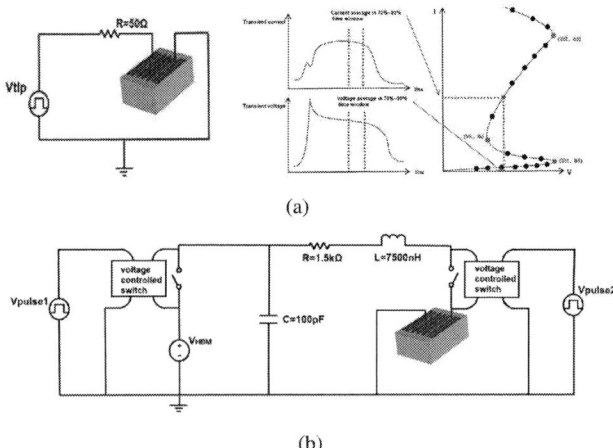

(a)

(b)

Fig. 1. The equivalent circuit for (a) TLP (b) HBM simulation.

III. 3D SIMULATION AND ANALYSIS

A. TLP-like simulation to evaluate SCRs' holding voltage

The fundamental operation of SCR devices is based on the positive feedback loop formed by two parasitic bipolar transistors (PNP and NPN). [7]While this mechanism enables high robustness characteristics, it simultaneously introduces inherent latch-up issues. Segmented SCR architecture mitigates this limitation through effective emitter area reduction, which effectively suppresses the bipolar gain (β) compared to conventional designs. Due to the significant reduction in the parasitic transistor's β value, a higher voltage is required to maintain the device in its conducting state. Consequently, the device's holding voltage increases.

As shown in Figure.2, three device configurations were modeled using 3D TCAD simulation: (a) Baseline SCR (no segmentation) (b) Segmented SCR with N+:P+ = 1:1 (c) Segmented SCR with N+:P+ = 1:2

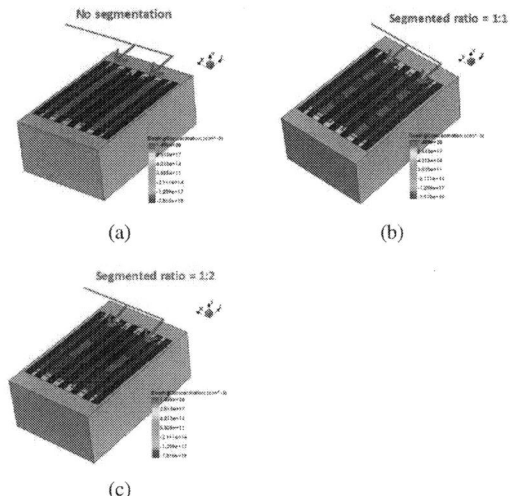

(a)   (b)

(c)

Fig. 2. Sample SCR structures with different segmentation ratio for simulation purpose.

Using the simulation model mentioned earlier, mixed-mode simulations incorporating transient thermal effects were conducted to replicate TLP testing conditions. High temperature simulations were performed on a segmented SCR with a 1:2 segmentation ratio to investigate its performance characteristics under automotive-grade high-temperature condition. The resulting current-voltage characteristics are demonstrated in Figure 3.

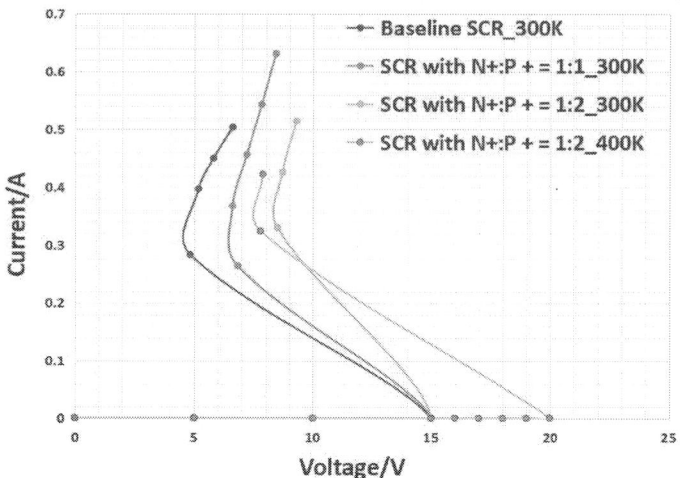

Fig. 3. The TLP characteristic curve of SCRs

Compared to the baseline SCR structure, the segmented configuration exhibits enhanced holding voltage. Additionally, the holding voltage of the segmented SCR was observed to decrease at elevated temperatures compared with its performance at room temperature. This is because increasing the temperature increases the intrinsic carrier concentration, which in turn decreases the barrier of PN junction in silicon. Thus, the higher temperature results in a lower holding voltage. The simulation results are consistent with both established physical principles and silicon proven reported in previous literature [8], thereby validating the credibility of the 3D simulation framework.

The unique feature of the 3D model successfully captures the interaction between vertical current distribution and thermal distribution, which is inaccessible in conventional 2D simulations.

B. HBM simulation to evaluate SCRs' Robustness

Previous studies [3] have demonstrated that the segmented SCR technique inevitably degrades the unit area robustness, which diminishes the competitive advantage of SCR structures renowned for their superior robustness per unit area. To address this limitation, we propose optimizing the current uniformity as a critical way to improve the robustness. Figure 4 presents 3D TCAD models of three SCR variants with different segmentation patterns: (a) non-uniform, (b) moderately uniform, and (c) optimally uniform layouts. To ensure comparative fairness, all devices maintain identical dimensions

979-8-3315-2209-4/25 $31.00 © 2025 IEEE

with a fixed N+/P+ segmentation ratio of 1:1. HBM stress pulses were applied to investigate significant differences in failure currents.

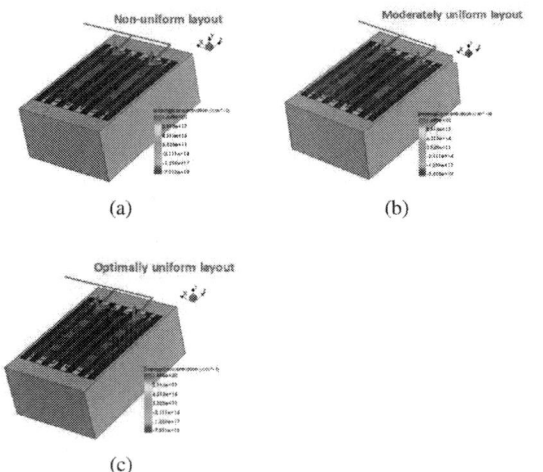

(a)　　　　　　　　(b)

(c)

Fig. 4. Sample SCR structures with different segmentation patterns for simulation purpose.

Typically, in TCAD ESD simulation, ESD thermal failure criterion is set to the Si melting temperature of 1688K and the corresponding maximum failure current (It2) reflects the HBM ESD protection capability. [9] As shown in Table 1, the failure current escalates from 0.53 A for the non-uniform layout to 1.13 A in the optimally uniform configuration at room temperature, establishing a direct correlation between geometric homogeneity and electrical robustness.As shown in Table 1, the failure current of the SCR operating at high temperature exhibits a significant decrease. This is attributed to the fact that the maximum device temperature reaching the silicon melting point (1688 K) was used as the failure criterion. Consequently, the device is more susceptible to thermal runaway at elevated temperatures, where an imbalance between heat generation and dissipation leads to rapid heat accumulation and subsequent failure. Therefore, enhancing the device's robustness through geometric optimization is crucial.

TABLE I
HBM LEVEL AND It2 OF EACH DEVICE FOR SIMULATION

Device	HBM Level _300K	Failure Current _300K	HBM Level _400K	Failure Current _400K
non-uniform	800V	0.53A	700V	0.46A
moderately uniform	1100V	0.73A	800V	0.53A
optimally uniform	1700V	1.13A	1000V	0.67A

As visualized in Figure 5, transient thermal analysis all under 750V HBM stress further demonstrates that the peak lattice temperature decreases from 1263 K in the non-uniform device to 937 K in the optimally uniform counterpart during the HBM pulse. This thermal improvement stems from enhanced

current dispersion characteristics, where uniform segmentation effectively alleviates localized current concentration. The evidence substantiates that layout-induced current uniformity constitutes the governing mechanism for achieving superior unit area robustness in segmented SCR structures, providing essential guidelines for ESD protection design optimization.

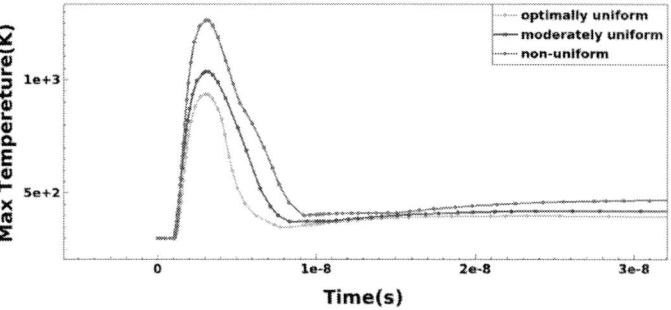

Fig. 5. Tmax-t curve of SCRs under 750V HBM stress

IV. CONCLUSION

This study presents a systematic investigation of segmented SCR structures through 3D TCAD simulations, focusing on two critical areas for ESD protection optimization. Firstly, the influence of segmentation ratios on holding voltage characteristics is comprehensively analyzed. Secondly, the layout optimization of segmented blocks is explored to enhance unit area robustness. Compared to conventional 2D simulation approaches that suffer from inherent limitations in modeling complex 3D effects, the proposed 3D methodology effectively captures vertical carrier transport dynamics and associated current distribution mechanisms. These capabilities enable accurate prediction of multidimensional electrical-thermal interactions in segmented SCRs. The simulation results provide critical physical insights for optimizing segmented SCR layouts, providing reliable guidance for practical fabrication processes in advanced power IC technologies. Because the Charged Device Model (CDM) test is indispensable in the automotive-grade chip certification standard AEC-Q100, we will expand our simulation approach to include CDM and VFTLP in future work to investigate the device performance under higher frequency electrical stress.

REFERENCES

[1] C.-W. Hsu, Y.-H. Li, and M.-D. Ker, "Optimization on bi-directional pnp esd protection device for high-voltage flexray applications," *IEEE Transactions on Electron Devices*, vol. 69, no. 10, pp. 5713–5721, 2022.

[2] X. Huang, J. J. Liou, Z. Liu, F. Liu, J. Liu, and H. Cheng, "A new high holding voltage dual-direction scr with optimized segmented topology," *IEEE Electron Device Letters*, vol. 37, no. 10, pp. 1311–1313, 2016.

[3] Z. Liu, J. J. Liou, and J. Vinson, "Novel silicon-controlled rectifier (scr) for high-voltage electrostatic discharge (esd) applications," *IEEE electron device letters*, vol. 29, no. 7, pp. 753–755, 2008.

[4] J. A. Salcedo, J.-J. Hajjar, S. Malobabic, and J. J. Liou, "Bidirectional devices for automotive-grade electrostatic discharge applications," *IEEE electron device letters*, vol. 33, no. 6, pp. 860–862, 2012.

[5] K.-I. Do, B.-B. Song, and Y.-S. Koo, "A novel dual-directional scr structure with high holding voltage for 12-v applications in 0.13-μm bcd process," *IEEE Transactions on Electron Devices*, vol. 67, no. 11, pp. 5020–5027, 2020.

[6] Q. Cui, S. Parthasarathy, J. A. Salcedo, J. J. Liou, J. J. Hajjar, and Y. Zhou, "Snapback and postsnapback saturation of pseudomorphic high-electron mobility transistor subject to transient overstress," *IEEE Electron Device Letters*, vol. 31, no. 5, pp. 425–427, 2010.

[7] M.-D. Ker and K.-C. Hsu, "Overview of on-chip electrostatic discharge protection design with scr-based devices in cmos integrated circuits," *IEEE Transactions on device and materials reliability*, vol. 5, no. 2, pp. 235–249, 2005.

[8] W. Liang, A. Dong, H. Li, M. Miao, C.-C. Kuo, M. Klebanov, and J. J. Liou, "Characteristics of esd protection devices operated under elevated temperatures," *Microelectronics Reliability*, vol. 66, pp. 46–51, 2016.

[9] Z. Pan, C. Li, M. Di, F. Zhang, and A. Wang, "3d tcad analysis enabling esd layout design optimization," *IEEE Journal of the Electron Devices Society*, vol. 8, pp. 1289–1296, 2020.

Degradation Behaviors of Bridged-Grain Polycrystalline Silicon Thin-Film Transistors under Dynamic Gate Voltage Stress with Fast Transition Time

Ming Guo[1], *Meng Zhang[1], Yunyang Wang[2], Lanrong Zou[3], Yan Yan[1], Lei Lu[4], Man Wong[5], Hoi-Sing Kwok[5]

[1]College of Electronics and Information Engineering, Shenzhen University, Shenzhen 518060, China
[2]Institute of Microscale Optoelectronics, Shenzhen University, Shenzhen, China
[3]ShenZhen Empyrean Technology Co., Ltd, Shenzhen, China
[4]School of Electronic and Computer Engineering, Shenzhen Graduate School, Peking University, Shenzhen 518055, China
[5]SKL of Advanced Displays and Optoelectronics Technologies, Hong Kong University of Science and Technology, Hong Kong, China
Phone: 0086 755 26925737, *Email: zhangmeng@connect.ust.hk

Abstract—This paper investigates the degradation of bridge-grain (BG) polycrystalline silicon (poly-Si) thin-film transistors (TFTs) under dynamic gate on-off switching stress (OOSS) with transition times as short as 1 nanosecond (ns). A larger gate voltage amplitude leads to more severe degradation. Compared to conventional poly-Si TFTs (ΔI_{on} = 99.98%), BG poly-Si TFTs exhibit smaller degradation (ΔI_{on} = 68.06%), demonstrating improved reliability. The study also reveals that hot-carrier (HC) degradation in BG poly-Si TFTs is influenced by the rising time (t_r) but not the falling time (t_f). As t_r decreases from 500 ns to 1 ns, ΔI_{on} after 10^4 s stress rises from 4.91% to 68.06%, while reducing t_f has no impact. Based on TCAD transient simulations the non-equilibrium PN junction degradation model is developed.

Keywords—Bridge-Grain (BG), Polycrystalline Silicon Thin-Film Transistor (TFT), Hot Carrier (HC), Non-Equilibrium PN Junction Degradation Model

I. INTRODUCTION

Polycrystalline silicon (Poly-Si) thin-film transistors (TFTs) are one of the preferred choices for developing superior active matrix (AM) displays [1], [2] and system-on-panel (SoP) applications[3]. The increasing requirements for faster refresh rates and higher resolutions in AM displays and SoP applications have led to poly-Si TFTs in pixel and driver circuits experiencing high-frequency (*f*) voltage pulse impacts. Over the years, a series of studies have been conducted on the reliability of poly-Si TFTs under various stress conditions, so that the associated degradation mechanisms are well understood [4], [5], [6], [7]. However, most studies have been limited to relatively low-*f* voltage stress, where the pulse rise time (t_r) and fall time (t_f) are at the level of approximately microsecond (~μs) [7], [8], [9], [10]. All experiments on dynamic gate on-off switching stress (OOSS) have been conducted under low-*f* voltage stress conditions, with t_r and t_f both at the level of approximately microsecond (~μs) [11], [12], [13]. So far, there have been only a few reports on poly-Si TFTs under dynamic transition times at the level of approximately nanosecond (~ns) [11],[12]. Moreover, experimental findings indicate that the degradation of poly-Si TFTs under high-*f* stress is not optimistic. To mitigate the degradation of poly-Si TFTs, we have systematically investigated the degradation of bridge-grain (BG) poly-Si TFTs under high-*f* gate on-off switching stress (OOSS) [16], [17], [18].

The study first examined degradation under fixed pulse rising (t_r) and falling (t_f) times while varying the voltage amplitude, showing that BG poly-Si TFTs mitigate degradation. Next, with a fixed voltage amplitude, the impact

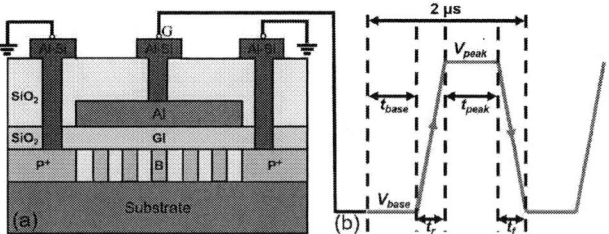

Fig. 1 The schematic of BG poly-Si TFTs and OOSS condition.

of t_r and t_f on BG poly-Si TFTs under dynamic gate OOSS was analyzed, particularly at a 1 ns transition time. Results show that at the nanosecond (~ns) scale, degradation depends on t_r but not t_f—a shorter t_r increases hot-carrier (HC) degradation, while t_f-related degradation remains around 40%. TCAD simulations further reveal carrier fluctuations and electric field formation, providing insights into HC degradation kinetics and guidance for improving device reliability in high-frequency applications.

II. EXPERIMENTAL

Fig. 1a illustrates the schematic of the BG poly-Si TFTs). The detailed fabrication process can be found elsewhere [7], [8], [17], [19]. For comparison, conventional poly-Si TFTs were also fabricated through the same process without the BG steps.

In this study, the channel width (*W*) and length (*L*) of both BG and conventional poly-Si TFTs were both set to 10 μm. The applied stress conditions are shown in Fig. 1b. Dynamic OOSS with a maximum amplitude of 24 V was applied to the gate electrode, while the source and drain remained at ground potential. The pulse period and duty cycle were fixed at 2 μs and 50%, respectively. The adjustable parameters included peak voltage (V_{peak}) (10 V to 12 V), base voltage (V_{base}) (−12 V to −10 V), t_r, and t_f. The t_r and t_f range from 1 ns to 500 ns. Device measurements were performed using a Keysight B1500A semiconductor parameter analyzer, while stress tests were conducted with a digital pulse generator. TCAD simulations were carried out using the Silvaco Atlas platform.

III. RESULTS AND DISCUSSION

The device transfer curve data are shown in Fig. 2. With $t_r = t_f$= 50 ns, three dynamic OOSS different with different amplitudes were applied in BG poly-Si TFTs (Fig. 2a, 2b, 2c). For the BG poly-Si TFT, while keeping other conditions constant, the I_{on} degradation rate was only 1.2% after 10^4 seconds when the voltage range was ± 10 V (Fig. 2a). In

979-8-3315-2209-4/25 $31.00 © 2025 IEEE

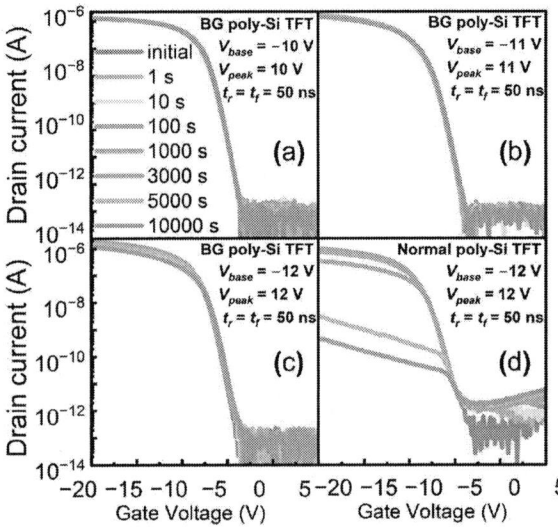

Fig. 2 The device transfer curves within 10^4 seconds under OOSS conditions.

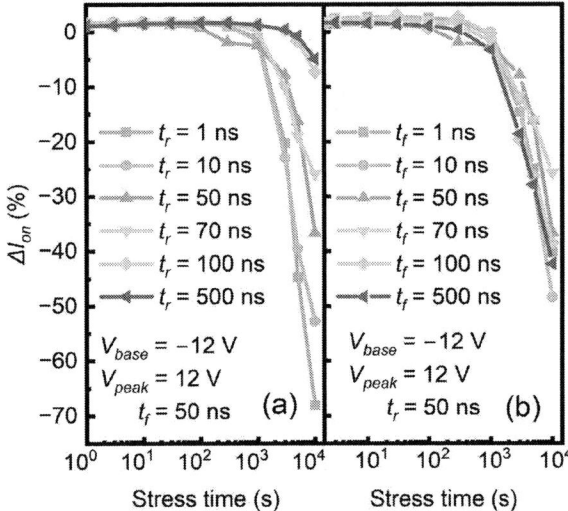

Fig. 3 Under the conditions of $V_{gs} = \pm 12V$ and a period of 2 μs (a)With t_f fixed at 50 ns and different t_r values. (b) With t_r fixed at 50 ns and different t_f values.

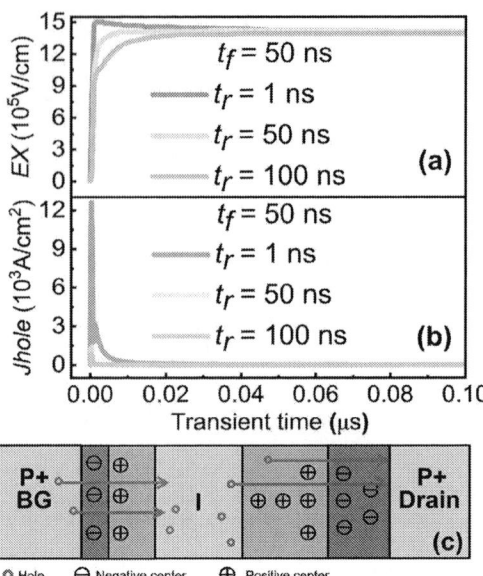

Fig. 4 Under the condition of $V_{gs} = \pm 12V$, (a) the lateral electric field in the channel (E_x) and (b) the hole current density (J_{hole}) obtained through TCAD simulation, (c) the carrier movement inside the device when t_r arrives.

contrast, when the voltage range was ±12 V (Fig. 2c), the I_{on} degradation rate reached 36.7% after 10^4 s OOSS (Fig. 2c). The experimental data shows that as the voltage amplitude increases, the degradation rate of I_{on} gradually increases. For the conventional poly-Si TFT, with the same voltage range as the BG poly-Si TFT (Fig. 2d), the subthreshold swing remained relatively unchanged, but the I_{on} degradation rate reached 99.98%. The BG poly-Si TFTs can significantly mitigate the degradation.

To further explore the relationship between device degradation and pulse transition times, the investigation of the I_{on} degradation under dynamic OOSS conditions with different t_r and t_f values was carried out. The test data are shown in Fig. 3.

Fig. 3a illustrates the effect of stress time on I_{on} degradation under dynamic OOSS with different t_r values while keeping t_f fixed at 50 ns. The t_r values ranged from 1 ns to 500 ns. When t_r was 1 ns, the I_{on} degradation rate reached 68.1% after 10^4 s of stress, whereas when t_r was 500 ns, the I_{on} degradation rate was only 4.9% after the same stress time. Shorter t_r values led to more significant device degradation. This observation is consistent with previous studies on both dynamic OOSS and dynamic OSS [5], [10], [19], [20], [21].

Fig. 3b shows the degradation of I_{on} over stress time under dynamic OOSS conditions with t_r fixed at 50 ns. The t_f values ranged from 1 ns to 500 ns. Interestingly, unlike t_r, the degree of device degradation was not affected by t_f. Shorter t_f values did not lead to more pronounced device degradation, which is consistent with previous reports on dynamic OSS and dynamic OOSS [11], [12], [13], [22], [23].

Regardless of the different t_r or t_f values, the device experienced a two-stage degradation process. In the first stage, I_{on} slightly increased with increasing stress time, and this stage generally ended after 300 s of applied stress. Subsequently, in the second stage, I_{on} dynamically decreased with increasing stress time. Similar degradation behavior was also observed in previous reports [4], [10], [14], [20]. According to previous studies, electron capture and injection into the gate oxide may primarily account for the initial stage of degradation, while hot-carrier (HC) effects may dominate the subsequent stage of degradation [10], [24].

To further clarify how BG poly-Si TFTs improve the degradation of conventional poly-Si TFTs and the degradation behavior under dynamic OOSS conditions at the gate, which is only related to the pulse t_r but not to t_f, TCAD transient simulations were carried out. These simulations were used to elucidate the mechanism by which BG poly-Si TFTs improve the degradation behavior of conventional poly-Si TFTs. Additionally, the non-equilibrium PN junction degradation model was integrated to explain the HC

979-8-3315-2209-4/25 $31.00 © 2025 IEEE

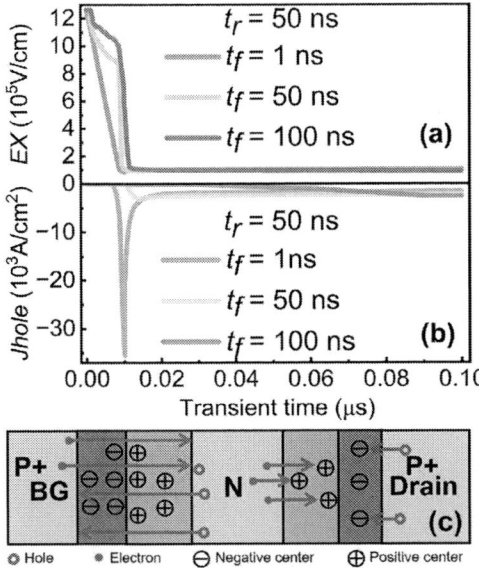

Fig. 5 Under the condition of V_{gs} = 12V, (a) the lateral electric field in the channel (E_x) and (b) the hole current density (J_{hole}) obtained through TCAD simulation, (c) the carrier movement inside the device when t_f arrives.

degradation behavior under rapid testing conditions, which is related to tr of dynamic gate OOSS but not to t_f.

Fig. 4 shows the simulation data of the lateral electric field in the channel (E_x) and hole current density (J_{hole}) for different t_r with a fixed t_f of 50 ns under a given voltage amplitude (the direction from the channel towards the drain is considered positive). To better explain the dynamic HC degradation under OOSS conditions, The non-equilibrium PN junction degradation model was utilized. During the t_r, two depletion regions are formed in the channel near the drain and the boundary of BG, as shown in Fig. 4c. During the t_r period from −12 V to 12 V, the PN junction formed between the BG and the channel will be forward-biased, reducing the barrier and facilitating the diffusion of holes from the BG line (P+) to the channel (I). Meanwhile, the PN junction formed between the drain and the channel will be reversely biased, causing the depletion region to extend towards the center of the channel. (Fig. 4c) This extension occurs sequentially, initially by emitting shallow traps. At this time, holes diffusing from the BG line (P+) to the channel (I) will also move towards the drain (Fig. 4b) to establish a larger E_x. Under this large E_x, deeper traps are emitted, and carriers emitted from deep trap states are injected into the strong built-in electric field, becoming HCs and causing dynamic HC degradation. Moreover, a shorter t_r will generate a larger built-in E_x during the pulse t_r (Fig. 4a). After the enhanced built-in E_x is established, a larger portion of carriers are emitted from the trap states, leading to more significant dynamic HC degradation (Fig. 3a) [14], [28]. After t_r ends, both the PN junctions formed between the drain and the channel and between the BG line and the channel reach a new equilibrium. Therefore, during the t_{peak} period, E_x and J_{hole} remain essentially unchanged.

When t_f arrives, the PN junction formed between the drain and the channel shortens its depletion region width and reduces the electric field in the depletion region (Fig. 5a). However, the E_x direction always points from the channel (I) to the drain (P+) (Fig. 5c). At this time, due to the capture of

holes by traps in the depletion region, the hole current direction is from the drain to the channel (Fig. 5b). The E_x direction is opposite to the hole current direction, resulting in a positive E_x and a negative J_{hole} (Fig. 5a,b). Therefore, even though there is a sufficiently large E_x during the t_f period and the hole current density remains high, E_x cannot accelerate holes to become HCs. Thus, device degradation is independent of t_f (Fig. 3b)[7], [15], [29]

Previous studies have already investigated the degradation mitigation of BG poly-Si TFTs compared to conventional poly-Si TFTs [8], [16], [17], [19], [30], [31], [32], [33]. This is because BG poly-Si TFTs incorporate multiple BG lines in the channel, and these BG lines have the same doping concentration as the source and drain terminals. These BG lines form PN junctions with the channel, which effectively share the applied voltage and reduce the lateral electric field near the source and drain. As a result, compared to conventional poly-Si TFTs, BG poly-Si TFTs can significantly reduce device degradation.

IV. CONCLUSION

This study presents a preliminary investigation of HC degradation in BG poly-Si TFTs under dynamic OOSS with rapid transition times. By integrating TCAD simulations, the study elucidates the device degradation mechanism that is related to t_r but not to t_f under dynamic OOSS with rapid transition times. Additionally, the mechanism by which BG poly-Si TFTs mitigate the degradation of conventional poly-Si TFTs is analyzed. This research offers profound insights into the degradation mechanism of HC in poly-Si TFTs. As a result, it furnishes valuable guidance for the reliable design of high-frequency applications.

V. ACKNOWLEDGMENTS

This work was financially supported by the Natural Science Foundation of China (62274111, 62374110), the Shenzhen Science and Technology Program (JCYJ20240813141301003, JCYJ20230808105806014), and the Independent Scientific Research Program from the State Key Laboratory of Radio Frequency Heterogeneous Integration (2024015).

REFERENCES

[1] Yu-Cheng Fan and Yi-Cheng Liu, "High-Speed Memory Cell Circuit Design Based on Low-Temperature Poly Silicon TFT Technology," *IEEE Trans. Magn.*, vol. 45, no. 5, pp. 2320–2323, May 2009, doi: 10.1109/TMAG.2009.2016495.

[2] Y. Nakajima, Y. Kida, M. Murase, Y. Toyoshima, and Y. Maki, "Latest developments for 'system-on-glass' displays using low-temperature poly-Si TFTs," *J Soc Info Display*, vol. 12, no. 4, pp. 361–365, Dec. 2004, doi: 10.1889/1.1847733.

[3] C.-L. Lin, P.-S. Chen, M.-Y. Deng, C.-E. Wu, W.-C. Chiu, and Y.-S. Lin, "UHD AMOLED Driving Scheme of Compensation Pixel and Gate Driver Circuits Achieving High-Speed Operation," *IEEE J. Electron Devices Soc.*, vol. 6, pp. 26–33, 2018, doi: 10.1109/JEDS.2017.2763601.

[4] M. Zhang and M. Wang, "An investigation of drain pulse induced hot carrier degradation in n-type low temperature polycrystalline silicon thin film transistors," *Microelectronics Reliability*, vol. 50, no. 5, pp. 713–716, May 2010, doi: 10.1016/j.microrel.2010.01.024.

[5] M. Zhang, M. Wang, X. Lu, M. Wong, and H.-S. Kwok, "Analysis of Degradation Mechanisms in Low-Temperature Polycrystalline Silicon Thin-Film Transistors under Dynamic Drain Stress," *IEEE Trans. Electron Devices*, vol. 59, no. 6, pp. 1730–1737, Jun. 2012, doi: 10.1109/TED.2012.2189218.

[6] M. Zhang, M. Wang, X. Lu, and M. Wong, "Characterization of hot carrier degradation in n-type poly-Si TFTs under dynamic drain pulse Stress with DC gate bias," in *2010 17th IEEE International Symposium on the Physical and Failure Analysis of Integrated Circuits*, Singapore, Singapore: IEEE, Jul. 2010, pp. 1–4. doi: 10.1109/IPFA.2010.5532002.

[7] M. Zhang, Z. Xia, W. Zhou, R. Chen, M. Wong, and H.-S. Kwok, "Dynamic-Gate-Stress-Induced Degradation in Bridged-Grain Polycrystalline Silicon Thin-Film Transistors," *IEEE Trans. Electron Devices*, vol. 63, no. 10, pp. 3964–3970, Oct. 2016, doi: 10.1109/TED.2016.2601218.

[8] Y. Yang, M. Zhang, L. Lu, M. Wong, and H.-S. Kwok, "Low-Frequency Noise in Bridged-Grain Polycrystalline Silicon Thin-Film Transistors," *IEEE Trans. Electron Devices*, vol. 69, no. 4, pp. 1984–1988, Apr. 2022, doi: 10.1109/TED.2022.3148697.

[9] M. Zhang. W. Zhou, S. Zhao, and H. S. Kwok, "Moisture Related Instability in p-Type Low Temperature Polycrystalline Silicon Thin Film transistors," 2011.

[10] X. Lu, M. Wang, M. Zhang, and M. Wong, "Negative drain pulse stress induced two-stage degradation of P-channel poly-Si thin-film transistors," in *18th IEEE International Symposium on the Physical and Failure Analysis of Integrated Circuits (IPFA)*, Incheon, Korea (South): IEEE, Jul. 2011, pp. 1–4. doi: 10.1109/IPFA.2011.5992756.

[11] L. Chen, Y. Wu, D. Zhang, H. Wang, and M. Wang, "Optimized Design of Carrier Injection Terminal for Reliable Low-Temperature Poly-Si TFTs Under Dynamic Hot-Carrier Stress," *IEEE Trans. Electron Devices*, vol. 67, no. 5, pp. 1987–1992, May 2020, doi: 10.1109/TED.2020.2977854.

[12] Huaisheng Wang, Mingxiang Wang, and Dongli Zhang, "Suppress Dynamic Hot-Carrier Induced Degradation in Polycrystalline Si Thin-Film Transistors by Using a Substrate Terminal," *IEEE Electron Device Lett.*, vol. 35, no. 5, pp. 551–553, May 2014, doi: 10.1109/LED.2014.2308987.

[13] J. S. Yoo, C. H. Kim, M. C. Lee, M. K. Han, and H. J. Kim, "Reliability of low temperature poly-Si TFT employing counter-doped lateral body terminal," in *International Electron Devices Meeting 2000. Technical Digest. IEDM (Cat. No.00CH37138)*, San Francisco, CA, USA: IEEE, 2000, pp. 217–220. doi: 10.1109/IEDM.2000.904296.

[14] Y. Wang, Z. Jiang, L. Zou, and M. Zhang, "Analysis of degradation mechanism in polycrystalline silicon thin-film transistors under dynamic off-state stress with fast transition time," *Phys. Scr.*, vol. 99, no. 12, p. 125906, Dec. 2024, doi: 10.1088/1402-4896/ad896a.

[15] M. Zhang *et al.*, "Dynamic Hot Carrier Degradation Behavior of Polycrystalline Silicon Thin-Film Transistors under Gate Voltage Pulse Stress with Fast Transition Time," in *2023 IEEE International Symposium on the Physical and Failure Analysis of Integrated Circuits (IPFA)*, Pulau Pinang, Malaysia: IEEE, Jul. 2023, pp. 1–5. doi: 10.1109/IPFA58228.2023.10249065.

[16] W. Zhou *et al.*, "Bridged-Grain Solid-Phase-Crystallized Polycrystalline-Silicon Thin-Film Transistors," *IEEE Electron Device Lett.*, vol. 33, no. 10, pp. 1414–1416, Oct. 2012, doi: 10.1109/LED.2012.2210019.

[17] M. Zhang *et al.*, "Degradation Induced by Forward Synchronized Stress in Poly-Si TFTs and Its Reduction by a Bridged-Grain Structure," *IEEE Electron Device Lett.*, vol. 40, no. 9, pp. 1467–1470, Sep. 2019, doi: 10.1109/LED.2019.2931007.

[18] W. Zhou *et al.*, "P-9: Bridged Grain MIC Poly-Si Thin-Film Transistors with Sputtered AlO$_x$ as Gate Dielectrics," *Symp Digest of Tech Papers*, vol. 43, no. 1, pp. 1079–1081, Jun. 2012, doi: 10.1002/j.2168-0159.2012.tb05978.x.

[19] M. Zhang *et al.*, "OFF-State-Stress-Induced Instability in Switching Polycrystalline Silicon Thin-Film Transistors and Its Improvement by a Bridged-Grain Structure," *IEEE Electron Device Lett.*, vol. 39, no. 11, pp. 1684–1687, Nov. 2018, doi: 10.1109/LED.2018.2872350.

[20] M. Zhang, Z. Jiang, L. Lu, M. Wong, and H.-S. Kwok, "Analysis of Degradation Mechanism in Poly-Si TFTs Under Dynamic Gate Voltage Stress With Short Pulse Width Duration," *IEEE Electron Device Lett.*, vol. 45, no. 2, pp. 204–207, Feb. 2024, doi: 10.1109/LED.2023.3345282.

[21] M. Zhang and M. Wang, "Observation of Hot Carrier Degradation in Poly-Si TFTs under Drain Pulse Stress".

[22] T. Gao, M. Wang, H. Wang, and D. Zhang, "TCAD Analysis of the Four-Terminal Poly-Si TFTs on Suppression Mechanisms of the DC and AC Hot-Carrier Degradation," *IEEE J. Electron Devices Soc.*, vol. 7, pp. 606–612, 2019, doi: 10.1109/JEDS.2019.2916619.

[23] D. Zhang, M. Wang, and X. Lu, "Two-Stage Degradation of p-Type Polycrystalline Silicon Thin-Film Transistors Under Dynamic Positive Bias Temperature Stress," *IEEE Trans. Electron Devices*, vol. 61, no. 11, pp. 3751–3756, Nov. 2014, doi: 10.1109/TED.2014.2359299.

[24] M. Zhang, Z. Jiang, L. Lu, M. Wong, and H.-S. Kwok, "Analysis of Degradation Mechanism in Poly-Si TFTs Under Dynamic Gate Voltage Stress With Short Pulse Width Duration," *IEEE Electron Device Lett.*, vol. 45, no. 2, pp. 204–207, Feb. 2024, doi: 10.1109/LED.2023.3345282.

[25] T. M. Brown and P. Migliorato, "Determination of the concentration of hot-carrier-induced bulk defects in laser-recrystallized polysilicon thin film transistors," *Applied Physics Letters*, vol. 76, no. 8, pp. 1024–1026, Feb. 2000, doi: 10.1063/1.125926.

[26] Y. Uraoka, Y. Morita, H. Yano, T. Hatayama, and T. Fuyuki, "Gate Length Dependence of Hot Carrier Reliability in Low-Temperature Polycrystalline-Silicon P-Channel Thin Film Transistors," *Jpn. J. Appl. Phys.*, vol. 41, no. Part 1, No. 10, pp. 5894–5899, Oct. 2002, doi: 10.1143/JJAP.41.5894.

[27] A. Valletta, L. Mariucci, G. Fortunato, S. D. Brotherton, and J. R. Ayres, "Hot carrier-induced degradation of gate overlapped lightly doped drain (GOLDD) polysilicon TFTs," *IEEE Trans. Electron Devices*, vol. 49, no. 4, pp. 636–642, Apr. 2002, doi: 10.1109/16.992873.

[28] Y. Zeng *et al.*, "Reliability of Poly-Si TFTs Under Voltage Pulse With Fast Transition Time," *IEEE Electron Device Lett.*, vol. 42, no. 12, pp. 1782–1785, Dec. 2021, doi: 10.1109/LED.2021.3124755.

[29] Y. Wang, Z. Jiang, L. Zou, and M. Zhang, "Analysis of degradation mechanism in polycrystalline silicon thin-film transistors under dynamic off-state stress with fast transition time," *Phys. Scr.*, vol. 99, no. 12, p. 125906, Dec. 2024, doi: 10.1088/1402-4896/ad896a.

[30] M. Zhang *et al.*, "High-Performance Polycrystalline Silicon Thin-Film Transistors without Source/Drain Doping by Utilizing Anisotropic Conductivity of Bridged-Grain Lines," *Adv Elect Materials*, vol. 6, no. 2, p. 1900961, Feb. 2020, doi: 10.1002/aelm.201900961.

[31] W. Zhou, R. Chen, S. Zhao, M. Zhang, M. Wong, and H. S. Kwok, "P.5: High Uniformity Solid Phase Crystallized Bridged-Grain Polycrystalline Silicon Thin Film Transistors," *Symp Digest of Tech Papers*, vol. 44, no. 1, pp. 1003–1006, Jun. 2013, doi: 10.1002/j.2168-0159.2013.tb06391.x.

[32] M. Zhang, S. Deng, W. Zhou, Y. Yan, M. Wong, and H.-S. Kwok, "Reversely-Synchronized-Stress-Induced Degradation in Polycrystalline Silicon Thin-Film Transistors and Its Suppression by a Bridged-Grain Structure," *IEEE Electron Device Lett.*, vol. 41, no. 8, pp. 1213–1216, Aug. 2020, doi: 10.1109/LED.2020.3005046.

[33] M. Zhang, Z. Xia, W. Zhou, R. Chen, and M. Wong, "Significant Reduction of Dynamic Negative Bias Stress-Induced Degradation in Bridged-Grain Poly-Si TFTs," *IEEE Electron Device Lett.*, vol. 36, no. 2, pp. 141–143, Feb. 2015, doi: 10.1109/LED.2014.2377040.

A Novel Super Junction SCR with High Holding Voltage for On-chip ESD Protection

Xinyu Zhu   Yipeng Chen   Shipeng Chen   Shurong Dong

Abstract—**A novel Super-junction SCR with high holding voltage is proposed and verified in 0.18-μm bipolar CMOS DMOS (BCD) process. Compared to traditional SCR, the integrated Super-junction structure can enhance the holding voltage by 30%. By adjusting the key parameters of the device, the characteristics of this device can be fine-tuned, while maintaining a robustness of 48 mA/μm. Additionally, the bidirectional devices exhibit minimal snapback, making them suitable for applications in negative voltage interfaces.**

Index Terms—**Electrostatic discharge (ESD), silicon controller rectifier (SCR), Superjunction**

I. INTRODUCTION

To enhance the device's voltage withstand capability, the Super-junction structure was introduced in power devices, widening the depletion region of the PN junction and enabling the device to withstand higher voltages [1]. In the design of ESD, for SCR devices, the focus often lies on enhancing the holding voltage of SCR devices to meet the requirements of the chip's ESD design window [2]–[4]. However, the traditional approach relies on serially connecting low-resistance structures within the SCR device to increase its holding voltage, which occupies a significant area. Devices embedding super-junction structures in LDMOS have long been used in ESD protection design [5], with subsequent structures undergoing further improvements and developments [6], [7]. Nevertheless, due to the deep STI structure beneath the gate in LDMOS, the device's conduction current capability is limited, resulting in higher trigger voltage, larger snapback, and lower unit area robustness when applied in ESD protection devices [8]. This paper proposed incorporating the Super-junction structure into the internal of the SCR-based ESD device, while adjusting the ESD characteristics by controlling the proportion of heavy doped P-type regions on the device surface in the width direction.

II. ESD DEVICE DESIGN

A. Device Structure and Description

Fig. 1 illustrates the structure and key parameters of the proposed device, including the widths D1 and lengths D2 of the strip P and strip N in the superjunction structure, as well as the proportion of P+ in the width direction of the

All Authors are with *College of Information Science & Electronic Engineering, Zhejiang University*, Hangzhou, China. This work is supported by STI2030-Major projects No.2021ZD0200401, Zhejiang Province high level talent special support plan No.2022R52042 and Zhejiang Province Key R&D programs No.2021C05004. Corresponding author: (Author 4, email: dongshurong@zju.edu.cn)

intermediate region. Parasitic SCR paths are formed by the P+/Nwell/Pwell near the anode and N+/Pwell/Nwell near the cathode, creating a superjunction structure between Pwell and Nwell. When an ESD event occurs, the PN junction in the well region becomes reverse-biased, widening the depletion region within the superjunction structure, thereby enhancing the holding voltage of the ESD device.

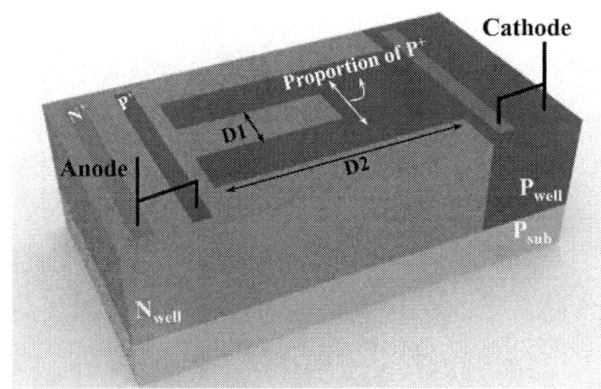

Fig. 1. The proposed device and key parameters

B. Mechanism of the Proposed Device

Fig. 2 illustrates the depletion region and electric field distribution at the superjunction. When the device is turned on, the PN junction is biased in the reverse direction, causing the depletion region to widen further until it fills the entire strip region. The interleaved strip P and strip N form a PN junction with linearly graded doping concentration, leading to a more uniform electric field distribution in the superjunction area, thereby enhancing the device's voltage sustainability capability.

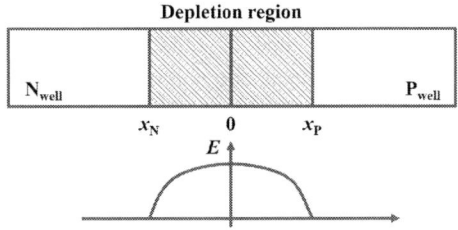

Fig. 2. Depletion region and electric field distribution in superjunction of proposed device

979-8-3315-2209-4/25 $31.00 © 2025 IEEE

For the linearly graded PN junction depicted in Fig. 2, the expression for the electric field can be derived as follows:

$$E(x) = \frac{q\alpha}{2\varepsilon_r\varepsilon_0}\left(x^2 - x_P^2\right), \quad (1)$$

where α represents the slope of the doping concentration at the PN junction, further leading to the expression for the voltage at that point:

$$V_{\text{bi}} + V_{\text{R}} = 2\int_0^{x_P} E(x)dx \quad (2)$$

Combining (1) and (2), we can obtain the expression for the width of the depletion region as:

$$2x_p = \sqrt[3]{\frac{12\varepsilon_r\varepsilon_0\left(V_{\text{bi}} + V_{\text{R}}\right)}{q\alpha}} \quad (3)$$

Fig. 3 presents the TCAD simulation results of the doping concentration and electric field distribution of the proposed device's superjunction structure. From (3), it is evident that the depletion region width is less than 1 μm, requiring a reduction in D1 to enable the superjunction to function effectively. However, due to constraints imposed by feature sizes and process design rules, D1 cannot be arbitrarily reduced.

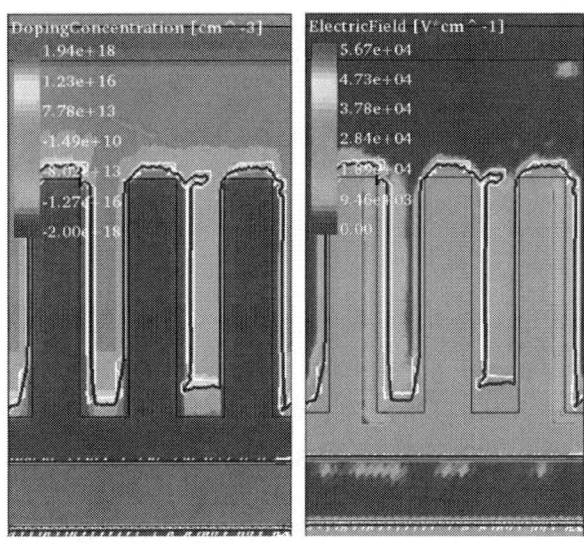

Fig. 3. TCAD simulation of doping concentration and electric field at the superjunction of proposed device

III. EXPERIMENTAL RESULTS AND DISCUSSIONS

Fig. 4 illustrates the TLP test results at different D1 values, where D1=0 implies the absence of a superjunction structure, and increasing D1 leads to a degradation in superjunction performance. The test results indicate a 30% increase in holding voltage due to the presence of the superjunction structure. Fig. 5 presents the TLP test outcomes for various D2 lengths, showing that increasing the D2 length effectively enhances the device's holding voltage while slightly reducing the trigger voltage, thereby mitigating the snapback characteristics of the SCR device.

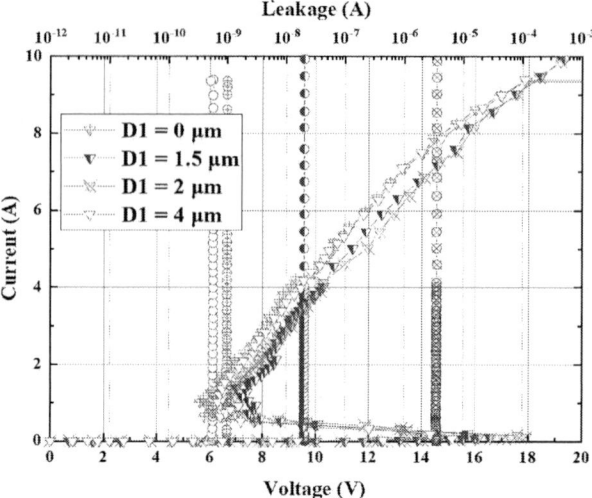

Fig. 4. TLP results for the proposed device at different D1 values

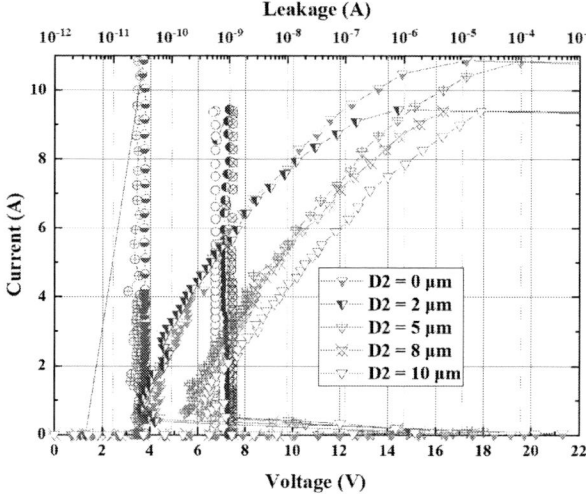

Fig. 5. TLP results for the proposed device at different D2 values

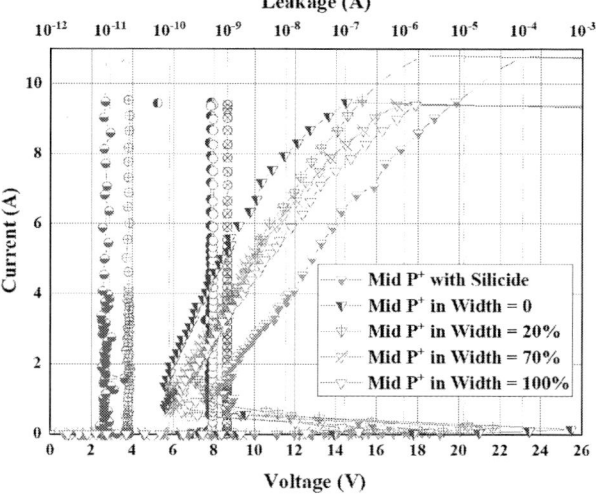

Fig. 6. TLP results for the proposed device at different proportion of P+

The TLP test results for different P+ concentrations are depicted in Fig. 6, indicating that the surface P+ can offer a current dissipation path upon device activation, resulting in minimal impact on the holding voltage but effectively reducing the trigger voltage. Fig. 7 displays the TLP test results for different widths, showing that increasing the device area does not significantly decrease the robustness per unit area. Fig. 8 demonstrates the TLP test outcomes for bidirectional devices, where a high-voltage N-well isolates the substrate and electrodes, serially connecting forward and reverse devices. The results reveal that the bidirectional proposed device exhibits small snapback and higher holding voltage.

simulations were employed to investigate the operational principles of the Super-junction SCR structure. In comparison to conventional SCR, the incorporation of the integrated Super-junction design resulted in a notable 30% increase in the holding voltage. By strategically adjusting the key parameters of the device, its characteristics can be finely tuned while upholding a robustness of 48 mA/μm. Moreover, the bidirectional devices exhibited minimal snapback, rendering them well-suited for applications involving negative voltage interfaces.

REFERENCES

[1] X. Chen, "Super-junction components," Dian Li Dian Zi Ji Shu, vol.42, no. 12, pp.2–7, 2008.

[2] F. Ma, B. Zhang, Y. Han, et al, "High holding voltage SCR-LDMOS stacking structure with ring-resistance-triggered technique," IEEE Electron Device Letters, vol. 34, no. 9, pp. 1178–1180, 2013.

[3] X. Huang, Z. Liu, F. Liu, et al, "High holding voltage SCRs with segmented layout for high-robust ESD protection," Electronics Letters, vol. 53, no. 18, pp. 1274–1275, 2017.

[4] Z. Wang, Z. Qi, L. Liang, et al, "Novel High Holding voltage SCR with embedded carrier recombination structure for latch-up immune and robust ESD protection," Nanoscale Research Letters, vol. 14, no. 1, pp. 175–177, 2019.

[5] V. A. Vashchenko and B. M. Ter, "ESD Protection Window Targeting Using LDMOS-SCR Devices with PWELL-NWELL Super-junction," 2005 IEEE International Reliability Physics Symposium, pp. 612–613, 2005.

[6] S. L. Chen and Y. T. Huang, "Drain Side Super Junctions Co-worked with NPN Arranged SCRs on ESD Robustness in the 45-V nLDMOS Devices," TENCON 2015 IEEE Region 10 Conference, 2015.

[7] H. W. Chen, S. L. Chen, Y. T. Huang, et al, "ESD Improvements on Power N-Channel LDMOS Devices by the Composite Structure of Super Junctions Integrated With SCRs in the Drain Side," IEEE Journal of the Electron Devices Society, vol. 8, pp. 864-872, 2020.

[8] L. Qian, M. Li, Y. Wang, et al, "A novel segmented LDMOS-SCR structure with 8-kV HBM ESD robustness in CMOS analog multiplexer," IEEE Transactions on Electron Devices, vol. 69, no. 12, pp. 1–6, 2022.

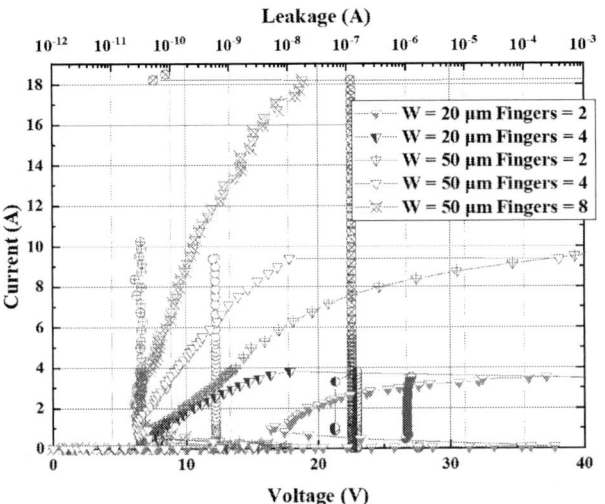

Fig. 7. TLP results for the proposed device at different width

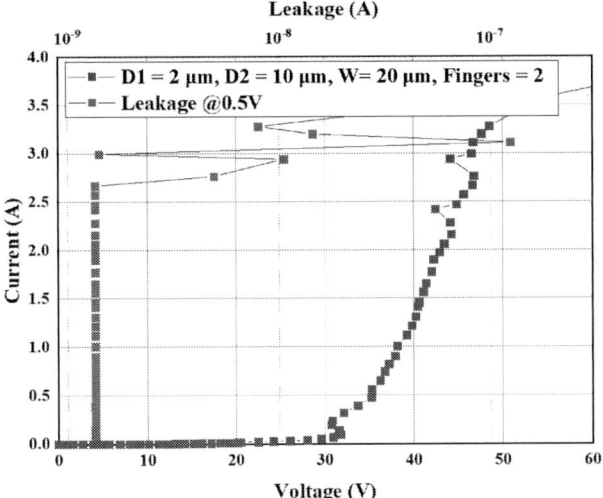

Fig. 8. TLP results for the bidirectional proposed device

IV. CONCLUSION

A novel Super-junction SCR with a high holding voltage is proposed and validated in a 0.18-μm BCD process. TCAD

979-8-3315-2209-4/25 $31.00 © 2025 IEEE

A novel fast-triggering DTSCR ESD protection device for low-voltage applications

1st Gongtang Yu   2nd Zhihua Zhu   3rd Shanglin Yang   4th Juin Jei Liou

Abstract—This paper proposes a novel diode-triggered silicon controlled (NDTSCR) for electrostatic discharge (ESD) protection, specifically designed for 1.8 V I/O interfaces. By optimizing the co-layout of the diode-triggering network and SCR structure, the design achieves bidirectional conduction capability under both positive and negative ESD events. The architecture separates the triggering path from the fully-conducting SCR path, enabling superior current discharge capacity and rapid device activation speed. The decoupling of the diode triggering path from the SCR's full conduction path reduces the trigger voltage and shortens the turn-on time.

Index Terms—Electrostatic Discharge (ESD), Diode-Triggered Silicon-Controlled Rectifier (DTSCR), Trigger Voltage, Turn-on Time

I. INTRODUCTION

The conventional diode-triggered silicon-controlled rectifier (SCR) primarily operates through substrate current injection via the triggering diode. When sufficient current accumulation induces a 0.7V potential difference between the PWELL along the SCR conduction path and the adjacent P+ region, the parasitic PNPN structure achieves full conductivity through sequential activation of its constituent junctions: the anode-connected P+ diffusion, underlying NWELL, adjacent PWELL, and terminal N+ region. While this architecture effectively reduces trigger voltage, its dependence on substrate current propagation introduces inherent limitations in turn-on speed. Furthermore, the unidirectional conduction characteristic necessitates parallel device configurations for bidirectional protection[2]–[4]. To address these constraints, A novel fast-triggering DTSCR ESD protection device for low-voltage applications (NDTSCR) has been developed, demonstrating enhanced performance metrics including sub-nanosecond turn-on response, sub triggering threshold, and inherent bidirectional current capability, making it particularly suitable for low-voltage interface ESD protection. Through comprehensive TCAD simulation analysis, this optimized topology demonstrates superior ESD current dissipation efficiency per unit layout area compared to conventional SCR implementations, while effectively resolving critical limitations such as excessive trigger voltage and delayed conduction response. The simulation results quantitatively validate significant improvements in both operational parameters and reliability indices under ESD stress conditions[5]–[7].

Gongtang Yu, Shanglin Yang, Juin Jei Liou are with *School of Electronic and Information Engineering, North Minzu University*, Ningxia 750021, China. Zhihua Zhu is with *School of Electrical Engineering, North China University of Water Resources and Electric Power*, Henan 450046, China. Corresponding author: (Shanglin Yang, email: ysl029@163.com, juin.liou@hotmail.com.)

II. DEVICE STRUCTURE AND OPERATION MECHANISM

Figures 1 and 2 illustrate the cross-sectional structures and equivalent devices of the conventional DTSCR and the novel NDTSCR, respectively. In the conventional DTSCR, ESD current is directed into the anode, causing the potential of the P-region to rise and establish a forward bias with the N-well.

During ESD events, partial current flowing through the diode string is diverted to the substrate via the N-well due to the Darlington amplification effect. Once the substrate current exceeds a critical threshold, the PN junction diode associated with the parasitic NPN transistor within the SCR structure becomes forward-biased, thereby initiating conductivity modulation in the SCR path. This mechanism ensures rapid ESD current discharge through the activated SCR conduction channel [8].

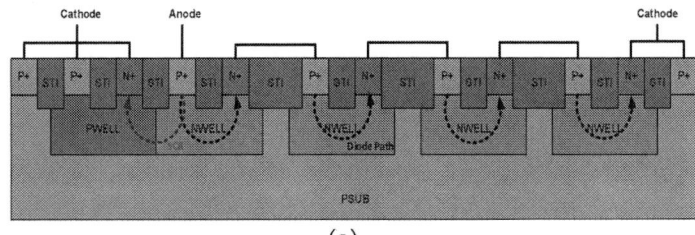

(a)

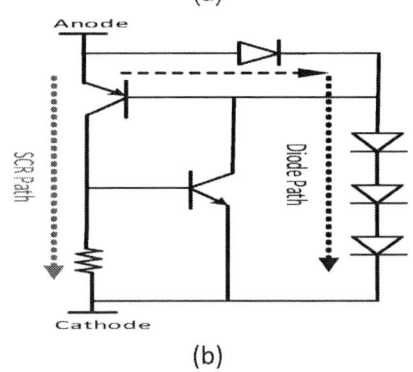

(b)

Fig. 1. (a)Cross-sectional view of the conventional DTSCR structure, and (b) the equivalent device diagram of DTSCR.

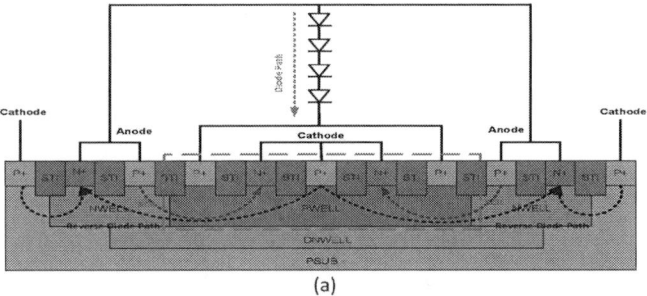

(a)

979-8-3315-2209-4/25 $31.00 © 2025 IEEE   516

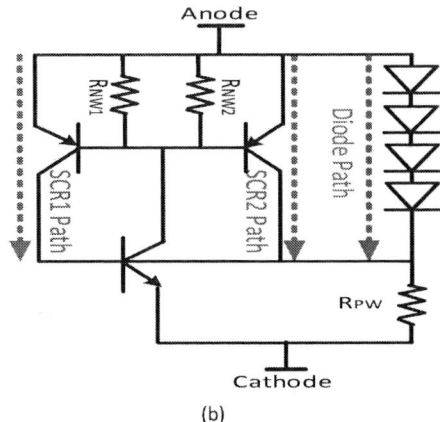

(b)

Fig. 2. (a) Cross-sectional view of the NDTSCR structure, and (b) the equivalent device diagram of NDTSCR.

The proposed NDTSCR architecture exhibits inherent bidirectional ESD current discharge capability through distinct conduction mechanisms. Under forward bias ESD stress, once the applied voltage surpasses the forward triggering threshold, cascade conduction initiates through the series diode network as illustrated in Figure 2(a) (red path). The current injection originates from the P+ anode diffusion into the PWELL substrate (red dashed box). Subsequent carrier accumulation induces forward biasing of the PWELL-to-cathode N+ junction, activating the parasitic NPN transistor within the SCR configuration[9]–[12]. This triggers the complete SCR conduction path (blue dashed path), establishing a sustained low-impedance conduction path for ESD current dissipation. During reverse ESD events, the integrated diode network (black dashed path) facilitates auxiliary discharge through its inherent forward conduction characteristics. Specifically, the reverse-polarity current achieves rapid shunting via ohmic conduction through the parallel diode elements, complementing the primary SCR-mediated protection mechanism. This dual-path topology ensures symmetrical transient response while maintaining layout efficiency.

III. SIMULATION RESULTS AND DISCUSSION

The TLP test was simulated using a current waveform characterized by a 10-ns rise time and 100-ns pulse width. A hybrid simulation methodology was adopted to achieve interconnect coupling between the diode string and SCR. Figures 3 and 4 illustrate the current density distribution within the diode string triggering path and SCR conduction path across different operational phases [13], [14].

As shown in Figure 3, the diode triggering path exhibits conduction under a current pulse of 7 mA. Simultaneously, a certain current component is observed in the SCR structure, primarily attributed to the initial conduction phase of the diode. During this phase, current flows through the P+ heavily doped region into the PWELL and subsequently exits via the P+ region connected to the cathode. This conduction current

establishes a localized potential gradient, which forward-biases the base-emitter junction of the parasitic NPN bipolar transistor. This mechanism effectively reduces the SCR's turn-on voltage, enabling its rapid activation.

I=7 mA

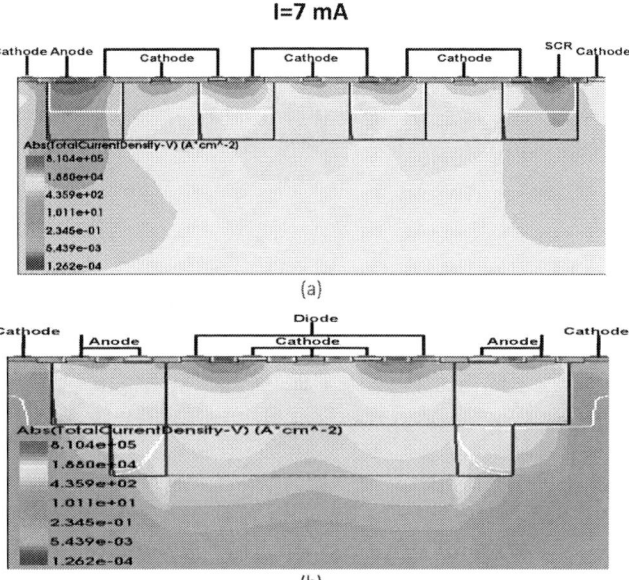

Fig. 3. (a) Diode-triggered path turns on, and (b) before the slave SCR triggers.

I=60 mA

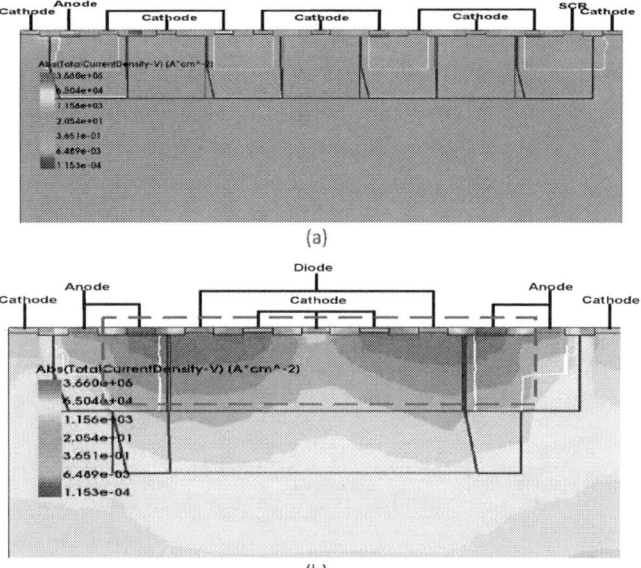

Fig. 4. (a) Diode current path under full activation, and (b) SCR conduction path in fully triggered state.

As shown in Figure 4(a), the current through the diode string in the NDTSCR is negligible based on the current distribution analysis. In contrast, Figure 4(b) demonstrates a significantly high current density in the SCR section of the NDTSCR. This phenomenon arises because once the SCR is activated, the

979-8-3315-2209-4/25 $31.00 © 2025 IEEE

517

positive feedback interaction between its internal PNP and NPN bipolar transistors causes a rapid surge in current through the SCR. This current increment progressively replaces the diode string triggering path, establishing the SCR as the dominant ESD current dissipation path.

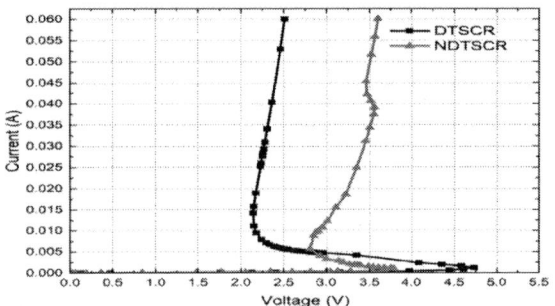

Fig. 5. TLP simulated results of NDTSCR and traditional DTSCR.

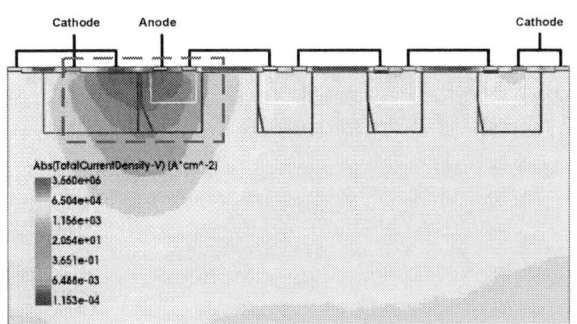

Fig. 6. DTSCR after full conduction.

Figure 5 presents the TLP simulation results of DTSCR and NDTSCR devices. TCAD simulation results reveal that the triggering voltage of the NDTSCR is reduced to 3.7 V, representing a 21.3% decrease compared to the conventional DTSCR's 4.7 V. This performance enhancement is primarily attributed to the direct injection of the triggering current into the base terminal of the internal NPN transistor within the SCR in the NDTSCR, which enables rapid and efficient SCR activation. In contrast, the triggering mechanism of the conventional DTSCR relies on minority current from the triggering path flowing into the substrate, resulting in a higher triggering voltage.

The holding voltage of the NDTSCR has been successfully elevated to 2.8V through two primary methodologies: (1) Anode-to-cathode spacing optimization: As illustrated in Figure 4(b) and Figure 6, the SCR conduction path in the NDTSCR exhibits an extended physical length compared to the DTSCR configuration. (2) The SCR conduction path typically resides near the surface. The incorporation of a P+ region above the PWELL increases the hole concentration in the base of the internal NPN transistor within the SCR. This enhances minority carrier recombination at the reverse-biased junction, thereby attenuating the positive feedback effect of the SCR. Consequently, the holding voltage is moderately elevated.

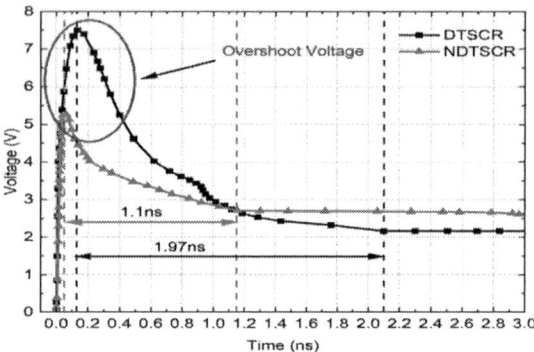

Fig. 7. The transient responses of of NDTSCR and DTSCR.

Current pulses with rise times of 0.1 ns and 5 ns were employed to analyze the transient waveform responses of DTSCR and NDTSCR devices. As illustrated in Figure 7, under single-pulse excitation, the DTSCR exhibits significantly higher voltage overshoot compared to the NDTSCR, a phenomenon predominantly caused by its triggering path characteristics. The NDTSCR not only demonstrates reduced overshoot voltage but also achieves 44.2% faster turn-on speed. This behavior aligns with the aforementioned operational mechanism of the NDTSCR, primarily attributed to the direct flow of triggering path current through the base terminal of the internal NPN transistor within the SCR.

IV. CONCLUSIONS

This study proposes a novel fast-triggering diode-activated SCR for ESD protection. The structure achieves direct injection of the diode string current into the base terminal of the internal NPN transistor within the SCR, effectively reducing the triggering voltage and turn-on time. Compared to conventional DTSCR devices, the NDTSCR demonstrates a reduced triggering voltage of 3.7 V, an elevated holding voltage of 2.8 V, and a 44.2% reduction in full conduction time. The device exhibits high compatibility with standard CMOS processes and demonstrates superior ESD protection capabilities, making it ideal for low-voltage interface applications.

REFERENCES

[1] C. Russ, "ESD Issues in Advanced CMOS bulk and FinFET Technologies: Processing, Protection Devices and Device Strategies," Microelectron. Rel, 2008, vol. 48, pp. 1403–1411.

[2] Boyang Ma, Shupeng Chen, Ruibo Chen et al., "Deep N-well DTSCR with Fast Turn-on Speed for Low-voltage ESD Protection Applications", Microelectronics Reliability, 2024, vol. 160, pp.115475.

[3] Feibo Du et al., "All-Directional Silicon-Controlled Rectifier with Improved Voltage Clamping Capability for FinFET ESD Protection", in IEEE Transactions on Electron Devices, 2024, vol. 71, pp.2247-2252.

[4] S. Park, Y. Choi, S. Lee et al., "A High Holding Voltage Diode-Triggered SCR for Low-Voltage ESD Application", 2024 International Conference on Electronics, Information, and Communication (ICEIC), Taipei, Taiwan, 2024.

979-8-3315-2209-4/25 $31.00 © 2025 IEEE

[5] Xiaofeng Gu, Jian Xu, Hailian Liang et al.," A Novel Dual-directional DTSCR in Twin-Well Process for Ultra-Low-Voltage ESD Protection", Solid-State Electronics, 2024, vol. 212, pp.108847.1-108847.6.

[6] Kyoung-Il Do, Bo-Bae Song, Yong-Seo Koo, "A Novel Dual-Directional SCR Structure with High Holding Voltage for 12-V Applications in 0.13μm BCD Process," IEEE Transactions on Electron Devices , 2020, vol. 67, pp. 5020 - 5027.

[7] Kyoung-Il Do, Yong-Seo Koo, "A New SCR Structure with High Holding Voltage and Low ON-Resistance for 5-V Applications," IEEE Transactions on Electron Devices, 2020,vol. 67, pp. 1052 – 1058.

[8] Long Chen, Feibo Du, Ruibo Chen et al., "Novel diode-triggered SCR with suppressed multiple triggering for ESD applications", 2018 IEEE International Conference on Electron Devices and Solid State Circuits (EDSSC), Shenzhen, China, 2018.

[9] Songyan Wang, Zhihua Zhu, Xikun Feng et al., A Modified Zener Diode-Triggered Silicon Controlled Rectifier Used for 5V Application in 0.18um BCD Process , 2022 10th International Symposium on Next-Generation Electronics (ISNE), Wuxi, China, 2023.

[10] Wen-Yi Chen, Elyse Rosenbaum, and Ming-Dou Ker, "Diode-Triggered Silicon-Controlled Rectifier With Reduced Voltage Overshoot for CDM ESD Protection, IEEE Transactions on Device and Materials Reliability,2011, vol. 12, pp. 10-14.

[11] Sang-Wook Kwon, Yong-Seo Koo, "Design of LDO Regulator With High Reliability ESD Protection Device Using Analog Current Switch Structure for 5-V Applications, IEEE Access", 2023, vol. 11. pp. 37472-37482.

[12] Meng Miao, Shurong Dong, Jian Wu et al., "Minimizing Multiple Triggering Effect in Diode-Triggered SiliconControlled Rectifiers for ESD Protection Applications",IEEE Transactions on Device and Materials Reliability, 2012, vol. 33, pp. 893-895.

[13] Robert G, Abou-Khalil M et al., "Investigation of Voltage Overshoots in Diode Triggered Silicon Controlled Rectifiers (DTSCRs) under Very Fast Transmission Line Pulsing (VFTLP)," EOS/ESD Symposium, 2009,pp. 1-10.

[14] Ulrich Glaser, Kai Esmark et al., "SCR Operation Mode of Diode Strings for ESD Protection," EOS/ESD Symposium,2009,vol.36, pp. 1-10.

Silver Film/CPSA ESD Failure Mechanism and Robustness

1st Shipeng Chen   2nd Zhencheng Xu   3rd Yipeng Chen   4th Jiabei Pan   5th Ling Zhang   6th Shurong Dong

Abstract—**In this paper, the antistatic ability of 50 nm silver film under electrostatic discharge (ESD) was analyzed and tested by transmission line pulse (TLP). The results show that the silver film has strong antistatic ability, which is mainly due to the continuity of its conductive path and good thermal diffusion performance. However, under the same ESD conditions, the performance of silver film / Conductive Pressure Sensitive Adhesive (CPSA) composite structure is not satisfactory. The main reason is that the surface carbonization of CPSA generates tensile stress, resulting in shrinkage and fracture of silver film. By analyzing its failure mechanism, this paper provides an important reference for the electrostatic protection design of related electronic devices.**

Index Terms—**silver film, electrostatic discharge (ESD), transmission line pulse (TLP), Conductive Pressure Sensitive Adhesive (CPSA)**

I. INTRODUCTION

With the development of integrated modules in electronic devices, system - level electrostatic discharge (ESD) has always been one of the key factors affecting the reliability and stability of equipment [1]. In integrated modules, the use of Conductive Pressure Sensitive Adhesive (CPSA) enables easy integration and interconnection of the modules. As a result, it is widely applied in various portable devices, such as mobile phones and laptops. This approach not only facilitates efficient assembly but also enhances the electrical performance and reliability of electronic components [2]. Although both silver film and CPSA possess strong ESD - resistance capabilities individually, when they are used in combination as a silver film/CPSA composite, it often leads to device failure.

For instance, the wireless charging coil on the backplane of a mobile phone is connected to the internal silver - containing devices through CPSA. However, it is precisely this connection structure that causes a sharp decline in charging efficiency or even complete failure of the device under ESD conditions. In the context of the rapid development of wireless charging technology, the need for reliable electrostatic protection in wireless charging backplanes has become more pressing. Wireless charging backplanes are complex structures integrating multiple components. ESD events can disrupt the normal operation of the charging circuit, resulting in issues such as reduced charging speed or complete charging failure [3]. A study found that voltage spikes induced by ESD can damage the sensitive components in the wireless charging module,

reducing its overall lifespan [4]. Moreover, as mobile devices become more compact and power - hungry, the density of components on the wireless charging backplane increases, making it more vulnerable to ESD threats [5].

Recent research has highlighted the significance of conductive materials like silver film in the field of flexible electronics [6]. Therefore, conducting in - depth research on silver film/CPSA is of great importance for ensuring the normal operation of electronic devices. This study aims to fill the research gaps in related fields through systematic experiments and theoretical analysis, providing strong support for the electrostatic protection design of electronic devices.

II. EXPERIMENTS

In order to perform TLP test on silver film and CPSA, three types of samples need to be set up : the first type is to prepare 50 nm silver film on PCB by sputtering process ; the second type of cutting CPSA is covered on the PCB ; the third type of silver film / CPSA has overlapping surfaces. The silver film part is marked as A, the CPSA part is marked as B, and the overlapping surface part is marked as C, as shown in Fig.1.

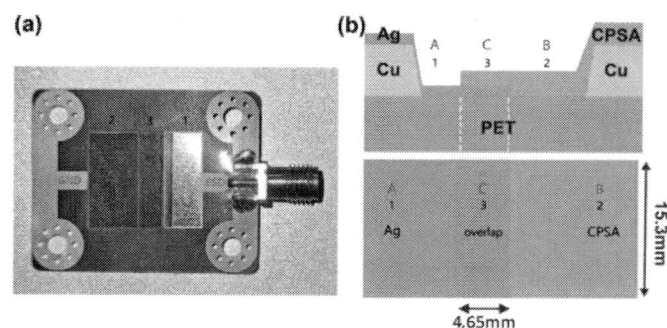

Fig. 1. Flexible PET circuit board covered with silver film/CPSA (a) object; (b) cross section.

A. silver film

When the TLP voltage is less than 500 V and the current is less than 5A, there is no visible deformation on the surface of silver film, and its resistance value remains stable. As the TLP voltage increased to 1000 V and the current increased to 10 A, there was still no crack or fracture on the surface of silver film. Scanning electron microscopy (SEM) showed that the surface of silver film was still smooth and continuous after TLP, as shown in Fig.2. Even when the pulse intensity is close to the limit value, such as 2500 V voltage and 40 A current, the surface of silver film has not undergone destructive

---

Shipeng Chen, Zhencheng Xu, Yipeng Chen, Jiabei Pan, Ling Zhang and Shurong Dong are with *College of Information Science & Electronic Engineering, Zhejiang University*, Hangzhou, China.This work is supported by xxx. Corresponding author: (Shurong Dong,email: dongshurong@zju.edu.cn)

979-8-3315-2209-4/25 $31.00 © 2025 IEEE

changes, and its conductivity remains above 98 % of the initial value. The results show that silver film has strong antistatic ability, and it has a highly continuous conductive path at the molecular scale, which avoids the local hot melt phenomenon caused by current concentration.

Fig. 2. Structure of Ag film (a) before TLP; (b) after TLP; (c) TLP curve.

In order to determine the antistatic ability of silver film under different areas, 50 nm silver film samples were divided into three categories : $130mm^2$, $75mm^2$ and $50mm^2$ for TLP test, as shown in Fig.3.

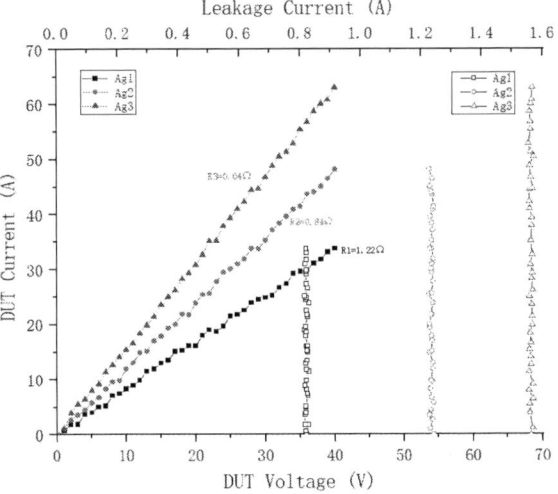

Fig. 3. TLP curves of silver films with different area.

At the same time, the resistance change of DUT before and after TLP was measured, as shown in Table 1. Obviously, the antistatic ability of silver film is still strong in a small area, and there is no sudden damage to the leakage current, which once again verifies the conclusion that silver film has strong antistatic ability.

TABLE I
SILVER FILM RESISTANCE

	area (mm^2)	before TLP (Ω)	after TLP (Ω)
Ag 1	50	1.19	1.22
Ag 2	75	0.82	0.84
Ag 3	150	0.65	0.64

B. CPSA

The same TLP test was performed on CPSA, and it was found that it also had strong antistatic ability and could withstand currents of up to 40 A.

C. silver film / CPSA

Although silver film has strong antistatic ability, silver film is usually used in combination with CPSA to achieve circuit interconnection. In the initial stage of TLP test, the silver film / CPSA overlap surface did not change significantly, and the overall structure remained stable. However, as the TLP voltage increases, especially when the voltage exceeds 1200 V, the edge area of the silver film / CPSA begins to appear tiny cracks, and as the voltage increases, the cracks are aggravated, and the measured failure current is only 13.8 A, as shown in Fig.4.

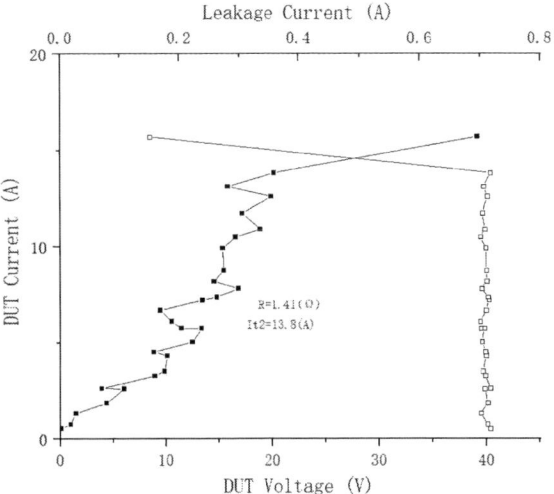

Fig. 4. TLP curve of silver film / CPSA.

III. SILVER FILM/CPSA FAILURE ANALYSIS

For the near-field scanning of the failure sample, since the continuous conductor electric field is also continuous, the discontinuous position of the electric field is on the overlapping surface of silver film / CPSA, as shown in Fig.5. Therefore, it can be determined that the failure position is on the silver film / CPSA overlap surface.

Fig. 5. Near-field scanning of failure samples.

Due to the previous analysis, silver film has strong antistatic ability, so the microscopic analysis of CPSA is carried out to find out the reason for the failure of silver film / CPSA overlap surface. After soaking CPSA in acetone and dissolving its viscous material, the microstructure of CPSA was observed by SEM. It can be seen that it is composed of conductive materials woven by fibers, as shown in Fig.6.

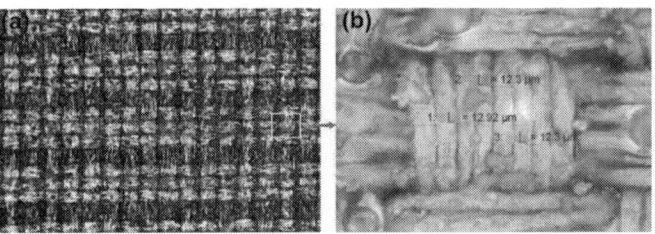

Fig. 6. CPSA structure after dissolution (a) 100 m; (b) 10 m.

TLP was performed on CPSA and it was found that it also had strong antistatic ability. However, the viscous substance on the surface shrinks after TLP, and the microstructure of CPSA is observed, as shown in Fig.7.

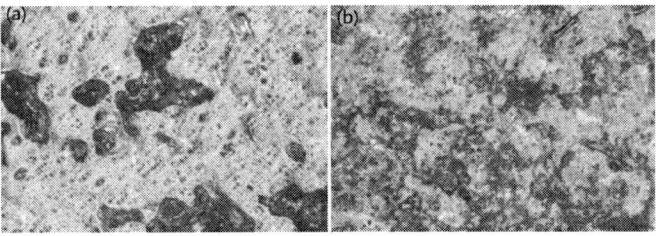

Fig. 7. CPSA surface viscous material (a) before TLP; (b) after TLP.

Since silver film and CPSA are tightly connected by a viscous substance, the shrinkage of the viscous substance will introduce tensile stress on the surface. Under the action of transient high voltage pulse, the heat of CPSA accumulates on the surface, and the viscous material softens and shrinks, so that the tensile stress is concentrated on the surface of silver film, which eventually leads to the large-scale shrinkage and fracture of the overlapping silver film, resulting in the failure of electrical connection, as shown in Fig.8.

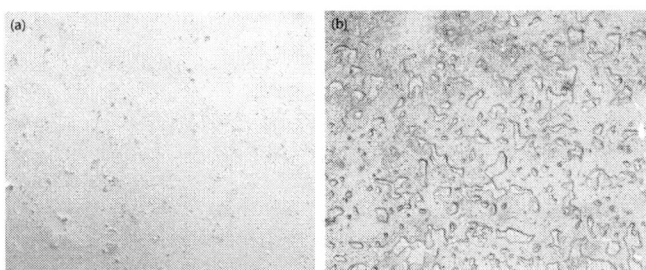

Fig. 8. Overlapped silver film structure (a) before TLP; (b) after TLP.

In order to verify the large-scale shrinkage and fracture of the overlapping silver film, the overlapping silver film of the failed sample was subjected to Energy Dispersive X-ray spectroscopy (EDX), as shown in Fig.9. It can be observed that the content of the intermediate carbon element is significantly increased, which means that the silver film has a hole in this part, which once again verifies the conclusion that the large-scale shrinkage and fracture of the silver film on the overlapping surface cause the failure of the electrical connection.

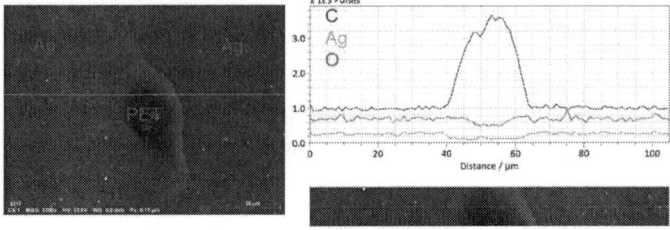

Fig. 9. Overlapping silver film element composition.

IV. DISCUSSIONS

50nm silver film has strong antistatic ability. However, under the transient high voltage pulse of TLP, the heat accumulates on the surface of silver film / CPSA, and the shrinkage of viscous material makes the tensile stress concentrate on the surface of silver film, which leads to the shrinkage and fracture of silver film under the overlapping surface, resulting in the failure of electrical connection. Future research can further optimize the material and structural design on this basis, and significantly improve the reliability of electronic devices in an electrostatic environment.

REFERENCES

[1] Smith J, Johnson M. The Impact of Electrostatic Discharge on Electronic Equipment Reliability[J]. IEEE Transactions on Electronics Packaging Manufacturing, 2015, 38(2): 78-85.
[2] Lee, S., Kim, H., & Park, K. (2024). Electrically Conductive Adhesives for Interconnection Technologies in Electronics Packaging. IEEE Transactions on Components, Packaging and Manufacturing Technology, 11(2), 112-123.
[3] Green A, Thompson B. Wireless Charging Backplane Integrity under ESD Stress. Journal of Wireless Power Transfer, 2023, 10(2): 111-120.
[4] Brown J, White S, Black M. ESD Impact on Wireless Charging Module Components. IEEE Transactions on Power Electronics, 2022, 37(9): 10102-10110.
[5] Johnson K, Miller D. ESD Challenges in High - Density Wireless Charging Backplanes. Electronics Packaging and Production, 2021, 61(4): 321-328.
[6] Li, C., Wang, Q., & Zhao, X. (2023). Advances in Conductive Materials for Flexible Electronics: A Comprehensive Review. Journal of Materials Science and Technology, 124, 156-179.

Design of Low-Leakage ESD Power Clamp Circuit with Multi-level Control Network

Gaoxiang Kai, Jiahao Xu, Liyao Wei, Hejiu Zhang, Yiqun Liu, Zhaonian Yang

Abstract—This paper proposes a low-leakage ESD clamp circuit design utilizing a silicon-controlled rectifier (SCR) as the clamping device for nanoscale process applications. The SCR triggering circuit consists of three structurally distinct RC networks with interactive control mechanisms that pull up or pull down the gate voltage of transistors. This configuration ensures transistors are thoroughly turned off during normal operation, thereby effectively reducing the overall leakage current of the system. Simulation results demonstrate that the SCR control module achieves rapid ESD event detection and maintains static leakage current at approximately 5.6 nA under normal operating conditions, exhibiting superior static low-power consumption characteristics.

Index Terms—Electrostatic Discharge (ESD), Low Leakage, SCR, Clamp Circuit

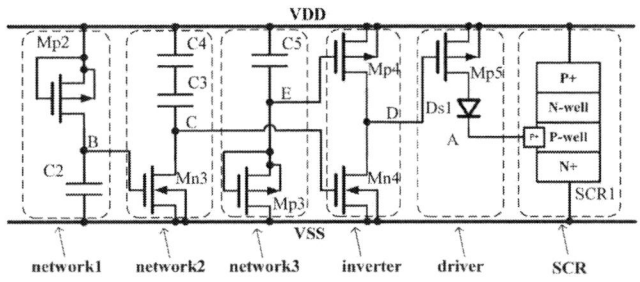

Fig. 1. Scheme of low leakage ESD power clamp circuit with multi-level control network

I. INTRODUCTION

Electrostatic discharge (ESD) refers to the phenomenon of rapid charge transfer occurring when accumulated static electricity reaches critical levels through objects or human bodies [1]. During the fabrication, transportation, and operation of integrated circuits, ESD events remain unavoidable [2]. The high-voltage, high-current transients generated by ESD can induce rapid dielectric breakdown in vulnerable electronic components, leading to catastrophic consequences including data corruption, performance degradation, and complete circuit failure [3]. Therefore, developing effective ESD protection solutions is critical for advancing integrated circuit technologies.

The ESD power-clamp circuit between power supply and ground serves dual functions: protecting power ports from ESD pulse damage and providing discharge paths for ESD energy between other ports. With the advancement of process technology and the growing demand for low-power design, the static leakage current in ESD power-clamp circuits under nanoscale processes has become increasingly problematic, necessitating innovative design solutions. In clamp circuit design, SCR-based clamping devices exhibit inherent advantages by maintaining non-conductive states during normal operation with low static leakage currents. Furthermore, through optimizing SCR triggering circuitry design, significant reduction in static leakage of discharge paths can be achieved while maintaining robust ESD protection performance[4].

Building upon the multi-stage trigger control methodology for ESD power-clamp circuits proposed in [5], this paper

Gaoxiang Kai, Jiahao Xu, Liyao Wei, Hejiu Zhang, Yiqun Liu and Zhaonian Yang are with *School of Automation and Information Engineering, Xi'an University of Technology*, Xi'an, China. This work is supported by the Scientific Research Program Funded by Shaanxi Provincial Education Department (Program No.24JP115). Corresponding author: (Zhaonian Yang, email: e_yangzhaonian@163.com)

further optimizes the trigger circuit. By designing three distinct RC networks with interactive control mechanisms based on SCR clamping devices, the gate voltages of pull-up/pull-down transistors are dynamically adjusted to ensure full activation or strict cutoff during normal operation, thereby significantly reducing the overall static leakage current of the ESD circuit to meet nanoscale process requirements.

II. CIRCUIT DESIGN DESCRIPTION

In traditional ESD power clamp circuit design, the structure primarily comprises three components: a detection circuit (typically implemented as an RC network), a driver circuit (composed of inverters), and a clamping device (a MOS transistor), which exhibits significant leakage current issues in nanoscale processes.

Based on the investigation of RC and CR network characteristics, this paper designs three RC control networks with distinct structures to optimize the SCR trigger circuit, regulating the transistor's static gate voltage to reduce leakage current. As shown in Fig. 1, the proposed low-leakage ESD power clamp circuit comprises the first, second, and third control networks, an inverter, a driver branch, and an SCR.

During ESD events, the first control network detects the rapid power-up surge, biasing node B at a low voltage level. This drives the NMOS transistor Mn3 into turned-off state, where it exhibits high equivalent resistance. The second control network (Mn3,C3 and C4) subsequently responds by elevating node C voltage to activate Mn4, which pulls down node D. This action activates the driver transistor Mp5, thereby injecting triggering current into the SCR through the substrate path.

Under normal operation, the normally-off PMOS transistor Mp2 with significant equivalent resistance and capacitor C2 form a voltage divider. Through dimensional optimization,

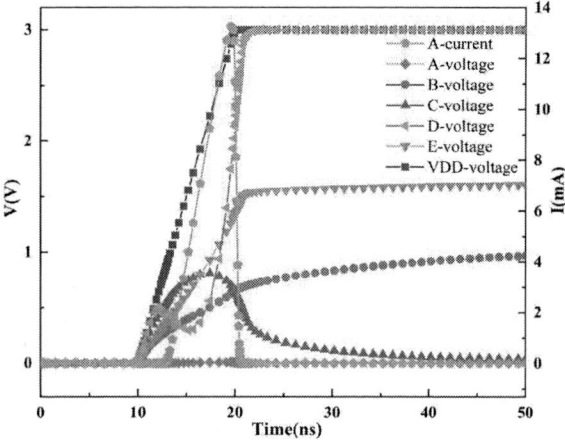

Fig. 2. Voltage-current simulation results of each node under ESD event

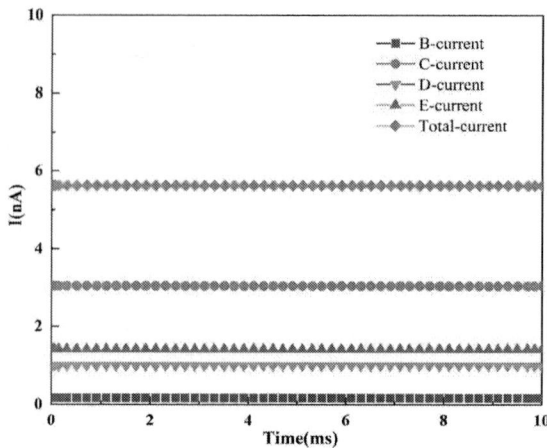

Fig. 4. Simulation results of leakage current characteristics under normal operation

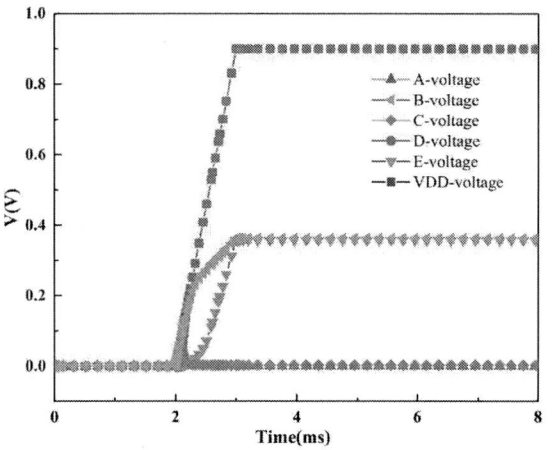

Fig. 3. Simulation results of voltage characteristics under normal operation

node B maintains a proper intermediate bias , ensuring full activation of Mn3. Consequently, node C voltage is pulled to ground potential, keeping NMOS Mn4 strictly in cut-off state. Similarly, node E settles at an intermediate level , activating PMOS Mp4 to pull node D to VDD rail. This configuration maintains driver transistor Mp5 in complete shutdown, effectively eliminating SCR triggering current generation.

This paper achieves leakage reduction in clamp circuits through circuit design optimization featuring three RC network interactions, enabling full activation or strict shutdown of transistors during normal operation. Furthermore,device and parametric co-optimization further minimizes overall static leakage: 1) The first and third control networks employ normally-off PMOS transistors with high equivalent resistance, reducing capacitor area while suppressing static leakage; 2) The second network implements series-connected capacitors to mitigate voltage-divider leakage; 3) A diode-added driver branch output provides additional leakage suppression.

III. RESULTS AND DISCUSSION

This design is implemented in a 45 nm CMOS technology with the following device parameters: where the normal operating voltage of the MOS transistors is 0.9V, the size of Mp2 is 1μm / 70 nm, Mp3 is 1μm / 70 nm, Mp4 is 1μm / 200 nm, Mp5 is 10μm / 200 nm, Mn3 is 10μm / 120 nm, Mn4 is 1μm / 100 nm. The capacitance is 2.4 fF. After the circuit design was completed, this paper carried out the ESD characteristic simulation and the normal working condition characteristic simulation.

A. ESD Pulse Characterization Simulation

An ESD voltage pulse with 10 ns rise time and 3 V amplitude was applied to verify nodal voltage/current variations. As shown in Fig. 2, the VDD voltage rises from 0 V to 3 V within 10 ns. Node C voltage is pulled high through VDD rising, activating NMOS Mn4 for 10 ns duration. Simultaneously, node D voltage is pulled low with PMOS Mp5 conducting for matching 10 ns period. This coordinated switching initiates the trigger path, injecting 13.24 mA peak current into node A to trigger the SCR.

B. Normal Operating Voltage Characteristics Simulation

A 1 ms rise time and a 0.9 V amplitude voltage waveform was applied to simulate normal operating conditions while evaluating the complete ESD characteristics of the circuit. As shown in Fig. 3, nodes B and E are biased at 0.36 V and 0.35 V respectively. This configuration forces NMOS Mn3 into the subthreshold region, yet sufficiently pulls node C voltage to 0 V to turn off Mn4. With node E maintained at 0.35 V, PMOS Mp4 operates in subthreshold region. The turned-off Mn4 allows node D to be pulled high to 0.9 V through Mp4, which consequently turns off PMOS Mp5. This results in node A being held at 0 V , maintaining the SCR in its non-conductive state.

979-8-3315-2209-4/25 $31.00 © 2025 IEEE

C. Normal operation leakage characteristics simulation

The leakage current of the first control network, second control network, third control network, drive branch, and total leakage current of SCR trigger circuit under normal operation is illustrated in Fig. 4. Simulation results demonstrate that the SCR trigger circuit in the proposed ESD power clamp design achieves a total leakage current of 5.6 nA. Compared with conventional circuits where the leakage current in detection-drive sections typically ranges from hundreds of nanoamperes to the microampere level, this design exhibits significantly reduced leakage characteristics.

IV. CONCLUSION

This paper presents a low-leakage ESD power clamp circuit utilizing SCR as the clamping device. Simulation results verify that the proposed design achieves rapid detection of ESD events and effectively reducing the overall static leakage current under normal working conditions, exhibiting excellent static low-power characteristics.This design is well-suited for low-power ESD protection solutions in nanoscale process technologies.

REFERENCES

[1] J. Ong, B. Chin and L. H. Koh, "Dummy Versus Live ESD Sensitive Devices Charge Analysis for Automated Handling Equipment ESD Qualification," 2019 41st Annual EOS/ESD Symposium (EOS/ESD), Riverside, CA, USA, 2019, pp. 1-5, doi: 10.23919/EOS/ESD.2019.8869997.

[2] J. Park, J. Lee, C. Jo, B. Seol and J. Kim, "A Proto-type ESD Generator for System Immunity Test of Wearable Devices," 2018 40th Electrical Overstress/Electrostatic Discharge Symposium (EOS/ESD), Reno, NV, USA, 2018, pp. 1-6, doi: 10.23919/EOS/ESD.2018.8509740.

[3] W. Stadler, J. Niemesheim, S. Seidl, R. Gaertner and T. Viheriaekoski, "The Risks of Electric Fields for ESD Sensitive Devices," 2018 40th Electrical Overstress/Electrostatic Discharge Symposium (EOS/ESD), Reno, NV, USA, 2018, pp. 1-9, doi: 10.23919/EOS/ESD.2018.8509774.

[4] P. -L. Peng et al., "Low-Capacitance SCR for On-Chip ESD Protection with High CDM Tolerance in 7nm Bulk FinFET Technology," 2019 41st Annual EOS/ESD Symposium (EOS/ESD), Riverside, CA, USA, 2019, pp. 1-5, doi: 10.23919/EOS/ESD.2019.8869982.

[5] M. Stockinger, "Low-Leakage NMOS Clamps with Gate-Assisted Bipolar Triggering," 2019 41st Annual EOS/ESD Symposium (EOS/ESD), Riverside, CA, USA, 2019, pp. 1-10, doi: 10.23919/EOS/ESD.2019.8869969.

Design of a Compact Clamp Circuit Based on Current Mirror Structure

Liyao Wei, Hejiu Zhang, Yiqun Liu, Zhaonian Yang

Abstract—**This paper presents a compact clamp circuit that optimizes the RC trigger network by incorporating a current mirror structure. The current mirror structure achieves equivalent amplification of the small resistor and capacitor, addressing the issue of excessive layout area caused by the resistor and capacitor. Through simulation of fast power-up and ESD events for the proposed clamp circuit, it is verified that the circuit can rapidly detect ESD events and dissipate ESD energy, effectively handling fast power-up conditions with 100ns, with excellent mistriggering immunity characteristics. Compared to traditional RC-triggered clamp circuits, this design achieves approximately 60% layout area optimization.**

Index Terms—**Electrostatic Discharge (ESD), Clamp Circuit, Current Mirror, Layout Area**

I. INTRODUCTION

Electrostatic Discharge (ESD) refers to the phenomenon of transient transfer of electrostatic charges between objects at different electrostatic potentials [1]. Throughout the processes of manufacturing, packaging, transportation, and usage of chips, there is a potential threat posed by ESD events. According to investigations, numerous failures in integrated circuit products are associated with ESD/EOS (Electrical Overstress), leading to losses amounting to billions of dollars. Therefore, incorporating ESD protection solutions is crucial [2, 3].

A clamp circuit provides protection between VDD and VSS. The clamp circuit which RC network-based serves as the detection unit, is the most commonly used design approach. However, traditional RC clamp circuits must balance the RC time constant between rapid detection and discharge delay. They need to trigger quickly during ESD events while also providing sufficient discharge time. This often results in larger values for R and C, leading to excessive layout area. In some processes, R and C can occupy more than half of the total layout area, which is unacceptable in high-integration and low-cost designs.

In [5], a circuit technique was proposed to achieve capacitive reactance amplification using a current mirror. A small capacitor is equivalently amplified through current amplification by a set of NMOS current mirrors, effectively reducing the area of an equivalent conventional capacitor. This approach significantly decreases the layout area required for the capacitance component while maintaining the necessary electrical characteristics.

Liyao Wei, Hejiu Zhang, Yiqun Liu and Zhaonian Yang are with *School of Automation and Information Engineering, Xi'an University of Technology,* Xi'an, China. This work is supported by the Scientific Research Program Funded by Shaanxi Provincial Education Department (Program No.24JP115). Corresponding author: (Zhaonian Yang, email: e_yangzhaonian@163.com)

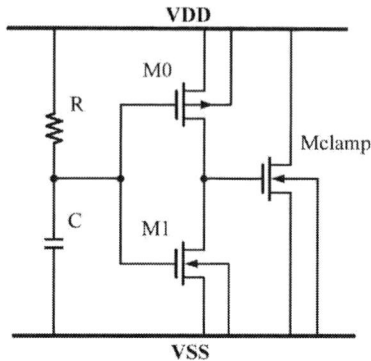

Fig. 1. Traditional RC clamp circuit

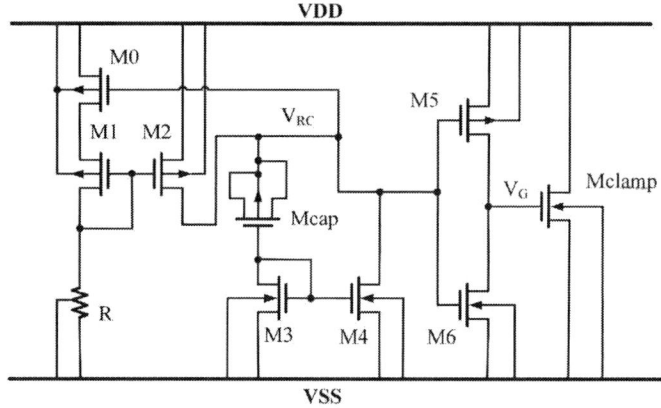

Fig. 2. Proposed RC clamp circuit incorporating two sets of current mirrors

In this paper, the RC network is further optimized by introducing two sets of current mirror structures to equivalently represent large resistors and large capacitors. Compared with traditional structures, this design significantly reduces the layout area and exhibits excellent performance in preventing false triggering, making it capable of handling fast power-up events.

II. CIRCUIT DESIGN DESCRIPTION

The traditional clamp circuit design is shown in Figure. 1, primarily consisting of a detection branch, driver branch, and clamp transistor. Based on the investigation of conventional clamp RC trigger networks, this paper optimizes both the R and C components by incorporating current mirror structures to form a new detection branch, as illustrated in Figure 2.

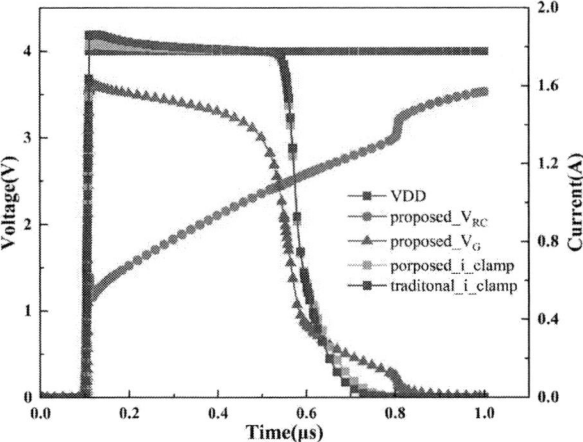

Fig. 3. Simulation results under ESD pulse

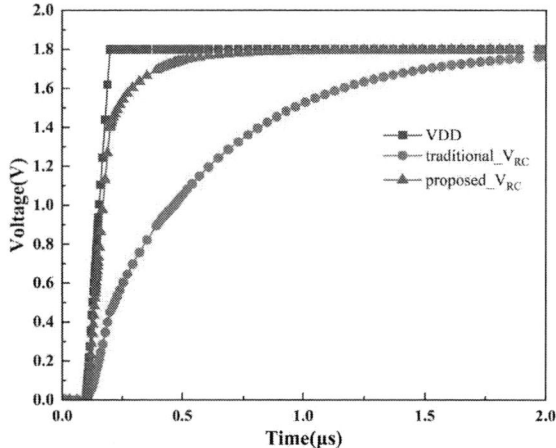

Fig. 5. Simulation of false triggering case under 100ns fast power-up

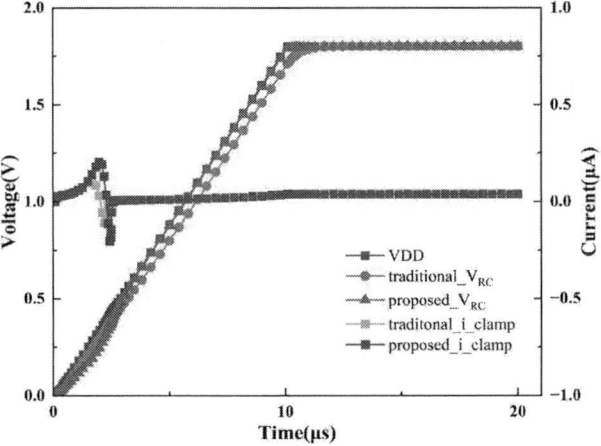

Fig. 4. Simulation results under normal power-up

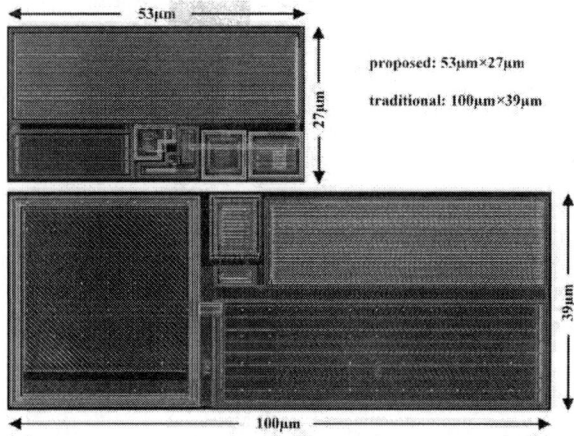

Fig. 6. Comparison of layout area between traditional scheme and proposed scheme

The proposed configuration employs a PMOS current mirror (M1 and M2) to achieve current reduction in the resistive path by a multiplicative factor, enabling low-current charging of the capacitor and effectively realizing the functionality of a large equivalent resistor. Simultaneously, an NMOS current mirror (M3 and M4) amplifies the capacitance value by a multiple, creating an equivalent large capacitor. This approach implements large resistance and capacitance characteristics through circuit techniques using small resistors and capacitors, effectively saving layout area.

When an ESD event occurs, the equivalent large-resistance path generates a small current to charge the large capacitor. This design effectively amplifies the RC time constant using small resistors and capacitors. Due to the inability of the capacitor voltage to change abruptly during charging or discharging, V_{RC} cannot follow VDD variations. The resulting voltage difference drives the activation of transistor M5, which clamps the gate voltage V_G of the clamp transistor M_{clamp} to a high level, enabling M_{clamp} to conduct and dissipate ESD energy. As the V_{RC} voltage increases, transistor M6 becomes activated, pulling V_G to a low level to deactivate the

clamp transistor, thereby completing the entire ESD protection process.

During normal power-up sequences where the rising edge duration of the voltage pulse typically exceeds the microsecond range, V_{RC} tracks VDD variations while maintaining transistor M6 in a non-conductive state. This configuration ensures the clamp transistor M_{clamp} remains persistently deactivated. The switching transistor M0 concurrently operates in its off-state, effectively blocking the leakage current path through the resistive branch, thereby minimizing leakage currents under standard operational conditions of internal circuitry.

Under fast power-up conditions with voltage pulse rising edges typically spanning tens to hundreds of nanoseconds, the transient characteristics exhibit temporal similarity to ESD pulses. This phenomenon introduces potential false triggering risks through waveform misidentification. In the proposed structure, during the detection process of fast power-up, owing to the supply voltage being 1.8 V, the M1 transistor within the first set of PMOS current mirror operates in the sub-threshold conduction state. The gate voltages of M1 and M2

979-8-3315-2209-4/25 $31.00 © 2025 IEEE   527

are only a few millivolts, leading to the left-hand resistive branch (consisting of M0, M1, and R) being essentially non-conductive.At this juncture, a new RC structure predominantly composed of M2 and the capacitor takes effect. Given that the on-resistance of M2 is significantly low, it leads to a diminished overall time constant, enabling V_{RC} to tightly follow VDD. Consequently, this can effectively circumvent mis-triggering under rapid power-up conditions.

III. RESULTS AND DISCUSSION

The proposed design is implemented using a 0.18 μm process technology, with all transistors operating at a supply voltage of 1.8 V. Following the completion of the circuit design, ESD characteristic simulations and analyses were conducted .Additionally, the layout design was finalized.

A. ESD Pulse Simulation

An ESD pulse with a 10 ns rise time and 4 V amplitude was simulated to verify the Human Body Model (HBM) protection level under this condition. As illustrated in Fig. 3, when the ESD pulse occurs, the circuit successfully detects the event and triggers M_{clamp} to discharge the energy, demonstrating that the design achieves an HBM protection level of 2500 V.

B. Normal power-up simulation

The protected internal circuit under normal power-up conditions was simulated using a voltage waveform with a 10 μs rise time and 1.8 V amplitude. As illustrated in Fig. 4, the detected branch voltage V_{RC} consistently follows VDD, while M_{clamp} remains inactive. The leakage current in the circuit is maintained at the nanosecond level throughout the normal working condition.

C. Fast power-up simulation

A voltage waveform with a rising edge of 100 ns and an amplitude of 1.8 V was applied to the proposed clamp circuit. The resulting VDD and V_{RC} voltage waveforms are shown in Fig. 5. In the proposed circuit, V_{RC} closely follows VDD with a voltage difference consistently below the transistor threshold voltage, thus keeping M_{clamp} in the off state. In contrast, the conventional circuit exhibits potential mis-triggering risks during the initial hundreds of nanoseconds. This comparison demonstrates that the proposed circuit achieves superior mis-triggering immunity characteristics.

D. Comparison of layout area

Figure. 6 illustrates the layout comparison between the conventional scheme and the proposed solution. Both designs maintain identical driver circuits and clamp transistor dimensions, differing only in RC network structures. In the implemented process technology, the conventional scheme's resistors R and capacitors C occupy over 50% of the layout area, resulting in significant area inefficiency. The proposed scheme optimizes the R and C configuration through structural improvements, achieving approximately 60% area reduction compared to the conventional approach.

IV. CONCLUSION

This paper proposes a compact clamp circuit scheme utilizing current-mirror optimized RC structure. Simulation results verify that the proposed solution effectively addresses fast power-on events, demonstrating superior mis-triggering immunity characteristics. Furthermore, it achieves approximately 60% layout area reduction compared to conventional implementations through optimized resistor and capacitor configuration. This design is particularly suitable for applications requiring enhanced integration density and cost efficiency in process technologies where resistor and capacitor unit impedances exhibit relatively large values.

REFERENCES

[1] Z. Yang et al., "Feedback Enhanced Area-Efficient ESD Power Clamp Circuit," in IEEE Transactions on Electron Devices, vol.71, no. 8, pp. 4504-4509, Aug. 2024.

[2] M. Park, J. Tseng, T. -y. Lee and D. Ripley, "Concurrent ESD and Surge Protection Clamps in RF Power Amplifier,"2019 41st Annual EOS/ESD Symposium (EOS/ESD), Riverside, CA, USA, 2019, pp. 1-6.

[3] M. Di, C. Li, Z. Pan and A. Wang, "Pad-Based CDM ESD Protection Methods Are Faulty," in IEEE Journal of the Electron Devices Society, vol. 8, pp. 1297-1304, 2020.

[4] S. Parthasarathy, J. A. Salcedo, S. Herrera and J. -J. Hajjar, "ESD protection clamp with active feedback and mis-trigger immunity in 28nm CMOS process,"2015 IEEE International Reliability Physics Symposium, Monterey, CA, USA, 2015, pp. EL.3.1-EL.3.5.

[5] F. A. Altolaguirre and M. -D. Ker, "Low-leakage power-rail ESD clamp circuit with gated current mirror in a 65-nm CMOS technology,"2013 IEEE International Symposium on Circuits and Systems (ISCAS), Beijing, China, 2013, pp. 2638-2641.

Microcontroller Unit ESD-Induced Soft Failure Study

Jiabei Pan   Yipeng Chen   Xinyu Zhu   Shipeng Chen   Ling Zhang   Shurong Dong

Abstract—Soft failures caused by system-level electrostatic discharges(ESD) can lead to functional abnormalities of chips in practical applications and have become one of the important forms of chip failures. However, there has always been a lack of a test method that can measure the power supply noise directly related to soft failures. This article presents the testing and analysis of power supply noise on actual microcontroller unit (MCU) products. Ground-Signal-Ground(GSG) probes are directly used to test the actual waveform of the power supply on the operating MCU when system-level ESD pulses are injected. The tested power supply noise waveform is utilized to analyze the soft-failure problems of the MCU chip and guide the design of the on-chip Low Dropout Regulator (LDO). By improving the design of the on-chip LDO, the system-level ESD soft-failure level of this MCU has been tripled.

Index Terms—electrostatic discharge, system level ESD, soft failures, supply noise, soft errors

I. INTRODUCTION

Soft failures caused by ESD are one of the main failure modes of IC failures.The system-level electrostatic discharge (ESD) pulse will generate a large transient current [1], resulting in clock flipping, power supply noise, logic flipping on the Microcontroller Unit (MCU) chip. These errors can be detected by software, but will not permanently damage the hardware. With the advancement of integrated circuit technology, new challenges have emerged: The reduction in operating voltage makes circuits highly susceptible to voltage fluctuations, leading to a significant increase in soft failures; The growth in circuit scale and the complexity of power domain structures make soft failures exceedingly difficult to locate and analyze. In order to analyze the soft failure caused by ESD on a MCU, Sandeep used a combination of hardware and software,the voltage monitor on the chip is used to detect the supply noise to determine whether the failure is related to the supply noise [2]. N.A.Thomson used a MCU including a glitch detector circuit that signals whether a logic-level glitch has occurred [3].A method of full-wave simulation of supply noise is shown in [4]. However, the above methods require additional detector circuits on the chip, which increases the cost of design and manufacture. In addition, the above methods cannot directly obtain the actual waveform of the power supply noise, which provides limited reference for improving the MCU design.

Jiabei Pan, Yipeng Chen, Xinyu Zhu, Shipeng Chen, Ling Zhang and Shurong Dong are with *College of Information Science & Electronic Engineering, Zhejiang University*, Hangzhou, China. Corresponding author: (Shurong Dong, email: dongshurong@zju.edu.cn)

This work shows a measurement method of MCU power supply noise without additional detector circuit, and proposes corresponding improvement measures to reduce supply noise. A radio frequency Ground-Signal-Ground (GSG) probe is used to directly measure the power supply noise on the chip,and the measured power supply noise provides a reference for improving the Low Dropout Regulator (LDO) in MCU.

II. TEST SYSTEM

The MCU product was fabricated in a 65-nm CMOS technology, packaged in a 48 pin LQFP package. A total of 3.3 V is supplied to the IO circuitry (VDD), while the core circuitry is connected to a 1.2 V supply (Vcore) provided by a LDO on chip.The ground of IO port is directly connected with the ground of digital circuit(VSS).Power clamp is placed on each IO port,and the primary ESD protection for each IO cell consists of dual-diodes.

Fig. 1. Test board.

The test board is shown in Fig. 1. In order to visually display whether the MCU is crashed, a water lamp program is burned to the MCU under test. When the function of the water lamp is wrong, it is proved that the MCU appears soft failure. Once the MCU appears soft failure, stop ESD pulse injection and press the reset button on the test board, the water lamp program re-enters the normal working state, indicating that the MCU does not occur hard failure.

979-8-3315-2209-4/25 $31.00 © 2025 IEEE

To directly test the the power supply noise, a decapped MCU is placed on the test board. ESD pulse is injected into each pin of the MCU.a GSG probe is used to directly test the voltage on the pad.The connection of the whole test system is shown in Fig. 2.

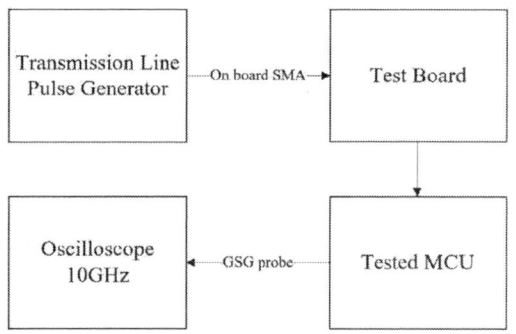

Fig. 2. Test system connection.

III. MEASUREMENT RESULTS AND DISCUSSION

The ESD gun was used to test the soft failure level of the MCU. The MCU has 35 IO ports, and each IO port is injected with a positive ESD gun pulse increasing from 0.5 kV to 2.5 kV until the MCU is dead. Similarly, the test results of negative ESD gun pulse are shown in TABLE I together with the positive ESD gun pulse result.In addition, very-fast transmission line pulse is used to test the soft failure level of MCU. The current pulses have a 100 ps rise-time and a 10 ns pulse width. Although the very-fast transmission line pulse is not completely consistent with the ESD gun pulse, measurement shows that very close agreement in the failure level to the two different pulses.

TABLE I
ESD GUN PULSE TEST OF THE MCU PRIOR TO IMPROVEMENT

Vgun(kV)	The number of IO ports cause crash at this level
0.5	17
1	10
1.5	3
2	2
2.5	3
-0.5	19
-1	7
-1.5	3
-2	4
-2.5	2

To detect the noise of the power supply, after the chip is decapped, the pad of the VDD and Vcore on die is directly tested using the GSG probe, as illustrated in Fig. 3. The probe is connected to a 10G bandwidth oscilloscope to read out the waveform of the Vcore. Once the probe is attached to the pad of the power supply, an incremental very-fast transmission line pulse is injected into the IO port until the MCU crashes. Since the oscilloscope is set to 50

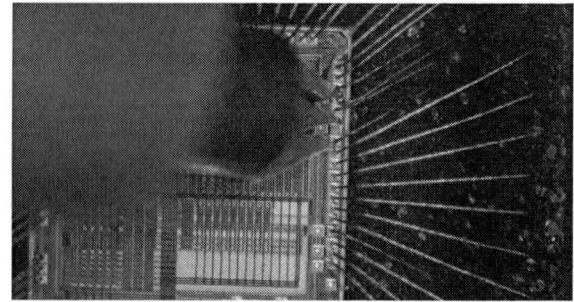

Fig. 3. On-chip Voltage measured by GSG probe.

Ω internal resistance, the impedance is matched throughout the measurement process, and the power supply noise can be correctly measured.Subsequently, the power supply noise waveform at the moment of crash is recorded, as shown in Fig.4.

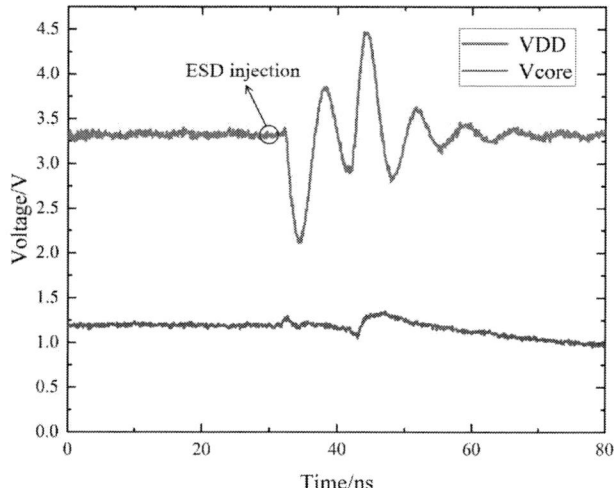

Fig. 4. measured VDD and Vcore of the tested MCU.

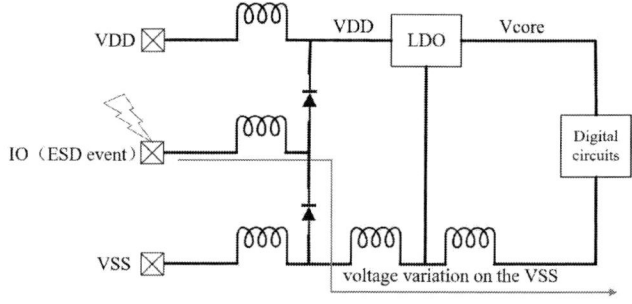

Fig. 5. ESD current causes ground bounce on VSS.

As shown in Fig. 5,since both the ESD gun pulse and the very-fast transmission line pulse have high frequency components, the parasitic inductance of the ground metal of the

MCU will cause a potential difference on the ground during the injection of these two pulses, so that the input voltage of the LDO is unstable [5]. Fourier transform is performed on the measured Vcore waveform, and it is found that there is supply noise component in the 10 kHz-300 MHz band. In this frequency band,the power supply rejection ratio (PSRR) of LDO is not enough to suppress the power supply noise on VDD, the voltage Vcore in the digital domain fluctuates, resulting in a logic error when the digital circuit clock flips.

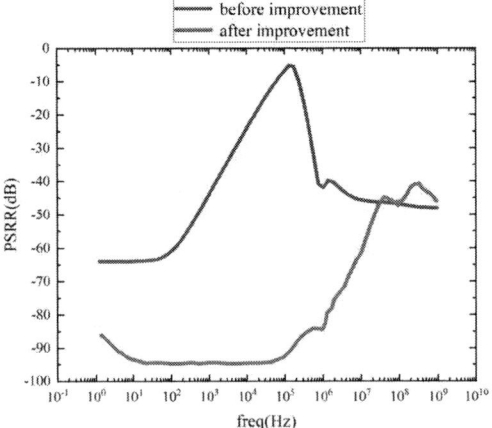

Fig. 6. PSRR of improved LDO.

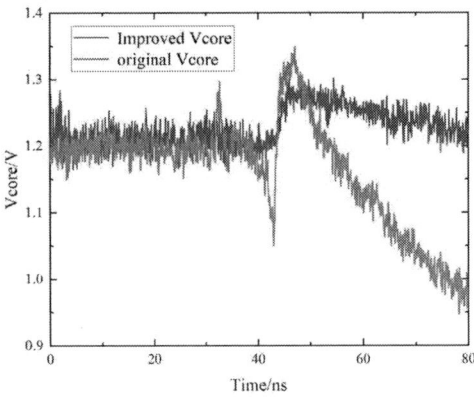

Fig. 7. Vcore comparison of original MCU and improved MCU.

In order to improve the soft failure level of the MCU, a LDO with a improved PSRR in the supply noise band is redesigned and fabricated,its PSRR is shown in Fig. 6. The improved MCU's ESD gun soft failure level is shown in TABLE II.Compared to TABLE I,it demonstrated that more IO ports can withstand high-intensity ESD pulses,The system-level ESD-induced soft failure problem of this MCU has been improved.The improved MCU was tested using the same method above. It was found that under the same ESD pulse

TABLE II
POSITIVE ESD GUN PULSE TEST AFTER IMPROVEMENT

Vgun(kV)	The number of IO ports cause crash at this level
0.5	0
1	0
1.5	7
2	6
2.5	22
-0.5	0
-1	0
-1.5	9
-2	3
-2.5	23

injection conditions, the Vcore of the improved MCU was more stable,as shown in Fig. 7.

IV. CONCLUTION

A system-level ESD test is performed on the designed MCU, and each pin will crash when a low-level ESD injection occurs. The measurement method of reading Vcore to IO port output through the program will lead to the parasitic bonding wire and IO port, so that the tested waveform is superimposed by the echo, which is not accurate. Therefore, after the chip is decapped, the GSG probe is directly used to measure the size of Vcore on die, and the waveform is correctly measured. According to the analysis, the PSRR of the LDO of the MCU is too small in the power supply noise frequency band, which leads to the fluctuation of Vcore. Excessive Vcore fluctuation causes the MCU to crash. After modifying the LDO design, the system-level ESD-induced soft failure level of the MCU is improved.After the improvements, the system-level ESD soft-failure level of this MCU has been tripled from 0.5kV to 1.5kV, a result validated through ESD gun testing and VFTLP testing.

REFERENCES

[1] K. Y. Kim and Y. Kim, "Systematic Analysis Methodology for Mobile Phone's Electrostatic Discharge Soft Failures," in IEEE Transactions on Electromagnetic Compatibility, vol. 53, no. 3,pp. 611-618,2011,doi: 10.1109/TEMC.2011.2143719.

[2] Vora, Sandeep and E. Rosenbaum, "Analysis of System-Level ESD-Induced Soft Failures in a CMOS Microcontroller," in IEEE Transactions on Electromagnetic Compatibility, vol. PP, no. 99,pp. 1-10,2020,doi: 10.1109/TEMC.2020.2986971.

[3] N. A. Thomson , Y. Xiu and E. Rosenbaum,"Soft-Failures Induced by System-Level ESD,"in IEEE Transactions on Device and Materials Reliability, 2017,doi:10.1109/TDMR.2017.2667712.

[4] Hosseinbeig et al. "Methodology for Analyzing ESD-Induced Soft Failure Using Full-Wave Simulation and Measurement." in IEEE Transactions on Electromagnetic Compatibility,2018,doi:10.1109/TEMC.2017.2787721.

[5] Y. S. Chang, S. K. Gupta and M . A . Breuer ,"Analysis of Ground Bounce in Deep Sub-Micron Circuits" in IEEE Vlsi Test Symposium,1997,doi:10.1109/VTEST.1997.599458.

Improved Snapback TVS Transient Behavior Model and Application to SEED

Yipeng Chen   Xinyu Zhu   Jiabei Pan   Shipeng Chen   Ling Zhang   Shurong Dong

Abstract—**With the rapid development of semiconductor technology, the electrostatic discharge (ESD) immunity of integrated circuits (IC) is decreasing due to the scaling down feature size of the semiconductor fabricated process. System-efficient ESD design (SEED) simulation, including ESD gun and printed circuit board (PCB) with components, can help predict the level of ESD stress seen by the IC when protected by a transient voltage suppressor (TVS). As a part of SEED simulation, TVS modeling is particularly critical to the simulation accuracy. Under the condition that both the structure and process parameters of most TVS products cannot be acquired, a novel snapback TVS behavioral model is proposed. The internal emission junction including its the diffusion capacitance and parallel resistance delays the turn on for TVS, and the on and off state transition is realized by controlling the ideal factor and the parasitic resistance of the variable diode. To adapt the model to different TVS quickly, a set of parameters can be tuned based on measurements by least square method. The novel TVS model successfully predicts the voltage overshoot and quasi-static voltage within 5% tolerance.**

Index Terms—**Electrostatic discharge (ESD), modeling, transient voltage suppressor (TVS), least square method**

I. INTRODUCTION

System-efficient ESD design (SEED) simulation, including ESD gun and printed circuit board (PCB) with components, can help predict the level of ESD stress seen by the IC when protected by a transient voltage suppressor (TVS). As a part of SEED simulation, TVS modeling is particularly critical to the simulation accuracy. Due to commercial confidentiality, both the structure and process parameters of most TVS products cannot be known and it is hard to establish a physical compact model for kinds of TVS. A variety of papers have addressed modeling of TVS devices and applied in SEED simulation. R. P. Santoro modeled quasi-static IV curve instead of the transient characteristic, which has convergence issue in transient solver [1]. L. Wei considered the snapback behavior of TVS, but conductivity modulation effect is not included in their model [2]. Some physics-based compact models are developed in [3], [4], which requires lots of model parameters and is hard to model. The novel behavioral TVS simulation framework was presented in [5] and further improved in [6] to investigate the interaction between a TVS and on-chip protection device during a transmission line pulse (TLP) event. Shown in Fig. 1, this model contains a linear small signal model and a non-linear section. The non-linear section realizes the conversion

Yipeng Chen, Xinyu Zhu, Jiabei Pan, Shipeng Chen, Ling Zhang and Shurong Dong are with *College of Information Science & Electronic Engineering, Zhejiang University*, Hangzhou, China. Corresponding author: (Shurong Dong, email: dongshurong@zju.edu.cn)

from a pre-clamping diode D5 to a quasi-static diode D3 through snapback delay and conductance modulation modules. However, the ideal switch in the snapback delay module will lead to a discontinuous point in the transient simulation and the parallel structure of pre-clamping diode and quasi-static diode results in the static IV curve to shift after the turn-on voltage.

In this paper, a novel snapback TVS behavioral model is proposed based on measured TVS responses. For faster behavioral TVS modeling, the least square method is used to extract the transient parameters of TVS model to fit the measured results. The model demonstrates good correlation with measurement for both the overshoot peak voltage as well as duration at different ESD level.

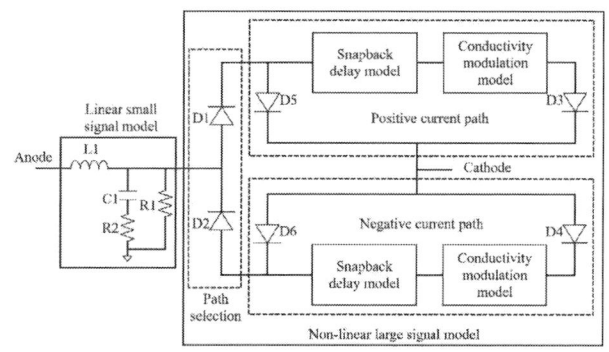

Fig. 1. Block diagram of the TVS transient behavior model in [5], [6].

II. NOVEL TVS MODEL FRAMEWORK

The circuit diagram of the proposed novel TVS model is shown in Fig. 2(a). It contains a linear small signal model and a non-linear section. The linear small signal model reflects the RF performance of the TVS device when it is not turned on, which is needed to simulate the effect of the TVS on signal integrity. Demonstrated in Fig. 2(b), the non-linear large signal model adjusts the diode ideal factor and its parasitic resistance according to snapback delay submodule and conductivity modulation submodule to replicate the dynamic characteristics of the TVS. The element equations and parameters list for TVS model are detailed in Fig. 3.

Compared with the model in [5], [6], the pre-clamping and quasi-static diodes are combined, and only one variable diode model D_1 in series with a variable resistor R_{CM} is used to characterize quasi-static IV curve of snapback TVS before and after triggering.

979-8-3315-2209-4/25 $31.00 © 2025 IEEE

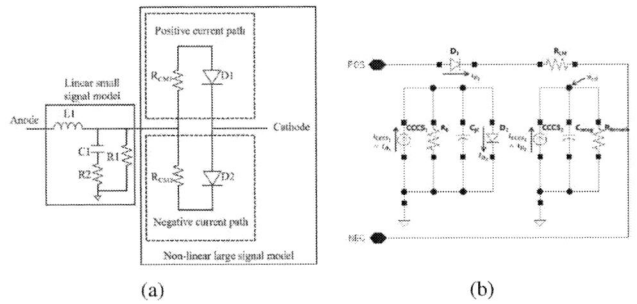

(a)   (b)

Fig. 2. (a) Block diagram of the proposed novel TVS model; (b) The circuit diagram of the non-linear large signal model.

Model Element	Model Expression	Model Parameter	Parameter Definition
D_1	$i_{D_1} = I_s[exp(\frac{v_{D_1}}{N \cdot V_T}) - 1]$	$N_0 \& R_0$	Diode ideal factor and parasitic resistance before triggering
	$N = N_1 + \frac{(N_0 - N_1) \cdot Q_{N_0}}{Q_{N_0} + v_{ctl}}$	$N_1 \& R_1$	Diode ideal factor and parasitic resistance after triggering
D_2	$i_{D_2} = I_s[exp(\frac{v_{D_2}}{V_T}) - 1]$	R_E	Emitter junction parallel resistance
		τ	Time constant of the forward recovery
R_{CM}	$R_{CM} = R_1 + \frac{(R_0 - R_1) \cdot Q_{R_0}}{Q_{R_0} + v_{ctl}}$	Q_{N_0}	Threshold voltage of variable diode ideal factor
C_{jE}	$Q_{jE} = \tau \cdot i_{D_2}$	Q_{R_0}	Threshold voltage of variable resistance

Fig. 3. The element equations and parameters list for the novel TVS model.

The initial value of the ideal factor and variable resistor are N_0 and R_0, which represents the avalanche IV characteristics before triggering. Both BJT and SCR type devices are triggered after the turn-on of the emission junction. The snapback delay effect can be realized by the parallel structure of the emitter junction D_2, the diffusion capacitor C_{jE} and the well resistance R_E [7]. At first, the emitter junction D_2 is not turn-on and its parallel resistor R_E discharges. Once the charge injected into the diffusion capacitor C_{jE} reaches the required turn-on condition for the emission junction, the variable diode D_1 begins to switch from its high-impedance off-state to its low-impedance on-state.

An equation for the conductivity modulation resistance is presented in Fig. 3 and was used to improve the conductivity modulation overshoot in [6]. Similar to variable resistance, the ideal factor of variable diode is also modulated by the amount of carriers in the neutral region near the depletion region edge. The emitter current of D_2 is transferred to the next conductivity modulation module by Current Controlled Current Source (CCCS). And the variable diode D_1 and its parasitic resistance R_{CM} are controlled by a first order RC network, where the emitter current charges the integrating capacitor C_{integ} and its voltage drop makes the ideal factor of variable diode D_1 from N_0 to N_1, and the variable resistance R_{CM} from R_0 to R_1. The dynamic transformation process will delay the transient voltage across the TVS into the steady state and cause the voltage overshoot beyond the value predicted by the quasi-static IV curve during the transition.

Besides, the variable diode in the novel TVS model can only be turned on forward without reverse breakdown characteristics. Therefore, no additional path selection diodes are needed for bidirectional TVS devices.

III. TRANSIENT MODEL EXTRACTION PROCEDURE

The TLP measurements were conducted with the Barth 4002 TLP standard test system by using the human body model parameters with a 10 ns rising time and a 100 ns pulse width. For each voltage pulse, current and voltage waveforms were captured by using a 10 GHz oscilloscope. By averaging over 70–90% time windows of the stress pulse waveform, a single data point of I-V characteristics was extracted. This was followed by a low 0.5 V bias DC test at room temperature to confirm device failure. In this section, the parameters extraction of one TVS sample (ESD18VU1B) according to its TLP test results and the detailed fitting process is described.

A. Small signal parameters

The linear small signal model characterization is the same as the method discussed in [5]. Shown in Table 1, The junction capacitance C, the parasitic inductance L and the effective series resistance R can be extracted by vector network analyzer (VNA).

Table 1
SMALL SIGNAL PARAMETERS EXTRACTED BY VNA.

C	L	R
0.231pF	0.403nH	10.76Ω

B. Variable diode parameters

The quasi-static behavior of TVS can be characterized by using the TLP measurement. Fig. 4 shows the measured quasi-static IV curve of the TVS sample compared with the fitted IV curve of the after-triggering diode model and its fitting parameters are extracted as illustrated in Table 2. According the TLP result, the trigger current is too low to fit the before-triggering diode so that the required parameters will be extracted by the least square method in next subsection.

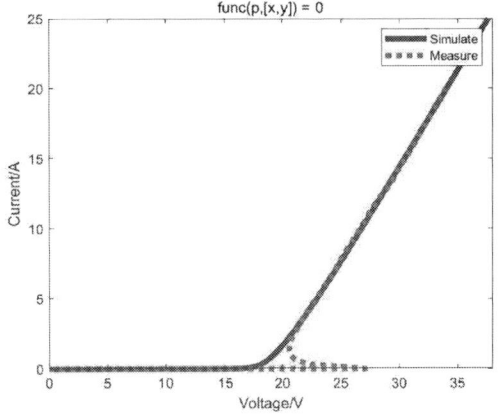

Fig. 4. Measured TVS quasi-static IV curve and the simulated after-triggering diode IV curve.

979-8-3315-2209-4/25 $31.00 © 2025 IEEE

Table 2
AFTER-TRIGGERING DIODE MODEL PARAMETERS.

I_S	R_1	N_1
1e-14A	0.687Ω	22.16

C. Other transient parameters

The fitting method of other transient parameters is shown in Fig. 5. The data processing and least squares fitting are implemented in Python, and the open source Ngspice is used for circuit simulation. It's worth noting that the variable diode model is implemented by Verilog-A and all other models use standard SPICE circuit elements.

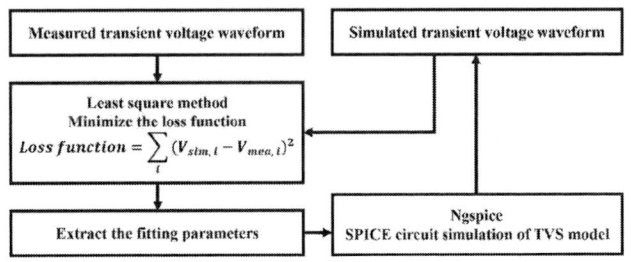

Fig. 5. Procedure of the proposed fitting method.

The fitting process is as follows:

1) Set the initial parameters of the TVS circuit model;
2) Run the Ngspice to get the simulated transient voltage waveform;
3) Calculate the loss function after data interpolation processing;
4) Fit the new parameters by least square method;
5) Turn to step 2) until the error is small enough or the number of iterations reaches the upper limit;

TVS in actual products generally withstand a higher level of ESD current, so the measured transient voltage waveforms of 100V, 300V and 1000V are selected for parameters fitting. The iterative process is shown in Fig. 6 and after the completion of the 150th iteration, the simulation and measurement have been almost consistent.

IV. TRANSIENT MODEL VALIDATION

The time domain simulation results of both the old TVS model in [5], [6] and the novel proposed model are fitted using 100V, 300V and 1000V TLP shown in Fig. 7 and validated using 200V, 600V and 800V forward voltage shown in Fig. 8.

The ideal switch in the snapback delay model in [5], [6] results in the strong discontinuity at voltage overshoot and it may worse the results of subsequent system-level simulations. The novel TVS model is delayed turn-on by the diffusion capacitance and the state transition is accomplished by controlling the ideal factor of the variable diode, which have better smooth transition properties and simulation accuracy. The quasi-static IV curve taken from the 100 ns TLP measurement result and the transient simulation result are also compared in Fig. 9. The average windows both for the measurement

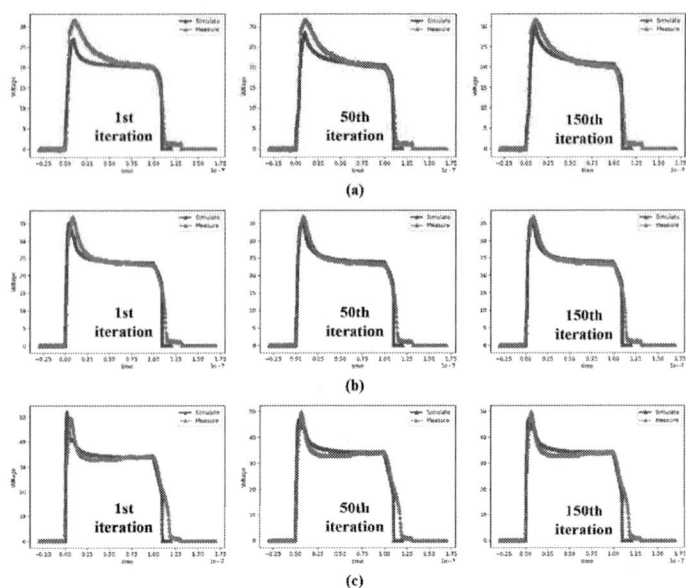

Fig. 6. Comparison between simulation and measurement of transient voltage waveform after each least square iteration (a)100V; (b)300V; (c)1000V.

and the simulation were found to be 70% to 90% of the time domain current and voltage waveforms. The novel TVS model successfully predicts the voltage overshoot and quasi-static voltage within 5% tolerance.

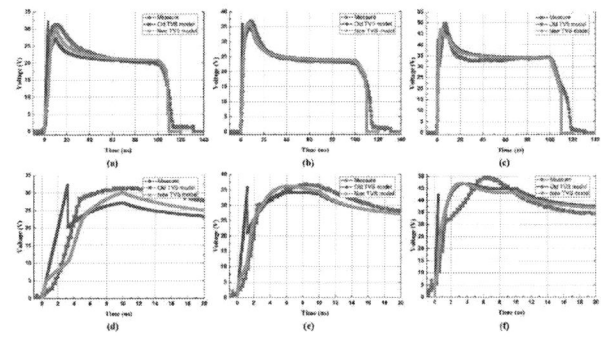

Fig. 7. The fitting results of both the old TVS model in [5], [6] and the novel proposed model (a) 100V; (b) 300V; (c) 1000V and their corresponding partial enlarged view (d-f).

V. DISCUSSION

A novel snapback TVS behavioral model is proposed based on measured TVS responses. It includes the variable diode, internal emission-junction portion and RC integral control circuit. The least square method is used to quickly extract the transient parameters of TVS model to fit the measured results. Compared with the model in [5], [6], the diffusion capacitance and its parallel resistance delay the turn on for TVS, and the on and off state transition is accomplished by controlling the ideal factor of the variable diode, which leads to better simulation accuracy in the voltage overshoot region and

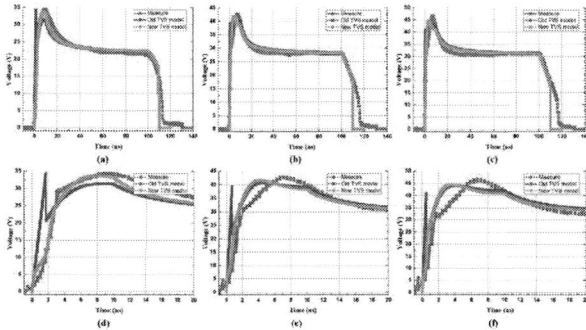

Fig. 8. The verification results of both the old TVS model in [5], [6] and the novel proposed model (a) 200V; (b) 600V; (c) 800V and their corresponding partial enlarged view (d-f).

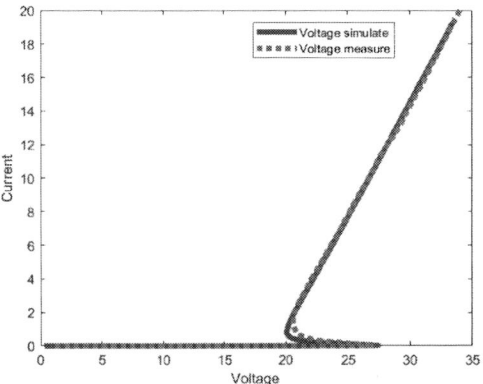

Fig. 9. Quasi-static VI curve comparison of the 100 ns TLP measurement result and the simulation result.

overall error between simulation and measurement is within 5%. This model can be applied to all kinds of snapback TVS.

REFERENCES

[1] R. P. Santoro, "Piecewise-linear modeling of I-V characteristics with SPICE," in IEEE Transactions on Education, vol. 38, no. 2, pp. 107-117, May 1995, doi: 10.1109/13.387211.

[2] L. Wei, C. E. Gill, W. Li, R. Wang and M. Zunino, "A convergence robust method to model snapback for ESD simulation," CAS 2011 Proceedings (2011 International Semiconductor Conference), Sinaia, Romania, 2011, pp. 369-372, doi: 10.1109/SMICND.2011.6095819.

[3] R. Mertens and E. Rosenbaum, "A physics-based compact model for SCR devices used in ESD protection circuits," 2013 IEEE International Reliability Physics Symposium (IRPS), Monterey, CA, USA, 2013, pp. 2B.2.1-2B.2.7, doi: 10.1109/IRPS.2013.6531947.

[4] S. Huang and E. Rosenbaum, "Physics-based Compact Model of N-Well ESD Diodes," 2023 45th Annual EOS/ESD Symposium (EOS/ESD), Riverside, CA, USA, 2023, pp. 1-6, doi: 10.23919/EOS/ESD58195.2023.10287734.

[5] P. Wei, G. Maghlakelidze, A. Patnaik, H. Gossner and D. Pommerenke, "TVS Transient Behavior Characterization and SPICEBased Behavior Model," 2018 40th Electrical Overstress/Electrostatic Discharge Symposium (EOS/ESD), Reno, NV, USA, 2018, pp. 1-10, doi: 10.23919/EOS/ESD.2018.8509780.

[6] Y. Xu et al., "Improved SEED Modeling of an ESD Discharge to a USB Cable," in IEEE Transactions on Electromagnetic Compatibility, vol. 65, no. 3, pp. 625-633, June 2023, doi: 10.1109/TEMC.2022.3232616.

[7] G. Notermans, H. -M. Ritter, S. Holland and D. Pogany, "Dynamic Voltage Overshoot During Triggering of an SCR-Type ESD Protection,"

in IEEE Transactions on Device and Materials Reliability, vol. 19, no. 4, pp. 583-590, Dec. 2019, doi: 10.1109/TDMR.2019.2952713.

Electrical Performance of Chiplet Interconnect Channels under thermal conditions

Tingyi Shi, Xiang Wang, Yuefeng He, Xiaofang Gao, Zhiqiang Zhu, Xujuan Wang and Xinnan Lin*

*Anhui Engineering Research Center of Vehicle Display Integrated Systems,
Joint Discipline Key Laboratory of Touch Display Materials and Devices, School of Integrated
Circuits, Anhui Polytechnic University, Wuhu 241000, China; xnlin@mail.aphu.edu.cn

Kunshan J.W. Micro Corp, Kunshan, Suzhou, Jiangsu, P.R.China

Abstract

This paper focuses on how thermal distributions in different directions, affect the electrical performance of chiplet interconnect. Through three distinct simulation experiments, this paper analyzes the performance impact under different temperature distributions, both perpendicular and parallel to the direction of signal transmission. Results reveal that thermal-induced temperature fields aligned parallel to the transmission lines exert the most substantial impact on channel performance, evidenced by more insertion loss, higher return loss which significantly impact eye diagram. These findings show the importance of chip system heat dissipation with high-speed data rate, especially in Chiplet system design.

Keywords: Chiplet, Signal Channel, Thermal distribution

Introduction

Chiplet technology, the most significant innovation in the semiconductor industry, focuses on the close interconnection of various functional small chips through interconnect channels to form an optimized system [1]. The data rate between chips is very critical for chiplet system, fundamentally affecting the system's communication efficiency and reliability [1]. The electrical performance metrics, such as signal integrity, transmission rate, and power consumption, are essential for evaluating the effectiveness of these channels [2].

There are a multitude of factors influencing the electrical performance of channels, including material properties, design layout, and power management but temperature distribution caused by system heat dissipation is always omitted due to EDA tools limitation. However, temperature fluctuations always impact both the physical properties of materials and the resistance and capacitance along the signal transmission path, subsequently affecting signal integrity and transmission speed [3]. This paper investigates the effects of temperature distribution on the electrical performance of chiplet interconnect channels, with the goal of offering theoretical and practical insights to quantify the impact. It will provide system heat dissipation design based on system electrical performance with temperature distribution.

Extraction of S-Parameters for Interconnect Signal Channel Models under Different Temperature Fields

S-parameters, or scattering parameters, are essential for characterizing signal reflection and transmission in RF and microwave circuits. They provide a quantitative measure of the reflection coefficient and gain at circuit ports, playing a vital role in the analysis and design of communication systems' electrical performance [4].

S-parameters for signal channel models can be efficiently extracted using electromagnetic simulation software, a critical process for analyzing and optimizing the performance of signal channels[5].

Figure 1. Trimmed Signal Channel Model

The initial step involves importing the designed layout into the electromagnetic simulation software RainbowStudio [6], with meticulous configuration of material properties and layer stack structure to replicate the actual application scenario within the simulation environment. Following this, a specific signal channel model is extracted from the layout in accordance with simulation requirements, as illustrated in Figure 1, to enable targeted analysis. Subsequently, a temperature field and excitation source are integrated into the transmission line model, with the appropriate frequency bandwidth established for simulation purposes. These procedures facilitate the simulation of signal transmission behavior under diverse conditions, thereby providing a solid data foundation for the design and optimization of signal channels.

To thoroughly examine how temperature fields from various orientations affect the electrical characteristics of signal channels, we conducted three tailored simulation experiments aimed at securing accurate data and insightful analysis. The first setup, serving as our baseline, involved no temperature variations, thus maintaining a stable thermal condition to establish a benchmark against which subsequent experimental outcomes could be compared.

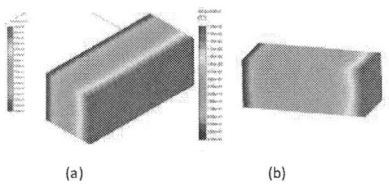

Figure 2. (a) Temperature field perpendicular to the transmission line direction, (b) Temperature field parallel to the transmission line direction

Figure 2(a) depicts the second experimental setup, where a linearly uniform temperature gradient ranging from 50 ° to 100 ° was applied perpendicularly to the signal channel model, mimicking the effects of vertical temperature gradients encountered in real-world scenarios. Figure 2(b) illustrates the third setup, wherein a similar temperature gradient was applied parallel to the signal channel model. This configuration was designed to assess how variations in the temperature field along the transmission line direction affect the electrical performance of the signal channel.

These experiments, encompassing three distinct configurations, are designed to thoroughly evaluate the directional impact of temperature fields on the electrical characteristics of signal channels, thereby laying a scientific foundation for the design and optimization processes of these channels.

Analysis of S-Parameters for Interconnect Signal Channel Models under Different Temperature Fields

Figure 3 illustrates the S-parameter comparisons: (a) between the first and second simulation sets, where the second set demonstrates increased insertion loss and decreased return loss relative to the first. (b) between the first and third sets, with the third set also exhibiting higher insertion loss and lower return loss compared to the first. (c) between the second and third sets, where the third set shows even greater insertion loss and smaller return loss than the second.

For signal channels, the optimal scenario is a return loss approaching zero, signifying minimal signal reflection, and an insertion loss approaching one, indicating minimal signal transmission loss. Consequently, an increase in insertion loss and a decrease in return loss both suggest a decline in signal channel performance. Comparative analysis of S-parameters reveals that the third set of simulation experiments, with the temperature field applied parallel to the direction of signal propagation, exerted the most substantial impact on the electrical performance of the signal channel.

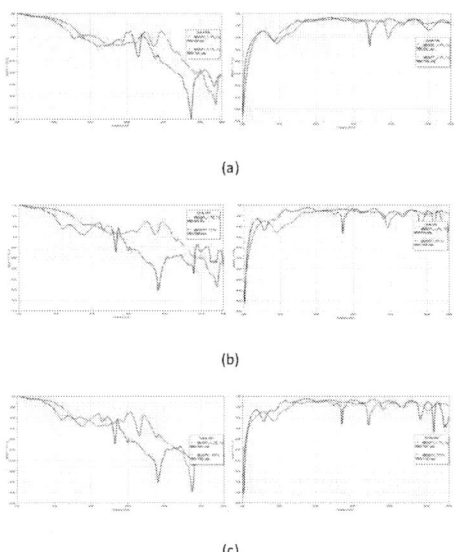

(a)

(b)

(c)

Figure 3. (a) Comparison between the first set (red) and the second set (blue), (b) Comparison between the first set (red) and the third set (blue), (c) Comparison between the second set (red) and the third set (blue).

The S-parameter model for signal channels enables eye diagram simulation [7]. During this process, the transmitter emits input signals at a 16 Gbps rate. These signals, once they have traversed the signal channel and are received, are transformed into eye diagrams for analysis.

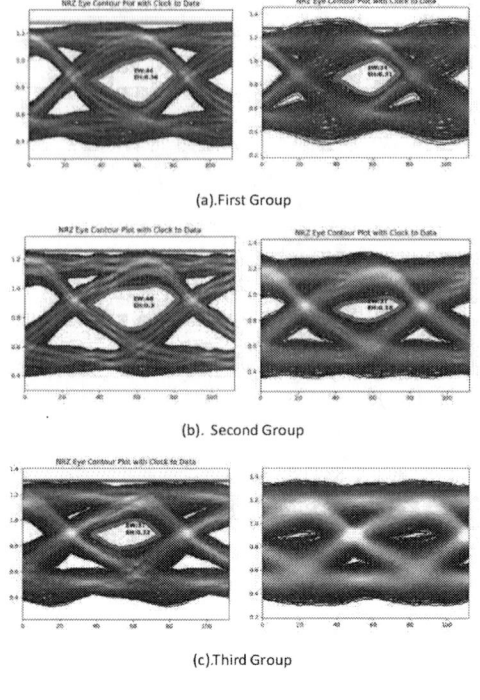

(a).First Group

(b). Second Group

(c).Third Group

name	no crosstalk				one crosstalk			
First group	EW:46	base	EH:0.36	base	EW:34	base	EH:0.31	base
Second group	EW:45	+4.36%	EH:0.20	-16.67%	EW:37	+8.82%	EH:0.18	-41.94%
Third group	EW:37	-19.56%	EH:0.22	-36.59%	EW:close	-100%	EH:close	-100%

(d).Comparison tables

Figure 4. (a), (b), (c) Eye diagram results for the three sets of experiments, with the left graph showing the results without crosstalk and the right graph showing the results with crosstalk, (d) Data comparison of the three sets of experiments

Figure 4 presents the eye diagram results from the three sets of simulation experiments as follows: (a) In the first set, the eye diagram remains open regardless of whether crosstalk is present. (b) For the second set, the eye diagram is slightly smaller than that of the first set in the absence of crosstalk, but it narrows significantly when crosstalk is introduced. (c) The eye diagram of the third set is considerably smaller than that of the first set even without crosstalk, and it closes completely when crosstalk is added.

Upon comparison of the eye diagram simulation results for all three sets, it is clear that there has been a notable decline in the performance of the signal channels in the second and third sets. Particularly, the third set stands out with the most severe degradation in performance.

Conclusion

Analysis of experimental data indicates that temperature fields oriented in various directions have a substantial impact on the electrical properties of signal channels. Notably, the experiments with temperature fields aligned parallel to the transmission lines (the third set) demonstrated the greatest negative impact, with increased insertion loss, decreased return loss, and pronounced eye diagram degradation. These findings underscore the critical role of temperature field orientation and distribution in maintaining signal integrity and efficiency, particularly in high-speed transmissions, highlighting the necessity of temperature management in the optimization of signal channels.

Acknowledgement

This work is supported by the "Sustained Support Project by Institutions Engaged In Fundamental Research" YG2406-1, in part by Key laboratory founding of Anhui Province 202205p12030001 and in part by the Start-Up Funding of Anhui Polytechnic University under Grant 2023YQQ003.

References

[1] Song R, Zhang J, Zhu Z, et al. Fault and self-repair for high reliability in die-to-die interconnection of 2.5 D/3D IC[J]. Microelectronics Reliability, 2024, 158: 115429.

[2] Wu K B, Kuo T Y, Hung C C, et al. Novel RDL design of wafer-level packaging for signal/power integrity in LPDDR4 application[J]. IEEE Transactions on Components, Packaging and Manufacturing Technology, 2018, 8(8): 1431-1439.

[3] Lee M, Jung D H, Kim H, et al. High-frequency temperature-dependent through-silicon-via (TSV) model and high-speed channel performance for 3-D ICs[J]. IEEE Design & Test, 2015, 33(2): 17-29.

[4] Jung D H, Kim Y, Kim J J, et al. Through silicon via (TSV) defect modeling, measurement, and analysis[J]. IEEE transactions on components, packaging and manufacturing technology, 2016, 7(1): 138-152.

[5] Alswat M, Alzahmi A. Energy efficient 5 Gb/s baseband transceiver for 3D memory interconnect[J]. Multimedia Tools and Applications, 2024: 1-16.

[6] Wuxi Flytrum Technology Co.,Ltd，Flytrum Multi-physics Electric thermal coupling Finite Element Method simulation software (RainbowStudio), Version 140798, WuXi,China,2024.

[7] Morales A, Agili S S, Meklachi T. S-parameter sampling in the frequency domain and its time-domain response[J]. IEEE Transactions on Instrumentation and Measurement, 2020, 70: 1-13.

A Low-power 16 Gb/s Single-Ended Voltage-Mode Transmitter With Two-Tap FFE in 55-nm CMOS

Haitao You   Qiang Cui   Xiaofang Gao   Dimitar Nikolov   Boris Dobrichkov   Vladimir Garistov

Abstract—**This paper introduces a low-power 16 Gb/s single-ended voltage-mode transmitter with a 2-tap feed-forward equalizer (FFE) using ICsprout's 55 nm CMOS technology. The transmitter consists of a CML-to-CMOS module, a 4:2 serializer, a 2-tap FFE, a source-series-terminated (SST) driver, and a clock module. The proposed SST driver calibrates the output impedance, achieving fine control of the impedance. Simulation results show that at a Nyquist frequency of 8 GHz, the insertion loss of the channel is 11 dB. A single-ended transmitter with a data transmission rate of 16 Gb/s and a supply voltage of 1.2 V, after passing through the channel, has an eye height of 146 mV and an eye width of 54 ps (0.864 UI). Compared to other works, this transmitter exhibits a lower power consumption of 31.2 mW and an energy efficiency of 1.95 pJ/bit.**

Index Terms—**transmitter, feed-forward equalizer (FFE), serializer, SST driver, impedance calibration**

I. INTRODUCTION

With the rapid development of fields such as Artificial Intelligence (AI) and High-Performance Computing (HPC), the demand for bandwidth has been continuously increasing, making traditional chip design methods unsuitable. In response, the industry has innovatively proposed the Chiplet design methodology. Among these, Die-to-Die single-ended interconnect technology is the core, with power consumption becoming the primary concern in interconnect design.

In differential interconnects, designers commonly use CML drivers, but they have the drawback of high power consumption. Therefore, in Die-to-Die single-ended interconnect designs, VM drivers are adopted, which consume only a quarter of the power compared to CML drivers. The main types of VM drivers include the N-over-N driver and the SST driver, as shown in Fig. 1(a) [1] and 1(b). Although the N-over-N driver significantly reduces area compared to the SST driver, it presents challenges in impedance matching and controlling the output voltage. Based on this, this paper selects the SST driver for improvement, aiming to calibrate the output impedance. Fig. 1(c) shows a common method for impedance calibration; however, the impedance tuning part consumes a certain amount of voltage margin [2]. The proposed novel SST single-ended transmitter not only effectively addresses the

above issues but also provides fine control of the impedance. At 16 Gb/s, the power consumption is as low as 31.2 mW, and the energy efficiency reaches as low as 1.95 pJ/bit.

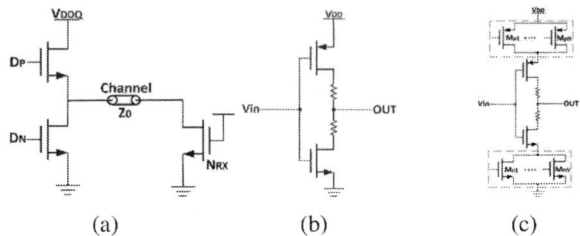

Fig. 1: VM drivers (a)N-over-N driver (b) SST driver (c) SST driver with a traditional impedance calibration.

II. THE PROPOSED DESIGN

As shown in Fig. 2, the overall architecture of the transmitter consists of a CML-to-CMOS module, a 4:2 serializer, a 2-tap FFE, an SST driver, and a clock module. First, the four 4 Gb/s parallel signals pass through the CML-to-CMOS module. These signals then enter the 4:2 serializer, where a multiplexer is used to convert the four signals into half-rate odd/even signals. Next, the half-rate odd/even signals undergo 2-tap FFE for equalization. Under the control of the half-rate clock, the odd/even signals are converted into a full-rate main tap signal. The odd/even signals are then pass through two additional latches, introducing a delay of half a half-rate clock cycle, resulting in a full-rate post-tap signal. Finally, the signals are transmitted through the SST driver to the channel for transmission.

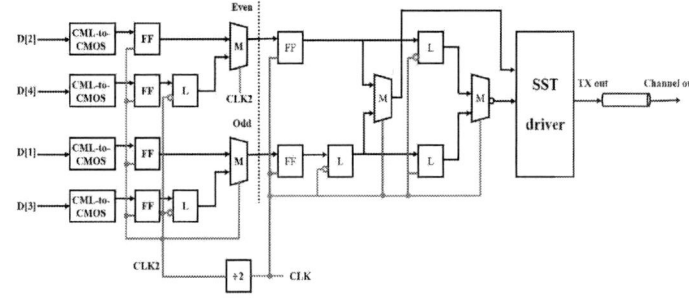

Fig. 2: The overall structure of the transmitter.

Considering the output signal swing of the test equipment, the main function of the CML-to-CMOS module is to increase

---

Author 1 and Author 2 are with *School of Integrated Circuit, Zhejiang University, Hangzhou, China.* Author 3 is with *Jiwei Microelectronics Technology Co., Ltd., Kunshan, China.* Author 4, Author 5 and Author 6 are with *Department of Electronics, Faculty of Electronic Engineering and Technologies, Technical University of Sofia, Sofia, Bulgaria.* This work is supported by the National Natural Science Foundation of China (Grant No. 92373106) and the process technology of Zhejiang ICsprout Semiconductor Co., Ltd.. Corresponding author: (Author 2, email: qiang_cui@zju.edu.cn)

979-8-3315-2209-4/25 $31.00 © 2025 IEEE

the signal swing and provide certain driving capability. The design of the DFF in the transmitter is shown in Fig. 3. This circuit is used to ensure that the two signals remain in phase. By using a single-phase clock, it can overcome the race conditions caused by the overlap of two-phase clocks, and it also has the advantage of low power consumption. When CLK is low, the D3 node is charged to a high level, and the level of Qb remains unchanged, meaning the level of Q also remains unchanged. When CLK transitions from low to high, the D2 node retains the value of $\overline{D}$ sampled when CLK was low, and the value of the D3 node becomes $\overline{D2}$, so Qb equals $\overline{D}$, and thus Q equals D.

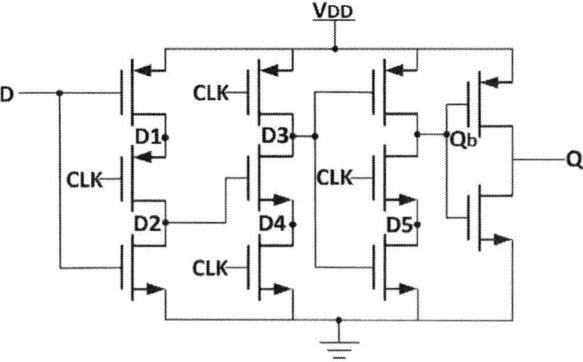

Fig. 3: The circuit of DFF.

The latch in the transmitter circuit functions to buffer data and introduce a phase delay between the two signals. The transmission gate latch designed in this paper is shown in Fig. 4. When CLK is low, the input data is stored in the latch. But when CLK is high, it passes the D input to the Q output, breaking the feedback loop.

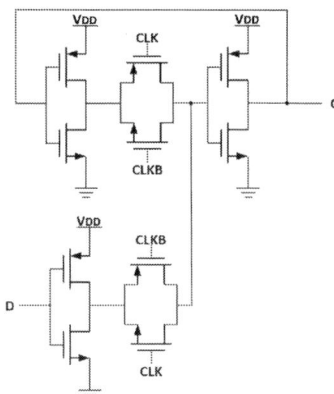

Fig. 4: The circuit of the transmission latch.

Due to fluctuations in the traditional SST driver caused by process variations, this paper proposes adding two MOS transistors in parallel with the pull-up PMOS and pull-down NMOS transistors for impedance calibration, as shown in Fig. 5. Take the main cursor as an example. The three PMOS transistors at the top are controlled by OR gates and digital control signals D1, D2, and D3, while the three NMOS transistors at the bottom are controlled by AND gates and digital control signals D4, D5, and D6. Since P1 and N1 are always in a conductive state, D1 is always 0, and D4 is always 1. Under the tt process corner, the output impedances of P1/N1, P2/N2, and P3/N3 are 45Ω, 90Ω, and 180Ω, respectively. Under normal conditions, D2/D6 are 0, and D3/D5 are 1. This keeps P1/N1 and P2/N2 conducting while P3/N3 is turned off. When the output impedance of P1/N1 decreases below 45Ω due to PVT variations, the P2/N2 is turned off, and P3/N3 is turned on to adjust the output impedance. When the output impedance of P1/N1 exceeds 45Ω, all the MOS transistors (both top and bottom) are turned on. This method achieves impedance calibration without consuming voltage margin.

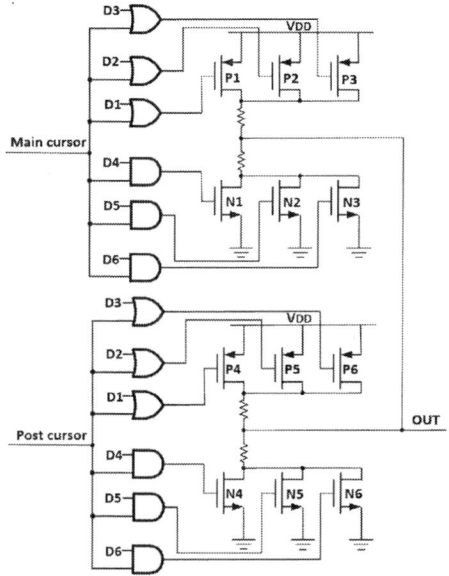

Fig. 5: The proposed SST driver with impedance calibration.

III. SIMULATION RESULTS

Since the signal transmission rate is 16 Gb/s, the Nyquist frequency is 8 GHz. At this point, the insertion loss S_{21} of the channel is 11 dB, as shown in Fig. 6.

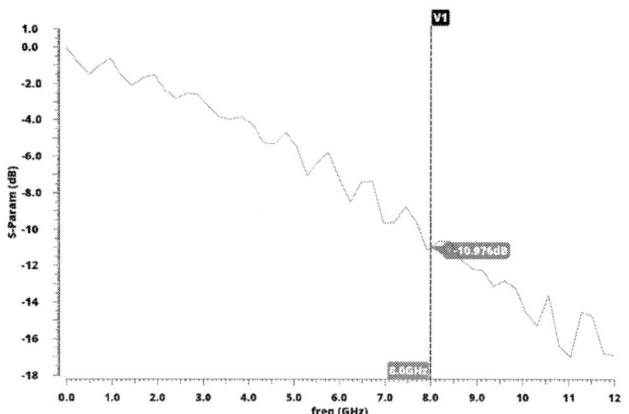

Fig. 6: The insertion loss S_{21} of the channel.

979-8-3315-2209-4/25 $31.00 © 2025 IEEE

Fig. 7 shows the output signal of the transmitter, demonstrating its de-emphasis capability.

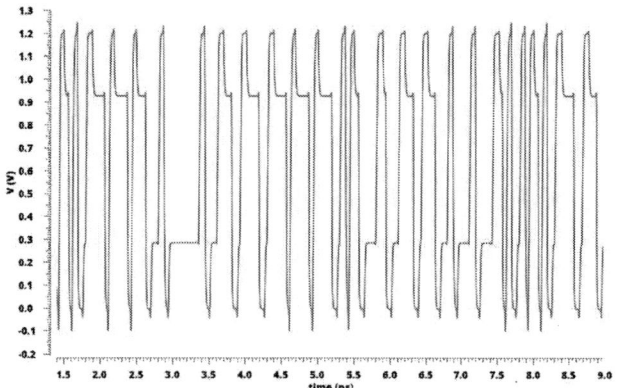

Fig. 7: The output signal of the transmitter.

Fig.8 (a) shows the eye diagram of the signal after transmission through the channel without FFE. The eye diagram is almost fully closed. Fig. 8(b) presents the eye diagram of the signal after transmission through the channel with FFE. The eye height is 146 mV, and the eye width is 54 ps (0.864 UI). The eye diagram shows good quality, indicating that the transmitter design effectively performs equalization, thereby enhancing the signal quality.

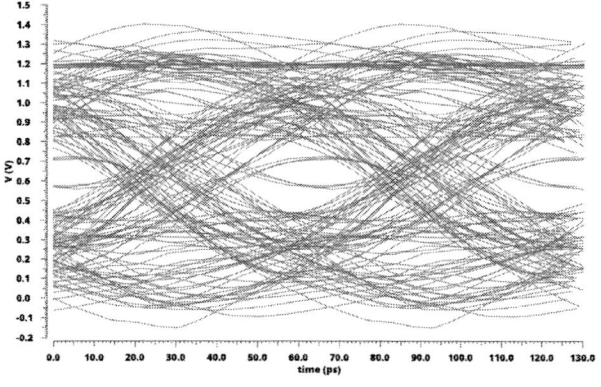

(a) The eye diagram without FFE.

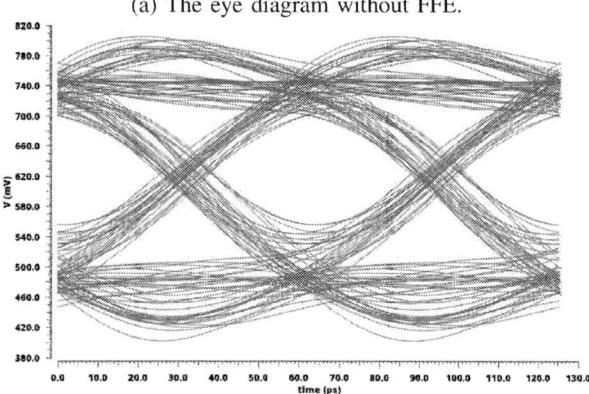

(b) The eye diagram with FFE.

Fig. 8: The eye diagram.

Compared to the transmitters designed in other works (TABLE I), the transmitter we designed offers lower power consumption and better energy efficiency, with a power consumption of 31.2 mW and an energy efficiency of 1.95 pJ/bit.

TABLE I: Performance Summary and Comparison

Parameters	[3]	[4]	[5]	This work
CMOS Technology	65nm	65nm	65nm	55nm
Signaling	Diff.	Diff.	Diff.	Single-ended
Data Rate(Gb/s)	6.4	8.5	20	16
Driver Type	SST	SST	SST	SST
Channel Loss(dB)	6.6	16	-	11
Eye Diagram Opening(mV)	750	1000	300	146
Power(mW)	26.24	90.4	167	31.2
Energy Efficiency(pJ/bit)	4.1	11.3	8.3	1.95

IV. CONCLUSION

This paper presents a low-power 16 Gb/s single-ended voltage-mode transmitter, which incorporates a 2-tap FFE. In VM drivers, traditional N-over-N drivers face challenges in impedance matching and output voltage control, so we chose to improve this with a SST driver. A novel SST driver is proposed, which enables precise output impedance calibration and offers better impedance matching with the channel compared to traditional SST drivers. This architecture also avoids the voltage margin loss that may occur with conventional calibration methods. Simulation results show that, using a 55 nm CMOS process, the signal from the 16 Gb/s transmitter after passing through the channel has an eye height of 146 mV and an eye width of 54 ps (0.864 UI). The overall power consumption of this transmitter is 31.2 mW, with an overall energy efficiency of 1.95 pJ/bit.

REFERENCES

[1] W. Bae, "Supply-scalable high-speed I/O interfaces," Electronics, 2020,9, 1315.

[2] D. Liu, Z. Wang, X. Bai, F. Lin, "A low power 10 Gbit/s transmitter based on SST driver," Microelectronics,2018,48(03):338-343.

[3] S. Chen, L. Yang, H. Jing, F. Zhang and Z. Gao, "A novel SST transmitter with mutually decoupled impedance self-calibration and equalization," 2011 IEEE International Symposium of Circuits and Systems (ISCAS), Rio de Janeiro, Brazil, 2011, pp. 173-176.

[4] M. Kossel et al., "A T-Coil-Enhanced 8.5 Gb/s high-swing SST transmitter in 65 nm bulk CMOS with ≪ −16 dB return loss over 10 GHz bandwidth," unpublished.

[5] R. A. Philpott, J. S. Humble, R. A. Kertis, K. E. Fritz, B. K. Gilbert and E. S. Daniel, "A 20Gb/s SerDes transmitter with adjustable source impedance and 4-tap feed-forward equalization in 65nm bulk CMOS," 2008 IEEE Custom Integrated Circuits Conference, San Jose, CA, USA, 2008, pp. 623-626.

AION-AMI： Chiplet System Simulation Platform

Qing Qian, Xiang Wang, Yuefeng He, Xiaofang Gao, Haofeng Zhu, Xujuan Wang,and Xinnan Lin*
*Anhui Engineering Research Center of Vehicle Display Integrated Systems,
Joint Discipline Key Laboratory of Touch Display Materials and Devices, School of Integrated Circuits, Anhui
Polytechnic University, Wuhu 241000, China; xnlin@mail.aphu.edu.cn
Kunshan J.W. Micro Corp, Kunshan, Suzhou, Jiangsu, P.R.China

Abstract

The evolution of chiplet design has emerged as a promising solution to overcome the limitation of traditional monolithic chip architecture. The performance of chiplet system has been focused all the time. From SPICE, known for the precision in circuit level test, to IBIS and its extension IBIS-AMI have made progress for simulation/validation in accuracy and efficiency step by step. However, with a variety of IP process emerging and the demands of high performance of system validation, it is imperative to provide a reliable solution to these challenges. In this paper, an alternative tool, the AION-AMI developed by JW Micro, has been introduced and illustrated the characters, simulation flow, output and the comparison in performance with other two commercial and matured tools, which are the ADS and Hspice, under the same simulation conditions. AION-AMI has shown its flexibly configurable for various IP process and foundry with general IO architecture and high performance of chiplet system evaluated in eye diagram.

INTRODUCTION

Chiplet system is a collection of small and modular chips from different process node, different foundries and then assembled in single package. It is unlike a conventional chip, which integrates all functional integrated circuits on a monolithic large die, and the chiplet is combined with multiple small dies as modules on a chip to perform specific functions. By this architecture, the chiplet is cost-efficiency, flexible in assembly and space [4]. However, due to unique characteristic of chiplet, chiplet system have different component resources as follows, which makes chiplet system performance evaluation difficult:

(a) Chiplet system includes different IPs type from different IP vendor from different foundries
(b) Chiplet system includes different IP process node

(same IP type) from the same IP vendor in same foundry
(c) Chiplet system includes different IP process node (same IP type) from the same IP vendor in different foundries

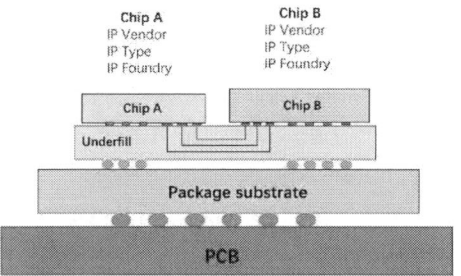

During the system design stage, the performance of system would be validated by realistic circuit models. Previously, these circuit models are supplied from venders by SPICE models that include detailed circuit and process information. However, those circuit information are intellectual properties (IP) and confidential for venders. In order to protect the venders' IP and also meet the demands of circuit models test, IBIS (I/O Buffer Information Specification) models are alternative solution for that. Traditional IBIS model is well defined below 1Gbs data rate. As data rate increases, Equalization needs to be used to compensate channel loss and IBIS modle is extended into IBIS-AMI.

IBIS_AMI INTRODUCTION

IBIS models provide a standardized way of the electrical characteristics for a circuit model, without revealing the underlying structure or process of a circuit. With the development of high speed interface needed in system, IBIS models no longer keep up with the performance evaluation requirement so that an extension solution of IBIS, the IBIS-AMI (Algorithmic Modeling Interface), has been generated. This new methodology could figure out most of link simulation requirements with large data stream and complex block, such as equalizers, FFE, CTLE, DFE, which performed output with eye diagram and bit error rate [2].

EDA tools would be utilized to conduct simulation for system performance with IBIS-AMI models, especially for chiplet system [3]. JW Micro developed AION-AMI simulator and this simulator has the features as follows:

(a) Support general IO architecture including Forward-clocking [1] and Embedded Clocking
(b) Support all kinds of TX/RX equalization
(c) System Power Supply Noise can be seamlessly integrated into simulation and impact on system performance.

AION-AMI Solution for IBIS-AMI

As one of simulators in AION families, AION_AMI is configurable to fit the IBIS-AMI models and package models, which are provided from one or various venders with the same or different processes, to get the result of system performance. To make different IP in different process node / foundry into unified simulation environment, AION Char is used to characterize IP into behavioral model.

The overall AION-AMI simulation flow is shown as follows:

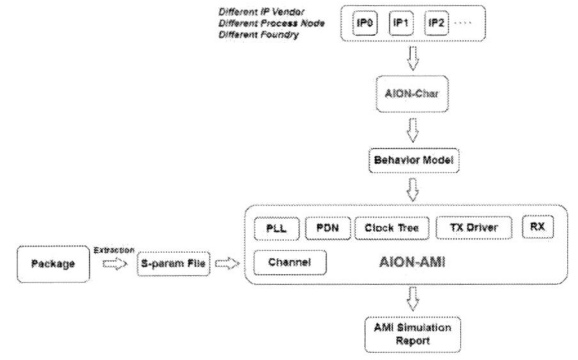

AION_AMI use behavior model file, the IBIS-AMI model file, and S-parameter file under the two general cases of clocking -- embedded or forwarded clock, to undertake simulation.

The configured interface of AION_AMI can be seen as below, with forwarded clock case as an example.

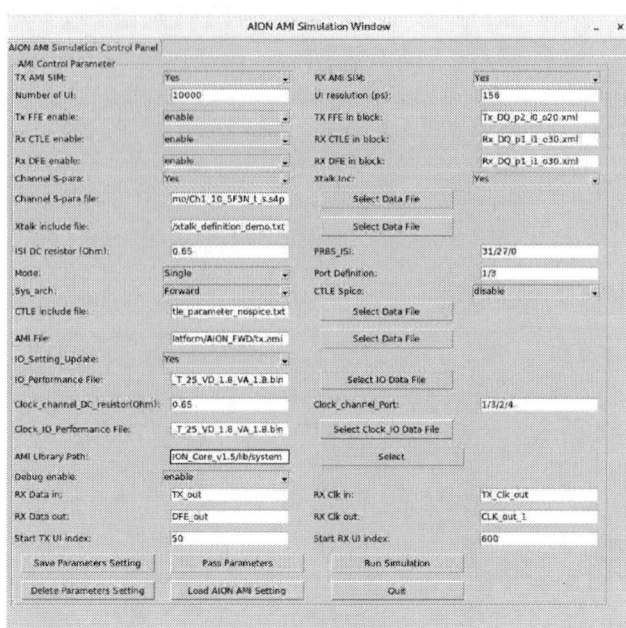

After configuration, AION_AMI will start the simulation and output will be demonstrated in two PDF file, the TX and RX result, respectively.

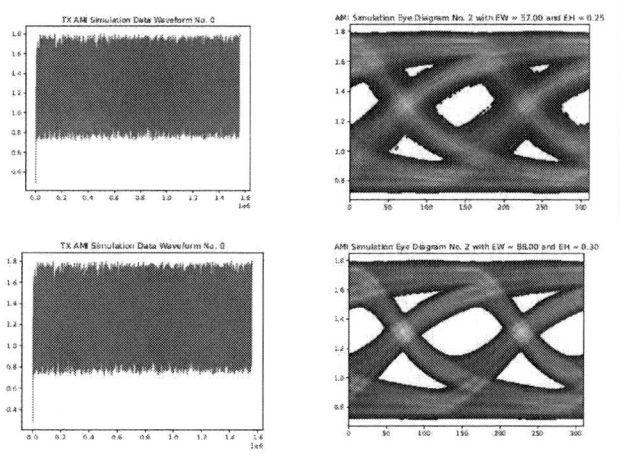

AION-AMI Simulation Accuracy

To investigate AION-AMI algorithm simulation accuracy, simulation results between ADS, HSPICE and AION-AMI are compared as shown below, which demonstrate AION-AMI algorithm is good for system performance evaluation.

979-8-3315-2209-4/25 $31.00 © 2025 IEEE

logs/en/tech/sim-des/2024/2/8/what-is-a-chiplet-and-why-should-you-care

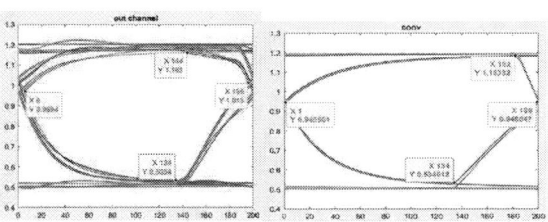

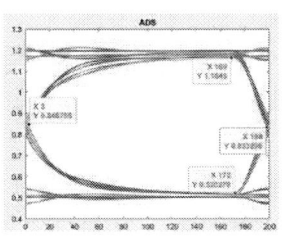

Simulator	Eye Height (V)	Eye Width (ps)
Hspice	0.6276	190
AION-AMI	0.6490	198
ADS	0.6442	195

Conclusion/Summary

AION-AMI simulation algorithm has been demonstrated. Comparing with ADS/HSPICE, good accuracy has been reached, which can make different IP process node, different IP vendor and different foundry simulate together.

Acknowledgement

This work is supported by the "Sustained Support Project by Institutions Engaged In Fundamental Research" YG2406-1, in part by Key laboratory founding of Anhui Province 202205p12030001 and in part by the Start-Up Funding of Anhui Polytechnic University under Grant 2023YQQ003

Reference

[1] Butterfield, J. (2019, February 1). *Modeling forwarded clock interfaces with IBIS-AMI*. Paper presented at the DesignCon 2019 IBIS Summit, Santa Clara, CA.

[2] Hongtao Zhang, et al (2017). *IBIS-AMI Modeling and Simulation of Link Systems using Duobinary Signaling*. In *Proceedings of* DesignCon 2017

[3] Sullivan, B., Rose, M., & Boh, J. (2015). *IBIS-AMI model simulations over six EDA platforms*. In *Proceedings of DesignCon 2015*.

[4] Keysight Technologies. (2024, February 8). *What is a chiplet and why should you care?* Retrieved October 23, 2023, from https://www.keysight.com/b

Optimization Algorithm for High-Speed IO Systems

Mengyuan Chu, Xiang Wang, Yuefeng He, Bo Zhang and Xinnan Lin*
*Anhui Engineering Research Center of Vehicle Display Integrated Systems,
Joint Discipline Key Laboratory of Touch Display Materials and Devices, School of Integrated Circuits, Anhui
Polytechnic University, Wuhu 241000, China; xnlin@mail.aphu.edu.cn
Kunshan J.W. Micro Corp, Kunshan, Suzhou, Jiangsu, P.R.China

Abstract

The data speed of chip IO interfaces accelerates as the demand for chip computing power increases and the operating speed of chip cores rises. This presents new challenges for high-speed chip design, necessitating the implementation of more advanced chip design processes, concepts, and technologies. In chip design optimization, equalizer optimization technology is a comprehensive approach that considers the entire system. This technology is related to the overall structure of the system, including factors such as chip packaging, signal transmission path lengths, materials, and connection methods. As a result, the design and optimization process can be quite complex.

This paper proposes a "Novel High-Speed IO Interconnect System Optimization Algorithm" as a solution to these challenges, offering an integrated application algorithm to address the issues at hand.

Fundamental theory

In the process of signal system optimization, the optimization is based on the impulse response curve of the communication channel, with the primary goal of assessing the impact of Inter-Symbol Interference (ISI) on the signal. The ISI level is calculated based on the channel's impulse response, which is used for subsequent worst-case peak analysis and eye diagram plotting, ultimately leading to optimized system design parameters.

First, a second-order filter is used to equalize the signal. The initial values of the filter coefficients are set, and the coefficients are continuously adjusted in steps. The impulse response of the channel output is filtered, using the maximum voltage of the impulse response as a reference. The lengths, counts, and positions of the pre-cursor and post-cursor are calculated, with the relationship between the cursor and ISI illustrated in Figure 1.

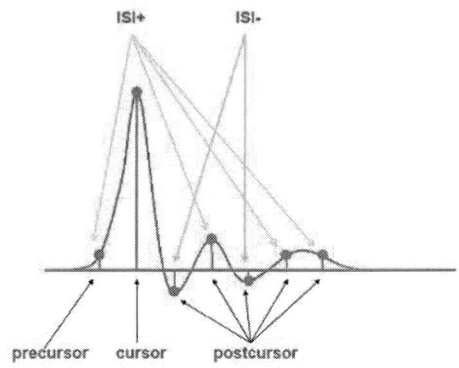

Figure 1. Schematic diagram of pre-cursor, cursor, post-cursor, and ISI

Calculate the ISI of the pre-cursor and post-cursor separately to obtain the absolute sum of the ISI. Based on the worst-case peak analysis method (PDA), the ISI (+/-) is superimposed on the current cursor to obtain the response curve of the current cursor under PDA within 1 UI (unit interval). The relationship before and after superposition is shown in Figure 2, where the orange curve represents the signal before superposition, and the green curve represents the signal after ISI superposition. As the ISI accumulates, the eye diagram becomes increasingly closed. By comparing the calculated PDA curve with the logical 0 level, the left and right intersection points can be obtained.

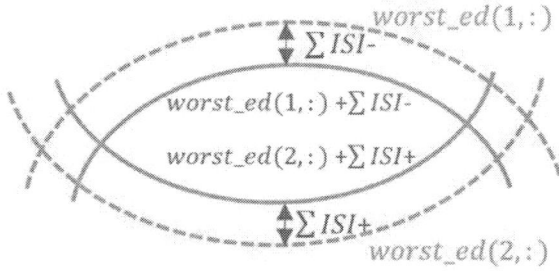

Figure 2. Schematic Diagram of the PDA Algorithm

After 15 iterations, the corresponding eye diagrams under N different PDAs are obtained. Based on the previously calculated PDA impulse response curves, the eye height and eye width of the curves can be determined. Using the eye height as a measurement standard, the filter coefficients corresponding to the highest peak after equalization, along with the eye height and eye width, can be identified.

Optimization Process and Algorithm for New High-Speed IO Interconnect Systems

The combination of equalizers for signal equalization and the selection of equalizers is shown in Figure 3. First, preprocessing is performed on the output signal of the single-bit impulse response to reduce the amount of data computation. The selection order of the equalizers is CTLE, FFE, and DFE. PDA analysis is conducted for each combination, providing the eye height and eye width for each combination. During each parameter selection process for the equalizers, it is necessary to optimize with the other equalizer combinations to ensure the overall optimization performance of the system during selection.

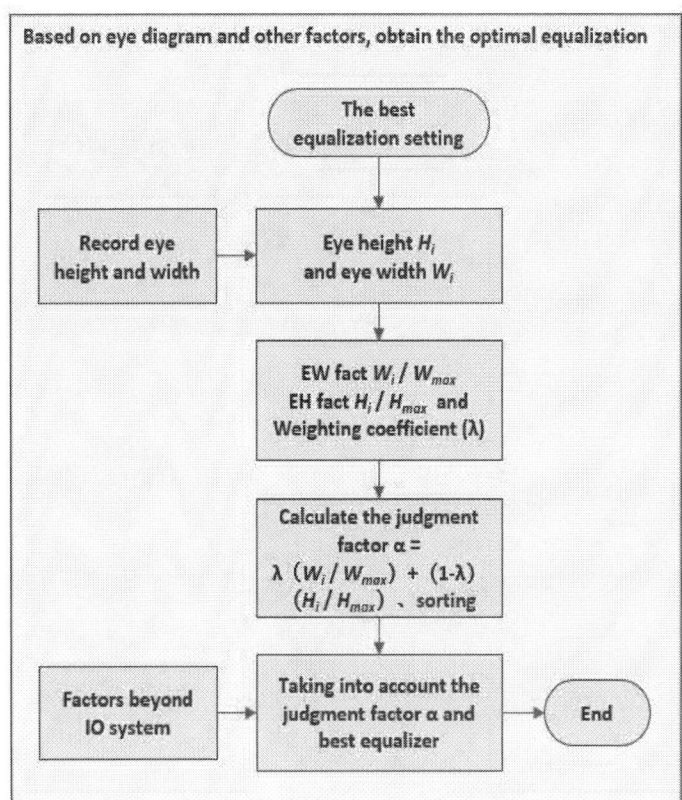

Figure 3 （b）

Application

Follow the simulation flowchart as Figure 3(a) and (b), simulation example has been done for a high speed IO system with different equalization technologies, a the blind selection of various combinations of equalizer, a (CTLE + FFE) equalization with optimization and a (CTLE + FFE+DFE) equalization with optimization. Based on the blind selection of various combinations of equalizers according to the output signal, a set of equalizer combinations with the best equalization effect and the best stability is obtained. As shown in Table 1, a comparative diagram of the effects of various equalizers is provided. The table indicates that the optimization of the CTLE+FFE equalizer has led to changes in the system eye diagram. On this basis, the optimized combination with the DFE equalizer greatly reduces inter-symbol interference (ISI), resulting in a clearer eye diagram, a wider eye opening, and an increased stability time of the signal on the transmission line.

Table1. Eye diagram comparison

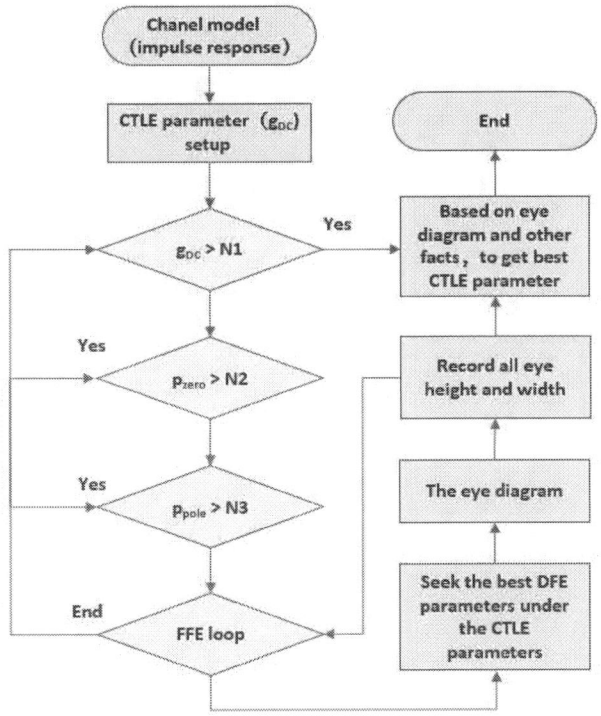

Figure 3 （a）

System configuration	Eye diagram
High-Speed IO System (Unoptimized)	
High-Speed IO System First Optimization (Including CTLE + FFE)	
High-Speed IO System Second Optimization (Including CTLE + FFE+DFE)	

Conclusion

This optimization algorithm is intended for High Speed IO system. Eventually, the methodology can be used for SI, power analysis and optimization in front-end design which is becoming a very critical constraint for high speed IO, chiplet and servers for AI. Several correlation activities are in place, especially to validate the algorithm for link BER and complete forwarded clock channel analysis.

Acknowledgement

This work is supported by the "Sustained Support Project by Institutions Engaged In Fundamental Research" YG2406-1, in part by Key laboratory founding of Anhui Province 202205p12030001 and in part by the Start-Up Funding of Anhui Polytechnic University under Grant 2023YQQ003

Reference

[1] Bo Zhang, Evelina F. Yeung, Santanu Chaudhuri, Sanjay Dabral, Adam J. Norman, "Integrated Link Analysis Tool – Using Equalization Design/Optimization Module as a Case Study", DTTC 2006, August 2006

[2] Jimmy Jackson, Bo Zhang "OEM/ODM SISTAI Tool as Bridge to Intel Leading Edge Technology", DTTC 2010 Demo9.

[3] Hall, S. H., Hall, G. W., & McCall, J. A. (2000). *High-speed digital system design: A handbook of interconnect theory and design practices*. Wiley-Interscience.

[4] S. Hwang, J. Song, Y. Lee and C. Kim, "A 1.62–5.4-Gb/s Receiver for DisplayPort Version 1.2a With Adaptive Equalization and Reference less Frequency Acquisition Techniques," in IEEE Transactions on Circuits and Systems I: Regular Papers, vol. 64, no. 10, pp. 2691-2702, Oct. 2017, doi: 10.1109/TCSI.2017. 2695612.

[5] Proakis J G. Digital Communications. McGraw-Hill. Fourth Edition, 2000.

[6] X. Zhang, W. Jiang and D. Li, "An ISI Transmission and the Optimal Detection/Decoding of the Coded ISI System," 2007 3rd International Workshop on Signal Design and Its Applications in Communications, 2007, pp. 283-287, doi: 10.1109/IWSDA.2007.4408378.

Author Index

Name	Page	Name	Page
Ailin Qiu	211	Boris Dobrichkov	540
Ailin Qiu	230	Bozhi Qiu	473
Amina Yasin	28	Carl Ma	272
Anqing Chen	111	Carl Ma	276
Anqing Chen	436	Chang Chen	215
Ao Han	356	Chang Chen	226
Ao Han	361	Chang Ding	313
Bharatha Kumar Thangarasu	238	Changjian Zhou	159
Bharatha Kumar Thangarasu	436	Changjian Zhou	488
Bharatha Kumar Thangarasu	452	Chao Gu	449
Bharatha Kumar Thangarasu	456	Chen Chen	174
Bharatha Kumar Thangarasu	463	Chen Tang	268
Bharatha Kumar Thangarasu	467	Cheng Li	185
Bharatha Kumar Thangarasu	485	Cheng Yuan	404
Bharatha Kumar Thangarasu	492	Chenguang Lv	203
Bharatha Thangarasu	104	Chengwu Pan	188
Bharatha Thangarasu	111	Chengyan Zhong	32
Bharatha Thangarasu	460	Chengyan Zhong	35
Bijiao Yang	300	Chenxu Zhao	396
Bin Li	291	Chi Zhang	203
Bin Li	419	Chia-Yen Li	272
Bin Wang	4	Chia-Yen Li	276
Binchen Wang	104	Chunhua He	280
Bing Chen	151	Chunhua He	283
Bing Chen	369	Chunxia Su	313
Bing Chen	392	Cloud Wang	272
Bingke Zhang	219	Cloud Wang	276
Bingke Zhang	242	Congwei Liao	163
Bo Qi	477	Congwei Liao	174
Bo Ran Zhang	500	Congwei Liao	373
Bo Yang	272	Congwei Liao	400
Bo Yang	276	Da Wang	191
Bo Yi	219	Da Wang	199
Bo Yi	222	Daiki Qin	272
Bo Yi	224	Daiki Qin	276
Bo Zhang	377	Daoguang Zhang	283
Bo Zhang	546	Decai Liu	55
Boran Cao	268	Dengfa Zhou	449
Boris Dobrichkov	505	Dimitar Nikolov	505

Name	Page	Name	Page
Dimitar Nikolov	540	Guangyin Feng	121
Dongyang Wang	230	Guiyue Mao	89
Dongyang Wang	234	Guiyue Mao	100
Dou Pei	350	Guiyue Mao	114
En Hong	433	Guiyue Mao	117
Fan Guo	163	Guofang Yu	495
Fang Wang	64	Guohua Liu	433
Fang Wang	75	Guohua Liu	443
Fanyi Meng	89	Guohua Liu	445
Fanyi Meng	93	Guohua Liu	477
Fanyi Meng	97	Guoqing Chen	443
Fanyi Meng	100	Guozheng Lin	38
Fanyi Meng	104	Hai Ye	111
Fanyi Meng	108	Haibing Xie	1
Fanyi Meng	111	Haigang Feng	309
Fanyi Meng	114	haigang feng	319
Fanyi Meng	117	Haigang Feng	323
Fanyi Meng	124	Haigang Feng	422
Fanyi Meng	128	Haigang Feng	473
Fanyi Meng	131	Haitao You	540
Fanyi Meng	135	Haizhen Guo	430
Fanyi meng	138	Haizhen Guo	470
fanyi meng	238	Han Yan	19
Fanyi Meng	436	Hang Li	22
Fanyi Meng	452	Hang Li	25
Fanyi Meng	456	Hang Li	306
Fanyi Meng	460	Hang Li	316
Fanyi Meng	463	Hang Xu	64
fanyi Meng	467	Hang Xu	85
Fanyi Meng	485	Haobo Jia	338
Fanyi Meng	492	Haobo Jia	342
Feilian Chen	4	Haobo Jia	346
Frank Schwierz	249	Haofeng Zhu	543
Fuyuan Wang	334	Haokai Jing	68
Fuyuan Wang	346	Haoliang Li	151
Gang Chen	350	Haoru Wang	207
Gaoxiang Kai	523	Haowen Wu	422
Gongtang Yu	516	Haoyu Dong	473
Guanglong Ding	22	Hayato Wang	272
Guanglong Ding	25	Hayato Wang	276
Guanglong Ding	49	Hejiu Zhang	523
Guangsheng Li	174	Hejiu Zhang	526
Guangyao Wang	257	Heng Wu	280

Name	Page	Name	Page
Heng Wu	283	Jiabei Pan	532
HengYue Gong	154	Jiahao Xu	523
Hetian Sun	415	Jiaji He	265
Hideki Hirayama	28	Jiaji He	412
Hiromitsu Sakai	28	Jialei Tan	207
Hiroshi Amano	52	Jiaming Zhao	138
Hiroyuki Yaguchi	28	Jian Wang	124
Hoi-Sing Kwok	509	Jian Zhan	280
Hong Zhang	384	Jian Zhou	350
Hongfei Su	452	Jianbing Liu	97
Hongqiang Yang	219	Jiangbo Hu	373
HongQiang Yang	222	Jian-gong Ni	350
HongQiang Yang	224	Jianing Li	392
Hongwei Shen	377	Jiawei Chen	121
Hongxing Ma	295	Jiaxing Yang	49
Hongxing Ma	334	JiaYuan Fan	449
Hongxing Ma	338	Jiayuan Wang	207
Hongxing Ma	342	Jichong Guo	313
Hongxing Ma	346	Jie Wei	207
Hongyi Li	502	Jigeng Sun	8
Hongyi Liu	439	Jigeng Sun	166
Huachao Xu	419	Jihong Yin	272
Huafei Cheng	430	Jihong Yin	276
Huafei Cheng	470	Jing Lin	280
Huan Li	211	Jing Lin	283
Huan Li	215	Jingwen Li	419
Huan Li	226	Jingyang Wen	295
Huan Li	230	Jingying Sun	377
Huan Ning	191	Jintai Chi	334
Huan Ning	199	Jintai Chi	338
Huan Yang	11	Jinwen Liu	11
Huanghao Ying	430	Jinxuan Wu	46
Huanghao Ying	470	Jinzhi Mu	166
Huanshi Guo	226	Jiping Hu	64
Huayi Wu	445	Jiping Hu	85
Hui Xia Yang	142	JiXiang Shang	502
Huilin Huang	439	Jixiang Zhang	408
HuiPing Zhu	154	Jixiang Zhang	412
HuiXia Yang	154	Jiyu Zhao	22
Jia Shen	388	Jiyu Zhao	25
Jia Wang	52	Jiyuan Huang	502
Jiabei Pan	520	JiZhi Liu	502
Jiabei Pan	529	Juin J.Liou	64

Name	Page	Name	Page
Juin Jei Liou	384	Kiat Seng Yeo	111
Juin Jei Liou	516	Kiat Seng Yeo	114
Juin Liou	75	Kiat Seng Yeo	436
Juinjei Liou	297	Kiat Seng Yeo	452
Juinjei Liou	300	Kiat Seng Yeo	456
Juinjei Liou	303	Kiat Seng Yeo	460
Jun Du	396	Kiat Seng Yeo	463
Jun Fu	495	Kiat Seng Yeo	492
Jun Zhang	64	kiatseng yeo	238
Jun Zhang	85	KiatSeng Yeo	485
Jun Zhang	207	Kohei Fujimoto	28
JunFeng Duan	222	Kuangyong Gao	430
JunFeng Duan	224	Kuangyong Gao	470
JunJi Cheng	222	Kui Xiao	245
JunJi Cheng	224	Kyle Xu	272
Junjie Liu	334	Kyle Xu	276
Junlang Yu	291	Lanrong Zou	509
Junun Zhu	272	Le Bian	46
Junun Zhu	276	Le Li	272
Junzhan Liu	257	Le Li	276
Kai Yuan	283	Lei Lu	11
Kaijun Ding	242	Lei Lu	509
Kaiqiang Qin	460	Lei Xu	169
Kaitseng Yeo	467	Lei Xu	172
Kaiwei Dai	207	Leo Kuai	272
Kaiwen Chen	342	Leo Kuai	276
Kaiwen Chen	346	Leran Huang	268
Kaixue Ma	104	Levi Chen	272
Kaixue Ma	111	Levi Chen	276
kaixue ma	238	Liangjing Bai	295
Kaixue Ma	436	LiJun Feng	488
Kaixue Ma	452	Lin Li	342
Kaixue Ma	456	Ling Li	295
Kaixue Ma	460	Ling Zhang	520
Kaixue Ma	463	Ling Zhang	529
Kaixue Ma	467	Ling Zhang	532
Kaixue Ma	485	Liyao Wei	523
Kaixue Ma	492	Liyao Wei	526
Kaiyun Deng	473	Lize Wang	108
Kangxiang Zhao	219	Lizhe Liu	306
Keer Chen	313	Lizhe Liu	316
Kiat Seng Yeo	89	Long Zhang	188
Kiat Seng Yeo	104	Lu Chang	163

Name	Page	Name	Page
Lu Chang	174	Nagarajan Mahalingam	436
M. Ajmal Khan	28	Nagarajan Mahalingam	452
M. Nawaz Sharif	28	Nagarajan Mahalingam	456
Mahalingam Nagarajan	104	Nagarajan Mahalingam	463
Mahalingam Nagarajan	460	Nagarajan Mahalingam	467
Man Wong	509	Nagarajan Mahalingam	485
Mansun Chan	159	Nagarajan Mahalingam	492
Mao Ye	265	Nelson Yang	272
Mao Ye	388	Nelson Yang	276
Mao Ye	396	Nengxu Zhu	89
Mao Ye	404	Nengxu Zhu	128
Mao Ye	408	Nuoya Yang	226
Mao Ye	412	Pao-Hsun Huang	327
Mao Ye	415	Pao-Hsun Huang	330
Maojin Liang	280	Pei Guo	207
Mark Zheng	272	Pengfei Li	138
Mark Zheng	276	Pengfei Ye	319
Meihua Liu	253	Pengwei Chen	215
Meng Zhang	4	Peter Liu	272
Meng Zhang	46	Peter Liu	276
Meng Zhang	509	Prabakaran Selvaraj	1
Mengran Chen	151	Qi Chen	297
Mengyuan Chu	546	Qi Chen	303
Mi Lin	477	Qi Liu	356
Miao Cui	195	Qi Liu	361
Miao Cui	426	Qiang Cui	505
Michail Michailow	144	Qiang Cui	540
Ming Guo	509	Qiang Yu	211
Mingchao Li	470	Qiang Yu	215
Mingcheng Shen	313	Qiang Yu	226
Minglin Zheng	25	Qiang Yu	230
Mingmin Huang	211	Qiaoyi Fu	261
Mingmin Huang	215	Qiji Huang	309
Mingmin Huang	226	Qijun Sun	15
Mingmin Huang	230	Qilin Hua	157
Mingmin Huang	234	Qimeng Jiang	234
Mingxin Sun	188	Qing Qian	543
Mitsuhiro Muta	28	Qinghua Guo	306
Moting Deng	316	Qinghua Guo	316
Moufu Kong	219	Qingsong Zhao	93
Moufu Kong	242	Qingsong Zhao	124
Nagarajan Mahalingam	111	Qinwen Huang	283
Nagarajan Mahalingam	238	Qiqi Wei	8

Name	Page	Name	Page
Qiuwei Wang	396	Shengdong Zhang	400
Qixian Zheng	306	Shi Zong	169
Qixian Zheng	316	Shi Zong	172
Ran Huo	159	Shijun Ou	159
Ren Wei Chen	500	Shipei Ji	64
Ronglin Yang	330	Shipei Ji	85
RR Zhu	272	Shipei Ji	181
RR Zhu	276	Shipeng Chang	188
Rui Jin	55	Shipeng Chen	513
Rui Liu	42	Shipeng Chen	520
Rui Liu	287	Shipeng Chen	529
Rui Wang	55	Shipeng Chen	532
Rui Yao	195	Shipeng Zhang	408
Rui Yao	426	Shipeng Zhang	412
Rui Zhang	443	Shiyu Zuo	327
Ruibo Chen	502	Shiyuan Fu	135
Rumeng Qiu	272	Shu Ming Qi	142
Rumeng Qiu	276	Shuai Li	111
Ruyu Liang	169	Shuangmei Xue	15
Saiya Wang	257	Shuangmei Xue	46
Sang Lam	195	Shuangmei Xue	49
Sang Lam	426	Shuhai Chen	203
Sha Zhang	356	Shumin You	330
Sha Zhang	361	Shunjing Lei	384
Shangle Ye	280	Shurong Dong	513
Shanglin Yang	380	Shurong Dong	520
Shanglin Yang	384	Shurong Dong	529
Shanglin Yang	516	Shurong Dong	532
Shanri Chen	38	Shuting Wang	234
Shaohui Pan	419	Shuxin Xu	439
Shaolin Zhou	8	Shuyan He	144
Shaolin Zhou	38	Sini Wu	238
Shaolin Zhou	42	Siyang Liu	188
Shaolin Zhou	166	Siyuan He	46
Shaolin Zhou	287	Songqing Deng	280
Shaowei Zhen	203	Stephen Taylor	195
Shaowei Zhen	377	Stephen Taylor	426
Shen Xu	245	Su-Ting Han	22
Sheng Zhang	268	Su-Ting Han	25
Shengdong Zhang	11	Takayuki Nakachi	297
Shengdong Zhang	163	Takayuki Nakachi	300
Shengdong Zhang	174	Takayuki Nakachi	303
Shengdong Zhang	373	Tao Zhu	55

Name	Page	Name	Page
Tao Zhu	222	Xiang Chen	433
Tao Zhu	224	Xiang LI	430
Tapa Arnauld Robert	1	Xiang Li	470
Tengyan Huang	400	Xiang Wang	32
Tianyi Zhang	505	Xiang Wang	35
Tianyong Ma	356	Xiang Wang	58
Tianyong Ma	361	Xiang Wang	62
Tinghua Chen	309	Xiang Wang	536
Tingyi Shi	536	Xiang Wang	543
Tongqing Liu	366	Xiang Wang	546
Vladimir Garistov	505	Xianyou Zeng	283
Vladimir Garistov	540	Xiao Long Xu	142
Wang Kang	257	Xiao Sen Liu	500
Wanghui Zou	439	Xiaobin Ren	338
Wanhong Luan	22	Xiaobin Ren	346
Wanhong Luan	25	Xiaobo She	32
Wansi Ge	463	Xiaobo She	35
Wei Sun	342	Xiaobo She	58
Wei Yao	245	Xiaobo She	62
Wei Zeng	22	Xiaodan Gu	265
Wei Zeng	25	Xiaodong Huang	366
Wei Zhang	295	Xiaofang Gao	536
Wei Zhang	353	Xiaofang Gao	540
Weibin Lin	15	Xiaofang Gao	543
Weifeng Qiao	323	Xiaolong Wei	481
Weifeng Sun	188	Xiaonan Yang	151
Weihui Liu	422	Xiaonan Yang	369
Weiping Li	366	Xiaonan Yang	392
Weiran Cao	353	Xiaoping Dong	211
Weiteng Hu	291	Xiaoping Dong	215
Wenbin Zhang	25	Xiaoping Dong	230
Wenbo Wang	380	Xiaotian Yang	64
Wenchang Zhang	151	Xiaotian Yang	85
WenDa Leng	253	Xiaozheng Guo	467
Wenhao Wu	272	Xien Sang	78
Wenhao Wu	276	Xin Wang	81
Wenlan Ma	71	Xin Yuan	174
Wentao Zhao	268	Xin Zheng	163
Wenting Wang	404	Xinghong Chen	377
Wenze Gao	297	Xingkun You	430
Wu Yutong	456	Xingkun You	470
XF Guan	272	Xinnan Lin	536
XF Guan	276	Xinnan Lin	543

Name	Page	Name	Page
Xinnan Lin	546	Yang Liu	114
Xinpeng Xing	309	Yang Liu	117
xinpeng xing	319	Yang Yang	157
Xinpeng Xing	323	Yang Yu	303
Xinpeng Xing	422	YangHui Xia	154
XinYi Wu	222	Yangyang Guan	306
XinYi Wu	224	Yangyang Guan	316
Xinyu Zhu	513	Yanlin Li	49
Xinyu Zhu	529	Yao Li	388
Xinyu Zhu	532	Yao Li	396
Xinyuan Lin	268	Yao Li	404
Xiufeng Zhou	174	Yao Li	408
Xiuyin Zhang	488	Yao Ma	211
Xixian Wang	330	Yao Ma	215
Xiyuan Wu	295	Yao Ma	226
Xu Chen Men	500	Yao Ma	230
Xu Liang Wang	500	Yao Ma	234
Xuan Li	100	Yao Wang	481
Xuan Li	114	Yaoli Guo	49
Xuan Li	117	Yaowen Zhang	219
Xuan Liu	334	Yaowen Zhang	242
Xuan Liu	338	Yaoxu An	388
Xuan Liu	342	Yashan Pang	306
Xudong Cai	495	Yashan Pang	316
Xugang Ma	42	Yasushi Iwaisako	28
Xugang Ma	287	Ye Zhou	22
Xujuan Wang	536	Ye Zhou	25
Xujuan Wang	543	Yejiang Lin	350
Xuying Zhou	300	Yeliang Wang	142
Ya Li	151	Yeliang Wang	154
Ya Li	369	Yi Wu	121
Ya Li	392	Yi Zou	8
YaDong Zhou	154	Yichu Qin	230
Yan Dong	388	Yi-Chuen Eng	272
Yan Liu	327	Yi-Chuen Eng	276
Yan Liu	330	Yidong Yuan	377
Yan Yan	4	Yifan Shu	207
Yan Yan	15	Yifan Zheng	49
Yan Yan	46	Yifei Huang	234
Yan Yan	49	Yihan Wang	353
Yan Yan	509	Yihong Sun	159
Yang Hui Xia	142	Yilidana Mamuti	502
Yang Liu	100	Yimeng Tang	242

Name	Page	Name	Page
Yimu Yang	502	Yuezhong Duan	400
Yingfei Wang	334	Yufei Wang	148
YingFeng He	55	Yufeng Guo	58
Yingqi Liu	473	Yufeng Guo	62
Yingxu Wang	62	Yufeng Jin	144
Yingzhi Luo	219	Yufeng Jin	253
Yingzhi Luo	242	Yuhan Zhang	11
Yipeng Chen	513	Yuhan Zhang	373
Yipeng Chen	520	Yuhuai Liu	28
Yipeng Chen	529	Yuhuai Liu	75
Yipeng Chen	532	Yuhuai Liu	85
Yiqiang Zhao	396	YuhuaiLiu	64
Yiqun Liu	523	Yujun Ye	15
Yiqun Liu	526	Yukun Fu	419
Yiting Zhang	128	YuMeng Wang	154
Yitu Wang	297	Yun Li	226
Yitu Wang	300	Yun Ma	338
Yitu Wang	303	Yun Ma	346
Yizhou Ye	366	Yun Xin Wang	500
Yongjie Zhao	148	Yunhao Liu	178
Yongjie Zhou	75	Yunyang Wang	509
Yongpan Liu	268	Yuqing Liu	492
Yongwang Ma	377	Yuri Ma	272
Yu Hang Zheng	142	Yuri Ma	276
Yu Hongshi	485	Yusuo Wang	384
Yu Li	369	Yuwen Long	121
Yu Li	392	Yuxuan Shen	46
Yu Liu	32	Yuya Nagata	28
Yu Liu	35	Zenglong Zhao	93
Yu Liu	58	Zengqing Liang	323
Yu Liu	62	Zexuan Chen	38
Yu Long	265	Zhao Wang	195
Yu Meng Wang	142	Zhao Wang	426
Yuan Xiao Ma	142	Zhaochen Huo	369
Yuanbo Sun	163	Zhaochen Huo	392
Yuanbo Sun	174	Zhaohui Wu	291
YuanXiao Ma	154	Zhaokai Qiu	42
Yuanzhan Sheng	327	Zhaokai Qiu	287
Yue Zhang	131	Zhaonian Yang	148
Yue Zhang	148	Zhaonian Yang	523
Yuefeng He	536	Zhaonian Yang	526
Yuefeng He	543	Zhen Huang	481
Yuefeng He	546	Zhen Lin	124

Name	Page	Name	Page
Zhen Shi	430	Zhipeng Liu	330
Zhen Shi	470	Zhiqiang Zhu	536
Zhencheng Xu	520	Zhiqun Cheng	306
Zheng Bian	245	Zhiqun Cheng	316
Zhenghao Lu	104	Zhiwei Liu	502
Zhenghao Lu	111	Zhiwei Ye	174
zhenghao lu	238	Zhiwei Zhang	449
Zhenghao Lu	436	Zhiyuan He	502
Zhenghao Lu	452	Zhonghai Zhang	430
Zhenghao Lu	456	Zhonghai Zhang	470
Zhenghao Lu	460	Zhongqiu Xing	75
Zhenghao Lu	463	Zhuang Miao	89
zhenghao Lu	467	Zhuojun Chen	261
Zhenghao Lu	485	Zi Chun Liu	142
Zhenghao Lu	492	Zichao Ma	19
Zhenxuan Zhang	481	Zichao Ma	159
Zherui Zhao	25	Zihan Zheng	502
Zherui Zhao	49	Zijin Jiang	195
ZheYang Li	55	ZiSheng Li	356
Zhi Lin	191	ZiSheng Li	361
Zhi Lin	199	Zishuo Li	124
Zhigang Zhou	350	Ziyi Cui	203
Zhihua Zhu	516	Zongyang Zhang	121
Zhijuan Zhao	384	Zupei Gu	377
Zhikang Lan	366	Zuyue Pang	203

IEEE
445 Hoes Lane
Piscataway, NJ 08854-4141

ISBN 979-8-3315-2209-4

2017 China Semiconductor Technology International Conference (CSTIC 2017)

Shanghai, China
12 – 13 March 2017

IEEE Catalog Number: CFP1760Y-POD
ISBN: 978-1-5090-6695-7